1999

47th Edition

NATIONAL CONSTRUCTION ESTIMATOR

**Edited by
Dave Ogershok
Doug Phillips**

Includes inside the back cover:
An estimating CD with all the costs in the book plus an interactive multimedia video that shows how to use this disk to compile estimates, and a billing program that can export your estimates to *QuickBooks Pro* for progressive billing and job costing.

Craftsman Book Company
6058 Corte del Cedro / P.O. Box 6500 / Carlsbad, CA 92018

Contents

A complete index begins on page 559

Copyright 1998 Craftsman Book Company ISBN 1-57218-060-9 1st printing November 1998 for the year 1999

This Book Is an Encyclopedia of 1999 Building Costs

The 1999 National Construction Estimator lists estimated construction costs to general contractors performing the work with their own crews, as of mid-1999. Overhead & profit are not included.

This Manual Has Two Parts; the Residential Construction Division begins on page 17. Use the figures in this division when estimating the cost of homes and apartments with a wood, steel or masonry frame. The Industrial and Commercial Division begins on page 250 and can be used to estimate costs for nearly all construction not covered by the Residential Division.

The Residential Construction Division is arranged in alphabetical order by construction trade and type of material. The Industrial and Commercial Division follows the 16 section Construction Specification Format. A complete index begins on page 559.

National Estimator (inside the back cover) is an electronic version of this book. Instructions for using this *Windows™* program begin on page 539.

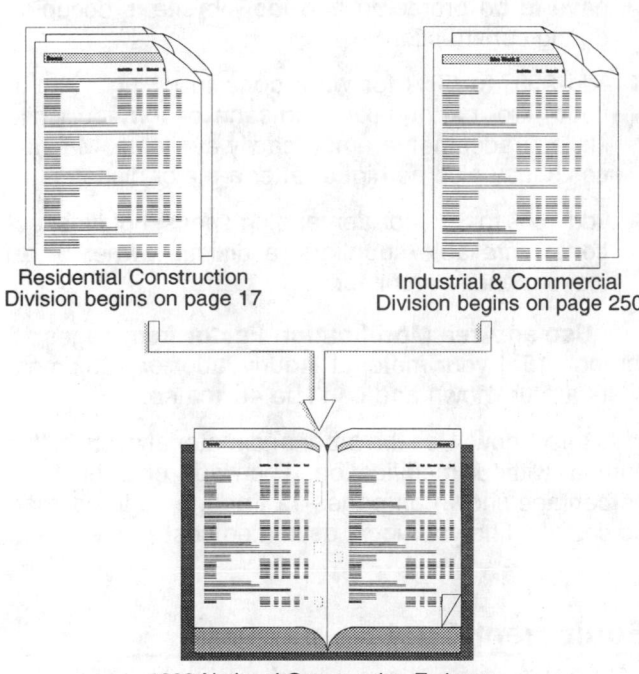

Residential Construction Division begins on page 17

Industrial & Commercial Division begins on page 250

1999 National Construction Estimator

1999 National Estimator is an electronic version of this book

Material Costs

Material Costs for each item are listed in the column headed "Material." These are neither retail nor wholesale prices. They are estimates of what most contractors who buy in moderate volume will pay suppliers as of mid-1999. Discounts may be available for purchases in larger volume.

Add Delivery Expense to the material cost for other than local delivery of reasonably large quantities. Cost of delivery varies with the distance from source of supply, method of transportation, and quantity to be delivered. But most material dealers absorb the delivery cost on local delivery (5 to 15 miles) of larger quantities to good customers. Add the expense of job site delivery when it is a significant part of the material cost.

Add Sales Tax when sales tax will be charged to the contractor buying the materials.

Waste and Coverage loss is included in the installed material cost. The cost of many materials per unit after installation is greater than the purchase price for the same unit because of waste, shrinkage or coverage loss during installation. For example, about 120 square feet of nominal 1" x 4" square edge boards will be needed to cover 100 square feet of floor or wall. There is no coverage loss with plywood sheathing, but waste due to cutting and fitting will average about 6%.

Costs in the "Material" column of this book assume normal waste and coverage loss. Small and irregular jobs may require a greater waste allowance. Materials priced without installation (with no labor cost) do not include an allowance for waste and coverage except as noted.

Labor Costs

Labor Costs for installing the material or doing the work described are listed in the column headed "Labor." The labor cost per unit is the labor cost per hour multiplied by the manhours per unit shown after the @ sign in the "Craft@Hours" column. Labor cost includes the basic wage, the employer's contribution to welfare, pension, vacation and apprentice funds and all tax and insurance charges based on wages. Hourly labor costs for the various crafts are listed on page 10 (for the Residential Division) and page 249 (for the Industrial and Commercial Division).

Hourly labor costs used in the Industrial and Commercial Division are higher than those used in the Residential Division, reflecting the fact that craftsmen on industrial and commercial jobs are often paid more than craftsmen on residential jobs.

Supervision Expense to the general contractor is not included in the labor cost. The cost of supervision and non-productive labor varies widely from job to job. Calculate the cost of supervision and non-productive labor and add this to the estimate.

Payroll Taxes and Insurance included in the labor cost are itemized in the sections beginning on pages 134 and 230.

Manhours per Unit and the Craft performing the work are listed in the "Craft@Hrs" column. Pages 7 through 9 explain the "Craft@Hrs" column. To find the units of work done per man in an 8-hour day, divide 8 by the manhours per unit. To find the units done by a crew in an 8-hour day, multiply the units per man per 8-hour day by the number of crew members.

Manhours Include all productive labor normally associated with installing the materials described. This will usually include tasks such as:

■ Unloading and storing construction materials, tools and equipment on site.

■ Moving tools and equipment from a storage area or truck on site at the beginning of the day.

■ Returning tools and equipment to a storage area or truck on site at the end of the day.

■ Normal time lost for work breaks.

■ Planning and discussing the work to be performed.

■ Normal handling, measuring, cutting and fitting.

■ Keeping a record of the time spent and work done.

■ Regular cleanup of construction debris.

■ Infrequent correction or repairs required because of faulty installation.

Adjust the Labor Cost to the job you are figuring when your actual hourly labor cost is known or can be estimated. The labor costs listed on pages 10 and 249 will apply within a few percent on many jobs. But labor costs may be much higher or much lower on the job you are estimating.

If the hourly wage rates listed on page 10 or page 249 are not accurate, divide your known or estimated cost per hour by the listed cost per hour. The result is your adjustment for any figure in the "Labor" column for that craft. See page 11 for more information on adjusting labor costs.

Adjust for Unusual Labor Productivity. Costs in the labor column are for normal conditions: experienced craftsmen working on reasonably well planned and managed new construction with fair to good productivity. Labor estimates assume that materials are standard grade, appropriate tools are on hand, work done by other crafts is adequate, layout and installation are relatively uncomplicated, and working conditions don't slow progress.

Working conditions at the job site have a major effect on labor cost. Estimating experience and careful analysis can help you predict the effect of most changes in working conditions. Obviously, no single adjustment will apply on all jobs. But the adjustments that follow should help you produce more accurate labor estimates. More than one condition may apply on a job.

■ Add 10% to 15% when working temperatures are below 40 degrees or above 95 degrees.

■ Add 15% to 25% for work on a ladder or a scaffold, in a crawl space, in a congested area or remote from the material storage point.

■ Deduct 10% when the work is in a large open area with excellent access and good light.

■ Add 1% for each 10 feet materials must be lifted above ground level.

■ Add 5% to 50% for tradesmen with below average skills. Deduct 5% to 25% for highly motivated, highly skilled tradesmen.

■ Deduct 10% to 20% when an identical task is repeated many times for several days at the same site.

■ Add 30% to 50% on small jobs where fitting and matching of materials is required, adjacent surfaces have to be protected and the job site is occupied during construction.

■ Add 25% to 50% for work done following a major flood, fire, earthquake, hurricane or tornado while skilled tradesmen are not readily available. Material costs may also be higher after a major disaster.

■ Add 10% to 35% for demanding specs, rigid inspections, unreliable suppliers, a difficult owner or an inexperienced architect.

Use an Area Modification Factor from pages 12 through 15 if your material, hourly labor or equipment costs are unknown and can't be estimated.

Here's how: Use the labor and material costs in this manual without modification. Then add or deduct the percentage shown on pages 12 through 15 to estimated costs to find your local estimated cost.

Equipment Costs

Equipment Costs for major equipment (such as cranes and tractors) are listed in the column headed "Equipment." Costs for small tools and expendable supplies (such as saws and tape) are usually considered overhead expense and do not appear in the Equipment cost column.

Equipment costs are based on rental rates listed in the section beginning on page 260 and assume that the equipment can be used productively for an entire 8-

hour day. Add the cost of moving equipment on and off the site. Allow for unproductive time when equipment can't be used for the full rental period. For example, the equipment costs per unit of work completed will be higher when a tractor is used for 4 hours during a day and sits idle for the remaining 4 hours. Generally, an 8-hour day is the minimum rental period for most heavy equipment. Many sections describe the equipment being used, the cost per hour and a suggested minimum job charge.

Subcontracted Work

Subcontractors do most of the work on construction projects. That's because specialty contractors can often get the work done at competitive cost, even after adding overhead and profit.

Many sections of this book cover work usually done by subcontractors. If you see the word "subcontract" in a section description, assume that costs are based on quotes by subcontractors and include typical subcontractor markup (about 30% on labor and 15% on material). Usually no material or labor costs will appear in these sections. The only costs shown will be in the "Total" column and will include all material, labor and equipment expense.

If you don't see the word "subcontract" in a section description, assume that costs are based on work done by a general contractor's crew. No markup is included in these costs. If the work is done by a subcontractor, the specialty contractor may be able to perform the work for the cost shown, even after adding overhead and profit.

Markup

The General Contractor's Markup is not included in any costs in this book. On page 154 we suggest a 20% markup on the contract price for general contractors handling residential construction. Apply this markup or some figure you select to all costs, including both subcontract items and work done by your own crews.

To realize a markup of 20% on the contract price, you'll have to add 25% to your costs. See page 154 for an example of how markup is calculated. Markup includes overhead and profit and may be the most difficult item to estimate.

Keep In Mind

Labor and Material Costs Change. These costs were compiled in the fall of 1998 and projected to mid-1999 by adding 2% to 4%. This estimate will be accurate for some materials but inaccurate for others. No one can predict material price changes accurately.

How Accurate Are These Figures? As accurate as possible considering that the estimators who wrote this book don't know your subcontractors or material suppliers, haven't seen the plans or specifications, don't know what building code applies or where the job is, had to project material costs at least 6 months into the future, and had no record of how much work the crew that will be assigned to the job can handle.

You wouldn't bid a job under those conditions. And we don't claim that all construction is done at these prices.

Estimating Is an Art, not a science. On many jobs the range between high and low bid will be 20% or more. There's room for legitimate disagreement on what the correct costs are, even when complete plans and specifications are available, the date and site are established, and labor and material costs are identical for all bidders.

No cost fits all jobs. Good estimates are custom made for a particular project and a single contractor through judgment, analysis and experience.

This book is not a substitute for judgment, analysis and sound estimating practice. It's an aid in developing an informed opinion of cost. If you're using this book as your sole cost authority for contract bids, you're reading more into these pages than the editors intend.

Use These Figures to compile preliminary estimates, to check your costs and subcontract bids and when no actual costs are available. This book will reduce the chance of error or omission on bid estimates, speed "ball park" estimates, and be a good guide when there's no time to get a quote.

Where Do We Get These Figures? From the same sources all professional estimators use: contractors and subcontractors, architectural and engineering firms, material suppliers, material price services, analysis of plans, specifications, estimates and completed project costs, and both published and unpublished cost studies. In addition, we conduct nationwide mail and phone surveys and have the use of several major national estimating databases.

 **We'll Answer Your Questions** about any part of this book and explain how to apply these costs. Free telephone assistance is available from 8 a.m. until 5 p.m. California time Monday through Friday except holidays. Phone 760-438-7828. We don't accept collect calls and won't estimate the job for you. But if you need clarification on something in this manual, we can help.

Abbreviations

AASHO	American Assn. of State Highway Officials	FICA	Federal Insurance Contributions Act (Social Security, Medicare tax)	oz	ounce
ABS	acrylonitrile butadiene styrene			perf	perforated
				Pr	pair
AC	alternating current	FOB	freight on board	PSF	pounds per square foot
AISC	American Institute of Steel Construction Inc.	FPM	feet per minute	PSI	pounds per square inch
		FRP	fiberglass reinforced plastic	PV	photovoltaic
APP	attactic polypropylene			PVC	polyvinyl chloride
ASHRAE	American Society of Heating, Refrigerating and Air Conditioning Engineers	FS	Federal Specification	Qt	quart
		FUTA	Federal Unemployment Compensation Act Tax	R	thermal resistance
		Gal	gallon	R/L	random length(s)
ASME	American Society of Mechanical Engineers	GFCI	ground fault circuit interruptor	R/W/L	random widths and lengths
ASTM	American Society for Testing Materials	GPH	gallon(s) per hour	RPM	revolutions per minute
		GPM	gallon(s) per minute	RSC	rigid steel conduit
AWPA	American Wood Products Association	H	height	S1S2E	surfaced 1 side, 2 edges
		HP	horsepower	S2S	surfaced 2 sides
AWWA	American Water Works Association	Hr(s)	hour(s)	S4S	surfaced 4 sides
		IMC	intermediate metal conduit	Sa	sack
Ba	bay	ID	Inside diameter	SBS	styrene butyl styrene
Bdle	bundle	KD	kiln dried or knocked down	SDR	size to diameter ratio
BF	board foot			SF	square foot
BHP	boiler horsepower	KSI	kips per square inch	SFCA	square feet of form in contact with concrete
Btr	better	KV	kilovolt(s)		
Btu	British thermal unit	KVA	1,000 volt amps	Sq	100 square feet
B & W	black & white	kw	kilowatt(s)	SSB	single strength B quality glass
C	thermal conductance	kwh	kilowatt hour		
C	one hundred	L	length	STC	sound transmission class
CF	cubic foot	Lb(s)	pound(s)	Std	standard
CFM	cubic feet per minute	LF	linear foot	SY	square yard
CLF	100 linear feet	LP	liquified propane	T	thick
cm	centimeter	LS	lump sum	T&G	tongue & groove edge
CPE	chlorinated polyethylene	M	one thousand	TV	television
CPM	cycles per minute	Mb	million bytes (characters)	UBC	Uniform Building Code
CPVC	chlorinated polyvinyl chloride	MBF	1,000 board feet	UL	Underwriter's Laboratory
		MBtu	1,000 British thermal units	USDA	United States Dept. of Agriculture
CSPE	chloro sulphinated polyethylene	MCM	1,000 circular mils		
		MDO	medium density overlaid	VLF	vertical linear foot
CSF	100 square feet	MH	manhour	W	width
CY	cubic yard	Mi	mile	Wk	week
d	penny	MLF	1,000 linear feet	W/	with
D	depth	MPH	miles per hour	x	by or times
DC	direct current	mm	millimeter(s)		
dia	diameter	Mo	month		
DSB	double strength B quality glass	MSF	1,000 square feet		
DWV	drain, waste, vent piping	NEMA	National Electrical Manufacturer's Association		
Ea	each				
EMT	electric metallic tube	NFPA	National Fire Protection Association		
EPDM	ethylene propylene diene monomer	No.	number		
equip.	equipment	NRC	noise reduction coefficient		
exp.	exposure	OC	spacing from center to center		
F	Fahrenheit	OD	outside diameter		
FAA	Federal Aviation Administration	OS & Y	outside screw & yoke		

Symbols

/	per
—	through or to
@	at
%	per 100 or percent
$	U.S. dollars
'	feet
"	inches
#	pound or number

Craft Codes, Hourly Costs and Crew Compositions

Both the Residential Division and Commercial and Industrial Division of this book include a column titled *Craft@Hrs*. Letters and numbers in this column show our estimates of:

- Who will do the work (the craft code)
- An @ symbol which means at
- How long the work will take (manhours).

For example, on page 45 you'll find estimates for installing BD plywood wall sheathing by the square foot. The *Craft@Hrs* column opposite 1/2" plywood wall sheathing shows:

B1@.016

That means we estimate the installation rate for crew B1 at .016 manhours per square foot. That's the same as 16 manhours per 1,000 square feet.

The table that follows defines each of the craft codes used in this book. Notice that crew B1 is composed of two craftsmen: one laborer and one carpenter. To install 1,000 square feet of 1/2" BD wall sheathing at .016 manhours per square foot, that crew would need 16 manhours (one 8-hour day for a crew of two).

Notice also in the table below that the cost per manhour for crew B1 is listed as $22.85. That's the average for a residential laborer (listed at $20.54 per hour on page 10) and a residential carpenter (listed at $25.17 per hour): $20.54 plus $25.17 is $45.71. Divide by 2 to get $22.85, the average cost per manhour for crew B1.

In the table below, the cost per manhour is the sum of hourly costs of all crew members divided by the number of crew members. That's the average cost per manhour.

Costs in the *Labor* column in this book are the installation time (in manhours) multiplied by the cost per manhour. For example, on page 45 the labor cost listed for 1/2" BD wall sheathing is $.37 per square foot. That's the installation time (.016 manhours per square foot) multiplied by $22.85, the average cost per manhour for crew B1.

Residential Division

Craft Code	Cost Per Manhour	Crew Composition	Craft Code	Cost Per Manhour	Crew Composition
B1	$22.85	1 laborer and 1 carpenter	BR	$25.11	1 lather
B2	$23.62	1 laborer, 2 carpenters	BS	$22.21	1 marble setter
B3	$22.08	2 laborers, 1 carpenter	CF	$24.57	1 cement mason
B4	$24.14	1 laborer 1 operating engineer 1 reinforcing iron worker	CT	$23.57	1 mosaic & terrazzo worker
			D1	$24.99	1 drywall installer 1 drywall taper
B5	$24.43	1 laborer, 1 carpenter 1 cement mason 1 operating engineer 1 reinforcing iron worker	DI	$24.99	1 drywall installer
			DT	$24.99	1 drywall taper
			HC	$19.46	1 plasterer helper
B6	$22.55	1 laborer, 1 cement mason	OE	$28.14	1 operating engineer
B7	$21.07	1 laborer, 1 truck driver	P1	$24.44	1 laborer, 1 plumber
B8	$24.34	1 laborer 1 operating engineer	PM	$28.34	1 plumber
			PP	$23.26	1 painter, 1 laborer
B9	$23.17	1 bricklayer 1 bricklayer's helper	PR	$25.51	1 plasterer
			PT	$25.98	1 painter
BB	$26.23	1 bricklayer	R1	$24.57	1 roofer, 1 laborer
BC	$25.17	1 carpenter	RI	$23.74	1 reinforcing iron worker
BE	$26.87	1 electrician	RR	$28.61	1 roofer
BF	$24.58	1 floor layer	SW	$27.76	1 sheet metal worker
BG	$23.06	1 glazier	T1	$22.53	1 tile layer, 1 laborer
BH	$20.10	1 bricklayer's helper	TL	$24.53	1 tile layer
BL	$20.54	1 laborer	TR	$21.60	1 truck driver

7

Commercial and Industrial Division

Craft Code	Cost Per Manhour	Crew Composition
A1	$36.65	1 asbestos worker 1 laborer
AT	$30.37	1 air tool operator
AW	$43.04	1 asbestos worker
BM	$41.51	1 boilermaker
BT	$29.23	1 bricklayer tender
C1	$30.54	4 laborers, 1 truck driver
C2	$34.32	1 laborer, 2 truck drivers 2 tractor operators
C3	$33.64	1 laborer, 1 truck driver 1 tractor operator
C4	$30.71	2 laborers, 1 truck driver
C5	$32.80	2 laborers, 1 truck driver 1 tractor operator
C6	$32.29	6 laborers, 2 truck drivers 2 tractor operators
C7	$33.15	2 laborers , 3 truck drivers 1 crane operator 1 tractor operator
C8	$36.27	1 laborer, 1 carpenter
C9	$33.95	1 laborer, 1 crane operator
CB	$37.98	1 bricklayer
CC	$42.27	1 carpenter
CD	$36.76	1 drywall Installer
CE	$40.07	1 electrician
CG	$36.43	1 glazier
CL	$30.27	1 laborer
CM	$37.51	1 cement mason
CO	$37.62	1 crane operator
CV	$37.40	1 elevator constructor
D2	$33.52	1 drywall installer 1 laborer
D3	$40.91	1 laborer, 1 iron worker (structural), 1 millwright
D4	$36.94	1 laborer, 1 millwright
D5	$35.89	1 boilermaker, 1 laborer
D6	$42.09	2 millwrights 1 tractor operator
D7	$33.64	1 painter, 1 laborer
D9	$39.17	2 millwrights, 1 laborer
E1	$35.94	2 electricians, 2 laborers 1 tractor operator
E2	$35.17	2 electricians, 2 laborers
E3	$37.53	2 electricians, 2 laborers 2 carpenters
E4	$35.17	1 electrician, 1 laborer
F5	$37.47	3 carpenters, 2 laborers

Craft Code	Cost Per Manhour	Crew Composition
F6	$36.83	2 carpenters, 2 laborers 1 tractor operator
F7	$38.46	2 carpenters, 1 laborer 1 tractor operator
F8	$36.16	2 plasterers 1 plasterer's helper
F9	$33.43	1 laborer, 1 floor layer
FL	$36.60	1 floor layer
G1	$33.35	1 glazier, 1 laborer
H1	$40.11	1 carpenter, 1 laborer 1 iron worker (structural) 1 tractor operator
H2	$34.62	1 crane operator 1 truck driver
H3	$33.72	1 carpenter, 3 laborers 1 crane operator 1 truck driver
H4	$45.28	1 crane operator 6 iron workers (structural) 1 truck driver
H5	$39.17	1 crane operator 2 iron workers (structural) 2 laborers
H6	$39.55	1 iron worker (structural) 1 laborer
H7	$45.10	1 crane operator 2 iron workers (structural)
H8	$44.10	1 crane operator 4 iron workers (structural) 1 truck driver
H9	$40.69	1 electrician 1 sheet metal worker
IW	$48.84	1 iron worker (structural)
LA	$33.23	1 lather
M1	$33.60	1 bricklayer 1 bricklayer's tender
M2	$34.27	1 carpenter, 2 laborers
M3	$34.98	1 plasterer 1 plasterer's helper
M4	$33.09	1 laborer, 1 marble setter
M5	$36.52	1 pipefitter, 1 laborer,
M6	$38.69	1 asbestos worker 1 laborer, 1 pipefitter
M8	$39.65	3 pipefitters, 1 laborer
M9	$41.42	1 electrician, 1 pipefitter
MI	$38.61	2 pipefitters, 1 laborer
MS	$35.92	marble setter
MT	$33.84	mosaic & terrazzo worker
MW	$43.62	millwright

Commercial and Industrial Division

Craft Code	Cost Per Manhour	Crew Composition	Craft Code	Cost Per Manhour	Crew Composition
P5	$32.29	3 laborers 1 tractor operator 1 truck driver	S8	$36.12	2 pile drivers, 2 laborers 1 truck driver 1 crane operator 1 tractor operator
P6	$36.11	1 laborer, 1 plumber			
P8	$33.89	1 laborer, 1 cement mason	S9	$34.64	1 pile driver, 2 laborers 1 tractor operator 1 truck driver
P9	$36.68	1 carpenter, 1 laborer 1 cement mason			
PA	$37.02	1 painter	SM	$41.31	1 sheet metal worker
PD	$42.01	1 pile driver	SP	$44.38	1 sprinkler fitter
PF	$42.78	1 pipefitter	SS	$36.12	1 laborer 2 tractor operators
PH	$31.46	1 plasterer's helper			
PL	$41.95	1 plumber	T2	$39.40	3 laborers, 3 carpenters 3 iron workers (structural) 1 crane operator 1 truck driver
PS	$38.50	1 plasterer			
R3	$35.38	2 roofers, 1 laborer			
RB	$38.92	1 reinforcing iron worker	T3	$34.59	1 laborer 1 reinforcing iron worker
RF	$37.94	1 roofer			
S1	$34.66	1 laborer 1 tractor operator	T4	$32.05	1 laborer, 1 mosaic worker
S3	$35.33	1 truck driver 1 tractor operator	T5	$35.79	1 sheet metal worker 1 laborer
S4	$30.27	3 laborers	T6	$37.63	2 sheet metal workers 1 laborer
S5	$31.51	5 laborers 1 crane operator 1 truck driver	TD	$31.61	1 truck driver
			TO	$39.05	1 tractor operator
S6	$33.20	2 laborers 1 tractor operator	U1	$35.38	1 plumber, 2 laborers 1 tractor operator
S7	$34.66	3 laborers 3 tractor operators	U2	$34.16	1 plumber, 2 laborers

Residential Division

Craft	1 Base wage per hour	2 Taxable fringe benefits (@5.15% of base wage)	3 Insurance and employer taxes (%)	4 Insurance and employer taxes ($)	5 Non-taxable fringe benefits (@4.55% of base wage)	6 Total hourly cost used in this book
Bricklayer	$18.34	$0.94	31.72%	$6.12	$0.83	$26.23
Bricklayer's Helper	14.05	0.72	31.72%	4.69	0.64	20.10
Building Laborer	14.22	0.73	33.05%	4.94	0.65	20.54
Carpenter	17.28	0.89	34.14%	6.21	0.79	25.17
Cement Mason	17.64	0.91	28.12%	5.22	0.80	24.57
Drywall installer	17.91	0.92	28.32%	5.33	0.82	24.98
Drywall Taper	17.91	0.92	28.32%	5.33	0.82	24.98
Electrician	20.25	1.04	21.88%	4.66	0.92	26.87
Floor Layer	17.08	0.88	32.55%	5.84	0.78	24.58
Glazier	16.44	0.85	29.03%	5.02	0.75	23.06
Lather	18.31	0.94	26.12%	5.03	0.83	25.11
Marble Setter	16.39	0.84	24.56%	4.23	0.75	22.21
Millwright	17.83	0.92	31.95%	5.99	0.81	25.55
Mosiac & Terrazzo Worker	17.39	0.90	24.56%	4.49	0.79	23.57
Operating Engineer	20.52	1.06	26.10%	5.63	0.93	28.14
Painter	18.28	0.94	30.84%	5.93	0.83	25.98
Plasterer	18.20	0.94	28.94%	5.54	0.83	25.51
Plasterer Helper	13.88	0.72	28.94%	4.23	0.63	19.46
Plumber	20.67	1.06	26.08%	5.67	0.94	28.34
Reinforcing Ironworker	17.18	0.88	27.15%	4.90	0.78	23.74
Roofer	17.83	0.92	48.29%	9.05	0.81	28.61
Sheet Metal Worker	19.97	1.03	27.88%	5.85	0.91	27.76
Sprinkler Fitter	20.32	1.05	26.99%	5.77	0.92	28.06
Tile Layer	18.11	0.93	24.56%	4.67	0.82	24.53
Truck Driver	14.94	0.77	33.15%	5.21	0.68	21.60

Hourly Labor Cost

The labor costs shown in Column 6 were used to compute the manhour costs for crews on pages 7 to 9 and the figures in the "Labor" column of the Residential Division of this manual. Figures in the "Labor" column of the Industrial and Commercial Division of this book were computed using the hourly costs shown on page 249. All labor costs are in U.S. dollars per manhour.

It's important that you understand what's included in the figures in each of the six columns above. Here's an explanation:

Column 1, the base wage per hour, is the craftsman's hourly wage. These figures are representative of what many contractors will be paying craftsmen working on residential construction in 1999.

Column 2, taxable fringe benefits, includes vacation pay, sick leave and other taxable benefits. These fringe benefits average 5.15% of the base wage for many construction contractors. This benefit is in addition to the base wage.

Column 3, insurance and employer taxes in percent, shows the insurance and tax rate for construction trades. The cost of insurance in this column includes workers' compensation and contractor's casualty and liability coverage. Insurance rates vary widely from state to state and depend on a contractor's loss experience. Typical rates are shown in the Insurance section of this manual beginning on page 134. Taxes are itemized in the section on page 230. Note that taxes and insurance increase the hourly labor cost by 30 to 35% for most trades. There is no legal way to avoid these costs.

Column 4, insurance and employer taxes in dollars, shows the hourly cost of taxes and insurance for each construction trade. Insurance and taxes are paid on the costs in both columns 1 and 2.

Column 5, non-taxable fringe benefits, includes employer paid non-taxable benefits such as medical coverage and tax-deferred pension and profit sharing plans. These fringe benefits average 4.55% of the base wage for many construction contractors. The employer pays no taxes or insurance on these benefits.

Column 6, the total hourly cost in dollars, is the sum of columns 1, 2, 4, and 5.

These hourly labor costs will apply within a few percent on many jobs. But wage rates may be much higher or lower in some areas. If the hourly costs shown in column 6 are not accurate for your work, develop modification factors that you can apply to the labor costs in this book. The following paragraphs explain the procedure.

Adjusting Labor Costs

Here's how to customize the labor costs in this book if your wage rates are different from the wage rates shown on page 10 or 249.

Start with the taxable benefits you offer. Assume craftsmen on your payroll get one week of vacation each year and one week of sick leave each year. Convert these benefits into hours. Your computation might look like this:

$$\begin{array}{r} 40 \text{ vacation hours} \\ + \ 40 \text{ sick leave hours} \\ \hline 80 \text{ taxable leave hours} \end{array}$$

Then add the regular work hours for the year:

$$\begin{array}{r} 2{,}000 \text{ regular hours} \\ + \ 80 \text{ taxable benefit hours} \\ \hline 2{,}080 \text{ total hours} \end{array}$$

Multiply these hours by the base wage per hour. If you pay carpenters $8.00 per hour, the calculation would be:

$$\begin{array}{r} 2{,}080 \text{ hours} \\ \times \ \$8.00 \text{ per hour} \\ \hline \$16{,}640 \text{ per year} \end{array}$$

Next determine the tax and insurance rate for each trade. If you know the rates that apply to your jobs, use those rates. If not, use the rates in column 3 on page

10. Continuing with our example, we'll use 34.14%, the rate for carpenters in column 3 on page 10. To increase the annual taxable wage by 34.14%, we'll multiply by 1.3414:

$$\begin{array}{r} \$16{,}640 \text{ per year} \\ \times \ 1.3414 \text{ tax \& insurance rate} \\ \hline \$22{,}321 \text{ annual cost} \end{array}$$

Then add the cost of non-taxable benefits. Suppose your company has no pension or profit sharing plan but does provide medical insurance for employees. Assume that the cost for your carpenter is $343.67 per month or $4,124 per year.

$$\begin{array}{r} \$4{,}124 \text{ medical plan} \\ + \ 22{,}321 \text{ annual cost} \\ \hline \$26{,}445 \text{ total annual cost} \end{array}$$

Divide this total annual cost by the actual hours worked in a year. This gives the contractor's total hourly labor cost including all benefits, taxes and insurance. Assume your carpenter will work 2,000 hours a year:

$$\frac{\$26{,}445}{2{,}000} = \$13.22 \text{ per hour}$$

Finally, find your modification factor for the labor costs in this book. Divide your total hourly labor cost by the total hourly cost shown on page 10. For the carpenter in our example, the figure in column 6 is $25.17.

$$\frac{\$13.22}{\$25.17} = .53$$

Your modification factor is 53%. Multiply any building carpenter (Craft Code BC) labor costs in the Residential Division of this book by .53 to find your estimated cost. For example, on page 23 the labor cost for installing an 18" long towel bar is $7.05 per each bar. If installed by your carpenter working at $8.00 per hour, your estimated cost would be 53% of $7.05 or $3.74. The manhours would remain the same @.280, assuming normal productivity.

If the Labor Rate Is Unknown

On some estimates you may not know what labor rates will apply. In that case, use both labor and material figures in this book without making any adjustment. When all labor, equipment and material costs have been compiled, add or deduct the percentage shown in the area modification table on pages 12 through 15.

Adjusting the labor costs in this book will make your estimates much more accurate.

Area Modification Factors

Construction costs are higher in some cities than in other cities. Add or deduct the percentage shown on pages 12 to 15 to adapt the costs in this book to your job site. Adjust your estimated total project cost by the percentage shown for the appropriate city in this table to find your total estimated cost. Where 0% is shown it means no modification is required.

These percentages were compiled by comparing the construction cost of buildings in nearly 700 communities throughout the U.S. Because these percentages are based on completed projects, they consider all construction cost variables, including labor, equip-ment and material cost, labor productivity, climate, job conditions and markup.

Modification factors are listed alphabetically by state. Areas within each state are listed by the first three digits of the postal zip code. For convenience, one representative city is identified in each zip code or range of zip codes.

These percentages are composites of many costs and will not necessarily be accurate when estimating the cost of any particular part of a building. But when used to modify costs for an entire structure, they should improve the accuracy of your estimates.

Location		Mat.	Lab.	Equip.	Total Wtd. Avg.
Alabama		-4	-23	-3	-14%
350-352	Birmingham	-4	-21	-3	-13%
354	Tuscaloosa	-4	-21	-3	-13%
355	Jasper	-4	-21	-3	-13%
356	Sheffield	-4	-30	-3	-16%
357	Scottsboro	-4	-30	-3	-16%
358	Huntsville	-5	-18	-3	-13%
359	Gadsden	-5	-18	-3	-13%
360-361	Montgomery	-4	-21	-3	-13%
362	Anniston	-5	-20	-3	-13%
363	Dothan	-4	-30	-3	-16%
364	Evergreen	-4	-30	-3	-16%
365-366	Mobile	-3	-17	-3	-11%
367	Selma	-4	-21	-3	-13%
368	Auburn	-4	-21	-3	-13%
369	Bellamy	-4	-21	-3	-13%
Alaska		41	46	18	43%
995	Anchorage	31	28	18	30%
996	King Salmon	50	53	18	51%
997	Fairbanks	46	43	18	45%
998	Juneau	38	54	18	44%
999	Ketchikan	40	54	18	46%
Arizona		1	-6	5	1%
850	Phoenix	1	-12	5	-1%
852-853	Mesa	1	-12	5	-1%
855	Douglas	1	-12	5	-1%
856-857	Tucson	2	-12	3	-2%
859	Show Low	2	1	5	4%
860	Flagstaff	2	-12	5	-1%
863	Prescott	1	1	5	3%
864	Kingman	1	1	5	3%
865	Chambers	1	1	5	3%
Arkansas		-3	-29	-3	-12%
716	Pine Bluff	-2	-29	-3	-11%
717	Camden	-4	-29	-3	-12%
718	Hope	-3	-29	-3	-12%
719	Hot Springs	-4	-29	-3	-12%
720-722	Little Rock	-3	-29	-3	-12%
723	West Memphis	-3	-29	-3	-12%
724	Jonesboro	-3	-29	-3	-12%
725	Batesville	-4	-29	-3	-12%
726	Harrison	-3	-29	-3	-12%
727	Fayetteville	-2	-29	-3	-11%
728	Russellville	-1	-25	-3	-9%
729	Fort Smith	-1	-25	-3	-9%

Location		Mat.	Lab.	Equip.	Total Wtd. Avg.
California		5	35	5	14%
900-901	Los Angeles	7	32	5	14%
902-905	Inglewood	7	28	5	13%
906	Whittier	7	28	5	13%
907-908	Long Beach	8	30	5	14%
910-912	Pasadena	7	28	5	13%
913-916	Van Nuys	7	28	5	13%
917-918	Alhambra	7	28	5	13%
919-921	San Diego	2	28	5	10%
922	El Centro	1	29	5	10%
923-925	San Bernardino	7	28	5	13%
926-928	Anaheim	7	28	5	13%
930	Oxnard	2	31	5	11%
931	Santa Barbara	2	31	5	11%
932-933	Bakersfield	7	28	5	13%
934	Lompoc	5	28	5	11%
935	Mojave	1	29	5	10%
936-938	Fresno	7	28	5	13%
939	Salinas	2	39	5	14%
940	Sunnyvale	5	42	5	17%
941	San Francisco	8	56	5	24%
942	Sacramento	-1	33	5	10%
943-944	San Mateo	4	42	5	16%
945-947	Oakland	4	42	5	16%
948	Richmond	8	51	5	22%
949	Novato	8	51	5	22%
950-951	San Jose	8	36	5	17%
952-953	Stockton	-1	37	5	11%
954	Santa Rosa	8	51	5	22%
955	Eureka	8	36	5	16%
956-958	R. Cordova	-1	37	5	11%
959	Marysville	-1	35	5	11%
960	Redding	5	30	5	12%
961	Herlong	1	51	5	18%
Colorado		1	-9	-2	-1%
800-801	Aurora	0	-13	-3	-3%
802-804	Denver	0	-3	-3	0%
805	Longmont	0	-13	-3	-3%
806	Greeley	0	-13	-3	-3%
807	Fort Morgan	0	-13	-3	-3%
808-809	Colorado Sp.	3	-7	-3	1%
810	Pueblo	-1	-14	-3	-4%
811	Pagosa Springs	1	-13	-3	-3%
812	Salida	1	-13	-3	-3%
813	Durango	1	-13	-3	-3%
814-815	Grand Junction	0	-13	-3	-3%
816	Glenwood Sp.	1	-13	-3	-3%
	Ski resorts	8	18	5	13%

Location		Mat.	Lab.	Equip.	Total Wtd. Avg.
Connecticut		4	24	4	9%
060-061	Hartford	3	23	4	9%
062	West Hartford	4	23	4	9%
063	Norwich	3	20	4	8%
064	Fairfield	3	25	4	10%
065	New Haven	3	25	4	10%
066	Bridgeport	3	20	4	8%
067	Waterbury	4	26	4	11%
068-069	Stamford	4	26	4	11%
Delaware		-2	15	0	3%
197	Newark	-3	14	0	3%
198	Wilmington	-1	16	0	4%
199	Dover	-3	14	0	3%
District of Columbia		-1	7	0	2%
200-205	Washington	-1	7	0	2%
Florida		-3	-18	-3	-8%
320	St. Augustine	-4	-22	-3	-10%
321	Daytona Bch.	-4	-22	-3	-10%
322	Jacksonville	-2	-6	-3	-3%
323	Tallahassee	-3	-34	-3	-14%
324	Panama City	-1	-34	-3	-13%
325	Pensacola	1	-38	-3	-13%
326	Gainesville	-4	-22	-3	-10%
327	Altamonte Sp.	-4	2	-3	-1%
328	Orlando	-4	2	-3	-1%
329	Melbourne	-4	2	-3	-1%
330-332	Miami	-6	-9	-3	-6%
333	Ft. Lauderdale	2	-9	-3	-1%
334	W. Palm Bch.	2	-9	-3	-1%
335-336	Tampa	-4	-26	-3	-11%
337	St Petersburg	-3	-26	-3	-11%
338	Lakeland	-4	-22	-3	-10%
339	Fort Myers	-5	-22	-3	-10%
342	Bradenton	-4	-22	-3	-10%
344	Ocala	-4	-22	-3	-10%
346	Brooksville	-4	-22	-3	-10%
347	Saint Cloud	-4	-22	-3	-10%
349	Fort Pierce	-5	-22	-3	-10%
Georgia		-1	-38	-3	-11%
300-302	Marietta	1	-15	-3	-1%
303	Atlanta	1	-15	-3	-1%
304	Statesboro	-1	-39	-3	-11%
305	Buford	-1	-39	-3	-11%
306	Athens	-1	-39	-3	-11%

Area Modification Factors

Location		Mat.	Lab.	Equip.	Total Wtd. Avg.
307	Calhoun	-1	-39	-3	-11%
308-309	Augusta	-6	-46	-3	-17%
310	Dublin	-1	-39	-3	-11%
311	Fort Valley	-1	-39	-3	-11%
312	Macon	-1	-39	-3	-11%
313	Hinesville	-2	-44	-3	-13%
314	Savannah	-2	-44	-3	-13%
315	Kings Bay	2	-45	-3	-12%
316	Valdosta	-1	-39	-3	-11%
317	Albany	-2	-44	-3	-13%
318-319	Columbus	-2	-49	-3	-16%
Hawaii		**26**	**52**	**5**	**34%**
967	Ewa	25	52	5	34%
967	Halawa Hts	25	52	5	34%
967	Hilo	25	52	5	34%
967	Lualualei	25	52	5	34%
967	Mililani Town	26	51	5	34%
967	Pearl City	26	51	5	34%
967	Wahiawa	25	52	5	34%
967	Waianae	25	52	5	34%
967	Wailuku (Maui)	25	52	5	34%
968	Alimanu	25	52	5	34%
068	Honolulu	26	51	6	34%
968	Kailua	27	52	5	35%
Idaho		**0**	**-11**	**1**	**-6%**
832	Pocatello	-1	-10	1	-6%
833	Sun Valley	2	-10	0	-5%
834	Idaho Falls	-1	-10	0	-6%
835	Lewiston	0	-16	1	-8%
836	Meridian	0	-16	1	-8%
837	Boise	0	-11	1	-6%
838	Coeur d'Alene	3	-6	1	-2%
Illinois		**-4**	**23**	**-1**	**6%**
600	Arlington Hts.	-3	22	-1	6%
601	Carol Stream	-4	34	0	10%
602	Quincy	-3	22	-1	6%
603	Oak Park	-3	22	-1	6%
604	Joliet	-3	22	-1	6%
605	Aurora	-3	22	-1	6%
606-608	Chicago	-3	34	0	10%
609	Kankakee	-4	22	-1	5%
610-611	Rockford	-4	14	-3	2%
612	Green River	-4	22	-1	5%
613	Peru	-4	22	-1	6%
614	Galesburg	-4	22	-1	5%
615-616	Peoria	-3	22	-1	6%
617	Bloomington	-4	22	-1	5%
618-619	Urbana	-3	22	-1	6%
620	Granite City	-4	22	-1	5%
622	Belleville	-4	17	-3	3%
623	Decatur	-4	22	-1	5%
624	Lawrenceville	-4	22	-1	6%
625-627	Springfield	-4	22	-1	6%
628	Centralia	-4	22	-1	5%
629	Carbondale	-4	22	-1	5%
Indiana		**-2**	**6**	**0**	**3%**
460-462	Indianapolis	-4	10	0	3%
463-464	Gary	-1	6	0	4%
465-466	South Bend	-1	6	0	4%
467-468	Fort Wayne	-1	6	0	4%
469	Kokomo	-1	6	0	3%
470	Aurora	-2	6	0	3%
471	Jeffersonville	-1	6	0	3%
472	Columbus	-1	6	0	4%
473	Muncie	-1	6	0	4%
474	Bloomington	0	6	0	4%
475	Jasper	-3	3	0	1%
476-477	Evansville	-3	3	0	1%
478	Terre Haute	-1	6	0	4%
479	Lafayette	-1	6	0	4%
Iowa		**1**	**-14**	**-3**	**-4%**
500-503	Des Moines	3	-5	-3	1%
504	Mason City	0	-14	-3	-4%
505	Fort Dodge	0	-14	-3	-4%
506-507	Waterloo	1	-14	-3	-4%
508	Creston	1	-14	-3	-4%
510	Cherokee	1	-14	-3	-4%
511	Sioux City	1	-14	-3	-4%
512	Sheldon	0	-14	-3	-4%
513	Spencer	1	-14	-3	-4%
514	Carroll	0	-14	-3	-5%
515	Council Bluffs	1	-14	-3	-4%
516	Shenandoah	0	-14	-3	-5%
520	Dubuque	1	-14	-3	-4%
522-524	Cedar Rapids	1	-14	-3	-4%
521	Decorah	1	-14	-3	-4%
525	Ottumwa	1	-14	-3	-4%
526	Burlington	0	-14	-3	-5%
527-528	Davenport	1	-9	-3	-2%
Kansas		**0**	**-16**	**-3**	**-5%**
660-662	Kansas City	0	-16	-3	-5%
664-666	Topeka	0	-16	-3	-5%
667	Fort Scott	0	-16	-3	-5%
668	Emporia	-1	-16	-3	-6%
669	Concordia	0	-16	-3	-6%
670-672	Wichita	3	-16	-3	-4%
673	Independence	0	-16	-3	-5%
674	Salina	0	-16	-3	-5%
675	Hutchinson	-1	-16	-3	-6%
676	Hays	0	-16	-3	-5%
677	Colby	0	-16	-3	-5%
678	Dodge City	0	-16	-3	-5%
679	Liberal	0	-16	-3	-5%
Kentucky		**-1**	**-11**	**0**	**-4%**
400-402	Louisville	-1	-11	0	-5%
403-406	Lexington	-1	-11	0	-5%
407-409	London	-1	-16	0	-7%
410	Covington	-1	-11	0	-5%
411-412	Ashland	-1	-11	0	-5%
413-414	Campton	-1	-11	0	-5%
415-416	Pikeville	-1	-11	0	-5%
417-418	Hazard	-1	-11	0	-5%
420	Paducah	-1	-11	0	-5%
421	Bowling Green	-1	-11	0	-5%
422	Hopkinsville	4	-16	0	-4%
423	Owensboro	-1	-11	0	-5%
424	White Plains	-2	-5	0	-3%
425-426	Somerset	-2	-5	0	-3%
427	Elizabethtown	-2	-5	0	-3%
Louisiana		**0**	**-17**	**-3**	**-4%**
700-701	New Orleans	5	-10	-3	1%
703	Houma	5	-10	-3	1%
704	Mandeville	-2	-18	-3	-6%
705	Lafayette	-1	-18	-3	-5%
706	Lake Charles	-1	-18	-3	-5%
707-708	Baton Rouge	-1	-18	-3	-5%
710	Minden	-3	-23	-3	-8%
711	Shreveport	-2	-18	-3	-6%
712	Monroe	-2	-18	-3	-6%
713-714	Alexandria	-2	-18	-3	-6%
Maine		**2**	**-16**	**4**	**-4%**
039-040	Brunswick	4	-13	4	-2%
041	Portland	3	-13	4	-2%
042	Auburn	2	-16	4	-4%
043	Augusta	2	-16	4	-4%
044	Bangor	1	-19	4	-6%
045	Bath	2	-16	4	-4%
046	Cutler	2	-16	4	-4%
047	Northern Area	2	-16	4	-4%
048	Camden	1	-16	4	-5%
049	Dexter	2	-16	4	-4%
Maryland		**-4**	**-1**	**0**	**0%**
206-207	Laurel	-1	-3	0	2%
208-209	Bethesda	-4	-1	0	0%
210-212	Baltimore	-4	-1	0	0%
214	Annapolis	-4	0	0	1%
215	Cumberland	-5	-1	0	0%
216	Church Hill	-5	-1	0	0%
217	Frederick	-5	-1	0	0%
218	Salisbury	-5	-1	0	0%
219	Elkton	-5	-1	0	0%
Massachusetts		**5**	**35**	**4**	**15%**
015-016	Ayer	5	35	4	15%
010	Chicopee	5	35	4	15%
011	Springfield	6	32	4	15%
012	Pittsfield	5	35	4	15%
013	Northfield	5	35	4	15%
014	Fitchburg	5	32	4	14%
017	Bedford	4	35	4	15%
018	Lawrence	6	35	4	16%
019	Dedham	5	35	4	15%
020	Hingham	6	35	4	16%
021-022	Boston	5	37	4	16%
023-024	Brockton	5	35	4	15%
025	Nantucket	4	35	4	15%
026	Centerville	5	35	4	15%
027	New Bedford	5	35	4	15%
Michigan		**-4**	**17**	**-2**	**6%**
480	Royal Oak	-2	19	0	7%
481-482	Detroit	-4	19	-2	6%
483	Pontiac	-4	19	-2	6%
484-485	Flint	-4	19	-2	6%
486-487	Saginaw	-5	6	-3	1%
488-489	Lansing	-4	19	-2	6%
490-491	Battle Creek	-3	19	-2	7%
492	Jackson	-4	19	-2	6%
493-495	Grand Rapids	-3	19	-2	7%
496	Traverse City	-4	19	-2	6%
497	Grayling	-4	19	-2	6%
498-499	Marquette	-4	6	-3	1%
Minnesota		**-3**	**20**	**-3**	**5%**
550-551	St Paul	-1	26	-3	9%
553-555	Minneapolis	-1	26	-3	9%
556-558	Duluth	-4	14	-3	3%
559	Rochester	-3	19	-3	5%
560	Mankato	-3	19	-3	5%
561	Magnolia	-4	19	-3	5%
562	Willmar	-4	19	-3	5%
563	St Cloud	-4	19	-3	5%
564	Brainerd	-3	19	-3	5%
565	Fergus Falls	-3	19	-3	5%
566	Bemidji	-3	19	-3	5%
567	Thief River Fl.	-3	19	-3	5%
Mississippi		**-2**	**-36**	**-3**	**-13%**
386	Clarksdale	-3	-38	-3	-14%
387	Greenville	2	-24	-3	-5%

Area Modification Factors

Location		Mat.	Lab.	Equip.	Total Wtd. Avg.
388	Tupelo	-3	-38	-3	-14%
389	Greenwood	-3	-38	-3	-14%
390-392	Jackson	-3	-38	-3	-14%
393	Meridian	-3	-38	-3	-14%
394	Laurel	-3	-38	-3	-14%
395	Gulfport	-2	-34	-3	-12%
396	McComb	-2	-34	-3	-12%
397	Columbus	-4	-42	-3	-16%
Missouri		**-2**	**6**	**-3**	**1%**
630-631	St Louis	-1	14	-3	5%
633	Saint Charles	-1	14	-3	5%
634	Hannibal	-3	6	-3	1%
635	Kirksville	-2	6	-3	1%
636	Farmington	-2	6	-3	1%
637	Cap. Girardeau	-2	6	-3	1%
638	Caruthersville	-3	6	-3	1%
639	Poplar Bluff	-3	6	-3	0%
640-641	Independence	-2	6	-3	1%
644-645	Saint Joseph	-2	6	-3	1%
646	Chillicothe	-3	6	-3	1%
647	East Lynne	-3	6	-3	1%
648	Joplin	-2	6	-3	1%
650-651	Jefferson City	-3	3	-3	-1%
652	Columbia	-3	3	-3	0%
653	Knob Noster	-3	3	-3	-1%
654-655	Lebanon	0	6	-3	2%
656-658	Springfield	-2	6	-3	1%
Montana		**0**	**4**	**-3**	**1%**
590-591	Billings	0	4	-3	1%
592	Fairview	0	4	-3	1%
593	Miles City	0	4	-3	1%
594	Great Falls	1	4	-3	2%
595	Havre	0	4	-3	1%
596	Helena	1	4	-3	1%
597	Butte	0	4	-3	1%
598	Missoula	0	4	-3	1%
599	Kalispell	0	4	-3	1%
Nebraska		**-1**	**-25**	**-3**	**-11%**
680-681	Omaha	0	-18	-3	-8%
683-685	Lincoln	-1	-26	-3	-11%
686	Columbus	-1	-26	-3	-11%
687	Norfolk	-1	-26	-3	-11%
688	Grand Island	-1	-26	-3	-11%
689	Hastings	-1	-26	-3	-11%
690	McCook	-1	-26	-3	-11%
691	North Platte	-1	-26	-3	-11%
692	Valentine	0	-26	-3	-11%
693	Alliance	0	-26	-3	-11%
Nevada		**0**	**14**	**5**	**5%**
889-891	Las Vegas	-2	15	5	4%
893	Ely	-2	15	5	4%
894	Fallon	4	12	5	7%
895	Reno	-2	15	5	4%
897	Carson City	1	13	5	6%
898	Elko	1	13	5	6%
New Hampshire		**0**	**-2**	**4**	**2%**
030-031	New Boston	-1	1	4	3%
032-033	Manchester	1	-3	4	2%
034	Concord	-1	1	4	3%
035	Littleton	1	-3	4	2%
036	Charlestown	0	-3	4	2%
037	Lebanon	0	-3	4	2%
038	Dover	1	-3	4	2%

Location		Mat.	Lab.	Equip.	Total Wtd. Avg.
New Jersey		**1**	**31**	**4**	**9%**
070-073	Newark	2	31	4	9%
074-075	Paterson	2	32	4	9%
076	Hackensack	1	31	4	9%
077	Monmouth	1	31	4	8%
078	Dover	1	32	4	9%
079	Summit	1	32	4	9%
080-084	Atlantic City	1	31	4	9%
085	Princeton	1	31	4	9%
086	Trenton	1	30	4	8%
087	Brick	1	31	4	9%
088-089	Edison	1	31	4	9%
New Mexico		**-2**	**-6**	**-3**	**-5%**
870-871	Albuquerque	-3	-6	-3	-4%
873	Gallup	-2	-6	-3	-4%
874	Farmington	-2	-6	-3	-4%
875	Santa Fe	-3	-6	-3	-4%
877	Holman	-3	-6	-3	-4%
878	Socorro	-2	-7	-3	-5%
879	Truth or Con.	-2	-7	-3	-5%
880	Las Cruces	-3	-7	-3	-5%
881	Clovis	-4	-7	-3	-6%
882	Fort Sumner	-2	-6	-3	-4%
883	Alamogordo	-2	-8	-3	-5%
884	Tucumcari	-2	-7	-3	-4%
New York		**0**	**35**	**4**	**9%**
100	NYC (Manhattan)	5	72	4	26%
100	New York City	2	41	4	13%
103	Staten Island	1	34	4	9%
104	Bronx	1	34	4	9%
105-108	White Plains	1	34	4	9%
109	West Point	-1	37	4	9%
110	Queens	3	48	4	16%
111	Long Island	4	48	4	16%
112	Brooklyn	1	34	4	9%
113	Flushing	1	34	4	9%
114	Jamaica	1	34	4	9%
115	Garden City	1	34	4	9%
116	Rockaway	1	34	4	9%
117	Amityville	1	34	4	9%
118	Hicksville	1	34	4	9%
119	Montauk	1	34	4	9%
120-123	Albany	-4	34	4	6%
124	Kingston	1	30	4	7%
125-126	Poughkeepsie	0	30	4	7%
127	Stewart	0	30	4	7%
128	Newcomb	-5	34	4	6%
129	Plattsburgh	-2	34	4	7%
130-132	Syracuse	0	34	4	8%
133-135	Utica	-1	34	4	8%
136	Watertown	-5	34	4	6%
137-139	Binghampton	0	34	4	8%
140	Batavia	0	32	4	8%
141	Tonawanda	0	32	4	8%
142	Buffalo	0	32	4	8%
143	Niagara Falls	0	34	4	8%
144-146	Rochester	0	30	4	7%
147	Jamestown	-1	26	4	5%
148-149	Ithaca	-1	26	4	5%
North Carolina		**-1**	**-40**	**-3**	**-16%**
270-274	Winston-Salem	-2	-40	-3	-17%
275-277	Raleigh	-2	-40	-3	-17%
278	Rocky Mount	-3	-40	-3	-17%
279	Elizabeth City	-3	-40	-3	-17%
280-282	Charlotte	-2	-40	-3	-17%
283	Fayetteville	6	-38	-3	-11%
284	Wilmington	6	-38	-3	-11%

Location		Mat.	Lab.	Equip.	Total Wtd. Avg.
285	Kinston	-3	-40	-3	-17%
286	Hickory	-3	-40	-3	-17%
287-289	Asheville	-3	-40	-3	-17%
North Dakota		**1**	**-16**	**-2**	**-4%**
580-581	Fargo	-1	-16	-2	-4%
582	Grand Forks	0	-16	-2	-4%
583	Nekoma	1	-16	-3	-4%
584	Jamestown	1	-16	-3	-4%
585	Bismarck	0	-16	-3	-4%
586	Dickinson	1	-16	-3	-4%
587	Minot	2	-16	-3	-3%
588	Williston	1	-16	-3	-4%
Ohio		**1**	**4**	**0**	**1%**
430-431	Newark	2	4	0	1%
432	Columbus	2	4	0	1%
433	Marion	1	4	0	0%
434-436	Toledo	2	6	0	2%
437-438	Zanesville	1	4	0	0%
439	Stubenville	1	4	0	1%
440-441	Cleveland	2	8	0	3%
442-443	Akron	2	4	0	1%
444-445	Youngstown	2	4	0	1%
446-447	Canton	2	4	0	1%
448-449	Sandusky	1	2	0	0%
450-452	Cincinnati	1	4	0	0%
453-455	Dayton	0	5	0	0%
456	Chillicothe	1	4	0	0%
457	Marietta	1	4	0	1%
458	Lima	2	8	0	2%
Oklahoma		**-2**	**-21**	**-3**	**-8%**
730-731	Oklahoma City	-2	-21	-3	-7%
734	Ardmore	-2	-21	-3	-8%
735	Lawton	-2	-21	-3	-8%
736	Clinton	-2	-21	-3	-8%
737	Enid	-2	-21	-3	-8%
738	Woodward	-2	-21	-3	-8%
739	Adams	-2	-21	-3	-8%
740-741	Tulsa	-2	-21	-3	-8%
743	Pryor	-2	-21	-3	-8%
744	Muskogee	-2	-16	-3	-6%
745	McAlester	-6	-26	-3	-12%
746	Ponca City	-2	-21	-3	-8%
747	Durant	-2	-21	-3	-8%
748	Shawnee	-2	-21	-3	-8%
749	Poteau	-2	-21	-3	-8%
Oregon		**4**	**19**	**0**	**6%**
970-972	Portland	3	19	0	6%
973	Salem	4	19	0	7%
974	Eugene	4	19	0	7%
975	Grants Pass	5	19	0	7%
976	Klamath Falls	5	19	0	7%
977	Bend	5	19	0	7%
978	Pendleton	3	18	0	6%
979	Adrian	4	18	0	6%
Pennsylvania		**-2**	**19**	**4**	**5%**
150-151	Warrendale	-3	13	4	2%
152	Pittsburgh	-3	13	4	2%
153	Washington	-3	13	4	2%
154	Uniontown	-3	13	4	2%
155	Somerset	-3	13	4	2%
156	Greensburg	-3	13	4	2%
157	Punxsutawney	-3	20	4	5%
158	DuBois	-2	20	4	5%
159	Johnstown	-2	20	4	5%
160	Butler	-3	20	4	5%

Area Modification Factors

Location		Mat.	Lab.	Equip.	Total Wtd. Avg.
161	New Castle	-3	20	4	5%
162	Kittanning	-3	20	4	5%
163	Meadville	-3	20	4	5%
164-165	Erie	-3	20	4	5%
166	Altoona	-3	20	4	5%
167	Bradford	-2	20	4	5%
168	Clearfield	-2	20	4	5%
169	Genesee	-2	20	4	5%
170-171	Harrisburg	-2	20	4	5%
172	Chambersburg	-5	17	4	2%
173-174	York	-2	20	4	5%
175-176	Lancaster	-2	20	4	5%
177	Williamsport	-3	20	4	5%
178	Beaver Springs	-2	20	4	5%
179	Pottsville	-3	20	4	5%
180-181	Allentown	-2	20	4	5%
182	Hazleton	-2	20	4	5%
183	E. Stroudsburg	-2	20	4	5%
184-185	Scranton	-1	15	4	4%
186-187	Wilkes Barre	-2	20	4	5%
188	Montrose	-3	20	4	5%
189	Warminster	-2	27	4	8%
190-191	Philadelphia	-1	27	4	9%
195-196	Reading	-2	20	4	5%
193	Southeastern	-2	20	4	5%
194	Valley Forge	-3	20	4	5%
Rhode Island		**3**	**26**	**4**	**10%**
028	Bristol	3	26	4	9%
028	Coventry	3	26	4	9%
028	Davisville	3	26	4	9%
028	Narragansett	0	20	4	9%
028	Newport	3	26	4	10%
028	Warwick	3	26	4	10%
029	Cranston	3	26	4	10%
029	Providence	3	26	4	10%
South Carolina		**-1**	**-37**	**-3**	**-13%**
290-292	Columbia	-3	-40	-3	-15%
293	Spartanburg	-3	-40	-3	-15%
294	Charleston	1	-39	-3	-13%
295	Myrtle Beach	0	-39	-3	-13%
296	Greenville	-3	-40	-3	-15%
297	Rock Hill	-3	-40	-3	-15%
298	Aiken	-1	-27	-3	-9%
299	Beaufort	-1	-27	-3	-9%
South Dakota		**1**	**-25**	**-3**	**-8%**
570-571	Sioux Falls	-4	-24	-3	-10%
572	Watertown	1	-25	-3	-8%
573	Mitchell	1	-25	-3	-8%
574	Aberdeen	2	-25	-3	-7%
575	Pierre	1	-25	-3	-7%
576	Mobridge	1	-25	-3	-8%
577	Rapid City	2	-26	-3	-8%
Tennessee		**-4**	**-20**	**-3**	**-8%**
370-372	Nashville	-6	-21	-3	-9%
373	Cleveland	-5	-21	-3	-9%
374	Chattanooga	-4	-20	-3	-8%
376	Kingsport	-4	-20	-3	-8%
377-379	Knoxville	-5	-20	-3	-8%
380-381	Memphis	-3	-19	-3	-7%
382	McKenzie	-4	-20	-3	-8%
383	Jackson	-4	-20	-3	-8%

Location		Mat.	Lab.	Equip.	Total Wtd. Avg.
384	Columbia	-5	-20	-3	-8%
385	Cookeville	-4	-20	-3	-8%
Texas		**-3**	**-24**	**-3**	**-9%**
750	Plano	2	-20	-3	-5%
751-753	Dallas	2	-20	-3	-5%
754	Greenville	-4	-25	-3	-11%
755	Texarkana	4	-32	-3	-8%
756	Longview	-5	-25	-3	-11%
757	Tyler	-4	-25	-3	-11%
758	Palestine	-5	-25	-3	-11%
759	Lufkin	-5	-25	-3	-11%
760	Arlington	-5	-25	-3	-11%
761-762	Ft Worth	1	-20	-3	-5%
763	Wichita Falls	-5	-25	-3	-11%
764	Woodson	-4	-26	-3	-11%
765-767	Waco	-4	-26	-3	-11%
768	Brownwood	-4	-30	-3	-12%
769	San Angelo	-4	-30	-3	-12%
770-772	Houston	-4	-25	-3	-10%
773	Huntsville	-1	-25	-3	-9%
774	Bay City	-4	-25	-3	-10%
775	Galveston	-5	-25	-3	-11%
776-777	Beaumont	-5	-25	-3	-11%
778	Bryan	-5	-25	-3	-11%
779	Victoria	-4	-25	-3	-10%
780-782	San Antonio	-4	-20	-3	-8%
783-784	Corpus Chrsti	-5	-25	-3	-11%
785	McAllen	-4	-25	-3	-11%
786-787	Austin	-5	-25	-3	-11%
788	Del Rio	4	-15	-3	-2%
789	Giddings	3	-15	-3	-2%
790-791	Amarillo	-5	-25	-3	-11%
792	Childress	-3	-22	-3	-9%
793-794	Lubbock	-2	-22	-3	-8%
795-796	Abilene	-5	-25	-3	-11%
797	Midland	-4	-25	-3	-10%
798-799	El Paso	-5	-25	-3	-11%
Utah		**-3**	**-16**	**5**	**-7%**
840	Clearfield	-3	-14	5	-7%
841	Salt Lake City	-2	-18	5	-8%
843-844	Ogden	-3	-14	5	-6%
845	Green River	-3	-16	5	-7%
846-847	Provo	-3	-16	5	-8%
Vermont		**1**	**-23**	**4**	**-4%**
050	White River Jct.	1	-22	4	-4%
051	Springfield	1	-22	4	-4%
052	Bennington	1	-22	4	-4%
053	Battleboro	1	-22	4	-4%
054	Burlington	1	-21	4	-3%
056	Montpelier	1	-24	4	-5%
057	Rutland	1	-22	4	-4%
058	Albany	1	-24	4	-4%
059	Beecher Falls	1	-24	4	-5%
Virgina		**-3**	**-23**	**0**	**-10%**
220-223	Alexandria	-2	-28	0	-11%
224-225	Fredericksburg	-2	-36	0	-15%
226	Winchester	-3	-22	0	-10%
227	Culpeper	-3	-22	0	-10%
228	Harrisonburg	-3	-22	0	-10%
229	Charlottesville	-3	-22	0	-10%
230-231	Williamsburg	-3	-16	0	-7%
232	Richmond	-3	-22	0	-10%

Location		Mat.	Lab.	Equip.	Total Wtd. Avg.
233-237	Norfolk	-3	-16	0	-7%
238	Petersburg	-2	-22	0	-9%
239	Farmville	-2	-20	0	-8%
240-241	Radford	-3	-22	0	-10%
242	Abingdon	-3	-22	0	-10%
243	Galax	-3	-22	0	-10%
244	Staunton	-3	-22	0	-10%
245	Lynchburg	-3	-22	0	-10%
246	Tazewell	-3	-22	0	-10%
Washington		**6**	**14**	**0**	**8%**
980-981	Seattle	6	13	0	8%
982	Everett	4	16	0	8%
983-984	Tacoma	6	13	0	8%
985	Olympia	6	13	0	8%
986	Vancouver	6	13	0	8%
988	Wenatchee	5	12	0	7%
989	Yakima	4	12	0	6%
990-992	Spokane	8	16	0	10%
993	Pasco	6	13	0	8%
994	Clarkston	6	13	0	8%
West Virginia		**-4**	**2**	**0**	**-1%**
247-248	Bluefield	-6	2	0	-3%
249	Lewisburg	-3	2	0	-1%
250-253	Charleston	-3	2	0	-1%
254	Martinsburg	-3	2	0	-1%
255-257	Huntington	-3	2	0	-1%
258-259	Beckley	-4	2	0	-1%
260	Wheeling	-3	2	0	-1%
261	Parkersburg	-3	2	0	-1%
262	New Martinsville	-3	2	0	-1%
263-264	Clarksburg	-3	2	0	-1%
265	Morgantown	-3	2	0	-1%
266	Fairmont	-3	2	0	-1%
267	Romney	-3	2	0	-1%
268	Sugar Grove	-3	-2	0	-3%
Wisconsin		**-4**	**5**	**-3**	**-2%**
530-534	Milwaukee	-1	17	-3	4%
535	Beloit	-3	5	-3	-2%
537	Madison	-7	0	0	-6%
538	Prairie du Chien	-8	0	0	-6%
539	Portage	-8	0	0	-7%
540	Amery	-2	5	-3	-2%
541-543	Green Bay	-2	5	-3	-1%
544	Wausau	-3	5	-3	-2%
545	Clam Lake	-3	5	-3	-2%
546	La Crosse	-3	5	-3	-2%
547	Eau Claire	-3	5	-3	-2%
548	Ladysmith	-3	5	-3	-2%
549	Oshkosh	-3	5	-3	-2%
Wyoming		**-1**	**-29**	**-3**	**-11%**
820	Cheyenne	0	-28	-3	-10%
821	Laramie	-1	-29	-3	-11%
822	Wheatland	-1	-29	-3	-11%
823	Rawlins	0	-29	-3	-10%
824	Powell	-1	-29	-3	-11%
825	Riverton	-1	-29	-3	-11%
826	Casper	-1	-29	-3	-11%
827	Gillette	-1	-29	-3	-11%
828	Sheridan	-1	-29	-3	-11%
829-831	Rock Springs	0	-29	-3	-10%

Credits and Acknowlegments

This book has over 30,000 cost estimates for 1999. To develop these estimates, the editors relied on information supplied by hundreds of construction cost authorities. We offer our sincere thanks to the contractors, engineers, design professionals, construction estimators, material suppliers and manufacturers who, in the spirit of cooperation, have assisted in the preparation of this Forty-seventh Edition of the National Construction Estimator. Many of the cost authorities who supplied for this volume are listed below.

A.A. Drafting Service., Andrew Gutt

ABC Construction Co., Wayne Czubernat

ABTCO, Inc., Tom Roe

Advance Lifts, Jim Oriowski

Advanced Stainless & Aluminum, Louie Sprague

AFCO Roofing Supply, John Ording

AGRA Earth and Environmental., Jack Eagan

Allied Barricade, Bob Johnson

Alply Incorporated., Alan E. Sleppy

Amerec Products, John Gunderson

Applied Waterproofing Technology, Inc., Mike Weinert

Architectural Illust., Bruce Wickern

Area Home Inspection, Anthony Galeota

Atascadero Glass, Jim Stevens

Atlas Mineral & Chemical Co., Glenn Kresge

Bell Blueprint, Ed Moore

Bessler Stairways, Morris Hudspeth

Beveled Glass Design, Lorna Koss

Brooks Products, Lynn Diaz

Brower Pease, Mark Brossert

Buy-Rite Decorating House, Hank Jordan

California Pneumatic Systems, Clayton Taylor

Cement Cutting, Inc., Greg Becker

Centennial Flag Pole Co., Eleanor Wiese

Classic Ceilings, Donna Morgan

Cold Spring Granite, Paul Dierkhising

Columbia Equipment Co., Arthur Cohen

Creative Ceilings, Bob Warren

C. R. Laurence, Inc, Michael Sedano

David Backer Limited, David D. Backer

Delta Pacific Builders, Jeff Andrews

Dial One, Charlie Gindele

Dryvit System, Inc., Roger Peevier

Eliason Corporation, John Duncan

Envel Design Corporation, Quinn Mayer

Escondido Roof Truss Co., David Truax

Finlandia Sauna Rooms, Reino J. Tarkianen

Fiorentine Company, Tom Forsyth

Flexi-Wall Systems, Nancy West

Fortifiber Corporation, Dave Olson

GAF Premium Products, Sandra Stern

Georgia Pacific, Travis Hemborg

Giles and Kendall, Inc., Todd M. Green

Glen-Gery Brick, David M. Reitz

Heintschel Plastering, Marvin J. Heintschel, Jr.

H&H Specialties, Reid Nesiage

I. C. C., Mike McInnis

Inclinator Company, Boc Crede

Int'l Conf. Of Bldg. Officials, Dennis McCreary

Iron-A-Way Company, Reg Smidt

J.A. Crawford Company, Jim Crawford

Jewett Refrigerator Co., Thomas D. Reider

J. H. Baxter Inc., Dean Rogers

Kentucky Wood Floors, John Stern

Kirsch Company, Dennis Kirby

Koffler Sales Corporation, Mary Holman

Landscape Structures, Inc., Mary Allen Grossman

Lasco Panels, Mike Bunke, Sales Manager

Lombardi, Diamond Core Drilling, Kit Sanders

Maccaferri Gabions, Inc., Ahmad Kasem

Marbleworks of San Diego, Charlene Butler

Mashburn Waste & Recycling, Bill Dean

M.C. Nottingham Company.

Mel-Nortley Co., Mel Nortley

Merli Concrete Pumping, John Caole

Microphor, Ross Beck

Morrow Equipment Co., John Morrow

Multi Water Systems, John Kratz

National Floor Products, Edsel Holden

National Schedule Masters, Steve Scholl

Nicolon Mirafi Group, Ted Pope

North County Interiors, Rod Gould

Nuclear Associates, Laura Rhein

Oregon Lumber Company, John Couch

Penhall Construction Services, Bruce Varney

Perma Grain Products, Thomas W. Gunuskey

Photocom, Ron Kenedi

Pittsburgh Corning Corp., Joe Zavada

Plastmo Raingutters, Ed Gunderson

Prof. Photographic Service, Larry Hoagland

Rally Rack, Shirlee Turnipseed

RCP Block & Brick Co., Bob Balderree

Rheem Manufacturing, Rick Miller

Ryan Roulette Photography, Ryan Roulette

R. W. Little Company, Inc., Ron Little

Safeway Steel Products, Patrick Whelan

Sauder Manufacturing Co., John Gascho

Sico, Inc., Charles Williamson

Simpler Solar Systems, Al Simpler

Slater Waterproofing, Merle Slater, Jr.

Southern Bleacher Co., Glen McNatt

Specialty Wood Company, Bob DiNello

SSHC, Inc., Richard Watson

Stark Ceramics, Inc., Paul Tauer

STGE, Don Altevers

Stebbins Company, Mrs. Pat Stebbins

Stone Products Corp., Robert Heath

Superior Fireplace Co., Mike Hammer

Superior Rock Products, John Knieff

Superior Stakes, George Kellett

Supradur Manufacturing, Marilyn Mueller

Ted's Quality Roofing, Ted Vanderhoof

The Parking Block Store, Jim Headley

Towill INC., Jim Smith

TRIMEX, Walter VanWickel

U. S. Gypsum Company, Roger Merchet

Ultraflo Corporation, Doug Didion

Unique Building Products, Charlie Gindele

United Panel Partnership, Donna Korp

U. S. Tile Company, Brian Austin

Velux-America, Inc., John Carigan

Vintage Lumber & Const. Co., Mark Harbold

Vital Services Co., Myrna Humphrey

Wasco Products, Estelle Gore

West Coast Lumber Inspect. Bur., Jim Kneaper

Western Colloid Products, Tim Griffin

Western Exterminators, Bill Sharp

Western Wood Products Assn., Frank Stewart

Westingconsul, Inc., Verne T. Kelling

Weyerhaeuser Company, Doug Barkee

York Spiral Stair, Jim A. Guerette

Zenith Aerial Photo, Lew Roberts

Zeus Lightning Protection, Dave Silvester

A special thanks to the following people. Special assistants to the editors: Edward Moselle and James Thomson. Production and Layout: Karen Sheffield. Software production: Jennifer Carro.

Cover Photography: Tom Page Photography, www.redrom.com/tom.

Adhesives See also Carpet, Caulking, Flooring, Glazing, Gypsum, and Tile in the Residential Division. Per C. R. Laurence Co., Inc

Panel adhesives Better quality, gun applied in continuous bead to wood or metal framing or furring members, material only, add labor below. Per 100 SF of wall, floor, or ceiling including 6% waste

			Bead diameter		
		1/8"	1/4"	3/8"	1/2"
Subfloor adhesive (at $.18 per oz), on floors					
12" OC members	CSF	7.26	16.30	18.30	29.20
16" OC members	CSF	5.67	13.20	14.50	23.20
20" OC members	CSF	5.17	11.20	13.00	19.90
Wall sheathing or shear panel adhesive (at $.17 per oz), on walls					
12" OC members	CSF	4.18	17.30	38.20	68.20
16" OC members	CSF	3.96	15.40	34.80	61.80
20" OC members	CSF	3.85	15.00	34.00	59.00
24" OC members	CSF	3.52	14.90	31.90	56.40
Polystyrene or polyurethane foam panel adhesive (at $.27 per oz), on walls					
12" OC members	CSF	3.85	16.20	36.30	64.90
16" OC members	CSF	3.63	15.00	33.10	59.10
20" OC members	CSF	3.41	14.30	31.80	56.50
24" OC members	CSF	3.25	13.80	30.40	53.70
Gypsum drywall adhesive (at $.24 per oz), on ceilings					
12" OC members	CSF	2.48	10.30	22.80	40.30
16" OC members	CSF	1.87	7.81	17.30	30.70
20" OC members	CSF	1.76	6.55	14.20	25.00
24" OC members	CSF	1.65	5.61	12.30	21.90
Gypsum drywall adhesive (at $.24 per oz), on walls					
12" OC members	CSF	3.47	13.80	30.80	55.20
16" OC members	CSF	3.14	12.70	28.00	49.60
20" OC members	CSF	2.97	12.30	27.90	47.40
24" OC members	CSF	2.81	11.90	25.50	45.70
Hardboard or plastic panel adhesive (at $.23 per oz), on walls					
12" OC members	CSF	3.19	13.50	30.40	54.60
16" OC members	CSF	3.08	12.30	27.60	48.20
20" OC members	CSF	2.92	11.80	26.50	46.90
24" OC members	CSF	2.70	11.30	25.30	45.20

	Craft@Hrs	Unit	Material	Labor	Total
Labor to apply adhesive to framing members, 1/8" to 1/2" bead diameter, no material included					
Floor or ceiling joists					
12" OC members	BC@.075	CSF	—	1.89	1.89
16" OC members	BC@.056	CSF	—	1.41	1.41
20" OC members	BC@.052	CSF	—	1.31	1.31
24" OC members	BC@.042	CSF	—	1.06	1.06
Interior and exterior wall members					
12" OC members	BC@.100	CSF	—	2.52	2.52
16" OC members	BC@.090	CSF	—	2.27	2.27
20" OC members	BC@.084	CSF	—	2.11	2.11
24" OC members	BC@.084	CSF	—	2.11	2.11

Adhesives

	29 oz cartridge	1 pint can	1 quart can	1 gallon can
General purpose adhesives				
Acoustic tile adhesive, solvent base, waterproof, sound deadening type	—	—	6.55	19.20
Aliphatic resin woodworking glue	—	6.75	9.75	27.10
Carpet cement, latex, multi-purpose	—	—	7.40	21.00
Contact cement, latex base, water cleanup, waterproof, bonds veneers to plywood, particleboard, wallboard	—	8.90	13.40	40.80
Contact cement, solvent base, solvent cleanup, waterproof, bonds metal, linoleum, rubber, veneers, plastic laminates, leather, plywood	—	7.10	11.20	30.50
Contact cement solvent	—	10.30	14.90	—
Foam panel insulation adhesive, waterproof, for polyurethane & polystyrene panels, bonds to wood, metal, masonry & concrete	10.20	—	—	32.20
Gypsum drywall adhesive, waterproof, bonds to wood, metal, masonry and concrete	7.50	—	—	28.20
Hardboard, fiberboard, and plastic panel adhesive, waterproof, bonds to wood, plaster, concrete, masonry and gypsum	—	—	9.80	29.00
Parquet tile adhesive, solvent base	—	—	—	37.00
Plastic resin glue, marine grade	—	7.30	—	—
Tile cement, solvent base, for installation of ceramic, quarry, slate and marble tiles	—	—	9.20	28.20
Resilient flooring adhesive, latex base, adheres to concrete, plywood, felt, sheet flooring	—	—	8.00	24.70
Resorcinol glue, waterproof, 2 part	—	25.00	40.00	—
Subfloor and general purpose construction adhesive, waterproof, specification grade, bonds to wood, metal, concrete and masonry	8.85	—	6.75	17.80

	Craft@Hrs	Unit	Material	Labor	Total
Aggregate Typical prices, 5 mile haul, 24 ton minimum. See also Roofing, Built-up					
Crushed stone (1.4 tons per CY)					
3/8" stone	—	Ton	14.10	—	14.10
3/4" (Number 3)	—	Ton	12.50	—	12.50
1-1/2" (Number 2)	—	Ton	12.50	—	12.50
Crushed slag, typical prices where available					
3/4" slag	—	Ton	9.50	—	9.50
1-1/2"	—	Ton	9.60	—	9.60
Gravel (1.4 tons per CY)					
3/4" gravel	—	Ton	15.00	—	15.00
1-1/2"	—	Ton	14.30	—	14.30
Sand (1.35 tons per CY)	—	Ton	14.00	—	14.00
Add per ton less than 24 tons	—	Ton	4.80	—	4.80
Add for delivery over 5 miles, one way	—	Mile	2.70	—	2.70

	Craft@Hrs	Unit	Material	Labor	Total

Appraisal Fees Costs for determining the value of existing buildings, land, and equipment. Actual fee charged is based on the level of difficulty and the time spent on appraisal plus travel to location and cost of support services, if any. Costs include research and report by a professional appraiser. Client may request an appraisal on a "fee not to exceed" basis. Fees shown are averages and are not quoted as a percentage of value. The Financial Institutions Reform, Recovery, and Enforcement Act of 1989 requires that federally regulated institutions order real estate appraisals when the appraisals are being used for underwriting loans, including construction loans. Costs shown are typical fees and will vary depending on the experience of the appraiser, the region of the country, and other factors.

Residences, single family dwellings (not including contents)

	Craft@Hrs	Unit	Material	Labor	Total
To $300,000 valuation	—	LS	—	—	300.00
Over $300,000 to $500,000 valuation	—	LS	—	—	400.00
Over $500,000 to $1,000,000 valuation	—	LS	—	—	600.00
Over $1,000,000 to $3,000,000 valuation	—	LS	—	—	1500.00

Apartment houses, commercial and industrial buildings

	Craft@Hrs	Unit	Material	Labor	Total
To $300,000 valuation	—	LS	—	—	3,000.00
Over $300,000 to $1,000,000 valuation	—	LS	—	—	4,000.00
Over $1,000,000 to $3,000,000 valuation	—	LS	—	—	4,500.00
Over $3,000,000 to $5,000,000 valuation	—	LS	—	—	5,500.00

Machinery (Fee is based on total value of equipment appraised.) Additional charges for travel and lodging may be required

	Craft@Hrs	Unit	Material	Labor	Total
To $30,000 valuation	—	LS	—	—	570.00
Over $30,000 to $100,000 valuation	—	LS	—	—	770.00
Over $100,000 to $500,000 valuation	—	LS	—	—	1,300.00
Over $500,000 to $1,000,000 valuation	—	LS	—	—	1,800.00
Over $1,000,000 to $5,000,000 valuation	—	LS	—	—	3,100.00
Appraisals per 1/2 day (typical)	—	Rate	—	—	500.00
Appraisals per day (typical)	—	Rate	—	—	800.00
Court testimony (excluding preparation)	—	Day	—	—	1,200.00

Arbitration These are administrative fees paid to the American Arbitration Association (AAA). Arbitrators are chosen from the National Roster of Construction Arbitrators and are paid a fee. Legal representation, if desired (although not necessary), is at the expense of each party. These fees do not include a hearing room rental.

Administrative AAA filing fee to initiate disclosed case is based on the amount of the claim. When no amount can be stated at the time of filing, the filing fee is $2,000, subject to adjustment when the claim or counterclaim is disclosed. If the claim or counterclaim is not for a monetary amount, an appropriate filing fee will be determined by the AAA.

The minimum filing fee for any case having three or more arbitrators is $2,000.

	Craft@Hrs	Unit	Material	Labor	Total
Up to $10,000	—	LS	—	—	500.00
$10,001 to $50,000	—	LS	—	—	750.00
$50,001 to $100,000	—	LS	—	—	1,250.00
$100,001 to $250,000	—	LS	—	—	2,000.00
$250,001 to $500,000	—	LS	—	—	3,500.00
$500,001 to $1,000,000	—	LS	—	—	5,000.00
$1,000,001 to 5,000,000	—	LS	—	—	7,000.00

Hearing fees, per day, payable by each party

	Craft@Hrs	Unit	Material	Labor	Total
Singe-arbitrator cases	—	LS	—	—	150.00
Multi-arbitrator cases	—	LS	—	—	250.00

Fee charged to party causing a postponement of any scheduled arbitration hearing

	Craft@Hrs	Unit	Material	Labor	Total
Single-arbitrator cases	—	LS	—	—	150.00
Multi-arbitrator cases	—	LS	—	—	250.00

Architectural Illustrations

	Craft@Hrs	Unit	Material	Labor	Total

Architectural Illustrations Full color painting on watercolor board with matted frame with title and credit on matte. Glass and elaborate framing are extra. Costs for pen and ink illustrations with color mylar overlay are similar to cost for watercolor illustrations. Typical fees

	Craft@Hrs	Unit	Material	Labor	Total
Custom home, eye level view					
Simple rendering	—	LS	—	—	800.00
Complex rendering	—	LS	—	—	1,150.00
Custom home, bird's eye view					
Simple rendering	—	LS	—	—	950.00
Complex rendering	—	LS	—	—	1,350.00
Tract homes in groups of five or more (single floor plans, multiple elevations), eye level view					
Simple rendering	—	LS	—	—	460.00
Complex rendering	—	LS	—	—	650.00
Tract homes in groups of five or more (single floor plans, multiple elevations), bird's eye view					
Simple rendering	—	LS	—	—	600.00
Complex rendering	—	LS	—	—	950.00
Tract homes or condominium project, overall bird's eye view					
10-25 homes or living units		LS	—	—	3,000.00
Typical commercial structure					
Eye level view	—	LS	—	—	1,250.00
Bird's eye view	—	LS	—	—	1,450.00
Complex commercial structure					
Eye level view	—	LS	—	—	1,900.00
Bird's eye view	—	LS	—	—	2,500.00
Deduct for pen and ink drawings (no color)	—	%	—	—	-33.0
Computer generated perspective drawings using CAD system for design studies					
Custom home	—	LS	—	—	500.00
Large condo or apartment projects	—	LS	—	—	1,250.00
Tract homes	—	LS	—	—	315.00
Commercial structure, line drawing	—	LS	—	—	950.00

Awnings and Canopies for Doors and Windows Costs for awnings include all hardware. All have adjustable support arms to control angle height and preferred amount of window coverage. For larger size aluminum awnings, price two awnings and add for splice kit below.

	Craft@Hrs	Unit	Material	Labor	Total
Natural aluminum ribbed awning with clear, weather-resistant finish and 26" arms					
36" wide x 30" long	SW@1.45	Ea	57.20	40.30	97.50
48" wide x 30" long	SW@1.86	Ea	73.80	51.60	125.40
60" wide x 30" long	SW@2.07	Ea	92.10	57.50	149.60
72" wide x 30" long	SW@2.27	Ea	105.00	63.00	168.00
Add for door canopy with 17" drop sides	—	%	50.0	—	—
Custom colored window awnings in stripes or solids, with baked enamel finish and ventilation panels					
30" wide x 24" high	SW@1.45	Ea	93.20	40.30	133.50
36" wide x 36" high	SW@1.45	Ea	140.00	40.30	180.30
48" wide x 48" high	SW@1.87	Ea	235.00	51.90	286.90
48" wide x 60" high	SW@1.87	Ea	333.00	51.90	384.90
60" wide x 72" high	SW@2.07	Ea	366.00	57.50	423.50
72" wide x 72" high	SW@2.26	Ea	500.00	62.70	562.70
Add for splice kit with overlap slats	SW@.218	Ea	19.40	6.05	25.45
Security roll-up awning with pull cord assembly and folding arms, clear weather-resistant finish. Awning rolls down to cover whole window. 48" long, 24" arms					
36" wide	SW@1.52	Ea	178.00	42.20	220.20
48" wide	SW@1.94	Ea	211.00	53.90	264.90

	Craft@Hrs	Unit	Material	Labor	Total

Plastic awning with baked-on acrylic finish, ventilated side panels, and reinforced metal frame, hardware included. 24" drop, 24" projection

	Craft@Hrs	Unit	Material	Labor	Total
36" wide	BC@1.58	Ea	96.30	39.80	136.10
42" wide	BC@1.75	Ea	107.00	44.00	151.00
48" wide	BC@2.02	Ea	122.00	50.80	172.80
60" wide	BC@2.26	Ea	138.00	56.90	194.90
72" wide	BC@2.47	Ea	149.00	62.20	211.20
96" wide	BC@2.79	Ea	188.00	70.20	258.20

Plastic door canopy with 36" projection

	Craft@Hrs	Unit	Material	Labor	Total
42" wide	BC@1.80	Ea	203.00	45.30	248.30

Traditional fabric awning, with waterproof, acrylic duck, colorfast fabric, double stitched seams, and tubular metal framing and pull cord assembly. 24" drop, 24" projection

	Craft@Hrs	Unit	Material	Labor	Total
30" wide	BC@1.35	Ea	42.30	34.00	76.30
36" wide	BC@1.58	Ea	55.60	39.80	95.40
42" wide	BC@1.80	Ea	51.90	45.30	97.20
48" wide	BC@2.02	Ea	61.00	50.80	111.80
Add for 30" drop, 30" projection	—	%	10.0	20.0	—

Cloth canopy patio cover, with front bar and tension support rafters, 9" valance and 8' projection

	Craft@Hrs	Unit	Material	Labor	Total
8' x 10'	BC@2.03	Ea	257.00	51.10	308.10
8' x 15'	BC@2.03	Ea	353.00	51.10	404.10

Barricades, Construction Safety Purchase prices except as noted

Plastic Type I barricade

		Unit	Material	Labor	Total
To 100 units	—	Ea	42.95	—	42.95
Over 100 units	—	Ea	38.75	—	38.75

Reflectorized plywood barricade, 8" to 12" wide rail, 4" to 6" wide stripes, 45" 14 gauge steel legs, no light

Type I, 2' wide, 3' high, 1 reflectorized rail each side

		Unit	Material	Labor	Total
To 100 units	—	Ea	21.50	—	21.50
Over 100 units	—	Ea	18.00	—	18.00

Type II, 2' wide, 3' high, 2 reflectorized rails each side

		Unit	Material	Labor	Total
To 100 units	—	Ea	27.80	—	27.80
Over 100 units	—	Ea	23.25	—	23.25

Type III, 4' wide, 5' high, 3 reflectorized rails each side, wood & steel legs

		Unit	Material	Labor	Total
To 100 units	—	Ea	99.00	—	99.00
Over 100 units	—	Ea	90.50	—	90.50

Utility barricade, 2" x 2" legs, 1" x 6" face, no light

		Unit	Material	Labor	Total
To 100 units	—	Ea	10.80	—	10.80
Over 100 units	—	Ea	9.50	—	9.50

Add for lighted units without batteries (batteries last 2 months)

		Unit	Material	Labor	Total
To 100 units	—	Ea	18.70	—	18.70
Over 100 units	—	Ea	13.00	—	13.00
Batteries, 6 volt (2 needed)	—	Ea	2.90	—	2.90
Lighted units, rental, per day	—	Ea	.48	—	.48
Unlighted units, rental per day	—	Ea	.35	—	.35
Add for pickup and delivery, per trip	—	Ea	15.00	—	15.00
"Road Closed", reflectorized, 30" x 48"	—	Ea	50.00	—	50.00
"Construction Zone", 4' x 4' high intensity grade	—	Ea	110.00	—	110.00
High-rise tripod with 3 orange flags	—	Ea	83.00	—	83.00
High-rise sign holder, wind resistant	—	Ea	145.00	—	145.00

	Craft@Hrs	Unit	Material	Labor	Total
Traffic cones, PVC					
Non-reflectorized type					
18" high	—	Ea	5.45	—	5.45
28" high	—	Ea	10.50	—	10.50
Reflectorized type					
18" high	—	Ea	9.90	—	9.90
28" high	—	Ea	18.80	—	18.80
Lane delineator, 42" orange plastic cylinder with 2 reflectors on a 12 pound rubber base	—	Ea	16.50	—	16.50
Mesh signs, orange, 48" x 48", includes brace and clamp	—	Ea	68.00	—	68.00
Hand-held traffic paddles, "Stop" and "Slow"	—	Ea	15.50	—	15.50
Typical labor cost, place and remove any barricade per use	BL@.160	Ea	—	3.29	3.29
Orange plastic safety fencing					
4' x 50' roll	—	Ea	40.00	—	40.00
5' x 50' roll	—	Ea	46.50	—	46.50
6' x 50' roll	—	Ea	58.00	—	58.00
4' x 150', light duty	—	Ea	79.00	—	79.00
4' x 200', light duty	—	Ea	86.00	—	86.00

Basement Doors Good quality 12 gauge primed steel, center opening basement doors. Costs include assembly and installation hardware. No concrete, masonry, anchor placement or finish painting included. Bilco Company

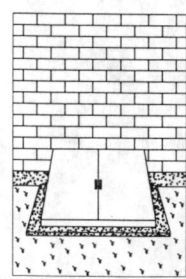

	Craft@Hrs	Unit	Material	Labor	Total
Doors (overall dimensions)					
30" H, 47" W, 58" L	BC@3.41	Ea	310.00	85.80	395.80
22" H, 51" W, 64" L	BC@3.41	Ea	314.00	85.80	399.80
19-1/2" H, 55" W, 72" L	BC@3.41	Ea	344.00	85.80	429.80
52" H, 51" W, 43-1/4" L	BC@3.41	Ea	369.00	85.80	454.80
Door extensions (available for 19-1/2" H, 55" W, 72" L door only)					
6" deep	BC@1.71	Ea	85.50	43.00	128.50
12" deep	BC@1.71	Ea	102.00	43.00	145.00
18" deep	BC@1.71	Ea	127.00	43.00	170.00
24" deep	BC@1.71	Ea	152.00	43.00	195.00
Stair stringers, steel, pre-cut for 2" x 10" wood treads (without treads)					
32" to 39" stair height	BC@1.71	Ea	72.20	43.00	115.20
48" to 55" stair height	BC@1.71	Ea	90.30	43.00	133.30
56" to 64" stair height	BC@1.71	Ea	102.00	43.00	145.00
65" to 72" stair height	BC@1.71	Ea	114.10	43.00	157.10
73" to 78" stair height	BC@1.71	Ea	157.00	43.00	200.00
81" to 88" stair height	BC@1.71	Ea	169.00	43.00	212.00
89" to 97" stair height	BC@1.71	Ea	181.00	43.00	224.00

Bathroom Accessories Average quality. Better quality brass accessories cost 75% to 100% more. See also Medicine Cabinets and Vanities

	Craft@Hrs	Unit	Material	Labor	Total
Cup and toothbrush holder, chrome	BC@.258	Ea	9.18	6.49	15.67
Cup holder, chrome, surface mounted	BC@.258	Ea	7.14	6.49	13.63
Cup, toothbrush & soap holder, recessed	BC@.258	Ea	23.50	6.49	29.99

	Craft@Hrs	Unit	Material	Labor	Total
Electrical plates, chrome plated					
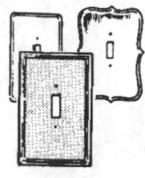 Switch plate, single	BE@.154	Ea	3.31	4.14	7.45
Switch plate, double	BE@.154	Ea	4.34	4.14	8.48
Duplex receptacle plate	BE@.154	Ea	3.31	4.14	7.45
Duplex receptacle and switch	BE@.154	Ea	4.34	4.14	8.48
Grab bars					
Tubular chrome plated, with anchor plates					
Straight bar, 16"	BC@.414	Ea	19.40	10.40	29.80
Straight bar, 24"	BC@.414	Ea	23.50	10.40	33.90
Straight bar, 32"	BC@.414	Ea	25.50	10.40	35.90
"L" shaped bar, 16" x 32"	BC@.620	Ea	59.20	15.60	74.80
Stainless steel, with anchor plates					
Straight bar, 16"	BC@.414	Ea	29.10	10.40	39.50
Straight bar, 24"	BC@.414	Ea	35.20	10.40	45.60
Straight bar, 32"	BC@.414	Ea	38.30	10.40	48.70
"L" shaped bar, 16" x 32"	BC@.620	Ea	82.60	15.60	98.20
Mirrors, stainless steel framed, surface mount, no light or cabinet					
16" high x 20" wide	BG@.420	Ea	49.00	9.69	58.69
18" high x 24" wide	BG@.420	Ea	57.10	9.69	66.79
18" high x 36" wide	BG@.420	Ea	87.70	9.69	97.39
24" high x 36" wide	BG@.420	Ea	103.00	9.69	112.69
48" high x 24" wide	BG@.420	Ea	127.50	9.69	137.19
Mirrors, wood framed, surface mount, better quality					
18" x 29" rectangular	BG@.420	Ea	61.20	9.69	70.89
20" x 27" oval, oak	BG@.420	Ea	86.70	9.69	96.39
Robe hook					
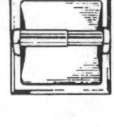 double, chrome	BC@.258	Ea	4.60	6.49	11.09
Shower curtain rods, chrome plated					
60", recessed	BC@.730	Ea	20.40	18.40	38.80
66", recessed	BC@.730	Ea	22.40	18.40	40.80
Soap holder, chrome, with drain holes					
surface mounted	BC@.258	Ea	6.40	6.49	12.89
Facial tissue holder, stainless steel, recessed	BC@.258	Ea	21.00	6.49	27.49
Toilet tissue roll holder, chrome, recessed	BC@.258	Ea	26.50	6.49	32.99
Toothbrush holder, chrome, surface mount	BC@.258	Ea	7.70	6.49	14.19
Towel bars, chrome, 3/4" round bar					
18" long	BC@.280	Ea	24.50	7.05	31.55
24" long	BC@.280	Ea	25.50	7.05	32.55
30" long	BC@.280	Ea	26.50	7.05	33.55
36" long	BC@.280	Ea	28.60	7.05	35.65
Towel rack, swing-arm, chrome, 3 bars, 12" L	BC@.280	Ea	14.30	7.05	21.35
Towel rings					
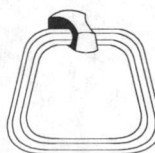 Chrome base, chrome ring	BC@.343	Ea	19.40	8.63	28.03
Chrome base, clear plastic ring	BC@.343	Ea	14.30	8.63	22.93
Towel shelf, chrome, 24" L with bar below	BC@.280	Ea	36.70	7.05	43.75

Beds, Folding

	Craft@Hrs	Unit	Material	Labor	Total

Beds, Folding Concealed-in-wall type. Steel framed, folding wall bed system. Bed requires 18-5/8" or 22" deep recess. Includes frame, lift mechanism, all hardware. Installed in framed opening. Padded vinyl headboard. Bed face panel accepts paint, wallpaper, vinyl or laminate up to 1/4" thick. Add for box spring and mattress. Sico Incorporated

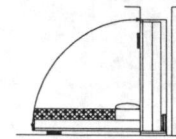

	Craft@Hrs	Unit	Material	Labor	Total
Twin, 41" W x 83-1/2" H x 18-5/8" D	B1@5.41	Ea	1,024.00	124.00	1,148.00
Box spring and mattress	—	Ea	321.00	—	321.00
Twin-long, 41" W x 88-5/8" H x 18-5/8" D	B1@5.41	Ea	1,057.00	124.00	1,181.00
Box spring and mattress	—	Ea	340.00	—	340.00
Double, 56" W x 83-1/2" H x 18-5/8" D	B1@5.41	Ea	1,024.00	124.00	1,148.00
Box spring and mattress	—	Ea	399.00	—	399.00
Double-long, 56" W x 88-5/8-1/2" H x 18-5/8" D	B1@5.41	Ea	1057.00	124.00	1,181.00
Box spring and mattress	—	Ea	422.00	—	422.00
Queen, 62-1/2" W x 88-5/8-1/2" H x 18-5/8" D	B1@5.41	Ea	1120.00	124.00	1,244.00
Box spring and mattress	—	Ea	481.00	—	481.00
King, 79" W x 88-5/8" H x 18-5/8" D	B1@5.41	Ea	1906.00	124.00	2,030.00
Box spring and mattress	—	Ea	693.00	—	693.00
Add for chrome frame, any bed	—	LS	216.00	—	216.00
Add for brass frame, any bed	—	LS	233.00	—	233.00

Blueprinting (Reproduction only) Assumes original is on semi-transparent drafting paper or film. See also Architectural Illustration and Drafting. Cost per square foot except where noted. Stapled edge and binder included. Diazo

	Craft@Hrs	Unit	Material	Labor	Total
Blueline / Blackline					
1-100 SF	—	SF	—	—	.13
101-1000 SF	—	SF	—	—	.12
1001-2000 SF	—	SF	—	—	.11
2001-to 3000 SF	—	SF	—	—	.10
3001-4000 SF	—	SF	—	—	.09
4001-UP	—	SF	—	—	.08
Pres. Blackline	—	SF	—	—	.60
Sepia	—	SF	—	—	.60
Mylar	—	SF	—	—	2.25
Xerographic					
Bond	—	SF	—	—	.50
Translucent bond	—	SF	—	—	.75
Vellum	—	SF	—	—	1.00
Erasable vellum	—	SF	—	—	1.25
Mylar	—	SF	—	—	2.25
Enlargements (Bond)	—	SF	—	—	.60
Reductions (Bond, per sheet)	—	Ea	—	—	2.50
Plotting					
Bond	—	SF	—	—	1.25
Translucent bond	—	SF	—	—	1.50
Vellum	—	SF	—	—	1.75
Erasable vellum	—	SF	—	—	2.50
Mylar	—	SF	—	—	2.50
Photo					
Mylar	—	SF	—	—	5.65
Add for local pickup and delivery, round trip	—	LS	—	—	5.00

	Craft@Hrs	Unit	Material	Labor	Total

Building Inspection Service (Home inspection service) Inspection of all parts of building by qualified engineer or certified building inspection technician. Includes written report covering all doors and windows, electrical system, foundation, heating and cooling system, insulation, interior and exterior surface conditions, landscaping, plumbing system, roofing, and structural integrity.

	Craft@Hrs	Unit	Material	Labor	Total
Single-family residence					
Base fee (up to 2,500 SF)	—	LS	—	—	250.00
Add for additional 1,000 SF or fraction	—	LS	—	—	70.00
Add for out buildings (each)	—	LS	—	—	50.00
Add for houses over 50 years old	—	LS	—	—	75.00
Add per room for houses with over 10 rooms	—	Ea	—	—	60.00
Add per room for houses with over 15 rooms	—	Ea	—	—	65.00
Add for swimming pool, spa or sauna	—	LS	—	—	200.00
Add for soil testing (expansive soil only)	—	LS	—	—	200.00
Add for water testing (coliform only)	—	LS	—	—	70.00
Add for warranty protection					
Houses to 10 rooms & 50 years old	—	LS	—	—	250.00
Houses over 50 years old	—	LS	—	—	270.00
Houses over 10 rooms	—	LS	—	—	270.00
Multi-family structures					
Two family residence base fee	—	LS	—	—	350.00
Apartment or condominium base fee	—	LS	—	—	205.00
Warranty protection (base cost)	—	LS	—	—	250.00
Add for each additional unit	—	LS	—	—	35.00
Add for each family living unit					
Standard inspection	—	LS	—	—	50.00
Detailed inspection	—	LS	—	—	75.00
Add for swimming pool, spa, sauna	—	LS	—	—	75.00
Add for potable water quality testing	—	LS	—	—	225.00
Add for water quantity test, per well	—	LS	—	—	150.00
Add for soil testing (EPA toxic)	—	LS	—	—	1,500.00
Add for soil testing (lead)	—	LS	—	—	125.00
Add for lead paint testing, full analysis, per room	—	LS	—	—	40.00
Hazards testing for single and multi-family dwellings					
Urea-formaldehyde insulation testing	—	LS	—	—	175.00
Asbestos testing	—	LS	—	—	175.00
Radon gas testing	—	LS	—	—	125.00
Geotechnical site examination, typical price	—	LS	—	—	400.00

Building Paper See also Roofing for roof applications and Polyethylene Film. Costs include 7% coverage allowance for 2" lap and 5% waste allowance. See installation costs at the end of this section.

	Craft@Hrs	Unit	Material	Labor	Total
Fortifiber products					
Asphalt felt					
15 lb., (432 SF roll at $24.00), 36" x 144'	—	SF	.05	—	.05
30 lb., (216 SF roll at $25.00), 36" x 72'	—	SF	.12	—	.12
Asphalt shake felt					
30 lb., (180 SF roll at $20.00), 18" x 120'	—	SF	.12	—	.12
Double Kraft (Aquabar)					
Class A, 36" wide, 30-50-30, (500 SF roll at $28.00)	—	SF	.06	—	.06
Class B, 36" wide, 30-30-30, (500 SF roll at $12.25)	—	SF	.03	—	.03

Building Paper

	Craft@Hrs	Unit	Material	Labor	Total
"Jumbo Tex" gun grade sheathing paper, 40" wide					
(324 SF roll at $6.65)	—	SF	.03	—	.03
"Jumbo Tex" black building paper, 36", 40" wide					
(500 SF roll at $15.00)	—	SF	.03	—	.03
Red rosin sized sheathing (duplex sheathing) 36" wide					
(500 SF roll at $9.95)	—	SF	.02	—	.02
Flashing paper					
Kraft paper, 6" wide x 150' long					
(75 SF roll at $6.95)	—	SF	.08	—	.08
Moistop flashing paper, 6" wide x 300' long					
(150 SF roll at $18.85)	—	SF	.14	—	.14
Copper armored Sisalkraft, 10" wide x 120' long					
1 oz per SF (at $120.00 per 120 LF roll)	—	LF	1.05	—	1.05
3 oz per SF (at $130.00 per 120 LF roll)	—	LF	1.14	—	1.14
Above grade vapor barriers, no waste or coverage allowance included.					
Vaporstop 298, (fiberglass reinforcing and asphaltic adhesive between 2 layers of Kraft),					
32" x 405' roll (1,080 SF roll at $83.40)	—	SF	.08	—	.08
Pyro-Kure 600, (2 layers of heavy Kraft with fire retardant adhesive edge reinforced with fiberglass)					
32" x 405' roll (1,080 SF roll at $156.00)	—	SF	.14	—	.14
Foil Barrier 718 (fiberglass reinforced aluminum foil and adhesive between 2 layers of Kraft)					
52" x 231' roll (1,000 SF roll at $243.00)	—	SF	.24	—	.24
Below grade vapor barriers					
Moistop (fiberglass reinforced Kraft between 2 layers of polyethylene)					
8' x 250' roll (2,000 SF roll at $155.00)	—	SF	.08	—	.08
Concrete curing papers					
Orange Label Sisalkraft (fiberglass and adhesive between 2 layers of Kraft), 4.8 lbs. per CSF					
48" x 125' roll, (900 SF roll at $81.00)	—	SF	.09	—	.09
Sisalkraft SK-10, economy papers, fiberglass and adhesive between 2 layers of Kraft, 4.2 lbs. per CSF, 48" x 300' roll (1,200 SF roll at $56.00)	—	SF	.05	—	.05
Protective papers					
Seekure (fiberglass reinforcing strands and non staining adhesive between 2 layers of Kraft)					
48" x 300' roll, (1,200 SF roll at $80.70)	—	SF	.07	—	.07
Tyvek™ house wrap by DuPont					
Air infiltration barrier (high-density polyethylene fibers in sheet form)					
3' x 195' rolls (585 SF @ $62.00) or 9' x 195' rolls (1755 SF @ $180.00)	—	SF	.11	—	.11
Labor to install building papers					
Felts, vapor barriers, infiltration barriers, building papers on walls, ceilings, roofs					
Heavy stapled, typical	BC@.003	SF	—	.08	.08
Tack stapled, typical	BC@.002	SF	—	.05	.05
Curing papers, protective papers and vapor barriers, minimal fasteners	BC@.001	SF	—	.03	.03
Flashing papers, 6" to 8" wide	BC@.010	LF	—	.25	.25

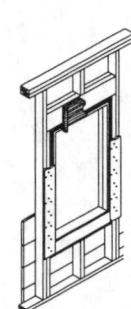

Building Permit Fees These fees are based on recommendations in the 1997 Uniform Building Code which was available May 1997 and will apply in many areas. Building Departments publish fee schedules. The permit fee will be doubled when work is started without getting a permit. When the valuation of the proposed construction exceeds $1,000, plans are often required. Estimate the plan check fee at 65% of the permit fee. The fee for reinspection is $47 per hour. Inspections outside normal business hours are $47 per hour with a two hour minimum. Additional plan review required by changes, additions, or revisions to approved plans are $47 per hour with a one-half hour minimum.

The minimum fee for construction values to $500 is $33.50

Over $500 to $2,000 the fee is $23.50 for the first $500
   plus $3.05 for each additional $100 or fraction thereof, up to $2,000.

Over $2,000 to $25,000 the fee is $69.25 for the first $2,000
   plus $14.00 for each additional $1,000 or fraction thereof to $25,000.

Over $25,000 to $50,000 the fee is $391.75 for the first $25,000
   plus $10.10 for each additional $1,000 or fraction thereof to $50,000.

Over $50,000 to $100,000 the fee is $643.75 for the first $50,000
   plus $7.00 for each additional $1,000 or fraction thereof to $100,000.

Over $100,000 to $500,000 the fee is $993.75 for the first $100,000
   plus $5.60 for each additional $1,000 or fraction thereof to $500,000.

Over $500,000 to $1,000,000 the fee is $3,233.75 for the first $500,000
   plus $4.75 for each additional $1,000 or fraction thereof to $1,000,000.

Over $1,000,000 the fee is $5,608.75 for the first $1,000,000
   plus $3.65 for each additional $1,000 or fraction thereof.

Cabinets, Kitchen See also Vanities. Good quality mill-made modular units with solid hardwood face frames, hardwood door frames and drawer fronts, hardwood veneer on raised door panels (front and back), glued mortise, dowel, and dado joint construction, full backs (1/8" vinyl laminated plywood), vinyl laminated cabinet interiors, vinyl laminated composition drawer bodies with nylon and metal guides. Includes self-closing hinges, door and drawer pulls, mounting hardware and adjustable shelves. See illustrations for unit types. See the price adjustments below for pricing of other units. No countertops included. See Countertops

Kitchen cabinet costs vary widely. The prices listed in this section are for standard grade cabinets. Add 65% to material costs for premium grade cabinets with solid hardwood fronts and frames, mitered corners and solid wood drawer bodies with steel guides and ball bearings. Deduct 45% from material costs for economy grade cabinets, melamine laminated to particleboard, in textured colors or woodgrain print finish.

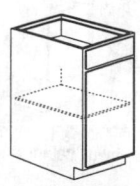

One door base

Drawer base

	Craft@Hrs	Unit	Material	Labor	Total
Cabinets, Kitchen (See the note above concerning cabinet costs.)					
One door base cabinets, 34-1/2" high, 24" deep					
9" wide, tray divider	BC@.461	Ea	109.00	11.60	120.60
12" wide, 1 door, 1 drawer	BC@.461	Ea	146.00	11.60	157.60
15" wide, 1 door, 1 drawer	BC@.638	Ea	156.00	16.10	172.10
18" wide, 1 door, 1 drawer	BC@.766	Ea	166.00	19.30	185.30
21" wide, 1 door, 1 drawer	BC@.766	Ea	178.00	19.30	197.30
24" wide, 1 door, 1 drawer	BC@.911	Ea	183.00	22.90	205.90
Drawer base cabinets, 34-1/2" high, 24" deep					
15" wide, 4 drawers	BC@.638	Ea	161.00	16.10	177.10
18" wide, 4 drawers	BC@.766	Ea	171.00	19.30	190.30
24" wide, 4 drawers	BC@.911	Ea	198.00	22.90	220.90

Cabinets, Kitchen

	Craft@Hrs	Unit	Material	Labor	Total
Sink base cabinets, 34-1/2" high, 24" deep					
24" wide, 1 door, 1 drawer front	BC@.740	Ea	156.00	18.60	174.60
30" wide, 2 doors, 2 drawer fronts	BC@.766	Ea	188.00	19.30	207.30
33" wide, 2 doors, 2 drawer fronts	BC@.766	Ea	198.00	19.30	217.30
36" wide, 2 doors, 2 drawer fronts	BC@.766	Ea	203.00	19.30	222.30
42" wide, 2 doors, 2 drawer fronts	BC@.911	Ea	224.00	22.90	246.90
48" wide, 2 doors, 2 drawer fronts	BC@.911	Ea	245.00	22.90	267.90

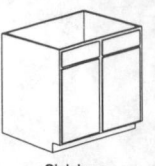
Sink base

	Craft@Hrs	Unit	Material	Labor	Total
Two door base cabinets, 34-1/2" high, 24" deep					
27" wide, 2 doors, 2 drawers	BC@1.25	Ea	245.00	31.50	276.50
30" wide, 2 doors, 2 drawers	BC@1.25	Ea	260.00	31.50	291.50
33" wide, 2 doors, 2 drawers	BC@1.25	Ea	270.00	31.50	301.50
36" wide, 2 doors, 2 drawers	BC@1.35	Ea	276.00	34.00	310.00
42" wide, 2 doors, 2 drawers	BC@1.50	Ea	297.00	37.80	334.80
48" wide, 2 doors, 2 drawers	BC@1.71	Ea	328.00	43.00	371.00

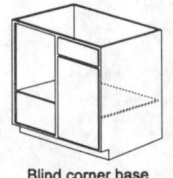
Blind corner base

	Craft@Hrs	Unit	Material	Labor	Total
Blind corner base cabinets, 34-1/2" high					
Minimum 36", maximum 39" at wall	BC@1.39	Ea	178.00	35.00	213.00
Minimum 39", maximum 42" at wall	BC@1.50	Ea	193.00	37.80	230.80
45-degree corner base, revolving, 34-1/2" high					
36" wide at each wall	BC@2.12	Ea	276.00	53.40	329.40
Corner sink front, 34-1/2" high					
40" wide at walls	BC@2.63	Ea	156.00	66.20	222.20

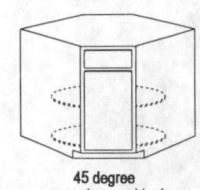
45 degree
corner base cabinet

	Craft@Hrs	Unit	Material	Labor	Total
Wall cabinets, adjustable shelves, 30" high, 12" deep					
9" wide, 1 door	BC@.461	Ea	99.00	11.60	110.60
12" wide or 15" wide, 1 door	BC@.461	Ea	104.00	11.60	115.60
18" wide, 1 door	BC@.638	Ea	126.00	16.10	142.10
21" wide, 1 door	BC@.638	Ea	131.00	16.10	147.10
24" wide, 1 door	BC@.766	Ea	141.00	19.30	160.30
27" wide, 2 doors	BC@.766	Ea	166.00	19.30	185.30
30" wide, 2 doors	BC@.911	Ea	166.00	22.90	188.90
33" wide, 2 doors	BC@.911	Ea	178.00	22.90	200.90
36" wide, 2 doors	BC@1.03	Ea	188.00	25.90	213.90
42" wide, 2 doors	BC@1.03	Ea	203.00	25.90	228.90
48" wide, 2 doors	BC@1.16	Ea	218.00	29.20	247.20

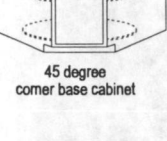
1 door
wall cabinet

	Craft@Hrs	Unit	Material	Labor	Total
Above-appliance wall cabinets, 12" deep					
12" high, 30" wide, 2 doors	BC@.461	Ea	104.00	11.60	115.60
15" high, 30" wide, 2 doors	BC@.461	Ea	119.00	11.60	130.60
15" high, 33" wide, 2 doors	BC@.537	Ea	126.00	13.50	139.50
15" high, 36" wide, 2 doors	BC@.638	Ea	131.00	16.10	147.10
18" high, 18" wide, 2 doors	BC@.537	Ea	99.00	13.50	112.50
18" high, 30" wide, 2 doors	BC@.766	Ea	131.00	19.30	150.30
18" high, 36" wide, 2 doors	BC@.911	Ea	141.00	22.90	163.90

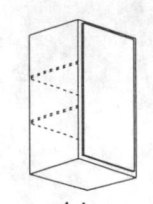
Above-appliance
wall cabinet

	Craft@Hrs	Unit	Material	Labor	Total
Corner wall cabinets, 30" high, 12" deep					
24" at each wall, fixed shelves	BC@1.03	Ea	178.00	25.90	203.90
24" at each wall, revolving shelves	BC@1.03	Ea	240.00	25.90	265.90
Blind corner wall cabinets, 30" high					
24" minimum, 1 door	BC@1.03	Ea	136.00	25.90	161.90
36" minimum, 1 door	BC@1.32	Ea	166.00	33.20	199.20
42" minimum, 2 doors	BC@1.20	Ea	208.00	30.20	238.20

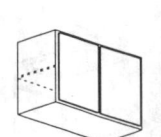

	Craft@Hrs	Unit	Material	Labor	Total
Utility cabinet, 66" high, 12" deep, no shelves					
18" wide	BC@1.32	Ea	208.00	33.20	241.20
24" wide	BC@1.71	Ea	235.00	43.00	278.00

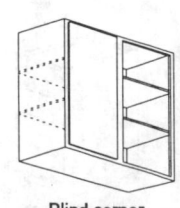
Blind corner
wall cabinet

	Craft@Hrs	Unit	Material	Labor	Total

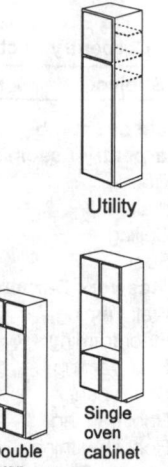

Utility

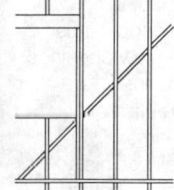

Double oven cabinet

Single oven cabinet

Utility cabinet, 66" high, 24" deep, add shelf cost below

	Craft@Hrs	Unit	Material	Labor	Total
18" wide	BC@1.24	Ea	218.00	31.20	249.20
24" wide	BC@1.71	Ea	260.00	43.00	303.00
Add for utility cabinet revolving shelves, includes mounting hardware					
18" wide x 24" deep	BC@.360	Ea	188.00	9.06	197.06
24" wide x 24" deep	BC@.360	Ea	218.00	9.06	227.06
Add for utility cabinet plain shelves					
18" wide x 24" deep	BC@.541	Ea	63.00	13.60	76.60
24" wide x 24" deep	BC@.541	Ea	66.00	13.60	79.60
Oven cabinets, 66" high, 24" deep					
27" wide, single oven	BC@2.19	Ea	260.00	55.10	315.10
27" wide, double oven	BC@2.19	Ea	201.00	55.10	256.10
Microwave cabinet, with trim,					
21" high, 20" deep, 30" wide	BC@.986	Ea	156.00	24.80	180.80

Cabinets, Rule of Thumb Cabinet cost per running foot of cabinets installed. These figures are useful only if a complete kitchen is being estimated. They are an average of the cabinets found in a typical 1,600 SF house. The base cabinet price assumes a sink base cabinet, 3 drawer bases and 6 base cabinets with doors.

	Craft@Hrs	Unit	Material	Labor	Total
Base cabinets, 34-1/2" high, 24" deep	BC@.521	LF	120.00	13.10	133.10
Wall cabinets, 30" high, 12" deep	BC@.340	LF	62.60	8.56	71.16

Carpentry See also the carpentry items in separate sections: Cabinets, Ceilings, Closet Door Systems, Countertops, Cupolas, Doors, Entrances, Flooring, Framing Connectors, Hardboard, Hardware, Lumber, Moulding, Paneling, Shutters, Siding, Skylights, Soffits, Stairs, Thresholds, and Weather-stripping.

Carpentry Rule of Thumb Rough carpentry (framing) cost per square foot of living area. These figures will apply on many types of residential jobs. The seven lines of cost data under "cost breakdown" below show the cost breakdown for framing a medium cost house. Labor on tract homes may run 50% less and complex framing jobs will cost more.

	Craft@Hrs	Unit	Material	Labor	Total
Low cost house	B1@.197	SF	4.04	4.50	8.54
Medium cost house	B1@.237	SF	4.56	5.42	9.98
Better quality house	B1@.314	SF	6.23	7.18	13.41

Carpentry, Rule of Thumb Cost Breakdown Costs per square foot of living area for single-story residences based on lumber at $544 per MBF. These figures were used to compile the framing cost for a medium cost home in the paragraph above.

	Craft@Hrs	Unit	Material	Labor	Total
Sills, pier blocks, floor beams					
(145 BF per 1,000 SF)	B1@.018	SF	.08	.41	.49
Floor joists, doublers, blocking, bridging					
(1,480 BF per 1,000 SF)	B1@.028	SF	.81	.64	1.45
Underlayment, 5/8" (at $571 per MSF)					
(1,150 SF per 1,000 SF)	B1@.011	SF	.66	.25	.91
Layout, studs, sole plates, top plates, header and					
end joists, backing, blocking, bracing and framing for openings					
(2,250 BF per 1,000 SF)	B1@.093	SF	1.22	2.13	3.35
Ceiling joists, header and end joists, backing, blocking and bracing					
(1,060 BF per 1,000 SF)	B1@.045	SF	.58	1.03	1.61
Rafters, braces, collar beams, ridge boards, 2" x 8" rafters					
16" OC, (1,340 BF per 1000 SF)	B1@.032	SF	.73	.73	1.46
Roof sheathing, 1/2" CDX plywood (at $415 per MSF)					
(1,150 SF per 1,000 SF)	B1@.010	SF	.48	.23	.71

Carpentry, Assemblies

	Craft@Hrs	Unit	Material	Labor	Total

Floor Assemblies Costs for wood framed floor joists with subflooring and R-19 insulation, based on performing the work at the construction site. These costs include the floor joists, subflooring as described, blocking, nails and 6-1/4" thick R-19 fiberglass insulation between the floor joists. Figures include box or band joists and typical double joists. No beams included. Floor joists and blocking are based on Std & Btr grade lumber at an average price of $750 per MBF. Underlayment is based on 5/8" thick at $16.50 per 4' x 8' sheet ($.52 per SF) and 3/4" thick at $19.80 per 4' x 8' sheet ($.62 per SF). 1" x 6" planked subflooring is based on 3 & Btr lumber at $760 per MBF with 1.24 BF per square foot of floor. Costs shown are per square foot of area covered and include normal waste. Deduct for openings over 25 SF.

Floor joists 16" OC, R-19 insulation and plywood subflooring

5/8" underlayment					
2" x 6" joists 16" o.c.	B1@.040	SF	1.46	.91	2.37
2" x 8" joists 16" o.c.	B1@.041	SF	1.64	.94	2.58
2" x 10" joists 16" o.c.	B1@.043	SF	2.32	.98	3.30
2" x 12" joists 16" o.c.	B1@.044	SF	3.77	1.01	4.78
3/4" underlayment					
2" x 6" joists 16" o.c.	B1@.042	SF	1.55	.96	2.51
2" x 8" joists 16" o.c.	B1@.043	SF	2.08	.98	3.06
2" x 10" joists 16" o.c.	B1@.045	SF	2.51	1.03	3.54
2" x 12" joists 16" o.c.	B1@.046	SF	2.88	1.05	3.93

Floor joists 16" OC, R-19 insulation and planked subflooring

1" x 6" plank flooring					
2" x 6" joists 16" o.c.	B1@.047	SF	1.81	1.07	2.88
2" x 8" joists 16" o.c.	B1@.048	SF	1.99	1.10	3.09
2" x 10" joists 16" o.c.	B1@.050	SF	2.67	1.14	3.81
2" x 12" joists 16" o.c.	B1@.051	SF	4.12	1.17	5.29

For different type insulation					
Fiberglass batts					
10" thick R-30, add	—	SF	.13	—	.13
12" thick R-38, add	—	SF	.28	—	.28

Wall Assemblies Costs for wood framed stud walls with wall finish treatment on both sides, based on performing the work at the construction site. These costs include wall studs at 16" center to center, double top plates, single bottom plates, fire blocking, nails and wall finish treatment as described. No headers or posts included. All lumber is Std & Btr. 2" x 4" walls have 1.12 BF (at $675 per MBF) per SF of wall and 2" x 6" walls have 1.68 BF (at $660 per MBF) per SF of wall. Costs shown are per SF or LF of wall measured on one face and include normal waste.

Interior wall assemblies

2" x 4" stud walls with 1/2" gypsum drywall both sides, ready for painting					
Cost per square foot	B1@.064	SF	.94	1.46	2.40
Cost per running foot, for 8' high walls	B1@.512	LF	7.52	11.70	19.22
2" x 4" stud walls with 5/8" gypsum drywall both sides, ready for painting					
Cost per square foot	B1@.068	SF	.96	1.55	2.51
Cost per running foot, for 8' high walls	B1@.544	LF	7.68	12.40	20.08
2" x 6" stud walls with 1/2" gypsum drywall both sides, ready for painting					
Cost per square foot	B1@.072	SF	1.16	1.65	2.81
Cost per running foot, for 8' high walls	B1@.576	LF	9.28	13.20	22.48
2" x 6" stud walls with 5/8" gypsum drywall both sides, ready for painting					
Cost per square foot	B1@.076	SF	1.18	1.74	2.92
Cost per running foot, for 8' high walls	B1@.608	LF	9.44	13.90	23.34

	Craft@Hrs	Unit	Material	Labor	Total

Exterior wall assemblies

2" x 4" stud walls with drywall interior, wood siding exterior, 1/2" gypsum drywall inside face ready for painting, over 3-1/2" R-11 insulation with 5/8" thick rough sawn T-1-11 (4 ply, 4' x 8' panels at $950 per MSF), exterior grade plywood siding on the outside face.

	Craft@Hrs	Unit	Material	Labor	Total
Cost per square foot	B1@.068	SF	1.95	1.55	3.50
Cost per running foot, for 8' high walls	B1@.544	LF	15.60	12.40	28.00

2" x 6" stud walls with drywall interior, wood siding exterior, same construction as above, except with 6-1/4" R-19 insulation

	Craft@Hrs	Unit	Material	Labor	Total
Cost per square foot	B1@.077	SF	2.05	1.76	3.81
Cost per running foot, for 8' high walls	B1@.616	LF	16.40	14.10	30.50

2" x 4" stud walls with drywall interior, 1" x 6" drop siding exterior, 1/2" gypsum drywall on inside face ready for painting, over 3-1/2" R-11 insulation with 1" x 6" southern yellow pine drop siding, D grade, (1.19 BF per SF with 5-1/4" exposure, at $940 per MBF) on the outside face.

	Craft@Hrs	Unit	Material	Labor	Total
Cost per square foot	B1@.074	SF	2.36	1.69	4.05
Cost per running foot, for 8' high wall	B1@.592	LF	18.88	13.50	32.38

2" x 6" stud walls with drywall interior, 1" x 6" drop siding exterior, same construction as above, except with 6-1/4" R-19 insulation

	Craft@Hrs	Unit	Material	Labor	Total
Cost per square foot	B1@.083	SF	2.71	1.90	4.61
Cost per running foot, for 8' high wall	B1@.664	LF	21.70	15.20	36.90

2" x 4" stud walls with drywall interior, stucco exterior, 1/2" gypsum drywall on inside face ready for painting, over 3-1/2" R-11 insulation and a three-coat exterior plaster (stucco) finish with integral color on the outside face

	Craft@Hrs	Unit	Material	Labor	Total
Cost per square foot	B1@.050	SF	2.95	1.14	4.09
Cost per running foot, for 8' high wall	B1@.400	LF	23.60	9.14	32.74

2" x 6" stud walls with drywall interior, stucco exterior, same construction as above, except with 6-1/4" R-19 insulation

	Craft@Hrs	Unit	Material	Labor	Total
Cost per square foot	B1@.059	SF	3.23	1.35	4.58
Cost per running foot, for 8' high wall	B1@.472	LF	25.84	10.80	36.64

Add for different type gypsum board
1/2" or 5/8" moisture resistant greenboard

	Craft@Hrs	Unit	Material	Labor	Total
Cost per SF, greenboard per side, add	—	SF	.12	—	.12

1/2" or 5/8" moisture resistant greenboard

	Craft@Hrs	Unit	Material	Labor	Total
Cost per running foot per side 8' high	—	LF	.96	—	.96

5/8" thick regular gypsum drywall

	Craft@Hrs	Unit	Material	Labor	Total
Cost per SF, per side, add	—	SF	.03	—	.03

5/8" thick regular gypsum drywall

	Craft@Hrs	Unit	Material	Labor	Total
Cost per running foot per side 8' high	—	LF	.20	—	.20

Ceiling Assemblies Costs for wood framed ceiling joists with ceiling finish and fiberglass insulation, based on performing the work at the construction site. These costs include the ceiling joists, ceiling finish as described, blocking, nails and 3-1/2" thick R-11 fiberglass insulation batts between the ceiling joists. Figures in parentheses indicate board feet per square foot of ceiling framing including end joists and typical header joists. No beams included. Ceiling joists and blocking are based on Std & Btr grade lumber. Costs shown are per square foot of area covered and include normal waste. Deduct for openings over 25 SF.

Ceiling joists with regular gypsum drywall taped and sanded smooth finish, ready for paint

2" x 4" ceiling joists at 16" on center (.59 BF per SF at $675 per MBF), with

	Craft@Hrs	Unit	Material	Labor	Total
insulation and 1/2" gypsum drywall	B1@.053	SF	.84	1.21	2.05

2" x 6" ceiling joists at 16" on center (.88 BF per SF at $660 per MBF), with

	Craft@Hrs	Unit	Material	Labor	Total
insulation and 1/2" gypsum drywall	B1@.055	SF	.98	1.26	2.24

2" x 8" ceiling joists at 16" on center (1.17 BF per SF at $690 per MBF), with

	Craft@Hrs	Unit	Material	Labor	Total
insulation and 1/2" gypsum drywall	B1@.057	SF	1.14	1.30	2.44

	Craft@Hrs	Unit	Material	Labor	Total
For spray applied plaster finish (sometimes called "popcorn" or "cottage cheese" texture)					
Add for ceiling texture	DT@.003	SF	.08	.07	.15
For different type gypsum drywall with taped and sanded smooth finish, ready for paint					
1/2" fire rated type X board, add	—	SF	.01	—	.01
1/2" moisture resistant greenboard, add	—	SF	.11	—	.11
5/8" thick plain board, add	—	SF	.02	—	.02
5/8" fire rated type X board, add	—	SF	.01	—	.01
For different ceiling joist center to center dimensions					
2" x 4" ceiling joists					
12" on center, add	B1@.004	SF	.10	.09	.19
20" on center, deduct	—	SF	-.06	-.06	-.12
24" on center, deduct	—	SF	-.09	-.12	-.21
2" x 6" ceiling joists					
12" on center, add	B1@.006	SF	.14	.14	.28
20" on center, deduct	—	SF	-.08	-.06	-.14
24" on center, deduct	—	SF	-.13	-.11	-.24
2" x 8" ceiling joists					
12" on center, add	B1@.006	SF	.19	.14	.33
20" on center, deduct	—	SF	-.11	-.08	-.19
24" on center, deduct	—	SF	-.17	-.13	-.30
For different type insulation					
Fiberglass batts					
6-1/4" thick R-19, add	—	SF	.13	—	.13
10" thick R-30, add	—	SF	.26	—	.26
12" thick R-38, add	—	SF	.38	—	.38
Blown-in fiberglass					
8" thick R-19, add	—	SF	.04	—	.04

Roofing Assemblies Costs for wood framed roof assemblies with roof finish material as shown based on performing the work at the construction site. Costs shown include all material and labor required above the top plate or ledger on the supporting walls. These costs assume the supporting wall structure is in-place and suitable for the assembly described.

Flat roof assembly Based on using 2" x 12" joists (at $587 per MBF) Std & Btr grade at 16" on center including blocking and ripped strips, 1/2" CDX (at $12.60 per 4' x 8' sheet) roof sheathing. Costs shown include an allowance for normal waste. The finished roof treatment is built-up 3 ply asphalt consisting of 2 plies of 15 lb. felt with 90 lb. cap sheet and 3 hot mop coats of asphalt.

	Craft@Hrs	Unit	Material	Labor	Total
Flat roof assembly as described above					
Framing, per Sq. (Sq. = 100 SF)	B1@5.00	Sq	188.00	114.00	302.00
Built-up roofing, per Sq.	B1@1.30	Sq	32.00	29.70	61.70
Total assembly cost	**B1@6.30**	**Sq**	**220.00**	**143.70**	**363.70**

Conventionally framed roof assemblies Based on straight gable type roof (no hips, valleys, or dormers) with 6" in 12" rise or less. Cost per 100 square feet (Sq.) of plan area under the roof, not actual roof surface area. Framing includes 2" x 8" common rafters (at $523 per MBF) Std & Btr grade at 24" on center. Cost includes blocking, ridge and normal bracing. Roof sheathing is 1/2" CDX plywood (at $12.60 per 4' x 8' sheet). Finish as shown.

	Craft@Hrs	Unit	Material	Labor	Total

Conventionally framed roof assembly with built-up rock finish roofing. The finished roof treatment is built-up 3 ply asphalt consisting of 1 ply 30 lb. felt, 2 plies of 15 lb. felt and 3 hot mop coats of asphalt with decorative crush rock spread at 180 lbs. per Sq.

	Craft@Hrs	Unit	Material	Labor	Total
Framing, per Sq (Sq = 100 SF)	B1@4.10	Sq	103.00	93.70	196.70
Built-up rock finish roofing, per Sq	B1@1.34	Sq	46.50	30.60	77.10
Total assembly cost	**B1@5.44**	**Sq**	**149.50**	**124.30**	**273.80**

Conventionally framed roof assembly with 25 year fiberglass shingle roofing.

	Craft@Hrs	Unit	Material	Labor	Total
Framing, per Sq (Sq = 100 SF)	B1@4.10	Sq	103.00	93.70	196.70
Fiberglass shingle roofing, per Sq	B1@1.83	Sq	44.00	41.80	85.80
Total assembly cost	**B1@5.93**	**Sq**	**147.00**	**135.50**	**282.50**

Conventionally framed roof assembly with 25 year asphalt shingle roofing.

	Craft@Hrs	Unit	Material	Labor	Total
Framing, per Sq (Sq = 100 SF)	B1@4.10	Sq	103.00	93.70	196.70
Asphalt shingle roofing, per Sq	B1@2.07	Sq	88.00	47.30	135.30
Total assembly cost	**B1@6.17**	**Sq**	**191.00**	**141.00**	**332.00**

Conventionally framed roof assembly, 28 gauge galvanized corrugated steel roofing, no plywood sheathing.

	Craft@Hrs	Unit	Material	Labor	Total
Framing, per Sq (Sq = 100 SF)	B1@2.80	Sq	103.00	64.00	167.00
28 gauge galvanized steel roofing, per Sq	B1@2.70	Sq	80.00	61.70	141.70
Total assembly cost	**B1@5.50**	**Sq**	**183.00**	**125.70**	**308.70**

Add for gable studs (at $523 per MBF)

	Craft@Hrs	Unit	Material	Labor	Total
2" x 4" spaced 16" OC (.54 BF per SF)	B1@.023	SF	.28	.53	.81

Add for purlins (purling), Std & Btr, installed below roof rafters. Figures in parentheses indicate board feet per LF including 5% waste

	Craft@Hrs	Unit	Material	Labor	Total
2" x 8" (at $571 per MBF, 1.40 BF per LF)	B1@.023	LF	.80	.53	1.33

Piecework Rough Carpentry Rough carpentry on residential tracts is usually done by framing subcontractors who bid at piecework rates (such as per square foot of floor). The figures below list typical piecework rates for repetitive framing work and assume all materials are supplied to the framing subcontractor. No figures appear in the Craft@Hrs column because the work is done for a fixed price per square foot and the labor productivity can be expected to vary widely.

Layout and plating Piecework rates
Lay out wall plates according to the plans (snap chalk lines for wall plates, mark location for studs, windows, doors and framing details), cut top and bottom plates and install bottom plates. Costs per square foot of floor (excluding garage).

	Craft@Hrs	Unit	Material	Labor	Total
Custom or more complex jobs	—	SF	—	.18	.18
Larger job, longer runs	—	SF	—	.09	.09

Wall framing. Piecework rates
Measure, cut, fit, assemble, and tip up walls, including studs, plates, cripples, let-in braces, trimmers and blocking. Costs per square foot of floor.

	Craft@Hrs	Unit	Material	Labor	Total
Complex job, radius walls, rake walls	—	SF	—	.52	.51
Larger job, 8' high walls, fewer partitions	—	SF	—	.18	.18

	Craft@Hrs	Unit	Material	Labor	Total

Plumb and align framed walls Piecework rates

Force walls into alignment, adjust walls to vertical and install temporary wood braces as needed, fasten nuts on anchor bolts or shoot power driven fasteners through wall plates into the slab. Based on accuracy to 3/16". Costs per square foot of floor are shown on next page.

Small or complex job, many braces	—	SF	—	.21	.21
Larger job, fewer braces	—	SF	—	.09	.09

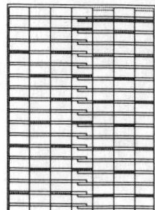

Floor joists or ceiling joists Piecework rates

Lay out, cut and install floor or ceiling joists, including rim joists, doubled joists, straps, joist hangers, blocking at 8' OC and ceiling backing for drywall. Based on larger jobs with simple joist layouts set 16" OC and pre-cut blocking supplied by the general contractor. Add the cost of floor beams, if required. Costs per square foot of horizontal joist area. More complex jobs with shorter runs may cost 50% more.

2" x 8" ceiling or floor joists	—	SF	—	.13	.13
2" x 10" ceiling or floor joists	—	SF	—	.15	.15
2" x 12" ceiling or floor joists	—	SF	—	.17	.17
2" x 14" ceiling or floor joists	—	SF	—	.21	.21
Add for 12" OC spacing	—	SF	—	.05	.05
Deduct for 20" OC spacing	—	SF	—	-.03	-.03
Deduct for 24" OC spacing	—	SF	—	-.05	-.05

Floor sheathing Piecework rates.

Lay out, cut, fit and install 5/8" or 3/4" tongue and groove plywood floor sheathing, including blocking as required. Based on nailing done with a pneumatic nailer and nails supplied by the general contractor. Costs per square foot of sheathing installed

Smaller, cut-up job	—	SF	—	.15	.15
Larger job, longer runs	—	SF	—	.11	.11
Add for 1-1/8" sheathing	—	SF	—	.03	.03

Stair framing Piecework rates

Lay out, cut, fit and install straight, U- or L-shaped 30" to 36" wide stairs made from plywood and 2" x 12" stringers set 16" OC. These costs include blocking in the adjacent stud wall and a 1" skirt board. Costs per 7-1/2" riser are shown on the next page. Framing more complex stairs may cost up to $500 per flight.

Small job, short runs	—	Ea	—	12.20	12.20
Larger job, longer runs	—	Ea	—	9.20	9.20
Add per 3' x 3' landing, including supports	—	Ea	—	31.60	31.60
Add for larger landings, including supports	—	Ea	—	78.50	78.50

Shear panels Piecework rates

Lay out, cut, fit and install structural 3/8" or 1/2" plywood wall panels. These figures assume shear studs were set correctly by others and that panel nails are driven at 4" OC with a pneumatic nailer. Not including hold-down straps, posts, shear blocking or extra studs. Costs per 4' x 9' panel installed.

Small job, many openings, 2nd floor	—	Ea	—	8.42	8.42
Larger job, few openings, 1st floor	—	Ea	—	5.25	5.25

Roof trusses Piecework rates

Setting and nailing engineered gable and hip roof trusses 24" OC on prepared wall plates. These figures assume that lifting equipment is provided by the general contractor and that the truss supplier provides a fill package, spreader blocks for each plate and the ridge and jack rafters (if required). Includes installation of ceiling backing where required and catwalks at the bottom chord. Costs per square foot of plan area under the truss.

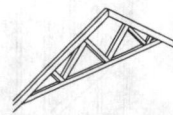

	Craft@Hrs	Unit	Material	Labor	Total

Small job assumes multiple California fill between roof surfaces and understacking

	Craft@Hrs	Unit	Material	Labor	Total
Small job, rake fill above a partition wall	—	SF	—	.47	.47
Larger job, little or no fill or understacking	—	SF	—	.27	.27

Conventional roof framing Piecework rates
Calculate lengths, lay out, cut and install 2" x 10" or 2" x 12" common, hip, valley and jack rafters on parallel and horizontal plates. Costs per square foot of plan area under the roof.

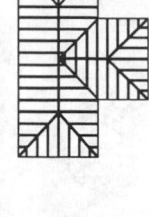

	Craft@Hrs	Unit	Material	Labor	Total
Small job, cut-up roof, few common rafters					
Rafters 12" OC	—	SF	—	.94	.94
Rafters 16" OC	—	SF	—	.79	.79
Rafters 20" OC	—	SF	—	.63	.63
Rafters 24" OC	—	SF	—	.47	.47
Larger job, longer runs, nearly all common rafters					
Rafters 12" OC	—	SF	—	.47	.47
Rafters 16" OC	—	SF	—	.42	.42
Rafters 20" OC	—	SF	—	.32	.32
Rafters 24" OC	—	SF	—	.27	.27
Add for slope over 6 in 12	—	SF	—	.15	.15
Deduct for 2" x 6" or 2" x 8" rafters	—	SF	—	.05	.05

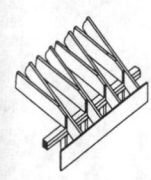

Fascia Piecework rates
Applied to rafter tails and as a barge rafter on gable ends. Includes trimming the rafter tails to the correct length and installing outlookers at gable ends. Costs per linear foot of 2" x 8" fascia installed.

	Craft@Hrs	Unit	Material	Labor	Total
Small job, short runs, with moulding	—	LF	—	2.04	2.04
Larger job, longer runs	—	LF	—	1.17	1.17

Roof sheathing Piecework rates
Lay out, cut, fit and install 1/2" or 5/8" plywood roof sheathing, including blocking and 1" x 8" starter board on overhangs as required. Based on nailing done with a pneumatic nailer and nails supplied by the general contractor. Costs per square foot of sheathing installed.

	Craft@Hrs	Unit	Material	Labor	Total
Smaller, cut-up hip and valley job	—	SF	—	.16	.16
Larger job, longer runs	—	SF	—	.11	.11
Add for slope over 6 in 12	—	SF	—	.06	.06

Carpentry Cost, Detailed Breakdown This section is arranged in the order of construction. Material costs shown here can be adjusted to reflect your actual lumber cost: divide your actual lumber cost (per MBF) by the cost listed in parentheses (per MBF). Then multiply the cost in the material column by this adjustment factor. No waste included.

Lally columns (Residential basement column) 3-1/2" diameter, steel tube

	Craft@Hrs	Unit	Material	Labor	Total
Material only	—	LF	6.68	—	6.68
Add for base or cap plate	—	Ea	2.70	—	2.70
Add for installation, per column to 12' high	B1@.458	Ea	—	10.50	10.50

Precast pier blocks Posts set on precast concrete pier block, including pier block with anchor (at $5.50) placed on existing grade, temporary 1" x 6" bracing (8 LF at $.35 = $2.80) and stakes (2 at $.40 = $.80). Cost is for each post set. Add for excavation if required

	Craft@Hrs	Unit	Material	Labor	Total
Heights to 8', cost of post not included	BL@.166	Ea	9.10	3.41	12.51

Pier pads 2" x 6", treated, #2 & Btr (lumber at $660 per MBF plus treatment at $120 per MBF)

	Craft@Hrs	Unit	Material	Labor	Total
1.10 BF per LF	B1@.034	LF	.86	.78	1.64

Carpentry, Detailed Breakdown

	Craft@Hrs	Unit	Material	Labor	Total

Posts 4" x 4" material costs including 10% waste (1.47 BF per LF). See also Lally columns and Posts (immediately following studding) in this section and Posts in Lumber section

	Craft@Hrs	Unit	Material	Labor	Total
Fir, rough Std & Btr, K.D. ($735 MBF)	—	LF	1.08	—	1.08
Red cedar, rough green constr ($1,140 MBF)	—	LF	1.67	—	1.67
Redwood, S4S constr heart ($1,420 MBF)	—	LF	3.08	—	3.08
Redwood, rough constr heart ($1,380 MBF)	—	LF	3.31	—	3.31
Southern yellow pine #2, treated ($740 MBF)	—	LF	1.08	—	1.08

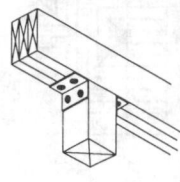

Girders Std & Btr, first floor work. Figures in parentheses show board feet per linear foot of girder, including 7% waste

	Craft@Hrs	Unit	Material	Labor	Total
4" x 6" (2.15 BF per LF, $748 per MBF)	B1@.034	LF	1.60	.78	2.38
4" x 8" (2.85 BF per LF, $741 per MBF)	B1@.045	LF	2.11	1.03	3.14
4" x 10" (3.58 BF per LF, $748 per MBF)	B1@.057	LF	2.68	1.30	3.98
4" x 12" (4.28 BF per LF, $735 per MBF)	B1@.067	LF	3.15	1.53	4.68
6" x 6" (3.21 BF per LF, $872 per MBF)	B1@.051	LF	2.80	1.17	3.97
6" x 8" (4.28 BF per LF, $872 per MBF)	B1@.067	LF	3.73	1.53	5.26
6" x 10" (5.35 BF per LF, $872 per MBF)	B1@.083	LF	4.67	1.90	6.57
6" x 12" (6.42 BF per LF, $872 per MBF)	B1@.098	LF	5.60	2.24	7.84
8" x 8" (5.71 BF per LF, $872 per MBF)	B1@.088	LF	4.98	2.01	6.99

Sill plates (At foundation), Std & Btr pressure treated lumber, drilled and installed with foundation bolts at 48" OC, no bolts, nuts or washers included. See also plates in this section. Figures in parentheses indicate board feet per LF of foundation, including 5% waste and wolmanized treatment at $120 per MBF.

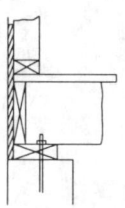

	Craft@Hrs	Unit	Material	Labor	Total
2" x 3" (.53 BF per LF at $635 per MBF)	B1@.020	LF	.34	.46	.80
2" x 4" (.70 BF per LF at $635 per MBF)	B1@.023	LF	.44	.53	.97
2" x 6" (1.05 BF per LF at $660 per MBF)	B1@.024	LF	.69	.55	1.24
2" x 8" (1.40 BF per LF at $740 per MBF)	B1@.031	LF	1.04	.71	1.75

Floor joists Per SF of area covered. Figures in parentheses indicate board feet per square foot of floor including box or band joist, typical double joists, and 6% waste. No beams, blocking or bridging included. Deduct for openings over 25 SF. Costs shown are based on a job with 1,000 SF of area covered. For scheduling purposes, estimate that a two-man crew can complete 750 SF of area per 8-hour day for 12" center to center framing; 925 SF for 16" OC; 1,100 SF for 20" OC; or 1,250 SF for 24" OC.

	Craft@Hrs	Unit	Material	Labor	Total
2" x 6" (at $515 per MBF), Std & Btr					
12" centers (1.28 BF per SF)	B1@.021	SF	.66	.48	1.14
16" centers (1.02 BF per SF)	B1@.017	SF	.53	.39	.92
20" centers (.88 BF per SF)	B1@.014	SF	.45	.32	.77
24" centers (.73 BF per SF)	B1@.012	SF	.38	.27	.65
2" x 8" (at $525 per MBF), Std & Btr					
12" centers (1.71 BF per SF)	B1@.023	SF	.90	.53	1.43
16" centers (1.36 BF per SF)	B1@.018	SF	.71	.41	1.12
20" centers (1.17 BF per SF)	B1@.015	SF	.61	.34	.95
24" centers (1.03 BF per SF)	B1@.013	SF	.54	.30	.84
2" x 10" (at $710 per MBF), Std & Btr					
12" centers (2.14 BF per SF)	B1@.025	SF	1.52	.57	2.09
16" centers (1.71 BF per SF)	B1@.020	SF	1.21	.46	1.67
20" centers (1.48 BF per SF)	B1@.016	SF	1.05	.37	1.42
24" centers (1.30 BF per SF)	B1@.014	SF	.92	.32	1.24
2" x 12" (at $965 per MBF), Std & Btr					
12" centers (2.56 BF per SF)	B1@.026	SF	2.47	.59	3.06
16" centers (2.05 BF per SF)	B1@.021	SF	1.98	.48	2.46
20" centers (1.77 BF per SF)	B1@.017	SF	1.71	.39	2.10
24" centers (1.56 BF per SF)	B1@.015	SF	1.51	.34	1.85

	Craft@Hrs	Unit	Material	Labor	Total

Floor joist wood, TJI truss type Suitable for residential use, 50 PSF floor load design. Costs shown are per square foot (SF) of floor area, based on joists at 16" OC, for a job with 1,000 SF of floor area. Figures in parentheses are the cost per linear foot for the joists. Beams, supports and blocking are not included. For scheduling purposes, estimate that a two-man crew can install 900 to 950 SF of joists in an 8-hour day.

	Craft@Hrs	Unit	Material	Labor	Total
9-1/2" TJI/15 ($1.75 LF)	B1@.017	SF	1.44	.39	1.83
11-7/8" TJI/15 ($1.90 LF)	B1@.017	SF	1.57	.39	1.96
14" TJI/35 ($2.75 LF)	B1@.018	SF	2.27	.41	2.68
16" TJI/35 ($3.00 LF)	B1@.018	SF	2.47	.41	2.88

Bridging or blocking Installed between 2" x 6" thru 2" x 12" joists. Costs shown are per each set of cross bridges or per each block for solid bridging, and include normal waste. The spacing between the bridging or blocking, sometimes called a "bay", depends on job requirements. Labor costs assume bridging is cut to size on site.

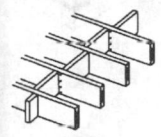
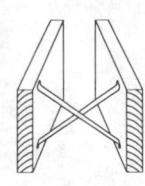

	Craft@Hrs	Unit	Material	Labor	Total
1" x 4" cross (at $557 per MBF)					
Joists on 12" centers	B1@.034	Ea	.28	.78	1.06
Joists on 16" centers	B1@.034	Ea	.38	.78	1.16
Joists on 20" centers	B1@.034	Ea	.48	.78	1.26
Joists on 24" centers	B1@.034	Ea	.58	.78	1.36
2" x 8" solid, Std & Btr (at $525 per MBF)					
Joists on 12" centers	B1@.042	Ea	.76	.96	1.72
Joists on 16" centers	B1@.042	Ea	1.01	.96	1.97
Joists on 20" centers	B1@.042	Ea	1.29	.96	2.25
Joists on 24" centers	B1@.042	Ea	1.54	.96	2.50
Deduct for 2" x 6" solid bridging at $660 per MBF)	—	Ea	-.34	—	-.34
2" x 10" solid, Std & Btr (at $571 per MBF)					
Joists on 12" centers	B1@.057	Ea	1.05	1.30	2.35
Joists on 16" centers	B1@.057	Ea	1.39	1.30	2.69
Joists on 20" centers	B1@.057	Ea	1.74	1.30	3.04
Joists on 24" centers	B1@.057	Ea	1.96	1.30	3.26
Add for 2" x 12" solid bridging (at $587 per MBF)	—	Ea	.23	—	.23
Steel, no nail type, cross					
Joists on 12" centers	B1@.020	Ea	1.54	.46	2.00
Joists on 16" centers	B1@.020	Ea	1.60	.46	2.06
Joists on 20" centers	B1@.020	Ea	1.75	.46	2.21
Joists on 24" centers	B1@.020	Ea	2.06	.46	2.52

Subflooring

Board sheathing, 1" x 6" #3 & Btr, (at $638 per MBF, 1.28 BF per SF)

	Craft@Hrs	Unit	Material	Labor	Total
includes 12% shrinkage, 5% waste & nails	B1@.020	SF	.82	.46	1.28
Add for diagonal patterns	B1@.001	SF	.05	.02	.07

Plywood sheathing, CD standard exterior grade. Material costs shown include 5% for waste & fasteners

	Craft@Hrs	Unit	Material	Labor	Total
5/16" (at $9.70 per 4' x 8' sheet)	B1@.011	SF	.32	.25	.57
3/8" (at $10.10 per 4' x 8' sheet)	B1@.011	SF	.33	.25	.58
1/2" (at $12.60 per 4' x 8' sheet)	B1@.012	SF	.41	.27	.68
5/8" (at $15.80 per 4' x 8' sheet)	B1@.012	SF	.52	.27	.79
3/4" (at $19.41 per 4' x 8' sheet)	B1@.013	SF	.64	.30	.94

Carpentry, Detailed Breakdown

	Craft@Hrs	Unit	Material	Labor	Total

Plates (Wall plates), Std & Btr, untreated. For pressure treated plates see also Sill Plates in this section. Figures in parentheses indicate board feet per LF. Costs shown include 10% for waste and nails

2" x 3" (at $525 per MBF, .55 BF per LF)	B1@.010	LF	.32	.23	.55
2" x 4" (at $525 per MBF, .73 BF per LF)	B1@.012	LF	.42	.27	.69
2" x 6" (at $515 per MBF, 1.10 BF per LF)	B1@.018	LF	.62	.41	1.03

Studding Per square foot of wall area. Do not subtract for openings less than 16' wide. Figures in parentheses indicate typical board feet per SF of wall area, measured on one side, and includes normal waste. Add for each corner and partition from below. Costs include studding, nails. Add for plates from the section above and also, door and window opening framing, backing, let-in bracing, fire blocking, and sheathing for shear walls from the sections that follow. Labor includes layout, plumb and align

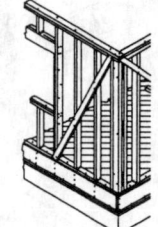

2" x 3" (at $530 per MBF), Std & Btr

12" center (.55 BF per SF)	B1@.020	SF	.29	.46	.75
16" center (.41 BF per SF)	B1@.018	SF	.22	.41	.63
20" center (.33 BF per SF)	B1@.018	SF	.18	.41	.59
24" center (.28 BF per SF)	B1@.017	SF	.15	.39	.54
Add for each corner or partition	B1@.083	Ea	6.40	1.90	8.30

2" x 4" (at $525 per MBF), Std & Btr

12" center (.73 BF per SF)	B1@.024	SF	.38	.55	.93
16" center (.54 BF per SF)	B1@.023	SF	.28	.53	.81
20" center (.45 BF per SF)	B1@.022	SF	.24	.50	.74
24" center (.37 BF per SF)	B1@.021	SF	.19	.48	.67
Add for each corner or partition	B1@.083	Ea	8.85	1.90	10.75

2" x 6" (at $515 per MBF), Std & Btr

12" center (1.10 BF per SF)	B1@.032	SF	.57	.73	1.30
16" center (.83 BF per SF)	B1@.029	SF	.43	.66	1.09
20" center (.67 BF per SF)	B1@.028	SF	.35	.64	.99
24" center (.55 BF per SF)	B1@.027	SF	.28	.62	.90
Add for each corner or partition	B1@.083	Ea	12.30	1.90	14.20

4" x 4" (at $915 per MBF), Std & Btr

installed in wall framing, with 10% waste	B1@.064	LF	1.35	1.46	2.81

Door opening framing In wall studs, based on walls 8' in height. Costs shown are per each door opening and include header of appropriate size, double vertical studs each side of the opening less than 8' wide (triple vertical studs each side of openings 8' wide or wider), cripples, blocking, nails and normal waste. Width shown is size of finished opening. Figures in parentheses indicate header size and cost.

2" x 4" wall studs (at $525 per MBF), Std & Btr, opening size as shown

To 3' wide (4" x 4" at $735 per MBF)	B1@.830	Ea	16.10	19.00	35.10
Over 3' to 4' wide (4" x 6" at $750 per MBF)	B1@1.11	Ea	19.90	25.40	45.30
Over 4' to 5' wide (4" x 6" at $750 per MBF)	B1@1.39	Ea	21.90	31.80	53.70
Over 5' to 6' wide (4" x 8" at $740 per MBF)	B1@1.66	Ea	26.50	37.90	64.40
Over 6' to 8' wide (4" x 10" at $750 per MBF)	B1@1.94	Ea	41.80	44.30	86.10
Over 8' to 10' wide (4" x 12" at $735 per MBF)	B1@1.94	Ea	52.00	44.30	96.30
Over 10' to 12' wide (4" x 14" at $800 per MBF)	B1@2.22	Ea	68.00	50.70	118.70
Add per foot of height for 4" walls over 8' in height	—	LF	2.10	—	2.10

	Craft@Hrs	Unit	Material	Labor	Total
2" x 6" wall studs (at $515 per MBF), Std & Btr, opening size as shown					
To 3' wide (6 x 4 at $750 per MBF)	B1@1.11	Ea	21.20	25.40	46.60
Over 3' to 4' wide (6 x 6 at $875 per MBF)	B1@1.39	Ea	28.30	31.80	60.10
Over 4' to 5' wide (6 x 6 at $875 per MBF)	B1@1.66	Ea	31.60	37.90	69.50
Over 5' to 6' wide (6 x 8 at $875 per MBF)	B1@1.94	Ea	39.70	44.30	84.00
Over 6' to 8' wide (6 x 10 at $875 per MBF)	B1@2.22	Ea	62.50	50.70	113.20
Over 8' to 10' wide (6 x 12 at $875 per MBF)	B1@2.22	Ea	81.00	50.70	131.70
Over 10' to 12' wide (6 x 14 at $875 per MBF)	B1@2.50	Ea	103.00	57.10	160.10
Add per foot of height for 6" walls over 8' in height	—	LF	3.09	—	3.09

Window opening framing In wall studs, based on walls 8' in height. Costs shown are per window opening and include header of appropriate size, sub-sill plate (double sub-sill if opening is 8' wide or wider), double vertical studs each side of openings less than 8' wide (triple vertical studs each side of openings 8' wide or wider), top and bottom cripples, blocking, nails and normal waste. Figures in parentheses indicate header size and cost.

	Craft@Hrs	Unit	Material	Labor	Total
2" x 4" wall studs (at $525 per MBF), Std & Btr, opening size shown is width of finished opening					
To 2' wide (4 x 4 at $735 per MBF)	B1@1.00	Ea	14.60	22.90	37.50
Over 2' to 3' wide (4 x 4 at $735 per MBF)	B1@1.17	Ea	18.20	26.70	44.90
Over 3' to 4' wide (4 x 6 at $750 per MBF)	B1@1.45	Ea	22.70	33.10	55.80
Over 4' to 5' wide (4 x 6 at $750 per MBF)	B1@1.73	Ea	25.50	39.50	65.00
Over 5' to 6' wide (4 x 8 at $740 per MBF)	B1@2.01	Ea	30.90	45.90	76.80
Over 6' to 7' wide (4 x 8 at $740 per MBF)	B1@2.29	Ea	39.20	52.30	91.50
Over 7' to 8' wide (4 x 10 at $750 per MBF)	B1@2.57	Ea	47.30	58.70	106.00
Over 8' to 10' wide (4 x 12 at $735 per MBF)	B1@2.57	Ea	59.00	58.70	117.70
Over 10' to 12' wide (4 x 14 at $800 per MBF)	B1@2.85	Ea	80.50	65.10	145.60
Add per foot of height for walls over 8' high		LF	2.10	—	2.10
2" x 6" wall studs (at $515 per MBF), Std & Btr, opening size as shown					
To 2' wide (4 x 4 at $735 per MBF)	B1@1.17	Ea	18.50	26.70	45.20
Over 2' to 3' wide (6 x 4 at $750 per MBF)	B1@1.45	Ea	24.40	33.10	57.50
Over 3' to 4' wide (6 x 6 at $875 per MBF)	B1@1.73	Ea	31.10	39.50	70.60
Over 4' to 5' wide (6 x 6 at $875 per MBF)	B1@2.01	Ea	36.70	45.90	82.60
Over 5' to 6' wide (6 x 8 at $875 per MBF)	B1@2.29	Ea	43.90	52.30	96.20
Over 6' to 8' wide (6 x 10 at $875 per MBF)	B1@2.85	Ea	68.00	65.10	133.10
Over 8' to 10' wide (6 x 12 at $875 per MBF)	B1@2.85	Ea	92.00	65.10	157.10
Over 8' to 12' wide (6 x 14 at $875 per MBF)	B1@3.13	Ea	115.00	71.50	186.50
Add per foot of height for 6" walls over 8' high	—	LF	3.09	—	3.09

Bracing See also sheathing for plywood bracing and shear panels

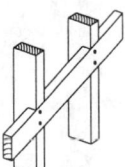

	Craft@Hrs	Unit	Material	Labor	Total
Let-in wall bracing, using Std & Btr lumber					
1" x 4" at $610 per MBF	B1@.021	LF	.20	.48	.68
1" x 6" at $660 per MBF	B1@.027	LF	.33	.62	.95
2" x 4" at $535 per MBF	B1@.035	LF	.36	.80	1.16
Steel strap bracing, 1-1/4" wide					
9'6" or 11'6" lengths	B1@.010	LF	.42	.23	.65
Steel "V" bracing, 3/4" x 3/4"					
9'6" or 11'6" lengths	B1@.010	LF	.58	.23	.81
Temporary wood frame wall bracing, assumes salvage at 50% and 3 uses					
1" x 4" Std & Btr (at $610 per MBF)	B1@.006	LF	.05	.14	.19
1" x 6" Std & Btr (at $660 per MBF)	B1@.010	LF	.08	.23	.31
2" x 4" utility (at $475 per MBF)	B1@.012	LF	.08	.27	.35
2" x 6" utility (at $480 per MBF)	B1@.018	LF	.12	.41	.53

Carpentry, Detailed Breakdown

	Craft@Hrs	Unit	Material	Labor	Total

Fireblocks Installed in wood frame walls, per LF of wall to be blocked. Figures in parentheses indicate board feet of fire blocking per linear foot of wall including 10% waste. See also Bridging and Backing and Nailers in this section

2" x 3" blocking (at $530 per MBF), Std & Btr

	Craft@Hrs	Unit	Material	Labor	Total
12" OC members (.48 BF per LF)	B1@.020	LF	.25	.46	.71
16" OC members (.50 BF per LF)	B1@.015	LF	.27	.34	.61
20" OC members (.51 BF per LF)	B1@.012	LF	.27	.27	.54
24" OC members (.52 BF per LF)	B1@.010	LF	.28	.23	.51

2" x 4" blocking (at $525 per MBF), Std & Btr

	Craft@Hrs	Unit	Material	Labor	Total
12" OC members (.64 BF per LF)	B1@.020	LF	.34	.46	.80
16" OC members (.67 BF per LF)	B1@.015	LF	.35	.34	.69
20" OC members (.68 BF per LF)	B1@.012	LF	.36	.27	.63
24" OC members (.69 BF per LF)	B1@.010	LF	.36	.23	.59

2" x 6" blocking (at $515 per MBF), Std & Btr

	Craft@Hrs	Unit	Material	Labor	Total
12" OC members (.96 BF per LF)	B1@.021	LF	.49	.48	.97
16" OC members (1.00 BF per LF)	B1@.016	LF	.52	.37	.89
20" OC members (1.02 BF per LF)	B1@.012	LF	.53	.27	.80
24" OC members (1.03 BF per LF)	B1@.010	LF	.53	.23	.76

Beams Std & Btr. Installed over wall openings and around floor, ceiling and roof openings or where a flush beam is called out in the plans, including 10% waste. Do not use these beams for door or window headers in framed walls, use door or window opening framing assemblies.

	Craft@Hrs	Unit	Material	Labor	Total
2" x 6" (at $515 per MBF, 1.10 BF per LF)	B1@.028	LF	.57	.64	1.21
2" x 8" (at $525 per MBF, 1.47 BF per LF)	B1@.037	LF	.77	.85	1.62
2" x 10" (at $570 per MBF, 1.83 BF per LF)	B1@.046	LF	1.04	1.05	2.09
2" x 12" (at $585 per MBF, 2.20 BF per LF)	B1@.057	LF	1.29	1.30	2.59
4" x 6" (at $750 per MBF, 2.20 BF per LF)	B1@.057	LF	1.65	1.30	2.95
4" x 8" (at $740 per MBF, 2.93 BF per LF)	B1@.073	LF	2.17	1.67	3.84
4" x 10" (at $750 per MBF, 3.67 BF per LF)	B1@.094	LF	2.75	2.15	4.90
4" x 12" (at $735 per MBF, 4.40 BF per LF)	B1@.112	LF	3.23	2.56	5.79
4" x 14" (at $800 per MBF, 5.13 BF per LF)	B1@.115	LF	4.10	2.63	6.73
6" x 6" (at $875 per MBF, 3.30 BF per LF)	B1@.060	LF	2.89	1.37	4.26
6" x 8" (at $875 per MBF, 4.40 BF per LF)	B1@.080	LF	3.85	1.83	5.68
6" x 10" (at $875 per MBF, 5.50 BF per LF)	B1@.105	LF	4.81	2.40	7.21
6" x 12" (at $875 per MBF, 6.60 BF per LF)	B1@.115	LF	5.78	2.63	8.41
6" x 14" (at $875 per MBF, 7.70 BF per LF)	B1@.120	LF	6.74	2.74	9.48

Posts S4S, green. Posts not in wall framing. Material costs include 10% waste. For scheduling purposes, estimate that a two-man crew can complete 100 to 125 LF per 8-hour day.

	Craft@Hrs	Unit	Material	Labor	Total
4" x 4" (at $735 per MBF)	B1@.110	LF	1.08	2.51	3.59
4" x 6" (at $750 per MBF)	B1@.120	LF	1.65	2.74	4.39
4" x 8" (at $740 per MBF)	B1@.140	LF	2.17	3.20	5.37
4" x 10" (at $750 per MBF)	B1@.143	LF	2.75	3.27	6.02
4" x 12" (at $735 per MBF)	B1@.145	LF	3.23	3.31	6.54
6" x 6" (at $875 per MBF)	B1@.145	LF	2.89	3.31	6.20
6" x 8" (at $875 per MBF)	B1@.145	LF	3.85	3.31	7.16
6" x 10" (at $875 per MBF)	B1@.147	LF	4.81	3.36	8.17
6" x 12" (at $875 per MBF)	B1@.150	LF	5.78	3.43	9.21
8" x 8" (at $875 per MBF)	B1@.150	LF	5.13	3.43	8.56
8" x 10" (at $875 per MBF)	B1@.155	LF	6.42	3.54	9.96
8" x 12" (at $875 per MBF)	B1@.166	LF	7.70	3.79	11.49

	Craft@Hrs	Unit	Material	Labor	Total

Ceiling joists and soffits Per SF of area covered. Figures in parentheses indicate board feet per square foot of ceiling including end joists, header joists, and 5% waste. No beams, bridging, blocking, or ledger strips included. Deduct for openings over 25 SF. Costs shown are based on a job with 1,000 SF of area covered. For scheduling purposes, estimate that a two-man crew can complete 650 SF of area per 8-hour day for 12" center to center framing; 800 SF for 16" OC; 950 SF for 20" OC; or 1,100 SF for 24" OC.

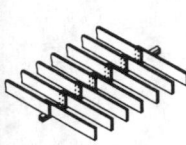

	Craft@Hrs	Unit	Material	Labor	Total
2" x 4" (at $525 per MBF), Std & Btr grade					
12" centers (.78 BF per SF)	B1@.023	SF	.41	.53	.94
16" centers (.59 BF per SF)	B1@.020	SF	.31	.46	.77
20" centers (.48 BF per SF)	B1@.016	SF	.25	.37	.62
24" centers (.42 BF per SF)	B1@.014	SF	.22	.32	.54
2" x 6" (at $515 per MBF), Std & Btr grade					
12" centers (1.15 BF per SF)	B1@.026	SF	.59	.59	1.18
16" centers (.88 BF per SF)	B1@.020	SF	.45	.46	.91
20" centers (.72 BF per SF)	B1@.017	SF	.37	.39	.76
24" centers (.63 BF per SF)	B1@.015	SF	.32	.34	.66
2" x 8" (at $525 per MBF), Std & Btr grade					
12" centers (1.53 BF per SF)	B1@.028	SF	.80	.64	1.44
16" centers (1.17 BF per SF)	B1@.022	SF	.61	.50	1.11
20" centers (.96 BF per SF)	B1@.018	SF	.50	.41	.91
24" centers (.84 BF per SF)	B1@.016	SF	.44	.37	.81
2" x 10" (at $570 per MBF), Std & Btr grade					
12" centers (1.94 BF per SF)	B1@.030	SF	1.11	.69	1.80
16" centers (1.47 BF per SF)	B1@.023	SF	.84	.53	1.37
20" centers (1.21 BF per SF)	B1@.020	SF	.69	.46	1.15
24" centers (1.04 BF per SF)	B1@.017	SF	.59	.39	.98
2" x 12" (at $585 per MBF), Std & Btr grade					
12" centers (2.30 BF per SF)	B1@.031	SF	1.35	.71	2.06
16" centers (1.76 BF per SF)	B1@.025	SF	1.03	.57	1.60
20" centers (1.44 BF per SF)	B1@.020	SF	.84	.46	1.30
24" centers (1.26 BF per SF)	B1@.018	SF	.74	.41	1.15

Catwalks 1" x 6" solid planking walkway, #2 & Btr grade

	Craft@Hrs	Unit	Material	Labor	Total
(at $755 per MBF, 0.50 BF per SF)	B1@.017	SF	.38	.39	.77

Backing and nailers Std & Btr, for wall finishes, "floating" backing for drywall ceilings, trim, z-bar, appliances and fixtures etc. See also Bridging and Fireblocking in this section. Figures in parentheses show board feet per LF including 10% waste.

	Craft@Hrs	Unit	Material	Labor	Total
1" x 4" (at $610 per MBF, .37 BF per LF)	B1@.011	LF	.23	.25	.48
1" x 6" (at $660 per MBF, .55 BF per LF)	B1@.017	LF	.36	.39	.75
1" x 8" (at $660 per MBF, .73 BF per LF)	B1@.022	LF	.48	.50	.98
2" x 4" (at $525 per MBF, .73 BF per LF)	B1@.023	LF	.38	.53	.91
2" x 6" (at $515 per MBF, 1.10 BF per LF)	B1@.034	LF	.57	.78	1.35
2" x 8" (at $525 per MBF, 1.47 BF per LF)	B1@.045	LF	.77	1.03	1.80
2" x 10" (at $570 per MBF, 1.83 BF per LF)	B1@.057	LF	1.04	1.30	2.34

Ledger strips Std & Btr, nailed to faces of studs, beams, girders, joists, etc. See also Ribbons in this section for let-in type. Figures in parentheses indicate board feet per LF including 10% waste.

	Craft@Hrs	Unit	Material	Labor	Total
1" x 2" (at $610 per MBF .18 BF per LF)	B1@.010	LF	.11	.23	.34
1" x 3" (at $610 per MBF .28 BF per LF)	B1@.010	LF	.17	.23	.40
1" x 4" (at $610 per MBF .37 BF per LF)	B1@.010	LF	.23	.23	.46
2" x 2" (at $525 per MBF .37 BF per LF)	B1@.010	LF	.19	.23	.42
2" x 3" (at $525 per MBF .55 BF per LF)	B1@.010	LF	.29	.23	.52
2" x 4" (at $525 per MBF .73 BF per LF)	B1@.010	LF	.38	.23	.61

Carpentry, Detailed Breakdown

	Craft@Hrs	Unit	Material	Labor	Total

Ribbons (Ribbands), let in to wall framing. See also Ledgers in this section. Figures in parentheses indicate board feet per LF including 10% waste.

	Craft@Hrs	Unit	Material	Labor	Total
1" x 3", Std & Btr (.28 BF per LF)					
(at $610 per MBF)	B1@.020	LF	.17	.46	.63
1" x 4", Std & Btr (.37 BF per LF)					
(at $610 per MBF)	B1@.020	LF	.23	.46	.69
1" x 6", Std & Btr (.55 BF per LF)					
(at $660 per MBF)	B1@.030	LF	.36	.69	1.05
2" x 3", Std & Btr (.55 BF per LF)					
(at $525 per MBF)	B1@.041	LF	.29	.94	1.23
2" x 4", Std & Btr (.73 BF per LF)					
(at $525 per MBF)	B1@.041	LF	.38	.94	1.32
2" x 6", Std & Btr (1.10 BF per LF)					
(at $515 per MBF)	B1@.045	LF	.57	1.03	1.60
2" x 8", Std & Btr (1.47 BF per LF)					
(at $525 per MBF)	B1@.045	LF	.77	1.03	1.80

Rafters Flat, shed, or gable roofs, up to 5 in 12 slope (5/24 pitch), maximum 25' span. Figures in parentheses indicate board feet per SF of actual roof surface area (not roof plan area), including rafters, ridge boards, collar beams and normal waste, but no blocking, bracing, purlins, curbs, or gable walls.

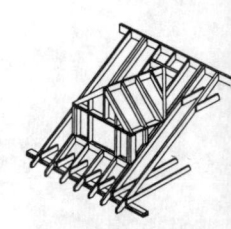

	Craft@Hrs	Unit	Material	Labor	Total
2" x 4" (at $525 per MBF), Std & Btr					
12" o.c. (.89 BF per SF)	B1@.021	SF	.47	.48	.95
16" o.c. (.71 BF per SF)	B1@.017	SF	.37	.39	.76
24" o.c. (.53 BF per SF)	B1@.014	SF	.28	.32	.60
2" x 6" (at $515 per MBF), Std & Btr					
12" o.c. (1.29 BF per SF)	B1@.028	SF	.66	.64	1.30
16" o.c. (1.02 BF per SF)	B1@.023	SF	.53	.53	1.06
24" o.c. (.75 BF per SF)	B1@.018	SF	.39	.41	.80
2" x 8" (at $525 per MBF), Std & Btr					
12" o.c. (1.71 BF per SF)	B1@.036	SF	.90	.82	1.72
16" o.c. (1.34 BF per SF)	B1@.030	SF	.70	.69	1.39
24" o.c. (1.12 BF per SF)	B1@.024	SF	.59	.55	1.14
2" x 10" (at $570 per MBF), Std & Btr					
12" o.c. (2.12 BF per SF)	B1@.045	SF	1.21	1.03	2.24
16" o.c. (1.97 BF per SF)	B1@.036	SF	1.12	.82	1.94
24" o.c. (1.21 BF per SF)	B1@.029	SF	.69	.66	1.35
2" x 12" (at $585 per MBF), Std & Btr					
12" o.c. (2.52 BF per SF)	B1@.050	SF	1.47	1.14	2.61
16" o.c. (1.97 BF per SF)	B1@.040	SF	1.15	.91	2.06
24" o.c. (1.43 BF per SF)	B1@.032	SF	.84	.73	1.57
Add for hip roof	B1@.007	SF	—	.16	.16
Add for slope over 5 in 12	B1@.015	SF	—	.34	.34

Trimmers and curbs At stairwells, skylights, dormers, etc. Figures in parentheses show board feet per LF, Std & Btr grade, including 10% waste

	Craft@Hrs	Unit	Material	Labor	Total
2" x 4" (.73 BF per LF at $525 per MBF)	B1@.018	LF	.38	.41	.79
2" x 6" (1.10 BF per LF at $515 per MBF)	B1@.028	LF	.57	.64	1.21
2" x 8" (1.47 BF per LF at $525 per MBF)	B1@.038	LF	.77	.87	1.64
2" x 10" (1.83 BF per LF at $570 per MBF)	B1@.047	LF	1.04	1.07	2.11
2" x 12" (2.20 BF per LF at $585 per MBF)	B1@.057	LF	1.29	1.30	2.59

	Craft@Hrs	Unit	Material	Labor	Total

Collar beams & collar ties Std & Btr grade, including 10% waste

	Craft@Hrs	Unit	Material	Labor	Total
Collar beams, 2" x 6" (at $515 per MBF)	B1@.013	LF	.57	.30	.87
Collar ties, 1" x 6" (at $660 per MBF)	B1@.006	LF	.36	.14	.50

Purlins (Purling), Std & Btr, installed below roof rafters. Figures in parentheses indicate board feet per LF including 5% waste

	Craft@Hrs	Unit	Material	Labor	Total
2" x 4" (at $525 per MBF, .70 BF per LF)	B1@.012	LF	.37	.27	.64
2" x 6" (at $515 per MBF, 1.05 BF per LF)	B1@.017	LF	.54	.39	.93
2" x 8" (at $525 per MBF, 1.40 BF per LF)	B1@.023	LF	.74	.53	1.27
2" x 10" (at $570 per MBF, 1.75 BF per LF)	B1@.027	LF	1.00	.62	1.62
2" x 12" (at $585 per MBF, 2.10 BF per LF)	B1@.030	LF	1.23	.69	1.92
4" x 6" (at $750 per MBF, 2.10 BF per LF)	B1@.034	LF	1.58	.78	2.36
4" x 8" (at $740 per MBF, 2.80 BF per LF)	B1@.045	LF	2.07	1.03	3.10

Dormer studs Std & Btr, per square foot of wall area, including 10% waste

	Craft@Hrs	Unit	Material	Labor	Total
2" x 4", (at $525 per MBF, .84 BF per SF)	B1@.033	SF	.44	.75	1.19

Roof trusses 24" OC, any slope from 3 in 12 to 12 in 12, total height not to exceed 12' high from bottom chord to highest point on truss. Prices for trusses over 12' high will be up to 100% higher. Square foot (SF) costs, where shown, are per square foot of roof area to be covered.

Scissor trusses
2" x 4" top and bottom chords

	Craft@Hrs	Unit	Material	Labor	Total
Up to 38' span	B1@.022	SF	1.86	.50	2.36
40' to 50' span	B1@.028	SF	2.35	.64	2.99

Fink truss "W" (conventional roof truss)
2" x 4" top and bottom chords

	Craft@Hrs	Unit	Material	Labor	Total
Up to 38' span	B1@.017	SF	1.63	.39	2.02
40' to 50' span	B1@.022	SF	1.92	.50	2.42

2" x 6" top and bottom chords

	Craft@Hrs	Unit	Material	Labor	Total
Up to 38' span	B1@.020	SF	1.97	.46	2.43
40' to 50' span	B1@.026	SF	2.37	.59	2.96

Truss with gable fill at 16" OC

	Craft@Hrs	Unit	Material	Labor	Total
28' span, 5 in 12 slope	B1@.958	Ea	133.00	21.90	154.90
32' span, 5 in 12 slope	B1@1.26	Ea	165.00	28.80	193.80
40' span, 5 in 12 slope	B1@1.73	Ea	230.00	39.50	269.50

Fascias Material column includes 10% waste

Engelmann spruce, #3 & Btr

	Craft@Hrs	Unit	Material	Labor	Total
1" x 4" (at $536 per MBF)	B1@.044	LF	.20	1.01	1.21
1" x 6" (at $585 per MBF)	B1@.044	LF	.32	1.01	1.33
1" x 8" (at $575 per MBF)	B1@.051	LF	.42	1.17	1.59

Hem-fir, S4S, dry, Std and Btr

	Craft@Hrs	Unit	Material	Labor	Total
2" x 6" (at $515 per MBF)	B1@.044	LF	.57	1.01	1.58
2" x 8" (at $530 per MBF)	B1@.051	LF	.78	1.17	1.95
2" x 10" (at $565 per MBF)	B1@.051	LF	1.04	1.17	2.21

Redwood, S4S kiln dried, clear

	Craft@Hrs	Unit	Material	Labor	Total
1" x 4" (at $1,600 per MBF)	B1@.044	LF	.59	1.01	1.60
1" x 6" (at $1,650 per MBF)	B1@.044	LF	.91	1.01	1.92
1" x 8" (at $1,820 per MBF)	B1@.051	LF	1.32	1.17	2.49
2" x 4" (at $1,800 per MBF)	B1@.044	LF	1.32	1.01	2.33
2" x 6" (at $2,000 per MBF)	B1@.044	LF	2.20	1.01	3.21
2" x 8" (at $2,000 per MBF)	B1@.051	LF	2.93	1.17	4.10

Carpentry, Detailed Breakdown

	Craft@Hrs	Unit	Material	Labor	Total
Sheathing, roof Per SF of roof surface including 10% waste					
CDX plywood sheathing, rough					
1/2" (at $12.60 per 4' x 8' sheet)	B1@.013	SF	.43	.30	.73
1/2" (4 ply at $12.70 per 4' x 8' sheet)	B1@.013	SF	.44	.30	.74
1/2" (5 ply at $12.70 per 4' x 8' sheet)	B1@.013	SF	.44	.30	.74
5/8" (at $15.80 per 4' x 8' sheet)	B1@.013	SF	.54	.30	.84
5/8" (4 ply at $15.80 per 4' x 8' sheet)	B1@.013	SF	.54	.30	.84
5/8" (5 ply at $16.30 per 4' x 8' sheet)	B1@.013	SF	.56	.30	.86
Add for hip roof	B1@.007	SF	—	.16	.16
Add for steep pitch or cut-up roof	B1@.015	SF	—	.34	.34
Board sheathing, 1" x 6" or 1" x 8" utility T&G laid diagonal (at $490 per MBF)					
(1.13 BF per SF)	B1@.026	SF	.45	.59	1.04
Add for hip roof	B1@.007	SF	—	.16	.16
Add for steep pitch or cut-up roof	B1@.015	SF	—	.34	.34

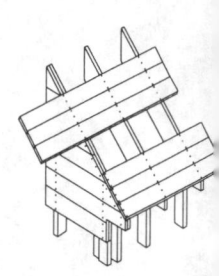

Roof decking Std & Btr grade. See also Sheathing in this section. Flat, shed, or gable roofs to 5 in 12 slope (5/24 pitch). Figures in parentheses indicate board feet per SF of actual roof area (not roof plan area), including 5% waste.

	Craft@Hrs	Unit	Material	Labor	Total
2" x 6" (2.28 BF per SF, $515 per MBF)	B1@.043	SF	1.17	.98	2.15
2" x 8" (2.25 BF per SF, $525 per MBF)	B1@.043	SF	1.18	.98	2.16
3" x 6" (3.43 BF per SF, $580 per MBF)	B1@.047	SF	1.99	1.07	3.06
Add for steep pitch or cut-up roof	B1@.019	SF	—	.43	.43

Cant strips Std & Btr grade lumber

	Craft@Hrs	Unit	Material	Labor	Total
2" x 4" beveled	B1@.020	LF	.58	.46	1.04
For short or cut-up runs	B1@.030	LF	.58	.69	1.27

Furring Per SF of surface area to be covered. Figures in parentheses show coverage including 7% waste, typical job.

	Craft@Hrs	Unit	Material	Labor	Total
Over masonry, lumber at $425 per MBF, Utility grade					
12" OC, 1" x 2" (.24 BF per SF)	B1@.025	SF	.10	.57	.67
16" OC, 1" x 2" (.20 BF per SF)	B1@.020	SF	.09	.46	.55
20" OC, 1" x 2" (.17 BF per SF)	B1@.018	SF	.07	.41	.48
24" OC, 1" x 2" (.15 BF per SF)	B1@.016	SF	.06	.37	.43
12" OC, 1" x 3" (.36 BF per SF)	B1@.025	SF	.15	.57	.72
16" OC, 1" x 3" (.29 BF per SF)	B1@.020	SF	.12	.46	.58
20" OC, 1" x 3" (.25 BF per SF)	B1@.018	SF	.11	.41	.52
24" OC, 1" x 3" (.22 BF per SF)	B1@.016	SF	.09	.37	.46
Over wood frame, lumber at $425 per MBF, Utility grade					
12" OC, 1" x 2" (.24 BF per SF)	B1@.016	SF	.10	.37	.47
16" OC, 1" x 2" (.20 BF per SF)	B1@.013	SF	.09	.30	.39
20" OC, 1" x 2" (.17 BF per SF)	B1@.011	SF	.07	.25	.32
24" OC, 1" x 2" (.15 BF per SF)	B1@.010	SF	.06	.23	.29
12" OC, 1" x 3" (.36 BF per SF)	B1@.016	SF	.15	.37	.52
16" OC, 1" x 3" (.29 BF per SF)	B1@.013	SF	.12	.30	.42
20" OC, 1" x 3" (.25 BF per SF)	B1@.011	SF	.11	.25	.36
24" OC, 1" x 3" (.22 BF per SF)	B1@.010	SF	.09	.23	.32
Over wood subfloor, 1" lumber at $425 per MBF, 2" lumber at $475 per MBF					
12" OC, 1" x 2" utility (.24 BF per SF)	B1@.033	SF	.11	.75	.86
16" OC, 1" x 2" utility (.20 BF per SF)	B1@.028	SF	.09	.64	.73
20" OC, 1" x 2" utility (.17 BF per SF)	B1@.024	SF	.07	.55	.62
24" OC, 1" x 2" utility (.15 BF per SF)	B1@.021	SF	.06	.48	.54

	Craft@Hrs	Unit	Material	Labor	Total
12" OC, 2" x 2" Std & Btr (.48 BF per SF)	B1@.033	SF	.23	.75	.98
16" OC, 2" x 2" Std & Btr (.39 BF per SF)	B1@.028	SF	.19	.64	.83
20" OC, 2" x 2" Std & Btr (.34 BF per SF)	B1@.024	SF	.16	.55	.71
24" OC, 2" x 2" Std & Btr (.30 BF per SF)	B1@.021	SF	.14	.48	.62

Grounds Screeds for plaster or stucco, #3, 1" x 2" lumber at $495 per MBF. Based on .18 BF per LF including 10% waste.

Over masonry	B1@.047	LF	.09	1.07	1.16
Over wood	B1@.038	LF	.09	.87	.96

Sheathing, wall Per SF of wall surface, including 5% waste

BD plywood sheathing, plugged and touch sanded, interior grade

1/4" (at $14.80 per 4' x 8' sheet)	B1@.013	SF	.49	.30	.79
3/8" (at $16.90 per 4' x 8' sheet)	B1@.015	SF	.55	.34	.89
1/2" (at $21.10 per 4' x 8' sheet)	B1@.016	SF	.69	.37	1.06
5/8" (at $25.20 per 4' x 8' sheet)	B1@.018	SF	.83	.41	1.24
3/4" (at $27.30 per 4' x 8' sheet)	B1@.020	SF	.90	.46	1.36

Board sheathing 1" x 6" or 1" x 8" utility T&G (at $490 per MBF)

(1.13 BF per SF)	B1@.020	SF	.55	.46	1.01
Add for diagonal patterns	B1@.006	SF	—	.14	.14

Sleepers Std & Btr pressure treated lumber at $658 per MBF and treatment at $120 per MBF, including 5% waste.

2" x 6" ($778 per MBF)	B1@.017	LF	.82	.39	1.21
Add for taper cuts on sleepers, per cut	B1@.009	Ea	—	.21	.21

Finish Carpentry These figures assume all materials to be installed are supplied by others. Labor costs assume good quality stain-grade work up to 9 feet above floor level. Paint-grade work will cost from 20% to 33% less. Add the cost of scaffolding, staining or painting, if required.

Blocking and backing Installation before drywall is hung. Cost per piece of blocking or backing set

2" x 4" backing or blocking	BC@.040	Ea	—	1.01	1.01
2" x 4" strongback 10' to 16' long					
screwed to ceiling joists	BC@.350	Ea	—	8.81	8.81

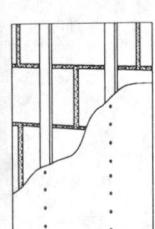

Furring Nailed over studs and adjusted with shims to bring the wall surface flush. Costs will be higher if the wall plane is very uneven or if the furred wall has to be exceptionally even.
Cost per linear foot of furring installed

1" x 4" furring, few shims required	BC@.010	LF	—	.25	.25

Straightening wall studs Costs per stud straightened
Cut saw kerf and drive shim or screw into the kerf

reinforce with a plywood scab	BC@.300	Ea	—	7.55	7.55
Remove and replace a badly warped stud	BC@.300	Ea	—	7.55	7.55

Cased openings Install jambs only. See costs for casing below. Costs per opening cased

Opening to 4' wide	BC@.800	Ea	—	20.10	20.10
Openings over 4' wide	BC@1.05	Ea	—	26.40	26.40

	Craft@Hrs	Unit	Material	Labor	Total

Exterior doors Based on hanging doors in prepared openings. No lockset, casing or stops included. See the sections that follow. Costs will be higher if the framing has to be adjusted before the door can be set. Hang a new door in an existing jamb

	Craft@Hrs	Unit	Material	Labor	Total
Cut door, set butt hinges, install door	BC@1.46	Ea	—	36.70	36.70

Prehung flush or panel 1-3/4" thick solid core single doors, with felt insulation, based on an adjustable metal sill

	Craft@Hrs	Unit	Material	Labor	Total
Up to 3' wide	BC@.750	Ea	—	18.90	18.90
Over 3' wide	BC@1.25	Ea	—	31.50	31.50
Add for door with sidelights	BC@.750	Ea	—	18.90	18.90
Sliding glass doors, opening to 8' wide	BC@1.75	Ea	—	44.00	44.00
French doors, 1-3/4" thick, to 6' wide	BC@1.75	Ea	—	44.00	44.00

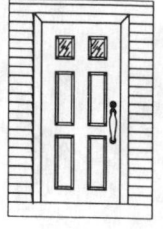

Interior doors Based on hanging 1-3/8" hollow core doors in prepared openings. No locksets or casing included. See the sections that follow.

	Craft@Hrs	Unit	Material	Labor	Total
Prehung interior doors	BC@.500	Ea	—	12.60	12.60

Add for assembling a factory-cut

	Craft@Hrs	Unit	Material	Labor	Total
mortised and rabbeted jamb	BC@.200	Ea	—	5.03	5.03

Pocket doors

	Craft@Hrs	Unit	Material	Labor	Total
Assemble and install the door frame	BC@.981	Ea	—	24.70	24.70
Install door and valances in existing frame	BC@.500	Ea	—	12.60	12.60

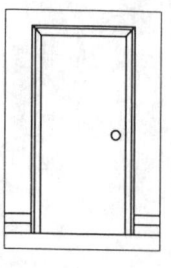

Bifold doors installed in cased openings. See costs for cased openings above

	Craft@Hrs	Unit	Material	Labor	Total
Single bifold door to 4' wide	BC@.700	Ea	—	17.60	17.60
Double bifold door to 8' wide, pair of doors	BC@1.20	Pr	—	30.20	30.20

Bypass closet doors installed on a track in a cased opening. See cased opening manhours above

	Craft@Hrs	Unit	Material	Labor	Total
Double doors, per opening	BC@.660	Ea	—	16.60	16.60
Triple doors, per opening	BC@1.00	Ea	—	25.20	25.20

Rehang most hollow core doors in an existing opening

	Craft@Hrs	Unit	Material	Labor	Total
re-shim jamb and plane door, per door	BC@1.50	Ea	—	37.80	37.80

Rehang large or heavy solid core doors in an existing opening

	Craft@Hrs	Unit	Material	Labor	Total
re-shim jamb and plane door, per door	BC@1.75	Ea	—	44.00	44.00

Correcting a sagging door hinge. Remove door from the jamb, remove hinge leaf

	Craft@Hrs	Unit	Material	Labor	Total
deepen the mortise and rehang the door	BC@.600	Ea	—	15.10	15.10

Hang a new door in an existing jamb. Mark the door to fit opening, cut the door, set butt hinges

	Craft@Hrs	Unit	Material	Labor	Total
install door in the jamb, install door hardware	BC@1.75	Ea	—	44.00	44.00

Prehanging doors off site. Costs per door hung, excluding door coast

	Craft@Hrs	Unit	Material	Labor	Total
Make and assemble frame, cost per frame	BC@.750	Ea	—	18.90	18.90

Set hinge on door and frame, panel or flush door, per hinge

	Craft@Hrs	Unit	Material	Labor	Total
Using jigs and router	BC@.160	Ea	—	4.03	4.03
Using hand tools	BC@.500	Ea	—	12.60	12.60
Add for French doors, per door	BC@.250	Ea	—	6.29	6.29

Lockset installation Set in either a solid core or hollow core door, includes setting the strike plate
Costs per lockset

	Craft@Hrs	Unit	Material	Labor	Total
Standard 2-1/8" bore, factory bored	BC@.160	Ea	—	4.03	4.03
Standard 2-1/8" bore, site bored	BC@1.00	Ea	—	25.20	25.20
Mortise type lockset, using a boring machine	BC@1.50	Ea	—	37.80	37.80
Mortise type lockset, using hand tools	BC@3.00	Ea	—	75.50	75.50

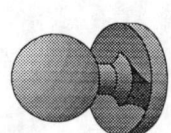

Pocket door hardware

	Craft@Hrs	Unit	Material	Labor	Total
Privacy latch	BC@.500	Ea	—	12.60	12.60
Passage latch	BC@.300	Ea	—	7.55	7.55
Edge pull	BC@.300	Ea	—	7.55	7.55

	Craft@Hrs	Unit	Material	Labor	Total

Setting door stop moulding Set at the head and on two sides. Costs per opening

Panel or flush door	BC@.250	Ea	—	6.29	6.29
French door	BC@.330	Ea	—	8.31	8.31
Pocket door	BC@.250	Ea	—	6.29	6.29

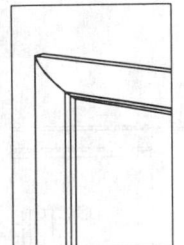

Set casing around a framed opening Mitered or butted moulding corners. Costs per opening, per side, using a pneumatic nail gun

Opening to 4' wide using casing to 2-1/4" wide	BC@.250	Ea	—	6.29	6.29
Opening over 4' wide to 8' wide, using casing up to 2-1/4" wide	BC@.300	Ea	—	7.55	7.55
Add for casing over 2-1/4" wide to 4" wide	BC@.160	Ea	—	4.03	4.03
Add for block and plinth door casing	BC@.300	Ea	—	7.55	7.55
Add for hand nailing	BC@.275	Ea	—	6.92	6.92

Window trim Based on trim set around prepared openings. Costs per window opening on square windows. Extra tall or extra wide windows will cost more.

Extension jambs cut, assembled, shimmed and installed

Up to 3/4" thick, 3 or 4 sides, per window	BC@.330	Ea	—	8.31	8.31
Over 3/4" thick, per LF of extension jamb	BC@.100	LF	—	2.52	2.52

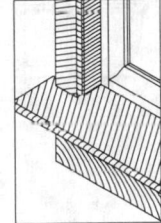

Window stool

Set to sheetrock wraps nailed directly on framing	BC@.500	Ea	—	12.60	12.60
When stool has to be shimmed up from framing Add per LF of stool	BC@.050	LF	—	1.26	1.26
Stool set to wooden jambs, per opening	BC@.400	Ea	—	10.10	10.10
When stool has to be shimmed up from jambs Add per LF of stool	BC@.050	LF	—	1.26	1.26
Add for mitered returns on horns per opening	BC@.500	Ea	—	12.60	12.60
Window apron moulding without mitered returns per opening	BC@.160	Ea	—	4.03	4.03
Window apron moulding with mitered returns per opening	BC@.320	Ea	—	8.05	8.05

Window casing, straight casing with square corners, per opening

3 sides (stool and apron), up to 5' on any side	BC@.330	Ea	—	8.31	8.31
4 sides (picture frame), up to 5' on any side	BC@.500	Ea	—	12.60	12.60
Add for any side over 5' long, per side	BC@.090	Ea	—	2.27	2.27
Add for angle corners (trapezoid or triangle) per corner	BC@.250	Ea	—	6.29	6.29

Installing a pre-manufactured arched top casing only, per opening

To 4' wide set against rosettes	BC@.330	Ea	—	8.31	8.31
To 4' wide, mitered corners	BC@.500	Ea	—	12.60	12.60
Over 4' wide set against rosettes	BC@.660	Ea	—	16.60	16.60
Over 4' wide, mitered corners	BC@1.00	Ea	—	25.20	25.20

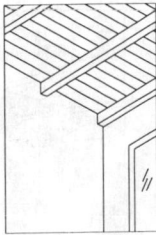

Ceiling treatments All figures assume a wall height not over 9' with work done from scaffolding. Add the cost of scaffolding

Set up and dismantle scaffolding, based on a typical 10' x 16' room, per room	BC@1.75	Ea	—	44.00	44.00

Wood strip ceiling, T&G or square edged, applied wall to wall with no miters

per square foot of ceiling	BC@.050	SF	—	1.26	1.26

	Craft@Hrs	Unit	Material	Labor	Total
Plywood panel ceiling					
Screwed in place, moulding strips nailed with a pneumatic nailer, including normal layout					
per square foot of ceiling covered	BC@.160	SF	—	4.03	4.03
Applied beams, including building up the beam, layout and installation on a flat ceiling					
per LF of beam	BC@3.10	LF	—	78.00	78.00
Add for each fitted joint or beam					
scribed to the wall	BC@.660	Ea	—	16.60	16.60
Add for each corbel	BC@.400	Ea	—	10.10	10.10
Add for each knee brace	BC@.600	Ea	—	15.10	15.10

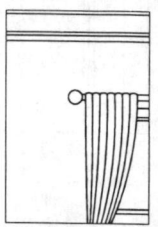

Cornice or crown moulding Including layout and coped joints, per LF of moulding

	Craft@Hrs	Unit	Material	Labor	Total
Up to 2-1/4" wide	BC@.050	LF	—	1.26	1.26
Over 2-1/4" wide to 3-1/2" wide	BC@.060	LF	—	1.51	1.51
Over 3-1/2" wide	BC@.070	LF	—	1.76	1.76

Wall paneling No moulding included. See moulding below. Costs per square foot of wall covered

	Craft@Hrs	Unit	Material	Labor	Total
Board wainscot or paneling, 3-1/2" to 5-1/2" wide tongue and groove					
per square foot of wall covered	BC@.080	SF	—	2.01	2.01
Frame and panel wainscot. Installation only. Add for fabricating the frames and panels					
per square foot of wall covered	BC@.090	SF	—	2.27	2.27
Add for frame attached with screws					
and covered with wood plugs	BC@.010	SF	—	.25	.25
Hardwood sheet paneling installed on wood frame or masonry walls					
Per square foot of wall covered	BC@.040	SF	—	1.01	1.01
Add for running moulding if required	BC@.040	SF	—	1.01	1.01
Add if vapor barrier must be installed					
with mastic over a masonry wall	BC@.010	SF	—	.25	.25
Add when 1" x 4" furring is nailed on a masonry wall with a powder-actuated nail gun					
No shimming or straightening included	BC@.020	SF	—	.50	.50
Add if sheet paneling is scribed around stone veneer					
per linear foot scribed	BC@.150	LF	—	3.78	3.78
Minimum time for scribing sheet paneling	BC@.500	Ea	—	12.60	12.60

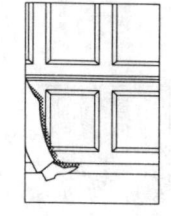

Running mouldings Single piece mitered installations nailed by hand. Add 100% for two-piece moulding. These estimates are based on stain-grade work. Subtract 33% for paint-grade work
Costs per LF of moulding.

	Craft@Hrs	Unit	Material	Labor	Total
Picture rail	BC@.050	LF	—	1.26	1.26
Chair rail	BC@.040	LF	—	1.01	1.01
Wainscot cap or bed moulding	BC@.040	LF	—	1.01	1.01
Baseboard	BC@.040	LF	—	1.01	1.01
Kneewall cap up to 8" wide and 1-1/2" thick. Complex job layout and bed moulding will increase costs					
Kneewall cap	BC@.160	LF	—	4.03	4.03

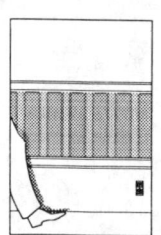

Closet poles and shelving Including layout and installation

	Craft@Hrs	Unit	Material	Labor	Total
1-1/4" closet pole supported with a metal bracket every 32"					
Per LF of pole	BC@.040	LF	—	1.01	1.01
Add for mitered corner joints, per corner	BC@.300	Ea	—	7.55	7.55
3/4" x 12" hat shelf supported with a metal shelf bracket every 32"					
Per LF of shelf	BC@.100	LF	—	2.52	2.52
Add for mitered corner joint, per corner	BC@.300	Ea	—	7.55	7.55

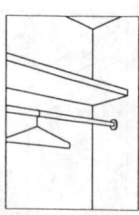

	Craft@Hrs	Unit	Material	Labor	Total

Site-built closet storage unit with drawers, using plate biscuit joiners and 3/4" board or plywood stock, including fabrication, layout, assembly and installation

	Craft@Hrs	Unit	Material	Labor	Total
Per square foot of storage unit face	BC@.680	SF	—	17.10	17.10
Add for each drawer	BC@.850	Ea	—	21.40	21.40

Cabinet installation Including layout, leveling and installation. No fabrication, assembly or countertops included. Costs per cabinet

	Craft@Hrs	Unit	Material	Labor	Total
Base cabinets					
Cabinets to 36" wide	BC@.766	Ea	—	19.30	19.30
Cabinets over 36" wide	BC@.911	Ea	—	22.90	22.90
Wall cabinets					
Cabinets to 18" wide	BC@.638	Ea	—	16.10	16.10
Cabinets over 18" to 36" wide	BC@.766	Ea	—	19.30	19.30
Cabinets over 36" wide	BC@1.03	Ea	—	25.90	25.90
Oven or utility cabinets, 70" high, to 30" wide	BC@1.71	Ea	—	43.00	43.00
Filler strip between cabinets, measured, cut, fit and installed					
Per strip installed	BC@.200	Ea	—	5.03	5.03

Fireplace surrounds and mantels Costs will be higher when a mantel or moulding must be scribed around irregular masonry surfaces.

Layout, cutting, fitting, assembling and installing a vertical pilaster on two sides and a horizontal frieze board above the pilasters. Additional mouldings will take more time.

	Craft@Hrs	Unit	Material	Labor	Total
Costs per linear foot of pilaster and frieze	BC@.350	LF	—	8.81	8.81
Mantel shelf attached to masonry, including cutting, fitting and installation					
Per mantel	BC@.750	Ea	—	18.90	18.90
Installing a prefabricated mantel and fireplace surround					
Per mantel and surround installed	BC@1.50	Ea	—	37.80	37.80

Wood flooring Including layout, vapor barrier and installation with a power nailer. Add for underlayment, sanding and finishing. Borders, feature strips and mitered corners will increase costs. Costs per square foot of floor covered

	Craft@Hrs	Unit	Material	Labor	Total
Tongue & groove strip flooring					
3/4" x 2-1/4" strips	BF@.036	SF	—	.88	.88
Straight edge plank flooring, 3/4" x 6" to 12" wide					
Set with biscuit joiner plates	BF@.055	SF	—	1.35	1.35
Add for plank flooring					
set with screws and butterfly plugs	BF@.010	SF	—	.25	.25
Filling and sanding a wood floor with drum and disc sanders,					
3 passes (60, 80, 100 grit), per square foot	BF@.023	SF	—	.57	.57
Finishing a wood floor, including filler, shellac					
polyurethane or waterborne urethane	BF@.024	SF	—	.59	.59

Skirt boards for stairways Including layout, fitting, assembly and installation. Based on a 9' floor-to-floor rise where the stair carriage has already been installed by others. Costs will be higher if re-cutting, shimming, or adjusting of an existing stair carriage is required.

Cutting and installing a skirt board on the closed side of a stairway. Costs per skirt board

	Craft@Hrs	Unit	Material	Labor	Total
Treads and risers					
Butted against skirt boards, no dadoes	BC@2.00	Ea	—	50.30	50.30
Treads & risers set in dadoes	BC@3.75	Ea	—	94.40	94.40
Cutting and installing a skirt board on the open side of a stairway					
Risers miter-jointed into the board	BC@5.50	Ea	—	138.00	138.00

Carpentry, Finish

	Craft@Hrs	Unit	Material	Labor	Total

Treads and risers for stairs Including layout, cutting, fitting, and installation of treads, risers and typical moulding under the tread nosing. Includes a miter return on one side of each tread. 9' rise is usually 14 treads and 15 risers

9' rise stairway	BC@2.50	Ea	—	62.90	62.90

Handrail for stairs Including layout, cutting, fitting and installation. Based on milled hardwood rail, posts and fittings

Setting newel posts, costs per post	BC@2.50	Ea	—	62.90	62.90
Setting post-to-post handrail					
Costs per 10 linear feet of rail	BC@1.50	Ea	—	37.80	37.80
Setting over-the-post handrail					
Costs per 10 linear feet of rail	BC@2.50	Ea	—	62.90	62.90
Add for each transition fitting					
Turn, easing or gooseneck	BC@1.75	Ea	—	44.00	44.00

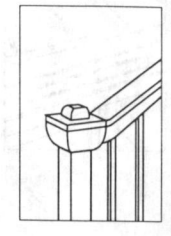

Balusters for stair railing Including layout, cutting, fitting and installation. Based on mill-made hardwood balusters. Cost per baluster

Balusters with round top section	BC@.330	Ea	—	8.31	8.31
Balusters with square top section	BC@.750	Ea	—	18.90	18.90

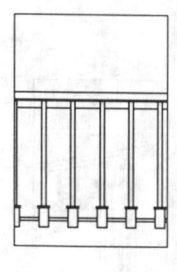

Carpeting, Subcontract Costs listed are average prices for complete residential jobs and include consultation, measurement, pad, carpet, and professional installation of pad and carpet using tack strips and hot melt tape on seams. Prices can be expected to vary, up or down, by 50% depending on quality and quantity of actual materials required.

Minimum quality, 25 to 35 oz face weight nylon carpet					
Including a 1/2" (4 lb density) rebond pad	—	SY	—	—	16.50
Medium quality, 35 to 50 oz face weight nylon carpet					
Including a 1/2" (6 lb density) rebond pad	—	SY	—	—	22.50
Better quality, 50 (plus) oz face weight nylon carpet					
Including a 1/2" (6 lb density) rebond pad	—	SY	—	—	30.50
Wool Berber carpet (large loop) installed	—	SY	—	—	46.00
Add for waterfall (box steps) stairways	—	Riser	—	—	4.00
Add for wrapped steps (open riser), sewn	—	Step	—	—	7.00
Add for sewn edge treatment on one side	—	Riser	—	—	8.50
Add for circular stair steps, costs vary depending on frame					
Add for circular stairs, typical cost	—	Step	—	—	12.00

Carports Typical prices. Material costs include all hardware. Labor is based on bolting posts to an existing concrete slab.

Single carport, 6 posts, 8' high

Natural aluminum finish, posts attach to galvanized steel beams					
8' wide x 16' long, 40 PSF roof load rating	B1@7.96	Ea	657.00	182.00	839.00
10' wide x 20' long, 20 PSF roof load rating	B1@7.96	Ea	850.00	182.00	1,032.00
White enamel finish, baked-on enamel roof panels, gutter-mounted posts					
10' wide x 20' long					
20 PSF roof load rating	B1@7.96	Ea	997.00	182.00	1,179.00
40 PSF roof load rating	B1@7.96	Ea	1,040.00	182.00	1,222.00
60 PSF roof load rating	B1@7.96	Ea	1,200.00	182.00	1,382.00
Double carport, 10' high, with steel posts and aluminum gutters, roof pans and supports					
20' x 20', 4 posts, 40 PSF roof load rating	B1@8.80	Ea	2,270.00	201.00	2,471.00
20' x 20', 6 posts, 60 PSF roof load rating	B1@9.34	Ea	2,950.00	213.00	3,163.00
24' x 24', 4 posts, 40 PSF roof load rating	B1@8.80	Ea	3,510.00	201.00	3,711.00
24' x 24', 6 posts, 60 PSF roof load rating	B1@9.34	Ea	3,910.00	213.00	4,123.00

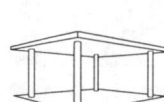

	Craft@Hrs	Unit	Material	Labor	Total

Caulking Material costs are typical costs for bead diameter listed. Figures in parentheses indicate approximate coverage including 5% waste. Labor costs are for good quality application on smooth to slightly irregular surfaces. Per LF bead length

Acoustical caulk, flexible, sound deadening (@$.33 per fluid oz)

1/4" (2.91 LF per fluid oz)	BC@.025	LF	.11	.63	.74
3/8" (1.29 LF per fluid oz)	BC@.030	LF	.26	.76	1.02
1/2" (.728 LF per fluid oz)	BC@.033	LF	.45	.83	1.28

Butyl caulk, premium quality, 20 year life expectancy (@$.35 per fluid oz)

1/8" (11.6 LF per fluid oz)	BC@.018	LF	.03	.45	..48
1/4" (2.91 LF per fluid oz)	BC@.025	LF	.12	.63	.75
3/8" (1.29 LF per fluid oz)	BC@.030	LF	.27	.76	1.03
1/2" (.728 LF per fluid oz)	BC@.033	LF	.48	.83	1.31

Latex, vinyl acrylic caulk, good quality, 10 year life (@$.45 per fluid oz)

1/8" (11.6 LF per fluid oz)	BC@.018	LF	.04	.45	.49
1/4" (2.91 LF per fluid oz)	BC@.025	LF	.15	.63	.78
3/8" (1.29 LF per fluid oz)	BC@.030	LF	.35	.76	1.11
1/2" (.728 LF per fluid oz)	BC@.033	LF	.62	.83	1.45

Latex caulk, economy grade (@$.42 per fluid oz)

1/8" (11.6 LF per fluid oz)	BC@.018	LF	.04	.45	.49
1/4" (2.91 LF per fluid oz)	BC@.025	LF	.14	.63	.77
3/8" (1.29 LF per fluid oz)	BC@.030	LF	.33	.76	1.09
1/2" (.728 LF per fluid oz)	BC@.033	LF	.58	.83	1.41

Oil base caulk, good quality (@$.36 per fluid oz)

1/8" (11.6 LF per fluid oz)	BC@.018	LF	.03	.45	.48
1/4" (2.91 LF per fluid oz)	BC@.025	LF	.12	.63	.75
3/8" (1.29 LF per fluid oz)	BC@.030	LF	.28	.76	1.04
1/2" (.728 LF per fluid oz)	BC@.033	LF	.49	.83	1.32

Oil base caulk, economy grade (@$.35 per fluid oz)

1/8" (11.6 LF per fluid oz)	BC@.018	LF	.03	.45	.48
1/4" (2.91 LF per fluid oz)	BC@.025	LF	.12	.63	.75
3/8" (1.29 LF per fluid oz)	BC@.030	LF	.27	.76	1.03
1/2" (.728 LF per fluid oz)	BC@.033	LF	.48	.83	1.31

Peel-off caulking for temporary use (@$.34 per fluid oz)

1/8" (11.6 LF per fluid oz)	BC@.018	LF	.03	.45	.48
1/4" (2.91 LF per fluid oz)	BC@.022	LF	.10	.55	.65
3/8" (1.29 LF per fluid oz)	BC@.027	LF	.26	.68	.94
1/2" (.728 LF per fluid oz)	BC@.030	LF	.47	.76	1.23

Siliconized acrylic caulk, premium quality, 20 year life (@$.33 per fluid oz)

1/8" (11.6 LF per fluid oz)	BC@.018	LF	.03	.45	.48
1/4" (2.91 LF per fluid oz)	BC@.025	LF	.12	.63	.75
3/8" (1.29 LF per fluid oz)	BC@.030	LF	.26	.76	1.02
1/2" (.728 LF per fluid oz)	BC@.033	LF	.45	.83	1.28

Silicon rubber sealant, premium quality, 20 year life (@$.36 per fluid oz)

1/8" (11.6 LF per fluid oz)	BC@.043	LF	.03	1.08	1.11
1/4" (2.91 LF per fluid oz)	BC@.048	LF	.12	1.21	1.33
3/8" (1.29 LF per fluid oz)	BC@.056	LF	.28	1.41	1.69
1/2" (.728 LF per fluid oz)	BC@.060	LF	.49	1.51	2.00

Tub caulk, white siliconized (@$.47 per fluid oz)

1/8" (11.6 LF per fluid oz)	BC@.018	LF	.04	.45	.49
1/4" (2.91 LF per fluid oz)	BC@.022	LF	.16	.55	.71
3/8" (1.29 LF per fluid oz)	BC@.027	LF	.36	.68	1.04
1/2" (.728 LF per fluid oz)	BC@.030	LF	.65	.76	1.41

Caulking

	Craft@Hrs	Unit	Material	Labor	Total
Anti-algae and mildew-resistant tub caulk, premium quality white or clear silicone (@$.36 per fluid oz)					
1/8" (11.6 LF per fluid oz)	BC@.018	LF	.03	.45	.48
1/4" (2.91 LF per fluid oz)	BC@.022	LF	.12	.55	.67
3/8" (1.29 LF per fluid oz)	BC@.027	LF	.28	.68	.96
1/2" (.728 LF per fluid oz)	BC@.030	LF	.49	.76	1.25
Urethane one-part caulk, good quality, 10 year life (@$.35 per fluid oz)					
1/8" (11.6 LF per fluid oz)	BC@.043	LF	.03	1.08	1.11
1/4" (2.91 LF per fluid oz)	BC@.048	LF	.12	1.21	1.33
3/8" (1.29 LF per fluid oz)	BC@.056	LF	.27	1.41	1.68
1/2" (.728 LF per fluid oz)	BC@.060	LF	.48	1.51	1.99
Urethane two-part caulk, top quality, 20 year life, 10 minute mixing time (@$.35 per fluid oz)					
1/8" (11.6 LF per fluid oz)	BC@.043	LF	.03	1.08	1.11
1/4" (2.91 LF per fluid oz)	BC@.052	LF	.12	1.31	1.43
3/8" (1.29 LF per fluid oz)	BC@.056	LF	.27	1.41	1.68
1/2" (.728 LF per fluid oz)	BC@.066	LF	.48	1.66	2.14
Caulking gun, professional type, heavy duty					
Bulk caulking gun	—	Ea	69.90	—	69.90
"Caulk Master" caulking gun	—	Ea	123.00	—	123.00
Caulking gun, economy grade					
11-oz cartridge	—	Ea	8.75	—	8.75

Cedar Closet Lining Material costs are per SF of floor, wall, ceiling or door of closet including 7% waste. Labor is for good quality installation in typical residential closet over studs or gypsum wallboard.

	Craft@Hrs	Unit	Material	Labor	Total
Compressed cedar chip panels, class C flame spread rating, 4' x 8' panels					
3/16" thick ($22.00 per 4' x 8' panel)	BC@.016	SF	.74	.40	1.14
1/4" thick ($24.00 per 4' x 8' panel)	BC@.016	SF	.80	.40	1.20
Cedar boards, 1/4" x 3-3/4", random lengths, T&G, end matched					
($20.00 per 15 SF bundle)	BC@.050	SF	1.42	1.26	2.68

Ceiling Domes Enveldomes™, complete ceiling dome kits, includes grid, hardware, panels and crating, from flat to 30 degree pitch in 5 degree increments, Envel Design Corp. Prices shown are for best quality. Prices may be as much as 40% lower depending on the requirements.

	Craft@Hrs	Unit	Material	Labor	Total
10' diameter, octagon	B1@10.7	Ea	3,400.00	245.00	3,645.00
16' diameter, octagon	B1@17.2	Ea	6,590.00	393.00	6,983.00
24' diameter, round	B1@24.7	Ea	15,040.00	564.00	15,604.00

Ceiling Lenses Envelite, including lens and border material. Labor costs shown do not allow for cutting panels. Envel Design Corp

	Craft@Hrs	Unit	Material	Labor	Total
UV stabilized commercial grade acrylic with decorative patterns (Costs vary from $3.90 to $6.50 per SF)					
Typical cost per SF	B1@.100	SF	5.15	2.29	7.44
Custom, dimensional Envelex™ Designs					
Standard sizes and patterns (2' x 2', 2' x 4')	B1@.006	SF	25.80	.14	25.94
Irregular shapes					
Oversize and custom designs	B1@.006	SF	38.10	.14	38.24
Add for cutting panels	B1@.033	Ea	—	.75	.75

Ceiling Panel Suspension Grids Typical costs, plain white. Add panel costs below.

	Craft@Hrs	Unit	Material	Labor	Total
Main runner, 12' long (@$4.62 Ea)	—	LF	.39	—	.39
Cross tee, 4' long (@$1.19 Ea)	—	LF	.30	—	.30
Cross tee, 2' long (@$.66 Ea)	—	LF	.33	—	.33

	Craft@Hrs	Unit	Material	Labor	Total
Wall mould, 12' long (@$2.52 Ea)	—	LF	.21	—	.21
Add for colored	—	%	10.0	—	—
Add for fire-rated white	—	%	6.0	—	—
Add for chrome or gold color grid	—	%	40.0	—	—
Suspension grid system including runners, tees, wall mould, hooks and hanging wire (but no tile or lighting fixtures)					
2' x 2' grid	BC@.017	SF	.54	.43	.97
2' x 4' grid	BC@.015	SF	.49	.38	.87
Add for under 400 SF job	BC@.007	SF	—	.18	.18
Lighting fixtures for 2' x 4' grid, includes installation, connection, lens and tubes					
2 tube fixtures	BE@1.00	Ea	49.00	26.90	75.90
4 tube fixtures	BE@1.00	Ea	73.50	26.90	100.40
Light-diffusing panels for installation in 2' x 4' suspension system under fluorescent lighting					
Acrylic, prismatic clear					
.095" thick	BC@.029	Ea	6.30	.73	7.03
.110" thick	BC@.029	Ea	8.30	.73	9.03
.125" thick	BC@.029	Ea	9.20	.73	9.93
Acrylic, dual white					
.080" thick	BC@.029	Ea	16.80	.73	17.53
.125" thick	BC@.029	Ea	24.10	.73	24.83
Acrylic frost overlay, clear or white					
.040" thick	BC@.029	Ea	8.30	.73	9.03
Egg crate louvers, 1/2" x 1/2" x 1/2" cell size					
White styrene	BC@.188	Ea	10.90	4.73	15.63
White acrylic	BC@.188	Ea	17.80	4.73	22.53
Mill aluminum	BC@.100	Ea	29.80	4.73	34.53
White aluminum	BC@.188	Ea	33.80	4.73	38.53
Anodized aluminum	BC@.188	Ea	38.80	4.73	43.53
Egg crate louvers, 1/2" x 1/2" x 3/8" cell size					
White styrene	BC@.188	Ea	8.50	4.73	13.23
White acrylic	BC@.188	Ea	16.00	4.73	20.73
Black styrene	BC@.188	Ea	10.50	4.73	15.23
Parabolic louvers, 5/8" x 5/8" x 7/16" cell size					
Silver styrene	BC@.188	Ea	24.70	4.73	29.43
Silver acrylic	BC@.188	Ea	41.50	4.73	46.23
Gold styrene	BC@.188	Ea	25.70	4.73	30.43
Gold acrylic	BC@.188	Ea	42.00	4.73	46.73
Polystyrene, clear or white					
.095", .070" thick	BC@.056	Ea	5.00	1.41	6.41
.040" thick, smooth white	BC@.056	Ea	5.00	1.41	6.41
Stained glass look, one dimension, 2' x 2'					
Standard colors	BC@.033	Ea	29.40	.83	30.23
Antiqued colors	BC@.035	Ea	39.40	.88	40.28

Ceiling Panels, Lay-in Type Costs per SF of ceiling area covered. No waste included. Add suspension grid system cost from preceding page

	Craft@Hrs	Unit	Material	Labor	Total
24" x 24" non-acoustic					
Smooth, plain white	BC@.007	SF	1.20	.18	1.38
Smooth, marbleized or simulated wood	BC@.007	SF	2.08	.18	2.26
Textured	BC@.005	SF	1.60	.13	1.73

Ceiling Panels

	Craft@Hrs	Unit	Material	Labor	Total
24" x 24" acoustic					
Deep textured, 3/4" reveal edge	BC@.010	SF	1.75	.25	2.00
Random pinhole, 5/8" square edge	BC@.004	SF	.42	.10	.52
24" x 48" acoustic					
Deep textured, 3/4" reveal edge	BC@.010	SF	1.53	.25	1.78
Fissured, 5/8" square edge	BC@.004	SF	.42	.10	.52
Random pinhole, 5/8" square edge	BC@.004	SF	.42	.10	.52
24" x 48" fire retardant, acoustic					
Deep textured, 3/4" reveal edge	BC@.012	SF	1.75	.30	2.05
Fissured, 5/8" square edge	BC@.004	SF	.49	.10	.59
Embossed parquet, pattern, grid hiding	BC@.004	SF	1.17	.10	1.27
Insulating (R-12), acoustic, textured	BC@.004	SF	2.20	.10	2.30
Fiberglass reinforced 24" x 24" gypsum panels for T-bar systems. Weight is 3 to 4 pounds each, depending on pattern					
White painted finish	BC@.015	SF	7.25	.38	7.63
Wood grain finish	BC@.020	SF	8.35	.50	8.85
Multi-dimensional faceted and beveled ceiling panels					
2' x 2' (Envel Design)	BC@.488	Ea	73.50	12.30	85.80
Ceiling Pans, Metal Flat, 2' x 2' pans					
Aluminum with concealed grid					
2' x 2' painted	BC@.083	SF	3.90	2.09	5.99
2' x 2' polished	BC@.083	SF	8.40	2.09	10.49
Stainless steel with concealed grid					
2' x 2', polished clear mirror	BC@.097	SF	12.60	2.44	15.04
Stainless steel with exposed grid					
2' x 2', polished clear mirror	BC@.074	SF	7.35	1.86	9.21
Ceiling Tile 12" x 12" tile nailed, glued or stapled to ceilings. Costs per SF of ceiling area covered including fasteners and trim. No waste included					
Non-acoustic					
Economy, embossed texture pattern	BC@.027	SF	.73	.68	1.41
Smooth	BC@.026	SF	.84	.65	1.49
Textured	BC@.027	SF	.89	.68	1.57
Acoustic					
Random pinhole	BC@.035	SF	1.04	.88	1.92
Heavily textured	BC@.035	SF	1.68	.88	2.56
Fire retardant					
Non-acoustic simulated cork	BC@.040	SF	2.15	1.01	3.16
Acoustic, textured	BC@.035	SF	1.62	.88	2.50
Ceilings, Tin Various plain and intricate patterns, unfinished					
Embossed panels					
2' x 2' square panels, cost each	B1@.147	Ea	14.50	3.36	17.86
Embossed plates					
12" x 12" plates, cost per SF	B1@.042	SF	7.50	.96	8.46
12" x 24" plates, cost per SF	B1@.040	SF	4.00	.91	4.91
24" x 24" plates, cost per SF	B1@.037	SF	3.50	.85	4.35
24" x 48" plates, cost per SF	B1@.035	SF	3.00	.80	3.80

	Craft@Hrs	Unit	Material	Labor	Total
Embossed moulding, 4' lengths					
3" and 4" widths, cost per linear foot	B1@.055	LF	2.00	1.26	3.26
6" widths, cost per linear foot	B1@.067	LF	2.75	1.53	4.28
12" widths, cost per linear foot	B1@.088	LF	3.88	2.01	5.89
Embossed moulding crosses, tees or ells					
For 3", 4" and 6" wide moulding, cost each	B1@.181	Ea	5.25	4.14	9.39
For 12" wide moulding, cost each	B1@.228	Ea	8.50	5.21	13.71
Perimeter and beam cornice moulding, 4' lengths					
2" projection x 2" deep, per LF	B1@.180	LF	2.50	4.11	6.61
3" projection x 4" deep, per LF	B1@.197	LF	2.50	4.50	7.00
4" projection x 6" deep, per LF	B1@.197	LF	2.63	4.50	7.13
6" projection x 8" deep, per LF	B1@.218	LF	3.13	4.98	8.11
6" projection x 10" deep, per LF	B1@.218	LF	3.13	4.98	8.11
10" projection x 12" deep, per LF	B1@.238	LF	4.25	5.44	9.69
12" projection x 18" deep, per LF	B1@.238	LF	5.50	5.44	10.94
Mitered corners for cornice moulding					
3" x 4"	B1@.184	Ea	9.50	4.20	13.70
4" x 6"	B1@.184	Ea	10.00	4.20	14.20
6" x 8"	B1@.230	Ea	10.50	5.26	15.76
6" x 10"	B1@.230	Ea	11.50	5.26	16.76
10" x 12"	B1@.280	Ea	12.50	6.40	18.90
12" x 18"	B1@.280	Ea	14.00	6.40	20.40
Frieze moulding, 4' lengths					
6" width, per LF	B1@.067	LF	2.75	1.53	4.28
12" and 14" widths per LF	B1@.088	LF	3.88	2.01	5.89
18" width per LF	B1@.110	LF	5.25	2.51	7.76
Foot moulding, 4' lengths					
4" width, per LF	B1@.055	LF	2.13	1.26	3.39
6" width, per LF	B1@.067	LF	2.75	1.53	4.28
Nosing, 4' lengths					
1-1/4" width, per LF	B1@.047	LF	2.00	1.07	3.07
4-1/2" width, per LF	B1@.055	LF	2.00	1.26	3.26

Cement See also Adhesives, Aggregates, and Concrete

		Unit	Material	Labor	Total
Portland cement, type II, 94 lb sack	—	Sa	5.50	—	5.50
Plastic cement, 94 lb sack	—	Sa	5.00	—	5.00
High early strength cement, 94 lb sack	—	Sa	7.30	—	7.30
Concrete mix, 90 lb sack	—	Sa	2.30	—	2.30
Concrete mix, 60 lb sack	—	Sa	1.60	—	1.60
Topping mix, 60 lb sack	—	Sa	2.60	—	2.60
Mortar mix, 60 lb sack	—	Sa	2.60	—	2.60
Exterior stucco mix, 60 lb sack	—	Sa	2.30	—	2.30
Plaster patch, 25 lb sack	—	Sa	6.90	—	6.90
Asphalt mix, 60 lb sack	—	Sa	3.10	—	3.10
White cement, 94 lb sack	—	Sa	12.50	—	12.50
Sand, 1/2 cubic foot sack	—	Sa	12.50	—	1.50
Deduct for pallet quantities	—	%	-15.0	—	—

Cleaning, Final

	Craft@Hrs	Unit	Material	Labor	Total

Cleaning, Final Interior surfaces, plumbing fixtures and windows. Includes removing construction debris and grime. Add for removing paint splatter or mastic. Work performed using hand tools and commercial non-toxic cleaning products or solvent. Use $80.00 as a minimum job charge.

	Craft@Hrs	Unit	Material	Labor	Total
Cabinets, per linear foot of cabinet face. Wall or base cabinet, (bathroom, kitchen or linen)					
Cleaned inside and outside	BL@.048	LF	.05	.99	1.04
Add for polish, exposed surface only	BL@.033	LF	.03	.68	.71
Countertops, per linear foot of front, bathroom, family room or kitchen					
Marble or granite	BL@.040	LF	.03	.82	.85
Tile, clean (including grout joints)	BL@.046	LF	.05	.94	.99
Vinyl or plastic	BL@.035	LF	.04	.72	.76
Painted surfaces, wipe down by hand					
Base board, trim or moulding	BL@.003	LF	.03	.06	.09
Cased openings, per side cleaned	BL@.060	Ea	.50	1.23	1.73
Ceilings, painted	BL@.003	SF	.03	.06	.09
Walls, painted or papered	BL@.002	SF	.03	.04	.07
Window sills	BL@.004	LF	.03	.08	.11
Doors, painted or stained					
Cleaned two sides, including casing	BL@.150	Ea	.05	3.08	3.13
Add for polishing two sides	BL@.077	Ea	.03	1.58	1.61
Floors					
Hardwood floor, clean, wax and polish	BL@.025	SF	.08	.51	.59
Mop floor	BL@.002	SF	.01	.04	.05
Vacuum carpet or sweep floor	BL@.025	SF	.01	.51	.52
Plumbing fixtures					
Bar sink, including faucet	BL@.380	Ea	.12	7.80	7.92
Bathtubs, including faucets & shower heads	BL@.750	Ea	.12	15.40	15.52
Kitchen sink, including faucets	BL@.500	Ea	.12	10.30	10.42
Lavatories, including faucets	BL@.500	Ea	.12	10.30	10.42
Toilets, including seat and tank	BL@.600	Ea	.12	12.30	12.42
Shower stalls, including walls, floor, shower rod and shower head					
Three-sided	BL@1.80	Ea	.21	37.00	37.21
Add for glass doors	BL@.500	Ea	.05	10.30	10.35
Sliding glass doors, clean and polish both sides including cleaning of track and screen					
Up to 8' wide	BL@.650	Ea	.05	13.30	13.35
Over 8' wide	BL@.750	Ea	.09	15.40	15.49
Windows, clean and polish, inside and outside, first or second floors					
Bay windows	BL@.750	Ea	.09	15.40	15.49
Sliding, casement or picture windows	BL@.300	Ea	.05	6.16	6.21
Jalousie type	BL@.350	Ea	.09	7.19	7.28
Remove window screen, hose wash, replace	BL@.500	Ea	—	10.30	10.30
Typical final cleanup for new residential construction					
per SF of floor (no glass cleaning)	BL@.004	SF	.11	.08	.19

Closet Door Systems Folding hollow core, flush wood doors, 1-3/8" thick, with hardware. Add labor from the section following bypassing closet doors.

	Hardboard	Lauan	Birch
6'8" high (2 door units)			
2'0" wide	48.30	72.80	76.00
2'6" wide	51.50	77.90	92.00
3'0" wide	55.70	85.60	92.00

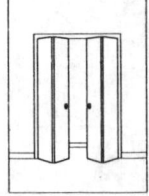

	Hardboard	Lauan	Birch
6'8" high (4 door units)			
4'0" wide	92.10	129.00	130.00
5'0" wide	98.70	142.00	148.00
6'0" wide	108.00	157.00	171.00
7'0" wide	160.00	198.00	219.00
8'0" wide	163.00	211.00	234.00
8'0" high (4 door units)			
4'0" wide	167.00	204.00	220.00
5'0" wide	183.00	230.00	250.00
6'0" wide	200.00	255.00	277.00
7'0" wide	279.00	320.00	349.00
8'0" wide	279.00	320.00	349.00

Folding pine closet doors, 6'8" high, 1-3/8" thick 2 door units (2 door units). With hardware. Add the labor cost from the section following bypassing closet doors.

	Full louver	Louver panel	2 panel design	3 panel design
2'0" wide	185.00	198.00	215.00	215.00
2'6" wide	193.00	215.00	240.00	240.00
3'0" wide	214.00	234.00	266.00	266.00
6'8" high (4 door units)				
4'0" wide	347.00	379.00	418.00	418.00
5'0" wide	372.00	314.00	469.00	469.00
6'0" wide	404.00	454.00	521.00	521.00
7'0" wide	454.00	518.00	577.00	577.00
8'0" wide	471.00	547.00	624.00	624.00

Folding doors, pine panel and louver, 1-1/8" thick, with hardware. Add the labor cost from the section following bypassing closet doors.

	Full louver	Louver panel	2 panel design	3 panel design
6'8" high (2 door units)				
2'0" wide	82.70	104.00	157.00	—
2'6" wide	99.30	119.00	179.00	—
2'8" wide	99.90	121.00	188.00	—
3'0" wide	108.00	137.00	210.00	—
Folding pine closet doors, 6'8" high, 1-3/8" thick 4 door units				
4'0" wide	165.00	203.00	302.00	—
5'0" wide	195.00	235.00	347.00	—
6'0" wide	214.00	267.00	386.00	—
7'0" wide	255.00	330.00	446.00	—
8'0" wide	266.00	345.00	481.00	—
8'0" high (2 door units)				
2'0" wide	133.00	148.00	—	—
3'0" wide	155.00	182.00	—	—
8'0" high (4 door units)				
4'0" wide	255.00	288.00	—	—
5'0" wide	280.00	328.00	—	—
6'0" wide	300.00	356.00	—	—
7'0" wide	362.00	441.00	—	—
8'0" wide	366.00	464.00	—	—

Closet Doors

	Hardboard	Lauan	Birch	Red oak
By-passing, hollow core wood closet doors, 1-3/8" thick. Includes hardware. Add labor from cost below.				
6'8" high				
4'0" (2 door unit)	76.00	135.00	151.00	163.00
5'0" (2 door unit)	83.00	145.00	162.00	185.00
6'0" (2 door unit)	93.10	160.00	185.00	200.00
7'0" (2 door unit)	141.00	211.00	298.00	308.00
8'0" (2 door unit)	169.00	230.00	312.00	330.00
9'0" (3 door unit)	145.00	246.00	279.00	305.00
10'0" (3 door unit)	210.00	316.00	444.00	461.00
12'0" (3 door unit)	257.00	340.00	469.00	498.00
4'0" (2 door unit)	121.00	221.00	242.00	281.00
5'0" (2 door unit)	132.00	235.00	246.00	298.00
6'0" (2 door unit)	145.00	256.00	254.00	308.00
7'0" (2 door unit)	179.00	312.00	341.00	418.00
8'0" (2 door unit)	193.00	327.00	365.00	424.00
9'0" (3 door unit)	217.00	382.00	378.00	457.00
10'0" (3 door unit)	267.00	409.00	512.00	628.00
12'0" (3 door unit)	292.00	495.00	551.00	626.00

	Craft@Hrs	Unit	Material	Labor	Total
Labor to install folding closet doors, 1-1/8" or 1-3/8" thick					
2'0", 2'6" or 3'0" wide x 6'8" high	BC@1.30	Ea	—	32.70	32.70
2'0", 2'6" or 3'0" wide x 8'0" high	BC@1.50	Ea	—	37.80	37.80
4'0" or 5'0" wide x 6'8" high	BC@1.60	Ea	—	40.30	40.30
4'0" or 5'0" wide x 8'0" high	BC@1.80	Ea	—	45.30	45.30
6'0", 7'0" or 8'0" wide x 6'8" high	BC@1.90	Ea	—	47.80	47.80
6'0", 7'0" or 8'0" wide x 8'0" high	BC@2.10	Ea	—	52.90	52.90
Labor to install bypassing closet doors, 1-3/8" thick					
4'0" or 5'0" wide x 6'8" high	BC@1.95	Set	—	49.10	49.10
4'0" or 5'0" wide x 8'0" high	BC@2.25	Set	—	56.60	56.60
6'0", 7'0", or 8'0" wide x 6'8" high	BC@2.42	Set	—	60.90	60.90
6'0", 7'0", or 8'0" wide x 8'0" high	BC@2.82	Set	—	71.00	71.00
9'0", 10'0", or 12'0" wide x 6'8" high	BC@2.69	Set	—	67.70	67.70
9'0", 10'0", or 12'0" wide x 8'0" high	BC@3.10	Set	—	78.00	78.00
Mirror panels for folding closet doors, polished edge					
12" x 6'8", for 24" bifold	BG@.406	Pr	99.70	9.36	109.06
15" x 6'8", for 30" bifold	BG@.406	Pr	117.00	9.36	126.36
18" x 6'8", for 36" bifold	BG@.406	Pr	136.00	9.36	145.36
Mirror panels for hinged doors, 3/16" float plate glass with 1/2" bevel on all edges. Includes mounting clips. Price per each					
14" x 50"	BG@.208	Ea	17.30	4.80	22.10
16" x 56" or 16" x 60"	BG@.208	Ea	27.50	4.80	32.30
18" x 60" or 18" x 68"	BG@.208	Ea	30.50	4.80	35.30
20" x 68" or 22" x 68"	BG@.208	Ea	41.70	4.80	46.50
24" x 68"	BG@.208	Ea	51.40	4.80	56.20
Louver-over-panel wood closet doors. Solid pine, unfinished, for 80" to 81" heights					
Two-section, folding door or two door set, hardware included					
24" wide	BC@1.20	Ea	57.80	30.20	88.00
30" wide	BC@1.20	Ea	73.50	30.20	103.70
32" wide	BC@1.50	Ea	83.00	37.80	120.80
36" wide	BC@1.50	Ea	88.20	37.80	126.00

	Craft@Hrs	Unit	Material	Labor	Total
Four-section (two 2-section doors)					
24" wide	BC@1.80	Ea	83.00	45.30	128.30
30" wide	BC@1.80	Ea	91.50	45.30	136.80
32" wide	BC@1.80	Ea	99.90	45.30	145.20
36" wide	BC@2.09	Ea	106.00	52.60	158.60
Add for colonial (raised panel) styles	—	LS	17.80	—	17.80

Sliding by-passing closet door units, steel frame, includes hardware and track, hollow core, opening sizes, 2 doors per set

	Craft@Hrs	Unit	Material	Labor	Total
Vinyl over hardboard doors, 6'8" high					
4'0"	BC@2.07	Set	90.20	52.10	142.30
6'0"	BC@2.58	Set	114.00	64.90	178.90
8'0"	BC@2.58	Set	129.00	64.90	193.90
10'0"	BC@3.01	Set	173.00	75.80	248.80

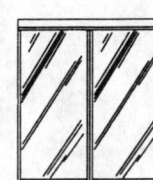

	Craft@Hrs	Unit	Material	Labor	Total
Mirror doors, 3/16" sheet glass over hardboard, 6'8" high					
4'0"	BC@2.08	Set	184.00	52.30	236.30
5'0"	BC@2.08	Set	218.00	52.30	270.30
6'0"	BC@2.58	Set	236.00	64.90	300.90
7'0"	BC@2.58	Set	299.00	64.90	363.90
8'0"	BC@2.58	Set	328.00	64.90	392.90
9'0"	BC@3.02	Set	393.00	76.00	469.00
10'0"	BC@3.02	Set	424.00	76.00	500.00
Mirror doors, 3/16" sheet glass over hardboard, 8'0" high					
5'0"	BC@2.08	Set	224.00	52.30	276.30
6'0"	BC@2.58	Set	266.00	64.90	330.90
7'0"	BC@2.58	Set	288.00	64.90	352.90
8'0"	BC@2.58	Set	348.00	64.90	412.90
9'0"	BC@3.02	Set	406.00	76.00	482.00
10'0"	BC@3.02	Set	442.00	76.00	518.00
Add for safety glass, any of above	—	%	50.0	—	—

Clothesline Units

Retractable type, indoor-outdoor, one end is wall-mounted, 5 lines, 34' long each, lines retract into metal case 34" L, 6-1/4" W, 5" H, mounting hardware included.

	Craft@Hrs	Unit	Material	Labor	Total
Cost per unit as described	B1@.860	Ea	42.00	19.70	61.70
Add for galvanized steel post, if required	B1@1.11	Ea	32.00	25.40	57.40
T-shaped galvanized steel pole type, including line, 6' x 1-1/2" post, 30" cross arm width, set in concrete					
5 hooks, two posts	BL@1.91	Ea	44.00	39.20	83.20
T-shaped galvanized steel pole type clothesline unit 6' x 2" post, 36" cross arm width, set in concrete					
7 hooks, two posts	BL@1.91	Ea	52.00	39.20	91.20
Umbrella type, 8' galvanized steel center post, requires 10' x 10' area, folds for storage,					
One post, set in concrete	BL@.885	Ea	67.20	18.20	85.40

Columns and Porch Posts Colonial style.

	Craft@Hrs	Unit	Material	Labor	Total
Round wood, standard cap and base					
8" x 8'	B1@2.00	Ea	132.00	45.70	177.70
10" x 10'	B1@2.00	Ea	198.00	45.70	243.70
Add for fluted wood, either of above	—	Ea	23.40	—	23.40
Round aluminum, standard cap and base					
8" x 9'	B1@2.00	Ea	135.00	45.70	180.70
10" x 10'	B1@2.00	Ea	179.00	45.70	224.70
12" x 12'	B1@3.00	Ea	305.00	68.60	373.60

Columns and Porch Posts

	Craft@Hrs	Unit	Material	Labor	Total
Add for ornamental wood cap, typical price	—	Ea	385.00	—	385.00
Add for ornamental aluminum cap, typical	—	Ea	369.00	—	369.00
Porch posts, clear laminated west coast hemlock, solid turned					
3-1/4" x 8'	B1@1.04	Ea	46.00	23.80	69.80
4-1/4" x 8'	B1@1.18	Ea	75.00	27.00	102.00
5-1/4" x 8'	B1@1.28	Ea	123.00	29.20	152.20

Concrete Ready-mix delivered by truck. Typical prices for most cities. Includes delivery up to 20 miles for 10 CY or more, 3" to 4" slump. Material cost only, no placing or pumping included. See forming and finishing costs on the following pages.

	Craft@Hrs	Unit	Material	Labor	Total
Footing and foundation concrete, 1-1/2" aggregate					
2,000 PSI, 4.8 sack mix	—	CY	57.50	—	57.50
2,500 PSI, 5.2 sack mix	—	CY	60.00	—	60.00
3,000 PSI, 5.7 sack mix	—	CY	62.50	—	62.50
3,500 PSI, 6.3 sack mix	—	CY	67.50	—	67.50
4,000 PSI, 6.9 sack mix	—	CY	70.70	—	70.70
Slab, sidewalk, and driveway concrete, 1" aggregate					
2,000 PSI, 5.0 sack mix	—	CY	58.50	—	58.50
2,500 PSI, 5.5 sack mix	—	CY	61.00	—	61.00
3,000 PSI, 6.0 sack mix	—	CY	64.00	—	64.00
3,500 PSI, 6.6 sack mix	—	CY	67.50	—	67.50
4,000 PSI, 7.1 sack mix	—	CY	70.00	—	70.00
Pea gravel pump mix / grout mix, 3/8" aggregate					
2,000 PSI, 6.0 sack mix	—	CY	63.70	—	63.70
2,500 PSI, 6.5 sack mix	—	CY	67.50	—	67.50
3,000 PSI, 7.2 sack mix	—	CY	71.00	—	71.00
3,500 PSI, 7.9 sack mix	—	CY	76.80	—	76.80
5,000 PSI, 8.5 sack mix	—	CY	79.80	—	79.80
Extra delivery costs for ready-mix concrete					
Add for delivery over 20 miles	—	Mile	.75	—	.75
Add for standby charge in excess of 5 minutes per CY delivered					
per minute of extra time	—	Ea	1.00	—	1.00
Add for less than 10 CY per load delivered	—	CY	10.00	—	10.00
Extra costs for non-standard aggregates					
Add for lightweight aggregate, typical	—	CY	34.00	—	34.00
Add for lightweight aggregate, pump mix	—	CY	34.00	—	34.00
Add for granite aggregate, typical	—	CY	3.25	—	3.25
Extra costs for non-standard mix additives					
High early strength 5 sack mix	—	CY	7.75	—	7.75
High early strength 6 sack mix	—	CY	9.75	—	9.75
Add for white cement (architectural)	—	CY	41.50	—	41.50
Add for 1% calcium chloride	—	CY	1.25	—	1.25
Add for chemical compensated shrinkage	—	CY	12.50	—	12.50
Add for super plasticized mix, 7" - 8" slump	—	%	8.0	—	—

Add for colored concrete, typical prices. The ready-mix supplier charges for cleanup of the ready-mix truck used for delivery of colored concrete. The usual practice is to provide one truck per day for delivery of all colored concrete for a particular job. Add the cost below to the cost per cubic yard for the design mix required. Also add the cost for truck cleanup.

	Craft@Hrs	Unit	Material	Labor	Total
Adobe	—	CY	14.50	—	14.50
Black	—	CY	26.00	—	26.00
Blended red	—	CY	13.50	—	13.50

	Craft@Hrs	Unit	Material	Labor	Total
Brown	—	CY	18.60	—	18.60
Green	—	CY	21.00	—	21.00
Yellow	—	CY	16.50	—	16.50
Colored concrete, truck cleanup, per day	—	LS	50.00	—	50.00

Concrete Expansion Joints Fiber or asphaltic-felt sided. Per linear foot

Wall (thickness)	1/4"	3/8"	1/2"
Depth			
3" deep	.18	.19	.20
3-1/2" deep	.19	.20	.22
4" deep	.21	.22	.24
6" deep	.29	.31	.34

	Craft@Hrs	Unit	Material	Labor	Total
Labor to install concrete expansion joints					
3" to 6" width, 1/4" to 1/2" thick	B1@.030	LF	—	.69	.69

Concrete Footing Form Ties Price each based on cartons of 200

		Unit	Material	Labor	Total
6" thru 12" form	—	Ea	.34	—	.34
14" form	—	Ea	.35	—	.35
16" form	—	Ea	.38	—	.38
18" form	—	Ea	.40	—	.40
20" form	—	Ea	.42	—	.42
22" form	—	Ea	.45	—	.45
24" form	—	Ea	.44	—	.44
Wedges, 3-3/4", 14 gauge	—	Ea	.07	—	.07

Concrete Form Excavation By hand using a pick and shovel, to 3'-0" deep. For wall footings, grade beams and column footings. Digging and one throw only, no disposal or back filling included. Per CY. See also, Excavation section

	Craft@Hrs	Unit	Material	Labor	Total
Light soil	BL@1.10	CY	—	22.60	22.60
Average soil	BL@1.65	CY	—	33.90	33.90
Heavy soil or loose rock	BL@2.14	CY	—	43.90	43.90

Concrete Form Stakes Costs per stake

		Unit	3/8 x 1-1/2" Material	Labor	Total
12" long	—	Ea	1.40	—	1.40
18" long	—	Ea	1.40	—	1.80
24" long	—	Ea	1.40	—	2.10
30" long	—	Ea	1.40	—	2.45
36" long	—	Ea	1.40	—	2.90

Concrete Formwork Multiple use of forms assumes that forms can be removed, partially disassembled and cleaned and that 75% of the form material can be reused. Costs include assembly, oiling, setting, stripping and cleaning. Material costs in this section can be adjusted to reflect your actual cost of lumber. Here's how: Divide your actual lumber cost per MBF by the assumed cost (listed in parentheses). Then multiply the cost in the material column by this adjustment factor.

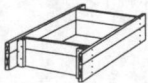

Board forming and stripping For wall footings, grade beams, column footings, site curbs and steps. Includes 5% waste. Per SF of contact area. When forms are required on both sides of the concrete, include the contact surface for each side.

Concrete Formwork

	Craft@Hrs	Unit	Material	Labor	Total

2" thick forms and bracing, using 2.85 BF of form lumber (@$525 per MBF) per SF. Includes nails, ties, and form oil (at $.20 per SF)

	Craft@Hrs	Unit	Material	Labor	Total
1 use	B2@.115	SF	1.77	2.72	4.49
3 use	B2@.115	SF	.91	2.72	3.63
5 use	B2@.115	SF	.72	2.72	3.44

Add for keyway beveled on two edges, one use. No stripping included

	Craft@Hrs	Unit	Material	Labor	Total
2" x 4" (@$525 per MBF)	B2@.027	LF	.37	.64	1.01
2" x 6" (@$515 per MBF)	B2@.027	LF	.54	.64	1.18

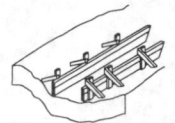

Driveway and walkway edge forms Material costs include stakes, nails, and form oil (@$.16 per LF) and 5% waste. Per LF of edge form. When forms are required on both sides of the concrete, include the length of each side plus the end widths.

2" x 4" form (@$525 per MBF, .7 BF per LF)

	Craft@Hrs	Unit	Material	Labor	Total
1 use	B2@.050	LF	.53	1.18	1.71
3 use	B2@.050	LF	.33	1.18	1.51
5 use	B2@.050	LF	.28	1.18	1.46

2" x 6" form (@$660 per MBF, 1.05 BF per LF)

	Craft@Hrs	Unit	Material	Labor	Total
1 use	B2@.050	LF	.71	1.18	1.89
3 use	B2@.050	LF	.41	1.18	1.59
5 use	B2@.050	LF	.34	1.18	1.52

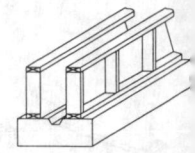

2" x 8" form (@$525 per MBF, 1.4 BF per LF)

	Craft@Hrs	Unit	Material	Labor	Total
1 use	B2@.055	LF	.90	1.30	2.20
3 use	B2@.055	LF	.49	1.30	1.79
5 use	B2@.055	LF	.40	1.30	1.70

2" x 10" form (@$570 per MBF, 1.75 BF per LF)

	Craft@Hrs	Unit	Material	Labor	Total
1 use	B2@.055	LF	1.63	1.30	2.93
3 use	B2@.055	LF	.61	1.30	1.91
5 use	B2@.055	LF	.49	1.30	1.79

2" x 12" form (@$585 per MBF, 2.1 BF per LF)

	Craft@Hrs	Unit	Material	Labor	Total
1 use	B2@.055	LF	1.39	1.30	2.69
3 use	B2@.055	LF	.71	1.30	2.01
5 use	B2@.055	LF	.57	1.30	1.87

Plywood forming For foundation walls, building walls and retaining walls, using 3/4" plyform (@$880 per MSF) with 10% waste, and 2" bracing (@$520 per MBF) with 20% waste. All material costs include nails, ties, clamps, form oil (@$.20 per SF). and stripping. Costs shown are per SF of contact area. When forms are required on both sides of the concrete, include the contact surface area for each side.

Walls up to 4' high (1.10 SF of plywood and .42 BF of bracing per SF of form area)

	Craft@Hrs	Unit	Material	Labor	Total
1 use	B2@.051	SF	1.43	1.20	2.63
3 use	B2@.051	SF	.75	1.20	1.95
5 use	B2@.051	SF	.61	1.20	1.81

Walls 4' to 8' high (1.10 SF of plywood and .60 BF of bracing per SF of form area)

	Craft@Hrs	Unit	Material	Labor	Total
1 use	B2@.060	SF	1.54	1.42	2.96
3 use	B2@.060	SF	.80	1.42	2.22
5 use	B2@.060	SF	.64	1.42	2.06

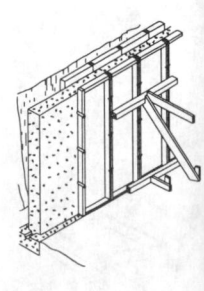

Walls 8' to 12' high (1.10 SF of plywood and .90 BF of bracing per SF of form area)

	Craft@Hrs	Unit	Material	Labor	Total
1 use	B2@.095	SF	1.73	2.24	3.97
3 use	B2@.095	SF	.98	2.24	3.22
5 use	B2@.095	SF	.77	2.24	3.01

Walls 12' to 16' high (1.10 SF of plywood and 1.05 BF of bracing per SF of form area)

	Craft@Hrs	Unit	Material	Labor	Total
1 use	B2@.128	SF	1.83	3.02	4.85
3 use	B2@.128	SF	.93	3.02	3.95
5 use	B2@.128	SF	.74	3.02	3.76

	Craft@Hrs	Unit	Material	Labor	Total

Walls over 16' high (1.10 SF of plywood and 1.20 BF of bracing per SF of form area)

	Craft@Hrs	Unit	Material	Labor	Total
1 use	B2@.153	SF	1.92	3.61	5.53
3 use	B2@.153	SF	.97	3.61	4.58
5 use	B2@.153	SF	.77	3.61	4.38

Form stripping Labor to remove forms and bracing, clean, and stack on job site.

Board forms at wall footings, grade beams and column footings. Per SF of contact area

	Craft@Hrs	Unit	Material	Labor	Total
1" thick lumber	BL@.010	SF	—	.21	.21
2" thick lumber	BL@.012	SF	—	.25	.25
Keyways, 2" x 4" and 2" x 6"	BL@.006	LF	—	.12	.12

Slab edge forms. Per LF of edge form

	Craft@Hrs	Unit	Material	Labor	Total
2" x 4" to 2" x 6"	BL@.012	LF	—	.25	.25
2" x 8" to 2" x 12"	BL@.013	LF	—	.27	.27

Walls, plywood forms. Per SF of contact area stripped

	Craft@Hrs	Unit	Material	Labor	Total
To 4' high	BL@.017	SF	—	.35	.35
Over 4' to 8' high	BL@.018	SF	—	.37	.37
Over 8' to 12' high	BL@.020	SF	—	.41	.41
Over 12' to 16' high	BL@.027	SF	—	.55	.55
Over 16' high	BL@.040	SF	—	.82	.82

Concrete Reinforcing Steel Steel reinforcing bars (rebar), ASTM A615 Grade 60. Material costs are for deformed steel reinforcing rebars, including 10% lap allowance, cutting and bending. These costs also include detailed shop drawings and delivery to jobsite with identity tags per shop drawings. Add for epoxy or galvanized coating of rebars, chairs, splicing, spiral caissons and round column reinforcing, if required, from the sections following the data below. Costs per pound (Lb) and per linear foot (LF) including tie wire and tying.

Reinforcing steel placed and tied in footings and grade beams

	Craft@Hrs	Unit	Material	Labor	Total
1/4" diameter, #2 rebar	RI@.015	Lb	.57	.36	.93
1/4" diameter, #2 rebar (.17 lb per LF)	RI@.003	LF	.10	.07	.17
3/8" diameter, #3 rebar	RI@.011	Lb	.42	.26	.68
3/8" diameter, #3 rebar (.38 lb per LF)	RI@.004	LF	.16	.09	.25
1/2" diameter, #4 rebar	RI@.010	Lb	.27	.24	.51
1/2" diameter, #4 rebar (.67 lb per LF)	RI@.007	LF	.18	.17	.35
5/8" diameter, #5 rebar	RI@.009	Lb	.27	.21	.48
5/8" diameter, #5 rebar (1.04 lb per LF)	RI@.009	LF	.28	.21	.49
3/4" diameter, #6 rebar	RI@.008	Lb	.27	.19	.46
3/4" diameter, #6 rebar (1.50 lb per LF)	RI@.012	LF	.41	.28	.69
7/8" diameter, #7 rebar	RI@.008	Lb	.27	.19	.46
7/8" diameter, #7 rebar (2.04 lb per LF)	RI@.016	LF	.55	.38	.93
1" diameter, #8 rebar	RI@.008	Lb	.27	.19	.46
1" diameter, #8 rebar (2.67 lb per LF)	RI@.021	LF	.72	.50	1.22

Reinforcing steel placed and tied in structural slabs

	Craft@Hrs	Unit	Material	Labor	Total
1/4" diameter, #2 rebar	RI@.014	Lb	.57	.33	.90
1/4" diameter, #2 rebar (.17 lb per LF)	RI@.002	LF	.10	.05	.15
3/8" diameter, #3 rebar	RI@.010	Lb	.42	.24	.66
3/8" diameter, #3 rebar (.38 lb per LF)	RI@.004	LF	.16	.09	.25
1/2" diameter, #4 rebar	RI@.009	Lb	.27	.21	.48
1/2" diameter, #4 rebar (.67 lb per LF)	RI@.006	LF	.18	.14	.32
5/8" diameter, #5 rebar	RI@.008	Lb	.27	.19	.46
5/8" diameter, #5 rebar (1.04 lb per LF)	RI@.008	LF	.28	.19	.47
3/4" diameter, #6 rebar	RI@.007	Lb	.27	.17	.44
3/4" diameter, #6 rebar (1.50 lb per LF)	RI@.011	LF	.41	.26	.67
7/8" diameter, #7 rebar	RI@.007	Lb	.27	.17	.44
7/8" diameter, #7 rebar (2.04 lb per LF)	RI@.014	LF	.55	.33	.88

Concrete Reinforcing Steel

	Craft@Hrs	Unit	Material	Labor	Total
1" diameter, #8 rebar	RI@.007	Lb	.27	.17	.44
1" diameter, #8 rebar (2.67 lb per LF)	RI@.019	LF	.72	.45	1.17
Reinforcing steel placed and tied in walls					
1/4" diameter, #2 rebar	RI@.017	Lb	.57	.40	.97
1/4" diameter, #2 rebar (.17 lb per LF)	RI@.003	LF	.10	.07	.17
3/8" diameter, #3 rebar	RI@.012	Lb	.42	.28	.70
3/8" diameter, #3 rebar (.38 lb per LF)	RI@.005	LF	.16	.12	.28
1/2" diameter, #4 rebar	RI@.011	Lb	.27	.26	.53
1/2" diameter, #4 rebar (.67 lb per LF)	RI@.007	LF	.18	.17	.35
5/8" diameter, #5 rebar	RI@.010	Lb	.27	.24	.51
5/8" diameter, #5 rebar (1.04 lb per LF)	RI@.010	LF	.28	.24	.52
3/4" diameter, #6 rebar	RI@.009	Lb	.27	.21	.48
3/4" diameter, #6 rebar (1.50 lb per LF)	RI@.014	LF	.41	.33	.74
7/8" diameter, #7 rebar	RI@.009	Lb	.27	.21	.48
7/8" diameter, #7 rebar (2.04 lb per LF)	RI@.018	LF	.55	.43	.98
1" diameter, #8 rebar	RI@.009	Lb	.27	.21	.48
1" diameter, #8 rebar (2.67 lb per LF)	RI@.024	LF	.72	.57	1.29

Add for reinforcing steel coating. Coating applied prior to shipment to job site. Labor for placing coated rebar is the same as that shown for plain rebar above. Cost per pound of reinforcing steel

Epoxy coating					
Any size rebar, add	—	Lb	.25	—	.25
Galvanized coating					
Any #2 rebar, #3 rebar or #4 rebar, add	—	Lb	.33	—	.33
Any #5 rebar or #6 rebar, add	—	Lb	.31	—	.31
Any #7 rebar or #8 rebar, add	—	Lb	.29	—	.29

Add for reinforcing steel chairs. Labor for placing rebar shown above includes time for setting chairs.

Chairs, individual type, cost each					
3" high					
Steel	—	Ea	.26	—	.26
Plastic	—	Ea	.36	—	.36
Galvanized steel	—	Ea	.40	—	.40
5" high					
Steel	—	Ea	.40	—	.40
Plastic	—	Ea	.56	—	.56
Galvanized steel	—	Ea	.52	—	.52
8" high					
Steel	—	Ea	.85	—	.85
Plastic	—	Ea	1.11	—	1.11
Galvanized steel	—	Ea	1.14	—	1.14
Chairs, continuous type, with legs 8" OC, cost per linear foot					
3" high					
Steel	—	LF	.36	—	.36
Plastic	—	LF	.43	—	.43
Galvanized steel	—	LF	.46	—	.46
6" high					
Steel	—	LF	.50	—	.50
Plastic	—	LF	.63	—	.63
Galvanized steel	—	LF	.66	—	.66
8" high					
Steel	—	LF	.70	—	.70
Plastic	—	LF	.89	—	.89
Galvanized steel	—	LF	.92	—	.92

	Craft@Hrs	Unit	Material	Labor	Total
Waterstop, 3/8"					
Rubber, 6"	B2@.059	LF	4.91	1.39	6.30
Rubber, 9"	B2@.059	LF	7.58	1.39	8.97
PVC, 9"	B2@.059	LF	4.39	1.39	5.78

Spiral caisson and round column reinforcing 3/8" diameter hot rolled steel spirals with main vertical bars as shown. Costs include typical engineering and shop drawings. Based on shop fabricated spirals delivered, tagged, ready to install. Cost per vertical linear foot (VLF)

	Craft@Hrs	Unit	Material	Labor	Total
16" diameter with 6 #6 bars, 16.8 lbs per VLF	RI@.041	VLF	16.20	.97	17.17
24" diameter with 6 #6 bars, 28.5 lbs per VLF	RI@.085	VLF	27.80	2.02	29.82
36" diameter with 8 #10 bars, 51.2 lbs per VLF	RI@.216	VLF	50.50	5.13	55.63

Welded wire mesh, steel, electric weld, including 15% waste and overlap.

	Craft@Hrs	Unit	Material	Labor	Total
2" x 2" W.9 x W.9 (#12 x #12), slabs	RI@.004	SF	.36	.09	.45
2" x 2" W.9 x W.9 (#12 x #12), beams and columns	RI@.020	SF	.36	.47	.83
4" x 4" W1.4 x W1.4 (#10 x #10), slabs	RI@.003	SF	.17	.07	.24
4" x 4" W2.0 x W2.0 (#8 x #8), slabs	RI@.004	SF	.20	.09	.29
4" x 4" W2.9 x W2.9 (#6 x #6), slabs	RI@.005	SF	.22	.12	.34
4" x 4" W4.0 x W4.0 (#4 x #4), slabs	RI@.006	SF	.28	.14	.42
6" x 6" W1.4 x W1.4 (#10 x #10), slabs	RI@.003	SF	.08	.07	.15
6" x 6" W2.0 x W2.0 (#8 x #8), slabs	RI@.004	SF	.13	.09	.22
6" x 6" W2.9 x W2.9 (#6 x #6), slabs	RI@.004	SF	.15	.09	.24
6" x 6" W4.0 x W4.0 (#4 x #4), slabs	RI@.005	SF	.21	.12	.33
Add for lengthwise cut, LF of cut	RI@.002	LF	—	.05	.05

Welded wire mesh, galvanized steel, electric weld, including 15% waste and overlap

	Craft@Hrs	Unit	Material	Labor	Total
2" x 2" W.9 x W.9 (#12 x #12), slabs	RI@.004	SF	.71	.09	.80
4" x 4" W1.4 x W1.4 (#10 x #10), slabs	RI@.003	SF	.33	.07	.40
4" x 4" W2.0 x W2.0 (#8 x #8), slabs	RI@.004	SF	.39	.09	.48
4" x 4" W2.9 x W2.9 (#6 x #6), slabs	RI@.005	SF	.44	.12	.56
4" x 4" W4.0 x W4.0 (#4 x #4), slabs	RI@.006	SF	.57	.14	.71
6" x 6" W1.4 x W1.4 (#10 x #10), slabs	RI@.003	SF	.18	.07	.25
6" x 6" W2.0 x W2.0 (#8 x #8), slabs	RI@.004	SF	.24	.09	.33
6" x 6" W2.9 x W2.9 (#6 x #6), slabs	RI@.004	SF	.28	.09	.37
6" x 6" W4.0 x W4.0 (#4 x #4), slabs	RI@.005	SF	.41	.12	.53
Add for lengthwise cut, LF of cut	RI@.002	LF	—	.05	.05

Welded wire mesh, epoxy coated steel, electric weld, including 15% waste and overlap

	Craft@Hrs	Unit	Material	Labor	Total
2" x 2" W.9 x W.9 (#12 x #12), slabs	RI@.004	SF	1.45	.09	1.54
4" x 4" W1.4 x W1.4 (#10 x #10), slabs	RI@.003	SF	1.25	.07	1.32
4" x 4" W2.0 x W2.0 (#8 x #8), slabs	RI@.004	SF	1.28	.09	1.37
4" x 4" W2.9 x W2.9 (#6 x #6), slabs	RI@.005	SF	1.30	.12	1.42
4" x 4" W4.0 x W4.0 (#4 x #4), slabs	RI@.006	SF	1.37	.14	1.51
6" x 6" W1.4 x W1.4 (#10 x #10), slabs	RI@.003	SF	1.18	.07	1.25
6" x 6" W2.0 x W2.0 (#8 x #8), slabs	RI@.004	SF	1.21	.09	1.30
6" x 6" W2.9 x W2.9 (#6 x #6), slabs	RI@.004	SF	1.23	.09	1.32
6" x 6" W4.0 x W4.0 (#4 x #4), slabs	RI@.005	SF	1.29	.12	1.41
Add for lengthwise cut, LF of cut	RI@.002	LF	—	.05	.05

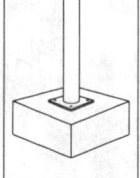

Column Footings Concrete material and placing only. Material costs are for 3,000 PSI, 5.7 sack mix, with 1-1/2" aggregate with 5% waste included ($59.70 per CY before waste allowance). Labor cost is for placing concrete in an existing form. Add for excavation, board forming and steel reinforcing

	Craft@Hrs	Unit	Material	Labor	Total
Typical cost per CY	B1@.875	CY	62.70	20.00	82.70

Concrete Walls

	Craft@Hrs	Unit	Material	Labor	Total

Concrete Walls Cast-in-place concrete walls for buildings or retaining walls. Material costs for concrete placed direct from chute are based on for 2,000 PSI, 5.0 sack mix, with 1" aggregate at $61.40 per CY including 5% waste ($58.50 per CY before waste allowance). Pump mix costs include an additive cost of $5.50 per CY. Labor costs are for placing only. Add the cost of excavation, formwork, steel reinforcing, finishes and curing. Square foot costs are based on SF of wall measured on one face only. Costs do not include engineering, design or foundations.

	Craft@Hrs	Unit	Material	Labor	Total
4" thick walls (1.23 CY per CSF)					
To 4' high, direct from chute	B1@.013	SF	.76	.30	1.06
4' to 8' high, pumped	B3@.015	SF	.82	.33	1.15
8' to 12' high, pumped	B3@.017	SF	.82	.38	1.20
12' to 16' high, pumped	B3@.018	SF	.82	.40	1.22
16' high, pumped	B3@.020	SF	.82	.44	1.26
6" thick walls (1.85 CY per CSF)					
To 4' high, direct from chute	B1@.020	SF	1.14	.46	1.60
4' to 8' high, pumped	B3@.022	SF	1.24	.49	1.73
8' to 12' high, pumped	B3@.025	SF	1.24	.55	1.79
12' to 16' high, pumped	B3@.027	SF	1.24	.60	1.84
16' high, pumped	B3@.030	SF	1.24	.66	1.90
8" thick walls (2.47 CY per CSF)					
To 4' high, direct from chute	B1@.026	SF	1.52	.59	2.11
4' to 8' high, pumped	B3@.030	SF	1.65	.66	2.31
8' to 12' high, pumped	B3@.033	SF	1.65	.73	2.38
12' to 16' high, pumped	B3@.036	SF	1.65	.79	2.44
16' high, pumped	B3@.040	SF	1.65	.88	2.53
10" thick walls (3.09 CY per CSF)					
To 4' high, direct from chute	B1@.032	SF	1.90	.73	2.63
4' to 8' high, pumped	B3@.037	SF	2.07	.82	2.89
8' to 12' high, pumped	B3@.041	SF	2.07	.91	2.98
12' to 16' high, pumped	B3@.046	SF	2.07	1.02	3.09
16' high, pumped	B3@.050	SF	2.07	1.10	3.17
12" thick walls (3.70 CY per CSF)					
To 4' high, placed direct from chute	B1@.040	SF	2.27	.91	3.18
4' to 8' high, pumped	B3@.045	SF	2.48	.99	3.47
8' to 12' high, pumped	B3@.050	SF	2.48	1.10	3.58
12' to 16' high, pumped	B3@.055	SF	2.48	1.21	3.69
16' high, pumped	B3@.060	SF	2.48	1.32	3.80

Concrete Footings, Grade Beams and Stem Walls Use the figures below for preliminary estimates. Concrete costs are based on 2,000 PSI, 5.0 sack mix with 1" aggregate placed directly from the chute of a ready-mix truck (at $58.50 per CY). Figures in parentheses show the cubic yards of concrete per linear foot of foundation (including 5% waste). Costs shown include concrete, 60 pounds of reinforcing per CY of concrete, and typical excavation using a 3/4 CY backhoe (at $25 per hour) with excess backfill spread on site.

	Craft@Hrs	Unit	Material	Labor	Equipment	Total

Footings and grade beams Cast directly against the earth, no forming or finishing required. These costs assume the top of the foundation will be at finished grade. For scheduling purposes estimate that a crew of 3 can lay out, excavate, place and tie the reinforcing steel and place 13 CY of concrete in an 8-hour day. Use $900.00 as a minimum job charge.

	Craft@Hrs	Unit	Material	Labor	Equipment	Total
Total cost per CY	B4@1.80	CY	76.50	43.50	15.40	135.40
12" W x 6" D (0.019CY per LF)	B4@.034	LF	1.45	.82	.29	2.56
12" W x 8" D (0.025CY per LF)	B4@.045	LF	1.91	1.09	.39	3.39

	Craft@Hrs	Unit	Material	Labor	Equipment	Total
12" W x 10" D (0.032 CY per LF)	B4@.058	LF	2.45	1.40	.49	4.34
12" W x 12" D (0.039 CY per LF)	B4@.070	LF	2.98	1.69	.60	5.27
12" W x 18" D (0.058 CY per LF)	B4@.104	LF	4.44	2.51	.89	7.84
16" W x 8" D (0.035 CY per LF)	B4@.063	LF	2.68	1.52	.54	4.74
16" W x 10" D (0.043 CY per LF)	B4@.077	LF	3.29	1.86	.66	5.81
16" W x 12" D (0.052 CY per LF)	B4@.094	LF	3.98	2.27	.80	7.05
18" W x 8" D (0.037 CY per LF)	B4@.067	LF	2.83	1.62	.57	5.02
18" W x 10" D (0.049 CY per LF)	B4@.088	LF	3.75	2.12	.75	6.62
18" W x 24" D (0.117 CY per LF)	B4@.216	LF	8.95	5.21	1.80	15.96
20" W x 12" D (0.065 CY per LF)	B4@.117	LF	4.97	2.82	1.00	8.79
24" W x 12" D (0.078 CY per LF)	B4@.140	LF	5.97	3.38	1.20	10.55
24" W x 24" D (0.156 CY per LF)	B4@.281	LF	11.90	6.78	2.40	21.08
24" W x 30" D (0.195 CY per LF)	B4@.351	LF	14.90	8.47	3.00	26.37
24" W x 36" D (0.234 CY per LF)	B4@.421	LF	17.90	10.20	3.60	31.70
30" W x 36" D (0.292 CY per LF)	B4@.526	LF	22.30	12.70	4.50	39.50
30" W x 42" D (0.341 CY per LF)	B4@.614	LF	26.10	14.80	5.25	46.15
36" W x 48" D (0.467 CY per LF)	B4@.841	LF	35.70	20.30	7.19	63.19

Tricks of the trade: To quickly estimate the cost of sizes not shown, multiply the width in inches by the depth in inches and divide the result by 3700. This is the CY of concrete per LF of footing including 5% waste. Multiply this quantity by the "Typical cost per CY" shown above to find the cost per LF.

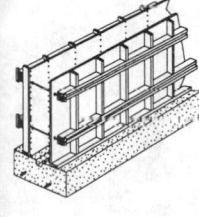

Continuous concrete footing with foundation stem wall These figures assume the foundation stem wall projects 24" above the finished grade and extends into the soil 18" to the top of the footing. Costs shown include typical excavation using a 3/4 CY backhoe (at an equipment cost of $25 per hour) with excess backfill spread on site, forming both sides of the foundation wall and the footing, based on three uses of the forms and 2 #4 rebar. Use $1,200.00 as a minimum cost for this type work.

	Craft@Hrs	Unit	Material	Labor	Equipment	Total
Typical cost per CY	B5@7.16	CY	86.00	175.00	35.80	296.80
Typical single story structure, footing 12" W x 8" D, wall 6" T x 42" D (.10 CY per LF)	B5@.716	LF	8.60	17.50	3.58	29.68
Typical two story structure, footing 18" W x 10" D, wall 8" T x 42" D (.14 CY per LF)	B5@1.00	LF	12.00	24.40	5.00	41.40
Typical three story structure, footing 24" W x 12" D, wall 10" T x 42" D (.19 CY per LF)	B5@1.36	LF	16.30	33.20	6.80	56.30

Slabs, Walks and Driveways Typical including typical excavation, slab base, forms, vapor barrier, wire mesh, 3000 PSI concrete, finishing and curing. For thickened edges, add SF of edge at thickness required.

	Craft@Hrs	Unit	Material	Labor	Equipment	Total
2" thick	B5@.022	SF	1.15	.54	.11	1.80
3" thick	B5@.023	SF	1.35	.56	.11	2.02
4" thick	B5@.024	SF	1.55	.59	.11	2.25
5" thick	B5@.025	SF	1.75	.61	.11	2.47
6" thick	B5@.026	SF	1.95	.64	.11	2.70

	Craft@Hrs	Unit	Material	Labor	Total
Slab Base Aggregate base for slabs. No waste included. Labor costs are for spreading aggregate from piles only. Add for fine grading, using hand tools.					
Crushed stone base (at $19.00 per cubic yard). 1.4 tons equal one cubic yard					
1" base (.309 CY per CSF)	BL@.001	SF	.06	.02	.08
2" base (.617 CY per CSF)	BL@.003	SF	.12	.06	.18
3" base (.926 CY per CSF)	BL@.004	SF	.18	.08	.26

Concrete, Slab Base

	Craft@Hrs	Unit	Material	Labor	Total
4" base (1.23 CY per CSF)	BL@.006	SF	.23	.12	.35
5" base (1.54 CY per CSF)	BL@.007	SF	.29	.14	.43
6" base (1.85 CY per CSF)	BL@.008	SF	.35	.16	.51

Sand fill base (at $14.80 per cubic yard). 1.35 tons equal one cubic yard

	Craft@Hrs	Unit	Material	Labor	Total
1" fill (.309 CY per CSF)	BL@.001	SF	.05	.02	.07
2" fill (.617 CY per CSF)	BL@.002	SF	.09	.04	.13
3" fill (.926 CY per CSF)	BL@.003	SF	.14	.06	.20
4" fill (1.23 CY per CSF)	BL@.004	SF	.18	.08	.26
5" fill (1.54 CY per CSF)	BL@.006	SF	.23	.12	.35
6" fill (1.85 CY per CSF)	BL@.007	SF	.27	.14	.41
Add for fine grading for slab on grade	BL@.008	SF	—	.16	.16

Slab Membrane Material costs are for 10' wide membrane with 6" laps. Labor costs are to lay membrane over a level slab base.

	Craft@Hrs	Unit	Material	Labor	Total
4 mil polyethylene (at $3.00 per CSF)	BL@.002	SF	.03	.04	.07
6 mil polyethylene (at $4.00 per CSF)	BL@.002	SF	.04	.04	.08
Reinforced kraft and polyethylene laminate (at $10.00 per CSF)	BL@.002	SF	.09	.04	.13
Add for adhesive cemented seams	BL@.001	SF	.01	.02	.03

Curbs and Gutters Small quantities. Material costs are for 2,000 PSI, 5.0 sack mix with 1" aggregate (at $58.50 per CY). No waste included. Labor cost is for placing and finishing concrete only. Add for excavation, wood forming, steel reinforcing.

	Craft@Hrs	Unit	Material	Labor	Total
Per CY of concrete	B6@1.61	CY	58.50	36.30	94.80

Steps on Grade Costs include layout, fabrication, and placing forms, setting and tying steel reinforcing, installation of steel nosing, placing 2,000 PSI concrete (at $58.50 per CY before waste) directly from the chute of a ready-mix truck, finishing, stripping forms, and curing. Costs assume excavation, back filling, and compacting have been completed. For scheduling purposes, estimate that a crew of 3 can set forms, place steel, and pour 4 to 5 CY of concrete cast-in-place steps in an 8 hour day. See also Cast-in-place Concrete Steps on Grade in the Industrial Section 3. Costs shown are for concrete steps with a 6" rise height and 12" tread depth supported by a 6" thick monolithic slab.

	Craft@Hrs	Unit	Material	Labor	Total
Total cost of steps per CY of concrete	B1@5.43	CY	177.00	124.00	301.00

Concrete Finishing

Slab finishes

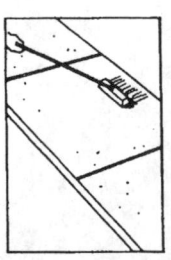

	Craft@Hrs	Unit	Material	Labor	Total
Broom finish	B6@.012	SF	—	.27	.27
Float finish	B6@.010	SF	—	.23	.23
Trowel finishing					
Steel, machine work	B6@.015	SF	—	.34	.34
Steel, hand work	B6@.018	SF	—	.41	.41
Finish treads and risers					
No abrasives, no plastering, per LF of tread	B6@.042	LF	—	.95	.95
With abrasives, plastered, per LF of tread	B6@.063	LF	.55	1.42	1.97
Scoring concrete surface, hand work	B6@.005	LF	—	.11	.11
Sweep, scrub and wash down	B6@.006	SF	.01	.14	.15
Liquid curing and sealing compound, spray-on,					
Mastercure, 400 SF and $20 per gallon,	B6@.003	SF	.05	.07	.12
Exposed aggregate (washed, including finishing),					
no disposal of slurry	B6@.017	SF	.25	.38	.63

Non-metallic color and hardener, troweled on, 2 applications, red, gray or black

	Craft@Hrs	Unit	Material	Labor	Total
60 pounds per 100 SF	B6@.021	SF	.35	.47	.82
100 pounds per 100 SF	B6@.025	SF	.58	.56	1.14
Add for other standard colors	—	%	50.0	—	

	Craft@Hrs	Unit	Material	Labor	Total
Wall finishes					
Cut back ties and patch	B6@.011	SF	.11	.25	.36
Remove fins	B6@.008	LF	.05	.18	.23
Grind smooth	B6@.021	SF	.08	.47	.55
Sack, burlap grout rub	B6@.013	SF	.05	.29	.34
Wire brush, green	B6@.015	SF	.05	.34	.39
Wash with acid and rinse	B6@.004	SF	.22	.09	.31
Break fins, patch voids, Carborundum rub	B6@.035	SF	.06	.79	.85
Break fins, patch voids, burlap grout rub	B6@.026	SF	.07	.59	.66
Specialty finishes					
Monolithic natural aggregate topping					
3/16"	B6@.020	SF	.20	.45	.65
1/2"	B6@.022	SF	.22	.50	.72
Integral colors,					
Typically 8 pounds per sack of cement	—	Lb	1.80	—	1.80
Stamped finish (embossed concrete)					
Diamond, square, octagonal patterns	B6@.047	SF	—	1.06	1.06
Spanish paver pattern	B6@.053	SF	—	1.20	1.20
Add for grouting of joints, if required	B6@.023	SF	.25	.52	.77

Countertops One-piece custom tops, 25"-36" deep, with 4" backsplash, shop assembled. Laminated plastic (such as Formica, Textolite or Wilsonare) on composition base. Costs are per LF of back edge and for solid colors.

	Craft@Hrs	Unit	Material	Labor	Total
Bar tops					
Square edge	B1@.181	LF	29.70	4.14	33.84
Bevel edge	B1@.181	LF	41.00	4.14	45.14
Rolled drip edge (post formed)	B1@.181	LF	20.50	4.14	24.64
Sink tops					
Square edge	B1@.181	LF	29.70	4.14	33.84
Square edge with woodgrain at edge	B1@.181	LF	25.60	4.14	29.74
Bevel edge with woodgrain at edge	B1@.181	LF	61.50	4.14	65.64
Rolled drip edge (post formed)	B1@.181	LF	20.50	4.14	24.64
Woodgrain vanity tops					
Square edge	B1@.181	LF	22.60	4.14	26.74
Square edge with woodgrain at edge	B1@.181	LF	25.60	4.14	29.74
Bevel edge with woodgrain at edge	B1@.181	LF	61.50	4.14	65.64
Rolled drip edge (post formed)	B1@.181	LF	12.80	4.14	16.94
Cultured stone vanity tops with integral back splash and sink, 22" deep					
Cultured marble					
24" long	B1@2.50	Ea	97.60	57.10	154.70
36" long	B1@2.50	Ea	126.80	57.10	183.90
48" long	B1@2.50	Ea	155.00	57.10	212.10
Cultured onyx					
24" long	B1@2.50	Ea	133.00	57.10	190.10
36" long	B1@2.50	Ea	171.00	57.10	228.10
48" long	B1@2.50	Ea	210.00	57.10	267.10
Corian vanity tops and integral sink, 3/4" thick, 22" deep					
25" long	B1@.654	Ea	295.00	14.90	309.90
31" long	B1@.742	Ea	395.00	17.00	412.00
37" long	B1@.830	Ea	460.00	19.00	479.00
43" long	B1@.924	Ea	445.00	21.10	466.10
49" long	B1@1.02	Ea	590.00	23.30	613.30
Add for square end splash, attached	—	Ea	15.00	—	15.00

Countertops

	Craft@Hrs	Unit	Material	Labor	Total
Add for square end splash, loose	—	Ea	14.00	—	14.00
Add for contour end splash, loose	—	Ea	15.00	—	15.00
Add for mitered corners	—	Ea	18.00	—	18.00
Add for seamless tops	—	Ea	26.00	—	26.00
Add for sink, range or vanity cutout	—	Ea	7.00	—	7.00
Add for drilling 3 plumbing fixture holes	—	LS	9.00	—	9.00
Add for quarter round corner	—	Ea	18.20	—	18.20
Add for half round corner	—	Ea	26.00	—	26.00

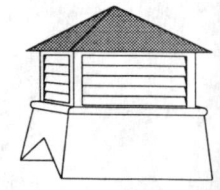

Cupolas Finished redwood. Add the cost of weather vanes.

Aluminum roof covering

	Craft@Hrs	Unit	Material	Labor	Total
22" x 22" x 9" base, 25" high	B1@.948	Ea	245.00	21.70	266.70
29" x 29" x 11-1/4" base, 30" high	B1@.948	Ea	306.00	21.70	327.70
35" x 35" x 11-1/4" base, 33" high	B1@1.43	Ea	418.00	32.70	450.70
47" x 47" x 11-1/4" base, 39" high	B1@2.02	Ea	724.00	46.20	770.20

Copper roof covering

	Craft@Hrs	Unit	Material	Labor	Total
22" x 22" x 9" base, 25" high	B1@.948	Ea	286.00	21.70	307.70
29" x 29" x 11-1/4" base, 30" high	B1@.948	Ea	347.00	21.70	368.70
35" x 35" x 11-1/4" base, 33" high	B1@1.43	Ea	490.00	32.70	522.70
47" x 47" x 11-1/4" base, 39" high	B1@2.02	Ea	765.00	46.20	811.20

Weather vanes for cupolas

Standard aluminum, baked black finish

	Craft@Hrs	Unit	Material	Labor	Total
26" height	B1@.275	Ea	56.00	6.28	62.28

Deluxe model, gold-bronze aluminum, black finish

	Craft@Hrs	Unit	Material	Labor	Total
36" height, bronze ornament	B1@.368	Ea	168.00	8.41	176.41

Decks These costs assume 2" x 8" decking supported by #2 & better 2" x 8" pressure treated joists spaced 24" on center over 4" x 8" pressure treated beams. Beams are supported by 4" x 4" pressure treated posts set 6' on center in concrete. Post length is assumed to be 2 feet. Fasteners are 16d galvanized nails, and galvanized steel joist hangers and post anchors. Decking is spaced at 1/8" to permit drainage. Costs are based on 150 SF rectangular deck, with one side fastened to an existing structure. Larger decks will reduce labor costs.

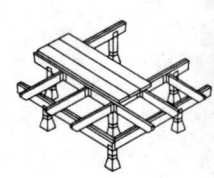

Deck, free standing, 3' above ground, with joists, beams, posts and hardware as described above, with decking material as shown.

10' x 45'.

	Craft@Hrs	Unit	Material	Labor	Total
Pine, pressure treated decking	B1@.170	SF	3.61	3.88	7.49
Redwood, select heart decking	B1@.170	SF	6.17	3.88	10.05

15' x 45'

	Craft@Hrs	Unit	Material	Labor	Total
Pine, pressure treated decking	B1@.170	SF	3.51	3.88	7.39
Redwood, select heart decking	B1@.170	SF	6.10	3.88	9.98

15' x 65'.

	Craft@Hrs	Unit	Material	Labor	Total
Pine, pressure treated decking	B1@.170	SF	3.25	3.88	7.13
Redwood, select heart decking	B1@.170	SF	5.90	3.88	9.78

20' x 35'

	Craft@Hrs	Unit	Material	Labor	Total
Pine, pressure treated decking	B1@.170	SF	3.82	3.88	7.70
Redwood, select heart decking	B1@.170	SF	7.06	3.88	10.94

20' x 55'

	Craft@Hrs	Unit	Material	Labor	Total
Pine, pressure treated decking	B1@.170	SF	3.19	3.88	7.07
Redwood, select heart decking	B1@.170	SF	5.69	3.88	9.57

Add for diagonal pattern decking

	Craft@Hrs	Unit	Material	Labor	Total
Pine decking	B1@.007	SF	.49	.16	.65
Redwood decking	B1@.007	SF	.83	.16	.99
Add for stain and sealer finish	B1@.006	SF	.10	.14	.24

	Craft@Hrs	Unit	Material	Labor	Total

Deck covers Conventionally framed wood cover joists. Per SF of area covered. Figures in parentheses indicate board feet per square foot of ceiling including end joists, header joists, and 5% waste. Costs shown are based on a job with 200 SF of area covered. Costs shown include normal waste. Costs for posts and finish roof are not included, add for same from costs that follow.

2" x 4" (at $525 per MBF), Std & Btr grade

	Craft@Hrs	Unit	Material	Labor	Total
16" centers (.59 BF per SF)	B1@.020	SF	.31	.46	.77
24" centers (.42 BF per SF)	B1@.014	SF	.22	.32	.54

2" x 6" (at $515 per MBF), Std & Btr grade

	Craft@Hrs	Unit	Material	Labor	Total
16" centers (.88 BF per SF)	B1@.020	SF	.45	.46	.91
24" centers (.63 BF per SF)	B1@.015	SF	.32	.34	.66

Add for posts to support conventionally framed wood cover joists. Posts are usually required at each corner.

	Craft@Hrs	Unit	Material	Labor	Total
3" x 3" Yellow pine, penta treated, 8' long	B1@.333	Ea	7.06	7.61	14.67

Deck finish roof Cover over conventionally framed wood ceiling joists.
Flat roof with built-up roofing. Based on using 1/2" CDX roof sheathing (at $12.70 per 4' x 8' sheet) with 3 ply asphalt consisting of 2 plies of 15 lb felt with 90 lb cap sheet and 3 hot mop coats of asphalt. Add the cost for framing from above.

	Craft@Hrs	Unit	Material	Labor	Total
Total flat roof assembly cost per SF	B1@.029	SF	.74	.66	1.40

Deck railing 42" high handrail. Consisting of: 2" x 2" boards 5" OC run vertically, lag bolted to edge of the deck, one 2" x 4" horizontal top rail and one 2" x 6" placed on edge directly below the top rail.

	Craft@Hrs	Unit	Material	Labor	Total
Pine, pressure treated, per running foot	B1@.333	LF	3.82	7.61	11.43
Redwood, select heart, per running foot	B1@.333	LF	6.49	7.61	14.10

Deck stairs All stairs assume 12" tread (step) composed of two 2" x 6" lumber and riser.
One or two steps high, box construction
Prices are per LF of step. A 4' wide stair with 2 steps has 8 LF of step. Use 8 LF as the quantity.

	Craft@Hrs	Unit	Material	Labor	Total
Pine treads, open risers	B1@.270	LF	3.15	6.17	9.32
Pine treads, with pine risers	B1@.350	LF	4.54	8.00	12.54
Redwood treads, open risers	B1@.270	LF	4.56	6.17	10.73
Redwood treads, with redwood risers	B1@.350	LF	8.76	8.00	16.76

Three or more steps high, 36" wide. 3 stringers cut from 2" x 12" pressure treated lumber.
Cost per riser

	Craft@Hrs	Unit	Material	Labor	Total
Pine treads, open risers	B1@.530	Ea	11.92	12.10	24.02
Pine treads, with pine risers	B1@.620	Ea	16.08	14.20	30.28
Redwood treads, open risers	B1@.530	Ea	16.14	12.10	28.24
Redwood treads, with redwood risers	B1@.620	Ea	28.74	14.20	42.94

Concrete footing to support stair. Cost includes brackets and fasteners to attach stair to footing.

	Craft@Hrs	Unit	Material	Labor	Total
18" x 18" x 12" footing	B1@.530	Ea	10.00	12.10	22.10

	Craft@Hrs	Unit	Material	Labor	Equipment	Total

Demolition Itemized costs for demolition of building components when building is being remodeled, repaired or rehabilitated, not completely demolished. Costs include protecting adjacent areas and normal clean-up. Costs are to break out the items listed and pile debris on site only. No hauling, dump fees or salvage value included. Equipment cost, where shown, is $7.00 per hour and includes one air compressor and a pneumatic breaker with jackhammer bits. (Figures in parentheses give the approximate "loose" volume and weight of the materials after being demolished.) Use $250.00 as a minimum charge.
Brick walls and walks, cost per SF removed, measured on one face, using pneumatic breaker
4" thick walls

	Craft@Hrs	Unit	Material	Labor	Equipment	Total
(60 SF per CY and 36 lbs. per SF)	BL@.061	SF	—	1.25	.43	1.68

Demolition

	Craft@Hrs	Unit	Material	Labor	Equipment	Total
8" thick walls						
(30 SF per CY and 72 lbs. per SF)	BL@.110	SF	—	2.26	.77	3.03
12" thick walls						
(20 SF per CY and 108 lbs. per SF)	BL@.133	SF	—	2.73	.93	3.66
Brick sidewalks, cost per SF removed						
2-1/2" thick bricks on sand base, no mortar bed, removed using hand tools						
(100 SF per CY and 28 lbs. per SF)	BL@.004	SF	—	.08	—	.08
Brick pavers up to 4-1/2" thick with mortar bed, removed using a pneumatic breaker						
(55 SF per CY and 50 lbs. per SF)	BL@.050	SF	—	1.03	.35	1.38
Ceiling tile, cost per in-place SF removed using hand tools						
Stapled or glued 12" x 12"						
(250 SF per CY and .25 lbs. per SF)	BL@.015	SF	—	.31	—	.31
Suspended panels 2' x 4'						
(250 SF per CY and .25 lbs. per SF)	BL@.010	SF	—	.21	—	.21
Concrete masonry walls, cost per in-place SF removed, using a pneumatic breaker						
4" thick walls						
(60 SF per CY and 19 lbs. per SF)	BL@.066	SF	—	1.36	.46	1.82
6" thick walls						
(40 SF per CY and 28 lbs. per SF)	BL@.075	SF	—	1.54	.53	2.07
8" thick walls						
(30 SF per CY and 34 lbs. per SF)	BL@.098	SF	—	2.01	.69	2.70
12" thick walls						
(20 SF per CY and 46 lbs. per SF)	BL@.140	SF	—	2.87	.98	3.85
Reinforced or grouted walls add	—	%	—	50.0	50.0	—
Concrete foundations (footings), steel reinforced, removed using a pneumatic breaker. Concrete at 3,900 pounds per in place cubic yard.						
Cost per CY (1.33 CY per CY)	BL@3.96	CY	—	81.30	27.70	109.00
Cost per LF with width and depth as shown						
6" W x 12" D (35 LF per CY)	BL@.075	LF	—	1.54	.53	2.07
8" W x 12" D (30 LF per CY)	BL@.098	LF	—	2.01	.69	2.70
8" W x 16" D (20 LF per CY)	BL@.133	LF	—	2.73	.93	3.66
8" W x 18" D (18 LF per CY)	BL@.147	LF	—	3.02	1.03	4.05
10" W x 12" D (21 LF per CY)	BL@.121	LF	—	2.48	.85	3.33
10" W x 16" D (16 LF per CY)	BL@.165	LF	—	3.39	1.16	4.55
10" W x 18" D (14 LF per CY)	BL@.185	LF	—	3.80	1.30	5.10
12" W x 12" D (20 LF per CY)	BL@.147	LF	—	3.02	1.03	4.05
12" W x 16" D (13 LF per CY)	BL@.196	LF	—	4.02	1.37	5.39
12" W x 20" D (11 LF per CY)	BL@.245	LF	—	5.03	1.72	6.75
12" W x 24" D (9 LF per CY)	BL@.294	LF	—	6.04	2.06	8.10
Concrete sidewalks, to 4" thick, cost per SF removed. Concrete at 3,900 pounds per in place cubic yard.						
Non-reinforced, removed by hand						
(60 SF per CY)	BL@.050	SF	—	1.03	—	1.03
Reinforced, removed using pneumatic breaker						
(55 SF per CY)	BL@.033	SF	—	.68	.23	.91
Concrete slabs, non-reinforced, removed using pneumatic breaker. Concrete at 3,900 pounds per in place cubic yard.						
Cost per CY (1.25 CY per CY)	BL@3.17	CY	—	65.10	22.20	87.30
Cost per SF with thickness as shown						
3" slab thickness (90 SF per CY)	BL@.030	SF	—	.62	.21	.83
4" slab thickness (60 SF per CY)	BL@.040	SF	—	.82	.28	1.10
6" slab thickness (45 SF per CY)	BL@.056	SF	—	1.15	.39	1.54
8" slab thickness (30 SF per CY)	BL@.092	SF	—	1.89	.64	2.53

	Craft@Hrs	Unit	Material	Labor	Equipment	Total
Concrete slabs, steel reinforced, removed using pneumatic breaker. Concrete at 3,900 pounds per in place cubic yard.						
Cost per CY (1.33 CY per CY)	BL@3.96	CY	—	81.30	27.70	109.00
Cost per SF with thickness as shown						
3" slab thickness (80 SF per CY)	BL@.036	SF	—	.74	.25	.99
4" slab thickness (55 SF per CY)	BL@.050	SF	—	1.03	.35	1.38
6" slab thickness (40 SF per CY)	BL@.075	SF	—	1.54	.53	2.07
8" slab thickness (30 SF per CY)	BL@.098	SF	—	2.01	.69	2.70
Concrete walls, steel reinforced, removed using pneumatic breaker. Concrete at 3,900 pounds per in place cubic yard.						
Cost per CY (1.33 CY per CY)	BL@4.76	CY	—	97.70	33.30	131.00
Cost per SF with thickness as shown (deduct openings over 25 SF)						
3" wall thickness (80 SF per CY)	BL@.045	SF	—	.92	.32	1.24
4" wall thickness (55 SF per CY)	BL@.058	SF	—	1.19	.41	1.60
5" wall thickness (45 SF per CY)	BL@.075	SF	—	1.54	.53	2.07
6" wall thickness (40 SF per CY)	BL@.090	SF	—	1.85	.63	2.48
8" wall thickness (30 SF per CY)	BL@.120	SF	—	2.46	.84	3.30
10" wall thickness (25 SF per CY)	BL@.147	SF	—	3.02	1.03	4.05
12" wall thickness (20 SF per CY)	BL@.176	SF	—	3.61	1.23	4.84
Curbs and 18" gutter, concrete, removed using pneumatic breaker. Concrete at 3,900 pounds per in place cubic yard.						
Cost per LF (8 LF per CY)	BL@.422	LF	—	8.67	2.95	11.62
Doors, typical cost to remove door, frame and hardware						
3' x 7' metal (2 doors per CY)	BL@1.00	Ea	—	20.50	—	20.50
3' x 7' wood (2 doors per CY)	BL@.497	Ea	—	10.20	—	10.20
Flooring, cost per in-place SY of floor area removed (1 square yard = 9 square feet)						
Ceramic tile						
(25 SY per CY and 34 lbs. per SY)	BL@.263	SY	—	5.40	—	5.40
Hardwood, nailed						
(25 SY per CY and 18 lbs. per SY)	BL@.290	SY	—	5.96	—	5.96
Hardwood, glued						
(25 SY per CY and 18 lbs. per SY)	BL@.503	SY	—	10.30	—	10.30
Linoleum, sheet						
(30 SY per CY and 3 lbs. per SY)	BL@.056	SY	—	1.15	—	1.15
Carpet						
(40 SY per CY and 5 lbs. per SY)	BL@.028	SY	—	.57	—	.57
Carpet pad						
(35 SY per CY and 1.7 lbs. per SY)	BL@.014	SY	—	.29	—	.29
Resilient tile						
(30 SY per CY and 3 lbs. per SY)	BL@.070	SY	—	1.44	—	1.44
Terrazzo						
(25 SY per CY and 34 lbs. per SY)	BL@.286	SY	—	5.87	—	5.87
Remove mastic, using power sander	BL@.125	SY	—	2.57	—	2.57
Insulation, fiberglass batts, cost per in-place SF removed						
R-11 to R-19						
(500 SF per CY and .3 lb per SF)	BL@.003	SF	—	.06	—	.06
Pavement, asphaltic concrete (bituminous) to 6" thick, removed using pneumatic breaker						
Cost per SY						
(4 SY per CY and 660 lbs. per SY)	BL@.250	SY	—	5.13	1.75	6.88
Plaster walls or ceilings, interior, cost per in-place SF removed by hand						
(150 SF per CY and 8 lbs. per SF)	BL@.015	SF	—	.31	—	.31

Demolition

	Craft@Hrs	Unit	Material	Labor	Equipment	Total

Roofing, cost per in-place Sq, removed using hand tools (1 square = 100 square feet)

Built-up, 5 ply

(2.5 Sq per CY and 250 lbs per Sq)	BL@1.50	Sq	—	30.80	—	30.80

Shingles, asphalt, single layer

(2.5 Sq per CY and 240 lbs per Sq)	BL@1.33	Sq	—	27.30	—	27.30

Shingles, asphalt, double layer

(1.2 Sq per CY and 480 lbs per Sq)	BL@1.33	Sq	—	27.30	—	27.30

Shingles, slate, Weight ranges from 600 lbs. to 1200 lbs. per Sq

(1 Sq per CY and 900 lbs per Sq).	BL@1.79	Sq	—	36.80	—	36.80

Shingles, wood

(1.7 Sq per CY and 400 lbs per Sq)	BL@2.02	Sq	—	41.50	—	41.50

Sheathing, up to 1" thick, cost per in-place SF, removed using hand tools

Gypsum

(250 SF per CY and 2.3 lbs. per SF)	BL@.017	SF	—	.35	—	.35

Plywood

(200 SF per CY and 2 lbs. per SF)	BL@.017	SF	—	.35	—	.35

Siding, up to 1" thick, cost per in-place SF, removed using hand tools

Metal

(200 SF per CY and 2 lbs. per SF)	BL@.027	SF	—	.55	—	.55

Plywood

(200 SF per CY and 2 lbs. per SF)	BL@.017	SF	—	.35	—	.35

Wood, boards

(250 SF per CY and 2 lbs. per SF)	BL@.030	SF	—	.62	—	.62

Stucco walls or soffits, cost per in-place SF, removed using hand tools

(150 SF per CY and 8 lbs. per SF)	BL@.036	SF	—	.74	—	.74

Trees. See Excavation

Wallboard up to 1" thick, cost per in-place SF, removed using hand tools

Gypsum, walls or ceilings

(250 SF per CY and 2.3 lbs. per SF)	BL@.018	SF	—	.37	—	.37

Plywood or insulation board

(200 SF per CY and 2 lbs. per SF)	BL@.018	SF	—	.37	—	.37

Windows, typical cost to remove window, frame and hardware, cost per SF of window

Metal (36 SF per CY)	BL@.058	SF	—	1.19	—	1.19
Wood (36 SF per CY)	BL@.063	SF	—	1.29	—	1.29

Wood stairs, cost per in-place SF or LF

Risers (25 SF per CY)	BL@.116	SF	—	2.38	—	2.38
Landings (50 SF per CY)	BL@.021	SF	—	.43	—	.43
Handrails (100 LF per CY)	BL@.044	LF	—	.90	—	.90
Posts (200 LF per CY)	BL@.075	LF	—	1.54	—	1.54

	Craft@Hrs	Unit	Material	Labor	Total

Wood Framing Demolition Typical costs for demolition of wood frame structural components using hand tools. Normal center to center spacing (12" thru 24" OC) is assumed. Costs include removal of Romex cable as necessary, protecting adjacent areas, and normal clean-up. Costs do not include temporary shoring to support structural elements above or adjacent to the work area. (Add for demolition of flooring, roofing, sheathing, etc.) Figures in parentheses give the approximate "loose" volume and weight of the materials after being demolished.

Ceiling joists, per in-place SF of ceiling area removed

2" x 4" (720 SF per CY and 1.18 lbs. per SF)	BL@.009	SF	—	.18	.18
2" x 6" (410 SF per CY and 1.76 lbs. per SF)	BL@.013	SF	—	.27	.27
2" x 8" (290 SF per CY and 2.34 lbs. per SF)	BL@.018	SF	—	.37	.37
2" x 10" (220 SF per CY and 2.94 lbs. per SF)	BL@.023	SF	—	.47	.47
2" x 12" (190 SF per CY and 3.52 lbs. per SF)	BL@.027	SF	—	.55	.55

	Craft@Hrs	Unit	Material	Labor	Total
Floor joists, per in-place SF of floor area removed					
2" x 6" (290 SF per CY and 2.04 lbs. per SF)	BL@.016	SF	—	.33	.33
2" x 8" (190 SF per CY and 2.72 lbs. per SF)	BL@.021	SF	—	.43	.43
2" x 10" (160 SF per CY and 3.42 lbs. per SF)	BL@.026	SF	—	.53	.53
2" x 12" (120 SF per CY and 4.10 lbs. per SF)	BL@.031	SF	—	.64	.64
Rafters, per in-place SF of actual roof area removed					
2" x 4" (610 SF per CY and 1.42 lbs. per SF)	BL@.011	SF	—	.23	.23
2" x 6" (360 SF per CY and 2.04 lbs. per SF)	BL@.016	SF	—	.33	.33
2" x 8" (270 SF per CY and 2.68 lbs. per SF)	BL@.020	SF	—	.41	.41
2" x 10" (210 SF per CY and 3.36 lbs. per SF)	BL@.026	SF	—	.53	.53
2" x 12" (180 SF per CY and 3.94 lbs. per SF)	BL@.030	SF	—	.62	.62
Stud walls, interior or exterior, includes allowance for plates and blocking, per in-place SF of wall area removed, measured on one face					
2" x 3" (430 SF per CY and 1.92 lbs. per SF)	BL@.013	SF	—	.27	.27
2" x 4" (310 SF per CY and 2.58 lbs. per SF)	BL@.017	SF	—	.35	.35
2" x 6" (190 SF per CY and 3.74 lbs. per SF)	BL@.025	SF	—	.51	.51

Wood Framed Assemblies Demolition Typical costs for demolition of wood framed structural assemblies using hand tools when building is being remodeled, repaired or rehabilitated but not completely demolished. Costs include removal of Romex cable as necessary, protecting adjacent areas, and normal clean-up. Costs do not include temporary shoring to support structural elements above or adjacent to the work area, loading, hauling, dump fees or salvage value. Costs shown are per in-place square foot (SF) of area being demolished. Normal center to center spacing (12" thru 24" OC) is assumed. Figures in parentheses give the approximate "loose" volume and weight of the materials (volume after being demolished and weight.)

	Craft@Hrs	Unit	Material	Labor	Total
Ceiling assemblies. Remove ceiling joists, fiberglass insulation and gypsum board up to 5/8" thick					
2" x 4" (135 SF per CY and 3.78 lbs. per SF)	BL@.030	SF	—	.62	.62
2" x 6" (118 SF per CY and 4.36 lbs. per SF)	BL@.034	SF	—	.70	.70
2" x 8" (106 SF per CY and 4.94 lbs. per SF)	BL@.039	SF	—	.80	.80
2" x 10" (95 SF per CY and 5.54 lbs. per SF)	BL@.043	SF	—	.88	.88
2" x 12" (89 SF per CY and 6.12 lbs. per SF)	BL@.048	SF	—	.99	.99
Floor assemblies. Remove floor joists and plywood subfloor up to 3/4" thick					
2" x 8" (97 SF per CY and 4.72 lbs. per SF)	BL@.035	SF	—	.72	.72
2" x 10" (88 SF per CY and 5.42 lbs. per SF)	BL@.040	SF	—	.82	.82
2" x 12" (75 SF per CY and 6.10 lbs. per SF)	BL@.043	SF	—	.88	.88
For gypsum board ceiling removed below floor joists, add to any of above					
(250 SF per CY and 2.3 lbs. per SF)	BL@.018	SF	—	.37	.37
Wall assemblies, interior. Remove studs, 2 top plates, 1 bottom plate and gypsum drywall up to 5/8" thick on both sides					
2" x 3" (97 SF per CY and 6.52 lbs. per SF)	BL@.048	SF	—	.99	.99
2" x 4" (89 SF per CY and 7.18 lbs. per SF)	BL@.052	SF	—	1.07	1.07
2" x 6" (75 SF per CY and 8.34 lbs. per SF)	BL@.059	SF	—	1.21	1.21
Wall assemblies, exterior. Remove studs, 2 top plates, 1 bottom plate, insulation, gypsum drywall up to 5/8" thick on one side, plywood siding up to 3/4" thick on other side					
2" x 4" (70 SF per CY and 7.18 lbs. per SF)	BL@.067	SF	—	1.38	1.38
2" x 6" (75 SF per CY and 8.34 lbs. per SF)	BL@.074	SF	—	1.52	1.52
For stucco in lieu of plywood, add to any of above					
(SF per CY is the same and add 6 lbs. per SF)	BL@.006	SF	—	.12	.12

Demolition, Subcontract Costs

	Craft@Hrs	Unit	Material	Labor	Total

Demolition, Subcontract Costs for demolishing and removing an entire building using a dozer and front-end loader. Costs shown include loading and hauling up to 5 miles, but no dump fees. Dump charges vary widely; see Disposal Fees in the Site Work Section of the Industrial and Commercial division of this book. Costs shown are for above ground structures only, per in-place SF of total floor area. Up to 10' high ceilings. No salvage value assumed. Add the cost of foundation demolition. Figures in parentheses give the approximate "loose" volume of the materials (volume after being demolished).

Light wood-frame structures, up to three stories in height. Based on 2,500 SF job. No basements included. Estimate each story separately. Cost per in-place SF floor area

	Craft@Hrs	Unit	Material	Labor	Total
First story (8 SF per CY)	—	SF	—	—	1.60
Second story (8 SF per CY)	—	SF	—	—	2.05
Third story (8 SF per CY)	—	SF	—	—	2.70
Masonry building, cost per in-place SF floor area					
based on 5,000 SF job (50 SF per CY)	—	SF	—	—	2.60
Concrete building, cost per in-place SF floor area					
based on 5,000 SF job (30 SF per CY)	—	SF	—	—	3.40
Reinforced concrete building, cost per in-place SF floor area					
based on 5,000 SF job (20 SF per CY)	—	SF	—	—	3.85

Door Chimes Electric, typical installation, add transformer and wire below.

Surface mounted, 2 notes

	Craft@Hrs	Unit	Material	Labor	Total
Simple white plastic chime, 4" x 7"	BE@3.50	Ea	25.00	94.10	119.10
Modern oak grained finish, plastic, 8" x 5"	BE@3.50	Ea	52.00	94.10	146.10
Walnut finish, plastic, cloth grille, 8" x 5"	BE@3.50	Ea	30.00	94.10	124.10
Maple finish, plastic, 8" x 5"	BE@3.50	Ea	36.00	94.10	130.10
Designer, wood-look plastic strips across a cloth grille, 9" x 7"	BE@3.50	Ea	48.00	94.10	142.10
Modern real oak wood, with brass plated chimes, 9" x 7"	BE@3.50	Ea	59.00	94.10	153.10

Surface mounted, 4 to 8 notes, solid wood door chime

	Craft@Hrs	Unit	Material	Labor	Total
Three 12" long tubes	BE@3.64	Ea	85.00	97.80	182.80
Four 55" long tubes	BE@3.64	Ea	85.00	97.80	182.80

Any of above

	Craft@Hrs	Unit	Material	Labor	Total
Add for chime transformer	—	Ea	7.00	—	7.00
Add for bell wire, 65' per coil	—	Ea	6.00	—	6.00
Add for lighted push buttons	—	Ea	6.50	—	6.50

Non-electric, push button type, door mounted

	Craft@Hrs	Unit	Material	Labor	Total
Modern design, 2-1/2" x 2-1/2"	BE@.497	Ea	14.00	13.40	27.40
With peep sight, 7" x 3-1/2"	BE@.497	Ea	16.00	13.40	29.40

Door Closers

Standard duty door closers, hydraulic, bronze or aluminum finish

	Craft@Hrs	Unit	Material	Labor	Total
To 4'0" interior door or 3'0" exterior door	BC@1.04	Ea	66.50	26.20	92.70
Add for adapter plate	—	Ea	11.40	—	11.40
Light duty model, pneumatic for interior doors					
Interior doors to 85 lbs	BC@1.04	Ea	36.00	26.20	62.20
Hinge mounted closer	BC@.390	Ea	10.30	9.82	20.12
Screen door closer, brass finish					
Hydraulic closer, medium duty	BC@.515	Ea	8.71	13.00	21.71
Pneumatic closer, heavy duty	BC@.546	Ea	12.80	13.70	26.50
Sliding door closer, pneumatic, aluminum finish					
Screen	BC@.250	Ea	12.20	6.29	18.49
Glass	BC@.250	Ea	27.20	6.29	33.49

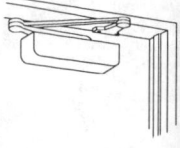

	Craft@Hrs	Unit	Material	Labor	Total

Door jambs and frames, exterior Includes brick or stucco mold casing. Exterior frames with 1-3/4" x 4", 4-3/8", or 4-3/4" head and side jambs. For doors up to 3'0" x 6'8"

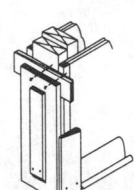

	Craft@Hrs	Unit	Material	Labor	Total
Select finger-jointed pine					
No sill	BC@1.08	Ea	56.50	27.20	83.70
Oak sill	BC@1.65	Ea	91.20	41.50	132.70
Aluminum sill	BC@1.46	Ea	76.60	36.70	113.30
Solid stock fir					
No sill	BC@1.08	Ea	75.10	27.20	102.30
Oak sill	BC@1.65	Ea	113.00	41.50	154.50
Aluminum sill	BC@1.46	Ea	96.10	36.70	132.80
Add for widths over 3'0"					
No sill	—	LF	4.21	—	4.21
Oak sill	—	LF	13.90	—	13.90
Aluminum sill	—	LF	9.50	—	9.50
Add for 7'0" high frame	—	Ea	7.82	—	7.82
Add for 8'0" high frame	—	Ea	26.20	—	26.20
For double frames, add the cost of two frames, plus	—	LS	11.30	—	11.30
For triple frames, add the cost of three frames, plus	—	LS	22.70	—	22.70
Add for solid mullions					
3" x 6", per mullion	—	Ea	65.10	—	65.10
4" x 6", per mullion	—	Ea	92.90	—	92.90

Door jambs only, exterior Jamb only, no sill, brick or stucco mould casing.

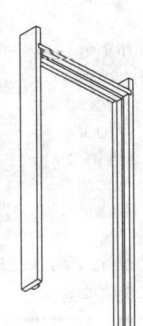

	Craft@Hrs	Unit	Material	Labor	Total
Select finger-joint, kiln dried pine, 6'9-1/2" high					
4-1/8" set or 4-1/2" set	BC@.800	Ea	39.00	20.10	59.10
4-3/4" set	BC@.800	Ea	44.30	20.10	64.40
5-1/4" set	BC@.850	Ea	50.80	21.40	72.20
Solid stock kiln dried fir, 6'9-1/2" high					
4-1/8" set or 4-1/2" set	BC@.800	Ea	54.10	20.10	74.20
4-3/4" set	BC@.800	Ea	54.10	20.10	74.20
5-1/4" set	BC@.850	Ea	57.30	21.40	78.70

Jambs, interior Flat jamb with square cut heads and rabbeted sides, pine, 6'9-1/2" high.

	Craft@Hrs	Unit	Material	Labor	Total
3-9/16" set	BC@.800	Ea	18.10	20.10	38.20
4-9/16" set	BC@.800	Ea	20.00	20.10	40.10
4-11/16" set	BC@.800	Ea	21.80	20.10	41.90
5-1/4" set	BC@.850	Ea	21.50	21.40	42.90
Add for solid jambs	—	Ea	23.80	—	22.70
Add for special jambs					
Add for extra wide jambs	—	Ea	14.00	—	14.00
Add for 7' high jambs	—	Ea	14.00	—	14.00
Add for 8' high jambs	—	Ea	27.80	—	27.80
Add for round jamb without stop	—	Ea	14.70	—	14.70
Deduct for unfinished lauan	—	Ea	-1.43	—	-1.43
Add for prefinished lauan	—	Ea	3.12	—	3.12
Add for fir or hemlock jamb sets	—	%	40.0	—	—

Door Jambs

	Craft@Hrs	Unit	Material	Labor	Total
Door jambs, pocket Complete pocket door frame and hardware, door and casing not included, 6'8" high, 4-5/8" or 4-3/8" jamb, for 2" x 4" studs.					
2'0" wide	BC@.981	Ea	61.70	24.70	86.40
2'4" wide	BC@.981	Ea	61.70	24.70	86.40
2'6" wide	BC@.981	Ea	61.70	24.70	86.40
2'8" wide	BC@.981	Ea	61.70	24.70	86.40
3'0" wide	BC@.981	Ea	61.70	24.70	86.40
3'6" wide	BC@.981	Ea	61.70	24.70	86.40
4'0" wide	BC@.981	Ea	61.70	24.70	86.40
Add for 7'0" height	—	Ea	11.90	—	11.90
Add for 8'0" height	—	Ea	23.90	—	23.90
Add for 2" x 6" stud wall	—	Ea	20.70	—	20.70
Add for solid jambs	—	Ea	21.70	—	21.70
Sliding door hardware set only	—	Ea	8.74	—	8.74
Door Trim Sets Gun nailed. Costs listed are per opening (two head casings and four side casings), 6'8" height, ponderosa pine. Costs include casing on both sides of door opening.					
Casing size 11/16" x 2-1/4"					
3'0" W, ranch style, unfinished	BC@.250	Ea	17.00	6.29	23.29
3'0" W, ranch style, primed	BC@.250	Ea	17.00	6.29	23.29
6'0" W, ranch style, unfinished	BC@.300	Ea	23.20	7.55	30.75
6'0" W, ranch style, primed	BC@.300	Ea	23.20	7.55	30.75
3'0" W, colonial style, unfinished	BC@.250	Ea	17.00	6.29	23.29
3'0" W, colonial style, primed	BC@.250	Ea	17.00	6.29	23.29
6'0" W, colonial style, unfinished	BC@.300	Ea	23.20	7.55	30.75
6'0" W, colonial style, primed	BC@.300	Ea	23.20	7.55	30.75
Casing size 1/2" x 2", embossed honey oak color					
3'0" W, ranch style, prefinished light	BC@.250	Ea	9.63	6.29	15.92
6'0" W, ranch style, prefinished light	BC@.300	Ea	10.80	7.55	18.35
3'0" W, colonial style, prefinished light	BC@.250	Ea	9.63	6.29	15.92
6'0" W, colonial style, prefinished light	BC@.300	Ea	12.00	7.55	19.55
Casing size 3/4" x 3-1/2", ponderosa pine, with corner and base blocks					
6'0" W, butterfly, prefinished dark	BC@.250	Ea	56.90	6.29	63.19
6'0" W, fluted, prefinished dark	BC@.300	Ea	62.40	7.55	69.95
Casing size 3/4" x 5-1/4", ponderosa pine, with corner and base blocks					
6'0" W, butterfly, prefinished dark	BC@.250	Ea	82.10	6.29	88.39
6'0" W, fluted, prefinished dark	BC@.300	Ea	82.10	7.55	89.65
Add for hand nailing	BC@.275	Ea	—	6.92	6.92

Doors See also Basement Doors, Closet Doors, Door Jambs and Frames, Door Closers, Door Trim, Entrances, Garage Doors, Hardware (door), Mouldings (astragal, sill, threshold), Shower and Tub Doors, Thresholds, and Weatherstripping listed alphabetically in the Residential Division of this book.

	Craft@Hrs	Unit	Material	Labor	Total
Bar doors (Cafe doors.)					
Louvered 1-1/8", 4' high, ponderosa pine, opening sizes, hardware included					
2'6" unfinished	BC@.686	Pr	48.10	17.30	65.40
2'8" unfinished	BC@.686	Pr	53.50	17.30	70.80
3'0" unfinished	BC@.686	Pr	57.90	17.30	75.20
2'6" prefinished	BC@.686	Pr	82.00	17.30	99.30
2'8" prefinished	BC@.686	Pr	86.30	17.30	103.60
3'0" prefinished	BC@.686	Pr	93.90	17.30	111.20

	Craft@Hrs	Unit	Material	Labor	Total

Spindle over raised panel design, 1-1/8", 4' high, hemlock, pair opening sizes

	Craft@Hrs	Unit	Material	Labor	Total
2'6" unfinished	BC@.686	Pr	110.00	17.30	127.30
2'8" unfinished	BC@.686	Pr	114.00	17.30	131.30
3'0" unfinished	BC@.686	Pr	119.00	17.30	136.30
Add for hinges					
Gravity pivot hinges	—	Pr	4.40	—	4.40
Spring hinges, polished brass	—	Pr	41.00	—	41.00

Beveled glass doors Spanish cedar or pine with beveled glass insert. Matching sidelites. Transom designed to fit over one 36" door and two 14" sidelites combination. Based on Beveled Glass Designs Symphony series. Door dimensions are 36" x 80" x 1 3/4". Labor cost includes hanging door only. Add the cost of hinges and lockset.

	Craft@Hrs	Unit	Material	Labor	Total
Bolero, oval insert					
Spanish cedar door	BC@1.45	Ea	1,070.00	36.50	1,106.50
Spanish cedar sidelight	BC@1.00	Ea	531.00	25.20	556.20
Pine door	BC@1.45	Ea	862.00	36.50	898.50
Pine sidelight	BC@1.00	Ea	453.00	25.20	478.20
Matching glass transom	BC@1.00	Ea	521.00	25.20	546.20

	Craft@Hrs	Unit	Material	Labor	Total
Crescendo, rectangular insert					
Spanish cedar door	BC@1.45	Ea	940.00	36.50	976.50
Spanish cedar sidelight	BC@1.00	Ea	527.00	25.20	552.20
Pine door	BC@1.45	Ea	804.00	36.50	840.50
Pine sidelight	BC@1.00	Ea	448.00	25.20	473.20
Matching glass transom	BC@1.00	Ea	436.00	25.20	461.20

	Craft@Hrs	Unit	Material	Labor	Total
Rhapsody, arched insert					
Spanish cedar door	BC@1.45	Ea	943.00	36.50	979.50
Spanish cedar sidelight	BC@1.00	Ea	551.00	25.20	576.20
Pine door	BC@1.45	Ea	804.00	36.50	840.50
Pine sidelight	BC@1.00	Ea	465.00	25.20	490.20
Matching glass transom	BC@1.00	Ea	480.00	25.20	505.20

	Craft@Hrs	Unit	Material	Labor	Total
Sonata, narrow arched insert					
Spanish cedar door	BC@1.45	Ea	1,230.00	36.50	1,266.50
Spanish cedar sidelight	BC@1.00	Ea	614.00	25.20	639.20
Pine door	BC@1.45	Ea	966.00	36.50	1,002.50
Pine sidelight	BC@1.00	Ea	518.00	25.20	543.20
Matching glass transom	BC@1.00	Ea	521.00	25.20	546.20

	Craft@Hrs	Unit	Material	Labor	Total
Tremolo, full rectangular insert					
Spanish cedar door	BC@1.45	Ea	928.00	36.50	964.50
Spanish cedar sidelight	BC@1.00	Ea	541.00	25.20	566.20
Pine door	BC@1.45	Ea	829.00	36.50	865.50
Pine sidelight	BC@1.00	Ea	475.00	25.20	500.20
Matching glass transom	BC@1.00	Ea	480.00	25.20	505.20

Combination doors (Storm and screen doors.) Wood construction, western hemlock, full view type, with aluminum screen cloth, tempered glass storm insert, 4-1/2" top rail and stiles, 9-1/4" bottom rail. Unfinished. No frame or hardware included

	Craft@Hrs	Unit	Material	Labor	Total
2'8" x 6'8"	BC@1.25	Ea	158.00	31.50	189.50
3'0" x 6'8"	BC@1.25	Ea	170.00	31.50	201.50

Wood construction, western hemlock, colonial style with half height crossbuck, aluminum screen cloth, tempered glass storm insert, 4-1/4" top rail and stiles, 9-1/4" bottom rail. Unfinished, no frame or hardware included

	Craft@Hrs	Unit	Material	Labor	Total
2'8" x 6'8"	BC@1.25	Ea	203.00	31.50	234.50
3'0" x 6'8"	BC@1.25	Ea	215.00	31.50	246.50

Doors, Combination

	Craft@Hrs	Unit	Material	Labor	Total

Aluminum frame, fiberglass screen cloth, latch, keyed lock, pneumatic closer, weather-stripping, and glass insert panel. Frame included. Better quality, half-height crossbuck with upper window or full length window styles, baked enamel finish and full weather-stripping

6'8" to 7'0" high, 2'8" to 3'0" wide	BC@1.01	Ea	257.00	25.40	282.40

Better quality with 3/4 length jalousie storm insert and foam-filled double wall kick plate

6'8" to 7'0" high, 2'8" to 3'0" wide	BC@1.05	Ea	340.00	26.40	366.40

Good quality with stationary upper glass panel, large aluminum kick plate, horizontal push bar, and baked enamel finish

6'8" to 7'0" high, 2'8" to 3'0" wide	BC@1.05	Ea	176.00	26.40	202.40

Dutch doors Fir, 1-3/4" thick stiles and rails, single pane tempered glazing. No frame, jambs, moulding, or hardware included. Add one top section and one bottom section to find the total door cost. See the labor cost to install two-section dutch doors below. 2'6", 2'8" or 3'0" wide by 6'8" overall height sizes.

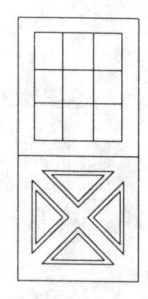

	Craft@Hrs	Unit	Material	Labor	Total
Bottom sections					
Two vertical or cross buck panel style	—	Ea	157.00	—	157.00
Top sections					
Full lite style	—	Ea	301.00	—	301.00
Four, six or nine lite style	—	Ea	323.00	—	323.00
Seven lite diagonal style	—	Ea	343.00	—	343.00
Twelve lite diagonal style	—	Ea	343.00	—	343.00
Add for dutch door shelf at lock rail	—	Ea	45.00	—	45.00
Labor to install two-section dutch door	BC@2.95	Ea	—	74.20	74.20

Fire doors Class B label, noncombustible mineral core, unfinished, labor hanging door only, 1-3/4" thick, natural birch.

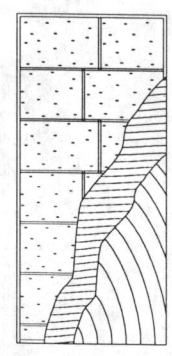

	Craft@Hrs	Unit	Material	Labor	Total
1 hour rating, 6'8" high					
2'0" wide	BC@1.36	Ea	169.00	34.20	203.20
2'4" wide	BC@1.41	Ea	184.00	35.50	219.50
2'6" wide or 2'8" wide	BC@1.41	Ea	194.00	35.50	229.50
3'0" wide	BC@1.46	Ea	223.00	36.70	259.70
3'6" wide	BC@1.46	Ea	242.00	36.70	278.70
4'0" wide	BC@1.46	Ea	267.00	36.70	303.70
1 hour rating, 7'0" high					
2'6" wide or 2'8" wide	BC@1.46	Ea	203.00	36.70	239.70
3'0" wide	BC@1.57	Ea	231.00	39.50	270.50
3'6" wide	BC@1.57	Ea	253.00	39.50	292.50
4'0" wide	BC@1.57	Ea	275.00	39.50	314.50
1-1/2 hour rating, 6'8" high					
2'6" wide	BC@1.36	Ea	275.00	34.20	309.20
2'8" wide	BC@1.46	Ea	291.00	36.70	327.70
3'0" wide or 3'6" wide	BC@1.46	Ea	308.00	36.70	344.70
4'0" wide	BC@1.46	Ea	375.00	36.70	411.70
1-1/2 hour rating, 7'0" high					
2'6" wide	BC@1.46	Ea	291.00	36.70	327.70
2'8" wide or 3'0" wide	BC@1.57	Ea	303.00	39.50	342.50
3'6" wide	BC@1.57	Ea	363.00	39.50	402.50
4'0" wide	BC@1.57	Ea	540.00	39.50	579.50
Add for 10" x 10" lite, wired glass	—	Ea	51.80	—	51.80

	Hardboard	Lauan	Birch	Red oak
Flush doors, hollow core Add labor cost from the section following Flush doors, additional costs				
6'8" high x 1-3/8" thick				
1'3" wide	21.60	35.30	43.00	48.50
1'4" wide	21.60	35.30	43.00	48.50
1'6" wide	21.60	35.30	43.00	48.50
1'8" wide	21.10	37.10	45.10	51.50
1'10" wide	21.10	37.10	45.10	51.50
2'0" wide	21.80	37.10	45.00	51.50
2'2" wide	23.00	38.50	49.80	57.50
2'4" wide	23.00	38.50	49.80	57.50
2'6" wide	23.10	40.50	49.00	59.00
2'8" wide	24.60	43.00	52.00	63.20
2'10" wide	25.70	46.00	57.50	65.20
3'0" wide	25.70	46.00	57.50	65.20
3'6" wide	43.40	69.30	112.00	117.00
4'0" wide	53.80	77.20	117.00	127.00
6'8" high x 1-3/4" thick				
2'6" or 2'8" wide	32.80	54.00	54.10	—
3'0" wide	33.90	59.00	59.20	—
6'8" high x 1-3/4" thick				
3'6" wide	57.10	—	121.00	—
7'0" high x 1-3/8" thick				
2'6" wide	31.30	47.00	61.50	—
2'8" wide	32.90	49.00	64.00	—
3'0" wide	34.30	52.00	67.10	—
7'0" high x 1-3/4" thick				
2'0" wide	42.40	—	56.00	—
2'8" wide	37.90	59.80	57.50	—
3'0" wide	39.50	63.00	62.00	—
3'6" wide	73.00	—	108.00	—
4'0" wide	76.90	—	116.50	—
8'0" high x 1-3/8" thick.				
1'6" wide	40.70	—	90.50	—
2'0" wide	40.70	—	90.50	—
2'2" wide	41.70	—	90.50	—
2'4" wide	41.70	—	90.50	—
2'6" wide	42.60	—	91.00	—
2'8" wide	43.80	—	92.50	—
3'0" wide	46.70	—	93.20	—
3'6" wide	59.10	—	134.70	—
4'0" wide	62.90	—	144.30	—
8'0" high x 1-3/4" thick				
2'6" wide	56.90	—	92.60	—
2'8" wide	57.80	—	109.00	—
3'0" wide	60.20	—	116.40	—
3'6" wide	74.70	—	—	—
4'0" wide	78.80	—	—	—
Flush doors, solid core Add the labor cost from the section following Flush doors, additional costs				
6'8" high x 1-3/8" thick				
2'0" wide	47.88	59.10	61.10	72.60
2'4" wide	48.62	59.10	62.30	74.30
2'6" wide	49.77	59.10	62.30	74.00
2'8" wide or 3'0" wide	51.87	62.70	64.40	75.30

Doors, Flush

	Hardboard	Lauan	Birch	Red oak
6'8" high x 1-3/4" thick				
2'6" wide	51.30	66.70	67.00	75.90
2'8" wide	53.10	69.30	69.60	78.90
3'0" wide	53.50	71.10	71.60	78.60
3'6" wide	103.00	—	117.60	120.70
7'0" high x 1-3/8" thick				
2'6" wide	56.00	64.90	73.90	—
2'8" or 3'0 wide	56.90	68.70	77.00	72.80
7'0" high x 1-3/4" thick				
2'0" wide	64.20	—	76.00	—
2'4" wide	64.20	—	77.30	—
2'6" wide	59.20	86.40	77.10	81.20
2'8" wide	40.70	89.80	81.70	83.70
3'0" wide	60.40	91.60	83.70	85.30
3'6" wide	109.00	—	119.70	—
4'0" wide	112.00	—	122.80	—
8'0" high x 1-3/4" thick				
2'6" or 2'8" wide	68.00	—	92.60	101.00
3'0" wide	71.00	—	98.80	104.00
3'6" wide	117.00	—	127.50	—
4'0" wide	119.00	—	142.00	—

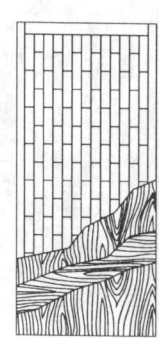

	Craft@Hrs	Unit	Material	Labor	Total
Additional costs for flush doors Hollow core and solid core					
Non-stock and odd sizes,					
Add to next larger size, hollow or solid core	—	Ea	36.40	—	36.40
Exterior glue, Type I waterproof					
Hardboard faces	—	Ea	3.20	—	3.20
Hardwood faces	—	Ea	15.60	—	15.60
Cut-outs, no stops included					
Hollow core, edge blocked	—	Ea	26.00	—	26.00
Solid core	—	Ea	20.80	—	20.80
Stops for cut-outs (openings in door), two sides					
Softwood	—	Ea	17.60	—	17.60
Birch or mahogany	—	Ea	22.50	—	22.50
Oak or walnut	—	Ea	37.50	—	37.50
Glazed openings, per SF or fraction of SF. No cut-outs or stops included					
Tempered clear, 3/16" thick	—	SF	14.10	—	14.10
Tempered bronze, 3/16" thick	—	SF	16.60	—	16.60
Tempered gray, 3/16" thick	—	SF	16.60	—	16.60
Tempered obscure, 7/32" thick	—	SF	16.60	—	16.60
Tempered Flemish amber	—	SF	20.80	—	20.80
Flemish acrylics, 3/16" thick	—	SF	11.70	—	11.70
Flemish acrylics, 1/8" thick	—	SF	8.20	—	8.20
Clear safety laminate 1/4" thick	—	SF	16.80	—	16.80
Polished wire plate 1/4" thick	—	SF	15.70	—	15.70
Vents, no cut-outs or stops included					
Fiber cane, 12" x 8"	—	Ea	13.10	—	13.10
Metal cane, 10" x 10"	—	Ea	38.90	—	38.90

	Craft@Hrs	Unit	Material	Labor	Total

Pre-fitting flush doors. Includes beveled lock stile, lockset bore, and hinge mortising. No hardware included

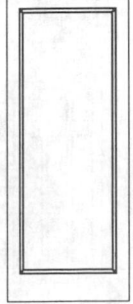

	Craft@Hrs	Unit	Material	Labor	Total
6'8" to 7'0" high x 1-3/8" or 1-3/4" thick					
To 3'0" wide, hollow core or solid core	—	Ea	2.73	—	2.73
Over 3'0" to 4'0", hollow core	—	Ea	5.35	—	5.35
Over 3'0" to 4'0" wide, solid core	—	Ea	8.61	—	8.61
8'0" high x 1-3/8" or 1-3/4" thick					
To 3'0" wide, hollow core	—	Ea	9.10	—	9.10
Over 3'0" to 4'0", hollow core	—	Ea	10.20	—	10.20
To 3'0" wide, solid core	—	Ea	10.20	—	10.20
Over 3'0" to 4'0" wide, solid core	—	Ea	15.00	—	15.00
Stave lumber cores for solid core doors					
To 3'0" x 7'0" sizes	—	Ea	56.20	—	56.20
Over 3'0" x 7'0 to 4'0" x 8'0" sizes	—	Ea	69.60	—	69.60
Special construction, 1-3/8" or 1-3/4" thick doors					
5" top and bottom rails	—	Ea	7.50	—	7.50
6" lock blocks	—	Ea	7.50	—	7.50
2-1/2" wide stiles	—	Ea	24.70	—	24.70
2-1/2" wide center stile	—	Ea	15.00	—	15.00

Labor to hang flush doors Typical cost per door. No frame, trim or finish. Add for hardware as required.

	Craft@Hrs	Unit	Material	Labor	Total
Hollow core, to 3'0" x 7'0"	BC@.941	Ea	—	23.70	23.70
Hollow core, over 3'0" x 7'0" to 4'0" x 8'0"	BC@1.08	Ea	—	27.20	27.20
Solid core, to 3'0" x 7'0"	BC@1.15	Ea	—	28.90	28.90
Solid core, over 3'0" x 7'0" to 4'0" x 8'0"	BC@1.32	Ea	—	33.20	33.20

Folding doors 4" wide hardwood panels with embossed, prefinished wood grain print, 3/8" thick, includes track and hardware. Add labor below.

	Walnut or oak 6'8"	Red oak 6'8"	Pine 8'0"	Birch 8'0"
2'8" wide	173.00	277.00	210.00	344.00
3'0" wide	188.00	299.00	232.00	372.00
3'6" wide	205.00	333.00	243.00	393.00
4'0" wide	228.00	366.00	275.00	438.00
4'6" wide	316.00	454.00	334.00	555.00
5'0" wide	293.00	477.00	350.00	582.00
5'6" wide	299.00	499.00	371.00	604.00
6'0" wide	360.00	582.00	431.00	607.00
7'0" wide	410.00	660.00	490.00	803.00
8'0" wide	477.00	742.00	549.00	920.00
9'0" wide	527.00	831.00	614.00	1,010.00
10'0" wide	576.00	843.00	673.00	1,100.00
11'0" wide	627.00	1,275.00	727.00	1,220.00
12'0" wide	721.00	1,310.00	840.00	1,410.00
13'0" wide	776.00	1,310.00	910.00	1,510.00
14'0" wide	831.00	1,340.00	970.00	1,610.00
15'0" wide	881.00	1,410.00	1,030.20	1,720.00

	Craft@Hrs	Unit	Material	Labor	Total
Labor to install wood panel folding doors					
Per LF of width	BC@.647	LF	—	16.30	16.30

	Craft@Hrs	Unit	Material	Labor	Total
French doors (sash doors) Douglas fir or hemlock with tempered glass set in putty, 6'8" high.					
1 lite, 1-3/8" thick sash					
2'0" to 2'8" wide	BC@1.45	Ea	179.00	36.50	215.50
3'0" wide	BC@1.45	Ea	186.00	36.50	222.50
5 lites, 5 high, 1-3/8" thick sash					
2'0" to 2'8" wide	BC@1.45	Ea	158.00	36.50	194.50
3'0" wide	BC@1.45	Ea	162.00	36.50	198.50
10 lites, 5 high, 1-3/8" thick sash					
2'0" to 2'8" wide	BC@1.45	Ea	175.00	36.50	211.50
3'0" wide	BC@1.45	Ea	179.00	36.50	215.50
1 lite, 1-3/4" thick sash					
2'0" to 2'8" wide	BC@1.45	Ea	181.00	36.50	217.50
3'0" wide	BC@1.45	Ea	185.00	36.50	221.50
5 lites, 5 high, 1-3/4" thick sash					
2'0" to 2'8" wide	BC@1.45	Ea	172.00	36.50	208.50
3'0" wide	BC@1.45	Ea	176.00	36.50	212.50
10 lites, 5 high, 1-3/4" thick sash					
2'0" to 2'8" wide	BC@1.45	Ea	169.30	36.50	205.80
3'0" wide	BC@1.45	Ea	175.00	36.50	211.50
Louver doors With full length slats, pine, 1-3/8", unfinished, 6'8" height					
1'0", 1'3" or 1'4"	BC@.750	Ea	66.20	18.90	85.10
1'6" or 1'8"	BC@.750	Ea	79.40	18.90	98.30
2'0"	BC@.750	Ea	91.60	18.90	110.50
2'4"	BC@.750	Ea	105.00	18.90	123.90
2'6" or 2'8"	BC@.750	Ea	110.00	18.90	128.90
2'8"	BC@.750	Ea	112.00	18.90	130.90
Add for prefinished doors	—	%	25.0	—	—
Add for half louver with raised lower panel	—	%	15.0	—	—
Add for false louvers	—	%	200.0	—	—
Add for 8'0" height, any design	—	%	145.0	—	—
Panel (colonial) doors 1-3/8" thick with raised panels 7/16" thick, 6'8" high, primed, labor hanging door only. No frame, trim, finish or hardware. Add for same as required.					
Primed hardboard, hollow core					
1'2"	BC@.827	Ea	39.40	20.80	60.20
1'4"	BC@.827	Ea	39.40	20.80	60.20
1'6"	BC@.827	Ea	39.40	20.80	60.20
2'0"	BC@.827	Ea	40.60	20.80	61.40
2'4"	BC@.827	Ea	46.00	20.80	66.80
2'6"	BC@.827	Ea	46.00	20.80	66.80
2'8"	BC@.827	Ea	47.10	20.80	67.90
3'0"	BC@.827	Ea	51.50	20.80	72.30
Ponderosa pine, stile and rail, 6 panel					
1'2"	BC@.827	Ea	78.80	20.80	99.60
1'3"	BC@.827	Ea	83.00	20.80	103.80
1'4"	BC@.827	Ea	87.70	20.80	108.50
1'6"	BC@.827	Ea	91.90	20.80	112.70
1'8"	BC@.827	Ea	96.30	20.80	117.10
1'10"	BC@.827	Ea	111.30	20.80	132.10
2'0"	BC@.827	Ea	115.60	20.80	136.40

	Craft@Hrs	Unit	Material	Labor	Total
2'2", 2'4" or 2'6"	BC@.827	Ea	120.80	20.80	141.60
2'8"	BC@.827	Ea	126.00	20.80	146.80
3'0"	BC@.827	Ea	137.00	20.80	157.80
Add for prefinished panel doors	—	%	20.0	—	—

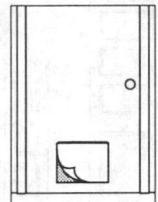

Pet doors

Entrance doors, lockable, swinging, aluminum frame with security panel

	Craft@Hrs	Unit	Material	Labor	Total
For cats and miniature dogs, 5" x 7-1/2"	BC@.800	Ea	48.50	20.10	68.60
For small dogs, 8-1/2" x 12-1/2"	BC@.800	Ea	69.30	20.10	89.40
For medium dogs, 11-1/2" x 16-1/2"	BC@.800	Ea	94.90	20.10	115.00
For large dogs, 15" x 20"	BC@.850	Ea	130.30	21.40	151.70

Sliding screen or patio doors, adjustable full length 1/2" panel, 80" high, "Lexan" plastic above lockable, swinging PVC door, aluminum frame

	Craft@Hrs	Unit	Material	Labor	Total
For cats and miniature dogs, panel is 11-1/2" wide with 5" x 7-1/2" door	BC@.500	Ea	194.40	12.60	207.00
For small dogs, panel is 15" wide with 8-1/2" x 12-1/2" door	BC@.500	Ea	220.00	12.60	232.60
For medium dogs, panel is 18" wide with 11-1/2" x 16-1/2" door	BC@.500	Ea	254.00	12.60	266.60

Prehung hollow core interior doors Primed split pine jamb, matching stop and casing, hinges, door sizes, 6'8" high, 1-3/8" thick. See costs on next page

	Prefinished embossed oak tone, flush	Natural birch flush	Unfinished panel pine
1'6"	80.60	101.00	172.40
2'0"	81.60	104.00	179.50
2'4"	86.10	108.10	186.70
2'6"	86.10	108.10	186.70
2'8" or 3'0"	88.20	117.30	206.00

	Craft@Hrs	Unit	Material	Labor	Total
Labor to install prehung hollow core doors					
To 2'0" wide	BC@.581	Ea	—	14.60	14.60
Over 2'0" wide	BC@.827	Ea	—	20.80	20.80

Prehung solid core exterior wood doors 1-3/4" thick, bored for lockset with finger-jointed pine frame, exterior casing, aluminum sill and 4-1/2" jamb. Assembled, primed and weather-stripped, 6'8" high. 5'4" and 6'0" widths are double doors. Add for interior trim and lockset.

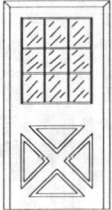

	Single door unit		Double door unit	
	2'8"	3'0"	5'4"	6'0"
Flush birch	275.00	280.00	595.00	575.00
6 panel, no glass, fir	404.00	415.00	845.00	850.00
French, 12 SF insulating glass	447.00	—	993.00	—
4 panel, 4 lite fan-shaped glass upper panels	—	447.00	—	963.00
Solid wood lower panel, 9 lite uninsulated glass upper panels	—	600.00	—	1,240.00
Mediterranean detail, no glass	—	532.00	—	1,110.00

Doors, Prehung

	Craft@Hrs	Unit	Material	Labor	Total
Labor to install prehung solid core doors					
2'8" or 3'0" wide single door units	BC@.750	Ea	—	18.90	18.90
5'4" or 6'0" wide double door units	BC@1.00	Ea	—	25.20	25.20

Sidelights for packaged prehung doors 1-3/4" thick, with insulated glass, assembled.

6'8" x 1'2", fir or hemlock	BC@1.36	Ea	289.00	34.20	323.20

Prehung steel-clad exterior doors 24 gauge steel cladding, 1-3/4" door bored for lockset, with wood frame, exterior casing only, polystyrene core, 1-1/2 hour (Class B) fire rating, weather-stripped, with hinges, assembled, primed, with aluminum sill, threshold. 6'8" high. Add for interior trim and lockset.

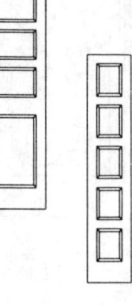

	Single doors			Double doors		
	2'8"	3'0"	3'6"	5'0"	5'4"	6'0"
Flush doors	195.00	198.00	266.00	488.00	488.00	500.00
Flush with up to 3 SF glass lite per door	247.00	250.00	266.00	610.00	610.00	615.00
Flush with 12 SF insulating glass per door	340.00	392.00	403.00	—	750.00	756.00
Doors with decorative insulating glass and moulding face	421.00	426.00	426.00	916.00	916.00	927.00

	Craft@Hrs	Unit	Material	Labor	Total
Labor to install prehung steel-clad doors					
2'8" to 3'6" wide single door units	BC@.750	Ea	—	18.90	18.90
5'0" to 6'0" wide double door units	BC@1.00	Ea	—	25.20	25.20
Add for:					
Single door steel frame	—	Ea	30.60	—	30.60
Double door steel frame	—	Ea	34.10	—	34.10
Sidelights, typical, 1'2" wide	BC@1.22	Ea	178.00	30.70	208.70
Interior trim sets					
2'6" to 2'8"	BC@.471	Ea	11.20	11.90	23.10
3'0" to 3'6"	BC@.471	Ea	12.30	11.90	24.20
5'0" to 6'0"	BC@.471	Ea	12.90	11.90	24.80
Mail slot (cut out only)	BC@.125	Ea	—	3.15	3.15
Peephole (hole only)	BC@.125	Ea	—	3.15	3.15
Extension jamb, 4-9/16" to 5-1/4"					
Single door	—	Ea	16.30	—	16.30
Double door	—	Ea	25.30	—	25.30

Screen doors No door frame or hardware included. Labor shown is for fitting and hanging the door.
Wood screen doors, fir, 1" thick, 6'9" high x 30", 32" or 36" wide ready for painting.
All doors include fiberglass screen cloth, 3-1/2" top rails and stiles, 5-1/2" bottom rails. No door frame or hardware included. Add for hardware and painting from below.

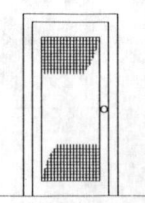

	Craft@Hrs	Unit	Material	Labor	Total
With one 5-1/2" solid panel at bottom	BC@1.25	Ea	63.60	31.50	95.10
With three 1" horizontal spindle bars and 2 corner blocks at top	BC@1.25	Ea	71.60	31.50	103.10
With one 3-1/2" center bar and two 5-1/2" solid panels at bottom	BC@1.25	Ea	84.80	31.50	116.30
With oval panel solid decorative blocks	BC@1.25	Ea	117.00	31.50	148.50
Add to any door above					
Hinges, latch, and pneumatic closer	BC@.500	LS	13.20	12.60	25.80
Paint or stain, 2 coats, both sides	BC@.500	LS	6.36	12.60	18.96

	Craft@Hrs	Unit	Material	Labor	Total

Aluminum screen doors, 1" thick, 6'8" to 6'9" high, 2'5" to 3'1" wide.
All doors include fiberglass screen cloth, latch, lock, pneumatic closer, and hinges. No door frame included.

	Craft@Hrs	Unit	Material	Labor	Total
With center pushbar, satin aluminum finish	BC@1.00	Ea	83.30	25.20	108.50
With half height grill, horizontal pushbar, kick plate, and baked enamel finish	BC@1.00	Ea	167.00	25.20	192.20
With ornamental grill, tempered glass, kick plate, and deadbolt lock	BC@1.00	Ea	236.00	25.20	261.20
With two decorative acrylic side panels, tempered glass, deadbolt lock	BC@1.00	Ea	241.00	25.20	266.20

Sliding 6'8" aluminum patio doors, baked on enamel finish Tempered safety glass, 5-1/2" frame, weather-stripped, adjustable locking device and screen included, 6'8 high. 3/16" single pane

	Craft@Hrs	Unit	Material	Labor	Total
1 sliding door, 1 stationary lite					
5'0" wide	BC@2.08	Ea	319.00	52.30	371.30
6'0" wide	BC@2.58	Ea	347.00	64.90	411.90
7'0" wide	BC@2.58	Ea	375.00	64.90	439.90
8'0" wide	BC@3.16	Ea	398.00	79.50	477.50
10'0" wide	BC@3.16	Ea	548.00	79.50	627.50
1 sliding door, 2 stationary lites					
9'0" wide	BC@3.16	Ea	504.00	79.50	583.50
12'0" wide	BC@3.65	Ea	570.00	91.90	661.90
15'0" wide	BC@4.14	Ea	749.00	104.00	853.00
2 sliding doors, 2 stationary lites					
12'0" wide	BC@3.65	Ea	651.00	91.90	742.90
16'0" wide	BC@4.14	Ea	770.00	104.00	874.00

Dual glazed, including hardware, screen included.

	Craft@Hrs	Unit	Material	Labor	Total
1 sliding door, 1 stationary lite					
5'0" x 6'8"	BC@2.36	Ea	442.00	59.40	501.40
6'0" x 6'8"	BC@2.80	Ea	493.00	70.50	563.50
7'0" x 6'8"	BC@2.80	Ea	604.00	70.50	674.50
8'0" x 6'8"	BC@3.42	Ea	621.00	86.10	707.10
10'0" x 6'8"	BC@3.42	Ea	794.00	86.10	880.10
1 sliding door, 2 stationary lites					
9'0" x 6'8" (OXO)	BC@3.85	Ea	648.00	96.90	744.90
12'0" x 6'8" (OXO)	BC@4.65	Ea	899.00	117.00	1,016.00
15'0" x 6'8" (OXO)	BC@5.15	Ea	981.00	130.00	1,111.00
Add to any of above					
Tinted gray glass, average	—	%	50.0	—	—
Key lock	—	Ea	20.40	—	20.40

Sliding aluminum patio doors, anodized finish Tempered safety glass, 5-1/2" frame, weather-stripped, adjustable locking device and screen included, 6'8 high.

3/16" single pane

	Craft@Hrs	Unit	Material	Labor	Total
1 sliding door, 1 stationary lite					
5'0" wide	BC@2.25	Ea	347.00	56.60	403.60
6'0" wide	BC@2.75	Ea	387.00	69.20	456.20
8'0" wide	BC@3.50	Ea	447.00	88.10	535.10
10'0" wide	BC@3.50	Ea	581.00	88.10	669.10

Doors, Sliding

	Craft@Hrs	Unit	Material	Labor	Total
1 sliding door, 2 stationary lites					
9'0" wide	BC@3.50	Ea	515.00	88.10	603.10
12'0" wide	BC@4.00	Ea	604.00	101.00	705.00
15'0" wide	BC@4.50	Ea	806.00	113.00	919.00
2 sliding doors, 2 stationary lites					
12'0" wide	BC@4.00	Ea	632.00	101.00	733.00
16'0" wide	BC@4.50	Ea	744.00	113.00	857.00
Add to any of above					
Tinted gray glass, average	—	%	50.0	—	—
Bronze finish frame	—	%	10.0	—	—
Key lock	—	Ea	19.20	—	19.20

Sliding aluminum-clad wood patio doors 5-1/2" frame, aluminum clad on exterior surfaces, ball bearing rollers, weather-stripped, with deadbolt security lock, and screen included, 6'8" high.

	Craft@Hrs	Unit	Material	Labor	Total
7/8" tempered insulating glass					
1 sliding door, 1 stationary lite					
5'0" wide	BC@2.00	Ea	886.00	50.30	936.30
6'0" wide	BC@2.75	Ea	925.00	69.20	994.20
8'0" wide	BC@3.25	Ea	1,100.00	81.80	1,181.80
1 sliding door, 2 stationary lites					
9'0" wide	BC@3.25	Ea	1,280.00	81.80	1,361.80
12'0" wide	BC@3.75	Ea	1,870.00	94.40	1,964.40
2 sliding doors, 2 stationary lites					
12'0" wide	BC@3.75	Ea	1,580.00	94.40	1,674.40
16'0" wide	BC@4.25	Ea	1,950.00	107.00	2,057.00
Add to any of above					
Tinted gray glass, average	—	%	50.0	—	—

Sliding wood patio doors Treated pine frame, double weather-stripped, adjustable locking device and screen included, 6'8 high.

	Craft@Hrs	Unit	Material	Labor	Total
5/8" tempered insulating glass					
1 sliding door, 1 stationary lite					
5' wide	BC@2.00	Ea	1,150.00	50.30	1,200.30
6' wide	BC@2.75	Ea	1,240.00	69.20	1,309.20
8' wide	BC@2.75	Ea	1,430.00	69.20	1,499.20
1 sliding door, 2 stationary lites					
9' wide	BC@3.50	Ea	1,540.00	88.10	1,628.10
12' wide	BC@3.75	Ea	2,190.00	94.40	2,284.40
Add to any of above					
Tinted gray glass, average	—	%	50.0	—	—
Colonial grille, per door section	BC@.500	LS	123.00	12.60	135.60
Key lock	—	Ea	19.20	—	19.20

Swinging steel patio doors Foam core, single lite, magnetic weather-stripping, with sliding screen, lock and deadbolt, 6'8" high.

	Craft@Hrs	Unit	Material	Labor	Total
1" tempered insulating glass					
5'4" wide, 1 swinging, 1 fixed lite	BC@2.00	Ea	749.00	50.30	799.30
6'0" wide, 1 swinging, 1 fixed lite	BC@2.75	Ea	749.00	69.20	818.20
7'0" wide, 2 swinging doors	BC@2.75	Ea	778.00	69.20	847.20
9'0" wide, 1 swinging door, 2 fixed lites	BC@3.25	Ea	1,140.00	81.80	1,221.80
Add for 15 lite door	—	Ea	45.00	—	45.00

	Craft@Hrs	Unit	Material	Labor	Total

Drafting Per SF of floor area, architectural only, not including engineering fees. See also Architectural Illustrations and Blueprinting. Typical prices.

	Craft@Hrs	Unit	Material	Labor	Total
Apartments	—	SF	—	—	1.50
Warehouses and storage buildings	—	SF	—	—	1.25
Office buildings	—	SF	—	—	2.50
Residences					
Minimum quality tract work	—	SF	—	—	2.00
Typical work	—	SF	—	—	2.50
Detailed jobs, exposed woods, hillsides	—	SF	—	—	3.00

Drainage piping Corrugated plastic drainage pipe, plain or perforated and snap-on fittings. Installed in trenches or at foundation footings. No excavation, gravel or backfill included. Polyethylene pipe.

	Craft@Hrs	Unit	Material	Labor	Total
3" pipe	BL@.010	LF	.34	.21	.55
4" pipe	BL@.010	LF	.43	.21	.64
5" pipe	BL@.010	LF	.86	.21	1.07
6" pipe	BL@.012	LF	1.04	.25	1.29
8" pipe	BL@.012	LF	1.57	.25	1.82
ABS snap-on fittings, reducers					
4" x 3"	BL@.050	Ea	2.24	1.03	3.27
5" x 4"	BL@.050	Ea	2.73	1.03	3.76
6" x 4"	BL@.050	Ea	4.04	1.03	5.07
8" x 6"	BL@.060	Ea	4.53	1.23	5.76
Adapters, end caps, split couplers					
3" fitting	BL@.050	Ea	1.29	1.03	2.32
4" fitting	BL@.050	Ea	1.70	1.03	2.73
5" fitting	BL@.050	Ea	2.29	1.03	3.32
6" fitting	BL@.060	Ea	2.87	1.23	4.10
Ells, wyes, tees					
3" fitting	BL@.050	Ea	4.13	1.03	5.16
4" fitting	BL@.050	Ea	6.22	1.03	7.25
5" fitting	BL@.050	Ea	7.60	1.03	8.63
6" fitting	BL@.060	Ea	9.26	1.23	10.49

Rock or sand fill for drainage systems. Labor cost is for spreading base and covering pipe. 3 mile haul dumped on site, and placed by wheelbarrow

	Craft@Hrs	Unit	Material	Labor	Total
Crushed stone (1.4 tons per CY)					
3/4" (Number 3) or 1-1/2" (Number 2)	BL@.700	CY	19.00	14.40	33.40
Crushed slag (1.86 tons per CY)					
3/4" or 1-1/2"	BL@.700	CY	9.50	14.40	23.90
Sand (1.35 tons per CY)	BL@.360	CY	14.80	7.39	22.19

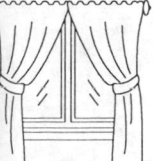

Draperies, Custom, Subcontract Custom draperies include 4" double hems and headers, 1-1/2" side hems, and weighted corners. Finished sizes allow 4" above, 4" below (except for sliding glass openings), and 5" on each side of wall openings. Prices are figured at one width, 48" of fabric, pleated to 16". Costs listed include fabric, full liner, manufacturing, all hardware, and installation. Add 50% per LF of width for jobs outside working range of metropolitan centers. Use $300 as a minimum job charge.

Minimum quality, 200% fullness, average is 6 panels of fabric, limited choice of fabric styles, colors, and textures

	Craft@Hrs	Unit	Material	Labor	Total
To 95" high	—	LF	—	—	32.60
To 68" high	—	LF	—	—	30.60
To 54" high	—	LF	—	—	29.60

	Craft@Hrs	Unit	Material	Labor	Total
Good quality, fully lined, 250% fullness, average is 5 panels of fabric, better selection of fabric styles, colors, and textures, weighted seams					
To 95" high	—	LF	—	—	77.50
To 84" high	—	LF	—	—	73.40
To 68" high	—	LF	—	—	69.40
To 54" high	—	LF	—	—	65.30
To 44" high	—	LF	—	—	62.20
Deduct for multi-unit jobs	—	%	—	—	-25.0
Add for insulated fabrics or liners					
Pleated shade liner	—	SY	—	—	21.40
Thermal liner	—	SY	—	—	24.50
Better quality, fully lined, choice of finest fabric styles, colors, and textures does not include special treatments such as elaborate swags, or custom fabrics					
To 95" high	—	LF	—	—	97.90
To 84" high	—	LF	—	—	92.80
To 68" high	—	LF	—	—	82.60
To 54" high	—	LF	—	—	75.50
To 44" high	—	LF	—	—	71.00

Electrical Work, Subcontract Costs listed below are for wiring new residential and light commercial buildings with Romex cable and assume circuit lengths averaging 40 feet. If flex cable is required, add $10.50 for 15 amp circuits and $30 for 20 amp circuits. No fixtures or appliances included except as noted. Work on second and higher floors may cost 25% more. Work performed by a qualified subcontractor.

Rule of thumb: Total cost for electrical work, performed by a qualified subcontractor, per SF of floor area. Includes service entrance, outlets, switches, basic lighting fixtures and connecting appliances only

	Craft@Hrs	Unit	Material	Labor	Total
All wiring and fixtures	—	SF	—	—	4.55
Lighting fixtures only (no wiring)	—	SF	—	—	1.85
Air conditioners (with thermostat)					
Central, 2 ton (220 volt)	—	LS	—	—	368.00
First floor room (115 volt)	—	LS	—	—	130.00
Second floor room (115 volt)	—	LS	—	—	292.00
Add for thermostat on second floor	—	LS	—	—	84.00
Alarms. See also Security Alarms					
Fire or smoke detector, wiring and outlet box only	—	LS	—	—	80.00
Add for detector unit	—	LS	—	—	31.90
Bathroom fixtures, wiring only, no fixtures or equipment included					
Mirror lighting (valance or side lighted mirrors)	—	LS	—	—	79.00
Sauna heater (40 amp branch circuit)	—	LS	—	—	256.00
Steam bath generator	—	LS	—	—	256.00
Sunlamps	—	LS	—	—	125.00
Whirlpool bath system/wall switch	—	LS	—	—	210.00
Clock outlets (recessed)	—	LS	—	—	58.00
Closet lighting					
Including ceramic "pull chain" fixture	—	LS	—	—	52.50
Including ceramic switch-operated fixture, add for wall switch from next page	—	LS	—	—	55.00

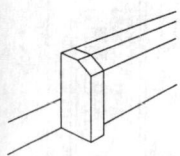

	Craft@Hrs	Unit	Material	Labor	Total
Clothes dryers					
Motor-driven gas dryer, receptacle only	—	LS	—	—	75.00
Direct connection for up to 5,760 watt (30 amp) electric dryer, 220 volt	—	LS	—	—	145.00
Direct connection for over 5,760 watt (40 amp) electric dryer, 220 volt	—	LS	—	—	160.00
Clothes washers (115 volt)	—	LS	—	—	82.50
Dishwashers	—	LS	—	—	90.00
Door bells. Rough wiring, front and rear with transformer.					
No chime or bell switch included	—	LS	—	—	84.00
Fans (exhaust)					
Attic fans (wiring only)	—	LS	—	—	100.00
Bathroom fans (includes 70 CFM fan)	—	LS	—	—	120.00
Garage fans (wiring only)	—	LS	—	—	95.00
Kitchen fans (includes 225 CFM fan)	—	LS	—	—	188.00
Freezers, branch circuit wiring only					
Upright (115 volt)	—	LS	—	—	80.00
Floor model	—	LS	—	—	65.00
Furnace wiring and blower hookup only	—	LS	—	—	90.00
Garage door opener, wiring and hookup only	—	LS		—	95.00
Garbage disposers. Wiring, switch and connection only					
No disposer included	—	LS	—	—	125.00
Grounding devices (see also Lightning protection systems on next page)					
Grounding entire electrical system	—	LS	—	—	105.00
Grounding single appliance	—	LS	—	—	62.50
Ground fault circuit interrupter (GFCI), rough and finish wiring,					
Includes outlet box and GFCI	—	LS	—	—	105.00
Heaters					
Baseboard (115 volt) per branch circuit	—	LS	—	—	95.00
Bathroom (ceiling type) wiring, switch connection (with GFCI) only	—	LS	—	—	120.00
Ceiling heat system (radiant-resistance type) 1,000 watt, 120 volt, per branch circuit, including thermostat	—	LS	—	—	230.00
Space heating (flush in-wall type) up to 2,000 watt, 220 volt	—	LS	—	—	210.00
Humidifiers, central	—	LS	—	—	80.00
Exterior lamp posts, wiring and hookup only, to 25'.					
Using buried wire and conduit	—	LS	—	—	180.00
Lighting fixture outlets (rough wiring and box only)					
Ceiling	—	Ea	—	—	42.50
Floor	—	Ea	—	—	62.50
Set in concrete	—	Ea	—	—	100.00
Set in masonry	—	Ea	—	—	100.00
Underground	—	Ea	—	—	130.00
Valance	—	Ea	—	—	52.50
Wall outlet	—	Ea	—	—	47.50

Electrical Work, Subcontract

	Craft@Hrs	Unit	Material	Labor	Total

Lightning protection systems (static electric grounding system)
  Residential, ridge protection including one chimney and connected garage (houses with cut-up roofs will cost more).

	Craft@Hrs	Unit	Material	Labor	Total
Typical home system	—	LS	—	—	1,900.00
Barns and light commercial buildings	—	LS	—	—	4,100.00

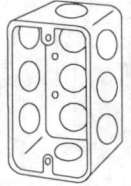

Ovens, wall type (up to 10' of wiring and hookup only), 220 volt

	Craft@Hrs	Unit	Material	Labor	Total
To 4,800 watts (20 amp)	—	LS	—	—	125.00
4,800 to 7,200 watts (30 amp)	—	LS	—	—	165.00
7,200 to 9,600 watts (40 amp)	—	LS	—	—	195.00

Ranges, (up to 10' of wiring and hookup only), 220 volt
  Countertop type

	Craft@Hrs	Unit	Material	Labor	Total
To 4,800 watts (20 amp)	—	LS	—	—	120.00
4,800 to 7,200 watts (30 amp)	—	LS	—	—	155.00
7,200 to 9,600 watts (40 amp)	—	LS	—	—	190.00
Freestanding type (50 amp)	—	LS	—	—	224.00

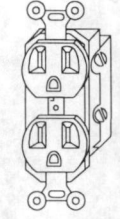

Receptacle outlets (rough wiring, box, receptacle and plate)

	Craft@Hrs	Unit	Material	Labor	Total
Ceiling	—	Ea	—	—	41.50
Countertop wall	—	Ea	—	—	52.50
Floor outlet	—	Ea	—	—	75.00
Split-wired	—	Ea	—	—	95.00
Standard indoor, duplex wall outlet	—	Ea	—	—	47.50
Waterproof, with ground fault circuit	—	Ea	—	—	80.00
Refrigerator or freezer wall outlet	—	Ea	—	—	57.50

Service entrance connections, complete (panel box hookup but no wiring)
  100 amp service including meter socket, main switch, GFCI and 5 single pole breakers

	Craft@Hrs	Unit	Material	Labor	Total
in 20 breaker space exterior panel box	—	LS	—	—	890.00

  200 amp service including meter socket, main switch, 2 GFCI and 15 single pole breakers

	Craft@Hrs	Unit	Material	Labor	Total
in 40 breaker space exterior panel box	—	LS	—	—	1,550.00

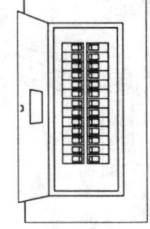

Sub panel connections (panel box hookup but no wiring)
  40 amp circuit panel including 12 single pole breakers

	Craft@Hrs	Unit	Material	Labor	Total
in 12 breaker indoor panel box	—	LS	—	—	290.00

  50 amp circuit panel including 16 single pole breakers

	Craft@Hrs	Unit	Material	Labor	Total
in 16 breaker indoor panel box	—	LS	—	—	310.00

Sump pump connection

	Craft@Hrs	Unit	Material	Labor	Total
including 15' of underground cable	—	Ea	—	—	160.00

Switches (includes rough wiring, box, switch and plate)

	Craft@Hrs	Unit	Material	Labor	Total
Dimmer	—	Ea	—	—	52.50
Lighted	—	Ea	—	—	52.50
Quiet	—	Ea	—	—	42.00
Mercury	—	Ea	—	—	52.50
Standard indoor wall switch	—	Ea	—	—	41.50
Stair-wired (3 way)	—	Ea	—	—	110.00
Waterproof	—	Ea	—	—	90.00
Television outlet wiring (300 ohm wire)	—	Ea	—	—	42.00

Trash compactor, 15" (includes compactor installed under counter and wiring)

	Craft@Hrs	Unit	Material	Labor	Total
Standard	—	LS	—	—	470.00
Deluxe	—	LS	—	—	550.00

Vacuuming system, central

	Craft@Hrs	Unit	Material	Labor	Total
Central unit hookup	—	LS	—	—	105.00

  Remote vacuum outlets.

	Craft@Hrs	Unit	Material	Labor	Total
Receptacle wiring only	—	Ea	—	—	44.00

	Craft@Hrs	Unit	Material	Labor	Total
Water heaters					
(up to 10' of wiring and hookup only), 220 volt	—	LS	—	—	155.00
Water pumps, domestic potable water.					
Connection only	—	LS	—	—	100.00

Elevators and Lifts, Subcontract Elevators for apartments or commercial buildings, hydraulic vertical cab, meets code requirements for public buildings, 2,500 pound capacity, to 13 passengers. Includes side opening sliding door, illuminated controls, emergency light, alarm, code-approved fire service operation, motor control unit, wiring in hoistway, cab and door design options.

	Craft@Hrs	Unit	Material	Labor	Total
Basic 2 stop typical prices.					
100 feet per minute	—	LS	—	—	40,500.00
125 feet per minute	—	LS	—	—	43,400.00
150 feet per minute	—	LS	—	—	46,800.00
Add per stop to 5 stops	—	LS	—	—	4,400.00
Add for infrared door protection	—	Ea	—	—	1,700.00
Add for hall position indicator	—	Ea	—	—	410.00
Add for car position indicator	—	Ea	—	—	410.00
Add for car direction light & tone	—	Ea	—	—	400.00
Add for hall lantern with audible tone	—	Ea	—	—	410.00

Elevators for private residences, electric cable, motor driven, 500 pound capacity, includes complete installation of elevator, controls and safety equipment. Meets code requirements for residences

	Craft@Hrs	Unit	Material	Labor	Total
2 stops (up to 10' rise)	—	LS	—	—	15,000.00
3 stops (up to 20' rise)	—	LS	—	—	17,000.00
4 stops (up to 30' rise)	—	LS	—	—	20,000.00
5 stops (up to 40' rise)	—	LS	—	—	22,000.00
Add for additional gate	—	Ea	—	—	750.00
Dumbwaiters, electric					
75 pound, 2 stop, 24" x 24" x 30"	—	LS	—	—	6,500.00
75 pound, add for each extra stop	—	Ea	—	—	1,500.00
100 pound, 2 stop, 24" x 30" x 36"	—	LS	—	—	7,500.00
100 pound, add for each extra stop	—	Ea	—	—	1,700.00
100 pound, 2 stop, 24" x 24" x 36", tray type	—	LS	—	—	12,000.00
100 pound tray type, add for each extra stop	—	Ea	—	—	1,870.00
300 pound, 2 stop, 30" x 30" x 36", tray type	—	LS	—	—	16,000.00
300 pound, add for each extra stop	—	Ea	—	—	3,000.00
500 pound, 2 stop, 23" x 56" x 48"	—	LS	—	—	18,000.00
500 pound, add for each extra stop	—	Ea	—	—	3,000.00

Hydraulic elevators, residential, 700 pound capacity. Includes controls and safety equipment with standard unfinished birch plywood cab and one gate.

	Craft@Hrs	Unit	Material	Labor	Total
2 stops (up to 15' rise)	—	LS	—	—	17,500.00
3 stops (up to 22' rise)	—	LS	—	—	19,000.00
4 stops (up to 28' rise)	—	LS	—	—	22,500.00
5 stops (up to 34' rise)	—	LS	—	—	25,200.00
Add for adjacent opening	—	Ea	—	—	350.00
Add for additional gate	—	Ea	—	—	800.00
Add for plastic laminate cab	—	Ea	—	—	600.00
Add for oak raised panel cab	—	Ea	—	—	2,800.00

Stairlifts (incline lifts), single passenger, indoor. Lift to 18' measured diagonally. 300 pound capacity. Includes top and bottom call button, safety chair, track, power unit, installation by a licensed elevator contractor. Single family residential use

	Craft@Hrs	Unit	Material	Labor	Total
Straight stairs	—	LS	—	—	3,800.00

	Craft@Hrs	Unit	Material	Labor	Total

Wheelchair lift (porch lift), screw driven, fully enclosed sides.
Typical electrical connections and control included. Up to 5' rise

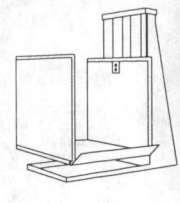

	Craft@Hrs	Unit	Material	Labor	Total
Residential wheelchair lift	—	LS	—	—	7,600.00

Engineering Fees Typical cost for consulting work.
Acoustical engineering (Based on an hourly wage of $65.00 to $125.00)

Environmental noise survey of a lot; includes technician with measuring equipment, data reduction and analysis, and written report with recommendations,

	Craft@Hrs	Unit	Material	Labor	Total
Written report with recommendations	—	LS	—	—	1,290.00
Exterior to interior noise analysis	—	LS	—	—	424.00

Measure "Impact Insulation Class" and "Sound Transmission Class" in existing buildings, (walls, floors, and ceilings)

	Craft@Hrs	Unit	Material	Labor	Total
Minimum cost	—	LS	—	—	848.00

Priced by wall, floor, or ceiling sections

	Craft@Hrs	Unit	Material	Labor	Total
Per section, approx. 100 SF	—	Ea	—	—	424.00

Analyze office, conference room, or examining room to assure acoustical privacy

	Craft@Hrs	Unit	Material	Labor	Total
Per room analyzed	—	LS	—	—	373.00

Prepare acoustical design for church, auditorium, or lecture hall, (fees vary greatly with complexity of acoustics desired)

	Craft@Hrs	Unit	Material	Labor	Total
Typical cost	—	LS	—	—	1,590.00

Evaluate noise problems from proposed building.

	Craft@Hrs	Unit	Material	Labor	Total
Impact on surrounding environment	—	LS	—	—	848.00

Evaluate noise problems from surrounding environment

	Craft@Hrs	Unit	Material	Labor	Total
Impact on a proposed building	—	LS	—	—	690.00

Front end scheduling. Coordinate architectural, engineering and other consultant services, plan check times and other approval times

	Craft@Hrs	Unit	Material	Labor	Total
For residential and commercial projects	—	LS	—	—	4,230.00

Project scheduling using CPM (Critical Path Method). Includes consultation, review of construction documents, development of construction logic, and graphic schedule. Comprehensive schedules to meet government or owner specifications will cost more.

	Craft@Hrs	Unit	Material	Labor	Total
Wood frame buildings, 1 or 2 stories	—	LS	—	—	2,120.00

Structural engineering plan check, typical 2-story, 8 to 10 unit apartment building

	Craft@Hrs	Unit	Material	Labor	Total
Typical price per apartment building	—	LS	—	—	828.00
Add for underground parking	—	LS	—	—	710.00

Entrances, Colonial
Single door entrance, pine, primed (no doors, frames, or sills included)

	Craft@Hrs	Unit	Material	Labor	Total
Without pediment, 3'0" x 6'8"	BC@1.19	Ea	177.00	29.90	206.90
With pediment, 3'0" x 6'8"	BC@1.19	Ea	216.00	29.90	245.90
With arch pediment, 3'0" x 6'8"	BC@1.19	Ea	353.00	29.90	382.90

Add for exterior door frames, 3'0" x 6'8" x 1-3/4"

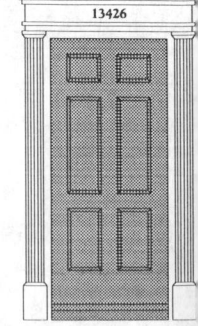

13426

	Craft@Hrs	Unit	Material	Labor	Total
4-9/16" jamb width	BC@1.25	Ea	53.40	31.50	84.90

Double door entrance, pine, primed (no doors, frames, or sills included)
Without pediment, two door (individual width)

	Craft@Hrs	Unit	Material	Labor	Total
2'6" x 6'8"	BC@1.45	Ea	237.00	36.50	273.50
2'8" x 6'8"	BC@1.45	Ea	222.00	36.50	258.50
3'0" x 6'8"	BC@1.45	Ea	232.00	36.50	268.50

With pediment, two doors (individual width)

	Craft@Hrs	Unit	Material	Labor	Total
2'6" x 6'8"	BC@1.45	Ea	237.00	36.50	273.50
2'8" x 6'8"	BC@1.45	Ea	262.00	36.50	298.50
3'0" x 6'8"	BC@1.45	Ea	363.00	36.50	399.50

	Craft@Hrs	Unit	Material	Labor	Total

With arched pediment, 7'6-1/2" x 1'9" (above door). Opening width of 5'7-1/4" will also fit 3'0" door with two 14" sidelights

	Craft@Hrs	Unit	Material	Labor	Total
3'0" x 6'8" (two doors)	BC@1.45	Ea	504.00	36.50	540.50

Add for exterior door frames, 1-3/4" thick, 5-1/4" jamb

	Craft@Hrs	Unit	Material	Labor	Total
2'8" x 6'8" or 3'0" x 6'8"	BC@1.25	Ea	57.40	31.50	88.90
6'0" x 6'8"	BC@2.75	Ea	68.60	69.20	137.80

Decorative split columns (with cap and base) alone, no pediment

	Craft@Hrs	Unit	Material	Labor	Total
8" x 8' to 12'	BC@.800	Ea	32.20	20.10	52.30
10" x 8' to 12'	BC@.800	Ea	35.30	20.10	55.40
12" x 8' to 12'	BC@.800	Ea	40.40	20.10	60.50
10" x 14' to 16'	BC@1.00	Ea	63.50	25.20	88.70

Excavation and Backfill by Hand Using hand tools.

General excavation, using a pick and shovel (loosening and one throw)

	Craft@Hrs	Unit	Material	Labor	Total
Light soil	BL@1.10	CY	—	22.60	22.60
Average soil	BL@1.70	CY	—	34.90	34.90
Heavy soil or loose rock	BL@2.25	CY	—	46.20	46.20

Backfilling (one shovel throw from stockpile)

	Craft@Hrs	Unit	Material	Labor	Total
Sand	BL@.367	CY	—	7.54	7.54
Average soil	BL@.467	CY	—	9.59	9.59
Rock or clay	BL@.625	CY	—	12.80	12.80
Add for compaction, average soil or sand	BL@.400	CY	—	8.21	8.21
Fine grading	BL@.008	SF	—	.16	.16

Footings, average soil, using a pick and shovel

	Craft@Hrs	Unit	Material	Labor	Total
6" deep x 12" wide (1.85 CY per CLF)	BL@.034	LF	—	.70	.70
8" deep x 12" wide (2.47 CY per CLF)	BL@.050	LF	—	1.03	1.03
8" deep x 16" wide (3.29 CY per CLF)	BL@.055	LF	—	1.13	1.13
8" deep x 18" wide (3.70 CY per CLF)	BL@.060	LF	—	1.23	1.23
10" deep x 12" wide (3.09 CY per CLF)	BL@.050	LF	—	1.03	1.03
10" deep x 16" wide (4.12 CY per CLF)	BL@.067	LF	—	1.38	1.38
10" deep x 18" wide (4.63 CY per CLF)	BL@.075	LF	—	1.54	1.54
12" deep x 12" wide (3.70 CY per CLF)	BL@.060	LF	—	1.23	1.23
12" deep x 16" wide (4.94 CY per CLF)	BL@.081	LF	—	1.66	1.66
12" deep x 20" wide (6.17 CY per CLF)	BL@.100	LF	—	2.05	2.05
16" deep x 16" wide (6.59 CY per CLF)	BL@.110	LF	—	2.26	2.26
12" deep x 24" wide (7.41 CY per CLF)	BL@.125	LF	—	2.57	2.57

Loading trucks (shoveling, one throw)

	Craft@Hrs	Unit	Material	Labor	Total
Average soil	BL@1.45	CY	—	29.80	29.80
Rock or clay	BL@2.68	CY	—	55.00	55.00

Pits to 5'

	Craft@Hrs	Unit	Material	Labor	Total
Light soil	BL@1.34	CY	—	27.50	27.50
Average soil	BL@2.00	CY	—	41.10	41.10
Heavy soil	BL@2.75	CY	—	56.50	56.50

Pits over 5' deep require special consideration: shoring, liners and method for removing soil.

Shaping trench bottom for pipe

	Craft@Hrs	Unit	Material	Labor	Total
To 10" pipe	BL@.024	LF	—	.49	.49
12" to 20" pipe	BL@.076	LF	—	1.56	1.56

Shaping embankment slopes

	Craft@Hrs	Unit	Material	Labor	Total
Up to 1 in 4 slope	BL@.060	SY	—	1.23	1.23
Over 1 in 4 slope	BL@.075	SY	—	1.54	1.54
Add for top crown or toe	—	%	—	50.0	—
Add for swales	—	%	—	90.0	—

Excavation

	Craft@Hrs	Unit	Material	Labor	Total
Spreading material piled on site					
Average soil	BL@.367	CY	—	7.54	7.54
Stone or clay	BL@.468	CY	—	9.61	9.61
Strip and pile top soil	BL@.024	SF	—	.49	.49
Tamping, hand tamp only	BL@.612	CY	—	12.60	12.60
Trenches to 5', soil piled beside trench					
Light soil	BL@1.13	CY	—	23.20	23.20
Average soil	BL@1.84	CY	—	37.80	37.80
Heavy soil or loose rock	BL@2.86	CY	—	58.70	58.70
Add for depth over 5' to 9'	BL@.250	CY	—	5.13	5.13

Trenching and Backfill with Heavy Equipment These costs and productivity are based on utility line trenches and continuous footings where the spoil is piled adjacent to the trench. Linear feet (LF) and cubic yards (CY) per hour shown are based on a 2-man crew. Reduce productivity by 10% to 25% when spoil is loaded in trucks. Shoring, dewatering or unusual conditions are not included.

	Craft@hrs	Unit	Material	Labor	Equipment	Total
Excavation, using equipment shown						
Wheel loader 55 HP, with integral backhoe, at $21 per hour						
12" wide bucket, for 12" wide trench. Depths 3' to 5'						
Light soil (60 LF per hour)	B8@.033	LF	—	.80	.35	1.15
Medium soil (55 LF per hour)	B8@.036	LF	—	.88	.38	1.26
Heavy or wet soil						
(35 LF per hour)	B8@.057	LF	—	1.39	.60	1.99
18" wide bucket, for 18" wide trench. Depths 3' to 5'						
Light soil (55 LF per hour)	B8@.036	LF	—	.88	.38	1.26
Medium soil (50 LF per hour)	B8@.040	LF	—	.97	.42	1.39
Heavy or wet soil (30 LF per hour)	B8@.067	LF	—	1.63	.70	2.33
24" wide bucket, for 24" wide trench. Depths 3' to 5'						
Light soil (50 LF per hour)	B8@.040	LF	—	.97	.42	1.39
Medium soil (45 LF per hour)	B8@.044	LF	—	1.07	.46	1.53
Heavy or wet soil (25 LF per hour)	B8@.080	LF	—	1.95	.84	2.79
Backfill trenches from loose material piled adjacent to trench. No compaction included.						
Soil, previously excavated						
Wheel loader 55 HP						
($15 & 50 CY per hour)	B8@.040	CY	—	.97	.29	1.26
D-3 crawler dozer						
($28 & 25 CY per hour)	B8@.080	CY	—	1.95	1.18	3.13
3/4 CY crawler loader						
($23 & 33 CY per hour)	B8@.061	CY	—	1.48	.58	2.06
D-7 crawler dozer						
($93 & 130 CY per hour)	B8@.015	CY	—	.37	.68	1.05
Sand or gravel bedding						
3/4 CY wheel loader						
($16 and 80 CY per hour)	B8@.025	CY	—	.61	.19	.80
Compaction of soil in trenches in 8" layers.						
Pneumatic tampers						
($5.70 & 40 CY per hour)	BL@.050	CY	—	1.03	.19	1.22
Vibrating rammers						
($5 & 20 CY per hour)	BL@.100	CY	—	2.05	2.05	4.10

	Craft@Hrs	Unit	Material	Labor	Total

Excavation with Heavy Equipment These figures assume a crew of one operator unless noted otherwise. Only labor costs are included here. See equipment rental costs at the end of this section. Use the productivity rates listed here to determine the number of hours or days that the equipment will be needed. For larger jobs and commercial work, see excavation costs under Site Work in the Commercial and Industrial division of this book.

	Craft@Hrs	Unit	Material	Labor	Total
Excavation rule of thumb for small jobs	—	CY	—	—	2.85
Backhoe, operator and one laborer, 3/4 CY bucket					
Light soil (13.2 CY per hour)	B8@.152	CY	—	3.70	3.70
Average soil (12.5 CY per hour)	B8@.160	CY	—	3.89	3.89
Heavy soil (10.3 CY per hour)	B8@.194	CY	—	4.72	4.72
Rock (16 CY per hour)	B8@.125	CY	—	3.04	3.04
Add when using 1/2 CY bucket	—	%	—	25.0	—
Bulldozer, 50 HP unit					
Backfill (36 CY per hour)	0E@.027	CY	—	.76	.76
Clearing brush (900 SF per hour)	0E@.001	SF	—	.03	.03
Add for thick brush	—	%	—	300.0	—
Spread dumped soil (42 CY per hour)	0E@.024	CY	—	.68	.68
Strip topsoil(17 CY per hour)	0E@.059	CY	—	1.66	1.66
Bulldozer, 66 HP unit					
Backfill (40 CY per hour)	0E@.025	CY	—	.70	.70
Clearing brush (1,000 SF per hour)	0E@.001	SF	—	.03	.03
Add for thick brush	—	%	—	300.0	—
Spread dumped soil (45 CY per hour)	0E@.023	CY	—	.65	.65
Strip top soil (20 CY per hour)	0E@.050	CY	—	1.41	1.41
Bulldozer, 120 HP unit					
Backfill (53 CY per hour)	0E@.019	CY	—	.53	.53
Clearing brush (1,250 SF per hour)	0E@.001	SF	—	.03	.03
Add for thick brush	—	%	—	200.0	—
Spread dumped soil (63 CY per hour)	0E@.016	CY	—	.45	.45
Strip top soil (25 CY per hour)	0E@.040	CY	—	1.13	1.13
Dump truck. Spot, load, unload, travel					
3 CY truck					
Short haul (9 CY per hour)	B7@.222	CY	—	4.68	4.68
2-3 mile haul (6 CY per hour)	B7@.333	CY	—	7.02	7.02
4 mile haul (4 CY per hour)	B7@.500	CY	—	10.50	10.50
5 mile haul (3 CY per hour)	B7@.666	CY	—	14.00	14.00
4 CY truck					
Short haul (11 CY per hour)	B7@.182	CY	—	3.83	3.83
2-3 mile haul (7 CY per hour)	B7@.286	CY	—	6.03	6.03
4 mile haul (5.5 CY per hour)	B7@.364	CY	—	7.67	7.67
5 mile haul (4.5 CY per hour)	B7@.444	CY	—	9.35	9.35
5 CY truck					
Short haul (14 CY per hour)	B7@.143	CY	—	3.01	3.01
2-3 mile haul (9.5 CY per hour)	B7@.211	CY	—	4.45	4.45
4 mile haul (6.5 CY per hour)	B7@.308	CY	—	6.49	6.49
5 mile haul (5.5 CY per hour)	B7@.364	CY	—	7.67	7.67
Jackhammer, one laborer (per CY of unloosened soil)					
Average soil (2-1/2 CY per hour)	BL@.400	CY	—	8.21	8.21
Heavy soil (2 CY per hour)	BL@.500	CY	—	10.30	10.30

Excavation

	Craft@Hrs	Unit	Material	Labor	Total
Igneous or dense rock (.7 CY per hour)	BL@1.43	CY	—	29.40	29.40
Most weathered rock (1.2 CY per hour)	BL@.833	CY	—	17.10	17.10
Soft sedimentary rock (2 CY per hour)	BL@.500	CY	—	10.30	10.30
Air tamp (3 CY per hour)	BL@.333	CY	—	6.84	6.84
Loader (tractor shovel)					
Piling earth on premises, 1 CY bucket					
Light soil (43 CY per hour)	OE@.023	CY	—	.65	.65
Heavy soil (35 CY per hour)	OE@.029	CY	—	.82	.82
Loading trucks, 1 CY bucket					
Light soil (53 CY per hour)	OE@.019	CY	—	.53	.53
Average soil (43 CY per hour)	OE@.023	CY	—	.65	.65
Heavy soil (40 CY per hour)	OE@.025	CY	—	.70	.70
Deduct for 2-1/4 CY bucket	—	%	—	-43.0	—
Sheeps foot roller (18.8 CSF per hour)	OE@.001	SF	—	.03	.03
Sprinkling with truck, (62.3 CSF per hour)	TR@.001	SF	—	.02	.02
Tree and brush removal, labor only (clear and grub, one operator and one laborer)					
Light brush	B8@10.0	Acre	—	243.00	243.00
Heavy brush	B8@12.0	Acre	—	292.00	292.00
Wooded	B8@64.0	Acre	—	1,560.00	1,560.00
Tree removal, cutting trees, removing branches, cutting into short lengths with chain saws and axes, by tree diameter, labor only					
8" to 12" (2.5 manhours per tree)	BL@2.50	Ea	—	51.30	51.30
13" to 18" (3.5 manhours per tree)	BL@3.50	Ea	—	71.90	71.90
19" to 24" (5.5 manhours per tree)	BL@5.50	Ea	—	113.00	113.00
25" to 36" (7.0 manhours per tree)	BL@7.00	Ea	—	144.00	144.00
Tree stump removal, using a small dozer, by tree diameter, operator only					
6" to 10" (1.6 manhours per stump)	OE@1.60	Ea	—	45.00	45.00
11" to 14" (2.1 manhours per stump)	OE@2.10	Ea	—	59.10	59.10
15" to 18" (2.6 manhours per stump)	OE@2.60	Ea	—	73.20	73.20
19" to 24" (3.1 manhours per stump)	OE@3.10	Ea	—	87.20	87.20
25" to 30" (3.3 manhours per stump)	OE@3.30	Ea	—	92.90	92.90
Trenching machine, crawler-mounted					
Light soil (27 CY per hour)	OE@.037	CY	—	1.04	1.04
Heavy soil (21 CY per hour)	OE@.048	CY	—	1.35	1.35
Add for pneumatic-tired machine	—	%	—	15.0	—

	1/2 day	Day	Week
Excavation Equipment Rental Costs Typical cost not including fuel, delivery or pickup charges. Add labor costs from the previous section. Half day minimums.			
Backhoe, wheeled mounted			
30 HP unit, 1/8 to 3/4 CY bucket	150.00	230.00	730.00
60 HP unit, 1/4 to 1 CY bucket	170.00	260.00	830.00
Add for each delivery or pickup	95.00	95.00	95.00
Bulldozer, crawler tractor			
65 HP unit, D-3	176.00	285.00	895.00
90 to 105 HP unit, D-4 or D-5	223.00	370.00	1,240.00
140 HP unit, D-6	370.00	580.00	1,890.00
Add for each delivery or pickup	95.00	95.00	95.00
Dump truck, on-highway type			
3 CY ($.40 per mile charge)	80.00	165.00	450.00
5 CY ($.44 per mile charge)	110.00	180.00	690.00
10 CY ($.48 per mile charge)	140.00	230.00	900.00

	1/2 day	Day	Week
Wheel loaders, front-end load and dump, diesel			
3/4 CY bucket, 4WD, articulated	115.00	190.00	630.00
1 CY bucket, 4WD, articulated	140.00	230.00	740.00
2 CY bucket, 4WD, articulated	186.00	310.00	1,060.00
3-1/4 CY bucket, 4WD, articulated	370.00	620.00	1,920.00
5 CY bucket, 4WD, articulated	570.00	950.00	2,370.00
Add for each delivery or pickup	94.00	94.00	94.00
Sheeps foot roller, towed type, 40" diameter, 48" wide, double drum	56.00	94.00	270.00
Trenching machine, to 5'6" depth, 6" to 16" width capacity, crawler-mounted type	140.00	230.00	750.00
Add for each delivery or pickup	95.00	95.00	95.00
Compactor, 32" plate width, manually guided, gasoline engine, vibratory plate type	72.00	120.00	400.00
Chain saw, 18" to 23" bar, gasoline engine	27.00	40.00	139.00
Breakers, pavement, medium duty, with blade. Including 50' of 1" hose	21.00	35.00	100.00
Air compressors, trailer-mounted, gas powered, silenced			
100 CFM to 150 CFM	36.00	60.00	215.00
175 CFM	42.00	70.00	248.00

	Craft@Hrs	Unit	Material	Labor	Total

Fans

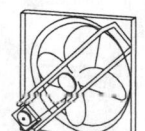

Attic fans, indoor type (see also Roof-mounted fans below), belt drive, overall dimensions, includes 10' of type NM cable, installation and connection.

	Craft@Hrs	Unit	Material	Labor	Total
Two speed, horizontal, 1/3 HP					
24" x 24" (5,100/3,400 CFM)	BE@3.80	Ea	414.00	102.00	516.00
30" x 30" (7,400/4,900 CFM)	BE@3.80	Ea	444.00	102.00	546.00
36" x 36' (9,750/6,500 CFM)	BE@3.80	Ea	526.00	102.00	628.00
42" x 42" (12,000/8,100 CFM)	BE@3.80	Ea	628.00	102.00	730.00
Variable speed, vertical or horizontal, 1/4 HP, with ceiling shutters					
24" x 24" (to 5,400 CFM)	BE@3.80	Ea	307.00	102.00	409.00
30" x 30" (to 8,150 CFM)	BE@3.80	Ea	327.00	102.00	429.00
Add for time switch	BE@.950	Ea	29.60	25.50	55.10

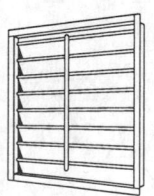

Automatic ceiling shutters for exhaust fans, white aluminum

	Craft@Hrs	Unit	Material	Labor	Total
24"	BE@.667	Ea	71.50	17.90	89.40
30"	BE@.667	Ea	92.00	17.90	109.90
36"	BE@.750	Ea	102.00	20.20	122.20
Window fans, 3 speed, white/gray finish					
12", 25" to 38" panel (2,000 CFM)	BE@1.50	Ea	81.80	40.30	122.10
16", 22-1/2" to 33" panel (6,000 CFM)	BE@1.50	Ea	92.00	40.30	132.30
20", 26-1/2" to 36" panel (6,500 CFM)	BE@1.50	Ea	112.00	40.30	152.30

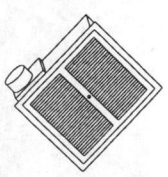

Bathroom exhaust fans, ceiling-mounted, includes 10' of type NM cable, installation and connection. See also Heaters

	Craft@Hrs	Unit	Material	Labor	Total
10" x 10" grille, 3" diameter duct (60 CFM)	BE@2.00	Ea	41.00	53.70	94.70
10" x 10" grille, With light, 4" diameter duct (50 CFM)	BE@2.00	Ea	66.50	53.70	120.20
Add for duct to 8', wall cap, and wire	BE@.417	LS	25.60	11.20	36.80

Bathroom wall-mounted direct exhaust fan, with plastic grille and damper. Includes 10' of type NM cable, installation, connection and wall vent.

	Craft@Hrs	Unit	Material	Labor	Total
8" x 8" (110 CFM)	BE@2.00	Ea	81.80	53.70	135.50

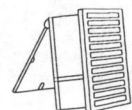

Fans

	Craft@Hrs	Unit	Material	Labor	Total

Kitchen exhaust fans, circular, aluminum. Includes 10' of type NM cable, installation, connection, and wall cap.

200 CFM ceiling fan (with duct)	BE@2.00	Ea	122.00	53.70	175.70
300 CFM direct vent wall fan	BE@2.00	Ea	102.00	53.70	155.70

Roof-mounted power ventilators, direct connected, with thermostat and exhaust hood. Includes 10' of type NM cable, installation and connection.

26" diameter roof vent flange, 1,200 CFM, 1/10 HP (for 4,800 to 7,200 CF attics)	BE@3.45	Ea	81.80	92.70	174.50
28" diameter vent, 1,600 CFM, 1/5 HP (for 7,200 to 9,600 CF attics)	BE@3.45	Ea	122.00	92.70	214.70
Add for automatic humidity sensor	—	Ea	15.20	—	15.20

Ceiling fans, decorative, 3-speed reversible motors, lighting kits optional (see below). Includes 10' of type NM cable, installation and connection.

42" four-blade plain fan, oak or white finish	BE@1.50	Ea	102.00	40.30	142.30
42" five-blade, modern styles	BE@1.50	Ea	102.00	40.30	142.30
52" four-blade plain fan	BE@1.50	Ea	112.00	40.30	152.30
52" five-blade plain fan	BE@1.50	Ea	122.00	40.30	162.30
52" five-blade premium fan	BE@1.50	Ea	307.00	40.30	347.30
Optional lighting fixtures for ceiling fans					
Standard globe dome light, 6" H, 10" Dia.	BE@.283	Ea	30.70	7.60	38.30
Schoolhouse or round globe, brass trim	BE@.283	Ea	16.40	7.60	24.00
Add for ball pull chain	BE@.017	Ea	2.04	.46	2.50

Fencing, Chain Link

Galvanized steel, 11.5 gauge 2" x 2" fence fabric, 2-3/8" terminal post with post caps, 1-5/8" line posts at 10' with loop line post caps, 1-3/8" top rail with rail ends, tension bands, tension bars and required bolts with nuts. Material includes 2/3 CF sack of concrete per post. Use 50 LF of fence as a minimum job cost

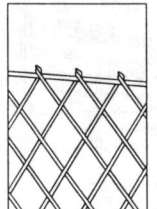

36" high	BL@.125	LF	2.75	2.57	5.32
42" high	BL@.125	LF	2.88	2.57	5.45
48" high	BL@.125	LF	2.94	2.57	5.51
60" high	BL@.125	LF	3.20	2.57	5.77
72" high	BL@.125	LF	3.44	2.57	6.01
Add for redwood filler strips	BL@.024	LF	2.70	.49	3.19
Add for aluminum filler strips	BL@.018	LF	2.99	.37	3.36
Add for plastic tubing filler strips	BL@.018	LF	3.11	.37	3.48

Gates, driveway or walkway. Same construction as fencing shown above, assembled units ready to install.
Add for gate hardware from below.

Driveway gates, costs per liner foot (LF)					
36" high	BL@.025	LF	4.23	.51	4.74
42" high	BL@.025	LF	4.45	.51	4.96
48" high	BL@.025	LF	4.56	.51	5.07
60" high	BL@.033	LF	4.95	.68	5.63
72" high	BL@.033	LF	5.33	.68	6.01
3'0" wide walkway gate, costs per gate					
36" high	BL@.500	Ea	12.70	10.30	23.00
42" high	BL@.500	Ea	13.30	10.30	23.60
48" high	BL@.500	Ea	13.50	10.30	23.80
60" high	BL@.666	Ea	14.80	13.70	28.50
72" high	BL@.666	Ea	16.00	13.70	29.70

	Craft@Hrs	Unit	Material	Labor	Total

Add for gate hardware, per gate. Labor is included with labor shown for gate above.

Driveway gate hardware

	Craft@Hrs	Unit	Material	Labor	Total
36" or 42" high	—	Ea	28.40	—	28.40
48" or 60" high	—	Ea	35.90	—	35.90
72" high	—	Ea	38.00	—	38.00

Walkway gate hardware

	Craft@Hrs	Unit	Material	Labor	Total
36" or 42" high	—	Ea	11.30	—	11.30
48" or 60" high	—	Ea	13.90	—	13.90
72" high	—	Ea	15.00	—	15.00

Pet enclosure fencing. Chain link fence, galvanized steel, 11.5 gauge, 2" x 2" mesh fence fabric, 1-3/8" tubular steel frame, 72" high, lockable gate. Posts set in concrete. Add for concrete slab, if required.

	Craft@Hrs	Unit	Material	Labor	Total
4' x 10'	BL@4.00	Ea	163.00	82.10	245.10
4' x 10', with chain link cover	BL@5.00	Ea	223.00	103.00	326.00
8' x 10'	BL@4.75	Ea	191.00	97.50	288.50
10' x 15'	BL@5.00	Ea	239.00	103.00	342.00

Fencing, Chain Link, Subcontract Work performed by a fence contractor. Costs shown include materials, installation and the subcontractor's overhead and profit. Materials assume galvanized steel, 11.5 gauge 2" x 2" fence fabric, 2-3/8" terminal post with post caps, 1-5/8" line posts at 10' with loop line post caps, 1-3/8" top rail with rail ends, tension bands, tension bars, with posts set in concrete. Installation assumes minimum hand labor, non-rocky soil and post locations accessible to a trailered mechanical post hole digger. Use 50 LF of fence as a minimum job cost.

	Craft@Hrs	Unit	Material	Labor	Total
36" high fence	—	LF	—	—	6.10
42" high fence	—	LF	—	—	6.20
48" high fence	—	LF	—	—	6.30
60" high fence	—	LF	—	—	6.40
72" high fence	—	LF	—	—	6.87
Add for vinyl coated fence and posts	—	%	—	—	10.0
Add for single strand barbed wire at top	—	LF	—	—	.25
Add for redwood filler strips	—	LF	—	—	3.35
Add for aluminum filler strips	—	LF	—	—	3.40
Add for plastic tubing filler strips	—	LF	—	—	3.45

Gates, driveway or walkway, installed in conjunction with fencing on subcontract basis. Same construction as fencing shown above, delivered ready to install. Add for gate hardware from below.

Driveway gates, costs per liner foot (LF)

	Craft@Hrs	Unit	Material	Labor	Total
36" high	—	LF	—	—	4.82
42" high	—	LF	—	—	5.07
48" high	—	LF	—	—	5.18
60" high	—	LF	—	—	5.69
72" high	—	LF	—	—	6.10

3'0" wide walkway gate, costs per gate

	Craft@Hrs	Unit	Material	Labor	Total
36" high	—	Ea	—	—	26.10
42" high	—	Ea	—	—	26.80
48" high	—	Ea	—	—	27.30
60" high	—	Ea	—	—	32.50
72" high	—	Ea	—	—	33.60

Add for gate hardware, per gate, material only. (Installation is included with gate above.)

Driveway gate hardware

	Craft@Hrs	Unit	Material	Labor	Total
36" or 42" high	—	Ea	—	—	32.70
48" or 60" high	—	Ea	—	—	41.20
72" high	—	Ea	—	—	43.90

	Craft@Hrs	Unit	Material	Labor	Total
Walkway gate hardware					
36" or 42" high	—	Ea	—	—	12.80
48" or 60" high	—	Ea	—	—	16.10
72" high	—	Ea	—	—	17.20

Fence, Galvanized Wire Mesh With 1-3/8" galvanized steel posts, set without concrete 10' OC.

	Craft@Hrs	Unit	Material	Labor	Total
12-1/2 gauge steel, 12" vertical wire spacing					
36" high, 6 horizontal wires	BL@.127	LF	.63	2.61	3.24
48" high, 8 horizontal wires	BL@.127	LF	.63	2.61	3.24
48" high, 8 horizontal wires (11 gauge)	BL@.127	LF	.83	2.61	3.44
11 gauge fence fabric, 9" vertical wire spacing					
42" high, 8 horizontal wires	BL@.127	LF	.83	2.61	3.44
48" high, 9 horizontal wires	BL@.127	LF	.90	2.61	3.51
14 gauge fence fabric, 6" vertical wire spacing					
36" high, 8 horizontal wires	BL@.127	LF	.56	2.61	3.17
48" high, 9 horizontal wires	BL@.127	LF	.83	2.61	3.44
60" high, 10 horizontal wires	BL@.127	LF	1.06	2.61	3.67
Add for components for electrified galvanized wire mesh fence					
15 gauge aluminum wire, single strand	BL@.050	LF	.05	1.03	1.08
12-1/2 gauge barbed wire, 4 barbs per LF	BL@.050	LF	1.07	1.03	2.10
Charging unit for electric fence, Solar powered,					
15 mile range, built in 12V battery	BL@.500	Ea	170.00	10.30	180.30
Charging unit, 110V,					
15 mile range, indoors or out	BL@.500	Ea	57.00	10.30	67.30
Charging unit, 110V,					
1 mile range, indoors or out	BL@.500	Ea	23.00	10.30	33.30
Fence stays for barbed wire, 48" high	BL@.250	Ea	.45	5.13	5.58
Add for electric fence warning sign					
Yellow with black lettering	BL@.125	Ea	.44	2.57	3.01
Add for polyethylene insulators	BL@.250	Ea	.11	5.13	5.24
Add for non-conductive gate fastener	BL@.500	Ea	2.24	10.30	12.54
Add for lightning arrestor kit	BL@.500	Ea	5.55	10.30	15.85

Fence Posts Setting only

Posts set with concrete, including temporary 1" x 6" bracing and stakes. Heights to 8'. Cost of post not included. Excavation and concrete assume post hole is 1' x 1' x 3' deep. Per hole dug and post set.

	Craft@Hrs	Unit	Material	Labor	Total
Excavate post hole by hand (light soil)	BL@.250	Ea	—	5.13	5.13
Set and brace post	BL@.125	Ea	3.45	2.57	6.02
Concrete for post, 60 lb. sack	BL@.125	Ea	3.05	2.57	5.62

Fence, Wood Costs shown include concrete and labor to set posts. Gates have 2" x 4" frame and are 3' wide. Add for digging post holes, gate hardware and painting as required. See additional costs at the end of this section.

	Craft@Hrs	Unit	Material	Labor	Total
Basket-weave fence, redwood, "B" grade, 1" x 6" boards,					
2" x 4" stringers or spreaders, 4" x 4" posts					
Tight weave, 4' high, posts @ 4' OC	B1@.090	LF	16.60	2.06	18.66
Tight weave, 6' high, posts @ 4' OC	B1@.095	LF	17.20	2.17	19.37
Wide span, 4' high, posts @ 8' OC	B1@.090	LF	14.00	2.06	16.06
Wide span, 8' high, posts @ 8' OC	B1@.095	LF	22.50	2.17	24.67
4' high gate, tight weave	B1@.846	Ea	73.20	19.30	92.50
4' high gate, wide span	B1@.846	Ea	75.80	19.30	95.10
6' high gate, tight weave	B1@1.05	Ea	61.60	24.00	85.60
8' high gate, wide span	B1@1.20	Ea	99.20	27.40	126.60

	Craft@Hrs	Unit	Material	Labor	Total
Board fence, 1" x 6" boards, 2" x 4" rails, 4" x 4" posts @ 8' OC					
Fir or larch, #4 & better					
4' high, 3 rail	B1@.110	LF	6.52	2.51	9.03
6' high, 3 rail	B1@.119	LF	8.79	2.72	11.51
4' high gate	B1@.846	Ea	28.70	19.30	48.00
6' high gate	B1@1.05	Ea	38.70	24.00	62.70
Red cedar, #3 & better					
4' high, 3 rail	B1@.112	LF	11.30	2.56	13.86
6' high, 3 rail	B1@.119	LF	17.70	2.72	20.42
4' high gate	B1@.846	Ea	49.50	19.30	68.80
6' high gate	B1@1.05	Ea	78.00	24.00	102.00
Redwood, B grade					
4' high, 3 rail	B1@.110	LF	13.20	2.51	15.71
6' high, 3 rail	B1@.119	LF	17.50	2.72	20.22
4' high gate	B1@.846	Ea	57.90	19.30	77.20
6' high gate	B1@1.05	Ea	77.10	24.00	101.10
Board and batten fence, 1" x 6" boards, 1" x 2" battens, 2" x 4" rails, 4" x 4" posts @ 8' OC					
Fir or larch, #4 & better					
4' high, 3 rail	B1@.124	LF	7.40	2.83	10.23
6' high, 3 rail	B1@.130	LF	10.10	2.97	13.07
4' high gate	B1@.846	Ea	32.60	19.30	51.90
6' high gate	B1@1.05	Ea	44.50	24.00	68.50
Red cedar, #3 & better					
4' high, 3 rail	B1@.124	LF	13.00	2.83	15.83
6' high, 3 rail	B1@.130	LF	20.30	2.97	23.27
4' high gate	B1@.846	Ea	57.20	19.30	76.50
6' high gate	B1@1.05	Ea	89.50	24.00	113.50
Redwood, B grade					
4' high, 3 rail	B1@.124	LF	15.60	2.83	18.43
6' high, 3 rail	B1@.130	LF	24.80	2.97	27.77
4' high gate	B1@.846	Ea	68.50	19.30	87.80
6' high gate	B1@1.05	Ea	109.00	24.00	133.00
Split rail fence, red cedar, 4" posts @ 8' OC					
Split 2" x 4" rails, 3' high, 2 rail	B1@.095	LF	3.18	2.17	5.35
Split 2" x 4" rails, 4' high, 3 rail	B1@.102	LF	4.07	2.33	6.40
Rustic 4" round, 3' high, 2 rail	B1@.095	LF	3.86	2.17	6.03
Rustic 4" round, 4' high, 3 rail	B1@.102	LF	4.94	2.33	7.27
Picket fence, 1" x 2" pickets, 2" x 4" rails, 4" x 4" posts @ 8' OC					
Fir or larch, #3 & better					
3' high, 2 rail	B1@.105	LF	2.94	2.40	5.34
5' high, 3 rail	B1@.113	LF	3.73	2.58	6.31
3' high gate	B1@.770	Ea	13.00	17.60	30.60
5' high gate	B1@.887	Ea	16.40	20.30	36.70
Red cedar, #3 & better					
3' high, 2 rail	B1@.105	LF	4.38	2.40	6.78
5' high, 3 rail	B1@.113	LF	5.90	2.58	8.48
3' high gate	B1@.770	Ea	19.30	17.60	36.90
5' high gate	B1@.887	Ea	25.90	20.30	46.20
Redwood, B grade					
3' high, 2 rail	B1@.105	LF	6.91	2.40	9.31
5' high, 3 rail	B1@.105	LF	8.88	2.40	11.28
3' high gate	B1@.770	Ea	30.40	17.60	48.00
5' high gate	B1@.887	Ea	39.10	20.30	59.40

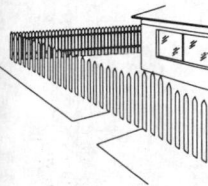

Fencing, Wood

	Craft@Hrs	Unit	Material	Labor	Total
Dig hole for 4" fence posts, by hand (light soil)					
8" diameter, 2' deep, typical	B1@.136	Ea	—	3.11	3.11
Gate hardware					
Latch, standard duty	B1@.281	Ea	4.08	6.42	10.50
Latch, heavy duty	B1@.281	Ea	7.65	6.42	14.07
Hinges, standard, per pair	B1@.374	Pr	6.63	8.55	15.18
Hinges, heavy duty, per pair	B1@.374	Pr	16.30	8.55	24.85
No-sag cable kit	B1@.281	Ea	8.16	6.42	14.58
Self closing spring	B1@.281	Ea	6.89	6.42	13.31

Fiberglass Panels Nailed or screwed onto wood frame.

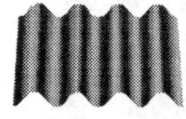

Corrugated, 8', 10', 12' panels, 2-1/2" corrugations, standard colors					
5 oz, 26" or 27-1/2" wide	BC@.012	SF	1.20	.30	1.50
6 oz, 26" or 51" wide	BC@.012	SF	1.40	.30	1.70
8 oz, 26" or 27" wide	BC@.012	SF	1.75	.30	2.05
Flat panels, clear, green or white					
.06" flat sheets, 4' x 8', 10', 12'	BC@.012	SF	1.70	.30	2.00
.03" rolls, 24", 36", 48" x 50'	BC@.012	SF	1.45	.30	1.75
.037" rolls, 24", 36", 48" x 50'	BC@.012	SF	1.65	.30	1.95
Accessories					
Nails, ring shank with rubber washer, 1-3/4" (labor included with panels)					
100 per box (covers 130 SF)	—	SF	.03	—	.03
Self-tapping screws, 1-1/4", rust-proof (labor included with panels)					
100 per box (covers 130 SF)	—	SF	.18	—	.18
Horizontal closure strips, corrugated (labor included with panels)					
Redwood, 2-1/2" x 1-1/2"	—	LF	.30	—	.30
Ploy-foam, 1" x 1"	—	LF	.23	—	.23
Rubber, 1" x 1"	—	LF	.42	—	.42
Vertical crown moulding					
Redwood, 1-1/2" x 1" or poly-foam, 1" x 1"	—	LF	.30	—	.30
Rubber, 1" x 1"	—	LF	.79	—	.79

Fire Sprinkler Systems Typical subcontract prices for systems installed in a single family residence not over two stories high in accord with NFPA Section 13D. Includes connection to domestic water line inside garage (1" pipe size minimum), exposed copper riser with shut-off and drain valve, CPVC concealed distribution piping and residential type sprinkler heads below finished ceilings. Crew is a sprinkler pipe installer and a helper at an average rate of $24.30 per manhour. These costs include hydraulic design calculations and inspection of completed system for compliance with design requirements. Costs shown are per SF of protected area. When estimating the square footage of protected area include non-inhabited areas such as bathrooms, closets and attached garages. For scheduling purposes, estimate that a crew of 2 men can install the rough-in piping for 1,500 to 1,600 SF of protected area in an 8-hour day and about the same amount of finish work in another 8-hour day.

Fabricate, install, and test system					
Single residence	—	SF	—	—	1.65
Tract work	—	SF	—	—	1.40
Condominiums or apartments	—	SF	—	—	2.20

Fireplace Components See also Masonry.

Angle iron (lintel support), lengths 25" to 96"					
3" x 3" x 3/16"	B9@.167	LF	2.50	3.87	6.37
Ash drops, cast iron top, galvanized container					
16-3/4" x 12-1/2" x 8"	B9@.750	Ea	56.50	17.40	73.90

	Craft@Hrs	Unit	Material	Labor	Total
Ash dumps, stamped steel					
4-1/2" x 9"	B9@.500	Ea	6.50	11.60	18.10
Chimney anchors, per pair					
48" x 1-1/2" x 3/16"	—	Pr	8.80	—	8.80
72" x 1-1/2" x 3/16"	—	Pr	11.70	—	11.70
Cleanout doors, 14 gauge stamped steel					
8" x 8"	B9@.500	Ea	11.30	11.60	22.90
8" x 10"	B9@.500	Ea	11.90	11.60	23.50
Combustion air inlet kit for masonry fireplaces	—	Ea	18.30	—	18.30
Dampers for single opening firebox, heavy steel with rockwool insulation					
14" high, 25" to 30" wide, 10" deep, 25 lbs	B9@1.33	Ea	37.80	30.80	68.60
14" high, 36" wide, 10" deep, 31 lbs	B9@1.33	Ea	40.80	30.80	71.60
14" high, 42" wide, 10" deep, 38 lbs	B9@1.33	Ea	49.90	30.80	80.70
14" high, 48" wide, 10" deep, 43 lbs	B9@1.75	Ea	65.10	40.50	105.60
16" high, 54" wide, 13" deep, 66 lbs	B9@2.50	Ea	128.00	57.90	185.90
16" high, 60" wide, 13" deep, 74 lbs	B9@2.50	Ea	132.00	57.90	189.90
16" high, 72" wide, 13" deep, 87 lbs	B9@2.50	Ea	159.00	57.90	216.90
Dampers for multiple opening or corner opening firebox, high capacity, steel downdraft shelf support					
35" wide	B9@1.33	Ea	155.00	30.80	185.80
41" wide	B9@1.33	Ea	176.00	30.80	206.80
47" wide	B9@1.75	Ea	226.00	40.50	266.50
Fuel grates					
21" x 16"	—	Ea	36.90	—	36.90
29" x 16"	—	Ea	47.40	—	47.40
40" x 30"	—	Ea	72.40	—	72.40
Gas valve, log lighter & key	—	LS	50.00	—	50.00

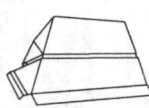

Fireplace Forms Heavy gauge steel heat circulating forms for masonry fireplaces. Based on Superior Fireplace Products. Finished opening dimensions as shown.

	Craft@Hrs	Unit	Material	Labor	Total
Single front opening					
31" x 25", 4,300 CF capacity	B9@3.50	Ea	444.00	81.10	525.10
34" x 25", 5,000 CF capacity	B9@3.50	Ea	472.00	81.10	553.10
37" x 29", 5,750 CF capacity	B9@3.50	Ea	528.00	81.10	609.10
42" x 29", 6,500 CF capacity	B9@3.50	Ea	697.00	81.10	778.10
48" x 31", 7,500 CF capacity	B9@4.00	Ea	881.00	92.70	973.70
60" x 31", 8,500 CF capacity	B9@4.00	Ea	1,339.00	92.70	1,431.70
72" x 31", 9,500 CF capacity	B9@5.50	Ea	1,836.00	127.00	1,963.00
Front and end opening (either right or left)					
38" wide, 27" high	B9@3.50	Ea	1,011.00	81.10	1,092.10
Open on both faces (see-through), including grate, 23" high					
33" wide	B9@3.50	Ea	1,011.00	81.10	1,092.10
43" wide	B9@4.00	Ea	1,145.00	92.70	1,237.70
Add for 2 air inlet grilles, 8" x 11"	—	LS	45.00	—	45.00
Add for 1 air outlet grille, 6" x 32"	—	LS	39.00	—	39.00
Add for 2 fan-forced air inlets					
With grilles and motor	—	LS	112.00	—	112.00

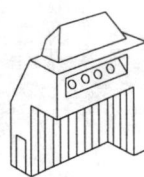

Fireplace, Prefabricated

	Craft@Hrs	Unit	Material	Labor	Total

Fireplaces, Prefabricated Zero clearance factory built fireplaces, including metal fireplace body, refractory interior and fuel grate. UL listed. Add for flues, doors and blower below. Listed by screen size. Based on Superior Fireplace Products.

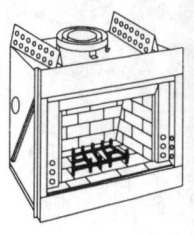

Radiant heating, open front only, with fuel grate

	Craft@Hrs	Unit	Material	Labor	Total
38" wide	B9@5.00	Ea	367.00	116.00	483.00
43" wide	B9@5.00	Ea	445.00	116.00	561.00

Radiant heating, open front and one end,

38" wide	B9@5.00	Ea	556.00	116.00	672.00

Radiant heating, three sides open, with brass and glass doors and refractory floor

36" wide (long side)	B9@6.00	Ea	890.00	139.00	1,029.00

Radiant heating, air circulating, open front only, recessed screen panels

38" screen size	B9@5.50	Ea	501.00	127.00	628.00
43" screen size	B9@5.50	Ea	579.00	127.00	706.00

Radiant heating traditional masonry look

45" wide	B9@5.00	Ea	835.00	116.00	951.00

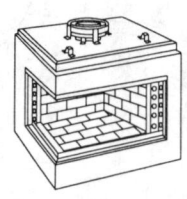

Convection-type heat circulating, open front only

38" screen size	B9@5.50	Ea	417.00	127.00	544.00
43" screen size	B9@5.50	Ea	512.00	127.00	639.00

Forced air heat circulating, open front only (add for blower below)

35" screen size	B9@5.50	Ea	367.00	127.00	494.00
45" screen size	B9@5.50	Ea	556.00	127.00	683.00

Radiant, see-through with brass and glass doors, designer model

38" screen size	B9@6.50	Ea	890.00	151.00	1,041.00

Header plate,

Connects venting system to gas fireplace	B9@.167	Ea	51.00	3.87	54.87

Fireplace doors, complete set including frame and hardware, polished brass

38", for radiant heat units	B9@1.25	Ea	140.00	29.00	169.00
45", for radiant heat units	B9@1.25	Ea	200.00	29.00	229.00
38", for radiant heat units, tall	B9@1.25	Ea	167.00	29.00	196.00
43", for radiant heat units, tall	B9@1.25	Ea	178.00	29.00	207.00
38", for forced air units	B9@1.25	Ea	140.00	29.00	169.00
43", for forced air units	B9@1.25	Ea	151.00	29.00	180.00
For corner open model	B9@1.25	Ea	221.00	29.00	250.00

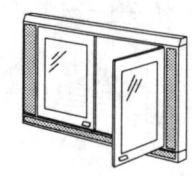

Blower for forced air fireplaces, requires 110 volt electrical receptacle outlet (electrical wiring not included)

Blower unit for connection to existing outlet	BE@.500	Ea	92.00	13.40	105.40

Fireplace flue (chimney), double wall, typical cost for straight vertical installation

8" inside diameter	B9@.167	LF	19.00	3.87	22.87
10" inside diameter	B9@.167	LF	27.00	3.87	30.87

Add for flue spacers, for passing through ceiling.

1" or 2" clearance	B9@.167	Ea	11.00	3.87	14.87
Add for typical flue offset	B9@.250	Ea	70.00	5.79	75.79

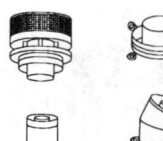 

Add for flashing and storm collar, one required per roof penetration

Flat to 6/12 pitch roofs	B9@3.25	Ea	44.00	75.30	119.30
6/12 to 12/12 pitch roofs	B9@3.25	Ea	64.00	75.30	139.30
Add for spark arrestor top	B9@.500	Ea	90.00	11.60	101.60

Flooring, Resilient Includes 10% waste and adhesive

Asphalt tile, 9" x 9" x 1/8" thick, marbleized

B grade colors, dark	BF@.021	SF	1.35	.52	1.87
C grade colors, medium	BF@.021	SF	1.51	.52	2.03
D grade colors, light	BF@.021	SF	1.51	.52	2.03

	Craft@Hrs	Unit	Material	Labor	Total

Armstrong Solarian sheet vinyl flooring, no wax, Armstrong Bonded with adhesive at perimeter and at seams. Ten year limited warranty. Performance Appearance Rating 4.1

	Craft@Hrs	Unit	Material	Labor	Total
Designer Solarian II	BF@.250	SY	40.80	6.14	46.94
Designer Solarian	BF@.250	SY	36.20	6.14	42.34
Etchings Solarian	BF@.250	SY	36.20	6.14	42.34
Rhythms Solarian	BF@.250	SY	32.94	6.14	39.08
Stencil Craft Solarian	BF@.250	SY	28.50	6.14	34.64
Starstep Solarian	BF@.250	SY	26.00	6.14	32.14
Traditions Solarian	BF@.250	SY	26.00	6.14	32.14

Armstrong sheet vinyl flooring. By Performance Appearance Rating (PAR). Bonded with adhesive at perimeter and at seams.

	Craft@Hrs	Unit	Material	Labor	Total
Fundamentals Successor, 5 year	BF@.250	SY	15.50	6.14	21.64
Caspian, PAR 3.8, urethane wear layer, 10 year	BF@.250	SY	14.40	6.14	20.54
Themes, PAR 2.9, urethane wear layer, 5 year	BF@.250	SY	12.10	6.14	18.24
Metro, PAR 2.3, vinyl wear layer, 3 year	BF@.250	SY	9.80	6.14	15.94
Cambray FHA, PAR 1.7, vinyl wear layer, 2 year	BF@.250	SY	8.50	6.14	14.64
Royelle, PAR 1.3, vinyl wear layer, 1 year	BF@.250	SY	5.90	6.14	12.04

Sheet vinyl, no wax, Congoleum Bonded with adhesive at perimeter and at seams.

	Craft@Hrs	Unit	Material	Labor	Total
Futura	BF@.250	SY	46.60	6.14	52.74
Celestial	BF@.250	SY	40.90	6.14	47.04
Forum Vinyl Plank, 3" or 4-1/2" x 36" x 1/8"	BF@.250	SY	38.50	6.14	44.64
Triumph	BF@.250	SY	34.00	6.14	40.14
Ovations Flex	BF@.250	SY	32.80	6.14	38.94
Ovations	BF@.250	SY	29.50	6.14	35.64
Starlight	BF@.250	SY	27.60	6.14	33.74
Endurance	BF@.250	SY	23.00	6.14	29.14
Exclamation	BF@.250	SY	23.60	6.14	29.74
Highlight Flex	BF@.250	SY	23.60	6.14	29.74
Highlight	BF@.250	SY	17.90	6.14	24.04

Sheet vinyl, no wax, Mannington. Bonded with adhesive at perimeter and at seams.

	Craft@Hrs	Unit	Material	Labor	Total
Gold sheet vinyl	BF@.250	SY	42.90	6.14	49.04
Aristocon or Classicon	BF@.250	SY	37.40	6.14	43.54
Lustrecon or Sterling	BF@.250	SY	29.40	6.14	35.54
Omnia, Resolution or Silverado	BF@.250	SY	23.60	6.14	29.74
Stardance	BF@.250	SY	21.40	6.14	27.54
Vega II	BF@.250	SY	13.90	6.14	20.04

Armstrong vinyl tile, 12" x 12", including 10% waste, self-adhesive

	Craft@Hrs	Unit	Material	Labor	Total
Solarian, .094"	BF@.023	SF	2.18	.57	2.75
Chelsea, .08"	BF@.023	SF	1.97	.57	2.54
Chesapeake, .08"	BF@.023	SF	1.75	.57	2.32
Harbour, .08"	BF@.023	SF	1.42	.57	1.99
Themes, .08", urethane	BF@.023	SF	1.09	.57	1.66
Stylistik, .065"	BF@.023	SF	.96	.57	1.53
Vernay, Metro, .045"	BF@.023	SF	.65	.57	1.22

Vinyl tile, 12" x 12", including 10% waste, self-adhesive

	Craft@Hrs	Unit	Material	Labor	Total
Congoleum Intrigue, .08"	BF@.023	SF	1.53	.57	2.10
Congoleum Majestic, .08"	BF@.023	SF	1.42	.57	1.99
Congoleum Advantage, .07"	BF@.023	SF	.98	.57	1.55
Magic Marble, .10"	BF@.023	SF	1.08	.57	1.65

Composition 12" x 12" tile, including adhesive, Armstrong

	Craft@Hrs	Unit	Material	Labor	Total
Excelon, 1/8"	BF@.023	SF	.68	.57	1.25
Civic Square, 1/8"	BF@.023	SF	.60	.57	1.17

	Craft@Hrs	Unit	Material	Labor	Total
Resilient flooring adhesive, $12 per gallon					
Non-porous subfloor (300 SF per gallon)	—	SF	.04	—	.04
Porous subfloor (150 SF per gallon)	—	SF	.08	—	.08
Vinyl seam sealer (50 LF and $9.78 per pint)	—	SF	.18	—	.18
Resilient flooring accessories					
Patching and leveling compound, 25 pound bag at $8.75, (levels uneven subfloor)					
1 pound per SF at .25" depth	BF@.016	SF	.35	.39	.74
Latex primer (used over dusting, porous or below grade concrete)					
($11 and 300 SF per gallon)	BF@.008	SF	.04	.20	.24
Liquid underlay (for application over existing vinyl floor),					
Gallon at $67 covers 300 SF	BF@.010	SF	.25	.25	.50
Lining felt, 15 Lb. asphalt felt (used over board subfloor), (432 SF roll					
at $24.00 covers 350 SF)	BF@.004	SF	.07	.10	.17
Tempered hardboard underlayment (levels uneven subfloor),					
1/8" x 4' x 8' sheet at $11.80	BF@.016	SF	.39	.39	.78
Cove base, colors,					
2-1/2" high, .08" thick	BF@.025	LF	.38	.61	.99
4" high, .08" thick	BF@.025	LF	.47	.61	1.08
4" high, self-adhesive	BF@.025	LF	1.10	.61	1.71
Cove base, colors,					
2-1/2" high, .08" thick	BF@.025	LF	.38	.61	.99
4" high, .08" thick	BF@.025	LF	.47	.61	1.08
4" high, self-adhesive	BF@.016	LF	1.10	.39	1.49
Cove base cement, at $20.65 per gallon					
200 LF of 4" base per gallon	—	LF	.10	—	.10
Accent strip, 1/8" x 24" long					
1/2" wide	BF@.025	LF	.42	.61	1.03
1" wide	BF@.025	LF	.52	.61	1.13
2" wide	BF@.025	LF	.83	.61	1.44
Vinyl edging strip, 1" wide					
1/16" thick	—	LF	.52	—	.52
1/8" thick	—	LF	.68	—	.68

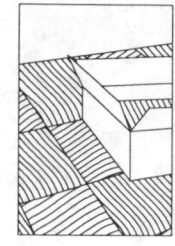

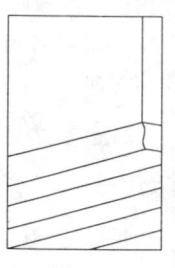

Flooring, Wood Plank and Strip Unfinished oak strip flooring, tongue and groove edge. Includes typical waste. Installed over a prepared subfloor. These are material costs per square foot based on a 1,000 SF job. (See installation, sanding and finishing cost below.)

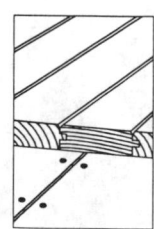

	Size and coverage			
	3/4" x 2-1/4" 138.3 BF/ 100 SF	3/4" x 1-1/2" 155 BF/ 100 SF	1/2" x 2" 130 BF/ 100 SF	1/2" x 1-1/2" 138 BF/ 100 SF
Select plain white or red	3.85	3.16	3.71	3.16
Number 1 common white	3.01	2.78	3.01	2.78
Number 1 common red	3.01	2.78	3.01	2.78
Number 2 common red & white	2.02	1.98	1.93	1.93
Number 1 common & better, short	2.12	2.02	2.10	2.02

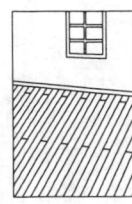

	Craft@Hrs	Unit	Material	Labor	Total
Labor installing, sanding and finishing hardwood strip flooring, nailed, including felt					
Install 1-1/2" width	BF@.066	SF	—	1.62	1.62
Install 2" or 2-1/4" width	BF@.062	SF	—	1.52	1.52
Sanding, 3 passes (60/80/100 grit)	BF@.023	SF	.05	.57	.62
2 coats of stain/sealer	BF@.010	SF	.10	.25	.35
2 coats of urethane	BF@.012	SF	.12	.29	.41
2 coats of lacquer	BF@.024	SF	.16	.59	.75

Unfinished antique tongue and groove plank flooring, 25/32" thick, random widths and lengths. Based on Aged Woods

	Craft@Hrs	Unit	Material	Labor	Total
Ash, select, 3" to 7"	BF@.057	SF	3.96	1.40	5.36
Chestnut, American, 3" to 7"	BF@.057	SF	8.56	1.40	9.96
Chestnut, distressed, 3" to 7"	BF@.057	SF	9.63	1.40	11.03
Chestnut, distressed, over 7"	BF@.057	SF	12.84	1.40	14.24
Cherry, #1, 3" to 7"	BF@.057	SF	8.56	1.40	9.96
Cherry, #1, 6" and up	BF@.057	SF	10.27	1.40	11.67
Hickory, #1, 3" to 7"	BF@.057	SF	3.85	1.40	5.25
Hemlock, 3" to 7"	BF@.057	SF	6.21	1.40	7.61
Hemlock, over 7"	BF@.057	SF	8.03	1.40	9.43
Oak, #1, red or white, 3" to 7"	BF@.057	SF	3.53	1.40	4.93
Oak, #1, red or white, over 7"	BF@.057	SF	5.56	1.40	6.96
Oak, 3" to 7"	BF@.057	SF	7.06	1.40	8.46
Oak, over 7"	BF@.057	SF	8.56	1.40	9.96
Oak, distressed, 3" to 7"	BF@.057	SF	8.03	1.40	9.43
Oak, distressed, over 7"	BF@.057	SF	10.17	1.40	11.57
Maple, FAS, 3" to 7"	BF@.057	SF	4.73	1.40	6.13
Poplar, 3" to 8"	BF@.057	SF	3.42	1.40	4.82
Poplar, over 8"	BF@.057	SF	4.60	1.40	6.00
White pine, 3" to 7"	BF@.057	SF	5.14	1.40	6.54
White pine, distressed, 3" to 7"	BF@.057	SF	5.89	1.40	7.29
White pine, distressed, over 7"	BF@.057	SF	8.03	1.40	9.43
Yellow pine, 3" to 5"	BF@.057	SF	5.99	1.40	7.39
Yellow pine, 6" to 10"	BF@.057	SF	8.13	1.40	9.53
Walnut, #3, 3" to 7"	BF@.057	SF	5.56	1.40	6.96
White pine, knotty grade, 12" and up	BF@.057	SF	8.24	1.40	9.64
Add for sanding and finishing					
Sanding, 3 passes (60/80/100 grit)	BF@.023	SF	.05	.57	.62
2 coats of stain/sealer	BF@.010	SF	.10	.25	.35
2 coats of urethane	BF@.012	SF	.12	.29	.41
2 coats of lacquer	BF@.024	SF	.16	.59	.75

Prefinished oak strip flooring, tongue and groove, 25/32" thick

	Craft@Hrs	Unit	Material	Labor	Total
2-1/4" wide, standard & better, Red and white oak	BF@.036	SF	5.21	.88	6.09
Random widths 3", 5" and 7", walnut plugs, Clear and select grade	BF@.042	SF	6.42	1.03	7.45
Random plank 2-1/4" to 3-1/4" wide, Dark number 1 to C and better	BF@.032	SF	4.44	.79	5.23
Adhesive for oak plank	—	SF	.80	—	.80

Flooring, Wood Plank and Strip

	Craft@Hrs	Unit	Material	Labor	Total
Unfinished northern hard maple strip flooring					
25/32" x 2-1/4" (138.3 BF per 100 SF)					
1st grade	BF@.036	SF	6.29	.88	7.17
2nd grade	BF@.036	SF	5.62	.88	6.50
3rd grade	BF@.036	SF	5.17	.88	6.05
25/32" x 1-1/2" (155 BF per 100 SF)					
1st grade	BF@.040	SF	7.51	.98	8.49
2nd grade	BF@.040	SF	6.73	.98	7.71
3rd grade	BF@.040	SF	6.21	.98	7.19
33/32" x 1-1/2" (170 BF per 100 SF)					
1st grade	BF@.036	SF	6.99	.88	7.87
2nd grade	BF@.036	SF	6.02	.88	6.90
3rd grade	BF@.036	SF	5.69	.88	6.57
Add for sanding and finishing					
Sanding, 3 passes (60/80/100 grit)	BF@.023	SF	.06	.57	.63
2 coats of stain/sealer	BF@.010	SF	.10	.25	.35
2 coats of urethane	BF@.012	SF	.13	.29	.42
2 coats of lacquer	BF@.024	SF	.17	.59	.76

Unfinished edge-wired "Worthwood" strip block flooring by Oregon Lumber Company. End grain unfinished kiln-dried fir, hemlock or teak in flexible strips. Costs include minimal cutting and fitting, installation in mastic but no subfloor preparation.

	Craft@Hrs	Unit	Material	Labor	Total
2-3/4, 3-5/8, or 4-5/8" wide strips 12" to 33" long					
1" thick	BF@.055	SF	4.28	1.35	5.63
1-1/2" thick	BF@.055	SF	5.08	1.35	6.43
2" thick	BF@.055	SF	5.89	1.35	7.24
2-1/2" thick	BF@.055	SF	6.69	1.35	8.04
Add for extensive cutting and fitting	—	%	—	100.0	—
Add for sanding and finishing					
Sanding, 3 passes (60/80/100 grit)	BF@.023	SF	.06	.57	.63
2 coats of stain/sealer	BF@.010	SF	.10	.25	.35
2 coats of urethane	BF@.012	SF	.12	.29	.41
2 coats of lacquer	BF@.024	SF	.18	.59	.77

Unfinished long plank flooring, random 3" to 8" widths, 4' to 16' lengths, 3/4" thick, including 8% waste, Specialty Wood Floors

	Craft@Hrs	Unit	Material	Labor	Total
Select grade, ash	BF@.062	SF	4.17	1.52	5.69
Common grade, cherry	BF@.062	SF	6.53	1.52	8.05
Select grade, cherry	BF@.062	SF	8.15	1.52	9.67
Select grade, rock maple	BF@.062	SF	6.63	1.52	8.15
Select grade, figured maple	BF@.062	SF	10.38	1.52	11.90
Select grade, red oak	BF@.062	SF	5.35	1.52	6.87
Select grade, white oak	BF@.062	SF	5.35	1.52	6.87
Add for sanding and finishing					
Sanding, 3 passes (60/80/100 grit)	BF@.023	SF	.06	.57	.63
2 coats of stain/sealer	BF@.010	SF	.10	.25	.35
2 coats of urethane	BF@.012	SF	.12	.29	.41
2 coats of lacquer	BF@.024	SF	.18	.59	.77

Unfinished wide plank flooring, random 7" to 14" widths, 4' to 16' lengths, 3/4" thick, including 8% waste, Specialty Wood Floors

	Craft@Hrs	Unit	Material	Labor	Total
Knotty grade, pine	BF@.074	SF	3.87	1.82	5.69
Select grade, cherry	BF@.074	SF	8.35	1.82	10.17

	Craft@Hrs	Unit	Material	Labor	Total
Select grade, maple	BF@.074	SF	8.99	1.82	10.81
Select grade, red oak	BF@.074	SF	5.53	1.82	7.35
Select grade, white oak	BF@.074	SF	5.53	1.82	7.35
Adhesive for 3/4" plank	—	SF	.48	—	.48
Add for sanding and finishing					
Sanding, 3 passes (60/80/100 grit)	BF@.023	SF	.06	.57	.63
2 coats of stain/sealer	BF@.010	SF	.12	.25	.37
2 coats of urethane	BF@.012	SF	.12	.29	.41
2 coats of lacquer	BF@.024	SF	.16	.59	.75

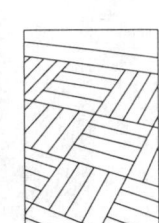

Flooring, Wood Parquet Blocks Includes typical waste and coverage loss, installed over a prepared subfloor.

Unfinished parquet block flooring, 3/4" thick, 10" x 10" to 39" x 39" blocks, includes 5% waste. Cost varies by pattern with larger blocks costing more. These are material costs only. Add labor cost below. Kentucky Wood Floors

		Unit	Material	Labor	Total
Ash or quartered oak, select	—	SF	16.80	—	16.80
Ash or quartered oak, natural	—	SF	11.70	—	11.70
Cherry or walnut select grade	—	SF	18.30	—	18.30
Cherry or walnut natural grade	—	SF	13.90	—	13.90
Oak, select plain	—	SF	13.10	—	13.10
Oak, natural plain	—	SF	11.40	—	11.40

Prefinished parquet block flooring, 5/16" thick, select grade, specified sizes, includes 5% waste. Cost varies by pattern and size, with larger blocks costing more. These are material costs only. Add labor cost below. Kentucky Wood Floors.

		Unit	Material	Labor	Total
Cherry or walnut					
9" x 9"	—	SF	6.00	—	6.00
13" x 13"	—	SF	10.10	—	10.10
Red oak					
9" x 9"	—	SF	4.55	—	4.55
13" x 13"	—	SF	8.50	—	8.50
Teak					
12" x 12"	—	SF	7.95	—	7.95

Labor and material installing parquet block flooring

	Craft@Hrs	Unit	Material	Labor	Total
Parquet laid in mastic	BF@.055	SF	.25	1.35	1.60
Edge-glued over high density foam	BF@.052	SF	.42	1.28	1.70
Custom and irregular-shaped tiles, with adhesive	BF@.074	SF	.43	1.82	2.25
Add for installations under 1,000 SF	BF@.025	SF	.04	.61	.65
Add for sanding and finishing for unfinished parquet block flooring					
Sanding, 3 passes (60/80/100 grit)	BF@.023	SF	.06	.57	.63
2 coats of stain/sealer	BF@.010	SF	.10	.25	.35
2 coats of urethane	BF@.012	SF	.12	.29	.41
2 coats of lacquer	BF@.047	SF	.16	1.16	1.32

Prefinished cross-band 3-ply birch parquet flooring, 6" to 12" wide by 12" to 48" blocks, Oregon Lumber. Costs include cutting, fitting and 5% waste, but no subfloor preparation.

	Craft@Hrs	Unit	Material	Labor	Total
"Saima", 9/16" thick, typical price	BF@.050	SF	3.86	1.23	5.09

Bonded blocks, 12" x 12", 1/8" thick, laid in mastic

	Craft@Hrs	Unit	Material	Labor	Total
Vinyl bonded cork	BF@.050	SF	3.22	1.23	4.45
Vinyl bonded domestic woods	BF@.062	SF	6.44	1.52	7.96
Vinyl bonded exotic woods	BF@.062	SF	12.90	1.52	14.42

Flooring, Wood Parquet

	Craft@Hrs	Unit	Material	Labor	Total
Acrylic (Permagrain) impregnated wood flooring. Material costs include mastic					
12" x 12" x 5/16" thick block laid in mastic					
Oak block	BF@.057	SF	8.45	1.40	9.85
Ash block	BF@.057	SF	9.00	1.40	10.40
Cherry block	BF@.057	SF	9.32	1.40	10.72
Maple block	BF@.057	SF	9.43	1.40	10.83
2-7/8" x RL plank laid in mastic					
Oak plank	BF@.057	SF	6.85	1.40	8.25
Maple plank	BF@.057	SF	9.00	1.40	10.40
Tupelo plank	BF@.057	SF	8.40	1.40	9.80
Lindenwood plank	BF@.057	SF	9.35	1.40	10.75
Cherry plank	BF@.057	SF	9.00	1.40	10.40

Flooring, Wood, Softwood Material costs include typical waste and coverage loss. Add for sanding and finishing from below.

	Craft@Hrs	Unit	Material	Labor	Total
Unfinished Douglas fir, "C" and better, dry, vertical grain, including 5% waste					
1" x 3"	BF@.037	SF	4.16	.91	5.07
1" x 4"	BF@.037	SF	4.10	.91	5.01
1/4" x 4"	BF@.037	SF	3.12	.91	4.03
Deduct for "D" grade	—	SF	-.70	—	-.70
Add for "B" grade and better	—	SF	.70	—	.70
Add for specific lengths	—	SF	.15	—	.15
Add for diagonal patterns	BF@.004	SF	—	.10	.10

Unfinished select kiln dried heart pine, 97% dense heartwood, random lengths, including 5% waste, 1" to 1-1/4" thick

	Craft@Hrs	Unit	Material	Labor	Total
4" wide	BF@.042	SF	6.35	1.03	7.38
5" wide	BF@.042	SF	6.56	1.03	7.59
6" wide	BF@.042	SF	6.61	1.03	7.64
8" wide	BF@.046	SF	7.11	1.13	8.24
10" wide	BF@.046	SF	8.26	1.13	9.39
Add for diagonal patterns	BF@.004	SF	—	.10	.10

	Craft@Hrs	Unit	Material	Labor	Total
Unfinished plank flooring, 3/4" thick, 3" to 8" widths, 2' to 12' lengths, Aged Woods					
Antique white pine	BF@.100	SF	7.00	2.46	9.46
Antique yellow pine	BF@.100	SF	6.46	2.46	8.92
Milled heart pine	BF@.100	SF	7.65	2.46	10.11
Distressed oak	BF@.100	SF	8.28	2.46	10.74
Milled oak	BF@.100	SF	7.51	2.46	9.97
Distressed American chestnut	BF@.100	SF	8.00	2.46	10.46
Milled American chestnut	BF@.100	SF	8.28	2.46	10.74
Aged hemlock/fir	BF@.100	SF	7.18	2.46	9.64
Antique poplar	BF@.100	SF	3.53	2.46	5.99
Aged cypress	BF@.100	SF	4.19	2.46	6.65
Add for 9" to 15" widths	—	%	15.0	—	15.0

	Craft@Hrs	Unit	Material	Labor	Total
Add for sanding and finishing.					
Sanding, 3 passes (60/80/100 grit)	BF@.023	SF	.06	.57	.63
2 coats of stain/sealer	BF@.010	SF	.10	.25	.35
2 coats of urethane	BF@.012	SF	.12	.29	.41
2 coats of lacquer	BF@.024	SF	.16	.59	.75

	Craft@Hrs	Unit	Material	Labor	Total
Flooring, Subcontract					
Hardwood flooring. See sanding and finishing costs below					
Strip flooring, T&G, unfinished, 2-1/4" wide, 25/32" thick oak					
Clear quartered	—	SF	—	—	9.47
Select quartered	—	SF	—	—	9.58
No. 1 common	—	SF	—	—	8.88
No. 2 common	—	SF	—	—	8.24
Deduct for 1/2" thick, typical cost	—	SF	—	—	-1.28
Deduct for tract work, typical cost	—	SF	—	—	-.64
Additions to hardwood flooring					
For sanding and finishing add					
Light finishes	—	SF	—	—	2.35
Dark finishes	—	SF	—	—	3.00
Custom fancy, bleached	—	SF	—	—	3.96
Parquet blocks, set in mastic, prefinished					
Oak, 3/16", select	—	SF	—	—	9.84
Oak, 5/16", select	—	SF	—	—	11.02
Oak, 3/8", select	—	SF	—	—	11.77
Oak, 25/32", select	—	SF	—	—	12.73
Maple, 5/16", select	—	SF	—	—	11.24
Teak, 5/16", typical	—	SF	—	—	21.40
Walnut, 5/16"	—	SF	—	—	14.34
Ranch plank flooring, finished including wooden pegs					
Red oak, 3", 5," 7" widths	—	SF	—	—	13.48
Resilient flooring					
Asphalt tile, 1/8", laid with felt underlayment, standard or wood designs					
B grade (dark colors)	—	SF	—	—	2.08
C grade (medium colors)	—	SF	—	—	2.29
D grade (light colors)	—	SF	—	—	2.29
Cork tile, standard					
3/16" thick	—	SF	—	—	5.20
5/16" thick	—	SF	—	—	6.24
Rubber tile,					
1/8" thick	—	SF	—	—	5.50
1/4" thick Dodge-Regupol All Purpose Rubber	—	SF	—	—	5.20
Vinyl composition tile					
1/16" standard	—	SF	—	—	1.56
3/32" standard	—	SF	—	—	1.66
1/8" standard	—	SF	—	—	1.87
1/16" metallic accent	—	SF	—	—	1.51
Vinyl tile, solid vinyl, Azrock					
.100" Marble look	—	SF	—	—	8.32
.100" Wood look	—	SF	—	—	7.80
Vinyl sheet flooring, Armstrong					
.085", "Classic Corlon"	—	SF	—	—	2.60
.085", "Possibilities"	—	SF	—	—	3.33
.080", "Translations"	—	SF	—	—	2.91
.080", "Medintech Tandem"	—	SF	—	—	2.86
Base for resilient flooring					
Rubber, 6"	—	LF	—	—	1.95
Rubber, 4"	—	LF	—	—	1.70
Rubber, 2-1/2"	—	LF	—	—	1.60

	Craft@Hrs	Unit	Material	Labor	Total
Stair treads, 12" width, Flexco					
Rubber, molded, 3/4"	—	LF	—	—	11.00
Rubber, molded, 1/8"	—	LF	—	—	8.70
Rubber, grit safety tread, 1/4"	—	LF	—	—	12.80
Rubber, grit safety tread, 1/8"	—	LF	—	—	10.50
Vinyl, molded, 1/8"	—	LF	—	—	5.70
Vinyl, molded, 1/4"	—	LF	—	—	6.95
Stair risers, 7" high, 1/8" thick					
Rubber riser	—	LF	—	—	3.70
Vinyl riser	—	LF	—	—	2.90
Slate, random rectangles					
Typical price	—	SF	—	—	19.80
Particleboard underlayment					
3/8" thick	—	SF	—	—	.70

Foundation Bolts With nuts attached. Costs per bolt based on full bag quantities.

	Craft@Hrs	Unit	Material	Labor	Total
1/2" x 6", .34 lb, bag of 300	B1@.107	Ea	.27	2.45	2.72
1/2" x 8", .44 lb, bag of 250	B1@.107	Ea	.29	2.45	2.74
1/2" x 10", .51 lb, bag of 200	B1@.107	Ea	.30	2.45	2.75
5/8" x 6", .62 lb, bag of 50	B1@.107	Ea	.53	2.45	2.98
5/8" x 8", .78 lb, bag of 50	B1@.107	Ea	.55	2.45	3.00
5/8" x 10", .88 lb, bag of 50	B1@.107	Ea	.53	2.45	2.98
5/8" x 12", 1.1 lb, bag of 50	B1@.121	Ea	.66	2.76	3.42
3/4" x 8", 1.2 lb, bag of 50	B1@.107	Ea	.85	2.45	3.30
3/4" x 10", 1.4 lb, bag of 50	B1@.107	Ea	.86	2.45	3.31
3/4" x 12", 1.7 lb, bag of 50	B1@.121	Ea	.99	2.76	3.75
Add for wrought, cut washers, plain					
1/2" washer	—	Ea	.13	—	.13
5/8" washer	—	Ea	.14	—	.14
3/4" washer	—	Ea	.15	—	.15

Framing Connectors All bolted connectors include nuts, bolts, and washers as needed. Equivalent Simpson Strong-Tie Co. model numbers are shown in parentheses.

	Craft@Hrs	Unit	Material	Labor	Total
Angle clips (A34 and A35), 18 gauge, galvanized					
A35 or A35F connector	BC@.030	Ea	.20	.76	.96
A34 connector	BC@.030	Ea	.14	.76	.90
Angle clips (L) for general utility use, 16 gauge, galvanized, no nails included					
3" (6 nails)	BC@.027	Ea	.28	.68	.96
5" (6 nails)	BC@.027	Ea	.53	.68	1.21
7" (8 nails)	BC@.027	Ea	.67	.68	1.35
9" (8 nails)	BC@.027	Ea	.84	.68	1.52
Brick wall ties (BT), 28 gauge, per 1,000					
3/4" x 6", corrugated	—	M	70.00	—	70.00

Bridging. See also Carpentry.

Cross bridging (NC), steel, per pair, no nails required.

	Craft@Hrs	Unit	Material	Labor	Total
12" joist spacing on center					
2" x 8" to 2" x 16" joists	BC@.022	Pr	.41	.55	.96
16" joist spacing on center					
2" x 8" to 2" x 10" joists	BC@.022	Pr	.42	.55	.97
2" x 12" to 2" x 16" joists	BC@.022	Pr	.43	.55	.98

	Craft@Hrs	Unit	Material	Labor	Total
24" joist spacing on center					
2" x 8" to 2" x 10" joists	BC@.022	Pr	.70	.55	1.25
2" x 12" to 2" x 14" joists	BC@.022	Pr	.71	.55	1.26
2" x 16" joists	BC@.022	Pr	.72	.55	1.27

Column bases, heavy duty (CB). Includes nuts and bolts. Labor shown is for placing column base before pouring concrete.

	Craft@Hrs	Unit	Material	Labor	Total
4" x 4", 4 lbs	BC@.121	Ea	6.93	3.05	9.98
4" x 6", 5 lbs	BC@.121	Ea	7.43	3.05	10.48
4" x 8", 6 lbs	BC@.121	Ea	7.87	3.05	10.92
6" x 6", 7 lbs	BC@.121	Ea	8.75	3.05	11.80
6" x 8", 8 lbs	BC@.121	Ea	9.74	3.05	12.79
6" x 10", 9 lbs	BC@.131	Ea	10.00	3.30	13.30
6" x 12", 10 lbs	BC@.131	Ea	10.80	3.30	14.10
8" x 8", 13 lbs	BC@.131	Ea	19.40	3.30	22.70
8" x 10", 14 lbs	BC@.131	Ea	21.90	3.30	25.20
8" x 12", 15 lbs	BC@.150	Ea	23.80	3.78	27.58
10" x 10", 16 lbs	BC@.150	Ea	25.40	3.78	29.18
10" x 12", 17 lbs	BC@.150	Ea	27.90	3.78	31.68
12" x 12", 20 lbs	BC@.150	Ea	35.20	3.78	38.98

Column caps or end column caps, heavy duty (CC) or (ECC). Includes nuts and bolts.

Column caps for solid lumber.

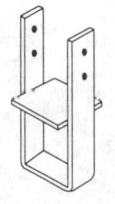

Beam size x post size	Craft@Hrs	Unit	Material	Labor	Total
4" x 4", 4 lbs	BC@.260	Ea	16.50	6.54	23.04
4" x 6", 5 lbs	BC@.300	Ea	22.00	7.55	29.55
6" x 4", 5 lbs	BC@.300	Ea	34.10	7.55	41.65
6" x 6", 7 lbs	BC@.400	Ea	34.10	10.10	44.20
6" x 8", 8 lbs	BC@.600	Ea	34.10	15.10	49.20
8" x 6", 14 lbs	BC@.600	Ea	39.90	15.10	55.00
8" x 8", 15 lbs	BC@.750	Ea	39.90	18.90	58.80
10" x 6", 20 lbs	BC@.750	Ea	53.40	18.90	72.30

Column caps for glu-lam beams

Glu-lam beam size x post size	Craft@Hrs	Unit	Material	Labor	Total
3-1/4" x 4", 4 lbs	BC@.260	Ea	30.90	6.54	37.44
3-1/4" x 6", 5 lbs	BC@.300	Ea	30.90	7.55	38.45
5-1/4" x 6", 4 lbs	BC@.400	Ea	31.50	10.10	41.60
5-1/4" x 8", 5 lbs	BC@.600	Ea	31.50	15.10	46.60
7" x 6", 9 lbs	BC@.600	Ea	35.90	15.10	51.00
7" x 7", 10 lbs	BC@.750	Ea	35.90	18.90	54.80
7" x 8", 13 lbs	BC@.750	Ea	35.90	18.90	54.80
9" x 6", 16 lbs	BC@.750	Ea	42.50	18.90	61.40
9" x 8", 17 lbs	BC@.750	Ea	42.50	18.90	61.40

Floor tie anchors (FTA)

	Craft@Hrs	Unit	Material	Labor	Total
FTA2	BC@.200	Ea	13.60	5.03	18.63
FTA5	BC@.200	Ea	16.00	5.03	21.03
FTA7	BC@.320	Ea	27.60	8.05	35.65

Header hangers (HH), 16 gauge, galvanized

	Craft@Hrs	Unit	Material	Labor	Total
4" header	BC@.046	Ea	2.49	1.16	3.65
6" header	BC@.046	Ea	4.60	1.16	5.76

Hold downs (HD), including typical nuts, bolts and washers for studs. Add for foundation bolts.

	Craft@Hrs	Unit	Material	Labor	Total
HD2A	BC@.310	Ea	3.28	7.80	11.08
HD5A	BC@.310	Ea	9.40	7.80	17.20
HD8A	BC@.420	Ea	9.85	10.60	20.45

Framing Connectors

	Craft@Hrs	Unit	Material	Labor	Total
HD10A	BC@.500	Ea	12.20	12.60	24.80
HD20A	BC@.500	Ea	23.80	12.60	36.40
Threaded rod (all thread) 24" long. Labor includes drilling through 6" of wood					
1/2" rod	BC@.185	Ea	2.19	4.66	6.85
5/8" rod	BC@.185	Ea	3.62	4.66	8.28
3/4" rod	BC@.200	Ea	4.94	5.03	9.97
1" rod	BC@.220	Ea	8.56	5.54	14.10
1-1/8" rod	BC@.235	Ea	12.10	5.91	18.01
1-1/4" rod	BC@.235	Ea	14.20	5.91	20.11
Hurricane and seismic ties (H)					
H1 tie	BC@.035	Ea	.38	.88	1.26
H2 tie	BC@.027	Ea	.51	.68	1.19
H2.5, H3 or H4	BC@.027	Ea	.17	.68	.85
H5 tie	BC@.027	Ea	.17	.68	.85
H6 tie	BC@.045	Ea	2.00	1.13	3.13
H7 tie	BC@.055	Ea	2.28	1.38	3.66
Mud sill anchors (MA)					
2" x 4"	BC@.046	Ea	.70	1.16	1.86
2" x 6"	BC@.046	Ea	.73	1.16	1.89
Panelized plywood roof hangers (F24N, F26N), for panelized plywood roof construction					
2" x 4"	—	Ea	.27	—	.27
2" x 6"	—	Ea	.50	—	.50
Purlin hangers (HHB or HW), without nails, 7 gauge					
4" x 6"	BC@.052	Ea	11.00	1.31	12.31
4" x 8"	BC@.052	Ea	11.80	1.31	13.11
4" x 10"	BC@.070	Ea	12.60	1.76	14.36
4" x 12"	BC@.070	Ea	13.50	1.76	15.26
4" x 14"	BC@.087	Ea	14.30	2.19	16.49
4" x 16"	BC@.087	Ea	15.10	2.19	17.29
6" x 6"	BC@.052	Ea	11.10	1.31	12.41
6" x 8"	BC@.070	Ea	11.90	1.76	13.66
6" x 10"	BC@.087	Ea	12.80	2.19	14.99
6" x 12"	BC@.087	Ea	13.60	2.19	15.79
6" x 14"	BC@.117	Ea	14.40	2.94	17.34
6" x 16"	BC@.117	Ea	15.30	2.94	18.24
8" x 10"	BC@.087	Ea	13.70	2.19	15.89
8" x 12"	BC@.117	Ea	14.50	2.94	17.44
8" x 14"	BC@.117	Ea	15.40	2.94	18.34
8" x 16"	BC@.117	Ea	16.20	2.94	19.14
Top flange hangers (JB or LB)					
2" x 6"	BC@.046	Ea	.71	1.16	1.87
2" x 8"	BC@.046	Ea	.89	1.16	2.05
2" x 10"	BC@.046	Ea	1.46	1.16	2.62
2" x 12"	BC@.046	Ea	1.60	1.16	2.76
2" x 14"	BC@.052	Ea	1.64	1.31	2.95
Top flange hangers, heavy (HW)					
4" x 6"	BC@.052	Ea	11.00	1.31	12.31
4" x 8"	BC@.052	Ea	11.80	1.31	13.11
4" x 10"	BC@.070	Ea	12.60	1.76	14.36
4" x 12"	BC@.070	Ea	13.50	1.76	15.26
4" x 14"	BC@.087	Ea	14.30	2.19	16.49
4" x 16"	BC@.087	Ea	15.10	2.19	17.29

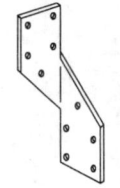

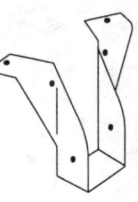

	Craft@Hrs	Unit	Material	Labor	Total
6" x 6"	BC@.052	Ea	11.10	1.31	12.41
6" x 8"	BC@.070	Ea	11.90	1.76	13.66
6" x 10"	BC@.087	Ea	12.80	2.19	14.99
6" x 12"	BC@.087	Ea	13.60	2.19	15.79
6" x 14"	BC@.117	Ea	14.40	2.94	17.34
6" x 16"	BC@.117	Ea	15.30	2.94	18.24
8" x 6"	BC@.070	Ea	12.00	1.76	13.76
8" x 8"	BC@.070	Ea	12.90	1.76	14.66
8" x 10"	BC@.087	Ea	13.70	2.19	15.89
8" x 12"	BC@.117	Ea	14.50	2.94	17.44
8" x 14"	BC@.117	Ea	15.40	2.94	18.34
8" x 16"	BC@.117	Ea	16.20	2.94	19.14

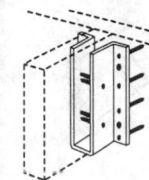

U type or formed seat type hangers (U). Standard duty, 16 gauge, galvanized

	Craft@Hrs	Unit	Material	Labor	Total
2" x 4"	BC@.040	Ea	.50	1.01	1.51
2" x 6" or 2" x 8"	BC@.046	Ea	.51	1.16	1.67
2" x 10" or 2" x 12"	BC@.046	Ea	.76	1.16	1.92
2" x 14"	BC@.052	Ea	1.46	1.31	2.77
3" x 4"	BC@.046	Ea	.83	1.16	1.99
3" x 6" or 3" x 8"	BC@.046	Ea	.89	1.16	2.05
3" x 10" or 3" x 12"	BC@.052	Ea	1.64	1.31	2.95
3" x 14" or 3" x 16"	BC@.070	Ea	1.77	1.76	3.53
4" x 4"	BC@.046	Ea	.86	1.16	2.02
4" x 6" or 4" x 8"	BC@.052	Ea	.92	1.31	2.23
4" x 10" or 4" x 12"	BC@.070	Ea	1.72	1.76	3.48
4" x 14" or 4" x 16"	BC@.087	Ea	2.43	2.19	4.62
6" x 6" or 6" x 8"	BC@.052	Ea	3.03	1.31	4.34
6" x 10" or 6" x 12"	BC@.087	Ea	3.32	2.19	5.51

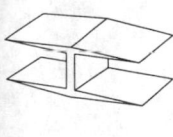

Plywood clips, extruded aluminum, purchased in cartons of 500.

	Craft@Hrs	Unit	Material	Labor	Total
3/8", 7/16", 1/2", or 5/8"	BC@.006	Ea	.07	.15	.22

Post base for decks (base with standoff, PBS), 12 gauge, galvanized

	Craft@Hrs	Unit	Material	Labor	Total
4" x 4"	BC@.046	Ea	3.31	1.16	4.47
4" x 6"	BC@.046	Ea	4.48	1.16	5.64
6" x 6"	BC@.070	Ea	6.63	1.76	8.39
Add for rough lumber sizes	—		10%	—	—

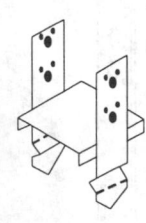

Post base (AB), adjustable, 16 gauge, galvanized

	Craft@Hrs	Unit	Material	Labor	Total
4" x 4"	BC@.046	Ea	3.54	1.16	4.70
4" x 6"	BC@.046	Ea	7.59	1.16	8.75
6" x 6"	BC@.070	Ea	8.79	1.76	10.55
Add for rough lumber sizes	—		10%	—	—

Post bases (PB), 12 gauge, galvanized

	Craft@Hrs	Unit	Material	Labor	Total
4" x 4"	BC@.046	Ea	3.03	1.16	4.19
4" x 6"	BC@.046	Ea	3.37	1.16	4.53
6" x 6"	BC@.070	Ea	6.10	1.76	7.86

Post cap and base combination (BC), 18 gauge, galvanized

	Craft@Hrs	Unit	Material	Labor	Total
4" x 4"	BC@.037	Ea	1.93	.93	2.86
4" x 6"	BC@.037	Ea	3.69	.93	4.62
6" x 6"	BC@.054	Ea	3.92	1.36	5.28
8" x 8"	BC@.054	Ea	14.00	1.36	15.36

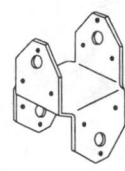

Post caps and end post caps (PC and EPC), 12 gauge, galvanized. Beam size x post size

	Craft@Hrs	Unit	Material	Labor	Total
4" x 4"	BC@.052	Ea	9.10	1.31	10.41
4" x 6"	BC@.052	Ea	12.00	1.31	13.31
4" x 8"	BC@.052	Ea	16.80	1.31	18.11

Framing Connectors

	Craft@Hrs	Unit	Material	Labor	Total
6" x 4"	BC@.052	Ea	11.70	1.31	13.01
6" x 6"	BC@.077	Ea	16.40	1.94	18.34
6" x 8"	BC@.077	Ea	14.00	1.94	15.94
8" x 8"	BC@.077	Ea	23.60	1.94	25.54
Post top tie or anchor (AC), 18 gauge, galvanized					
4" x 4"	BC@.040	Pr	1.27	1.01	2.28
6" x 6"	BC@.040	Pr	1.40	1.01	2.41
Purlin anchors (PA), 12 gauge, 2" wide					
18" anchor	BC@.165	Ea	2.81	4.15	6.96
23" anchor	BC@.165	Ea	3.30	4.15	7.45
28" anchor	BC@.185	Ea	3.37	4.66	8.03
35" anchor	BC@.185	Ea	4.63	4.66	9.29
Split ring connectors					
2-5/8" ring	BC@.047	Ea	1.39	1.18	2.57
4" ring	BC@.054	Ea	2.16	1.36	3.52
Straps					
"L" straps, 12 gauge					
5" x 5", 1" wide	BC@.132	Ea	.51	3.32	3.83
6" x 6", 1-1/2" wide	BC@.132	Ea	1.19	3.32	4.51
8" x 8", 2" wide	BC@.132	Ea	1.78	3.32	5.10
12" x 12", 2" wide	BC@.165	Ea	2.41	4.15	6.56
Add for heavy duty straps, 7 gauge	—	Ea	1.80	—	1.80
Long straps (MST), 12 gauge					
27"	BC@.165	Ea	2.18	4.15	6.33
36"	BC@.165	Ea	3.12	4.15	7.27
48"	BC@.265	Ea	4.30	6.67	10.97
60"	BC@.265	Ea	6.66	6.67	13.33
72"	BC@.285	Ea	8.82	7.17	15.99
Plate and stud straps (FHA), galvanized, 1-1/2" wide, 12 gauge					
6"	BC@.100	Ea	.68	2.52	3.20
9"	BC@.132	Ea	.80	3.32	4.12
12"	BC@.132	Ea	.83	3.32	4.15
18"	BC@.132	Ea	1.11	3.32	4.43
24"	BC@.165	Ea	1.55	4.15	5.70
30"	BC@.165	Ea	2.23	4.15	6.38
Tie straps (ST292, ST2122, ST2215, and ST6200 series)					
2-1/16" x 9-5/16"	BC@.132	Ea	.51	3.32	3.83
3/4" x 16-5/16"	BC@.046	Ea	.47	1.16	1.63
2-1/16" x 16-5/16"	BC@.132	Ea	.76	3.32	4.08
2-1/16" x 23-5/16"	BC@.165	Ea	1.24	4.15	5.39
2-1/16" x 33-13/16"	BC@.165	Ea	1.62	4.15	5.77
"T" straps, 14 gauge					
6" x 6", 1-1/2" wide	BC@.132	Ea	1.19	3.32	4.51
12" x 8", 2" wide	BC@.132	Ea	2.77	3.32	6.09
12" x 12", 2" wide	BC@.132	Ea	4.14	3.32	7.46
Add for heavy-duty straps, 7 gauge	—	Ea	2.37	—	2.37
Twist straps (TS), 16 gauge, galvanized, 1-1/4" wide					
9-5/8"	BC@.132	Ea	.35	3.32	3.67
13-1/8"	BC@.132	Ea	.48	3.32	3.80
16-5/8"	BC@.132	Ea	.69	3.32	4.01
23-5/8"	BC@.165	Ea	.78	4.15	4.93

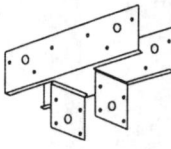

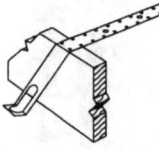

	Craft@Hrs	Unit	Material	Labor	Total
Wall braces (WB), diagonal, 16 gauge, 1-1/4" wide					
10' long, corrugated or flat	BC@.101	Ea	3.37	2.54	5.91
12' long, flat	BC@.101	Ea	3.97	2.54	6.51
14' long, flat	BC@.101	Ea	6.81	2.54	9.35
Wall braces (CWB), diagonal, 18 gauge, 1" wide					
10' long, L shaped	BC@.125	Ea	3.49	3.15	6.64
12' long, L shaped	BC@.125	Ea	4.12	3.15	7.27

Garage Door Openers Radio controlled, electric, single or double door, includes typical electric hookup to existing adjacent 110 volt outlet. Screw worm drive, 1/3 HP opener, safety stop device, receiver and one transmitter

	Craft@Hrs	Unit	Material	Labor	Total
Multiple frequency, lighted	BC@3.90	Ea	175.00	98.20	273.20

Deluxe model, includes 1/2 HP opener, receiver, transmitter, automatic light, built-in time delay, safety top device

	Craft@Hrs	Unit	Material	Labor	Total
Chain drive, not for vault-type garages	BC@4.04	Ea	260.00	102.00	362.00
Add for additional transmitter	—	Ea	30.00	—	30.00
Add for additional key switch	—	Ea	20.00	—	20.00

Garage Doors Unfinished, with hardware, Add labor from below.

	Type	Single		Double
	Size	8'x 7'	9' x 7'	16' x 7'
Aluminum framed fiberglass, sectional				
White, with key lock	Ea	440.00	480.00	800.00
Steel garage doors				
One piece, primed	Ea	200.00	210.00	—
Sectional, continuous hinges	Ea	325.00	350.00	600.00
Sectional, with glass panels	Ea	375.00	400.00	700.00
Wood garage doors, sectional, without insulation.				
Sectional door	Ea	275.00	297.00	479.00
Door with two small lites	Ea	292.00	330.00	550.00
Door with plain panels	Ea	303.00	325.00	605.00
Door with raised redwood panels	Ea	550.00	704.00	1,210.00
Door with detailed face	Ea	550.00	660.00	990.00
Honeycomb core wood door	Ea	297.00	308.00	—
Styrofoam core hardboard door	Ea	297.00	314.00	523.00
Styrofoam core textured wood door	Ea	303.00	319.00	528.00
Wood doors with from 4 to 12 better-quality glass panel inserts.				
Cathedral style	Ea	385.00	380.00	820.00
Sunburst style	Ea	445.00	375.00	860.00
Add for low headroom hardware	Ea	29.00	30.00	40.00

	Craft@Hrs	Unit	Material	Labor	Total
Labor installing garage doors					
8' or 9' wide doors	B1@4.42	Ea	—	101.00	101.00
16' wide doors	B1@5.90	Ea	—	135.00	135.00

Garage Doors, Subcontract Complete installed costs including hardware.

Aluminum (.019 gauge) on wood frame, jamb hinge

		Unit	Material	Labor	Total
8' x 7'	—	Ea	—	—	438.00
16' x 7'	—	Ea	—	—	609.00

Garage Doors, Subcontract

	Craft@Hrs	Unit	Material	Labor	Total
Plywood (3/8") on wood frame, flush face, jamb hinge					
8' x 7'	—	Ea	—	—	318.00
16' x 7'	—	Ea	—	—	439.00
Add for moulding on face	—	LF	—	—	2.00
Board on wood frame, jamb hinge					
Plain face, 8' x 7'	—	Ea	—	—	489.00
Plain face, 16' x 7'	—	Ea	—	—	881.00
Pattern face, 8' x 7'	—	Ea	—	—	500.00
Pattern face, 16' x 7'	—	Ea	—	—	905.00
Sectional upward acting garage doors					
Aluminum, 8' x 7'	—	Ea	—	—	614.00
Aluminum, 16' x 7'	—	Ea	—	—	1,183.00
Steel, 8' x 7'	—	Ea	—	—	557.00
Steel, 16' x 7'	—	Ea	—	—	910.00
Wood, 8' x 7'	—	Ea	—	—	501.00
Wood, 16' x 7'	—	Ea	—	—	1,265.00
Add for garage door operators, radio controlled					
Single door, 1 transmitter	—	LS	—	—	434.00
Double door, 2 transmitters	—	LS	—	—	471.00
Glass Material only. See labor costs below					
SSB glass (single strength, "B" grade)					
60" width plus length	—	SF	2.18	—	2.18
70" width plus length	—	SF	2.37	—	2.37
80" or more width plus length	—	SF	2.67	—	2.67
DSB glass (double strength, "B" grade)					
60" width plus length	—	SF	3.02	—	3.02
70" width plus length	—	SF	3.07	—	3.07
80" width plus length	—	SF	3.12	—	3.12
100" width plus length	—	SF	3.33	—	3.33
Non-glare glass	—	SF	5.15	—	5.15
Float glass, 1/4" thick					
To 25 SF	—	SF	3.38	—	3.38
25 to 35 SF	—	SF	3.64	—	3.64
35 to 50 SF	—	SF	3.95	—	3.95
Insulating glass, clear					
1/2" thick, 1/4" airspace, up to 25 SF	—	SF	7.28	—	7.28
5/8" thick, 1/4" airspace, up to 35 SF	—	SF	8.53	—	8.53
1" thick, 1/2" airspace, up to 50 SF	—	SF	10.09	—	10.09
Tempered glass, clear, 1/4" thick					
To 25 SF	—	SF	7.28	—	7.28
25 to 50 SF	—	SF	8.74	—	8.74
50 to 75 SF	—	SF	9.67	—	9.67
Tempered glass, clear, 3/16" thick					
To 25 SF	—	SF	7.18	—	7.18
25 to 50 SF	—	SF	8.22	—	8.22
50 to 75 SF	—	SF	9.10	—	9.10
Laminated safety glass, 1/4" thick, ASI safety rating					
Up to 25 SF	—	SF	9.36	—	9.36
25 to 35 SF	—	SF	9.78	—	9.78

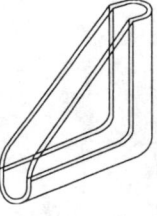

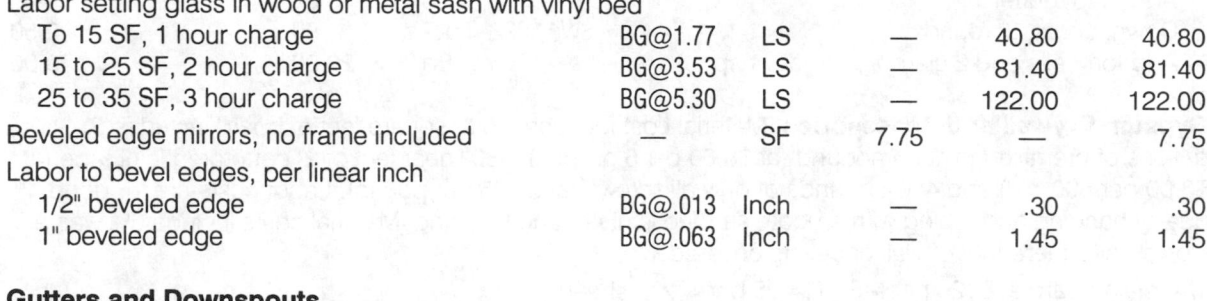

	Craft@Hrs	Unit	Material	Labor	Total
Labor setting glass in wood or metal sash with vinyl bed					
To 15 SF, 1 hour charge	BG@1.77	LS	—	40.80	40.80
15 to 25 SF, 2 hour charge	BG@3.53	LS	—	81.40	81.40
25 to 35 SF, 3 hour charge	BG@5.30	LS	—	122.00	122.00
Beveled edge mirrors, no frame included	—	SF	7.75	—	7.75
Labor to bevel edges, per linear inch					
1/2" beveled edge	BG@.013	Inch	—	.30	.30
1" beveled edge	BG@.063	Inch	—	1.45	1.45

Gutters and Downspouts

	Craft@Hrs	Unit	Material	Labor	Total
Galvanized steel, "K" style box (10' lengths)					
Gutter, 4", 28 gauge	SW@.065	LF	.99	1.80	2.79
Gutter, 4", 26 gauge	SW@.065	LF	1.65	1.80	3.45
Gutter, 5", 28 gauge	SW@.070	LF	1.13	1.94	3.07
Gutter, 5", 26 gauge	SW@.070	LF	2.23	1.94	4.17
Add for fittings (connectors, hangers, etc)	—	LF	1.14	—	1.14
Add for inside or outside corner, 4"	SW@.143	Ea	5.94	3.97	9.91
Add for inside or outside corner, 5"	SW@.143	Ea	13.68	3.97	17.65
Downspouts, 2" x 3", 28 gauge,					
Square, corrugated	SW@.054	LF	1.25	1.50	2.75
Add for drop outlet, 3 ells and strap,					
Per downspout	—	Ea	18.77	—	18.77

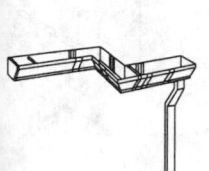

Galvanized steel, half round gutter (10' lengths), single bead					
28 gauge, 4"	SW@.065	LF	2.07	1.80	3.87
28 gauge, 5"	SW@.070	LF	2.23	1.94	4.17
28 gauge, 6"	SW@.078	LF	2.87	2.17	5.04
26 gauge, 4"	SW@.065	LF	2.44	1.80	4.24
26 gauge, 5"	SW@.070	LF	2.49	1.94	4.43
26 gauge, 6"	SW@.078	LF	3.50	2.17	5.67
Add for double bead gutter	—	%	25.0	—	—
Add for fittings (connectors, hangers, etc)	—	LF	1.11	—	1.11
Add for inside or outside corner	SW@.143	Ea	10.82	3.97	14.79
Downspouts, square, plain					
2" x 3", 28 gauge	SW@.054	LF	2.02	1.50	3.52
2" x 3", 26 gauge	SW@.054	LF	2.12	1.50	3.62
2" x 4", 28 gauge	SW@.054	LF	2.33	1.50	3.83
2" x 4", 26 gauge	SW@.054	LF	2.12	1.50	3.62
3" x 4", 28 gauge	SW@.054	LF	2.76	1.50	4.26
3" x 4", 26 gauge	SW@.054	LF	3.02	1.50	4.52
Downspouts, round					
2", 28 gauge	SW@.054	LF	2.44	1.50	3.94
2", 26 gauge	SW@.054	LF	3.08	1.50	4.58
3", 28 gauge	SW@.054	LF	1.48	1.50	2.98
3", 26 gauge	SW@.054	LF	1.96	1.50	3.46
4", 28 gauge	SW@.054	LF	1.86	1.50	3.36
4", 26 gauge	SW@.054	LF	2.44	1.50	3.94
Add for downspout fittings, per downspout	—	Ea	17.50	—	17.50
Vinyl rain gutter, white PVC, (Plastmo Vinyl), half round gutter, (10' lengths)					
Gutter, 4"	SW@.054	LF	.88	1.50	2.38
Add for fittings (connectors, hangers, etc)	—	LF	.93	—	.93
Add for inside or outside corner	SW@.091	Ea	4.69	2.53	7.22
Add for bonding kit (covers 150 LF)	—	Ea	3.65	—	3.65

Gutters and Downspouts

	Craft@Hrs	Unit	Material	Labor	Total
Add for 5" gutter	—	%	15.0	—	—
Downspouts, 3" round	SW@.022	LF	1.89	.61	2.50
Add for 3 ells and 2 clamps, per downspout	—	Ea	20.00	—	20.00

Gypsum Drywall and Accessories Material cost for each 1,000 square feet of board includes 5 gallons of premixed joint compound (at $8.00 per 5 gallons), 380 linear feet of 2" perforated joint tape (at $3.00 per 500' roll) and 4-1/2 pounds of drywall screws (at $2.25 per pound). Labor costs are for good quality hanging and taping with smooth sanded joints but no texturing. Material costs include 6% waste. Costs per square foot of wall or ceiling covered.

1/4" plain board at $126 per MSF ($4.05 per 4' x 8' sheet)

	Craft@Hrs	Unit	Material	Labor	Total
Walls	D1@.016	SF	.16	.40	.56
Ceilings	D1@.021	SF	.16	.52	.68

3/8" plain board at $126 per MSF ($4.05 per 4' x 8' sheet)

Walls	D1@.017	SF	.16	.42	.58
Ceilings	D1@.023	SF	.16	.57	.73

1/2" plain board at $211 per MSF ($6.75 per 4' x 8' sheet)

Walls	D1@.018	SF	.23	.45	.68
Ceilings	D1@.024	SF	.23	.60	.83

5/8" plain board at $228 per MSF ($7.30 per 4' x 8' sheet)

Walls	D1@.020	SF	.26	.50	.76
Ceilings	D1@.025	SF	.26	.62	.88

3/8" foil backed board at $176 per MSF ($5.63 per 4' x 8' sheet)

Walls	D1@.017	SF	.21	.42	.63
Ceilings	D1@.023	SF	.21	.57	.78

1/2" foil backed board at $276 per MSF ($8.83 per 4' x 8' sheet)

Walls	D1@.018	SF	.31	.45	.76
Ceilings	D1@.024	SF	.31	.60	.91

5/8" foil backed board at $300 per MSF ($9.60 per 4' x 8' sheet)

Walls	D1@.020	SF	.34	.50	.84
Ceilings	D1@.025	SF	.34	.62	.96

1/2" moisture resistant greenboard at $305 per MSF ($9.75 per 4' x 8' sheet)

Walls	D1@.018	SF	.36	.45	.81
Ceilings	D1@.024	SF	.36	.60	.96

1/2" fire rated type C board at $220 per MSF ($7.05 per 4' x 8' sheet)

Walls	D1@.018	SF	.27	.45	.72
Ceilings	D1@.024	SF	.27	.60	.87

5/8" fire rated type X board at $220 per MSF ($7.05 per 4' x 8' sheet)

Walls	D1@.020	SF	.27	.50	.77
Ceilings	D1@.025	SF	.27	.62	.89

1/2" fire rated type X foil backed insulating board at $310 per MSF ($10.00 per 4' x 8' sheet)

Walls	D1@.018	SF	.36	.45	.81
Ceilings	D1@.024	SF	.36	.60	.96

5/8" fire rated type X foil backed insulating board at $310 per MSF ($10.00 per 4' x 8' sheet)

Walls	D1@.020	SF	.36	.50	.86
Ceilings	D1@.025	SF	.36	.62	.98

3/8" plain backerboard at $240 per MSF ($7.70 per 4' x 8' sheet)

Walls	D1@.017	SF	.28	.42	.70
Ceilings	D1@.023	SF	.28	.57	.85

1/2" plain backerboard at $240 per MSF ($7.70 per 4' x 8' sheet)

Walls	D1@.018	SF	.28	.45	.73
Ceilings	D1@.024	SF	.28	.60	.88

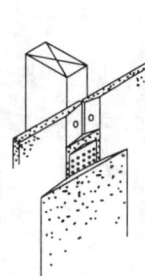

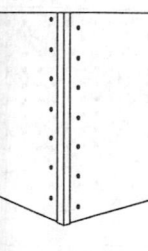

	Craft@Hrs	Unit	Material	Labor	Total
5/8" plain backerboard at $240 per MSF ($7.70 per 4' x 8' sheet)					
Walls	D1@.020	SF	.28	.50	.78
Ceilings	D1@.025	SF	.28	.62	.90
Sheathing board, (greenboard)1/2" thickness	D1@.010	SF	.52	.25	.77
Sheathing board, (greenboard) 5/8" thickness	D1@.010	SF	.58	.25	.83
Drywall casing and channel trim					
Stop or casing or jamb casing	D1@.020	LF	.25	.50	.75
RC-1 channel or 7/8" hat channel	D1@.011	LF	.21	.27	.48
Drywall corner trim					
Beadex outside corner	D1@.030	LF	.21	.75	.96
Beadex inside corner	D1@.030	LF	.21	.75	.96
Corner guard	D1@.030	LF	.21	.75	.96
Drywall edge trim					
"L" casing	D1@.016	LF	.22	.40	.62
Clip on casing	D1@.016	LF	.28	.40	.68
Beadex casing B-8	D1@.016	LF	.31	.40	.71
Beadex casing B-9	D1@.016	LF	.25	.40	.65
Drywall joint finishing compound (figure 5 gallons per 1,000 SF of board)					
Premixed joint compound (3.59 gallons at $5.75)	—	Gal	1.60	—	1.60
Perforated paper tape (figure 380 LF of tape per 1,000 SF of board)					
500 LF roll	—	Ea	3.00	—	3.00
250 LF roll	—	Ea	2.10	—	2.10
Deduct for no taping or joint finishing					
Walls	—	SF	-.02	-.22	-.24
Ceilings	—	SF	-.02	-.24	-.26
Add for spray applied plaster finish (sometimes called "popcorn" or "cottage cheese" texture)					
Textured or acoustic	DT@.003	SF	.10	.07	.17

Hardboard See also Siding for hardboard siding.

	Craft@Hrs	Unit	Material	Labor	Total
Plastic coated decorative tileboard, Bestile. Solid colors or patterns					
1/8" x 4' x 8' sheets at $39.00	B1@.034	SF	1.25	.78	2.03
Adhesive, 1 gallon covers 80 SF					
80 SF per gallon at $25.00	—	SF	.30	—	.30
Sealer for edges, 11 oz cartridge covers 130 LF					
130 LF per 11 oz cartridge at $4.50	—	LF	.04	—	.04
Tub kits, various colors, Bestile (includes two 5' x 5' panels, adhesive, caulking, and moulding)					
Per kit with accessories	—	Ea	175.00	—	175.00
Labor applying tub kit panels with adhesive					
In bathrooms	B1@.043	SF	—	.98	.98
Mouldings for bathroom tileboard panels, cost per LF					
Aluminum moulding	B1@.034	LF	1.35	.78	2.13
PVC moulding	B1@.034	LF	.82	.78	1.60
Standard hardboard, untempered, surfaced one side					
1/8" x 4' x 8' sheets at $7.20	B1@.034	SF	.24	.78	1.02
1/4" x 4' x 8' sheets at $10.50	B1@.034	SF	.34	.78	1.12
Tempered hardboard					
1/8" x 4' x 8' sheets at $11.80	B1@.034	SF	.39	.78	1.16
1/8" x 4' x 10' sheets at $16.60	B1@.034	SF	.54	.78	1.32
1/4" x 4' x 8' sheets at $14.80	B1@.034	SF	.49	.78	1.27
1/4" x 4' x 10' sheets at $18.90	B1@.034	SF	.62	.78	1.40

Hardboard

	Craft@Hrs	Unit	Material	Labor	Total
Surfaced two sides					
1/8" x 4' x 8' sheets at $11.80	B1@.034	SF	.38	.78	1.16
1/8" x 4' x 10' sheets at $16.60	B1@.034	SF	.43	.78	1.21
1/4" x 4' x 8' sheets at $14.80	B1@.034	SF	.48	.78	1.26
1/4" x 4' x 10' sheets at $18.90	B1@.034	SF	.49	.78	1.27
Pegboard, 4' x 8' sheets with holes 1" on centers					
1/8" standard sheets at $11.50	B1@.034	SF	.37	.78	1.15
1/4" standard sheets at $13.50	B1@.034	SF	.43	.78	1.21
1/8" tempered sheets at $16.20	B1@.034	SF	.52	.78	1.30
1/4" tempered sheets at $17.50	B1@.034	SF	.56	.78	1.34
Underlayment, particleboard, standard grade					
3/8" x 4' x 8' sheets at $7.50	B1@.011	SF	.23	.25	.48
1/2" x 4' x 8' sheets at $9.20	B1@.011	SF	.29	.25	.54
5/8" x 4' x 8' sheets at $11.50	B1@.011	SF	.36	.25	.61
3/4" x 4' x 8' sheets at $13.90	B1@.011	SF	.43	.25	.68
Marlite plastic coated panels, High gloss or satin gloss					
1/8" x 4' x 8' sheets at $70.00	B1@.043	SF	2.15	.98	3.13

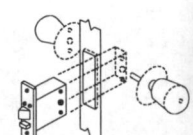

Hardboard, Masonite

Decorative hardboard panels for bathroom applications, adhesive applied. Add for adhesive below.

	Craft@Hrs	Unit	Material	Labor	Total
Royalclad, solid colors or patterns,					
1/4" x 4' x 8' sheets at $17.80	B1@.043	SF	.55	.98	1.53

Interior hardboard paneling, nail applied. Add for adhesive below

	Craft@Hrs	Unit	Material	Labor	Total
Standard Presdwood, untempered, surfaced one side					
1/8" x 4' x 8' sheets at $11.20	B1@.034	SF	.34	.78	1.12
1/4" x 4' x 8' sheets at $15.80	B1@.034	SF	.48	.78	1.26
Royalcote panels, smooth and grooved, Standard woodgrain, Somerset or Hanover finish					
1/4" x 4' x 8' sheets at $16.80	B1@.034	SF	.51	.78	1.29
Royalcote panels, textured, Woodfield or Natural finish					
1/4" x 4' x 8' sheets at $17.80	B1@.034	SF	.55	.78	1.33
Add for panel adhesive (@$24.40 per gal)					
(80 SF per gal)	B1@.008	SF	.30	.18	.48
Add for edge sealer (@$6.60 per 11 oz cartridge)					
(130 LF per cartridge)	B1@.004	LF	.05	.09	.14

Hardware See also Bath Accessories, Door Chimes, Door Closers, Foundation Bolts, Garage Door Openers, Mailboxes, Nails, Screen Wire, Sheet Metal, Framing Connectors, and Weatherstripping.

Rule of thumb: Cost of finish hardware (including door hardware, catches, stops, racks, brackets and pulls) per house.

	Craft@Hrs	Unit	Material	Labor	Total
Low to medium cost house (1600 SF)	BC@16.0	LS	440.00	403.00	843.00
Better quality house (2000 SF)	BC@20.0	LS	527.00	503.00	1,030.00

Door exterior locksets

Cabinet locks (commercial quality), keyed, 6 pin, including strike,

	Craft@Hrs	Unit	Material	Labor	Total
1/2" or 1" throw	BC@.572	Ea	33.70	14.40	48.10

Deadbolt locks, keyed cylinder outside, thumb turn inside, including strike, strike reinforcer, and strike box

	Craft@Hrs	Unit	Material	Labor	Total
Standard duty, 5 pin, 1" throw	BC@1.04	Ea	31.50	26.20	57.70
Heavy duty (commercial quality), 5 pin, 1" throw					
Steel reinforced	BC@1.04	Ea	37.80	26.20	64.00
Extra heavy duty, 6 pin, 1" throw	BC@1.04	Ea	65.20	26.20	91.40

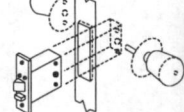

	Craft@Hrs	Unit	Material	Labor	Total
Entry locks, bored type, including trim, strike, and strike box with escutcheon trim and strike reinforcer					
Heavy duty, 6 pin tumbler, 1/2" throw, knob	BC@1.04	Ea	43.00	26.20	69.20
Heavy duty, 6 pin tumbler, 1/2" throw, lever	BC@1.04	Ea	49.40	26.20	75.60
Standard, 5 pin tumbler, 1/2" throw, knob	BC@1.04	Ea	22.00	26.20	48.20
Standard, 5 pin tumbler, 1/2" throw, lever	BC@1.10	Ea	25.20	27.70	52.90
Economy quality, escutcheon trim, no strike reinforcer					
5 pin tumbler, 1/2" throw, knob	BC@1.00	Ea	14.70	25.20	39.90
Add for construction keying system (minimum order 20 keyed entry locks),					
Per lock	—	Ea	2.00	—	2.00
Add for electrically operated latch release, no switches or wiring included					
Per latch release	—	Ea	75.50	—	75.50

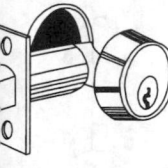

	Craft@Hrs	Unit	Material	Labor	Total
Grip handle entry lock, deluxe trim, 5 pin cylinder, includes strike, strike reinforcer, and strike box					
Per entry lock	BC@1.04	Ea	97.70	26.20	123.90

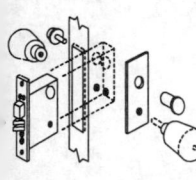

	Craft@Hrs	Unit	Material	Labor	Total
Mortise type locks, heavy duty, 6 pin tumbler, with integral deadbolt, strike, strike reinforcer, strike box and escutcheon trim					
Residential, knob type	BC@1.30	Ea	120.00	32.70	152.70
Residential, lever type	BC@1.30	Ea	127.00	32.70	159.70
Apartment, lever type	BC@1.30	Ea	127.00	32.70	159.70
Retail store, lever type	BC@1.30	Ea	127.00	32.70	159.70
Add for construction keying system (minimum order of 20 keyed entry locks),					
Per lock	—	Ea	2.63	—	2.63
Add for armored fronts	—	Ea	65.20	—	65.20
Add for electrically operated locks, no switches or wiring included					
Per lock	—	Ea	352.00	—	352.00
Add for monitor switch for electrically operated locks					
Per lock	—	Ea	59.90	—	59.90
Security entrance (cylinder/deadbolt combination), 5 pin cylinder and 1" throw,					
5 pin deadbolt	BC@1.30	Ea	87.20	32.70	119.90

Door passage locksets Keyless

	Craft@Hrs	Unit	Material	Labor	Total
Bored type, including knob trim, non-locking					
Standard residential latchset	BC@.572	Ea	15.80	14.40	30.20
Apartment/commercial latchset	BC@.572	Ea	32.50	14.40	46.90
Heavy duty commercial latchset	BC@.572	Ea	92.40	14.40	106.80
Standard residential button locking	BC@.572	Ea	16.80	14.40	31.20
Apartment/commercial button locking	BC@.572	Ea	39.80	14.40	54.20
Heavy duty commercial button locking	BC@.572	Ea	108.00	14.40	122.40
Dummy knobs, one side only					
Residential dummy knob	BC@.428	Ea	6.30	10.80	17.10
Apartment/commercial dummy knob	BC@.428	Ea	12.60	10.80	23.40
Heavy duty commercial dummy knob	BC@.428	Ea	35.70	10.80	46.50
Add for lever trim	—	Ea	8.40	—	8.40
Mortise type latchsets					
Passage (non-locking) knob trim	BC@1.04	Ea	163.00	26.20	189.20
Passage (non-locking) lever trim	BC@1.04	Ea	220.00	26.20	246.20
Privacy (thumb turn lock), knob trim	BC@1.04	Ea	191.00	26.20	217.20
Privacy (thumb turn lock), lever trim	BC@1.04	Ea	220.00	26.20	246.20
Office privacy lock, keyed,					
With lever and escutcheon trim	BC@1.04	Ea	216.00	26.20	242.20
Add for lead X-ray shielding	—	Ea	87.20	—	87.20
Add for electricity operated locks, no switches					
or wiring included	—	Ea	352.00	—	352.00

Hardware

	Craft@Hrs	Unit	Material	Labor	Total
Miscellaneous door hardware					
Hasps, safety type, heavy duty zinc plated					
3-1/2" hasp	—	Ea	2.48	—	2.48
4-1/2" hasp	—	Ea	3.17	—	3.17
6" hasp	—	Ea	5.34	—	5.34
Hinges					
Full mortise, heavy steel, satin brass finish					
3" x 3", loose pin	—	Ea	3.05	—	3.05
4" x 4", loose pin	—	Ea	4.30	—	4.30
4" x 4", ball bearing	—	Ea	9.87	—	9.87
Full mortise, light steel, satin brass finish					
3-1/2" x 3-1/2", loose pin	—	Ea	2.46	—	2.46
4" x 4", loose pin	—	Ea	3.50	—	3.50
Screen door					
Bright zinc	—	Ea	2.66	—	2.66
Galvanized	—	Ea	6.14	—	6.14
Door knockers, good quality polished brass					
3-1/2" x 7", traditional style	BC@.258	Ea	33.70	6.49	40.19
4" x 8-1/2", ornate style	BC@.258	Ea	45.90	6.49	52.39
Door pull, brass finish, 4" x 16"	BC@.258	Ea	35.00	6.49	41.49
Door push plates					
Brass finish, 3" x 12"	BC@.258	Ea	7.31	6.49	13.80
Brass finish, 4" x 16"	BC@.258	Ea	8.58	6.49	15.07
Door kick plate					
Brass					
6" x 34"	BC@.258	Ea	19.00	6.49	25.49
8" x 30"	BC@.258	Ea	22.40	6.49	28.89
8" x 34"	BC@.258	Ea	25.40	6.49	31.89
8" x 40"	BC@.258	Ea	29.80	6.49	36.29
Stainless					
6" x 34"	BC@.258	Ea	15.70	6.49	22.19
8" x 30"	BC@.258	Ea	15.70	6.49	22.19
8" x 34"	BC@.258	Ea	20.90	6.49	27.39
8" x 40"	BC@.258	Ea	24.60	6.49	31.09
Stops					
Door stop	BC@.084	Ea	14.10	2.11	16.21
Dome stop 2" x 1"	BC@.084	Ea	3.00	2.11	5.11
Dome stop 2" x 1 3/8"	BC@.084	Ea	3.60	2.11	5.71
Floor stop	BC@.084	Ea	23.10	2.11	25.21
Wall bumper	BC@.084	Ea	2.60	2.11	4.71
Wall stop	BC@.084	Ea	20.40	2.11	22.51
Door peep sights, 5/8" diameter	BC@.258	Ea	4.20	6.49	10.69
House numbers, 4" high, brass or black coated	BC@.103	Ea	1.90	2.59	4.49
Garage door hardware, jamb type, 7' height, with springs and slide bolt					
8' to 10' wide door	BC@4.36	Ea	125.50	110.00	235.50
10' to 16' wide door, with truss	BC@5.50	Ea	196.90	138.00	334.90
Truss assembly for 16' door	BC@.493	Ea	32.40	12.40	44.80
Garage door hardware, track type, 7' height, with springs and lock					
7' to 9' wide door	BC@4.36	Ea	186.70	110.00	296.70
10' to 16' double track	BC@5.50	Ea	330.50	138.00	468.50

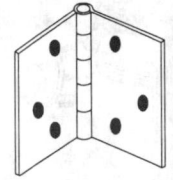

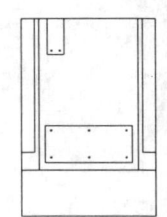

	Craft@Hrs	Unit	Material	Labor	Total

Hydronic Heating and Controls A complete system will consist of a Hydronic Boiler, Cast Iron Radiators or Hydronic Radiation Baseboard Heaters, Boiler Controls and Accessories. This type system is engineered and designed to meet specific heating requirements. Once these design requirements are known, the following components may be used to estimate the cost of a complete installed system. See the Plumbing and Heating Piping section and add for pipe, fittings, hangers and insulation as required.

Hydronic Boilers Packaged type units with standard equipment, including circulating pump and controls. Based on Burnham

Gas Fired

	Craft@Hrs	Unit	Material	Labor	Total
37,500 Btu with standard gas pilot	P1@3.00	Ea	940.00	73.30	1,013.30
62,000 Btu with standard gas pilot	P1@3.50	Ea	1,110.00	85.50	1,195.50
96,000 Btu with standard gas pilot	P1@3.50	Ea	1,220.00	85.50	1,305.50
130,000 Btu with standard gas pilot	P1@4.00	Ea	1,400.00	97.70	1,497.70
164,000 Btu with standard gas pilot	P1@4.00	Ea	1,620.00	97.70	1,717.70
198,000 Btu with standard gas pilot	P1@4.00	Ea	1,840.00	97.70	1,937.70
Add for electronic ignition, any above	—	Ea	202.00	—	202.00

Hydronic Heating Specialties

Air eliminator-purger threaded connections

	Craft@Hrs	Unit	Material	Labor	Total
3/4", cast iron, steam & water, 15 PSI	P1@.250	Ea	285.00	6.11	291.11
3/4", cast iron, 150 PSI water	P1@.250	Ea	92.00	6.11	98.11
1", cast iron, 150 PSI water	P1@.250	Ea	105.00	6.11	111.11
3/8", brass, 125 PSI steam	P1@.250	Ea	75.00	6.11	81.11
1/2", cast iron, 250 PSI steam	P1@.250	Ea	155.00	6.11	161.11
3/4", cast iron, 250 PSI steam	P1@.250	Ea	195.00	6.11	201.11
Airtrol fitting, 3/4"	P1@.300	Ea	33.00	7.33	40.33
Air eliminator vents, 1/4"	P1@.150	Ea	11.60	3.67	15.27

Atmospheric (Anti-siphon) vacuum breakers, rough brass, threaded

	Craft@Hrs	Unit	Material	Labor	Total
1/2" vacuum breaker	P1@.210	Ea	16.00	5.13	21.13
3/4" vacuum breaker	P1@.250	Ea	17.00	6.11	23.11
1" vacuum breaker	P1@.300	Ea	27.00	7.33	34.33
1-1/4" vacuum breaker	P1@.400	Ea	45.00	9.77	54.77
1-1/2" vacuum breaker	P1@.450	Ea	53.00	11.00	64.00
2" vacuum breaker	P1@.500	Ea	79.00	12.20	91.20

Circuit Balancing valves, brass, threaded

	Craft@Hrs	Unit	Material	Labor	Total
1/2" circuit balancing valve	P1@.210	Ea	19.50	5.13	24.63
3/4" circuit balancing valve	P1@.250	Ea	25.00	6.11	31.11
1" circuit balancing valve	P1@.300	Ea	32.00	7.33	39.33
1-1/2" circuit balancing valve	P1@.450	Ea	42.00	11.00	53.00

Baseboard fin tube radiation, per linear foot, copper tube with aluminum fins, wall mounted

	Craft@Hrs	Unit	Material	Labor	Total
1/2" element and 8" cover	P1@.280	LF	5.75	6.84	12.59
3/4" element and 9"cover	P1@.280	LF	9.50	6.84	16.34
3/4" element, 10" cover, high capacity	P1@.320	LF	14.75	7.82	22.57
3/4" fin tube element only	P1@.140	LF	3.75	3.42	7.17
1" fin tube element only	P1@.150	LF	4.50	3.67	8.17
8" cover only	P1@.140	LF	3.50	3.42	6.92
9" cover only	P1@.140	LF	4.25	3.42	7.67
10" cover only	P1@.160	LF	4.65	3.91	8.56
Add for pipe connection and control valve	P1@2.15	Ea	95.00	52.50	147.50
Add for corners, fillers and caps, average per foot	—	%	10.0	—	—

	Craft@Hrs	Unit	Material	Labor	Total
Flow check valves, brass, threaded, horizontal type					
3/4" check valve	P1@.250	Ea	16.00	6.11	22.11
1" check valve	P1@.300	Ea	23.00	7.33	30.33
Thermostatic mixing valves					
1/2", soldered	P1@.240	Ea	35.00	5.86	40.86
3/4", threaded	P1@.250	Ea	45.00	6.11	51.11
3/4", soldered	P1@.300	Ea	38.50	7.33	45.83
1", threaded	P1@.300	Ea	175.00	7.33	182.33
Liquid level gauges					
1/2", 175 PSI bronze	P1@.210	Ea	42.50	5.13	47.63
Wye pattern strainers, threaded, 250 PSI, cast iron body valves					
3/4" strainer	P1@.260	Ea	7.50	6.35	13.85
1" strainer	P1@.330	Ea	9.50	8.06	17.56
1-1/4" strainer	P1@.440	Ea	13.75	10.80	24.55
1-1/2" strainer	P1@.495	Ea	16.00	12.10	28.10
2" strainer	P1@.550	Ea	25.00	13.40	38.40

Baseboard heaters Electric baseboard heaters, convection type, surface mounted, 20 gauge steel, labor includes connecting and installation, 7" high, 3-1/4" deep, 3.41 Btu/hr/watt, 240 volt. Low or medium density.

	Craft@Hrs	Unit	Material	Labor	Total
2'6" long, 500 watt (1,700 Btu)	BE@1.70	Ea	31.20	45.70	76.90
3'0" long, 750 watt (2,225 Btu)	BE@1.81	Ea	39.10	48.60	87.70
4'0" long, 1,000 watt (3,400 Btu)	BE@1.81	Ea	47.60	48.60	96.20
5'0" long, 1,250 watt (4,250 Btu)	BE@1.81	Ea	54.10	48.60	102.70
6'0" long, 1,500 watt (5,100 Btu)	BE@1.89	Ea	60.60	50.80	111.40
8'0" long, 2,000 watt (6,800 Btu)	BE@2.04	Ea	77.30	54.80	132.10
10'0" long, 2,500 watt (10,200 Btu)	BE@2.32	Ea	98.90	62.30	161.20
Add for integral thermostat for any baseboard heater	BE@.094	Ea	21.60	2.53	24.13
Add for wall mounted thermostat, for line voltage for any of baseboard heater	BE@1.62	Ea	28.80	43.50	72.30

Hot water circulating baseboard heaters, 240 volts, 9-1/2" high, 3-1/4" deep. Labor includes connecting and installation.

	Craft@Hrs	Unit	Material	Labor	Total
39" long, 500 watts	BE@1.62	Ea	110.00	43.50	153.50
47" long, 750 watts	BE@1.70	Ea	125.00	45.70	170.70
59" long, 1,000 watts	BE@1.77	Ea	145.00	47.60	192.60
71" long, 1,380 watts	BE@1.79	Ea	165.00	48.10	213.10
83" long, 1,500 watts	BE@1.88	Ea	175.00	50.50	225.50
107" long, 2,000 watts	BE@2.03	Ea	210.00	54.60	264.60
Add for wall mounted thermostat, for line voltage any baseboard heater	BE@1.62	Ea	26.00	43.50	69.50

Recessed water circulating baseboard heaters, 240 volt floor mounted, 10" high, 6" wide. Labor includes connecting and installation.

	Craft@Hrs	Unit	Material	Labor	Total
2'4" long, 500 watts	BE@2.27	Ea	124.00	61.00	185.00
3'10" long, 1,000 watts	BE@2.27	Ea	160.00	61.00	221.00
5'10" long, 1,500 watts	BE@2.36	Ea	202.00	63.40	265.40
Add for wall mounted line voltage thermostat for any recessed heater	BE@1.62	Ea	26.90	43.50	70.40

	Craft@Hrs	Unit	Material	Labor	Total

Bathroom electric heaters

Ceiling blower type, snap-on grille, ventilator with switch, light, 14" long, 8" wide, 7" high,

| 1,500 watts, 120 volts (with vent kit at $12.00) | BE@2.00 | Ea | 134.00 | 53.70 | 187.70 |

Wall mount forced-air chrome wall unit, 11" x 8" x 4" rough-in.

| Includes rough-in box, 1,500 watts | BE@2.00 | Ea | 131.00 | 53.70 | 184.70 |

Circular ceiling unit, fan-forced, 8-1/2" diameter, 3-1/2" deep, chrome, surface mount,

| 1,250 watts, with switch | BE@2.00 | Ea | 67.50 | 53.70 | 121.20 |

Radiant ceiling unit

| 1,500 watts, 120 volts | BE@2.25 | Ea | 88.00 | 60.50 | 148.50 |
| Add for switch (4 position) | — | Ea | 20.00 | — | 20.00 |

Radiant wall unit, steel grille, built-in switch, 14" wide, 25" high, 4" deep, 120 volts

1,000 watt, 3,400 Btu, with blower	BE@2.75	Ea	78.30	73.90	152.20
1,200 watt, 4,095 Btu	BE@1.95	Ea	78.30	52.40	130.70
4,500 watt, with blower	BE@2.23	Ea	119.00	59.90	178.90

Radiant ceiling panels

Surface mounted, 1" thick, 120/208/240/277 volts AC.

200 watts, 23-3/4" x 23-3/4"	BE@.850	Ea	84.70	22.80	107.50
300 watts, 23-3/4" x 35-3/4"	BE@.900	Ea	101.00	24.20	125.20
400 watts, 23-3/4" x 47-3/4"	BE@.950	Ea	122.00	25.50	147.50
500 watts, 23-3/4" x 59-3/4"	BE@1.00	Ea	143.00	26.90	169.90
610 watts, 23-3/4" x 71-3/4"	BE@1.11	Ea	163.00	29.80	192.80
820 watts, 23-3/4" x 95-3/4"	BE@1.31	Ea	199.00	35.20	234.20
600 watts, 47-3/4" x 47-3/4"	BE@1.11	Ea	163.00	29.80	192.80
800 watts, 47-3/4" x 47-3/4"	BE@1.31	Ea	199.00	35.20	234.20
1,000 watts, 47-3/4" x 59-3/4"	BE@1.46	Ea	245.00	39.20	284.20
1,220 watts, 47-3/4" x 71-3/4"	BE@1.61	Ea	281.00	43.30	324.30
1,640 watts, 47-3/4" x 95-3/4"	BE@1.91	Ea	362.00	51.30	413.30
Add for radiant line voltage wall thermostat	BE@.450	Ea	24.50	12.10	36.60
Add for motion sensing setback thermostat	BE@.450	Ea	71.40	12.10	83.50

T-bar mounted, 1" thick, 120/208/240/277 volts AC.

200/250/375 watts, 23-3/4" x 23-3/4"	BE@.470	Ea	89.80	12.60	102.40
400/500/750 watts, 23-3/4" x 47-3/4"	BE@.500	Ea	136.00	13.40	149.40
610/750/1,000 watts, 23-3/4" x 71-3/4"	BE@.550	Ea	181.00	14.80	195.80
Add for radiant line voltage wall thermostat	BE@.450	Ea	24.50	12.10	36.60
Add for motion sensing setback thermostat	BE@.450	Ea	89.80	12.10	101.90

Surface mounted, 2-1/2" thick, 120/208/240/277 volts AC.

750/1000 watts, 23-3/4" x 47-3/4"	BE@.950	Ea	275.00	25.50	300.50
1125/1500 watts, 23-3/4" x 71-3/4"	BE@1.31	Ea	372.00	35.20	407.20
Add for radiant line voltage wall thermostat	BE@.450	Ea	24.50	12.10	36.60
Add for motion sensing setback thermostat	BE@.450	Ea	89.80	12.10	101.90

Wall and ceiling electric heaters

Wall heater, forced-air type, with built-in thermostat, and automatic safety shut-off switch,
11" high, 8" wide, 4" deep, recessed

| 2,000 watt, 240 volt | BE@1.78 | Ea | 115.00 | 47.80 | 162.80 |
| 1,500 watt, 120 volt | BE@1.78 | Ea | 115.00 | 47.80 | 162.80 |

Heating

	Craft@Hrs	Unit	Material	Labor	Total
Floor units, drop-in type, 14" long x 7-1/4" wide, includes fan motor assembly, housing and grille					
120 volts, 375 or 750 watts	BE@1.76	Ea	210.00	47.30	257.30
277 volts, 750 watts	BE@1.76	Ea	258.00	47.30	305.30
Add for wall thermostat kit	BE@1.62	Ea	49.40	43.50	92.90
Add for concrete accessory kit and housing	—	LS	18.50	—	18.50
Infrared quartz tube heaters, indoor or outdoor, chromed guard, polished aluminum reflector. Includes modulating control					
1,500 watts, 33" long, 120 or 240 volts	BE@4.14	Ea	235.00	111.00	346.00
2,000 watts, 33" long, 240 volts	BE@4.14	Ea	245.00	111.00	356.00
3,000 watts, 58-1/2" long, 240 volts	BE@4.14	Ea	311.00	111.00	422.00
4,000 watts, 58-1/2" long, 240 volts	BE@4.14	Ea	316.00	111.00	427.00
Add for wall or ceiling brackets (pair)	—	LS	10.20	—	10.20
Deduct for indoor use only	—	%	-30.0	—	—
Suspension blower heaters, 208 or 240 volts, single phase, propeller type					
3,000 watts	BE@4.80	Ea	219.00	129.00	348.00
5,600 watts	BE@4.80	Ea	240.00	129.00	369.00
7,500 watts	BE@4.80	Ea	362.00	129.00	491.00
10,000 watts	BE@4.80	Ea	383.00	129.00	512.00
Add for wall mounting bracket	—	LS	16.30	—	16.30
Add for line voltage wall thermostat					
Single pole	—	LS	26.00	—	26.00
Double pole	—	LS	37.00	—	37.00

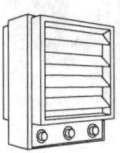

	Craft@Hrs	Unit	Material	Labor	Total
Wall mount fan-forced downflow insert heaters, heavy duty, built-in thermostat, 240 volts, 14" wide, 20" high, 4" deep rough-in, 3.41 Btu per hr per watt					
1,500 watts	BE@3.95	Ea	120.00	106.00	226.00
2,000 watts	BE@3.95	Ea	130.00	106.00	236.00
3,000 watts	BE@3.95	Ea	150.00	106.00	256.00
4,500 watts	BE@3.95	Ea	165.00	106.00	271.00
Add for surface mounting kit	—	LS	16.00	—	16.00

Gas and oil furnaces Complete costs including installation, connection and vent.

	Craft@Hrs	Unit	Material	Labor	Total
Floor gas furnaces, with vent at $80, valves and wall thermostat					
30 MBtu input, 24 MBtu output	PM@7.13	Ea	694.00	202.00	896.00
50 MBtu input, 35 MBtu output	PM@8.01	Ea	774.00	227.00	1,001.00
Add for dual room floor furnaces	—	%	30.0	—	—
Add for spark ignition furnaces	—	LS	50.00	—	50.00
Wall gas furnaces, 65" high, 14" wide, includes typical piping and valve at $40 and built-in remote dial thermostat, pilot ignition, blower, and vent at $80, modulating bulb, safety pilot					
25 MBtu input, 17.5 MBtu output	PM@4.67	Ea	492.00	132.00	624.00
35 MBtu input, 24.5 MBtu output	PM@4.67	Ea	472.00	132.00	604.00
50 MBtu input, 40 MBtu output	PM@4.67	Ea	697.00	132.00	829.00
Add for wall thermostat	PM@.972	Ea	—	27.50	27.50
Add for automatic blower kit	—	LS	118.00	—	118.00
Add for free standing vent trim kit	PM@.254	LS	144.00	7.20	151.20
Add for 2 wall unit (rear registers)	PM@.455	LS	41.00	12.90	53.90
Direct vent gas wall furnaces, including valve at $40, pilot ignition, wall-mounted thermostat, wall vent cap and guard					
12 MBtu output	PM@2.98	Ea	594.00	84.50	678.50
20 MBtu output	PM@2.98	Ea	716.00	84.50	800.50
30 MBtu output	PM@2.98	Ea	763.00	84.50	847.50
Add for vent extender kit	PM@.455	Ea	64.00	12.90	76.90

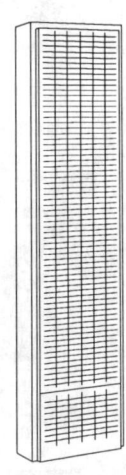

	Craft@Hrs	Unit	Material	Labor	Total

Forced air, upflow gas furnaces with electronic ignition. Includes vent at $100, typical piping at $50, thermostat at $40 and distribution plenum at $100. No ducting included

	Craft@Hrs	Unit	Material	Labor	Total
50 MBtu input, 48 MBtu output	PM@7.13	Ea	2,150.00	202.00	2,352.00
75 MBtu input, 70 MBtu output	PM@7.13	Ea	2,260.00	202.00	2,462.00
100 MBtu input, 92 MBtu output	PM@7.80	Ea	2,410.00	221.00	2,631.00
125 MBtu input, 113 MBtu output	PM@7.80	Ea	2,560.00	221.00	2,781.00
Add for liquid propane models	—	LS	236.00	—	236.00
Add for time control thermostat	—	Ea	92.30	—	92.30

Ductwork for forced air furnaces

Sheet metal, 6 duct and 2 air return residential system, with typical elbows and attaching boots

	Craft@Hrs	Unit	Material	Labor	Total
4" x 12" registers, and 2" R-5 insulation	SW@22.1	LS	493.00	614.00	1,107.00
Add for each additional duct run	SW@4.20	Ea	106.00	117.00	223.00
Fiberglass flex duct, insulated, 8" dia.	SW@.012	LF	2.52	.33	2.85

Heating and cooling equipment

Evaporative coolers, no wiring or duct work, roof installation, with switch, electrical and water connection at $100, recirculation pump, and mounting hardware at $75.

	Craft@Hrs	Unit	Material	Labor	Total
2,000 CFM, 1/3 HP	PM@11.1	Ea	459.00	315.00	774.00
4,300 CFM, 1/3 HP	PM@11.1	Ea	673.00	315.00	988.00
4,700 CFM, 1/2 HP	PM@11.l	Ea	729.00	315.00	1,044.00
5,600 CFM, 3/4 HP	PM@11.1	Ea	780.00	315.00	1,095.00

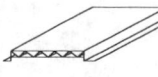

Window type air conditioning/cooling units, high-efficiency units, adjustable thermostat, labor is for installation in an existing opening, connected to existing electrical outlet.

	Craft@Hrs	Unit	Material	Labor	Total
5,200 Btu/hr	PM@.751	Ea	536.00	21.30	557.30
6,000 Btu/hr	PM@.75l	Ea	546.00	21.30	567.30
8,100 Btu/hr	PM@.751	Ea	551.00	21.30	572.30
10,000 Btu/hr	PM@1.00	Ea	653.00	28.30	681.30

Split system heat pump, one outdoor unit, with indoor evaporator coil, direct drive multi-speed blower, filter, and fiberglass cabinet insulation. No electrical or concrete work included

	Craft@Hrs	Unit	Material	Labor	Total
18 MBtu cooling, 17.8 MBtu heating	PM@9.59	Ea	780.00	272.00	1,052.00
21 MBtu cooling, 20.8 MBtu heating	PM@11.3	Ea	995.00	320.00	1,315.00
25 MBtu cooling, 24.7 MBtu heating	PM@12.3	Ea	1,050.00	349.00	1,399.00
29 MBtu cooling, 28.7 MBtu heating	PM@13.1	Ea	1,420.00	371.00	1,791.00
Add for electrical connection	PM@7.34	LS	61.20	208.00	269.20
Add for drain pan	—	LS	45.90	—	45.90

Packaged split system electric heat pumps, galvanized steel cabinets, window-wall units. 35' of tubing, two thermostats and control package. Installed on concrete pad. No electrical, concrete, or duct work included.

	Craft@Hrs	Unit	Material	Labor	Total
8 MBtu cooling, 5 Mbtu heating	PM@10.9	Ea	622.00	309.00	931.00
15 MBtu cooling, 13 Mbtu heating	PM@10.9	Ea	872.00	309.00	1,181.00
18 Mbtu cooling, 16 Mbtu heating	PM@12.9	Ea	984.00	366.00	1,350.00
Add for electrical connections	PM@14.0	LS	128.00	397.00	525.00

Insulation See also Building Paper and Polyethylene Film. Coverage allows for studs and joists.

Aluminum foil insulation, wall and ceiling applications, including 10% waste and overlap, 36" width on 40 lb kraft paper, 250 SF per roll

	Craft@Hrs	Unit	Material	Labor	Total
1 side foil ($11.50 per roll)	BC@.006	SF	.05	.15	.20
2 sides foil ($18.50 per roll)	BC@.006	SF	.07	.15	.22

Insulation

	Craft@Hrs	Unit	Material	Labor	Total
Fiberglass insulation, wall and ceiling application, no waste included					
Kraft paper-faced, roll and batt insulation					
16" OC framing members					
3-1/2" (R-11)	BC@.007	SF	.14	.18	.32
6-1/4" (R-19)	BC@.007	SF	.27	.18	.45
10" (R-30)	BC@.008	SF	.40	.20	.60
12" (R-38)	BC@.008	SF	.55	.20	.75
24" OC framing members					
3-1/2" (R-11)	BC@.004	SF	.14	.10	.24
6-1/4" (R-19)	BC@.005	SF	.27	.13	.40
10" (R-30)	BC@.005	SF	.38	.13	.51
12" (R-38)	BC@.005	SF	.50	.13	.63
Foil faced, roll and batt insulation					
16" OC framing members					
3-1/2" (R-11)	BC@.007	SF	.15	.18	.33
6-1/2" (R-19)	BC@.007	SF	.28	.18	.46
24" OC framing members					
3-1/2" (R-11)	BC@.004	SF	.15	.10	.25
6-1/4" (R-19)	BC@.005	SF	.28	.13	.41
Unfaced, roll and batt insulation					
16" OC framing members					
3-1/2" (R-11)	BC@.004	SF	.14	.10	.24
6-1/4" (R-19)	BC@.005	SF	.27	.13	.40
10" (R-30)	BC@.005	SF	.40	.13	.53
12" (R-38)	BC@.006	SF	.52	.15	.67
24" OC framing members					
3-1/2" (R-11)	BC@.003	SF	.14	.08	.22
6-1/4" (R-19)	BC@.003	SF	.27	.08	.35
10" (R-30)	BC@.003	SF	.40	.08	.48
12" (R-38)	BC@.004	SF	.52	.10	.62
Polystyrene insulation board, No. 100 regular, white, applied over walls with adhesive					
2' x 8' panels, including 5% waste					
1" thick (R-5)	BC@.010	SF	.44	.25	.69
2" thick (R-8)	BC@.011	SF	.83	.28	1.11
Add for adhesive	—	SF	.08	—	.08
Rigid foam (isocyanurate) insulation board, foil faced, two sides,					
4' x 8', including 5% waste					
1/2" (R-4)	BC@.010	SF	.33	.25	.58
3/4" (R-4)	BC@.010	SF	.43	.25	.68
1-1/2" (R-8)	BC@.010	SF	.65	.25	.90
Rigid urethane insulation board, foil faced no paper backing,					
4' x 8' panels, including 5% waste					
1-1/2" (R-10)	BC@.011	SF	.76	.28	1.04
2" (R-14)	BC@.011	SF	.93	.28	1.21
Add for nails, 50 lb cartons (large square heads)					
2-1/2", for 1-1/2" boards	—	LS	96.00	—	96.00
3", for 2" boards	—	LS	96.00	—	96.00
Vermiculite insulation, poured over ceilings					
Vermiculite, 25 lb sack (3 CF)	—	Ea	9.29	—	9.29
At 3" depth (60 sacks per 1,000 SF)	BL@.007	SF	.55	.14	.69
At 4" depth (72 sacks per 1,000 SF)	BL@.007	SF	.67	.14	.81

	Craft@Hrs	Unit	Material	Labor	Total
Roof insulation, 24" x 48", perlite board					
3/4" thick (R-2.08)	RR@.007	SF	.32	.20	.52
1" thick (R-2.78)	RR@.007	SF	.43	.20	.63
1-1/2" thick (R-4.17)	RR@.007	SF	.60	.20	.80
2" thick (R-5.26)	RR@.007	SF	.73	.20	.93
Cant strips for insulated roof board, one piece wood, tapered					
2" strip	RR@.018	LF	.21	.51	.72
3" strip	RR@.020	LF	.43	.57	1.00
4" strip	RR@.018	LF	.74	.51	1.25
Sill sealer, 1" x 50', 6" wide	RR@.003	LF	.32	.09	.41
Sound control batts, unfaced, 96" long					
4" x 16", 24"	BC@.007	SF	.49	.18	.67
2-3/4" x 16", 24"	BC@.007	SF	.42	.18	.60
Masonry batts, 1-1/2" x 15" x 48"	B9@.008	SF	.23	.19	.42
Masonry fill, 3 CF bag @$6.20, perlite poured in concrete block cores					
4" wall, 8.1 SF per CF	B9@.006	SF	.27	.14	.41
6" wall, 5.4 SF per CF	B9@.006	SF	.40	.14	.54
8" wall, 3.6 SF per CF	B9@.006	SF	.59	.14	.73
Asphalt sheathing board on walls, nailed, 4' x 8' panels					
1/2", standard	BC@.013	SF	.30	.33	.63
1/2", intermediate	BC@.013	SF	.36	.33	.69
1/2", nail base	BC@.013	SF	.44	.33	.77
25/32", standard	BC@.014	SF	.43	.35	.78
Sound board on wall, 1/2"	BC@.013	SF	.30	.33	.63
Sound board on ceilings, 1/2"	BC@.017	SF	.30	.43	.73

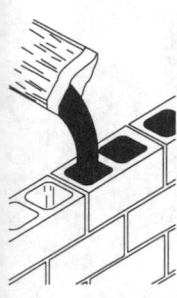

Insulation, Subcontract New construction, coverage includes framing.

	Craft@Hrs	Unit	Material	Labor	Total
Batts, walls and ceilings to 10' high, fiberglass, kraft faced					
3-1/2" (R-11) in walls	—	MSF	—	—	371.00
3-1/2" (R-11) in ceilings	—	MSF	—	—	413.00
6-1/4" (R-19)	—	MSF	—	—	551.00
10" (R-30)	—	MSF	—	—	847.00
Blown-in rockwool					
3-1/2" (R-11)	—	MSF	—	—	515.00
6" (R-19)	—	MSF	—	—	709.00
Blown-in cellulose					
3-1/2" (R-12)	—	MSF	—	—	505.00
6" (R-21)	—	MSF	—	—	719.00
Blown-in fiberglass, ceilings					
5" (R-11)	—	MSF	—	—	434.00
8" (R-19)	—	MSF	—	—	581.00
Sprayed-on urethane foam (rigid)					
1" thick	—	MSF	—	—	1,244.00
Add for acrylic waterproofing	—	MSF	—	—	895.00
Add for urethane rubber waterproofing	—	MSF	—	—	1,448.00
Add for cleaning rock off existing roof	—	MSF	—	—	90.00

Installed in existing structures

Fiberglass blown into walls in an existing structure, includes coring hole in exterior or interior of wall, filling cavity, sealing hole and paint to match existing surface.

	Craft@Hrs	Unit	Material	Labor	Total
R-12 to R-14 rating depending on cavity thickness					
Stucco (exterior application)	—	MSF	—	—	1,470.00
Wallboard (interior application)	—	MSF	—	—	1,470.00
Wood siding (exterior application)	—	MSF	—	—	1,950.00

Insulation

	Craft@Hrs	Unit	Material	Labor	Total

Thermal Analysis (Infrared thermography). Includes infrared video inspection of entire building, written report by qualified technician or engineer, and thermograms (infrared photographs) of all problem areas of building. Inspection is done using high-quality thermal sensitive instrumentation. Costs listed are for exterior facade shots, or interior (room by room) shots. All travel expenses for inspector are extra

	Craft@Hrs	Unit	Material	Labor	Total
Residence, typically 4 hours required	—	Hr	—	—	193.00
Commercial structure ($800.00 minimum)	—	Hr	—	—	115.00

Add per day lost due to weather, lack of preparation by building occupants, etc., when inspector and equipment are at building site prepared to commence inspection

	Craft@Hrs	Unit	Material	Labor	Total
Typical cost per day	—	LS	—	—	320.00

Add per night for accommodations when overnight stay is required. This is common when inspection must be performed at night to get accurate results.

	Craft@Hrs	Unit	Material	Labor	Total
Typical cost, per night	—	LS	—	—	110.00

Insurance and Bonding Typical rates. Costs vary by state and class of construction.

Rule of thumb: Employer's cost for payroll taxes, and insurance.

	Craft@Hrs	Unit	Material	Labor	Total
Per $100 (C$) of payroll	—	C$	—	—	29.90

Rule of thumb: Complete insurance program (comprehensive general liability policy, truck, automobile, and equipment floaters and fidelity bond), contractor's typical cost

	Craft@Hrs	Unit	Material	Labor	Total
Per $100 of payroll	—	C$	—	—	6.75

Liability insurance Comprehensive contractor's liability insurance, including operations, completed operations, bodily injury and property damage, protective and contractual coverage's, $1,000,000 policy limit. Minimum annual premium will be between $2,500 and $10,000. Rates vary by state and with the contractor's loss experience. Typical costs per $100 (C$) of payroll for each trade employed. (Calculate and add for each category separately.)

	Craft@Hrs	Unit	Material	Labor	Total
General contractors	—	C$	—	—	2.47
Carpentry	—	C$	—	—	3.71
Concrete, formed or flat	—	C$	—	—	4.75
Drywall hanging and finishing	—	C$	—	—	3.20
Electrical wiring	—	C$	—	—	3.23
Floor covering installation	—	C$	—	—	4.28
Glaziers	—	C$	—	—	3.20
Heating, ventilating, air conditioning	—	C$	—	—	4.47
Insulation	—	C$	—	—	4.09
Masonry, tile	—	C$	—	—	3.52
Painting	—	C$	—	—	4.85
Plastering and stucco	—	C$	—	—	3.42
Plumbing	—	C$	—	—	5.75

Workers compensation coverage Rates vary from state to state and by contractor's loss history. Coverage cost per $100 (C$) of base payroll excluding fringe benefits.

	Craft@Hrs	Unit	Material	Labor	Total
Bricklayer	—	C$	—	—	15.87
Carpenter					
1 and 2 family dwellings	—	C$	—	—	18.09
Multiple units and commercial	—	C$	—	—	32.10
Clerical (office worker)	—	C$	—	—	.99
Concrete					
1 and 2 family dwellings	—	C$	—	—	10.97
Other concrete	—	C$	—	—	19.80
Construction laborer	—	C$	—	—	15.90
Drywall taper	—	C$	—	—	12.80
Electrical wiring	—	C$	—	—	6.33
Elevator erectors	—	C$	—	—	21.05

	Craft@Hrs	Unit	Material	Labor	Total
Executive supervisors	—	C$	—	—	3.50
Excavation, grading	—	C$	—	—	8.20
Excavation, rock (no tunneling)	—	C$	—	—	12.10
Glazing	—	C$	—	—	13.51
Insulation work	—	C$	—	—	20.00
Iron or steel erection work	—	C$	—	—	38.41
Lathing	—	C$	—	—	10.07
Operating engineers	—	C$	—	—	9.95
Painting and paperhanging	—	C$	—	—	13.59
Pile driving	—	C$	—	—	27.26
Plastering and stucco	—	C$	—	—	13.19
Plumbing	—	C$	—	—	7.88
Reinforcing steel installation (concrete)	—	C$	—	—	10.00
Roofing	—	C$	—	—	31.14
Sewer construction	—	C$	—	—	9.10
Sheet metal work (on site)	—	C$	—	—	11.03
Steam and boiler work	—	C$	—	—	10.00
Tile, stone, and terrazzo work	—	C$	—	—	8.71
Truck driver	—	C$	—	—	16.00
Tunneling	—	C$	—	—	20.00

Payment and performance bonds Cost per $1,000 (M$) of final contract price. Rates depend on the experience, credit, and net worth of the applicant.

	Craft@Hrs	Unit	Material	Labor	Total
Preferred rates, contract coverage for:					
First $100,000	—	M$	—	—	13.70
Next $2,400,000	—	M$	—	—	8.27
Next $2,500,000	—	M$	—	—	6.55
Standard rates, contract coverage for:					
First $100,000	—	M$	—	—	28.50
Next $400,000	—	M$	—	—	17.10
Next $2,000,000	—	M$	—	—	11.40
Next $2,500,000	—	M$	—	—	8.55
Next $2,500,000	—	M$	—	—	7.98
Over $7,500,000	—	M$	—	—	7.41
Substandard rates, contract coverage for:					
First $100,000	—	M$	—	—	34.20
Balance of contract	—	M$	—	—	22.80

Intercom System, Subcontract Costs include all labor and equipment but assume that wiring is done before interior wall and ceiling finish is applied. Central system with AM/FM music, hands-free talk back, privacy feature and room monitoring. Includes 1 master station, 5 selective call remote stations and 2 door stations

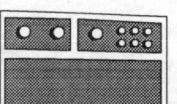

	Craft@Hrs	Unit	Material	Labor	Total
Basic residential system	—	LS	—	—	820.00
Add cassette tape player	—	LS	—	—	165.00
Add door and gate release	—	Ea	—	—	150.00
Add telephone answering feature, per station	—	Ea	—	—	61.00
Add telephone handset stations, per station	—	Ea	—	—	36.00

Ironing Centers

	Craft@Hrs	Unit	Material	Labor	Total

Ironing Centers, Built-In Wood frame with hinged door, can be recessed or surface mounted. Includes one 110 volt control panel where shown, receptacle outlet, timed shutoff switch, spotlight, storage shelf, and steel ironing board. No electrical work included. Iron-A-Way Products

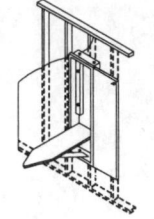

	Craft@Hrs	Unit	Material	Labor	Total
Unit type 1, no electric control panel, non-adjustable 42" board, 48" high, 15" wide, 6" deep	BC@.994	Ea	144.00	25.00	169.00
Unit type 2, with electric control panel, non-adjustable 42" board, 48" high, 15" wide, 6" deep	BC@.994	Ea	206.00	25.00	231.00
Unit type 3, with electric control panel, adjustable 46" board, 61" high, 15" wide, 6" deep	BC@.994	Ea	273.00	25.00	298.00
Added cost for any unit					
Add for sleeve board with holder	—	Ea	18.50	—	18.50
Add for door mirror	—	Ea	36.10	—	36.10
Add for oak paneled door front	—	Ea	46.50	—	46.50

Jackposts, Steel, Adjustable Permanent or temporary adjustable steel posts, 3" diameter.

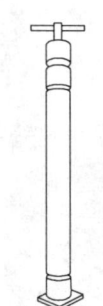

	Craft@Hrs	Unit	Material	Labor	Total
12" to 15" adjustment, 20,000 lb capacity	B1@.130	Ea	28.80	2.97	31.77
20" to 36" adjustment, 16,000 lb capacity	B1@.130	Ea	27.70	2.97	30.67
51" to 90" adjustment, 13,000 lb capacity	B1@.211	Ea	36.30	4.82	41.12
37" to 60" adjustment, 17,500 lb capacity	B1@.211	Ea	34.10	4.82	38.92
48" to 100" adjustment, 16,000 lb capacity	B1@.330	Ea	42.60	7.54	50.14
56" to 96" adjustment, 25,000 lb capacity	B1@.330	Ea	53.30	7.54	60.84

Landscaping

Ground cover. Costs include plants and hand labor to plant

	Craft@Hrs	Unit	Material	Labor	Total
English Ivy, 2-3/4" pots	BL@.098	Ea	.41	2.01	2.42
Liriope Muscari, 1 quart	BL@.133	Ea	1.25	2.73	3.98
Periwinkle, 2-3/4" pots	BL@.095	Ea	.50	1.95	2.45
Purple-leafed Wintercreeper, 1 quart	BL@.128	Ea	1.25	2.63	3.88

Plants and shrubs. Costs include plants, planting by hand, fertilization, small quantity

	Craft@Hrs	Unit	Material	Labor	Total
Abelia, glossy, 18" - 24"	BL@.835	Ea	8.40	17.10	25.50
Acacia, Fern leaf, 5 gal.	BL@.868	Ea	10.50	17.80	28.30
Australian Brush Cherry, 5 gal.	BL@.868	Ea	10.50	17.80	28.30
Azalea, Indica or Kaepferi, 18" - 24"	BL@.868	Ea	7.80	17.80	25.60
Azalea, Kurume types, 18" - 24"	BL@.868	Ea	7.80	17.80	25.60
Begonia, Angel Wing	BL@.835	Ea	9.20	17.10	26.30
Blue Hibiscus, 5 gal.	BL@.835	Ea	10.00	17.10	27.10
Bougainvillea, all varieties, 5 gal.	BL@.868	Ea	10.70	17.80	28.50
Bridal Wreath, 30" - 36"	BL@1.44	Ea	8.50	29.60	38.10
Burford Holly, dwarf, 30" - 36"	BL@1.44	Ea	22.00	29.60	51.60
Camellia, Sasanqua, 2 1/2' - 3'	BL@.835	Ea	10.80	17.10	27.90
Cherry Laurel, 18" - 24"	BL@1.53	Ea	12.80	31.40	44.20
Cleyera Japonica, 24" - 30"	BL@1.14	Ea	11.00	23.40	34.40
Compact Japanese Holly, 15" - 18"	BL@.696	Ea	8.60	14.30	22.90
Dwarf Lily of the Nile, 5 gal.	BL@.835	Ea	9.25	17.10	26.35
Dwarf Yaupon Holly, 15" - 18"	BL@.668	Ea	8.25	13.70	21.95
Gardenia, all varieties, 15" - 18"	BL@.835	Ea	8.50	17.10	25.60
Geranium, all varieties, 5 gal.	BL@.835	Ea	9.00	17.10	26.10
Gold Dust Aucuba, 24" - 30"	BL@1.19	Ea	10.80	24.40	35.20
Grey Lavender Cotton, 12" - 15"	BL@.601	Ea	3.00	12.30	15.30
Juniper, San Jose 18" - 24"	BL@.868	Ea	8.25	17.80	26.05
Lily of the Nile, 5 gal.	BL@.835	Ea	9.20	17.10	26.30
Night Blooming Jessamine, 5 gal.	BL@.835	Ea	9.20	17.10	26.30

	Craft@Hrs	Unit	Material	Labor	Total
Pampas Grass, 2 gal.	BL@.417	Ea	4.60	8.56	13.16
Pink Escallonia, 5 gal.	BL@.835	Ea	9.20	17.10	26.30
Poinciana Paradise Shrub, 5 gal.	BL@.868	Ea	10.20	17.80	28.00
Prickly Pear Cactus, 2 gal.	BL@.417	Ea	7.20	8.56	15.76
Redleaved Jap. Barberry, 18" - 24"	BL@.868	Ea	8.00	17.80	25.80
Rhododendron, 24" - 30"	BL@.835	Ea	27.00	17.10	44.10
Rubber Plant, 5 gal.	BL@.835	Ea	13.40	17.10	30.50
Sprengers Ash, 5 gal.	BL@.907	Ea	10.20	18.60	28.80
Shore Juniper, 18" - 24"	BL@.868	Ea	8.25	17.80	26.05
Tamarix Juniper, 18" - 24"	BL@.835	Ea	8.25	17.10	25.35
Thunbergi Spiraea, 24" - 30"	BL@1.14	Ea	8.10	23.40	31.50
Unedo- Strawberry Tree, 5 gal.	BL@.835	Ea	9.15	17.10	26.25
Vanhoutte Spiraea, 24" - 36"	BL@1.31	Ea	8.25	26.90	35.15
Veronica, all varieties, 5 gal.	BL@.835	Ea	9.15	17.10	26.25
Viburnum, all varieties, 3' - 4'	BL@.835	Ea	12.80	17.10	29.90
Willow Leaf Japanese Holly, 24" - 30"	BL@.442	Ea	12.40	9.08	21.48
Wintergreen Barberry, 18" - 24"	BL@.868	Ea	9.60	17.80	27.40
Yew, Hicks, 24" - 30"	BL@1.31	Ea	17.00	26.90	43.90

Trees and shrubs. Costs include planting, fertilization, backfill, and support as required. Caliper means trunk diameter

	Craft@Hrs	Unit	Material	Labor	Total
American Holly, 5' 6'	BL@1.75	Ea	56.00	35.90	91.90
American Elm, 12' - 14'	BL@3.66	Ea	97.00	75.20	172.20
Armstrong Juniperus, 24" - 30"	BL@.835	Ea	8.50	17.10	25.60
Australian Tree Fern, 5 gal.	BL@.835	Ea	10.30	17.10	27.40
Austrian Pine, 4' - 5'	BL@1.99	Ea	25.00	40.90	65.90
Bamboo Golden, 10' - 12'	BL@2.08	Ea	55.00	42.70	97.70
Bradford Pear, 8' - 10',	BL@2.25	Ea	80.00	46.20	126.20
Camellia, common type, #3 container	BL@1.10	Ea	10.80	22.60	33.40
Cape Myrtle, 8' - 10'	BL@2.25	Ea	65.00	46.20	111.20
Deodara Cedar, 6' - 8'	BL@3.66	Ea	52.00	75.20	127.20
Eastern Redbud, 6' - 8'	BL@2.00	Ea	40.00	41.10	81.10
Foster's American Holly, 5' - 6'	BL@1.75	Ea	64.00	35.90	99.90
Flowering Crabapple, Japanese, 2-1/2" - 3" caliper	BL@2.25	Ea	56.00	46.20	102.20
Fuchsia, all varieties, 5 gal.	BL@.835	Ea	8.70	17.10	25.80
Goldenrain Tree, 8' - 10'	BL@2.25	Ea	68.00	46.20	114.20
Green Ash, 2 gal.	BL@2.00	Ea	7.60	41.10	48.70
Greenspire Linden, 2-1/2" - 3" caliper	BL@3.05	Ea	100.00	62.60	162.60
Hibiscus, all varieties, 5 gal.	BL@.835	Ea	10.00	17.10	27.10
Japanese Privet, 4' - 5'	BL@1.46	Ea	10.80	30.00	40.80
Kwanzan Cherry, 1-3/4" - 2" caliper	BL@3.05	Ea	60.00	62.60	122.60
Live Oak, 3" - 3-1/2" caliper	BL@2.25	Ea	136.00	46.20	182.20
London Planetree Sycamore, 3" - 3-1/2" caliper	BL@3.66	Ea	115.00	75.20	190.20
Maidenhair Tree, 2 1/2" - 3" caliper	BL@3.66	Ea	110.00	75.20	185.20
Northern Red Oak, 3" - 3-1/2" caliper	BL@2.00	Ea	124.00	41.10	165.10
Norway Blue Spruce, 4' - 5'	BL@1.99	Ea	38.00	40.90	78.90
Pin Oak, 3" - 3-1/2" caliper	BL@3.48	Ea	12.40	71.50	83.90
Prince of Wales Juniperus, 5 gal.	BL@.835	Ea	8.25	17.10	25.35
Psidium (guava), 5 gal.	BL@.835	Ea	8.70	17.10	25.80
Pyracantha Firethorn, 3' - 4'	BL@1.10	Ea	7.20	22.60	29.80
Purpleleaf Flowering Plum, 6' - 8'	BL@2.00	Ea	40.00	41.10	81.10
Red Maple, 2" - 2-1/2" caliper	BL@3.05	Ea	85.00	62.60	147.60
Saucer Magnolia, 6' - 8'	BL@2.00	Ea	62.00	41.10	103.10

Landscaping

	Craft@Hrs	Unit	Material	Labor	Total
Star Magnolia, 4' - 5'	BL@1.46	Ea	40.00	30.00	70.00
Sugar Maple, 2-1/2" - 3" caliper	BL@2.00	Ea	90.00	41.10	131.10
Thornless Honeylocust, 10' - 12'	BL@2.25	Ea	98.00	46.20	144.20
Washington Hawthorne, 8' - 10'	BL@2.25	Ea	118.00	46.20	164.20
Weeping Fig, 5 gal.	BL@.835	Ea	13.00	17.10	30.10
Weeping Forsythia, 3' - 4'	BL@1.10	Ea	8.25	22.60	30.85
White Flowering Dogwood, 5' - 6'	BL@1.75	Ea	35.00	35.90	70.90
White Pine, 6' - 8'	BL@2.00	Ea	46.00	41.10	87.10
Willow Oak, 3" - 3-1/2" caliper	BL@2.74	Ea	125.00	56.30	181.30
Wisconsin Weeping Willow, 2-1/2" - 3" caliper	BL@2.00	Ea	65.70	41.10	106.80

Palm trees, based on West Coast prices

	Craft@Hrs	Unit	Material	Labor	Total
Mexican Fan Palm, 5 gal.	BL@1.50	Ea	9.99	30.80	40.79
Mediterranean Fan Palm, 5 gal.	BL@1.50	Ea	12.00	30.80	42.80
Pigmy Date Palm, 5 gal.	BL@1.50	Ea	29.00	30.80	59.80
Common King Palm, 7 gal.	BL@2.25	Ea	21.99	46.20	68.19

Seeding, level area (general mixture at $3.20 per lb)

	Craft@Hrs	Unit	Material	Labor	Total
Seeding preparation, grade, rake, clean	BL@.003	SY	.10	.06	.16
Fine grade, seed (with grass), and lime	BL@.018	SY	.78	.37	1.15
Mechanical seeding (1 lb per 22 SY)	BL@.005	SY	.16	.10	.26
Hand seeding (1 lb per 10 SY)	BL@.006	SY	.35	.12	.47

Fertilizer, pre-planting,
12-nitrogen, 4-phosphorus, 6-potassium

	Craft@Hrs	Unit	Material	Labor	Total
(1 lb per 12 SY)	BL@.002	SY	.34	.04	.38
Liming (1 lb per 1.6 SY)	BL@.002	SY	.12	.04	.16

Placing topsoil, delivered to site, 1 CY covers 81 SF at 4" depth. Add cost for equipment, where needed, from Excavation Equipment Costs section. Labor shown below is for operation of equipment.

	Craft@Hrs	Unit	Material	Labor	Total
With equipment, level site	BL@.092	CY	19.00	1.89	20.89
With equipment, sloped site	BL@.107	CY	19.20	2.20	21.40
By hand, level site	BL@.735	CY	18.70	15.10	33.80
By hand, sloped site	BL@.946	CY	18.70	19.40	38.10

Placing soil amendments, by hand spreading

	Craft@Hrs	Unit	Material	Labor	Total
Wood chip mulch, pine bark	BL@1.27	CY	20.00	26.10	46.10
Bale of straw, 25 bails per ton	BL@.200	Ea	3.00	4.11	7.11

Landscape stepping stones, concrete

	Craft@Hrs	Unit	Material	Labor	Total
12" round	BL@.133	Ea	2.45	2.73	5.18
14" round	BL@.133	Ea	3.30	2.73	6.03
18" round	BL@.153	Ea	5.70	3.14	8.84
24" round	BL@.153	Ea	9.60	3.14	12.74
Add for square shapes	—	%	20.0	—	—
18" or 24" diameter natural tree ring	BL@.133	Ea	2.08	2.73	4.81

Redwood benderboard, staked redwood

	Craft@Hrs	Unit	Material	Labor	Total
5/16" x 4"	BL@.011	LF	.45	.23	.68
1" x 6"	BL@.016	LF	1.42	.33	1.75
1" x 8"	BL@.016	LF	1.85	.33	2.18
2" x 8" rough	BL@.020	LF	6.20	.41	6.61

Roto-tilling light soil

	Craft@Hrs	Unit	Material	Labor	Total
To 4" depth	BL@.584	CSY	7.07	12.00	19.07

Sodding, no soil preparation, nursery sod, Bermuda or Blue Rye Mix

	Craft@Hrs	Unit	Material	Labor	Total
Per SF (500 SF minimum charge)	BL@.013	SF	.38	.27	.65

Lath

	Craft@Hrs	Unit	Material	Labor	Total
Lath					
Gypsum lath, perforated or plain, nailed to walls and ceilings					
3/8" x 16" x 48"	BR@.076	SY	2.68	1.91	4.59
1/2" x 16" x 48"	BR@.083	SY	2.93	2.08	5.01
Add for foil back insulating lath	—	SY	.64	—	.64
Steel lath, diamond pattern (junior mesh), 27" x 96" sheets, nailed to walls					
2.5 lb black painted	BR@.076	SY	1.96	1.91	3.87
2.5 lb galvanized	BR@.076	SY	2.13	1.91	4.04
1.75 lb black painted	BR@.076	SY	1.92	1.91	3.83
3.4 lb black painted	BR@.076	SY	2.34	1.91	4.25
3.4 lb galvanized	BR@.076	SY	2.45	1.91	4.36
Wood lath. See Lumber, Redwood					
Z-Riblath, 1/8" to wood frame (flat rib)					
2.75 lb painted, 24", 27" x 96" sheets	BR@.076	SY	3.21	1.91	5.12
3.4 lb painted, 24", 27" x 96" sheets	BR@.088	SY	3.90	2.21	6.11
Riblath, 3/8" to wood frame (high rib)					
3.4 lb painted, 24", 27" x 96" sheets	BR@.088	SY	2.62	2.21	4.83
3.4 lb galvanized, 24", 27" x 96" sheets	BR@.088	SY	3.10	2.21	5.31
3.4 lb painted paperback, 27" x 96" sheets	BR@.088	SY	3.03	2.21	5.24
Add for ceiling applications	—	%	—	25.0	—
Corner bead, 26 gauge, 8' to 12' lengths. Use 100 LF as a minimum job charge					
Flexible all-purpose bead	BR@.027	LF	.48	.68	1.16
Expansion, 2-1/2" flange	BR@.027	LF	.50	.68	1.18
Cornerite or cornalath					
2" x 2" x 4", steel-painted copper alloy	BR@.027	LF	.16	.68	.84
Corneraid					
2-1/2" smooth wire	BR@.027	LF	.26	.68	.94
Casing beads					
Square nose, short flange, 1/2" or 3/4"	BR@.025	LF	.38	.63	1.01
Quarter round, short flange, 1/2" or 3/4"	BR@.025	LF	.36	.63	.99
Square nose, expansion, 1/2" x 3/4"	BR@.025	LF	.43	.63	1.06
Quarter round, expansion, 1/2" x 3/4"	BR@.025	LF	.43	.63	1.06
Archbead (flexible plastic nose) for curves	BR@.025	LF	.58	.63	1.21
Base screed, flush or curved point	BR@.018	LF	.42	.45	.87
Expansion joint, 26 gauge, 1/2" ground	BR@.025	LF	1.06	.63	1.69
Archaid (for curves and arches), 8' length	BR@.028	LF	.48	.70	1.18
Galvanized stucco netting					
1" x 20 gauge x 48", 450 SF/roll	—	Roll	99.80	—	99.80
1" x 18 gauge x 48", 450 SF/roll	—	Roll	154.00	—	154.00
1" x 18 gauge x 48", 600 SF/roll	—	Roll	194.00	—	194.00
1-1/2" x 17 gauge x 36", 150 SF/roll	—	Roll	55.10	—	55.10
Self-furring	—	Roll	55.10	—	55.10
Self-furring, paperback, 100 SF/roll	—	Roll	60.90	—	60.90
Lathing, Subcontract					
Metal lath, nailed on wood studs					
2.5 lb diamond mesh, copper alloy	—	SY	—	—	4.96
2.5 lb galvanized	—	SY	—	—	5.54
3.4 lb painted copper alloy	—	SY	—	—	5.92
Add for ceiling applications	—	%	—	—	15.0

Lathing, Subcontract

	Craft@Hrs	Unit	Material	Labor	Total
Gypsum lath, perforated or plain, nailed on wood studs					
3/8" thick	—	SY	—	—	5.70
1/2" thick	—	SY	—	—	6.12
Add for ceiling applications	—	%	—	—	15.0
Add for foil facing	—	SY	—	—	.95
15 lb. felt and stucco netting	—	SY	—	—	3.17

Lawn Sprinkler Systems PVC pipe, Schedule 40, including typical hand trenching, plastic fittings (but no valves or sprinkler heads), installed

	Craft@Hrs	Unit	Material	Labor	Total
1/2" pipe at $.12/LF, with 1 fitting each					
10 LF at $.35	BL@.040	LF	.16	.82	.98
3/4" pipe at $.14/LF, with 1 fitting each					
10 LF at $.45	BL@.040	LF	.19	.82	1.01
1" pipe at $.20/LF, with 1 fitting each					
10 LF at $.55	BL@.040	LF	.26	.82	1.08
Deduct for Class 315 pipe	—	%	-40.0	—	—

Lawn sprinklers Individual components.

	Craft@Hrs	Unit	Material	Labor	Total
Trenching for pipe installation, by hand					
Light soil, 8" wide					
12" deep	BL@.027	LF	—	.55	.55
18" deep	BL@.040	LF	—	.82	.82
24" deep	BL@.053	LF	—	1.09	1.09
Average soil, 8" wide					
12" deep	BL@.043	LF	—	.88	.88
18" deep	BL@.065	LF	—	1.33	1.33
24" deep	BL@.087	LF	—	1.79	1.79
Heavy soil or loose rock, 8" wide					
12" deep	BL@.090	LF	—	1.85	1.85
18" deep	BL@.100	LF	—	2.05	2.05
24" deep	BL@.134	LF	—	2.75	2.75
Impact sprinkler heads, riser mounted, with nozzle, full circle					
Brass, with diffuser	BL@.422	Ea	15.00	8.67	23.67
Plastic, with diffuser	BL@.422	Ea	6.00	8.67	14.67
Impact sprinkler, pop-up type					
Brass and stainless steel	BL@.422	Ea	16.00	8.67	24.67
Plastic	BL@.422	Ea	11.00	8.67	19.67
Rotor pop-up, sprinkler heads					
Full circle	BL@.422	Ea	80.00	8.67	88.67
Rotor pop-up, with electric valve in head,					
Full circle	BL@2.52	Ea	110.00	51.70	161.70
Spray head pop-up type, plastic, with nozzle					
2" height	BL@.422	Ea	1.00	8.67	9.67
4" height	BL@.422	Ea	1.25	8.67	9.92
6" height	BL@.422	Ea	4.00	8.67	12.67
Spray head, brass, for riser mounting					
Flat pattern	BL@.191	Ea	2.00	3.92	5.92
Standard	BL@.191	Ea	2.25	3.92	6.17
Stream bubbler, all patterns	BL@.191	Ea	3.00	3.92	6.92

	Craft@Hrs	Unit	Material	Labor	Total
Sprinkler head risers, 1/2" or 3/4", installed to main or branch supply					
6" riser with three fittings	BL@.056	Ea	7.25	1.15	8.40
8" riser with three fittings	BL@.056	Ea	7.25	1.15	8.40
12" riser with three fittings	BL@.056	Ea	8.50	1.15	9.65
18" riser with three fittings	BL@.056	Ea	12.00	1.15	13.15
24" riser with three fittings	BL@.056	Ea	14.00	1.15	15.15

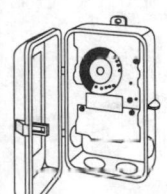

Valves, non-siphon, manual, brass

	Craft@Hrs	Unit	Material	Labor	Total
3/4" valve	BL@1.03	Ea	22.50	21.20	43.70
1" valve	BL@1.38	Ea	26.50	28.30	54.80
Valves, electric solenoid, brass, with siphon breaker, hookup but no wire included					
3/4" valve	BL@2.52	Ea	37.00	51.70	88.70
1" valve	BL@3.53	Ea	40.00	72.50	112.50

Valve control wire, 16 gauge low voltage direct burial cable, laid with pipe. No trenching or end connections included

	Craft@Hrs	Unit	Material	Labor	Total
2 conductor	BL@.003	LF	.12	.06	.18
3 conductor	BL@.003	LF	.13	.06	.19
4 conductor	BL@.003	LF	.15	.06	.21
6 conductor	BL@.003	LF	.21	.06	.27
8 conductor	BL@.003	LF	.27	.06	.33

Automatic irrigation control stations, no valve or wiring included, add for electrical service connection below.

	Craft@Hrs	Unit	Material	Labor	Total
Electro-mechanical type, controls up to					
6 separate electric valves	BL@1.83	LS	115.00	37.60	152.60
Programmable computerized type.					
Controls up to 8 electric valves	BL@1.83	LS	225.00	37.60	262.60
Add for 120 volt power source	—	LS	—	—	85.00

Residential system, typical contract price for 1,700 SF area using PVC Schedule 40 pipe, 340 LF of 3/4" pipe, 35 LF of 1" pipe, four 1" valves, 30 brass sprinkler heads and 6" risers on each head, based on heads spaced at 10'.

	Craft@Hrs	Unit	Material	Labor	Total
Large, regular shaped areas	—	SF	—	—	.55
Narrow and irregular areas	—	SF	—	—	.62
Add for freezing zones	—	SF	—	—	.06

Layout Foundation layout, medium to large size residence, 6 to 8 outside corners, includes shooting elevation from nearby reference, stakes and batterboards as required. No surveying or clearing included.

	Craft@Hrs	Unit	Material	Labor	Total
Typical cost per residence	B1@7.86	LS	35.00	180.00	215.00

Lighting Fixtures See also Electrical Work. Costs are to hang and connect fixtures only. No wiring included.

Fluorescent lighting fixtures

Surface mounted fluorescent fixtures Including ballasts but no lamps.

	Craft@Hrs	Unit	Material	Labor	Total
Wraparound acrylic diffuser					
24", two 20 watt tubes	BE@1.60	Ea	57.20	43.00	100.20
48", two 40 watt tubes	BE@1.60	Ea	71.00	43.00	114.00
48", four 40 watt	BE@1.60	Ea	112.00	43.00	155.00
96" tandem, four 40 watt tubes	BE@1.60	Ea	140.00	43.00	183.00

Decorative model with vinyl trim on metal frame and acrylic diffuser.

	Craft@Hrs	Unit	Material	Labor	Total
Available in oak, walnut or butcher block woodgrained vinyl					
10" x 48" x 4", two 20 watt tubes	BE@1.60	Ea	53.60	43.00	96.60
15" x 48" x 4", four 40 watt tubes	BE@1.60	Ea	90.60	43.00	133.60
15" x 96" x 4", eight 40 watt tubes	BE@1.60	Ea	182.00	43.00	225.00

Lighting Fixtures

	Craft@Hrs	Unit	Material	Labor	Total
14" x 48" x 4", four 40 watt tubes	BE@1.60	Ea	90.60	43.00	133.60
19" x 19" x 4", one 32/40 watt circline tube	BE@1.60	Ea	58.70	43.00	101.70
24" x 24" x 4", two 40 watt U/6 tubes	BE@1.60	Ea	71.10	43.00	114.10
Basic wraparound acrylic diffusers, solid oak wood ends					
10" x 24" x 3", two 20 watt tubes	BE@1.60	Ea	37.10	43.00	80.10
10" x 48" x 3", two 40 watt tubes	BE@1.60	Ea	45.30	43.00	88.30
Economy, wraparound acrylic diffuser, basic metal ends					
14" x 48" x 2", four 40 watt tubes	BE@1.60	Ea	68.00	43.00	111.00
8" x 48" x 2", two 40 watt tubes	BE@1.60	Ea	37.10	43.00	80.10
Basic circular fixtures for circline tubes					
Simulated glass crystal with antiqued metal band,					
11" round x 3" high, one 22 watt tube	BE@.696	Ea	27.30	18.70	46.00
Circular fixture					
12" round x 3 high, 22 and 32 watt tubes	BE@.696	Ea	26.30	18.70	45.00
Fixture with frosted glass diffuser					
19" round x 4" high, 32 and 40 watt tubes	BE@.696	Ea	60.90	18.70	79.60
Low profile rectangular wall or ceiling mount fixture, metal housing, white acrylic diffuser					
15" x 5", two 9 watt energy saver bulbs	BE@.696	Ea	37.80	18.70	56.50
Rectangular indoor/outdoor model with discoloration resistant/impact resistant acrylic diffuser and cold weather ballast					
10" x 10" x 5" deep, one 22 watt tube	BE@.696	Ea	56.70	18.70	75.40
12" x 12" x 5" deep, one 34 watt tube	BE@.696	Ea	73.50	18.70	92.20
Hall lantern type, including tubes					
4-1/4" wide x 12-1/8" long x 4" deep	BE@.696	Ea	35.70	18.70	54.40
Decorative valance type, with solid oak frame and acrylic diffuser					
12" x 25" x 4", two 20 watt tubes	BE@2.06	Ea	95.60	55.40	151.00
25" x 25" x 4", two 40 watt U/6 tubes	BE@1.59	Ea	120.00	42.70	162.70
12" x 49" x 4", two 40 watt tubes	BE@1.59	Ea	110.00	42.70	152.70
Economy valance type, with metal ends. Round or square models available.					
All models are 2-5/8" wide x 4-1/4" deep					
24-3/4" L (requires one 15 watt tube)	BE@1.59	Ea	26.30	42.70	69.00
24-3/4" L (requires one 20 watt tube)	BE@1.59	Ea	28.40	42.70	71.10
36-3/4" L (requires one 30 watt tube)	BE@1.59	Ea	36.70	42.70	79.40
36-3/4" L (requires two 30 watt tubes)	BE@1.59	Ea	44.70	42.70	87.40
48-3/4" L (requires one 40 watt tube)	BE@1.59	Ea	50.40	42.70	93.10
48-3/4" L (requires two 40 watt tubes)	BE@1.59	Ea	56.70	42.70	99.40
Under shelf/cabinet type for wiring to wall switch					
18" x 5" x 2", one 15 watt tube	BE@.696	Ea	14.20	18.70	32.90
24" x 5" x 2", one 20 watt tube	BE@.696	Ea	18.90	18.70	37.60
36" x 5" x 2", one 30 watt tube	BE@.696	Ea	29.40	18.70	48.10
48" x 5" x 2", one 40 watt tube	BE@.696	Ea	32.50	18.70	51.20

	Cool white	Deluxe cool white	Energy Saving cool white	Warm white
Fluorescent tubes				
8 watt (12" long)	5.92	9.27	—	8.76
15 watt (18" long)	6.28	8.03	—	7.00
20 watt (24" long)	5.67	8.55	8.24	6.10
30 watt (36" long)	6.60	10.70	5.67	9.17
40 watt (48" long)	2.06	—	—	—
Slimline straight tubes				
30 watt (24" or 36" long)	14.70	—	11.70	16.00
40 watt (48" long)	9.68	13.40	18.50	12.00

	Cool white	Deluxe cool white	Energy Saving cool white	Warm white
Circular (circline) tubes				
20 watt (6-1/2" diameter)	9.27	—	—	10.30
22 watt (8-1/4" diameter)	7.98	9.80	—	10.00
32 watt (12" diameter)	8.55	11.80	—	11.30
40 watt (16" diameter)	12.40	—	—	15.00
"U" shaped tubes (3" spacing x 12" long)				
U-3 40 watt	12.70	17.00	—	12.70
U-6 40 watt	12.70	17.00	—	12.70
Labor to install each tube	1.16	1.16	1.16	1.16

	Craft@Hrs	Unit	Material	Labor	Total
Incandescent lighting fixtures					
Outdoor post lantern fixtures, antique satin brass finish, no digging included					
Post lantern type fixture, cast aluminum	BE@2.00	Ea	56.00	53.70	109.70
Modern porch ceiling fixture					
7-1/2" x 7-1/2" x 3" drop, single light	BE@2.15	Ea	21.80	57.80	79.60
Flood light lampholders					
Cadmium steel plate	BE@2.15	Ea	26.00	57.80	83.80
1/2" male swivel 7-5/16" long, aluminum	BE@2.15	Ea	25.00	57.80	82.80
Path lighting, kits with cable and timer					
20 light low voltage	BE@2.63	LS	51.00	70.70	121.70
6 flood lights, 11 watt	BE@2.98	LS	54.00	80.10	134.10
12 tier lights, 7 watt, programmable	BE@2.63	LS	72.70	70.70	143.40
Recessed fixtures, including housing, incandescent					
Thermally protected shower/utility lights					
Round 75 watt shower light	BE@1.72	Ea	93.60	46.20	139.80
Utility light with fresnel lens diffuser	BE@1.72	Ea	23.00	46.20	69.20
Utility light with molded opal diffuser	BE@1.72	Ea	25.50	46.20	71.70
75 watt utility light, 7" frame, flush lens	BE@1.72	Ea	31.00	46.20	77.20
7" clear reflector lights, large area	BE@1.72	Ea	41.50	46.20	87.70
Eyeball spotlight, 75 watt	BE@1.72	Ea	43.50	46.20	89.70
Utility 12" x 10", diffusing glass lens	BE@1.72	Ea	43.50	46.20	89.70
Ceiling fixtures, using 60 watt standard bulbs, pre-wired housing white canopy finish					
Cloud 9-3/4" x 5-1/4" (2 bulbs)	BE@1.59	Ea	23.20	42.70	65.90
Cloud 8-1/4" x 5-1/2" (1 bulb)	BE@1.59	Ea	29.40	42.70	72.10
Standard globe 8-3/4" x 5" (2 bulbs)	BE@1.59	Ea	20.30	42.70	63.00
Cloud globe, brass canopy 10" x 5"	BE@1.59	Ea	25.00	42.70	67.70

Lumber. Prices for lumber vary with changes in supply and demand and increase with increases in distance from the mill or source of supply. Lumber costs in this section are based on average prices as of the fourth quarter of 1998. Carload lots will cost 15% to 20% less and retail purchases will cost from 20% to 25% more. Generally, lumber prices will be 10% to 20% lower in the Southern U.S. and slightly higher in the Northeast. See also Lumber Preservation Treatments.

Prices shown per MBF mean per 1,000 Board Feet. To determine the board feet (BF) per linear foot (LF) of lumber, multiply the width of the lumber by its depth, in inches, and divide the answer by 12".

		Std. & Btr K.D.	Utility K.D.
Douglas Fir S4S, random lengths, kiln dried.			
2" x 4" (Std & Btr = $0.37 per LF Utility = $0.32 per LF)	MBF	535.00	475.00
2" x 6" (Std & Btr = $0.54 per LF Utility = $0.48 per LF)	MBF	540.00	480.00

Lumber

				Std. & Btr K.D.	Utility K.D.
2" x 8"	(Std & Btr = $0.74 per LF Utility = $0.55 per LF)		MBF	555.00	415.00
2" x 10"	(Std & Btr = $1.00 per LF Utility = $0.71 per LF)		MBF	600.00	425.00
2" x 12"	(Std & Btr = $1.20 per LF Utility = $.88 per LF)		MBF	600.00	440.00

		Craft@Hrs	Unit	Material	Labor	Total
Fir & Larch S4S, random lengths, standard and better, kiln dried						
2" x 4"	($.35 per LF)	—	MBF	530.00	—	530.00
2" x 6"	($.54 per LF)	—	MBF	535.00	—	535.00
2" x 8"	($.72 per LF)	—	MBF	540.00	—	540.00
2" x 10"	($.98 per LF)	—	MBF	590.00	—	590.00
2" x 12"	($1.18 per LF)	—	MBF	590.00	—	590.00
Hemlock - Fir S4S, random lengths, standard and better, kiln dried						
2" x 4"	($.35 per LF)	—	MBF	530.00	—	530.00
2" x 6"	($.52 per LF)	—	MBF	515.00	—	515.00
2" x 8"	($.71 per LF)	—	MBF	530.00	—	530.00
2" x 10"	($.94 per LF)	—	MBF	565.00	—	565.00
2" x 12"	($1.13 per LF)	—	MBF	565.00	—	565.00
Ponderosa Pine S4S, random lengths, standard and better, kiln dried						
2" x 4"	($.36 per LF)	—	MBF	535.00	—	535.00
2" x 6"	($.53 per LF)	—	MBF	530.00	—	530.00
2" x 8"	($.71 per LF)	—	MBF	535.00	—	535.00
2" x 10"	($.95 per LF)	—	MBF	570.00	—	570.00
2" x 12"	($1.27 per LF)	—	MBF	635.00	—	635.00
Red Cedar S4S, random lengths, #2 and better						
2" x 4"	($.64 per LF)	—	MBF	960.00	—	960.00
2" x 6"	($1.07 per LF)	—	MBF	1,070.00	—	1,070.00
2" x 8"	($1.47 per LF)	—	MBF	1,100.00	—	1,100.00
2" x 10"	($2.07 per LF)	—	MBF	1,240.00	—	1,240.00
2" x 12	($2.48 per LF)	—	MBF	1,240.00	—	1,240.00

				#1	#2
Southern Yellow Pine S4S, random lengths					
2" x 4"	($.43 per LF)		MBF	645.00	685.00
2" x 6"	($.67 per LF)		MBF	665.00	665.00
2" x 8"	($.91 per LF)		MBF	680.00	675.00
2" x 10"	($1.09 per LF)		MBF	655.00	765.00
2" x 12"	($1.59 per LF)		MBF	795.00	885.00

		Craft@Hrs	Unit	Material	Labor	Total
Spruce & Pine S4S, random lengths, standard & better, kiln dried.						
2" x 4"	($.32 per LF)	—	MBF	485.00	—	485.00
2" x 6"	($.46 per LF)	—	MBF	455.00	—	455.00
2" x 8"	($.61 per LF)	—	MBF	455.00	—	455.00
2" x 10"	($.88 per LF)	—	MBF	530.00	—	530.00
2" x 12"	($.91 per LF)	—	MBF	545.00	—	545.00
Studs S4S, random lengths, standard & better, kiln dried.						
Hemlock - Fir						
(2" x 4" = $.38 per LF 2" x 6" = $.57 per LF)		—	MBF	550.00		550.00

	Craft@Hrs	Unit	Material	Labor	Total
Douglas Fir					
(2" x 4" = $.38 per LF 2" x 6" = $.57 per LF)	—	MBF	570.00	—	570.00
Fir & Larch					
(2" x 4" = $.37 per LF 2" x 6" = $.56 per LF)	—	MBF	560.00	—	560.00
White Woods					
(2" x 4" = $.37 per LF 2" x 6" = $.56 per LF)	—	MBF	560.00	—	560.00
Spruce or Pine					
(2" x 4" = $.34 per LF 2" x 6" = $.51 per LF)	—	MBF	510.00	—	510.00
Southern Yellow Pine					
(2" x 4" = $.38 per LF 2" x 6" = $.57 per LF)	—	MBF	570.00	—	570.00

Studs S4S, precut lengths for 8' high walls, standard or better, kiln dried.

	Craft@Hrs	Unit	Material	Labor	Total
Hemlock - Fir					
(2" x 4" = $2.74 Ea, 2" x 6" = $4.12 Ea)	—	MBF	515.00	—	515.00
Douglas Fir					
(2" x 4" = $2.98 Ea, 2" x 6" = $4.48 Ea)	—	MBF	560.00	—	560.00
Fir & Larch					
(2" x 4" = $2.90 Ea, 2" x 6" = $4.36 Ea)	—	MBF	545.00	—	545.00
White Woods					
(2" x 4" = $2.93 Ea, 2" x 6" = $4.40 Ea)	—	MBF	550.00	—	550.00
Spruce or Pine					
(2" x 4" = $2.64 Ea, 2" x 6" = $3.96 Ea)	—	MBF	495.00	—	495.00
Southern Pine					
(2" x 4" = $2.96 Ea, 2" x 6" = $4.44 Ea)	—	MBF	555.00	—	555.00

White Woods Spruce Lodgepole S4S, random lengths, standard and better, kiln dried.

	Craft@Hrs	Unit	Material	Labor	Total
2" x 4" ($.36 por LF)	—	MBF	535.00	—	535.00
2" x 6" ($.52 per LF)	—	MBF	520.00	—	520.00

Douglas Fir timbers Rough, random lengths, standard and better, kiln dried.

	Craft@Hrs	Unit	Material	Labor	Total
4" x 4" ($.98 per LF)	—	MBF	735.00	—	735.00
4" x 6" ($1.50 per LF)	—	MBF	750.00	—	750.00
4" x 8" ($1.97 per LF)	—	MBF	740.00	—	740.00
4" x 10" ($2.50 per LF)	—	MBF	750.00	—	750.00
4" x 12" ($2.94 per LF)	—	MBF	735.00	—	735.00
6" x 6" ($2.61 per LF)	—	MBF	870.00	—	870.00

Red Cedar timbers Rough, random lengths, construction grade, green.

	Craft@Hrs	Unit	Material	Labor	Total
4" x 4" ($1.52 per LF)	—	MBF	1,140.00	—	1,140.00
4" x 6" ($2.80 per LF)	—	MBF	1,400.00	—	1,400.00
4" x 8" ($3.57 per LF)	—	MBF	1,340.00	—	1,340.00
4" x 10" ($4.93 per LF)	—	MBF	1,480.00	—	1,480.00
4" x 12" ($5.80 per LF)	—	MBF	1,450.00	—	1,450.00
6" x 6" ($4.56 per LF)	—	MBF	1,520.00	—	1,520.00

Ponderosa Pine boards Random lengths.

	Unit	C & Btr	D & Btr
1" x 4" (C & Btr = $.85 per LF D & Btr = $.42 per LF)	MBF	2,560.00	1,250.00
1" x 6" (C & Btr = $1.64 per LF D & Btr = $.83 per LF)	MBF	3,270.00	1,650.00
1" x 8" (C & Btr = $2.10 per LF D & Btr = $1.15 per LF)	MBF	3,150.00	1,720.00
1" x 10" (C & Btr = $2.42 per LF D & Btr = $1.52 per LF)	MBF	2,900.00	1,820.00
1" x 12" (C & Btr = $3.16 per LF D & Btr = $2.30 per LF)	MBF	3,430.00	2,410.00

Lumber

		#2 & Btr	#3 & Btr
Ponderosa Pine boards Random lengths.			
1" x 4" (#2 & Btr = $.24 per LF #3 & Btr = $.19 per LF)	MBF	730.00	560.00
1" x 6" (#2 & Btr = $.39 per LF #3 & Btr = $.30 per LF)	MBF	770.00	605.00
1" x 8" (#2 & Btr = $.49 per LF #3 & Btr = $.39 per LF)	MBF	735.00	590.00
1" x 10" (#2 & Btr = $.87 per LF #3 & Btr = $.43 per LF)	MBF	1,050.00	520.00
1" x 12" (#2 & Btr = $1.12 per LF #3 & Btr = $.66 per LF)	MBF	1,120.00	655.00

		Choice & Btr	Quality
Idaho White Pine boards Random lengths.			
1" x 4" (Choice & Btr = $.89 per LF Quality = $.48 per LF)	MBF	2,670.00	1,440.00
1" x 6" (Choice & Btr = $1.69 per LF Quality = $.90 per LF)	MBF	3,370.00	1,800.00
1" x 8" (Choice & Btr = $2.10 per LF Quality = $1.23 per LF)	MBF	3,150.00	1,840.00
1" x 10" (Choice & Btr = $2.50 per LF Quality = $1.62 per LF)	MBF	3,000.00	1,940.00
1" x 12" (Choice & Btr = $3.40 per LF Quality = $2.46 per LF)	MBF	3,400.00	2,460.00

		Standard	Utility
Idaho White Pine boards Random lengths.			
1" x 4" (Standard = $.20 per LF Utility = $.14 per LF)	MBF	610.00	425.00
1" x 6" (Standard = $.33 per LF Utility = $.22 per LF)	MBF	660.00	445.00
1" x 8" (Standard = $.44 per LF Utility = $.32 per LF)	MBF	655.00	475.00
1" x 10" (Standard = $.55 per LF Utility = $.40 per LF)	MBF	655.00	475.00
1" x 12" (Standard = $.74 per LF Utility = $.48 per LF)	MBF	735.00	480.00

		D & Btr	#2 & Btr
Engelmann Spruce boards Random lengths.			
1" x 4" (D & Btr = $.48 per LF #2 & Btr = $.24 per LF)	MBF	1,450.00	720.00
1" x 6" (D & Btr = $.97 per LF #2 & Btr = $.37 per LF)	MBF	1,940.00	745.00
1" x 8" (D & Btr = $1.29 per LF #2 & Btr = $.48 per LF)	MBF	1,940.00	720.00
1" x 10" (D & Btr = $1.67 per LF #2 & Btr = $.88 per LF)	MBF	2,010.00	1,060.00
1" x 12" (D & Btr = $2.35 per LF #2 & Btr = $1.14 per LF)	MBF	2,350.00	1,140.00

		#3 & Btr	#4 & Btr
Engelmann Spruce boards Random lengths.			
1" x 4" (#3 & Btr = $.18 per LF #4 & Btr = $.14 per LF)	MBF	535.00	425.00
1" x 6" (#3 & Btr = $.29 per LF #4 & Btr = $.22 per LF)	MBF	585.00	440.00
1" x 8" (#3 & Btr = $.38 per LF #4 & Btr = $.27 per LF)	MBF	575.00	430.00
1" x 10" (#3 & Btr = $.42 per LF #4 & Btr = $.36 per LF)	MBF	500.00	430.00
1" x 12" (#3 & Btr = $.62 per LF #4 & Btr = $.43 per LF)	MBF	620.00	430.00

		D & Btr	#4 & Btr
Fir & Larch boards Random lengths.			
1" x 4" (D & Btr = $.50 per LF #4 & Btr = $.20 per LF)	MBF	1,450.00	550.00
1" x 6" (D & Btr = $.91 per LF #4 & Btr = $.41 per LF)	MBF	1,750.00	772.00
1" x 8" (D & Btr = $1.37 per LF #4 & Btr = $.49 per LF)	MBF	1,960.00	638.00
1" x 10" (D & Btr = $1.62 per LF #4 & Btr = $.75 per LF)	MBF	1,860.00	519.00
1" x 12" (D & Btr = $2.50 per LF #4 & Btr = $1.05 per LF)	MBF	2,390.00	778.00

	Craft@Hrs	Unit	Material	Labor	Total
Red Cedar boards S1S2E, random lengths, #3 and better.					
1" x 4" ($.28 per LF)	—	MBF	835.00	—	835.00
1" x 6" ($.45 per LF)	—	MBF	900.00	—	900.00
1" x 8" ($.74 per LF)	—	MBF	1,110.00	—	1,110.00
1" x 10" ($.92 per LF)	—	MBF	1,100.00	—	1,100.00
1" x 12" ($1.25 per LF)	—	MBF	1,250.00	—	1,250.00

		Clear Select	B Grade

Redwood Random lengths 6' to 20', S4S kiln dried. West Coast prices. Add 10% to 30% for transportation to other areas.

	Unit	Clear Select	B Grade
1" x 2" (Clear Select = $.27per LF B Grade = $.23 per LF)	MBF	1,598.00	1,363.00
1" x 3" (Clear Select = $.40 per LF B Grade = $.34per LF)	MBF	1,598.00	1,363.00
1" x 4" (Clear Select = $.53 per LF B Grade = $.45 per LF)	MBF	1,598.00	1,363.00
1" x 6" (Clear Select = $.83 per LF B Grade = $.71 per LF)	MBF	1,651.00	1,416.00
1" x 8" (Clear Select = $1.20 per LF B Grade = $1.04 per LF)	MBF	1,821.00	1,576.00
1" x 10" (Clear Select = $1.52 per LF B Grade = $1.31 per LF)	MBF	1,821.00	1,576.00
1" x 12" (Clear Select = $1.98 per LF B Grade = $1.74 per LF)	MBF	1,981.00	1,736.00
2" x 4" (Clear Select = $1.20 per LF B Grade = $1.04 per LF)	MBF	1,800.00	1,555.00
2" x 6" (Clear Select = $2.00 per LF B Grade = $1.73 per LF)	MBF	2,002.00	1,736.00
2" x 8" (Clear Select = $2.67 per LF B Grade = $2.31 per LF)	MBF	2,002.00	1,736.00
2" x 10" (Clear Select = $3.53 per LF B Grade = $3.12per LF)	MBF	2,119.00	1,874.00
2" x 12" (Clear Select = $4.86 per LF B Grade = $4.32 per LF)	MBF	2,428.00	2,162.00
4" x 4" (Clear Select = $3.39 per LF B Grade = $2.93 per LF)	MBF	2,545.00	2,194.00
4" x 6" (Clear Select = $5.09 per LF B Grade = $4.43per LF)	MBF	2,545.00	2,215.00

		C & Btr	D

Southern Pine boards Random lengths.

	Unit	C & Btr	D
1" x 4" (C & Btr = $.44 per LF D = $.41 per LF)	MBF	1,320.00	1,220.00
1" x 6" (C & Btr = $.76 per LF D = $.67 per I F)	MBF	1,520.00	1,330.00
1" x 8" (C & Btr = $.99 per LF D = $.89 per LF)	MBF	1,480.00	1,330.00
1" x 10" (C & Btr = $1.29 per LF D – $1.02 per LF)	MBF	1,550.00	1,220.00
1" x 12" (C & Btr = $1.73 per LF D = $1.29 per LF)	MBF	1,730.00	1,290.00

		#2 & Btr	#3

Southern Pine boards Random lengths.

	Unit	#2 & Btr	#3
1" x 4" (#2 & Btr = $.19 per LF #3 = $.16 per LF)	MBF	585.00	495.00
1" x 6" (#2 & Btr = $.35 per LF #3 = $.23 per LF)	MBF	690.00	460.00
1" x 8" (#2 & Btr = $.57 per LF #3 = $.32 per LF)	MBF	850.00	480.00
1" x 10" (#2 & Btr = $.77 per LF #3 = $.50 per LF)	MBF	925.00	595.00
1" x 12" (#2 & Btr = $1.07 per LF #3 = $.61 per LF)	MBF	1,070.00	610.00

		C & Btr	D Grade

Fir Flooring 4' to 20' random lengths, T&G, Vertical Grain or Flat Grain

	Unit	C & Btr	D Grade
1" x 3" flooring, V/G (C & Btr = $.68 per LF D grade = $.49)	MBF	2,720.00	1,950.00
1" x 4" flooring, V/G (C & Btr = $.91 per LF D grade = $.65)	MBF	2,720.00	1,950.00
1" x 3" flooring, F/G (C & Btr = $.41 per LF D grade = $.32)	MBF	1,640.00	1,280.00
1" x 4" flooring, F/G (C & Btr = $.55 per LF D grade = $.43)	MBF	1,640.00	1,280.00

	Craft@Hrs	Unit	Material	Labor	Total
Southern Yellow Pine decking Random lengths, premium grade.					
5/4" x 6" ($.88 per LF)	—	MBF	1,400.00		1,400.00
Southern Yellow Pine flooring Random lengths, D grade.					
1" x 4" ($.41 per LF)	—	MBF	1,220.00	—	1,220.00
1" x 6" ($.67 per LF)	—	MBF	1,330.00	—	1,330.00
1" x 8" ($.89 per LF)	—	MBF	1,330.00	—	1,330.00
1" x 10" ($1.02 per LF)	—	MBF	1,220.00	—	1,200.00
1" x 12" ($1.29 per LF)	—	MBF	1,290.00	—	1,290.00

Lumber

Douglas fir finish and clear S4S, dry, 8' to 20' specified lengths, vertical or flat grain. West Coast prices. Add 10% to 30% for transportation to other areas.

			C & Btr (V/G)	C & Btr (F/G)
1" x 2" (V/G = $.49 per LF F/G = $.42 per LF)		MBF	2,950.00	2,540.00
1" x 3" (V/G = $.73 per LF F/G = $.64 per LF)		MBF	2,950.00	2,540.00
1" x 4" (V/G = $.98 per LF F/G = $.84 per LF)		MBF	2,950.00	2,540.00
1" x 6" (V/G = $1.67 per LF F/G = $1.32 per LF)		MBF	3,350.00	2,660.00
1" x 8" (V/G = $2.72 per LF F/G = $2.26 per LF)		MBF	4,090.00	3,420.00
1" x 10" (V/G = $3.79 per LF F/G = $3.53 per LF)		MBF	4,550.00	4,280.00
1" x 12" (V/G = $4.87 per LF F/G = $4.33 per LF)		MBF	4,870.00	4,370.00
2" x 2" (V/G = $.98 per LF F/G = $.77 per LF)		MBF	2,950.00	2,320.00
2" x 3" (V/G = $1.48 per LF F/G = $1.16 per LF)		MBF	2,950.00	2,320.00
2" x 4" (V/G = $2.01 per LF F/G = $1.55 per LF)		MBF	3,010.00	2,350.00
2" x 6" (V/G = $3.35 per LF F/G = $2.70 per LF)		MBF	3,350.00	2,720.00
2" x 8" (V/G = $5.04 per LF F/G = $3.63 per LF)		MBF	3,790.00	2,740.00
2" x 10" (V/G = $7.85 per LF F/G = $4.53 per LF)		MBF	4,710.00	2,740.00
2" x 12" (V/G = $12.20 per LF F/G = $7.15 per LF)		MBF	6,100.00	3,610.00

Particleboard, underlayment

		Standard	Industrial
3/8" thick	MSF	255.00	310.00
1/2" thick	MSF	290.00	350.00
5/8" thick	MSF	340.00	405.00
3/4" thick	MSF	415.00	455.00

Sanded exterior grade plywood

		AA	AB	AC	BC
1/4" thick	MSF	755.00	730.00	545.00	485.00
3/8" thick	MSF	830.00	805.00	620.00	555.00
1/2" thick	MSF	960.00	940.00	750.00	690.00
5/8" thick	MSF	1,090.00	1,070.00	885.00	820.00
3/4" thick	MSF	1,170.00	1,150.00	950.00	890.00

Sanded interior grade plywood

		AA	AB	AD	BD
1/4" thick	MSF	740.00	720.00	530.00	470.00
3/8" thick	MSF	815.00	790.00	605.00	540.00
1/2" thick	MSF	950.00	925.00	735.00	675.00
5/8" thick	MSF	1,080.00	1,060.00	870.00	805.00
3/4" thick	MSF	1,150.00	1,130.00	935.00	875.00

Exterior grade
Sheathing, mill grade

		CD	CC	Mill Grade
5/16" thick	MSF	310.00	340.00	265.00
3/8" thick	MSF	325.00	355.00	300.00
1/2" thick	MSF	405.00	—	335.00
1/2" thick, 4 ply	MSF	405.00	490.00	—
1/2" thick, 5 ply	MSF	415.00	—	345.00
5/8" thick, 4 ply	MSF	505.00	595.00	420.00
3/4" thick	MSF	610.00	720.00	570.00

Identification Index: A set of two numbers separated by a slash that appears in the grade trademarks on standard sheathing, Structural I, Structural II and C-C grades. Number on left indicates recommended maximum spacing in inches for supports when panel is used for roof decking. Number on right shows maximum recommended spacing in inches for supports when the panel is used for subflooring.

A-A EXT-DFPA
Designed for use in exposed applications where the appearance of both sides is important. Fences, wind screens, exterior cabinets and built-ins, signs, boats, commercial refrigerators, shipping containers, tanks, and ducts

A-A G-2 EXT-DFPA PS 1-6
Typical edge-mark

A-B EXT-DFPA
For uses similar to A-A EXT where the appearance of one side is a little less important than the face.

A-B G-3 EXT-DFPA PS 1-6
Typical edge-mark

		Sq. Edged	T & G
Strand board Oriented grain (OSB).			
1/4" thick	MSF	225.00	—
3/8" thick	MSF	275.00	—
7/16" thick	MSF	300.00	—
1/2" thick	MSF	330.00	—
5/8" thick	MSF	—	450.00
3/4" thick	MSF	—	530.00

	Craft@Hrs	Unit	Material	Labor	Total
Underlayment. X-Band, C Grade, tongue and groove.					
5/8" thick, 4 ply	—	MSF	490.00	—	490.00
3/4" thick, 4 ply	—	MSF	570.00	—	570.00
Concrete form (Plyform), square edged.					
5/8" thick, 4 ply	—	MSF	805.00	—	805.00
3/4" thick, 4 ply	—	MSF	880.00	—	880.00

			T-1-11	RBB
Siding, rough sawn, 4 ply.				
3/8" thick, 4' x 8' panels		MSF	575.00	—
5/8" thick, 4' x 8' panels		MSF	820.00	895.00
3/8" thick, 4' x 9' panels		MSF	780.00	—
5/8" thick, 4' x 9' panels		MSF	1,030.00	1,100.00

	Craft@Hrs	Unit	Material	Labor	Total
Lath Redwood "A" & better, 50 per bundle, nailed to walls for architectural treatment.					
5/16" x 1-1/2" x 4'	BC@.500	Bdle	14.90	12.60	27.50
5/16" x 1-1/2" x 6'	BC@.625	Bdle	29.10	15.70	44.80
5/16" x 1-1/2" x 8'	BC@.750	Bdle	34.90	18.90	53.80

Poles Select appearance grade building poles. Poles have an average taper of 1" in diameter per 10' of length. Ponderosa pine treated with Penta-Dow or ACZA (.60 lbs/CF), meets AWPA-C 23-77. Costs based on quantity of 8 to 10 poles. For delivery of smaller quantities, add $50.00 to the total cost. Add for labor to install on the next page. Costs are per linear foot.

		8" butt size	10" butt size	12" butt size
To 15' long pole	LF	7.42	9.68	12.80
To 20' long pole	LF	7.20	9.14	11.94
To 25' long pole	LF	6.99	7.63	10.86
Add for Douglas fir	LF	1.51	1.51	1.51
Add for delivery, typical per trip	Ea	39.78	39.78	39.78

		6' long	7' long	8' long	10' long
Posts, sawn posts, merchantable, redwood.					
4" x 4"	Ea	5.94	6.68	9.41	11.82
Yellow pine penta treated posts					
3" x 3"	Ea	5.00	6.09	7.06	8.15
3-1/2" x 3-1/2"	Ea	5.63	6.61	7.48	8.56
4" x 4"	Ea	6.97	8.15	8.93	10.00
4-1/2" x 4-1/2"	Ea	8.45	8.36	9.23	10.32
5" x 5"	Ea	8.31	10.21	10.57	12.38
6" x 6"	Ea	9.96	11.29	11.84	13.23

Lumber

	Craft@Hrs	Unit	Material	Labor	Total

Labor to install poles and posts Costs shown are per post and include measuring, cutting one end, drilling two holes at the top and bottom and installation. Add for hardware and bolts as needed.

	Craft@Hrs	Unit	Material	Labor	Total
3" x 3", 4" x 4" or 5" x 5" posts	B1@.333	Ea	—	7.61	7.61
6" x 6" or 7" x 7" posts	B1@.500	Ea	—	11.40	11.40
8" or 10" butt diameter poles	B1@.666	Ea	—	15.20	15.20
12" butt diameter poles	B1@.750	Ea	—	17.10	17.10

Board siding See also Siding and Hardboard.

Douglas fir drop siding, flat grain, coverage and 5% waste included, rustic or shiplap patterns

C grade and better siding

	Craft@Hrs	Unit	Material	Labor	Total
1" x 6" ($2.87/BF, 1.20 BF per SF)	BC@3.16	Sq	362.00	79.50	441.50
1" x 8" ($3.28/BF, 1.18 BF per SF)	BC@3.16	Sq	406.00	79.50	485.50

D grade siding

	Craft@Hrs	Unit	Material	Labor	Total
1" x 6" ($2.56/BF, 1.20 BF per SF)	BC@3.16	Sq	323.00	79.50	402.50
1" x 8" ($2.74/BF, 1.18 BF per SF)	BC@3.16	Sq	340.00	79.50	419.50

Redwood siding Clear all heart, resawn, VG KD, random lengths, includes 5% waste and coverage

	Craft@Hrs	Unit	Material	Labor	Total
1/2" x 4", 1.51 BF per SF at 2-3/4" exposure	BC@2.77	Sq	337.00	69.70	406.70
5/8" x 6", 1.31 BF per SF at 4-3/4" exposure	BC@2.68	Sq	293.00	67.40	360.40
5/8" x 8", 1.23 BF per SF at 6-3/4" exposure	BC@2.68	Sq	309.00	67.40	376.40
3/4" x 6", 1.31 BF per SF at 4-3/4" exposure	BC@2.68	Sq	304.00	67.40	371.40
3/4" x 8", 1.23 BF per SF at 6-3/4" exposure	BC@2.68	Sq	379.00	67.40	446.40
3/4" x 10", 1.19 BF per SF at 8-3/4" exposure	BC@2.58	Sq	416.00	64.90	480.90
Deduct for clear all heart, flat grain	—	%	-10.0	—	—
Deduct for clear flat grain	—	%	-15.0	—	—
Deduct for green	—	Sq	-50.00	—	-50.00
Deduct for clear vertical grain	—	%	-10.0	—	—

Rustic, anzac or drop pattern redwood siding

	Craft@Hrs	Unit	Material	Labor	Total
1" x 4" ($3,140/MBF, 1.25 BF per SF)	BC@2.48	Sq	431.00	62.40	493.40
1" x 6" ($3,360/MBF, 1.17 BF per SF)	BC@2.30	Sq	437.00	57.90	494.90
1" x 8" ($3,440/MBF, 1.12 BF per SF)	BC@2.30	Sq	426.00	57.90	483.90
1" x 10" ($3,060/MBF, 1.11 BF per SF)	BC@2.30	Sq	375.00	57.90	432.90
1" x 12" ($3,620/MBF, 1.09 BF per SF)	BC@2.30	Sq	437.00	57.90	494.90
Deduct for clear all-heart flat grain	—	%	-10.0	—	—
Deduct for clear vertical grain	—	%	-15.0	—	—
Add for resawn face	—	Sq	10.00	—	10.00
Add for specified lengths	—	Sq	10.00	—	10.00

Western red cedar clear bevel siding Smooth face, includes waste and coverage allowance

	Craft@Hrs	Unit	Material	Labor	Total
1/2" x 4", 1.51 BF per SF at 2-3/4" exposure	BC@2.77	Sq	371.00	69.70	440.70
1/2" x 6", 1.31 BF per SF at 4-3/4" exposure	BC@2.68	Sq	353.00	67.40	420.40
1/2" x 8", 1.23 BF per SF at 6-3/4" exposure	BC@2.68	Sq	322.00	67.40	389.40
3/4" x 8", 1.23 BF per SF at 6-3/4" exposure	BC@2.68	Sq	322.00	67.40	389.40
3/4" x 10", 1.19 BF per SF at 8-3/4" exposure	BC@2.68	Sq	327.00	67.40	394.40

Western red cedar "A" grade bevel siding Smooth face, dry, includes waste and coverage allowance

	Craft@Hrs	Unit	Material	Labor	Total
1/2" x 4", 1.51 BF per SF at 2-3/4" exposure	BC@2.77	Sq	439.00	69.70	508.70
1/2" x 6", 1.31 BF per SF at 4-3/4" exposure	BC@2.68	Sq	349.00	67.40	416.40
1/2" x 8", 1.23 BF per SF at 6-3/4" exposure	BC@2.68	Sq	334.00	67.40	401.40
3/4" x 8", 1.23 BF per SF at 6-3/4" exposure	BC@2.68	Sq	355.00	67.40	422.40
3/4" x 10", 1.19 BF per SF at 8-3/4 "exposure	BC@2.58	Sq	323.00	64.90	387.90

	Craft@Hrs	Unit	Material	Labor	Total
Western red cedar "B" grade bevel siding Smooth face, dry, includes waste and coverage allowance					
1/2" x 6", 1.29 BF per SF at 4-3/4" exposure	BC@2.68	Sq	298.00	67.40	365.40
1/2" x 8", 1.32 BF per SF at 6-3/4" exposure	BC@2.68	Sq	294.00	67.40	361.40
3/4" x 8", 1.32 BF per SF at 6-3/4" exposure	BC@2.68	Sq	294.00	67.40	361.40
3/4" x 10", 1.27 BF per SF at 8-3/4" exposure	BC@2.58	Sq	290.00	64.90	354.90
Spruce shiplap siding No. 2 and better, dry, random lengths					
1" x 6", 1.27 BF per SF at 5-1/4" exposure	BC@2.30	Sq	184.00	57.90	241.90
1" x 8", 1.22 BF per SF at 7-1/4" exposure	BC@2.30	Sq	184.00	57.90	241.90
Southern yellow pine, drop siding D grade, random lengths					
1" x 6", 1.27 BF per SF at 5-1/4" exposure	BC@2.39	Sq	153.00	60.10	213.10
1" x 8", 1.22 BF per SF at 7-1/4" exposure	BC@2.39	Sq	163.00	60.10	223.10

Stakes

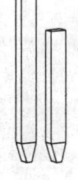

	Craft@Hrs	Unit	Material	Labor	Total
Survey stakes, pine or fir, 1" x 2", per 1,000					
24"	—	M	307.00	—	307.00
36"	—	M	466.00	—	466.00
48"	—	M	636.00	—	636.00
Ginnies, pine or fir, 1" x 1", per 1,000					
6"	—	M	71.00	—	71.00
8"	—	M	76.00	—	76.00
Pointed lath, pine or fir, per 1,000, 1/4" x 1-1/2", S4S					
18"	—	M	170.00	—	170.00
24"	—	M	212.00	—	212.00
36"	—	M	339.00	—	339.00
48"	—	M	403.00	—	403.00
Foundation stakes, 1" x 3", per 1,000					
12" or 14"	—	M	300.00	—	300.00
18"	—	M	352.00	—	352.00
24"	—	M	421.00	—	421.00
36"	—	M	544.00	—	544.00
48"	—	M	737.00	—	737.00
Hubs, pine or fir, 2" x 2", per 1,000					
8"	—	M	382.00	—	382.00
10"	—	M	382.00	—	382.00
12"	—	M	509.00	—	509.00
Form stakes (wedges), pine or fir, 5" point, per 1,000					
2" x 2" x 6"	—	M	180.00	—	180.00
2" x 2" x 8"	—	M	196.00	—	196.00
2" x 2" x 10"	—	M	223.00	—	223.00
2" x 4" x 6"	—	M	445.00	—	445.00
2" x 4" x 8"	—	M	477.00	—	477.00
2" x 4" x 10"	—	M	498.00	—	498.00
2" x 4" x 12"	—	M	562.00	—	562.00
Feather wedges	—	M	117.00	—	117.00

Lumber Grading

	Craft@Hrs	Unit	Material	Labor	Total

Lumber Grading Quality inspector daily rate: $375.00. Half day rate: $220.00 plus actual expenses. For lumber in remote areas, the daily rate applied is $525.00 plus actual travel costs. Rates assume adequate lifting equipment and a helper are available.

	Craft@Hrs	Unit	Material	Labor	Total
1" lumber (15 MBF per day)	—	MBF	—	—	14.10
2" x 4", 6" lumber (20 MBF per day)	—	MBF	—	—	13.00
2" x 8", 10", 12" (28 MBF per day)	—	MBF	—	—	11.30
4" x 4", 6" (30 MBF per day)	—	MBF	—	—	11.30
4" x 8" and over (28 MBF per day)	—	MBF	—	—	13.00
Lumber inventory verification, excluding travel	—	Day	—	—	237.00
Court appearance, plus preparation and travel	—	Day	—	—	237.00
Lumber grading training (on site)	—	Day	—	—	430.00

Lumber Preservation Treatments, Subcontract See also Soil Treatments. Costs assume treatment at a treatment plant, not on the construction site. Treatments listed below are designed to protect wood and wood products from decay and destruction caused by moisture (fungi), wood boring insects and organisms. All retention rates listed (in parentheses) are in pounds per cubic foot and are the minimum recommended by the Society of American Wood Preservers. Costs are for treatment only and assume treating of 10,000 board feet or more. No lumber or transportation included. Add $70.00 per MBF for kiln-drying after treatment.

	Craft@Hrs	Unit	Material	Labor	Total
Lumber submerged in or frequently exposed to salt water					
ACZA treatment (2.5)	—	MBF	—	—	687.00
ACZA treatment for structural lumber					
Piles and building poles (2.5)	—	CF	—	—	7.70
Board lumber and timbers (2.5)	—	MBF	—	—	687.00
Creosote treatment for structural lumber					
Piles and building poles (20.0)	—	CF	—	—	6.88
Board lumber and timbers (25.0)	—	MBF	—	—	407.00
Lumber in contact with fresh water or soil, ACZA treatment					
Building poles, structural (.60)	—	CF	—	—	4.29
Boards, timbers, structural (.60)	—	MBF	—	—	198.00
Boards, timbers, non-structural (.40)	—	MBF	—	—	176.00
Plywood, structural (.40), 1/2"	—	MSF	—	—	198.00
Posts, guardrail, signs (.60)	—	MBF	—	—	245.00
All weather foundation KD lumber (.60)	—	MBF	—	—	275.00
All weather foundation, 1/2" plywood (.60)	—	MSF	—	—	220.00
Lumber in contact with fresh water or soil, creosote treatment					
Piles, building poles, structural (12.0)	—	CF	—	—	4.84
Boards, timbers, structural (12.0)	—	MBF	—	—	297.00
Boards, timbers, non-structural (10.0)	—	MBF	—	—	275.00
Plywood, structural, 1/2" (10.0)	—	MSF	—	—	237.00
Posts (fence, guardrail, sign) (12.0)	—	MBF	—	—	297.00
Lumber used above ground, ACZA treatment					
Boards, timbers, structural (.25)	—	MBF	—	—	160.00
Boards, timbers, non-structural (.25)	—	MBF	—	—	160.00
Plywood, structural 1/2" (.25)	—	MSF	—	—	182.00
Lumber used above ground, creosote treatment					
Boards, timbers, structural (8.0)	—	MBF	—	—	248.00
Boards, timbers, non-structural (8.0)	—	MBF	—	—	248.00
Plywood, structural (8.0), 1/2"	—	MSF	—	—	220.00

	Craft@Hrs	Unit	Material	Labor	Total
Lumber used for decking, ACQ treatment					
Boards, structural (.41)	—	MBF	—	—	215.00
Boards, timbers, non-structural (.41)	—	MSF	—	—	215.00
Plywood and other 1/2" sheets (.41)	—	MSF	—	—	231.00
Millwork (.26)	—	MBF	—	—	193.00
Fire retardant treatments, structural use (interior only), "D-Blaze"					
Surfaced lumber	—	MBF	—	—	260.00
Rough lumber	—	MBF	—	—	260.00
Architectural use (interior and exterior), "NCX"					
Surfaced lumber	—	MBF	—	—	606.00
Rough lumber	—	MBF	—	—	606.00
Wolmanized treatment					
Lumber used above ground (.25)	—	MBF	—	—	92.00
Lumber in ground or fresh water (.40)	—	MBF	—	—	116.00
Lumber used as foundation (.60)	—	MBF	—	—	126.00

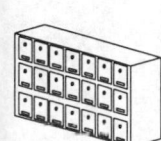

Mailboxes Apartment type, meets current postal regulations, tumbler locks, aluminum or gold finish, price per unit, recessed

	Craft@Hrs	Unit	Material	Labor	Total
3 box unit, 5-1/2" wide x 16" high	B1@1.88	Ea	68.00	43.00	111.00
4 box unit, 5-1/2" wide x 16" high	B1@1.88	Ea	101.00	43.00	144.00
5 box unit, 5-1/2" wide x 16" high	B1@1.88	Ea	161.00	43.00	204.00
6 box unit, 5-1/2" wide x 16" high	B1@1.88	Ea	135.00	43.00	178.00
7 box unit, 5-1/2" wide x 16" high	B1@1.88	Ea	158.00	43.00	201.00
Add to any above for surface mounted	—	Ea	12.00	—	12.00

Office type, horizontal, meets postal regulations, tumbler locks, aluminum or brass finish
For front or rear load, per box

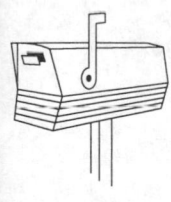

	Craft@Hrs	Unit	Material	Labor	Total
6" wide x 5" high x 15" deep	B1@1.02	Ea	30.00	23.30	53.30
Rural, aluminum finish, 19" long x 6-1/2" wide x 9" high, with flag, Post mount	B1@.776	Ea	12.90	17.70	30.60
Rural, #1-1/2 size, aluminum finish, 19" long x 6-1/2" wide x 11" high, with flag, Post mount	B1@.776	Ea	30.10	17.70	47.80
Suburban style, finished, 16" long x 3" wide x 6" deep, With flag, post mount	B1@.776	Ea	24.70	17.70	42.40
Add for redwood or wrought iron post	—	Ea	14.00	—	14.00
Thru-the-wall type, 13" long x 3" high, Gold or aluminum finish	B1@1.25	Ea	31.20	28.60	59.80
Wall-mounting security type, lockable, 15" wide x 14" high x 8" deep, Black plastic	B1@1.25	Ea	76.30	28.60	104.90

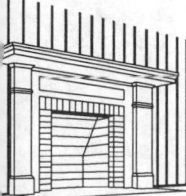

Mantels, Ponderosa Pine, Unfinished

	Craft@Hrs	Unit	Material	Labor	Total
Modern design, 50" x 37" opening, paint grade, 11" x 77" shelf	B1@3.81	Ea	447.00	87.10	534.10
Colonial design, 50" x 39" opening, stain grade, 7" x 72" shelf	B1@3.81	Ea	714.00	87.10	801.10
Ornate design, 51-1/2" x 39" opening, with detailing and scroll work, stain grade, 11" x 68" shelf	B1@3.81	Ea	1,330.00	87.10	1,417.10
Mantel moulding (simple trim for masonry fireplace openings), 15 LF, for opening 6'0" x 3'6" or less	B1@1.07	Set	156.00	24.50	180.50

Mantels

	Craft@Hrs	Unit	Material	Labor	Total
Mantel shelf only, with iron or wood support brackets, prefinished					
4" x 8" x 8"	B1@.541	Ea	40.90	12.40	53.30
4" x 8" x 6"	B1@.541	Ea	31.20	12.40	43.60
4" x 10" x 10"	B1@.541	Ea	70.00	12.40	82.40

Marble Setting, Subcontract See also Tile section. Marble prices vary with quality, quantity, and availability of materials. Check prices with local suppliers. These prices are based on 200 SF quantities.

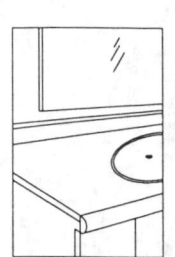

Marble tile flooring, tiles or slabs

3/8" x 12" x 12", thin set over concrete slab, based on $5.00 to $8.00 per SF material cost and $7.00 to $10.00 per SF labor cost. (Includes cutting)

Typical cost per SF of area covered	—	SF	—	—	15.00

3/4" slabs set in mortar bed, cut to size, based on $30.00 to $40.00 per SF material cost and $15.00 per SF labor cost

Typical cost per SF of area covered	—	SF	—	—	50.00

Marble bathroom vanity top, 6' x 2' wide, with 1 sink cutout and 4" splash on 3 sides

One per home, based on $560.00 to $860.00 per unit material cost and $150.00 per unit labor cost

Single top, typical cost	—	Ea	—	—	860.00

Four per home, based on $560.00 to $860.00 per unit material cost and $50.00 per unit labor cost

Multiple tops, typical cost	—	Ea	—	—	760.00

Marble fireplace hearth and facing, one per home, based on $400.00 to $550.00 per unit material cost and $260.00 to $460.00 per unit labor cost

Typical cost per unit	—	Ea	—	—	835.00

Markup Typical markup for light construction, percentage of gross contract price. No insurance or taxes included. See also Insurance and Taxes sections. Markup is the amount added to an estimate after all job costs have been accounted for: labor and materials, job fees, permits, bonds and insurance, job site supervision and similar job-related expenses. Profit is the return on money invested in the construction business.

Contingency (allowance for unknown and unforeseen conditions)

Varies widely from job to job, typically	—	%	—	—	2.0

Overhead (office, administrative and general business expense)

Typical overhead expense	—	%	—	—	10.0

Profit (return on the money invested in the business, varies with competitive conditions)

Typical profit for residential construction	—	%	—	—	8.0

Total markup (add 25% to the total job cost to attain 20% of gross contract price)

Total for Contingency, Overhead & Profit	—	%	—	—	20.0

Note that costs listed in this book do not include the general contractor's markup. However, sections identified as "subcontract" include the subcontractor's markup. Here's how to calculate markup:

Project Costs	
Material	$50,000
Labor (with taxes & insurance)	40,000
Subcontract	10,000
Total job costs	100,000
Markup (25% of $100,000)	**25,000**
Contract price	$125,000

Markup is $25,000 which is 25% of the total job costs ($100,000) but only 20% of the contract price ($25,000 divided by $125,000 is .20).

Markup For General Contractors Handling larger commercial and industrial projects, a markup of 24.8% on the total job cost is suggested. See the paragraph General Contractor Markup at the beginning of the section General Requirements in the Industrial and Commercial Division of this manual. For example, note how markup is calculated on a project with these costs:

Material	$3,000,000
Labor (with taxes and insurance)	4,500,000
Equipment	1,000,000
Subcontracts	1,500,000
Total job cost	$10,000,000
Indirect overhead (8%)	**800,000**
Direct overhead (7.3%)	**730,000**
Contingency (2%)	**200,000**
Profit (7.5%)	**750,000**
Contract price	$12,480,000

The total job cost shown above includes work done with the general contractor's crew and work done by subcontractor-specialists. Note that markup for the general contractor (overhead, including supervision expense, contingency and profit) is $2,480,000. This amounts to 24.8% of the total job costs but only 19.87% of the contract price (2,480,000 divided by 12,480,000 is 0.1987).

Masonry, Brick Wall Assemblies Typical costs for smooth red clay brick walls, laid in running bond with 3/8" concave joints. These costs include the bricks (cost of bricks only shown in parentheses), mortar for bricks and cavities, typical ladder type reinforcing, wall ties and normal waste. Foundations are not included. Wall thickness shown is the nominal size based on using the type bricks described. "Wythe" means the quantity of bricks in the thickness of the wall. Costs shown are per square foot (SF) of wall measured on one face. Deduct for openings over 10 SF in size. The names and dimensions of bricks can be expected to vary, depending on the manufacturer.

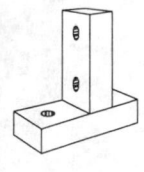

	Craft@Hrs	Unit	Material	Labor	Total
Standard bricks 3-3/4" wide x 2-1/4" high x 8" long (at $292 per M)					
4" thick wall, single wythe, veneer facing	B9@.211	SF	2.48	4.89	7.37
8" thick wall, double wythe, cavity filled	B9@.464	SF	4.45	10.70	15.15
12" thick wall, triple wythe, cavity filled	B9@.696	SF	6.42	16.10	22.52
Norman bricks 3-1/2" wide x 2-1/2" high x 11-1/2" long (at $468 per M)					
4" thick wall, single wythe, veneer facing	B9@.211	SF	2.59	4.89	7.48
8" thick wall, double wythe, cavity filled	B9@.464	SF	4.67	10.70	15.37
12" thick wall, triple wythe, cavity filled	B9@.696	SF	6.74	16.10	22.84
Modular bricks, 3" wide x 3-1/2" x 11-1/2" long (at $586 per M)					
3-1/2" thick wall, single wythe, veneer face	B9@.156	SF	2.43	3.61	6.04
7-1/2" thick wall, double wythe, cavity filled	B9@.343	SF	4.35	7.95	12.30
11-1/2" thick wall, triple wythe, cavity filled	B9@.515	SF	6.28	11.90	18.18
Colonial bricks, 3" wide x 3-1/2" x 10" long (at $267 per M)					
3-1/2" thick wall, single wythe, veneer face	B9@.177	SF	1.51	4.10	5.61
7-1/2" thick wall, double wythe, cavity filled	B9@.389	SF	2.52	9.01	11.53
11-1/2" thick wall, triple wythe, cavity filled	B9@.584	SF	3.53	13.50	17.03

Masonry, Face Brick Laid in running bond with 3/8" concave joints. Mortar, wall ties and foundations not included. Includes normal waste and delivery up to 30 miles with a 5 ton minimum. Coverage (units per SF) includes 3/8" mortar joints but the SF cost does not include mortar cost. Add for mortar below.

	Craft@Hrs	Unit	Material	Labor	Total
Standard brick, 2-1/4" high x 3-3/4" deep x 8" long (6.45 units per SF)					
Brown (at $281 per M)	B9@.211	SF	1.81	4.89	6.70
Burnt oak (at $354 per M)	B9@.211	SF	2.28	4.89	7.17
Flashed (at $297 per M)	B9@.211	SF	1.92	4.89	6.81

	Craft@Hrs	Unit	Material	Labor	Total
Old English (at $297 per M)	B9@.211	SF	1.92	4.89	6.81
Red (at $271 per M)	B9@.211	SF	1.75	4.89	6.64
Norman brick, 2-1/2" high x 3-1/2" deep x 11-1/2" long (4.22 units per SF)					
Brown (at $469 per M)	B9@.211	SF	1.99	4.89	6.88
Burnt Oak (at $479 per M)	B9@.211	SF	2.02	4.89	6.91
Flashed (at $469 per M)	B9@.211	SF	1.99	4.89	6.88
Red (at $469 per M)	B9@.211	SF	1.99	4.89	6.88
Modular, 3-1/2" high x 3" deep x 11-1/2" long (3.13 units per SF)					
Brown (at $573 per M)	B9@.156	SF	1.79	3.61	5.40
Burnt oak (at $666 per M)	B9@.156	SF	2.08	3.61	5.69
Flashed (at $573per M)	B9@.156	SF	1.79	3.61	5.40
Red (at $573 per M)	B9@.156	SF	1.79	3.61	5.40
Jumbo (filler brick) 3-1/2" high x 3" deep x 11-1/2" long (3.13 units per SF)					
Fireplace, cored (at $312 per M)	B9@.156	SF	.98	3.61	4.59
Fireplace, solid (at $322 per M)	B9@.156	SF	1.01	3.61	4.62
Mission, cored (at $380 per M)	B9@.156	SF	1.19	3.61	4.80
Mission, solid (at $390 per M)	B9@.156	SF	1.22	3.61	4.83
Padre brick, 4" high x 3" deep					
7-1/2" long cored, 4.2 per SF, $.66 each	B9@.156	SF	2.77	3.61	6.38
7-1/2" long solid, 4.2 per SF, $.83 each	B9@.156	SF	3.49	3.61	7.10
11-1/2" long cored, 2.8 per SF, $.83 each	B9@.177	SF	2.32	4.10	6.42
11-1/2" long solid, 2.8 per SF, $1.04 each	B9@.177	SF	2.91	4.10	7.01
15-1/2" long cored, 2.1 per SF, $1.10 each	B9@.211	SF	2.31	4.89	7.20
15-1/2" long solid, 2.1 per SF, $1.38 each	B9@.211	SF	2.90	4.89	7.79
Commercial, 3-1/4" high x 3-1/4" deep					
Smooth, 8" long, 4.4 per SF, $.38 each	B9@.156	SF	1.67	3.61	5.28
Colonial, 10" long, 3.6 per SF, $.46 each	B9@.177	SF	1.66	4.10	5.76

Mortar for brick Based on an all-purpose mortar mix costing $5.70 per cubic foot (60 lb. bag at $2.85) used in single wythe walls laid with running bond. Includes normal waste. Costs are per SF of wall face using the brick sizes (width x height x length) listed.

	Craft@Hrs	Unit	Material	Labor	Total
4" x 2-1/2" to 2-5/8" x 8" bricks					
3/8" joints (5.5 CF per CSF)	—	SF	.32	—	.32
1/2" joints (7.0 CF per CSF)	—	SF	.40	—	.40
4" x 3-1/2" to 3-5/8" x 8" bricks					
3/8" joints (4.8 CF per CSF)	—	SF	.28	—	.28
1/2" joints (6.1 CF per CSF)	—	SF	.35	—	.35
4" x 4" x 8" bricks					
3/8" joints (4.2 CF per CSF)	—	SF	.23	—	.23
1/2" joints (5.3 CF per CSF)	—	SF	.30	—	.30
4" x 5-1/2" to 5-5/8" x 8" bricks					
3/8" joints (3.5 CF per CSF)	—	SF	.20	—	.20
1/2" joints (4.4 CF per CSF)	—	SF	.25	—	.25
4" x 2" x 12" bricks					
3/8" joints (6.5 CF per CSF)	—	SF	.37	—	.37
1/2" joints (8.2 CF per CSF)	—	SF	.29	—	.29
4" x 2-1/2" to 2-5/8" x 12" bricks					
3/8" joints (5.1 CF per CSF)	—	SF	.29	—	.29
1/2" joints (6.5 CF per CSF)	—	SF	.37	—	.37
4" x 3-1/2" to 3-5/8" x 12" brick					
3/8" joints (4.4 CF per CSF)	—	SF	.25	—	.25
1/2" joints (5.6 CF per CSF)	—	SF	.32	—	.32

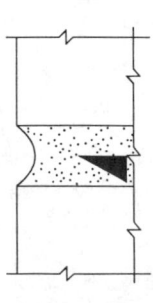

	Craft@Hrs	Unit	Material	Labor	Total
4" x 4" x 12" bricks					
3/8" joints (3.7 CF per CSF)	—	SF	.21	—	.21
1/2" joints (4.8 CF per CSF)	—	SF	.30	—	.30
4" x 5-1/2" to 5-5/8" x 12" brick					
3/8" joints (3.0 CF per CSF)	—	SF	.17	—	.17
1/2" joints (3.9 CF per CSF)	—	SF	.22	—	.22
6" x 2-1/2" to 2-5/8" x 12" bricks					
3/8" joints (7.9 CF per CSF)	—	SF	.45	—	.45
1/2" joints (10.2 CF per CSF)	—	SF	.58	—	.58
6" x 3-1/2" to 3-5/8" x 12" bricks					
3/8" joints (6.8 CF per CSF)	—	SF	.39	—	.39
1/2" joints (8.8 CF per CSF)	—	SF	.50	—	.50
6" x 4" x 12" bricks					
3/8" joints (5.6 CF per CSF)	—	SF	.32	—	.32
1/2" joints (7.4 CF per CSF)	—	SF	.42	—	.42
Add for 8" 22 gauge galvanized wall					
50 ties per 100 SF	B9@.011	SF	.05	.25	.30
Add for raked joints	B9@.027	SF	—	.63	.63
Add for flush joints	B9@.072	SF	—	1.67	1.67
Add for header course every 6th course	B9@.009	SF	.53	.21	.74
Add for Flemish bond					
With header course every 6th course	B9@.036	SF	.32	.83	1.15

Mini-brick veneer strips Figures in parentheses indicate coverage using 1/2" grouted joints. No waste included.

	Craft@Hrs	Unit	Material	Labor	Total
2-3/16" high x 7/16" thick x 7-1/2" long(6.75 units per SF)					
Unglazed, standard colors, $.24 each	B9@.300	SF	1.67	6.95	8.62
Unglazed, flashed, $.40 each	B9@.300	SF	2.78	6.95	9.73
Glazed, all colors, $.59 each	B9@.300	SF	4.10	6.95	11.05
2-3/16" high x 7/16" thick x 11-1/2" long(4.50 units per SF)					
Unglazed, standard colors, $.52 each	B9@.200	SF	2.41	4.63	7.04
Unglazed, flashed, $.86 each	B9@.200	SF	3.93	4.63	8.56
Glazed, all colors, $1.03 each	B9@.200	SF	4.77	4.63	9.40
Mini-brick veneer strip corners					
Unglazed, standard colors	B9@.077	Ea	.91	1.78	2.69
Unglazed, flashed	B9@.077	Ea	1.21	1.78	2.99
Glazed, all colors	B9@.077	Ea	1.49	1.78	3.27

Firebrick Used in chimneys and boilers, delivered to 30 miles, typical costs.

	Craft@Hrs	Unit	Material	Labor	Total
Standard backs, 9" x 4-1/2" x 2-1/2"	B9@30.9	M	844.00	716.00	1,560.00
Split backs, 9" x 4-1/2" x 1-1/2"	B9@23.3	M	844.00	540.00	1,384.00
Fireclay, Dosch clay					
10 lb sack	—	Ea	1.22	—	1.22
50 lb sack	—	Ea	3.98	—	3.98

Brick pavers Costs assume simple pattern with 1/2" mortar joints. Delivered to 30 miles, 10 ton minimum. Costs in material column include 8% for waste and breakage.

	Craft@Hrs	Unit	Material	Labor	Total
Adobe paver (Fresno)					
6" x 12" (2 units per SF, $.49 each)	B9@.193	SF	1.06	4.47	5.53
12" x 12" (1 unit per SF, $.77 each)	B9@.135	SF	.83	3.13	3.96

Masonry

	Craft@Hrs	Unit	Material	Labor	Total
California paver, 1-1/4" thick, brown					
3-5/8" x 7-5/8" (4.5 units per SF, $.45 each)	B9@.340	SF	2.19	7.88	10.07
3-5/8" x 11-5/8" (3 units per SF, $.56 each)	B9@.222	SF	1.81	5.14	6.95
Giant mission paver, 2-3/8" thick x 5-5/8" x 11-5/8"					
Red (2 units per SF, $.69 each)	B9@.193	SF	1.49	4.47	5.96
Redskin (2 units per SF, $.78 each)	B9@.193	SF	1.68	4.47	6.15
Mini-brick paver, 7/16" thick x 3-5/8" x 7-5/8"					
Unglazed, standard colors, (4.5 units per SF, $.39 each)	B9@.340	SF	1.89	7.88	9.77
Unglazed, flashed (4.5 units per SF $.70 each)	B9@.340	SF	3.40	7.88	11.28
Glazed, all colors (4.5 units per SF $.69 each)	B9@.340	SF	3.35	7.88	11.23
Norman paver, 2-1/2" thick x 3-1/2" x 11-1/2" (3 units per SF, $.80 each)	B9@.222	SF	2.40	5.14	7.54
Padre paver, 1-1/4" thick x 7-1/2" x 7-1/2" (2.5 units per SF, $1.12 each)	B9@.211	SF	3.02	4.89	7.91
Mortar joints for pavers, 1/2" typical	B9@.010	SF	.45	.23	.68

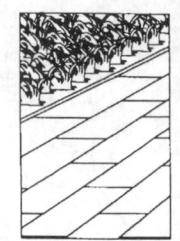

Concrete block wall assemblies Typical costs for standard natural gray medium weight masonry block walls including blocks, mortar, typical reinforcing and normal waste. Foundations are not included. See also individual component material and labor costs for blocks, mortar and reinforcing on the pages that follow.

	Craft@Hrs	Unit	Material	Labor	Total
Walls constructed with 8" x 16" blocks laid in running bond					
4" thick wall	B9@.090	SF	.99	2.08	3.07
6" thick wall	B9@.100	SF	1.18	2.32	3.50
8" thick wall	B9@.120	SF	1.43	2.78	4.21
12" thick wall	B9@.150	SF	2.10	3.47	5.57
Add for grouting cores at 36" intervals, poured by hand					
4" thick wall	B9@.007	SF	.14	.16	.30
6" thick wall	B9@.007	SF	.22	.16	.38
8" thick wall	B9@.007	SF	.29	.16	.45
12" thick wall	B9@.100	SF	.43	2.32	2.75
Add for 2" thick caps, natural gray concrete					
4" thick wall	B9@.027	LF	.60	.63	1.23
6" thick wall	B9@.038	LF	.72	.88	1.60
8" thick wall	B9@.046	LF	.84	1.07	1.91
12" thick wall	B9@.065	LF	1.01	1.51	2.52
Add for detailed block, 3/8" score					
Single score, one side	—	SF	.46	—	.46
Single score, two sides	—	SF	.50	—	.50
Multi-scores, one side	—	SF	.65	—	.65
Multi-scores, two sides	—	SF	.94	—	.94
Add for color block, any wall assembly					
Light colors	—	SF	.12	—	.12
Medium colors	—	SF	.18	—	.18
Dark colors	—	SF	.25	—	.25
Add for other than running bond	B9@.021	SF	—	.49	.49

	Craft@Hrs	Unit	Material	Labor	Total

Concrete block Foundations not included. Includes local delivery, mortar and 8% waste. Natural gray, medium weight, no reinforcing included. See also concrete block wall assemblies. Prices shown in parenthesis are for the block only. Costs shown per SF include 8% for waste and cost for mortar. Costs shown as Ea are for the block only, the mortar from the adjacent block will be sufficient for these blocks.

	Craft@Hrs	Unit	Material	Labor	Total
4" wide units					
8" x 16" ($.65), 1.125 per SF	B9@.085	SF	.98	1.97	2.95
8" x 12", corner	B9@.054	Ea	.65	1.25	1.90
8" x 8" ($.62 ea), 2.25 per SF	B9@.093	SF	1.70	2.15	3.85
4" x 16" ($.55 ea), 2.25 per SF	B9@.093	SF	1.53	2.15	3.68
4" x 12", corner	B9@.052	Ea	.65	1.20	1.85
4" x 8" ($.47 ea), 4.5 per SF, (half block)	B9@.106	SF	2.50	2.46	4.96
6" wide units					
8" x 16" ($.78 ea), 1.125 per SF	B9@.095	SF	1.21	2.20	3.41
8" x 14" corner	B9@.058	Ea	.77	1.34	2.11
8" x 8" ($.76 ea), 2.25 per SF, (half block)	B9@.103	SF	2.11	2.39	4.50
4" x 16" ($.66 ea), 2.25 per SF	B9@.103	SF	1.86	2.39	4.25
4" x 14" corner	B9@.070	Ea	.72	1.62	2.34
4" x 8" ($.63 ea), 4.5 per SF, (half block)	B9@.114	SF	3.32	2.64	5.96
4" x 12" (3/4 block)	B9@.052	Ea	.65	1.20	1.85
8" wide units					
4" x 16" ($.79 ea), 2.25 per SF	B9@.141	SF	2.23	3.27	5.50
8" x 16" ($.98 ea), 1.125 per SF	B9@.111	SF	1.50	2.57	4.07
6" x 8" ($.99 ea), 3 per SF	B9@.137	SF	3.52	3.17	6.69
8" x 12" (3/4 block)	B9@.074	Ea	.98	1.71	2.69
8" x 8" (half)	B9@.052	Ea	1.01	1.20	2.21
8" x 8" (lintel)	B9@.052	Ea	1.31	1.20	2.51
6" x 16" ($.97 ea), 1.5 per SF	B9@.117	SF	1.88	2.71	4.59
4" x 8" (half block)	B9@.053	Ea	.84	1.23	2.07
12" wide units					
8" x 16" ($1.54 ea), 1.125 per SF	B9@.148	SF	2.49	3.43	5.92
8" x 8" (half) ($1.35 ea)	B9@.082	Ea	1.50	1.90	3.40
8" x 8" (lintel) ($1.40 ea)	B9@.082	Ea	1.54	1.90	3.44
Add for high strength units	—	%	27.0	—	—
Add for lightweight block, typical	—	Ea	.16	—	.16
Deduct for lightweight block, labor	—	%	—	-7.0	—
Add for ladder type reinforcement	B9@.004	SF	.12	.09	.21
Add for 1/2" rebars 24" OC	B9@.004	SF	.12	.09	.21
Add for other than running bond	—	%	—	20.0	—
Add for 2" wall caps, natural color					
4" x 16" ($.72 ea)	B9@.027	LF	.54	.63	1.17
6" x 16" ($.85 ea)	B9@.038	LF	.64	.88	1.52
8" x 16" ($.98 ea)	B9@.046	LF	.73	1.07	1.80
Add for detailed block, 3/8" score, typical prices					
Single score, one side	—	SF	.46	—	.46
Single score, two sides	—	SF	.50	—	.50
Multi-scores, one side	—	SF	.65	—	.65
Multi-scores, two sides	—	SF	.94	—	.94
Add for color block					
Light colors	—	SF	.08	—	.08
Medium colors	—	SF	.12	—	.12
Dark colors	—	SF	.17	—	.17

Masonry

	Craft@Hrs	Unit	Material	Labor	Total
Add for grouting cores at (at $2.30 per CF), poured by hand, normal waste included					
36" interval					
4" wide wall (16.1 SF/CF grout)	B9@.007	SF	.14	.16	.30
6" wide wall (10.7 SF/CF grout)	B9@.007	SF	.22	.16	.38
8" wide wall (8.0 SF/CF grout)	B9@.007	SF	.29	.16	.45
12" wide wall (5.4 SF/CF grout)	B9@.010	SF	.43	.23	.66
24" interval					
4" wide wall (16.1 SF/CF grout)	B9@.009	SF	.27	.21	.48
6" wide wall (10.7 SF/CF grout)	B9@.009	SF	.22	.21	.43
8" wide wall (8.0 SF/CF grout)	B9@.009	SF	.41	.21	.62
12" wide wall (5.4 SF/CF grout)	B9@.013	SF	.61	.30	.91

Split face concrete block Textured block, medium weight, split one side, structural. Includes 8% waste and local delivery.

	Craft@Hrs	Unit	Material	Labor	Total
8" wide, 4" x 16", ($1.12 ea)	B9@.111	SF	2.72	2.57	5.29
8" wide, 8" x 16", ($1.23 ea)	B9@.111	SF	2.25	2.57	4.82
12" wide, 8" x 8", ($1.72 ea)	B9@.117	SF	4.18	2.71	6.89
12" wide, 8" x 16", ($1.78 ea)	B9@.117	SF	2.16	2.71	4.87
Add for lightweight block	—	%	12.0	—	—
Add for color	—	SF	.27	—	.27
Add for mortar (at $5.50 per cubic foot and 7 cubic feet,					
Per 100 square feet of wall)	—	SF	.39	—	.39

Concrete pavers Natural concrete 3/4" thick, including local delivery. Concrete bedding not included.

	Craft@Hrs	Unit	Material	Labor	Total
6" x 6", $3.05 ea, 4 per SF	—	SF	12.20	—	12.20
6" x 12", $3.10 ea, 2 per SF	—	SF	6.20	—	6.20
12" x 12", $2.90 ea, 1 per SF	—	SF	2.90	—	2.90
12" x 18", $3.05 ea, .666 per SF	—	SF	2.03	—	2.03
18" x 18", $3.15 ea, .444 per SF	—	SF	1.40	—	1.40
Add for light colors	—	SF	.40	—	.40
Add for dark colors	—	SF	1.20	—	1.20
Add for beach pebble surface	—	SF	.80	—	.80
Labor installing pavers in concrete	B9@.150	SF	—	3.47	3.47

Concrete screen block Natural color. Foundations and supports not included. Material costs include 8% waste, (cost per block and quantity per SF are shown in parentheses)

	Craft@Hrs	Unit	Material	Labor	Total
Textured screen (no corebar defacements)					
4" x 12" x 12", (at $1.60 Ea and 1 per SF)	B9@.211	SF	1.73	4.89	6.62
4" x 6" x 6", (at $.52 Ea and 3.68 per SF)	B9@.407	SF	2.07	9.43	11.50
4" x 8" x 8", (at $1.29 Ea and 2.15 per SF)	B9@.285	SF	3.00	6.60	9.60
Four sided sculpture screens					
4" x 12" x 12", (at $1.60 Ea and 1 per SF)	B9@.312	SF	1.73	7.23	8.96
Sculpture screens (patterned face, rug texture reverse)					
4" x 12" x 12", (at $2.17 Ea and 1 per SF)	B9@.211	SF	2.34	4.89	7.23
4" x 16" x 16", (at $3.77 Ea and .57 per SF)	B9@.285	SF	2.32	6.60	8.92
4" x 16" x 12", (at $3.05 Ea and .75 per SF)	B9@.285	SF	2.47	6.60	9.07
Add for channel steel support (high rise)					
2" x 2" x 3/16"	—	LF	1.85	—	1.85
3" x 3" x 3/16"	—	LF	2.20	—	2.20
4" x 4" x 1/4"	—	LF	4.35	—	4.35
6" x 4" x 3/8"	—	LF	8.45	—	8.45

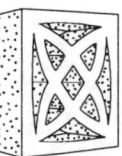

	Craft@Hrs	Unit	Material	Labor	Total

Concrete slump block Natural concrete color, 4" widths are veneer, no foundation or reinforcing included. Cost in material column include 8% waste, (cost per block and quantity per SF are shown in parentheses).

	Craft@Hrs	Unit	Material	Labor	Total
4" x 4" x 8", (at $.63 Ea and 4.5 per SF)	B9@.267	SF	3.06	6.19	9.25
4" x 4" x 12", (at $.62 Ea and 3 per SF)	B9@.267	SF	2.01	6.19	8.20
4" x 4" x 16", (at $.80 Ea and 2.25 per SF)	B9@.267	SF	1.94	6.19	8.13
6" x 4" x 16", (at $.80 Ea and 2.25 per SF)	B9@.243	SF	1.94	5.63	7.57
6" x 6" x 16", (at $.87 Ea and 1.5 per SF)	B9@.217	SF	1.41	5.03	6.44
8" x 4" x 16", (at $1.04 Ea and 2.25 per SF)	B9@.264	SF	2.53	6.12	8.65
8" x 6" x 16", (at $1.15 Ea and 1.5 per SF)	B9@.243	SF	1.86	5.63	7.49
12" x 4" x 16", (at $1.78 Ea and 2.25 per SF)	B9@.267	SF	4.33	6.19	10.52
12" x 6" x 16", (at $1.89 Ea and 1.5 per SF)	B9@.260	SF	3.06	6.02	9.08

Cap or slab block, 2" thick, 16" long, .75 units per LF. No waste included

	Craft@Hrs	Unit	Material	Labor	Total
4" width ($.64 Ea)	B9@.160	LF	.48	3.71	4.19
6" width ($.92 Ea)	B9@.160	LF	.69	3.71	4.40
8" width ($1.18 Ea)	B9@.160	LF	.89	3.71	4.60

Add for curved slump block

	Craft@Hrs	Unit	Material	Labor	Total
4" width, 2' or 4' diameter circle	—	%	125.0	25.0	—
8" width, 2' or 4' diameter circle	—	%	150.0	20.0	—
Add for slump one side block	—	SF	.05	—	.05
Add for light colors for slump block	—	%	13.0	—	—
Add for darker colors for slump block	—	%	25.0	—	—

Add for mortar for slump block($5.50 per CF). Cost includes normal waste

	Craft@Hrs	Unit	Material	Labor	Total
4" x 4" x 8" (4 CF per CSF)	—	SF	.24	—	.24
6" x 6" x 16" (4 CF per CSF)	—	SF	.24	—	.24
8" x 6" x 16" (5 CF per CSF)	—	SF	.30	—	.30
12" x 6" x 16" (7.5 CF per CSF)	—	SF	.45	—	.45

Adobe block Walls to 8', not including reinforcement or foundations, plus delivery, 4" high, 16" long, 2.25 blocks per SF, including 10% waste.

	Craft@Hrs	Unit	Material	Labor	Total
4" wide, ($.65 Ea)	B9@.206	SF	1.46	4.77	6.23
6" wide, ($.92 Ea)	B9@.233	SF	2.07	5.40	7.47
8" wide, ($1.15 Ea)	B9@.238	SF	2.59	5.51	8.10
12" wide, ($1.87 Ea)	B9@.267	SF	4.21	6.19	10.40
Add for reinforcement (24" OC both ways)	—	SF	.24	—	.24

Flagstone Including concrete bed.

	Craft@Hrs	Unit	Material	Labor	Total
Veneer on walls, 3" thick	B9@.480	SF	3.45	11.10	14.55
Walks and porches, 2" thick	B9@.248	SF	1.70	5.74	7.44
Coping, 4" x 12"	B9@.272	SF	5.45	6.30	11.75
Steps, 6" risers	B9@.434	LF	1.95	10.10	12.05
Steps, 12" treads	B9@.455	LF	3.90	10.50	14.40

Flue lining Pumice, 1' lengths, delivered to 30 miles, including 8% waste.

	Craft@Hrs	Unit	Material	Labor	Total
8-1/2" round	B9@.148	Ea	6.80	3.43	10.23
8" x 13" oval	B9@.206	Ea	7.65	4.77	12.42
8" x 17" oval	B9@.228	Ea	10.20	5.28	15.48
10" x 17 oval"	B9@.243	Ea	11.90	5.63	17.53
13" x 13" square	B9@.193	Ea	11.90	4.47	16.37
13" x 17" oval	B9@.193	Ea	9.35	4.47	13.82
13" x 21" oval	B9@.228	Ea	13.60	5.28	18.88

Masonry

	Craft@Hrs	Unit	Material	Labor	Total
17" x 17" square	B9@.228	Ea	15.30	5.28	20.58
17" x 21" oval	B9@.254	Ea	17.90	5.88	23.78
21" x 21" square	B9@.265	Ea	31.50	6.14	37.64

Glass block Costs include 6% for waste and local delivery. Add mortar and reinforcing costs below. Labor cost includes caulking, cleaning with sponge and fiber brush, and continuous tuckpointing. Based on Pittsburgh Corning Glass, costs are for a 100 SF job.

3-7/8" thick block, smooth or irregular faces (Argus, Decora, Essex AA or Vue)

	Craft@Hrs	Unit	Material	Labor	Total
4" x 8", 4.5 per SF, $2.70 each	B9@.392	SF	12.90	9.08	21.98
6" x 6", 4 per SF, $2.80 each	B9@.376	SF	11.90	8.71	20.61
8" x 8", 2.25 per SF, $3.67 each	B9@.296	SF	8.75	6.86	15.61
12" x 12", 1 per SF, $9.80 each	B9@.211	SF	10.40	4.89	15.29
Add for heat and glare reducing insert		%	75.0	—	—
3-7/8" thick corner units	B9@.125	Ea	16.00	2.90	18.90
3-7/8" thick end units	B9@.125	Ea	16.00	2.90	18.90
3-7/8" thick 45 degree block	B9@.125	Ea	13.00	2.90	15.90
3-7/8" thick double finish end	B9@.125	Ea	21.40	2.90	24.30
4" x 8" x 3-1/8" thick end finishing unit	B9@.125	Ea	12.80	2.90	15.70

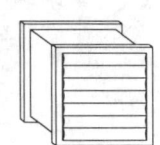

3" thick block, solid glass, coated or uncoated, smooth faces (Vistabrick)

	Craft@Hrs	Unit	Material	Labor	Total
8" x 8", 2.25 per SF, $23.00 each	B9@.376	SF	51.80	8.71	60.51

3-1/8" thick block, smooth, irregular, or diamond relief faces (Decora, or Delphi thinline)

	Craft@Hrs	Unit	Material	Labor	Total
6" x 6", 4 per SF, $2.45 each	B9@.376	SF	10.40	8.71	19.11
8" x 8", 2.25 per SF, $2.62 each	B9@.296	SF	6.25	6.86	13.11
4" x 8", 4.5 per SF, $2.36 each	B9@.391	SF	11.30	9.06	20.36
6" x 8", 3 per SF, $2.45 each	B9@.344	SF	7.79	7.97	15.76
Add for heat and glare reducing insert	—	%	75.0	—	—
Deduct for over 100 SF to 500 SF job	—	%	-5.0	—	—
Deduct for over 500 SF job to 1,000 SF job	—	%	-10.0	—	—
Deduct for over 1,000 SF job	—	%	-12.0	—	—
Add for wall ties	—	Ea	2.00	—	2.00
Add for expansion strips, fiberglass	—	Ea	.70	—	.70
Add for panel anchors, 20 gauge perforated steel, 1-3/4" x 24"	—	LF	1.20	—	1.20
Add for panel reinforcing, 10' lengths, 3" wide	—	LF	.25	—	.25

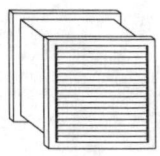

Mortar for glass block White cement mix (1 part white cement, 1/2 part hydrated lime, 2-1/4 to 3 parts #20 or #30 silica sand)

	Craft@Hrs	Unit	Material	Labor	Total
Needed with white cement mix	—	CF	7.00	—	7.00
1/4" mortar joints, 3-7/8" thick block, white cement mix					
6" x 6" block (4.3 CF/CSF)	—	SF	.30	—	.30
8" x 8" block (3.2 CF/CSF)	—	SF	.22	—	.22
12" x 12" block (2.2 CF/CSF)	—	SF	.15	—	.15
1/4" mortar joints, 3" and 3-1/8" thick block, white cement mix					
6" x 6" block (3.5 CF/CSF)	—	SF	.25	—	.25
8" x 8" block (2.6 CF/CSF)	—	SF	.18	—	.18

Masonry Accessories Asphalt emulsion for below grade waterproofing

	Craft@Hrs	Unit	Material	Labor	Total
5 gallon can (50 SF and $19.00 per gallon)	—	SF	.38	—	.38
1 gallon can (50 SF and $ 5.00 per gallon)	—	SF	.10	—	.10
Glass reinforcing mesh	—	SF	.11	—	.11

	Craft@Hrs	Unit	Material	Labor	Total
Lime					
Hydrated (builders)					
50 lb sack	—	Sack	6.40	—	6.40
25 lb sack	—	Sack	3.45	—	3.45
10 lb sack	—	Sack	1.80	—	1.80
Pebbled lime, 100 lb sack	—	Sack	11.10	—	11.10
Processed or ground lime, 60 lb sack	—	Sack	7.55	—	7.55
Silica sand, #30, 100 lb sack	—	Sack	3.07	—	3.07
White cement, 94 lb sack	—	Sack	10.40	—	10.40
Portland cement, 94 lb sack	—	Sack	4.98	—	4.98
Mortar colors					
Red, yellow, brown or black	—	Lb	1.90	—	1.90
Green	—	Lb	4.10	—	4.10
Trowel-ready mortar (factory prepared, ready to use), 30 hour life.					
Deposit required on container, local delivery included.					
Type "S" natural mortar, 1/3 CY	—	Ea	53.60	—	53.60
Type "M" natural mortar, 1/3 CY	—	Ea	58.70	—	58.70
Add to natural mortar for colored, 1/3 CY	—	Ea	9.00	—	9.00
Type "S" white mortar, 1/3 CY	—	Ea	70.60	—	70.60
Trowel-ready mortar, cost per square foot of wall, 3/8" joint					
4" x 4" x 8" brick wall	—	SF	.30	—	.30
8" x 8" x 16" block wall	—	SF	.27	—	.27
Hi-rise channel reinforcing steel (Hat Channel)					
2" x 2" x 3/16"	—	LF	1.75	—	1.75
3" x 3" x 3/16"	—	LF	2.10	—	2.10
4" x 4" x 1/4"	—	LF	4.15	—	4.15
6" x 4" x 3/8"	—	LF	8.10	—	8.10
Welding	—	Ea	5.10	—	5.10
Cutting	—	Ea	2.55	—	2.55
Flanges, drill one hole, weld	—	LS	7.90	—	7.90
Additional holes	—	Ea	.82	—	.82
Expansion joints					
1/4" x 2"	—	LF	.08	—	.08
1/4" x 3-1/2"	—	LF	.11	—	.11
1/2" x 3-1/2"	—	LF	.13	—	.13
Wall anchors, 1-3/4" x 24"	—	Ea	1.38	—	1.38
Dur-O-Wal reinforcing, Ladur or standard					
3" wide or 4" wide	—	LF	.10	—	.10
6" wide	—	LF	.10	—	.10
8" wide	—	LF	.11	—	.11
10" wide or 12" wide	—	LF	.14	—	.14
Wired adobe blocks					
2" x 2"	—	Ea	.11	—	.11
3" x 3"	—	Ea	.23	—	.23
Tie wire, 16 gauge					
Roll, 400'	—	Ea	2.13	—	2.13
Box, 20 rolls	—	Ea	46.80	—	46.80
Wall ties, 22 gauge	—	M	29.70	—	29.70
Wall ties, 28 gauge	—	M	23.00	—	23.00

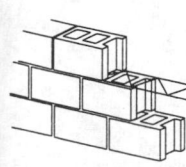

Masonry Accessories

	Craft@Hrs	Unit	Material	Labor	Total
Angle iron					
3" x 3" x 3/16"	—	LF	1.68	—	1.68
4" x 4" x 1/4"	—	LF	3.83	—	3.83
6" x 4" x 3/8"	—	LF	7.19	—	7.19
Post anchors (straps)					
4" x 4" x 10"	—	Ea	2.00	—	2.00
4" x 4" x 16"	—	Ea	2.75	—	2.75

Masonry, Subcontract Typical prices.

Fireplace chimneys, masonry only. Height above firebox shown.

	Craft@Hrs	Unit	Material	Labor	Total
5' high chimney (3' opening)	—	Ea	—	—	1,840.00
13' high chimney (3' opening)	—	Ea	—	—	2,170.00
Floors					
Flagstone	—	SF	—	—	8.80
Patio tile	—	SF	—	—	6.73
Pavers	—	SF	—	—	7.34
Quarry tile, unglazed	—	SF	—	—	9.64

Mats, Runners and Treads Koffler Products.

	Craft@Hrs	Unit	Material	Labor	Total
Mats					
One piece, corrugated rubber mats, black					
1/4" thick, stock sizes up to 3' x 6'	B1@.125	SF	5.70	2.86	8.56
Vinyl link mats, open link, 1/2" thick, including nosing					
Solid black	B1@.167	SF	15.00	3.82	18.82
Solid colors	B1@.167	SF	16.50	3.82	20.32
Any two colors	B1@.167	SF	16.50	3.82	20.32
For 9" to 12" letters add	—	Ea	16.60	—	16.60
For closed link add	—	%	25.0	—	—
Steel link mats, 3/8" deep, 1" mesh					
16" x 24"	B1@.501	Ea	18.40	11.40	29.80
22" x 36"	B1@.501	Ea	35.75	11.40	47.15
30" x 48"	B1@.501	Ea	62.05	11.40	73.45
36" x 54"	B1@.501	Ea	75.95	11.40	87.35
36" x 72"	B1@.501	Ea	97.00	11.40	108.40
Runners					
Corrugated rubber runner, cut lengths, 2', 3', 4' widths					
1/8" thick, black	B1@.060	SF	2.20	1.37	3.57
3/16" thick, black	B1@.060	SF	4.85	1.37	6.22
1/4" thick, black	B1@.060	SF	5.70	1.37	7.07
Corrugated or round ribbed vinyl runner, 1/8" thick, 2', 3', or 4' wide, cut lengths, flame resistant					
Black or brown	B1@.060	SF	2.20	1.37	3.57
Terra cotta, green or gray	B1@.060	SF	2.40	1.37	3.77
White, 3' widths only	B1@.060	SF	2.70	1.37	4.07
Deduct for full rolls (100 LF)	—	%	-10.0	—	—
Safety strips, non-slip strips, press-on abrasive, black, 1/16" thick					
3/4" x 24", box of 50 at $36.20	B1@.080	Ea	.73	1.83	2.56
6" x 24" cleats, box of 50 at $225.00	B1@.080	Ea	4.95	1.83	6.78
Treads					
Stair treads, homogeneous rubber, 1/4" thick, 1-1/2" nosing, 12-3/8" deep					
Diamond design, marbleized	B1@.080	LF	12.70	1.83	14.53
Diamond design, black	B1@.080	LF	12.70	1.83	14.53
Plain, with abrasive strips	B1@.080	LF	12.34	1.83	14.17

	Craft@Hrs	Unit	Material	Labor	Total

Medicine Cabinets No electrical work included.

Recessed cabinets, hinged door, no lighting

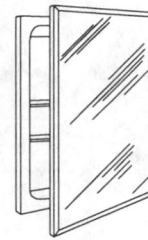

    Basic chrome frame cabinet with mirror

14" x 18"	BC@.850	Ea	36.00	21.40	57.40

    Better quality stainless steel or brass frame cabinet with mirror

14" x 18"	BC@.850	Ea	83.00	21.40	104.40

    Oak framed cabinet

24" x 26", tri-view mirrors	BC@.850	Ea	135.00	21.40	156.40
30" x 26", tri-view mirrors	BC@.941	Ea	140.00	23.70	163.70

    Vertical beveled mirror, frameless swing door

14" x 24"	BC@.850	Ea	89.00	21.40	110.40

    Oval smoked bevel mirror, frameless swing door

14" x 18"	BC@.850	Ea	92.00	21.40	113.40

Recessed cabinets with lighting

    Rectangular beveled, frameless swing door. Add the cost of electrical work.

14" x 34"	BC@.941	Ea	72.00	23.70	95.70
14" x 18"	BC@.850	Ea	62.00	21.40	83.40

Surface mounted cabinets, no lighting

    Corner cabinet, single swing door

16" x 36"	BC@.941	Ea	90.00	23.70	113.70

    Double slide door, lighted

24" x 20"	BC@.941	Ea	90.00	23.70	113.70

    Single swing doors

Frameless beveled mirror 16" x 26"	BC@.850	Ea	72.00	21.40	93.40
Chrome, brass picture frame 17" x 27"	BC@.850	Ea	83.00	21.40	104.40

Surface mounted cabinets with lighting

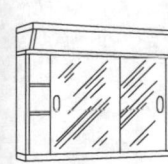

    Beveled tri-view, frameless

24" x 26"	BC@.941	Ea	149.00	23.70	172.70
30" x 30"	BC@1.18	Ea	176.00	29.70	205.70
36" x 30"	BC@1.18	Ea	203.00	29.70	232.70
48" x 30"	BC@1.18	Ea	268.00	29.70	297.70

Matching light fixtures

2 lights	BC@1.00	Ea	50.00	25.20	75.20
3 lights	BC@1.00	Ea	70.00	25.20	95.20
4 lights	BC@1.00	Ea	80.00	25.20	105.20
6 lights	BC@1.00	Ea	110.00	25.20	135.20

Add for hardwood and glass valance lighting units,

decorator style, no electrical work	BC@.377	Ea	91.00	9.49	100.49
Add for wiring and connecting lighted units	BE@1.00	Ea	21.00	26.90	47.90

Meter Boxes Precast concrete, FOB manufacturer, inside dimension, approximately 100 to 190 lbs each

    13-1/4" x 19-1/8" x 11" deep

With 1 piece concrete cover	BL@.334	Ea	15.90	6.86	22.76
With metal cover (cast iron)	BL@.334	Ea	22.10	6.86	28.96

    14-5/8" x 19-3/4" x 11" deep

With 1 piece concrete cover	BL@.334	Ea	12.70	6.86	19.56
With metal cover (cast iron)	BL@.334	Ea	23.20	6.86	30.06

    15-3/4" x 22-3/4" x 11" deep

With 1 piece concrete cover	BL@.334	Ea	23.20	6.86	30.06
With metal cover (cast iron)	BL@.334	Ea	41.00	6.86	47.86
Add hinged or reading lid, concrete cover	—	Ea	5.28	—	5.28

Meter Boxes

	Craft@Hrs	Unit	Material	Labor	Total
Plastic meter and valve boxes, bolting plastic covers					
Standard 16" x 10-3/4" x 12" deep	BL@.167	Ea	17.70	3.43	21.13
Jumbo 20-1/4" x 14-1/2" x 12" deep	BL@.167	Ea	31.50	3.43	34.93
Round pit box, 6" diameter, 9-1/16"	BL@.167	Ea	6.66	3.43	10.09

Mirrors See also Bathroom Accessories and Glass.

	Craft@Hrs	Unit	Material	Labor	Total
Wall mirrors, distortion-free float glass, 1/8" x 3/16" thick					
Cut and polished edges	—	SF	3.25	—	3.25
Beveled edges	—	SF	5.89	—	5.89
Labor installing wall mirrors					
To 5 SF	BG@.142	SF	—	3.27	3.27
5 to 15 SF	BG@.127	SF	—	2.93	2.93
15 to 25 SF	BG@.113	SF	—	2.61	2.61
25 to 50 SF	BG@.093	SF	—	2.14	2.14
Decorative mirrors					
Mirror strips, 8" x 72" (4 SF)					
Clear	BG@.568	Ea	16.70	13.10	29.80
Clear with beveled edge	BG@.568	Ea	24.90	13.10	38.00
Bronze with beveled edge	BG@.568	Ea	26.90	13.10	40.00
Mirror arches (radius, top edge), 68" high					
16" wide, clear (8.25 SF)	BG@1.05	Ea	23.90	24.20	48.10
20" wide, clear (10.9 SF)	BG@1.38	Ea	28.90	31.80	60.70
24" wide, clear (10.9 SF)	BG@1.38	Ea	77.10	31.80	108.90
Framed mirrors					
Traditional rectangle with urethane corner and side ornamentation					
31" x 42"	BG@.331	Ea	75.10	7.63	82.73
35" x 49"	BG@.331	Ea	98.50	7.63	106.13
Traditional hexagon mirror with urethane frame					
32" x 42"	BG@.331	Ea	103.00	7.63	110.63
38" x 48"	BG@.331	Ea	111.00	7.63	118.63
Solid oak frame, rectangle plate glass					
18" x 29"	BG@.331	Ea	82.20	7.63	89.83
24" x 24"	BG@.331	Ea	93.40	7.63	101.03
30" x 24"	BG@.331	Ea	117.00	7.63	124.63
36" x 26"	BG@.450	Ea	132.00	10.40	142.40
48" x 26"	BG@.450	Ea	167.00	10.40	177.40
Door mirrors					
16" x 56"	BG@.251	Ea	34.50	5.79	40.29
18" x 60"	BG@.301	Ea	40.60	6.94	47.54
24" x 68"	BG@.331	Ea	52.30	7.63	59.93

Mouldings Unfinished. Material costs include 5% waste. Labor costs are for typical installation only. (For installation costs of Hardwood Moulding, see Softwood Mouldings starting on next page.)

	Unit	Oak	Poplar	Birch
Hardwood moulding				
Base				
1/2" x 1-1/2"	LF	.90	.69	.77
1/2" x 2-1/2"	LF	1.12	.83	.91
Casing				
1/2" x 1-1/2"	LF	.99	.75	.81
5/8" x 1-5/8"	LF	1.12	.83	.91

	Unit	Oak	Poplar	Birch
Cap moulding, 1/2" x 1-1/2"	LF	.90	.67	.71
Crown, 1/2" x 2-1/4"	LF	1.04	.81	.82
Cove				
1/2" x 1/2"	LF	.48	.40	.44
3/4" x 3/4	LF	.53	.44	.47
Corner mould				
3/4" x 3/4"	LF	.54	.43	.46
1" x 1"	LF	.70	.54	.60
Round edge stop, 3/8" x 1-1/4"	LF	.48	.37	.46
Chair rail, 1/2" x 2"	LF	1.10	.83	.92
Quarter round				
1/4" x 1/4"	LF	.43	.37	.41
1/2" x 1/2"	LF	.47	.44	.45
3/4" x 3/4"	LF	.53	.44	.46
Base shoe, 3/8" x 3/4"	LF	.41	.35	.38
Battens, 1/4" x 3/4"	LF	.37	.28	.30
Door trim sets, 16'8" total length				
1/2" x 1-5/8" casing	Ea	21.30	18.90	18.90
5/8" x 1-5/8" casing	Ea	22.40	19.90	21.30
3/8" x 1-1/4" round edge stop	Ea	14.90	12.80	12.80

	Craft@Hrs	Unit	Material	Labor	Total
Oak threshold, LF of threshold					
5/8" x 3-1/2" or 3/4" x 3-1/2"	BC@.174	LF	3.77	4.38	8.15
3/4" x 5-1/2"	BC@.174	LF	4.20	4.38	8.58
7/8" x 3-1/2"	BC@.174	LF	3.87	4.38	8.25
Redwood moulding					
Band moulding, 3/4" x 5/16"	BC@.030	LF	.63	.76	1.39
Battens and lattice					
5/16" x 1-1/4"	BC@.026	LF	.33	.65	.98
5/16" x 1-5/8"	BC@.026	LF	.34	.65	.99
5/16" x 2-1/2"	BC@.026	LF	.54	.65	1.19
5/16" x 3-1/2"	BC@.026	LF	.54	.65	1.19
3/8" x 2-1/2"	BC@.026	LF	.56	.65	1.21
Brick mould, 1-1/2" x 1-1/2"	BC@.026	LF	1.29	.65	1.94
Drip cap-water table					
1-5/8" x 2-1/2" (no lip)	BC@.060	LF	1.59	1.51	3.10
1-5/8" x 2-1/2"	BC@.060	LF	1.59	1.51	3.10
Quarter round, 11/16" x 11/16"	BC@.026	LF	.43	.65	1.08
Rabbeted siding mould, 1-1/8" x 1-5/8"	BC@.026	LF	.92	.65	1.57
S4S (rectangular)					
11/16" x 11/16"	BC@.026	LF	.46	.65	1.11
1" x 2"	BC@.026	LF	.68	.65	1.33
1" x 3"	BC@.026	LF	1.02	.65	1.67
1" x 4"	BC@.026	LF	1.30	.65	1.95
Stucco mould					
13/16" x 1-5/16"	BC@.026	LF	.72	.65	1.37
13/16" x 1/2"	BC@.026	LF	1.04	.65	1.69
Window sill, 2" x 8"	BC@.043	LF	5.84	1.08	6.92

Mouldings, Softwood

	Craft@Hrs	Unit	Material	Labor	Total
Softwood moulding, pine (These labor costs can also be used for the hardwood mouldings on the previous page.)					
Astragal moulding					
For 1-3/8" x 7' doors	BC@.032	LF	3.00	.81	3.81
For 1-3/4" x 7' doors	BC@.032	LF	3.23	.81	4.04
Base (all patterns)					
7/16" x 7/16"	BC@.016	LF	.46	.40	.86
3/8" x 2-1/4"	BC@.016	LF	.71	.40	1.11
7/16" x 1-5/8"	BC@.016	LF	.71	.40	1.11
1/2" x 2-1/4"	BC@.016	LF	1.06	.40	1.46
1/2" x 2-1/2"	BC@.016	LF	1.06	.40	1.46
1/2" x 3-1/2"	BC@.016	LF	1.29	.40	1.69
Base (combination or cove)					
7/16" x 1-5/8"	BC@.016	LF	.67	.40	1.07
1/2" x 2-1/2"	BC@.016	LF	1.10	.40	1.50
1/2" x 3-1/2"	BC@.016	LF	1.39	.40	1.79
Base shoe					
3/8" x 11/16"	BC@.016	LF	.34	.40	.74
7/16" x 3/4"	BC@.016	LF	.42	.40	.82
Blind stop					
3/8" x 1/2"	BC@.015	LF	.29	.38	.67
3/8" x 3/4"	BC@.015	LF	.31	.38	.69
3/8" x 1-1/4"	BC@.015	LF	.50	.38	.88
1/2" x 3/4"	BC@.015	LF	.50	.38	.88
1/2" x 1-1/4"	BC@.015	LF	.56	.38	.94
1/2" x 1-1/2"	BC@.015	LF	.66	.38	1.04
Casing (all patterns)					
7/16" x 1-1/2"	BC@.023	LF	.46	.58	1.04
9/16" x 1-5/8"	BC@.023	LF	.72	.58	1.30
1/2" x 1-5/8"	BC@.023	LF	.80	.58	1.38
5/8" x 2-1/2"	BC@.023	LF	.96	.58	1.54
5/8" x 3-1/2"	BC@.023	LF	1.77	.58	2.35
Chair rail					
1/2" x 1-5/8"	BC@.021	LF	.71	.53	1.24
5/8" x 2-1/2"	BC@.021	LF	.96	.53	1.49
Chamfer strip, 3/4" x 3/4"	BC@.016	LF	.46	.40	.86
Corner bead (outside corner moulding)					
3/4" x 3/4"	BC@.016	LF	.47	.40	.87
15/16" x 15/16"	BC@.016	LF	.96	.40	1.36
1-5/16" x 1-5/16"	BC@.016	LF	1.08	.40	1.48
1" x 2"	BC@.016	LF	1.10	.40	1.50
Cove moulding, solid					
3/8" x 3/8"	BC@.020	LF	.29	.50	.79
1/2" x 1/2"	BC@.020	LF	.37	.50	.87
5/8" x 5/8"	BC@.020	LF	.56	.50	1.06
3/4" x 3/4"	BC@.020	LF	.56	.50	1.06
15/16" x 15/16"	BC@.020	LF	.89	.50	1.39
Cove moulding, sprung					
3/4" x 1-5/8"	BC@.020	LF	.78	.50	1.28
3/4" x 3-1/2"	BC@.020	LF	1.51	.50	2.01

	Craft@Hrs	Unit	Material	Labor	Total
Crown or bed moulding					
3/4" x 3/4"	BC@.044	LF	.58	1.11	1.69
9/16" x 1-5/8"	BC@.044	LF	.79	1.11	1.90
9/16" x 2-1/4"	BC@.044	LF	1.10	1.11	2.21
3/4" x 3-1/2"	BC@.044	LF	1.45	1.11	2.56
5/8" x 4-1/4"	BC@.044	LF	1.85	1.11	2.96
Add for special cutting, any crown moulding shown:					
Inside corner, using coping saw	BC@.167	Ea	—	4.20	4.20
Outside corners, using miter saw	BC@.083	Ea	—	2.09	2.09
Drip cap (water table moulding or bar nosing)					
1-1/2" x 2-5/16" with lip	BC@.060	LF	1.85	1.51	3.36
1-1/2" x 2-5/16" no lip	BC@.060	LF	1.85	1.51	3.36
Drip moulding, 3/4" x 1-1/4"	BC@.030	LF	.52	.76	1.28
Glass bead, 5/8" x 3/4", 3/8" x 5/8"	BC@.016	LF	.43	.40	.83
Handrail, fir					
2" x 2" (oval)	BC@.060	LF	2.49	1.51	4.00
1" x 2-1/2" (detailed)	BC@.060	LF	2.31	1.51	3.82
Lattice					
1/4" x 1-1/4"	BC@.016	LF	.31	.40	.71
1/4" x 1-5/8"	BC@.016	LF	.35	.40	.75
1/4" x 2-1/2"	BC@.016	LF	.56	.40	.96
1/4" x 3-1/2"	BC@.016	LF	.68	.40	1.08
Mullion casing					
1/4" x 3-1/2"	BC@.016	LF	.89	.40	1.29
Panel moulding					
3/8" x 5/8"	BC@.030	LF	.24	.76	1.00
5/8" x 3/4"	BC@.030	LF	.40	.76	1.16
3/4" x 1"	BC@.030	LF	.56	.76	1.32
3/4" x 1-5/8"	BC@.030	LF	.59	.76	1.35
1/2" x 7/8"	BC@.030	LF	.47	.76	1.23
Parting bead, 3/8" x 3/4"	BC@.016	LF	.31	.40	.71
Picture frame mould					
11/16" x 13/16"	BC@.040	LF	.74	1.01	1.75
3/4" x 1-5/8"	BC@.040	LF	1.10	1.01	2.11
7/8" x 2-1/2"	BC@.040	LF	2.13	1.01	3.14
Plaster ground					
3/4" x 7/8"	BC@.010	LF	.46	.25	.71
Quarter round					
1/4" x 1/4"	BC@.016	LF	.27	.40	.67
3/8" x 3/8"	BC@.016	LF	.29	.40	.69
1/2" x 1/2"	BC@.016	LF	.37	.40	.77
5/8" x 5/8"	BC@.016	LF	.35	.40	.75
3/4" x 3/4"	BC@.016	LF	.41	.40	.81
1" x 1"	BC@.016	LF	.79	.40	1.19
Half round					
1/4" x 1/2" or 5/16" x 5/8"	BC@.016	LF	.25	.40	.65
3/8" x 3/4" or 1/2" x 1"	BC@.016	LF	.33	.40	.73
3/4" x 1-1/2"	BC@.016	LF	.64	.40	1.04

Mouldings, Softwood

	Craft@Hrs	Unit	Material	Labor	Total
Full round					
1/2" or 3/4" diameter	BC@.016	LF	.56	.40	.96
1-1/4" or 1-3/8" diameter	BC@.016	LF	.59	.40	.99
1" diameter	BC@.016	LF	.85	.40	1.25
1-5/8" diameter	BC@.016	LF	.99	.40	1.39
2" diameter	BC@.016	LF	1.97	.40	2.37
S4S (rectangular)					
1/2" x 2-1/2"	BC@.016	LF	.80	.40	1.20
1/2" x 3-1/2"	BC@.016	LF	1.28	.40	1.68
1/2" x 5-1/2"	BC@.016	LF	1.91	.40	2.31
1/2" x 7-1/4"	BC@.016	LF	2.70	.40	3.10
1/2" x 3/4"	BC@.016	LF	.31	.40	.71
3/4" x 3/4"	BC@.016	LF	.42	.40	.82
3/4" x 1-1/4"	BC@.016	LF	.58	.40	.98
3/4" x 1-5/8"	BC@.016	LF	.64	.40	1.04
3/4" x 2-1/2"	BC@.016	LF	1.10	.40	1.50
3/4" x 3-1/2"	BC@.016	LF	1.61	.40	2.01
1" x 1"	BC@.016	LF	.82	.40	1.22
1-1/2" x 1-1/2"	BC@.016	LF	1.38	.40	1.78
Sash bar, 7/8" x 1-3/8"	BC@.016	LF	.80	.40	1.20
Screen moulding					
1/4" x 3/4" beaded	BC@.016	LF	.34	.40	.74
1/4" x 3/4" flat or insert	BC@.016	LF	.34	.40	.74
3/8" x 3/4" clover leaf	BC@.016	LF	.35	.40	.75
Square moulding					
1/2" x 1/2"	BC@.016	LF	.35	.40	.75
3/4" x 3/4"	BC@.016	LF	.48	.40	.88
1" x 1"	BC@.016	LF	.74	.40	1.14
2" x 2"	BC@.016	LF	1.55	.40	1.95
Stops, round edge					
3/8" x 1/2"	BC@.025	LF	.29	.63	.92
3/8" x 3/4"	BC@.025	LF	.30	.63	.93
3/8" x 1"	BC@.025	LF	.38	.63	1.01
3/8" x 1-1/4"	BC@.025	LF	.47	.63	1.10
3/8" x 1-5/8" or 1/2" x 3/4"	BC@.025	LF	.46	.63	1.09
1/2" x 1-1/4"	BC@.025	LF	.55	.63	1.18
1/2" x 1-5/8"	BC@.025	LF	.64	.63	1.27
Wainscot cap					
5/8" x 1-1/4" or 1/2" x 1-1/2"	BC@.016	LF	.73	.40	1.13
Window stool, beveled					
7/8" x 2-1/4"	BC@.037	LF	1.15	.93	2.08
15/16" x 2-1/2"	BC@.037	LF	1.67	.93	2.60
1" x 3-1/2"	BC@.037	LF	2.70	.93	3.63
1" x 5-1/2"	BC@.037	LF	4.15	.93	5.08
Window stool, flat					
11/16" x 4-5/8"	BC@.037	LF	1.67	.93	2.60
1" x 5-1/2"	BC@.037	LF	3.81	.93	4.74

Polystyrene moulding Primed, ready for painting.

	Craft@Hrs	Unit	Material	Labor	Total
Base, 9/16" x 3-5/16"	BC@.012	LF	1.39	.30	1.69
Casing, fluted, 5/8" x 3-1/4"	BC@.016	LF	1.42	.40	1.82
Cove, 13/16" x 3-3/4"	BC@.025	LF	1.39	.63	2.02

	Craft@Hrs	Unit	Material	Labor	Total
Chair rail, 5/8" x 3-1/8"	BC@.015	LF	1.25	.38	1.63
Crown, 3/4" x 3-13/16"	BC@.025	LF	1.39	.63	2.02
Casing corner block, 3/4" x 3-1/2" x 3-1/2"	BC@.025	Ea	4.49	.63	5.12
Plinth block, 3/4" x 3-1/2" x 5-3/4"	BC@.025	Ea	5.01	.63	5.64
Base corner block					
Inside, 1-1/16" x 1-1/16" x 5-3/4"	BC@.025	Ea	3.09	.63	3.72
Outside, 1-5/16" x 1-5/16" x 5-3/4"	BC@.025	Ea	3.71	.63	4.34
Ceiling corner block					
Inside, 2-7/8" x 2-7/8" x 3-3/8"	BC@.025	Ea	4.94	.63	5.57
Outside, 3-1/2" x 3-1/2" x 3-3/8"	BC@.025	Ea	5.93	.63	6.56

Nails Cost per pound in 50 lb cartons.
Add 33% for 5 lb packages
Add 50% for 1 lb packages

Size & Kind of Material	Size of Nail	Quantity Required
1 x 2 fur. on brick walls	20d common cut nails	5-1/4 lbs. / 100 LF
Base nailed 16" centers	8d finish	1 lb. / 100 LF
Battens	4d finish	1/2 lb. / 100 LF
Sides of trim	4d, 6d, 8d finish	1/2 lb. / side of trim
48" wood lath	3d fine	7 lb. / 100 lath
Metal lath	7/16" hd. 1-1/2" Roof nail	17-1/2 lb. / 100 SY
3/8" rock lath	Plaster board nail	9 lb. / 100 SY
1/4" plaster board	4d cement coated	4-1/2 lbs. / 1000 SF
3/8" plaster board	4d cement coated	4-1/2 lbs. / 1000 SF
1/2" plaster board	6d cement coated	6 lbs. / 1000 SF
3/16" wallboard	3d cement coated	5 lbs. / 1000 SF
1/4" wallboard	4d cement coated	9 lbs. / 1000 SF
3/8" wallboard	4d cement coated	9 lbs. / 1000 SF

	Sinkers cement coated	Common bright	Common galvanized
3 penny	.63	.66	.83
4 penny	.62	.63	.81
5 penny	.52	.56	.74
6 penny	.47	.48	.70
7 penny	.47	.51	.70
8 penny	.45	.47	.66
10 penny	.45	.47	.66
12 penny	.45	.48	.66
16 penny	.45	.48	.65
20 penny	.45	.48	.66
30 penny	.47	.48	.70
40 penny	.49	.48	.70
50 penny	—	.48	.66
60 penny	—	.48	.66

Nails for Particle Board Underlayment

3/8", 1/2" and 5/8" 6d cement coated box nails – 10-1/2 lbs. / 1000 SF

3/4" 8d cement coated box nails – 17 lbs. / 1000 SF

Space nails at 6" centers around edges and 10" centers in interior area along line of floor joists.

Edge nailing to be at a minimum of 1/2" from edge of panel.

(Approx. 100 nails per 4' x 8' panel)

	Bright	Galvanized	Electro galv.	Cement coated
Box nails				
3 penny	.47	.70	.72	.66
4 penny	.47	.66	.68	.64
5 penny	.47	.66	.66	.57
6 penny	.43	.63	.64	.53
7 penny	.41	.63	.65	.56
8 penny	.39	.63	.61	.53
10 penny	.41	.65	.63	.54
12 penny	.41	.65	.63	.53
16 penny	.39	.61	.59	.51
20 penny	.41	.63	.62	.54

Cement Coated Nails for Plywood		
Thickness		Lbs. / 1000 SF
1/4	4d	9
5/16	6d	11
3/8	6d	11
1/2	7d	12
5/8	8d	17
3/4	10d	21

Note: 1/4" plywood will require about 5 lbs. per 1000 SF of 4d finish nails or 9 lbs. of 4d flat head nails when battens are not used. Figure 4-1/2 lbs. of 4d flat head nails together with 3 lbs. of 4d finish nails when battens are used to cover joints spaced 4'0" apart.

	Craft@Hrs	Unit	Material	Labor	Total
Quantity of nails per pound is in parentheses.					
Box nails, blued					
2 penny (890) or 3 penny (614)	—	Lb	.63	—	.63
4 penny (473)	—	Lb	.65	—	.65
Casing nails, bright					
3 penny (635)	—	Lb	.66	—	.66
4 penny (473)	—	Lb	.60	—	.60

Nails

	Craft@Hrs	Unit	Material	Labor	Total
5 penny (406) or 6 penny (236)	—	Lb	.57	—	.57
8 penny (145) or 10 penny (94)	—	Lb	.56	—	.56
16 penny (71)	—	Lb	.56	—	.56
Concrete nails (priced per pound in 25 lb carton)					
1/2" x 9 gauge (375)	—	Lb	.89	—	.89
3/4" x 9 gauge (238)	—	Lb	.84	—	.84
1" x 9 gauge (185) or 1-1/2" x 9 gauge (128)	—	Lb	.77	—	.77
2" x 9 gauge (91) or 2-1/2" x 9 gauge (67)	—	Lb	.83	—	.83
3" x 9 gauge (44)	—	Lb	.83	—	.83
Duplex head nails (scaffold and form nails)					
6 penny (150) or 8 penny (88)	—	Lb	.57	—	.57
10 penny (62)	—	Lb	.57	—	.57
16 penny (44) or 20 penny (29)	—	Lb	.57	—	.57
Finishing nails, bright or galvanized					
3 penny (775)	—	Lb	.77	—	.77
4 penny (560) or 6 penny (295)	—	Lb	.75	—	.75
8 penny (180)	—	Lb	.74	—	.74
16 penny (86)	—	Lb	.71	—	.71
20 penny (60)	—	Lb	.72	—	.72
Furring nails, 3/8" pads, 5,000 nails per 50 lb case					
1-1/4"	—	Case	42.30	—	42.30
1-1/2"	—	Case	45.10	—	45.10
1-3/4"	—	Case	53.70	—	53.70
Joist hanger nails					
1-1/2" x 10-1/4 gauge	—	Lb	.46	—	.46
2-1/2" x 10-1/4 gauge or 1-1/2" x 9 gauge	—	Lb	.45	—	.45
3" x 9 gauge	—	Lb	.52	—	.52
3-1/2" x 8 gauge	—	Lb	.63	—	.63
1-3/4" or 2-1/8" x .192" annular ring	—	Lb	.90	—	.90
2-1/2" x .250" annular ring	—	Lb	.92	—	.92
Lath nails, blued					
2 penny (1351) or 3 penny (778)	—	Lb	.84	—	.84
3 penny (light) (1015)	—	Lb	.84	—	.84
Plasterboard nails, 13 gauge, blued					
3/8" head, 1-1/8" (449) or 1-1/4" (405)	—	Lb	.78	—	.78
3/8" head, 1-1/2" (342)	—	Lb	.76	—	.76
19/64" head, 1-1/8" (473)	—	Lb	.69	—	.69
19/64" head, 1-1/4" (425)	—	Lb	.65	—	.65
Ring shank nails, bright, 19/64" head, 12-1/2 gauge					
1-1/4", 1-3/8", 1-1/2", 1-5/8"	—	Lb	.69	—	.69
Dritite nails, bright, 1/4" head					
1-3/8", 14 gauge or 1-5/8", 13-1/2 gauge	—	Lb	.65	—	.65
1-7/8", 13 gauge or 2-3/8", 11-1/2 gauge	—	Lb	.65	—	.65
Roofing nails, steel, large flat head, barbed, 7/16" head, 11 gauge					
Hot dipped galvanized					
1/2" (355)	—	Lb	.83	—	.83
3/4" (315) or 7/8" (280)	—	Lb	.74	—	.74
1" (255)or 1-1/4" (210)	—	Lb	.70	—	.70
1-1/2" (180), 1-3/4" (150) or 2" (138)	—	Lb	.69	—	.69
Electro galvanized					
1/2" (380)	—	Lb	.65	—	.65
3/4" (315)	—	Lb	.55	—	.55

	Craft@Hrs	Unit	Material	Labor	Total
7/8" (280)	—	Lb	.53	—	.53
1" (255) or 1-1/4" (210)	—	Lb	.53	—	.53
1-1/2" (180)	—	Lb	.52	—	.52
1-3/4" (150) or 2" (138)	—	Lb	.50	—	.50
Aluminum roofing nails, 7/16" head with 3/8" neoprene washer attached					
Screw thread, 10 gauge					
1-3/4", 525 per can, 1.87 lbs	—	Can	10.40	—	10.40
2", 525 per can, 2.60 lbs	—	Can	11.10	—	11.10
2-1/2", 315 per can, 3.40 lbs	—	Can	8.50	—	8.50
Screw shank, twin thread, 10 gauge					
1-3/4", 525 per can, 2.00 lbs	—	Can	10.40	—	10.40
2", 525 per can, 2.25 lbs	—	Can	11.10	—	11.10
2-1/2", 315 per can, 1.75 lbs	—	Can	9.30	—	9.30
Siding nails, board siding, plain shank					
6 penny (283), 2" or 7 penny (248), 2-1/4"	—	Lb	1.28	—	1.28
8 penny (189), 2-1/2"	—	Lb	1.23	—	1.23
9 penny (171), 2-3/4"	—	Lb	1.23	—	1.23
10 penny (153), 3" or 12 penny (113), 3-1/4"	—	Lb	1.23	—	1.23
16 penny (104), 3-1/2"	—	Lb	1.19	—	1.19
Siding nails, hardboard siding, plain shank					
6 penny (181), 2"	—	Lb	1.12	—	1.12
8 penny (146), 2-1/2"	—	Lb	1.10	—	1.10
Siding nails, for aluminum, steel and vinyl siding, plain shank					
1-1/4" (270) or 1-1/2" (228)	—	Lb	1.15	—	1.15
2" (178) or 2-1/2" (123)	—	Lb	1.07	—	1.07
3" (103)	—	Lb	1.17	—	1.17
Add for screw shank	—	Lb	.28	—	.28
Sinkers, green vinyl coated					
8 penny (90)	—	Lb	.28	—	.28
16 penny (42)	—	Lb	.28	—	.28
Spikes					
7" x 5/16" (6)	—	Lb	.21	—	.21
8" x 5/16" (4) or 8" x 3/8" (4)	—	Lb	.20	—	.20
10" x 3/8" (3)	—	Lb	.22	—	.22
12" x 3/8" (2-1/2)	—	Lb	.32	—	.32

Nails for Asbestos Roofing Sheathing

Extra-heavy American Colonial and American Colonial.

New work requires 1-3/4 lbs., 1-1/4" galvanized roofing nails per square.

Re-roofing requires 2-1/2 lbs. 2" galvanized roofing nails per square.

Ranch Design

New work requires 1 lb. 1-1/4" galvanized nails and 80 storm anchors per square.

Re-roofing requires 1-1/2 lbs. 2" galvanized nails and 80 storm anchors per square.

Nails for Metal Roofing

Corrugated and 5-V crimp roofing - approx. 105 nails per square. Galvanized - approx. 1 lb. of spring head or approx. 1-1/2 lbs. of lead head galvanized nails. Aluminum - approx. 1/3 lb. of aluminum roofing nails with neoprene washers.

Kaiser aluminim diamond-rib and twin-rib roofing - approx. 90 nails per square.

Nuts, Bolts and Washers Bolts, standard hex head, material cost per each bolt.

	Length				
	4"	6"	8"	10"	12"
Lag bolts					
1/2"	.71	1.09	1.99	2.61	2.71
5/8"	1.38	2.04	3.70	3.99	4.18
3/4"	2.33	3.37	6.27	6.70	6.89
Machine bolts					
1/2"	.65	1.04	1.62	1.99	2.18
5/8"	.95	1.62	2.66	3.13	3.28
3/4"	1.90	2.85	3.80	4.84	4.94
7/8"	2.38	3.66	4.84	5.98	6.17
1"	2.85	4.37	5.70	7.22	7.36

Nuts, Bolts and Washers

	Craft@Hrs	Unit	Material	Labor	Total
Labor to install machine bolts or lag bolts Includes installation of nuts and washers and drilling as needed.					
4" to 6" long	BC@.100	Ea	—	2.52	2.52
8" to 10" long	BC@.150	Ea	—	3.78	3.78
12" long	BC@.200	Ea	—	5.03	5.03
Nuts, standard hex					
1/2"	—	Ea	.14	—	.14
5/8"	—	Ea	.28	—	.28
3/4"	—	Ea	.41	—	.41
7/8"	—	Ea	.87	—	.87
1"	—	Ea	1.49	—	1.49
1-1/8"	—	Ea	1.66	—	1.66
1-1/4"	—	Ea	2.38	—	2.38
Washers, standard cut steel					
1/2"	—	Ea	.07	—	.07
5/8"	—	Ea	.11	—	.11
3/4", 7/8" or 1"	—	Ea	.19	—	.19
1-1/8" or 1-1/4"	—	Ea	.24	—	.24

Ornamental Iron, Railings and Fence, Subcontract Typical prices, fabricated, delivered and installed.

		Unit			Total
Handrail, 1-1/2" x 1" rectangular top rail tubing, 1" square bottom rail tubing, pickets, 1/2" square with 5-1/2" space between.					
42" high apartment style railing	—	LF	—	—	17.20
36" high railing for single family dwelling	—	LF	—	—	21.50
Add for inclined runs	—	LF	—	—	3.95
Pool fence					
5' high, 4-1/2" centers	—	LF	—	—	21.50
Add for gate, with self-closer and latch	—	LS	—	—	102.00

Paint Removal, Lead-Based and Surface Preparation Dustless removal of lead-based paints. The removal system is a mechanical process and no water, chemicals or abrasive grit are used. Containment or special ventilation is not required. Based on PENTEK Inc. Costs shown are for a 5,000 SF or larger job. For estimating purposes figure one 55-gallon drum of contaminated waste per 2,500 square feet of coatings removed from surfaces with an average coating thickness of 12 mils. Waste disposal costs are not included. Use $5,000 as a minimum charge for work of this type.

Ceilings, doors, door & window frames, roofs, siding Equipment is based on using three pneumatic needle scaler units attached to one air-powered vacuum-waste packing unit. Equipment cost is $46.70 per hour and includes one vacuum-waste packing unit (at $4,500 per month), three needle scalers (at $975 each per month) and one 150 CFM gas or diesel powered air compressor (at $600 per month) including all hoses and connectors. Add the one-time charge for move-on, move-off and refurbishing of equipment as shown.

	Craft@Hrs	Unit	Material	Labor	Equipment	Total
Ceilings						
Plaster, 70 SF per hour	PT@.042	SF	—	1.09	.65	1.74
Doors, wood or metal. Remove and re-hang doors, no carpentry included						
To 3'0" x 6'8"	PT@1.00	Ea	—	26.00	—	26.00
Over 3'0" x 6'8" to 6'0" x 8'0"	PT@1.50	Ea	—	39.00	—	39.00
Over 6'0" x 8'0" to 12'0" x 20'0"	PT@2.00	Ea	—	52.00	—	52.00
Paint removal (assumes doors are laying flat)						
80 SF per hour	PT@.038	SF	—	.99	.59	1.58

	Craft@Hrs	Unit	Material	Labor	Total
Door or window frames					
To 3'0" x 6'8"	PT@1.50	Ea	—	39.00	23.30	62.30
Over 3'0" x 6'8" to 6'0" x 8'0"	PT@2.00	Ea	—	52.00	31.10	83.10
Over 6'0" x 8'0" to 12'0" x 20'0"	PT@3.00	Ea	—	77.90	46.70	124.60
Lintels					
Per SF of surface area	PT@.033	SF	—	.86	.51	1.37
Roofs					
Metal roofs to 3 in 12 pitch					
75 SF of surface per hour	PT@.040	SF	—	1.04	.62	1.66
Metal roofs over 3 in 12 pitch					
65 SF of surface per hour	PT@.046	SF	—	1.20	.72	1.92
Wood roofs to 3 in 12 pitch					
80 SF of surface per hour	PT@.038	SF	—	.99	.59	1.58
Wood roofs over 3 in 12 pitch					
70 SF of surface per hour	PT@.042	SF	—	1.09	.65	1.74
Siding					
Metal siding, 80 SF per hour	PT@.038	SF	—	.99	.59	1.58
Wood siding, 85 SF per hour	PT@.035	SF	—	.91	.54	1.45
Walls measured on one face of wall					
Concrete, 80 SF per hour	PT@.038	SF	—	.99	.59	1.58
Concrete block, 70 SF per hour	PT@.042	SF	—	1.09	.65	1.74
Plaster, 85 SF per hour	PT@.035	SF	—	.91	.54	1.45
Add for mobilization and demobilization, per job					
Move-on & off and refurbish	—	LS	—	—	—	2,500.00

Concrete floor and slab surfaces Equipment is based on using three pneumatic manually-operated wheel-mounted scabblers attached to one air-powered vacuum-waste packing unit to scarify concrete floors and slabs. Equipment cost is $89 per hour and includes one vacuum-waste packing unit (at $4500 per month), three scabblers (at $3,400 each per month) and one 150 CFM gas or diesel powered air compressor (at $600 per month) including all hoses and connectors. Add the one-time charge for move-on, move-off and refurbishing of equipment as shown.

	Craft@Hrs	Unit	Material	Labor	Total
Large unobstructed areas; floor slabs, driveways, sidewalks					
90 SF of surface per hour	PT@.033	SF	—	.86	1.77	2.63
Obstructed areas; narrow aisles, areas around equipment					
60 SF of surface per hour	PT@.050	SF	—	1.30	2.68	3.98
Small areas; around decks					
40 SF of surface per hour	PT@.075	SF	—	1.95	4.02	5.97
Add for mobilization and demobilization, per job					
Move-on & off and refurbish	—	LS	—	—	—	2,500.00

Paints, Coatings, and Supplies Costs listed are for good to better quality coatings purchased in 1 to 4 gallon retail quantities. General contractor prices are generally discounted 10% to 15% on residential job quantities. Painting contractors receive discounts of 20% to 25% on steady volume accounts. Figures in parentheses are typical coverage per coat. Labor costs are listed at the end of this section. Add 30% for premium quality paints. Deduct 30% for minimum quality paints.

	Craft@Hrs	Unit	Material	Labor	Total
Oil base primer					
Alkyd primer/sealer, interior					
(500 SF per Gal at $14.00 per Gal)	—	SF	.03	—	.03
Alkyd wood primer, exterior					
(400 SF per Gal at $13.80 per Gal)	—	SF	.03	—	.03
Prime coat oil wood primer					
(310 SF per Gal at $17.97 per Gal)	—	SF	.06	—	.06

Paint, Materials

	Craft@Hrs	Unit	Material	Labor	Total
Epoxy primer					
Catalyzed epoxy primer, two component type					
(425 SF per Gal at $32.10 per Gal)	—	SF	.08	—	.08
Cement base primer, per 25 lb sack, dry mix	—	Sack	16.00	—	16.00
Alkali resistant concrete and masonry primer, interior or exterior					
walls (225 SF per Gal at $25.10 per Gal)	—	SF	.11	—	.11
Latex primers					
Acrylic flat exterior, fast drying wood undercoater and backprimer					
(400 SF per Gal at $15 per Gal)	—	SF	.04	—	.04
Acrylic (vinyl acrylic) block filler					
(50 SF per Gal at $12.50 per Gal)	—	SF	.25	—	.25
Acrylic (vinyl acrylic) rust inhibiting metal primer					
(500 SF per Gal at $15.50 per Gal)	—	SF	.04	—	.04
Acrylic heavy duty wall primer/sealer, low odor type, alkali-resistant					
(800 SF per Gal at $18.40 per Gal)	—	SF	.02	—	.02
Acrylic (vinyl acrylic) undercoater, all purpose, interior/exterior					
(500 SF per Gal at $18.40 per Gal)	—	SF	.04	—	.04
Trimex™ primer acrylic undercoater, primes and masks glass, interior/exterior					
(250 SF per Gal at $58.50 per Gal)		SF	.23	—	.23
Latex wood primer, exterior					
(300 SF per Gal at $20.70 per Gal)	—	SF	.07	—	.07
Linseed primers					
Linseed (oil base) exterior/ interior primer					
(300 SF per Gal at $13.50 per Gal)	—	SF	.05	—	.05
Red alkyd primer					
(500 SF per Gal at $24.00 per Gal)	—	SF	.05	—	.05
Wood sealer/primer, interior, clear					
(600 SF per Gal at $14.40 per Gal)	—	SF	.02	—	.02
Zinc chromate primer					
(300 SF per Gal at $11.50 per Gal)	—	SF	.04	—	.04
Zinc dust/zinc oxide primer					
(300 SF per Gal at $29.80 per Gal)	—	SF	.10	—	.10
Aluminum fiber roof coat on composition roof					
(50 SF per Gal at $10.10 per Gal)	—	SF	.20	—	.20
Anti-rust enamel					
(500 SF/Qt at $8.10 per Qt)	—	SF	.02	—	.02
Asphalt-fiber roof and foundation coating at $5.40 per gallon, solvent type					
75 SF per gallon on roof or concrete	—	SF	.07	—	.07
100 SF per gallon on metal	—	SF	.05	—	.05
Concrete enamel epoxy					
(350 SF per Gal at $31.20 per Gal)	—	SF	.09	—	.09
Concrete acrylic sealer, rolled on concrete slab					
(400 SF per Gal at $16.40 per Gal)	—	SF	.05	—	.05
Driveway coating					
Tar emulsion (120 SF per Gal at $5.40/Gal)	—	SF	.05	—	.05
Acrylic latex (250 SF per Gal at $8.20/Gal)	—	SF	.03	—	.03
Gutter paint					
(100 LF per Qt at $5.70 per Qt)	—	LF	.06	—	.06
House paint, exterior					
Oil base, gloss enamel					
(400 SF per Gal at $16.60 per Gal)	—	SF	.04	—	.04

	Craft@Hrs	Unit	Material	Labor	Total
One coat, gloss					
(400 SF per Gal at $22.10 per Gal)	—	SF	.06	—	.06
Latex, semi-gloss,					
(400 SF per Gal at $15.70 per Gal)	—	SF	.04	—	.04
Trim paint, flat latex,					
(500 SF per Gal at $18.50 per Gal)	—	SF	.04	—	.04
Latex flat paint,					
(300 SF per Gal at $13.00 per Gal)	—	SF	.04	—	.04
House paint, interior					
Latex flat paint					
(300 SF per Gal at $12.97 per Gal)	—	SF	.05	—	.05
Latex semi-gloss enamel,					
(300 SF per Gal at $18.78 per Gal)	—	SF	.06	—	.06
Latex fire-retardant flat paint					
(200 SF per Gal at $33.80 per Gal)	—	SF	.17	—	.17
Ceiling white latex flat,					
(350 SF per Gal at $10.50 per Gal)	—	SF	.03	—	.03
Linseed oil, boiled or raw at $12.00 per Gal					
New exterior wood (400 SF per Gal)	—	SF	.03	—	.03
Exterior brick, common, (200 SF per Gal)	—	SF	.06	—	.06
Exterior concrete, stucco, brick, and shingle paint					
(300 SF per Gal at $14 per Gal)	—	SF	.05	—	.05
Masonry paint, two component type at $15.60 per Gal					
Concrete floors or walls (200 SF per Gal)	—	SF	.08	—	.08
Masonry block walls (120 SF per Gal)	—	SF	.13	—	.13
Masonry waterproofing					
(100 SF per Gal at $21 per Gal)	—	SF	.21	—	.21
Paint remover, methylene chloride,					
(200 SF per Gal at $8.00 per Gal)	—	SF	.04	—	.04
Patio and floor paint					
Latex (300 SF per Gal at $20.00 per Gal)	—	SF	.05	—	.05
Acrylic (400 SF per Gal at $19.00 per Gal)	—	SF	.05	—	.05
Sealer					
Thompson's Water Seal					
(750 SF and $46.80 for 5 Gallon can)	—	SF	.06	—	.06
Spackling compounds (vinyl-paste), DAP					
Exterior	—	Qt	5.67	—	5.67
Exterior	—	Gal	14.20	—	14.20
Interior	—	Qt	5.22	—	5.22
Stain					
Latex redwood stain (water base)					
(300 SF per Gal at $17.50 per Gal)	—	SF	.06	—	.06
Redwood oil stain					
(300 SF per Gal at $15.60 per Gal)	—	SF	.05	—	.05
Swimming pool enamel					
Sau-Sea Type R (rubber base)					
(350 SF per Gal at $29.00 per Gal)	—	SF	.08	—	.08
Sau-Sea Plaster Pool Coating					
(350 SF per Gal at $29.00 per Gal)	—	SF	.08	—	.08
Sau-Sea Thinner					
For Plaster or Rubber	—	Gal	9.45	—	9.45

Paint, Materials

	Craft@Hrs	Unit	Material	Labor	Total
Sau-Sea Patching Compound for Plastic					
(150 LF per Gal at $38.60 per Gal)	—	LF	.26	—	.26
Texture paint, sand texture latex, armor coat					
(100 SF per Gal at $7.50 per Gal)	—	SF	.08	—	.08
Thinners					
Acetone	—	Gal	8.00	—	8.00
Paint thinner	—	Gal	1.76	—	1.76
Shellac or lacquer thinner	—	Gal	7.50	—	7.50
Turpentine	—	Gal	11.00	—	11.00
Traffic paint, alkyd latex					
Black	—	Gal	15.30	—	15.30
White	—	Gal	15.30	—	15.30
Colors	—	Gal	18.50	—	15.30
Varnish, marine, clear	—	Gal	28.50	—	28.50
Wood filler paste	—	Gal	16.90	—	16.90
Wood preservative					
Pentachlorophenol, general purpose, 40%					
(150 SF per Gal at $14.50 Gal)	—	SF	.10	—	.10

	Quart	1 Gal	5 Gal
Stains and wood preservatives			
Original P.A.R., penetrating water-repellent			
Redwood hue or clear	5.80	18.60	72.00
Woodlife wood preservative, clear	5.80	16.70	63.00
Patiolife redwood stain	6.90	19.50	78.30
Cuprolignum green wood preservative	7.80	17.60	71.10
Cuprinol stains, semi-transparent and solid colors	8.80	19.50	81.90
Wrought iron paint, black	20.70	—	—

Painting, Labor Single coat applications except as noted. Not including equipment rental costs. These figures are based on hand work (roller, and brush where required) on residential jobs. Where spray equipment can be used to good advantage, reduce these costs 30% to 40%. Spray painting will increase paint requirements by 30% to 60%. Protection of adjacent materials is included in these costs but little or no surface preparation is assumed. Per SF of surface area to be painted. Figures in parenthesese show how much work should be done in an hour. For more complete coverage of painting costs, see *National Painting Cost Estimator*. An order form is at the back of this manual.

	Craft@Hrs	Unit	Material	Labor	Total
Exterior surfaces, per coat, paint grade					
Door and frame, 6 sides (2.5 doors/hour)	PT@.400	Ea	—	10.40	10.40
Siding, smooth, wood (250 SF/hour)	PT@.004	SF	—	.10	.13
Siding, rough or shingle (200 SF/hour)	PT@.005	SF	—	.13	.10
Shutters (50 SF/hour)	PT@.020	SF	—	.52	.52
Stucco (300 SF/hour)	PT@.003	SF	—	.08	.08
Trim, posts, rails (20 LF/hour)	PT@.050	LF	—	1.30	1.30
Windows, including mullions, per SF of opening,					
one side only(67 SF/hour)	PT@.015	SF	—	.39	.39
Add for light sanding (330 SF/hour)	PT@.003	SF	—	.08	.08
Interior surfaces, per coat					
Ceiling, flat latex (250 SF/hour)	PT@.004	SF	—	.10	.10
Ceiling, enamel (200 SF/hour)	PT@.005	SF	—	.13	.13
Walls, flat latex (500 SF/hour)	PT@.002	SF	—	.05	.16
Walls, enamel (400 SF/hour)	PT@.003	SF	—	.08	.08
Baseboard (120 LF/hour)	PT@.008	LF	—	.20	.20

	Craft@Hrs	Unit	Material	Labor	Total
Book cases, cabinets (90 SF/hour)	PT@.011	SF	—	.29	.29
Windows, including mullions and frame, per SF of opening, one side only(67 SF/hour)	PT@.015	SF	—	.39	.39

Floors, wood

	Craft@Hrs	Unit	Material	Labor	Total
Filling	PT@.007	SF	—	.18	.18
Staining	PT@.006	SF	—	.16	.16
Shellacking	PT@.006	SF	—	.16	.16
Varnishing	PT@.006	SF	—	.16	.16
Waxing, machine	PT@.006	SF	—	.16	.16
Polishing, machine	PT@.008	SF	—	.21	.21
Sanding, hand	PT@.012	SF	—	.31	.31
Sanding, machine	PT@.007	SF	—	.18	.18

Walls, brick, concrete and masonry

	Craft@Hrs	Unit	Material	Labor	Total
Oiling and sizing (140 SF/hour)	PT@.007	SF	—	.18	.18
Sealer coat (150 SF/hour)	PT@.007	SF	—	.18	.18
Brick or masonry (165 SF/hour)	PT@.006	SF	—	.16	.16
Concrete (200 SF/hour)	PT@.005	SF	—	.13	.13
Concrete steps, 3 coats (38 SF/hour)	PT@.026	SF	—	.68	.68

Window frames and sash, using Trimex™ primer to mask glass and prime sash, topcoats as shown

	Craft@Hrs	Unit	Material	Labor	Total
1 topcoat (100 SF/hour)	PT@.010	SF	—	.26	.26
2 topcoats (67 SF/hour)	PT@.015	SF	—	.39	.39
Score and peel Trimex primer from glass (peels from glass only)					
Per lite	PT@.020	Ea	—	.52	.52

Paint Spraying Equipment Rental Complete spraying outfit including paint cup, hose, spray gun

	Day	Week	Month
8 CFM electric compressor	41.00	103.00	315.00
17 CFM gas-driven compressor	51.00	144.00	435.00
Professional airless, complete outfit	82.00	212.00	630.00

	Craft@Hrs	Unit	Material	Labor	Total

Painting, Subcontract, Rule of Thumb Including material, labor, equipment and the subcontractor's overhead and profit.

Typical subcontract costs per square foot of floor area for painting the interior and exterior of residential buildings, including only minimum surface preparation

	Craft@Hrs	Unit	Material	Labor	Total
Economy, 1 or 2 coats, little brushwork	—	SF	—	—	1.63
Good quality, 2 or 3 coats	—	SF	—	—	3.72

Painting, subcontract, cost breakdown The costs that follow are based on the area of the surface being prepared or painted. Typical subcontract costs, including labor, material, equipment and the subcontractor's overhead and profit.

Surface preparation, subcontract Work performed by hand.

Light cleaning to remove surface dust and stains

	Craft@Hrs	Unit	Material	Labor	Total
Concrete or masonry surfaces	—	SF	—	—	.22
Gypsum or plaster surfaces	—	SF	—	—	.22
Wood surfaces	—	SF	—	—	.21

Normal preparation including scraping, patching and puttying

Concrete or masonry surface

	Craft@Hrs	Unit	Material	Labor	Total
Painted	—	SF	—	—	.22
Unpainted	—	SF	—	—	.17

Gypsum or plaster surfaces

	Craft@Hrs	Unit	Material	Labor	Total
Painted	—	SF	—	—	.22
Unpainted	—	SF	—	—	.08

Painting, Subcontract

	Craft@Hrs	Unit	Material	Labor	Total
Metal surfaces, light sanding	—	SF	—	—	.08
Wood surfaces, surface preparation, subcontract					
Painted	—	SF	—	—	.22
Unpainted	—	SF	—	—	.08
Puttying or reglazing windows	—	LF	—	—	1.16

Exterior painting, subcontract Hand work, costs will be 15% to 30% lower when spray equipment can be used to good advantage. These costs include only minimum surface preparation. Add additional surface preparation, if required, from above. Typical subcontract costs

	Craft@Hrs	Unit	Material	Labor	Total
Brick or concrete, including acrylic latex masonry primer and latex paint					
1 coat latex	—	SF	—	—	.50
2 coats latex	—	SF	—	—	.69
Concrete floors, etch and epoxy enamel	—	SF	—	—	.33
Columns and pilasters per coat	—	SF	—	—	.40
Cornices, per coat	—	SF	—	—	1.02
Doors, including trim, exterior only					
3 coats	—	Ea	—	—	32.60
2 coats	—	Ea	—	—	23.50
Downspouts and gutters					
Per coat	—	LF	—	—	.83
Eaves					
No rafters, per coat	—	SF	—	—	.39
With rafters, per coat, brush	—	SF	—	—	.69
Fences (gross area)					
Plain, 2 coats (per side)	—	SF	—	—	.48
Plain, 1 coat (per side)	—	SF	—	—	.27
Picket, 2 coats (per side)	—	SF	—	—	.44
Picket, 1 coat (per side)	—	SF	—	—	.24
Lattice work					
1 coat 1 side, gross area	—	SF	—	—	.44
Metal, typical					
1 coat	—	SF	—	—	.34
2 coats	—	SF	—	—	.61
Porch rail and balusters, per coat					
Gross area	—	SF	—	—	.89
Handrail only	—	LF	—	—	.59
Roofs, wood shingle					
1 coat stain plus 1 coat sealer					
Flat roof	—	SF	—	—	.41
4 in 12 pitch	—	SF	—	—	.52
8 in 12 pitch	—	SF	—	—	.55
12 in 12 pitch	—	SF	—	—	.67
1 coat stain plus 2 coats sealer					
Flat roof	—	SF	—	—	.67
4 in 12 pitch	—	SF	—	—	.79
8 in 12 pitch	—	SF	—	—	.87
12 in 12 pitch	—	SF	—	—	.93
Siding, plain					
Sanding and puttying, typical	—	SF	—	—	.26
Siding and trim, 3 coats	—	SF	—	—	.84
Siding and trim, 2 coats	—	SF	—	—	.61

	Craft@Hrs	Unit	Material	Labor	Total
Trim only, 2 coats	—	SF	—	—	.66
Trim only, 3 coats	—	SF	—	—	.90
Siding, shingle, including trim					
2 coats oil paint	—	SF	—	—	.63
1 coat oil paint	—	SF	—	—	.38
2 coats stain	—	SF	—	—	.59
1 coat stain	—	SF	—	—	.33
Steel sash					
3 coats (per side)	—	SF	—	—	.80
Stucco					
1 coat, smooth surface	—	SF	—	—	.39
2 coats, smooth surface	—	SF	—	—	.60
1 coat, rough surface	—	SF	—	—	.44
2 coats, rough surface	—	SF	—	—	.72
Window frames and sash, 3 coats					
Per side, one lite	—	Ea	—	—	25.00
Per side, add for each additional lite	—	Ea	—	—	1.38
Window frames and sash, using Trimex™ primer to mask glass and prime, topcoats as shown					
1 topcoat, per SF	—	SF	—	—	.71
2 topcoats, per SF	—	SF	—	—	.96
Score and peel Trimex™ primer from glass,					
Per lite	—	Ea	—	—	.68

Interior painting, subcontract Hand work (roller with some brush cut-in), costs will be 15% to 30% lower when spray equipment can be used to good advantage. These costs include only minimum surface preparation. Add for additional surface preparation, if required, from preceding page. Typical subcontract costs

	Craft@Hrs	Unit	Material	Labor	Total
Cabinets, bookcases, cupboards, per SF of total surface area					
3 coats, paint	—	SF	—	—	2.00
Stain, shellac, varnish or plastic	—	SF	—	—	2.04
Concrete block or brick					
Primer and 1 coat latex	—	SF	—	—	.46
Doors, including trim					
3 coats (per side)	—	SF	—	—	1.23
2 coats (per side)	—	SF	—	—	.86
Plaster or drywall walls, latex. Add for ceilings from below					
Prime coat or sealer	—	SF	—	—	.24
1 coat smooth surface	—	SF	—	—	.29
1 coat rough surface	—	SF	—	—	.32
1 coat sealer, 1 coat flat					
Smooth surface	—	SF	—	—	.40
Rough surface	—	SF	—	—	.53
1 coat sealer, 2 coats flat					
Smooth surface	—	SF	—	—	.54
Rough surface	—	SF	—	—	.72
1 coat sealer, 2 coats gloss or semi-gloss					
Smooth surface	—	SF	—	—	.52
Rough surface	—	SF	—	—	.71
Add for ceilings	—	%	—	—	45.0
Stairs, including risers					
3 coats paint	—	SF	—	—	1.38
Stain, shellac, varnish	—	SF	—	—	1.43

	Craft@Hrs	Unit	Material	Labor	Total
Woodwork, painting					
Priming	—	SF	—	—	.31
2 coats	—	SF	—	—	.52
2 coats, top workmanship	—	SF	—	—	.66
3 coats	—	SF	—	—	.86
3 coats, top workmanship	—	SF	—	—	1.12
Woodwork, staining					
Apply stain	—	SF	—	—	.26
Filling (paste wood filler)	—	SF	—	—	.57
Sanding	—	SF	—	—	.32
Apply wax and polish	—	SF	—	—	.32

Paneling See also Hardwood under Lumber and Hardboard.

Solid plank paneling, prefinished V joint or T&G, costs include nails and waste.

	Craft@Hrs	Unit	Material	Labor	Total
Tennessee cedar					
Clear 3/8" x 4" or 6" wide	BC@.050	SF	2.07	1.26	3.33
Select tight knot, 3/4" x 4" or 6" wide	BC@.050	SF	2.18	1.26	3.44
Knotty pine, 5/16" x 4" or 6" wide	BC@.050	SF	2.02	1.26	3.28
Redwood, clear, 5/16" x 4" wide	BC@.050	SF	4.09	1.26	5.35

Plywood and hardboard paneling Prefinished, 4' x 8' x 1/4" panels. Costs include nails, adhesive (at $.02 per SF) and 5% waste. Georgia-Pacific.

	Craft@Hrs	Unit	Material	Labor	Total
Front Street (paper overlay)	BC@.030	SF	.73	.76	1.49
Concepts (direct print)	BC@.030	SF	.73	.76	1.49
Hillside birch, Lionite, Styleboard, Woodgrains, Terrace, or McKenzie birch	BC@.030	SF	.83	.76	1.59
Valley Forge, Barnplank, Millplank, Renaissance, Firelight, or Woodglen	BC@.030	SF	1.24	.76	2.00
Timber Ridge	BC@.030	SF	1.41	.76	2.17
Bridgeport	BC@.030	SF	1.47	.76	2.23
Estate Oak	BC@.030	SF	1.70	.76	2.46
Firelight walnut	BC@.030	SF	1.80	.76	2.56
Estate cherry	BC@.030	SF	2.03	.76	2.79

Hardwood plywood paneling Unfinished, 4' x 8' panels
Costs include nails, adhesive (at $.02 per SF) and 5% waste. Georgia-Pacific

	Craft@Hrs	Unit	Material	Labor	Total
Rotary cut ash, imported, flush face					
1/4"	BC@.032	SF	.79	.81	1.60
1/2"	BC@.040	SF	1.80	1.01	2.81
3/4", veneer core	BC@.042	SF	2.09	1.06	3.15
Rotary cut birch imported,					
1/4"	BC@.032	SF	.79	.81	1.60
Plain slice white oak, imported, natural, flush face					
1/2"	BC@.040	SF	1.41	1.01	2.42
3/4"	BC@.042	SF	1.80	1.06	2.86
1/4"	BC@.032	SF	.82	.81	1.63
1/2"	BC@.040	SF	1.92	1.01	2.93
3/4", veneer core	BC@.042	SF	2.20	1.06	3.26
Rotary cut maple, natural					
1/4"	BC@.032	SF	1.15	.81	1.96
1/2"	BC@.040	SF	1.33	1.01	2.34
3/4"	BC@.042	SF	1.96	1.06	3.02

	Craft@Hrs	Unit	Material	Labor	Total
Plain slice walnut, flush face					
1/4"	BC@.032	SF	1.24	.81	2.05
1/2"	BC@.040	SF	2.70	1.01	3.71
3/4", veneer core	BC@.042	SF	3.21	1.06	4.27
Rotary cut lauan, Philippine mahogany					
1/4"	BC@.032	SF	.39	.81	1.20
1/2"	BC@.040	SF	.94	1.01	1.95
3/4"	BC@.042	SF	1.25	1.06	2.31
Plain slice Honduras mahogany					
1/4"	BC@.032	SF	2.09	.81	2.90
1/2"	BC@.040	SF	2.82	1.01	3.83
3/4", veneer core	BC@.042	SF	3.67	1.06	4.73

Vinyl-clad hardboard paneling Vinyl-clad film over hardboard, 4' x 8' x 1/4" panels. Costs include nails, adhesive (at $.02 per SF) and 5% waste.

	Craft@Hrs	Unit	Material	Labor	Total
White or Aegean Gold	BC@.032	SF	1.23	.81	2.04
Blue marble	BC@.032	SF	.47	.81	1.28
Harvest pattern	BC@.032	SF	.95	.81	1.76

Grooved tambour paneling 5/32" thick, grooves 1/2" OC, 4' x 8' panels. Costs include nails, adhesive (at $.02 per SF) and 5% waste. Walker and Zanger

	Craft@Hrs	Unit	Material	Labor	Total
Oak, teak, ash					
Unfinished veneer on plywood	BC@.032	SF	4.19	.81	5.00
Prefinished veneer on plywood	BC@.032	SF	4.95	.81	5.76
Walnut					
Unfinished veneer on plywood	BC@.032	SF	5.35	.81	6.16
Prefinished veneer on plywood	BC@.032	SF	5.58	.81	6.39
Bright or brushed aluminum					
On hardboard	BC@.032	SF	6.10	.81	6.91
Bright or brushed brass					
On hardboard	BC@.032	SF	6.10	.81	6.91
White or almond enameled aluminum					
On hardboard	BC@.032	SF	7.02	.81	7.83
Plain unfinished hardboard	BC@.032	SF	2.79	.81	3.60

Z-brick paneling Non-ceramic mineral brick-like veneer. Interior or exterior. 2-3/8" x 8-1/8" x 5/16" face pattern. Including adhesive, sealer and 5% waste

	Craft@Hrs	Unit	Material	Labor	Total
4' x 8' panels	BC@.035	SF	3.80	.88	4.68

Paving, Subcontract Small areas such as walks and driveways around residential buildings. Typical costs including material, labor, equipment and subcontractor's overhead and profit. No substrate preparation included. Use $1,800 as a minimum job charge.

Asphalt paving, including oil seal coat

	Craft@Hrs	Unit	Material	Labor	Total
2" asphalt	—	SF	—	—	1.27
3" asphalt	—	SF	—	—	1.43
4" asphalt	—	SF	—	—	1.84
4" base and fine grading	—	SF	—	—	.77
6" base and fine grading	—	SF	—	—	.87

Paving

	Craft@Hrs	Unit	Material	Labor	Total
Asphalt seal coats					
Applied with broom or squeegee	—	SF	—	—	.20
Fog seal, thin layer of diluted SS1H oil	—	SF	—	—	.09
Seal and sand					
Thin layer of oil with sand laid on top	—	SF	—	—	.16
Simco "walk-top" or other quick-dry applications					
Creamy oil	—	SF	—	—	.18
Brick paving					
On concrete, grouted, laid flat, including concrete base	—	SF	—	—	9.08
On concrete, laid solid on edge, 8" x 2-1/4" exposure, no grouting, including concrete base	—	SF	—	—	10.20
On sand bed, laid flat	—	SF	—	—	7.14
On sand bed, 8" x 2-1/4" exposure	—	SF	—	—	8.67
Concrete (including fine grading) check					
3" concrete, unreinforced	—	SF	—	—	2.04
4" concrete, unreinforced	—	SF	—	—	2.45
4" concrete, mesh reinforcing	—	SF	—	—	2.55
4" concrete, seeded aggregate finish	—	SF	—	—	4.74
Add for colored concrete, most colors	—	SF	—	—	.31

Plastering, Subcontract See also Lathing. Typical costs including material, labor, equipment and subcontractor's overhead and profit.

	Craft@Hrs	Unit	Material	Labor	Total
Acoustical plaster					
Including brown coat, 1-1/4" total thickness	—	SY	—	—	19.80
Plaster on concrete or masonry, with bonding agent					
Gypsum plaster, 1/2" sanded	—	SY	—	—	10.40
Plaster on interior wood frame walls, no lath included					
Gypsum plaster, 7/8", 3 coats	—	SY	—	—	15.60
Keene's cement finish, 3 coats	—	SY	—	—	16.50
Cement plaster finish, 3 coats	—	SY	—	—	16.10
Gypsum plaster, 2 coats, tract work	—	SY	—	—	13.60
Add for ceiling work	—	%	—	—	15.0
Plaster for tile					
Scratch coat	—	SY	—	—	7.75
Brown coat and scratch coat	—	SY	—	—	11.50
Exterior plaster (stucco)					
Walls, 3 coats, 7/8", no lath included					
Textured finish	—	SY	—	—	11.40
Float finish	—	SY	—	—	13.60
Dash finish	—	SY	—	—	15.10
Add for soffits	—	%	—	—	15.0
Portland cement plaster, including metal lath					
Exterior, 3 coats, 7/8" on wood frame,					
Walls	—	SY	—	—	18.80
Soffits	—	SY	—	—	19.00
Deduct for paper back lath	—	SY	—	—	-.16

Plumbing Fixtures and Equipment Costs are for good to better quality fixtures and trim installed with the accessories listed. All vitreous china, enameled steel cast iron, and fiberglass fixtures are white. Colored acrylic fiberglass fixtures will cost about 6% more. Colored enameled steel, cast iron, or vitreous fixtures will cost about 25% more. No rough plumbing included. Rough plumbing is the drain, waste and vent (DWV) piping and water supply piping that isn't exposed when construction is complete. See the costs for rough plumbing at the end of this section. Minimum quality fixtures and trim will cost 10% to 20% less and deluxe fixtures will cost much more. Installation times assume standard quality work and uncomplicated layouts.

	Craft@Hrs	Unit	Material	Labor	Total

Bathtubs

Acrylic fiberglass bathtub, 5' L, 32" W, 17" H, (at $300). Includes polished chrome pop-up drain/overflow (at $60). Mixing valve, tub filler, and 2.5 GPM shower head (at $130)

	Craft@Hrs	Unit	Material	Labor	Total
Recessed acrylic tub and trim	P1@3.00	Ea	490.00	73.30	563.30

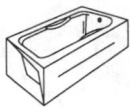

Acrylic fiberglass drop-in whirlpool bath, 5' L, 32" W, 21-½ " H (at $1390), includes chrome pop- up drain/overflow (at $60), mixing valve and tub filler (at $130)

Whirlpool acrylic tub and trim	P1@4.00	Ea	1,580.00	97.70	1,677.70

Enameled cast iron bathtub, 5' L, 32" W, 15" H (at $450). Includes polished chrome pop-up drain/overflow (at $60), mixing valve, tub filler and 2.5 GPM shower head (at $130)

Cast iron tub and trim	P1@4.50	Ea	640.00	110.00	750.00

Enameled cast iron bathtub with chrome grab rails, 5' L, 32" W, 16" H, (at $700). Includes polished chrome pop-up drain/overflow (at $60), mixing valve, tub filler, and 2.5 GPM shower head at $130)

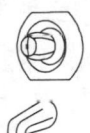

Cast Iron tub with handrail and trim	P1@5.25	Ea	890.00	128.00	1,018.00

Enameled steel bathtub, 5' L, 30" W, 15" H (at $150) includes polished chrome pop-up drain/overflow (at $60), mixing valve, tub filler, and 2.5 GPM shower head (at $130)

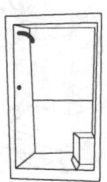

Enameled steel tub and trim	P1@2.50	Ea	340.00	61.10	401.10
Add for shower curtain rod	P1@.572	Ea	25.00	14.00	39.00

Enameled cast iron bathtub with built-in whirlpool, 5' L, 42" W, 16" H (at $2,700) including integral drain. Trim includes mixing valve and tub filler (at $130)

Cast iron whirlpool tub and trim	P1@5.50	Ea	2,830.00	134.00	2,964.00
Add for electric connection	—	Ea	—	—	200.00

Fiberglass whirlpool bath, 72" L x 37" W x 17" H (at $1,250), mixing valve, tub filler, 2.5 GPM head and drain kit (at $60)

Fiberglass whirlpool tub and trim	P1@6.75	Ea	1,440.00	165.00	1,605.00
Add for electrical connection	—	Ea	—	—	200.00

Showers

Acrylic fiberglass shower stall, 32" W, 32" D, 72" H (at $250), includes mixing valve 2.5 GPM shower head (at $115) and door (at $195)

Fiberglass 32" shower stall and trim	P1@4.15	Ea	560.00	101.00	661.00

Acrylic fiberglass shower stall, 48" W, 35" D, 72" H with integral soap dishes, grab bar and drain (at $360), mixing valve and 2.5 GPM shower head (at $115) and door (at $195)

Fiberglass 48" shower stall and trim	P1@4.65	Ea	670.00	114.00	784.00

Acrylic fiberglass shower stall, 60" W, 35" D, 72" H with integral soap dishes, grab bar, drain and seat, (at $395), mixing valve and 2.5 GPM shower head (at $115) and door (at $195)

Fiberglass 60" shower stall and trim	P1@4.80	Ea	705.00	117.00	822.00
Roll-in wheelchair stall and trim	P1@5.40	Ea	1,210.00	132.00	1,342.00
Add for shower cap (roof)	P1@.500	Ea	125.00	12.20	137.20

Plumbing Fixtures and Equipment

	Craft@Hrs	Unit	Material	Labor	Total

Tub and shower combinations

Acrylic one-piece fiberglass tub and shower, 5' W, 32" D, 72" H with integral soap dishes, high and low grab bars (at $435). Includes polished chrome pop-up drain/overflow (at $60), mixing valve, tub filler, and 2.5 GPM shower head (at $130)

	Craft@Hrs	Unit	Material	Labor	Total
Acrylic tub-shower and trim	P1@4.75	Ea	625.00	116.00	741.00

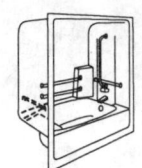

Handicapped and elderly access fiberglass tub and shower, 64" L, 66" D, 82" H with integral soap dishes, seat and four grab bars (at $975), body washer shower (at $45), polished chrome pop-up drain/overflow (at $60), mixing valve and tub filler (at $120)

	Craft@Hrs	Unit	Material	Labor	Total
Handicap shower-tub and trim	P1@4.75	Ea	1,200.00	116.00	1,316.00
Add for shower curtain rod	P1@.572	Ea	25.00	14.00	39.00
Add for aluminum framed sliding door	P1@.750	Ea	225.00	18.30	243.30

Water closets

Floor-mounted, residential tank type vitreous china toilet (at $110), valved closet supply (at $10), and toilet seat with cover (at $20)

	Craft@Hrs	Unit	Material	Labor	Total
Floor-mount tank type closet and trim	P1@2.10	Ea	140.00	51.30	191.30

Wall-hung, residential tank type vitreous china toilet (at $280), valved closet supply (at $10), and toilet seat with cover (at $20)

	Craft@Hrs	Unit	Material	Labor	Total
Wall hung tank type closet and trim	P1@2.50	Ea	310.00	61.10	371.10

One piece, floor-mounted vitreous china closet with elongated bowl (at $480), including toilet seat and valved closet supply (at $30)

	Craft@Hrs	Unit	Material	Labor	Total
One-piece water closet and trim	P1@2.35	Ea	510.00	57.40	567.40
Add for elongated bowl models	—	Ea	60.00	—	60.00

Bidets

Standard vitreous china bidet with floor-mounted hardware (at $300), hot and cold valved supplies (at $15) and polished chrome faucet (at $175)

	Craft@Hrs	Unit	Material	Labor	Total
Vitreous china bidet and trim	P1@2.50	Ea	490.00	61.10	551.10

Deluxe, designer type, vitreous china with floor mounting hardware (at $430), hot and cold valved supplies (at $15) and designer faucet (at $225)

	Craft@Hrs	Unit	Material	Labor	Total
Designer bidet and trim	P1@2.65	Ea	670.00	64.80	734.80

Lavatories

Oval self-rimming enameled steel lavatory (at $55), hot and cold valved supplies (at $15), single lever faucet and pop-up drain fitting (at $110) and trap assembly (at $25)

	Craft@Hrs	Unit	Material	Labor	Total
Oval self-rim steel lavatory and trim	P1@2.50	Ea	315.00	61.10	376.10

Oval, self-rimming vitreous china lavatory (at $90), hot and cold valved supplies (at $15), single lever faucet and pop-up drain fitting (at $110) and trap assembly (at $25)

	Craft@Hrs	Unit	Material	Labor	Total
Oval self-rim china lavatory and trim	P1@2.00	Ea	240.00	48.90	288.90

Oval, pedestal-mounted vitreous china lavatory (at $125), hot and cold valved supplies (at $15), single lever faucet & pop-up drain fitting (at $110) and trap (at $25)

	Craft@Hrs	Unit	Material	Labor	Total
Oval pedestal mount china lavatory and trim	P1@2.50	Ea	275.00	61.10	336.10

Rectangular, pedestal-mount vitreous china lavatory (at $165), hot and cold valved supplies (at $15), single lever faucet and pop-up drain fitting (at $110) and trap (at $25)

	Craft@Hrs	Unit	Material	Labor	Total
Rectangular pedestal china lavatory and trim	P1@2.50	Ea	315.00	61.10	376.10

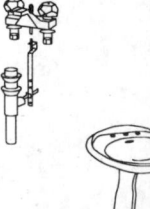

Rectangular, self-rimming vitreous china lavatory (at $130), hot and cold valved supplies (at $15), single lever faucet & pop-up drain fitting (at $110) and trap assembly (at $25)

	Craft@Hrs	Unit	Material	Labor	Total
Counter type rectangular china lavatory & trim	P1@2.00	Ea	280.00	48.90	328.90

Rectangular, wall-hung vitreous china lavatory (at $100), hot and cold valved supplies (at $15), single lever faucet & pop-up drain fitting (at $110) and trap assembly (at $25)

	Craft@Hrs	Unit	Material	Labor	Total
Wall-hung china lavatory and trim	P1@2.50	Ea	260.00	61.10	321.10

	Craft@Hrs	Unit	Material	Labor	Total

Wheelchair access lavatory, wall-hung vitreous china, with carrier (at $260)
insulated hot and cold valved supplies (at $20), gooseneck faucet with wrist handles and
pop-up drain (at $150) and insulated offset trap (at $30)

	Craft@Hrs	Unit	Material	Labor	Total
Wall-hung wheelchair lavatory and trim	P1@2.75	Ea	460.00	67.20	527.20

Bar sinks

Acrylic bar sink, 15" L, 15" W, 6" D, self-rimming (at $55), with chrome faucet (at $70),
hot and cold water valved supplies (at $15) and P trap assembly (at $25)

	Craft@Hrs	Unit	Material	Labor	Total
Acrylic bar sink and trim	P1@2.00	Ea	105.00	48.90	153.90

Enameled cast iron bar sink, 16" L, 19" W, 6" D with chrome faucet (at $ 275),
hot and cold water valved supplies (at $15) and P trap assembly (at $25)

	Craft@Hrs	Unit	Material	Labor	Total
Enameled cast iron bar sink and trim	P1@2.25	Ea	315.00	55.00	370.00

Stainless steel bar sink 15" L, 15" W, 6" D (at $75), chrome faucet
(at $70), hot and cold valved supplies (at $15), and trap assembly (at $25)

	Craft@Hrs	Unit	Material	Labor	Total
Stainless steel bar sink and trim	P1@2.00	Ea	160.00	48.90	208.90

Kitchen sinks

Single bowl, enameled cast iron, self-rimming, 25" L, 22" W, 8" D with chrome faucet
and sprayer, P trap assembly and valved supplies

	Craft@Hrs	Unit	Material	Labor	Total
Single bowl cast iron sink and trim	P1@1.75	Ea	210.00	42.80	252.80

Single bowl, stainless steel, self-rimming sink, 20" L, 20" W, 8" D with chrome faucet
and sprayer, strainer, P-trap assembly and valved supplies

	Craft@Hrs	Unit	Material	Labor	Total
Single bowl stainless sink and trim	P1@2.00	Ea	180.00	48.90	228.90

Double bowl, enameled cast iron, self-rimming sink, 33" L, 22" W, 8" D, chrome faucet
and sprayer, two strainers, P-trap assembly and valved supplies

	Craft@Hrs	Unit	Material	Labor	Total
Double bowl cast iron sink and trim	P1@2.00	Ea	280.00	48.90	328.90

Double bowl, stainless steel, self-rimming sink, 32" L, 22" W, 8" D with chrome faucet
and sprayer, two strainers, P-trap assembly and valved supplies

	Craft@Hrs	Unit	Material	Labor	Total
Double bowl stainless sink and trim	P1@2.25	Ea	215.00	55.00	270.00

Triple bowl, stainless steel, self-rimming sink, 43" L, 22" W (at $215) with chrome faucet
and sprayer (at $145), three strainers, P-trap assembly (at $40) and valved supplies (at $15)

	Craft@Hrs	Unit	Material	Labor	Total
Triple bowl stainless sink and trim	P1@2.75	Ea	415.00	67.20	482.20

Laundry sinks

Enameled cast iron laundry sink, 24" L, 20" D, 14" D, wall-mounted faucet with integral stops and
vacuum breaker, strainer and P-trap assembly

	Craft@Hrs	Unit	Material	Labor	Total
Cast iron laundry sink and trim	P1@2.76	Ea	310.00	67.40	377.40

Acrylic laundry sink, 24" L, 20" D, 14" D , deck-mounted faucet valved supplies, and P-trap assembly

	Craft@Hrs	Unit	Material	Labor	Total
Acrylic laundry sink and trim	P1@2.00	Ea	135.00	48.90	183.90

Service sinks

Enameled cast iron service sink, 22" L, 18" D (at $290), wall mounted faucet with integral stops
and vacuum breaker (at $140), strainer (at $20) and cast iron P-trap (at $170)

	Craft@Hrs	Unit	Material	Labor	Total
Cast iron service sink and trim	P1@3.00	Ea	620.00	73.30	693.30

Drinking fountains and electric water coolers.

Wall-hung vitreous china fountain, 14" W, 13" H, complete with fittings and 1-1/4" trap

	Craft@Hrs	Unit	Material	Labor	Total
Wall hung vitreous fountain and trim	P1@2.00	Ea	275.00	48.90	323.90

Semi-recessed vitreous china, 15" W, 27" H, with fittings, wall cleanout and 1-1/4" trap

	Craft@Hrs	Unit	Material	Labor	Total
Semi-recessed vitreous fountain and trim	P1@2.50	Ea	460.00	61.10	521.10

Plumbing Fixtures and Equipment

	Craft@Hrs	Unit	Material	Labor	Total

Wall-hung electric water cooler, 13 gallon capacity, complete with trap and fittings
(requires electrical outlet for plug-in connection, not included)

	Craft@Hrs	Unit	Material	Labor	Total
Electric water cooler and trim	P1@2.75	Ea	730.00	67.20	797.20

Wall-hung wheelchair access electric water cooler, front push bar, 7.5 gallon capacity, complete
with trap and fittings (requires electrical outlet for plug-in connection, not included)

Wheelchair access water cooler and trim	P1@2.50	Ea	750.00	61.10	811.10

Fully recessed electric water cooler, 11.5 gallon capacity, complete
with trap and fittings (requires electrical outlet for plug-in connection, not included)

Recessed electric water cooler and trim	P1@3.00	Ea	1,275.00	73.30	1,348.30

Garbage disposers

Small, 1/3 HP kitchen disposer (at $70), including fittings (at $30), no waste pipe included

1/3 HP kitchen disposer	P1@2.00	Ea	100.00	48.90	148.90

Standard, 1/2 HP kitchen disposer (at $85), including fittings (at $30), no waste pipe

1/2 HP kitchen disposer	P1@2.00	Ea	115.00	48.90	163.90

Standard, 3/4 HP kitchen disposer (at $165), including fittings ($30), no waste pipe included

3/4 HP kitchen disposer	P1@2.00	Ea	195.00	48.90	243.90

Large capacity, 1 HP kitchen disposer (at $235), including fittings (at $30), no waste pipe

1 HP kitchen disposer	P1@2.15	Ea	265.00	52.50	317.50
Add for electrical connection and switch	BE@1.00	Ea	25.00	26.90	51.90

Dishwashers. Under counter installation, connected to existing drain.

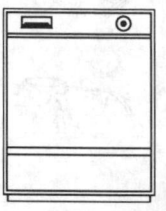

Deluxe	P1@2.40	Ea	635.00	58.60	693.60
Standard	P1@2.40	Ea	535.00	58.60	593.60
Economy	P1@2.40	Ea	320.00	58.60	378.60
Add for electrical connection	BE@1.00	Ea	22.00	26.90	48.90

Ultraflo water systems. Single line, solid state push button fixture controls with LED's, centralized water distribution system, includes centralized mixing and metering valves, control wire, spouts, and push button fixture controls.

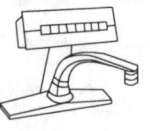

Kitchen and one full bath system	P1@7.63	Ea	1,160.00	186.00	1,346.00
Kitchen and two full bath system	P1@11.5	Ea	1,840.00	281.00	2,121.00

Electric water heaters Includes labor and material to connect to existing water supply piping. 4,500 watt elements, 6 year warranty, with temperature and pressure relief valve. All meet NAECA-1990 efficiency standards.

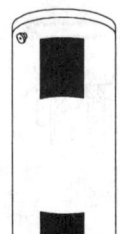

30 gallon (at $170)	P1@2.50	LS	220.00	61.10	281.10
40 gallon (at $180)	P1@2.75	LS	230.00	67.20	297.20
52 gallon (at $255)	P1@3.25	LS	305.00	79.40	384.40
82 gallon (at $680)	P1@4.00	LS	780.00	97.70	877.70

Gas water heaters Includes labor and material to connect to existing water supply piping and gas line. 6 year warranty, with temperature and pressure relief valve (at $70) and flue (at $90). Labor column includes cost of installing heater, making the gas connection and installing the flue. No gas piping included.

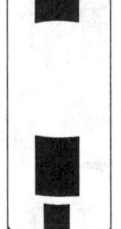

30 gallon (at $260)	P1@3.50	LS	420.00	85.50	505.50
40 gallon (at $270)	P1@3.75	LS	430.00	91.60	521.60
50 gallon (at $340)	P1@4.25	LS	500.00	104.00	604.00

Solar Water Heating System Water storage tanks compatible with most closed-loop solar water heating systems, heat transfer by insulated all-copper heat exchanger, double wall construction and back-up electric heating element. Rheem

80 gallon, 59" high x 25" deep	P1@8.44	Ea	745.00	206.00	951.00

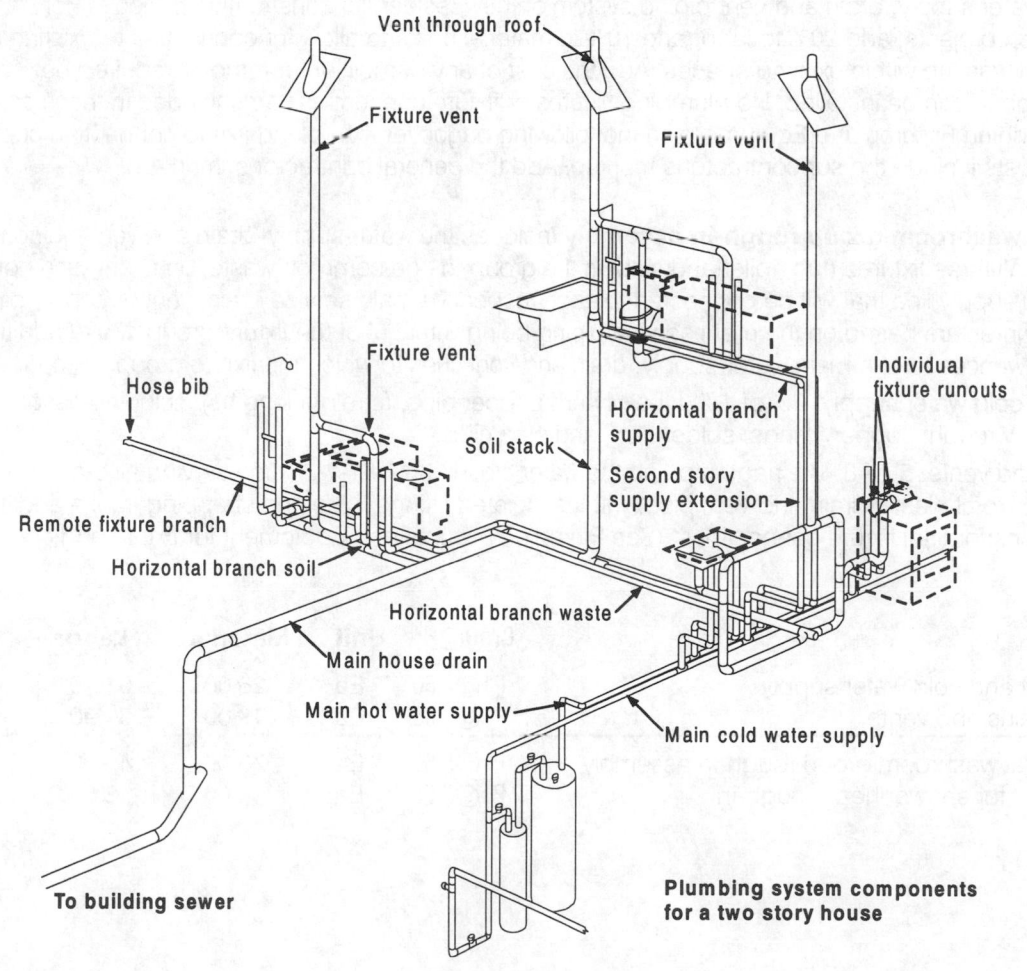

	Craft@Hrs	Unit	Material	Labor	Total
Water heaters, installed on a subcontract basis					
Passive thermosiphon domestic hot water system with integral 80 gallon					
water tank, retrofitted to an existing single story structure, with 5 year limited warranty					
Passive solar water heater system (retrofit)	—	Ea	—	—	3,260.00
Active solar hot water system for 100 or 120 gallon water supply tank, fitted with automatic					
drain retrofitted to existing single story, single family unit, 5 year warranty					
Active solar water heater system (retrofit)	—	Ea	—	—	3,670.00
Add for strengthening existing roof structure under solar collectors					
Per SF of roof strengthened	—	SF	—	—	1.04
Active solar hot water system for new one-story home					
Including design and engineering (new work)	—	Ea	—	—	3,160.00
For designs requiring special cosmetic treatment					
Add for concealment of pipe, collectors, etc.	—	%	—	—	15.0
Solar pool heater, active heater for pools with freeze tolerant collector, roof mounted					
Per SF of collector area	—	SF	—	—	5.00
Water softeners Installed on a subcontract basis					
9 gallons per minute, automatic	—	Ea	—	—	950.00
12 gallons per minute, automatic	—	Ea	—	—	1,000.00

Plumbing System Components

Plumbing system components
for a two story house

Plumbing Rough-in Assemblies

Whole House Plumbing Rough-In
Rough plumbing is the drain, waste and vent (DWV) piping and water supply piping that isn't visible when construction is complete. Rough plumbing costs vary with the number of fixtures and length of the pipe runs. The figures in this section show rough plumbing costs for a typical single family home using plastic drainage and copper water supply materials. No plumbing fixtures, equipment or their final connections are included in these costs. See Plumbing Fixtures and Equipment for fixture and fixture connection costs. These costs include the subcontractor's markup. Add the general contractor's markup when applicable. For more complete information on plumbing and piping costs, see *National Plumbing & HVAC Estimator*. An order form is at the back of this manual.

	Craft@Hrs	Unit	Material	Labor	Total
Single story homes					
Hot and cold potable water rough-in	P1@15.0	Ea	360.00	367.00	727.00
Drainage and venting rough-in	P1@20.0	Ea	320.00	489.00	809.00
Single story, total plumbing rough-in	P1@35.0	Ea	680.00	856.00	1,536.00
Two story homes					
Hot and cold potable water rough-in	P1@20.0	Ea	531.00	489.00	1,020.00
Drainage and venting rough-in	P1@25.0	Ea	566.00	611.00	1,177.00
Two story, total plumbing rough-in	P1@45.0	Ea	1,097.00	1,100.00	2,197.00

Plumbing Fixture Rough-In Assemblies Includes labor and material costs for the installation of a typical water supply, drain and vent piping system on new residential construction projects. For renovation projects, add 20% to labor and 10% to material costs to allow for connection to existing service lines and working within existing spaces. Add the cost of any demolition or removal required before the new rough-in can be installed. No plumbing fixtures or fixture connections are included in these costs. See Plumbing Fixtures and Equipment on the following pages for fixture and fixture connection costs. These costs include the subcontractor's markup. Add the general contractor's mark-up.

3-piece washroom group rough-in assembly Includes the water supply, drain and vent piping to connect all three fixtures (tub, toilet and sink) in the group to one common waste, one vent, one hot and one cold supply line that will be connected to each respective main service drain, vent or supply pipe. These figures are based on the main service piping being within 5' of the fixture group. Make additional cost allowances for the main water supply, drain and vent lines to which the fixture group will connect.

Hot and cold water supply: 26' of 1/2" type M hard copper pipe, (straight lengths) including associated soldered wrought copper fittings, solder, flux, and pipe clips.

Drains and vents: 5' of 3" drainage pipe on the water closet, 38' of 1-1/2" pipe serving all other drains and vents. 43' total plastic drain and vent pipe, with associated fittings, glue and strapping. Make additional allowances for final fixture connections. (See Plumbing Fixtures and Equipment for fixture and connection costs)

	Craft@Hrs	Unit	Material	Labor	Total
Hot and cold water supply	P1@2.50	Ea	26.00	61.10	87.10
Drains and vents	P1@3.35	Ea	47.00	81.90	128.90
Total washroom group rough-in assembly	P1@5.85	Ea	73.00	143.00	216.00
Add for shower head rough-in	P1@.350	Ea	4.00	8.55	12.55

	Craft@Hrs	Unit	Material	Labor	Total

Kitchen sink rough-in assembly

Hot and cold water supply: 10' of 1/2" type M hard copper pipe, (straight lengths) including associated soldered wrought copper fittings, solder, flux, and pipe clips. Drains and vents: 15' of 1-1/2" plastic drain and vent pipe with associated fittings, glue and strapping. Make additional allowance for final fixture connection. (See Plumbing Fixtures and Equipment for fixture and connection costs)

	Craft@Hrs	Unit	Material	Labor	Total
Hot and cold water supply	P1@.750	Ea	9.60	18.30	27.90
Drains and vents	P1@1.00	Ea	14.40	24.40	38.80
Total Kitchen sink rough-in assembly	P1@1.75	Ea	24.00	42.70	66.70

Laundry tub and washing machine rough-in assembly

Hot and cold water supply: 16' of 1/2" type M hard copper pipe, (straight lengths) including associated fittings, solder, flux, and pipe clips. Drains and vents: 21' of 1-1/2" plastic drain and vent pipe with associated fittings, glue and strapping.

	Craft@Hrs	Unit	Material	Labor	Total
Hot and cold water supply	P1@1.00	Ea	15.36	24.40	39.76
Drains and vents	P1@1.35	Ea	20.16	33.00	53.16
Total Laundry tub & washer rough-in assembly	P1@2.35	Ea	35.52	57.40	92.92

Hot water tank rough-in assembly

Hot and cold water supply: 10' of 3/4" type M hard copper pipe, (straight lengths) including associated soldered wrought copper fittings, solder, flux, and pipe clips

	Craft@Hrs	Unit	Material	Labor	Total
Hot and cold water supply	P1@.500	Ea	20.09	12.20	32.29
Add for 3/4" isolation valve	P1@.250	Ea	5.90	6.11	12.01

Water softener rough-in assembly

Hot and cold water supply: 10' of 3/4" type M hard copper pipe, (straight lengths) including associated soldered wrought copper fittings, solder, flux, and pipe clips

	Craft@Hrs	Unit	Material	Labor	Total
Cold water supply and return	P1@.500	Ea	20.09	12.20	32.29
Regeneration drain	P1@.500	Ea	10.00	12.20	22.20

For fixture rough-in assemblies not listed here, see plumbing assemblies, Mechanical Section 15

	Craft@Hrs	Unit	Material	Labor	Total

Natural gas appliance rough-in assembly

Includes labor and material to install and test 12' of 3/4" schedule 40 black steel piping, associated 150 lb malleable iron fittings and a plug valve to serve a typical residential gas fired appliance. Make additional allowances for the natural gas main supply line running through the building, the meter connection and the final connection to the appliance. Assembly assumes the appliance is within 5' of the main supply line.

	Craft@Hrs	Unit	Material	Labor	Total
Gas fired domestic hot water heater	P1@.650	Ea	18.50	15.90	34.40
Gas fired hot water heating boiler	P1@.650	Ea	18.50	15.90	34.40
Gas fired forced air furnace	P1@.650	Ea	18.50	15.90	34.40
Gas fired stove and oven	P1@.650	Ea	18.50	15.90	34.40
Gas fired clothes dryer	P1@.650	Ea	18.50	15.90	34.40
Gas main, 1"dia.with fittings and hangers	P1@.120	LF	2.80	2.93	5.73
Gas meter connection (house side only)	P1@.750	Ea	15.00	18.30	33.30

Plumbing Rough-in Assemblies

	Craft@Hrs	Unit	Material	Labor	Total

Natural gas appliance connections

Includes labor and material to connect and test a typical residential natural gas fired appliance. Make additional allowances for the natural gas main supply line running through the building, the appliance gas supply rough-in and labor to set the appliance in place if it has not been set in place by others.

	Craft@Hrs	Unit	Material	Labor	Total
Gas fired domestic hot water heater	P1@.650	Ea	10.00	15.90	25.90
Gas fired hot water heating boiler	P1@.650	Ea	10.00	15.90	25.90
Gas fired forced air furnace	P1@.650	Ea	10.00	15.90	25.90
Gas fired grill, stove or oven	P1@.650	Ea	25.00	15.90	40.90
Gas fired clothes dryer	P1@.650	Ea	25.00	15.90	40.90

Hot Water Heating Systems

Whole house hot water heating system assembly

Includes 150' of 1-1/2" insulated supply and return mains, 150' of ¾" insulated branch piping, 10 radiant baseboard convectors (50'), a 110 mbh high efficiency gas fired hot water boiler, a circulation pump, electric controls, installed and tested. (Based on 1,200 SF of heated space)

	Craft@Hrs	Unit	Material	Labor	Total
Distribution system	P1@46.0	Ea	580.00	1,120.00	1,700.00
Distribution system (second floor)	P1@16.0	Ea	250.00	391.00	641.00
Boiler, pump and controls	P1@16.0	Ea	2,450.00	391.00	2,841.00
Radiant baseboard and valves	P1@16.0	Ea	1,800.00	391.00	2,191.00
Radiant baseboard and valves (2nd floor)	P1@12.0	Ea	1,200.00	293.00	1,493.00
Total hot water heating system, bungalow	P1@78.0	Ea	4,830.00	1,902.00	6,732.00
Total hot water heating system, 2 story	P1@106	Ea	6,280.00	2,586.00	8,866.00

Gate Valves

1/2" bronze gate valves

	Craft@Hrs	Unit	Material	Labor	Total
125 lb, threaded	P1@.210	Ea	17.30	5.13	22.43
125 lb, soldered	P1@.200	Ea	14.10	4.89	18.99
150 lb, threaded	P1@.210	Ea	21.70	5.13	26.83
200 lb, threaded	P1@.210	Ea	50.70	5.13	55.83

3/4" bronze gate valves

	Craft@Hrs	Unit	Material	Labor	Total
125 lb, threaded	P1@.250	Ea	21.30	6.11	27.41
125 lb, soldered	P1@.240	Ea	17.30	5.86	23.16
150 lb, threaded	P1@.250	Ea	25.40	6.11	31.51
200 lb, threaded	P1@.250	Ea	60.20	6.11	66.31

1" bronze gate valves

	Craft@Hrs	Unit	Material	Labor	Total
125 lb, threaded	P1@.300	Ea	28.90	7.33	36.23
125 lb, soldered	P1@.290	Ea	22.00	7.09	29.09
150 lb, threaded	P1@.300	Ea	33.30	7.33	40.63
200 lb, threaded	P1@.300	Ea	85.80	7.33	93.13

1-1/2" bronze gate valves

	Craft@Hrs	Unit	Material	Labor	Total
125 lb, threaded	P1@.450	Ea	51.50	11.00	62.50
125 lb, soldered	P1@.440	Ea	38.40	10.80	49.20
150 lb, threaded	P1@.450	Ea	56.50	11.00	67.50
200 lb, threaded	P1@.450	Ea	133.00	11.00	144.00

2" bronze gate valves

	Craft@Hrs	Unit	Material	Labor	Total
125 lb, threaded	P1@.500	Ea	62.30	12.20	74.50
125 lb, soldered	P1@.480	Ea	50.30	11.70	62.00
150 lb, threaded	P1@.500	Ea	76.50	12.20	88.70
200 lb, threaded	P1@.500	Ea	193.00	12.20	205.20

	Craft@Hrs	Unit	Material	Labor	Total
Forged steel Class 800 bolted bonnet OS&Y gate valves, threaded ends					
1/2" valve	P1@.210	Ea	62.20	5.13	67.33
3/4" valve	P1@.250	Ea	85.00	6.11	91.11
1" valve	P1@.300	Ea	153.00	7.33	160.33
1-1/2" valve	P1@.450	Ea	193.00	11.00	204.00
2" valve	P1@.500	Ea	295.00	12.20	307.20

Globe Valves   125 lb and 150 lb ratings are saturated steam pressure. 200 lb rating is working steam pressure, union bonnet. 125 lb valves have bronze disks. 150 lb valves have composition disks. 300 lb valves have regrinding disks.

	Craft@Hrs	Unit	Material	Labor	Total
1/2" threaded bronze globe valves					
125 lb	P1@.210	Ea	25.80	5.13	30.93
150 lb	P1@.210	Ea	32.20	5.13	37.33
300 lb	P1@.210	Ea	57.90	5.13	63.03
3/4" threaded bronze globe valves					
125 lb	P1@.250	Ea	32.70	6.11	38.81
150 lb	P1@.250	Ea	43.70	6.11	49.81
300 lb	P1@.250	Ea	80.20	6.11	86.31
1" threaded bronze globe valves					
125 lb	P1@.300	Ea	43.80	7.33	51.13
150 lb	P1@.300	Ea	67.90	7.33	75.23
300 lb	P1@.300	Ea	116.00	7.33	123.33
1-1/2" threaded bronze globe valves					
125 lb	P1@.450	Ea	75.20	11.00	86.20
150 lb.	P1@.450	Ea	129.00	11.00	140.00
300 lb.	P1@.450	Ea	200.00	11.00	211.00
2" threaded bronze globe valves					
125 lb	P1@.500	Ea	113.00	12.20	125.20
150 lb	P1@.500	Ea	194.00	12.20	206.20
300 lb	P1@.500	Ea	306.60	12.20	318.80

Swing Check Valves

	Craft@Hrs	Unit	Material	Labor	Total
1/2" threaded bronze check valves					
125 lb	P1@.210	Ea	20.50	5.13	25.63
125 lb, soldered	P1@.200	Ea	16.10	4.89	20.99
125 lb, vertical lift check	P1@.210	Ea	47.10	5.13	52.23
150 lb, bronze disk	P1@.210	Ea	31.70	5.13	36.83
200 lb, swing, regrinding bronze disk	P1@.210	Ea	31.20	5.13	36.33
3/4" threaded bronze check valves					
125 lb	P1@.250	Ea	24.60	6.11	30.71
125 lb, soldered	P1@.240	Ea	19.70	5.86	25.56
125 lb, vertical check	P1@.250	Ea	64.10	6.11	70.21
150 lb, bronze disk	P1@.250	Ea	39.50	6.11	45.61
200 lb, swing, regrinding bronze disk	P1@.250	Ea	39.00	6.11	45.11
1" threaded bronze check valves					
125 lb	P1@.300	Ea	33.50	7.33	40.83
125 lb, soldered	P1@.290	Ea	27.00	7.09	34.09
125 lb, vertical check	P1@.300	Ea	80.60	7.33	87.93
150 lb, bronze disk	P1@.300	Ea	53.30	7.33	60.63
200 lb, swing, regrinding bronze disk	P1@.300	Ea	52.60	7.33	59.93

Plumbing and Heating

	Craft@Hrs	Unit	Material	Labor	Total
1-1/2" threaded bronze check valves					
125 lb	P1@.450	Ea	55.20	11.00	66.20
125 lb, soldered	P1@.440	Ea	44.50	10.80	55.30
125 lb, vertical check	P1@.450	Ea	136.00	11.00	147.00
150 lb, bronze disk	P1@.450	Ea	91.70	11.00	102.70
200 lb, swing, regrinding bronze disk	P1@.450	Ea	90.80	11.00	101.80
2" threaded bronze check valves					
125 lb	P1@.500	Ea	79.50	12.20	91.70
125 lb, soldered	P1@.480	Ea	64.80	11.70	76.50
125 lb, vertical check	P1@.500	Ea	202.00	12.20	214.20
150 lb, bronze disk	P1@.500	Ea	135.00	12.20	147.20
200 lb, swing, regrinding bronze disk	P1@.500	Ea	134.00	12.20	146.20

Miscellaneous Valves and Regulators

	Craft@Hrs	Unit	Material	Labor	Total
Angle valves, bronze, 150 lb, threaded, Teflon disk					
1/2" valve	P1@.210	Ea	41.00	5.13	46.13
3/4" valve	P1@.250	Ea	55.70	6.11	61.81
1" valve	P1@.300	Ea	80.50	7.33	87.83
1-1/4" valve	P1@.400	Ea	105.00	9.77	114.77
1-1/2" valve	P1@.450	Ea	140.00	11.00	151.00
Backflow preventers reduced pressure					
3/4", threaded, bronze, with 2 gate valves	P1@1.00	Ea	172.00	24.40	196.40
1", threaded, bronze, with 2 gate valves	P1@1.25	Ea	209.00	30.50	239.50
2", threaded, with 2 gate valves	P1@2.13	Ea	309.00	52.10	361.10
Ball valves, bronze, 150 lb threaded, Teflon seat					
1/2" valve	P1@.210	Ea	5.38	5.13	10.51
3/4" valve	P1@.250	Ea	7.95	6.11	14.06
1" valve	P1@.300	Ea	11.30	7.33	18.63
1-1/4" valve	P1@.400	Ea	19.50	9.77	29.27
1-1/2" valve	P1@.450	Ea	23.80	11.00	34.80
2" valve	P1@.500	Ea	32.10	12.20	44.30

Lavatory supply valve with hand wheel shutoff, flexible compression connection tube with length as shown

	Craft@Hrs	Unit	Material	Labor	Total
1/2" angle or straight stops					
12" tube	P1@.294	Ea	15.75	7.18	22.93
15" tube	P1@.294	Ea	16.80	7.18	23.98
20" tube	P1@.473	Ea	19.30	11.60	30.90
Gas stops, lever handle					
1/2" stop	P1@.210	Ea	7.57	5.13	12.70
3/4" stop	P1@.250	Ea	10.20	6.11	16.31
1" stop	P1@.300	Ea	15.60	7.33	22.93
1-1/4" stop	P1@.450	Ea	23.40	11.00	34.40
1-1/2" stop	P1@.500	Ea	40.60	12.20	52.80
2" stop	P1@.750	Ea	62.70	18.30	81.00
Hose bibbs, threaded					
3/4", brass	P1@.350	Ea	3.25	8.55	11.80
Pressure regulator valves, bronze, 300 lb threaded, 25 to 75 PSI, with "Y" strainer					
3/4" valve	P1@.250	Ea	92.70	6.11	98.81
1" valve	P1@.300	Ea	123.00	7.33	130.33
1-1/4" valve	P1@.400	Ea	192.00	9.77	201.77
1-1/2" valve	P1@.450	Ea	295.00	11.00	306.00
2" valve	P1@.500	Ea	372.00	12.20	384.20

	Craft@Hrs	Unit	Material	Labor	Total
Meters, Gauges and Indicators Water meters					
Disc type, 1-1/2"	P1@1.00	Ea	129.00	24.40	153.40
Turbine type, 1-1/2"	P1@1.00	Ea	194.00	24.40	218.40
Displacement type, AWWA C7000, 1"	P1@.941	Ea	199.00	23.00	222.00
Displacement type, AWWA C7000, 1-1/2"	P1@1.23	Ea	684.00	30.10	714.10
Displacement type, AWWA C7000, 2"	P1@1.57	Ea	1,020.00	38.40	1,058.40
Curb stop and waste valve, 1"	P1@1.00	Ea	71.50	24.40	95.90
Thermoflow indicators, soldered					
1-1/4" indicator	P1@2.01	Ea	298.00	49.10	347.10
2" indicator	P1@2.01	Ea	340.00	49.10	389.10
Plumbing Specialties					
Access doors for plumbing, painted steel, with cam lock, for masonry, drywall, tile, etc. Milcor					
8" x 8"	SW@.400	Ea	18.00	11.10	29.10
12" x 12"	SW@.500	Ea	35.60	13.90	49.50
18" x 18"	SW@.800	Ea	44.00	22.20	66.20
24" x 24"	SW@1.25	Ea	63.00	34.70	97.70
36" x 36"	SW@1.60	Ea	130.00	44.40	174.40
Add for fire rating	—	%	400.0	—	—
Add for stainless steel	—	%	300.0	—	—
Add for cylinder lock	—	Ea	8.38	—	8.38
Hangers and Supports					
Angle bracket hangers, steel, by rod size					
3/8" angle bracket hanger	P1@.070	Ea	1.41	1.71	3.12
1/2" angle bracket hanger	P1@.070	Ea	1.58	1.71	3.29
5/8" angle bracket hanger	P1@.070	Ea	3.48	1.71	5.19
3/4" angle bracket hanger	P1@.070	Ea	5.35	1.71	7.06
7/8" angle bracket hanger	P1@.070	Ea	9.66	1.71	11.37
Top beam clamps, steel, by rod size					
3/8" top beam clamp	P1@.070	Ea	1.83	1.71	3.54
1/2" top beam clamp	P1@.070	Ea	2.98	1.71	4.69
5/8" top beam clamp	P1@.070	Ea	3.65	1.71	5.36
3/4" top beam clamp	P1@.070	Ea	5.34	1.71	7.05
7/8" top beam clamp	P1@.070	Ea	7.00	1.71	8.71
Clevis hangers, standard, by intended pipe size, no hanger rod or angle bracket included					
1/2" clevis hanger	P1@.156	Ea	1.44	3.81	5.25
3/4" clevis hanger	P1@.156	Ea	1.44	3.81	5.25
1" clevis hanger	P1@.156	Ea	1.53	3.81	5.34
1-1/4" clevis hanger	P1@.173	Ea	1.74	4.23	5.97
1-1/2" clevis hanger	P1@.173	Ea	1.91	4.23	6.14
2" clevis hanger	P1@.173	Ea	2.13	4.23	6.36
Copper tube riser clamps, by intended tube size					
1/2" riser clamp	P1@.100	Ea	4.68	2.44	7.12
3/4"riser clamp	P1@.100	Ea	4.91	2.44	7.35
1" riser clamp	P1@.100	Ea	6.32	2.44	8.76
1-1/4" riser clamp	P1@.105	Ea	6.55	2.57	9.12
1-1/2" riser clamp	P1@.110	Ea	7.95	2.69	10.64
2" riser clamp	P1@.115	Ea	8.42	2.81	11.23

Plumbing and Heating

	Craft@Hrs	Unit	Material	Labor	Total
Pipe clamps, medium duty, by intended pipe size					
1/2" or 3/4" pipe clamp	P1@.156	Ea	3.15	3.81	6.96
1" pipe clamp	P1@.156	Ea	3.32	3.81	7.13
1-1/4" or 1-1/2" pipe clamp	P1@.173	Ea	4.13	4.23	8.36
2" pipe clamp	P1@.173	Ea	5.10	4.23	9.33
Pipe straps, galvanized, two hole, lightweight					
1/2" pipe strap	P1@.072	Ea	.07	1.76	1.83
3/4" pipe strap	P1@.080	Ea	.11	1.95	2.06
1" pipe strap	P1@.088	Ea	.13	2.15	2.28
1-1/4" pipe strap	P1@.096	Ea	.21	2.35	2.56
1-1/2" pipe strap	P1@.104	Ea	.24	2.54	2.78
2" pipe strap	P1@.112	Ea	.30	2.74	3.04
Plumber's tape, galvanized					
26 gauge 3/4" x 10' roll	—	Ea	1.05	—	1.05
22 gauge 3/4" x 10' roll	—	Ea	1.40	—	1.40
Threaded hanger rod, by rod size, in 6' to 10' lengths, per LF					
3/8" rod	—	LF	1.00	—	1.00
1/2" rod	—	LF	1.56	—	1.56
5/8" rod	—	LF	2.23	—	2.23
3/4" rod	—	LF	3.90	—	3.90
U bolts, with hex nuts, light duty					
1/2" bolt	P1@.089	Ea	1.83	2.17	4.00
1" bolt	P1@.089	Ea	2.04	2.17	4.21
2" bolt	P1@.089	Ea	2.81	2.17	4.98

Closed Cell Elastomeric Pipe and Tubing Insulation Semi split, 1/2" wall thickness, no cover, by nominal pipe or tube diameter. R factor equals 3.58 at 220 degrees F. Manufactured by Rubatex. These costs do not include scaffolding. Also see fittings, flanges and valves at the end of this section.

	Craft@Hrs	Unit	Material	Labor	Total
1/4" pipe	P1@.039	LF	.31	.95	1.26
3/8" pipe	P1@.039	LF	.33	.95	1.28
1/2 " pipe	P1@.039	LF	.37	.95	1.32
3/4" pipe	P1@.039	LF	.42	.95	1.37
1" pipe	P1@.042	LF	.47	1.03	1.50
1-1/4" pipe	P1@.042	LF	.53	1.03	1.56
1-1/2" pipe	P1@.042	LF	.62	1.03	1.65
2" pipe	P1@.046	LF	.78	1.12	1.90

Pipe fittings and flanges:

For each fitting or flange use the cost for 3 LF of pipe or tube of the same size.

Valves:

For each valve body use the cost for 5 LF of pipe or tube of the same size.

Fiberglass Pipe Insulation with AP-T Plus (all purpose self sealing) Jacket By nominal pipe diameter. R factor equals 2.56 at 300 degrees F. These costs do not include scaffolding, add for same, if required. Also see fittings, flanges and valves at the end of this section. Available in 3 foot lengths, price is per linear foot.

	Craft@Hrs	Unit	Material	Labor	Total
1/2" diameter pipe					
1/2" thick insulation	P1@.038	LF	1.18	.93	2.11
1" thick insulation	P1@.038	LF	1.40	.93	2.33
1-1/2" thick insulation	P1@.040	LF	2.80	.98	3.78

	Craft@Hrs	Unit	Material	Labor	Total
3/4" diameter pipe					
1/2" thick insulation	P1@.038	LF	1.30	.93	2.23
1" thick insulation	P1@.038	LF	1.60	.93	2.53
1-1/2" thick insulation	P1@.040	LF	2.95	.98	3.93
1" diameter pipe					
1/2" thick insulation	P1@.040	LF	1.37	.98	2.35
1" thick insulation	P1@.040	LF	1.40	.98	2.38
1-1/2" thick insulation	P1@.042	LF	3.10	1.03	4.13
1-1/4" diameter pipe					
1/2" thick insulation	P1@.040	LF	1.85	.98	2.83
1" thick insulation	P1@.040	LF	1.85	.98	2.83
1-1/2" thick insulation	P1@.042	LF	3.33	1.03	4.36
1-1/2" diameter pipe					
1/2" thick insulation	P1@.042	LF	1.63	1.03	2.66
1" thick insulation	P1@.042	LF	2.10	1.03	3.13
1-1/2" thick insulation	P1@.044	LF	3.59	1.08	4.67
2" diameter pipe					
1/2" thick insulation	P1@.044	LF	1.74	1.08	2.82
1" thick insulation	P1@.044	LF	2.39	1.08	3.47
1-1/2" thick insulation	P1@.046	LF	3.85	1.12	4.97

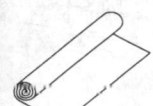

Polyethylene Film, Clear or Black Material costs include 5% for waste and 10% for laps. Labor shown is for installation on grade.

	Craft@Hrs	Unit	Material	Labor	Total
4 mil (.004" thick)					
50 LF rolls, 3' to 20' wide	BL@.002	SF	.04	.04	.08
6 mil (.006" thick)					
50 LF rolls, 3' to 40' wide	BL@.002	SF	.06	.04	.10
Add for installation on walls, ceilings and roofs, tack stapled, typical	BL@.001	SF	—	.02	.02

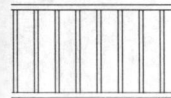

Railings, Steel, Prefabricated See also Ornamental Iron. Porch or step rail, site fitted.

	Craft@Hrs	Unit	Material	Labor	Total
2'-10" high, 6" OC 1/2" solid twisted pickets, 4' or 6' sections					
1" x 1/2" rails	RI@.171	LF	2.68	4.06	6.74
1" x 1" rails	RI@.180	LF	3.71	4.27	7.98
Add for newel post 3' H with hardware	—	Ea	10.20	—	10.20
Add lamb's tongue for end post	—	Ea	4.02	—	4.02

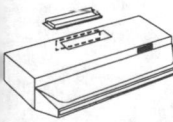

Range Hoods Material costs are for ducted or ductless range hood only, no electrical wiring included. Add for ducting or charcoal filters below. Labor costs are for installation of range hood only. No carpentry work included. All models are steel construction with baked enamel finish in various colors.

	Craft@Hrs	Unit	Material	Labor	Total
Minimum quality, 160 CFM, 2 speed fan, 75 watt light and .64 SF grease filter, 5.5 sone, 110 V, UL listed, front-mounted controls, top or back ducting					
24" x 17-1/2" x 6"	SW@1.42	Ea	38.50	39.40	77.90
30" x 17-1/2" x 6"	SW@1.42	Ea	42.60	39.40	82.00
36" x 17-1/2" x 6"	SW@1.42	Ea	44.60	39.40	84.00
42" x 17-1/2" x 6"	SW@1.42	Ea	50.80	39.40	90.20
Good quality, 200 CFM, variable speed fan, 75 watt light and a 1.04 SF grease filter, 5.5 sone, 110 V, UL listed, front-mounted controls					
24" x 17-1/2" x 6"	SW@1.42	Ea	142.00	39.40	181.40
30" x 17-1/2" x 6"	SW@1.42	Ea	152.00	39.40	191.40
36" x 17-1/2" x 6"	SW@1.42	Ea	182.00	39.40	221.40

Range Hoods

	Craft@Hrs	Unit	Material	Labor	Total
Better quality, 360 CFM, variable speed fan, two 75 watt lights and a 1.56 SF grease filter, 5.5 sone, 110 V, UL listed, front-mounted controls					
30" x 20" x 9"	SW@1.42	Ea	213.00	39.40	252.40
36" x 20" x 9"	SW@1.42	Ea	259.00	39.40	298.40
30" x 20" x 9", stainless steel	SW@1.42	Ea	238.00	39.40	277.40
36" x 20" x 9", stainless steel	SW@1.42	Ea	279.00	39.40	318.40
42" x 20" x 9", stainless steel	SW@1.42	Ea	299.00	39.40	338.40
Add for ducting, including cap with back draft preventer at exterior wall or roof					
Back ducted straight through wall	SW@.526	Ea	15.20	14.60	29.80
Top or back ducted through wall with single bend, to 4'	SW@.785	Ea	51.00	21.80	72.80
Top or back ducted through roof straight through roof, to 4'	SW@.657	Ea	43.00	18.20	61.20
Top or back ducted through roof with single bend, to 4'	SW@.801	Ea	56.00	22.20	78.20
Add for each additional LF of straight duct over 4'	SW@.136	LF	6.34	3.78	10.12
Add for each additional bend	SW@.136	Ea	15.20	3.78	18.98
Add for charcoal filters, ductless installation	—	Ea	10.30	—	10.30

Ranges, Built-In Material costs are for good to better quality built-in appliances. Labor costs are for carpentry installation only. Add for gas and electric runs below. Note that all cooktops with grilles require exhaust venting. See Range Hoods above.

	Craft@Hrs	Unit	Material	Labor	Total
Counter-mounted cooktops					
Gas, pilot free ignition, 30" x 21" x 4-1/2"					
Baked enamel top	P1@1.92	Ea	238.00	46.90	284.90
Chrome plated top	P1@1.92	Ea	259.00	46.90	305.90
Electric					
Smooth ceramic cooktop with one 9" element, one 8" element and two 6" elements					
30" x 21" x 3-3/8"	BE@.903	Ea	360.00	24.30	384.30
Radiant heat type with smooth ceramic cooktop, one 9", two 8", and one 6" element					
29-3/8" x 20-1/2" x 3-3/8"	BE@.903	Ea	532.00	24.30	556.30
Conventional coil type cooktop with two 8" and two 6" elements, 30-1/4" x 21-1/4" x 3-3/8"					
Baked enamel finish, various colors	BE@.903	Ea	213.00	24.30	237.30
Chrome plated finish	BE@.903	Ea	233.00	24.30	257.30
Conventional coil type cooktop, with two 8" and two 6" elements, built-in griddle, 35-1/2" x 20-1/2" x 3-3/8"					
Baked enamel finish, various colors	BE@.903	Ea	269.00	24.30	293.30
Chrome plated finish	BE@.903	Ea	289.00	24.30	313.30
Deluxe quality kitchen countertop range/grill combination, one 8" and one 6" element, grill module, chrome finish, built-in exhaust system below and 2 speed fan					
30" x 23" x 6"	BE@1.44	Ea	513.00	38.70	551.70
Wall ovens, with light and clock.					
Gas, electronic ignition, continuous cleaning, with broiler below, 39" high, 23-3/4" wide, 26-3/8" deep					
Porcelain enamel black glass door	P1@1.92	Ea	517.00	46.90	563.90
Electric, single oven 29" high, 27" deep, black glass door, with broiler					
24" wide, continuous cleaning	BE@1.67	Ea	543.00	44.90	587.90
24" wide, self-cleaning	BE@1.67	Ea	669.00	44.90	713.90
24" wide, porcelain enameled	BE@1.67	Ea	431.00	44.90	475.90

	Craft@Hrs	Unit	Material	Labor	Total
Electric, double oven, 50" high, 27" deep, black glass doors					
24" wide, continuous cleaning	BE@1.67	Ea	735.00	44.90	779.90
24" wide, self-cleaning	BE@1.67	Ea	953.00	44.90	997.90
27" wide, self-cleaning	BE@1.67	Ea	963.00	44.90	1,007.90
27" wide, microwave upper, with					
electric self-cleaning lower	BE@1.67	Ea	1,313.20	44.90	1,358.10

Range and oven combinations, automatic controls, with light and clock
 Drop-in models, 4 unit burners, black glass doors

	Craft@Hrs	Unit	Material	Labor	Total
Gas, pilot-free ignition, continuous cleaning, timer, baked enamel finish					
30" wide x 27" deep x 36" high	P1@1.92	Ea	1,088.00	46.90	1,134.90
Electric, self-cleaning, chrome-plated top					
30" wide x 27" deep x 31" high	BE@2.00	Ea	725.00	53.70	778.70
Add for electric circuit for electric ovens or ranges,					
typical cost	BE@2.00	Ea	102.00	53.70	155.70
Add for electrical circuit for clocks on gas ovens or oven & range combinations,					
typical cost	BE@.500	Ea	22.30	13.40	35.70
Add for gas piping run for gas ovens or ranges,					
typical cost	P1@2.00	Ea	35.00	48.90	83.90

Roof Coatings and Adhesives

	Craft@Hrs	Unit	Material	Labor	Total
Roofing asphalt, 100 lb carton (160 SF per carton)					
145 to 165 degree (low melt)	—	Ea	14.20	—	14.20
185 degree, certified (high melt)	—	Ea	14.20	—	14.20
Asphalt emulsion, Henry #107, (35 SF per gallon)					
1 gallon	—	Ea	5.25	—	5.25
5 gallons	—	Ea	18.90	—	18.90
55 gallons	—	Ea	163.80	—	163.80
Asphalt primer, Henry #105, (150 to 300 SF per gallon)					
1 gallon	—	Ea	8.93	—	8.93
5 gallons	—	Ea	36.80	—	36.80
55 gallons	—	Ea	373.00	—	373.00
Asphalt roof coating, Henry #101, plain, economy, (75 SF per gallon)					
1 gallon	—	Ea	5.46	—	5.46
5 gallons	—	Ea	22.90	—	22.90
55 gallons	—	Ea	205.00	—	205.00
Coal tar wet patch, Western Colloid #101, (12.5 SF per gallon at 1/8" thick)					
3 gallons	—	Ea	23.60	—	23.60
5 gallons	—	Ea	33.60	—	33.60
Cold application cement, Henry #203, (67 SF per gallon)					
5 gallons	—	Ea	29.40	—	29.40
55 gallons	—	Ea	263.00	—	263.00
Elastomeric mastic, Henry #209, (12.5 SF per gallon at 1/8" thick)					
11 oz cartridge for caulking gun	—	Ea	2.94	—	2.94
1 gallon	—	Ea	16.30	—	16.30
3.5 gallons	—	Ea	46.20	—	46.20
5 gallons	—	Ea	73.50	—	73.50
Fiber roof coating, Henry #201, non-asbestos fiber, (50 SF per gallon)					
1 gallon	—	Ea	4.50	—	4.50
5 gallons	—	Ea	22.00	—	22.00
55 gallons	—	Ea	221.00	—	221.00

Roof Coatings and Adhesives

	Craft@Hrs	Unit	Material	Labor	Total
Flashing compound, Western Colloid #106, (12.5 SF per gallon at 1/8" thick)					
3 gallons	—	Ea	16.80	—	16.80
5 gallons	—	Ea	24.90	—	24.90
Lap cement, Henry #108, (280 LF at 2" lap per gallon)					
1 gallon	—	Ea	5.93	—	5.93
5 gallons	—	Ea	25.80	—	25.80
Plastic roof cement, Henry #204, (12-1/2 SF at 1/8" thick per gallon)					
1 gallon	—	Ea	4.73	—	4.73
3.5 gallons	—	Ea	17.90	—	17.90
5 gallons	—	Ea	22.00	—	22.00
55 gallons	—	Ea	205.00	—	205.00
Reflective non-fibered aluminum coating, Henry #120, (200 to 400 SF per gallon)					
1 gallon	—	Ea	15.00	—	15.00
5 gallons	—	Ea	64.10	—	64.10
50 gallons	—	Ea	604.00	—	604.00
Reflective fibered aluminum coating, Henry #220, (60 to 85 SF per gallon)					
1 gallon	—	Ea	13.70	—	13.70
5 gallons	—	Ea	62.90	—	62.90
50 gallons	—	Ea	599.00	—	599.00
Reflective white acrylic roof coating, Henry #280, (65 to 100 SF per gallon)					
1 gallon	—	Ea	20.50	—	20.50
5 gallons	—	Ea	92.80	—	92.80
55 gallons	—	Ea	971.00	—	971.00
Wet patch, Henry #208, (12-1/2 SF at 1/8" thick per gallon)					
11 oz cartridge for caulking gun	—	Ea	1.52	—	1.52
1 gallon	—	Ea	6.30	—	6.30
3.5 gallons	—	Ea	20.00	—	20.00
5 gallons	—	Ea	24.20	—	24.20
Polyester fabric, Henry					
3 oz per SY, 36" x 375'	—	Sq	11.90	—	11.90

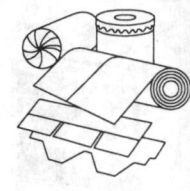

Roofing Rule of Thumb Cost per square to install a roof on a roof 1,500 to 2,000 SF roof not over two stories above ground. Typical subcontract prices. See detailed labor and material costs in the following sections. Many communities restrict use of wood roofing products.

	Craft@Hrs	Unit	Material	Labor	Total
Composition shingles (class A, minimum)	—	Sq	—	—	90.00
Composition shingles (class C, premium)	—	Sq	—	—	130.00
Built-up roof, 3 ply and gravel	—	Sq	—	—	100.00
Cedar shingles, #3, 16"	—	Sq	—	—	275.00
Cedar shingles, #1, 18", fire treated	—	Sq	—	—	375.00
Clay mission tile	—	Sq	—	—	285.00
Concrete tile	—	Sq	—	—	235.00
Metal tile	—	Sq	—	—	300.00
Perlite shakes	—	Sq	—	—	275.00

Obtaining Roof Area from Plan Area

Rise	Factor	Rise	Factor
3"	1.031	8"	1.202
3½"	1.042	8½"	1.225
4"	1.054	9"	1.250
4½"	1.068	9½"	1.275
5"	1.083	10"	1.302
5½"	1.100	10½"	1.329
6"	1.118	11"	1.357
6½"	1.137	11½"	1.385
7"	1.158	12"	1.414
7½"	1.179		

When a roof has to be figured from a plan only, and the roof pitch is known, the roof area may be fairly accurately computed from the table above. The horizontal or plan area (including overhangs) should be multiplied by the factor shown in the table opposite the rise, which is given in inches per horizontal foot. The result will be the roof area.

Roofing, Built-Up Installed over existing suitable substrate.

Type 1: 2 ply felt and 1 ply 90 lb cap sheet, including 3 coats hot mop asphalt

	Craft@Hrs	Unit	Material	Labor	Total
Asphalt felt, 15 lb, 2 plies, roll and mop	R1@.720	Sq	12.00	17.70	29.70
Cap sheet, 90 lb	R1@.720	Sq	11.60	17.70	29.30
Asphalt, 100 lbs per Sq, at $.11 per lb	—	Sq	11.60	—	11.60
Type 1, total per 100 square feet	R1@1.44	Sq	35.30	35.40	70.70

200

	Craft@Hrs	Unit	Material	Labor	Total
Type 2: 3 ply asphalt, 1 ply 30 lb felt, 2 plies 15 lb felt, 3 coats hot mop asphalt and gravel					
Asphalt felt, 30 lb, 1 ply, roll and mop	R1@.495	Sq	8.98	12.20	21.18
Asphalt felt, 15 lb, 2 plies, roll and mop	R1@.720	Sq	11.60	17.70	29.30
Asphalt, 100 lbs per Sq, at $.11 per lb	—	Sq	11.60	—	11.60
Gravel, 3 bags, 60 lbs Ea at $4.50	R1@.125	Sq	14.00	3.07	17.07
Type 2, total per 100 square feet	R1@1.34	Sq	46.50	32.90	79.40
Type 3: 4 ply asphalt, 4 plies 15 lb felt including 4 coats hot mop and gravel					
Asphalt felt, 15 lb, 4 plies, roll and mop	R1@1.44	Sq	23.50	35.40	58.90
Asphalt, 133 lbs per Sq, at $.11 per lb	—	Sq	15.40	—	15.40
Gravel, 3 bags, 60 lbs Ea at $4.50	R1@.125	Sq	14.00	3.07	17.07
Type 3, total per 100 square feet	R1@1.57	Sq	53.10	38.60	91.70

Cold process built-up roofing, subcontract

3 plies 25 lb fiberglass base sheet, 2 coats #203 cold application cement, one coat #107 asphalt emulsion, and #120 aluminum coating

	Craft@Hrs	Unit	Material	Labor	Total
Over suitable existing substrate	—	Sq	—	—	126.00

1 ply 25 lb fiberglass base sheet, 2 plies #184 Rufon polyester fabric embedded and top coated with #106 asphalt emulsion, and Henry #120 aluminum coating

	Craft@Hrs	Unit	Material	Labor	Total
Over suitable existing substrate	—	Sq	—	—	138.00

Roofing, Composition Shingle Material costs include 5% waste. Labor costs include roof loading and typical cutting and fitting for roofs of average complexity.

	Craft@Hrs	Unit	Material	Labor	Total
Asphalt shingles					
Celotex Dimensional 4 (355 lb, 25 year)	R1@2.05	Sq	88.00	50.40	138.40
Certainteed Hallmark (340 lb, 30 year)	R1@2.05	Sq	78.00	50.40	128.40
Fiberglass roofing shingles. Add for hip and ridge below					
Celotex Big D (225 lb, 20 year)	R1@1.83	Sq	25.00	45.00	70.00
Celotex Big D 25 (300 lb, 25 year)	R1@1.83	Sq	44.00	45.00	89.00
Celotex Presidential (40 year)	R1@2.60	Sq	78.00	63.90	141.90
GAF Sentinel (225 lb, 20 year class A)	R1@1.83	Sq	25.00	45.00	70.00
GAF Royal Sovereign (25 year)	R1@1.83	Sq	30.00	45.00	76.00
GAF Woodline (25 year)	R1@1.83	Sq	41.60	45.00	86.60
GAF Timberline (25 year)	R1@1.83	Sq	35.00	45.00	98.00
GAF Timberline (30 year)	R1@1.83	Sq	53.00	45.00	98.00
GS Firescreen (20 year)	R1@1.83	Sq	25.00	45.00	70.00
GS Firescreen Plus (25 year)	R1@1.83	Sq	30.70	45.00	75.70
GS Firehalt (25 year)	R1@1.83	Sq	38.50	45.00	83.50
GS Architect 80 (30 year class A)	R1@1.83	Sq	45.00	45.00	90.00
GS High Sierra (40 year)	R1@1.83	Sq	55.50	45.00	100.50
Elk Prestique I (320 lb, 30 year class A)	R1@1.83	Sq	49.00	45.00	94.00
Elk Prestique II (240 lb, 25 year class A)	R1@1.83	Sq	38.50	45.00	83.50
Elk Prestique Plus (40 year, class A)	R1@1.83	Sq	57.20	45.00	102.20
Masonite Woodruf Shingles					
Traditional	R1@1.83	Sq	85.30	45.00	130.30
Add for hip and ridge units	R1@.020	LF	.58	.49	1.07
Allowance for felt, flashing, fasteners, and vents					
Typical price	—	Sq	—	—	4.00

Roofing Papers

	Craft@Hrs	Unit	Material	Labor	Total

Roofing Papers See also Building Paper. Labor laying paper on 3 in 12 pitch roof.
Asphalt roofing felt (108 SF covers 100 SF), labor to roll and mop

	Craft@Hrs	Unit	Material	Labor	Total
15 lb (432 SF roll at $15.00) 36" x 144'	R1@.360	Sq	3.50	8.85	12.35
30 lb (216 SF roll at $12.00) 36" x 72'	R1@.495	Sq	5.60	12.20	17.80
Roofing asphalt, 100 lbs/Sq	—	Sq	11.00	—	11.00

Mineral surfaced roofing felt(108 SF covers 100 SF), with fasteners (at $12.00)

	Craft@Hrs	Unit	Material	Labor	Total
72 lb (108 SF roll at $13.50) 36" x 36'	R1@1.03	Sq	15.00	25.30	40.30
72 lb (96 SF roll at $12.50) 36" x 32'	R1@1.16	Sq	25.00	28.50	53.50

Smooth roofing felt (roll covers 100 SF), with fasteners (at $12.00)

	Craft@Hrs	Unit	Material	Labor	Total
50 lb medium ($16.00 per roll)	R1@.738	Sq	28.00	18.10	46.10
65 lb extra heavy ($18.00 per roll)	R1@.738	Sq	30.00	18.10	48.10

Roofing Shakes and Shingles Red cedar. No pressure treating. Labor includes flashing.

Shakes
Shake hip and ridge units, 20 per bundle, covers 16.7 LF at 10" and 20 LF at 12" exposure,

	Craft@Hrs	Unit	Material	Labor	Total
Per bundle	R1@1.00	Ea	31.90	24.60	56.50

Sawn shakes, sawn 1 side, class "C" fire retardant

	Craft@Hrs	Unit	Material	Labor	Total
1/2" to 3/4" x 24" (4 bdle/sq at 10" exp.)	R1@3.52	Sq	143.00	86.50	229.50
3/4" to 5/4" x 24" (5 bdle/sq at 10" exp.)	R1@4.16	Sq	198.00	102.00	300.00

Taper split shakes, 1/2" x 24"

	Craft@Hrs	Unit	Material	Labor	Total
(5 bdle/sq at 10" exp.)	R1@3.52	Sq	110.00	86.50	196.50
Shake felt, 30 lb (100 SF roll at $8.25)	R1@.160	Sq	9.90	3.93	13.83
Add for pressure treated shakes	—	%	30.0	—	—

Shingles (red cedar)
Perfects 5X, green, 16", 5" exposure, 5 shingles are 2" thick (5/2)

	Craft@Hrs	Unit	Material	Labor	Total
#1 (for houses)	R1@3.52	Sq	171.00	86.50	257.50
#2, red label (houses or garages)	R1@3.52	Sq	138.00	86.50	224.50
#3 (for garages)	R1@3.52	Sq	99.00	86.50	185.50
#4 (undercourse grade)	R1@1.09	Sq	41.80	26.80	68.60

Perfections, dry, 18", 5-1/2" exposure, 5 shingles are 2-1/4" thick, (5/2-1/4), fire treated

	Craft@Hrs	Unit	Material	Labor	Total
#1 (houses)	R1@3.52	Sq	193.00	86.50	279.50
#2, red label (houses or garages)	R1@3.52	Sq	160.00	86.50	246.50
#3 (garages)	R1@3.52	Sq	138.00	86.50	224.50

Hip and ridge units, 40 pieces per bundle (20 LF at 6" exposure, 16-2/3 LF at 5" exposure and 15 LF at 4-1/2" exposure) per bundle

	Craft@Hrs	Unit	Material	Labor	Total
#1	R1@.500	Ea	34.10	12.30	46.40
#2	R1@.500	Ea	30.80	12.30	43.10
Add for pitch over 6" in 12"	—	%	—	40.0	—

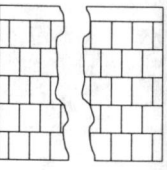

Roofing Sheets See also Fiberglass Panels.

Aluminum roofing and siding, includes 15% waste
4-V corrugated, 2-1/2" corrugations, 5-V crimp, plain or embossed finish
On wood frame, includes 15% loss for coverage

	Craft@Hrs	Unit	Material	Labor	Total
.017" x 26" x 6' to 24'	R1@.022	SF	1.62	.54	2.16
.019" x 26" x 6' to 24'	R1@.022	SF	1.68	.54	2.22

Ridge cap

	Craft@Hrs	Unit	Material	Labor	Total
12" x 10', plain or 10" x 28' formed	R1@.024	LF	1.51	.59	2.10

Flashing, corrugated aluminum, embossed

	Craft@Hrs	Unit	Material	Labor	Total
End wall, 10" x 52"	R1@.047	Ea	2.87	1.15	4.02
Plain side wall, 7-1/2" x 10'	R1@.014	LF	1.26	.34	1.60
Rubber filler strip, 3/4" x 7/8" x 6'	—	LF	.26	—	.26

	Craft@Hrs	Unit	Material	Labor	Total
Galvanized corrugated steel sheet roofing					
27-1/2" wide (includes 20% coverage loss)					
28 gauge, 6' to 12' lengths	R1@.027	SF	.78	.66	1.44
26" wide (include 15% coverage loss)					
26 gauge, 6' to 12' lengths	R1@.027	SF	.82	.66	1.48
Ridge roll, plain					
28 gauge, 10" wide	R1@.030	LF	.83	.74	1.57
Ridge cap, formed, plain					
28 gauge, 2-1/2"	R1@.061	LF	1.33	1.50	2.83
Sidewall flashing, plain					
28 gauge, 3" x 4" x 10'	R1@.035	LF	.71	.86	1.57
Endwall flashing, 2-1/2" corrugated					
28 gauge, 10" x 28"	R1@.035	LF	1.36	.86	2.22
Twin rib, 10" x 52"	R1@.035	LF	1.63	.86	2.49
Wood filler strip, corrugated					
7/8" x 7/8" x 6' (at $2.50 each)	R1@.035	LF	.44	.86	1.30
Galvanized flat sheets					
26 gauge	R1@.027	SF	.99	.66	1.65

Roofing Slate Local delivery included. Costs will be higher where slate is not mined. Add freight cost at 800 to 1,000 pounds per 100 SF. Includes 20" long random width slate, 3/16" thick, 7-1/2" exposure. Meets Standard SS-S-451.

	Craft@Hrs	Unit	Material	Labor	Total
Semi-weathering green and gray	R1@11.3	Sq	315.00	278.00	593.00
Vermont black and gray black	R1@11.3	Sq	373.00	278.00	651.00
China black or gray	R1@11.3	Sq	315.00	278.00	593.00
Unfading and variegated purple	R1@11.3	Sq	425.00	278.00	703.00
Unfading mottled green and purple	R1@11.3	Sq	347.00	278.00	625.00
Unfading green	R1@11.3	Sq	347.00	278.00	625.00
Red slate	R1@13.6	Sq	1,520.00	334.00	1,854.00
Add for other specified widths and lengths	—	%	20.0	—	—

Roofing tile, clay Clay roof tile. Material costs include felts and flashing. No freight or waste included. Costs will be higher where clay tile is not manufactured locally. U.S. Tile Co.

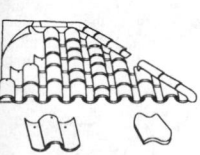

Spanish tile, "S" shaped, 88 pieces per square,
(800 lbs per square at 11" centers and 15" exposure)

	Craft@Hrs	Unit	Material	Labor	Total
Red clay tile ($57 per square)	R1@3.46	Sq	74.00	85.00	159.00
Add for coloring	—	Sq	10.00	—	10.00
Red hip and ridge units	R1@.047	LF	.65	1.15	1.80
Color hip and ridge units	R1@.047	LF	.88	1.15	2.03
Red rake units	R1@.047	LF	1.17	1.15	2.32
Color rake units	R1@.047	LF	1.28	1.15	2.43

Red clay mission tile, 2-piece, 86 pans and 86 tops per square,
7-1/2" x 18" x 8-1/2" tiles at 11" centers and 15" exposure

	Craft@Hrs	Unit	Material	Labor	Total
Red clay tile ($107 per square)	R1@5.84	Sq	138.00	143.00	281.00
Add for coloring (costs vary widely)	—	Sq	36.00	—	36.00
Red hip and ridge units	R1@.047	LF	.65	1.15	1.80
Color hip and ridge units	R1@.047	LF	.88	1.15	2.03
Red rake units	R1@.047	LF	1.17	1.15	2.32
Color rake units	R1@.047	LF	1.28	1.15	2.43

Roofing Tile, Clay

	Craft@Hrs	Unit	Material	Labor	Total
Concrete roof tile, material includes felt, nails, and flashing. Approximately 90 pieces per square. Monier, Inc.					
Homestead, slurry coat (at $48.00/Sq)	R1@3.25	Sq	78.00	79.90	157.90
Homestead, thru color (at $72.00/Sq)	R1@3.25	Sq	102.00	79.90	181.90
Mission "S", slurry coated (at $67.00/Sq)	R1@3.25	Sq	97.00	79.90	176.90
Mission "S", thru color (at $71.00/Sq)	R1@3.25	Sq	101.00	79.90	180.90
Normandie slate (at $80.00/Sq)	R1@3.25	Sq	110.00	79.90	189.90
Shake, thru color (at $70.00/Sq)	R1@3.25	Sq	100.00	79.90	179.90
Slate, slurry coated (at $66.00/Sq)	R1@3.25	Sq	96.00	79.90	175.90
Split shake (at $75.00/Sq)	R1@3.25	Sq	105.00	79.90	184.90
Villa, slurry coat (at $66.00/Sq)	R1@3.25	Sq	96.00	79.90	175.90
Villa, Roma, thru color (at $70.00/Sq)	R1@3.25	Sq	100.00	79.90	179.90
Trim tile					
Mansard "V" ridge or rake, slurry coated	—	Ea	1.20	—	1.20
Mansard, ridge or rake, thru color	—	Ea	1.25	—	1.25
Hipstarters, slurry coated	—	Ea	10.00	—	10.00
Hipstarters, thru color	—	Ea	12.00	—	12.00
Accessories					
Eave closure or birdstop	R1@.030	LF	.90	.74	1.64
Hurricane or wind clips	R1@.030	Ea	.30	.74	1.04
Underlayment or felt	R1@.050	Sq	12.00	1.23	13.23
Metal flashing and nails	R1@.306	Sq	7.00	7.52	14.52
Pre-formed plastic flashings					
Hip (13" long)	R1@.020	Ea	24.00	.49	24.49
Ridge (39" long)	R1@.030	Ea	24.00	.74	24.74
Anti-ponding foam	R1@.030	LF	.20	.74	.94
Batten extenders	R1@.030	Ea	.80	.74	1.54
Roof loading					
Add to load tile and accessories on roof	R1@.822	Sq	—	20.20	20.20

Roofing, Tin Shingles Material costs shown include 5% for waste and laps and 2% for fasteners.

	Craft@Hrs	Unit	Material	Labor	Total
Painted					
5" x 7" (411 per Sq)	R1@3.40	Sq	66.00	83.50	149.50
8" x 12" (150 per Sq)	R1@1.55	Sq	78.00	38.10	116.10
Galvanized					
5" x 7" (411 per Sq)	R1@3.40	Sq	115.00	83.50	198.50
8" x 12" (150 per Sq)	R1@1.55	Sq	135.00	38.10	173.10

Safes, Residential "Safe-N-Sekure" wall storage safe, 14-1/2' wide, 9" high, 4" deep, with heavy gauge steel body, and dead bolt combination lock, self-trimming.

	Craft@Hrs	Unit	Material	Labor	Total
Installed during new construction	BC@.788	Ea	86.00	19.80	105.80

Sandblasting and Waterblasting, Subcontract The minimum charge for on-site (truck-mounted) sandblasting and waterblasting will usually be about $500.00. Add the cost of extra liability insurance and physical barriers to protect adjacent swimming pools, residences, process plants, etc. No scaffolding included. Based on sandblasting with 450 CFM, 125 PSI compressor. Portable sandblasting equipment will increase costs about 35%.

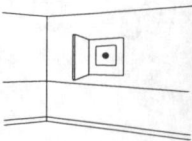

	Craft@Hrs	Unit	Material	Labor	Total
Sandblasting most surfaces					
Water soluble paints	—	SF	—	—	.89
Oil paints	—	SF	—	—	1.00
Heavy mastic	—	SF	—	—	1.22

	Craft@Hrs	Unit	Material	Labor	Total
Sandblasting masonry					
Brick	—	SF	—	—	.70
Block walls, most work	—	SF	—	—	.70
Remove heavy surface grime	—	SF	—	—	.95
Heavy, exposing aggregate	—	SF	—	—	1.05
Sandblasting concrete tilt-up panels					
Light blast	—	SF	—	—	.66
Medium, exposing aggregate	—	SF	—	—	1.05
Heavy, exposing aggregate	—	SF	—	—	1.25
Sandblasting steel					
New, uncoated (commercial grade)	—	SF	—	—	1.00
New, uncoated (near white grade)	—	SF	—	—	1.20
Epoxy coated (near white grade)	—	SF	—	—	1.70
Sandblasting wood					
Medium blast, clean and texture	—	SF	—	—	.90
Waterblast (hydroblast) with mild detergent					
To 5,000 PSI blast (4 hour minimum)	—	Hr	—	—	168.00
5,000 to 10,000 PSI blast (8 hour min.)	—	Hr	—	—	145.00
Over 10,000 PSI blast (8 hour minimum)	—	Hr	—	—	180.00
Wet sandblasting					
(4 hour minimum)	—	Hr	—	—	180.00

Sauna Rooms Costs listed are for labor and materials to install sauna rooms as described. Units are shipped as a disassembled kit complete with kiln dried softwood paneling boards, sauna heater, temperature and humidity controls and monitors, pre-assembled benches, prehung door, water bucket and dipper, and interior light. No electrical work included.

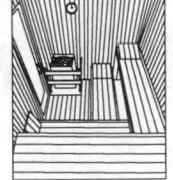

Pre-cut western red cedar sauna room package with heater, controls, duckboard floor and accessories, ready for nailing to existing framed and insulated walls with exterior finish. Amerec.

	Craft@Hrs	Unit	Material	Labor	Total
4' x 4' room	B1@8.01	Ea	1,830.00	183.00	2,013.00
4' x 6' room	B1@8.01	Ea	2,205.00	183.00	2,388.00
6' x 6' room	B1@10.0	Ea	2,550.00	229.00	2,779.00
6' x 8' room	B1@12.0	Ea	3,112.00	274.00	3,386.00
8' x 8' room	B1@14.0	Ea	3,575.00	320.00	3,895.00
8' x 10' room	B1@15.0	Ea	4,103.00	343.00	4,446.00
8' x 12' room	B1@16.0	Ea	5,082.00	366.00	5,448.00
Add for electrical connection wiring	—	LS	—	—	293.00

Modular sauna room assembly. Includes the wired and insulated panels to construct a free-standing sauna on a waterproof floor. With heater, rocks, door, benches and accessories. Amerec.

	Craft@Hrs	Unit	Material	Labor	Total
4' x 4' room	B1@4.00	Ea	2,501.00	91.40	2,592.40
4' x 6' room	B1@4.00	Ea	3,079.00	91.40	3,170.40
6' x 6' room	B1@6.00	Ea	3,586.00	137.00	3,723.00
6' x 8' room	B1@8.01	Ea	4,297.00	183.00	4,480.00
6' x 10' room	B1@8.01	Ea	5,026.00	183.00	5,209.00
8' x 8' room	B1@10.0	Ea	5,077.00	229.00	5,306.00
8' x 10' room	B1@11.0	Ea	5,797.00	251.00	6,048.00
8' x 12' room	B1@12.0	Ea	6,751.00	274.00	7,025.00
Add for glass panel, 22-1/2" or 24"	—	Ea	70.00	—	70.00
Add for electrical connection wiring	—	LS	—	—	225.00

Sauna Rooms

	Craft@Hrs	Unit	Material	Labor	Total

Pre-cut 100% vertical grain redwood or clear western red cedar sauna room package with heater, controls and accessories (including non-skid removable rubberized flooring). Ready for nailing to existing framed and insulated walls with exterior finish. Finlandia.

	Craft@Hrs	Unit	Material	Labor	Total
3' x 5' or 4' x 4' room	B1@6.68	Ea	1,650.00	153.00	1,803.00
4' x 6' room	B1@6.68	Ea	1,900.00	153.00	2,053.00
5' x 6' room	B1@6.68	Ea	2,100.00	153.00	2,253.00
6' x 6' room	B1@8.01	Ea	2,250.00	183.00	2,433.00
6' x 8' room	B1@10.7	Ea	2,750.00	245.00	2,995.00
8' x 8' room	B1@10.7	Ea	3,200.00	245.00	3,445.00
8' x 10' room	B1@10.7	Ea	3,700.00	245.00	3,945.00
8' x 12' room	B1@10.7	Ea	4,400.00	245.00	4,645.00
10' x 10' room	B1@10.7	Ea	4,300.00	245.00	4,545.00
10' x 12' room	B1@10.7	Ea	4,750.00	245.00	4,995.00
Add for electrical connection wiring	—	LS	—	—	293.00

Modular sauna room assembly. Completely pre-wired pre-insulated panels to construct a free-standing sauna on waterproof floor. Includes heater, rocks, door, benches and accessories. Finlandia.

	Craft@Hrs	Unit	Material	Labor	Total
4' x 4' room	B1@1.34	Ea	2,150.00	30.60	2,180.60
4' x 6' room	B1@1.34	Ea	2,700.00	30.60	2,730.60
5' x 6' room	B1@2.68	Ea	2,850.00	61.20	2,911.20
6' x 6' room	B1@2.68	Ea	3,300.00	61.20	3,361.20
6' x 8' room	B1@2.68	Ea	3,700.00	61.20	3,761.20
8' x 8' room	B1@4.02	Ea	4,500.00	91.90	4,591.90
8' x 10' room	B1@4.02	Ea	5,200.00	91.90	5,291.90
8' x 12' room	B1@4.02	Ea	5,800.00	91.90	5,891.90
10' x 10' room	B1@4.02	Ea	5,700.00	91.90	5,791.90
10' x 12' room	B1@4.02	Ea	6,300.00	91.90	6,391.90
Add for electrical connection wiring	—	LS	—	—	230.00
Add for thermometer	—	Ea	18.00	—	18.00
Add for sand timer	—	Ea	26.00	—	26.00
Add for delivery (typical)	—	%	—	—	5.0

	Alum mesh	Galv mesh	Bronze mesh	Fiber mesh
Screen Wire				
Cost per 100 LF roll				
18" wide roll	39.00	42.50	—	—
24" wide roll	48.00	58.90	170.00	38.00
30" wide roll	61.00	73.00	215.00	41.00
36" wide roll	79.00	82.00	255.00	51.50
42" wide roll	89.00	103.00	298.00	57.00
48" wide roll	105.00	117.00	340.00	62.00
54" wide roll	119.00	—	—	70.00
60" wide roll	134.00	—	—	78.00
72" wide roll	159.00	—	—	95.00
84" wide roll	—	—	—	109.00

	Craft@Hrs	Unit	Material	Labor	Total
Labor to install screen wire					
Includes measure, layout, cut and attach screenwire to an existing metal or wood frame					
Per linear foot of perimeter of frame	BC@.017	LF	—	.43	.43

	Craft@Hrs	Unit	Material	Labor	Total

Security Alarms, Subcontract Costs are for securing building perimeter and all accessible openings with electronic alarm system, including all labor and materials. Costs assume that no interior wall, floor, or ceiling finishes are in place. All equipment is good to better quality.

	Craft@Hrs	Unit	Material	Labor	Total
Alarm control panel	—	Ea	—	—	405.00
Wiring, detectors per opening, with switch	—	Ea	—	—	70.00
Monthly monitoring charge	—	Mo	—	—	28.80
Alarm options (installed)					
Audio detectors	—	Ea	—	—	100.00
Communicator (central station service)	—	Ea	—	—	225.00
Digital touch pad control	—	Ea	—	—	123.00
Entry/exit delay	—	Ea	—	—	87.00
Fire/smoke detectors	—	Ea	—	—	150.00
Interior/exterior sirens	—	Ea	—	—	67.00
Motion detector	—	Ea	—	—	205.00
Panic button	—	Ea	—	—	50.00
Passive infrared detector	—	Ea	—	—	280.00
Pressure mat	—	LF	—	—	80.00
Security light control	—	Ea	—	—	492.00
Add for wiring detectors and controls in existing buildings					
With wall and ceiling finishes already in place	—	%	—	—	25.0

Security Guards, Subcontract Construction site security guards, unarmed, in uniform with two-way radio, backup patrol car, bond, and liability insurance. Per manhour.

	Craft@Hrs	Unit	Material	Labor	Total
Short term (1 night to 1 week)	—	Hr	—	—	21.50
Medium duration (1 week to 1 month)	—	Hr	—	—	17.30
Long term (1 to 6 months)	—	Hr	—	—	15.30
Add for licensed armed guard	—	Hr	—	—	2.50
Add for holidays	—	%	—	—	50.0

Construction site guard dog service for sites with totally enclosed perimeter, 24 hour per day, 7 day per week service. Local business codes may restrict or prohibit unattended guard dogs. Includes dog handling training session, liability insurance, on-site kennel. Cost per month.

	Craft@Hrs	Unit	Material	Labor	Total
Contractor feeding and tending dog	—	Mo	—	—	525.00

Daily delivery of guard dog at quitting time and pickup at starting time by guard dog service (special hours require advance notice)

	Craft@Hrs	Unit	Material	Labor	Total
Service feeding and tending dog	—	Mo	—	—	550.00

Construction site man and dog guard team, includes two-way radio communication with guard service office, backup patrol car, and liability insurance. Guard service may require that a telephone and guard shack be provided by the contractor. Cost for man and dog per hour.

	Craft@Hrs	Unit	Material	Labor	Total
Short term (1 week to 1 month)	—	Hr	—	—	18.50
Long term (1 to 6 months)	—	Hr	—	—	16.50

Septic Sewer Systems, Subcontract Soils testing (percolation test) by qualified engineer (required by some communities when applying for septic system permit). Does not include application fee.

	Craft@Hrs	Unit	Material	Labor	Total
Minimum cost	—	Ea	—	—	700.00

Residential septic sewer tanks (costs include excavation for tank with good site conditions, placing of tank, inlet and outlet fittings, and backfill after hookup).

	Craft@Hrs	Unit	Material	Labor	Total
Steel reinforced concrete tanks					
1,250 gallons (3 or 4 bedroom house)	—	Ea	—	—	1,680.00
1,500 gallons (5 or 6 bedroom house)	—	Ea	—	—	1,780.00

Septic Sewer Systems, Subcontract

	Craft@Hrs	Unit	Material	Labor	Total
Fiberglass tanks					
1,000 gallons (3 bedroom house)	—	Ea	—	—	1,680.00
1,250 gallons (4 bedroom house)	—	Ea	—	—	1,730.00
1,500 gallons (5 or 6 bedroom house)	—	Ea	—	—	1,840.00
Polyethylene tanks, 4" lines, 20" manhole diameter					
500 gallons, 42" H x 51" W x 100" L	—	Ea	—	—	790.00
750 gallons, 57" H x 52" W x 95" L	—	Ea	—	—	880.00
1,000 gallons, 66" H x 52" W x 100" L	—	Ea	—	—	1,075.00
1,250 gallons, 68" H x 56" W x 112" L	—	Ea	—	—	1,170.00
1,500 gallons, 65" H x 55" W x 143" L	—	Ea	—	—	1,450.00

Residential septic sewer drain fields (leach lines). Costs include labor, unsaturated paper, piping, gravel, excavation with good site conditions, backfill, and disposal of excess soil. 4" PVC pipe (ASTM D2729) laid in 3' deep by 1' wide trench.

	Craft@Hrs	Unit	Material	Labor	Total
With 12" gravel base	—	LF	—	—	6.30
With 24" gravel base	—	LF	—	—	7.40
With 36" gravel base	—	LF	—	—	8.40
Add for pipe laid 6' deep	—	LF	—	—	2.10

Add for piping from house to septic tank, and from septic tank to remote drain fields,

	Craft@Hrs	Unit	Material	Labor	Total
4" PVC Schedule 40	—	LF	—	—	4.50

Low cost pumping systems for residential wastes, including fiberglass basin, pump, installation from septic tank or sewer line, 40' pipe run, and automatic float switch (no electric work or pipe included).

	Craft@Hrs	Unit	Material	Labor	Total
To 15' head	—	LS	—	—	2,090.00
To 25' head	—	LS	—	—	2,460.00
To 30' head	—	LS	—	—	3,210.00
Add for high water/pump failure alarm	—	LS	—	—	454.00

Better quality pump system with two alternating pumps, 700 to 800 gallon concrete or fiberglass basin, automatic float switch, indoor control panel, high water/pump failure alarm, explosion proof electrical system but no electric work or pipe .

	Craft@Hrs	Unit	Material	Labor	Total
Per pump system	—	LS	—	—	6,490.00

Pipe locating service to detect sewage and water leaks, breaks, and stoppages in pipes with diameter of 2" or more. Equipment can't be used where pipe has 90 degree turns. Maximum depth is 18' for cast iron and 30' for vitreous clay.

	Craft@Hrs	Unit	Material	Labor	Total
Residential 1/2 day rate	—	LS	—	—	310.00
Commercial 1/2 day rate	—	LS	—	—	410.00
Add for each additional pipe entered	—	LS	—	—	40.00
Plus per LF of pipe length	—	LF	—	—	1.50
Add for travel over 15 miles, per mile	—	Ea	—	—	.30

Sewer Connections, Subcontract Including typical excavation and backfill.

	Craft@Hrs	Unit	Material	Labor	Total
4" vitrified clay pipeline, house to property line					
Long runs	—	LF	—	—	16.50
Short runs	—	LF	—	—	17.50
6" vitrified clay pipeline					
Long runs	—	LF	—	—	18.50
Short runs	—	LF	—	—	22.00
Street work	—	LF	—	—	46.00
4" PVC sewer pipe, city installed, property line to main line.					
Up to 40 feet (connect in street).	—	LS	—	—	1,670.00
Up to 40 feet (connect in alley).	—	LS	—	—	1,210.00
Each foot over 40 feet	—	LF	—	—	36.00

	Craft@Hrs	Unit	Material	Labor	Total
6" PVC sewer pipe, city installed, property line to main line					
Up to 40 feet (connect in street)	—	LS	—	—	2,735.00
Up to 40 feet (connect in alley)	—	LS	—	—	1,960.00
Each foot over 40 feet	—	LF	—	—	60.00

Sheet Metal Access Doors

	Craft@Hrs	Unit	Material	Labor	Total
Attic access doors					
22" x 22"	SW@.363	Ea	42.20	10.10	52.30
22" x 30"	SW@.363	Ea	45.40	10.10	55.50
24" x 36"	SW@.363	Ea	45.40	10.10	55.50
30" x 30"	SW@.363	Ea	68.60	10.10	78.70
Wall access doors, nail-on type, galvanized or painted					
24" x 18"	SW@.363	Ea	41.20	10.10	51.30
24" x 24"	SW@.363	Ea	48.60	10.10	58.70
24" x 18", screen only	—	Ea	12.10	—	12.10
24" x 24", screen only	—	Ea	13.80	—	13.80
Tub access doors					
14" x 14", plaster type	SW@.363	Ea	21.10	10.10	31.20
14" x 14", drywall type	SW@.363	Ea	26.40	10.10	36.50

Sheet Metal Area Walls Provides light well for basement windows, galvanized, corrugated.

	Craft@Hrs	Unit	Material	Labor	Total
12" deep, 6" projection, 37" wide	SW@.410	Ea	15.30	11.40	26.70
18" deep, 6" projection, 37" wide	SW@.410	Ea	20.40	11.40	31.80
24" deep, 6" projection, 37" wide	SW@.410	Ea	26.50	11.40	37.90
30" deep, 6" projection, 37" wide	SW@.410	Ea	31.60	11.40	43.00
Add for steel grille cover	—	Ea	35.70	—	35.70

Sheet Metal Flashing

	Craft@Hrs	Unit	Material	Labor	Total
Counter flashing, in 200' pack					
1/2" x 2"	SW@.020	LF	.37	.56	.93
1/2" x 3"	SW@.020	LF	.43	.56	.99
1/2" x 4"	SW@.020	LF	.55	.56	1.11
1/2" x 5"	SW@.020	LF	.62	.56	1.18
1/2" x 6"	SW@.020	LF	.72	.56	1.28

	Galvanized	FHA painted
Flashing, all pitch, for plumbing vent pipes, see labor cost below.		
1/2", 3/4", 1	4.90	6.83
1-1/4", 1-1/2", 2"	3.88	5.92
3" conductor	6.12	9.03
3" amerivent, transite, soil	7.24	9.13
4" conductor	7.14	10.20
4" transite	9.59	11.70
4" amerivent, soil	7.96	9.49
5" conductor	9.89	14.50
5" transite	15.50	15.20
5" amerivent	12.00	14.80
6" conductor	14.20	18.00
6" transite	18.20	19.90
6" amerivent	17.30	19.90
7" conductor	18.00	19.90
7" transite	22.30	23.30

Sheet Metal Flashing

	Galvan-ized	FHA painted
8" conductor	24.40	25.00
8" transite	28.40	25.00
9" conductor	26.00	25.00
10" conductor or transite	34.20	34.20
12" conductor or transite	38.30	38.40
14" conductor or transite	69.40	67.80
Flashing, all pitch, oval, galvanized		
4" to 6" opening	14.30	14.30
Flashing, all pitch, tee top		
3" opening	10.10	10.10
4" opening	10.10	10.10
5" opening	18.50	18.50
6" opening	21.90	21.90

	Craft@Hrs	Unit	Material	Labor	Total
Vent pipe flashing labor					
To 3" diameter	SW@.136	Ea	—	3.78	3.78
4" to 8" diameter	SW@.270	Ea	—	7.50	7.50
Gravel guard					
4-1/2" gravel guard	SW@.030	LF	.38	.83	1.21
6" gravel guard	SW@.030	LF	.56	.83	1.39
7-1/4" gravel guard	SW@.030	LF	.62	.83	1.45
Nosing, roof edging, at 90 or 105 degree angle, galvanized, 10' lengths					
3/4" x 3/4" or 1" x 1"	SW@.040	LF	.22	1.11	1.33
1" x 2" or 1-1/2" x 1-1/2"	SW@.040	LF	.26	1.11	1.37
2" x 2"	SW@.040	LF	.33	1.11	1.44
2" x 3"	SW@.040	LF	.43	1.11	1.54
2" x 4"	SW@.040	LF	.56	1.11	1.67
3" x 3"	SW@.040	LF	.56	1.11	1.67
3" x 4"	SW@.040	LF	.72	1.11	1.83
3" x 5"	SW@.040	LF	.85	1.11	1.96
4" x 4"	SW@.043	LF	.85	1.19	2.04
4" x 5"	SW@.043	LF	1.02	1.19	2.21
4" x 6"	SW@.043	LF	1.12	1.19	2.31
5" x 5"	SW@.043	LF	1.12	1.19	2.31
6" x 6"	SW@.043	LF	1.33	1.19	2.52
Add for 90 degree kickout nosing	—	%	5.0	—	—
Rain diverter, 1" x 3", galvanized, unpainted					
4' length	SW@.136	Ea	3.32	3.78	7.10
5' length	SW@.136	Ea	3.67	3.78	7.45
10' length	SW@.265	Ea	6.12	7.36	13.48
Roof edging (eave drip, with 1/4" kickout, galvanized)					
1" x 2"	SW@.040	LF	.37	1.11	1.48
1-1/2" x 1-1/2"	SW@.040	LF	.39	1.11	1.50
2" x 2"	SW@.040	LF	.43	1.11	1.54
Sill pans, galvanized, 5", 5-1/2" or 6",					
per pair	SW@.136	Pr	.77	3.78	4.55
Valley flashing, tin roll type, 50' lengths, 28 gauge, galvanized					
8" wide	SW@.020	LF	.75	.56	1.31
14" wide	SW@.020	LF	1.28	.56	1.84
20" wide	SW@.020	LF	1.79	.56	2.35

	Craft@Hrs	Unit	Material	Labor	Total
Tin seamless flashing, painted					
8" wide	SW@.020	LF	.63	.56	1.19
14" wide	SW@.020	LF	.97	.56	1.53
20" wide	SW@.020	LF	1.38	.56	1.94
Aluminum seamless flashing, .016 gauge					
4" wide	SW@.020	LF	.46	.56	1.02
6" or 8" wide	SW@.020	LF	.74	.56	1.30
10" wide	SW@.020	LF	.86	.56	1.42
12" wide	SW@.020	LF	1.07	.56	1.63
14" wide	SW@.020	LF	1.28	.56	1.84
20" wide	SW@.020	LF	1.79	.56	2.35
Valley flashing, "W" type					
18", 26 gauge	SW@.030	LF	1.60	.83	2.43
24", 26 gauge	SW@.030	LF	2.09	.83	2.92
Vertical chimney flashing					
2" x 3", galvanized	SW@.050	LF	.58	1.39	1.97
Window flashing, painted tin					
3/4" x 1" x 28" long	SW@.020	LF	.23	.56	.79
1" x 1" x 28" long	SW@.020	LF	.26	.56	.82
"Z" bar flashing, galvanized					
Standard	SW@.015	LF	.52	.42	.94
Old style	SW@.015	LF	.52	.42	.94
For plywood siding	SW@.015	LF	.37	.42	.79

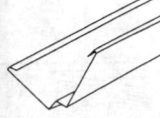

Sheet Metal Vents

	Craft@Hrs	Unit	Material	Labor	Total
Attic and gable vents, opening sizes, louvers					
8" x 12"	SW@.448	Ea	10.70	12.40	23.10
14" x 12"	SW@.448	Ea	11.20	12.40	23.60
14" x 18"	SW@.448	Ea	16.80	12.40	29.20
14" x 24"	SW@.448	Ea	21.40	12.40	33.80
Chimney caps					
3" diameter	SW@.324	Ea	5.87	8.99	14.86
4" diameter	SW@.324	Ea	6.07	8.99	15.06
5" diameter	SW@.324	Ea	7.70	8.99	16.69
6" diameter	SW@.324	Ea	9.33	8.99	18.32
7" diameter	SW@.324	Ea	12.10	8.99	21.09
8" diameter	SW@.324	Ea	19.80	8.99	28.79
Clothes dryer vent set, aluminum, with hood and inside plate					
3" or 4" diameter	SW@.440	Ea	6.12	12.20	18.32
Flex hose, 4" diameter	—	LF	1.02	—	1.02
Flex hose clamps, 4" diameter	—	Ea	1.43	—	1.43
Crawl hole vents, screen, painted, 24" x 24"	SW@.448	Ea	24.50	12.40	36.90

	Craft@Hrs	Unit	Material	Labor	Total
Cowl caps					
3" diameter	SW@.324	Ea	10.20	8.99	19.19
4" diameter	SW@.324	Ea	10.20	8.99	19.19
5" diameter	SW@.324	Ea	12.30	8.99	21.29
6" diameter	SW@.324	Ea	13.80	8.99	22.79
7" diameter	SW@.324	Ea	17.00	8.99	25.99
8" diameter	SW@.493	Ea	24.90	13.70	38.60
9" diameter	SW@.493	Ea	40.20	13.70	53.90
10" diameter	SW@.493	Ea	44.60	13.70	58.30
12" diameter	SW@.493	Ea	51.00	13.70	64.70

Sheet Metal Vents

	Craft@Hrs	Unit	Material	Labor	Total
Dormer louvers, half round, galvanized, 1/4" mesh screen					
18" x 9", 3/12 pitch	SW@.871	Ea	43.70	24.20	67.90
24" x 12", 3/12 pitch	SW@.871	Ea	52.80	24.20	77.00
18" x 9", 5/12 pitch	SW@.871	Ea	34.70	24.20	58.90
24" x 12", 5/12 pitch	SW@.871	Ea	42.80	24.20	67.00
Add 1/8" mesh	—	%	3.0	—	—
Foundation vents, galvanized, with screen, no louvers					
6" x 14" stucco	SW@.255	Ea	2.55	7.08	9.63
8" x 14" stucco	SW@.255	Ea	3.11	7.08	10.19
6" x 14" 2-way	SW@.255	Ea	3.01	7.08	10.09
6" x 14" for siding (flat type)	SW@.255	Ea	2.55	7.08	9.63
6" x 14" louver type	SW@.255	Ea	3.57	7.08	10.65
6" x 14" louver type with mesh	SW@.255	Ea	3.77	7.08	10.85
6" x 14" foundation insert	SW@.255	Ea	2.55	7.08	9.63
Ornamental grill & louvers, 6" x 14"	SW@.255	Ea	6.20	7.08	13.28
Heater closet door vents, louver					
70 square inches, louvers only	SW@.220	Ea	3.72	6.11	9.83
70 square inches, louvers and screen	SW@.255	Ea	6.89	7.08	13.97
38 square inches, louvers only	SW@.190	Ea	3.62	5.27	8.89
38 square inches, louvers and screen	SW@.210	Ea	5.46	5.83	11.29
Midget louvers, aluminum, pack of 50					
1" louvers	SW@.255	Ea	1.73	7.08	8.81
2" louvers	SW@.255	Ea	2.55	7.08	9.63
3" louvers	SW@.255	Ea	3.37	7.08	10.45
Rafter vents, 1/8" mesh, galvanized, louver					
3" x 14"	SW@.220	Ea	1.94	6.11	8.05
3" x 22"	SW@.220	Ea	3.37	6.11	9.48
Plaster ground rafter vent four sides					
3" x 14"	SW@.220	Ea	1.94	6.11	8.05
3" x 22"	SW@.220	Ea	3.37	6.11	9.48
Flat type for siding					
3" x 14"	SW@.220	Ea	2.24	6.11	8.35
3" x 22"	SW@.220	Ea	3.37	6.11	9.48
Rafter inserts					
3" x 14"	SW@.220	Ea	1.94	6.11	8.05
3" x 22"	SW@.220	Ea	3.37	6.11	9.48
Rafter vent with ears					
3" x 14"	SW@.220	Ea	2.70	6.11	8.81
3" x 22"	SW@.220	Ea	3.21	6.11	9.32
6" x 22"	SW@.220	Ea	3.88	6.11	9.99
For FHA painted, add	—	%	5.0	—	—
For galvanized 1/4" mesh, deduct	—	%	-5.0	—	—
Ridge ventilators 1/8" louver opening, aluminum					
Mill finish	SW@.086	LF	3.21	2.39	5.60
Black or brown	SW@.086	LF	3.77	2.39	6.16
Add for end fitting	—	Ea	3.47	—	3.47
Roof jacks, round					
7" diameter	SW@.186	Ea	10.20	5.16	15.36
8" diameter	SW@.186	Ea	18.40	5.16	23.56
9" diameter	SW@.186	Ea	24.50	5.16	29.66

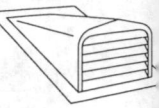

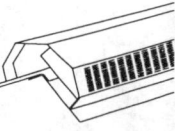

	Craft@Hrs	Unit	Material	Labor	Total
Roof vents, with mesh screen					
Square cap type, 26" L x 23" D x 5" H	SW@.448	Ea	41.80	12.40	54.20
Dormer type, 20" L x 10" D x 6" H	SW@.448	Ea	27.50	12.40	39.90
Dormer type, louvers, 19" L x 27" D x 18" H	SW@.448	Ea	64.30	12.40	76.70
Rotary roof ventilators, with base, by turbine diameter					
6" diameter	SW@.694	Ea	25.00	19.30	44.30
8" diameter	SW@.694	Ea	27.10	19.30	46.40
10" diameter	SW@.694	Ea	32.50	19.30	51.80
12" diameter	SW@.694	Ea	33.50	19.30	52.80
14" diameter	SW@.694	Ea	56.40	19.30	75.70
16" diameter	SW@.930	Ea	82.50	25.80	108.30
18" diameter	SW@1.21	Ea	100.00	33.60	133.60
20" diameter	SW@1.21	Ea	128.00	33.60	161.60
Round louver vents, 1/8" mesh, galvanized					
12" or 14" diameter	SW@.440	Ea	33.70	12.20	45.90
16" diameter	SW@.440	Ea	38.80	12.20	51.00
18" diameter	SW@.440	Ea	42.80	12.20	55.00
24" diameter	SW@.661	Ea	77.50	18.40	95.90

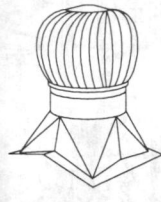

Sheet Metal Hoods

	Craft@Hrs	Unit	Material	Labor	Total
Entrance hoods, galvanized					
48" x 22" x 12", double mold	SW@1.00	Ea	83.60	27.80	111.40
60" x 24" x 12", double mold	SW@1.00	Ea	79.60	27.80	107.40
48" x 22" x 12", scalloped edge	SW@1.00	Ea	102.00	27.80	129.80
60" x 24" x 12", scalloped edge	SW@1.00	Ea	108.00	27.80	135.80
Side hoods, galvanized, square					
4" x 4"	SW@.190	Ea	7.91	5.27	13.18
6" x 6"	SW@.216	Ea	13.30	6.00	19.30
7" x 7"	SW@.272	Ea	15.00	7.55	22.55

Shelving Wood shelving, based on using 1" thick #3 & Btr paint grade board lumber. Prices shown in parentheses are before allowance for accessories and waste. Costs shown include an allowance for ledger boards, nails and normal waste. Painting not included.

	Craft@Hrs	Unit	Material	Labor	Total
Closet shelves, with 1" diameter clothes pole with brackets 3' OC					
10" wide shelf ($550 per MBF = $.46 per LF)	BC@.080	LF	.86	2.01	2.87
12" wide shelf ($680 per MBF = $.68 per LF)	BC@.080	LF	1.10	2.01	3.11
18" wide shelf ($540 per MBF = $.81 per LF)	BC@.100	LF	1.24	2.52	3.76
Linen cabinet shelves, cost of linen closet not included					
18" wide shelf ($540 per MBF = $.81 per LF)	BC@.125	LF	.89	3.15	4.04
24" wide shelf ($680 per MBF = $1.36 per LF)	BC@.125	LF	1.50	3.15	4.65
Utility shelves, laundry room walls, garage walls, etc.					
18" wide shelf ($473 per MBF = $.71 per LF)	BC@.100	LF	.78	2.52	3.30
24" wide shelf ($480 per MBF = $.96 per LF)	BC@.112	LF	1.06	2.82	3.88

Shower and Tub Doors.

	Craft@Hrs	Unit	Material	Labor	Total
Swinging shower doors, with hardware, anodized aluminum frame, tempered safety glass					
Door 64" high, silver or gold tone					
24" to 26" wide	BG@1.57	Ea	124.00	36.20	160.20
26" to 28" wide	BG@1.57	Ea	128.00	36.20	164.20
Door 66" high, silver or brass frame					
27" to 31" wide	BG@1.57	Ea	151.00	36.20	187.20
31" to 36" wide	BG@1.57	Ea	165.00	36.20	201.20

Shower and Tub Doors

	Craft@Hrs	Unit	Material	Labor	Total
Door 73" high, silver or gold frame					
32" to 36" wide	BG@1.57	Ea	200.00	36.20	236.20
42" to 48" wide	BG@1.57	Ea	228.00	36.20	264.20
Bi-folding doors, 70" high,					
46" to 48" total width	BG@2.41	Ea	171.00	55.60	226.60

Sliding shower doors, 2 panels, 5/32" tempered glass, anodized aluminum frame, 70" high, outside towel bar

	Craft@Hrs	Unit	Material	Labor	Total
Non-textured glass, 48" wide	BG@1.57	Ea	202.00	36.20	238.20
Non-textured glass, 56" wide	BG@1.57	Ea	210.00	36.20	246.20
Non-textured glass, 64" wide	BG@1.57	Ea	214.00	36.20	250.20
Texture, opaque, 48" wide	BG@1.57	Ea	231.00	36.20	268.20
Textured, opaque, 56" wide	BG@1.57	Ea	244.00	36.20	280.20
Texture, opaque, 64" wide	BG@1.57	Ea	252.00	36.20	288.20
1 panel double mirrored, 1 bronze tinted	BG@1.57	Ea	261.00	36.20	297.20
Snap-on plastic trim kit, per set of doors	BG@.151	Ea	23.80	3.48	27.28

Sliding shower doors, 2 panels, 1/4" tempered glass, anodized aluminum frame, 70" high, outside towel bar

	Craft@Hrs	Unit	Material	Labor	Total
Non-textured glass, 48" wide	BG@1.57	Ea	231.00	36.20	238.20
Non-textured glass, 56" wide	BG@1.57	Ea	244.00	36.20	246.20
Non-textured glass, 64" wide	BG@1.57	Ea	252.00	36.20	250.20
Frosted opaque design, 48" wide	BG@1.57	Ea	302.00	36.20	268.20
Frosted opaque design, 56" wide	BG@1.57	Ea	320.00	36.20	280.20
Frosted, opaque design, 64" wide	BG@1.57	Ea	337.00	36.20	288.20

Tub doors, sliding, tempered safety glass, anodized aluminum frame

	Craft@Hrs	Unit	Material	Labor	Total
Three panels, 57" H x 60" W opening					
Mirrored center panel	BG@2.02	Ea	257.00	46.60	303.60
Reflective glass panels	BG@2.02	Ea	323.00	46.60	369.60
Two panels, 56" H x 60" W opening					
Shatter-resistant plastic panels	BG@1.77	Ea	71.30	40.80	112.10
Semi-clear, tempered pebble pattern	BG@1.77	Ea	105.00	40.80	145.80
Semi-clear, with design	BG@1.77	Ea	176.00	40.80	216.80
One semi-clear, one mirrored	BG@1.77	Ea	143.00	40.80	183.80
Reflective glass panels	BG@1.77	Ea	190.00	40.80	230.80
Accordion folding tub and shower doors, vinyl, chrome frame					
60" x 57" high, tub enclosure	BG@1.57	Ea	71.30	36.20	107.50
30" to 32", 69" high, shower enclosure	BG@1.57	Ea	66.50	36.20	102.70
33" to 36", 69" high, shower enclosure	BG@1.57	Ea	71.30	36.20	107.50
46" to 48", 69" high, shower enclosure	BG@1.57	Ea	76.00	36.20	112.20

Shower and Tub Enclosures, Subcontract No plumbing or plumbing fixtures included.
Shower enclosures, 48" W x 34" D x 66" H, tempered glass, installed, with composition.
Receptor, door and towel bar

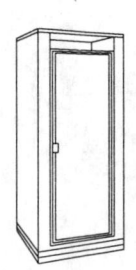

	Unit	Material	Labor	Total
Aluminum frame, opaque glass	Ea	—	—	423.00
Gold frame, clear glass	Ea	—	—	446.00
Bronze frame, bronze tint	Ea	—	—	486.00
Corner shower enclosures, 2-wall, aluminum frame, plastic panels,				
installed, with composition receptor, door and towel bar				
30" to 34"W, 72"H	Ea	—	—	490.00
35" to 45"W, 72"H	Ea	—	—	410.00
45" to 60"W, 72"H	Ea	—	—	446.00
Add for tempered glass panels	Ea	—	—	115.00

	Craft@Hrs	Unit	Material	Labor	Total
"Neo-Angle" corner shower cabinets, tempered glass					
3'W x 3'D x 72"H, hammered glass	—	Ea	—	—	464.00
3'4"W x 3'4"D x 72"H, bronze frame	—	Ea	—	—	612.00
Shower cabinet door, tempered opaque glass, installed					
Aluminum frame, 27"W x 66"H	—	Ea	—	—	212.00
Add for extra towel bars, installed	—	Ea	—	—	16.00
Add for clear or tinted glass	—	Ea	—	—	40.00
Tub enclosures, cultured marble					
60" high, 2 panels 30" wide and 1 panel 60" wide	—	Ea	—	—	869.00
Tub enclosures, fiberglass, installed, 70"H, 60"W Plexiglass sliding door					
Semi-clear	—	Ea	—	—	212.00
Colors, tints, designs	—	Ea	—	—	243.00
Tub enclosures, tempered reflective glass, installed, aluminum frame					
70" high, two panel frame	—	Ea	—	—	347.00

Shower Receptors

Molded fiberglass, corner style, with drain assembly.

	Craft@Hrs	Unit	Material	Labor	Total
34" x 34"	PM@.480	Ea	97.70	13.60	111.30
36" x 34"	PM@.480	Ea	100.00	13.60	113.60
42" x 34"	PM@.480	Ea	118.00	13.60	131.60
48" x 34"	PM@.480	Ea	122.00	13.60	135.60
54" x 34"	PM@.480	Ea	147.00	13.60	160.60
60" x 34"	PM@.480	Ea	156.00	13.60	169.60
36" x 36" neo corner	PM@.480	Ea	235.00	13.60	248.60
Cultured marble					
36" x 48"	PM@2.00	Ea	367.00	56.70	423.70

Shutters, Exterior Louvered, price per pair. Add hinges and anchors from the installation section below.

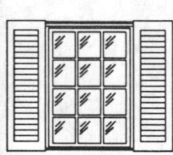

	Polypropylene		
	Unfinished 7/8"	Painted 1-1/8"	Painted 1-1/8"
Height	12" wide	16" wide	15" wide
25"	18.10	33.40	51.60
35" or 39"	25.70	44.20	69.80
43" or 47"	33.40	53.80	87.90
51"	34.60	56.10	91.80
55"	36.80	58.30	95.80
59"	39.10	63.50	102.00
63"	41.90	65.70	107.00
67"	43.00	68.10	111.00

	Craft@Hrs	Unit	Material	Labor	Total
Shutters, Exterior. Installation (Figures in the material column show the cost of anchors and hinges).					
Installing shutters on hinges.					
Wood frame construction, 1-3/8" throw					
25" to 51" high	BC@.400	Pr	23.20	10.10	33.30
52" to 80" high	BC@.634	Pr	34.80	16.00	50.80
Masonry construction, 4-1/4" throw					
25" to 51" high	BC@.598	Pr	37.10	15.00	52.10
52" to 80" high	BC@.827	Pr	49.40	20.80	70.20
Installing fixed shutters on wood frame or masonry					
Any size shown above	BC@.390	Pr	3.60	9.82	13.42

Shutters, Interior

	Craft@Hrs	Unit	Material	Labor	Total

Shutters, Interior Movable 1-1/4" louver, 3/4" thick, kiln dried pine, with knobs and hooks, panels attached with hanging strips hinged to panels, finished, priced per set of either two or four panels.

Width x Height

	Craft@Hrs	Unit	Material	Labor	Total
23" x 20", 2 panels	BC@.250	Ea	58.50	6.29	64.79
23" x 24", 2 panels	BC@.300	Ea	64.60	7.55	72.15
23" x 28", 2 panels	BC@.300	Ea	76.20	7.55	83.75
23" x 32", 2 panels	BC@.333	Ea	81.80	8.38	90.18
23" x 36", 2 panels	BC@.367	Ea	96.80	9.24	106.04
27" x 20", 4 panels	BC@.250	Ea	77.90	6.29	84.19
27" x 24", 4 panels	BC@.300	Ea	87.90	7.55	95.45
27" x 28", 4 panels	BC@.300	Ea	105.00	7.55	112.55
27" x 32", 4 panels	BC@.333	Ea	110.00	8.38	118.38
27" x 36", 4 panels	BC@.367	Ea	117.00	9.24	126.24
31" x 20", 4 panels	BC@.250	Ea	81.80	6.29	88.09
31" x 24", 4 panels	BC@.300	Ea	93.50	7.55	101.05
31" x 28", 4 panels	BC@.300	Ea	105.00	7.55	112.55
31" x 32", 4 panels	BC@.367	Ea	117.00	9.24	126.24
31" x 36", 4 panels	BC@.367	Ea	129.00	9.24	138.24
33" x 20", 4 panels	BC@.250	Ea	93.50	6.29	99.79
33" x 24", 4 panels	BC@.300	Ea	99.60	7.55	107.15
33" x 28", 4 panels	BC@.300	Ea	110.00	7.55	117.55
33" x 32", 4 panels	BC@.333	Ea	129.00	8.38	137.38
33" x 36", 4 panels	BC@.367	Ea	141.00	9.24	150.24
35" x 20", 4 panels	BC@.250	Ea	99.60	6.29	105.89
35" x 24", 4 panels	BC@.300	Ea	117.00	7.55	124.55
35" x 28", 4 panels	BC@.300	Ea	123.00	7.55	130.55
35" x 32", 4 panels	BC@.367	Ea	134.00	9.24	143.24
35" x 36", 4 panels	BC@.367	Ea	146.00	9.24	155.24
39" x 20", 4 panels	BC@.250	Ea	117.00	6.29	123.29
39" x 24", 4 panels	BC@.300	Ea	123.00	7.55	130.55
39" x 28", 4 panels	BC@.300	Ea	134.00	7.55	141.55
39" x 32", 4 panels	BC@.315	Ea	146.00	7.93	153.93
39" x 36", 4 panels	BC@.315	Ea	152.00	7.93	159.93
47" x 24", 4 panels	BC@.300	Ea	129.00	7.55	136.55
47" x 28", 4 panels	BC@.300	Ea	146.00	7.55	153.55
47" x 32", 4 panels	BC@.367	Ea	152.00	9.24	161.24
47" x 36", 4 panels	BC@.367	Ea	159.00	9.24	168.24

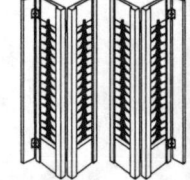

Siding See also Siding in the Lumber Section, Plywood and Hardboard.

Fiber-cement siding 1/8" to 1/4" thick, prefinished. Includes caulking and 5% waste, Supradur products.

	Craft@Hrs	Unit	Material	Labor	Total
12" x 24", 11" exposure, 3 bundles per square					
Various patterns	B1@1.88	Sq	87.70	43.00	130.70
14-1/2" x 25", 13" exposure, 4 bundles per square, thatched edge					
"Cedatex"	B1@1.88	Sq	95.90	43.00	138.90
"Pocono"	B1@1.88	Sq	117.00	43.00	160.00
14-1/2" x 32 x 13-1/2" exposure, 3 bundles per square					
"Magna-Tones"	B1@1.88	Sq	87.70	43.00	130.70
9" x 32", 8" exposure, 3 bundles per square					
"Stride"	B1@1.88	Sq	100.00	43.00	143.00

	Craft@Hrs	Unit	Material	Labor	Total
Hardboard siding, ABTCO, Inc.					
Siding, 7/16" thick. Includes caulking and 10% waste.					
Bevelside, 12" x 16', primed	B1@.032	SF	.75	.73	1.48
Cedar lap, 8" or 12" x 16' primed	B1@.032	SF	.61	.73	1.34
"Great Random Shakes"					
12" x 4' unprimed	B1@.032	SF	.69	.73	1.42
Lockside, 8" x 16', primed	B1@.032	SF	.64	.73	1.37
Smooth Lap, 8" or 12" x 16', primed	B1@.032	SF	.60	.73	1.33
Smooth beaded lap, 8" x 16' primed	B1@.032	SF	.61	.73	1.34
Textured beaded lap, 8" x 16' primed	B1@.032	SF	.61	.73	1.34
Panels, 7/16" thick. Includes caulking and 10% waste.					
Cedar panels					
4' x 8' or 4' x 9' primed	B1@.032	SF	.59	.73	1.32
Cross sawn fir panels					
4' x 8' or 4' x 9' primed	B1@.032	SF	.61	.73	1.34
Stucco panels					
4' x 8' or 4' x 9' primed	B1@.032	SF	.73	.73	1.46
Siding, 1/2" thick. Includes caulking and 10% waste					
Bevelside, 12" x 16', primed	B1@.032	SF	.78	.73	1.51
Lockside, 8" x 16', primed	B1@.032	SF	.83	.73	1.56
Hardboard siding, Georgia-Pacific 7/16" thick, primed. Includes caulking, nails and 10% waste.					
Cadence lap, 8" or 12" x 16', smooth	B1@.032	SF	.79	.73	1.52
Fairfax 1/2", 4" or 6" multilap x 16'	B1@.032	SF	1.18	.73	1.91
Provincetown beaded lap, 8" x 16'	B1@.032	SF	.86	.73	1.59
Summerwood lap, 8" x 16',					
beaded or sq. edge	B1@.032	SF	.97	.73	1.70
Summerwood, 4' x 8'	B1@.025	SF	1.16	.57	1.73
Sturbridge 1/2", 4" or 6" multilap x 16'	B1@.032	SF	1.18	.73	1.91
Sundance lap, 8" x 16'	B1@.032	SF	.86	.73	1.59
Sundance lap, 12" x 16'	B1@.032	SF	.81	.73	1.54
Sundance (grooved) 4' x 8', or 9'	B1@.025	SF	.86	.57	1.43
V-8 Cadence, smooth, 4' x 8'	B1@.025	SF	.86	.57	1.43
Yorktown textured beaded lap, 8" x 16'	B1@.032	SF	.92	.73	1.65
Windridge lap, 8" x 16', deep texture	B1@.032	SF	.94	.73	1.67
Windridge, 4' x 8', deep texture	B1@.025	SF	.97	.57	1.54
Hardboard siding, Masonite 7/16" thick. Includes caulking, nails and 10% waste.					
Colorlok, various styles, smooth or textured					
8" x 16'	B1@.032	SF	1.52	.73	2.25
Stuccato panel, 4' x 8'	B1@.032	SF	.96	.73	1.69
Woodsman, textured, 12" x 16'	B1@.032	SF	.75	.73	1.48
Hardboard siding, Masonite Superside 1/2" thick. Includes caulking, nails and 10% waste.					
Cedarside lap, 12" x 16'	B1@.032	SF	1.52	.73	2.25
Channelside lap, textured, 16" x 16'	B1@.032	SF	1.22	.73	1.95
Drop-side lap, smooth or textured					
12" x 16'	B1@.032	SF	1.22	.73	1.95
16" x 16'	B1@.032	SF	1.22	.73	1.95
Pineridge lap, textured					
12" x 16'	B1@.032	SF	1.22	.73	1.95
16" x 16'	B1@.032	SF	1.22	.73	1.95
Pinewood beaded lap, 8" x 16'	B1@.032	SF	1.16	.73	1.89

Siding

	Craft@Hrs	Unit	Material	Labor	Total
Self-aligning lap, plain or textured					
6" x 16'	B1@.032	SF	1.11	.73	1.84
8" x 16'	B1@.032	SF	1.11	.73	1.84
12" x 16'	B1@.032	SF	1.11	.73	1.84
V-side lap, textured, 16" x 16'	B1@.032	SF	1.22	.73	1.95
Woodridge, smooth or textured					
12" x 16'	B1@.032	SF	1.22	.73	1.95
16" x 16'	B1@.032	SF	1.22	.73	1.95
Woodsman, 8" OC, textured, 4' x 8'	B1@.025	SF	1.05	.57	1.62
Woodsman, plain, textured, 4' x 8'	B1@.025	SF	1.06	.57	1.63
Woodsman, splined, textured					
8" x 16'	B1@.032	SF	1.42	.73	2.15
X-90 splined siding, smooth, 8" x 16'	B1@.032	SF	1.42	.73	2.15

Hardboard siding, Weyerhaeuser Includes caulking, nails and 10% waste.

	Craft@Hrs	Unit	Material	Labor	Total
Adobe panels, shiplap edge, 7/16" thick					
4' x 8' primed	B1@.025	SF	.70	.57	1.27
4' x 9' primed	B1@.025	SF	.78	.57	1.35
Adobe panels, shiplap edge, 1/2" thick					
4' x 8' primed	B1@.025	SF	.78	.57	1.35
4' x 9' primed	B1@.025	SF	.90	.57	1.47
Old Mill texture lap, 7/16" thick					
6" x 16' primed	B1@.032	SF	.69	.73	1.42
8" x 16' primed	B1@.032	SF	.63	.73	1.36
9-1/2" x 16' primed	B1@.032	SF	.63	.73	1.36
12" x 16' primed	B1@.032	SF	.63	.73	1.36
Old Mill texture panel, 8" OC, 3/4" channel, 404 or 808, 7/16" thick					
4' x 7' primed	B1@.025	SF	.53	.57	1.10
4' x 8' primed	B1@.025	SF	.63	.57	1.20
4' x 9' primed	B1@.025	SF	.63	.57	1.20
Cedar Shake lap, 1/2" thick, self-aligning					
10-1/2" x 16' primed	B1@.025	SF	1.04	.57	1.61
Designer Shake lap, 1/2" thick, various patterns					
9-1/2" x 8' primed	B1@.032	SF	1.64	.73	2.37
Old Mill texture panel, 8" OC, 3/4" channel, 404 or 808, 1/2" thick					
4' x 7' primed	B1@.025	SF	.61	.57	1.18
4' x 8' primed	B1@.025	SF	.61	.57	1.18
4' x 9' primed	B1@.025	SF	.71	.57	1.28
Old Mill texture lap, 1/2" thick, various patterns and groove sizes					
16" x 16' primed	B1@.032	SF	.78	.73	1.51
Old Mill texture lap, bevel edge or self-aligning, 1/2" thick					
6" x 16" primed	B1@.032	SF	.76	.73	1.49
6" x 16' Sure-lock, concealed nail	B1@.032	SF	.78	.73	1.51
8" x 16' primed	B1@.032	SF	.72	.73	1.45
8" x 16' Sure-lock, concealed nail	B1@.032	SF	.70	.73	1.43
9-1/2" x 16' primed	B1@.032	SF	.72	.73	1.45
12" x 16' primed	B1@.032	SF	.70	.73	1.43
Old Mill texture panel					
4' x 8' plain square edge	B1@.032	SF	.71	.73	1.44

	Craft@Hrs	Unit	Material	Labor	Total

Oriented strand board (OSB),Louisiana-Pacific, Inner-Seal OSB Siding and trim, (figures in parentheses show quantity required including 8% waste allowance)

Lap siding, smooth finish, square edge or lap, 7/16" thick

	Craft@Hrs	Unit	Material	Labor	Total
8" x 16' (190 LF per Sq at $.37 per LF)	B1@.032	SF	.97	.73	1.70
12" x 16' (114 LF per Sq at $.59 per LF)	B1@.032 .	SF	.92	.73	1.65

Lap siding, rough sawn cedar finish, square edge or lap, 7/16" thick

6" x 16' (250 LF per Sq at $.33 per LF)	B1@.032	SF	1.14	.73	1.87
8" x 16' (190 LF per Sq at $.43 per LF)	B1@.032	SF	1.13	.73	1.86

Panel siding, rough sawn cedar finish, ungrooved face, square edge or T&G, 7/16" thick

4' x 8' (108 SF per Sq at $21.80 per panel)	B1@.025	SF	1.02	.57	1.59

Panel siding, rough sawn cedar finish, grooved face, square edge or T&G, 7/16" thick

4' x 8' (108 SF per Sq at $22.00 per panel)	B1@.025	SF	1.02	.57	1.59

Pre-stained finish, Olympic Machine coat, any of 85 colors, factory applied.

Add to any of the above siding costs

One-coat application, 5-year warranty	—	SF	.11	—	.11
Two-coat application, 10-year warranty	—	SF	.22	—	.22

Exterior trim and soffit panels, (figures in parentheses show cost per LF before 10% waste allowance)

Fascia, band boards, corner boards or rake boards, smooth finished, primed

3/4" x 4" (at $.39 per LF)	B1@.026	LF	.52	.59	1.11
3/4" x 6" (at $.59 per LF)	B1@.026	LF	78	.59	1.37
3/4" x 8" (at $.79 per LF)	B1@.026	LF	1.04	.59	1.63
3/4" x 10" (at $.97 per LF)	B1@.026	LF	1.28	.59	1.87
3/4" x 12" (at $1.18 per LF)	B1@.026	LF	1.56	.59	2.15
15/16" x 4" (at $.43 per LF)	B1@.026	LF	.56	.59	1.15
15/16" x 6" (at $.64 per LF)	B1@.026	LF	.84	.59	1.43
15/16" x 8" (at $.84 per LF)	B1@.026	LF	1.10	.59	1.69
15/16" x 10" (at $1.07 per LF)	B1@.026	LF	1.42	.59	2.01
15/16" x 12" (at $1.27 per LF)	B1@.026	LF	1.68	.59	2.27

Accessories, pre-finished to match OSB siding stain

Maze "splitless" corrosion resistant nails

Siding (1.7 Lb per Sq at $1.90 per pound)	—	SF	.03	—	.03
Trim (.85 Lb per LF at $1.90 per pound)	—	LF	.02	—	.02

Geocel "stain match" acrylic latex caulking, 1/8" to 1/2" bead

Siding (12 tubes per Sq at $3.35 per tube)	B1@.016	SF	.40	.37	.77
Trim (1 tube per CLF at $3.35 per tube)	B1@.001	LF	.03	.02	.05

Sheet metal flashing

7/8", 5/4", 1-5/8" x 10' (at $5.00 each)	B1@.040	LF	.50	.91	1.41
Brick mold, 10' lengths (at $8.50 each)	B1@.040	LF	.85	.91	1.76

Under eave soffit vents, screen type

4" or 8" x 16"	B1@.500	Ea	4.50	11.40	15.90

Continuous aluminum soffit vents

2" x 8' (at $5.00 each)	B1@.050	LF	.63	1.14	1.77

Touch-up stain, Olympic, per coat, applied with brush or roller

Prime coat (300 SF and $16 per gal)	PT@.006	SF	.05	.16	.21
Oil stain (400 SF and $16 per gal)	PT@.006	SF	.04	.16	.20
Penofin Oil Finish (400 SF and $19.00 per gal)	PT@.006	SF	.05	.16	.21

Hammerhead covers, plastic

Per SF of siding (1 per Sq at $1.50 each)	—	SF	.02	—	.02

Siding

	Craft@Hrs	Unit	Material	Labor	Total
Plywood siding, Douglas fir Plain or patterns, rough sawn. Includes caulking, nails and 6% waste.					
3/8" premium grade					
4' x 8'	B1@2.18	Sq	76.00	49.80	125.80
4' x 9'	B1@2.18	Sq	111.00	49.80	160.80
4' x 10'	B1@2.18	Sq	146.00	49.80	195.80
3/8" shop grade					
4' x 8'	B1@2.18	Sq	61.00	49.80	110.80
4' x 9'	B1@2.18	Sq	71.00	49.80	120.80
4' x 10'	B1@2.18	Sq	81.00	49.80	130.80
5/8" premium grade					
4' x 8'	B1@2.45	Sq	86.00	56.00	142.00
4' x 9'	B1@2.45	Sq	121.00	56.00	177.00
4' x 10'	B1@2.45	Sq	152.00	56.00	208.00
5/8" shop grade					
4' x 8'	B1@2.45	Sq	66.00	56.00	122.00
4' x 9'	B1@2.45	Sq	71.00	56.00	127.00
4' x 10'	B1@2.45	Sq	81.00	56.00	137.00
Add for inverted batten patterns	—	Sq	10.00	—	10.00
Add for water repellent treated	—	Sq	16.60	—	16.60
Plywood siding, with batten boards fiber bonded to one face. Costs include 3/8" thick x 4' x 8' rough sawn texture exterior grade plywood, caulking, nails and 6% waste allowance. Batten boards are 5/8" thick x 2" wide at 4", 6", or 8" OC					
Flat batten boards	B1@2.45	Sq	79.30	56.00	135.30
Inverted batten boards	B1@2.45	Sq	84.50	56.00	140.50
Plywood siding, redwood Water repellent treated. Costs include caulking, nails and 6% waste.					
Premium grade, plain, rough sawn					
3/8" thick x 4' x 8'	B1@2.18	Sq	88.10	49.80	137.90
5/8" thick x 4' x 8'	B1@2.45	Sq	93.70	56.00	149.70
Premium grade, batten boards are 5/8" thick x 2" wide at 4", 6" or 8" OC					
3/8" thick x 4' x 8', flat batten	B1@2.45	Sq	95.40	56.00	151.40
Premium grade, batten boards are 5/8" thick x 2" wide at 8" or 12" OC					
3/8" thick x 4' x 8', inverted batten	B1@2.45	Sq	103.00	56.00	159.00
Select grade, plain, rough sawn					
3/8" thick x 4' x 8'	B1@2.18	Sq	63.50	49.80	113.30
5/8" thick 4' x 8'	B1@2.45	Sq	70.50	56.00	126.50
Select grade, rough sawn with batten boards 5/8" thick x 2" wide at 4", 6" or 8" OC					
3/8" thick x 4' x 8', flat batten	B1@2.45	Sq	63.50	56.00	119.50
Select grade, rough sawn with batten boards 5/8" thick x 2" wide at 4", 6" or 8" OC					
3/8" thick x 4' x 8', inverted batten	B1@2.45	Sq	67.30	56.00	123.30
Plywood siding, western red cedar 4' x 8' sheets. Costs include caulking, nails and 6% waste.					
3/8" thick, select grade, plain or patterns	B1@2.18	Sq	72.10	49.80	121.90
1/2" thick, select grade, plain or patterns	B1@2.30	Sq	82.40	52.60	135.00
5/8" thick, select grade, plain or patterns	B1@2.45	Sq	87.60	56.00	143.60
3/8" thick, cabin grade	B1@2.18	Sq	52.50	49.80	102.30
1/2" thick, cabin grade	B1@2.30	Sq	59.70	52.60	112.30
5/8" thick, cabin grade	B1@2.45	Sq	75.20	56.00	131.20

	Craft@Hrs	Unit	Material	Labor	Total
Sheet metal siding					
Aluminum corrugated 4-V x 2-1/2", plain or embossed finish. Includes 15% waste.					
17 gauge, 26" x 6' to 24'	B1@.034	SF	1.76	.78	2.54
19 gauge, 26" x 6' to 24'	B1@.034	SF	1.82	.78	2.60
Rubber filler strip, 3/4" x 7/8" x 6'	—	LF	.28	—	.28
Flashing for corrugated aluminum siding, embossed					
End wall, 10" x 52"	B1@.048	Ea	3.15	1.10	4.25
Side wall, 7-1/2" x 10'	B1@.015	LF	1.35	.34	1.69
Aluminum smooth 24 gauge, horizontal patterns, non-insulated.					
8" or double 4" widths, acrylic finish	B1@2.77	Sq	125.00	63.30	188.30
12" widths, bonded vinyl finish	B1@2.77	Sq	128.00	63.30	191.30
Add for foam backing	—	Sq	25.30	—	24.70
Starter strip	B1@.030	LF	.29	.69	.98
Inside and outside corners	B1@.033	LF	.89	.75	1.64
Casing and trim	B1@.033	LF	.29	.75	1.04
Drip cap	B1@.044	LF	.30	1.01	1.31
Galvanized steel siding					
28 gauge, 27-1/2" wide (includes 20% coverage loss)					
6' to 12' standard lengths	B1@.034	SF	.83	.78	1.61
26 gauge, 26" wide (includes 15% coverage loss)					
6' to 12' standard lengths	B1@.034	SF	1.01	.78	1.79
Shingle siding					
Red cedar perfects 5X green, 16" long, 7-1/2" exposure, 5 shingles are 2" thick (5/2)					
3 bundles cover 1 Sq including waste. Cost shown includes fasteners.					
#1 ($50.00 per bundle)	B1@3.86	Sq	150.00	88.20	238.20
#2 red label ($28.40 per bundle)	B1@3.86	Sq	85.20	88.20	173.40
#3 ($20.30 per bundle)	B1@3.86	Sq	60.90	88.20	149.10
Red cedar perfects, dry, 18" long, 8" exposure, 5 shingles are 2-1/4" thick (5/2-1/4).					
3 bundles cover 1 Sq including waste. Cost shown includes fasteners.					
#1 ($60.50 per bundle)	B1@3.75	Sq	182.00	85.70	267.70
#2 red label ($29.50 per bundle)	B1@3.75	Sq	88.50	85.70	174.20
#3 ($37.50 per bundle)	B1@3.75	Sq	113.00	85.70	198.70
Panelized shingles, Shakertown, KD, regraded shingle on plywood backing, no waste.					
8' long x 7" wide single course					
Colonial 1	B1@.028	SF	3.01	.64	3.65
Cascade	B1@.028	SF	2.91	.64	3.55
8' long x 14" wide, single course					
Colonial 1	B1@.028	SF	3.01	.64	3.65
8' long x 14" wide, double course					
Colonial 2	B1@.028	SF	3.28	.64	3.92
Cascade Classic	B1@.028	SF	2.91	.64	3.55
Decorative shingles, 5" wide x 18" long, Shakertown, fancy cuts, 9 hand-shaped patterns.					
7-1/2" exposure	B1@.030	SF	3.51	.69	4.20
Deduct for 10" exposure	—	SF	-.42	—	-.42
Vinyl siding Solid vinyl .044 gauge, woodgrained or colors. Costs include caulking, nails and 5% waste.					
10" or double 5" panels	B1@3.34	Sq	70.30	76.30	146.60
Starter strip	B1@.030	LF	.25	.69	.94
J channel for corners	B1@.033	LF	.26	.75	1.01
Outside corner	B1@.033	LF	.95	.75	1.70
Casing and trim	B1@.033	LF	.26	.75	1.01
Add for Amocor insulation (R-3)	B1@.350	Sq	22.60	8.00	30.60

Skylights

Skylights Polycarbonate dome, clear transparent or tinted, roof opening sizes, price each, Crestline Skylights, curb or flush mount, self-flashing

	22" x 22"	30" x 30"	22" x 46"	30" x 46"	46" x 46"
Single dome	42.80	71.30	85.5	124.00	214.00
Double dome	57.00	109.00	124.00	162.00	261.00
Triple dome	76.00	143.00	171.00	214.00	328.00

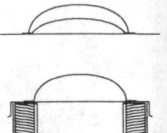

	Craft@Hrs	Unit	Material	Labor	Total

Labor installing self-flashing skylights, installation in new construction. No carpentry or roofing included.

	Craft@Hrs	Unit	Material	Labor	Total
Less than 4 SF	BG@1.56	Ea	—	36.00	36.00
4 SF to 10 SF	BG@2.58	Ea	—	59.50	59.50
Over 11 SF	BG@2.84	Ea	—	65.50	65.50

Tempered over laminated low-E argon gas filled doubled insulated fixed glass skylights. Includes costs for flashing. Conforms to all building codes. For installation on roofs with 15 to 85 degree slope. Labor costs are for installation of skylight and flashing. Roll shade costs are for factory attached shades. No carpentry or roofing included.

	Craft@Hrs	Unit	Material	Labor	Total
Operable skylight, VELUX, VS models, including flashing					
21-1/2" x 27-1/2", #101	B1@3.42	Ea	334.00	78.20	412.20
21-1/2" x 38-1/2", #104	B1@3.42	Ea	366.00	78.20	444.20
21-1/2" x 46-3/8", #106	B1@3.42	Ea	394.00	78.20	472.20
21-1/2" x 55", #108	B1@3.42	Ea	424.00	78.20	502.20
30-5/8" x 38-1/2", #304	B1@3.42	Ea	416.00	78.20	494.20
30-5/8" x 46-3/8", #306	B1@3.42	Ea	457.00	78.20	535.20
30-5/8" x 55", #308	B1@3.52	Ea	488.00	80.40	568.40
44-3/4" x 27-1/2", #601	B1@3.52	Ea	440.00	80.40	520.40
44-3/4" x 46-1/2", #606	B1@3.52	Ea	542.00	80.40	622.40
Add for roll shade, natural					
21-1/2" x 27-1/2" to 55" width	—	Ea	48.50	—	48.50
30-5/8" x 38-1/2" to 55" width	—	Ea	56.00	—	56.00
Fixed skylight, VELUX, FS models					
21-1/2" x 27-1/2", #101	B1@3.42	Ea	183.00	78.20	261.20
21-1/2" x 38-1/2", #104	B1@3.42	Ea	218.00	78.20	296.20
21-1/2" x 46-3/8", #106	B1@3.42	Ea	241.00	78.20	319.20
21-1/2" x 55", #108	B1@3.42	Ea	265.00	78.20	343.20
21-1/2" x 70-7/8", #112	B1@3.42	Ea	320.00	78.20	398.20
30-5/8" x 38-1/2", #304	B1@3.42	Ea	258.00	78.20	336.20
30-5/8" x 46-3/8", #306	B1@3.42	Ea	287.00	78.20	365.20
30-5/8" x 55", #308	B1@3.52	Ea	312.00	80.40	392.40
44-3/4" x 27-1/2", #601	B1@3.52	Ea	256.00	80.40	336.40
44-3/4" x 46-1/2", #606	B1@3.52	Ea	357.00	80.40	437.40
Add for roll shade, natural					
21-1/2" x 27-1/2" to 70-7/8" width	—	Ea	48.50	—	48.50
30-5/8" x 38-1/2" to 55" width	—	Ea	56.00	—	56.00
Fixed ventilation, with removable filter on ventilation flap, VELUX model FSF					
21-1/2" x 27-1/2", #101	B1@3.14	Ea	209.00	71.80	280.80
21-1/2" x 38-1/2", #104	B1@3.14	Ea	244.00	71.80	315.80
21-1/2" x 46-1/2", #106	B1@3.14	Ea	272.00	71.80	343.80
21-1/2" x 70-7/8", #108	B1@3.42	Ea	296.00	78.20	374.20
30-5/8" x 38-1/2", #304	B1@3.14	Ea	360.80	71.80	350.70
30-5/8" x 46-3/8", #306	B1@3.14	Ea	326.00	71.80	397.80
30-5/8" x 55", #308	B1@3.39	Ea	350.00	77.50	427.50
44-3/4" x 27-1/2", #601	B1@3.36	Ea	312.00	76.80	388.80
44-3/4" x 46-1/2", #606	B1@3.39	Ea	403.00	77.50	480.50

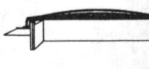

	Craft@Hrs	Unit	Material	Labor	Total
Add for roll shade, natural					
21-1/2" x 27-1/2" to 70-7/8" width	—	Ea	48.50	—	48.50
30-5/8" x 38-1/2" to 55" width	—	Ea	56.00	—	56.00
Production builder's skylight, VELUX model FS-156 lay-in type for 24" C to C rafters					
23-5/16" x 46-1/2"	B1@3.14	Ea	236.00	71.80	307.80
23-5/16" x 23-9/16"	B1@3.14	Ea	181.00	71.80	252.80

Skywindows

Skywindow, E-Class. Ultraseal Self-Flashing Units. Clear insulated glass, flexible flange, no mastic or step flashing required. Other glazings available. Integral skyshade optional. Wasco.

	Craft@Hrs	Unit	Material	Labor	Total
22-1/2" x 22-1/2", fixed	B1@0.75	Ea	190.00	17.10	207.10
22-1/2" x 38-1/2", fixed	B1@0.75	Ea	219.00	17.10	236.10
22-1/2" x 46-1/2", fixed	B1@0.75	Ea	257.00	17.10	274.10
30-1/2" x 30-1/2", fixed	B1@0.75	Ea	257.00	17.10	274.10
30-1/2" x 46-1/2", fixed	B1@0.75	Ea	314.00	17.10	331.10
46-1/2" x 46-1/2", fixed	B1@0.75	Ea	352.00	17.10	369.10
22-1/2" x 22-1/2", vented	B1@0.75	Ea	361.00	17.10	378.10
22-1/2" x 38-1/2", vented	B1@0.75	Ea	390.00	17.10	407.10
22-1/2" x 46-1/2", vented	B1@0.75	Ea	442.00	17.10	459.10
30-1/2" x 30-1/2", vented	B1@0.75	Ea	418.00	17.10	435.10
30-1/2" x 46-1/2", vented	B1@0.75	Ea	480.00	17.10	497.10

Skywindow, Excel-10. Step Flash/Pan Flashed Units. Insulated clear tempered glass. Deck mounted step flash unit. Other glazings available. Wasco.

	Craft@Hrs	Unit	Material	Labor	Total
22-1/2" x 22-1/2", fixed	B1@2.35	Ea	176.00	53.70	229.70
22-1/2" x 38-1/2", fixed	B1@2.35	Ea	185.00	53.70	238.70
22-1/2" x 46-1/2", fixed	B1@2.35	Ea	195.00	53.70	248.70
30-1/2" x 30-1/2", fixed	B1@2.35	Ea	209.00	53.70	262.70
30-1/2" x 46-1/2", fixed	B1@2.35	Ea	252.00	53.70	305.70
22-1/2" x 22-1/2", vented	B1@2.35	Ea	295.00	53.70	348.70
22-1/2" x 30-1/2", vented	B1@2.35	Ea	318.00	53.70	371.70
22-1/2" x 46-1/2", vented	B1@2.35	Ea	380.00	53.70	433.70
30-1/2" x 30-1/2", vented	B1@2.35	Ea	366.00	53.70	419.70
30-1/2" x 46-1/2", vented	B1@2.35	Ea	409.00	53.70	462.70
Add for flashing kit	—	Ea	60.00	—	60.00

Skywindow, Genra-1. Low profile insulated glass skywindow.
Deck mount, clear tempered glass, self-flashing models. Other glazings available. Wasco.

	Craft@Hrs	Unit	Material	Labor	Total
22-1/2" x 22-1/2", fixed	B1@2.35	Ea	157.00	53.70	210.70
22-1/2" x 38-1/2", fixed	B1@2.35	Ea	185.00	53.70	238.70
22-1/2" x 46-1/2", fixed	B1@2.35	Ea	228.00	53.70	281.70
30-1/2" x 30-1/2", fixed	B1@2.35	Ea	214.00	53.70	267.70
30-1/2" x 46-1/2", fixed	B1@2.35	Ea	261.00	53.70	314.70
46-1/2" x 46-1/2", fixed	B1@2.35	Ea	333.00	53.70	386.70
22-1/2" x 22-1/2", vented	B1@2.35	Ea	323.00	53.70	376.70

Roof windows Aluminum-clad wood frame, double insulated tempered over laminated low E gas filled double insulated glass, including prefabricated flashing, exterior awning, interior roller blind and insect screen. Sash rotates 180 degrees. For installation on roofs with 10 to 85 degree slope. Labor costs are for installation of roof window, flashing and sun screening accessories. No carpentry or roofing work other than curb included. Add for interior trim. Listed by actual unit dimensions, top hung, VELUX model TPS.

	Craft@Hrs	Unit	Material	Labor	Total
21-9/16" wide x 38-1/2" high	B1@3.42	Ea	492.00	78.20	570.20
27-9/16" wide x 46-3/8" high	B1@3.52	Ea	516.00	80.40	596.40

	Craft@Hrs	Unit	Material	Labor	Total
30-5/8" wide x 38-1/2" high	B1@3.42	Ea	551.00	78.20	629.20
30-5/8" wide x 55" high	B1@3.52	Ea	619.00	80.40	699.40
44-3/4" wide x 46-3/8" high	B1@3.52	Ea	689.00	80.40	769.40

Soffit Systems

Baked enamel finish, 6" fascia, J-channel, 8" perforated or solid soffit, .019 gauge aluminum

	Craft@Hrs	Unit	Material	Labor	Total
12" soffit	B1@.047	LF	2.85	1.07	3.92
18" soffit	B1@.055	LF	3.30	1.26	4.56
24" soffit	B1@.060	LF	3.75	1.37	5.12

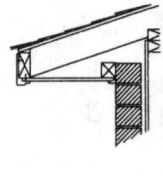

Vinyl soffit systems

	Craft@Hrs	Unit	Material	Labor	Total
12" soffit	B1@.030	LF	2.45	.69	3.14
18" soffit	B1@.030	LF	2.80	.69	3.49
24" soffit	B1@.030	LF	3.15	.69	3.84

Aluminum fascia alone

	Craft@Hrs	Unit	Material	Labor	Total
4" fascia	B1@.030	LF	.76	.69	1.45
6" fascia	B1@.030	LF	.92	.69	1.61
8" fascia	B1@.030	LF	1.09	.69	1.78

Soil Testing

Field observation and testing

	Craft@Hrs	Unit	Material	Labor	Total
Soil technician (average)		Hr	—	—	60.00

Maximum density/optimum moisture test

	Craft@Hrs	Unit	Material	Labor	Total
4" mold	—	Ea	—	—	126.00
6" mold	—	Ea	—	—	145.00
Expansion index test	—	Ea	—	—	90.00
Report on compacted fill (costs vary widely)	—	Ea	—	—	1,500.00

Foundation investigation (bearing capacity, lateral loads, piles, seismic, stability, settlement for design purposes) and preliminary soils investigation (depth of fill, classification and profile of soil, expansion, shrinkage, grading, drainage recommendations). Including drilling and sampling costs

	Craft@Hrs	Unit	Material	Labor	Total
Drill rig (driller and helper)	—	Hr	—	—	140.00
Staff engineer	—	Hr	—	—	70.00
Project engineer, RCE	—	Hr	—	—	79.00
Principal engineer	—	Hr	—	—	132.00
Transportation	—	Mile	—	—	.50
Transportation	—	Hr	—	—	6.90
Laboratory testing (general)	—	Hr	—	—	60.00
Laboratory testing (varies on individual test basis)	—	Ea	—	—	100.00

Report preparation

	Craft@Hrs	Unit	Material	Labor	Total
Analysis, conclusions, recommendations	—	Ea	—	—	2,000.00
Court preparation (Principal Engineer)	—	Hr	—	—	150.00
Expert witness court appearance	—	Hr	—	—	250.00
Add for hazardous studies	—	%	—	—	20.0

Soil Treatments, Subcontract Costs based on total square footage inside of, and including, foundation. Includes inside and outside perimeter of foundation, interior foundation walls, and underneath all slabs. Three-plus year re-application guarantee against existence of termites. Use $300 as a minimum job charge.

Dursban T.C. Mix applied per manufacturers' and state and federal specifications

	Craft@Hrs	Unit	Material	Labor	Total
Complete application	—	SF	—	—	.22

Solar Photovoltaic Electrical Systems, Subcontract

	Craft@Hrs	Unit	Material	Labor	Total

Solar Photovoltaic Electrical Systems, Subcontract Costs are for solar electric generating systems for use where commercial power is not available. These estimates assume 8 clear sunlight hours per day. Costs will be higher where less sun time is available. Photocomm with Kyocera modules.

Small remote vacation home system (12 volt DC). 64 watt solar electric module (98 amp hours), digital voltage regulator, mounting rack and wiring.

384 watt-hours per day	—	LS	—	—	1,150.00

Larger remote primary home system (120 volt AC). Includes eight L16 storage batteries (1,050 amp hours), voltage regulator, mounting rack, 2,400 watt DC to AC inverter.

2,880 watt-hours per day	—	LS	—	—	6,900.00

Basic two room solar power package Simpler Solar Systems.
Includes three 48 watt PV panels, one 12 amp voltage regulator, one 220 Ah battery with hydrocaps, one 600 watt inverter, mounting frame and hardware, and two receptacles.

12 amp regulator, with 600 watt inverter	—	LS	—	—	2,800.00

Small sized home solar power package Simpler Solar Systems.
90% solar powered system for 1,500 SF home includes six 60 watt PV panels, one 750 Ah battery, 20 amp voltage regulator and DC distribution box, and one 2.2 kW inverter. Costs also include one solar hot water system, one water source heat pump, one 19 cubic foot refrigerator freezer, one ringer type washing machine, six 12 volt lights and three 12 volt ceiling fans.

20 amp regulator, 12 volt system	—	LS	—	—	14,500.00

Medium sized home solar power package Simpler Solar Systems. 80% solar powered system for 2,500 SF home includes ten 45 watt PV panels, one 750 Ah battery, one 30 amp voltage regulator, and one 2 kW inverter with charger. Costs also include one solar domestic hot water system and 22 assorted 12 volt light fixtures.

30 amp regulator, 12 volt system	—	LS	—	—	8,800.00

Solar power pool heating system Simpler Solar Systems.

Includes eight panels, valves, and piping.	—	LS	—	—	3,300.00

Solar power spa heating system Simpler Solar Systems.

Includes one panel, valves and piping.	—	LS	—	—	980.00

Stairs

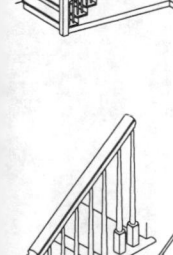

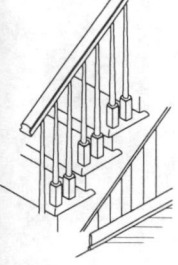

Factory cut and assembled straight closed box stairs, unfinished, 36" width

	Craft@Hrs	Unit	Material	Labor	Total
Oak treads with 7-1/2" risers, price per riser					
3'0" wide	B1@.324	Ea	101.00	7.40	108.40
3'6" wide	B1@.324	Ea	130.00	7.40	137.40
4'0" wide	B1@.357	Ea	140.00	8.16	148.16
Deduct for pine treads and plywood risers	—	%	-22.0	—	—
Add for prefinished assembled stair rail with balusters and newel, per riser	B1@.257	Ea	51.50	5.87	57.37
Add for prefinished handrail, with brackets and balusters	B1@.135	LF	15.10	3.08	18.18
Basement stairs, open riser, delivered to job site assembled, 12 treads, 2'11-3/4" wide.					
13 risers at 7-13/16" riser height	B1@4.00	Ea	224.00	91.40	315.40
13 risers at 8" riser height	B1@4.00	Ea	229.00	91.40	320.40
13 risers at 8-3/16" riser height	B1@4.00	Ea	234.00	91.40	325.40
Add for handrail	B1@.161	LF	9.36	3.68	13.04

Stairs

	Craft@Hrs	Unit	Material	Labor	Total
Curved stair, clear oak, self-supporting, 8'6" radii, factory cut, unfinished, unassembled					
Open one side (1 side against wall)					
8'9" to 9'4" rise	B1@24.6	Ea	5,200.00	562.00	5,762.00
9'5" to 10'0" rise	B1@24.6	Ea	5,620.00	562.00	6,182.00
10'1" to 10'8" rise	B1@26.5	Ea	6,030.00	606.00	6,636.00
10'9" to 11'4" rise	B1@26.5	Ea	6,450.00	606.00	7,056.00
Open two sides					
8'9" to 9'4" rise	B1@30.0	Ea	9,360.00	686.00	10,046.00
9'5" to 10'0" rise	B1@30.0	Ea	10,100.00	686.00	10,786.00
10'1" to 10'8" rise	B1@31.9	Ea	10,700.00	729.00	11,429.00
10'9" to 11'4" rise	B1@31.9	Ea	11,400.00	729.00	12,129.00
Add for newel posts	B1@.334	Ea	65.00	7.63	72.63

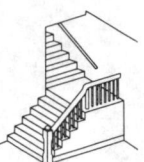

	Craft@Hrs	Unit	Material	Labor	Total
Spiral stairs, aluminum, delivered to job site unassembled, 7-3/4" or 8-3/4" riser heights, 5' diameter					
85-1/4" to 96-1/4" with 10 treads	B1@9.98	Ea	1,650.00	228.00	1,878.00
93" to 105" with 11 treads	B1@10.7	Ea	1,800.00	245.00	2,045.00
100-3/4" to 113-3/4" with 12 treads	B1@11.3	Ea	1,950.00	258.00	2,208.00
108-1/2" to 122-1/2" with 13 treads	B1@12.0	Ea	2,100.00	274.00	2,374.00
116-1/4" to 131-1/4" with 14 risers	B1@12.6	Ea	2,240.00	288.00	2,528.00
124" to 140" with 15 risers	B1@12.8	Ea	2,410.00	292.00	2,702.00
131-3/4" to 148-3/4" with 16 risers	B1@13.5	Ea	2,540.00	308.00	2,848.00
138-1/2" to 157-1/2" with 17 risers	B1@14.2	Ea	2,690.00	324.00	3,014.00
147-1/4" to 166-1/4" with 18 risers	B1@14.8	Ea	2,850.00	338.00	3,188.00
Deduct for 4' diameter	—	%	-20.0	—	—
Deduct for 4'6" diameter	—	%	-10.0	—	—
Add for additional aluminum treads	—	Ea	153.00	—	153.00
Add for oak tread inserts	—	Ea	45.90	—	45.90

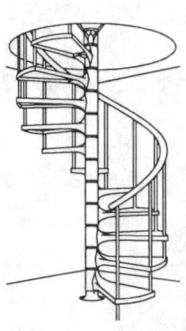

	Craft@Hrs	Unit	Material	Labor	Total
Spiral stairs, red oak, FOB factory, double handrails, 5' diameter					
74" to 82" with 9 risers	B1@16.0	Ea	2,838.00	366.00	3,204.00
82" to 90" with 10 risers	B1@16.0	Ea	3,044.00	366.00	3,410.00
90" to 98" with 11 risers	B1@16.0	Ea	3,100.00	366.00	3,466.00
98" to 106" with 12 risers	B1@16.0	Ea	3,156.00	366.00	3,522.00
106" to 114" with 13 risers	B1@20.0	Ea	3,232.00	457.00	3,689.00
114" to 122" with 14 risers	B1@20.0	Ea	3,344.00	457.00	3,801.00
122" to 130" with 15 risers	B1@20.0	Ea	3,514.00	457.00	3,971.00
Add for 6' diameter	—	%	20.0	—	—
Add for 8'6" diameter	—	%	25.0	—	—
Add for mahogany, any of above	—	%	10.0	—	—
Add for straight rail, unbored	B1@.060	LF	10.00	1.37	11.37
Add for straight nosing, unbored	B1@.060	LF	6.25	1.37	7.62
Add for balusters, 7/8" x 42"	B1@.334	Ea	4.00	7.63	11.63
Add for newel posts 1-7/8" x 44"	B1@.334	Ea	26.00	7.63	33.63
Spiral stairs, steel, tubular steel handrail, composition board treads, delivered unassembled, 5' diameter					
86" to 95" with 10 risers	B1@9.98	Ea	895.00	228.00	1,123.00
95" to 105" with 11 risers	B1@10.7	Ea	988.00	245.00	1,233.00
105" to 114" with 12 risers	B1@11.3	Ea	1,040.00	258.00	1,298.00
114" to 124" with 13 risers	B1@12.0	Ea	1,140.00	274.00	1,414.00
124" to 133" with 14 risers	B1@12.7	Ea	1,310.00	290.00	1,600.00
Deduct for 4' diameter	—	%	-20.0	—	—
Add for 6' diameter	—	%	24.0	—	—
Add for oak treads	—	%	25.0	—	—
Add for oak handrail	—	%	45.0	—	—

	Craft@Hrs	Unit	Material	Labor	Total

Straight stairs, red oak, factory cut, unassembled , 36" tread, includes two 1-1/2" x 3-1/2" handrails and 7/8" round balusters

	Craft@Hrs	Unit	Material	Labor	Total
70" to 77" with 10 risers	B1@12.0	Ea	1,560.00	274.00	1,834.00
77" to 85" with 11 risers	B1@12.0	Ea	1,770.00	274.00	2,044.00
85" to 93" with 12 risers	B1@12.0	Ea	1,920.00	274.00	2,194.00
93" to 100" with 13 risers	B1@14.0	Ea	2,080.00	320.00	2,400.00
100" to 108" with 14 risers	B1@14.0	Ea	2,290.00	320.00	2,610.00
108" to 116" with 15 risers	B1@14.0	Ea	2,390.00	320.00	2,710.00
116" to 124" with 16 risers	B1@16.0	Ea	2,600.00	366.00	2,966.00
124" to 131" with 17 risers	B1@16.0	Ea	2,700.00	366.00	3,066.00
131" to 139" with 18 risers	B1@16.0	Ea	3,020.00	366.00	3,386.00
Add for 44" tread, any of above	—	Ea	90.00	—	90.00
Add for straight rail, unbored	B1@.060	LF	11.50	1.37	12.87
Add for straight nosing	B1@.060	LF	6.25	1.37	7.62
Add for round balusters, 7/8" x 42"	B1@.334	Ea	4.25	7.63	11.88
Add for square balusters, 1-1/8" x 1-1/8" x 44"	B1@.334	Ea	8.00	7.63	15.63
Add for newel posts, 3" x 3" x 44"	B1@.334	Ea	40.00	7.63	47.63

Job-built stairways 3 stringers cut from 2" x 12" material (at $585 per MBF), treads and risers from 3/4" CDX plywood (at $550 per MSF). Cost per 7-1/2" rise, 36" wide.

	Craft@Hrs	Unit	Material	Labor	Total
Straight run, 8'0" to 10'0" rise, per riser	B1@.530	Ea	9.14	12.10	21.24
"L" or "U" shape, per riser, add for landings	B1@.625	Ea	10.40	14.30	24.70
Semi-circular, repetitive tract work, per riser	B1@.795	Ea	10.40	18.20	28.60
Landings, framed and 3/4" CDX plywood surfaced, per SF of landing surface	B1@.270	SF	2.24	6.17	8.41

Stairs, Disappearing Bessler Stairways.

Better quality folding stairway, full width main hinge, fir door panel, molded handrail, preassembled, 1" x 6" treads, 1" x 5" stringer and frame of select yellow pine, "space saver" for narrow openings, 22", 25-1/2", or 30" wide by 54" long,

	Craft@Hrs	Unit	Material	Labor	Total
8'9" or 10' height	B1@2.20	Ea	220.00	50.30	270.30

Superior quality solid one-piece stringer sliding stairway, stringer and treads of clear yellow pine, unassembled, full-width piano hinge at head of stairs, with springs and cables.

	Craft@Hrs	Unit	Material	Labor	Total
Compact unit, for a 2' x 4' opening, 7'7" to 9'3" height, 1" x 4" stringers and treads	B1@3.33	Ea	400.00	76.10	476.10
Standard unit, for openings from 2'0" x 5'6" to 2'6" x 6'0", 7'7" to 10'10" heights, 1" x 6" stringers and treads	B1@3.86	Ea	450.00	88.20	538.20
Heavy-duty unit, for openings from 2'6" x 5'10" to 2'6" x 8'0", 7'7" to 12'10" heights, 1" x 8" stringers and treads	B1@4.56	Ea	885.00	104.00	989.00

Steam Bath Generators Steam bath equipment, 240 Volts AC (Amperage shown in parentheses), single phase, UL approved. Costs include steam generator, timer switch and chrome-plated steam outlet. Labor costs include installation of steam unit and connections to enclosure, no rough plumbing or electrical included. Installed in existing, adequately sealed, tub enclosure. Maximum enclosure size as shown.

	Craft@Hrs	Unit	Material	Labor	Total
160 cubic feet (25 Amps)	P1@10.4	Ea	685.00	254.00	939.00
400 cubic feet (41.5 Amps)	P1@10.4	Ea	800.00	254.00	1,054.00
Add for thermostat	—	Ea	150.00	—	150.00
Add for brass steam head with escutcheon	—	Ea	24.00	—	24.00

Steel Stud Framing

	Craft@Hrs	Unit	Material	Labor	Total

Steel stud wall framing Installed using screws. Costs shown include galvanized metal studs, top and bottom track, screws and normal waste. Do not subtract for openings less than 16' wide. Add door and window opening framing, backing, let-in bracing, and sheathing for shear walls. Costs are per square foot of wall area, measured on one side. The figures in parentheses show production per day for a 2-man crew. Use one day as a minimum job charge.

Non-load bearing steel studding, 20 gauge

2-1/2" wide studs

	Craft@Hrs	Unit	Material	Labor	Total
16" on center spacing (1,070 SF per day)	B1@.015	SF	.19	.34	.53
24" on center spacing (1,250 SF per day)	B1@.013	SF	.15	.30	.45
Add per corner or partition intersection	—	Ea	2.00	—	2.00

3-5/8" wide studs

	Craft@Hrs	Unit	Material	Labor	Total
16" on center spacing (1,000 SF per day)	B1@.016	SF	.22	.37	.59
24" on center spacing (1,150 SF per day)	B1@.014	SF	.17	.32	.49
Add per corner or partition intersection	—	Ea	2.92	—	2.92

6" wide studs

	Craft@Hrs	Unit	Material	Labor	Total
16" on center spacing (890 SF per day)	B1@.018	SF	.31	.41	.72
24" on center spacing (1,000 SF per day)	B1@.016	SF	.24	.37	.61
Add per corner or partition intersection	—	Ea	4.97	—	4.97

Load bearing steel studding, 16 gauge

2-1/2" wide studs

	Craft@Hrs	Unit	Material	Labor	Total
16" on center spacing (700 SF per day)	B1@.023	SF	.29	.53	.82
24" on center spacing (850 SF per day)	B1@.019	SF	.23	.43	.66
Add per corner or partition intersection	—	Ea	3.54	—	3.54

3-5/8" wide studs

	Craft@Hrs	Unit	Material	Labor	Total
16" on center spacing (650 SF per day)	B1@.025	SF	.34	.57	.91
24" on center spacing (800 SF per day)	B1@.020	SF	.27	.46	.73
Add per corner or partition intersection	—	Ea	4.61	—	4.61

6" wide studs

	Craft@Hrs	Unit	Material	Labor	Total
16" on center spacing (600 SF per day)	B1@.027	SF	.44	.62	1.06
24" on center spacing (760 SF per day)	B1@.021	SF	.35	.48	.83
Add per corner or partition intersection	—	Ea	7.28	—	7.28

Door opening steel framing Using 16 gauge load bearing steel studs (hollow metal studs), based on walls 8' high. Figures in parentheses indicate typical built-up header size. Costs shown are per door opening and include header (built up using steel studs), double vertical steel studs each side of the opening less than 8' wide (triple vertical steel studs each side of openings 8' wide or wider), with steel stud cripples and blocking installed using screws. These costs include the studs, screws and normal waste. Width shown is size of finished opening.

3-5/8" wall studs, opening size as shown

	Craft@Hrs	Unit	Material	Labor	Total
Up to 3'0" wide (4" x 4" header)	B1@1.50	Ea	12.10	34.30	46.40
Up to 4'0" wide (4" x 6" header)	B1@1.60	Ea	13.80	36.60	50.40
Up to 5'0" wide (4" x 6" header)	B1@1.80	Ea	15.20	41.10	56.30
Up to 6'0" wide (4" x 8" header)	B1@2.00	Ea	18.10	45.70	63.80
Up to 8'0" wide (4" x 10" header)	B1@2.30	Ea	36.10	52.60	88.70
Up to 10'0" wide (4" x 12" header)	B1@2.40	Ea	46.70	54.80	101.50
Up to 12'0" wide (4" x 14" header)	B1@2.50	Ea	59.10	57.10	116.20
Add per foot of height, walls over 8' high	—	LF	.94	—	.94
Deduct for 2-1/2" wall studs	—	%	-15.0	-10.0	—
Add for 6" wall studs	—	%	30.0	10.0	—

	Craft@Hrs	Unit	Material	Labor	Total

Window opening framing Using 16 gauge load bearing steel studs (hollow metal studs), based on walls 8' high. Figures in parentheses indicate typical built-up header size. Costs shown are per window opening and include header (built up using steel studs), sub-sill plate (double sub-sill if opening is 8' wide or wider), double vertical steel studs each side of openings less than 8' wide (triple vertical steel studs each side of openings 8' wide or wider), with steel stud top and bottom cripples and blocking installed using screws. These costs include the studs, screws and normal waste.

3-5/8" wall studs, opening size shown is width of finished opening

	Craft@Hrs	Unit	Material	Labor	Total
Up to 2'0" wide (4" x 4" header)	B1@1.90	Ea	15.70	43.40	59.10
Up to 3'0" wide (4" x 4" header)	B1@2.20	Ea	19.70	50.30	70.00
Up to 4'0" wide (4" x 6" header)	B1@2.50	Ea	20.70	57.10	77.80
Up to 5'0" wide (4" x 6" header)	B1@2.70	Ea	24.30	61.70	86.00
Up to 6'0" wide (4" x 8" header)	B1@3.00	Ea	30.80	68.60	99.40
Up to 7'0" wide (4" x 8" header)	B1@3.30	Ea	34.60	75.40	110.00
Up to 8'0" wide (4" x 10" header)	B1@3.50	Ea	56.10	80.00	136.10
Up to 10'0" wide (4" x 12" header)	B1@3.80	Ea	67.70	86.80	154.50
Up to 12'0" wide (4" x 14" header)	B1@4.00	Ea	84.60	91.40	176.00
Add per foot of height, walls over 8' high	—	LF	1.52	—	1.52
Deduct for 2-1/2" wall studs	—	%	-15.0	-10.0	—
Add for 6" wall studs	—	%	30.0	10.0	—

Bracing for steel framing See also sheathing for plywood bracing and shear panels.

Let-in wall bracing, using 16 gauge load bearing steel studs installed using screws. These costs include screws and normal waste

	Craft@Hrs	Unit	Material	Labor	Total
2-1/2"	B1@.027	LF	.22	.62	.84
3-5/8"	B1@.035	LF	.24	.80	1.04
6"	B1@.035	LF	.36	.80	1.16
Steel strap bracing, 1-1/4" wide, 9'6" or 11'6" lengths	B1@.010	LF	.42	.23	.65
Steel "V" bracing, 3/4" x 3/4", 9'6" or 11'6" lengths	B1@.010	LF	.58	.23	.81

Temporary steel stud frame wall bracing, assumes salvage at 50% and 3 uses

	Craft@Hrs	Unit	Material	Labor	Total
2-1/2", 16 gauge load bearing steel studs	B1@.010	LF	.11	.23	.34
3-5/8", 16 gauge load bearing steel studs	B1@.012	LF	.12	.27	.39
6", 16 gauge load bearing steel studs	B1@.018	LF	.18	.41	.59

Tarpaulins

Water resistant, with grommets, cut sizes

	Unit	Vinyl-coated nylon	Cotton duck	Canvas	Poly-ethylene
6'6" x 8'6"	Ea	6.95	—	—	—
8' x 10'	Ea	8.00	—	31.00	—
10' x 12'	Ea	9.90	—	45.00	—
11'6" x 18'	Ea	16.95	—	—	—
16' x 20'	Ea	24.00	—	—	—
4'8" x 6'8" or 5'8" x 7'8"	Ea	—	35.00	23.00	10.00
7'6" x 9'6"	Ea	—	52.00	38.00	14.00
9'6" x 11'6"	Ea	—	68.00	57.00	18.90
9'6" x 15'6"	Ea	—	—	73.00	27.30
15'6" x 19'6"	Ea	—	—	—	51.00

Taxes

	Craft@Hrs	Unit	Material	Labor	Total

Taxes

Employer's cost for payroll taxes and insurance, expressed as a percent of total payroll.

Rule of thumb: When actual payroll taxes and insurance are not known, estimate that for each $1 paid in wages the employer must pay an additional $.30.

	Craft@Hrs	Unit	Material	Labor	Total
Payroll taxes and insurance	—	%	—	—	30.0

See also, Insurance for specific costs.

Specific taxes, percent of total payroll

	Craft@Hrs	Unit	Material	Labor	Total
Typical SUI, State unemployment insurance	—	%	—	—	3.70
FICA Social Security 6.20%, Medicare 1.45%	—	%	—	—	7.65
FUTA, Federal Unemployment Insurance	—	%	—	—	.80

Thresholds See the Mouldings section for oak threshold.

Aluminum with vinyl insert

1-1/8" high, low rug, 3-1/4" wide

	Craft@Hrs	Unit	Material	Labor	Total
36" long	BC@.515	Ea	14.50	13.00	27.50
48" long	BC@.515	Ea	19.70	13.00	32.70
60" long	BC@.730	Ea	25.40	18.40	43.80
72" long	BC@.730	Ea	29.00	18.40	47.40

1-1/8" high, 1-3/4" wide

	Craft@Hrs	Unit	Material	Labor	Total
30" long	BC@.515	Ea	16.50	13.00	29.50
48" long	BC@.515	Ea	23.30	13.00	36.30
60" long	BC@.730	Ea	29.00	18.40	47.40
72" long	BC@.730	Ea	33.20	18.40	51.60

Combination threshold, vinyl door shoe and drip cap

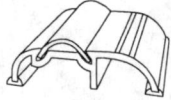

Low rug

	Craft@Hrs	Unit	Material	Labor	Total
32" or 36" long	BC@.730	Ea	14.90	18.40	33.30
42" long	BC@.730	Ea	19.60	18.40	38.00
48" long	BC@.730	Ea	22.20	18.40	40.60
60" long	BC@.941	Ea	28.20	23.70	51.90
72" long	BC@.941	Ea	34.00	23.70	57.70

High rug

	Craft@Hrs	Unit	Material	Labor	Total
32" long	BC@.730	Ea	18.60	18.40	37.00
36" long	BC@.730	Ea	20.70	18.40	39.10
42" long	BC@.730	Ea	24.90	18.40	43.30
48" long	BC@.730	Ea	28.70	18.40	47.10
60" long	BC@.941	Ea	36.10	23.70	59.80
72" long	BC@.941	Ea	42.90	23.70	66.60

Aluminum sill and interlocking threshold low rug

	Craft@Hrs	Unit	Material	Labor	Total
32" or 36"	BC@1.05	Ea	14.50	26.40	40.90
42"	BC@1.27	Ea	17.00	32.00	49.00
72"	BC@1.27	Ea	29.50	32.00	61.50

Door bottom caps, surface type, aluminum and vinyl

	Craft@Hrs	Unit	Material	Labor	Total
32" or 36"	BC@.515	Ea	6.21	13.00	19.21
42"	BC@.515	Ea	9.32	13.00	22.32
48"	BC@.515	Ea	9.32	13.00	22.32

Rain drip

	Craft@Hrs	Unit	Material	Labor	Total
32" or 36"	BC@.230	Ea	4.14	5.79	9.93
42"	BC@.230	Ea	5.69	5.79	11.48
48"	BC@.230	Ea	6.21	5.79	12.00

	Craft@Hrs	Unit	Material	Labor	Total

Tile Costs shown are typical costs for the tile only, including 10% allowance for breakage and cutting waste. Add costs for grout and labor to install tile from the Installation of Tile section below.

Glazed ceramic wall tile (5/16" thickness)

	Craft@Hrs	Unit	Material	Labor	Total
2" x 2"					
Minimum quality	—	SF	2.04	—	2.04
Good quality	—	SF	4.50	—	4.50
Better quality	—	SF	7.89	—	7.89
3" x 3"					
Minimum quality	—	SF	2.02	—	2.02
Good quality	—	SF	3.94	—	3.94
Better quality	—	SF	7.04	—	7.04
4-1/4" x 4-1/4"					
Minimum quality	—	SF	2.04	—	2.04
Good quality	—	SF	3.93	—	3.93
Better quality	—	SF	5.75	—	5.75
6" x 6"					
Minimum quality	—	SF	2.02	—	2.02
Good quality	—	SF	3.83	—	3.83
Better quality	—	SF	6.43	—	6.43
8" x 8"					
Minimum quality	—	SF	2.30	—	2.30
Good quality	—	SF	4.29	—	4.29
Better quality	—	SF	6.99	—	6.99
Ceramic mosaic tile (1/4" thickness)					
1" hexagonal					
Minimum quality	—	SF	2.02	—	2.02
Good quality	—	SF	2.74	—	2.74
Better quality	—	SF	4.69	—	4.69
2" hexagonal					
Minimum quality	—	SF	2.02	—	2.02
Good quality	—	SF	3.50	—	3.50
Better quality	—	SF	6.10	—	6.10
3/4" penny round					
Minimum quality	—	SF	2.53	—	2.53
Good quality	—	SF	2.61	—	2.61
Better quality	—	SF	4.08	—	4.08
7/8" x 7/8" square					
Minimum quality	—	SF	2.27	—	2.27
Good quality	—	SF	3.99	—	3.99
Better quality	—	SF	6.96	—	6.96
1-1/2" x 1-1/2" square					
Minimum quality	—	SF	2.02	—	2.02
Good quality	—	SF	3.39	—	3.39
Better quality	—	SF	5.92	—	5.92
2" x 2" square					
Minimum quality	—	SF	2.00	—	2.00
Good quality	—	SF	4.26	—	4.26
Better quality	—	SF	7.30	—	7.30
Quarry tile (1/2" or 3/4" thickness)					
6" x 6" or 8" x 8"					
Minimum quality	—	SF	2.12	—	2.12
Good quality	—	SF	4.14	—	4.14
Better quality	—	SF	8.69	—	8.69
Add for abrasive grain surfaces	—	SF	.66	—	.66

Tile

	Craft@Hrs	Unit	Material	Labor	Total
Paver tile (1" thickness)					
4" x 4" or 6" x 6"					
Minimum quality	—	SF	2.67	—	2.67
Good quality	—	SF	5.04	—	5.04
Better quality	—	SF	10.80	—	10.80
12" x 12"					
Minimum quality	—	SF	2.02	—	2.02
Good quality	—	SF	4.19	—	4.19
Better quality	—	SF	9.09	—	9.09
Mexican red pavers, 12" x 12"	—	SF	1.21	—	1.21
Saltillo paver, 12" x 12"	—	SF	1.86	—	1.86
Marble and granite tile, "Martile." Prices for marble and granite tile vary widely.					
Marble, 3/8" x 12" x 12", polished one face					
Low price	—	SF	6.05	—	6.05
Medium price	—	SF	9.25	—	9.25
High price	—	SF	18.10	—	18.10
Granite tile, natural, 3/8" x 12" x 12", polished on one face					
Low price	—	SF	6.05	—	6.05
Medium price	—	SF	11.10	—	11.10
High price	—	SF	20.20	—	20.20
Granite, 3/4", cut to size, imported	—	SF	25.20	—	25.20
Marble, 3/4", cut to size, imported	—	SF	18.10	—	18.10
Travertine beige marble tile,					
3/8" x 12" x 12", imported	—	SF	4.54	—	4.54
Carrera white marble tile,					
3/8" x 12" x 12", imported	—	SF	4.99	—	4.99

Installation of tile Using organic adhesive and grout. The costs below are for installation only. Cost of tile or surface preparation not included.

	Craft@Hrs	Unit	Material	Labor	Total
Countertops					
Ceramic mosaic	TL@.203	SF	1.00	4.98	5.98
4-1/4" x 4-1/4" to 6" x 6" glazed	TL@.180	SF	1.00	4.42	5.42
Floors					
Ceramic mosaic	TL@.121	SF	1.00	2.97	3.97
4-1/4" x 4-1/4" to 6" x 6" glazed	TL@.110	SF	1.00	2.70	3.70
Walls					
Ceramic mosaic	TL@.143	SF	1.00	3.51	4.51
4-1/4" x 4-1/4" to 6" x 6" glazed	TL@.131	SF	1.00	3.21	4.21

Installation of tile in a conventional mortar bed, with grout The costs below are for mortar bed, grout and installation only. Cost of tile or surface preparation not included, add for same.

	Craft@Hrs	Unit	Material	Labor	Total
Countertops					
Ceramic mosaic	TL@.407	SF	1.75	9.98	11.73
4-1/4" x 4-1/4" to 6" x 6" glazed	TL@.352	SF	1.85	8.63	10.48
Floors					
Ceramic mosaic	TL@.238	SF	1.95	5.84	7.79
4-1/4" x 4-1/4" to 6" x 6" glazed	TL@.210	SF	1.65	5.15	6.80
Quarry or paver	TL@.167	SF	1.65	4.10	5.75
Marble or granite, 3/8" thick	TL@.354	SF	2.30	8.68	10.98
Marble or granite, 3/4" thick	TL@.591	SF	2.30	14.50	16.80

	Craft@Hrs	Unit	Material	Labor	Total
Walls					
Ceramic mosaic	TL@.315	SF	1.60	7.73	9.33
4-1/4" x 4-1/4" to 6" x 6" glazed	TL@.270	SF	1.70	6.62	8.32
Quarter round trim	TL@.020	LF	—	.49	.49

Tile Backer Board (Durock™ or Wonderboard™). Water-resistant underlayment for ceramic tile on floors, walls, countertops and other interior wet areas. Material cost for 100 square feet (CSF) of board includes the backer board (at $100), 50 pounds of job mixed latex mortar for the joints and surface skim coat (at $20 per 50 pound sack), 75 linear feet of joint tape (at $3.00 per 75' roll), 1/2 pound of 1-1/2" galvanized roofing nails (at $0.70 per pound) and 10% for waste. For scheduling purposes, estimate that a crew of 2 can install, tape and apply the skim coat on the following quantity of backer board in an 8-hour day: countertops 180 SOF, floors 525 SF and walls 350 SF. Use $100.00 as a minimum cost for this type work.

1/2" thick backer board at $0.65 per SF plus mortar, tape and nails					
Floors	T1@.030	SF	1.36	.68	2.04
Walls	T1@.045	SF	1.36	1.01	2.37
Countertops	T1@.090	SF	1.36	2.03	3.39

Vacuum Cleaning Systems, Central, Subcontract Costs listed are for better quality central vacuum power unit, piping to 4 remote outlets, hose and attachments (adequate for typical 2,000 SF home). Assumes that all vacuum tubes are installed before any interior finishes are applied.

Basic package with 23' hose	—	LS	—	—	1,170.00
Additional remote outlets	—	Ea	—	—	102.00
Add for electric or turbo power head	—	Ea	—	—	157.00

Vanities, Good Quality No countertops, lavatory bowls or plumbing included. See also Countertops and Marble Setting.

Lavatory sink bases					
21" deep x 31-1/2" high					
24" wide, 2 doors	B1@1.00	Ea	155.00	22.90	177.90
30" wide, 2 doors	B1@1.37	Ea	175.00	31.30	206.30
36" wide, 2 doors	B1@1.49	Ea	196.00	34.00	230.00
42" wide, 3 drawers over 3 doors	B1@1.65	Ea	280.00	37.70	317.70
48" wide, 3 drawers over 3 doors	B1@1.65	Ea	295.00	37.70	332.70
18" deep x 31-1/2" high					
24" wide, 2 doors	B1@1.00	Ea	150.00	22.90	172.90
30" wide, 2 doors	B1@1.37	Ea	170.00	31.30	201.30
16" deep x 31-1/2" high x 18" wide, 1 door	B1@1.00	Ea	135.00	22.90	157.90
Drawer bases					
21" deep x 31-1/2" high					
12" wide, 3 drawers	B1@.703	Ea	140.00	16.10	156.10
15" wide, 3 drawers	B1@.703	Ea	145.00	16.10	161.10
18" wide, 3 drawers	B1@.845	Ea	150.00	19.30	169.30

Wallboard partitions, subcontract Costs for interior partitions in one and two story buildings. 1/4" gypsum wallboard on 2 sides with "V-grooved" edges and plastic moulding at corners fastened to 25 gauge metal framing at 24" OC, including framing, per LF of wall.

Vinyl covered 8' high	—	LF	—	—	41.50
Vinyl covered 10' high	—	LF	—	—	55.60
Paintable paper-covered 8' high	—	LF	—	—	30.20
Paintable paper-covered 10' high	—	LF	—	—	44.80
Add for glued vinyl base on two sides	—	LF	—	—	2.35

Wallcoverings

	Craft@Hrs	Unit	Material	Labor	Total

Wallcoverings Costs listed are for installation on a clean, smooth surface, and include adhesive, edge trimming, pattern matching, and normal amount of cutting and fitting and waste. Surface preparation such as patching and priming are extra. Typical roll is 8 yards long and 18" wide (approximately 36 square feet). Good to better material styles and quality.

	Craft@Hrs	Unit	Material	Labor	Total
Scrape off old wallpaper, up to 2 layers	PP@.015	SF	—	.35	.35
Patch holes, scrape marks	PP@.010	SF	.02	.23	.25
Blank stock (underliner)					
Bath, kitchen, and laundry room	PP@.668	Roll	9.54	15.50	25.04
Other rooms	PP@.558	Roll	9.54	13.00	22.54
Grasscloth or strings, vertical pattern					
Bath, kitchen, and laundry rooms	PP@.995	Roll	53.00	23.10	76.10
Other rooms	PP@.830	Roll	53.00	19.30	72.30
Papers, vinyl-coated papers, trimmed, and pre-pasted					
Bath, kitchen, and laundry rooms	PP@.884	Roll	22.30	20.60	42.90
Other rooms	PP@.780	Roll	22.30	18.10	40.40
Foil with or without pattern					
Bath, kitchen, and laundry rooms	PP@1.12	Roll	30.70	26.00	56.70
Other rooms	PP@.844	Roll	30.70	19.60	50.30
Designer paper and hand-painted prints					
Bath, kitchen, and laundry rooms	PP@1.68	Roll	74.20	39.10	113.30
Other rooms	PP@1.41	Roll	74.20	32.80	107.00
Vinyls, patterned					
Bath, kitchen, and laundry rooms	PP@.844	Roll	31.80	19.60	51.40
Other rooms	PP@.730	Roll	31.80	17.00	48.80
Vinyls, textured, no pattern					
Double cut seams, cloth backing	PP@.844	Roll	37.10	19.60	56.70
Coordinating borders, vinyl coated					
roll is 4" to 6" x 12' to 15'	PP@.017	LF	.68	.40	1.08
Wallpaper adhesive and sizing					
Adhesive (300 SF and $9.00 per Gallon)	—	SF	.03	—	.03
Sizing (800 SF and $16.00 per Gallon)	—	SF	.02	—	.02

Waterproofing

Below grade waterproofing, applied to concrete or masonry walls, no surface preparation, excavation, or backfill included, typical prices, average job

	Craft@Hrs	Unit	Material	Labor	Total
Bentonite waterproofing					
One layer, nailed in place	BL@.079	SF	.55	1.62	2.17
Cementitious waterproofing with protection board					
Spray or brush applied, 2 coats	BL@.077	SF	1.05	1.58	2.63
Crystalline waterproofing					
Spray or brush applied, 3 coats	BL@.110	SF	1.15	2.26	3.41
Elastomeric waterproofing 30 mm per coat					
Sprayed, troweled, or rolled, 2 coats,	BL@.070	SF	1.05	1.44	2.49

Above grade waterproofing, applied to smooth concrete or plywood deck, no surface preparation included, typical prices, average job

	Craft@Hrs	Unit	Material	Labor	Total
Between slabs membrane					
Sprayed, troweled or rolled, 2 coats	PT@.047	SF	.85	1.22	2.07
Pedestrian walking surface, elastomeric membrane,					
Applied 4 coats thick with aggregate	PT@.086	SF	1.20	2.23	3.43
Chloroprene/Neoprene latex deck					
Surfacing	PT@.144	SF	1.55	3.74	5.29

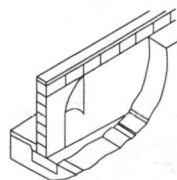

	Craft@Hrs	Unit	Material	Labor	Total
Weatherstripping Materials					
Door bottom seals					
Spring brass in pinch grip carrier (KEL-EEZ)					
32" x 1-3/4" brass	BC@.250	Ea	6.95	6.29	13.24
36" x 1-3/4"	BC@.250	Ea	7.90	6.29	14.19
Brush seal door sweep gold anodized frame					
with screws	BC@.250	Ea	6.60	6.29	12.89
Door frame seal sets					
Spring bronze, nailed every 1 inch					
36" x 6'8" door	BC@.500	Ea	9.90	12.60	22.50
36" x 7'0" door	BC@.500	Ea	10.30	12.60	22.90
Rigid PVC frame	BC@.008	Ea	8.00	.20	8.20
Header and jamb side stripping, adhesive					
36" x 6'8" door, brass stripping	BC@.500	Ea	17.50	12.60	30.10
Window seal sets, spring brass, nailed every 1 inch, linear feet per side					
Double hung window, head and sill	BC@.008	LF	.70	.20	.90
Add to remove & replace window sash	BC@.250	Ea	—	6.29	6.29
Casement window, head and sill	BC@.008	LF	.65	.20	.85
Gray or black pile gasket	BC@.008	LF	.51	.20	.71
Gray of black pile gasket with adhesive	BC@.006	LF	.71	.15	.86
Garage door top and side seal					
Kry-O-Gem fiber gasket in aluminum or vinyl frame (KEL-EEZ)					
9' x 7' door	BC@.500	Ea	38.50	12.60	51.10
16' x 7' door	BC@.600	Ea	50.00	15.10	65.10
Flexible PVC	BC@.050	LF	1.06	1.26	2.32
Garage door bottom seal					
Soft rubber, set with adhesive					
9', neoprene or EPDM	BC@.383	Ea	16.00	9.64	25.64
16', neoprene or EPDM	BC@.500	Ea	28.00	12.60	40.60
Flexible PVC	BC@.250	LF	.90	6.29	7.19
EPDM Rubber	BC@.250	LF	1.00	6.29	7.29

Well Drilling, Subcontract Typical costs, based on well depth including welded steel casing (or continuous PVC casing). The amount of casing needed depends on soil conditions. Some types of rock require casings at the well surface only. Your local health department can probably provide information on the average well depth and subsurface conditions in the community. Note that unforeseen subsurface conditions can increase drilling costs substantially. Most subcontractors will estimate costs on an hourly basis.

	Craft@Hrs	Unit	Material	Labor	Total
Well hole with 4" ID PVC casing	—	LF	—	—	22.40
Well hole with 6" ID steel casing	—	LF	—	—	40.00
Well hole with 8" ID steel casing	—	LF	—	—	56.00
6" well hole in sturdy rock (no casing)	—	LF	—	—	13.30
Add per well for 6" diameter bottom filter screen if needed due to site conditions					
Stainless steel	—	LF	—	—	105.00
Low carbon steel	—	LF	—	—	62.20
Add per well for drive shoe	—	LS	—	—	83.60
Add for surface seal (20' to 50' depth) if needed due to local code or site conditions					
Per well	—	LS	—	—	620.00
Add per well for well drilling permit					
Typical fee in most municipalities	—	LS	—	—	306.00

	Craft@Hrs	Unit	Material	Labor	Total

Add for automatic electric pumping system (domestic use), no electrical work included

| 87 gallon pressure tank, typical cost | — | LS | — | — | 3,670.00 |

Add for electrical service drop pole

| Typical cost | — | LS | — | — | 970.00 |

Window Sills Cultured marble, 1/2" radius front edge.

	Craft@Hrs	Unit	Material	Labor	Total
3-1/2" sill depth, 2" x 4" stud walls	BS@.118	LF	3.25	2.62	5.87
4-7/8" sill depth, 2" x 6" stud walls	BS@.131	LF	3.60	2.91	6.51
5-1/2" sill depth	BS@.147	LF	4.35	3.26	7.61

Window Treatments Blinds and shades, stock sizes except as noted. See also Draperies.

Mini-blinds

	Craft@Hrs	Unit	Material	Labor	Total
Aluminum, 1" standard colors	BC@.050	SF	4.45	1.26	5.71
Custom aluminum sizes and colors	BC@.050	SF	6.46	1.26	7.72
Fabric, vertical, 3-1/2" wide, fire retarding					
Custom sizes and colors	BC@.083	SF	5.84	2.09	7.93
PVC mini-blinds, vertical, 3-1/2" wide, fire retarding					
Custom sizes and colors	BC@.075	SF	4.12	1.89	6.01
Polyester fabric blinds, continuous pleated	BC@.066	SF	4.35	1.66	6.01
Wood blinds, 1" prefinished	BC@.075	SF	8.56	1.89	10.45
Vertical blinds, 3-1/2" vanes	BC@.083	SF	6.68	2.09	8.77
Steel blinds					
1" enamel finish	BC@.050	SF	4.79	1.26	6.05
2" white enamel only	BC@.045	SF	3.34	1.13	4.47
Traverse rods for draperies					
Steel	BC@.075	LF	5.68	1.89	7.57
Extruded aluminum, hand operated	BC@.083	LF	8.90	2.09	10.99
Extruded aluminum, cord operated	BC@.083	LF	10.00	2.09	12.09

Windows See also Skylights and Sky Windows

Awning windows, aluminum Dual glazed , with hardware, no trim or surround needed, weatherstripped, natural finish, with fiberglass screens

		Width		
Height	**19"**	**26"**	**37"**	**53"**
26", 2 lites high	50.40	62.70	72.20	86.50
38-1/2", 3 lites high	60.80	73.20	83.60	97.90
50-1/2", 4 lites high	72.20	95.00	96.90	126.40
63", 5 lites high	85.50	111.00	130.00	150.00

	Craft@Hrs	Unit	Material	Labor	Total
Add for storm sash, per square foot of glass	—	SF	6.30	—	6.30
Deduct for single glazing, per SF of glass	—	SF	-3.60	—	-3.60
Add for bronze finish	—	%	15.0	—	—
Add for enamel finish	—	%	10.0	—	—

Labor setting aluminum awning windows, no carpentry included

To 8 SF opening size	BG@1.00	Ea	—	23.10	23.10
Over 8 SF to 16 SF opening size	BG@1.50	Ea	—	34.60	34.60
Over 16 SF to 32 SF opening size	BG@2.00	Ea	—	46.10	46.10
Over 32 SF opening size	BG@3.00	Ea	—	69.20	69.20

Double hung aluminum windows Dual glazed, bronze or white finish, vertical slide with screen.

Height	Width 2'0"	2'6"	3'0"	3'6"	4'0"
Double hung, two lites					
2'0" or 2'6"	105.00	—	108.00	—	—
3'0"	113.00	125.00	138.00	—	164.00
3'6"	121.00	133.00	146.00	159.00	175.00
4'0"	129.00	144.00	157.00	172.00	188.00
4'6"	138.00	152.00	167.00	182.00	198.00
5'0"	146.00	162.00	175.00	193.00	210.00
6'0"	162.00	195.00	195.00	—	—

	Craft@Hrs	Unit	Material	Labor	Total
Deduct for single glazing, per SF of glass	—	SF	-3.65	—	-3.65
Add for storm sash, per square foot of glass	—	SF	6.30	—	6.30
Add for bronze finish	—	%	15.0	—	—
Add for enamel finish	—	%	10.0	—	—
Labor setting aluminum double hung windows, no carpentry included					
To 8 SF opening size	BG@1.00	Ea	—	23.10	23.10
Over 8 SF to 16 SF opening size	BG@1.50	Ea	—	34.60	34.60
Over 16 SF to 32 SF opening size	BG@2.00	Ea	—	46.10	46.10
Over 32 SF opening size	BG@3.00	Ea	—	69.20	69.20

Fixed aluminum windows

Single glazed, including bronze or white finish and hardware.

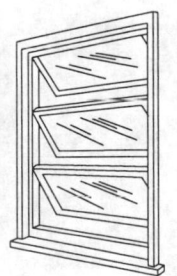

Height	Width 1'4"	1'8"	2'2"	2'6"	3'0"
1'4" high	29.70	32.90	37.10	41.30	45.60
1'8" high	32.90	36.00	41.30	46.10	50.90
2'2" high	36.00	40.30	45.60	47.50	55.10
2'6" high	39.20	45.00	49.80	55.10	60.40
3'0" high	44.00	48.80	54.60	60.40	65.70

Height	Width 1'4"	1'8"	2'2"	2'6"	3'0"
3'4" high	47.00	52.20	59.10	64.40	73.90
3'10" high	—	58.00	63.80	73.90	—
4'2" high	55.00	61.20	68.60	80.70	84.40
5'0" high	65.40	72.80	80.70	89.70	96.00
5'10" high	—	80.70	90.70	98.10	—

Height	Width 4'2"	5'0"	5'10"	6'0"
1'4" high	49.60	—	—	—
1'8" high	54.90	64.90	78.10	86.50
2'2" high	60.10	74.40	91.80	103.00
2'6" high	65.40	80.70	—	—
3'0" high	74.90	—	—	—
3'4" high	82.30	93.90	—	—
4'2" high	92.80	—	—	—

Windows, Aluminum

Dual glazed fixed aluminum windows, including bronze or white enamel finish and hardware.

Height	Width 2'0"	2'6"	3'0"	3'6"	4'0"
1'8" high	68.60	83.30	99.20	119.21	141.37
2'6" high	80.20	98.10	119.00	147.70	168.80
3'4" high	95.00	116.00	149.00	173.02	194.12
4'2" high	—	141.00	127.00	—	—
5'0" high	—	162.00	209.00	—	—

	Craft@Hrs	Unit	Material	Labor	Total
Labor setting aluminum fixed windows, no carpentry included					
To 8 SF opening size	BG@1.00	Ea	—	23.10	23.10
Over 8 SF to 16 SF opening size	BG@1.50	Ea	—	34.60	34.60
Over 16 SF to 32 SF opening size	BG@2.00	Ea	—	46.10	46.10
Over 32 SF opening size	BG@3.00	Ea	—	69.20	69.20
Add for storm sash, per square foot of glass	—	SF	6.00	—	6.00

Louvered aluminum windows Single glazed, 4" clear glass louvers, including hardware.

Height	Width 1'6"	2'0"	2'6"	3'0"	3'6"
1'0" high	19.50	22.20	25.30	27.40	30.60
1'4" high	26.40	28.50	32.70	34.80	39.00
1'6", or 1'8" high	28.50	32.70	38.00	41.70	47.00
1'10" high	33.80	38.00	45.90	50.10	55.90
2'0" high	38.00	42.70	51.20	55.40	64.40
2'4", or 2'6" high	41.10	48.00	56.40	63.30	70.70
2'8", or 2'10" high	51.70	57.00	67.50	73.90	84.40
3'0" high	54.90	62.20	72.80	80.70	91.80
3'4", or 3'6" high	59.10	67.50	80.70	88.60	100.20
4'0" high	67.50	78.60	92.30	101.00	116.00
4'6" high	79.10	89.70	105.50	116.00	132.00
5'0" high	88.60	101.30	119.20	132.00	148.00
5'6" high	93.90	107.00	128.70	141.00	158.00
6'0" high	104.00	119.00	141.40	155.00	174.00

	Craft@Hrs	Unit	Material	Labor	Total
Labor setting aluminum louvered windows, no carpentry included					
To 8 SF opening size	BG@1.00	Ea	—	23.10	23.10
Over 8 SF to 32 SF opening size	BG@1.50	Ea	—	34.60	34.60
Over 32 SF opening size	BG@2.00	Ea	—	46.10	46.10

Single hung aluminum windows Dual glazed, bronze or white finish, vertical slide with screen.

Height	Width 2'0"	2'6"	3'0"	3'6"	4'0"
Single hung, two lites					
2'0"	55.40	—	67.50	—	—
2'6"	59.60	—	—	—	—
3'0"	63.80	71.70	79.10	—	93.90
3'6"	68.60	76.00	83.30	91.30	101.00
4'0"	72.80	81.20	89.20	97.10	104.00
4'6"	77.00	86.00	93.90	102.30	103.00
5'0"	82.30	91.30	100.20	107.60	119.00
6'0"	92.80	101.30	109.00	—	—

	Craft@Hrs	Unit	Material	Labor	Total
Additional costs for aluminum windows					
Add for storm sash, per square foot of glass	—	SF	6.30	—	6.30
Deduct for single glazing, per SF of glass	—	SF	-3.60	—	-3.60
Labor setting single hung aluminum windows, no carpentry included					
To 8 SF opening size	BG@1.00	Ea	—	23.10	23.10
Over 8 SF to 16 SF opening size	BG@1.50	Ea	—	34.60	34.60
Over 16 SF to 32 SF opening size	BG@2.00	Ea	—	46.10	46.10
Over 32 SF opening size	BG@3.00	Ea	—	69.20	69.20

Sliding aluminum windows Dual glazed, including bronze or white finish and hardware. One horizontal rolling sash and one stationary sash. See installation cost below.

	Width				
Height	3'0"	3'4"	4'0"	5'0"	6'0"
1'0"	45.40	—	52.80	—	—
1'6"	51.70	—	58.00	—	—
2'0"	57.00	60.70	65.90	74.90	83.90
2'6"	65.40	67.50	73.90	82.80	92.80
3'0"	70.70	74.90	81.80	91.30	101.00
3'6"	78.10	82.30	89.20	100.00	111.00
4'0"	84.90	89.20	95.00	108.00	119.00
5'0"	106.00	—	120.00	142.00	156.00

Two horizontal rolling sashes and one stationary center sash, dual glazed

	Width				
Height	6'0"	7'0"	8'0"	9'0"	10'0"
2'0"	102.00	111.00	119.00	—	196.00
2'6"	115.00	—	133.00	—	—
3'0"	123.00	136.00	147.00	165.00	173.00
3'6"	—	154.00	168.00	188.00	197.00
4'0"	156.00	168.00	179.00	202.00	209.00
5'0"	196.00	210.00	223.00	—	280.00

	Craft@Hrs	Unit	Material	Labor	Total
Labor setting sliding aluminum windows, no carpentry included.					
To 8 SF opening size	BG@1.00	Ea	—	23.10	23.10
Over 8 SF to 16 SF opening size	BG@1.50	Ea	—	34.60	34.60
Over 16 SF to 32 SF opening size	BG@2.00	Ea	—	46.10	46.10
Over 32 SF opening size	BG@3.00	Ea	—	69.20	69.20

Windows, Wood

Awning windows, pine Treated and primed, includes weatherstripping and exterior trim, glazed with 3/4" insulating glass, no drip cap, 1-3/4" sash, frame for 4-9/16" wall, each unit has 1 ventilating lite for each lite of width, overall dimension.

Height	One 48" lite wide	Two 48" lites wide	Three 48" lites wide
20" or 24", 1 lite high	225.00	455.00	669.00
40", 2 lites high	401.00	803.00	1,220.00
60", 3 lites high	589.00	1,120.00	1,660.00
48", 2 lites high	417.00	821.00	1,180.00
72", 3 lites high	605.00	1,180.00	1,710.00
30", 1 lite high	262.00	509.00	770.00
60", 2 lites high	471.00	920.00	1,340.00
90", 3 lites high	679.00	1,290.00	1,930.00

Windows, Wood

	Craft@Hrs	Unit	Material	Labor	Total
Labor installing wood awning windows, no carpentry included					
To 8 SF opening size	BG@1.00	Ea	—	23.10	23.10
Over 8 SF to 16 SF opening size	BG@1.50	Ea	—	34.60	34.60
Over 16 SF to 32 SF opening size	BG@2.00	Ea	—	46.10	46.10
Over 32 SF to 48 SF opening size	BG@3.00	Ea	—	69.20	69.20
Over 48 SF opening size	BG@5.00	Ea	—	115.00	115.00
Modifiers for wood awning windows with single glazing.					
Deduct for DSB glass, per SF of glass	—	SF	-5.00	—	-5.00
Add for storm panels, per SF of glass	—	SF	8.00		8.00
Add for operating sash	—	Ea	40.00	—	40.00

Bay window Angle bay double hung units, 30 degree angle, pine, assembled, aluminum cladding, hardware, 2 ventilating sash, weatherstripped, double hung sides with picture center sash, insulated glass, screens, white or bronze finish, 4-9/16" wall thickness, overall unit dimensions, 4'8" high. Labor is for installing the window, no carpentry included.

	Craft@Hrs	Unit	Material	Labor	Total
6'4"	BG@4.00	Ea	979.00	92.20	1,071.20
7'2"	BG@4.00	Ea	985.00	92.20	1,077.20
7'10"	BG@4.50	Ea	1,070.00	104.00	1,174.00
8'6"	BG@4.50	Ea	1,220.00	104.00	1,324.00
8'10"	BG@4.50	Ea	1,250.00	104.00	1,354.00
9'2"	BG@5.00	Ea	1,270.00	115.00	1,385.00
9'10"	BG@5.00	Ea	1,140.00	115.00	1,255.00
Add for 45 degree angle windows	—	Ea	50.00	—	50.00
3, 4 or 5 lites	BG@2.00	Set	24.00	46.10	70.10
Add for roof framing kit, 5 lites	—	Ea	100.00	—	100.00

Bow windows Casement bow window unit, glazed with 1/2" insulated glass, assembled, screens, pine, fixed sash, 22" x 56" lites, weatherstripped, unfinished, 16" projection, 4-7/16" jambs, 19" glass, listed by actual unit dimensions. Labor is for installing the window, no carpentry included.

	Craft@Hrs	Unit	Material	Labor	Total
8'1" wide x 4'8" (4 lites, 2 venting)	BG@4.75	Ea	1,100.00	110.00	1,210.00
8'1" wide x 5'4" (4 lites, 2 venting)	BG@4.75	Ea	1,190.00	110.00	1,300.00
8'1" wide x 6'0" (4 lites, 2 venting)	BG@4.75	Ea	1,290.00	110.00	1,400.00
10' wide x 4'8" (5 lites, 2 venting)	BG@4.75	Ea	1,340.00	110.00	1,450.00
10' wide x 5'4" (5 lites, 2 venting)	BG@4.75	Ea	1,460.00	110.00	1,570.00
10' wide x 6'0" (5 lites, 2 venting)	BG@4.75	Ea	1,590.00	110.00	1,700.00
Bow windows, add for interior and exterior trim					
3, 4 or 5 lites	BG@2.00	Set	26.40	46.10	72.50
Add for roof framing kit, 5 lites	—	Ea	100.00	—	100.00

Casement windows, wood Dual glazed 3/4" glass, with hardware, primed wood exterior frames and sash, screens, weatherstripped, with drip cap.

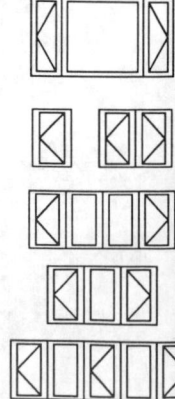

		Height			
Width of glass	**2'8"**	**3'4"**	**4'0"**	**5'0"**	**6'0"**
20" (1 ventilating lite)	—	194.00	222.00	260.00	296.00
40" (2 ventilating lites)	—	372.00	442.00	523.00	594.00
60" (2 ventilating, 1 fixed)	—	534.00	626.00	723.00	852.00
80" (2 ventilating, 2 fixed)	—	674.0	809.00	934.00	1,100.00
100" (3 ventilating, 2 fixed)	—	826.00	993.00	1,130.00	1,350.00
24" (1 ventilating lite)	194.00	205.00	222.00	260.00	302.00
48" (2 ventilating lites)	392.00	421.00	459.00	518.00	615.00
72" (2 ventilating, 1 fixed)	540.00	594.00	648.00	734.00	880.00
96" (2 ventilating, 2 fixed)	691.00	755.00	831.00	960.00	1,130.00
120" (3 ventilating, 2 fixed)	842.00	928.00	1,020.00	1,180.00	1,400.00
30" (1 ventilating lite)	—	237.00	275.00	313.00	345.00

Width of glass	2'8"	3'4"	Height 4'0"	5'0"	6'0"
60" (2 ventilating lites)	—	454.00	508.00	605.00	1,190.00
90" (2 ventilating, 1 fixed)	—	643.00	723.00	852.00	880.00
120" (2 ventilating, 2 fixed)	—	826.00	939.00	1,100.00	1,190.00
150" (3 ventilating, 2 fixed)	—	1,010.00	1,150.00	1,350.00	1,460.00
Add for storm panels, per lite	8.00	8.00	8.00	8.00	8.00
Deduct for single sheet glazing, SF of glass	-2.50	-2.50	-2.50	-2.50	-2.50

	Craft@Hrs	Unit	Material	Labor	Total
Labor setting wood casement windows, no carpentry included.					
To 8 SF	BG@1.00	Ea	—	23.10	23.10
Over 8 SF to 16 SF	BG@1.50	Ea	—	34.60	34.60
Over 16 SF	BG@2.00	Ea	—	46.10	46.10

Double hung wood windows With primed exterior aluminum cladding, screens, white or bronze finish, 5/8" insulating glass, weatherstripped, two lites per sash, listed by overall size (width x length). Labor is for installing the window, no carpentry included.

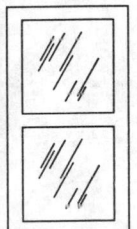

	Craft@Hrs	Unit	Material	Labor	Total
1'10" x 4'8" (16" x 24" lite)	BG@1.00	Ea	181.00	23.10	204.10
2'2" x 3'4" (20" x 16" lite)	BG@1.00	Ea	172.00	23.10	195.10
2'2" x 4'4" (20" x 24" lite)	BG@1.00	Ea	202.00	23.10	225.10
2'2" x 5'4" (20" x 28" lite)	BG@1.50	Ea	226.00	34.60	260.60
2'2" x 6'4" (20" x 34" lite)	BG@1.50	Ea	248.00	34.60	282.60
2'6" x 3'4" (24" x 16" lite)	BG@1.00	Ea	183.00	23.10	206.10
2'6" x 4'0" (24" x 20" lite)	BG@1.50	Ea	202.00	34.60	236.60
2'6" x 4'8" (24" x 24" lite)	BG@1.50	Ea	207.00	34.60	241.60
2'10" x 3'4" (28" x 16" lite)	BG@1.00	Ea	198.00	23.10	221.10
2'10" x 4'0" (28" x 20" lite)	BG@1.50	Ea	213.00	34.60	247.60
2'10" x 4'8" (28" x 24" lite)	BG@1.50	Ea	237.00	34.60	271.60
2'10" x 5'4" (28" x 28" lite)	BG@1.50	Ea	260.00	34.60	294.60
3'7" x 3'4" (32" x 16" lite)	BG@1.50	Ea	339.00	34.60	373.60
3'7" x 4'0" (32" x 20" lite)	BG@1.50	Ea	355.00	34.60	389.60
3'7" x 4'8" (32" x 24" lite)	BG@1.50	Ea	407.00	34.60	441.60
3'7" x 5'4" (32" x 28" lite)	BG@1.50	Ea	429.00	34.60	463.60
4'3" x 4'0" (36" x 20" lite)	BG@1.50	Ea	395.00	34.60	429.60
4'3" x 4'8" (36" x 24" lite)	BG@2.00	Ea	422.00	46.10	468.10
4'3" x 5'4" (36" x 28" lite)	BG@2.25	Ea	463.00	51.90	514.90
Add for interior trim set, per SF of opening	—	SF	1.25	—	1.25
Install interior trim set, any size opening listed	BG@1.00	Ea	—	23.10	23.10
Add for removable grille, per SF of opening	—	SF	3.00	—	3.00
Deduct for single sheet glazing, SF of glass	—	SF	-2.70	—	-2.70
Add for white aluminum screen with fiberglass screen cloth					
Per SF of glass	—	SF	2.00	—	2.00

Picture windows, wood Pine, listed by overall rough opening sizes. Labor is for installing the window, no carpentry included.

Picture windows, with two double hung flankers, includes hardware and screens primed, weatherstripped. Glazed with 1/2" insulated glass

	Craft@Hrs	Unit	Material	Labor	Total
8'1" x 4'9"	BG@2.75	Ea	746.00	63.40	809.40
8'9" x 4'9"	BG@2.75	Ea	778.00	63.40	841.40
9'5" x 4'9"	BG@2.75	Ea	799.00	63.40	862.40

Windows, Wood

Picture windows, with two casement flankers, weatherstripped, with hardware, casing, screens, 15" glass, and moulding, unit dimension sizes. Glazed with 3/4" insulated center sash and 1/2" insulated casements

Height	Width				
	7'4"	8'0"	8'4"	8'8"	9'4"
3'4"	635.00	668.00	678.00	690.00	745.00
4'0"	745.00	766.00	789.00	810.00	865.00
5'0"	854.00	931.00	974.00	1,010.00	1,070.00
6'0"	1,020.00	1,090.00	1,140.00	1,180.00	1,230.00

	Craft@Hrs	Unit	Material	Labor	Total
Labor to hang and fit wood picture casement windows, no carpentry included					
To 8 SF opening size	BG@1.00	Ea	—	23.10	23.10
Over 8 SF to 16 SF opening size	BG@1.50	Ea	—	34.60	34.60
Over 16 SF to 32 SF opening size	BG@2.00	Ea	—	46.10	46.10
Over 32 SF opening size	BG@2.50	Ea	—	57.70	57.70

Windows, Vinyl

Awning windows, vinyl Dual glazed, white frame, clear glass. Glass is 1" dual overall.

Height	Width				
	1'8"	2'2"	2'6"	3'4"	4'2"
1'4"	277.00	288.00	299.00	321.00	343.00
1'8"	288.00	305.00	315.00	343.00	371.00
2'2"	305.00	315.00	254.00	403.00	393.00
2'6"	300.00	332.00	349.00	382.00	426.00

	Craft@Hrs	Unit	Material	Labor	Total
Labor setting vinyl awning windows, no carpentry included					
To 8 SF opening size	BG@1.00	Ea	—	23.10	23.10
Over 8 SF to 16 SF opening size	BG@1.50	Ea	—	34.60	34.60

Bay windows, vinyl Dual glazed, white frames, clear glass.

Height	Width		
	5'0"	5'10"	6'8"
2'6"	607.00	632.00	654.00
3'0"	654.00	682.00	699.00
3'4"	709.00	731.00	748.00
4'2"	793.00	827.00	916.00

	Craft@Hrs	Unit	Material	Labor	Total
Labor setting vinyl bay windows, no carpentry included					
To 8 SF opening size	BG@4.00	Ea	—	92.20	92.20
Over 8 SF to 16 SF opening size	BG@4.00	Ea	—	92.20	92.20
Over 16 SF to 32 SF opening size	BG@4.50	Ea	—	104.00	104.00

Casement windows, vinyl Dual glazed, white frame, clear glass. Glass is 1" dual overall.

	Width		
Height	**1'8"**	**2'2"**	**2'6"**
1'4"	262.00	279.00	289.00
1'8"	284.00	294.00	311.00
2'2"	294.00	311.00	328.00
2'6"	311.00	328.00	345.00
3'0"	323.00	349.00	367.00
3'4"	345.00	367.00	384.00
3'10"	350.00	384.00	406.00
4'2"	372.00	395.00	428.00

	Craft@Hrs	Unit	Material	Labor	Total
Labor setting vinyl casement windows, no carpentry included					
To 8 SF	BG@1.00	Ea	—	23.10	23.10
Over 8 SF to 16 SF	BG@1.50	Ea	—	34.60	34.60
Add for storm panels, per lite	—	SF	15.00	—	15.00
Deduct for single sheet glazing, SF of glass	—	SF	-2.75	—	-2.75

Full circle windows, vinyl Dual glazed, white frame. Minimum diameter for a full circle is 2'0". Maximum diameter for a full circle is 60". If for a larger full circle, stack two half circles.

	Craft@Hrs	Unit	Material	Labor	Total
20" x 20"	BG@2.75	Ea	405.00	63.40	468.40
30" x 30"	BG@2.75	Ea	405.00	63.40	468.40
40" x 40"	BG@2.75	Ea	542.00	63.40	605.40
50" x 50"	BG@2.75	Ea	696.00	63.40	759.40

Half circle windows, vinyl Dual glazed, white frame. Standard fixed frame and bead are used for all circles and rakes.

	Craft@Hrs	Unit	Material	Labor	Total
20" x 10"	BG@2.75	Ea	289.00	63.40	352.40
30" x 16"	BG@2.75	Ea	289.00	63.40	352.40
40" x 20"	BG@2.75	Ea	384.00	63.40	447.40
50" x 26"	BG@2.75	Ea	454.00	63.40	517.40
60" x 30"	BG@2.75	Ea	516.00	63.40	579.40
80" x 40"	BG@2.75	Ea	747.00	63.40	810.40

Fixed windows, vinyl Dual glazed, white frame. Glass is 1" dual overall. See below for installation labor.

	Width			
Height	**1'8"**	**2'2"**	**2'6"**	**3'4"**
0'10", 1'4" or 1'8"	95.00	102.00	109.00	122.00
2'6" or 3'0"	117.00	128.00	139.00	156.00
3'4"	122.00	133.00	150.00	172.00
3'10"	128.00	145.00	156.00	178.00
4'2"	139.00	161.00	167.00	194.00
5'0"	150.00	167.00	184.00	272.00

	Width			
Height	**4'2"**	**5'0"**	**6'8"**	**8'4"**	**10'0"**
0'10", 1'4", or 1'8"	133.00	156.00	217.00	262.00	344.00
2'2"	156.00	167.00	245.00	289.00	—
2'6"	167.00	183.00	279.00	328.00	—
3'0"	178.00	199.00	306.00	364.00	—
3'4"	194.00	272.00	333.00	400.00	—
3'10"	223.00	294.00	378.00	—	—
4'2"	250.00	317.00	483.00	—	—
5'0"	323.00	362.00	—	—	—

Windows, Wood

	Craft@Hrs	Unit	Material	Labor	Total
Labor setting fixed vinyl windows, no carpentry included					
To 8 SF opening size	BG@1.00	Ea	—	23.10	23.10
Over 8 SF to 16 SF opening size	BG@1.50	Ea	—	34.60	34.60
Over 16 SF to 32 SF opening size	BG@2.00	Ea	—	46.10	46.10
Over 32 SF opening size	BG@3.00	Ea	—	69.20	69.20

Single sliding/rolling windows, vinyl Dual glazed, white frame, clear glass. Add labor from below.

	Width			
Height	**1'8"**	**2'2"**	**2'6"**	**3'4"**
0'10", 1'4" or 1'8"	145.00	156.00	161.00	178.00
2'2"	156.00	172.00	172.00	194.00
2'6"	167.00	178.00	189.00	211.00
3'0"	172.00	178.00	200.00	223.00
3'4"	184.00	184.00	211.00	239.00
3'10"	—	—	228.00	250.00
4'2"	—	—	239.00	267.00

	Width		
	4'2"	**5'0"**	**6'8"**
0'10", 1'4", or 1'8"	194.00	211.00	250.00
2'2"	211.00	233.00	272.00
2'6"	233.00	250.00	301.00
3'0"	250.00	279.00	328.00
3'4"	267.00	294.00	356.00
3'10"	283.00	323.00	384.00
4'2"	311.00	345.00	411.00

	Craft@Hrs	Unit	Material	Labor	Total
Labor setting vinyl single sliding/rolling windows, no carpentry included					
To 8 SF opening size	BG@1.00	Ea	—	23.10	23.10
Over 8 SF to 16 SF opening size	BG@1.50	Ea	—	34.60	34.60
Over 16 SF to 32 SF opening size	BG@2.00	Ea	—	46.10	46.10
Over 32 SF opening size	BG@3.00	Ea	—	69.20	69.20

Dual rolling windows, vinyl Dual glazed, white frame, clear glass. Add labor from below

	Width		
Height	**5'0"**	**6'8"**	**8'4"**
1'8"	294.00	333.00	372.00
2'2"	323.00	362.00	406.00
2'6"	345.00	389.00	445.00
3'0"	367.00	428.00	483.00
3'4"	395.00	473.00	517.00
3'10"	428.00	490.00	551.00
4'2"	445.00	517.00	700.00

	Craft@Hrs	Unit	Material	Labor	Total
Labor setting vinyl dual sliding/rolling windows, no carpentry included.					
To 8 SF opening size	BG@1.00	Ea	—	23.10	23.10
Over 8 SF to 16 SF opening size	BG@1.50	Ea	—	34.60	34.60
Over 16 SF to 32 SF opening size	BG@2.00	Ea	—	46.10	46.10
Over 32 SF opening size	BG@3.00	Ea	—	69.20	69.20

Single hung windows, vinyl Dual glazed, white frame, clear windows. See installation costs below

Height		Width		
		1'4"	**1'8"**	**2'2"**
2'2"		156.00	166.69	—
2'6"		161.00	172.00	189.00
3'0"		172.00	184.00	194.00
3'4"		178.00	189.00	206.00
3'10"		184.00	200.00	217.00
4'2"		189.00	206.00	223.00
5'0"		200.00	228.00	245.00
		Width		
		2'6"	**3'0"**	**3'4"**
2'6"		200.45	211.00	222.60
3'0"		205.72	222.60	233.15
3'4"		217.33	233.15	244.76
3'10"		233.15	244.76	261.64
4'2"		239.48	255.31	272.19
5'0"		266.91	283.79	300.67
6'8"		305.95	328.10	355.53

	Craft@Hrs	Unit	Material	Labor	Total
Labor setting vinyl single hung windows, no carpentry included					
To 8 SF opening size	BG@1.00	Ea	—	23.10	23.10
Over 8 SF to 16 SF opening size	BG@1.50	Ea	—	34.60	34.60
Over 16 SF to 23 SF opening size	BG@2.00	Ea	—	46.10	46.10

Single Family Home Costs

Use the costs below to check your estimates and to verify bids submitted by subcontractors. These costs are based on a 1,600 square foot home (as described on page 247) built during the third quarter of 1998 at a cost of $70.22 per square foot. Figures in the column "% of Total" are the same as those listed in the Residential Rule of Thumb Construction Costs on page 247. These are averages and will vary widely with the type and quality of construction. (Costs below have been rounded so columns and rows may not add correctly.)

Direct Costs	Manpower 1600 SF	% of Total	Unit	$ Material	$ Labor	$ Equipment	Total $ per SF	Total $ 1600 SF
Excavation	3 men 2 days	1.2%	SF	—	$.63	$.21	$.84	$1,348
Foundation, slab, piers	3 men 4 days	3.7%	SF	$1.04	1.30	.25	2.60	4,157
Flatwork, drive and walk	3 men 3 days	2.4%	SF	.67	.84	.17	1.69	2,696
Brick hearth & veneer	2 men 1 day	0.7%	SF	.20	.25	.05	.49	786
Rough hardware	with carpentry	0.6%	SF	.17	.21	.04	.42	674
Finish hardware	with carpentry	0.2%	SF	.08	.06	—	.14	225
Rough lumber	with carpentry	8.4%	SF	5.90	—	—	5.90	9,438
Finish lumber	with carpentry	0.5%	SF	.35	—	—	.35	562
Rough carpentry labor	2 men 23 days	8.9%	SF	—	6.25	—	6.25	9,999
Finish carpentry labor	1 man 7 days	1.7%	SF	—	1.19	—	1.19	1,910
Countertops	1 man 2 days	1.5%	SF	.63	.42	—	1.05	1,685
Cabinets	2 men 2 days	3.7%	SF	2.08	.52	—	2.60	4,157
Insulation (R19 ceiling)	1 man 3 days	2.3%	SF	1.04	.57	—	1.62	2,584
Roofing	4 men 3 days	5.5%	SF	2.32	1.54	—	3.86	6,179
Painting	2 men 3 days	3.6%	SF	.88	1.64	—	2.53	4,045
Shower and tub enclosure	1 man 1 day	0.5%	SF	.21	.14	—	.35	562
Prefabricated fireplace	1 man 1 day	0.9%	SF	.47	.13	.03	.63	1,011
Bath accessories	1 man 1 day	0.7%	SF	.33	.16	—	.49	786
Built-In appliances	1 man 1 day	1.6%	SF	1.01	.11	—	1.12	1,798
Heating and ducting	2 men 5 days	2.9%	SF	.81	1.22	—	2.04	3,258
Plumbing & sewer, connect	3 men 5 days	7.3%	SF	2.03	2.56	.53	5.13	8,202
Doors	2 men 2 days	1.9%	SF	.87	.47	—	1.33	2,135
Garage door	1 man 1 day	0.4%	SF	.21	.07	—	.28	449
Alum windows & sliding doors	1 man 2 days	1.2%	SF	.51	.34	—	.84	1,348
Exterior stucco	3 men 3 days	6.4%	SF	2.90	1.37	.22	4.49	7,191
Gypsum wallboard	2 men 6 days	4.7%	SF	1.48	1.82	—	3.30	5,281
Resilient flooring	1 man 1 day	2.0%	SF	.66	.74	—	1.40	2,247
Carpeting	2 men 1 day	2.4%	SF	1.31	.36	—	1.69	2,696
Wiring (Romex)	1 man 10 days	3.2%	SF	.90	1.34	—	2.25	3,595
Lighting fixtures	with wiring	1.2%	SF	.67	.17	—	.84	1,348
Subtotal, Direct Job Costs		**82.2%**	**SF**	**$29.75**	**$26.44**	**$1.50**	**$57.72**	**$92,353**
Amounts above, as a percent of total direct costs				52%	46%	3%	100.0%	

Indirect Costs	% of Total	Unit	$ Material	$ Labor	$ Equipment	Total $ per SF	Total $ 1600 SF
Insurance, payroll tax	2.8%	SF	$1.98	—	—	$1.98	$3,162
Plans & specs	0.4%	SF	.28	—	—	.28	449
Permits & utilities	1.7%	SF	1.19	—	—	1.19	1,910
Final cleanup	0.4%	SF	—	.28	—	.28	449
Subtotal, Indirect Job Costs	**5.3%**	**SF**	**$3.45**	**$.28**	**—**	**$3.73**	**$5,971**
Overhead & profit	**12.5%**	**SF**	**$8.78**	**—**	**—**	**$8.78**	**$14,044**
Total all costs, rounded	**100.0%**	**SF**	**$41.98**	**$26.72**	**$1.50**	**$70.22**	**$112,368**
Amounts above, as a percent of total costs			59.8%	38.1%	2.2%	100.0%	

The "Manpower" column assumes work is done in phases. Elapsed time, from breaking ground to final cleanup, will be about 8 weeks.

Construction Economics Division
Construction Cost Index for New Single Family Homes

Period	% of 1967 Cost	$ per SF of Floor	Period	% of 1967 Cost	$ per SF of Floor	Period	% of 1967 Cost	$ per SF of Floor
1981			**1987**			**1993**		
1st quarter	318.1	44.51	1st quarter	387.1	54.17	1st quarter	443.9	62.15
2nd quarter	325.7	45.58	2nd quarter	388.2	54.32	2nd quarter	450.7	63.10
3rd quarter	331.5	46.30	3rd quarter	392.3	54.89	3rd quarter	456.4	63.90
4th quarter	338.8	47.41	4th quarter	398.6	55.78	4th quarter	458.0	64.12
1982			**1988**			**1994**		
1st quarter	336.9	47.14	1st quarter	398.6	55.78	1st quarter	459.5	64.33
2nd quarter	339.3	47.48	2nd quarter	399.6	55.92	2nd quarter	460.7	64.50
3rd quarter	342.3	47.90	3rd quarter	402.8	56.37	3rd quarter	461.4	64.60
4th quarter	344.3	48.18	4th quarter	407.1	56.97	4th quarter	461.6	64.62
1983			**1989**			**1995**		
1st quarter	350.8	49.09	1st quarter	407.5	57.02	1st quarter	462.1	64.68
2nd quarter	354.3	49.57	2nd quarter	411.9	57.64	2nd quarter	463.6	64.90
3rd quarter	363.9	50.92	3rd quarter	422.6	59.13	3rd quarter	464.3	65.00
4th quarter	370.1	51.79	4th quarter	429.0	60.02	4th quarter	464.8	65.07
1984			**1990**			**1996**		
1st quarter	372.0	52.06	1st quarter	432.4	60.49	1st quarter	469.4	65.72
2nd quarter	373.0	52.20	2nd quarter	434.2	60.74	2nd quarter	474.1	66.37
3rd quarter	375.1	52.49	3rd quarter	435.3	60.89	3rd quarter	478.6	67.00
4th quarter	372.6	52.14	4th quarter	435.2	60.74	4th quarter	480.6	672.5
1985			**1991**			**1997**		
1st quarter	367.8	51.47	1st quarter	433.0	60.62	1st quarter	482.0	67.53
2nd quarter	374.1	52.35	2nd quarter	432.6	60.56	2nd quarter	487.4	68.20
3rd quarter	376.2	52.64	3rd quarter	433.2	60.65	3rd quarter	491.9	68.84
4th quarter	374.0	52.34	4th quarter	433.0	60.62	4th quarter	493.5	69.09
1986			**1992**			**1998**		
1st quarter	376.3	52.34	1st quarter	432.6	60.56	1st quarter	495.2	69.30
2nd quarter	383.5	53.67	2nd quarter	432.4	60.54	2nd quarter	498.4	69.76
3rd quarter	385.5	53.95	3rd quarter	432.3	60.52	3rd quarter	501.7	70.22
4th quarter	386.1	54.03	4th quarter	436.9	61.17	4th quarter	—	—

The figures under the column "$ per SF of Floor" show construction costs for building a good quality home in a suburban area under competitive conditions in each calendar quarter since 1981. This home is described below under the section "Residential Rule of Thumb." These costs include the builder's overhead and profit and a 450 square foot garage but no basement. The cost of a finished basement per square foot will be approximately 40% of the square foot cost of living area. If the garage area is more than 450 square feet, use 50% of the living area cost to adjust for the larger or smaller garage. To find the total construction cost of the home, multiply the living area (excluding the garage) by the cost in the column "$ per SF of Floor."

Deduct for rural areas	5.0%
Add for 1,800 SF house (better quality)	4.0
Add for 2,000 SF house (better quality)	3.0
Deduct for over 2,400 SF house	3.0
Add for split level house	3.0
Add for 3-story house	10.0
Add for masonry construction	9.0

Construction costs are higher in some cities and lower in others. Square foot costs listed in the table above are national averages. To modify these costs to your job site, apply the appropriate area modification factor from pages 12 through 15 of this manual. But note that area modifications on pages 12 through 15 are based on recent construction and may not apply to work that was completed many years ago.

Residential Rule of Thumb

Construction Costs

The following figures are percentages of total construction cost for a good quality single family residence: 1,600 square foot single story non-tract 3-bedroom, 1¾ bath home plus attached two car (450 SF) garage. Costs assume a conventional floor, stucco exterior, wallboard interior, shake roof, single fireplace, forced air heat, copper water supply, ABS drain lines, range with double oven, disposer, dishwasher and 900 SF of concrete flatwork.

Item	Percent	Item	Percent	Item	Percent	Item	Percent	Item	Percent
Excavation	1.2	Finish lumber	.5	Painting	3.6	Doors	1.9	Wiring (Romex)	3.2
Flatwork (drive & walk)	2.4	Rough carpentry labor	8.9	Shower & tub enclosure	.5	Garage door	.4	Lighting fixtures	1.2
Foundation, slab, piers	3.7	Finish carpentry labor	1.7	Prefabricated fireplace	.9	Alum. windows, door	1.2	Insurance, payroll tax	2.8
Brick hearth & veneer	.7	Countertops	1.5	Bath accessories	.7	Exterior stucco	6.4	Plans and specs	.4
Rough hardware	.6	Cabinets	3.7	Built-in appliances	1.6	Gypsum wallboard	4.7	Permits & utilities	1.7
Finish hardware	.2	Insulation (R19 ceiling)	2.3	Heating and ducting	2.9	Resilient flooring	2.0	Final cleanup	.4
Rough lumber	8.4	Roofing	5.5	Plumb. & sewer conn.	7.3	Carpeting	2.4	Overhead & profit	12.5

Industrial and Commercial Division Contents

A complete index begins on page 559

Industrial and Commercial Division
Hourly Labor Costs

The hourly labor costs shown in the column headed "Hourly Cost" have been used to compute the manhour costs for the crews on pages 7 to 9 and the costs in the "Labor" column on pages 255 to 532 of this book. All figures are in U.S. dollars per hour.

"Hourly Wage and Benefits" includes the wage, welfare, pension, vacation, apprentice and other mandatory contributions. The "Employer's Burden" is the cost to the contractor for Unemployment Insurance (FUTA), Social Security and Medicare (FICA), state unemployment insurance, workers' compensation insurance and liability insurance. Tax and insurance expense included in these labor-hour costs are itemized in the sections beginning on pages 133 and 235.

These hourly labor costs will apply within a few percent on many jobs. But wages may be much higher or lower on the job you are estimating. If the hourly cost on this page is not accurate, use the labor cost adjustment procedure on page 11.

If your hourly labor cost is not known and can't be estimated, use both labor and material figures in this book without adjustment. When all material and labor costs have been compiled, multiply the total by the appropriate figure in the area modification table on pages 12 through 15.

Craft	Hourly Wage and Benefits ($)	Typical Employer Burden (%)	Employer's Burden Per Hour ($)	Hourly Cost ($)
Air Tool Operator	22.83	33.05%	7.54	30.37
Asbestos Worker	31.54	36.45%	11.50	43.04
Boilermaker	32.38	28.20%	9.13	41.51
Bricklayer	28.83	31.72%	9.15	37.98
Bricklayer Tender	22.19	31.72%	7.04	29.23
Building Laborer	22.75	33.05%	7.52	30.27
Carpenter	28.53	48.15%	13.74	42.27
Cement Mason	27.39	36.95%	10.12	37.51
Crane Operator	29.83	26.10%	7.79	37.62
Drywall Installer	28.65	28.32%	8.11	36.76
Electrician	32.88	21.88%	7.19	40.07
Elevator Constructor	28.56	30.95%	8.84	37.40
Floor Layer	27.61	32.55%	8.99	36.60
Glazier	28.23	29.03%	8.20	36.43
Iron Worker (Structural)	31.40	55.56%	17.44	48.84
Lather	26.35	26.12%	6.88	33.23
Marble Setter	28.83	24.56%	7.08	35.91
Millwright	29.44	48.15%	14.18	43.62
Mosaic & Terrazzo Worker	27.17	24.56%	6.67	33.84
Painter	28.29	30.84%	8.72	37.01
Pile Driver	28.88	45.46%	13.13	42.01
Pipefitter	33.10	29.20%	9.67	42.77
Plasterer	29.86	28.94%	8.64	38.50
Plasterer Helper	24.40	28.94%	7.06	31.46
Plumber	33.27	26.08%	8.68	41.95
Reinforcing Ironworker	30.61	27.15%	8.31	38.92
Roofer	25.59	48.29%	12.35	37.94
Sheet Metal Worker	32.30	27.88%	9.01	41.31
Sprinkler Fitter	34.35	29.20%	10.03	44.38
Tractor Operator	29.33	33.15%	9.72	39.05
Truck Driver	23.74	33.15%	7.87	31.61

General Contractor's Markup Costs in this manual do not include the general contractor's markup.

Costs in sections identified as *subcontract* are based on quotes from subcontractors and include the subcontractor's markup. See pages 3 through 5 for more on subcontracted work and markup. Typical markup for general contractors handling commercial and industrial projects is shown at the bottom of this page. The two sections that follow, Indirect Overhead and Direct Overhead give a more detailed breakdown.

Indirect Overhead (home office overhead).
This cost, usually estimated as a percentage, is calculated by dividing average annual receipts by average annual indirect overhead costs. Indirect overhead usually doesn't increase or decrease as quickly as gross volume increases or decreases. Home office expense may be about the same even if annual sales double or drop by one-half or more.

The figures below show typical indirect overhead <u>costs per $1,000</u> of estimated project costs for a general contractor handling $500,000 to $1,000,000 commercial and industrial projects. Indirect overhead costs vary widely but will be about 8% of gross for most profitable firms. Total indirect overhead cost ($20.00 per $1,000 for materials and $60.00 per $1,000 for labor) appears as 2.0% and 6.0% under *Total General Contractor's Markup* at the bottom of this page.

	Craft@Hrs	Unit	Material	Labor	Total
Rent, office supplies, utilities, equipment, advertising, etc.	—	M$	20.00	—	20.00
Office salaries and professional fees	—	M$	—	60.00	60.00
General Contractor's indirect overhead	**—**	**M$**	**20.00**	**60.00**	**80.00**

Direct Overhead (job site overhead).
The figures below show typical direct overhead costs per $1,000 of a project's total estimated costs for a general contractor handling a $500,000 to $1,000,000 job that requires 6 to 9 months for completion. Use these figures for preliminary estimates and to check final bids. Add the cost of mobilization, watchmen, fencing, hoisting, permits, bonds, scaffolding and testing. Total direct overhead cost ($14.30 per $1,000 for materials and $58.70 per $1,000 for labor) appears as 1.4% and 5.9% under *Total General Contractor's Markup* at the bottom of this page. These costs can be expected to vary and will usually be lower on larger projects.

	Craft@Hrs	Unit	Material	Labor	Total
Job superintendent ($4,400 per month)	—	M$	—	44.00	44.00
Pickup truck for superintendent ($400 per month)	—	M$	4.00	—	4.00
Temporary power, light and heat ($225 per month)	—	M$	2.50	—	2.50
Temporary water ($100 per month)	—	M$	1.00	—	1.00
Temporary phone ($100 per month)	—	M$	1.00	—	1.00
Job site toilets ($100 per month)	—	M$	1.00	—	1.00
Job site office trailer 8' x 30' ($150 per month)	—	M$	1.50	—	1.50
Storage bin and tool shed 8' x 24' ($80 per month)	—	M$	.80	—	.80
Job site cleanup & debris removal ($300 per month)	—	M$	—	3.00	3.00
Job signs & first aid equipment ($50 per month)	—	M$	.50	—	.50
Small tools, supplies ($200 per month)	—	M$	2.00	—	2.00
Taxes and insurance on wages (at 24.9%)	—	M$	—	11.70	11.70
General Contractor's direct overhead	—	M$	14.30	58.70	73.00

Total General Contractor's Markup Typical costs for commercial and industrial projects.

	Craft@Hrs	Unit	Material	Labor	Total
Indirect overhead (home office overhead)	—	%	2.0	6.0	8.0
Direct overhead (job site overhead)	—	%	1.4	5.9	7.3
Contingency (allowance for unknown conditions)	—	%	—	—	2.0
Profit (varies widely, mid-range shown)	—	%	—	—	7.5
Total General Contractor's Markup	**—**	**%**	**—**	**—**	**24.8**

	Craft@Hrs	Unit	Material	Labor	Total

Project Financing Construction loans are usually made for a term of up to 18 months. The maximum loan will be based on the value of the land and building when completed. Typically this maximum is 75% for commercial, industrial and apartment buildings and 77.5% for tract housing.

The initial loan disbursement will be about 50% of the land cost. Periodic disbursements are based on the percentage of completion as verified by voucher or inspection. The last 20% of loan proceeds is disbursed when an occupancy permit is issued. Fund control fees cover the cost of monitoring disbursements. Typical loan fees, loan rates and fund control fees are listed below.Loan origination fee, based on amount of loan

	Craft@Hrs	Unit	Material	Labor	Total
Tracts	—	%	—	—	2.5
Apartments	—	%	—	—	2.0
Commercial, industrial buildings	—	%	—	—	3.0
Loan interest rate, prime rate plus					
Tracts, commercial, industrial	—	%	—	—	2.0
Apartments	—	%	—	—	1.5
Fund control fees (costs per $1,000 disbursed)					
To $200,000	—	LS	—	—	1,000.00
Over $200,00 to $500,000, $1,000 plus	—	M$	—	—	1.50
Over $500,000 to $1,000,000, $1,450 plus	—	M$	—	—	1.00
Over $1,000,000 to $3,000,000, $1,250 plus	—	M$	—	—	.85
Over $3,000,000 to $5,000,000, $1,150 plus	—	M$	—	—	.75
Over $5,000,000, $1,000 plus	—	M$	—	—	.55

Project Scheduling by CPM (Critical Path Method). Includes consultation, review of construction documents, development of construction logic, and graphic schedule.

	Craft@Hrs	Unit	Material	Labor	Total
Wood frame buildings, one or two stories					
Simple schedule	—	LS	—	—	690.00
Complex schedule	—	LS	—	—	1,330.00
Tilt-up concrete buildings, 10,000 to 50,000 SF					
Simple schedule	—	LS	—	—	1,020.00
Complex schedule	—	LS	—	—	1,530.00
Two to five story buildings (mid rise)					
Simple schedule	—	LS	—	—	1,530.00
Complex schedule	—	LS	—	—	2,040.00
Five to ten story buildings (high rise)					
Simple schedule	—	LS	—	—	2,550.00
Complex schedule	—	LS	—	—	6,380.00
Manufacturing plants and specialized use low rise buildings					
Simple schedule	—	LS	—	—	1,330.00
Complex schedule	—	LS	—	—	1,840.00

Comprehensive schedules to meet government or owner specifications may cost 20% to 40% more. Daily schedule updates can increase costs by 40% to 50%.

Sewer Connection Fees Check with the local sanitation district for actual charges. These costs are typical for work done by city or county crews and include excavation, recompaction and repairs to the street. Note that these costs will vary widely depending on local government policy. These costs do not include capacity fees or other charges which are often levied on connection to new construction. Cost per living unit may be $2,000 or more. For commercial buildings, figure 20 fixture units equal one living unit. A 6" sewer connection will be required for buildings that include more than 216 fixture units. (Bathtubs, showers and sinks are 2 fixture units, water closets are 4, clothes washers are 3 and lavatories are 1.) The costs shown assume that work will be done by sanitation district crews. Similar work done by a private contractor under a public improvement permit may cost 50% less.

	Craft@Hrs	Unit	Material	Labor	Total
Typical charges based on main up to 8' deep and lateral up to 5' deep.					
4" connection and 40' run to main in street	—	LS	—	—	1,900.00
4" pipe runs over 40' to street	—	LF	—	—	45.00
6" connection and 40' run to main in street	—	LS	—	—	2,980.00
6" pipe runs over 40' to street	—	LF	—	—	70.00
4" connection and 15' run to main in alley	—	LS	—	—	1,350.00
4" pipe runs over 15' in alley	—	LF	—	—	45.00
6" connection and 15' run to main in alley	—	LS	—	—	2,240.00
6" pipe runs over 15' in alley	—	LF	—	—	70.00
Manhole cut-in	—	Ea	—	—	390.00
Add for main depths over 5'					
Over 5' to 8'	—	%	—	—	30.0
Over 8' to 11'	—	%	—	—	60.0
Over 11'	—	%	—	—	100.0

Water Meters Check with the local water district for actual charges. These costs are typical for work done by city or county crews and include excavation and pipe to 50' from the main, meter, vault, recompaction and repairs to the street. These costs do not include capacity fees or other charges which are often levied on connection to new construction. The capacity fee per living unit may be $1,000 or more. For commercial buildings, figure 20 fixture units equal one living unit. (Bathtubs, showers and sinks are 2 fixture units, water closets are 4, clothes washers are 3 and lavatories are 1.) The cost for discontinuing service will usually be about the same as the cost for starting new service, but no capacity fee will be charged. Add the backflow device and capacity fee, if required.

	Craft@Hrs	Unit	Material	Labor	Total
1" service, 3/4" or 1" meter	—	LS	—	—	980.00
Add for 1" service per LF over 50'	—	LF	—	—	36.20
2" service, 1-1/2" or 2" meter	—	LS	—	—	1,490.00
Add for 2" service per LF over 50'	—	LF	—	—	51.50
Two 2" service lines and meter manifold	—	LS	—	—	5,660.00
Add for two 2" lines per LF over 50'	—	LF	—	—	83.10

Backflow preventer with single check valve. Includes pressure vacuum breaker (PVB), two ball valves, one air inlet and one check valve

	Craft@Hrs	Unit	Material	Labor	Total
5/8" or 3/4" pipe	—	LS	—	—	236.00
1" pipe	—	LS	—	—	235.00
1-1/2" pipe	—	LS	—	—	343.00
2" pipe	—	LS	—	—	465.00

Backflow preventer with double check valves. Includes pressure vacuum breaker (PVB), two ball or gate valves, one air inlet and two check valves

	Craft@Hrs	Unit	Material	Labor	Total
3/4" pipe	—	LS	—	—	308.00
1" pipe	—	LS	—	—	309.00
1-1/2" pipe	—	LS	—	—	375.00
2" pipe	—	LS	—	—	507.00

Reduced pressure backflow preventer. Includes pressure reducing vacuum breaker (RPVB), two ball or gate valves, one air inlet and two check valves and a relief valve

	Craft@Hrs	Unit	Material	Labor	Total
3/4" pipe	—	LS	—	—	607.00
1" pipe	—	LS	—	—	665.00
1-1/2" pipe	—	LS	—	—	750.00
2" pipe	—	LS	—	—	984.00

Surveying

Surveying party, 2 or 3 technicians

	Craft@Hrs	Unit	Material	Labor	Total
Typical cost	—	Day	—	—	1,050.00
Higher cost	—	Day	—	—	1,350.00
Data reduction and drafting	—	Hr	—	—	56.00

	Craft@Hrs	Unit	Material	Labor	Total

Surveys Including wood hubs or pipe markers as needed. These figures assume that recorded monuments are available adjacent to the site.

Residential lot, tract work, 4 corners	—	LS	—	—	635.00
Residential lot, individual, 4 corners	—	LS	—	—	1,080.00
Commercial lot, based on 30,000 SF lot	—	LS	—	—	1,160.00
Over 30,000 SF, add per acre	—	Acre	—	—	175.00

Lots without recorded markers cost more, depending on distance to the nearest recorded monument.

Add up to	—	%	—	—	100.0

Some states require that a corner record be prepared and filed with the county surveyor's office when new monuments are set. The cost of preparing and filing a corner record will be about $350.00 to $400.00.

Aerial Mapping Typical costs based on a scale of 1" to 40'.

Back up ground survey including flagging model	—	Ea	—	—	2,850.00

(Typical "model" is approximately 750' by 1,260' or 25 acres)

Aerial photo flight, plane and crew, per local flight. One flight can cover several adjacent models

Per local flight	—	Ea	—	—	355.00
Vertical photos (2 required per model)	—	Ea	—	—	40.00
Data reduction and drafting, per hour	—	Hr	—	—	55.00

Complete survey with finished drawing, per model, including both field work and flying

Typical cost per model	—	Ea	—	—	2,950.00
Deduct if ground control is marked on the ground by others,					
Per model	—	LS	—	—	-950.00

Typical mapping cost per acre including aerial control, flight and compilation

Based on 25 acre coverage	—	Acre	—	—	135.00

Aerial photos, oblique (minimum 4 per job), one 8" x 10" print

Black and white	—	Ea	—	—	40.00
Color	—	Ea	—	—	82.00

Quantitative studies for earthmoving operations to evaluate soil quantities moved

Ground control and one flight	—	Ea	—	—	2,450.00
Each additional flight	—	Ea	—	—	1,660.00

Engineering Fees Typical billing rates.

Registered engineers (calculations, design and drafting)

Assistant engineer	—	Hr	—	—	68.00
Senior engineer	—	Hr	—	—	105.00

Principal engineers (cost control, project management, client contact

Assistant principal engineer	—	Hr	—	—	120.00
Senior principal engineer	—	Hr	—	—	165.00
Job site engineer (specification compliance)	—	Hr	—	—	60.00

Estimating Cost Compiling detailed labor and material cost estimate for commercial and industrial construction, in percent of job cost. Actual fees are based on time spent plus cost of support services.

Most types of buildings, typical fees

Under $100,000 (range of .75 to 1.5%)	—	%	—	—	1.1
Over $100,000 to $500,000 (range of .5 to 1.25%)	—	%	—	—	0.9
Over $500,000 to $1,000,000 (range of .5 to 1.0%)	—	%	—	—	0.8
Over $1,000,000 (range of .25 to .75%)	—	%	—	—	0.5

Complex, high-tech, research or manufacturing facilities or other buildings with many unique design features, typical fees

Under $100,000 (range of 2.3 to 3%)	—	%	—	—	2.7
Over 100,000 to $500,000 (range of 1.5 to 2.5%)	—	%	—	—	2.0

	Craft@Hrs	Unit	Material	Labor	Total
Over $500,000 to $1,000,000					
(range of 1.0 to 2.0%)	—	%	—	—	1.5
Over $1,000,000 (range of .5 to 1.0%)	—	%	—	—	0.8
Typical estimator's fee per hour	—	Hr	—	—	50.00

Specifications Fee Preparing detailed construction specifications to meet design requirements for commercial and industrial buildings, in percent of job cost. Actual fees are based on time spent plus cost of support services, if any.

	Craft@Hrs	Unit	Material	Labor	Total
Typical fees for most jobs	—	%	—	—	0.5
Small or complex jobs					
Typical cost	—	%	—	—	1.0
High cost	—	%	—	—	1.5
Large jobs or repetitive work					
Typical cost	—	%	—	—	0.2
High cost	—	%	—	—	0.5

Job Layout Setting layout stakes for the foundation from monuments identified on the plans and existing on site. Per square foot of building first floor area. Typical costs. Actual costs will be based on time spent.

	Craft@Hrs	Unit	Material	Labor	Total
Smaller or more complex jobs	—	SF	—	—	.57
Large job, easy work	—	SF	—	—	.25

General Non-Distributable Supervision and Expense

	Craft@Hrs	Unit	Material	Labor	Total
Project manager, typical	—	Mo	—	—	5,200.00
Job superintendent	—	Mo	—	—	4,430.00
Factory representative	—	Day	—	—	600.00
Timekeeper	—	Mo	—	—	1,650.00
Job site engineer for layout	—	Mo	—	—	3,100.00
Office manager	—	Mo	—	—	1,850.00
Office clerk	—	Mo	—	—	1,380.00

Mobilization Typical costs for mobilization and demobilization within a 50 mile radius as a percent of total contract price. Includes cost of moving general contractor-owned equipment and job office to the site, setting up a fenced material yard, and removing equipment, office and fencing at job completion.

	Craft@Hrs	Unit	Material	Labor	Total
Allow, as a percentage of total contract price	—	%	—	—	0.5

Aggregate Testing Add $45.00 per hour for sample preparation.

	Craft@Hrs	Unit	Material	Labor	Total
Abrasion, L.A. Rattler, 100 and 500 cycles,					
ASTM C 131	—	Ea	—	—	131.00
Absorption, ASTM C 127, 128	—	Ea	—	—	54.00
Acid solubility	—	Ea	—	—	32.00
Aggregate test for mix design, including sieve analysis, specific gravity, number 200 wash,					
organic impurities and weight per cubic foot	—	Ea	—	—	73.00
Clay lumps and friable particles, ASTM C 142	—	Ea	—	—	63.00
Coal and lignite, ASTM C 123	—	Ea	—	—	95.00
Material finer than #200 sieve, ASTM C 117	—	Ea	—	—	37.00
Organic impurities, ASTM C 40	—	Ea	—	—	47.00
Percent crushed particles, CAL 205	—	Ea	—	—	63.00
Percent flat or elongated particles, CRD C 119	—	Ea	—	—	105.00
Potential reactivity, chemical method, 3 determinations per series,					
ASTM C 289	—	Ea	—	—	315.00
Potential reactivity, mortar bar method, ASTM C 227	—	Ea	—	—	420.00
Sieve analysis, pit run aggregate	—	Ea	—	—	63.00
Sieve analysis, processed (each size), ASTM C 136	—	Ea	—	—	73.00
Soft particles, ASTM C 235	—	Ea	—	—	63.00

	Craft@Hrs	Unit	Material	Labor	Total
Soundness, sodium or magnesium, 5 cycle, per series, ASTM C 88	—	Ea	—	—	135.00
Specific gravity, coarse, ASTM C 127	—	Ea	—	—	58.00
Specific gravity, fine, ASTM C 127	—	Ea	—	—	79.00
Unit weight per cubic foot, ASTM C 29	—	Ea	—	—	32.00

Asphaltic Concrete Testing Minimum fee is usually $40.00.

	Craft@Hrs	Unit	Material	Labor	Total
Asphaltic concrete core drilling, field or laboratory					
Per linear inch	—	Ea	—	—	5.00
Add for equipment and technician for core drilling	—	Hr	—	—	70.00
Asphaltic core density	—	Ea	—	—	25.00
Extraction, percent asphalt, (Method B) excluding ash correction, ASTM D 2172	—	Ea	—	—	60.00
Gradation on extracted sample (including wash), ASTM C 136	—	Ea	—	—	75.00
Maximum density, lab mixed, Marshall, 1559	—	Ea	—	—	250.00
Maximum density, premixed, Marshall, 1559	—	Ea	—	—	150.00
Maximum theoretical unit weight (Rice Gravity), ASTM 2041	—	Ea	—	—	75.00
Penetration, ASTM D 5	—	Ea	—	—	60.00
Stability, lab mixed, Marshall, ASTM D 1559	—	Ea	—	—	250.00
Stability, premixed, Marshall, ASTM D 1559	—	Ea	—	—	150.00

Concrete Testing Minimum fee is usually $40.00.

	Craft@Hrs	Unit	Material	Labor	Total
Aggregate tests for concrete mix designs only, including sieve analysis, specific gravity, number 200 wash, organic impurities, weight per cubic foot					
Per aggregate size	—	Ea	—	—	72.00
Amend or re-type existing mix designs					
Not involving calculations	—	Ea	—	—	42.00
Compression test, 2", 4", or 6" cores (excludes sample preparation)					
ASTM C 42	—	Ea	—	—	40.00
Compression test, 6" x 12" cylinders, including mold					
ASTM C 39	—	Ea	—	—	13.00
Core cutting in laboratory	—	Ea	—	—	40.00
Cylinder pickup within 40 miles of laboratory					
Cost per cylinder (minimum of 3)	—	Ea	—	—	9.00
Flexure test, 6" x 6" beams, ASTM C 78	—	Ea	—	—	53.00
Laboratory trial, concrete batch, ASTM C 192	—	Ea	—	—	345.00
Length change (3 bars, 4 readings, up to 90 days)					
ASTM C 157 modified	—	LS	—	—	240.00
Additional readings, per set of 3 bars	—	LS	—	—	37.00
Storage over 90 days, per set of 3 bars, per month	—	Mo	—	—	37.00
Modulus of elasticity test, static, ASTM C 469	—	Ea	—	—	73.00
Mix design, determination of proportions	—	Ea	—	—	79.00
Pick up aggregate sample					
Within 40 mile radius of laboratory, per trip	—	Ea	—	—	47.00
Pickup or delivery of shrinkage molds, per trip	—	Ea	—	—	32.00
Prepare special strength documentation for mix design	—	Ea	—	—	89.00
Proportional analysis, cement factor and percent of aggregate					
Per analysis	—	Ea	—	—	230.00
Review mix design prepared by others	—	Ea	—	—	68.00
Splitting tensile, 6" x 12" cylinder, ASTM C 496	—	Ea	—	—	32.00
Weight per cubic foot determination of lightweight concrete					
Cylinders	—	Ea	—	—	6.00

	Craft@Hrs	Unit	Material	Labor	Total
Masonry and Tile Testing					
Brick, ASTM C 67					
Modulus of rupture (flexure) or compressive strength	—	Ea	—	—	27.00
Absorption initial rate, 5 hour or 24 hour	—	Ea	—	—	22.00
Boil, 1, 2, or 5 hours	—	Ea	—	—	22.00
Efflorescence	—	Ea	—	—	27.00
Dimensions, overall, coring, shell and web	—	Ea	—	—	22.00
Compression of core	—	Ea	—	—	37.00
Cores, shear, 6" and 8" diameter, 2 faces, per core	—	Ea	—	—	47.00
Concrete block, ASTM C 140					
Moisture content as received	—	Ea	—	—	22.00
Absorption	—	Ea	—	—	27.00
Compression	—	Ea	—	—	32.00
Shrinkage, modified British, ASTM C 426	—	Ea	—	—	68.00
Compression, 4", 6", 8" cores	—	Ea	—	—	37.00
Fireproofing, oven dry density	—	Ea	—	—	32.00
Gunite					
Compression, 2", 4", 6" cores, ASTM C 42	—	Ea	—	—	37.00
Pickup gunite field sample (40 mile trip maximum)	—	Ea	—	—	53.00
Masonry prisms, ASTM E 447					
Compression test, grouted prisms	—	Ea	—	—	136.00
Pickup prisms, within 40 miles of lab	—	Ea	—	—	37.00
Mortar and grout, UBC Standard 24-22 & 24-28					
Compression test, 2" x 4" mortar cylinder	—	Ea	—	—	18.00
Compression test, 3" x 6" grout prisms	—	Ea	—	—	24.00
Compression test, 2" cubes, ASTM C 109	—	Ea	—	—	47.00
Roof fill, lightweight ASTM C 495					
Compression test	—	Ea	—	—	22.00
Density	—	Ea	—	—	16.00
Roofing tile					
Roofing tile breaking strength, UBC	—	Ea	—	—	22.00
Roofing tile absorption	—	Ea	—	—	22.00
Slate					
Modulus of rupture	—	Ea	—	—	37.00
Modulus of elasticity	—	Ea	—	—	58.00
Water absorption	—	Ea	—	—	32.00
Reinforcement Testing					
Bend test, number 11 bar or smaller, ASTM 615	—	Ea	—	—	19.00
Modulus of elasticity for prestressing wire	—	Ea	—	—	105.00
Nick-break test, welded bar	—	Ea	—	—	58.00
Sampling at fabricator's plant (within 40 miles of lab), $50.00 minimum					
During normal business hours, per sample	—	Ea	—	—	18.00
Other than normal business hours, per sample	—	Ea	—	—	27.00
Tensile test, mechanically spliced bar	—	Ea	—	—	100.00
Tensile test, number 11 bar or smaller, ASTM 615	—	Ea	—	—	26.00
Tensile test, number 14 bar or smaller, ASTM 615	—	Ea	—	—	58.00
Tensile test, number 18 bar, ASTM 615	—	Ea	—	—	68.00
Tensile test, welded number 11 bar or smaller	—	Ea	—	—	37.00

	Craft@Hrs	Unit	Material	Labor	Total
Tensile test, welded number 14 bar	—	Ea	—	—	68.00
Tensile test, welded number 18 bar	—	Ea	—	—	84.00
Tensile and elongation test for prestress strands					
24" test, ASTM A 416	—	Ea	—	—	84.00
10" test, ASTM A 421	—	Ea	—	—	47.00

Soils and Aggregate Base Testing Minimum fee is usually $400.

	Craft@Hrs	Unit	Material	Labor	Total
Bearing ratio (excluding moisture-density curve)	—	Ea	—	—	110.00
Consolidation test (single point)	—	Ea	—	—	75.00
Consolidation test (without rate data)	—	Ea	—	—	95.00
Direct shear test					
At natural moisture - strain rate 0.04 inch/min	—	Ea	—	—	50.00
Saturated - strain rate 0.04 inch/min	—	Ea	—	—	55.00
ASTM 3080 - consolidated, drained 3 points/test	—	Ea	—	—	290.00
Durability index - coarse and fine	—	Ea	—	—	140.00
Expansion index test, UBC 29-2	—	Ea	—	—	110.00
Liquid limit or plastic limit, Atterberg limits	—	Ea	—	—	75.00
Mechanical analysis - ASTM D1140 (wash 200 sieve)	—	Ea	—	—	41.00
Mechanical analysis - sand and gravel (wash sieve)	—	Ea	—	—	116.00
Moisture content	—	Ea	—	—	15.00
Moisture-density curve for compacted fill, ASTM 1557					
4-inch mold	—	Ea	—	—	130.00
6-inch mold	—	Ea	—	—	140.00
Permeability (falling head)	—	Ea	—	—	175.00
Permeability (constant head)	—	Ea	—	—	225.00
Resistance value	—	Ea	—	—	220.00
Sand equivalent	—	Ea	—	—	65.00
Specific gravity - fine-grained soils	—	Ea	—	—	60.00
Unconfined compression test (undisturbed sample)	—	Ea	—	—	70.00
Unit dry weight and moisture content					
(undisturbed sample)	—	Ea	—	—	21.00

Steel Testing $100.00 minimum for machine time.

	Craft@Hrs	Unit	Material	Labor	Total
Charpy impact test, reduced temperature	—	Ea	—	—	26.00
Charpy impact test, room temperature	—	Ea	—	—	13.00
Electric extensometer	—	Ea	—	—	63.00
Hardness tests, Brinell or Rockwell	—	Ea	—	—	22.00
Photo micrographs					
20x to 2000x	—	Ea	—	—	22.00
With negative	—	Ea	—	—	26.00
Tensile test, under 100,000 lbs	—	Ea	—	—	37.00
Welder Qualification Testing. Add $36 minimum per report					
Tensile test	—	Ea	—	—	37.00
Bend test	—	Ea	—	—	26.00
Macro etch	—	Ea	—	—	53.00
Fracture test	—	Ea	—	—	32.00
Machining for weld tests, 1/2" and less	—	Ea	—	—	37.00
Machining for weld tests, over 1/2"	—	Ea	—	—	53.00

Construction Photography Professional quality work, 2-1/4" x 2-1/4" negatives, hand printed, labeled, numbered and dated.

Job progress photos, includes shots of all building corners, two 8" x 10" black and white prints of each shot, per visit. See prices on next page.

	Craft@Hrs	Unit	Material	Labor	Total
Typical cost	—	LS	—	—	370.00
High cost	—	LS	—	—	530.00

Periodic survey visits to job site to figure angles (for shots of all building corners),

Minimum two photos, per visit	—	Ea	—	—	270.00

Exterior completion photographs, two 8" x 10" prints using 4" x 5" negatives (two angles)

Black and white	—	Ea	—	—	350.00
Color	—	Ea	—	—	390.00

Interior completion photographs, with lighting setup, two 8" x 10" prints of each shot (minimum three shots)

Black and white	—	Ea	—	—	350.00
Color	—	Ea	—	—	390.00

Reprints or enlargements from existing negative. Hand (custom) prints

8" x 10", black and white	—	Ea	—	—	10.00
11" x 14", black and white	—	Ea	—	—	20.00
8" x 10", color	—	Ea	—	—	20.00
11" x 14", color	—	Ea	—	—	35.00

Signs and Markers Pipe markers, self-stick vinyl, by overall outside diameter of pipe or covering, meets OSHA and ANSI requirements. Costs are based on standard wording and colors.

	Craft@Hrs	Unit	Material	Labor	Total
3/4" to 2-1/2" diameter, 3/4" letters, 7" long	PA@.077	Ea	1.75	2.85	4.60
2-1/2" to 8" diameter, 1-1/4" letters, 12" long	PA@.077	Ea	2.40	2.85	5.25
8" to 10" diameter, 2-1/2" letters, 24" long	PA@.077	Ea	4.00	2.85	6.85
10" and larger diameter, 3-1/2" letters, 32" long	PA@.095	Ea	4.50	3.52	8.02
Flow arrows, 2" wide tape x 108' roll, perforated every 8"					
(use $50.00 as minimum material cost)	PA@.095	LF	.52	3.52	4.04

Barrier tapes, 3" wide x 1,000' long. "Caution," "Danger," "Open Trench," and other stock wording (Use $35.00 as minimum material cost). Labor shown is based on one man installing 500 LF in 1 hour.

	Craft@Hrs	Unit	Material	Labor	Total
2 mil thick, short-term or indoor use	PA@.002	LF	.02	.07	.09
4 mil thick, long-term or outdoor use	PA@.002	LF	.04	.07	.11

Traffic control signage, "No Parking Between Signs," "Customer Parking," and other stock traffic control signs, rectangular, 30 gauge, baked enamel

	Craft@Hrs	Unit	Material	Labor	Total
6" x 12", non-reflective	PA@.301	Ea	6.30	11.10	17.40
6" x 12", reflective	PA@.301	Ea	14.50	11.10	25.60
12" x 18", non-reflective	PA@.455	Ea	17.00	16.80	33.80
12" x 18", reflective	PA@.455	Ea	25.00	16.80	41.80
18" x 24", non-reflective	PA@.455	Ea	32.40	16.80	49.20
18" x 24", reflective	PA@.455	Ea	24.20	16.80	41.00
Add for pole-mounted signs	CL@1.00	Ea	13.00	30.30	43.30

Temporary Utilities No permit fees included.

Drop pole for electric service

Power pole installation, wiring, 6 months rent and pickup

100 amp meter pole	—	Ea	—	—	427.00
200 amp meter pole	—	Ea	—	—	556.00

Electric drop line and run to contractor's panel (assuming proper phase and voltage and service are available within 100' of site)

Overhead run to contractor's "T" pole (including removal)

Single phase, 100 amps	—	LS	—	—	190.00
Single phase, over 100 to 200 amps	—	LS	—	—	284.00
Three phase, to 200 amps	—	LS	—	—	341.00

	Craft@Hrs	Unit	Material	Labor	Total
Underground run in contractor's trench and conduit (including removal)					
Single phase, 100 amps, "USA" wire	—	LS	—	—	325.00
Single phase, 101-200 amps, "USA" wire	—	LS	—	—	427.00
Three phase, to 200 amps, "USA" wire	—	LS	—	—	486.00
Typical power cost					
Per kilowatt hour ($5.00 minimum)	—	Ea	—	—	.11
Per 1000 SF of building per month of construction	—	MSF	—	—	1.85
Temporary water service, meter installed on fire hydrant. Add the cost of distribution hose or piping					
Meter fee, 2" meter (excluding $400 deposit)	—	LS	—	—	68.00
Water cost, per 1,000 gallons	—	M	—	—	1.60
Job phone, typical charge for 1 line, excluding message unit charges					
Installation and 3 months service	—	LS	—	—	425.00
Portable job site toilets, including delivery, weekly service, and pickup, per 28 day month, six month contract					
Good quality (polyethylene)	—	Mo	—	—	55.00
Deluxe quality,					
with sink, soap dispenser, towel dispenser, mirror	—	Mo	—	—	85.00
Trailer-mounted male and female compartment double, 2 stalls in each, sink, soap dispenser, mirror and towel dispenser in each compartment					
Rental per week	—	Wk	—	—	250.00

Construction Elevators and Hoists

	Craft@Hrs	Unit	Material	Labor	Total
Hoisting towers to 200', monthly rent					
Material hoist	CO@176.	Mo	2,120.00	6,620.00	8,740.00
Personnel hoist	CO@176.	Mo	4,850.00	6,620.00	11,470.00
Tower cranes, self-climbing (electric), monthly rent					
5 ton lift at 180' radius, one operator	CO@176.	Mo	12,100.00	6,620.00	18,720.00
With two operators	CO@352.	Mo	12,100.00	13,200.00	25,300.00
7 ton lift at 170' radius, one operator	CO@176.	Mo	16,700.00	6,620.00	23,320.00
With two operators	CO@352.	Mo	16,700.00	13,200.00	29,900.00
Mobilize, erect and dismantle self-climbing tower cranes					
Good access	—	LS	—	—	36,000.00
Average access	—	LS	—	—	42,000.00
Poor access	—	LS	—	—	47,500.00

Hydraulic Truck Cranes Includes equipment rental and operator. Four hour minimum. Time charge begins at the rental yard. By rated crane capacity.

	Craft@Hrs	Unit	Material	Labor	Total
12 1/2 ton	—	Hr	—	—	97.20
15 ton	—	Hr	—	—	105.00
25 ton	—	Hr	—	—	128.00
50 ton	—	Hr	—	—	160.00
70 ton	—	Hr	—	—	193.30
100 ton	—	Hr	—	—	232.00
130 ton	—	Hr	—	—	275.00
150 ton	—	Hr	—	—	295.00
Add for clamshell or breaker ball work	—	Hr	—	—	30.00

Conventional Cable Cranes Including equipment rental and operator. Four hour minimum. Time charge begins at the rental yard. By rated crane capacity with 80' boom.

	Craft@Hrs	Unit	Material	Labor	Total
50 tons	—	Hr	—	—	160.00
70 tons	—	Hr	—	—	180.00
90 tons	—	Hr	—	—	195.00
115 tons	—	Hr	—	—	205.00
140 tons	—	Hr	—	—	215.00

Equipment Rental Rates Costs are for equipment in good condition but exclude accessories. Includes the costs of damage waiver (about 10% of the rental cost) insurance, fuel, oil. Add the cost of an equipment operator, pickup, return to yard, and repairs if necessary. Delivery and pickup within 30 miles of the rental yard are usually offered at an additional cost of one day's rent. A rental "day" is assumed to be one 8-hour shift and begins when the equipment leaves the rental yard. A "week" is 40 hours in five consecutive days. A "month" is 176 hours in 30 consecutive days.

	Day	Week	Month
Air equipment rental			
Air compressors, wheel-mounted			
15 CFM, shop type, electric	35.00	119.00	360.00
50 CFM, shop type, electric	41.00	140.00	410.00
90 CFM, gasoline unit	50.00	174.00	510.00
150 CFM, gasoline unit	60.00	215.00	580.00
175 CFM, gasoline unit	70.00	248.00	701.00
190 CFM, gasoline unit	80.00	264.00	762.00
Air compressors wheel-mounted diesel units			
to 159 CFM	60.00	200.00	600.00
160 - 249 CFM	80.00	266.00	750.00
250 - 449 CFM	135.00	470.00	1,240.00
450 - 749 CFM	210.00	775.00	1,820.00
750 - 1149 CFM	270.00	890.00	2,530.00
1150 CFM and over	440.00	1,400.00	3,960.00
Paving breakers (no bits included) hand-held, pneumatic			
To 40 lb	22.00	71.00	195.00
41 - 55 lb	24.00	80.00	225.00
56 - 70 lb	28.00	85.00	230.00
71 - 90 lb	30.00	89.00	240.00
Paving breakers jackhammer bits			
Moil points, 15" to 18"	6.00	13.00	27.00
Chisels, 3"	6.00	14.00	33.00
Clay spades, 5-1/2"	9.00	21.00	48.00
Asphalt cutters, 5"	8.00	22.00	49.00
Rock drills, jackhammers, hand-held, pneumatic			
to 29 lb	30.00	94.00	270.00
30 - 39 lb	31.00	100.00	272.00
40 - 49 lb	34.00	111.00	320.00
50 lb and over	36.00	120.00	328.00
Rock drill bits, hexagonal			
7/8", 2' long	8.00	21.00	48.00
1", 6' long	8.00	25.00	60.00
Pneumatic chippers, medium weight, 10 lb	25.00	77.00	211.00
Pneumatic tampers, medium weight, 30 lb	30.00	90.00	240.00
Pneumatic grinders, hand-held, 7 lb (no stone)	28.00	110.00	260.00
Air hose, 50 LF section			
5/8" air hose	4.00	12.00	30.00
3/4" air hose	6.00	18.00	44.00
1" air hose	7.00	23.00	55.00
1-1/2" air hose	17.00	51.00	125.00

	Day	Week	Month
Compaction equipment rental			
Vibro plate, 300 lb, 24" plate width, gas	68.00	220.00	595.00
Vibro plate, 600 lb, 32" plate width, gas	120.00	400.00	1,100.00
Rammer, 60 CPM, 200 lb, gas powered	60.00	193.00	540.00
Rollers, two axle steel drum, road type			
1 ton, gasoline	95.00	290.00	830.00
4 ton, gasoline	210.00	575.00	1,810.00
8 ton, diesel	250.00	845.00	2,590.00
Rollers, rubber tired, self propelled			
12 ton, 9 wheel, diesel	254.00	790.00	2,280.00
15 ton, 9 wheel, diesel	251.00	877.00	2,580.00
30 ton, 11 wheel, diesel	450.00	1,270.00	3,930.00
Rollers, towed, sheepsfoot type			
40" diameter, double drum, 4' wide	94.00	270.00	800.00
60" diameter, double drum, 5' wide	115.00	325.00	1,400.00
Rollers, self-propelled vibratory dual sheepsfoot drum, gas			
25 HP, 26" wide roll	295.00	1,000.00	3,020.00
100 HP, 36" wide roll	480.00	1,600.00	4,930.00
Rollers, vibrating steel drum, gas powered, walk behind			
7 HP, 1,000 lb, one drum, 2' wide	100.00	320.00	920.00
10 HP, 2,000 lb, two drum, 2'-6" wide	160.00	530.00	1,500.00
Rollers, self-propelled riding type, vibrating steel drum			
12 HP, 4,000 lb, gasoline, two drum, 3' wide	190.00	605.00	1,700.00
35 HP, 10,000 lb, diesel, two drum, 4' wide	340.00	1,100.00	3,300.00
100 HP, 20,000 lb, diesel, two drum, 8' wide	447.00	1,400.00	4,140.00
Concrete equipment rental			
Buggies, push type, 7 CF	20.00	65.00	165.00
Buggies, walking type, 12 CF	66.00	208.00	580.00
Buggies, riding type, 14 CF	70.00	230.00	630.00
Buggies, riding type, 18 CF	100.00	320.00	920.00
Vibrator, electric, 3 HP, flexible shaft	37.00	125.00	333.00
Vibrator, gasoline, 6 HP, flexible shaft	48.00	160.00	440.00
Troweling machine, 36", 4 paddle	49.00	160.00	440.00
Troweling machine, 48", riding type	134.00	412.00	1,220.00
Concrete saws, gas powered, excluding blade cost			
10 HP, push type	48.00	175.00	488.00
20 HP, self propelled	50.00	164.00	457.00
30 HP, self propelled	75.00	370.00	841.00
Concrete bucket, bottom dump 1 CY	45.00	130.00	352.00
Concrete mixer, gas, trailer mount, 6 CF	52.00	170.00	475.00
Plaster mixer, gas, portable, 5 CF	50.00	160.00	445.00
Concrete conveyor, belt type, portable, gas powered			
all sizes	248.00	745.00	2,030.00
Vibrating screed, 16', single beam	82.00	205.00	570.00
Column clamps, to 48", per set	—	—	5.00
Crane rental Add for clamshell, dragline, or pile driver attachments from below.			
Cable controlled crawler-mounted lifting cranes, diesel			
45 ton	—	—	5,400.00
60 ton	—	2,200.00	6,400.00
100 ton (125 ton)	1,180.00	3,200.00	7,700.00
150 ton	1,200.00	3,800.00	12,500.00

	Day	Week	Month
Cable controlled, truck-mounted lifting cranes, diesel			
45 ton	—	—	5,500.00
75 ton	780.00	2,440.00	7,150.00
100 ton	1,250.00	3,600.00	9,200.00
Hydraulic - rough terrain			
10 ton, gasoline	380.00	1,090.00	3,200.00
15 ton, diesel	340.00	1,100.00	3,300.00
35 ton, diesel	675.00	2,150.00	6,150.00
50 ton, diesel	1,100.00	2,800.00	8,220.00

Clamshell, dragline, and pile driver attachments Add to crane rental costs shown above.

	Day	Week	Month
Clamshell buckets			
Rehandling type, 1 CY	—	320.00	850.00
General purpose, 2 CY	75.00	480.00	1,300.00
Dragline buckets			
1 CY bucket	40.00	100.00	280.00
2 CY bucket	75.00	190.00	560.00
Pile drivers, hammers			
MKT diesel, model DA 35	655.00	1,840.00	4,500.00
MKT steam, double acting, model 7	—	626.00	1,550.00
MKT vibratory, model V-20	2,100	3,800.00	9,500.00
MKT extractor, model E4	—	610.00	1,640.00

Excavation equipment rental

	Day	Week	Month
Backhoe/loaders, crawler mounted, diesel			
3/8 CY backhoe	200.00	750.00	2,200.00
1/2 CY backhoe	210.00	780.00	2,300.00
3/4 CY backhoe	225.00	810.00	2,370.00
1 CY backhoe	275.00	930.00	2,700.00
1-1/2 CY backhoe	350.00	1,200.00	3,500.00
2 CY backhoe	530.00	1,740.00	5,150.00
Backhoe/loaders, wheel mounted, diesel or gasoline			
1/2 CY bucket capacity, 55 HP	230.00	730.00	2,160.00
1 CY bucket capacity, 65 HP	260.00	830.00	2,310.00
1-1/4 CY bucket capacity, 75 HP	282.00	900.00	2,670.00
1-1/2 CY bucket capacity, 100 HP	330.00	1,120.00	3,280.00
Crawler dozers, with angle dozer or bulldozer blade, diesel			
65 HP, D-3	285.00	895.00	2,600.00
90-105 HP, D-4 or D-5	370.00	1,240.00	3,600.00
140 HP, D-6	580.00	1,890.00	5,570.00
200 HP, D-7	880.00	2,900.00	8,600.00
335 HP, D-8	1,250.00	4,070.00	12,000.00
460 HP, D-9	1,700.00	5,800.00	17,000.00
Crawler loaders, diesel			
3/4 CY loader	230.00	800.00	2,350.00
1-1/2 CY loader	350.00	1,200.00	3,470.00
2 CY loader	530.00	1,740.00	5,160.00
Hydraulic excavators, crawler mounted with backhoe type arm, diesel			
.2 - .4 CY bucket	2,700.00	925.00	290.00
.4 - .6 CY bucket	3,200.00	1,090.00	365.00
.6 - .8 CY bucket	3,580.00	1,240.00	390.00

	Day	Week	Month
Gradall, truck mounted, diesel			
3/4 CY bucket	783.00	2,410.00	6,900.00
1 CY bucket	—	3,170.00	8,900.00
Wheel loaders, front-end load and dump, diesel			
3/4 CY bucket, 4WD, articulated	190.00	630.00	1,750.00
1 CY bucket, 4WD, articulated	230.00	740.00	2,220.00
2 CY bucket, 4WD, articulated	310.00	1,060.00	2,900.00
3-1/4 CY bucket, 4WD, articulated	620.00	1,920.00	5,620.00
5 CY bucket, 4WD, articulated	950.00	2,370.00	7,000.00
Motor scraper-hauler, single engine drive, 2 wheel tractor			
200 HP, 15 CY capacity	1,050.00	3,850.00	11,400.00
250 HP, 18 CY capacity	1,470.00	5,400.00	16,000.00
350 HP, 24 CY capacity	1,920.00	6,320.00	18,600.00
550 HP, 32 CY capacity	3,790.00	11,300.00	30,000.00
Graders, diesel, pneumatic tired, articulated			
100 HP	370.00	1,190.00	3,500.00
150 HP	590.00	1,890.00	5,700.00
200 HP	820.00	2,400.00	7,500.00
Trenchers, inclined chain boom type, pneumatic tired			
15 HP, 12" wide, 48" max. depth, walking	200.00	550.00	1,850.00
20 HP, 12" wide, 60" max. depth, riding	230.00	750.00	2,200.00
55 HP, 18" wide, 96" max. depth, riding	340.00	1,100.00	3,150.00
100 HP, 24" x 96" deep, crawler mount	590.00	1,800.00	5,700.00
Trench boxes, steel, 8' high x 12' long, single wall	133.00	500.00	1,480.00
Hydraulic trench braces, 8' long, to 42" wide	17.00	50.00	155.00
Track-mounted wagon drills, self propelled, crawler mounted			
4" drifter, straight boom	436.00	1,150.00	3,220.00
4-1/2" drifter, hydraulic swing boom	699.00	2,130.00	6,300.00
Dump trucks, rental rate plus mileage			
3 CY ($.40 mileage charge)	165.00	450.00	1,350.00
5 CY ($.44 mileage charge)	180.00	690.00	2,200.00
10 CY ($.48 mileage charge)	230.00	900.00	2,600.00
25 CY off highway	—	3,100.00	9,300.00

Forklifts and fork loaders

	Day	Week	Month
Rough terrain, towable, gas powered			
4,000 lb capacity	188.00	555.00	1,630.00
8,000 lb capacity	190.00	630.00	1,850.00
Extension boom, 2 wheel steering			
4,000 lb, 30' lift	220.00	740.00	2,140.00
6,000 lb, 35' lift	270.00	880.00	2,460.00
8,000 lb, 40' lift, 4 wheel steering	315.00	990.00	2,800.00

Platforms, aerial

	Day	Week	Month
Rolling scissor lifts, 2' x 3' platform, 650 lb capacity, push around, electric powered			
30' high	60.00	172.00	516.00
40' high	114.00	300.00	879.00
50' high	135.00	440.00	1,250.00
Rolling scissor lifts, self-propelled, hydraulic, electric powered			
to 20'	85.00	230.00	625.00
21' - 30'	107.00	320.00	879.00
31' - 40'	150.00	460.00	1,360.00

	Day	Week	Month
Rolling scissor lifts, self-propelled, hydraulic, diesel powered			
to 20'	105.00	285.00	780.00
21' - 30'	135.00	385.00	1,110.00
31' - 40'	180.00	510.00	1,450.00
Boomlifts, telescoping and articulating booms, self-propelled, gas or diesel powered, 2-wheel drive			
21' - 30'	160.00	512.00	1,400.00
31' - 40'	243.00	660.00	1,900.00
41' - 50'	242.00	740.00	2,180.00
51' - 60'	376.00	1,100.00	3,1500.00
Pump rental Hoses not included.			
Submersible pumps			
1-1/2" - 2"	33.00	105.00	295.00
3" - 4"	75.00	240.00	666.00
Trash pumps, self-priming, gas			
1-1/2" connection	34.00	116.00	334.00
3" connection	57.00	190.00	525.00
4" connection	150.00	532.00	1,590.00
Diaphragm pumps			
2", gas, single action	39.00	125.00	345.00
4", gas, double action	76.00	220.00	621.00
Hose, coupled, 25 LF sections			
2" suction line	10.00	28.00	72.00
1-1/2" discharge line	15.00	39.00	96.00
3" discharge line	10.00	28.00	72.00
Sandblasting equipment rental			
Compressor and hopper			
To 250 PSI	48.00	178.00	440.00
Over 250 to 300 PSI	75.00	275.00	715.00
Over 600 PSI to 1000 PSI	170.00	550.00	1,600.00
Sandblasting accessories			
Hoses, 50', coupled			
3/8" sandblast hose or air hose (whip line)	8.00	20.00	50.00
3/4" air hose	9.00	30.00	71.00
1" air hose or 1" sandblast hose	16.00	50.00	120.00
Nozzles, all types	15.00	50.00	98.00
Valve, remote control (deadman valve), all sizes	30.00	45.00	130.00
Air-fed hood	25.00	72.00	194.00
Roadway and pavement equipment rental			
Boom mounted pavement breaker, hydraulic 1,500 lb	330.00	1,085.00	3,300.00
Paving breaker, backhoe mount, pneumatic, 1,000 lb	270.00	870.00	1,850.00
Paving machine, diesel, 10' width, self propelled	1,300.00	3,900.00	12,000.00
Distribution truck for asphalt prime coat, 3,000 gallon	840.00	2,510.00	7,420.00
Pavement striper, 1 line, walk behind	50.00	200.00	700.00
Water blast truck for 1,000 PSI pavement cleaning	1,260.00	3,650.00	11,300.00
Router-groover for joint or crack sealing in pavement	191.00	650.00	1,960.00

	Day	Week	Month
Sealant pot for joint or crack sealing	230.00	720.00	2,150.00
Water truck, 4,000 gallon	565.00	1,800.00	5,200.00
Self-propelled street sweeper, vacuum (7 CY max.)	255.00	1,850.00	5,400.00

Miscellaneous construction equipment rental

	Day	Week	Month
Brush chipper, trailer mounted 12" diameter capacity	256.00	779.00	2,240.00
Chain saws, 18", gasoline	40.00	139.00	410.00
Circular hand saws, electric	21.00	70.00	200.00
Generators, electric, portable			
2.5 kw, gasoline	45.00	144.00	409.00
5 kw, gasoline	66.00	205.00	580.00
15 kw, diesel	108.00	305.00	861.00
60 kw, diesel	214.00	555.00	1,620.00
Grinders, hand held, electric	23.00	77.00	210.00
Heaters, portable, oil or gasoline, fan forced			
To 200 MBtu	38.00	120.00	355.00
Over 200 to 300 MBtu	42.00	133.00	387.00
Over 500 to 1,000 MBtu	77.00	260.00	695.00
Salamanders, LP gas (no powered fan)	53.00	176.00	470.00
Laser pipe levels	77.00	260.00	695.00
Light towers, trailer mounted, gas powered			
through 7,000 watt	85.00	280.00	780.00
8,000 watt and over	100.00	325.00	890.00
Masonry saws, excluding blade cost			
Portable stand saw	55.00	175.00	550.00
Table saw, 2 HP	55.00	180.00	535.00
Post hole digger, hand held	50.00	170.00	480.00
Power washer (water blaster), gas powered	66.00	236.00	649.00
Sander, belt, 3" width	15.00	53.00	139.00
Sander, disc, 8" diameter	17.00	60.00	170.00
Welding machine, with helmet			
to 200 amp	51.00	161.00	450.00
201 - 300 amp	59.00	190.00	535.00
Wheelbarrows, contractor style	8.00	26.00	64.00

	Craft@Hrs	Unit	Material	Labor	Total
Concrete Pumping, Subcontract Small jobs, including truck rent and operator. Based on 6 sack concrete mix designed for pumping. No concrete included.					
3/8" aggregate mix (pea gravel), using hose to 200'					
First 7 CY (7 CY minimum)	—	CY	—	—	16.80
Additional cubic yards over 7	—	CY	—	—	6.88
Add for hose over 200', per LF	—	LF	—	—	1.00
3/4" aggregate mix, using hose to 200'					
First 10 CY (10 CY minimum)	—	CY	—	—	27.50
Additional cubic yards over 10	—	CY	—	—	6.88
Add for hose over 200', per LF	—	LF	—	—	1.00

	Craft@Hrs	Unit	Material	Labor	Total

Concrete Pumping with a Boom Truck, Subcontract Includes truck rent, operator, local travel but no concrete. Add costs equal to 1 hour for equipment setup and 1 hour for cleanup. Use 4 hours as the minimum cost for 23, 28 and 32 meter boom trucks and 5 hours as the minimum cost for 36 through 52 meter boom trucks. Estimate the actual pour rate at 70% of the rated capacity on thicker slabs and 50% of the capacity on most other work.

Boom lengths over 23 meters include both an operator and an oiler. Costs shown include subcontractor's markup.

	Craft@Hrs	Unit	Material	Labor	Total
23 meter boom (75'), 70 CY per hour rating	—	Hr	—	—	55.00
Add per CY pumped with 23 meter boom	—	CY	—	—	2.00
28 meter boom (92'), 70 CY per hour rating	—	Hr	—	—	65.00
Add per CY pumped with 28 meter boom	—	CY	—	—	2.00
32 meter boom (105'), 90 CY per hour rating	—	Hr	—	—	75.00
Add per CY pumped with 32 meter boom	—	CY	—	—	2.00
36 meter boom (120'), 90 CY per hour rating	—	Hr	—	—	95.00
Add per CY pumped with 36 meter boom	—	CY	—	—	2.00
42 meter boom (138'), 100 CY per hour rating	—	Hr	—	—	125.00
Add per CY pumped with 42 meter boom	—	CY	—	—	2.50
52 meter boom (170'), 100 CY per hour rating	—	Hr	—	—	150.00
Add per CY pumped with 52 meter boom	—	CY	—	—	3.00

Temporary Enclosures Rented chain link fence and accessories. Costs are a one-time charge for up to six months usage on a rental basis. Costs include installation and one trip for removal and assume level site with truck access along pre-marked fence line. Add for gates and barbed wire as shown. Minimum charge is $295.

	Craft@Hrs	Unit	Material	Labor	Total
Chain link fence, 6' high					
Less than 250 feet	—	LF	—	—	1.00
250 to 500 feet	—	LF	—	—	.90
501 to 750 feet	—	LF	—	—	.80
751 to 1,000 feet	—	LF	—	—	.70
Over 1,000 feet	—	LF	—	—	.65
Add for gates					
6' x 10' single	—	Ea	—	—	77.50
6' x 12' single	—	Ea	—	—	100.00
6' x 15' single	—	Ea	—	—	130.00
6' x 20' double	—	Ea	—	—	150.00
6' x 24' double	—	Ea	—	—	200.00
6' x 30' double	—	Ea	—	—	250.00
Add for barbed wire					
per strand, per lineal foot	—	LF	—	—	.15
Contractor furnished items, installed and removed, based on single use and no salvage value					
Railing on stairway, two sides, 2" x 4"	CL@.121	LF	.78	3.66	4.44
Guardrail at second and higher floors					
toe rail, mid rail and top rail	CL@.124	LF	2.08	3.75	5.83
Plywood barricade fence					
Bolted to pavement, 8' high	CL@.430	LF	7.79	13.00	20.79
Post in ground, 8' high	CL@.202	LF	8.68	6.11	14.79
8' high with 4' wide sidewalk cover	CL@.643	LF	22.20	19.50	41.70

Scaffolding Rental Exterior scaffolding, tubular steel, 60" wide, with 6'4" open end frames set at 7', with cross braces, base plates, mud sills, adjustable legs, post-mounted guardrail, climbing ladders and landings, brackets, clamps and building ties. Including two 2" x 10" scaffold planks on all side brackets and six 2" x 10" scaffold planks on scaffold where indicated. Add local delivery and pickup. Minimum rental is one month.

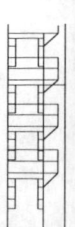

	Craft@Hrs	Unit	Material	Labor	Total
Costs are per square foot of building wall covered, per month.					
With plank on scaffold only	—	SF	.65	—	.65
With plank on side brackets and scaffold	—	SF	.90	—	.90

	Craft@Hrs	Unit	Material	Labor	Total
Caster-mounted scaffolding, 30" wide by 7' or 10' long, rental. Minimum charge is one month.					
6' high	—	Mo	70.00	—	70.00
7' high	—	Mo	90.00	—	90.00
10' high	—	Mo	95.00	—	95.00
13' high	—	Mo	100.00	—	100.00
15' high	—	Mo	115.00	—	115.00
20' high	—	Mo	200.00	—	200.00
Hook-end scaffold plank, rental, each	—	Mo	8.00	—	8.00
Plain-end micro-lam scaffold plank, rental each					
8' length	—	Mo	5.00	—	5.00
16' length	—	Mo	6.50	—	6.50
Add for erection and dismantling, level ground, truck accessible					
With plank on scaffold only	CL@.008	SF	—	.24	.24
With planks on side brackets and scaffold	CL@.012	SF	—	.36	.36
Safety nets nylon, 4" mesh, purchase	CL@.004	SF	1.60	.12	1.72
Safety nets nylon, 1/2" mesh, purchase	CL@.004	SF	1.80	.12	1.92
Swinging stage, 10', complete, motor operated, purchase	CL@4.00	Ea	1,220.00	121.00	1,341.00

Heavy Duty Shoring Adjustable vertical tower type shoring. Rated capacity 5.5 tons per leg. Base frames are 4' wide by 6' high or 2' wide by 5' high. Extension frames are 4' wide by 5'4" high or 2' wide by 4'4" high. Frames are erected in pairs using cross-bracing to form a tower. Combinations of base frames and extension frames are used to reach the height required. Screw jacks with base plates are installed in the bottom of each leg. Similar screw jacks with "U" bracket heads are installed in the top of each leg. Material costs shown are rental rates based on a 1-month minimum. Add the cost of delivery and pickup. For scheduling purposes, estimate that a crew of 2 can unload, handle and erect 8 to 10 frames (including typical jacks and heads) per hour. Dismantling, handling and moving or loading will require nearly the same time.

	Craft@Hrs	Unit	Material	Labor	Total
Frames (any size), including braces and assembly hardware, rental per frame per month	—	Mo	8.40	—	8.40
Screw jacks (four required per frame with base plate or "U" head),					
Rental per jack per month	—	Mo	3.15	—	3.15
Add for erecting each frame	CL@.245	Ea	—	7.42	7.42
Add for dismantling each frame	CL@.197	Ea	—	5.96	5.96

Temporary Structures Portable job site office trailers with electric air conditioners and baseboard heat, built-in desks, plan tables, insulated walls, tiled floors, paneled walls, locking doors and windows with screens, closet, fluorescent lights, and electrical receptacles. Add $30 per month for units with toilet and lavatory. Monthly rental based on 6 month minimum. Dimensions show length, width and height overall and floor area (excluding hitch area).

	Craft@Hrs	Unit	Material	Labor	Total
16' x 8' x 7' high (108 SF)	—	Mo	—	—	113.00
20' x 8' x 7' high (160 SF)	—	Mo	—	—	123.00
24' x 8' x 7' high (192 SF)	—	Mo	—	—	138.00
30' x 8' x 7' high (240 SF)	—	Mo	—	—	179.00
36' x 10' x 10' high (360 SF)	—	Mo	—	—	200.00
44' x 10' x 10' high (440 SF)	—	Mo	—	—	217.00
50' x 10' x 10' high (500 SF)	—	Mo	—	—	246.00
50' x 12' x 10' high (552 SF)	—	Mo	—	—	251.00
60' x 12' x 10' high (720 SF)	—	Mo	—	—	287.00
60' x 14' x 10' high (840 SF)	—	Mo	—	—	308.00
Add for skirting, per LF of perimeter	—	LF	—	—	9.00
Deduct for unit without heating & cooling	—	Mo	—	—	-25.00
Add for delivery and setup of office trailers, within 15 miles					
Typical cost	—	LS	—	—	169.00
Add for typical pickup, within 15 miles	—	LS	—	—	179.00

	Craft@Hrs	Unit	Material	Labor	Total
Add for delivery or pickup over 15 miles, per mile					
8' wide units	—	Mile	—	—	1.65
10' wide units	—	Mile	—	—	2.50
12' wide units	—	Mile	—	—	2.80
14' wide units	—	Mile	—	—	3.30

Portable steel storage containers (lockable), suitable for job site storage of tools and materials. Rental per month, based on 6 month rental

	Craft@Hrs	Unit	Material	Labor	Total
50' x 12' x 8'	—	Mo	—	—	250.00
26' x 8' x 8'	—	Mo	—	—	101.00
22' x 8' x 7'	—	Mo	—	—	91.70
16' x 8' x 7'	—	Mo	—	—	60.00
Add for delivery and pickup, within 15 miles					
One time charge	—	LS	—	—	61.50

Portable job site shacks with lights, power receptacles, locking door and window. Rental per month, based on 6 month rental

	Craft@Hrs	Unit	Material	Labor	Total
12' x 8' x 8'	—	Mo	—	—	67.70
8' x 8' x 8'	—	Mo	—	—	56.40
Add for typical delivery and pickup, within 15 miles					
One time charge	—	LS	—	—	56.40

Cleanup

Progressive "broom clean" cleanup, as necessary, per 1,000 SF of floor per month

	Craft@Hrs	Unit	Material	Labor	Total
Typical cost	CL@.183	MSF	—	5.54	5.54
Final, total floor area (no glass cleaning)	CL@1.65	MSF	—	49.90	49.90

Glass cleaning, per 1,000 square feet of glass cleaned on one side. (Double these figures when both sides are cleaned.) Add the cost of staging or scaffolding when needed

	Craft@Hrs	Unit	Material	Labor	Total
Cleaning glass with sponge and squeegee	CL@1.34	MSF	10.20	40.60	50.80
Cleaning glass and window trim					
With solvent, towel, sponge and squeegee	CL@5.55	MSF	13.90	168.00	181.90
Removing paint and concrete splatter from windows and trim					
By scraping and solvent	CL@15.9	MSF	40.00	481.00	521.00
Mop resilient floor by hand	CL@1.20	MSF	1.25	36.30	37.55

Waste Disposal

Dump fees Tippage charges for solid waste disposal at the dump vary from $30 to $120 per ton. For planning purposes, estimate waste disposal at $75 per ton plus the hauling cost. Call the dump or trash disposal company for actual charges. Typical costs are shown below

	Craft@Hrs	Unit	Material	Labor	Total
Dumpster, 3 CY trash bin, emptied weekly	—	Mo	—	—	100.00
Dumpster, 40 CY solid waste bin (lumber, drywall, roofing)					
Hauling cost, per load	—	Ea	—	—	195.00
Add to per load charge, per ton	—	Ton	—	—	50.00
Low-boy, 14 CY solid waste container (asphalt, dirt, masonry, concrete)					
Hauling cost, per load (use 7 CY as maximum load)	—	Ea	—	—	205.00
Add to per load charge, per ton	—	Ton	—	—	45.00

Recycler fees The use of companies that recycle construction waste materials can substantially reduce disposal costs. Recycling charges vary from $95 to $120 per load, depending on the type material and the size of the load. For planning purposes, estimate waste recycling charges at $10 per CY, or approximately $7 per ton. Call the recycling company for actual charges. Typical Recycler fees are shown below. Add the cost for hauling.

	Craft@Hrs	Unit	Material	Labor	Total
Asphalt, per load (10 to 12 CY)	—	Ea	—	—	100.00
Concrete, masonry or rock, per load (10 to 12 CY)	—	Ea	—	—	75.00
Dirt, per load (10 to 12 CY)	—	Ea	—	—	60.00
Mixed loads, per load (10 to 12 CY)	—	Ea	—	—	135.00

	Craft@Hrs	Unit	Material	Labor	Equipment	Total

Structure Moving Up to 5 mile haul, not including new foundation costs, utility hookup or finishing. These figures assume two stories maximum, no obstruction from trees or utility lines and that adequate right of way is available. Fees and charges imposed by government are not included. Costs shown are per SF of total floor area. Equipment cost per hour is $160 and includes one tow-truck equipped with dollies, jacks, support beams and hand tools. For estimating purpose, use 1,000 SF as a minimum job charge.

Concrete or masonry structures, 12" maximum thick walls or floors

	Craft@Hrs	Unit	Material	Labor	Equipment	Total
1,000 to 2,000 SF	C1@.168	SF	—	5.13	5.34	10.47
2,000 to 4,000 SF	C1@.149	SF	—	4.55	4.74	9.29

Wood frame

	Craft@Hrs	Unit	Material	Labor	Equipment	Total
1,000 to 2,000 SF	C1@.142	SF	—	4.34	4.52	8.86
2,000 to 4,000 SF	C1@.131	SF	—	4.00	4.17	8.17

Steel frame

	Craft@Hrs	Unit	Material	Labor	Equipment	Total
1,000 to 2,000 SF	C1@.215	SF	—	6.57	6.84	13.41
2,000 to 4,000 SF	C1@.199	SF	—	6.08	6.33	12.41

Site Clearing Using a dozer as described below, a 3/4 CY wheel loader and two 10 CY dump trucks. Clearing costs will be higher on smaller jobs and where the terrain limits production. These costs include hauling 6 miles to a legal dump. Dump fees are not included. See Waste Disposal.
Quantity shown in parentheses gives the approximate "loose" volume of the materials removed (volume being hauled to the dump). For estimating purposes, add the cost equal to 1 acre of the appropriate clearing operation to allow for moving the equipment on and off the job.

	Craft@Hrs	Unit	Material	Labor	Equipment	Total
Clear light brush and grub roots						
Using a 105 HP D-5 dozer (350 CY)	C2@34.8	Acre	—	1,190.00	732.50	1,922.50
Clear medium brush and small trees, grub roots						
Using a 130 HP D-6 dozer (420 CY)	C2@41.8	Acre	—	1,430.00	1,030.00	2,460.00
Clear brush and trees to 6" trunk diameter						
Using a 335 HP D-8 dozer (455 CY)	C2@45.2	Acre	—	1,550.00	1,661.00	3,211.00
Clear wooded area, pull stumps						
Using a 460 HP D-9 dozer (490 CY)	C2@48.7	Acre	—	1,670.00	2,102.00	3,772.00

Paving and Curb Demolition No salvage of materials. These costs include the cost of loading and hauling to a legal dump within 6 miles. Dump fees are not included. See Waste Disposal. Equipment cost per hour is $52.00 and includes one wheel-mounted air compressor, one paving breaker and jackhammer bits, one 55 HP wheel loader with integral backhoe and one 5 CY dump truck. The figures in parentheses give the approximate "loose" volume of the materials (volume after being demolished). Use $400.00 as a minimum charge.

	Craft@Hrs	Unit	Material	Labor	Equipment	Total
Bituminous paving, depths to 3" (85 SF per CY)						
Large area, with a wheel loader	C3@.002	SF	—	.07	.03	.10
Strips 24" wide for utility lines	C3@.003	SF	—	.10	.05	.15
Add for jobs under 500 SF	C3@.001	SF	—	.03	.02	.05
Bituminous curbs,						
to 12" width (65 LF per CY)	C3@.002	LF	—	.07	.03	.10
Concrete paving and slabs on grade						
4" concrete without rebars						
(.80 CY per CY)	C3@.325	CY	—	10.90	5.60	16.50
4" concrete without rebars						
(65 SF per CY)	C3@.004	SF	—	.13	.07	.20
6" concrete without rebars						
(.80 CY per CY)	C3@.421	CY	—	14.20	7.30	21.50
6" concrete without rebars						
(45 SY per CY)	C3@.007	SF	—	.24	.12	.36

	Craft@Hrs	Unit	Material	Labor	Equipment	Total
8" concrete without rebars						
(.80 CY per CY)	C3@.666	CY	—	22.40	11.54	33.94
8" concrete without rebars						
(32 SF per CY)	C3@.009	SF	—	.30	.16	.46
Over 8" to 12" thick concrete						
Per CY without rebars						
(.80 CY per CY)	C3@1.00	CY	—	33.60	17.33	50.93
Per CY with rebars						
(.68 CY per CY)	C3@1.20	CY	—	40.40	20.80	61.20
Per SF without rebars						
(20 SF per CY)	C3@.020	SF	—	.67	.35	1.02
Per SF with rebars						
(20 SF per CY)	C3@.025	SF	—	.84	.43	1.27
Concrete curbs, demolition						
Curb and 24" monolithic						
gutter (7 LF per CY)	C3@.100	LF	—	3.36	1.73	5.09
Planter and batter type curbs,						
6" width (30 LF per CY)	C3@.043	LF	—	1.45	.75	2.20

Removal of pavement markings by water blasting. Equipment cost is $7.00 per hour. Use $100.00 as a minimum charge

	Craft@Hrs	Unit	Material	Labor	Equipment	Total
4" wide strips	CL@.032	LF	—	.97	.22	1.19
Per square foot	CL@.098	SF	—	2.97	.69	3.66

Fence and Guardrail Demolition No salvage except as noted. These costs include the cost of loading and hauling to a legal dump within 6 miles. Dump fees are not included. See Waste Disposal.

Fencing demolition Equipment cost per hour is $20.00 and includes one 5 CY dump truck. Use $250.00 as a minimum charge.

	Craft@Hrs	Unit	Material	Labor	Equipment	Total
Remove and dispose chain link,						
6' high	C4@.015	LF	—	.46	.10	.56
Remove and salvage chain link,						
6' high	C4@.037	LF	—	1.14	.25	1.39
Remove and dispose board fence,						
6' high	C4@.016	LF	—	.49	.11	.60

Highway type guardrail demolition Equipment cost per hour is $63.80 and includes one wheel-mounted air compressor, one paving breaker and jackhammer bits, one 55 HP wheel loader with integral backhoe and one 5 CY dump truck. Use $400.00 as a minimum charge

	Craft@Hrs	Unit	Material	Labor	Equipment	Total
Remove and dispose guardrail	C4@.043	LF	—	1.32	.91	2.23
Remove guardrail in salvage condition	C4@.074	LF	—	2.27	1.58	3.85
Remove and dispose guardrail posts	C3@.245	Ea	—	8.24	5.22	13.46

Manhole, Piping and Underground Tank Demolition No salvage of materials except as noted. These costs include the cost of loading and hauling to a legal dump within 6 miles. Dump fees are not included. See Waste Disposal. Equipment cost per hour is $50.00 and includes one wheel-mounted air compressor, one paving breaker and jackhammer bits, one 55 HP wheel loader with integral backhoe and one 5 CY dump truck. Use $400.00 as a minimum charge.

Manholes and catch basins, demolition, to 10' deep. Break below collar and plug

	Craft@Hrs	Unit	Material	Labor	Equipment	Total
Brick	C3@5.16	Ea	—	174.00	86.17	260.17
Masonry	C3@5.33	Ea	—	179.00	89.00	268.00
Precast concrete	C3@6.64	Ea	—	223.00	110.90	333.90
Add for sand fill, any of above	C3@.125	CY	12.00	4.21	2.09	18.30

	Craft@Hrs	Unit	Material	Labor	Equipment	Total
Frame and cover from manhole or catch basin						
Remove in salvage condition	C3@1.45	Ea	—	48.80	24.22	73.02
Remove and reset	C3@3.99	Ea	—	134.00	66.63	200.63
Fire hydrant demolition						
Remove and dispose	C3@5.57	Ea	—	187.00	93.02	280.02
Break out storm or sewer pipe, non-salvageable. Excavation or backfill not included						
Up to 12" diameter	C3@.114	LF	—	3.84	1.90	5.74
15" to 18"	C3@.134	LF	—	4.51	2.24	6.75
21" to 24"	C3@.154	LF	—	5.18	2.57	7.75
27" to 36"	C3@.202	LF	—	6.80	3.37	10.17
Remove welded steel pipe for salvage. Excavation or backfill not included						
4" diameter or smaller	C3@.111	LF	—	3.73	60.96	64.69
6" to 10" diameter	C3@.202	LF	—	6.80	3.37	10.17

Remove and haul away empty underground liquid storage tanks. (Draining and disposing of hazardous liquid in a tank may require special waste handling equipment.) Cost of draining not included. Includes excavation and backfill to 6' deep.

	Craft@Hrs	Unit	Material	Labor	Equipment	Total
50 to 250 gallon tank	C3@4.24	Ea	—	143.00	70.80	213.80
Over 250 to 600 gallon tank	C3@11.7	Ea	—	394.00	195.39	589.39
Over 600 to 1,000 gallon tank	C3@22.4	Ea	—	754.00	374.00	1,128.00
Add for sand fill, any of above	C3@.125	CY	12.00	4.21	2.09	18.30

Miscellaneous Sitework Demolition No salvage of materials except as noted. These costs include the cost of loading and hauling to a legal dump within 6 miles. Dump fees are not included. See Waste Disposal.

Railroad demolition, siding quantities. Equipment cost per hour is $58.20 and includes one wheel-mounted air compressor, one paving breaker and jackhammer bits, one pneumatic breaker and jackhammer bits, one 55 HP wheel loader with integral backhoe and one 5 CY dump truck. Use $400.00 as a minimum charge

	Craft@Hrs	Unit	Material	Labor	Equipment	Total
Remove track and ties for scrap	C5@.508	LF	—	16.70	7.36	24.06
Remove and dispose of ballast stone	C5@.111	CY	—	3.64	1.61	5.25
Remove wood ties alone	C5@.143	Ea	—	4.69	2.10	6.79

Remove light standards, flagpoles, playground poles, up to 30' high, including foundations. Equipment cost per hour is $29.30 and includes one wheel-mounted air compressor, one pneumatic breaker and jackhammer bits and one 5 CY dump truck. Use $400.00 as a minimum charge

	Craft@Hrs	Unit	Material	Labor	Equipment	Total
Remove item, in salvage condition	C4@6.24	Ea	—	192.00	57.60	249.60
Remove item, no salvage	C4@.746	Ea	—	22.90	6.90	29.80

Torch cutting steel plate. Equipment cost per hour is $5.00 for an oxygen-acetylene manual welding and cutting torch with gases, regulator and goggles. Use $200.00 as a minimum charge

	Craft@Hrs	Unit	Material	Labor	Equipment	Total
To 3/8" thick	CL@.073	LF	—	2.21	.37	2.58

Building Demolition Costs for demolishing an entire building. Includes loading and hauling up to 6 miles but no dump fees. See Waste Disposal. Costs are by square foot of floor area based on 8' ceiling height. No salvage value assumed. Figures in parentheses give the approximate "loose" volume of the materials (volume after being demolished).

Light wood-frame structures Up to three stories in height. Based on 2,500 SF job. No basements included. Estimate each story separately. Equipment cost per hour is $40.00 and includes one 55 HP wheel loader with integral backhoe and one 5 CY dump truck. Use $3,200.00 as a minimum charge

	Craft@Hrs	Unit	Material	Labor	Equipment	Total
First story (8 SF per CY)	C5@.038	SF	—	1.25	.39	1.64
Second story (8 SF per CY)	C5@.053	SF	—	1.74	.54	2.28
Third story (8 SF per CY)	C5@.070	SF	—	2.30	.71	3.01

	Craft@Hrs	Unit	Material	Labor	Equipment	Total

Building demolition with pneumatic tools. Equipment cost per hour is $99.30 and includes one wheel-mounted air compressor, three breakers and jackhammer bits, two 55 HP wheel loaders with integral backhoes and two 5 CY dump trucks. Use $10,000.00 as a minimum charge

	Craft@Hrs	Unit	Material	Labor	Equipment	Total
Concrete building (30 SF per CY)	C6@.089	SF	—	2.87	.80	3.67
Reinforced concrete building (20 SF per CY)	C6@.101	SF	—	3.26	.91	4.17
Masonry building (50 SF per CY)	C6@.074	SF	—	2.39	.67	3.06

Building demolition with crane and headache ball. Equipment cost per hour is $112.00 and includes one wheel-mounted air compressor, one pneumatic paving breaker and jackhammer bits, one 55 HP wheel loader with integral backhoe, one 15-ton hydraulic crane with headache ball and three 5 CY dump trucks. Use $10,000.00 as a minimum charge

	Craft@Hrs	Unit	Material	Labor	Equipment	Total
Concrete building (30 SF per CY)	C7@.031	SF	—	1.03	.52	1.55
Reinforced concrete building (20 SF per CY)	C7@.037	SF	—	1.23	.62	1.85
Masonry building (50 SF per CY)	C7@.026	SF	—	.86	.44	1.30

Concrete foundation and footing demolition with D-6 crawler dozer and pneumatic tools. Equipment cost per hour is $79.90 and includes one wheel-mounted air compressor, one pneumatic paving breaker and jackhammer bits, one D-6 crawler dozer with attachments and one 5 CY dump truck. Use $10,000.00 as a minimum charge
Non-reinforced

	Craft@Hrs	Unit	Material	Labor	Equipment	Total
concrete (.75 CY per CY)	C3@1.10	CY	—	37.00	29.30	66.30
Reinforced concrete (.60 CY per CY)	C3@1.58	CY	—	53.20	42.10	95.30

Chip out concrete using paving breaker

	Craft@Hrs	Unit	Material	Labor	Equipment	Total
No dozer used, (.8 CF per CF)	CL@.203	CF	—	6.14	5.41	11.55

Gutting a building Interior finishes stripped back to the structural walls. Building structure to remain. No allowance for salvage value. These costs include loading and hauling up to 6 miles. Dump fees are not included. See Waste Disposal. Costs shown are per square foot of floor area based on 8' ceiling height. Costs will be about 50% less if a small tractor can be used and up to 25% higher if debris must be carried to ground level in an elevator. Equipment cost per hour is $32.50 and includes one air compressor, two pneumatic breakers with jackhammer bits and one 5 CY dump truck. Figures in parentheses give the approximate "loose" volume of the materials (volume after being demolished). Use $2,500.00 as a minimum charge.

	Craft@Hrs	Unit	Material	Labor	Equipment	Total
Residential buildings (125 SF per CY)	C4@.265	SF	—	8.14	2.86	11.00
Commercial buildings (140 SF per CY)	C4@.243	SF	—	7.46	2.62	10.08

Partition wall demolition Building structure to remain. No allowance for salvage value. Dump fees are not included. See Waste Disposal. Costs shown are per square foot of wall removed (as measured on one side). Knock down with pneumatic jackhammers and pile adjacent to site but no removal. Equipment cost per hour is $12.30 and includes one air compressor and two pneumatic breakers with jackhammer bits. Figures in parentheses give the approximate "loose" volume of the materials (volume after being demolished). Use $350.00 as a minimum charge.

Brick or block partition demolition

	Craft@Hrs	Unit	Material	Labor	Equipment	Total
4" thick partition (60 SF per CY)	CL@.040	SF	—	1.21	.25	1.46
8" thick partition (30 SF per CY)	CL@.057	SF	—	1.73	.35	2.08
12" thick partition (20 SF per CY)	CL@.073	SF	—	2.21	.45	2.66

Concrete partition demolition

	Craft@Hrs	Unit	Material	Labor	Equipment	Total
Non-reinforced (18 CF per CY)	CL@.273	CF	—	8.26	1.68	9.94
Reinforced (20 CF per CY)	CL@.363	CF	—	11.00	2.23	13.23

	Craft@Hrs	Unit	Material	Labor	Equipment	Total

Knock down with hand tools and pile adjacent to site but no removal. Stud partition demolition (typically 75 to 100 SF per CY)

	Craft@Hrs	Unit	Material	Labor	Equipment	Total
Gypsum or terra cotta on metal lath	CL@.027	SF	—	.82	—	.82
Drywall on metal or wood studs	CL@.028	SF	—	.85	—	.85
Plaster on metal studs	CL@.026	SF	—	.79	—	.79

Ceiling demolition Building structure to remain. No allowance for salvage value. Dump fees are not included. See Waste Disposal. Costs shown are per square foot of ceiling (as measured on one side). Figures in parentheses give the approximate "loose" volume of the materials (volume after being demolished). Use $300.00 as a minimum charge. Knock down with hand tools to remove ceiling finish on framing at heights to 9'.

	Craft@Hrs	Unit	Material	Labor	Equipment	Total
Plaster ceiling (typically 175 to 200 SF per CY)						
Including lath and furring	CL@.025	SF	—	.76	—	.76
Including suspended grid	CL@.020	SF	—	.61	—	.61
Acoustic tile ceiling (typically 200 to 250 SF per CY)						
Including suspended grid	CL@.010	SF	—	.30	—	.30
Including grid in salvage condition	CL@.019	SF	—	.58	—	.58
Including strip furring	CL@.014	SF	—	.42	—	.42
Tile glued to ceiling	CL@.015	SF	—	.45	—	.45
Drywall ceiling (typically 250 to 300 SF per CY)						
Nailed or attached with screws to joists	CL@.018	SF	—	.54	—	.54
Including strip furring	CL@.023	SF	—	.70	—	.70

Roof demolition Building structure to remain. No allowance for salvage value. Dump fees are not included, refer to construction materials disposal section and add for same. Costs shown are per square foot of area removed (as measured on one side).

Roof surface removal, using hand tools. Figures in parentheses give the approximate "loose" volume of the materials (volume after being demolished). Use $900.00 as a minimum charge.

	Craft@Hrs	Unit	Material	Labor	Equipment	Total
Asphalt shingles (2.50 Sq per CY)	CL@1.33	Sq	—	40.30	—	40.30
Built-up roofing, including sheathing and gravel (1.25 Sq per CY)	CL@2.91	Sq	—	88.10	—	88.10
Clay or concrete tile (.70 Sq per CY)	CL@1.03	Sq	—	31.20	—	31.20
Concrete plank, no covering (.80 Sq per CY)	CL@1.10	Sq	—	33.30	—	33.30
Gypsum plank, no covering (.70 Sq per CY)	CL@.820	Sq	—	24.80	—	24.80
Metal deck, no covering (.50 Sq per CY)	CL@2.27	Sq	—	68.70	—	68.70
Wood shingles (1.66 Sq per CY)	CL@.756	Sq	—	22.90	—	22.90
Remove gravel stop or flashing	CL@.070	LF	—	2.12	—	2.12

Floor slab demolition using pneumatic jackhammers and piled adjacent to site but no removal. Equipment cost per hour is $12.30 and includes one wheel-mounted air compressor and two pneumatic breakers with jackhammer bits. Use $350.00 as a minimum charge.

	Craft@Hrs	Unit	Material	Labor	Equipment	Total
Slab on grade, 4" to 6" thick, reinforced						
With wire mesh (55 SF per CY)	CL@.057	SF	—	1.73	.35	2.08
With number 4 bars (45 SF per CY)	CL@.063	SF	—	1.91	.39	2.30

	Craft@Hrs	Unit	Material	Labor	Equipment	Total
Slab, suspended, 6" to 8" thick, free fall						
(35 SF per CY)	CL@.092	SF	—	2.78	.57	3.35
Slab fill, lightweight concrete fill						
On metal deck (45 SF per CY)	CL@.029	SF	—	.88	.18	1.06
Topping, insulating (150 SF per CY)	CL@.023	SF	—	.70	.14	.84

Floor covering demolition using pneumatic jackhammers and piled adjacent to site but no removal. Equipment cost per hour is $11.00 and includes one air compressor and two pneumatic breakers with jackhammer bits. Figures in parentheses give the approximate "loose" volume of the materials (volume after being demolished). Use $350.00 as a minimum charge.

	Craft@Hrs	Unit	Material	Labor	Equipment	Total
Ceramic or quarry tile, brick						
(200 SF per CY)	CL@.029	SF	—	.88	.18	1.06
Resilient materials only						
(270 SF per CY)	CL@.008	SF	—	.24	.05	.29
Terrazzo						
(225 SF per CY)	CL@.032	SF	—	.97	.20	1.17
Wood blocks						
(200 SF per CY)	CL@.051	SF	—	1.54	.31	1.85
Wood, residential strip floor						
(225 SF per CY)	CL@.032	SF	—	.97	.20	1.17

Cutting openings in frame walls Using hand tools. Cost per square foot of wall measured on one face.

	Craft@Hrs	Unit	Material	Labor	Equipment	Total
Metal stud wall with stucco or plaster	CL@.094	SF	—	2.85	—	2.85
Wood stud wall with drywall	CL@.051	SF	—	1.54	—	1.54
Dust control partitions, 6 mil plastic	C8@.011	SF	.30	.40	—	.70

Debris removal Break demolition debris into manageable size with hand tools, load into a 5 CF wheelbarrow, move to chute and dump. Costs shown are per cubic foot of material dumped.

	Craft@Hrs	Unit	Material	Labor	Equipment	Total
Wheelbarrow, 50'						
to trash chute and dump	CL@.018	CF	—	.54	—	.54
Wheelbarrow, 100'						
to trash chute and dump	CL@.022	CF	—	.67	—	.67
Wheelbarrow, 50' to elevator, descend 10 floors						
to trash chute and dump	CL@.024	CF	—	.73	—	.73

Trash chutes. Prefabricated steel chute installed and removed in a multi-story building. One-time charge for up to six-months usage. Use 50 LF as minimum charge.

	Craft@Hrs	Unit	Material	Labor	Equipment	Total
18" diameter	--	LS	—	—	—	1,160.00
36" diameter	--	LS	—	—	—	1,735.00

Load demolition debris on truck and haul 6 miles to a dump site. Includes truck cost but dump fees are not included. See Waste Disposal. Equipment cost per hour is based on $20.00 per hour for a 5 CY dump truck plus $22.00 for a 55 HP wheel loader with integral backhoe where shown. Per cubic yard of debris

	Craft@Hrs	Unit	Material	Labor	Equipment	Total
Truck loaded by hand	C4@.828	CY	—	25.40	5.55	30.95
Truck loaded using a wheel loader	C5@.367	CY	—	12.00	4.03	16.03

Removal of interior items Based on removal using hand tools. Items removed in salvageable condition do not include an allowance for salvage value. Figures in parentheses give the approximate "loose" volume of the materials (volume after being demolished.) Removal, in non-salvageable condition (demolished).

Hollow metal door and frame in a masonry wall. (2 doors per CY)

	Craft@Hrs	Unit	Material	Labor	Equipment	Total
Single door to 4' x 7'	CL@1.00	Ea	—	30.30	—	30.30
Two doors, per opening to 8' x 7'	CL@1.50	Ea	—	45.40	—	45.40

	Craft@Hrs	Unit	Material	Labor	Equipment	Total
Wood door and frame in a wood-frame wall (2 doors per CY)						
Single door to 4' x 7'	CL@.500	Ea	—	15.10	—	15.10
Two doors, per opening to 8' x 7'	CL@.750	Ea	—	22.70	—	22.70
Remove door and frame in masonry wall, salvage condition						
Hollow metal door to 4' x 7'	C8@2.00	Ea	—	72.50	—	72.50
Wood door to 4' x 7'	C8@1.00	Ea	—	36.30	—	36.30
Frame for resilient mat, metal						
per SF of mat area	C8@.054	SF	—	1.96	—	1.96
Lockers, metal 12" W, 60" H, 15" D	C8@.500	Ea	—	18.10	—	18.10
Sink and soap dispenser, wall-hung	C8@.500	Ea	—	18.10	—	18.10
Toilet partitions, wood or metal,						
per partition	C8@.750	Ea	—	27.20	—	27.20
Urinal screens, wood or metal,						
per partition	C8@.500	Ea	—	18.10	—	18.10
Window and frame, wood or metal	C8@.076	SF	—	2.76	—	2.76
Remove and reset airlock doors	C8@5.56	Ea	—	202.00	—	202.00

Asbestos Hazard Surveys, Subcontract Building inspection, hazard identification, sampling and ranking of asbestos risk.

	Craft@Hrs	Unit	Material	Labor	Equipment	Total
Asbestos hazard survey and sample collection (10,000 SF per hour),						
four hour minimum	—	Hour	—	—	—	60.00
Sample analysis (usually one sample per 1,000 SF),						
ten sample minimum	—	Ea	—	—	—	31.00
Report writing (per 1,000 SF of floor),						
$200 minimum	—	SF	—	—	—	.02

Asbestos Removal, Subcontract Typical costs including site preparation, monitoring, equipment, and removal of waste. Disposal of hazardous waste materials vary widely, consult your local waste disposal facility concerning prices and procedures.

	Craft@Hrs	Unit	Material	Labor	Equipment	Total
Ceiling insulation in containment structure						
500 to 5,000 SF job	—	SF	—	—	—	34.00
5,000 to 20,000 SF job	—	SF	—	—	—	23.00
Pipe insulation in containment structure						
100 to 1,000 LF of 6" pipe	—	LF	—	—	—	61.00
1,000 to 3,000 LF of 6" pipe	—	LF	—	—	—	39.00
Pipe insulation using glove bags						
100 to 1,000 LF of 6" pipe	—	LF	—	—	—	51.00
1,000 to 3,000 LF of 6" pipe	—	LF	—	—	—	39.00

Wall Sawing, Subcontract Using an electric, hydraulic or air saw. Per LF of cut including overcuts at each corner equal to the depth of the cut. Costs shown assume electric power is available Minimum cost will be $380.

	Unit	Brick or block	Concrete w/#4 bar	Concrete w/#7 bar
To 4" depth	LF	8.25	12.00	12.00
To 5" depth	LF	9.50	14.00	14.00
To 6" depth	LF	10.50	10.50	12.25
To 7" depth	LF	14.00	14.00	16.00
To 8" depth	LF	16.00	16.00	18.00
To 10" depth	LF	22.50	22.50	26.00
To 12" depth	LF	27.00	27.00	31.00
To 14" depth	LF	31.50	31.50	36.00
To 16" depth	LF	36.00	36.00	41.00
To 18" depth	LF	54.00	54.00	62.00

	Craft@Hrs	Unit	Material	Labor	Total

Concrete Core Drilling, Subcontract Using an electric, hydraulic or air drill. Costs shown are per LF of depth for diameter shown. Prices are based on one mat reinforcing with 3/8" to 5/8" bars and include cleanup. Difficult drill setups are higher. Prices assume availability of 110 volt electricity. Figure travel time at $85 per hour. Minimum cost will be $190.

	Craft@Hrs	Unit	Material	Labor	Total
1" or 1-1/2" diameter	—	LF	—	—	33.60
2" or 2-1/2" diameter	—	LF	—	—	34.80
3" or 3-1/2"diameter	—	LF	—	—	42.00
4" or 5" diameter	—	LF	—	—	51.00
6" or 8" diameter	—	LF	—	—	70.00
10" diameter	—	LF	—	—	96.00
12" diameter	—	LF	—	—	156.00
14" diameter	—	LF	—	—	180.00
16" diameter	—	LF	—	—	204.00
18" diameter	—	LF	—	—	228.00
20" diameter	—	LF	—	—	252.00
24" diameter	—	LF	—	—	300.00
30" diameter	—	LF	—	—	372.00
36" diameter	—	LF	—	—	444.00
Hourly rate, 2 hour minimum	—	Hr	—	—	85.00

Concrete Slab Sawing, Subcontract Using a gasoline powered saw. Costs per linear foot for cured concrete assuming a level surface with good access and saw cut lines laid out and pre-marked by others. Costs include local travel time. Minimum cost will be $190. Electric powered slab sawing will be approximately 40% higher for the same depth.

Depth	Unit	Under 200'	200'-1000'	Over 1000'
1" deep	LF	.40	.35	.30
1-1/2" deep	LF	.60	.55	.45
2" deep	LF	.80	.70	.60
2-1/2" deep	LF	1.00	.90	.75
3" deep	LF	1.20	1.05	.90
3-1/2" deep	LF	1.40	1.25	1.05
4" deep	LF	1.60	1.40	1.20
5" deep	LF	2.00	1.75	1.50
6" deep	LF	2.40	2.10	1.80
7" deep	LF	2.85	2.50	2.15
8" deep	LF	3.30	2.90	2.50
9" deep	LF	3.75	3.30	2.85
10" deep	LF	4.25	3.75	3.25
11" deep	LF	4.75	4.20	3.65
12" deep	LF	5.25	4.65	4.05

Green concrete (2 days old) sawing will usually cost 15 to 20% less. Work done on an hourly basis will cost $98 per hour for slabs up to 4" thick and $108 per hour for 5" or 6" thick slabs. A two hour minimum charge will apply on work done on an hourly basis.

Asphalt Sawing, Subcontract Using a gasoline powered saw. Minimum cost will be $170. Cost per linear foot of green or cured asphalt, assuming level surface with good access and saw cut lines laid out and pre-marked by others. Costs include local travel time. Work done on an hourly basis will cost $98 per hour for asphalt up to 4" thick and $108 per hour for 5" or 6" thick asphalt. A two hour minimum charge will apply on work done on an hourly basis.

Depth	Unit	Under 450'	450'-1000'	Over 1000'
1" or 1-1/2" deep	LF	.25	.20	.15
2" deep	LF	.40	.30	.25
2-1/2" deep	LF	.50	.40	.30
3" deep	LF	.60	.45	.35
3-1/2" deep	LF	.70	.55	.40
4" deep	LF	.80	.60	.45
5" deep	LF	1.00	.75	.55
6" deep	LF	1.20	.90	.65

	Craft@Hrs	Unit	Material	Labor	Equipment	Total

Rock Excavation Rock drilling costs. Based on using a pneumatic track-mounted wagon drill. Equipment cost per hour is $65 and includes one wheel-mounted air compressor and one 4" pneumatic wagon drill with drill bits. Costs shown are per linear foot of hole drilled. Add blasting costs below. Use $2,500 as a minimum job charge. LF per hour shown is for a 2-man crew

	Craft@Hrs	Unit	Material	Labor	Equipment	Total
Easy work, 38 LF per hour	S1@.052	LF	—	1.80	1.69	3.49
Moderate work, 30 LF per hour	S1@.067	LF	—	2.32	2.18	4.50
Most work, 25 LF per hour	S1@.080	LF	—	2.77	2.60	5.37
Hard work, 20 LF per hour	S1@.100	LF	—	3.47	3.25	6.72
Dense rock, 15 LF per hour	S1@.132	LF	—	4.57	4.29	8.86
Drilling 2-1/2" hole for						
rock bolts, 24 LF/hour	S1@.086	LF	—	2.98	2.80	5.78

Blasting costs Based on two cycles per hour and 20 linear foot lifts. These costs assume 75% fill of holes with explosives costing $2.70 per pound. Equipment cost per hour is $20.00 and includes one flatbed truck. Costs shown are per cubic yard of area blasted. Loading or hauling of blasted material not included. Use $3,500 as a minimum job charge. Load explosives in holes and detonate, by pattern spacing. CY per hour shown is for a 5-man crew.

	Craft@Hrs	Unit	Material	Labor	Equipment	Total
6' x 6' pattern (26 CY per hour)	C1@.192	CY	7.90	5.86	.81	14.57
7' x 7' pattern (35 CY per hour)	C1@.144	CY	6.00	4.40	.61	11.01
7' x 8' pattern (47 CY per hour)	C1@.106	CY	4.65	3.24	.44	8.33
9' x 9' pattern (60 CY per hour)	C1@.084	CY	4.05	2.57	.36	6.98
10' x 10' pattern (72 CY per hour)	C1@.069	CY	3.30	2.11	.29	5.70
12' x 12' pattern (108 CY per hour)	C1@.046	CY	2.25	1.40	.19	3.84
14' x 14' pattern (147 CY per hour)	C1@.034	CY	1.60	1.04	.15	2.79

Ripping rock Based on using a tractor-mounted ripper. Equipment cost is $113 per hour for a D-8 tractor and $160 per hour for a D-9 tractor. Costs shown are per CY of area ripped. Loading or hauling of material not included. Use $2,000 as a minimum job charge.

	Craft@Hrs	Unit	Material	Labor	Equipment	Total
Clay or glacial tills, D-9 cat tractor with						
2-shank ripper (500 CY per hour)	T0@.002	CY	—	.08	.32	.40
Clay or glacial tills, D-8 cat tractor with						
1-shank ripper (335 CY per hour)	T0@.003	CY	—	.12	.34	.46
Shale, sandstone, or limestone, D-9 cat tractor with						
2-shank ripper (125 CY per hour)	T0@.008	CY	—	.31	1.28	1.59
Shale, sandstone, or limestone, D-8 cat tractor with						
1-shank ripper (95 CY per hour)	T0@.011	CY	—	.43	1.19	1.62
Slate, metamorphic rock, D-9 cat tractor with						
2-shank ripper (95 CY per hour)	T0@.011	CY	—	.43	1.68	2.11

	Craft@Hrs	Unit	Material	Labor	Equipment	Total
Slate, metamorphic rock, D-8 cat tractor with 1-shank ripper (63 CY per hour)	T0@.016	CY	—	.62	1.79	2.41
Granite or basalt, D-9 cat tractor with 2-shank ripper (78 CY per hour)	T0@.013	CY	—	.51	2.05	2.56
Granite or basalt, D-8 cat tractor with 1-shank ripper (38 CY per hour)	T0@.027	CY	—	1.05	2.97	4.02

Rock loosening Based on using pneumatic jackhammers. Equipment cost per hour is $13.50 and includes one compressor and two jackhammers with drill points. Costs shown are per CY of area loosened. Loading and hauling not included. CY per hour shown is for a 2-man crew. Use $800 as a minimum job charge.

	Craft@Hrs	Unit	Material	Labor	Equipment	Total
Igneous or dense rock (1.4 CY per hour)	CL@1.43	CY	—	43.30	9.64	52.94
Most weathered rock (2.4 CY per hour)	CL@.835	CY	—	25.30	5.63	30.93
Soft sedimentary (4 CY per hour)	CL@.500	CY	—	15.10	3.38	18.48

Site Grading Layout, staking, flagmen, lights, watering, ripping, rock breaking, loading or hauling not included. Use $3,000.00 as a minimum job charge. (One acre is 43,560 SF or 4,840 SY.) Site preparation grading based on using one crawler tractor. Equipment cost per hour is as shown.

	Craft@Hrs	Unit	Material	Labor	Equipment	Total
General area rough grading with 100 HP D-4 tractor at $35.00 per hour (.5 acres per hour)	S1@4.00	Acre	—	139.00	70.20	209.20
General area rough grading with 335 HP D-8 tractor at $113 per hour (1 acre per hour)	S1@2.00	Acre	—	69.30	113.00	182.30

Site grading based on using one 100 HP (10,000 lb) motor grader. Equipment cost is $34 per hour.

	Craft@Hrs	Unit	Material	Labor	Equipment	Total
General area grading (.7 acres per hour)	S1@2.86	Acre	—	99.10	48.57	147.67
Fine grading subgrade to 1/10' (.45 acres per hour)	S1@4.44	Acre	—	154.00	75.60	229.56
Cut slope or shape embankment to 2' high (.25 acres per hour)	S1@8.00	Acre	—	277.00	136.00	413.00
Rough grade sub-base course on roadway (.2 acres per hour)	S1@10.0	Acre	—	347.00	170.00	517.00
Finish grade base or leveling course on roadway (.17 acres per hour)	S1@11.8	Acre	—	409.00	200.00	609.00
Finish grading for building slab, windrow excess (1,000 SF per hour)	S1@.002	SF	—	.07	.03	.10

Embankment Grading Earth embankment spreading, shaping, compacting and watering, and finish shaping. Use $15,000 as a minimum job charge, loading or hauling of earth fill not included.

Spreading and shaping Spread and shape earth from loose piles, based on using a D-8 tractor at $113 per hour

	Craft@Hrs	Unit	Material	Labor	Equipment	Total
6" to 10" lifts (164 CY per hour)	T0@.006	CY	—	.23	.68	.91

Compacting and watering (Add for cost of water from next page.) CY per hour shown is for a 3-man crew. Based on using a self-propelled 100 HP vibrating roller and a 4,000 gallon truck, equipment cost is $93 per hour. Productivity assumes 3 passes at 7' wide

	Craft@Hrs	Unit	Material	Labor	Equipment	Total
6" lifts (500 CY per hour)	C3@.006	CY	—	.20	.18	.38
8" lifts (750 CY per hour)	C3@.004	CY	—	.13	.12	.25
10" lifts (1,000 CY per hour)	C3@.003	CY	—	.10	.09	.19

	Craft@Hrs	Unit	Material	Labor	Equipment	Total

Compacting with a sheepsfoot roller towed behind a D-7 tractor and a 4,000 gallon truck, equipment cost is $131 per hour. Productivity assumes 3 passes at 5' wide.

	Craft@Hrs	Unit	Material	Labor	Equipment	Total
6" lifts (185 CY per hour)	C3@.016	CY	—	.54	.70	1.24
8" lifts (245 CY per hour)	C3@.012	CY	—	.40	.53	.93

Cost of water for compacted earth embankments. Based on water at $2.50 per 1,000 gallons and 66 gallons per cubic yard of compacted material. Assumes optimum moisture at 10%, natural moisture of 2% and evaporation of 2%. Placed in conjunction with compaction shown above.

	Craft@Hrs	Unit	Material	Labor	Equipment	Total
Cost per CY of compacted embankment	—	CY	.17	—	—	.17

Finish shaping Earth embankment slopes and swales up to 1 in 4 incline.

Based on using a 10,000 pound grader and a 15 ton self-propelled rubber tired roller, equipment cost is $58.80 per hour.

	Craft@Hrs	Unit	Material	Labor	Equipment	Total
(200 SY per hour based on a 3-man crew)	SS@.016	SY	—	.58	.29	.87

Based on using a D-8 tractor, equipment cost is $116 per hour.

	Craft@Hrs	Unit	Material	Labor	Equipment	Total
(150 SY per hour)	T0@.007	SY	—	.27	.77	1.04

Finish shaping of embankment slopes and swales by hand. SY per hour shown is for 1 man.

	Craft@Hrs	Unit	Material	Labor	Equipment	Total
Slopes up to 1 in 4 (16 SY per hour)	CL@.063	SY	—	1.91	—	1.91
Slopes over 1 in 4 (12.5 SY per hour)	CL@.080	SY	—	2.42	—	2.42

Trench Excavation and Backfill These costs and productivity are based on utility line trenches and continuous footings where the spoil is piled adjacent to the trench. Linear feet (LF), cubic yards (CY), square feet (SF), or square yards (SY) per hour shown are based on a 2-man crew. Increase costs by 10% to 25% when spoil is loaded in trucks. Hauling, shoring, dewatering or unusual conditions are not included.

12" wide bucket, for 12" wide trench. Depths 3' to 5'. Equipment is a wheel loader with 1 CY bucket and integral backhoe, at $21 per hour. Use $400 as a minimum job charge.

	Craft@Hrs	Unit	Material	Labor	Equipment	Total
Light soil (60 LF per hour)	S1@.033	LF	—	1.14	.35	1.49
Medium soil (55 LF per hour)	S1@.036	LF	—	1.25	.38	1.63
Heavy or wet soil (35 LF per hour)	S1@.057	LF	—	1.98	.61	2.59

18" wide bucket, for 18" wide trench. Depths 3' to 5'. Equipment is a wheel loader with 1 CY bucket and integral backhoe, at $21 per hour. Use $400 as a minimum job charge.

	Craft@Hrs	Unit	Material	Labor	Equipment	Total
Light soil (55 LF per hour)	S1@.036	LF	—	1.25	.38	1.63
Medium soil (50 LF per hour)	S1@.040	LF	—	1.39	.43	1.82
Heavy or wet soil (30 LF per hour)	S1@.067	LF	—	2.32	.72	3.04

24" wide bucket, for 24" wide trench. Depths 3' to 5'. Equipment is a wheel loader with 1 CY bucket and integral backhoe, at $21 per hour. Use $400 as a minimum job charge.

	Craft@Hrs	Unit	Material	Labor	Equipment	Total
Light soil (50 LF per hour)	S1@.040	LF	—	1.39	.43	1.82
Medium soil (45 LF per hour)	S1@.044	LF	—	1.52	.47	1.99
Heavy or wet soil (25 LF per hour)	S1@.080	LF	—	2.77	.85	3.62

Truck-mounted Gradall with 1 CY bucket at $88 per hour. Use $700 as a minimum job charge.

	Craft@Hrs	Unit	Material	Labor	Equipment	Total
Light soil (40 CY per hour)	S3@.050	CY	—	1.77	2.20	3.97
Medium soil (34 CY per hour)	S3@.059	CY	—	2.08	2.60	4.68
Heavy or wet soil (27 CY per hour)	S3@.074	CY	—	2.61	3.26	5.87
Loose rock (23 CY per hour)	S3@.087	CY	—	3.07	3.83	6.90

Crawler-mounted hydraulic backhoe with 3/4 CY bucket at $41 per hour. Use $500 as a minimum job charge.

	Craft@Hrs	Unit	Material	Labor	Equipment	Total
Light soil (33 CY per hour)	S1@.061	CY	—	2.11	1.25	3.36
Medium soil (27 CY per hour)	S1@.074	CY	—	2.56	1.52	4.08
Heavy or wet soil (22 CY per hour)	S1@.091	CY	—	3.15	1.87	5.02
Loose rock (18 CY per hour)	S1@.112	CY	—	3.88	2.30	6.18

	Craft@Hrs	Unit	Material	Labor	Equipment	Total
Crawler-mounted hydraulic backhoe with 1 CY bucket at $46 per hour. Use $500 as a minimum job charge.						
Light soil (65 CY per hour)	S1@.031	CY	—	1.07	.71	1.78
Medium soil (53 CY per hour)	S1@.038	CY	—	1.32	.87	2.19
Heavy or wet soil (43 CY per hour)	S1@.047	CY	—	1.63	1.08	2.71
Loose rock (37 CY per hour)	S1@.055	CY	—	1.91	1.27	3.18
Blasted rock (34 CY per hour)	S1@.058	CY	—	2.01	1.33	3.34
Crawler-mounted hydraulic backhoe with 1-1/2 CY bucket at $56 per hour. Use $500 as a minimum job charge						
Light soil (83 CY per hour)	S1@.024	CY	—	.83	.67	1.50
Medium soil (70 CY per hour)	S1@.029	CY	—	1.01	.81	1.82
Heavy or wet soil (57 CY per hour)	S1@.035	CY	—	1.21	.98	2.19
Loose rock (48 CY per hour)	S1@.042	CY	—	1.46	1.18	2.64
Blasted rock (43 CY per hour)	S1@.047	CY	—	1.63	1.32	2.95
Crawler-mounted hydraulic backhoe with 2 CY bucket at $86 per hour. Use $700 as a minimum job charge						
Light soil (97 CY per hour)	S1@.021	CY	—	.73	.91	1.64
Medium soil (80 CY per hour)	S1@.025	CY	—	.87	1.08	1.95
Heavy or wet soil (65 CY per hour)	S1@.031	CY	—	1.07	1.33	2.40
Loose rock (55 CY per hour)	S1@.036	CY	—	1.25	1.55	2.80
Blasted rock (50 CY per hour)	S1@.040	CY	—	1.39	1.72	3.11
Crawler-mounted hydraulic backhoe with 2-1/2 CY bucket at $105 per hour. Use $750 as a minimum job charge						
Light soil (122 CY per hour)	S1@.016	CY	—	.55	.84	1.39
Medium soil (100 CY per hour)	S1@.020	CY	—	.69	1.05	1.74
Heavy or wet soil (82 CY per hour)	S1@.024	CY	—	.83	1.26	2.09
Loose rock (68 CY per hour)	S1@.029	CY	—	1.01	1.52	2.53
Blasted rock (62 CY per hour)	S1@.032	CY	—	1.11	1.68	2.79
Chain-boom Ditch Witch digging trench to 12" wide and 5' deep at $21 per hour. Use $400 as a minimum job charge						
Light soil (10 CY per hour)	S1@.200	CY	—	6.93	2.10	9.03
Most soils (8.5 CY per hour)	S1@.235	CY	—	8.14	2.47	10.61
Heavy soil (7 CY per hour)	S1@.286	CY	—	9.91	3.00	12.91
Trenchers, chain boom, 55 HP, digging trench to 18" wide and 8' deep at $31 per hour. Use $500 as a minimum job charge						
Light soil (62 CY per hour)	S1@.032	CY	—	1.11	.50	1.61
Most soils (49 CY per hour)	S1@.041	CY	—	1.42	.64	2.06
Heavy soil (40 CY per hour)	S1@.050	CY	—	1.73	.78	2.51
Trenchers, chain boom, 100 HP, digging trench to 24" wide and 8' deep at $50 per hour. Use $500 as a minimum job charge						
Light soil (155 CY per hour)	S1@.013	CY	—	.45	.33	.78
Most soils (125 CY per hour)	S1@.016	CY	—	.55	.40	.95
Heavy soil (100 CY per hour)	S1@.020	CY	—	.69	.51	1.20
Trim trench bottom to 1/10'						
By hand (200 SF per hour)	CL@.010	SF	—	.30	—	.30

Backfill trenches from loose material piled adjacent to trench. No compaction included. Soil previously excavated

	Craft@Hrs	Unit	Material	Labor	Equipment	Total
Backfill trenches by hand.						
By hand (8 CY per hour)	CL@.250	CY	—	7.57	—	7.57
Wheel loader 55 HP at $21 per hour. Use $400 as a minimum job charge						
Typical soils (50 CY per hour)	S1@.040	CY	—	1.39	.43	1.82
D-3 crawler dozer at $30 per hour. Use $475 as a minimum job charge						
Typical soils (25 CY per hour)	S1@.080	CY	—	2.77	1.20	3.97
3/4 CY crawler loader at $23 per hour. Use $450 as a minimum job charge						
Typical soils (33 CY per hr)	S1@.061	CY	—	2.11	.71	2.82
D-7 crawler dozer at $81 per hour. Use $700 as a minimum job charge						
Typical soils (130 CY per hour)	S1@.015	CY	—	.52	.61	1.13
Wheel loader 55 HP at $21 per hour. Sand or gravel bedding. Use $400 as a minimum job charge						
Typical soils (80 CY per hour)	S1@.025	CY	16.50	.87	.27	17.64
Fine grade bedding by hand						
(22 SY per hr)	CL@.090	SY	—	2.72	—	2.72

	Craft@Hrs	Unit	Material	Labor	Equipment	Total
Compaction of soil in trenches in 8" layers. Use $200 as a minimum job charge						
Pneumatic tampers						
($9 & 20 CY per hour)	CL@.050	CY	—	1.51	.45	1.96
Vibrating rammers						
($6 & 10 CY per hour)	CL@.100	CY	—	3.03	.60	3.63

Dragline Excavation For mass excavation, footings or foundations. Costs shown include casting the excavated soil adjacent to the excavation or loading it into trucks. Hauling costs are not included. Cubic yards (CY) per hour are as measured in an undisturbed condition (bank measure) and are based on a 2-man crew. Use $3,000 as a minimum job charge.

	Craft@Hrs	Unit	Material	Labor	Equipment	Total
1-1/2 CY dragline with a 45 ton crawler crane at $53 per hour						
Loam or light clay (73 CY per hour)	H2@.027	CY	—	.93	.72	1.65
Sand or gravel (67 CY per hour)	H2@.030	CY	—	1.04	.79	1.83
Heavy clay (37 CY per hour)	H2@.054	CY	—	1.87	1.43	3.30
Unclassified soil (28 CY per hour)	H2@.071	CY	—	2.46	1.88	4.34
2 CY dragline with a 60 ton crawler crane at $66 per hour						
Loam or light clay (88 CY per hour)	H2@.023	CY	—	.80	.76	1.56
Sand or gravel (85 CY per hour)	H2@.024	CY	—	.83	.79	1.62
Heavy clay (48 CY per hour)	H2@.042	CY	—	1.45	1.38	2.83
Unclassified soil (36 CY per hour)	H2@.056	CY	—	1.94	1.84	3.78
2-1/2 CY dragline with a 100 ton crawler crane at $82 per hour						
Loam or light clay (102 CY per hour)	H2@.019	CY	—	.66	.80	1.46
Sand or gravel (98 CY per hour)	H2@.021	CY	—	.73	.84	1.57
Heavy clay (58 CY per hour)	H2@.035	CY	—	1.21	1.41	2.62
Unclassified soil (44 CY per hour)	H2@.045	CY	—	1.56	1.86	3.42
3 CY dragline with a 150 ton crawler crane at $117 per hour						
Loam or light clay (116 CY per hour)	H2@.017	CY	—	.59	1.00	1.59
Sand or gravel (113 CY per hour)	H2@.018	CY	—	.62	1.05	1.67
Heavy clay (70 CY per hour)	H2@.029	CY	—	1.00	1.67	2.67
Unclassified soil (53 CY per hour)	H2@.038	CY	—	1.32	2.23	3.55

Backhoe Excavation For mass excavation, footings or foundations. Costs shown include casting the excavated soil adjacent to the excavation or loading it into trucks. Hauling costs are not included. Cubic yards (CY) per hour are as measured in an undisturbed condition (bank measure) and are based on a 2-man crew. Equipment costs are based on using a crawler-mounted hydraulic backhoe. Use $2,500 as a minimum job charge

	Craft@Hrs	Unit	Material	Labor	Equipment	Total
1 CY backhoe at $26 per hour						
Light soil (69 CY per hour)	S1@.029	CY	—	1.01	.39	1.40
Most soils (57 CY per hour)	S1@.035	CY	—	1.21	.47	1.68
Wet soil, loose rock (46 CY per hour)	S1@.043	CY	—	1.49	.57	2.06
1-1/2 CY backhoe at $34 per hour						
Light soil (90 CY per hour)	S1@.022	CY	—	.76	.37	1.13
Most soils (77 CY per hour)	S1@.026	CY	—	.90	.44	1.34
Wet soil, loose rock (60 CY per hour)	S1@.033	CY	—	1.14	.56	1.70
Blasted rock (54 CY per hour)	S1@.044	CY	—	1.52	.63	2.15
2 CY backhoe at $49 per hour						
Light soil (100 CY per hour)	S1@.020	CY	—	.69	.49	1.18
Most soils (85 CY per hour)	S1@.023	CY	—	.80	.56	1.36
Wet soil, loose rock (70 CY per hour)	S1@.029	CY	—	1.01	.71	1.72

	Craft@Hrs	Unit	Material	Labor	Equipment	Total

Moving and Loading Excavated Materials Costs shown include moving the material 50' and dumping it into piles or loading it into trucks. Add 25% for each 50' of travel beyond the first 50'. Hauling costs are not included. Cubic yards (CY) per hour are for material in a loose condition (previously excavated) and are based on a 2-man crew performing the work. Equipment costs are based on using a wheel-mounted front end loader. Use $2,000 as a minimum job charge

3/4 CY loader

	Craft@Hrs	Unit	Material	Labor	Equipment	Total
($18 and 32 CY per hour)	S1@.062	CY	—	2.15	.57	2.72

1 CY loader

($21 and 55 CY per hour)	S1@.036	CY	—	1.25	.38	1.63

2 CY loader

($26 and 90 CY per hour)	S1@.022	CY	—	.76	.29	1.05

3-1/4 CY loader

($54 and 125 CY per hour)	S1@.016	CY	—	.55	.43	.98

5 CY loader

($66 and 218 CY per hour)	S1@.009	CY	—	.31	.30	.61

Hauling excavated material Using trucks. Costs shown include 4 minutes for loading, 3 minutes for dumping and travel time based on the one-way distance, speed and cycles per hour as noted. Costs for equipment to excavate and load the material are not included. Truck capacity shown is based on loose cubic yards of material. Allow the following percentage amounts when estimating quantities based on undisturbed (bank measure) materials for swell when exporting from the excavation site or shrinkage when importing from borrow site: clay (33%), common earth (25%), granite (65%), mud (21%), sand or gravel (12%).

3 CY dump truck at $13 per hour. Use $550 as a minimum job charge

1 mile haul at 20 MPH

	Craft@Hrs	Unit	Material	Labor	Equipment	Total
(4.2 cycles and 34 CY per hour)	TD@.029	CY	—	.92	.39	1.31

3 mile haul at 30 MPH

(2.9 cycles and 23 CY per hour)	TD@.044	CY	—	1.39	.59	1.98

6 mile haul at 40 MPH

(2.1 cycles and 17 CY per hour)	TD@.058	CY	—	1.83	.78	2.61

5 CY dump truck at $20 per hour. Use $600 as a minimum job charge

1 mile haul at 20 MPH

(4.2 cycles and 50 CY per hour)	TD@.020	CY	—	.63	.40	1.03

3 mile haul at 30 MPH

(2.9 cycles and 35 CY per hour)	TD@.028	CY	—	.88	.56	1.44

6 mile haul at 40 MPH

(2.1 cycles and 25 CY per hour)	TD@.041	CY	—	1.30	.80	2.10

25 CY off-highway dump truck at $86.20 per hour. Use $1,600 as a minimum job charge

1 mile haul at 20 MPH

(4.2 cycles and 67 CY per hour)	TD@.015	CY	—	.47	1.28	1.75

3 mile haul at 30 MPH

(2.9 cycles and 46 CY per hour)	TD@.022	CY	—	.70	1.87	2.57

6 mile haul at 40 MPH

(2.1 cycles and 34 CY per hour)	TD@.029	CY	—	.92	2.54	3.46

25 CY off-highway dump truck at $86 per hour. Use $2,000 as a minimum job charge

1 mile haul at 20 MPH

(4.2 cycles and 105 CY per hour)	TD@.009	CY	—	.28	.82	1.10

3 mile haul at 30 MPH

(2.9 cycles and 73 CY per hour)	TD@.014	CY	—	.44	1.18	1.62

6 mile haul at 40 MPH

(2.1 cycles and 53 CY per hour)	TD@.019	CY	—	.60	1.62	2.22

	Craft@Hrs	Unit	Material	Labor	Equipment	Total

Dozer Excavation Mass excavation using a crawler tractor with a dozing blade attached. Costs shown include excavation and pushing soil 150' to stockpile based on good access and a maximum of 5% grade. Increase or reduce costs 20% for each 50' of push more or less than 150'. Cubic yards (CY) per hour shown are undisturbed bank measure and are based on a 1-man crew. Use the following as minimum job charges: $850 if using a D-4 or a D-6 dozer; $1,350 if using a D-7 or D-8 dozer; or $2,000 if using a D-9 dozer.

	Craft@Hrs	Unit	Material	Labor	Equipment	Total
Gravel or loose sand						
100 HP D-4 dozer with "S" blade at $35 per hour						
(29 CY per hour)	TO@.034	CY	—	1.33	1.19	2.52
140 HP D-6 dozer with "S" blade at $53 per hour						
(56 CY per hour)	TO@.018	CY	—	.70	.95	1.65
200 HP D-7 dozer with "S" blade at $89.80 per hour						
(78 CY per hour)	TO@.013	CY	—	.51	1.15	1.66
335 HP D-8 dozer with "U" blade at $113 per hour						
(140 CY per hour)	TO@.007	CY	—	.27	.82	1.09
460 HP D-9 dozer with "U" blade at $160 per hour						
(205 CY per hour)	TO@.005	CY	—	.20	.78	.98
Loam or soft clay						
100 HP D-4 dozer with "S" blade at $45 per hour						
(26 CY per hour)	TO@.038	CY	—	1.48	1.73	3.21
140 HP D-6 dozer with "S" blade at $58 per hour						
(50 CY per hour)	TO@.020	CY	—	.78	1.16	1.94
200 HP D-7 dozer with "S" blade at $93 per hour						
(70 CY per hour)	TO@.014	CY	—	.55	1.33	2.43
335 HP D-8 dozer with "U" blade at $117 per hour						
(125 CY per hour)	TO@.008	CY	—	.31	.94	1.25
460 HP D-9 dozer with "U" blade at $195 per hour						
(185 CY per hour)	TO@.005	CY	—	.20	1.05	1.25
Shale, sandstone or blasted rock						
100 HP D-4 dozer with "S" blade at $45 per hour						
(20 CY per hour)	TO@.051	CY	—	1.99	2.25	4.24
140 HP D-6 dozer with "S" blade at $58 per hour						
(38 CY per hour)	TO@.026	CY	—	1.02	1.53	2.55
200 HP D-7 dozer with "S" blade at $93 per hour						
(53 CY per hour)	TO@.019	CY	—	.74	1.75	2.49
335 HP D-8 dozer with "U" blade at $117 per hour						
(95 CY per hour)	TO@.011	CY	—	.43	1.23	1.66
460 HP D-9 dozer with "U" blade at $195 per hour						
(140 CY per hour)	TO@.007	CY	—	.27	1.39	1.66

Scraper-hauler Excavation Mass excavation using a self-propelled scraper-hauler. Equipment costs include the scraper-hauler and a crawler tractor with a dozer blade attached pushing it 10 minutes each hour. Cubic yards (CY) per hour shown are undisturbed bank measure. Work done in clay, shale or soft rock will cost 10% to 25% more.

15 CY self-propelled 200 HP scraper-hauler & D-8 335 HP tractor pushing ($218 per hour).

	Craft@Hrs	Unit	Material	Labor	Equipment	Total
Use $6,000 as a minimum job charge						
1,000' haul						
(9 cycles and 135 CY per hr)	SS@.016	CY	—	.58	1.62	2.20
2,500' haul						
(6 cycles and 90 CY per hr)	SS@.024	CY	—	.87	2.42	3.29
4,000' haul						
(4.5 cycles and 68 CY per hr)	SS@.032	CY	—	1.16	3.21	4.37

	Craft@Hrs	Unit	Material	Labor	Equipment	Total
24 CY self-propelled 350 HP scraper-hauler & D-9 460 HP tractor pushing ($335 per hour).						
Use $8,000 as a minimum job charge						
1,000' haul						
(9 cycles and 225 CY per hr)	SS@.010	CY	—	.36	1.49	1.85
2,500' haul						
(6 cycles and 150 CY per hr)	SS@.014	CY	—	.51	2.23	2.74
4,000' haul						
(4.5 cycles and 113 CY per hr)	SS@.019	CY	—	.69	2.96	3.65
32 CY self-propelled 550 HP scraper-hauler & D-9 460 HP tractor pushing ($471 per hour).						
Use $10,000 as a minimum job charge						
1,000' haul						
(9 cycles and 315 CY per hr)	SS@.007	CY	—	.25	1.50	1.75
2,500' haul						
(6 cycles and 219 CY per hr)	SS@.010	CY	—	.36	2.15	2.51
4,000' haul						
(4.5 cycles and 158 CY per hr)	SS@.014	CY	—	.51	2.98	3.49

Roadway and Embankment Earthwork Cut & Fill Mass excavating, hauling, placing and compacting (cut & fill) using a 200 HP, 15 CY capacity self-propelled scraper-hauler with attachments. Equipment cost is $106 per hour. Productivity is based on a 1,500' haul with 6" lifts compacted to 95% per AASHO requirements. Cubic yards (CY) or square yards (SY) per hour shown are bank measurement. Use $3,000 as a minimum job charge.

	Craft@Hrs	Unit	Material	Labor	Equipment	Total
Cut, fill and compact, clearing or finishing not included						
Most soil types (45 CY per hour)	TO@.022	CY	—	.86	2.33	3.19
Rock and earth mixed (40 CY per hour)	TO@.025	CY	—	.98	2.65	3.63
Rippable rock (25 CY per hour)	TO@.040	CY	—	1.56	4.24	5.80
Earth banks and levees, clearing or finishing not included						
Cut, fill and compact (85 CY per hour)	TO@.012	CY	—	.47	1.27	1.74
Channelize compacted fill						
(55 CY per hour)	TO@.018	CY	—	.70	1.91	2.61
Roadway subgrade preparation						
Scarify & compact (91 CY per hour)	TO@.011	CY	—	.43	1.17	1.60
Roll & compact base course						
(83 CY per hour)	TO@.012	CY	—	.47	1.27	1.74
Excavation, open ditches						
(38 CY per hour)	TO@.026	CY	—	1.02	2.76	3.78
Trimming and finishing						
Banks, swales or ditches						
(140 CY per hour)	TO@.007	CY	—	.27	.74	1.01
Scarify asphalt pavement						
(48 SY per hour)	TO@.021	SY	—	.82	2.23	3.05

Slope Protection See also, Soil Stabilization. Gabions as manufactured by Maccaferri Inc. Typical prices for gabions as described, placed by hand. Prices can be expected to vary based on quantity. Cost for stone is not included. Add for stone as required. Costs shown per cubic yard of stone capacity.

	Craft@Hrs	Unit	Material	Labor	Equipment	Total
8 x 10 type (3-1/4" x 4-1/2") double twisted hexagonal wire mesh, galvanized						
3'0" depth	CL@.600	CY	22.50	18.20	—	40.70
1'6" depth	CL@.500	CY	30.60	15.10	—	45.70
1'0" depth	CL@.400	CY	39.40	12.10	—	51.50
8 x 10 type (3-1/4" x 4-1/2") double twisted hexagonal wire mesh, PVC coated						
3'0" depth	CL@.600	CY	28.80	18.20	—	47.00
1'6" depth	CL@.500	CY	38.50	15.10	—	53.60
1'0" depth	CL@.400	CY	49.20	12.10	—	61.30
Erosion and drainage control polypropylene geotextile fabric, (Mirafi FW700)						
For use under riprap, hand placed	CL@.011	SY	1.20	.33	—	1.53

	Craft@Hrs	Unit	Material	Labor	Equipment	Total

Riprap Dumped from trucks and placed using a 15 ton hydraulic crane and a 10 cubic yard dump truck. Typical riprap cost is $18.00 per cubic yard FOB the quarry and will vary widely. Add $.75 per CY per mile for trucking to the site. Cubic (CY) yards per hour shown in parentheses are based on a 7-man crew. Equipment cost is $56.80 per hour. Use $2500 as a minimum job charge

	Craft@Hrs	Unit	Material	Labor	Equipment	Total
5 to 7 CF pieces (9.4 CY per hour)	S5@.745	CY	18.00	23.50	6.04	47.54
5 to 7 CF pieces, sacked and placed, excluding bag cost (3 CY per hour)	S5@2.33	CY	18.00	73.40	18.90	110.30
PVC coated nylon riprap bags, Quantities of 50 to 100						
5' x 7', 1.50 CY capacity	—	Ea	50.80	—	—	50.80
5' x 10', 2.25 CY capacity	—	Ea	68.00	—	—	68.00
5 'x 13', 3.00 CY capacity	—	Ea	83.00	—	—	83.00
Loose small riprap stone, under 30 lbs each						
Hand placed (2 CY per hour)	CL@.500	CY	50.00	15.10	—	65.10

Ornamental large rock Rock prices include local delivery of 10 CY minimum. Cubic (CY) yards per hour shown in parentheses are based on a 7-man crew. Equipment cost is $56.80 per hour based on using a 15 ton hydraulic crane and a 10 cubic yard dump truck. Use $2,500 as a minimum job charge
Volcanic cinder, 1,200 lbs per CY at $.21 per lb.

	Craft@Hrs	Unit	Material	Labor	Equipment	Total
(8 CY per hour)	S5@.875	CY	252.00	27.60	7.10	286.70

Featherock, 1,000 lbs per CY at $.25 per lb.

	Craft@Hrs	Unit	Material	Labor	Equipment	Total
(12 CY per hour)	S5@.583	CY	250.00	18.40	4.72	273.12

Rock fill Dumped from trucks and placed with a 100 HP D-4 tractor. Rock prices include local delivery of 10 CY minimum. Cubic (CY) yards per hour shown in parentheses are based on a 3-man crew. Equipment cost is $35.10 per hour. Use $1,350 as a minimum job charge

	Craft@Hrs	Unit	Material	Labor	Equipment	Total
Drain rock, 3/4" to 1-1/2"						
(12 CY per hour)	S6@.250	CY	18.30	8.30	2.95	29.55
Bank run gravel (12 CY per hour)	S6@.250	CY	16.30	8.09	2.95	28.14
Pea gravel (12 CY per hour)	S6@.250	CY	17.30	8.30	2.95	28.55

Straw bales secured to ground with reinforcing bars, bale is 3'6" long, 2' wide, 1'6" high. Straw at $100 per ton (25 bales) and rebars at $1.00

	Craft@Hrs	Unit	Material	Labor	Equipment	Total
Per bale of straw, placed by hand	CL@.166	Ea	4.00	5.02	—	9.02

Sedimentation control fence, installed vertically at bottom of slope, without stakes attached

	Craft@Hrs	Unit	Material	Labor	Equipment	Total
36" high woven fabric (Mirafi 100X)	CL@.003	LF	.28	.09	—	.37

Add to fence above for #6 rebar stakes 6'0" long at 10'0" OC hand-driven 36" into the ground

	Craft@Hrs	Unit	Material	Labor	Equipment	Total
Per LF of fence	CL@.050	LF	.28	1.51	—	1.79

Manual Excavation Based on one throw (except where noted) of shoveled earth from trench or pit. Shoring or dewatering not included

	Craft@Hrs	Unit	Material	Labor	Equipment	Total
Trench or pit in light soil (silty sand or loess, etc.)						
Up to 4' deep (.82 CY per hour)	CL@1.22	CY	—	36.90	—	36.90
Over 4' to 6' deep (.68 CY per hour)	CL@1.46	CY	—	44.20	—	44.20
Over 6' deep, two throws						
(.32 CY per hour)	CL@3.16	CY	—	95.60	—	95.60
Trench or pit in medium soil (sandy clay, clayey sand, etc.)						
Up to 4' deep (.68 CY per hour)	CL@1.46	CY	—	44.20	—	44.20
Over 4' to 6' deep (.54 CY per hour)	CL@1.86	CY	—	56.30	—	56.30
Over 6' deep, two throws						
(.27 CY per hour)	CL@3.66	CY	—	111.00	—	111.00

	Craft@Hrs	Unit	Material	Labor	Equipment	Total
Trench or pit in heavy soil (clayey materials, shales, caliche, etc.)						
Up to 4' deep (.54 CY per hour)	CL@1.86	CY	—	56.30	—	56.30
Over 4' to 6' deep (.46 CY per hour)	CL@2.16	CY	—	65.40	—	65.40
Over 6' deep, 2 throws						
(.25 CY per hour)	CL@3.90	CY	—	118.00	—	118.00
Wheelbarrow (5 CF) on firm ground, load by shovel from pile, wheel 300 feet and dump						
Light soil (.74 CY per hour)	CL@1.35	CY	—	40.90	—	40.90
Medium soil (.65 CY per hour)	CL@1.54	CY	—	46.60	—	46.60
Heavy soil (.56 CY per hour)	CL@1.79	CY	—	54.20	—	54.20
Loose rock (.37 CY per hour)	CL@2.70	CY	—	81.70	—	81.70
Hand trim and shape						
Around utility lines (15 CF hour)	CL@.065	CF	—	1.97	—	1.97
For slab on grade (63 SY per hour)	CL@.016	SY	—	.48	—	.48
Trench bottom (100 SF per hour)	CL@.010	SF	—	.30	—	.30

Soil Stabilization See also, Slope Protection. Lime slurry injection treatment. Equipment cost is $82.20 per hour based on using a 100 HP grader, a 1 ton self-propelled steel roller and a 10 ton 100 HP vibratory steel roller. Use $3,000 as a minimum job charge. Cost per CY of soil treated

	Craft@Hrs	Unit	Material	Labor	Equipment	Total
(25 CY per hour)	S7@.240	CY	9.80	8.32	3.30	21.42

Vibroflotation treatment Equipment cost is $185 per hour based on using a 50 ton hydraulic crane, a 1 CY wheel loader for 15 minutes of each hour, and a vibroflotation machine with pumps. Use $3,500 as a minimum job charge Typical cost per CY of soil treated

	Craft@Hrs	Unit	Material	Labor	Equipment	Total
Low cost (92 CY per hour)	S7@.065	CY	—	2.25	2.00	4.25
High cost (46 CY per hour)	S7@.130	CY	—	4.51	4.00	8.51

Soil cement treatment Includes materials and placement for 7% cement mix. Equipment cost is $105 per hour based on using a cross shaft mixer, a 2 ton 3 wheel steel roller and a vibratory roller. Cement priced at $100 per ton. Costs shown are per CY of soil treated. Use $3,000 as a minimum job charge

	Craft@Hrs	Unit	Material	Labor	Equipment	Total
On level ground						
(5 CY per hour)	S7@1.20	CY	7.69	41.60	22.24	71.53
On slopes under 3 in 1						
(4.3 CY per hour)	S7@1.40	CY	7.69	48.50	25.91	82.10
On slopes over 3 in 1						
(4 CY per hour)	S7@1.50	CY	7.69	52.00	27.85	87.54

Construction fabrics (geotextiles) Mirafi products. Costs shown include 10% for lapover. Drainage fabric (Mirafi 140NC), used to line trenches, allows water passage but prevents soil migration into drains

	Craft@Hrs	Unit	Material	Labor	Equipment	Total
Per SY of fabric placed by hand	CL@.016	SY	.70	.48	—	1.18

Stabilization fabric (Mirafi 500X), used to prevent mixing of unstable soils with road base aggregate

	Craft@Hrs	Unit	Material	Labor	Equipment	Total
Per SY of fabric placed by hand	CL@.008	SY	.70	.24	—	.94

Prefabricated drainage fabric (Miradrain), used to drain subsurface water from structural walls

	Craft@Hrs	Unit	Material	Labor	Equipment	Total
Per SY of fabric placed by hand	CL@.006	SF	.90	.18	—	1.08

Erosion control mat (Miramat TM8), used to reduce surface soil erosion on slopes or in ditches while promoting seed growth

	Craft@Hrs	Unit	Material	Labor	Equipment	Total
Per SY of mat placed by hand	CL@.020	SY	6.00	.61	—	6.61

Pile Foundations These costs assume solid ground for support of the pile driving equipment with all site work performed by others at no cost prior to pile driving being performed. Costs shown per linear foot (LF) are per vertical foot of pile depth. Standby or idle time, access roads, rig mats, test piles or special engineering required by unusual conditions are not included.

	Craft@Hrs	Unit	Material	Labor	Equipment	Total

Pile testing Testing costs will vary with the soil type, method of testing, and seismic requirements for any given job site. Use the following as guide line costs for estimating purposes. Typical costs including cost for moving test equipment on and off job site Add cost below for various types of piling.

	Craft@Hrs	Unit	Material	Labor	Equipment	Total
50 to 100 ton range	S8@224.	Ea	—	8,090.00	6,800.00	14,890.00
Over 100 ton to 200 ton range	S8@280.	Ea	—	10,100.00	8,500.00	18,600.00
Over 200 ton to 300 ton range	S8@336.	Ea	—	12,100.00	10,200.00	22,300.00

Prestressed concrete piles Equipment cost is $181 per hour and includes a 60 ton crawler mounted crane equipped with pile driving attachments, a 4-ton forklift, a truck with hand tools, a portable air compressor complete with hoses and pneumatic operated tools, a welding machine and an oxygen-acetylene cutting torch. Use $28,000 as minimum job charge, including 1,000 LF of piling and the cost for moving the pile driving equipment on and off the job site

	Craft@Hrs	Unit	Material	Labor	Equipment	Total
Minimum job charge	—	LS	—	—	—	28,000.00

For jobs requiring over 1,000 LF of piling, add to minimum charge as follows. (Figures in parentheses show approximate production per day based on a 7-man crew)

	Craft@Hrs	Unit	Material	Labor	Equipment	Total
12" square						
30' or 40' long piles (550 LF per day)	S8@.102	LF	10.50	3.68	2.64	16.82
50' or 60' long piles (690 LF per day)	S8@.082	LF	10.50	2.96	2.12	15.58
14" square						
30' or 40' long piles (500 LF per day)	S8@.113	LF	12.80	4.08	2.92	19.80
50' or 60' long piles (645 LF per day)	S8@.087	LF	12.80	3.14	2.25	18.19
16" square						
30' or 40' long piles (420 LF per day)	S8@.134	LF	18.10	4.84	3.47	26.41
50' or 60' long piles (550 LF per day)	S8@.102	LF	18.10	3.68	2.64	24.42
18" square						
30' or 40' long piles (400 LF per day)	S8@.143	LF	22.70	5.17	3.70	31.57
50' or 60' long piles (525 LF per day)	S8@.107	LF	22.70	3.86	2.77	29.33

Additional costs for prestressed concrete piles Pre-drilling holes for piles. Equipment cost per hour is $22.50 and includes a flatbed truck with a boom and a power operated auger. Use $2,000 as a minimum job charge, including drilling 100 LF of holes and the cost to move the drilling equipment on and off the job site

	Craft@Hrs	Unit	Material	Labor	Equipment	Total
Minimum job charge	—	LS	—	—	—	2,000.00

For jobs requiring more than 100 LF of holes drilled, add to minimum cost as shown below

	Craft@Hrs	Unit	Material	Labor	Equipment	Total
Pre-drilling first 10 LF, per hole	C9@1.50	LS	—	50.90	99.50	150.40
Add per LF of drilling,						
after first 10 LF	C9@.177	LF	—	6.01	11.80	17.81

Steel "HP" shape piles Equipment cost per hour is $180 and includes a 60 ton crawler mounted crane equipped with pile driving attachments, a 4-ton forklift, a truck with hand tools, a portable air compressor complete with hoses and pneumatic operated tools, a welding machine and an oxygen-acetylene cutting torch. Use $25,000 as minimum job charge, including 1,000 LF of piling and cost for moving pile driving equipment on and off the job site.

	Craft@Hrs	Unit	Material	Labor	Equipment	Total
Minimum job charge	—	LS	—	—	—	25,000.00

For jobs requiring over 1,000 LF of piling add to minimum charge as follows. (Figures in parentheses show approximate production per day based on a 7-man crew)

	Craft@Hrs	Unit	Material	Labor	Equipment	Total
HP8 8" x 8", 36 lbs per LF at $.32 per pound.						
30' or 40' long piles (800 LF per day)	S8@.070	LF	12.10	2.53	1.80	16.43
50' or 60' long piles (880 LF per day)	S8@.064	LF	12.10	2.31	1.65	16.06
HP10 10" x 10", 42 lbs per LF at $.30 per pound						
30' or 40' long piles (760 LF per day)	S8@.074	LF	13.20	2.67	1.90	17.77
50' or 60' long piles (800 LF per day)	S8@.070	LF	13.20	2.53	1.80	17.53
HP10 10" x 10", 48 lbs per LF at $.30 per pound						
30' or 40' long piles (760 LF per day)	S8@.074	LF	15.10	2.67	1.90	19.67
50' or 60' long piles (800 LF per day)	S8@.070	LF	15.10	2.53	1.80	19.43

	Craft@Hrs	Unit	Material	Labor	Equipment	Total
HP12 12" x 12", 53 lbs per LF at $.29 per pound						
30' or 40' long piles (730 LF per day)	S8@.077	LF	16.10	2.78	1.98	20.86
50' or 60' long piles (760 LF per day)	S8@.074	LF	16.10	2.67	1.90	20.67
HP12 12" x 12", 74 lbs per LF at $.29 per pound						
30' or 40' long piles (730 LF per day)	S8@.077	LF	22.50	2.78	1.98	27.26
50' or 60' long piles (760 LF per day)	S8@.074	LF	22.50	2.67	1.90	27.07
HP14 14" x 14", 73 lbs per LF at $.28 per pound						
30' or 40' long piles (690 LF per day)	S8@.081	LF	21.50	2.93	2.08	26.51
50' or 60' long piles (730 LF per day)	S8@.077	LF	21.50	2.78	1.98	26.26
HP14 14" x 14", 89 lbs per LF at $.28 per pound						
30' or 40' long piles (630 LF per day)	S8@.089	LF	26.20	3.21	2.29	31.70
50' or 60' long piles (690 LF per day)	S8@.081	LF	26.20	2.93	2.08	31.21
HP14 14" x 14", 102 lbs per LF at $.28 per pound						
30' or 40' long piles (630 LF per day)	S8@.089	LF	30.00	3.21	2.29	35.50
50' or 60' long piles (690 LF per day)	S8@.081	LF	30.00	2.93	2.08	35.01
HP14 14" x 14", 117 lbs per LF at $.28 per pound						
30' or 40' long piles (630 LF per day)	S8@.089	LF	34.40	3.21	2.29	39.90
50' or 60' long piles (690 LF per day)	S8@.081	LF	34.40	2.93	2.08	39.41
Additional costs related to any of above						
Standard driving points or splices						
HP8	S8@1.50	Ea	40.20	54.20	16.90	111.30
HP10	S8@1.75	Ea	43.70	63.20	19.80	126.70
HP12	S8@2.00	Ea	54.90	72.20	22.60	149.70
HP14	S8@2.50	Ea	64.60	90.30	28.20	183.10
Cutting pile off to required elevation						
HP8 or HP10	S8@1.00	Ea	5.25	36.10	11.30	52.65
HP12 or HP14	S8@1.50	Ea	7.90	54.20	16.90	79.00

Steel pipe piles Equipment cost per hour is $152 and includes a 45 ton crawler mounted crane equipped with pile driving attachments, a 2-ton forklift, a truck with hand tools, a portable air compressor complete with hoses and pneumatic operated tools, a welding machine and an oxygen-acetylene cutting torch. Use $20,000 as minimum job charge, including 500 LF of piling and cost for moving pile driving equipment on and off the job site

	Craft@Hrs	Unit	Material	Labor	Equipment	Total
Minimum job charge	—	LS	—	—	—	20,000.00

For jobs requiring over 500 LF of piling add to minimum charge as follows. (Figures in parentheses show approximate production per day based on a 7-man crew)

	Craft@Hrs	Unit	Material	Labor	Equipment	Total
Steel pipe piles, non-filled						
8" (660 LF per day)	S8@.085	LF	11.10	3.07	1.83	16.00
10" (635 LF per day)	S8@.088	LF	15.80	3.18	1.90	20.88
12" (580 LF per day)	S8@.096	LF	19.50	3.47	2.10	25.07
14" (500 LF per day)	S8@.112	LF	21.50	4.05	2.42	27.97
16" (465 LF per day)	S8@.120	LF	24.80	4.33	2.60	31.73
18" (440 LF per day)	S8@.128	LF	28.20	4.62	2.76	35.58
Steel pipe piles, concrete filled						
8" (560 LF per day)	S8@.100	LF	13.00	3.61	2.16	18.77
10" (550 LF per day)	S8@.101	LF	16.40	3.65	2.19	22.24
12" (500 LF per day)	S8@.112	LF	21.70	4.05	2.42	28.17
14" (425 LF per day)	S8@.131	LF	23.30	4.73	2.83	30.86
16" (400 LF per day)	S8@.139	LF	27.70	5.02	3.00	35.72
18" (375 LF per day)	S8@.149	LF	32.00	5.38	3.22	40.60
Splices for steel pipe piles						
8" pile	S8@1.12	Ea	51.60	40.50	33.88	125.98
10" pile	S8@1.55	Ea	58.00	56.00	46.89	160.89
12" pile	S8@2.13	Ea	62.50	76.90	64.43	203.83

	Craft@Hrs	Unit	Material	Labor	Equipment	Total
14" pile	S8@2.25	Ea	82.60	81.30	57.00	220.90
16" pile	S8@2.39	Ea	131.00	86.30	68.06	285.36
18" pile	S8@2.52	Ea	157.00	91.00	76.23	324.23
Standard points for steel pipe piles						
8" point	S8@1.49	Ea	79.20	53.80	45.10	178.10
10" point	S8@1.68	Ea	107.00	60.70	50.82	218.52
12" point	S8@1.84	Ea	144.00	66.50	55.66	266.16
14" point	S8@2.03	Ea	181.00	73.30	61.41	315.71
16" point	S8@2.23	Ea	265.00	80.50	67.46	412.96
18" point	S8@2.52	Ea	351.00	91.00	76.23	518.23

Wood piles End bearing SYP, treated with 12 lb creosote. Piles up to 40' long are 8" minimum tip diameter. Over 45' long, 7" is the minimum tip diameter. Equipment cost per hour is $115 and includes a 45 ton crawler mounted crane equipped with pile driving attachments, a 2-ton forklift, a truck with hand tools, a portable air compressor complete with hoses and pneumatic operated tools. Use $18,000 as minimum job charge, including 1,000 LF of piling and cost for moving pile driving equipment on and off the job site

	Craft@Hrs	Unit	Material	Labor	Equipment	Total
Minimum job charge	—	LS	—	—	—	18,000.00

For jobs requiring over 1,000 LF of piling, add to minimum charge as follows. (Figures in parentheses show approximate production per day based on a 7-man crew)

	Craft@Hrs	Unit	Material	Labor	Equipment	Total
To 30' long (725 LF per day)	S8@.077	LF	5.83	2.78	1.26	9.87
Over 30' to 40' (800 LF per day)	S8@.069	LF	5.93	2.49	1.13	9.55
Over 40' to 50' (840 LF per day)	S8@.067	LF	6.51	2.42	1.10	10.03
Over 50' to 60' (920 LF per day)	S8@.061	LF	6.97	2.20	1.00	10.17
Over 60' to 80' (950 LF per day)	S8@.059	LF	8.85	2.13	0.97	11.95
Add for drive shoe, per pile	—	Ea	30.90	—	—	30.90

Caisson Foundations Drilled in stable soil, no shoring required, filled with reinforced concrete. Add for belled bottoms, if required, from next page. Equipment cost per hour is $175.00 per hour and includes a truck-mounted A-frame hoist and a power operated auger and a concrete pump. Concrete is based on using 3,000 PSI pump mix at $71.00 per CY, before 5% allowance for waste, delivered to the job site in ready-mix trucks. Spiral caisson reinforcing is based on using 3/8" diameter hot rolled steel spirals with main vertical bars as shown, shop fabricated delivered ready to install, with typical engineering and drawings. Costs for waste disposal are not included, refer to construction materials disposal section and add for same. Use $15,000 as a minimum charge, including 200 LF of reinforced concrete filled caissons and cost for moving the equipment on and off the job

	Craft@Hrs	Unit	Material	Labor	Equipment	Total
Minimum job charge	—	LS	—	—	—	15,000.00

For jobs requiring over 200 LF of reinforced concrete filled caissons add to minimum charge as follows. (Figures in parentheses show approximate production per day based on a 3-man crew)

16" diameter, drilled caisson filled with reinforced concrete

	Craft@Hrs	Unit	Material	Labor	Equipment	Total
Drilling						
(680 LF per day)	S6@.035	LF	—	1.16	1.20	2.36
Reinforcing with 6 #6 bars						
(590 LF per day)	S6@.041	LF	16.25	1.36	2.56	20.17
Concrete filling						
(615 LF per day)	S6@.039	LF	3.88	1.29	2.46	7.63
Total for 16" caisson	S6@.115	LF	20.13	3.81	6.22	30.16

24" diameter, drilled caisson filled with reinforced concrete

	Craft@Hrs	Unit	Material	Labor	Equipment	Total
Drilling						
(570 LF per day)	S6@.042	LF	—	1.39	2.65	4.04
Reinforcing 6 #6 bars						
(280 LF per day	S6@.085	LF	27.80	2.82	5.40	36.02
Concrete filling						
(280 LF per day)	S6@.086	LF	8.65	2.85	5.40	16.90
Total for 24" caisson	S6@.213	LF	36.45	7.06	13.45	56.96

Site Work 2

	Craft@Hrs	Unit	Material	Labor	Equipment	Total
36" diameter, drilled caisson filled with reinforced concrete						
Drilling						
(440 LF per day)	S6@.055	LF	—	1.83	3.43	5.26
Reinforcing 8 #10 bars						
(135 LF per day)	S6@.176	LF	50.50	5.84	11.20	67.54
Concrete filling						
(125 LF per day)	S6@.194	LF	19.60	6.44	12.10	38.14
Total for 36" caisson	S6@.425	LF	70.10	14.10	26.70	110.90

Bell-bottom footings concrete filled, for pre-drilled caissons. Add to the cost of any of the caissons above. (Figures shown in parentheses show approximate production per day based on a 3-man crew)

	Craft@Hrs	Unit	Material	Labor	Equipment	Total
6' diameter bell bottom						
concrete filled (8 per day)	S6@3.00	Ea	234.00	99.60	176.50	510.10
8' diameter bell bottom						
concrete filled (4 per day)	S6@6.00	Ea	567.10	199.00	346.00	1,112.10

Shoring, Bulkheads and Underpinning Steel sheet piling, driven and pulled, based on good driving conditions and using sheets 30' to 50' in length with 32 lbs of 38.5 KSI steel per SF. Material costs are based on steel at $870 per ton. Equipment cost for driving and pulling averages $200 per ton. Material cost assumes the piling is salvaged for $650 per ton. Use $16,000 as a minimum job charge

	Craft@Hrs	Unit	Material	Labor	Equipment	Total
Up to 20' deep	S8@.096	SF	3.52	3.47	3.55	10.54
Over 20' to 35' deep	S8@.088	SF	3.52	3.18	3.26	9.96
Over 35' to 50' deep	S8@.087	SF	3.52	3.14	3.22	9.88
Over 50' deep	S8@.085	SF	3.52	3.07	3.15	9.74

	Craft@Hrs	Unit	Material	Labor	Equipment	Total
Extra costs for sheet steel piling						
Add for:						
50' to 65' lengths	—	SF	.07	—	—	.07
65' to 100' lengths	—	SF	.18	—	—	.18
Epoxy coal tar 16 mil factory finish	—	SF	2.00	—	—	2.00
32 lb per SF piling material						
left in place	—	SF	3.67	—	—	3.67
38 lbs per SF (MZ 38) steel	—	SF	.77	—	—	.77
50 KSI high-strength steel	—	SF	.42	—	—	.42
Marine grade 50 KSI						
high strength steel	—	SF	.83	—	—	.83
Deductive costs for sheet steel piling						
Deduct for:						
28 lbs per SF (PS 28) steel	—	SF	-.38	—	—	-.38
23 lbs per SF (PS 23) steel	—	SF	-.80	—	—	-.80
Deduct labor & equipment cost						
when piling is left in place	—	SF	—	-1.45	-.52	-1.97

Timber trench sheeting and bracing per square foot of trench wall measured on one side of trench. Costs include pulling and salvage, depths to 12'. Equipment cost is $40.60 per hour for a 10 ton hydraulic crane and a 1 ton flat-bed truck. Use $4,000 as a minimum job charge.

	Craft@Hrs	Unit	Material	Labor	Equipment	Total
Less than 6' wide, open bracing	S5@.053	SF	1.95	1.67	.31	3.93
6' to 10' wide, open bracing	S5@.064	SF	2.05	2.02	.37	4.44
11' to 16' wide, open bracing	S5@.079	SF	2.42	2.49	.45	5.36
Over 16' wide, open bracing	S5@.091	SF	2.65	2.87	.49	6.01
Less than 6' wide, closed sheeting	S5@.114	SF	2.25	3.59	.53	6.37
6' to 10' wide, closed sheeting	S5@.143	SF	2.00	4.51	.83	7.34
11' to 16' wide, closed sheeting	S5@.163	SF	2.38	5.14	.95	8.47
Over 16' wide, closed sheeting	S5@.199	SF	3.00	6.27	1.16	10.43

	Craft@Hrs	Unit	Material	Labor	Equipment	Total
Additional costs for bracing on trenches over 12' deep. Add to costs on previous page.						
Normal bracing, to 15' deep	S5@.050	SF	.41	1.58	.29	2.28
One line of bracing, over 15' to 22' deep	S5@.053	SF	.50	1.67	.31	2.48
Two lines of bracing, over 22' to 35' deep	S5@.061	SF	.82	1.92	.35	3.09
Three lines of bracing, over 35' to 45' deep	S5@.064	SF	.91	2.02	.37	3.30

Wellpoint Dewatering Based on header pipe connecting 2" diameter jetted wellpoints 5' on center. Includes header pipe, wellpoints, filter sand, pumps and swing joints. The header pipe length is usually equal to the perimeter of the area excavated.

Typical cost for installation and removal of wellpoint system with wellpoints 14' deep and placed 5' on center along 6" header pipe. Based on equipment rented for 1 month. Use 100 LF as a minimum job charge and add cost for operator as required. Also add (if required) for additional months of rental, fuel for the pumps, a standby pump, water truck for jetting, outflow pipe, permits and consultant costs.

6" header pipe and accessories, system per LF installed, rented 1 month and removed

	Craft@Hrs	Unit	Material	Labor	Equipment	Total
100 LF system	C5@.535	LF	1.30	17.50	43.50	62.30
200 LF system	C5@.535	LF	1.30	17.50	26.00	44.80
500 LF system	C5@.535	LF	1.30	17.50	22.50	41.30
1,000 LF system	C5@.535	LF	1.30	17.50	19.00	37.80
Add for 8" header pipe	C5@.010	LF	.06	.33	1.50	1.89
Add for 18' wellpoint depth	C5@.096	LF	.11	3.15	4.75	8.01
Add for second month	—	LF	—	—	14.00	14.00
Add for each additional month	—	LF	—	—	8.35	8.35
Add for operator, per hour	T0@1.00	Hr	—	39.10	—	39.10

Pipe Jacking Typical costs for jacking .50" thick wall pipe casing under an existing roadway. Costs include leaving casing in place. Add 15% when ground water is present. Add 100% for light rock conditions. Includes jacking pits on both sides. Equipment cost shown is for a 1 CY wheel mounted backhoe at $23.80 per hour plus a 2-ton truck at $13.10 per hour equipped for this type work. Size shown is casing diameter. Use $2,200 as a minimum job charge.

	Craft@Hrs	Unit	Material	Labor	Equipment	Total
2" casing	C5@.289	LF	5.80	9.48	2.67	17.95
3" casing	C5@.351	LF	6.86	11.50	3.24	21.60
4" casing	C5@.477	LF	8.98	15.60	4.40	28.98
6" casing	C5@.623	LF	11.60	20.40	5.75	37.75
8" casing	C5@.847	LF	15.30	27.80	7.82	50.92
10" casing	C5@1.24	LF	22.70	40.70	11.45	74.85
12" casing	C5@1.49	LF	27.40	48.90	13.75	90.05
16" casing	C5@1.78	LF	32.70	58.40	16.43	107.53
17" casing	C5@2.09	LF	37.40	68.50	19.30	125.20
24" casing	C5@7.27	LF	132.00	238.00	67.12	437.12
30" casing	C5@7.88	LF	143.00	258.00	72.75	473.75
36" casing	C5@8.61	LF	158.00	282.00	79.49	519.49
42" casing	C5@9.29	LF	169.00	305.00	85.76	559.76
48" casing	C5@10.2	LF	185.00	335.00	94.17	614.17

Asbestos Cement Pipe (Transite) Class 2400 sewer pipe or Class 3000 storm drain pipe, standard 13' lengths (Certain Teed Products). Installed in an open trench. Costs include couplers, inspection and test and are based on truck load quantities. Equipment cost shown is for a 1 CY wheel mounted backhoe at $23.80 per hour for lifting and placing the pipe. Excavation, bedding material or backfill are not included. Use $4,000 as a minimum job charge.

	Craft@Hrs	Unit	Material	Labor	Equipment	Total
6" pipe	U1@.077	LF	2.87	2.72	.45	6.04
8" pipe	U1@.077	LF	4.64	2.72	.45	7.81
10" pipe	U1@.080	LF	7.31	2.83	.48	10.62

	Craft@Hrs	Unit	Material	Labor	Equipment	Total
12" pipe	U1@.083	LF	9.39	2.94	.49	12.82
14" pipe	U1@.116	LF	13.10	4.10	.69	17.89
16" pipe	U1@.123	LF	15.70	4.35	.73	20.78

Gas Distribution Lines Installed in an open trench. Equipment cost shown is for a wheel mounted 1/2 CY backhoe at $21.10 per hour for lifting and placing the pipe. Excavation, bedding material or backfill are not included. Use $2,000 as a minimum job charge.

Polyethylene pipe, 60 PSI

	Craft@Hrs	Unit	Material	Labor	Equipment	Total
1-1/4" diameter coils	U1@.064	LF	.47	2.26	.34	3.07
1-1/2" diameter coils	U1@.064	LF	.66	2.26	.34	3.26
2" diameter coils	U1@.071	LF	.85	2.51	.37	3.73
3" diameter coils	U1@.086	LF	1.71	3.04	.45	5.20
3" diameter in 38' lengths with couplings	U1@.133	LF	1.42	4.71	.70	6.83
4" diameter in 38' lengths with couplings	U1@.168	LF	2.84	5.94	.89	9.67
6" diameter in 38' lengths with couplings	U1@.182	LF	6.05	6.44	.96	13.45
8" diameter in 38' lengths with couplings	U1@.218	LF	11.00	7.71	1.15	19.86

Meters and pressure regulators Natural gas diaphragm meters, direct digital reading, not including temperature and pressure compensation. For use with pressure regulators below

	Craft@Hrs	Unit	Material	Labor	Equipment	Total
425 CFH at 10 PSI	P6@4.38	Ea	225.00	158.00	—	383.00
425 CFH at 25 PSI	P6@4.38	Ea	297.00	158.00	—	455.00

Natural gas pressure regulators with screwed connections

	Craft@Hrs	Unit	Material	Labor	Equipment	Total
3/4" or 1" connection	P6@.772	Ea	23.00	27.90	—	50.90
1-1/4" or 1-1/2" connection	P6@1.04	Ea	121.00	37.60	—	158.60
2" connection	P6@1.55	Ea	989.00	56.00	—	1,045.00

Polyethylene Pipe Belled ends with rubber ring joints. Installed in an open trench. Equipment cost shown is for a wheel mounted 1/2 CY backhoe at $21.10 per hour for lifting and placing the pipe. Excavation, bedding material, shoring, backfill or dewatering are not included. Use $2,000 as a minimum job charge.

160 PSI polyethylene pipe, Standard Dimension Ratio (SDR) 11 to 1

	Craft@Hrs	Unit	Material	Labor	Equipment	Total
3" pipe	U1@.108	LF	1.42	3.82	.57	5.81
4" pipe	U1@.108	LF	2.36	3.82	.57	6.75
5" pipe	U1@.108	LF	3.69	3.82	.57	8.08
6" pipe	U1@.108	LF	5.12	3.82	.57	9.51
8" pipe	U1@.145	LF	8.67	5.13	.76	14.56
10" pipe	U1@.145	LF	13.60	5.13	.76	19.49
12" pipe	U1@.145	LF	19.00	5.13	.76	24.89

100 PSI polyethylene pipe, Standard Dimension Ratio (SDR) 17 to 1

	Craft@Hrs	Unit	Material	Labor	Equipment	Total
4" pipe	U1@.108	LF	1.61	3.82	.57	6.00
6" pipe	U1@.108	LF	3.45	3.82	.57	7.84
8" pipe	U1@.145	LF	5.82	5.13	.76	11.71
10" pipe	U1@.145	LF	9.03	5.13	.76	14.92
12" pipe	U1@.145	LF	12.60	5.13	.76	18.49

Low pressure polyethylene pipe, Standard Dimension Ratio (SDR) 21 to 1

	Craft@Hrs	Unit	Material	Labor	Equipment	Total
4" pipe	U1@.108	LF	1.33	3.82	.57	5.72
5" pipe	U1@.108	LF	1.99	3.82	.57	6.38
6" pipe	U1@.108	LF	2.89	3.82	.57	7.28
7" pipe	U1@.145	LF	3.26	5.13	.76	9.15
8" pipe	U1@.145	LF	4.83	5.13	.76	10.72
10" pipe	U1@.145	LF	7.24	5.13	.76	13.13
12" pipe	U1@.145	LF	10.40	5.13	.76	16.29
14" pipe	U1@.169	LF	12.60	5.98	.89	19.47

	Craft@Hrs	Unit	Material	Labor	Equipment	Total
16" pipe	U1@.169	LF	16.40	5.98	.89	23.27
18" pipe	U1@.169	LF	20.60	5.98	.89	27.47
20" pipe	U1@.212	LF	25.70	7.50	1.12	34.32
22" pipe	U1@.212	LF	31.00	7.50	1.12	39.62
24" pipe	U1@.212	LF	36.70	7.50	1.12	45.32
Fittings, molded, bell ends						
4" 90 degree ell, 160 PSI	U1@2.00	Ea	31.20	70.80	10.56	112.56
4" 45 degree ell, 160 PSI	U1@2.00	Ea	31.20	70.80	10.56	112.56
6" 90 degree ell, 160 PSI	U1@2.00	Ea	64.30	70.80	10.56	145.66
6" 45 degree ell, 160 PSI	U1@2.00	Ea	64.30	70.80	10.56	145.66
4" to 3" reducer, 160 PSI	U1@2.00	Ea	18.70	70.80	10.56	100.06
6" to 4" reducer, 160 PSI	U1@2.00	Ea	44.70	70.80	10.56	126.06
3" transition section, 160 PSI	U1@2.00	Ea	39.30	70.80	10.56	120.66
6" transition section, 160 PSI	U1@2.00	Ea	174.00	70.80	10.56	255.36
10" by 4" branch saddle, 160 PSI	U1@3.00	Ea	56.60	106.00	15.84	178.44
3" flanged adapter, low pressure	U1@1.00	Ea	29.40	35.40	5.28	70.08
4" flanged adapter, low pressure	U1@1.00	Ea	42.00	35.40	5.28	82.68
6" flanged adapter, low pressure	U1@1.00	Ea	55.10	35.40	5.28	95.78
8" flanged adapter, low pressure	U1@1.50	Ea	79.10	53.10	7.92	140.12
10" flanged adapter, low pressure	U1@1.50	Ea	113.00	53.10	7.92	174.02
4" flanged adapter, 160 PSI	U1@1.00	Ea	45.80	35.40	5.28	86.48
6" flanged adapter, 160 PSI	U1@1.00	Ea	64.80	35.40	5.28	105.48
Fittings, fabricated, belled ends						
10" 90 degree ell, 100 PSI	U1@3.00	Ea	323.00	106.00	15.84	444.84
10" 45 degree ell, 100 PSI	U1@3.00	Ea	179.00	106.00	15.84	300.84
4" 30 degree ell, 100 PSI	U1@2.00	Ea	52.30	70.80	10.56	133.66
8" to 6" reducer, 100 PSI	U1@3.00	Ea	95.80	106.00	15.84	217.64
10" to 8" reducer, 100 PSI	U1@3.00	Ea	113.00	106.00	15.84	234.84
4" 45 degree wye, 100 PSI	U1@2.00	Ea	139.00	70.80	10.56	220.36

Polyethylene Rigid Drainage Pipe Standard 10' lengths, 1500 lb crush rating, plain or perforated and high impact ABS glued-on fittings. Installed in an open trench. No excavation, shoring, backfill or dewatering included.

	Craft@Hrs	Unit	Material	Labor	Equipment	Total
3" pipe	CL@.070	LF	.33	2.12	—	2.45
4" pipe	CL@.073	LF	.47	2.21	—	2.68
6" pipe	CL@.079	LF	1.06	2.39	—	3.45
Elbows fittings (1/4 bend)						
3" elbow	CL@.168	Ea	1.45	5.09	—	6.54
4" elbow	CL@.250	Ea	2.05	7.57	—	9.62
6" elbow	CL@.332	Ea	8.48	10.00	—	18.48
Tee fittings						
3" elbow	CL@.251	Ea	1.63	7.60	—	9.23
4" elbow	CL@.375	Ea	2.18	11.40	—	13.58
6" elbow	CL@.500	Ea	9.49	15.10	—	24.59

Polyethylene Flexible Drainage Tubing Corrugated drainage tubing, plain or perforated and snap-on ABS fittings. Installed in an open trench. Excavation, bedding material or backfill are not included.

	Craft@Hrs	Unit	Material	Labor	Equipment	Total
Tubing						
3" (10' length or 100' coil)	CL@.009	LF	.33	.27	—	.60
4" (10' length or 250' coil)	CL@.011	LF	.38	.33	—	.71
6" (100' coil)	CL@.014	LF	.83	.42	—	1.25
8" (20' or 40' length)	CL@.017	LF	1.40	.51	—	1.91
10" (20' length)	CL@.021	LF	3.07	.64	—	3.71
12" (20' length)	CL@.022	LF	4.37	.67	—	5.04

	Craft@Hrs	Unit	Material	Labor	Equipment	Total
Snap fittings, elbows or tees						
3" fitting	CL@.025	Ea	2.90	.76	—	3.66
4" fitting	CL@.025	Ea	3.47	.76	—	4.23
6" fitting	CL@.025	Ea	5.80	.76	—	6.56

Cast Iron Flanged Pipe AWWA C151 pipe with cast iron flanges on both ends. Pipe is laid and connected in an open trench. Equipment cost shown is for a wheel mounted 1 CY backhoe at $23.80 per hour for lifting and placing the pipe and fittings. Trench excavation, dewatering, backfill and compaction are not included. Use $3,000 as a minimum job charge. See next section for fittings, valves and accessories.

	Craft@Hrs	Unit	Material	Labor	Equipment	Total
4" pipe, 260 pounds per 18' section	U1@.057	LF	9.66	2.02	.34	12.02
6" pipe, 405 pounds per 18' section	U1@.075	LF	11.20	2.65	.45	14.30
8" pipe, 570 pounds per 18' section	U1@.075	LF	16.00	2.65	.45	19.10
10" pipe, 740 pounds per 18' section	U1@.094	LF	21.90	3.33	.56	25.79
12" pipe, 930 pounds per 18' section	U1@.094	LF	29.60	3.33	.56	33.49

Cast Iron Fittings, Valves and Accessories Based on type ASA A21.10-A21-11-6 and 4C110-64, C111-64 fittings. Installed in an open trench and connected. Equipment cost shown is for a wheel mounted 1 CY backhoe at $23.80 per hour for lifting and placing. Trench excavation, dewatering, backfill and compaction are not included. These items are also suitable for use with the Ductile Iron Pipe in the section that follows. Costs estimated using this section may be combined with the costs estimated from the Cast Iron Flanged Pipe section or the Ductile Iron Pipe section for arriving at the minimum job charge shown in those sections.

	Craft@Hrs	Unit	Material	Labor	Equipment	Total
Cast iron mechanical joint 90 degree ells						
4" ell	U1@.808	Ea	72.00	28.60	4.82	105.42
6" ell	U1@1.06	Ea	109.00	37.50	6.32	152.82
8" ell	U1@1.06	Ea	158.00	37.50	6.32	201.82
10" ell	U1@1.32	Ea	238.00	46.70	7.86	292.56
12" ell	U1@1.32	Ea	297.00	46.70	7.86	351.56
14" ell	U1@1.32	Ea	683.00	46.70	7.86	737.56
16" ell	U1@1.65	Ea	780.00	58.40	9.83	848.23
Deduct for 1/8 or 1/16 bends	—	%	-20.0	—	—	—
Cast iron mechanical joint tees						
4" x 4"	U1@1.22	Ea	109.00	43.20	7.27	159.47
6" x 4"	U1@1.59	Ea	140.00	56.30	9.47	205.77
6" x 6"	U1@1.59	Ea	153.00	56.30	9.47	218.77
8" x 8"	U1@1.59	Ea	220.00	56.30	9.47	285.77
10" x 10"	U1@1.59	Ea	319.00	56.30	9.47	384.77
12" x 12"	U1@2.48	Ea	414.00	87.80	14.78	516.58
Add for wyes	—	%	40.0	—	—	—
Add for crosses	—	%	30.0	33.0	33.0	—
Cast iron mechanical joint reducers						
6" x 4"	U1@1.06	Ea	81.00	37.50	6.31	124.81
8" x 6"	U1@1.06	Ea	118.00	37.50	6.31	161.81
10" x 8"	U1@1.32	Ea	156.00	46.70	7.87	210.57
12" x 10"	U1@1.32	Ea	211.00	46.70	7.87	265.57
AWWA mechanical joint gate valves						
3" valve	U1@3.10	Ea	264.00	110.00	18.46	392.46
4" valve	U1@4.04	Ea	353.00	143.00	24.08	520.08
6" valve	U1@4.80	Ea	563.00	170.00	28.61	761.61
8" valve	U1@5.14	Ea	952.00	182.00	30.63	1,164.63
10" valve	U1@5.67	Ea	1509.00	201.00	33.79	1,743.79
12" valve	U1@6.19	Ea	2560.00	219.00	36.89	2,815.89
Indicator post valve						
Upright, adjustable	U1@2.10	Ea	633.00	74.30	11.70	719.00

Site Work 2

	Craft@Hrs	Unit	Material	Labor	Equipment	Total
Fire hydrant with spool						
6"	U1@5.00	Ea	872.00	177.00	27.80	1,076.80
Backflow preventers, installed at ground level						
3" backflow preventer	U1@3.50	Ea	1,100.00	124.00	20.86	1,244.86
4" backflow preventer	U1@4.00	Ea	1,450.00	142.00	23.84	1,615.84
6" backflow preventer	U1@5.00	Ea	2,380.00	177.00	29.80	2,586.80
8" backflow preventer	U1@6.00	Ea	4,580.00	212.00	35.76	4,827.76
Block structure for backflow preventer,						
5' x 7' x 3'4"	M1@32.0	Ea	630.00	1,080.00	—	1,710.00
Sheer gate						
8", with adjustable release handle	U1@2.44	Ea	1,100.00	86.30	14.54	1,200.84
Adjustable length valve boxes, for up to 20" valves,						
5' depth	U1@1.75	Ea	68.00	61.90	10.43	140.33
Concrete thrust blocks, minimum formwork						
1/4 CY thrust block	C8@1.20	Ea	47.90	43.50	—	91.40
1/2 CY thrust block	C8@2.47	Ea	96.30	89.60	—	185.90
3/4 CY thrust block	C8@3.80	Ea	144.00	138.00	—	282.00
1 CY thrust block	C8@4.86	Ea	191.00	176.00	—	367.00
Tapping saddles, double strap, iron body, tap size to 2"						
Pipe to 4"	U1@2.38	Ea	27.00	84.20	14.18	125.38
8" pipe	I1@2.89	Ea	35.20	102.00	17.22	154.42
10" pipe	U1@3.15	Ea	35.90	111.00	18.77	165.67
12" pipe	U1@3.52	Ea	47.40	125.00	20.98	193.38
14" pipe, bronze body	U1@3.78	Ea	56.90	134.00	22.53	213.43
Tapping valves, mechanical joint						
4" valve	I1@6.93	Ea	273.00	245.00	41.30	559.30
6" valve	U1@8.66	Ea	368.00	306.00	51.61	725.61
8" valve	U1@9.32	Ea	541.00	330.00	55.55	926.55
10" valve	U1@10.4	Ea	819.00	368.00	61.98	1,248.98
12" valve	U1@11.5	Ea	1,250.00	407.00	68.54	1,725.54
Labor and equipment tapping pipe, by hole size						
4" pipe	U1@5.90	Ea	—	209.00	35.16	244.16
6" pipe	U1@9.32	Ea	—	330.00	55.55	385.55
8" pipe	U1@11.7	Ea	—	414.00	69.73	483.73
10" pipe	U1@15.5	Ea	—	548.00	92.38	640.38
12" pipe	U1@23.4	Ea	—	828.00	139.50	967.50

Ductile Iron Pressure Pipe Mechanical joint 150 lb Class 52 ductile iron pipe. Installed in an open trench to 6' deep. Equipment cost shown is for a wheel mounted 1 CY backhoe at $23.80 per hour for lifting and placing the pipe. Trench excavation, dewatering, backfill and compaction are not included. Use the Cast Iron Fittings and Valves from the preceding section as required. Use $4,000 as a minimum job charge.

	Craft@Hrs	Unit	Material	Labor	Equipment	Total
4" pipe	U1@.140	LF	10.00	4.95	.83	15.78
6" pipe	U1@.180	LF	14.50	6.37	1.07	21.94
8" pipe	U1@.200	LF	20.10	7.08	1.19	28.37
10" pipe	U1@.230	LF	29.60	8.14	1.37	39.11
12" pipe	U1@.260	LF	40.60	9.20	1.55	51.35
14" pipe	U1@.300	LF	62.10	10.60	1.79	74.49
16" pipe	U1@.340	LF	79.20	12.00	2.03	93.23

	Craft@Hrs	Unit	Material	Labor	Equipment	Total

Non-Reinforced Concrete Pipe Smooth wall, standard strength, 3' lengths with tongue and groove mortar joint ends. Installed in an open trench. Equipment cost shown is for a wheel mounted 1 CY backhoe at $23.80per hour for lifting and placing the pipe. Trench excavation, bedding material and backfill are not included. Use $3,500 as a minimum job charge.

	Craft@Hrs	Unit	Material	Labor	Equipment	Total
6" pipe	C5@.076	LF	2.56	2.49	.36	5.41
8" pipe	C5@.088	LF	3.96	2.89	.42	7.27
10" pipe	C5@.088	LF	4.71	2.89	.42	8.02
12" pipe	C5@.093	LF	6.57	3.05	.44	10.06
15" pipe	C5@.105	LF	8.17	3.44	.50	12.11
18" pipe	C5@.117	LF	10.30	3.84	.56	14.70
21" pipe	C5@.117	LF	13.10	3.84	.56	17.50
24" pipe	C5@.130	LF	19.30	4.26	.62	24.18
30" pipe	C5@.143	LF	33.60	4.69	.68	38.97
36" pipe	C5@.158	LF	36.40	5.18	.75	42.33

Non-Reinforced Perforated Concrete Underdrain Pipe Perforated smooth wall, standard strength, 3' lengths with tongue and groove mortar joint ends. Installed in an open trench. Equipment cost shown is for a wheel mounted 1 CY backhoe at $23.80 per hour for lifting and placing the pipe. Trench excavation, bedding material and backfill are not included. Use $2,200 as a minimum job charge.

	Craft@Hrs	Unit	Material	Labor	Equipment	Total
6" pipe	C5@.076	LF	3.05	2.49	.36	5.90
8" pipe	C5@.104	LF	4.60	3.41	.50	8.51

Reinforced Concrete Pipe Class III, 1350 "D" load, ASTM C-76, 8' lengths. Tongue and groove mortar joint ends. Installed in an open trench. Equipment cost shown is for a wheel mounted 1 CY backhoe at $23.80 per hour for lifting and placing the pipe. Trench excavation, dewatering, backfill and compaction are not included. Use $3,000 as a minimum job charge.

	Craft@Hrs	Unit	Material	Labor	Equipment	Total
18" pipe	C5@.272	LF	13.00	8.92	1.30	23.22
24" pipe	C5@.365	LF	16.00	12.00	1.74	29.74
30" pipe	C5@.426	LF	22.00	14.00	2.03	38.03
36" pipe	C5@.511	LF	31.00	16.80	2.44	50.24
42" pipe	C5@.567	LF	38.00	18.60	2.70	59.30
48" pipe	C5@.623	LF	46.00	20.40	2.98	69.38
54" pipe	C5@.707	LF	55.00	23.20	3.37	81.57
60" pipe	C5@.785	LF	67.00	25.70	3.74	96.44
66" pipe	C5@.914	LF	83.00	30.00	4.36	117.36
72" pipe	C5@.914	LF	96.00	30.00	4.36	130.36
78" pipe	C5@.914	LF	110.00	30.00	4.36	144.36
84" pipe	C5@1.04	LF	127.00	34.10	4.96	166.06
90" pipe	C5@1.04	LF	137.00	34.10	4.96	176.06
96" pipe	C5@1.18	LF	153.00	38.70	5.63	197.33

Reinforced Elliptical Concrete Pipe Class III, 1350 "D", C-502-72, 8' lengths, tongue and groove mortar joint ends. Installed in an open trench. Equipment cost shown is for a wheel mounted 1 CY backhoe at $23.80 per hour for lifting and placing the pipe. Trench excavation, dewatering, backfill and compaction are not included. Use $3,500 as a minimum job charge.

	Craft@Hrs	Unit	Material	Labor	Equipment	Total
19" x 30" pipe, 24" pipe equivalent	C5@.356	LF	24.80	11.70	1.68	38.18
24" x 38" pipe, 30" pipe equivalent	C5@.424	LF	33.30	13.90	2.02	49.22
29" x 45" pipe, 36" pipe equivalent	C5@.516	LF	53.80	16.90	2.46	73.16
34" x 53" pipe, 42" pipe equivalent	C5@.581	LF	66.60	19.10	2.77	88.47
38" x 60" pipe, 48" pipe equivalent	C5@.637	LF	79.40	20.90	3.04	103.34
48" x 76" pipe, 60" pipe equivalent	C5@.797	LF	132.00	26.10	3.80	161.90
53" x 83" pipe, 66" pipe equivalent	C5@.923	LF	162.00	30.30	4.40	196.70
58" x 91" pipe, 72" pipe equivalent	C5@.923	LF	192.00	30.30	4.40	226.70

	Craft@Hrs	Unit	Material	Labor	Equipment	Total

Flared Concrete End Sections For round concrete drain pipe, precast. Equipment cost shown is for a wheel mounted 1 CY backhoe at $23.80 per hour for lifting and placing the items. Trench excavation, dewatering, backfill and compaction are not included. Use $2,500 as a minimum job charge.

	Craft@Hrs	Unit	Material	Labor	Equipment	Total
12" opening, 530 lbs	C5@2.41	Ea	244.00	79.00	11.49	334.49
15" opening, 740 lbs	C5@2.57	Ea	287.00	84.30	12.25	383.55
18" opening, 990 lbs	C5@3.45	Ea	319.00	113.00	16.45	448.45
24" opening, 1,520 lbs	C5@3.45	Ea	392.00	113.00	16.45	521.45
30" opening, 2,190 lbs	C5@3.96	Ea	537.00	130.00	18.88	685.88
36" opening, 4,100 lbs	C5@3.96	Ea	681.00	130.00	18.88	829.88
42" opening, 5,380 lbs	C5@4.80	Ea	958.00	157.00	22.89	1,137.89
48" opening, 6,550 lbs	C5@5.75	Ea	1,130.00	189.00	27.42	1,346.42
54" opening, 8,000 lbs	C5@7.10	Ea	1,440.00	233.00	33.85	1,706.85

Precast Reinforced Concrete Box Culvert Pipe ASTM C-850 with tongue and groove mortar joint ends. Installed in an open trench. Equipment cost shown is for a wheel mounted 1 CY backhoe at $23.80 per hour for lifting and placing the pipe. Trench excavation, dewatering, backfill and compaction are not included. Laying length is 8' for smaller cross sections and 6' or 4' for larger cross sections. Use $8,500 as a minimum job charge.

	Craft@Hrs	Unit	Material	Labor	Equipment	Total
4' x 3'	C5@.273	LF	102.00	8.95	1.30	112.25
5' x 5'	C5@.336	LF	151.00	11.00	1.60	163.60
6' x 4'	C5@.365	LF	176.00	12.00	1.74	189.74
6' x 6'	C5@.425	LF	205.00	13.90	2.03	220.93
7' x 4'	C5@.442	LF	215.00	14.50	2.11	231.61
7' x 7'	C5@.547	LF	263.00	17.90	2.61	283.51
8' x 4'	C5@.579	LF	229.00	19.00	2.76	250.76
8' x 8'	C5@.613	LF	298.00	20.10	2.92	321.02
9' x 6'	C5@.658	LF	318.00	21.60	3.14	342.74
9' x 9'	C5@.773	LF	376.00	25.40	3.69	405.09
10' x 6'	C5@.787	LF	380.00	25.80	3.75	409.55
10' x 10'	C5@.961	LF	464.00	31.50	4.58	500.08

Corrugated Metal Pipe, Galvanized 20' lengths, installed in an open trench. Costs include couplers. Equipment cost shown is for a wheel mounted 1 CY backhoe at $23.80 per hour for lifting and placing the pipe. Trench excavation, dewatering, backfill and compaction are not included. Use $3,500 as a minimum job charge.

Round pipe

	Craft@Hrs	Unit	Material	Labor	Equipment	Total
8", 16 gauge (.064)	C5@.102	LF	8.06	3.35	.49	11.90
10", 16 gauge (.064)	C5@.102	LF	8.44	3.35	.49	12.28
12", 16 gauge (.064)	C5@.120	LF	8.76	3.94	.57	13.27
15", 16 gauge (.064)	C5@.120	LF	10.80	3.94	.57	15.31
18", 16 gauge (.064)	C5@.152	LF	13.00	4.99	.72	18.71
24", 14 gauge (.079)	C5@.196	LF	20.30	6.43	.93	27.66
30", 14 gauge (.079)	C5@.196	LF	25.00	6.43	.93	32.36
36", 14 gauge (.079)	C5@.231	LF	30.40	7.58	1.10	39.08
42", 14 gauge (.079)	C5@.231	LF	35.30	7.58	1.10	43.98
48", 14 gauge (.079)	C5@.272	LF	40.10	8.92	1.30	50.32
60", 12 gauge (.109)	C5@.311	LF	68.40	10.20	1.48	80.08

Oval pipe

	Craft@Hrs	Unit	Material	Labor	Equipment	Total
18" x 11", 16 gauge (.064)	C5@.133	LF	12.80	4.36	.63	17.79
22" x 13", 16 gauge (.064)	C5@.167	LF	15.30	5.48	.80	21.58
29" x 18", 14 gauge (.079)	C5@.222	LF	23.90	7.28	1.06	32.24
36" x 22", 14 gauge (.079)	C5@.228	LF	29.70	7.48	1.09	38.27
43" x 27", 14 gauge (.079)	C5@.284	LF	35.90	9.31	1.35	46.56

	Craft@Hrs	Unit	Material	Labor	Equipment	Total
Flared end sections for oval pipe						
18" x 11"	C5@1.59	Ea	69.50	52.10	7.58	129.18
22" x 13"	C5@1.75	Ea	82.70	57.40	8.34	148.44
29" x 18"	C5@1.85	Ea	124.60	60.70	8.82	194.12
36" x 22"	C5@2.38	Ea	200.80	78.10	11.34	290.24
43" x 27"	C5@2.85	Ea	337.80	93.50	13.59	444.89
Add for bituminous coating, with paved invert						
On round pipe	—	%	20.0	—	—	—
On oval pipe or flared ends	—	%	24.0	—	—	—
Deduct for aluminum pipe, round						
oval or flared ends	—	%	-12.0	—	—	—

Corrugated Metal Nestable Pipe Split-in-half, type I, 16 gauge. Trench excavation, backfill, compaction and grading are not included. Equipment cost shown is for a wheel mounted 1 CY backhoe at $23.80 per hour for lifting and placing the pipe. Use $5,500 as a minimum job charge.

	Craft@Hrs	Unit	Material	Labor	Equipment	Total
12" diameter, galvanized	C5@.056	LF	14.80	1.84	.27	16.91
18" diameter, galvanized	C5@.064	LF	19.00	2.10	.31	21.41
Add for bituminous coating, with paved invert	—	%	20.0	—	—	—

Drainage Tile Vitrified clay drainage tile pipe. Standard weight, 6' joint length with 1 coupler per joint. Installed in an open trench. Couplers consist of a rubber sleeve with stainless steel compression bands. Equipment cost shown is for a wheel mounted 1/2 CY backhoe at $23.80 per hour for lifting and placing the pipe. Excavation, bedding material or backfill are not included. Use $2,000 as a minimum job charge.

	Craft@Hrs	Unit	Material	Labor	Equipment	Total
4" tile	S6@.095	LF	3.64	3.15	.75	7.54
6" tile	S6@.119	LF	7.63	3.95	.94	12.52
8" tile	S6@.143	LF	10.50	4.75	1.13	16.38
10" tile	S6@.165	LF	17.50	5.48	1.31	24.29
Add for covering 4" to 10" tile, 30 lb felt building paper	CL@.048	LF	.10	1.45	—	1.55
Elbows (1/4 bends), costs include one coupler						
4" elbow	S6@.286	Ea	22.40	9.49	2.27	34.16
6" elbow	S6@.360	Ea	38.80	12.00	2.86	53.66
8" elbow	S6@.430	Ea	51.80	14.30	3.41	69.51
10" elbow	S6@.494	Ea	88.90	16.40	3.92	109.22
Tees, costs include two couplers						
4" tee	S6@.430	Ea	28.80	14.30	3.41	46.51
6" tee	S6@.540	Ea	48.80	17.90	4.28	70.98
8" tee	S6@.643	Ea	72.20	21.30	5.10	98.60
10" tee	S6@.744	Ea	155.00	24.70	5.90	185.60

Corrugated Polyethylene Culvert Pipe Heavy duty, 20' lengths. Installed in an open trench. Equipment cost shown is for a wheel mounted 1 CY backhoe at $23.80 per hour for lifting and placing the pipe. Trench excavation, dewatering, backfill and compaction are not included. Based on 500' minimum job. Various sizes may be combined for total footage. Couplings are split type that snap onto the pipe. Use $2,500 as a minimum job charge.

	Craft@Hrs	Unit	Material	Labor	Equipment	Total
8" diameter pipe	C5@.035	LF	1.47	1.15	.17	2.79
10" diameter pipe	C5@.038	LF	3.24	1.25	.18	4.67
12" diameter pipe	C5@.053	LF	4.61	1.74	.25	6.60
18" diameter pipe	C5@.111	LF	9.32	3.64	.53	13.49
8" coupling	C5@.105	Ea	3.93	3.44	.50	7.87
10" coupling	C5@.114	Ea	3.80	3.74	.54	8.08
12" coupling	C5@.158	Ea	5.92	5.18	.75	11.85
18" coupling	C5@.331	Ea	12.20	10.90	1.58	24.68

	Craft@Hrs	Unit	Material	Labor	Equipment	Total

PVC Sewer Pipe 13' lengths, bell and spigot ends with rubber ring gasket in bell. Installed in open trench. Equipment cost shown is for a wheel mounted 1 CY backhoe at $23.80 per hour for lifting and placing the pipe. Trench excavation, dewatering, backfill and compaction are not included. Use $2,500 as a minimum job charge.

	Craft@Hrs	Unit	Material	Labor	Equipment	Total
4" pipe	C5@.047	LF	.87	1.54	.22	2.63
6" pipe	C5@.056	LF	1.92	1.84	.27	4.03
8" pipe	C5@.061	LF	3.36	2.00	.29	5.65
10" pipe	C5@.079	LF	5.37	2.59	.38	8.34
12" pipe	C5@.114	LF	7.59	3.74	.54	11.87

PVC Water Pipe Installed in an open trench. Equipment cost shown is for a wheel mounted 1 CY backhoe at $23.80 per hour for lifting and placing the pipe. Trench excavation, bedding material and backfill are not included. Use $2,500 as a minimum job charge.

Class 200 PVC cold water pressure pipe

	Craft@Hrs	Unit	Material	Labor	Equipment	Total
1-1/2" pipe	U1@.029	LF	.27	1.03	.17	1.47
2" pipe	U1@.036	LF	.42	1.27	.21	1.90
2-1/2" pipe	U1@.037	LF	.62	1.31	.22	2.15
3" pipe	U1@.065	LF	.90	2.30	.39	3.59
4" pipe	U1@.068	LF	1.50	2.41	.40	4.31
6" pipe	U1@.082	LF	3.25	2.90	.49	6.64

90 degree ells Schedule 40

	Craft@Hrs	Unit	Material	Labor	Equipment	Total
1-1/2" ells	P6@.385	Ea	.90	13.90	—	14.80
2" ells	P6@.489	Ea	1.45	17.70	—	19.15
2-1/2" ells	P6@.724	Ea	4.40	26.10	—	30.50
3" ells	P6@1.32	Ea	5.20	47.70	—	52.90
4" ells	P6@1.57	Ea	9.30	56.70	—	66.00
6" ells	P6@1.79	Ea	30.00	64.60	—	94.60
Deduct for 45 degree ells	—	%	-5.0	—	—	—

Tees

	Craft@Hrs	Unit	Material	Labor	Equipment	Total
1-1/2" tees	P6@.581	Ea	1.20	21.00	—	22.20
2" tees	P6@.724	Ea	1.80	26.10	—	27.90
2-1/2" tees	P6@.951	Ea	5.75	34.30	—	40.05
3" tees	P6@1.57	Ea	7.60	56.70	—	64.30
4" tees	P6@2.09	Ea	13.80	75.50	—	89.30
6" tees	P6@2.70	Ea	46.00	97.50	—	143.50

Vitrified Clay Pipe C-200, extra strength, installed in open trench. 4" through 12" are 6'0" joint length, over 12" are either 7' or 7'6" joint length. Costs shown include one coupler per joint. Couplers consist of a rubber sleeve with stainless steel compression bands. Equipment cost shown is for a wheel mounted 1 CY backhoe at $23.80 per hour for lifting and placing the pipe. Trench excavation, dewatering, backfill and compaction are not included. Use $3,500 as a minimum job charge.

	Craft@Hrs	Unit	Material	Labor	Equipment	Total
4" pipe	C5@.081	LF	3.80	2.66	.39	6.85
6" pipe	C5@.175	LF	7.90	5.74	.83	14.47
8" pipe	C5@.183	LF	11.00	6.00	.87	17.87
10" pipe	C5@.203	LF	18.00	6.66	.97	25.63
12" pipe	C5@.225	LF	24.20	7.38	1.07	32.65
15" pipe	C5@.233	LF	24.40	7.64	1.11	33.15
18" pipe	C5@.254	LF	33.00	8.33	1.21	42.54
21" pipe	C5@.283	LF	45.20	9.28	1.35	55.83
24" pipe	C5@.334	LF	59.90	11.00	1.59	72.49
27" pipe	C5@.396	LF	68.40	13.00	1.89	83.29
30" pipe	C5@.452	LF	83.30	14.80	2.15	100.25

	Craft@Hrs	Unit	Material	Labor	Equipment	Total
1/4 bends including couplers						
4" bend	C5@.494	Ea	22.90	16.20	2.35	41.45
6" bend	C5@.524	Ea	39.70	17.20	2.50	59.40
8" bend	C5@.549	Ea	53.10	18.00	2.62	73.72
10" bend	C5@.610	Ea	91.30	20.00	2.91	114.21
12" bend	C5@.677	Ea	117.00	22.20	3.23	142.43
15" bend	C5@.715	Ea	308.00	23.50	3.41	334.91
18" bend	C5@.844	Ea	423.00	27.70	4.02	454.72
21" bend	C5@.993	Ea	575.00	32.60	4.73	612.33
24" bend	C5@1.17	Ea	753.00	38.40	5.58	796.98
30" bend	C5@1.37	Ea	1,070.00	44.90	6.53	1,121.43
Wyes and tees including couplers						
4" wye or tee	C5@.744	Ea	29.50	24.40	3.55	57.45
6" wye or tee	C5@.787	Ea	50.20	25.80	3.75	79.75
8" wye or tee	C5@.823	Ea	74.30	27.00	3.92	105.22
10" wye or tee	C5@.915	Ea	160.00	30.00	4.36	194.36
12" wye or tee	C5@1.02	Ea	211.00	33.50	4.86	249.36
15" wye or tee	C5@1.10	Ea	345.00	36.10	5.24	386.34
18" wye or tee	C5@1.30	Ea	468.00	42.60	6.19	516.79
21" wye or tee	C5@1.53	Ea	614.00	50.20	7.29	671.49
24" wye or tee	C5@1.80	Ea	730.00	59.00	8.58	797.58
27" wye or tee	C5@2.13	Ea	888.00	69.90	10.15	968.05
30" wye or tee	C5@2.50	Ea	870.00	82.00	11.91	963.91

Sewer Main Cleaning Ball method, typical costs for cleaning new sewer mains prior to inspection, includes operator and equipment. Add $200 to total costs for setting up and removing equipment. Minimum cost will be $500. These costs include the subcontractor's overhead and profit. Typical costs per 1,000 LF of main diameter listed.

		Unit	Material	Labor	Equipment	Total
6" diameter	—	MLF	—	—	—	70.00
8" diameter	—	MLF	—	—	—	90.00
10" diameter	—	MLF	—	—	—	105.00
12" diameter	—	MLF	—	—	—	135.00
14" diameter	—	MLF	—	—	—	160.00
16" diameter	—	MLF	—	—	—	180.00
18" diameter	—	MLF	—	—	—	210.00
24" diameter	—	MLF	—	—	—	275.00
36" diameter	—	MLF	—	—	—	425.00
48" diameter	—	MLF	—	—	—	540.00

Water Main Sterilization Gas chlorination method, typical costs for sterilizing new water mains prior to inspection, includes operator and equipment. Add $200 to total costs for setting up and removing equipment. Minimum cost will be $500. These costs include the subcontractor's overhead and profit. Typical cost per 1,000 LF of main with diameter as shown.

		Unit	Material	Labor	Equipment	Total
3" or 4" diameter	—	MLF	—	—	—	40.00
6" diameter	—	MLF	—	—	—	70.00
8" diameter	—	MLF	—	—	—	95.00
10" diameter	—	MLF	—	—	—	108.00
12" diameter	—	MLF	—	—	—	135.00
14" diameter	—	MLF	—	—	—	160.00
16" diameter	—	MLF	—	—	—	170.00
18" diameter	—	MLF	—	—	—	210.00
24" diameter	—	MLF	—	—	—	275.00
36" diameter	—	MLF	—	—	—	395.00
48" diameter	—	MLF	—	—	—	540.00

	Craft@Hrs	Unit	Material	Labor	Equipment	Total
Soil Covers						
Using staked burlap and tar over straw	CL@.019	SY	.75	.58	—	1.33
Using copolymer-base sprayed-on liquid soil sealant						
On slopes for erosion control	CL@.012	SY	.40	.36	—	.76
On level ground for dust abatement	CL@.012	SY	.35	.36	—	.71

Catch Basins Catch basin including 6" concrete top and base. Equipment cost shown is for a wheel mounted 1 CY backhoe at $23.80 per hour for lifting and placing the items. Trench excavation, bedding material and backfill are not included. Use $3,000 as a minimum job charge.

	Craft@Hrs	Unit	Material	Labor	Equipment	Total
4' diameter precast concrete basin						
4' deep	S6@19.0	Ea	1,170.00	631.00	151.00	1,952.00
6' deep	S6@23.7	Ea	1,470.00	787.00	188.00	2,445.00
8' deep	S6@28.9	Ea	1,790.00	959.00	229.00	2,978.00
Light duty grate, gray iron, asphalt coated						
Grate on pipe bell						
6" diameter, 13 lb	S6@.205	Ea	10.30	6.80	1.63	18.73
8" diameter, 25 lb	S6@.247	Ea	28.00	8.20	1.96	38.16
Frame and grate						
8" diameter, 55 lb	S6@.424	Ea	60.00	14.10	3.36	77.46
17" diameter, 135 lb	S6@.823	Ea	145.00	27.30	6.53	178.83
Medium duty, frame and grate						
11" diameter, 70 lb	S6@.441	Ea	72.50	14.60	3.50	90.60
15" diameter, 120 lb	S6@.770	Ea	129.00	25.60	6.11	160.71
Radial grate, 20" diameter, 140 lb	S6@1.68	Ea	150.00	55.80	13.30	219.10
Heavy duty frame and grate						
Flat grate						
11 1/2" diameter, 85 lb	S6@.537	Ea	95.50	17.80	4.26	117.56
20" diameter, 235 lb	S6@2.01	Ea	255.00	66.70	15.95	337.65
21" diameter, 315 lb	S6@2.51	Ea	310.00	83.30	19.91	413.21
24" diameter, 350 lb	S6@2.51	Ea	345.00	83.30	19.91	448.21
30" diameter, 555 lb	S6@4.02	Ea	505.00	133.00	31.89	669.89
Convex or concave grate						
20" diameter, 200 lb	S6@1.83	Ea	180.00	60.70	14.52	255.22
20" diameter, 325 lb	S6@2.51	Ea	295.00	83.30	19.91	398.21
Beehive grate and frame						
11" diameter, 80 lb	S6@.495	Ea	129.00	16.40	3.93	149.33
15" diameter, 120 lb	S6@1.83	Ea	124.00	60.70	14.52	199.22
21" diameter, 285 lb	S6@2.51	Ea	260.00	83.30	19.91	363.21
24" diameter, 375 lb	S6@2.87	Ea	360.00	95.30	22.77	478.07

Manholes Equipment cost shown is for a wheel mounted 1 CY backhoe at $23.80 per hour for lifting and placing the items. Trench excavation, bedding material and backfill are not included. Use $3,000 as a minimum job charge.

Precast concrete manholes, including concrete top and base. No excavation or backfill included. Add for steps, frames and lids from the costs listed on the next page.

	Craft@Hrs	Unit	Material	Labor	Equipment	Total
3' x 6' to 8' deep	S6@17.5	Ea	1,320.00	581.00	139.00	2,040.00
3' x 9' to 12' deep	S6@19.4	Ea	1,540.00	644.00	154.00	2,338.00
3' x 13' to 16' deep	S6@23.1	Ea	1,990.00	767.00	183.00	2,940.00
3' diameter, 4' deep	S6@12.1	Ea	773.00	402.00	96.00	1,271.00
4' diameter, 5' deep	S6@22.8	Ea	938.00	757.00	181.00	1,876.00
4' diameter, 6' deep	S6@26.3	Ea	1,050.00	873.00	209.00	2,132.00
4' diameter, 7' deep	S6@30.3	Ea	1,160.00	1,010.00	240.00	2,410.00
4' diameter, 8' deep	S6@33.7	Ea	1,290.00	1,120.00	267.00	2,677.00
4' diameter, 9' deep	S6@38.2	Ea	1,370.00	1,270.00	303.00	2,943.00
4' diameter, 10' deep	S6@43.3	Ea	1,500.00	1,440.00	344.00	3,284.00

	Craft@Hrs	Unit	Material	Labor	Equipment	Total
Concrete block radial manholes, 4' inside diameter, no excavation or backfill included						
4' deep	M1@8.01	Ea	305.00	269.00	—	574.00
6' deep	M1@13.4	Ea	462.00	450.00	—	912.00
8' deep	M1@20.0	Ea	680.00	672.00	—	1,352.00
10' deep	M1@26.7	Ea	786.00	897.00	—	1,683.00
Depth over 10', add per LF	M1@2.29	LF	90.00	76.90	—	166.90
2' depth cone block for 30" grate	M1@5.34	LF	133.00	179.00	—	312.00
2'6" depth cone block for 24" grate	M1@6.43	LF	173.00	216.00	—	389.00
Manhole steps, cast iron, heavy type, asphalt coated						
10" x 14-1/2" in job-built manhole	M1@.254	Ea	14.60	8.53		23.13
12" x 10" in precast manhole	—	Ea	13.10	—	—	13.10
Gray iron manhole frames, asphalt coated, standard sizes						
Light duty frame and lid						
15" diameter, 65 lb	S6@.660	Ea	70.10	21.90	5.24	97.24
22" diameter, 140 lb	S6@1.42	Ea	154.00	47.10	11.27	212.37
Medium duty frame and lid						
11" diameter, 75 lb	S6@1.51	Ea	81.60	50.10	11.98	143.68
20" diameter, 185 lb	S6@3.01	Ea	188.00	99.90	23.90	311.80
Heavy duty frame and lid						
17" diameter, 135 lb	S6@2.51	Ea	143.00	83.30	19.91	246.21
21" diameter, 315 lb	S6@3.77	Ea	320.00	125.00	29.90	474.90
24" diameter, 375 lb	S6@4.30	Ea	369.00	143.00	34.10	546.10
Connect new drain line to existing manhole, no excavation or backfill included. Typical cost						
Per connection	S6@6.83	Ea	99.80	227.00	54.18	380.98
Connect existing drain line to new manhole, no excavation or backfill included. Typical cost						
Per connection	S6@3.54	Ea	60.70	118.00	28.10	206.80
Manhole repairs and alterations, typical costs						
Repair manhole leak with grout	S6@15.8	Ea	314.00	524.00	125.00	963.00
Repair inlet leak with grout	S6@15.8	Ea	369.00	524.00	125.00	1,018.00
Grout under manhole frame	S6@1.29	Ea	6.25	42.80	10.23	59.28
Drill and grout pump setup cost	S6@17.4	LS	369.00	578.00	138.00	1,085.00
Grout for pressure grouting	—	Gal	5.90	—	—	5.90
Replace brick in manhole wall	M1@8.79	SF	9.00	295.00	—	304.00
Replace brick under manhole frame	M1@8.79	LS	30.10	295.00	—	325.10
Raise existing frame and cover 2"	CL@4.76	LS	167.00	144.00	—	311.00
Raise existing frame and cover more than 2", per each 1" added	CL@.250	LS	57.60	7.57	—	65.17
TV inspection of pipe interior, subcontract		LF	—	—	—	2.00
Grouting concrete pipe joints						
6" to 30" diameter	S6@.315	Ea	17.20	10.50	2.50	30.20
33" to 60" diameter	S6@1.12	Ea	33.90	37.20	8.89	79.99
66" to 72" diameter	S6@2.05	Ea	68.40	68.00	16.26	152.66

Accessories for Site Utilities Equipment cost shown is for a wheel mounted 1 CY backhoe at $23.80 per hour for lifting and placing the items. Trench excavation, bedding material and backfill are not included. Costs estimated using this section may be combined with costs estimated from other site work utility sections for arriving at a minimum job charge.

	Craft@Hrs	Unit	Material	Labor	Equipment	Total
Curb inlets, gray iron, asphalt coated, heavy duty frame, grate and curb box						
20" x 11", 260 lb	S6@2.75	Ea	260.00	91.30	21.82	373.12
20" x 16.5", 300 lb	S6@3.04	Ea	285.00	101.00	24.12	410.12
20" x 17", 400 lb	S6@4.30	Ea	360.00	143.00	34.11	537.11
19" x 18", 500 lb	S6@5.06	Ea	430.00	168.00	40.14	638.14
30" x 17", 600 lb	S6@6.04	Ea	465.00	200.00	47.90	712.90

	Craft@Hrs	Unit	Material	Labor	Equipment	Total
Gutter inlets, gray iron, asphalt coated, heavy duty frame and grate						
8" x 11.5", 85 lb	S6@.444	Ea	88.00	14.70	3.52	106.22
22" x 17", 260 lb, concave	S6@1.84	Ea	250.00	61.10	14.59	325.69
22.3" x 22.3", 475 lb	S6@2.87	Ea	270.00	95.30	22.77	388.07
29.8" x 17.8", 750 lb	S6@3.37	Ea	415.00	112.00	26.74	553.74
Trench inlets, ductile iron, light duty frame and grate for pedestrian traffic, 53.5" x 8.3", 100 lb	S6@.495	Ea	93.00	16.40	3.93	113.33
Trench inlets, gray iron, asphalt coated frame and grate or solid cover						
Light duty, 1-1/4"						
8" wide grate	S6@.247	LF	36.00	8.20	1.96	46.16
12" wide grate	S6@.275	LF	52.00	9.13	2.18	63.31
8" wide solid cover	S6@.247	LF	40.00	8.20	1.96	50.16
12" wide solid cover	S6@.275	LF	57.00	9.13	2.18	68.31
Heavy duty, 1-3/4"						
8" wide grate	S6@.275	LF	39.00	9.13	2.18	50.31
12" wide grate	S6@.309	LF	58.00	10.30	2.45	70.75
8" wide solid cover	S6@.275	LF	45.00	9.13	2.18	56.31
12" wide solid cover	S6@.309	LF	62.00	10.30	2.45	74.75

Asphalt Paving To arrive at the total cost for new asphalt paving jobs, add the cost of the subbase preparation and aggregate to the costs for the asphalt binder course, the rubberized interliner and the wear course. Cost for site preparation is not included. These costs are intended to be used for on-site roadways and parking lots. Use $7,500 as a minimum charge.

Subbase and aggregate Equipment cost is $101 per hour and includes a pneumatic tired articulated 10,000 pound motor grader, a 4 ton two axle steel drum roller and a 4,000 gallon water truck.

	Craft@Hrs	Unit	Material	Labor	Equipment	Total
Fine grading and compacting existing subbase.						
To within plus or minus 1/10'	P5@.066	SY	.01	2.13	1.33	3.47
Stone aggregate base, Class 2, (at $14.10 per CY or $9.00 ton). Includes grading and compacting of base material						
1" thick base (36 SY per CY)	P5@.008	SY	.39	.26	.16	.81
4" thick base (9 SY per CY)	P5@.020	SY	1.57	.65	.40	2.62
6" thick base (6 SY per CY)	P5@.033	SY	2.35	1.07	.66	4.08
8" thick base (4.5 SY per CY)	P5@.039	SY	3.13	1.26	.79	5.18
10" thick base (3.6 SY per CY)	P5@.042	SY	3.92	1.36	.85	6.13
12" thick base (3 SY per CY)	P5@.048	SY	4.70	1.55	.97	7.22

Asphalt materials Equipment cost is $125 per hour and includes a 10' wide self-propelled paving machine and a 4 ton two axle steel drum roller.

	Craft@Hrs	Unit	Material	Labor	Equipment	Total
Binder course (at $38.10 per ton), spread and rolled						
1" thick (18 SY per ton)	P5@.028	SY	2.11	.90	.70	3.71
1-1/2" thick (12 SY per ton)	P5@.036	SY	3.52	1.16	.90	5.58
2" thick (9 SY per ton)	P5@.039	SY	4.69	1.26	.98	6.93
3" thick (6 SY per ton)	P5@.050	SY	7.03	1.61	1.25	9.89
Rubberized asphalt interlayer with tack coat. Based on using 10' wide rolls 3/16" thick x 500' long at $100 per roll ($.18 per SY) and one coat of mopped-on bituminous tack coat, MC70 at $1.50 per gallon (with .15 gallon per SY) before allowance for waste.						
Laid, cut, mopped and rolled	P5@.005	SY	.48	.16	.13	.77
Wear course (at $42.20 per ton before markup)						
1" thick (18 SY per ton)	P5@.028	SY	2.34	.90	.70	3.94
1-1/2" thick (12 SY per ton)	P5@.036	SY	3.52	1.16	.90	5.58
2" thick (9 SY per ton)	P5@.039	SY	4.69	1.26	.98	6.93
3" thick (6 SY per ton)	P5@.050	SY	7.03	1.61	1.25	9.89

	Craft@Hrs	Unit	Material	Labor	Equipment	Total
Add to total installed cost of paving as described on previous page, if required.						
Bituminous prime coat	P5@.001	SY	.42	.03	.02	.47
Tar emulsion protective seal coat	P5@.034	SY	1.50	1.10	.85	3.45
Asphalt slurry seal	P5@.043	SY	1.89	1.39	1.08	4.36
Asphaltic concrete curb, straight, placed during paving						
8" x 6", placed by hand	P5@.141	LF	.94	4.55	—	5.49
8" x 6", machine extruded	P5@.067	LF	.86	2.16	1.54	4.56
8" x 8", placed by hand	P5@.141	LF	1.50	4.55	—	6.05
8" x 8", machine extruded	P5@.068	LF	1.14	2.20	1.56	4.90
Add for curved sections	P5@.034	LF	—	1.10	.81	1.91
Asphaltic concrete speed bumps, 3" high, 24" wide (135 LF and 2 tons per CY), placed on parking lot surface and rolled						
100 LF job	P5@.198	LF	.69	6.39	4.55	11.63
Painting speed bumps, 24" wide, white	P5@.022	LF	.24	.71	—	.95

See also the section on Speed Bumps and Parking Blocks following Pavement Striping and Marking.

Asphalt Paving for Roadways, Assembly costs On-site roadway, with 4" thick aggregate base, 2" thick asphalt binder course, rubberized asphalt interliner, and 2" thick wear course. This is often called 4" paving over 4" aggregate base.

	Craft@Hrs	Unit	Material	Labor	Equipment	Total
Fine grading and compact						
existing subbase	P5@.066	SY	.01	2.13	1.25	3.39
Stone aggregate base,						
Class 2, 4" thick	P5@.020	SY	1.63	.65	.38	2.66
Binder course 2" thick (9 SY per ton)	P5@.039	SY	4.23	1.26	.90	6.39
Rubberized asphalt interlayer	P5@.005	SY	.48	.16	.12	.76
Wear course 2" thick (9 SY per ton)	P5@.039	SY	4.70	1.26	.90	6.86
Tar emulsion protective seal coat	P5@.034	SY	1.50	1.10	.79	3.39
Total estimated cost for typical 10' wide roadway, 4" paving over 4" aggregate base						
Cost per SY of roadway	P5@.203	SY	12.55	6.56	4.34	23.45
Cost per SF roadway	P5@.023	SF	1.39	.74	.48	2.61
Cost per LF, 10' wide roadway	P5@.226	LF	13.94	7.30	4.82	26.06

Asphalt Paving for Parking Areas, Assembly costs On-site parking lot with 6" thick aggregate base, 3" thick asphalt binder course, rubberized asphalt interliner, and 3" thick wear course. This is often called 6" paving over 6" aggregate base.

	Craft@Hrs	Unit	Material	Labor	Equipment	Total
Fine grading and compact						
existing subbase	P5@.066	SY	.01	2.13	1.25	3.39
Stone aggregate base,						
Class 2, 6" thick	P5@.033	SY	2.43	1.07	.62	4.12
Binder course 3" thick	P5@.050	SY	6.35	1.61	1.16	9.12
Rubberized asphalt interlayer	P5@.005	SY	.48	.16	.12	.76
Wear course 3" thick	P5@.050	SY	7.04	1.61	1.16	9.81
Tar emulsion protective seal coat	P5@.034	SY	1.50	1.10	.79	3.39
Total estimated cost of typical 1,000 SY parking lot, 6" paving over 6" aggregate base						
Cost per SY of parking lot	P5@.238	SY	17.81	7.68	5.10	30.59
Cost per SF of parking lot	P5@.026	SF	1.98	.85	.57	3.40

Cold Milling of Asphalt Surface to 1" Depth Equipment cost is $155 per hour for a motorized street broom, a 20" pavement miller, two 5 CY dump trucks and a pickup truck. Hauling and disposal not included. Use $3,000 as a minimum charge.

	Craft@Hrs	Unit	Material	Labor	Equipment	Total
Milling asphalt to 1" depth	C6@.017	SY	—	.55	.27	.82

	Craft@Hrs	Unit	Material	Labor	Equipment	Total

Asphalt Pavement Repairs to conform to existing street, strips of 100 to 500 SF. Equipment cost is $66 per hour and includes one wheel-mounted 150 CFM air compressor, one paving breaker and jackhammer bits, one 55 HP wheel loader with integral backhoe, one medium weight pneumatic tamper, one 1 ton two axle steel roller and one 5 CY dump truck. These costs include the cost of loading and hauling to a legal dump within 6 miles. Dump fees are not included. See Waste Disposal.

Asphalt Pavement Repairs, Assembly costs Use $3,600 as a minimum job charge.

	Craft@Hrs	Unit	Material	Labor	Equipment	Total
Cut and remove 2" asphalt and 36" subbase	P5@.250	SY	—	8.07	3.27	11.34
Place 30" fill compacted to 90%, 6" aggregate base, 2" thick asphalt and tack coat						
30" fill (30 SY per CY at $15 CY)	P5@.099	SY	.53	3.20	1.29	5.02
6" thick base (6 SY per CY)	P5@.033	SY	2.46	1.07	.43	3.96
2" thick asphalt (9 SY per ton)	P5@.039	SY	4.27	1.26	.51	6.04
Tar emulsion protective seal coat	P5@.034	SY	1.50	1.10	.44	3.04
Total per SY	P5@.455	SY	8.76	14.70	5.94	29.40

Asphalt Overlay on existing asphalt pavement. Equipment cost is $125 per hour and includes a 10' wide self-propelled paving machine and a 4 ton two axle steel drum roller. Costs shown assume unobstructed access to area receiving overlay. Use $6,000 as a minimum job charge.

Rubberized asphalt interlayer with tack coat. Based on using 10' wide rolls 3/16" thick x 500' long at $100 per roll ($.18 per SY) and one coat of mopped-on bituminous tack coat, MC70 at $1.50 per gallon (with .15 gallon per SY) before allowance for waste.

	Craft@Hrs	Unit	Material	Labor	Equipment	Total
Laid, cut, mopped and rolled	P5@.005	SY	.44	.16	.13	.73
Wear course (at $42.20 per ton)						
1" thick (18 SY per ton)	P5@.028	SY	2.34	.90	.70	3.94
1-1/2" thick (12 SY per ton)	P5@.036	SY	3.52	1.16	.90	5.58
2" thick (9 SY per ton)	P5@.039	SY	4.69	1.26	.98	6.93
Add to overlaid paving as described above, if required.						
Bituminous prime coat, MC70	P5@.001	SY	.41	.03	.03	.47
Tar emulsion protective seal coat	P5@.034	SY	1.50	1.10	.85	3.45
Asphalt slurry seal	P5@.043	SY	1.89	1.39	1.08	4.36

Concrete Paving No grading or base preparation included.

	Craft@Hrs	Unit	Material	Labor	Equipment	Total
Steel edge forms to 15", rented	C8@.028	LF	—	1.02	.53	1.55

2,000 PSI concrete (at $58.50 per CY before allowance for waste or markup) placed directly from the chute of a ready-mix truck, vibrated and broom finished. Costs include 4% allowance for waste but no forms, reinforcing, or joints are included.

	Craft@Hrs	Unit	Material	Labor	Equipment	Total
4" thick (80 SF per CY)	P8@.013	SF	.76	.44	.01	1.21
6" thick (54 SF per CY)	P8@.016	SF	1.13	.54	.02	1.69
8" thick (40 SF per CY)	P8@.019	SF	1.52	.64	.03	2.19
10" thick (32 SF per CY)	P8@.021	SF	1.90	.71	.03	2.64
12" thick (27 SF per CY)	P8@.024	SF	2.25	.81	.04	3.10
15" thick (21 SF per CY)	P8@.028	SF	2.90	.95	.05	3.90

Clean and Sweep Pavement Equipment cost is $20.00 per hour for a self-propelled vacuum street sweeper, 7 CY maximum capacity. Use $300 as a minimum charge for sweeping using equipment.

	Craft@Hrs	Unit	Material	Labor	Equipment	Total
Pavement sweeping, using equipment	TO@.200	MSF	—	7.81	4.00	11.81
Clean and sweep pavement, by hand	CL@1.44	MSF	—	43.60	—	43.60

Joint Treatment for Concrete Pavement Seal joint in concrete pavement with ASTM D-3569 PVC coal tar (at $13.50 per gallon before markup or 5% waste). Cost per linear foot (LF), by joint size

	Craft@Hrs	Unit	Material	Labor	Equipment	Total
3/8" x 1/2" (128 LF per gallon)	PA@.002	LF	.11	.07	—	.18
1/2" x 1/2" (96 LF per gallon)	PA@.002	LF	.14	.07	—	.21
1/2" x 3/4" (68 LF per gallon)	PA@.003	LF	.20	.11	—	.31

	Craft@Hrs	Unit	Material	Labor	Equipment	Total
1/2" x 1" (50 LF per gallon)	PA@.003	LF	.27	.11	—	.38
1/2" x 1-1/4" (40 LF per gallon)	PA@.004	LF	.34	.15	—	.49
1/2" x 1-1/2" (33 LF per gallon)	PA@.006	LF	.41	.22	—	.63
3/4" x 3/4" (46 LF per gallon)	PA@.004	LF	.29	.15	—	.44
3/4" x 1-1/4" (33 LF per gallon)	PA@.006	LF	.41	.22	—	.63
3/4" x 1-1/2" (27 LF per gallon)	PA@.007	LF	.50	.26	—	.76
3/4" x 2" (22 LF per gallon)	PA@.008	LF	.61	.30	—	.91
1" x 1" (25 LF per gallon)	PA@.008	LF	.54	.30	—	.84

Backer rod for joints in concrete pavement

	Craft@Hrs	Unit	Material	Labor	Equipment	Total
1/2" backer rod	PA@.002	LF	.02	.07	—	.09
3/4" or 1" backer rod	PA@.002	LF	.03	.07	—	.10

Clean out old joint sealer in concrete pavement by sandblasting and reseal with PVC coal tar sealer, including new backer rod. Equipment cost is $95 per day. Use $400 as a minimum job charge.

	Craft@Hrs	Unit	Material	Labor	Equipment	Total
1/2" x 1/2"	PA@.009	LF	.17	.33	.12	.62
1/2" x 3/4"	PA@.013	LF	.25	.48	.16	.89
1/2" x 1"	PA@.013	LF	.33	.48	.16	.97
1/2" x 1-1/4"	PA@.018	LF	.41	.67	.22	1.30
1/2" x 1-1/2"	PA@.022	LF	.49	.81	.27	1.57

Pavement Striping and Marking Use $150 as a minimum job charge. Pavement line markings

	Craft@Hrs	Unit	Material	Labor	Equipment	Total
Single line striping	PA@.005	LF	.06	.19	.02	.27
Pedestrian or bike lane striping	PA@.007	LF	.08	.26	.02	.36
Red curb painting	PA@.006	LF	.07	.22	.02	.31

Parking lot spaces. For estimating purposes figure one space per 300 SF of parking pavement area

	Craft@Hrs	Unit	Material	Labor	Equipment	Total
Single line striping, per space	PA@.115	Ea	1.19	4.26	.37	5.82
Dual line striping, per space	PA@.175	Ea	2.09	6.48	.57	9.14

Traffic symbols

	Craft@Hrs	Unit	Material	Labor	Equipment	Total
Arrows	PA@.500	Ea	5.60	18.50	.29	24.39
Lettering, 2'-0" x 8'-0" template	PA@2.50	Ea	35.90	92.50	2.00	130.40
Handicapped symbol						
One color	PA@.250	Ea	4.55	9.25	.20	14.00
Two color	PA@.500	Ea	5.85	18.50	.39	24.74

Speed Bumps and Parking Blocks Prefabricated solid plastic, pre-drilled. Based on Saver Bumps or Saver Blocks as manufactured by The Parking Block Store.
Costs shown include installation hardware. Labor includes drilling holes in substrate for anchors

	Craft@Hrs	Unit	Material	Labor	Total
Speed bumps, 2" high x 12" wide with lengths and integral color as shown. Based on 16 bumps					
4'-0", yellow (35 Lbs each)	CL@.500	Ea	36.70	15.10	51.80
6'-0", yellow (51 Lbs each)	CL@.750	Ea	51.00	22.70	73.70
8'-0", yellow (68 Lbs each)	CL@1.00	Ea	66.30	30.30	96.60
Parking blocks, 4" high x 6" wide with lengths and integral color as shown. Based on 25 blocks					
4'-0", gray (30 Lbs each)	CL@.500	Ea	19.40	15.10	34.50
6'-0", gray (44 Lbs each)	CL@.750	Ea	22.40	22.70	45.10
4'-0", yellow (30 Lbs each)	CL@.500	Ea	19.90	15.10	35.00
6'-0", yellow (44 Lbs each)	CL@.750	Ea	22.40	22.70	45.10
4'-0", blue (30 Lbs each)	CL@.500	Ea	19.40	15.10	34.50
6'-0", blue (44 Lbs each)	CL@.750	Ea	22.40	22.70	45.10

Site Work 2

	Craft@Hrs	Unit	Material	Labor	Total

Traffic Signs Reflectorized steel, with a high strength U-channel galvanized steel pipe post 10' long set 2' into the ground. Includes digging of hole with a manual auger and backfill.

	Craft@Hrs	Unit	Material	Labor	Total
Stop, 24" x 24"	CL@1.25	Ea	118.00	37.80	155.80
Speed limit, Exit, etc, 18" x 24"	CL@1.25	Ea	81.60	37.80	119.40
Warning, 24" x 24"	CL@1.25	Ea	104.00	37.80	141.80
Yield, 30" triangle	CL@1.25	Ea	153.00	37.80	190.80

Brick Paving Extruded hard red brick. These costs do not include site preparation. Material costs shown include 2% for waste. (Figures in parentheses show quantity and cost for brick before allowance for waste.)

Brick laid flat without mortar
Running bond

	Craft@Hrs	Unit	Material	Labor	Total
4" x 8" x 2-1/4" (4.5 per SF at $490 per M)	M1@.114	SF	2.24	3.83	6.07
4" x 8" x 1-5/8" (4.5 per SF at $490 per M)	M1@.112	SF	2.24	3.76	6.00
3-5/8" x 7-5/8" x 2-1/4" (5.2 per SF at $570 per M)	M1@.130	SF	2.99	4.37	7.36
3-5/8" x 7-5/8" x 1-5/8" (5.2 per SF at $480 per M)	M1@.127	SF	2.55	4.27	6.82
Add for herringbone pattern	M1@.033	SF	—	1.11	1.11

Basketweave

	Craft@Hrs	Unit	Material	Labor	Total
4" x 8" x 2-1/4" (4.5 per SF at $490 per M)	M1@.133	SF	2.24	4.47	6.71
4" x 8" x 1-5/8" (4.5 per SF at $490 per M)	M1@.130	SF	2.24	4.37	6.61

Brick laid on edge without mortar
Running bond

	Craft@Hrs	Unit	Material	Labor	Total
4" x 8" x 2-1/4" (8 per SF at $490 per M)	M1@.251	SF	4.00	8.43	12.43
3-5/8" x 7-5/8" x 2-1/4" (8.4 per SF at $570 per M)	M1@.265	SF	4.88	8.90	13.78

Brick laid flat with 3/8" mortar joints and mortar setting bed. Mortar included in material cost at $.50 per SF.
Running bond

	Craft@Hrs	Unit	Material	Labor	Total
4" x 8" x 2-1/4" (3.9 per SF at $490 per M)	M1@.127	SF	2.41	4.27	6.68
4" x 8" x 1-5/8" (3.9 per SF at $490 per M)	M1@.124	SF	2.41	4.17	6.58
3-5/8" x 7-5/8" x 2-1/4" (4.5 per SF at $570 per M)	M1@.144	SF	3.07	4.84	7.91
3-5/8" x 7-5/8" x 1-5/8" (4.5 per SF at $480 per M)	M1@.140	SF	2.66	4.70	7.36
Add for herringbone pattern	M1@.033	SF	—	1.11	1.11

Basketweave

	Craft@Hrs	Unit	Material	Labor	Total
4" x 8" x 2-1/4" (3.9 per SF at $490 per M)	M1@.147	SF	2.41	4.94	7.35
4" x 8" x 1-5/8" (3.9 per SF at $490 per M)	M1@.144	SF	2.41	4.84	7.25
3-5/8" x 7-5/8" x 2-1/4" (4.5 per SF at $570 per M)	M1@.147	SF	3.07	4.94	8.01
3-5/8" x 7-5/8" x 1-5/8" (4.5 per SF at $480 per M)	M1@.144	SF	2.66	4.84	7.50

Brick laid on edge w/ 3/8" mortar joints and setting bed, running bond. Mortar included in material cost at $.60 per SF.

	Craft@Hrs	Unit	Material	Labor	Total
4" x 8" x 2-1/4" (6.5 per SF at $490 per M)	M1@.280	SF	3.79	9.41	13.20
3-5/8" x 7-5/8" x 2-1/4" (6.9 per SF at $570 per M)	M1@.314	SF	4.53	10.60	15.13

Mortar color for brick laid in mortar, add per SF of paved area

		Unit	Material	Labor	Total
Red or yellow	—	SF	.09	—	.09
Black or brown	—	SF	.14	—	.14
Green	—	SF	.37	—	.37

Base underlayment for paving brick

	Craft@Hrs	Unit	Material	Labor	Total
Add for 2" sand base (124 SF at $15 per ton)	M1@.009	SF	.12	.30	.42
Add for 15 lb felt underlayment	M1@.002	SF	.10	.07	.17
Add for asphalt tack coat (40 SF at $5 per gallon)	M1@.005	SF	.13	.17	.30
Add for 2% neoprene tack coat (40 SF at $7 per gal.)	M1@.005	SF	.19	.17	.36

Brick edging, 8" deep, set in concrete with dry joints

	Craft@Hrs	Unit	Material	Labor	Total
Headers 4" x 8" x 2-1/4" (3/LF at $490/M)	M1@.147	LF	2.07	4.94	7.01
Headers 3-5/8" x 7-5/8" x 2-1/4" (3.3/LF at $570/M)	M1@.159	LF	2.48	5.34	7.82
Rowlocks 3-5/8" x 7-5/8" x 2-1/4" (4.5/LF at $570/M)	M1@.251	LF	3.16	8.43	11.59
Rowlocks 4" x 8" x 2-1/4" (4.7/LF at $490/M)	M1@.265	LF	2.90	8.90	11.80

	Craft@Hrs	Unit	Material	Labor	Total
Brick edging, 8" deep, set in concrete with 3/8" mortar joints					
Headers 4" x 8" x 2-1/4" (2.8/LF at $490/M)	M1@.159	LF	1.97	5.34	7.31
Headers 3-5/8" x 7-5/8" x 2-1/4" (3/LF at $570/M)	M1@.173	LF	2.31	5.81	8.12
Rowlocks 3-5/8" x 7-5/8" x 2-1/4" (4.2/LF at $570/M)	M1@.298	LF	2.99	10.00	12.99
Rowlocks 4" x 8" x 2-1/4" (4.5/LF at $490/M)	M1@.314	LF	2.80	10.60	13.40

Masonry, Stone and Slate Paving These costs do not include site preparation. Costs for stone and slate vary widely. (Costs shown in parentheses are before waste allowance.) Use $500 as a minimum job charge.

	Craft@Hrs	Unit	Material	Labor	Total
Interlocking 9" x 4-1/2" concrete pavers, dry joints, natural gray					
2-3/8" thick (3.5 per SF at $.43 each)	M1@.058	SF	1.51	1.95	3.46
3-1/8" thick (3.5 per SF at $.48 each)	M1@.063	SF	1.68	2.12	3.80
Add for standard colors (3.5 per SF at $.05 each)	—	SF	.18	—	.18
Add for custom colors (3.5 per SF at $.10 each)	—	SF	.37	—	.37
Patio blocks, 8" x 16" x 2", dry joints					
Natural gray (1.125 per SF at $.96 each)	M1@.052	SF	1.08	1.75	2.83
Standard colors (1.125 per SF at $1.07 each)	M1@.052	SF	1.20	1.75	2.95
Flagstone pavers in sand bed with dry joints					
Random ashlar, 1-1/4"	M1@.193	SF	2.27	6.49	8.76
Random ashlar, 2-1/2"	M1@.228	SF	3.45	7.66	11.11
Irregular fitted, 1-1/4"	M1@.208	SF	1.91	6.99	8.90
Irregular fitted, 2-1/2"	M1@.251	SF	2.94	8.43	11.37
Flagstone pavers in mortar bed with mortar joints					
Random ashlar, 1-1/4"	M1@.208	SF	2.37	6.99	9.36
Random ashlar, 2-1/2"	M1@.251	SF	3.61	8.43	12.04
Irregular fitted, 1-1/4"	M1@.228	SF	2.01	7.66	9.67
Irregular fitted, 2-1/2"	M1@.277	SF	3.04	9.31	12.35
Granite pavers, grey, sawn, thermal finish, 3/4" joint. Add for bedding below.					
4" x 4" x 2" thick (2.83 Ea)	M1@.184	SF	18.10	6.18	24.28
4" x 4" x 3" thick (3.41 Ea)	M1@.184	SF	21.80	6.18	27.98
4" x 4" x 4" thick (3.86 Ea)	M1@.184	SF	24.60	6.18	30.78
4" x 8" x 2" thick (4.54 Ea)	M1@.171	SF	15.70	5.75	21.45
4" x 8" x 3" thick (5.10 Ea)	M1@.184	SF	17.70	6.18	23.88
4" x 8" x 4" thick (5.68 Ea)	M1@.184	SF	19.70	6.18	25.88
8" x 8" x 2" thick (6.80 Ea)	M1@.171	SF	12.80	5.75	18.55
8" x 8" x 3" thick (7.37 Ea)	M1@.171	SF	13.90	5.75	19.65
8" x 8" x 4" thick (7.94 Ea)	M1@.184	SF	14.90	6.18	21.08
12" x 12" x 2" thick (7.16 Ea)	M1@.171	SF	6.34	5.75	12.09
12" x 12" x 3" thick (8.51 Ea)	M1@.171	SF	7.54	5.75	13.29
12" x 12" x 4" thick (9.07 Ea)	M1@.171	SF	8.03	5.75	13.78
Mortar bed, 1/2" thick, based on mortar at $6 per cubic foot	M1@.039	SF	.25	1.31	1.56
Sand bed, 2" thick, based on sand at $20 per CY	M1@.024	SF	.13	.81	.94
Limestone flag paving					
Sand bed, dry joints					
Random ashlar, 1-1/4"	M1@.193	SF	2.37	6.49	8.86
Random ashlar, 2-1/2"	M1@.228	SF	4.53	7.66	12.19
Irregular fitted, 1-1/2"	M1@.208	SF	2.06	6.99	9.05
Irregular fitted, 2-1/2"	M1@.251	SF	3.50	8.43	11.93
Mortar bed and joints					
Random ashlar, 1"	M1@.208	SF	2.11	6.99	9.10
Random ashlar, 2-1/2"	M1@.251	SF	4.64	8.43	13.07

	Craft@Hrs	Unit	Material	Labor	Total
Irregular fitted, 1"	M1@.228	SF	1.91	7.66	9.57
Irregular fitted, 2-1/2"	M1@.277	SF	3.55	9.31	12.86
Slate flag paving, natural cleft					
Sand bed, dry joints					
Random ashlar, 1-1/4"	M1@.173	SF	7.57	5.81	13.38
Irregular fitted, 1-1/4"	M1@.193	SF	7.00	6.49	13.49
Mortar bed and joints					
Random ashlar, 3/4"	M1@.173	SF	5.97	5.81	11.78
Random ashlar, 1"	M1@.208	SF	7.00	6.99	13.99
Irregular fitted, 3/4"	M1@.193	SF	5.56	6.49	12.05
Irregular fitted, 1"	M1@.228	SF	6.49	7.66	14.15
Add for sand rubbed slate	—	SF	1.49	—	1.49

Curbs, Gutters and Driveway Aprons Costs assume 3 uses of the forms and 2,000 PSI concrete (at $58.50 per CY before markup or allowance for waste) placed directly from the chute of a ready-mix truck. Concrete cast-in-place curb. Excavation or backfill not included. Use $750 as a minimum job charge.

	Craft@Hrs	Unit	Material	Labor	Equipment	Total
Curbs						
Vertical curb						
6" x 12" straight curb	P9@.109	LF	2.19	4.00	1.00	7.19
6" x 12" curved curb	P9@.150	LF	2.23	5.50	1.38	9.11
6" x 18" straight curb	P9@.119	LF	3.25	4.37	1.09	8.71
6" x 18" curved curb	P9@.166	LF	3.31	6.09	1.52	10.92
6" x 24" straight curb	P9@.123	LF	4.32	4.51	1.13	9.96
6" x 24" curved curb	P9@.179	LF	4.37	6.57	1.64	12.58
Rolled curb and gutter, 6" roll						
18" x 6" base, straight curb	P9@.143	LF	4.32	5.25	1.31	10.88
18" x 6" base, curved curb	P9@.199	LF	4.37	7.30	1.82	13.49
24" x 6" base, straight curb	P9@.168	LF	5.43	6.16	1.54	13.13
24" x 6" base, curved curb	P9@.237	LF	5.49	8.69	2.17	16.35

Driveway Aprons Concrete cast-in-place driveway aprons, including 4% waste, wire mesh, forms and finishing. Per square foot of apron, no site preparation included. Use $200 as a minimum job charge.

	Craft@Hrs	Unit	Material	Labor	Equipment	Total
4" thick (80 SF per CY)	P9@.024	SF	.98	.88	.22	2.08
6" thick (54 SF per CY)	P9@.032	SF	1.32	1.17	.29	2.78

Walkways Concrete, 2,000 PSI (at $58.50 per CY before allowance for waste or markup) placed directly from the chute of a ready mix truck, vibrated, plain scored, and broom finished. Costs include 2% allowance for waste but no forms and reinforcing.

	Craft@Hrs	Unit	Material	Labor	Equipment	Total
4" thick (80 SF per CY)	P8@.013	SF	.71	.44	.01	1.16
Add for integral colors, most pastels	—	SF	1.00	—	—	1.00
Add for 1/2" color top course	P8@.012	SF	1.10	.41	.09	1.60
Add for steel trowel finish	CM@.011	SF	.02	.41	.12	.55
Add for seeded aggregate finish	CM@.011	SF	.10	.41	.12	.63
Add for exposed aggregate wash process	CM@.006	SF	.05	.23	.06	.34

Asphalt walkway, temporary, including installation and breakout, but no hauling. Use $500 as a minimum job charge.

	Craft@Hrs	Unit	Material	Labor	Equipment	Total
2" to 2-1/2" thick	P5@.015	SF	.78	.48	.35	1.61

Headers and Dividers

	Craft@Hrs	Unit	Material	Labor	Equipment	Total
2" x 4", treated pine	C8@.027	LF	.64	.98	—	1.62
2" x 6", treated pine	C8@.028	LF	.98	1.02	—	2.00
2" x 6", redwood, B grade	C8@.028	LF	1.73	1.02	—	2.75
2" x 4", patio type dividers, untreated	C8@.030	LF	.41	1.09	—	1.50

	Craft@Hrs	Unit	Material	Labor	Equipment	Total

Post Holes Drilling 2' deep by 10" to 12" diameter fence post holes. Costs will be higher on slopes or where access is limited. These costs do not include layout, setting of fence posts or disposal of excavated material.

Light to medium soil, laborer working with a hand auger

	Craft@Hrs	Unit	Material	Labor	Equipment	Total
(at $10 per day), 4 holes per hour	CL@.250	Ea	—	7.57	.31	7.88

Medium to heavy soil, laborer working with a breaking bar and hand auger

(at $10 per day), 2 holes per hour	CL@.500	Ea	—	15.10	.62	15.72

Medium to heavy soil, 2 laborers working with a gas powered auger

(at $60 per day), 15 holes per hour	CL@.133	Ea	—	4.03	.50	4.53

Broken rock, laborer and equipment operator working with a compressor and jackhammer

(at $100 per day) 2 holes per hour	S1@1.00	Ea	—	34.70	6.25	40.95

Rock, laborer and an equipment operator working with a truck-mounted drill

(at $700 per day) 12 holes per hour	S1@.166	Ea	—	5.75	7.29	13.04

Fencing, Chain Link Fence, industrial grade 9 gauge 2" x 2" galvanized steel chain link fabric and framework, including 2" or 2-3/8" line posts as indicated set at 10' intervals, set in concrete in augured post holes and 1-5/8" top rail, tension panels at corners and abrupt grade changes. Add for gates, gate posts and corner posts as required. Equipment is a gasoline powered 12" auger at $6.00 an hour. Use $600 as a minimum job charge. For scheduling purposes, estimate that a 2-man crew will install 125 to 130 LF of fence per 8-hour day

	Craft@Hrs	Unit	Material	Labor	Equipment	Total
4' high galvanized fence, 2" line posts	C4@.073	LF	5.47	2.24	.38	8.09
5' high galvanized fence, 2" line posts	C4@.091	LF	6.10	2.79	.38	9.27
6' high galvanized fence, 2" line posts	C4@.106	LF	6.69	3.26	.38	10.33
6' high galvanized fence, 2-3/8" line posts	C4@.109	LF	7.44	3.35	.38	11.17
7' high galvanized fence, 2-3/8" line posts	C4@.127	LF	7.59	3.90	.38	11.87
8' high galvanized fence, 2-3/8" line posts	C4@.146	LF	8.78	4.48	.38	13.64
10' high galvanized fence, 2-3/8" line posts	C4@.182	LF	10.30	5.59	.47	16.36
10' high galvanized fence, 2-3/8" line posts	C4@.218	LF	11.90	6.70	.59	19.19
Add for vertical aluminum privacy slats	C4@.016	SF	.62	.49	—	1.11
Add for vinyl coated fabric and posts	—	%	50.0	—	—	—
Add for under 200 LF quantities	C4@.006	LF	.33	.18	.02	.53
Deduct for highway quantities	—	LF	-.70	—	—	-.70
Deduct for 11 gauge chain link fabric	—	SF	-.70	—	—	-.70
Deduct for "C" section line posts	—	LF	-1.00	—	—	-1.00

Gates Driveway or walkway. 9 gauge 2" x 2" galvanized steel chain link with frame and tension bars. Costs shown are per square foot of frame area. Add for hardware and gate posts, from below.

	Craft@Hrs	Unit	Material	Labor	Equipment	Total
With 1-3/8" diameter pipe frame	C4@.025	SF	3.00	.77	—	3.77
With 1-5/8" diameter pipe frame	C4@.025	SF	3.45	.77	—	4.22

Gate hardware, per gate

Walkway gate	C4@.250	Ea	14.00	7.68	—	21.68
Driveway gate	C4@.250	Ea	34.20	7.68	—	41.88
Add for vinyl coated fabric and frame	—	SF	.34	—	—	.34

Corner, end or gate posts, complete, heavyweight, 2-1/2" outside diameter galvanized, with fittings and brace. Costs include 2/3 CF sack of concrete per post.

	Craft@Hrs	Unit	Material	Labor	Equipment	Total
4' high fence	C4@.612	Ea	18.50	18.80	1.49	38.79
5' high fence	C4@.668	Ea	20.90	20.50	1.63	43.03
6' high fence	C4@.735	Ea	21.20	22.60	1.80	45.60
7' high fence	C4@.800	Ea	25.80	24.60	1.95	52.35
8' high fence	C4@.882	Ea	28.10	27.10	2.16	57.36
10' high fence	C4@.962	Ea	31.80	29.50	2.35	63.65
12' high fence, with middle rail	C4@1.00	Ea	34.90	30.70	2.45	68.05

	Craft@Hrs	Unit	Material	Labor	Equipment	Total
Add for vinyl coated posts,						
per vertical foot	—	VLF	.35	—	—	.35
Deduct for 2" outside diameter	—	VLF	-.50	—	—	-.50
Barbed wire topper for galvanized chain link fence						
Single strand barbed wire	C4@.004	LF	.10	.12	—	.22
3 strands on one side,						
with support arm	C4@.008	LF	.64	.25	—	.89
3 strand on each side,						
with support arm	C4@.013	LF	1.30	.40	—	1.70
Double coil of 24" to 30" barbed tape	C4@.038	LF	5.10	1.17	—	6.27

Guardrails and Bumpers Equipment is a 4000 lb forklift at $16.00 per hour. Use $3,500 as a minimum job charge.

	Craft@Hrs	Unit	Material	Labor	Equipment	Total
Steel guardrail, standard beam	H1@.250	LF	13.30	10.00	1.00	24.30
Posts, wood, treated, 4" x 4" x 36" high	H1@1.00	Ea	9.90	40.10	4.00	54.00
Guardrail two sides,						
channel rail & side blocks	H1@1.00	LF	24.60	40.10	4.00	68.70
Sight screen	H1@.025	LF	1.29	1.00	.10	2.39
Highway median barricade, precast concrete						
2'8" high x 10' long	H1@.500	Ea	242.00	20.10	2.00	264.10
Add for reflectors, 2 per 10' section	—	LS	11.10	—	—	11.10
Parking bumper, precast concrete, including dowels						
3' wide	CL@.381	Ea	19.20	11.50	1.52	32.22
6' wide	CL@.454	Ea	33.80	13.70	1.82	49.32

Bollards Posts mounted in walkways to limit vehicle entry. Equipment is a 4000 lb capacity forklift at $16.00 per hour. Add for concrete footings, if required, from costs shown at the end of this section. Use $1,000 as a minimum job charge.

	Craft@Hrs	Unit	Material	Labor	Equipment	Total
Cast iron bollards, ornamental, surface mounted						
12" diameter base, 42" high	S1@1.00	Ea	677.00	34.70	8.00	719.70
17" diameter base, 42" high	S1@1.00	Ea	1,140.00	34.70	8.00	1,182.70
Lighted, electrical work not included,						
17" diameter base, 42" high	S1@1.00	Ea	1,260.00	34.70	8.00	1,302.70
Concrete bollards, precast, exposed aggregate, surface mounted						
Square, 12" x 12" x 30" high	S1@1.00	Ea	244.00	34.70	8.00	286.70
Round, 12" diameter x 30" high	S1@1.00	Ea	249.00	34.70	8.00	291.70
Granite bollards, doweled to concrete foundation or slab (foundation not included.)						
Square, smooth matte finish, flat top						
16" x 16" x 30", 756 lbs	S1@1.00	Ea	827.00	34.70	8.00	869.70
16" x 16" x 54", 1,320 lbs	S1@1.50	Ea	1,050.00	52.00	12.00	1,114.00
Square, rough finish, pyramid top						
12" x 12" x 24", 330 lbs	S1@1.00	Ea	370.00	34.70	8.00	412.70
Round, smooth matte finish, flat top						
12" diameter, 18" high	S1@1.00	Ea	801.00	34.70	8.00	843.70
Octagonal, smooth matte finish, flat top,						
24" x 24" x 20", 740 lbs	S1@1.00	Ea	1,200.00	34.70	8.00	1,242.70
Pipe bollards, concrete filled steel pipe, painted yellow, 8' long, set 4' in concrete, includes digging hole.						
6" diameter pipe	S1@.750	Ea	77.80	26.00	6.00	109.80
8" diameter pipe	S1@.750	Ea	124.00	26.00	6.00	156.00
12" diameter pipe	S1@.750	Ea	208.00	26.00	6.00	240.00
Wood bollards, pressure treated timber, includes hand-digging and backfill of hole but no disposal of excess soil. Costs per MBF, shown in parentheses, are before markup. Length shown is portion exposed, costs include 3' bury depth.						
6" x 6" x 36" high (at $1,315 MBF)	S1@.750	Ea	24.00	26.00	6.00	56.00
8" x 8" x 30" high (at $1,400 MBF)	S1@.750	Ea	41.00	26.00	6.00	73.00

	Craft@Hrs	Unit	Material	Labor	Equipment	Total
8" x 8" x 36" high (at $1,400 MBF)	S1@.750	Ea	44.80	26.00	6.00	76.80
8" x 8" x 42" high (at $1,400 MBF)	S1@.750	Ea	48.50	26.00	6.00	80.50
12" x 12" x 24" high (at $1,400 MBF)	S1@.750	Ea	84.00	26.00	6.00	116.00
12" x 12" x 36" high (at $1,400 MBF)	S1@.750	Ea	100.00	26.00	6.00	132.00
12" x 12" x 42" high (at $1,400 MBF)	S1@.750	Ea	110.00	26.00	6.00	142.00
Footings for bollards, add if required						
Permanent concrete footing	C8@.500	Ea	20.60	18.10	—	38.70
Footing for removable bollards	C8@.500	Ea	124.00	18.10	—	142.10

Baseball Backstops Galvanized 9 gauge mesh chain link backstop with 2-5/8" OD galvanized vertical end posts, 2-3/8" OD galvanized vertical back posts and 1-5/8" OD galvanized framing. Includes 4' deep overhang, 20 LF behind home plate, 16 LF on both first and third base lines, rear and wing planks 3' high and two 12' long benches. Equipment is based on using a 4000 lb forklift at $16.00 per hour.

	Craft@Hrs	Unit	Material	Labor	Equipment	Total
Backstop 19'-6" high x 35' wide x 9' deep	S6@24.0	Ea	3,820.00	797.00	128.00	4,745.00

Add for concrete footings, 12" diameter x 24" deep, includes digging by hand and spreading excavated material adjacent to hole, using fence post mix bagged concrete

	Craft@Hrs	Unit	Material	Labor	Equipment	Total
Total 18 footings required	S6@4.00	LS	90.00	133.00	—	223.00

Playing Fields Typical costs. These costs include the subcontractor's overhead and profit. Use 3,000 SF as a minimum job charge.

	Craft@Hrs	Unit	Material	Labor	Equipment	Total
Playing fields, synthetic surface						
Turf, including base and turf	—	SF	—	—	—	9.00
Turf, 3 layer, base foam, turf	—	SF	—	—	—	10.00
Uniturf,						
embossed running surface, 3/8"	—	SF	—	—	—	6.60
Running track						
Volcanic cinder, 7"	—	SF	—	—	—	1.70
Bituminous and cork, 2"	—	SF	—	—	—	2.75

Playground Equipment Equipment is a 4000 lb capacity forklift (at $16.00 per hour.) Use $2,500 as a minimum job charge.

	Craft@Hrs	Unit	Material	Labor	Equipment	Total
Balance beams. All are supported 12" above the ground by 2-3/8" OD galvanized pipe uprights						
10' long Z-shaped beam	S6@6.00	Ea	259.00	199.00	32.00	490.00
15' long straight beam	S6@6.00	Ea	204.00	199.00	32.00	435.00
30' long Z-shaped beam	S6@9.00	Ea	398.00	299.00	48.00	745.00
Geodesic dome beehive climber, 1" galvanized pipe rungs						
8' diameter, 4' high	S6@24.0	Ea	361.00	797.00	128.00	1,286.00
13' diameter, 5' high	S6@36.0	Ea	1,090.00	1,200.00	192.00	2,482.00
18' diameter, 7' high	S6@48.0	Ea	1,360.00	1,590.00	256.00	3,206.00
Horizontal ladders, 20" wide with 1-1/16" OD galvanized pipe rungs 12" OC and 2-3/8" OD galvanized pipe headers and top rails						
12' long, 6' high	S6@6.00	Ea	460.00	199.00	32.00	691.00
18' long, 7' high	S6@9.00	Ea	565.00	299.00	48.00	912.00
Rope and pole climbs. Horizontad vertical end posts are 3-1/2" OD galvanized steel tubing						
10' H x 10' W						
with three 1-5/8" diameter poles	S6@24.0	Ea	678.00	797.00	128.00	1,603.00
10' H x 10' W						
with three 1" diameter ropes	S6@24.0	Ea	734.00	797.00	128.00	1,659.00

Site Work 2

	Craft@Hrs	Unit	Material	Labor	Equipment	Total
16' H x 12' W						
with three 1-5/8" diameter poles	S6@24.0	Ea	900.00	797.00	130.00	1,827.00
16' H x 12' W						
with three 1-1/2" diameter ropes	S6@24.0	Ea	1,030.00	797.00	130.00	1,957.00
Sandboxes, prefabricated. Painted galvanized steel, 12" high, pine seats, sand not included.						
6' x 6'	S6@9.00	Ea	300.00	299.00	48.80	647.80
12' x 12'	S6@12.0	Ea	425.00	398.00	65.00	888.00
Seesaws, galvanized pipe beam with polyethylene saddle or animal seats. Seesaw beam pivots between two coil springs. Double units have seesaw beams spaced 90 degrees apart.						
Single unit, 2 saddle seats	S6@12.0	Ea	770.00	398.00	65.00	1,233.00
Double unit, 4 saddle seats	S6@24.0	Ea	949.00	797.00	130.0O	1,876.00
Single unit, 2 animal seats	S6@12.0	Ea	1,220.00	398.00	65.00	1,683.00
Double unit, 2 animal						
& 2 saddle seats	S6@24.0	Ea	1,380.00	797.00	130.00	2,307.00
Double unit, 4 animal seats	S6@24.0	Ea	1,800.00	797.00	130.00	2,727.00
Slides						
Traditional slide with 18" wide 16 gauge stainless steel slide and galvanized steel stairway						
Chute 5' H x 10' L,						
overall length is 13'	S6@12.0	Ea	872.00	398.00	65.00	1,335.00
Chute 6' H x 12' L,						
overall length is 15'	S6@12.0	Ea	945.00	398.00	65.00	1,408.00
Chute 8' H x 16' L,						
overall length is 20'	S6@12.0	Ea	1,280.00	398.00	65.00	1,743.00
Swings, belt seats, galvanized pipe frame						
8' high two-leg ends						
2 seats	S6@12.0	Ea	419.00	398.00	65.00	882.00
4 seats	S6@18.0	Ea	719.00	598.00	97.50	1,414.50
6 seats	S6@24.0	Ea	997.00	797.00	130.00	1,924.00
10' high two-leg ends						
2 seats	S6@18.0	Ea	457.00	598.00	97.50	1,152.50
4 seats	S6@24.0	Ea	781.00	797.00	130.00	1,708.00
6 seats	S6@30.0	Ea	1,100.00	996.00	162.60	2,258.60
10' high three-leg ends						
4 seats	S6@24.0	Ea	972.00	797.00	130.00	1,899.00
6 seats	S6@30.0	Ea	1,200.00	996.00	162.60	2,358.60
8 seats	S6@36.0	Ea	1,720.00	1,200.00	195.12	3,115.12
12' high three-leg ends						
3 seats	S6@30.0	Ea	820.00	996.00	162.60	1,978.60
6 seats	S6@36.0	Ea	1,230.00	1,200.00	195.12	2,625.12
8 seats	S6@42.0	Ea	1,700.00	1,390.00	227.64	3,317.64
Add for platform for wheelchair,						
any of above	S6@12.0	Ea	475.00	398.00	65.00	938.00
Tetherball sets, galvanized pipe post, ground sleeve and chain,						
complete with nylon rope and ball	S6@3.00	Ea	121.00	99.60	16.25	236.85

Tower climber. Complete with 8' high fireman's slide pole, "loft" platform, cargo net, horizontal ladders and vertical ladders. Frame is 1-1/4" square steel tubing with painted finish, bracing, ladders and slide pole are 1" OD steel pipe with painted finish. The platform consists of two 30" x 96" fiberglass planks with integral color located 5'0" above the ground. Available in mixed or matched colors.

	Craft@Hrs	Unit	Material	Labor	Equipment	Total
8' L x 6' W x 8' H	S6@36.0	Ea	2,300.00	1,200.00	195.12	3,695.12

	Craft@Hrs	Unit	Material	Labor	Equipment	Total

Athletic Equipment Equipment is a 4000 lb capacity forklift (at $16.00 per hour).
Use $1,600 as a minimum job charge.

Football goals, costs shown are for two goals for one field

	Craft@Hrs	Unit	Material	Labor	Equipment	Total
Regulation goal, galvanized steel pipe						
4 ½" center support post	S6@22.0	Pr	2,060.00	730.00	156.00	2,946.00
6 5/8" center support post	S6@24.0	Pr	3,710.00	797.00	171.00	4,678.00
Combination football/soccer goal, regulation size, galvanized						
steel pipe, dual uprights	S6@24.0	Pr	1,840.00	797.00	171.00	2,808.00
Soccer goals, regulation size						
aluminum frame with net	S6@24.0	Pr	2,210.00	797.00	171.00	3,178.00
Basketball goals, single support, 6' backboard extension aluminum fan backboard, hoop and net						
Single goal	S6@12.0	Ea	755.00	398.00	85.44	1,238.44
Double goal, back to back	S6@15.0	Ea	1,060.00	498.00	106.50	1,664.50
Volleyball nets, with galvanized posts, net, and ground sleeves						
Per net	S6@18.0	Ea	477.00	598.00	128.00	1,203.00

Tennis Courts Typical costs for championship size single playing court installation including normal fine grading. These costs do not include site preparation, excavation, drainage, or retaining walls. Equipment, where shown, is a 4000 lb capacity forklift at $16.00 per hour. Use $22,000 as a minimum job charge. Court, 60' x 120' asphaltic concrete or portland cement slab (4" thickness built up to 6" thickness at perimeter). Includes compacted base material and wire mesh reinforcing in concrete

	Craft@Hrs	Unit	Material	Labor	Equipment	Total
Typical cost per court, slab-on-grade	—	LS	—	—	—	19,400.00
Court surface seal coat (7,200 SF)						
One color plus 2" wide boundary lines	—	LS	—	—	—	2,290.00
Fence, 10' high galvanized chain link (360 LF) at court perimeter						
Fencing including gates	—	LS	—	—	—	4,700.00
Lighting						
Typical installation	—	LS	—	—	—	7,500.00
Net posts, with tension reel						
Galvanized steel, 3-1/2" diameter	S6@6.00	Pr	303.00	199.00	42.72	544.72
Galvanized steel, 4-1/2" diameter	S6@6.00	Pr	389.00	199.00	42.72	630.72
Tennis nets						
Nylon	S6@1.50	Ea	208.00	49.80	10.68	268.48
Steel	S6@1.50	Ea	404.00	49.80	10.68	464.48
Ball wall, fiberglass, 8' x 5'						
Attached to existing wall	S6@1.50	Ea	525.00	49.80	10.68	585.48
Windbreak, 9' high, closed mesh polypropylene						
Lashed to existing chain link fence	S6@.003	SF	.10	.10	.02	.22

	Craft@Hrs	Unit	Material	Labor	Total

Athletic Benches Backed athletic benches, galvanized pipe frame, stationary

	Craft@Hrs	Unit	Material	Labor	Total
Aluminum seat and back					
7' long	C8@.930	Ea	222.00	33.70	255.70
8' long	C8@1.16	Ea	256.00	42.10	298.10
15' long	C8@1.31	Ea	410.00	47.50	457.50
Fiberglass seat and back					
10' long	C8@.930	Ea	319.00	33.70	352.70
12' long	C8@1.16	Ea	367.00	42.10	409.10
16' long	C8@1.31	Ea	490.00	47.50	537.50
Galvanized steel seat and back					
15' long	C8@1.31	Ea	293.00	47.50	340.50

Site Work 2

	Craft@Hrs	Unit	Material	Labor	Total
Backless athletic bench, galvanized pipe frame, stationary, 12" wide x 10' long					
Wood seat	C8@.776	Ea	155.00	28.10	183.10
Steel seat	C8@.776	Ea	172.00	28.10	200.10
Aluminum seat	C8@.776	Ea	197.00	28.10	225.10

Surface Preparation and Striping for Athletic Courts Use $150 as a minimum job charge.

	Craft@Hrs	Unit	Material	Labor	Total
Surface preparation					
Acid etch concrete surface	PA@.013	SY	.16	.48	.64
Primer on asphalt	PA@.006	SY	.21	.22	.43
Primer on concrete	PA@.006	SY	.33	.22	.55
Asphalt emulsion filler, 1 coat	PA@.008	SY	.38	.30	.68
Acrylic filler, 1 coat	PA@.008	SY	.65	.30	.95
Rubber/acrylic compound, per coat	PA@.012	SY	1.63	.44	2.07
Acrylic emulsion texture, per coat					
Sand filled	PA@.006	SY	.98	.22	1.20
Rubber filled	PA@.006	SY	1.03	.22	1.25
Acrylic emulsion color coat, per coat					
Tan, brown, or red	PA@.006	SY	.33	.22	.55
Greens	PA@.006	SY	.38	.22	.60
Blue	PA@.006	SY	.43	.22	.65
Rubber cushioned concrete pavers					
18" x 18", dry joints	FL@.039	SF	13.50	1.43	14.93
Striping for athletic courts					
Marked with white or yellow striping paint					
Lines 3" wide	PA@.005	LF	.05	.19	.24
Lines 4" wide	PA@.005	LF	.07	.19	.26
Lines 4" wide, reflectorized	PA@.002	LF	.13	.07	.20
Lines 8" wide, reflectorized	PA@.004	LF	.26	.15	.41
Marked with reflectorized thermoplastic tape					
Lines 4" wide	PA@.004	LF	1.08	.15	1.23
Lines 6" wide	PA@.006	LF	1.63	.22	1.85
Symbols, lettering					
Within 2'-0" x 8'-0" template	PA@2.50	Ea	37.90	92.50	130.40

Bicycle, Moped and Motorcycle Security Racks Installation kit includes tamper proof bolts and quick setting cement. Locks supplied by users. Labor is to install units on concrete surface. Cost per bicycle or motorcycle of capacity. Add cost for drilling holes in concrete as shown. These costs are based on a minimum purchase of six racks.

	Craft@Hrs	Unit	Material	Labor	Total
Basic bicycle rack					
for locking rear wheels (RR-100)	C8@.575	Ea	70.00	20.90	90.90
for locking rear wheels and front wheels (RR-200)	C8@.575	Ea	76.00	20.90	96.90
for locking wheels, and chassis housing (RR-300)	C8@.862	Ea	183.00	31.30	214.30
Moped or motorcycle rack					
With 24" security chain (RR-700)	C8@.575	Ea	106.00	20.90	126.90

Add for mounting track for use on asphalt or non-concrete surfaces. No foundation work included. Each 10' track section holds 6 bicycle racks or 4 moped or motorcycle racks

	Craft@Hrs	Unit	Material	Labor	Total
Per 10' track section, including drilling holes	C8@2.50	Ea	100.00	90.70	190.70
Add for purchasing less than six mounting racks	—	Ea	14.20	—	14.20
Add for drilling mounting holes in concrete,					
2 required per rack, cost per rack	C8@.125	Ea	—	4.53	4.53

Bicycle Racks. Galvanized pipe rack, embedded or surface mounted to concrete, cost per rack

	Craft@Hrs	Unit	Material	Labor	Total
Double entry					
5' long, 7 bikes	C8@3.31	Ea	240.00	120.00	360.00
10' long, 14 bikes	C8@3.31	Ea	345.00	120.00	465.00

	Craft@Hrs	Unit	Material	Labor	Total
20' long, 28 bikes	C8@4.88	Ea	630.00	177.00	807.00
30' long, 42 bikes	C8@6.46	Ea	920.00	234.00	1,154.00
Single side entry					
5' long, 4 bikes	C8@2.41	Ea	230.00	87.40	317.40
10' long, 8 bikes	C8@2.41	Ea	260.00	87.40	347.40
20' long, 16 bikes	C8@3.55	Ea	470.00	129.00	599.00

Wood and metal rack, embedded or surface mounted

	Craft@Hrs	Unit	Material	Labor	Total
Double entry					
10' long, 14 bikes	C8@3.31	Ea	376.00	120.00	496.00
20' long, 28 bikes	C8@4.88	Ea	705.00	177.00	882.00
Single entry					
10' long, 8 bikes	C8@2.41	Ea	314.00	87.40	401.40
20' long, 10 bikes	C8@3.55	Ea	564.00	129.00	693.00

Bike post bollard,

	Craft@Hrs	Unit	Material	Labor	Total
dimensional wood, concrete footing	C8@.223	Ea	188.00	8.09	196.09

Single tube galvanized steel, looped wave pattern, 30" high, surface mounted

	Craft@Hrs	Unit	Material	Labor	Total
3' long, 5 bikes	C8@3.45	Ea	616.00	125.00	741.00
5' long, 7 bikes	C8@3.45	Ea	705.00	125.00	830.00
7' long, 9 bikes	C8@4.61	Ea	920.00	167.00	1,087.00
9' long, 11 bikes	C8@4.61	Ea	1,090.00	167.00	1,257.00
Bonded color, add	—	Ea	94.00	—	94.00
Add for embedded mounting	C8@1.77	Ea	23.40	64.20	87.60

Sprinkler Irrigation Systems Typical complete system costs, including PVC pipe, heads, valves, fittings, trenching and backfill. Per SF of area watered. Add 10% for irrigation systems installed in areas subject to freezing hazard.

	Craft@Hrs	Unit	Material	Labor	Equipment	Total
Large areas using pop-up impact heads, 200 to 300 SF per head, manual system						
To 5,000 SF job	CL@.005	SF	.14	.15	.05	.34
Over 5,000 SF to 10,000 SF	CL@.004	SF	.12	.12	.04	.28
Over 10,000 SF	CL@.003	SF	.10	.09	.03	.22
Add for automatic system	CL@.001	SF	.04	.03	.01	.08
Small areas, 5,000 SF and under, spray heads						
Strip, automatic, shrub type	CL@.008	SF	.24	.24	.08	.56
Commercial type, manual	CL@.007	SF	.22	.21	.07	.50
Residential type, manual	CL@.007	SF	.22	.21	.07	.50
Add for automatic system	CL@.001	SF	.05	.03	.01	.09
Trenching for sprinkler systems, in normal non-compacted topsoil						
Main lines with a 55 HP riding type trencher, 18" wide (at $29.00 per hour)						
12" deep, 143 LF per hour	CL@.007	LF	—	.21	.22	.43
18" deep, 125 LF per hour	CL@.008	LF	—	.24	.25	.49
24" deep, 100 LF per hour	CL@.010	LF	—	.30	.30	.60
Lateral lines with a 20 HP riding type trencher, 12" wide (at $17.00 per hour)						
8" deep, 200 LF per hour	CL@.005	LF	—	.15	.11	.26
12" deep, 125 LF per hour	CL@.008	LF	—	.24	.17	.41
Add for hard soil						
(shelf, rock field, hardpan)	CL@.006	LF	—	.18	.13	.31
Boring under walkways and pavement						
To 8' wide, by hand, 8 LF per hour	CL@.125	LF	—	3.78	—	3.78
Backfill and compact trench, by hand						
Mains to 24" deep, 125 LF per hour	CL@.008	LF	—	.24	—	.24
Laterals to 12" deep, 190 LF per hour	CL@.005	LF	—	.15	—	.15

	Craft@Hrs	Unit	Material	Labor	Total
Restore turf above pipe after sprinkler installation					
Replace and roll sod	CL@.019	SY	.04	.58	.62
Reseed by hand, cover, water and mulch	CL@.010	SY	.11	.30	.41
Connection to existing water line					
Residential or small commercial tap	CL@.939	Ea	18.00	28.40	46.40
Residential or small commercial stub	CL@.683	Ea	14.00	20.70	34.70
Medium commercial stub	CL@1.25	Ea	38.00	37.80	75.80
Atmospheric vacuum breaker, brass					
3/4"	CL@.470	Ea	10.60	14.20	24.80
1"	CL@.500	Ea	14.20	15.10	29.30
1-1/2"	CL@.575	Ea	36.20	17.40	53.60
2"	CL@.625	Ea	63.00	18.90	81.90
Pressure type vacuum breaker, with threaded ball valves					
3/4"	CL@.700	Ea	155.00	21.20	176.20
1"	CL@.750	Ea	161.00	22.70	183.70
1-1/2"	CL@.850	Ea	369.00	25.70	394.70
2"	CL@.939	Ea	420.00	28.40	448.40
Double check valve assembly with two ball valves					
1-1/2" valve	CL@.750	Ea	387.00	22.70	409.70
2" valve	CL@.939	Ea	460.00	28.40	488.40
3" valve	CL@1.25	Ea	1,620.00	37.80	1,657.80
4" valve	CL@1.50	Ea	2,510.00	45.40	2,555.40
Reduced pressure backflow preventer with gate valves					
1" backflow preventer	CL@.750	Ea	177.00	22.70	199.70
1-1/2" backflow preventer	CL@.939	Ea	332.00	28.40	360.40
2" backflow preventer	CL@.939	Ea	374.00	28.40	402.40
3" backflow preventer	CL@1.25	Ea	1,300.00	37.80	1,337.80
4" backflow preventer	CL@1.50	Ea	1,960.00	45.40	2,005.40
6" backflow preventer	CL@1.88	Ea	3,000.00	56.90	3,056.90
Automatic control valves, Brass, inline globe valves, no vacuum breaker or feeder wire included.					
3/4" valve	CL@.341	Ea	39.00	10.30	49.30
1" valve	CL@.341	Ea	50.00	10.30	60.30
1-1/4" valve	CL@.418	Ea	78.00	12.70	90.70
1-1/2" valve	CL@.418	Ea	82.00	12.70	94.70
2" valve	CL@.470	Ea	92.00	14.20	106.20
Valve boxes, plastic box with plastic lid					
7" round	CL@.314	Ea	4.80	9.50	14.30
10" round	CL@.375	Ea	10.40	11.40	21.80
Sprinkler controllers, solid state electronics, add electrical connection and feeder wire to valves					
Indoor mount, 4 station	CL@.960	Ea	96.00	29.10	125.10
Indoor mount, 6 station	CL@.960	Ea	112.00	29.10	141.10
Indoor mount, 8 station	CL@1.10	Ea	155.00	33.30	188.30
Outdoor mount, add pedestal cost below					
4 station	CL@2.77	Ea	112.00	83.80	195.80
6 station	CL@2.77	Ea	155.00	83.80	238.80
8 station	CL@2.77	Ea	182.00	83.80	265.80
10-12 station	CL@3.72	Ea	235.00	113.00	348.00
16-18 station	CL@3.72	Ea	661.00	113.00	774.00
20-24 station	CL@4.45	Ea	863.00	135.00	998.00
Add for pedestal for outdoor units	CL@5.73	Ea	47.90	173.00	220.90

	Craft@Hrs	Unit	Material	Labor	Total

Underground control wire for automatic valves, includes waterproof connectors, wire laid in an open trench, two conductors per valve, costs are per pair of conductors

	Craft@Hrs	Unit	Material	Labor	Total
14 gauge	CL@.001	LF	.10	.03	.13
12 gauge	CL@.001	LF	.19	.03	.22
10 gauge	CL@.002	LF	.31	.06	.37
8 gauge	CL@.003	LF	.49	.09	.58

Manual sprinkler control valves, anti-siphon, with union

	Craft@Hrs	Unit	Material	Labor	Total
3/4" valve	CL@.518	Ea	26.00	15.70	41.70
1" valve	CL@.610	Ea	33.30	18.50	51.80
1-1/4" valve	CL@.677	Ea	53.40	20.50	73.90
1-1/2" valve	CL@.796	Ea	58.80	24.10	82.90
2" valve	CL@.866	Ea	89.00	26.20	115.20

Angle valves, with union

	Craft@Hrs	Unit	Material	Labor	Total
3/4" valve	CL@.518	Ea	18.50	15.70	34.20
1" valve	CL@.610	Ea	21.50	18.50	40.00
1-1/4" valve	CL@.677	Ea	60.00	20.50	80.50
1-1/2" valve	CL@.796	Ea	70.00	24.10	94.10
2" valve	CL@.866	Ea	89.00	26.20	115.20

Sprinkler pipe, PVC, solvent welded, including typical fittings but no excavation

Class 200 PVC pipe

	Craft@Hrs	Unit	Material	Labor	Total
3/4" diameter	CL@.005	LF	.11	.15	.26
1" diameter	CL@.005	LF	.14	.15	.29
1-1/4" diameter	CL@.006	LF	.21	.18	.39
1-1/2" diameter	CL@.006	LF	.28	.18	.46
2" diameter	CL@.008	LF	.44	.24	.68
Deduct for Class 160 PVC pipe	—	%	-30.0	—	—

Class 315 PVC pipe

	Craft@Hrs	Unit	Material	Labor	Total
1/2" diameter	CL@.005	LF	.08	.15	.23
3/4" diameter	CL@.005	LF	.14	.15	.29
1" diameter	CL@.005	LF	.20	.15	.35
1-1/4" diameter	CL@.006	LF	.29	.18	.47
1-1/2" diameter	CL@.006	LF	.36	.18	.54
2" diameter	CL@.008	LF	.66	.24	.90

Schedule 40 PVC pipe

	Craft@Hrs	Unit	Material	Labor	Total
1/2" diameter	CL@.005	LF	.14	.15	.29
3/4" diameter	CL@.005	LF	.19	.15	.34
1" diameter	CL@.006	LF	.27	.18	.45
1-1/4" diameter	CL@.006	LF	.37	.18	.55
1-1/2" diameter	CL@.007	LF	.44	.21	.65
2" diameter	CL@.008	LF	.31	.24	.55

Polyethylene pipe, 80 PSI

	Craft@Hrs	Unit	Material	Labor	Total
3/4" diameter	CL@.006	LF	.08	.18	.26
1" diameter	CL@.006	LF	.10	.18	.28
1-1/4" diameter	CL@.007	LF	.18	.21	.39
1-1/2" diameter	CL@.008	LF	.21	.24	.45
2" diameter	CL@.009	LF	.26	.27	.53

Polyethylene pipe, 100 PSI

	Craft@Hrs	Unit	Material	Labor	Total
1-1/2" diameter	CL@.008	LF	.29	.24	.53
2" diameter	CL@.009	LF	.48	.27	.75
Add for 125 PSI polyethylene pipe	—	%	30.0	—	—

	Craft@Hrs	Unit	Material	Labor	Total
Shrub Sprinklers and Bubblers These costs include a plastic pipe riser as noted at each head.					
Bubblers (5' spacing, 12" riser), plastic	CL@.032	Ea	1.58	.97	2.55
Flex riser	CL@.035	Ea	2.08	1.06	3.14
Double swing joint	CL@.035	Ea	2.50	1.06	3.56
Triple swing joint	CL@.035	Ea	3.75	1.06	4.81
Add for brass head and metal riser	—	Ea	4.69	—	4.69
Shrub spray (8' to 18' spacing, 12" riser)	CL@.038	Ea	1.78	1.15	2.93
Flex riser	CL@.038	Ea	2.18	1.15	3.33
Double swing joint	CL@.038	Ea	2.50	1.15	3.65
Triple swing joint	CL@.041	Ea	3.96	1.24	5.20
Add for brass head and metal riser	—	Ea	4.69	—	4.69
Stream spray (12' to 22' spacing, 12" riser)					
Plastic riser	CL@.041	Ea	2.94	1.24	4.18
Flex riser	CL@.041	Ea	3.31	1.24	4.55
Double swing joint	CL@.045	Ea	3.71	1.36	5.07
Triple swing joint	CL@.045	Ea	4.21	1.36	5.57
Add for brass head and metal riser	—	Ea	4.72	—	4.72
Rotary head, plastic (20' to 40' spacing, 12" riser)					
Plastic riser	CL@.063	Ea	10.50	1.91	12.41
Flex riser	CL@.063	Ea	11.20	1.91	13.11
Double swing joint	CL@.067	Ea	11.50	2.03	13.53
Triple swing joint	CL@.067	Ea	12.00	2.03	14.03
Rotary head, metal (20' to 40' spacing, 12" riser)					
Plastic riser	CL@.063	Ea	15.70	1.91	17.61
Flex riser	CL@.063	Ea	16.10	1.91	18.01
Double swing joint	CL@.067	Ea	16.50	2.03	18.53
Triple swing joint	CL@.067	Ea	17.00	2.03	19.03
Hi-pop spray, 6" pop-up (6" riser)					
Plastic riser	CL@.041	Ea	6.50	1.24	7.74
Flex riser	CL@.043	Ea	6.97	1.30	8.27
Double swing joint	CL@.045	Ea	7.34	1.36	8.70
Triple swing joint	CL@.045	Ea	7.86	1.36	9.22
Rotary, hi-pop, 10" to 12" pop-up					
Plastic riser	CL@.076	Ea	27.80	2.30	30.10
Flex riser	CL@.076	Ea	28.30	2.30	30.60
Double swing joint	CL@.079	Ea	29.20	2.39	31.59
Triple swing joint	CL@.079	Ea	29.50	2.39	31.89
Add for shrub sprinklers and bubblers					
Installed in existing shrub and flower beds	—	%	—	25.0	—
Lawn Sprinklers These costs include a plastic pipe riser as noted at each head.					
1" pop-up spray, plastic head (10' to 12' spacing)					
Single riser	CL@.045	Ea	3.26	1.36	4.62
Double swing joint	CL@.048	Ea	4.20	1.45	5.65
Triple swing joint	CL@.048	Ea	4.72	1.45	6.17
2" pop-up spray, plastic head (10' to 12' spacing)					
Single riser	CL@.045	Ea	4.46	1.36	5.82
Double swing joint	CL@.048	Ea	5.08	1.45	6.53
Triple swing joint	CL@.048	Ea	5.90	1.45	7.35
Add for brass head and metal riser	—	Ea	9.63	—	9.63

	Craft@Hrs	Unit	Material	Labor	Total
Rotary pop-up spray (20' to 30' spacing)					
Regular	CL@.094	Ea	15.70	2.85	18.55
Double swing joint	CL@.098	Ea	17.60	2.97	20.57
Triple swing joint	CL@.102	Ea	18.80	3.09	21.89
Rotary pop-up spray (30' to 50' spacing)					
Regular	CL@.124	Ea	18.60	3.75	22.35
Double swing joint	CL@.132	Ea	20.50	4.00	24.50
Triple swing joint	CL@.133	Ea	21.70	4.03	25.73
Rotary head (50' to 70' spacing), full or part circle					
Without valve	CL@.130	Ea	72.90	3.93	76.83
With integral check valve	CL@.130	Ea	94.30	3.93	98.23
With integral automatic electric valve	CL@.137	Ea	98.50	4.15	102.65
Rotary head (70' to 80' spacing), full or part circle					
Without valve	CL@.149	Ea	94.30	4.51	98.81
With integral check valve	CL@.149	Ea	114.00	4.51	118.51
With integral automatic electric valve	CL@.159	Ea	136.00	4.81	140.81
With metal cover	—	Ea	6.16	—	6.16
Quick coupling valves, add swing joint costs below					
3/4" regular	CL@.089	Ea	19.20	2.69	21.89
3/4" 2-piece	CL@.089	Ea	22.60	2.69	25.29
1" regular	CL@.094	Ea	27.40	2.85	30.25
1" 2-piece	CL@.094	Ea	32.80	2.85	35.65
1-1/4" regular	CL@.108	Ea	34.50	3.27	37.77
1-1/2" regular	CL@.124	Ea	41.50	3.75	45.25
Add for double swing joint	CL@.132	Ea	3.59	4.00	7.59
Add for triple swing joint	CL@.133	Ea	5.94	4.03	9.97
Add for locking vinyl cover	—	Ea	8.90	—	8.90
Hose bibb, 3/4" on 12" galvanized riser	CL@.041	Ea	8.85	1.24	10.09

	Craft@Hrs	Unit	Material	Labor	Equipment	Total
Planting Bed Preparation						

Spreading topsoil from piles on site, based on topsoil delivered to the site. Topsoil prices can be expected to vary widely. Typical price is $25 per CY.

	Craft@Hrs	Unit	Material	Labor	Equipment	Total
Hand work, 10 CY job,						
level site, 10' throw	CL@.500	CY	25.00	15.10	—	40.10
Move 25' in wheelbarrow	CL@1.00	CY	25.00	30.30	—	55.30
With 3/4 CY loader (at $16.00 per hour)						
100 CY job, level site	TO@.080	CY	25.00	3.12	1.46	29.58

Spreading granular or powdered soil conditioners, based on fertilizer at $.20 per pound. Spread 20 pounds per 1,000 square feet (MSF). Add cost for mixing soil and fertilizer from below.

	Craft@Hrs	Unit	Material	Labor	Equipment	Total
By hand, 2,250 SF per hour	CL@.445	MSF	4.00	13.50	—	17.50
Hand broadcast spreader,						
9,000 SF per hour	CL@.111	MSF	4.00	3.36	.07	7.43
Push gravity spreader,						
3,000 SF per hour	CL@.333	MSF	4.00	10.10	.41	14.51
Push broadcast spreader,						
20,000 SF per hour	CL@.050	MSF	4.00	1.51	.08	5.59
Tractor drawn broadcast spreader						
87 MSF per hour	CL@.011	MSF	4.00	.33	.21	4.54

	Craft@Hrs	Unit	Material	Labor	Equipment	Total

Spreading organic soil conditioners, based on peat humus at $.15 per pound. Spread 200 pounds per 1,000 square feet (MSF). Add mixing cost from below

	Craft@Hrs	Unit	Material	Labor	Equipment	Total
By hand,						
1,000 SF per hour	CL@1.00	MSF	40.00	30.30	—	70.30
Push gravity spreader,						
2,750 SF per hour	CL@.363	MSF	40.00	11.00	.44	51.44
Manure spreader,						
34,000 SF per hour	CL@.029	MSF	40.00	.88	.35	41.23
Fertilizer and soil conditioners,						
Most nitrate fertilizers,						
ammonium sulfate	—	Lb	.10	—	—	.10
Lawn and garden fertilizer,						
calcium nitrate	—	Lb	.35	—	—	.35
Hydrated limesodium nitrate	—	Lb	.25	—	—	.25
Ground limestone	—	Ton	140.00	—	—	140.00
Ground dolomitic limestone	—	Ton	145.00	—	—	145.00
Composted manure	—	Lb	.07	—	—	.07
Peat humus	—	Lb	.18	—	—	.18
Vermiculite, perlite	—	CF	2.90	—	—	2.90

Spreading lime at 70 pounds per 1,000 SF. Based on an 8 acre job using a tractor-drawn spreader at $24.20 per hour and 1 acre per hour. Add mixing cost from below. Note: 1 acre equals 43,560 square feet.

	Craft@Hrs	Unit	Material	Labor	Equipment	Total
Lime at $7.10 per 100 pounds	TO@1.00	Acre	216.00	39.10	24.20	279.30
Lime at $7.10 per 100 pounds	TO@.656	Ton	142.00	25.60	15.86	183.46

Mixing soil and fertilizer in place. Medium soil. Light soil (sand or soft loam) will decrease costs shown by 20% to 30%. Heavy soil (clay, wet or rocky soil) will increase costs shown by 20% to 40%. Add soil conditioner cost from sections preceding this section.

	Craft@Hrs	Unit	Material	Labor	Equipment	Total
Mixing by hand						
2" deep, 9 SY per hour	CL@.111	SY	—	3.36	—	3.36
4" deep, 7.5 SY per hour	CL@.133	SY	—	4.03	—	4.03
6" deep, 5.5 SY per hour	CL@.181	SY	—	5.48	—	5.48
Mixing with a 26" tiller rented for $32 per day						
2" deep, 110 SY per hour	CL@.009	SY	—	.27	.03	.30
4" deep, 90 SY per hour	CL@.011	SY	—	.33	.04	.37
6" deep, 65 SY per hour	CL@.015	SY	—	.45	.06	.51
8" deep, 45 SY per hour	CL@.022	SY	—	.67	.08	.75
Mixing with a 6' disk and 20 HP garden tractor rented for $100 per day						
2" deep, 50,000 SF per hour	TO@.020	MSF	—	.78	.25	1.03
4" deep, 40,000 SF per hour	TO@.025	MSF	—	.98	.32	1.30
6" deep, 30,000 SF per hour	TO@.033	MSF	—	1.29	.41	1.70
8" deep, 20,000 SF per hour	TO@.050	MSF	—	1.95	.62	2.57
Mixing with a 12' disk and 40 HP utility tractor rented for $126 per day						
2" deep, 90,000 SF per hour	TO@.011	MSF	—	.43	.18	.61
4" deep, 70,000 SF per hour	TO@.014	MSF	—	.55	.22	.77
6" deep, 53,000 SF per hour	TO@.019	MSF	—	.74	.30	1.04
8" deep, 35,000 SF per hour	TO@.028	MSF	—	1.09	.44	1.53
Leveling the surface for lawn or planting bed						
By hand, 110 SY per hour	CL@.009	SY	—	.27	—	.27
With 6' drag harrow, at $95 per day						
60,000 SF per hour	TO@.017	MSF	—	.66	.21	.87
With 12' drag harrow, at $120 per day						
115,000 SF per hour	TO@.009	MSF	—	.35	.15	.50

	Craft@Hrs	Unit	Material	Labor	Equipment	Total
Plant bed preparation						
With a 3/4 CY wheel loader, at $126 per day						
12.5 CY per hour	TO@.080	CY	—	3.12	1.26	4.38
By hand to 18" deep, 8.5 SY/hr	CL@.118	SY	—	3.57	—	3.57
Mixing planting soil						
By hand, .9 CY per hour	CL@1.11	CY	—	33.60	—	33.60
With 1/2 CY wheel loader, at $126 per day,						
7 CY per hour	TO@.143	CY	—	5.58	2.25	7.83
With a soil shredder, at $126 per day,						
11 CY per hour	TO@.090	CY	—	3.51	1.42	4.93

Seeding and Planting

	Craft@Hrs	Unit	Material	Labor	Equipment	Total
Seeding with a hand broadcast spreader, 5 pounds per 1,000 SF						
4,000 SF per hour,						
seed at $3.70 per pound	CL@.250	MSF	18.50	7.57	.16	26.23
Seeding with push broadcast spreader, 5 pounds per 1,000 SF						
10,000 SF per hour,						
seed at $3.70 per pound	CL@.100	MSF	18.50	3.03	.19	21.72
Seeding with a mechanical seeder at $20 per hour and 175 pounds per acre, 5 acre job						
Seed at $3.75 per pound	TO@13.1	Acre	656.00	512.00	261.00	1,429.00
Seed at $3.75 per pound	TO@.300	MSF	14.20	11.70	5.99	31.89

Hydroseeding (spray application of seed, binder and fertilizer slurry), 3 pounds of Alta Fescue seed applied in a 12 pound mix per 1,000 SF. Equipment cost is $399 per day. Use $1,500 as a minimum job charge.

	Craft@Hrs	Unit	Material	Labor	Equipment	Total
10,000 SF job	C4@1.12	MSF	21.00	34.40	18.60	74.00
25,000 SF job	C4@.645	MSF	21.00	19.80	10.70	51.50
50,000 SF job	C4@.399	MSF	21.00	12.30	6.62	39.92
Add for work in congested areas	—	%	—	—	—	50.0

Sealing soil, application by a hydroseeding unit. Use $750 as a minimum job charge.

	Craft@Hrs	Unit	Material	Labor	Equipment	Total
Copolymer based liquid						
for erosion control	C4@.020	SY	.25	.61	.34	1.20
Grass seed						
Most Kentucky bluegrass	—	Lb	4.80	—	—	4.80
Most fescue	—	Lb	2.50	—	—	2.50
Most ryegrass, timothy, orchard grass	—	Lb	1.90	—	—	1.90
Annual ryegrass	—	Lb	.76	—	—	.76
Bentgrass, certified	—	Lb	11.00	—	—	11.00
Native grass PLS,						
(bluestream, blue grama)	—	Lb	17.00	—	—	17.00

Sodding, placed on level ground at 25 SY per hour. Add delivery cost below. Use 50 SY as a minimum job charge.

	Craft@Hrs	Unit	Material	Labor	Equipment	Total
Northeast, Midwest, bluegrass blend	CL@.040	SY	1.55	1.21	—	2.76
Great Plains, South, bluegrass	CL@.040	SY	1.85	1.21	—	3.06
West, bluegrass, tall fescue	CL@.040	SY	2.90	1.21	—	4.11
South, Bermuda grass,						
centipede grass	CL@.040	SY	3.25	1.21	—	4.46
South, Zoysia grass, St. Augustine	CL@.040	SY	5.00	1.21	—	6.21
Typical delivery cost, to 90 miles	—	SY	.50	—	—	.50
Add for staking sod on slopes	CL@.010	SY	.10	.30	—	.40
Work grass seed into soil, no seed included						
By hand, 220 SY per hour	CL@.005	SY	—	.15	—	.15
6' harrow and 20 HP tractor, at $100 per day						
60 MSF per hour	TO@.017	MSF	—	.66	.21	.87
12' harrow and 40 HP tractor, at $125 per day						
100 MSF per hour	TO@.010	MSF	—	.39	.16	.55

	Craft@Hrs	Unit	Material	Labor	Equipment	Total
Apply top dressing over seed or stolons						
300 pounds per 1,000 SF (2.7 pounds per SY), dressing $.16 per pound						
By hand, 65 SY per hour	CL@.015	SY	.43	.45	—	.88
With manure spreader,						
10,000 SF per hour	CL@.125	MSF	48.00	3.78	—	51.78
Roll sod or soil surface with a roller						
With a push roller, 400 SY per hour	CL@.002	SY	—	.06	—	.06
With a 20 HP tractor, at $100 per day						
25,000 SF per hour	TO@.040	MSF	—	1.56	.50	2.06
Sprigging, by hand						
6" spacing, 50 SY per hour	CL@.020	SY	.21	.61	—	.82
9" spacing, 100 SY per hour	CL@.010	SY	.14	.30	—	.44
12" spacing, 150 SY per hour	CL@.007	SY	.11	.21	—	.32
Hybrid Bermuda grass stolons,						
per bushel	—	Ea	3.65	—	—	3.65
Ground cover, large areas, no surface preparation						
Ice plant (80 SF per flat)	CL@.008	SF	.19	.24	—	.43
Strawberry (75 SF per flat)	CL@.008	SF	.22	.24	—	.46
Ivy and similar (70 SF per flat)	CL@.008	SF	.23	.24	—	.47
Gravel bed, pea gravel, at $22.50 a ton (1.4 tons per CY)						
Spread by hand	CL@.555	CY	31.50	16.80	—	48.30
Trees and shrubs. See also Landscaping in the Residential Division						
Shrubs, most varieties						
1 gallon, 100 units	CL@.085	Ea	8.00	2.57	—	10.57
1 gallon, over 100 units	CL@.076	Ea	5.15	2.30	—	7.45
5 gallons, 100 units	CL@.443	Ea	18.90	13.40	—	32.30
Trees, most varieties, complete, staked, typical costs						
5 gallon, 4' to 6' high	CL@.716	Ea	24.00	21.70	—	45.70
15 gallon, 8' to 10' high	CL@1.37	Ea	78.80	41.50	—	120.30
20" x 24" box, 10' to 12' high	CL@4.54	Ea	300.00	137.00	—	437.00
Specimen size 36" box,						
14' to 20' high	CL@11.6	Ea	775.00	351.00	—	1,126.00
Guaranteed establishment of plants						
Add, as a % of total contract price	—	%	—	—	—	12.0
Edging						
Redwood benderboard,						
6" x 3/8", staked	CL@.013	LF	.75	.39	—	1.14
Redwood benderboard,						
4" x 5/16", staked	CL@.013	LF	.49	.39	—	.88
Redwood headerboard, 2" x 4", staked						
Specification grade	CL@.127	LF	4.80	3.84	—	8.64
Decomposed granite edging,						
Saturated and rolled, 3" deep	CL@.011	SF	.35	.33	—	.68
Landscape stepping stones (concrete pavers)						
Laid on level ground	CL@.024	SF	4.00	.73	—	4.73
Pressure treated piles in 2' to 5' sections,						
9" to 12" diameter	CL@.045	LF	3.10	1.36	—	4.46

	Craft@Hrs	Unit	Material	Labor	Equipment	Total

Excavation for Concrete Work These costs do not include shoring or disposal. Typical soil conditions.

Batterboards, lay out for footings,

per corner	C8@1.24	Ea	6.50	45.00	—	51.50

Trenching with a 1/2 CY utility backhoe/loader, small jobs, good soil conditions, no backfill included. Equipment cost is $22 per hour and 15 CY per hour

18" x 24" depth, .111 CY per LF (135 LF/Hr)	S1@.015	LF	—	.52	.17	.69
24" x 36" depth, .222 CY per LF (68 LF/Hr)	S1@.029	LF	—	1.01	.33	1.34
36" x 48" depth, .444 CY per LF (34 LF/Hr)	S1@.059	LF	—	2.04	.66	2.70
48" x 60" depth, .741 CY per LF (20 LF/Hr)	S1@.099	LF	—	3.43	1.11	4.54
Per cubic yard, 15 CY per hour	S1@.133	CY	—	4.61	1.49	6.10

Hand labor

Clearing trench shoulders of obstructions	CL@.017	LF	—	.51	—	.51
Hard pan or rock outcropping Breaking and stacking	CL@3.23	CY	—	97.80	—	97.80
Column pads and small piers, average soil	CL@1.32	CY	—	40.00	—	40.00
Backfilling against foundations, no import or export of materials						
Hand, including hand tamp	CL@.860	CY	—	26.00	—	26.00

Backfill with a 1/2 CY utility backhoe/loader (at $22 per hour)

Minimum compaction, wheel-rolled	S1@.055	CY	—	1.91	.62	2.53

Backfill with 1/2 CY utility backhoe/loader, 150CFM compressor & pneumatic tamper (at $32 per hour),

Spec grade compaction	S1@.141	CY	—	4.89	2.24	7.13

Disposal of excess backfill material with a 1/2 CY backhoe (at $22 per hour), no compaction

Spot spread, 100' haul	S1@.034	CY	—	1.18	.38	1.56
Area spread, 4" deep (covers 81 SF per CY)	S1@.054	CY	—	1.87	.61	2.48

Grading for slabs

Using a D-4 crawler tractor with dozer blade (1000 SF and $44.00 per hour)	S1@.001	SF	—	.03	.02	.05
Fine grading, by hand, light to medium soil (165 to 170 SF per hour)	CL@.006	SF	—	.18	—	.18

Capillary fill, at $20.00 per CY, one CY covers 81 SF 4" deep

4", hand graded and rolled, small job	CL@.014	SF	.25	.42	.09	.76
4", machine graded and rolled, larger job	S1@.006	SF	.25	.21	.01	.47
Add for each extra 1" hand graded	CL@.004	SF	.07	.12	—	.19
Add for each extra 1" machine graded	S1@.001	SF	.07	.03	.02	.12

Sand fill, at $22.00 per CY, 1 CY covers 162 SF at 2" deep

2" sand cushion, hand spread, 1 throw	CL@.003	SF	.14	.09	—	.23
Add for each additional 1"	CL@.002	SF	.07	.06	—	.13

Waterproof membrane, polyethylene, over sand bed, including 20% lap and waste

.004" (4 mil) clear or black	CL@.001	SF	.02	.03	—	.05
.006" (6 mil) clear or black	CL@.001	SF	.04	.03	—	.07

Formwork for Concrete Labor costs include the time needed to prepare formwork sketches at the job site, measure for the forms, fabricate, erect, align, brace, strip, clean and stack the forms. Costs for reinforcing and concrete are not included. Multiple use of forms shows the cost per use when a form is used several times on the same job without being totally disassembled or discarded. Normal cleaning and repairs are included in the labor cost on lines that show multiple use of forms. Cost will be higher if plans are not detailed enough to identify the work to be performed. No salvage value is assumed except as noted. Costs listed are per square foot of form in contact with the concrete (SFCA). These costs assume a standard and better lumber price of $520 per MBF and $880 per MSF for 3/4" plyform before waste allowance.

	Craft@Hrs	Unit	Material	Labor	Total

Wall footing, grade beam or tie beam forms These figures assume nails, stakes and form oil costing $.25 per square foot and 2.5 board feet of lumber are used for each square foot of form in contact with the concrete (SFCA). To calculate the quantity of formwork required, multiply the depth of footing in feet by the length of footing in feet. Then double the results if two sides will be formed. For scheduling purposes, estimate that a crew of 5 can lay out, fabricate and erect 600 to 700 SF of footing, grade beam or tie beam forms in an 8-hour day.

	Craft@Hrs	Unit	Material	Labor	Total
1 use	F5@.070	SFCA	1.55	2.62	4.17
3 uses	F5@.050	SFCA	.90	1.87	2.77
5 uses	F5@.040	SFCA	.77	1.50	2.27
Add for stepped footings	F5@.028	SFCA	.45	1.05	1.50
Add for keyed joint, 1 use	F5@.020	LF	.62	.75	1.37

Reinforcing bar supports for footing, grade beam and tie beam forms.
Bars suspended from 2" x 4" lumber, including stakes. Based on .86 BF of lumber per LF.

	Craft@Hrs	Unit	Material	Labor	Total
1 use	F5@.060	LF	.70	2.25	2.95
3 uses	F5@.050	LF	.48	1.87	2.35
5 uses	F5@.040	LF	.43	1.50	1.93

Integral starter wall forms (stem walls) formed monolithic with footings, heights up to 4'0", with 3 BF of lumber per SFCA plus an allowance of $.25 per SFCA for nails, stakes and form oil.

	Craft@Hrs	Unit	Material	Labor	Total
1 use	F5@.100	SFCA	1.01	3.75	5.56
3 uses	F5@.075	SFCA	1.03	2.81	3.84
5 uses	F5@.067	SFCA	.87	2.51	3.38

Bulkheads or pour-stops. When integral starter walls are formed, forms will usually be required at the ends of each wall or footing. These are usually called bulkheads or pour-stops. These figures assume the use of 1.1 SF of 3/4" plyform per SFCA plus an allowance of $.10 per SFCA for nails, stakes and bracing.

	Craft@Hrs	Unit	Material	Labor	Total
1 use	F5@.150	SFCA	1.07	5.62	6.69
3 uses	F5@.120	SFCA	.59	4.50	5.09
5 uses	F5@.100	SFCA	.49	3.75	4.24

Column footing or pile cap forms These figures assume nails, stakes and form oil costing $.25 per square foot and 3.5 board feet of lumber are used for each square foot of form in contact with the concrete. For scheduling purposes, estimate that a crew of 5 can lay out, fabricate and erect 500 to 600 SF of square or rectangular footing forms in an 8-hour day or 350 to 450 SF of octagonal, hexagonal or triangular footing forms.

Square or rectangular column forms

	Craft@Hrs	Unit	Material	Labor	Total
1 use	F5@.090	SFCA	2.07	3.37	5.44
3 uses	F5@.070	SFCA	1.16	2.62	3.78
5 uses	F5@.060	SFCA	.98	2.25	3.23

Octagonal, hexagonal or triangular forms

	Craft@Hrs	Unit	Material	Labor	Total
1 use	F5@.117	SFCA	2.07	4.38	6.45
3 uses	F5@.091	SFCA	1.16	3.41	4.57
5 uses	F5@.078	SFCA	.98	2.92	3.90

Reinforcing bar supports for column forms. Bars suspended from 2" x 4" lumber, including stakes. Based on .86 BF of lumber per LF

	Craft@Hrs	Unit	Material	Labor	Total
1 use	F5@.060	LF	.45	2.25	2.70

	Craft@Hrs	Unit	Material	Labor	Total

Anchor bolt templates or dowel supports for column forms. Based on using nails, stakes, bracing, and form oil costing $.10 per SF and 1.1 SF of 3/4" plyform for each SF of column base or dowel support. No anchor bolts included

1 use	F5@.150	SF	1.07	5.62	6.69
3 uses	F5@.100	SF	.59	3.75	4.34
5 uses	F5@.090	SF	.49	3.37	3.86

Slab-on-grade forms These costs assume nails, stakes, form oil, and accessories costing $.15 are used for each board foot of lumber. Note that costs listed below are per linear foot (LF) of form. Figures in parentheses show the average square feet of contact area (SFCA) per linear foot of form. Use this figure to convert costs and manhours per linear foot to costs and manhours per square foot. For example, 12" to 24" high edge forms average 1.5 SFCA per LF of form. The manhours and costs per SF for 3 uses would be: .079 manhours (.119 divided by 1.5), material $.93 ($1.40 divided by 1.5) and labor $3.06 ($4.46 divided by 1.5). For scheduling purposes, estimate that a crew of 5 will lay out, fabricate and erect the following quantities of slab-on-grade edge forms in an 8-hour day:

Slabs to 6" high at 615 to 720 LF. Slabs over 6" to 12" high at 450 to 500 LF

Slabs over 12" to 24" high at 320 to 350 LF. Slabs over 24" to 36" high, 240 to 260 LF

Edge forms to 6" high, 1.5 BF per LF of form (.5 SFCA per LF).

1 use	F5@.065	LF	.93	2.44	3.37
3 uses	F5@.061	LF	.54	2.29	2.83
5 uses	F5@.055	LF	.46	2.06	2.52

Edge forms over 6" high to 12" high, 2.25 BF per LF of form (average .75 SFCA per LF)

1 use	F5@.090	LF	1.28	3.37	4.65
3 uses	F5@.086	LF	.70	3.22	3.92
5 uses	F5@.080	LF	.58	3.00	3.58

Edge forms over 12" high to 24" high, 4.5 BF per LF of form (average 1.5 SFCA per LF)

1 use	F5@.124	LF	2.57	4.65	7.22
3 uses	F5@.119	LF	1.40	4.46	5.86
5 uses	F5@.115	LF	1.17	4.31	5.48

Edge forms over 24" high to 36" high, 7.5 BF per LF of forms (average 2.5 SFCA per LF)

1 use	F5@.166	LF	4.28	6.22	10.50
3 uses	F5@.160	LF	2.33	6.00	8.33
5 uses	F5@.155	LF	1.94	5.81	7.75

Add for 2" x 2" tapered keyed joint, one-piece, .38 BF per LF

1 use	F5@.044	LF	.20	1.65	1.85
3 uses	F5@.040	LF	.10	1.50	1.60
5 uses	F5@.035	LF	.08	1.31	1.39

Blockout and slab depression forms. Figure the linear feet required. Use the linear foot costs for slab edge forms for the appropriate height and then add per linear foot

Blockouts	F5@.030	LF	.50	1.12	1.62

Wall forms Formwork over 6' high includes an allowance for a work platform and handrail built on one side of the form for use by the concrete placing crew.

Heights to 4', includes 1.1 SF of plyform, 1.5 BF of lumber and $.29 for nails, ties and oil per SFCA

1 use	F5@.119	SFCA	2.04	4.46	6.50
3 uses	F5@.080	SFCA	1.17	3.00	4.17
5 uses	F5@.069	SFCA	.99	2.59	3.58

Heights over 4' to 6', includes 1.1 SF of plyform, 2.0 BF of lumber and $.37 per SFCA for nails, ties and oil

1 use	F5@.140	SFCA	2.38	5.25	7.63
3 uses	F5@.100	SFCA	1.38	3.75	5.13
5 uses	F5@.080	SFCA	1.17	3.00	4.17

	Craft@Hrs	Unit	Material	Labor	Total
Heights over 6' to 12', includes 1.2 SF of plyform, 2.5 BF of lumber and $.43 per SFCA for nails, ties and oil					
1 use	F5@.160	SFCA	2.79	6.00	8.79
3 uses	F5@.110	SFCA	1.61	4.12	5.73
5 uses	F5@.100	SFCA	1.37	3.75	5.12
Heights over 12' to 16', includes 1.2 SF of plyform, 3.0 BF of lumber and $.44 per SFCA for nails, ties and oil					
1 use	F5@.180	SFCA	3.71	6.74	10.45
3 uses	F5@.130	SFCA	2.10	4.87	6.97
5 uses	F5@.110	SFCA	1.90	4.12	6.02
Heights over 16', includes 1.3 SF of plyform, 3.5 BF of lumber and $.50 per SFCA for nails, ties and oil					
1 use	F5@.199	SFCA	3.51	7.46	10.97
3 uses	F5@.140	SFCA	2.03	5.25	7.28
5 uses	F5@.119	SFCA	1.73	4.46	6.19
Architectural form liner for wall forms					
Low cost liners	F5@.020	SFCA	1.20	.75	1.95
Average cost liners	F5@.100	SFCA	1.95	3.75	5.70
Reveal strips 1" deep by 2" wide, one-piece, wood					
1 use	F5@.069	LF	.11	2.59	2.70
3 uses	F5@.050	LF	.07	1.87	1.94
5 uses	F5@.020	LF	.06	.75	.81
Battered wall forms (wall form is inclined from the vertical)					
Add for 1 side battered	F5@.016	SFCA	.21	.60	.81
Add for 2 sides battered	F5@.024	SFCA	.42	.90	1.32

Blockouts for openings. Form area is the opening perimeter times the depth. These figures assume that nails and form oil costing $.08 per square foot, 1.2 SF of plyform and .5 board feet of lumber are used per SF of form. Blockouts usually can be used only once.

	Craft@Hrs	Unit	Material	Labor	Total
1 use	F5@.250	SF	1.40	9.37	10.77

Bulkheads or pour-stops. These costs assume that nails, form oil and accessories costing $.17 per square foot, 1.1 SF of plyform and 1 board foot of lumber are used per SF of form

	Craft@Hrs	Unit	Material	Labor	Total
1 use	F5@.220	SF	1.67	8.24	9.91
3 uses	F5@.149	SF	.93	5.58	6.51
5 uses	F5@.130	SF	.78	4.87	5.65
Add for keyed wall joint, two-piece, tapered, 1.2 BF per SF					
1 use	F5@.100	SF	.79	3.75	4.54
3 uses	F5@.069	SF	.48	2.59	3.07
5 uses	F5@.061	SF	.42	2.29	2.71
Curved wall forms					
Smooth radius, add to straight wall cost	F5@.032	SFCA	.70	1.20	1.90
8' chord sections, add to straight wall cost	F5@.032	SFCA	.42	1.20	1.62

Haunches or ledges. Area is the width of the ledge times the length. These costs assume nails, form oil and accessories costing $.12 per square foot, 2 SF of plyform, and .5 board foot of lumber are used per SF of ledge

	Craft@Hrs	Unit	Material	Labor	Total
1 use	F5@.300	SF	2.14	11.20	13.34
3 uses	F5@.210	SF	1.13	7.87	9.00
5 uses	F5@.180	SF	.93	6.74	7.67

Steel framed plywood forms Rented steel framed plywood forms can reduce forming costs on many jobs. Where rented forms are used 3 times a month, use the wall forming costs shown for 3 uses at the appropriate height but deduct 25% from the material cost and 50% from the labor manhours and labor costs. Savings will be smaller where layouts change from one use to the next, when form penetrations must be made and repaired before returning a form, where form delivery costs are high, and where non-standard form sizes are needed.

	Craft@Hrs	Unit	Material	Labor	Total

Column forms for square or rectangular columns Quantities shown in parenthesis give cross section area of a square column in square inches. When estimating a rectangular column, use the costs for the square column with closest cross section area. For scheduling purposes estimate that a crew of 5 can lay out, fabricate and erect about 550 to 600 SFCA of square or rectangular column forms in an 8-hour day.

Up to 12" x 12" (144 square inches), using nails, snap ties, oil and column clamps costing $1.00, with 1.15 SF of plyform and 2 BF of lumber per SFCA

	Craft@Hrs	Unit	Material	Labor	Total
1 use	F5@.130	SFCA	3.05	4.87	7.92
3 uses	F5@.090	SFCA	2.03	3.37	5.40
5 uses	F5@.080	SFCA	1.82	3.00	4.82

Over 12" x 12" to 16" x 16" (256 square inches), using nails, snap ties, oil and column clamps costing $.81, with 1.125 SF of plyform and 2.1 BF of lumber per SFCA

	Craft@Hrs	Unit	Material	Labor	Total
1 use	F5@.100	SFCA	3.49	3.75	7.24
3 uses	F5@.069	SFCA	1.91	2.59	4.50
5 uses	F5@.061	SFCA	1.69	2.29	3.98

Over 16" x 16" to 20" x 20" (400 square inches), using nails, snap ties, oil and column clamps costing $.78, with 1.1 SF of plyform and 2.2 BF of lumber per SFCA

	Craft@Hrs	Unit	Material	Labor	Total
1 use	F5@.080	SFCA	2.89	3.00	5.89
3 uses	F5@.061	SFCA	1.84	2.29	4.13
5 uses	F5@.050	SFCA	1.62	1.87	3.49

Over 20" x 20" to 24" x 24" (576 square inches), using nails, snap ties, oil and column clamps costing $.55, with 1.1 SF of plyform and 2.4 BF of lumber per SFCA

	Craft@Hrs	Unit	Material	Labor	Total
1 use	F5@.069	SFCA	2.77	2.59	5.36
3 uses	F5@.050	SFCA	1.66	1.87	3.53
5 uses	F5@.040	SFCA	1.44	1.50	2.94

Over 24" x 24" to 30" x 30" (900 square inches), using nails, snap ties, oil and column clamps costing $.55, with 1.1 SF of plyform and 2.4 BF of lumber per SFCA

	Craft@Hrs	Unit	Material	Labor	Total
1 use	F5@.061	SFCA	2.77	2.29	5.06
3 uses	F5@.040	SFCA	1.66	1.50	3.16
5 uses	F5@.031	SFCA	1.44	1.16	2.60

Over 30" x 30" to 36" x 36" (1,296 square inches), using nails, snap ties, oil and column clamps costing $.55, with 1.1 SF of plyform and 2.4 BF of lumber per SFCA

	Craft@Hrs	Unit	Material	Labor	Total
1 use	F5@.050	SFCA	3.28	1.87	5.15
3 uses	F5@.040	SFCA	1.66	1.50	3.16
5 uses	F5@.031	SFCA	1.44	1.16	2.60

Over 36" x 36" to 48" x 48" (2,304 square inches), using nails, snap ties, oil and column clamps costing $.42, with 1.1 SF of plyform and 2.5 BF of lumber per SFCA

	Craft@Hrs	Unit	Material	Labor	Total
1 use	F5@.040	SFCA	2.69	1.50	4.19
3 uses	F5@.031	SFCA	1.56	1.16	2.72
5 uses	F5@.020	SFCA	1.33	.75	2.08

Column capitals for square columns Capital forms for square columns usually have four symmetrical sides. Length and width at the top of the capital are usually twice the length and width at the bottom of the capital. Height is usually the same as the width at the capital base. These costs assume capitals installed not over 12' above floor level, use of nails, form oil, shores and accessories costing $3.02, with 1.5 SF of plyform and 2 BF of lumber per SF of contact area (SFCA). Complexity of these forms usually makes more than 3 uses impractical. For scheduling purposes estimate that a crew of 5 can lay out, fabricate and erect the following quantities of capital formwork in an 8-hour day:

70 to 80 SF for 12" to 16" columns	90 to 100 SF for 20" to 24" columns
110 to 120 SF for 30" to 36" columns	160 to 200 SF for 48" columns

Up to 12" x 12" column, 6.0 SFCA

	Craft@Hrs	Unit	Material	Labor	Total
1 use of forms	F5@4.01	Ea	17.18	150.00	167.18
3 uses of forms	F5@3.00	Ea	10.01	112.00	122.01

	Craft@Hrs	Unit	Material	Labor	Total
Over 12" x 12" to 16" x 16" column, 10.7 SFCA					
1 use of forms	F5@6.00	Ea	28.27	225.00	253.27
3 uses of forms	F5@4.01	Ea	15.65	150.00	165.65
Over 16" x 16" to 20" x 20" column, 16.6 SFCA					
1 use of forms	F5@8.31	Ea	42.20	311.00	353.20
3 uses of forms	F5@6.00	Ea	22.61	225.00	247.61
Over 20" x 20" to 24" x 24" column, 24.0 SFCA					
1 use of forms	F5@12.0	Ea	59.66	450.00	509.66
3 uses of forms	F5@9.00	Ea	31.34	337.00	368.34
Over 24" x 24" to 30" x 30" column, 37.5 SFCA					
1 use of forms	F5@16.0	Ea	91.52	600.00	691.52
3 uses of forms	F5@12.0	Ea	47.27	450.00	497.27
Over 30" x 30" to 36" x 36" column, 54.0 SFCA					
1 use of forms	F5@20.0	Ea	130.46	749.00	879.46
3 uses of forms	F5@16.0	Ea	66.74	600.00	666.74
Over 36" x 36" to 48" x 48", 96.0 SFCA					
1 use of forms	F5@24.0	Ea	229.58	899.00	1,128.58
3 uses of forms	F5@18.0	Ea	116.30	674.00	790.30

Column forms for round columns Use the costs below to estimate round fiber tube forms (Sonotube is one manufacturer). These costs do not include column footings or foundations. These forms are peeled off when the concrete has cured. Costs shown include setting, aligning, bracing and stripping and assume that bracing and collars at top and bottom can be used 3 times. Column forms over 12'0" long will cost more per linear foot. Costs are per linear foot for standard weight column forms. Plastic lined forms have one vertical seam. The figures in parentheses show the tube cost per LF. For scheduling purposes, estimate that a crew of 5 can lay out, erect and brace the following quantities of 10' to 12' high round fiber tube forms in an 8-hour day: twenty 8" to 12" diameter columns; sixteen 14" to 20" columns and twelve 24" to 48" columns.

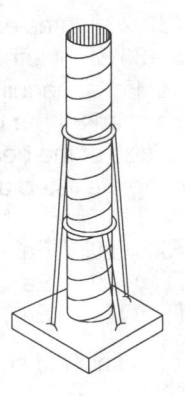

	Craft@Hrs	Unit	Material	Labor	Total
8" spiral type, ($1.25 per LF)	F5@.166	LF	2.49	6.22	8.71
8" plastic line, ($5.90 per LF)	F5@.166	LF	7.19	6.22	13.41
10" spiral type, ($1.70 per LF)	F5@.182	LF	2.89	6.82	9.71
10" plastic lined, ($6.20 per LF)	F5@.177	LF	7.49	6.63	14.12
12" spiral type, ($1.90 per LF)	F5@.207	LF	3.09	7.76	10.85
12" plastic lined, ($8.30 per LF)	F5@.207	LF	9.59	7.76	17.35
14" spiral type, ($2.50 per LF)	F5@.224	LF	3.79	8.39	12.18
14" plastic lined, ($9.10 per LF)	F5@.224	LF	10.39	8.39	18.78
16" spiral type, ($2.90 per LF)	F5@.232	LF	4.19	8.69	12.88
16" plastic lined, ($10.70 per LF)	F5@.232	LF	12.00	8.69	20.69
20" spiral type, ($4.30 per LF)	F5@.250	LF	5.59	9.37	14.96
20" plastic lined, ($14.20 per LF)	F5@.250	LF	15.50	9.37	24.87
24" spiral type, ($5.00 per LF)	F5@.275	LF	6.29	10.30	16.59
24" plastic lined, ($17.00 per LF)	F5@.275	LF	18.30	10.30	28.60
30" spiral type, ($8.90 per LF)	F5@.289	LF	10.20	10.80	21.00
30" plastic lined, ($20.00 per LF)	F5@.289	LF	21.40	10.80	32.20
36" spiral type, ($10.00 per LF)	F5@.308	LF	11.30	11.50	22.80
36" plastic lined, ($21.80 per LF)	F5@.308	LF	23.20	11.50	34.70
48" spiral type, ($24.50 per LF)	F5@.334	LF	25.90	12.50	38.40
48" plastic lined, ($33.00 per LF)	F5@.334	LF	34.50	12.50	47.00

	Craft@Hrs	Unit	Material	Labor	Total

Column capitals for round columns Use these figures to estimate costs for rented prefabricated steel conical shaped capital forms. Costs assume a minimum rental period of 1 month and include an allowance for nails and form oil. Labor shown assumes the capital will be supported from the deck formwork above the capital and includes laying out and cutting the hole in the deck to receive the capital. Capital bottom diameter is sized to fit into the column tube form. The column form size (diameter) does not significantly affect the capital form rental cost. Dimensions shown below are the top diameter. Form weights are shown in parentheses. Only 3 uses of these forms are usually possible in a 30-day period. Setting forms and installing reinforcing steel takes one day, concrete placing takes another day, curing requires 7 to 10 days, removing and cleaning forms takes another day. For scheduling purposes estimate that a crew of 5 can lay out, cut openings in floor deck formwork and install an average of 8 or 9 round column capital forms in an 8-hour day.

	Craft@Hrs	Unit	Material	Labor	Total
3'6" (100 pounds) 1 use	F5@4.01	Ea	175.00	150.00	325.00
3'6" (100 pounds) 3 uses	F5@4.01	Ea	58.50	150.00	208.50
4'0" (125 pounds) 1 use	F5@4.01	Ea	188.00	150.00	338.00
4'0" (125 pounds) 3 uses	F5@4.01	Ea	63.00	150.00	213.00
4'6" (150 pounds) 1 use	F5@4.30	Ea	196.00	161.00	357.00
4'6" (150 pounds) 3 uses	F5@4.30	Ea	65.40	161.00	226.40
5'0" (175 pounds) 1 use	F5@4.51	Ea	206.00	169.00	375.00
5'0" (175 pounds) 3 uses	F5@4.51	Ea	73.40	169.00	242.40
5'6" (200 pounds) 1 use	F5@4.51	Ea	237.00	169.00	406.00
5'6" (200 pounds) 3 uses	F5@4.51	Ea	79.20	169.00	248.20
6'0" (225 pounds) 1 use	F5@4.70	Ea	247.00	176.00	423.00
6'0" (225 pounds) 3 uses	F5@4.70	Ea	78.20	176.00	254.20

Beam and girder forms Using nails, snap ties and oil costing $.55, 1.3 SF of 3/4" plyform and 2.1 BF of lumber per SFCA. For scheduling purposes, estimate that a crew of 5 can lay out, fabricate and erect 250 to 300 SF of beam and girder forms in an 8-hour day.

	Craft@Hrs	Unit	Material	Labor	Total
1 use	F5@.149	SFCA	2.79	5.58	8.37
3 uses	F5@.141	SFCA	1.67	5.28	6.95
5 uses	F5@.120	SFCA	1.45	4.50	5.95

Shores for beams and girders Using 4" x 4" wooden posts. These figures assume an average shore height of 12'0" with a 2'0" long horizontal 4" x 4" head and 2" x 6" diagonal brace (30 BF of lumber per shore) and include nails, adjustable post clamps and accessories costing $2.50 per shore. For scheduling purposes estimate that a crew of 5 can lay out, fabricate and install 30 to 35 adjustable wood shores in an 8-hour day when shores have an average height of 12'0". Quantity and spacing of shores is dictated by the size of the beams and girders. Generally, allow one shore for each 6 LF of beam or girder. Costs for heavy duty shoring are listed under General Requirements, Heavy Duty Shoring.

	Craft@Hrs	Unit	Material	Labor	Total
1 use	F5@1.50	Ea	18.10	56.20	74.30
3 uses	F5@1.00	Ea	10.30	37.50	47.80
5 uses	F5@.748	Ea	8.74	28.00	36.74

Elevated slab forms These are costs for forms per square foot of finished slab area and include shoring and stripping at heights to 12'. Bays are assumed to be 20'0" x 20'0" bays.

Waffle slab-joist pans, monolithic Using rented forms (30-day rental period minimum). For scheduling purposes, estimate that a crew of 5 can lay out, fabricate and install 500 to 600 SF of waffle slab-joist pan forms and shoring in an 8-hour day. Two uses of the forms in a 30-day period is usually the maximum.

	Craft@Hrs	Unit	Material	Labor	Total
Metal pans, 10,000 to 20,000 SF					
1 use	F5@.100	SF	3.99	3.75	7.74
2 uses	F5@.066	SF	2.23	2.47	4.70
Fiberglass pans, 10,000 to 20,000 SF					
1 use	F5@.077	SF	4.01	2.89	6.90
2 uses	F5@.061	SF	2.28	2.29	4.57

	Craft@Hrs	Unit	Material	Labor	Total

Flat slab with beams, monolithic Using nails, braces, stiffeners and oil costing $.50, 1.2 SF of plyform and 3.2 BF of lumber per SFCA for one-way beam systems. Two-way beam systems require approximately 40% more forming materials than one-way beam systems. For scheduling purposes, estimate that a crew of 5 can lay out, fabricate and erect 350 to 400 SF of one-way beam systems and 250 to 300 SF of two-way beam systems in an 8-hour day, including shoring.

	Craft@Hrs	Unit	Material	Labor	Total
One-way beam 1 use	F5@.110	SFCA	3.22	4.12	7.34
One-way beam 2 uses	F5@.080	SFCA	2.00	3.00	5.00
Two-way beam 1 use	F5@.152	SFCA	4.31	5.70	10.01
Two-way beam 2 uses	F5@.100	SFCA	2.21	3.75	5.96

Structural flat slab Using nails, braces, stiffeners and oil costing $.50, 1.2 SF of plyform and 2.25 BF of lumber per SFCA. For scheduling purposes, estimate that a crew of 5 can lay out, fabricate and erect 500 to 600 SFCA of structural flat slab forms in an 8-hour day, including shoring.

	Craft@Hrs	Unit	Material	Labor	Total
Edge forms	F5@.072	LF	1.67	2.70	4.37
Slab forms 1 use	F5@.069	SFCA	2.73	2.59	5.32
Slab forms 3 uses	F5@.048	SFCA	1.62	1.80	3.42
Slab forms 5 uses	F5@.043	SFCA	1.39	1.61	3.00

Additional forming costs for elevated slabs, if required Based on single use of plyform and lumber. For scheduling purposes estimate that a 5 man crew can install 500 SFCA per 8-hour day.

	Craft@Hrs	Unit	Material	Labor	Total
Control joints	F5@.031	LF	1.79	1.16	2.95
Curbs & pads	F5@.083	SFCA	2.73	3.11	5.84
Depressions	F5@.083	SFCA	.82	3.11	3.93
Keyed joints	F5@.031	LF	.75	1.16	1.91

Miscellaneous formwork Based on single use of plyform and lumber. For scheduling purposes, estimate that a crew of 5 can lay out, fabricate and erect 225 to 250 SFCA of these types of forms in an 8-hour day.

	Craft@Hrs	Unit	Material	Labor	Total
Flat soffits & landings	F5@.166	SFCA	4.23	6.22	10.45
Sloping soffits	F5@.166	SFCA	4.37	6.22	10.59
Stair risers (steps)	F5@.120	SFCA	1.64	4.50	6.14

Driveway, curb and sidewalk forms See Site Work section.

Reinforcing for Cast-in-Place Concrete Steel reinforcing bars (rebar), ASTM A615 Grade 60. Material costs are for deformed steel reinforcing rebars, including 10% lap allowance, cutting and bending. These costs also include detailed shop drawings and delivery to jobsite with identity tags per shop drawings. Add for epoxy or galvanized coating of rebars, chairs, splicing, spiral caissons and round column reinforcing, if required, from the sections following the data below. Costs per pound (Lb) and per linear foot (LF) including tie wire and tying.

Reinforcing steel placed and tied in footings, foundations and grade beams

	Craft@Hrs	Unit	Material	Labor	Total
1/4" diameter, #2 rebar	RB@.015	Lb	.55	.58	1.13
1/4" diameter, #2 rebar (.17 lb per LF)	RB@.003	LF	.09	.12	.21
3/8" diameter, #3 rebar	RB@.011	Lb	.41	.43	.84
3/8" diameter, #3 rebar (.38 lb per LF)	RB@.004	LF	.16	.16	.32
1/2" diameter, #4 rebar	RB@.010	Lb	.26	.39	.65
1/2" diameter, #4 rebar (.67 lb per LF)	RB@.007	LF	.18	.27	.45
5/8" diameter, #5 rebar	RB@.009	Lb	.26	.35	.61
5/8" diameter, #5 rebar (1.04 lb per LF)	RB@.009	LF	.27	.35	.62
3/4" diameter, #6 rebar	RB@.008	Lb	.26	.31	.57
3/4" diameter, #6 rebar (1.50 lb per LF)	RB@.012	LF	.39	.47	.86
7/8" diameter, #7 rebar	RB@.008	Lb	.26	.31	.57
7/8" diameter, #7 rebar (2.04 lb per LF)	RB@.016	LF	.53	.62	1.15
1" diameter, #8 rebar	RB@.008	Lb	.26	.31	.57
1" diameter, #8 rebar (2.67 lb per LF)	RB@.021	LF	.69	.82	1.51
1-1/8" diameter, #9 rebar	RB@.008	Lb	.26	.31	.57
1-1/8" diameter, #9 rebar (3.40 lb per LF)	RB@.027	LF	.88	1.05	1.93

	Craft@Hrs	Unit	Material	Labor	Total
1-1/4" diameter, #10 rebar	RB@.007	Lb	.26	.27	.53
1-1/4" diameter, #10 rebar (4.30 lb per LF)	RB@.030	LF	1.12	1.17	2.29
1-3/8" diameter, #11 rebar	RB@.007	Lb	.26	.27	.53
1-3/8" diameter, #11 rebar (5.31 lb per LF)	RB@.037	LF	1.38	1.44	2.82
Reinforcing steel placed and tied in structural slabs					
1/4" diameter, #2 rebar	RB@.014	Lb	.57	.54	1.11
1/4" diameter, #2 rebar (.17 lb per LF)	RB@.002	LF	.09	.08	.17
3/8" diameter, #3 rebar	RB@.010	Lb	.42	.39	.81
3/8" diameter, #3 rebar (.38 lb per LF)	RB@.004	LF	.17	.16	.33
1/2" diameter, #4 rebar	RB@.009	Lb	.27	.35	.62
1/2" diameter, #4 rebar (.67 lb per LF)	RB@.006	LF	.19	.23	.42
5/8" diameter, #5 rebar	RB@.008	Lb	.27	.31	.58
5/8" diameter, #5 rebar (1.04 lb per LF)	RB@.008	LF	.28	.31	.59
3/4" diameter, #6 rebar	RB@.007	Lb	.27	.27	.54
3/4" diameter, #6 rebar (1.50 lb per LF)	RB@.011	LF	.40	.43	.83
7/8" diameter, #7 rebar	RB@.007	Lb	.27	.27	.54
7/8" diameter, #7 rebar (2.04 lb per LF)	RB@.014	LF	.55	.54	1.09
1" diameter, #8 rebar	RB@.007	Lb	.27	.27	.54
1" diameter, #8 rebar (2.67 lb per LF)	RB@.019	LF	.71	.74	1.45
1-1/8" diameter, #9 rebar	RB@.007	Lb	.27	.27	.54
1-1/8" diameter, #9 rebar (3.40 lb per LF)	RB@.024	LF	.91	.93	1.84
1-1/4" diameter, #10 rebar	RB@.006	Lb	.27	.23	.50
1-1/4" diameter, #10 rebar (4.30 lb per LF)	RB@.026	LF	1.16	1.01	2.17
1-3/8" diameter, #11 rebar	RB@.006	Lb	.27	.23	.50
1-3/8" diameter, #11 rebar (5.31 lb per LF)	RB@.032	LF	1.43	1.25	2.68
Reinforcing steel placed and tied in columns, stairs and walls					
1/4" diameter, #2 rebar	RB@.017	Lb	.57	.66	1.23
1/4" diameter, #2 rebar (.17 lb per LF)	RB@.003	LF	.09	.12	.21
3/8" diameter, #3 rebar	RB@.012	Lb	.42	.47	.89
3/8" diameter, #3 rebar (.38 lb per LF)	RB@.005	LF	.17	.19	.36
1/2" diameter, #4 rebar	RB@.011	Lb	.27	.43	.70
1/2" diameter, #4 rebar (.67 lb per LF)	RB@.007	LF	.19	.27	.46
5/8" diameter, #5 rebar	RB@.010	Lb	.27	.39	.66
5/8" diameter, #5 rebar (1.04 lb per LF)	RB@.010	LF	.28	.39	.67
3/4" diameter, #6 rebar	RB@.009	Lb	.27	.35	.62
3/4" diameter, #6 rebar (1.50 lb per LF)	RB@.014	LF	.40	.54	.94
7/8" diameter, #7 rebar	RB@.009	Lb	.27	.35	.62
7/8" diameter, #7 rebar (2.04 lb per LF)	RB@.018	LF	.55	.70	1.25
1" diameter, #8 rebar	RB@.009	Lb	.27	.35	.62
1" diameter, #8 rebar (2.67 lb per LF)	RB@.024	LF	.71	.93	1.64
1-1/8" diameter, #9 rebar	RB@.009	Lb	.27	.35	.62
1-1/8" diameter, #9 rebar (3.40 lb per LF)	RB@.031	LF	.91	1.21	2.12
1-1/4" diameter, #10 rebar	RB@.008	Lb	.27	.31	.58
1-1/4" diameter, #10 rebar (4.30 lb per LF)	RB@.034	LF	1.16	1.32	2.48
1-3/8" diameter, #11 rebar	RB@.008	Lb	.27	.31	.58
1-3/8" diameter, #11 rebar (5.31 lb per LF)	RB@.037	LF	1.43	1.44	2.87
Any size rebar, add	—	Lb	.26	—	.26
Galvanized coating					
Any #2 rebar, #3 rebar or #4 rebar, add	—	Lb	.33	—	.33
Any #5 rebar or #6 rebar, add	—	Lb	.31	—	.31
Any #7 rebar or #8 rebar, add	—	Lb	.29	—	.29
Any #9 rebar, #10 rebar or #11 rebar, add	—	Lb	.27	—	.27

	Craft@Hrs	Unit	Material	Labor	Total

Add for reinforcing steel chairs. Labor for placing rebar shown above includes time for setting chairs.

Chairs, individual type, cost each

3" high

	Craft@Hrs	Unit	Material	Labor	Total
steel	—	Ea	.26	—	.26
plastic	—	Ea	.36	—	.36
galvanized steel	—	Ea	.40	—	.40

5" high

	Craft@Hrs	Unit	Material	Labor	Total
steel	—	Ea	.40	—	.40
plastic	—	Ea	.56	—	.56
galvanized steel	—	Ea	.52	—	.52

8" high

	Craft@Hrs	Unit	Material	Labor	Total
steel	—	Ea	.85	—	.85
plastic	—	Ea	1.11	—	1.11
galvanized steel	—	Ea	1.14	—	1.14

12" high

	Craft@Hrs	Unit	Material	Labor	Total
steel	—	Ea	1.66	—	1.66
plastic	—	Ea	1.91	—	1.91
galvanized steel	—	Ea	2.04	—	2.04

Chairs, continuous type, with legs 8" OC, cost per linear foot

3" high

	Craft@Hrs	Unit	Material	Labor	Total
steel	—	LF	.36	—	.36
plastic	—	LF	.44	—	.44
galvanized steel	—	LF	.47	—	.47

6" high

	Craft@Hrs	Unit	Material	Labor	Total
steel	—	LF	.50	—	.50
plastic	—	LF	.63	—	.63
galvanized steel	—	LF	.66	—	.66

8" high

	Craft@Hrs	Unit	Material	Labor	Total
steel	—	LF	.70	—	.70
plastic	—	LF	.89	—	.89
galvanized steel	—	LF	.92	—	.92

12" high

	Craft@Hrs	Unit	Material	Labor	Total
steel	—	LF	1.68	—	1.68
plastic	—	LF	1.88	—	1.88
galvanized steel	—	LF	1.91	—	1.91

Reinforcing bar weld splicing. Cost per splice

	Craft@Hrs	Unit	Material	Labor	Total
#8 bars, #9 bars or #10 bars	RB@.527	Ea	2.07	20.50	22.57

Reinforcing bar clip splicing, sleeve and wedge, with hand-held hydraulic ram. Cost per splice

	Craft@Hrs	Unit	Material	Labor	Total
Number 4 bars, 1/2"	RB@.212	Ea	3.66	8.25	11.91
Number 5 bars, 5/8"	RB@.236	Ea	8.61	9.18	17.79
Number 6 bars, 3/4"	RB@.258	Ea	16.10	10.00	26.10

Waterstop, 3/8"

	Craft@Hrs	Unit	Material	Labor	Total
Rubber, 6"	B2@.059	LF	5.08	1.39	6.47
Rubber, 9"	B2@.059	LF	7.85	1.39	9.24
PVC, 9"	B2@.059	LF	4.54	1.39	5.93

Spiral caisson and round column reinforcing, 3/8" diameter hot rolled steel spirals with main vertical bars as shown. Costs include typical engineering and shop drawings. Shop fabricated spirals delivered, tagged, ready to install. Cost per vertical linear foot (VLF)

	Craft@Hrs	Unit	Material	Labor	Total
16" diameter with 6 #6 bars, 16.8 lbs per VLF	RB@.041	VLF	16.25	1.60	17.85
24" diameter with 6 #6 bars, 28.5 lbs per VLF	RB@.085	VLF	27.80	3.31	31.11
36" diameter with 8 #10 bars, 51.2 lbs per VLF	RB@.216	VLF	50.50	8.41	58.91

	Craft@Hrs	Unit	Material	Labor	Total
Welded wire mesh steel, electric weld, including 15% waste and overlap.					
2" x 2" W.9 x W.9 (#12 x #12), slabs	RB@.004	SF	.36	.16	.52
2" x 2" W.9 x W.9 (#12 x #12), beams and columns	RB@.020	SF	.36	.78	1.14
4" x 4" W1.4 x W1.4 (#10 x #10), slabs	RB@.003	SF	.17	.12	.29
4" x 4" W2.0 x W2.0 (#8 x #8), slabs	RB@.004	SF	.20	.16	.36
4" x 4" W2.9 x W2.9 (#6 x #6), slabs	RB@.005	SF	.22	.19	.41
4" x 4" W4.0 x W4.0 (#4 x #4), slabs	RB@.006	SF	.28	.23	.51
6" x 6" W1.4 x W1.4 (#10 x #10), slabs	RB@.003	SF	.08	.12	.20
6" x 6" W2.0 x W2.0 (#8 x #8), slabs	RB@.004	SF	.12	.16	.28
6" x 6" W2.9 x W2.9 (#6 x #6), slabs	RB@.004	SF	.14	.16	.30
6" x 6" W4.0 x W4.0 (#4 x #4), slabs	RB@.005	SF	.21	.19	.40
Add for lengthwise cut, LF of cut	RB@.002	LF	—	.08	.08
Welded wire mesh, galvanized steel, electric weld, including 15% waste and overlap.					
2" x 2" W.9 x W.9 (#12 x #12), slabs	RB@.004	SF	.71	.16	.87
2" x 2" W.9 x W.9 (#12 x #12), beams and columns	RB@.020	SF	.71	.78	1.49
4" x 4" W1.4 x W1.4 (#10 x #10), slabs	RB@.003	SF	.33	.12	.45
4" x 4" W2.0 x W2.0 (#8 x #8), slabs	RB@.004	SF	.39	.16	.55
4" x 4" W2.9 x W2.9 (#6 x #6), slabs	RB@.005	SF	.44	.19	.63
4" x 4" W4.0 x W4.0 (#4 x #4), slabs	RB@.006	SF	.57	.23	.80
6" x 6" W1.4 x W1.4 (#10 x #10), slabs	RB@.003	SF	.18	.12	.30
6" x 6" W2.0 x W2.0 (#8 x #8), slabs	RB@.004	SF	.24	.16	.40
6" x 6" W2.9 x W2.9 (#6 x #6), slabs	RB@.004	SF	.28	.16	.44
6" x 6" W4.0 x W4.0 (#4 x #4), slabs	RB@.005	SF	.41	.19	.60
Add for lengthwise cut, LF of cut	RB@.002	LF	—	.08	.08
Welded wire mesh, epoxy coated steel, electric weld, including 15% waste and overlap.					
2" x 2" W.9 x W.9 (#12 x #12), slabs	RB@.004	SF	1.45	.16	1.61
2" x 2" W.9 x W.9 (#12 x #12), beams and columns	RB@.020	SF	1.45	.78	2.23
4" x 4" W1.4 x W1.4 (#10 x #10), slabs	RB@.003	SF	1.25	.12	1.37
4" x 4" W2.0 x W2.0 (#8 x #8), slabs	RB@.004	SF	1.28	.16	1.44
4" x 4" W2.9 x W2.9 (#6 x #6), slabs	RB@.005	SF	1.30	.19	1.49
4" x 4" W4.0 x W4.0 (#4 x #4), slabs	RB@.006	SF	1.37	.23	1.60
6" x 6" W1.4 x W1.4 (#10 x #10), slabs	RB@.003	SF	1.18	.12	1.30
6" x 6" W2.0 x W2.0 (#8 x #8), slabs	RB@.004	SF	1.21	.16	1.37
6" x 6" W2.9 x W2.9 (#6 x #6), slabs	RB@.004	SF	1.23	.16	1.39
6" x 6" W4.0 x W4.0 (#4 x #4), slabs	RB@.005	SF	1.29	.19	1.48
Add for lengthwise cut, LF of cut	RB@.002	LF	—	.08	.08

Ready-Mix Concrete Ready-mix delivered by truck. Typical prices for most cities. Includes delivery up to 20 miles for 10 CY or more, 3" to 4" slump. Material cost only, no placing or pumping included. All material costs which include concrete are based on these figures.

	Craft@Hrs	Unit	Material	Labor	Total
5.0 sack mix, 2,000 PSI	—	CY	58.50	—	58.50
6.0 sack mix, 3,000 PSI	—	CY	64.00	—	64.00
6.6 sack mix, 3,500 PSI	—	CY	67.50	—	67.50
7.1 sack mix, 4,000 PSI	—	CY	70.00	—	70.00
8.5 sack mix, 5,000 PSI	—	CY	78.80	—	78.80
Extra costs for ready-mix concrete					
Add for less than 10 CY per load	—	CY	10.00	—	10.00
Add for delivery over 20 miles	—	Mile	.75	—	.75
Add for standby charge in excess of 5 minutes per CY delivered, per minute of extra time	—	Ea	1.00	—	1.00

	Craft@Hrs	Unit	Material	Labor	Total
Add for super plasticized mix, 7"-8" slump	—	%	8.0	—	—
Add for high early strength concrete					
5 sack mix	—	CY	7.75	—	7.75
6 sack mix	—	CY	9.75	—	9.75
Add for lightweight aggregate, typical	—	CY	34.00	—	34.00
Add for pump mix (pea-gravel aggregate)	—	CY	7.00	—	7.00
Add for granite aggregate, typical	—	CY	3.25	—	3.25
Add for white cement (architectural)	—	CY	41.50	—	41.50
Add for 1% calcium chloride	—	CY	1.25	—	1.25
Add for chemical compensated shrinkage	—	CY	12.50	—	12.50

Add for colored concrete, typical prices. The ready-mix supplier charges for cleanup of the ready-mix truck used for delivery of colored concrete. The usual practice is to provide one truck per day for delivery of all colored concrete for a particular job. Add the cost below to the cost per cubic yard for the design mix required. Also add the cost for truck cleanup.

	Craft@Hrs	Unit	Material	Labor	Total
Colored concrete, truck cleanup, per day	—	LS	50.00	—	50.00
Adobe	—	CY	14.50	—	14.50
Black	—	CY	26.00	—	26.00
Blended red	—	CY	13.50	—	13.50
Brown	—	CY	18.60	—	18.60
Green	—	CY	21.00	—	21.00
Yellow	—	CY	16.50	—	16.50

Ready-Mix Concrete and Placing No forms, finishing or reinforcing included. Material cost is based on 3,000 PSI concrete at $64 per CY, pump mix is $71.00.

	Craft@Hrs	Unit	Material	Labor	Equipment	Total
Columns						
By crane	H3@.873	CY	64.00	29.40	25.17	118.57
By pump	M2@.738	CY	71.00	25.30	17.14	113.44
Slabs-on-grade						
Direct from chute	CL@.430	CY	64.00	13.00	1.12	78.12
By crane	H3@.594	CY	64.00	20.00	17.03	101.03
By pump	M2@.456	CY	71.00	15.60	10.61	97.21
With buggy	CL@.543	CY	64.00	16.40	3.64	84.04
Elevated slabs						
By crane	H3@.868	CY	64.00	29.30	25.70	119.00
By pump	M2@.690	CY	71.00	23.60	16.12	110.72
Footings, pile caps, foundations						
Direct from chute	CL@.564	CY	64.00	17.10	1.50	82.60
By pump	M2@.546	CY	71.00	18.70	12.65	102.35
With buggy	CL@.701	CY	64.00	21.20	4.71	89.91
Beams and girders						
By pump	M2@.608	CY	71.00	20.80	14.18	105.98
By crane	H3@.807	CY	64.00	27.20	23.36	114.56
Stairs						
Direct from chute	CL@.814	CY	64.00	24.60	2.19	90.79
By crane	H3@1.05	CY	64.00	35.40	30.40	129.80
By pump	M2@.817	CY	71.00	28.00	18.97	117.97
With buggy	CL@.948	CY	64.00	28.70	6.31	99.01
Walls and grade beams to 4' high						
Direct from chute	CL@.479	CY	64.00	14.50	1.29	79.79
By pump	M2@.476	CY	71.00	16.30	11.04	98.34
With buggy	CL@.582	CY	64.00	17.60	3.91	85.51

	Craft@Hrs	Unit	Material	Labor	Equipment	Total
Walls over 4' to 8' high						
By crane	H3@.821	CY	61.00	27.70	23.67	112.37
By pump	M2@.608	CY	71.00	20.80	14.04	105.84
Walls over 8' to 16' high						
By crane	H3@.895	CY	64.00	30.20	25.91	120.11
By pump	M2@.667	CY	71.00	22.90	15.40	109.30
Walls over 16' high						
By crane	H3@.986	CY	64.00	33.20	28.66	125.86
By pump	M2@.704	CY	71.00	24.10	16.42	111.52

Concrete Slab-on-Grade Assemblies Typical costs including fine grading, edge forms, 3,000 PSI concrete, wire mesh, finishing and curing.

	Craft@Hrs	Unit	Material	Labor	Equipment	Total
4" thick 4' x 6' equipment pad						
Hand trimming and shaping, 24 SF	CL@.512	LS	—	15.50	—	15.50
Lay out, set and strip edge forms, 20 LF, 5 uses	C8@1.30	LS	9.20	47.10	—	56.30
Place W2.9 x W2.9 x 6" x 6" mesh, 24 SF	RB@.097	LS	5.04	3.77	—	8.81
Place ready-mix concrete, from chute .3 CY at $64.00 per CY	CL@.127	LS	19.20	3.84	—	23.04
Finish concrete, broom finish	CM@.591	LS	—	22.20	—	22.20
Cure concrete, curing paper	CL@.124	LS	1.00	3.75	—	4.75
Total job cost for 4" thick 24 SF pad	**—@2.75**	**LS**	**34.44**	**96.16**	**—**	**130.60**
Cost per SF for 4" thick 24 SF pad	—@.115	SF	1.44	4.01	—	5.45
Cost per CY of 4" thick concrete, .3 CY job	—@9.17	CY	114.80	320.53	—	435.33
8" thick 10' x 10' equipment pad						
Hand trimming and shaping, 100 SF	CL@2.79	LS	—	84.40	—	84.40
Lay out, set and strip edge forms, 40 LF, 5 uses	C8@3.28	LS	23.20	119.00	—	142.20
Place W2.9 x W2.9 x 6" x 6" mesh, 100 SF	RB@.400	LS	21.00	15.60	—	36.60
Place ready-mix concrete, from chute 2.5 CY at $64.00 per CY	CL@1.08	LS	160.50	32.70	—	193.20
Finish concrete, broom finish	CM@2.11	LS	—	79.10	—	79.10
Cure concrete, curing paper	CL@.515	LS	4.00	15.60	—	19.60
Total job cost for 8" thick 100 SF pad	**—@10.2**	**LS**	**208.70**	**346.40**	**—**	**555.10**
Cost per SF for 8" thick 100 SF job	—@.102	SF	2.09	3.46	—	5.55
Cost per CY of 8" thick concrete, 2.5 CY job	—@4.08	CY	83.48	138.56	—	222.04

Note: Ready-mix concrete unit price at $61.00 per CY, used in both examples, assumes a minimum of 10 cubic yards will be delivered (per load) with the excess used elsewhere on the same job.

	Craft@Hrs	Unit	Material	Labor	Equipment	Total
Add for each CY less than 10 delivered	—	CY	10.00	—	—	10.00

Floor Slab Assemblies Typical reinforced concrete slab-on-grade including excavation, gravel fill, forms, vapor barrier, wire mesh, 3000 PSI concrete at $64.00 per CY, finishing and curing.
Based on 100' x 75' slab (7,500 SF)

	Craft@Hrs	Unit	Material	Labor	Equipment	Total
4" thick slab	—@.024	SF	1.55	.88	.16	2.59
5" thick slab	—@.025	SF	1.75	.92	.16	2.83
6" thick slab	—@.026	SF	1.95	.95	.16	3.06

	Craft@Hrs	Unit	Material	Labor	Equipment	Total

Detailed cost breakdown slab as described above:

4" thick 100' x 75' slab

	Craft@Hrs	Unit	Material	Labor	Equipment	Total
Grade sandy loam site using a D-4 tractor (at $35.00 per hour), 140 CY, balanced job, (no import or export)	S1@2.66	LS	—	92.20	46.60	138.80
Buy and spread 6" crushed rock base using a D-4 tractor, 140 CY at $19.00 per CY	S1@15.8	LS	2,660.00	548.00	277.00	3,485.00
Lay out, set and strip edge forms, 350 LF, 5 uses	C8@19.6	LS	161.00	711.00	—	872.00
Place .006" polyethylene vapor barrier 7,500 SF	CL@8.84	LS	300.00	268.00	—	568.00
Place W2.9 x W2.9 x 6" x 6" mesh, 7,500 SF	RB@30.0	LS	2,100.00	1,170.00	—	3,270.00
Place and remove 2" x 4" keyway, 200 LF, 1 use	C8@8.80	LS	74.00	319.00	—	393.00
Place 4" concrete, 93 CY, from chute	CL@39.9	LS	5,952.00	1,210.00	—	7,162.00
Float finish 7,500 SF	CM@52.4	LS	—	1,970.00	750.00	2,720.00
Cure with slab curing paper	CM@5.58	LS	375.00	209.00	—	584.00
Total job cost 4" thick 100' x 75' floor slab	—@180	LS	**11,622.00**	**6,497.20**	**1,073.60**	**19,192.80**
Cost per SF for 4" thick 7,500 SF job	—@.024	SF	1.55	.87	.14	2.57
Cost per CY for 4" thick 93 CY job	—@1.94	CY	124.99	69.86	11.54	206.39
Cost per each additional 1" of concrete	—@.001	SF	.20	.04	—	.24

Cast-in-Place Concrete, Subcontract Typical costs including forms, concrete, reinforcing, finishing and curing. These costs include the subcontractor's overhead and profit but no excavation, shoring or backfill. Use these figures only for preliminary estimates and on jobs that require a minimum of 1,000 CY of concrete.

	Craft@Hrs	Unit	Material	Labor	Equipment	Total
Foundations						
Institutional or office buildings	—	CY	—	—	—	340.00
Heavy engineered structures	—	CY	—	—	—	370.00
Structural concrete walls						
Single story, 8" wall	—	CY	—	—	—	451.00
Single story, 10" wall	—	CY	—	—	—	514.00
Multi-story, 8" wall	—	CY	—	—	—	614.00
Multi-story, 10"	—	CY	—	—	—	651.00
Slip formed 8" wall	—	CY	—	—	—	554.00
Structural slabs, including shoring						
6", 1 way beams, 100 pounds reinforcing	—	CY	—	—	—	554.00
6", 2 way beams, 225 pounds reinforcing	—	CY	—	—	—	511.00
8", flat, with double mat reinforcing over steel frame by others, 100 pounds reinforcing	—	CY	—	—	—	461.00
12" flat, 250 pounds reinforcing bars per CY	—	CY	—	—	—	418.00
7", post-tensioned, to 1 pound tempered bars	—	CY	—	—	—	521.00
6" to 8" including beam jacketing	—	CY	—	—	—	495.00
6" to 8" on permanent metal form	—	CY	—	—	—	446.00
8" concurrent with slipform construction	—	CY	—	—	—	478.00
Lift slab construction, 6" to 8"	—	SF	—	—	—	10.30

	Craft@Hrs	Unit	Material	Labor	Equipment	Total
Pan and joist slabs 3" thick, 30" pan, including shoring						
6" x 16" joist	—	CY	—	—	—	527.00
6" x 12" joist	—	CY	—	—	—	466.00
6" x 10" joist	—	CY	—	—	—	435.00
Pan and joist slabs 4-1/2" thick, 20" pan, including shoring						
6" x 20" joist	—	CY	—	—	—	438.00
6" x 16" joist	—	CY	—	—	—	428.00
6" x 12" joist	—	CY	—	—	—	410.00
Parapet and fascia, 6" thick	—	CY	—	—	—	667.00
Loading docks,						
based on 8" walls and 8" slab	—	CY	—	—	—	297.00
Beams and girders, not slab integrated						
12" x 24", typical	—	CY	—	—	—	906.00
18" x 24", typical	—	CY	—	—	—	702.00
Columns, average reinforcing						
Square, wood formed, with chamfer						
12" column	—	CY	—	—	—	1,150.00
16" column	—	CY	—	—	—	1,060.00
18" column	—	CY	—	—	—	950.00
20" column	—	CY	—	—	—	863.00
24" column	—	CY	—	—	—	742.00
Round, fiber tube formed, 12" to 18"	—	CY	—	—	—	769.00

Cast-in-Place Concrete Steps on Grade Assemblies Typical in-place costs, including layout, fabrication and placing forms, setting and tying steel reinforcing, installation of steel nosing, placing 2,000 PSI concrete directly from the chute of a ready-mix truck, finishing, stripping forms and curing. Costs assume excavation, back filling and compaction have been completed by others. Crew is a carpenter, laborer and finisher. Formwork is based on using standard and better lumber at $525 per MBF before waste allowance, and concrete at $58.50 per CY before waste allowance. For scheduling purposes, estimate that a crew of 3 can set forms and place steel for and pour 4 to 5 CY of concrete cast-in-place steps in an 8-hour day. Costs shown are for concrete steps with a 6" riser height and 12" tread depth supported by a 6" thick monolithic slab. The "tricks of the trade" below explain how to use these unit prices to estimate stairways of various widths and heights. Note: Total cost per CY has been rounded.

	Craft@Hrs	Unit	Material	Labor	Total
Fabricate and erect forms, 2 uses of forms, includes stakes, braces, form oil and nails at $.13 per LF of forms					
Side forms, 1.7 BF per LF of tread	P9@.080	LF	.66	2.93	3.59
Riser forms, 1.5 BF per LF of tread	P9@.040	LF	.59	1.47	2.06
Rebars, #3, Grade 60, 1 pound per LF of tread	P9@.009	LF	.27	.33	.60
Embedded steel step nosing, 2-1/2" x 2-1/2" x 1/4" black mild steel					
Angle iron, 4.5 pounds per LF of tread at $.44 per Lb	P9@.004	LF	1.98	.15	2.13
Concrete, .03 CY per LF of tread	P9@.030	LF	1.83	1.10	2.93
Total cost per LF of risers	**P9@.163**	**LF**	**5.33**	**5.98**	**11.31**
Total cost per CY of concrete	**P9@5.43**	**CY**	**178.00**	**199.00**	**377.00**

Tricks of the trade:

To quickly estimate the cost of cast-in-place concrete steps on grade, do this:

Multiply the width of the steps (in feet) times their height in feet and multiply the result times 2.

This will give the total linear feet (LF) of risers in the steps. Multiply the total LF by the costs per LF.

Sample estimates for cast-in-place concrete steps on grade using the procedure and costs from above:

2'6" wide by 3' high (6 steps) would be: 2.5' x 3' x 2 = 15 LF x $11.31 = $170

3' wide by 3' high (6 steps) would be: 3' x 3' x 2 = 18 LF x $11.31 = $204

4' wide by 4' high (8 steps) would be: 4' x 4' x 2 = 32 LF x $11.31 = $362

7' wide by 4'6" high (9 steps) would be: 7' x 4.5' x 2 = 63 LF x $11.31 = $713

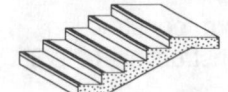

	Craft@Hrs	Unit	Material	Labor	Total

Tip for using the previous page with The National Estimator computer estimating program: Calculate the linear feet of stair, copy and paste the line "Total cost per LF of risers" and type the quantity.

Concrete Accessories

	Craft@Hrs	Unit	Material	Labor	Total
Bentonite granules,					
50 pound bag	—	Ea	15.00	—	15.00
Bond breaker and form release at $10.00 per gallon, Thompson's CBA, spray on					
Lumber & plywood form release, 200 SF/gal	CL@.006	SF	.05	.18	.23
Metal form release, 600 SF per gallon	CL@.003	SF	.02	.09	.11
Casting bed form breaker, 300 SF per gallon	CL@.005	SF	.03	.15	.18
Cure coat, 300 SF per gallon	CL@.005	SF	.03	.15	.18
Chamfer strip, reusable polyethylene					
1/2" radius	CC@.015	LF	.36	.63	.99
3/4" radius	CC@.015	LF	.57	.63	1.20
1" radius	CC@.015	LF	.62	.63	1.25
Column clamps for plywood forms, per set, by column size, Econ-o-clamp					
8" x 8" to 24" x 24", purchase	—	Set	65.00	—	65.00
8" x 8" to 30" x 30", purchase	—	Set	75.00	—	75.00
10" x 10" to 36" x 36", purchase	—	Set	85.00	—	85.00
Monthly rental cost as % of purchase price	—	%	5.0	—	—
Column clamps for lumber forms, per set, by column size, Symons					
10" x 10" to 28" x 28", purchase	—	Set	120.00	—	120.00
16" x 16" to 40" x 40", purchase	—	Set	150.00	—	150.00
21" x 21" to 50" x 50", purchase	—	Set	250.00	—	250.00
10" x 10" to 28" x 28", rent per month, per set	—	Mo	4.00	—	4.00
16" x 16" to 40" x 40", rent per month, per set	—	Mo	4.50	—	4.50
21" x 21" to 50" x 50", rent per month, per set	—	Mo	7.50	—	7.50
Concrete sleeves, circular polyethylene liner, any slab thickness					
1-1/2" to 3" diameter	—	Ea	1.70	—	1.70
4" to 6" diameter	—	Ea	3.35	—	3.35
Embedded iron, installation only	C8@.045	Lb	—	1.63	1.63
Expansion joints, cane fiber, asphalt impregnated, pre-molded					
1/2" x 4", slabs	C8@.028	LF	.24	1.02	1.26
1/2" x 6", in walls	C8@.035	LF	.34	1.27	1.61
1/2" x 8", in walls	C8@.052	LF	.43	1.89	2.32
Footing ties, 6" to 12" wide form	—	Ea	.34	—	.34
Form spreaders, 4" to 10"	—	Ea	.29	—	.29
Form oil, 800 to 1,200 SF per gallon	—	Gal	3.75	—	3.75
Grout for machine bases, mixed and placed (add for edge forms if required, from below)					
Five Star epoxy (at $125 per CF)					
per SF for each 1" of thickness	C8@.170	SF	10.40	6.17	16.57
Five Star NBEC non-shrink, non-metallic (at $28.00 per CF)					
per SF for each 1" of thickness	C8@.170	SF	2.33	6.17	8.50
Embeco 636, metallic aggregate (at $72 per CF)					
per SF for each 1" of thickness	C8@.163	SF	6.00	5.91	11.91
Forms for grout, based on using 2" x 4" lumber, 1 use	C8@.195	LF	.92	7.07	7.99
Inserts, Unistrut, average	C8@.068	Ea	7.21	2.47	9.68

	Craft@Hrs	Unit	Material	Labor	Total
She bolt form clamps, with cathead and bolt, purchase					
17" rod	—	Ea	3.90	—	3.90
21" rod	—	Ea	6.50	—	6.50
24" rod	—	Ea	7.00	—	7.00
Monthly rental as % of purchase price	—	%	15.0	—	—
Snap ties, average, 4,000 pound, 6" to 10"	—	Ea	.41	—	.41
Water stop, 3/8"					
Rubber, 6"	C8@.059	LF	4.84	2.14	6.98
Rubber, 9"	C8@.059	LF	7.47	2.14	9.61
PVC, 9"	C8@.059	LF	4.33	2.14	6.47
Concrete waler brackets					
2" x 4"	—	Ea	2.75	—	2.75
2" x 6"	—	Ea	3.60	—	3.60
Waler jacks (scaffold support on form), purchase	—	Ea	38.50	—	38.50
Rental per month, each	—	Mo	4.25	—	5.40

Concrete Slab Finishes

	Craft@Hrs	Unit	Material	Labor	Total
Float finish	CM@.009	SF	—	.34	.34
Trowel finishing					
Steel, machine work	CM@.014	SF	—	.53	.53
Steel, hand work	CM@.017	SF	—	.64	.64
Broom finish	CM@.012	SF	—	.45	.45
Scoring concrete surface, hand work	CM@.005	LF	—	.19	.19
Exposed aggregate (washed, including finishing), no disposal of slurry	CM@.015	SF	.25	.56	.81
Color hardener, nonmetallic, trowel application					
Standard colors	—	Lb	.55	—	.55
Non standard colors	—	Lb	.75	—	.75
Light duty floors (40 pounds per 100 SF)	CM@.008	SF	—	.30	.30
Pastel shades (60 pounds per 100 SF)	CM@.012	SF	—	.45	.45
Heavy duty (90 pounds per 100 SF)	CM@.015	SF	—	.56	.56
Add for colored wax, 800 SF and $25.75 per gal	CM@.003	SF	.03	.11	.14
Sidewalk grits, abrasive, 25 pounds per 100 SF					
Aluminum oxide, $67 per 100 pound bag	CM@.008	SF	.18	.30	.48
Silicon carbide, $129 per 100 pound bag	CM@.008	SF	.32	.30	.62
Metallic surface hardener, dry shake, Masterplate 200, at $.62 per pound					
Moderate duty, 1 pound per SF	CM@.008	SF	.62	.30	.92
Heavy duty, 1.5 pounds per SF	CM@.009	SF	.93	.34	1.27
Concentrated traffic, 2 pounds per SF	CM@.010	SF	1.25	.38	1.63
Liquid curing and sealing compound, sprayed-on, Mastercure, 400 SF and $26 per gallon	CL@.004	SF	.06	.12	.18
Sweep, scrub and wash down	CL@.006	SF	.02	.18	.20
Finish treads and risers					
No abrasives, no plastering, per SF of tread	CM@.038	SF	—	1.43	1.43
With abrasives, plastered, per SF of tread	CM@.058	SF	.65	2.18	2.83
Wax coating	CL@.002	SF	.04	.06	.10

	Craft@Hrs	Unit	Material	Labor	Total
Concrete Wall Finishes					
Cut back ties and patch	CM@.010	SF	.12	.38	.50
Remove fins	CM@.007	LF	.05	.26	.31
Grind smooth	CM@.020	SF	.09	.75	.84
Sack, simple	CM@.012	SF	.06	.45	.51
Bush hammer light, green concrete	CM@.020	SF	.13	.75	.88
Bush hammer standard, cured concrete	CM@.029	SF	.13	1.09	1.22
Bush hammer, heavy, cured concrete	CM@.077	SF	.13	2.89	3.02
Needle gun treatment, large areas	CM@.058	SF	.13	2.18	2.31
Wire brush, green	CM@.014	SF	.05	.53	.58
Wash with acid and rinse	CM@.004	SF	.28	.15	.43
Break fins, patch voids, Carborundum rub	CM@.032	SF	.07	1.20	1.27
Break fins, patch voids, burlap grout rub	CM@.024	SF	.07	.90	.97
Sandblasting, Subcontract. No scaffolding included, flat vertical surfaces					
Light sandblast	—	SF	—	—	1.00
Medium, to expose aggregate	—	SF	—	—	1.30
Heavy, with dust control requirements	—	SF	—	—	2.00
Note: Minimum charge for sandblasting	—	LS	—	—	300.00
Miscellaneous Concrete Finishes					
Monolithic natural aggregate topping					
1/16" topping	P9@.006	SF	.14	.22	.36
3/16" topping	P9@.018	SF	.36	.66	1.02
1/2" topping	P9@.021	SF	.92	.77	1.69
Integral colors,					
Typically 8 pounds per sack of cement	—	Lb	1.80	—	1.80
Mono rock, 3/8" wear course only, typical cost	—	SF	—	—	1.05
Kalman 3/4", wear course only, typical cost	—	SF	—	—	1.60
Acid etching, 5% muriatic acid, typical cost	P9@.005	SF	.30	.18	.48
Felton sand, for use with white cement	—	CY	33.00	—	33.00
Specially Placed Concrete Typical subcontract prices.					
Gunite, no forms and no reinforcing included					
Flat plane, vertical areas, 1" thick	—	SF	—	—	2.63
Curved arch, barrel, per 1" thickness	—	SF	—	—	3.54
Pool or retaining walls, per 1" thickness	—	SF	—	—	2.73
Add for architectural finish	—	SF	—	—	1.09
Wire mesh for gunite 2" x 2", W.9 x W.9	—	SF	—	—	1.64
Pressure grouting, large quantity, 50/50 mix	—	CY	—	—	515.00

Precast Concrete Panel costs are based on 3,000 PSI natural gray concrete with a smooth or board finish on one side. Placing (lifting) costs include handling, hoisting, alignment, bracing and permanent connections. Erection costs vary with the type of crane. Heavier lifts over a longer radius (reach) require larger cranes. Equipment costs include the crane and a 300 amp gas powered welding machine. Add the cost of grouting or caulking joints (about $2 per LF). Costs will be higher on small jobs (less than 35,000 SF).

	Craft@Hrs	Unit	Material	Labor	Equipment	Total
Precast wall panels Costs include delivery to 40 miles, typical reinforcing steel and embedded items but no lifting. Use 2,500 SF as a minimum job charge. See placing costs below.						
4" thick wall panels (52 lbs per SF)	—	SF	8.31	—	—	8.31
5" thick wall panels (65 lbs per SF)	—	SF	9.11	—	—	9.11
6" thick wall panels (78 lbs per SF)	—	SF	9.87	—	—	9.87
8" thick wall panels (105 lbs per SF)	—	SF	10.50	—	—	10.50

	Craft@Hrs	Unit	Material	Labor	Equipment	Total
Add for insulated sandwich wall panels						
2-1/2" of cladding and 2" of fiberboard	—	SF	3.28	—	—	3.28
Add for high strength concrete						
3,500 lb concrete mix	—	%	2.0	—	—	—
4,000 lb concrete mix	—	%	3.0	—	—	—
4,500 lb concrete mix	—	%	4.0	—	—	—
5,000 lb concrete mix	—	%	6.0	—	—	—
Add for white facing or integral white mix	—	%	20.0	—	—	—
Add for broken rib						
or fluted surface design	—	%	20.0	—	—	—
Add for exposed aggregate finish	—	%	17.0	—	—	—
Add for sandblasted surface finish	—	%	12.0	—	—	—

Placing precast structural wall panels Figures in parentheses show the panels placed per day with a 100 ton crawler-mounted crane and welding machine at $900 per day. Use $8,500 as a minimum job charge.

	Craft@Hrs	Unit	Material	Labor	Equipment	Total
1 ton panels to 180' reach (23 per day)	H4@2.79	Ea	—	126.00	44.64	170.64
1 to 3 tons, to 121' reach (21 per day)	H4@3.05	Ea	—	138.00	48.80	186.80
3 to 5 tons, to 89' reach (18 per day)	H4@3.54	Ea	—	160.00	56.64	216.64
5 to 7 tons, to 75' reach (17 per day)	H4@3.76	Ea	—	170.00	60.16	230.16
7 to 9 tons, to 65' reach (16 per day)	H4@4.00	Ea	—	181.00	64.00	245.00
9 to 11 tons, to 55' reach (15 per day)	H4@4.28	Ea	—	194.00	68.48	262.48
11 to 13 tons, to 50' reach (14 per day)	H4@4.56	Ea	—	206.00	72.96	278.96
13 to 15 tons, to 43' reach (13 per day)	H4@4.93	Ea	—	223.00	78.88	301.88
15 to 20 tons, to 36' reach (11 per day)	H4@5.81	Ea	—	263.00	92.96	355.96
20 to 25 tons to 31' reach (10 per day)	H4@6.40	Ea	—	290.00	102.4	392.40
For placing panels with cladding, add	—	%	—	5.0	5.0	—
For placing sandwich wall panels, add	—	%	—	10.0	10.0	—
If a 45 ton crane can be used, deduct	—	%	—	—	-22.0	—

Placing precast partition wall panels Figures in parentheses show the panels placed per day with a 45 ton truck-mounted crane and welding machine at $550 per day. Use $6,000 as a minimum job charge.

	Craft@Hrs	Unit	Material	Labor	Equipment	Total
1 ton panels to 100' reach (23 per day)	H4@2.79	Ea	—	126.00	19.90	145.90
1 to 2 tons, to 82' reach (21 per day)	H4@3.05	Ea	—	138.00	21.75	159.75
2 to 3 tons, to 65' reach (18 per day)	H4@3.54	Ea	—	160.00	25.25	185.25
3 to 4 tons, to 57' reach (16 per day)	H4@4.00	Ea	—	181.00	28.50	209.50
4 to 5 tons, to 50' reach (15 per day)	H4@4.28	Ea	—	194.00	30.50	224.50
5 to 6 tons, to 44' reach (14 per day)	H4@4.56	Ea	—	206.00	32.50	238.50
6 to 7 tons, to 40' reach (13 per day)	H4@4.93	Ea	—	223.00	35.15	258.15
7 to 8 tons, to 36' reach (11 per day)	H4@5.81	Ea	—	263.00	41.40	304.40
8 to 10 tons, to 31' reach (10 per day)	H4@6.40	Ea	—	290.00	45.60	335.60
10 to 12 tons to 26' reach (9 per day)	H4@7.11	Ea	—	322.00	50.70	372.70
12 to 14 tons, to 25' reach (8 per day)	H4@8.02	Ea	—	363.00	57.20	420.20
14 to 16 tons, to 21' reach (7 per day)	H4@9.14	Ea	—	414.00	65.20	479.20
If a 35 ton crane can be used, deduct	—	%	—	—	-5.0	—

Precast flat floor slab or floor plank Costs include delivery to 40 miles, 3,000 PSI concrete, typical reinforcing steel and embedded items. Use $2,000 as a minimum job charge. See placing costs below.

	Craft@Hrs	Unit	Material	Labor	Equipment	Total
4" thick floor panels (52 lbs per SF)	—	SF	5.88	—	—	5.88
5" thick floor panels (65 lbs per SF)	—	SF	7.00	—	—	7.00
6" thick floor panels (78 lbs per SF)	—	SF	9.01	—	—	9.01
8" thick floor panels (105 lbs per SF)	—	SF	9.35	—	—	9.35

	Craft@Hrs	Unit	Material	Labor	Equipment	Total

Placing panels with a 100 ton crawler-mounted crane and welding machine at $900 per day. Figures in parentheses show the number of panels placed per day. Use $8,500 as a minimum job charge

	Craft@Hrs	Unit	Material	Labor	Equipment	Total
2 ton panels to 150' reach						
(23 per day)	H4@2.79	Ea	—	126.00	61.38	187.38
2 to 3 tons, to 121' reach						
(21 per day)	H4@3.05	Ea	—	138.00	67.10	205.10
3 to 4 tons, to 105' reach						
(19 per day)	H4@3.37	Ea	—	153.00	74.14	227.14
4 to 5 tons, to 88' reach (18 per day)	H4@3.56	Ea	—	161.00	78.30	239.30
5 to 6 tons, to 82' reach (17 per day)	H4@3.76	Ea	—	170.00	82.70	252.70
6 to 7 tons, to 75' reach (16 per day)	H4@4.00	Ea	—	181.00	88.00	269.00
7 to 8 tons, to 70' reach (15 per day)	H4@4.28	Ea	—	194.00	94.16	288.16
8 to 9 tons, to 65' reach (14 per day)	H4@4.56	Ea	—	206.00	100.32	306.32
9 to 10 tons, to 60' reach (11 per day)	H4@5.82	Ea	—	264.00	128.00	392.00
20 to 25 tons to 31' reach (10 per day)	H4@6.40	Ea	—	290.00	140.80	430.80

Precast combination beam and slab units Beam and slab units with beam on edge of slab. Weight of beams and slabs will be about 4,120 lbs per CY. Use 40 CY as a minimum job charge. See placing costs below.

	Craft@Hrs	Unit	Material	Labor	Equipment	Total
Slabs with no intermediate beam	—	CY	509.00	—	—	509.00
Slabs with one intermediate beam	—	CY	540.00	—	—	540.00
Slabs with two intermediate beams	—	CY	588.00	—	—	588.00

Placing deck units with a 100 ton crawler-mounted crane and a welding machine at $900 per day. Figures in parentheses show the number of units placed per day. Use $8,500 as a minimum job charge.

	Craft@Hrs	Unit	Material	Labor	Equipment	Total
5 ton units to 88' reach (21 per day)	H4@3.05	Ea	—	138.00	67.10	205.10
5 to 10 tons, to 60' reach (19 per day)	H4@3.37	Ea	—	153.00	74.14	227.14
10 to 15 tons, to 43' reach (17 per day)	H4@3.76	Ea	—	170.00	82.72	252.72
15 to 20 tons, to 36' reach (15 per day)	H4@4.28	Ea	—	194.00	94.16	288.16
20 to 25 tons, to 31' reach (14 per day)	H4@4.57	Ea	—	207.00	100.54	307.54
If a 150 ton crane is required, add	—	%	—	—	16.0	—

Precast beams, girders and joists Beams, girders and joists, costs per linear foot of span, including 3,000 PSI concrete, reinforcing steel, embedded items and delivery to 40 miles. Use 400 LF as a minimum job charge. See placing costs below.

	Craft@Hrs	Unit	Material	Labor	Equipment	Total
1,000 lb load per foot, 15' span	—	LF	56.30	—	—	56.30
1,000 lb load per foot, 20' span	—	LF	62.60	—	—	62.60
1,000 lb load per foot, 30' span	—	LF	72.10	—	—	72.10
3,000 lb load per foot, 10' span	—	LF	56.30	—	—	56.30
3,000 lb load per foot, 20' span	—	LF	72.10	—	—	72.10
3,000 lb load per foot, 30' span	—	LF	86.90	—	—	86.90
5,000 lb load per foot, 10' span	—	LF	56.30	—	—	56.30
5,000 lb load per foot, 20' span	—	LF	79.50	—	—	79.50
5,000 lb load per foot, 30' span	—	LF	93.30	—	—	93.30

Placing beams, girders and joists with a 100 ton crawler-mounted crane and a welding machine at $1240 per day. Figures in parentheses show the number of units placed per day. Use $8,500 as a minimum job charge.

	Craft@Hrs	Unit	Material	Labor	Equipment	Total
5 ton units to 88' reach (18 per day)	H4@3.56	Ea	—	161.00	83.00	244.00
5 to 10 tons, to 60' reach (17 per day)	H4@3.76	Ea	—	170.00	87.60	257.60
10 to 15 tons, to 43' reach (16 per day)	H4@4.00	Ea	—	181.00	93.20	274.20
15 to 20 tons, to 36' reach (15 per day)	H4@4.27	Ea	—	193.00	99.50	292.50
If a 150 ton crane is required, add	—	%	—	—	16.0	—

	Craft@Hrs	Unit	Material	Labor	Equipment	Total

Precast concrete columns Precast concrete columns, 12" x 12" to 36" x 36" including concrete, reinforcing steel, embedded items and delivery to 40 miles. Use 40 CY as a minimum job charge. See placing costs below.

	Craft@Hrs	Unit	Material	Labor	Equipment	Total
Columns with 500 lbs of reinforcement per CY	—	CY	530.00	—	—	530.00
Columns with 350 lbs of reinforcement per CY	—	CY	520.00	—	—	520.00
Columns with 200 lbs of reinforcement per CY	—	CY	795.00	—	—	795.00

Placing concrete columns with a 100 ton crawler-mounted crane and a welding machine at $1240 per day Figures in parentheses show the number of units placed per day. Use $8,500 as a minimum job charge.

	Craft@Hrs	Unit	Material	Labor	Equipment	Total
3 ton columns to 121' reach (29 per day)	H4@2.21	Ea	—	100.00	51.50	151.50
3 to 5 tons, to 89' reach (27 per day)	H4@2.37	Ea	—	107.00	55.20	162.20
5 to 7 tons, to 75' reach (25 per day)	H4@2.55	Ea	—	115.00	59.40	174.40
7 to 9 tons, to 65' reach (23 per day)	H4@2.79	Ea	—	126.00	65.00	191.00
9 to 11 tons, to 55' reach (22 per day)	H4@2.92	Ea	—	132.00	68.00	200.00
11 to 13 tons, to 50' reach (20 per day)	H4@3.20	Ea	—	145.00	74.60	219.60
13 to 15 tons, to 43' reach (18 per day)	H4@3.56	Ea	—	161.00	83.00	244.00

Precast concrete stairs Costs include concrete, reinforcing, embedded steel, nosing, and delivery to 40 miles. Use $500 as a minimum job charge. See placing costs below.

Stairs, 44" to 48" wide, 10' to 12' rise, "U" or "L" shape, including typical landings,

	Craft@Hrs	Unit	Material	Labor	Equipment	Total
Cost per each step (7" rise)	—	Ea	27.60	—	—	27.60

Placing precast concrete stairs with a 45 ton truck-mounted crane and a welding machine at $329 per day. Figures in parentheses show the number of stair flights placed per day. Use $6,000 as a minimum job charge.

	Craft@Hrs	Unit	Material	Labor	Equipment	Total
2 ton stairs to 82' reach (21 per day)	H4@3.05	Ea	—	138.00	22.30	160.30
2 to 3 tons, to 65' reach (19 per day)	H4@3.37	Ea	—	153.00	24.70	177.70
3 to 4 tons, to 57' reach (18 per day)	H4@3.54	Ea	—	160.00	25.90	185.90
4 to 5 tons, to 50' reach (17 per day)	H4@3.76	Ea	—	170.00	27.60	197.60
5 to 6 tons, to 44' reach (16 per day)	H4@4.00	Ea	—	181.00	29.30	210.30
6 to 7 tons, to 40' reach (15 per day)	H4@4.28	Ea	—	194.00	31.40	225.40
7 to 8 tons, to 36' reach (14 per day)	H4@4.56	Ea	—	206.00	33.40	239.40
8 to 9 tons, to 33' reach (10 per day)	H4@6.40	Ea	—	290.00	47.00	337.00
If a 35 ton crane can be used, deduct	—	%	—	—	-10.0	—

Precast double tees 8' wide, 24" deep. Design load is shown in pounds per square foot (PSF). Costs include lifting and placing but no slab topping. Equipment cost is $550 per day for a 45 ton truck crane and a welding machine. Use $6,000 as a minimum job charge.

	Craft@Hrs	Unit	Material	Labor	Equipment	Total
30' to 35' span, 115 PSF	H4@.019	SF	6.33	.86	.15	7.34
30' to 35' span, 141 PSF	H4@.019	SF	6.93	.86	.15	7.94
35' to 40' span, 78 PSF	H4@.018	SF	6.33	.82	.15	7.30
35' to 40' span, 98 to 143 PSF	H4@.018	SF	6.93	.82	.15	7.90
40' to 45' span, 53 PSF	H4@.017	SF	6.33	.77	.14	7.24
40' to 45' span, 69 to 104 PSF	H4@.017	SF	6.93	.77	.14	7.84
40' to 45' span, 134 PSF	H4@.017	SF	7.19	.77	.14	8.10
45' to 50' span, 77 PSF	H4@.016	SF	6.93	.72	.13	7.78
45' to 50' span, 101 PSF	H4@.016	SF	7.19	.72	.13	8.04

Concrete topping, based on 350 CY job 3,000 PSI design mix with lightweight aggregate, no forms or finishing included.

	Craft@Hrs	Unit	Material	Labor	Equipment	Total
By crane	H3@.868	CY	117.00	29.30	22.03	168.33
By pump	M2@.690	CY	117.00	23.60	13.67	154.27

	Craft@Hrs	Unit	Material	Labor	Equipment	Total
Hollow core roof planks, not including topping, 35 PSF live load, 10' to 15' span						
4" thick	H4@.018	SF	3.84	.82	.15	4.81
6" thick	H4@.018	SF	4.71	.82	.15	5.68
8" thick	H4@.020	SF	5.51	.91	.16	6.58
10" thick	H4@.026	SF	5.74	1.18	.21	7.13

Prestressed concrete support poles Costs shown are per linear foot for sizes and lengths shown. Equipment cost is $500 per day for a 45 ton truck crane. These costs do not include base or excavation. Lengths other than as shown will cost more per foot. Use $6,000 as a minimum job charge.

	Craft@Hrs	Unit	Material	Labor	Equipment	Total
10" x 10" poles with four 7/16" 270 KSI ultimate tendon strand strength						
10', 14', 20', 25' or 30' long	H4@.028	LF	10.00	1.27	.18	11.45
12" x 12" poles with six 7/16" 270 KSI ultimate tendon strand strength						
10', 14' or 20' long	H4@.028	LF	12.20	1.27	.18	13.65
25' or 30' long	H4@.028	LF	13.30	1.27	.18	14.75

Tilt-Up Concrete Construction Costs will be higher if access is not available from all sides or obstructions delay the work. Except as noted, no grading, compacting, engineering, design, permit or inspection fees are included.

Foundations and footings for tilt-up Costs assume normal soil conditions with no forming or shoring required and are based on concrete poured directly from the chute of a ready-mix truck. Soil excavated and not needed for backfill can be stored on site and used by others. For scheduling purposes, estimate that a crew of 5 can set forms for and pour 40 to 50 CY of concrete footings and foundations in an 8-hour day.

	Craft@Hrs	Unit	Material	Labor	Equipment	Total
Lay out foundation from known point on site, set stakes and mark for excavation.						
Per SF of floor slab	C8@.001	SF	—	.04	—	.04
Excavation with a 1/2 CY utility backhoe at $22.50 per hour						
27 CY per hour	S1@.075	CY	—	2.60	.84	3.44
Set and tie grade A60 reinforcing bars, typical foundation has 50 to 60 pounds of						
reinforcing bar per CY of concrete	RB@.008	Lb	.36	.31	—	.67
Place stringer supports for reinforcing bars (at 1.1 LF of support per LF of rebar), using 2" x 4" or 2" x 6" lumber at $525 per MBF before waste allowance. Includes typical stakes and hardware						
Based on 3 uses of supports	C8@.017	LF	.46	.62	—	1.08
(Inspection of reinforcing may be required before concrete is placed.)						
Make and place steel column anchor bolt templates (using 1.1 SF of plywood per SF of column base). Based on 2 uses of 3/4" plyform at $28.00 per 4' x 8' sheet ($.88 per SF) before waste allowance. If templates are furnished by the steel column fabricator, include only the labor cost to place the templates as shown						
Make and place templates	C8@.160	SF	.54	5.80	—	6.34
Place templates furnished by others	C8@.004	SF	—	.15	—	.15
Blockout form at foundation, step footings, columns, etc., per SF of contact area, using 2" x 6" lumber at $525 per MBF before allowance for waste. Includes typical stakes and hardware.						
Based on 3 uses of forms	C8@.018	SF	.38	.65	—	1.03
Column anchor bolts, usually 4 per column. J-hook bolt includes flat washer and two nuts						
1/2" x 6"	C8@.037	Ea	.55	1.34	—	1.89
3/4" x 12"	C8@.037	Ea	2.00	1.34	—	3.34
Concrete, 2,000 PSI design mix, at $58.50 per CY before 3% allowance for waste,						
Placed directly from chute	CL@.552	CY	60.30	16.70	.97	77.97
Remove, clean and stack						
reinforcing supports	CL@.020	LF	—	.61	—	.61
Dry pack non-shrink grout under column bases, at $18.50 per CF. Typical grout area is						
2' x 2' x 4" (1.33 CF of grout)	CM@.625	Ea	24.60	23.40	—	48.00
Panel support pads. (Supports panel when grout is placed.) One needed per panel. Based on						
3 CF of 2,000 PSI concrete per pad	C8@.141	Ea	6.50	5.11	—	11.61

	Craft@Hrs	Unit	Material	Labor	Equipment	Total
Backfill at foundation with 1/2 CY utility backhoe at $22.50 per hour						
90 CY per hour	S1@.022	CY	—	.76	.24	1.00
Dispose of excess soil on site,						
using backhoe	S1@.011	CY	—	.38	.12	.50
Concrete test cylinders, at $10 each, typically 5 required						
per 100 CY of concrete	—	CY	.50	—	—	.50

Floor slabs for tilt-up These costs assume that the site has been graded and compacted by others before slab work begins. The costs below include fine grading the site, placing a 2" layer of sand, a vapor barrier and a second 2" layer of sand over the barrier. Each layer of sand should be screeded level. Reinforcing may be either wire mesh or bars. When sand and vapor barrier are used, concrete may have to be pumped in place. If inspection of the reinforcing is required, the cost will usually be included in the building permit fee. Floor slabs are usually poured 6" deep in 14' to 18' wide strips with edge forms on both sides and ends. Edge forms are used as supports for the mechanical screed which consolidates the concrete with pneumatic vibrators as the surface is leveled. The screed is pulled forward by a winch and cable. Edge forms should have an offset edge to provide a groove which is filled by the adjacent slab. The slab perimeter should end 2' to 3' inside the finished building walls. This "pour strip" is filled in with concrete when the walls are up and braces have been removed. About 2 days after pouring, the slab should be saw-cut 1-1/2" to 2" deep and 1/4" wide each 20' both ways. This controls cracking as the slab cures. For scheduling purposes, estimate that a finishing crew of 10 workers can place and finish 10,000 to 12,000 square feet of slab in 8 hours.

	Craft@Hrs	Unit	Material	Labor	Equipment	Total
Fine grade the building pad with a utility tractor with blade and bucket, at $36.00 per hour						
2,500 SF per hour	S6@.001	SF	—	.03	.01	.04
Spread sand 4" deep in 2" layers, (1.3 tons is 1 cubic yard and covers 81 SF at 4" depth), using a utility tractor at $36.00 per hour.						
With sand at $15.00 per ton,						
60 SF per hour	S6@.006	SF	.24	.20	.07	.51
Vapor barrier, 6 mil polyethylene	CL@.001	SF	.04	.03	—	.07
Welded wire mesh, 6" x 6", W1.4 x W1.4						
Including 10% for overlap and waste	RB@.004	SF	.15	.16	—	.31
Reinforcing bars, grade 60, typically #4 bars on 2' centers each way						
With 1.05 LF bars per SF						
at $.18 per LF	RB@.004	SF	.19	.16	—	.35
Dowel out for #4 bars						
at construction joints	RB@.004	LF	.20	.16	—	.36
Edge forms with keyed joint, rented	F5@.013	LF	—	.49	2.00	2.49
Ready-mix concrete for the slab						
2,000 PSI, chute mix at $58.50 per CY and pump mix at $67.50 per CY before 3% allowance for waste						
Placed from the chute						
of the ready-mix truck	CL@.421	CY	60.30	12.70	.98	73.98
Pump mix, including pumping cost	M2@.442	CY	69.50	15.10	9.03	93.63
Finish the floor slab, using power trowels	CM@.003	SF	—	.11	.09	.20
Vibrating screed and accessories, (includes screeds, fuel, air compressor and hoses) rented						
Per SF of slab	—	SF	—	—	.13	.13
Curing compound, spray applied,						
$5.00 and 250 SF per gallon	CL@.001	SF	.02	.03	—	.05
Saw cut green concrete,						
1-1/2" to 2" deep	C8@.010	LF	—	.36	.25	.61
Strip, clean and stack edge forms	CL@.011	LF	—	.33	—	.33

	Craft@Hrs	Unit	Material	Labor	Equipment	Total

Ready-mix concrete for the pour strip,
2,000 PSI, chute mix at $58.50 per CY and pump mix at $67.50 per CY before 3% allowance for waste
Placed from the chute

	Craft@Hrs	Unit	Material	Labor	Equipment	Total
of the ready-mix truck	CL@.421	CY	60.30	12.70	.98	73.98
Pump mix, including pumping cost	M2@.442	CY	69.50	15.10	9.03	93.63
Finish the pour strip	CM@.004	SF	—	.15	.11	.26

Special inspection of the concrete pour (if required) at $50 per hour, (Use $200 as a minimum job charge.)

	Craft@Hrs	Unit	Material	Labor	Equipment	Total
Based on 10,000 SF in 8 hours	—	SF	—	—	—	.04

Clean and caulk the saw joints, based on 50 LF per hour and

	Craft@Hrs	Unit	Material	Labor	Equipment	Total
caulk at $30 and 1,500 LF per gallon	CL@.020	LF	.02	.61	—	.63
Clean slab (broom clean)	CL@.001	SF	—	.03	—	.03

Tilt-up wall panels Wall thickness is usually a nominal 6" (5-1/2" actual) or a nominal 8" (7-1/2"). Panel heights over 40' are uncommon. Typical panel width is 20'. When the floor area isn't large enough to form all wall panels on the floor slab, stacking will be required. Construction period will be longer and the costs higher than the figures shown below when panels must be poured and cured in stacked multiple layers. When calculating the volume of concrete required, the usual practice is to multiply overall panel dimensions (width, length and thickness) and deduct only for panel openings that exceed 25 square feet. When pilasters (thickened wall sections) are formed as part of the panel, add the cost of extra formwork and the additional concrete. Use the production rates that follow for scheduling a job.

Production rates for tilt-up

Lay out and form panels, place reinforcing bars and install embedded lifting devices:
2,000 to 3,000 SF of panel face (measured one side) per 8-hour day for a crew of 5. Note that reinforcing may have to be inspected before concrete is poured.

Place and finish concrete: 4,000 to 5,000 SF of panel per 8-hour day for a crew of 5.
Continuous inspection of the concrete pour by a licensed inspector may be required.
The minimum charge will usually be 4 hours at $50 per hour plus travel cost.

Install ledgers with embedded anchor bolts on panels before the concrete hardens:
150 to 200 LF of ledger per 8-hour day for a crew of 2.

Sack and patch exposed panel face before panel is erected:
2,000 to 3,000 SF of panel face per 8-hour day for a crew of 2.
Install panel braces before lifting panels:
30 to 40 braces per 8-hour day for a crew of 2. Panels usually need 3 or 4 braces.
Braces are usually rented for 30 days.

Tilt up and set panels 20 to 24 panels per 8-hour day for a crew of 11 (6 workers, the crane operator, oiler and 3 riggers). If the crew sets 22 panels averaging 400 SF each in an 8-hour day, the crew is setting 1,100 SF per hour. These figures include time for setting and aligning panels, drilling the floor slab for brace bolts, and securing the panel braces. Panel tilt-up will go slower if the crane can't work from the floor slab when lifting panels. Good planning of the panel lifting sequence and easy exit of the crane from the building interior will increase productivity. Size and weight of the panel has little influence on the setting time if the crane is capable of making the lift and is equipped with the right rigging equipment, spreader bars and strongbacks. Panels will usually have to cure for 14 days before lifting unless high early strength concrete is used.

Weld or bolt steel embedded in panels: 16 to 18 connections per 8-hour day for a crew of 2 working from ladders. Work more than 30 feet above the floor may take more time.

Sack and patch panel faces after erection: 400 SF of face per 8-hour day for a crew of 2.

	Craft@Hrs	Unit	Material	Labor	Equipment	Total

Tilt-up panel costs

Engineering fee, per panel, typical
| | — | Ea | — | — | — | 36.00 |

Fabricate and lay out forms for panel edges and blockouts, assumes 2 uses and includes typical hardware and chamfer

	Craft@Hrs	Unit	Material	Labor	Equipment	Total
2" x 6" at $515 per MBF	F5@.040	LF	.52	1.50	—	2.02
2" x 8" at $525 per MBF	F5@.040	LF	.70	1.50	—	2.20

Reveals (1" x 4" accent strips on exterior panel face), based on 1 use of lumber, nailed to slab, at $390 per MBF
| | C8@.018 | LF | .13 | .65 | — | .78 |

Clean and prepare the slab prior to placing the concrete for the panels, 1,250 SF per hour
| | CL@.001 | SF | — | .03 | — | .03 |

Releasing agent (bond breaker), sprayed on, 250 SF and $7.50 per gallon
| | CL@.001 | SF | .03 | .03 | — | .06 |

Place and tie grade 60 reinforcing bars and supports. Based on using #3 and #4 bars, including waste and laps. A typical panel has 80 to 100 pounds of reinforcing steel per CY of concrete in the panel

	Craft@Hrs	Unit	Material	Labor	Equipment	Total
Reinforcing per CY of concrete, at $.26 per lb	RB@.500	CY	26.00	19.50	—	45.50
Reinforcing per SF of 6" thick panel	RB@.009	SF	.48	.35	—	.83
Reinforcing per SF of 8" thick panel	RB@.012	SF	.65	.47	—	1.12

Ledger boards with anchor bolts. Based on 4" x 12" treated lumber at $790 per MBF and 3/4" x 12" bolts at $2.00 each placed each 2 feet, installed prior to lifting panel
| Typical ledgers and anchor bolts | C8@.353 | LF | 4.25 | 12.80 | — | 17.05 |

Embedded steel for pickup and brace point hardware. Usually 10 to 12 points per panel, quantity and type depend on panel size
| Embedded pickup & brace hardware, typical | C8@.001 | SF | .32 | .04 | — | .36 |

Note: Inspection of reinforcing may be required before concrete is placed.

Concrete, 3,000 PSI, placed direct from chute (at $64 per CY before 3% allowance for waste)

	Craft@Hrs	Unit	Material	Labor	Equipment	Total
6" thick panels (54 SF per CY)	CL@.008	SF	1.22	.24	.02	1.48
8" thick panels (40.5 SF per CY)	CL@.010	SF	1.63	.30	.02	1.95

Concrete, 3,000 PSI, pumped in place (at $71.00 per CY before 3% allowance for waste)

	Craft@Hrs	Unit	Material	Labor	Equipment	Total
6" thick panels (54 SF per CY)	M2@.008	SF	1.35	.27	.18	1.80
8" thick panels (40.5 SF per CY)	M2@.011	SF	1.81	.38	.22	2.41

Finish concrete panels, one face before lifting
| | P8@.004 | SF | — | .14 | .10 | .24 |

Continuous inspection of pour (at $50 per hour), if required, based on 5,000 to 6,000 SF being placed per 8-hour day
| Typical cost for continuous inspection of pour | — | SF | — | — | — | .07 |

Curing wall panels with spray-on curing compound, per SF of face, based on 250 SF and $5 per gallon
| | CL@.001 | SF | .02 | .03 | — | .05 |

Strip, clean and stack 2" edge forms
| | CL@.030 | LF | — | .91 | — | .91 |

Braces for panels

Install rented panel braces, based on an average of three braces per panel for panels averaging 400 to 500 SF each and rental cost of $12.00 per month each. See also, Remove panel braces.
| Typical cost for braces, per SF of panel | F5@.005 | SF | — | .19 | .08 | .27 |

Miscellaneous equipment rental, typical costs per SF of wall panel

	Craft@Hrs	Unit	Material	Labor	Equipment	Total
Rent power trowels and vibrators	—	SF	—	—	.04	.04
Rent compressors, hoses, rotohammers, fuel	—	SF	—	—	.05	.05

	Craft@Hrs	Unit	Material	Labor	Equipment	Total
Lifting panels into place, with crane and riggers, using a 140 ton truck crane						
Move crane on and off site, typical	—	LS	—	—	2,200.00	2,200.00
Hourly cost (4 hr minimum applies), per hour	T2@11.0	Hr	—	433.00	275.00	708.00
Set, brace and align panels, per SF of panel measured on one face, Based on 1,100 SF per hour	T2@.010	SF	—	.39	.25	.64
Allowance for shims, bolts for panel braces and miscellaneous hardware, labor cost is included with associated items						
Typical cost per SF of panel	—	SF	.25	—	—	.25
Weld or bolt panels together, typically 2 connections are required per panel, equipment is a welding machine at $58 per day						
Typical cost per panel connection	T3@.667	Ea	70.00	23.10	2.42	95.52
Continuous inspection of welding, if required, at $50 per hour ($200 minimum applies)						
Inspection of welds, per connection (3 per hour)	—	Ea	—	—	—	16.70
Set grout forms at exterior wall perimeter. Based on using 2" x 12" boards at $590 per MBF before allowance for waste. Costs shown include stakes and assume 3 uses of the forms						
Forms per linear foot (LF)	F5@.019	LF	.77	.71		1.48
Pour grout under panel bottoms. Place 2,000 PSI concrete (grout) directly from the chute of ready-mix truck (at $58.50 per CY before 3% allowance for waste),						
Typically 1.85 CY per 100 LF of wall	M2@.005	LF	1.11	.17	—	1.28
Sack and patch wall, per SF of area (measure area to be sacked and patched, both inside and outside),						
Per SF of wall face(s)	P8@.007	SF	.06	.24	—	.30
Caulking and backing strip for panel joints. Cost per LF for caulking both inside & outside face of joint,						
per linear foot (LF) of joint	PA@.055	LF	.20	2.04	—	2.24
Remove panel braces,						
stack & load onto trucks	CL@.168	Ea	—	5.09	—	5.09
Final cleanup & patch floor,						
per SF of floor	P8@.003	SF	.01	.10	—	.11
Additional costs for tilt-up panels, if required						
Architectural form liner, minimum cost	F5@.001	SF	1.17	.04	—	1.21
Architectural form liner, typical cost	F5@.003	SF	1.53	.11	—	1.64
Sandblast, light and medium, subcontract	—	SF	—	—	—	1.50
Sandblast, heavy, subcontract	—	SF	—	—	—	2.25
Truck dock bumpers, installation only	C8@2.00	Ea	—	72.50	—	72.50
Dock levelers, installation only	C8@7.46	Ea	—	271.00	—	271.00
Angle iron guard around door opening, 2" x 2" x 1/4", mild steel (4' lengths)	C8@.120	LF	2.00	4.35	—	6.35
Pipe guard at door opening, 6" black steel, concrete filled	C8@.281	LF	21.00	10.20	—	31.20
Embedded reglet at parapet wall, aluminum	F5@.019	LF	1.00	.71	—	1.71
Integral color in concrete, light colors, add to concrete Price per CY of concrete to be colored	—	CY	20.00	—	—	20.00
Backfill outside perimeter wall with a utility tractor at $29.00 per hour 90 CY per hour	S1@.022	CY	—	.76	.32	1.08

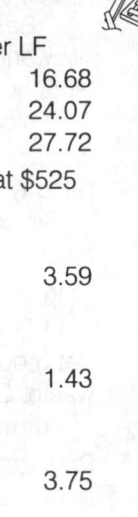

	Craft@Hrs	Unit	Material	Labor	Equipment	Total

Cast-in-place columns for tilt-up Concrete columns are often designed as part of the entryway of a tilt-up building or to support lintel panels over a glass storefront. Both round and square freestanding columns are used. They must be formed and poured before the lintel panels are erected. Round columns are formed with plastic-treated fiber tubes (Sonotube is one manufacturer) that are peeled off when the concrete has cured. Square columns are usually formed with plywood and 2" x 4" wood braces or adjustable steel column clamps.

Round plastic-lined fiber tube forms, set and aligned, based on 12' long tubes. Longer tubes will cost more per LF

	Craft@Hrs	Unit	Material	Labor	Equipment	Total
12" diameter tubes	M2@.207	LF	9.59	7.09	—	16.68
20" diameter tubes	M2@.250	LF	15.50	8.57	—	24.07
24" diameter tubes	M2@.275	LF	18.30	9.42	—	27.72

Rectangular plywood forms, using 1.1 SF of 3/4" plyform at $27.50 per 4' x 8' sheet ($.86 per SF), and 2" x 4" (at $525 per MBF) bracing 3' OC
Per SF of contact area

	Craft@Hrs	Unit	Material	Labor	Equipment	Total
based on 3 uses	F5@.083	SF	.48	3.11	—	3.59

Column braces, 2" x 4" at $575 per MBF, with 5 BF per LF of column height
Per LF of column height

	Craft@Hrs	Unit	Material	Labor	Equipment	Total
based on 3 uses	F5@.005	LF	1.24	.19	—	1.43

Column clamps rented at $4.50 for 30 days, 1 per LF of column height, cost per LF of column height
Per LF of column

	Craft@Hrs	Unit	Material	Labor	Equipment	Total
based on 2 uses a month	F5@.040	LF	2.25	1.50	—	3.75

Reinforcing bars, grade 60, placed and tied, typically 150 to 200 pounds per CY of concrete

	Craft@Hrs	Unit	Material	Labor	Equipment	Total
Cages, round or rectangular (at $.26/lb)	RB@1.50	CY	52.00	58.40	—	110.40
Vertical bars						
attached to cages (at $.26/lb)	RB@1.50	CY	52.00	58.40	—	110.40

Ready-mix concrete, 3,000 PSI pump mix (at $71.00 before 3% allowance for waste)

	Craft@Hrs	Unit	Material	Labor	Equipment	Total
Placed by pump	M2@.670	CY	73.10	23.00	14.90	111.00

Strip off fiber forms and dismantle bracing, per LF of height

	Craft@Hrs	Unit	Material	Labor	Equipment	Total
12" or 16" diameter	CL@.075	LF	—	2.27	—	2.27
24" diameter	CL@.095	LF	—	2.78	—	2.78

Strip wood forms and dismantle bracing

	Craft@Hrs	Unit	Material	Labor	Equipment	Total
Per SF of contact area, no salvage	CL@.011	SF	—	.33	—	.33
Clean and stack column clamps	CL@.098	Ea	—	2.97	—	2.97
Sack and patch concrete column face	P8@.011	SF	.10	.37	—	.47

Allowance for miscellaneous equipment rental (vibrators, compressors, hoses, etc.)

	Craft@Hrs	Unit	Material	Labor	Equipment	Total
Per CY of concrete in column	—	CY	—	—	2.50	2.50

Trash enclosures for tilt-up These usually consist of a 6" concrete slab measuring 5' x 10' and precast concrete tilt-up panels 6' high. A 6" ledger usually runs along the wall interior. Overall wall length is usually 20 feet. Cost for the trash enclosure foundation, floor slab and wall will be approximately the same as for the tilt-up foundation, floor slab and wall. The installed cost of a two-leaf steel gate measuring 6' high and 8' long overall will be about $350 including hardware.

Equipment Foundations Except as noted, no grading, compacting, engineering, design, permit or inspection fees are included. Costs assume normal soil conditions with no shoring or dewatering required and are based on concrete poured directly from the chute of a ready-mix truck. Costs will be higher if access is not available from all sides or if obstructions delay the work.

Labor costs include the time needed to prepare formwork sketches at the job site, measure for the forms, fabricate, erect, align and brace the forms, place and tie the reinforcing steel and embedded steel items, pour and finish the concrete and strip, clean and stack the forms.

Reinforcing steel costs assume an average of #3 and #4 bars at 12" on center each way. Weights shown include an allowance for waste and laps.

Form costs assume 3 uses of the forms and show the cost per use when a form is used on the same job without being totally disassembled. Normal cleaning and repairs are included. No salvage value is assumed. Costs listed are per square foot of form in contact with the concrete, assume a number #2 & Btr lumber price of $525 per MBF and a price of $880 per MSF for 3/4" plyform.

For scheduling purposes, estimate that a crew of 5 can set the forms and place and tie the reinforcing steel for and pour 8 to 10 CY of concrete equipment foundations in an 8-hour day.

Rule-of-thumb estimates The costs below are based on the volume of concrete in the foundation. If plans for the foundation have not been prepared, estimate the quantity of concrete required from the weight of the equipment using the rule-of-thumb estimating data below. Design considerations may alter the actual quantities required. Based on the type foundation required, use the cost per CY given in the detailed cost breakdown section which follows the table on the next page.

Estimating concrete quantity from the weight of equipment:

1. Rectangular pad type foundations —
   Boilers and similar equipment mounted at grade:
   Allow 1 CY per ton

   Centrifugal pumps and compressors, including weight of motor:
   Up to 2 HP, allow .5 CY per ton
   Over 2 to 10 HP, allow 1 CY per ton
   Over 10 to 30 HP, allow 2 CY per ton
   Over 30 HP, allow 3 CY per ton

   Reciprocating pumps and compressors, including weight of motor:
   Up to 2 HP, allow .75 CY per ton
   Over 2 to 10 HP, allow 1.5 CY per ton
   Over 10 to 30 HP, allow 2.5 CY per ton
   Over 30 HP, allow 4.5 CY per ton

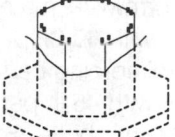

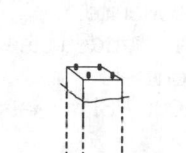

2. Rectangular pad and stem wall type foundations —
   Horizontal tanks or shell and tube heat exchangers, allow 6 CY per ton

3. Square pad and pedestal type foundations —
   Vertical tanks not over 3' in diameter, allow 6 CY per ton

4. Octagonal pad and pedestal type foundations —
   Vertical tanks over 3' in diameter, allow 4 CY per ton of empty weight

Rectangular pad type foundations For pumps, compressors, boilers and similar equipment mounted at grade. Form costs assume nails, stakes and form oil costing $.50 per square foot and 1.1 SF of 3/4" plyform plus 2.2 BF of lumber per square foot of contact surface. Costs shown are per cubic yard (CY) of concrete within the foundation.

	Craft@Hrs	Unit	Material	Labor	Equipment	Total
Lay out foundation from known point on site, set stakes and mark for excavation,						
with 1 CY per CY of concrete	C8@.147	CY	.05	5.33	—	5.38
Excavation with utility backhoe, at 28 CY and $29.00 per hour,						
with 1.5 CY per CY of concrete	S1@.108	CY	—	3.74	1.57	5.31
Formwork, material at $1.63 SF and labor at .150 manhours per SF						
with 15 SF per CY of concrete	F5@2.25	CY	24.50	84.30	—	108.80
Reinforcing bars, Grade 60, set and tied, material at $.26 per lb and labor .006 manhours per lb						
with 25 lb per CY of concrete	RB@.150	CY	6.50	5.84	—	12.34
Anchor bolts, sleeves, embedded steel, material at $1.67per lb and labor at .040 manhours per lb						
with 5 lb per CY of concrete	C8@.200	CY	8.35	7.25	—	15.60
Ready-mix concrete, 3,000 PSI mix at $64.00 per CY before allowance for 10% waste						
with 1 CY per CY of concrete	C8@.385	CY	70.40	14.00	—	84.40

	Craft@Hrs	Unit	Material	Labor	Equipment	Total
Concrete test cylinders including test reports, at $10 each						
with 5 per 100 CY of concrete	—	CY	.50	—	—	.50
Finish, sack and patch concrete, material at $.03 SF and labor at .010 manhours per SF						
with 15 SF per CY of concrete	P8@.150	CY	.45	5.08	—	5.53
Backfill around foundation with utility backhoe at 40 CY and $29.00 per hour,						
with .5 CY per CY of concrete	S1@.025	CY	—	.87	.36	1.23
Dispose of excess soil on site with utility backhoe at 80 CY and $29.00 per hour,						
with 1 CY per CY of concrete	S1@.025	CY	—	.87	.36	1.23
Rectangular pad type foundation						
Total cost per CY of concrete	**—@3.44**	**CY**	**110.75**	**127.28**	**2.29**	**240.32**

Rectangular pad and stem wall type foundations For horizontal cylindrical or shell and tube heat exchangers mounted above grade. Form costs assume nails, stakes and form oil costing $.60 per square foot and 1.1 SF of 3/4" plyform plus 4.5 BF of lumber per square foot of contact surface. Costs shown are per cubic yard (CY) of concrete within the foundations.

	Craft@Hrs	Unit	Material	Labor	Equipment	Total
Lay out foundation from known point on site, set stakes and mark for excavation,						
with 1 CY per CY of concrete	C8@.147	CY	.04	5.33	—	5.37
Excavation with utility backhoe, at 28 CY and $29.00 per hour,						
with 1 CY per CY of concrete	S1@.072	CY	—	2.50	1.04	3.54
Formwork, material at $2.66 SF and labor at .200 manhours per SF						
with 60 SF per CY of concrete	F5@12.0	CY	160.00	450.00	—	610.00
Reinforcing bars, Grade 60, set and tied, material at $.26 per lb and labor at .006 manhours per lb						
with 45 lb per CY of concrete	RB@.270	CY	11.70	10.50	—	22.20
Anchor bolts, sleeves, embedded steel, material at $1.67 per lb and labor at .040 manhours per lb						
with 5 lb per CY of concrete	C8@.200	CY	8.35	7.25	—	15.60
Ready-mix concrete, 3,000 PSI mix, at $64 before 10% waste allowance						
with 1 CY per CY	C8@.385	CY	70.40	14.00	—	84.40
Concrete test cylinders including test reports, at $10 each						
with 5 per 100 CY	—	CY	.50	—	—	.50
Finish, sack and patch concrete, material at $.03 per SF and labor at .010 manhours per SF						
with 60 SF per CY of concrete	P8@.600	CY	1.80	20.30	—	22.10
Backfill around foundation with utility backhoe, at 40 CY and $29.00 per hour,						
with .5 CY per CY of concrete	S1@.025	CY	—	.87	.36	1.23
Dispose of excess soil on site with utility backhoe, at 80 CY and $29.00 per hour,						
with .5 CY per CY of concrete	S1@.025	CY	—	.87	.36	1.23
Rectangular pad and stem wall type foundation						
Total cost per CY of concrete	**—@13.7**	**CY**	**252.79**	**511.62**	**1.76**	**766.17**

Square pad and pedestal type foundations For vertical cylindrical tanks not over 3' in diameter. Form costs assume nails, stakes and form oil costing $.50 per square foot and 1.1 SF of 3/4" plyform plus 3 BF of lumber per square foot of contact surface. Costs shown are per cubic yard (CY) of concrete within the foundations.

	Craft@Hrs	Unit	Material	Labor	Equipment	Total
Lay out foundation from known point on site, set stakes and mark for excavation,						
with 1 CY per CY of concrete	C8@.147	CY	.05	5.33	—	5.38
Excavation with utility backhoe, at 28 CY and $29.00 per hour,						
with 1 CY per CY of concrete	S1@.072	CY	—	2.50	1.04	3.54
Formwork, material at $1.64 SF and labor at .192 manhours per SF						
with 13 SF per CY of concrete	F5@2.50	CY	23.27	93.70	—	116.97
Reinforcing bars, Grade 60, set and tied, material at $.26 per lb and labor at .006 manhours per lb						
with 20 lb per CY of concrete	RB@.120	CY	5.20	4.67	—	9.87
Anchor bolts, sleeves, embedded steel, material at $1.67 per lb and labor at .040 manhours per lb						
with 2 lb per CY of concrete	C8@.080	CY	3.34	2.90	—	6.24
Ready-mix concrete, 3,000 PSI mix, at $64 before 10% waste allowance						
with 1 CY per CY	C8@.385	CY	70.40	14.00	—	84.40

	Craft@Hrs	Unit	Material	Labor	Equipment	Total
Concrete test cylinders including test reports, at $10 each						
with 5 per 100 CY	—	CY	.50	—	—	.50
Finish, sack and patch concrete, material at $.03 per SF and labor at .010 manhours per SF						
with 60 SF per CY of concrete	P8@.600	CY	1.80	20.30	—	22.10
Backfill around foundation with utility backhoe, at 40 CY and $29.00 per hour,						
with .5 CY per CY of concrete	S1@.025	CY	—	.87	.36	1.23
Dispose of excess soil on site with utility backhoe, at 80 CY and $29.00 per hour,						
with .5 CY per CY of concrete	S1@.025	CY	—	.87	.36	1.23
Square pad and pedestal type foundation						
Total cost per CY of concrete	**—@3.95**	**CY**	**104.56**	**145.14**	**1.76**	**251.46**

Octagonal pad and pedestal type foundations For vertical cylindrical tanks over 3' in diameter. Form costs assume nails, stakes and form oil costing $.50 per square foot and 1.1 SF of 3/4" plyform plus 4.5 BF of lumber per square foot of contact surface. Costs shown are per cubic yard (CY) of concrete within the foundation.

	Craft@Hrs	Unit	Material	Labor	Equipment	Total
Lay out foundation from known point on site, set stakes and mark for excavation,						
with 1 CY per CY of concrete	C8@.147	CY	.05	5.33	—	5.38
Excavation with utility backhoe, at 28 CY and $29.00 per hour,						
with 1 CY per CY of concrete	S1@.072	CY	—	2.50	1.04	3.54
Formwork, material at $1.85 SF and labor at .250 manhours per SF						
with 10 SF per CY of concrete	F5@2.50	CY	18.50	93.70	—	112.20
Reinforcing bars, Grade 60, set and tied, material at $.26 per lb and labor at .006 manhours per lb						
with 20 lb per CY of concrete	RB@.120	CY	5.20	4.67	—	9.87
Anchor bolts, sleeves, embedded steel, material at $1.70 per lb and labor at .040 manhours per lb						
with 2 lb per CY of concrete	C8@.080	CY	3.40	2.90	—	6.30
Ready-mix concrete, 3,000 PSI mix, at $64 before 10% waste allowance						
with 1 CY per CY of concrete	C8@.385	CY	70.40	14.00	—	84.40
Concrete test cylinders including test reports, at $10 ea						
with 5 per 100 CY of concrete	—	CY	.50	—	—	.50
Finish, sack and patch concrete, material at $.03 SF and labor at .010 manhours per SF						
with 60 SF per CY of concrete	P8@.600	CY	1.80	20.30	—	22.10
Backfill around foundation with utility backhoe, at 40 CY and $29.00 per hour,						
with .5 CY per CY of concrete	S1@.025	CY	—	.87	.36	1.23
Dispose of excess soil on site with utility backhoe, at 80 CY and $29.00 per hour,						
with .5 CY per CY of concrete	S1@.025	CY	—	.87	.36	1.23
Octagonal pad and pedestal type foundation						
Total cost per CY of concrete	**—@3.95**	**CY**	**99.85**	**145.14**	**1.76**	**246.75**

Cementitious Decks, Subcontract Typical costs including the subcontractor's overhead and profit.

	Craft@Hrs	Unit	Material	Labor	Equipment	Total
Insulating concrete deck, poured on existing substrate 1-1/2" thick						
Under 3,000 SF	—	SF	—	—	—	1.29
Over 3,000 SF to 10,000 SF	—	SF	—	—	—	1.12
Over 10,000 SF	—	SF	—	—	—	.94
Poured on existing metal deck, 2-5/8" thick, lightweight						
Concrete fill, trowel finish	—	SF	—	—	—	1.52
Gypsum decking,						
poured over form board, screed and float	—	SF	—	—	—	1.37
Vermiculite,						
screed and float, field mixed	—	SF	—	—	—	1.43

	Craft@Hrs	Unit	Material	Labor	Equipment	Total
Exterior concrete deck for walking service, troweled						
Lightweight, 2-5/8", 110 lbs per CF	—	SF	—	—	—	2.00
Add for 4,000 PSI mix	—	SF	—	—	—	.15
Add for 5,000 PSI mix	—	SF	—	—	—	.23
Pool deck at pool side,						
1/8" to 3/16" thick, colors, finished	—	SF	—	—	—	1.80
Fiber deck, tongue and groove, cementitious planks, complete						
2" thick	—	SF	—	—	—	1.90
2-1/2" thick	—	SF	—	—	—	2.31
3" thick	—	SF	—	—	—	2.52

Masonry 4

	Craft@Hrs	Unit	Material	Labor	Total

Brick Wall Assemblies Typical costs for smooth red clay brick walls, laid in running bond with 3/8" concave joints. These costs include the bricks, mortar for bricks and cavities, typical ladder type reinforcing, wall ties and normal waste. Foundations are not included. Wall thickness shown is the nominal size based on using the type bricks described. "Wythe" means the quantity of bricks in the thickness of the wall. Costs shown are per square foot (SF) of wall measured on one face. Deduct for openings over 10 SF in size. The names and dimensions of bricks can be expected to vary, depending on the manufacturer.

	Craft@Hrs	Unit	Material	Labor	Total
Standard bricks, 3-3/4" wide x 2-1/4" high x 8" long (at $292 per M)					
4" thick wall, single wythe, veneer facing	M1@.211	SF	2.48	7.09	9.57
8" thick wall, double wythe, cavity filled	M1@.464	SF	4.45	15.60	20.05
12" thick wall, triple wythe, cavity filled	M1@.696	SF	6.42	23.40	29.82
Modular bricks, 3" wide x 3-1/2" x 11-1/2" long (at $586 per M)					
3-1/2" thick wall, single wythe, veneer facing	M1@.156	SF	2.43	5.24	7.67
7-1/2" thick wall, double wythe, cavity filled	M1@.343	SF	4.35	11.50	15.85
11-1/2" thick wall, triple wythe, cavity filled	M1@.515	SF	6.28	17.30	23.58
Colonial bricks, 3" wide x 3-1/2" x 10" long (at $267 per M)					
3-1/2" thick wall, single wythe, veneer facing	M1@.177	SF	1.51	5.95	7.46
7-1/2" thick wall, double wythe, cavity filled	M1@.389	SF	2.52	13.10	15.62
11-1/2" thick wall, triple wythe, cavity filled	M1@.584	SF	3.53	19.60	23.13
Add to any brick wall assembly above for:					
American bond or stacked bond	—	%	5.0	10.0	—
Basketweave bond or soldier bond	—	%	10.0	10.0	—
Flemish bond	—	%	10.0	15.0	—
Herringbone pattern	—	%	15.0	15.0	—
Institutional inspection	—	%	—	20.0	—
Short runs or cut-up work	—	%	5.0	15.0	—

Chemical Resistant Brick or Tile, Subcontract (also known as acid-proof brick).Typical prices for chemical resistant brick or tile meeting ASTM C-279, types I, II, and III requirements. Installed with 1/8" thick bed and mortar joints over a membrane of either elastomeric bonded to the floors or walls, sheet membrane with a special seam treatment, squeegee hot-applied asphalt, cold-applied adhesive troweled on, or special spray-applied material. The setting bed and joint materials can be either 100% carbon-filled furan or phenolic, sodium or potassium silicates, silica-filled epoxy, polyester, vinyl ester or other special mortars. Expansion joints are required in floor construction, particularly where construction joints exist in the concrete. Additional expansion joints may be required, depending on the service conditions to which the floor will be subjected. Brick or tile will be either 3-3/4" x 8" face, 1-1/8" thick (splits); 3-3/4" x 8" face, 2-1/4" thick (singles); or 4" x 8" face, 1-3/8" thick (pavers). Costs are based on a 2,000 SF job. For smaller projects, increase the cost 5% for each 500 SF or portion of 500 SF less than 2,000 SF.

	Craft@Hrs	Unit	Material	Labor	Total

For scheduling purposes estimate that a crew of 4 can install 75 to 100 SF of membrane and brick in an 8-hour day. These costs assume that the membrane is laid on a suitable surface prepared by others and include the subcontractor's overhead and profit.

Typical installed costs for chemical resistant brick or tile (acid-proof brick), installed

	Craft@Hrs	Unit	Material	Labor	Total
Less complex jobs	—	SF	—	—	22.50
More complex jobs	—	SF	—	—	27.50
Minimum job cost	—	LS	—	—	15,300.00

Quarry Tile Unglazed natural red tile set in a portland cement bed with mortar joints.

	Craft@Hrs	Unit	Material	Labor	Total
Quarry floor tile					
4" x 4" x 1/2", 1/8" straight joints	M1@.112	SF	2.86	3.76	6.62
6" x 6" x 1/2", 1/4" straight joints	M1@.101	SF	2.70	3.39	6.09
6" x 6" x 3/4", 1/4" straight joints	M1@.105	SF	3.52	3.53	7.05
8" x 8" x 3/4", 3/8" straight joints	M1@.095	SF	3.32	3.19	6.51
8" x 8" x 1/2", 1/4" hexagon joints	M1@.112	SF	3.77	3.76	7.53
Quarry wall tile					
4" x 4" x 1/2", 1/8" straight joints	M1@.125	SF	2.86	4.20	7.06
6" x 6" x 3/4", 1/4" straight joints	M1@.122	SF	3.52	4.10	7.62
Quarry tile stair treads					
6" x 6" x 3/4", 12" wide tread	M1@.122	LF	3.88	4.10	7.98
Quarry tile window sill					
6" x 6" x 3/4" tile on 6" wide sill	M1@.078	LF	3.72	2.62	6.34
Quarry tile trim or cove base					
5" x 6" x 1/2" straight top	M1@.083	LF	2.86	2.79	5.65
6" x 6" x 3/4" round top	M1@.087	LF	3.32	2.92	6.24
Deduct for tile set in epoxy bed without grout joints	—	SF	—	-1.87	-1.87
Add for tile set in epoxy bed with grout joints	—	SF	1.12	—	1.12

Pavers and Floor Tile

	Craft@Hrs	Unit	Material	Labor	Total
Brick, excluding platform cost					
Plate, glazed	M1@.114	SF	2.35	3.83	6.18
Laid in sand	M1@.089	SF	1.79	2.99	4.78
Pavers, on concrete, grouted	M1@.107	SF	3.16	3.60	6.76
Add for special patterns	—	%	25.0	35.0	—
Slate	M1@.099	SF	2.75	3.33	6.08
Terrazzo tiles, 1/4" thick					
Standard	M1@.078	SF	4.44	2.62	7.06
Granite	M1@.085	SF	7.24	2.86	10.10
Brick steps	M1@.186	SF	2.81	6.25	9.06

Interlocking Paving Stones These costs include fine sand to fill joints and machine vibration, but no excavation.

	Craft@Hrs	Unit	Material	Labor	Total
Interlocking pavers, rectangular, 60mm thick, Orion	M1@.083	SF	1.21	2.79	4.00
Interlocking pavers, hexagonal, 80mm thick, Omni	M1@.085	SF	1.47	2.86	4.33
Interlocking pavers, multi-angle, 80mm thick, Venus	M1@.081	SF	1.26	2.72	3.98
Concrete masonry grid pavers (erosion control)	M1@.061	SF	2.11	2.05	4.16
Add for a 1" sand cushion	M1@.001	SF	.12	.03	.15
Add for a 2" sand cushion	M1@.002	SF	.17	.07	.24
Deduct for over 5,000 SF	—	%	-3.0	-7.0	—

	Craft@Hrs	Unit	Material	Labor	Total

Concrete Block Wall Assemblies Typical costs for standard natural gray medium weight masonry block walls including blocks, mortar, typical reinforcing and normal waste. Foundations are not included.

Walls constructed with 8" x 16" blocks laid in running bond

	Craft@Hrs	Unit	Material	Labor	Total
4" thick wall	M1@.090	SF	.99	3.02	4.01
6" thick wall	M1@.100	SF	1.18	3.36	4.54
8" thick wall	M1@.120	SF	1.43	4.03	5.46
12" thick wall	M1@.150	SF	2.10	5.04	7.14

Additional costs for concrete block wall assemblies

Add for grouting concrete block cores at 36" intervals, poured by hand

	Craft@Hrs	Unit	Material	Labor	Total
4" thick wall	M1@.007	SF	.14	.24	.38
6" thick wall	M1@.007	SF	.22	.24	.46
8" thick wall	M1@.007	SF	.29	.24	.53
12" thick wall	M1@.010	SF	.43	.34	.77

Add for 2" thick caps, natural gray concrete

	Craft@Hrs	Unit	Material	Labor	Total
4" thick wall	M1@.027	LF	.60	.91	1.51
6" thick wall	M1@.038	LF	.72	1.28	2.00
8" thick wall	M1@.046	LF	.84	1.55	2.39
12" thick wall	M1@.065	LF	1.01	2.18	3.19

Add for detailed block, 3/8" score

	Craft@Hrs	Unit	Material	Labor	Total
Single score, one side	—	SF	.47	—	.47
Single score, two sides	—	SF	.51	—	.51
Multi-scores, one side	—	SF	.66	—	.66
Multi-scores, two sides	—	SF	.95	—	.95

Add for color block, any of above

	Craft@Hrs	Unit	Material	Labor	Total
Light colors	—	%	12.0	—	—
Medium colors	—	%	18.0	—	—
Dark colors	—	%	25.0	—	—
Add for other than running bond	—	%	—	20.0	—

Glazed Concrete Block Prices shown in parentheses are per block. Costs per square foot include allowance for mortar and waste. 8" high x 16" long, 3/8" joints, no reinforcing, grout or foundations included. Based on 1.05 blocks per square foot (SF).

Glazed 1 side

	Craft@Hrs	Unit	Material	Labor	Total
2" wide ($6.50 per block)	M1@.102	SF	7.50	3.43	10.93
4" wide ($6.60 per block)	M1@.107	SF	7.62	3.60	11.22
6" wide ($6.80 per block)	M1@.111	SF	7.85	3.73	11.58
8" wide ($7.00 per block)	M1@.118	SF	8.08	3.97	12.05

Glazed 2 sides

	Craft@Hrs	Unit	Material	Labor	Total
4" wide ($9.90 per block)	M1@.122	SF	11.40	4.10	15.50
6" wide ($10.50 per block)	M1@.127	SF	12.10	4.27	16.37
8" wide ($10.70 per block)	M1@.136	SF	12.40	4.57	16.97

Glazed cove base, glazed 1 side, 8" high

	Craft@Hrs	Unit	Material	Labor	Total
4" length ($7.50 per block)	M1@.127	LF	22.50	4.27	26.77
6" length ($8.50 per block)	M1@.137	LF	17.00	4.60	21.60
8" length ($8.70 per block)	M1@.147	LF	13.10	4.94	18.04

Decorative Concrete Block Prices shown in parentheses are per block. Costs per square foot include allowance for mortar and waste.

Regular weight units, 1.05 blocks per SF.

Split face hollow block, 8" high x 16" long, 3/8" joints. No foundations, grout or reinforcing included

	Craft@Hrs	Unit	Material	Labor	Total
4" wide ($.99 per block)	M1@.110	SF	1.11	3.70	4.81
6" wide ($1.15 per block)	M1@.121	SF	1.29	4.07	5.36
8" wide ($1.32 per block)	M1@.133	SF	1.48	4.47	5.95

	Craft@Hrs	Unit	Material	Labor	Total
10" wide ($2.04 per block)	M1@.140	SF	2.28	4.70	6.98
12" wide ($1.98 per block)	M1@.153	SF	2.22	5.14	7.36

Split rib (combed)hollow block, 8" high x 16" long, 3/8" joints. No foundations, grout or reinforcing included

	Craft@Hrs	Unit	Material	Labor	Total
4" wide ($.94 per block)	M1@.107	SF	1.05	3.60	4.65
6" wide ($1.00 per block)	M1@.115	SF	1.12	3.86	4.98
8" wide ($1.27 per block)	M1@.121	SF	1.42	4.07	5.49
10" wide ($1.82 per block)	M1@.130	SF	2.03	4.37	6.40
12" wide ($1.76 per block)	M1@.146	SF	1.97	4.91	6.88

Screen block, 3/8" joints, no foundations or reinforcing included

	Craft@Hrs	Unit	Material	Labor	Total
6" x 6" x 4" wide (4 per SF at $.52 each)	M1@.210	SF	2.23	7.06	9.29
8" x 8" x 4" wide (2.25 per SF at $1.29)	M1@.140	SF	3.11	4.70	7.81
12" x 12" x 4" wide (1 per SF at $1.60 each)	M1@.127	SF	1.71	4.27	5.98
8" x 16" x 8" wide (1.11 per SF at $1.50 each)	M1@.115	SF	1.78	3.86	5.64

Acoustical slotted block, by rated Noise Reduction Coefficient (NRC), no foundation or reinforcing included

	Craft@Hrs	Unit	Material	Labor	Total
4" wide, with metal baffle, NRC, .45 to .55	M1@.121	SF	2.78	4.07	6.85
6" wide, with metal baffle, NRC, .45 to .55	M1@.130	SF	3.30	4.37	7.67
8" wide, with metal baffle, NRC, .45 to .55	M1@.121	SF	4.10	4.07	8.17
8" wide, with fiber filler, NRC, .65 to .75	M1@.143	SF	5.30	4.81	10.11

Concrete Block Bond Beams For door and window openings, no grout or reinforcing included.
Cost per linear foot of bond beam block

	Craft@Hrs	Unit	Material	Labor	Total
6" wide, 8" high, 16" long	M1@.068	LF	.57	2.28	2.85
8" wide, 8" high, 16" long	M1@.072	LF	.71	2.42	3.13
10" wide, 8" high, 16" long	M1@.079	LF	1.23	2.65	3.88
12" wide, 8" high, 16" long	M1@.089	LF	1.19	2.99	4.18
12" wide, 8" high double bond beam, 16" long	M1@.102	LF	1.19	3.43	4.62
8" wide, 16" high, 16" long	M1@.127	LF	1.99	4.27	6.26

Grout for Concrete Block Grout for block cores, type "S" at $2.30 per cubic foot, before 3% allowance for waste.
Costs shown are based on 2 cores per 16" block.
Grout per SF of wall, all cells filled, pumped in place and rodded

	Craft@Hrs	Unit	Material	Labor	Total
4" wide block, .11 CF per SF	M1@.018	SF	.26	.60	.86
6" wide block, .22 CF per SF	M1@.025	SF	.51	.84	1.35
8" wide block, .32 CF per SF	M1@.035	SF	.78	1.18	1.96
10" wide block, .45 CF per SF	M1@.035	SF	1.10	1.18	2.28
12" wide block, .59 CF per SF	M1@.039	SF	1.45	1.31	2.76

Grout per LF of cell filled measured vertically

	Craft@Hrs	Unit	Material	Labor	Total
4" wide block, .06 CF per SF	M1@.010	LF	.14	.34	.48
6" wide block, .11 CF per SF	M1@.015	LF	.26	.50	.76
8" wide block, .16 CF per SF	M1@.021	LF	.39	.71	1.10
10" wide block, .23 CF per SF	M1@.022	LF	.56	.74	1.30
12" wide block, .30 CF per SF	M1@.024	LF	.73	.81	1.54
Add for type M mortar (2,500 PSI)	—	%	10.0	—	—

Grout for bond beams, 3,000 PSI concrete, by block size and linear foot of beam, no block or reinforcing included, based on block cavity at 50% of block volume

	Craft@Hrs	Unit	Material	Labor	Total
8" high x 4" wide, .10 CF per LF	M1@.005	LF	.28	.17	.45
8" high x 6" wide, .15 CF per LF	M1@.010	LF	.41	.34	.75
8" high x 8" wide, .22 CF per LF	M1@.017	LF	.59	.57	1.16
8" high x 10" wide, .30 CF per LF	M1@.023	LF	.82	.77	1.59
8" high x 12" wide, .40 CF per LF	M1@.032	LF	1.08	1.08	2.16
16" high x 8" wide, .44 CF per LF	M1@.061	LF	1.20	2.05	3.25
16" high x 10" wide, .60 CF per LF	M1@.058	LF	1.60	1.95	3.55
16" high x 12" wide, .80 CF per LF	M1@.079	LF	2.15	2.65	4.80

	Craft@Hrs	Unit	Material	Labor	Total
Grout hollow metal door frame in a masonry wall					
Single door	M1@.668	Ea	6.00	22.40	28.40
Double door	M1@.851	Ea	8.00	28.60	36.60
Additional Costs for Concrete Block					
Sill blocks	M1@.087	LF	1.40	2.92	4.32
Cutting blocks	M1@.137	LF	—	4.60	4.60
Stack bond	—	%	5.0	10.0	—
Add for cut-up jobs	—	%	5.0	5.0	—
Add for institutional grade work	—	%	5.0	15.0	—
Add for colored block	—	%	15.0	—	—

Masonry Reinforcing and Flashing Reinforcing bars for concrete block, ASTM A615 grade 60 bars, cost per linear foot including 10% overlap and waste.

	Craft@Hrs	Unit	Material	Labor	Total
#3 bars (3/8", 5,319 LF per ton), horizontal	M1@.005	LF	.16	.17	.33
#3 bars placed vertically	M1@.006	LF	.16	.20	.36
#3 galvanized bars, horizontal	M1@.005	LF	.28	.17	.45
#3 galvanized bars placed vertically	M1@.006	LF	.28	.20	.48
#4 bars (1/2", 2,994 LF per ton), horizontal	M1@.008	LF	.18	.27	.45
#4 bars placed vertically	M1@.010	LF	.18	.34	.52
#4 galvanized bars, horizontal	M1@.008	LF	.39	.27	.66
#4 galvanized bars placed vertically	M1@.010	LF	.39	.34	.73
#5 bars (5/8", 1,918 LF per ton), horizontal	M1@.009	LF	.27	.30	.57
#5 bars placed vertically	M1@.012	LF	.27	.40	.67
#5 galvanized bars, horizontal	M1@.009	LF	.59	.30	.89
#5 galvanized bars placed vertically	M1@.012	LF	.59	.40	.99
#6 bars (3/4", 1,332 LF per ton), horizontal	M1@.010	LF	.39	.34	.73
#6 bars placed vertically	M1@.013	LF	.39	.44	.83
#6 galvanized bars, horizontal	M1@.014	LF	.86	.47	1.33
#6 galvanized bars placed vertically	M1@.014	LF	.86	.47	1.33
Wall ties					
Rectangular, galvanized	M1@.005	Ea	.17	.17	.34
Rectangular, copper coated	M1@.005	Ea	.19	.17	.36
Cavity "Z" type, galvanized	M1@.005	Ea	.16	.17	.33
Cavity "Z" type, copper coated	M1@.005	Ea	.17	.17	.34
Masonry anchors, steel, including bolts					
3/16" x 18" long	M1@.173	Ea	1.92	5.81	7.73
3/16" x 24" long	M1@.192	Ea	2.42	6.45	8.87
1/4" x 18" long	M1@.216	Ea	2.47	7.26	9.73
1/4" x 24" long	M1@.249	Ea	3.02	8.37	11.39
Wall reinforcing					
Truss type, plain					
4" wall	M1@.002	LF	.16	.07	.23
6" wall	M1@.002	LF	.17	.07	.24
8" wall	M1@.003	LF	.21	.10	.31
10" wall	M1@.003	LF	.23	.10	.33
12" wall	M1@.004	LF	.24	.13	.37
Truss type, galvanized					
4" wall	M1@.002	LF	.20	.07	.27
6" wall	M1@.002	LF	.22	.07	.29
8" wall	M1@.003	LF	.26	.10	.36
10" wall	M1@.003	LF	.28	.10	.38
12" wall	M1@.004	LF	.31	.13	.44
Ladder type, galvanized, 8" wide	M1@.003	LF	.18	.10	.28

	Craft@Hrs	Unit	Material	Labor	Total
Control joints					
Cross-shaped PVC	M1@.024	LF	2.20	.81	3.01
8" wide PVC	M1@.047	LF	2.78	1.58	4.36
Closed cell 1/2" joint filler	M1@.012	LF	.28	.40	.68
Closed cell 3/4" joint filler	M1@.012	LF	.56	.40	.96
Through the wall flashing					
5 ounce copper	M1@.023	SF	2.18	.77	2.95
.030" elastomeric sheeting	M1@.038	SF	.48	1.28	1.76

Expansion Shields Masonry anchors set in block or concrete including drilling. With nut and washer. Length indicates drilled depth.

	Craft@Hrs	Unit	Material	Labor	Total
One wedge anchors, 1/4" diameter					
1-3/4" long	CL@.079	Ea	.53	2.39	2.92
2-1/4" long	CL@.114	Ea	.63	3.45	4.08
3-1/4" long	CL@.171	Ea	.85	5.18	6.03
One wedge anchors, 1/2" diameter					
4-1/4" long	CL@.227	Ea	1.38	6.87	8.25
5-1/2" long	CL@.295	Ea	1.81	8.93	10.74
7" long	CL@.409	Ea	2.22	12.40	14.62
One wedge anchors, 3/4" diameter					
5-1/2" long	CL@.286	Ea	3.71	8.66	12.37
8-1/2" long	CL@.497	Ea	5.30	15.00	20.30
10" long	CL@.692	Ea	6.26	20.90	27.16
One wedge anchors, 1" diameter					
6" long	CL@.295	Ea	9.12	8.93	18.05
9" long	CL@.515	Ea	11.20	15.60	26.80
12" long	CL@.689	Ea	13.00	20.90	33.90
Flush self-drilling anchors					
1/4" x 1-1/4"	CL@.094	Ea	.53	2.85	3.38
5/16" x 1-1/4"	CL@.108	Ea	.63	3.27	3.90
3/8" x 1-1/2"	CL@.121	Ea	.85	3.66	4.51
1/2" x 1-1/2"	CL@.159	Ea	1.28	4.81	6.09
5/8" x 2-3/8"	CL@.205	Ea	2.02	6.20	8.22
3/4" x 2-3/8"	CL@.257	Ea	3.28	7.78	11.06

Clay Backing Tile Load bearing, 12" x 12".

	Craft@Hrs	Unit	Material	Labor	Total
4" thick	M1@.066	SF	3.16	2.22	5.38
6" thick	M1@.079	SF	3.57	2.65	6.22
8" thick	M1@.091	SF	4.18	3.06	7.24
Deduct for non-load bearing tile	—	%	-8.0	-10.0	—

Structural Glazed Tile Heavy-duty, fire-safe, structural glazed facing tile for interior surfaces. No foundations or reinforcing included.

	Craft@Hrs	Unit	Material	Labor	Total
Glazed one side, all colors, 5-1/3" x 12"					
2" thick	M1@.125	SF	3.59	4.20	7.79
4" thick	M1@.136	SF	4.32	4.57	8.89
6" thick	M1@.141	SF	6.40	4.74	11.14
8" thick	M1@.172	SF	8.14	5.78	13.92
Glazed one side, all colors, 8" x 16" block					
2" thick	M1@.072	SF	4.04	2.42	6.46
4" thick	M1@.072	SF	4.32	2.42	6.74
6" thick	M1@.079	SF	5.95	2.65	8.60
8" thick	M1@.085	SF	6.96	2.86	9.82

	Craft@Hrs	Unit	Material	Labor	Total

Acoustical glazed facing tile, with small holes on a smooth ceramic glazed facing, and one mineral fiberglass sound batt located in the hollow core behind perforated face

	Craft@Hrs	Unit	Material	Labor	Total
8" x 8" face size unit, 4" thick	M1@.136	SF	7.07	4.57	11.64
8" x 16" face size unit, 4" thick	M1@.141	SF	6.28	4.74	11.02

Reinforced glazed security tile, with vertical cores and knockouts for horizontal reinforcing, 8" x 16" block

	Craft@Hrs	Unit	Material	Labor	Total
4" thick	M1@.092	SF	5.67	3.09	8.76
6" thick	M1@.095	SF	7.35	3.19	10.54
8" thick	M1@.101	SF	8.86	3.39	12.25

Add to 8" security tile for horizontal and vertical reinforcing and grout cores

	Craft@Hrs	Unit	Material	Labor	Total
to meet high security standards	M1@.047	SF	2.70	1.58	4.28
Add for glazed 2 sides	—	%	55.0	10.0	—
Base, glazed 1 side	M1@.190	LF	5.50	6.38	11.88
Cap, glazed 2 sides	M1@.193	LF	8.86	6.49	15.35
Deduct for large areas	—	%	-5.0	-5.0	—
Add for institutional inspection	—	%	—	10.0	—

Glass Block Clear 3-7/8" thick block installed in walls to 8' high. Costs shown include 1/4" mortar joints, caulking, wall ties and 6% waste. Heights over 8' to 10' require additional scaffolding and handling and may increase labor costs up to 40%. Mortar is 1 part Portland cement, 1/2 part lime, and 4 parts sand, plus integral waterproofer. Costs for continuous panel reinforcing in horizontal joints are listed below.

Flat glass block, under 1,000 SF job

	Craft@Hrs	Unit	Material	Labor	Total
6" x 6", 4 per SF, $3.40 each	M1@.323	SF	15.80	10.90	26.70
8" x 8", 2.25 per SF, $3.90 each	M1@.245	SF	10.20	8.23	18.43
12" x 12", 1 per SF, $10.40 each	M1@.202	SF	12.10	6.79	18.89
Deduct for jobs over 1,000 SF to 5,000 SF	—	%	-4.0	-4.0	—
Deduct for jobs over 5,000 SF	—	%	-10.0	-10.0	—
Deduct for Thinline interior glass block	—	%	-10.0	—	—
Add for thermal control fiberglass inserts	—	%	5.0	5.0	—

Replacing 8" x 8" x 4" clear glass block, 10 unit minimum,

	Craft@Hrs	Unit	Material	Labor	Total
Per block replaced	M1@.851	Ea	15.00	28.60	43.60

Additional costs for glass block

	Craft@Hrs	Unit	Material	Labor	Total
Fiberglass or polymeric expansion joints	—	LF	.67	—	.67
Reinforcing steel mesh	—	LF	.46	—	.46
Wall anchors, 2' long					
Used where blocks aren't set into a wall	—	Ea	2.50	—	2.50
Asphalt emulsion, 600 LF per gallon	—	Gal	6.00	—	6.00
Waterproof sealant, 180 LF per gallon	—	Gal	8.00	—	8.00
Cleaning block after installation, using sponge and water only,					
Per SF cleaned	CL@.023	SF	—	.70	.70

Stone Work The delivered price of stone can be expected to vary widely. These costs include mortar, based on 1/2" joints, but no scaffolding. Figures in parentheses show typical quantities installed per 8-hour day by a crew of two.

Rough stone veneer, 4", placed over stud wall

	Craft@Hrs	Unit	Material	Labor	Total
Lava stone (79 SF)	M4@.202	SF	5.65	6.68	12.33
Rubble stone (40 SF)	M4@.216	SF	5.85	7.15	13.00
Most common veneer (60 SF)	M4@.210	SF	5.25	6.95	12.20

Cut stone,

	Craft@Hrs	Unit	Material	Labor	Total
Thinset granite tile, 1/2" thick (120 SF)	M4@.333	SF	12.50	11.00	23.50
Granite, 7/8" thick, interior (70 SF)	M4@.333	SF	17.00	11.00	28.00
Granite, 1-1/4" thick, exterior (60 SF)	M4@.333	SF	20.00	11.00	31.00
Limestone, 2" thick (70 SF)	M4@.333	SF	12.00	11.00	23.00
Limestone, 3" thick (60 SF)	M4@.333	SF	16.00	11.00	27.00
Marble, 7/8" thick, interior (70 SF)	M4@.333	SF	17.00	11.00	28.00

	Craft@Hrs	Unit	Material	Labor	Total
Marble, 1-1/4" thick, interior (70 SF)	M4@.333	SF	20.00	11.00	31.00
Sandstone, 2" thick (70 SF)	M4@.444	SF	11.20	14.70	25.90
Sandstone, 3" thick (60 SF)	M4@.444	SF	12.30	14.70	27.00
Add for honed finish	—	SF	2.00	—	2.00
Add for polished finish	—	SF	3.50	—	3.50
Marble specialties, typical prices					
Floors, 7/8" thick, exterior (70 SF)	M1@.333	SF	15.30	11.20	26.50
Thresholds, 1-1/4" thick (70 LF)	M1@.239	LF	9.18	8.03	17.21
Bases, 7/8" x 6" (51 LF)	M1@.317	LF	8.16	10.70	18.86
Columns, plain (15 CF)	M1@1.10	CF	94.90	37.00	131.90
Columns, fluted (13 CF)	M1@1.26	CF	133.00	42.30	175.30
Window stools, 5" x 1" (60 LF)	M1@.268	LF	10.00	9.01	19.01
Toilet partitions (1.5 Ea)	M1@10.9	Ea	530.00	366.00	896.00
Stair treads, 1-1/4" x 11" (50 LF)	M1@.323	LF	22.40	10.90	33.30
Limestone, rough cut large blocks (39 CF)	M1@.409	CF	27.50	13.70	41.20
Manufactured stone veneer, including mortar, Cultured Stone®					
Prefitted types (160 SF)	M1@.100	SF	4.35	3.36	7.71
Brick veneer type (120 SF)	M1@.133	SF	3.90	4.47	8.37
Random cast type (120 SF)	M1@.133	SF	4.05	4.47	8.52

Masonry Lintels and Coping

	Craft@Hrs	Unit	Material	Labor	Total
Reinforced precast concrete lintels					
4" wide x 8" high, to 6'6" long	M1@.073	LF	4.68	2.45	7.13
6" wide x 8" high, to 6'6" long	M1@.078	LF	8.37	2.62	10.99
8" wide x 8" high, to 6'6" long	M1@.081	LF	8.71	2.72	11.43
10" wide x 8" high, to 6'6" long	M1@.091	LF	11.30	3.06	14.36
4" wide x 8" high, over 6'6" to 10' long	M1@.073	LF	8.81	2.45	11.26
6" wide x 8" high, over 6'6" to 10' long	M1@.077	LF	10.20	2.59	12.79
8" wide x 8" high, over 6'6" to 10' long	M1@.081	LF	10.80	2.72	13.52
10" wide x 8" high, over 6'6" to 10' long	M1@.091	LF	14.90	3.06	17.96
Precast concrete coping, 5" average thickness, 4' to 8' long					
10" wide, gray concrete	M1@.190	LF	7.81	6.38	14.19
12" wide, gray concrete	M1@.200	LF	8.81	6.72	15.53
12" wide, white concrete	M1@.226	LF	21.60	7.59	29.19

Cleaning and Pointing Masonry These costs assume the masonry surface is in fair to good condition with no unusual damage. Add the cost of protecting adjacent surfaces such as trim or the base of the wall and the cost of scaffolding, if required. Labor required to presoak or saturate the area cleaned is included in the labor cost. Work more than 12' above floor level will increase costs.

	Craft@Hrs	Unit	Material	Labor	Total
Brushing (hand cleaning) masonry, includes the cost of detergent or chemical solution					
Light cleanup (100 SF per manhour)	M1@.010	SF	.02	.34	.36
Medium scrub (75 SF per manhour)	M1@.013	SF	.03	.44	.47
Heavy (50 SF per manhour)	M1@.020	SF	.04	.67	.71

Water blasting masonry, using rented 400 to 700 PSI power washer with 3 to 8 gallon per minute flow rate, includes blaster rental at $50 per day

	Craft@Hrs	Unit	Material	Labor	Total
Smooth face (250 SF per manhour)	M1@.004	SF	.05	.13	.18
Rough face (200 SF per manhour)	M1@.005	SF	.07	.17	.24

Sandblasting masonry, using 150 PSI compressor (with accessories and sand) at $150.00 per day

	Craft@Hrs	Unit	Material	Labor	Total
Smooth face (50 SF per manhour)	M1@.020	SF	.22	.67	.89
Rough face (40 SF per manhour)	M1@.025	SF	.25	.84	1.09

	Craft@Hrs	Unit	Material	Labor	Total
Steam cleaning masonry, using rented steam cleaning rig (with accessories) at $52.50 per day					
Smooth face (75 SF per manhour)	M1@.013	SF	.04	.44	.48
Rough face (55 SF per manhour)	M1@.018	SF	.06	.60	.66
Add for masking adjacent surfaces	—	%	—	5.0	—
Add for difficult stain removal	—	%	50.0	50.0	—
Add for working from scaffold	—	%	—	20.0	—
Repointing brick, cut out joint, mask (blend-in), and regrout (tuck pointing)					
30 SF per manhour	M1@.033	SF	.05	1.11	1.16

Masonry Cleaning Costs to clean masonry surfaces using commercial cleaning agents, including washing and scaffolding equipment.

	Craft@Hrs	Unit	Material	Labor	Total
Typical cleaning of surfaces					
Granite, sandstone, terra cotta, brick	CL@.015	SF	.25	.45	.70
Cleaning surfaces of heavily carbonated					
Limestone or cast stone	CL@.045	SF	.35	1.36	1.71
Typical wash with acid and rinse	CL@.004	SF	.40	.12	.52

Fireplaces Common brick, including the chimney foundation, damper, flue lining and stack to 15'

	Craft@Hrs	Unit	Material	Labor	Total
30" box, brick to 5', large quantity	M1@30.3	Ea	612.00	1,020.00	1,632.00
36" box, brick to 5', custom	M1@33.7	Ea	740.00	1,130.00	1,870.00
42" box, brick to 8', custom	M1@48.1	Ea	867.00	1,620.00	2,487.00
48" box, brick to 8', custom	M1@53.1	Ea	995.00	1,780.00	2,775.00
Flue lining, 1' lengths					
8-1/2" x 17"	M1@.148	LF	4.28	4.97	9.25
10" x 14"	M1@.206	LF	4.64	6.92	11.56
13" x 13"	M1@.193	LF	5.36	6.49	11.85
Prefabricated cast concrete					
30" box and 15' chimney,	M1@5.43	Ea	1,040.00	182.00	1,222.00
Face brick, mantle to ceiling	M1@.451	SF	6.25	15.20	21.45

Pargeting, Waterproofing and Dampproofing

	Craft@Hrs	Unit	Material	Labor	Total
Pargeting, 2 coats, 1/2" thick	M3@.070	SF	.16	2.45	2.61
Pargeting, 2 coats, waterproof, 3/4" thick	M3@.085	SF	.17	2.97	3.14
Waterproofing, per coat of asphalt	CL@.007	SF	.13	.21	.34
Dampproofing, asphalt primer at 1 gallon per 100 SF,					
Dampproof coat at 30 pounds per 100 SF	CL@.029	SF	.68	.88	1.56

Pre-Engineered Steel Buildings 26 gauge colored galvanized steel roof and siding with 4 in 12 (20 pound live load) roof. Cost per SF of floor area. These costs do not include foundation or floor slab. Add delivery cost to the site. Equipment cost shown is $44.50 for a 15 ton truck crane at $33 per hour and a 2-ton truck at $11.50 per hour equipped for work of this type.

	Craft@Hrs	Unit	Material	Labor	Equipment	Total
40' x 100' (4,000 SF)						
14' eave height	H5@.074	SF	6.28	2.90	.66	9.84
16' eave height	H5@.081	SF	7.04	3.17	.72	10.93
20' eave height	H5@.093	SF	7.96	3.64	.83	12.43
60' x 100' (6,000 SF)						
14' eave height	H5@.069	SF	6.06	2.70	.61	9.37
16' eave height	H5@.071	SF	6.20	2.78	.63	9.61
20' eave height	H5@.081	SF	7.17	3.17	.72	11.06

	Craft@Hrs	Unit	Material	Labor	Equipment	Total
80' x 100' (8,000 SF)						
14' eave height	H5@.069	SF	5.93	2.70	.61	9.24
16' eave height	H5@.071	SF	6.18	2.78	.63	9.59
20' eave height	H5@.080	SF	7.17	3.13	.71	11.01
100' x 100' (10,000 SF)						
14' eave height	H5@.067	SF	5.90	2.62	.60	9.12
16' eave height	H5@.071	SF	6.18	2.78	.63	9.59
20' eave height	H5@.080	SF	6.93	3.13	.71	10.77
100' x 150' (15,000 SF)						
14' eave height	H5@.065	SF	5.04	2.55	.58	8.17
16' eave height	H5@.069	SF	5.36	2.70	.61	8.67
20' eave height	H5@.074	SF	5.95	2.90	.66	9.51
100' x 200' (20,000 SF)						
14' eave height	H5@.063	SF	4.95	2.47	.56	7.98
16' eave height	H5@.066	SF	5.17	2.59	.59	8.35
20' eave height	H5@.071	SF	5.56	2.78	.63	8.97
140' x 150' (21,000 SF)						
14' eave height	H5@.058	SF	4.57	2.27	.52	7.36
16' eave height	H5@.059	SF	4.73	2.31	.53	7.57
20' eave height	H5@.066	SF	5.27	2.59	.59	8.45
140' x 175' (24,500 SF)						
14' eave height	H5@.057	SF	4.41	2.23	.51	7.15
16' eave height	H5@.058	SF	4.57	2.27	.52	7.36
20' eave height	H5@.065	SF	5.03	2.55	.59	8.17
160' x 200' (32,000 SF)						
14' eave height	H5@.055	SF	4.28	2.15	.50	6.93
16' eave height	H5@.056	SF	4.41	2.19	.50	7.10
20' eave height	H5@.061	SF	4.85	2.39	.54	7.78
200' x 200' (40,000 SF)						
14' eave height	H5@.052	SF	4.14	2.04	.46	6.64
16' eave height	H5@.053	SF	4.24	2.08	.47	6.79
20' eave height	H5@.058	SF	4.57	2.27	.52	7.36
Additive costs:						
Personnel door, 3'0" x 6'8"	H5@2.83	Ea	399.00	111.00	25.20	535.20
Overhead door, 8'0" x 10'0"	H5@8.02	Ea	977.00	314.00	71.40	1,362.40
Interior liner panel, 26 gauge, per SF of wall	H5@.012	SF	1.31	.47	.11	1.89
Foam insulated sandwich panel roofing and siding Per SF of roof and sidewall	H5@.005	SF	3.81	.20	.04	4.05
Plastic skylight roof panel, per SF of panel	H5@.013	SF	3.29	.51	.12	3.92
R-11 blanket insulation, 3-1/2", wall or floor	H5@.005	SF	.64	.20	.04	.88
Ridge ventilator, 9" throat	H5@.193	LF	31.50	7.56	1.72	40.78
24 gauge colored eave gutter	H5@.028	LF	5.04	1.10	.25	6.39
4" x 4" colored downspouts	H5@.014	LF	4.50	.55	.12	5.17
30 pound live load roof design	—	%	12.0	—	—	—

	Craft@Hrs	Unit	Material	Labor	Equipment	Total

Railings, Aluminum, Brass and Wrought Iron See also, Steel Railings.

Aluminum railings. 3'6" high with 1-1/2" square horizontal rails top and bottom, 5/8" square vertical bars at 4" OC, and 1-1/2" square posts spaced 4' OC. Welded construction with standard mill finish.

	Craft@Hrs	Unit	Material	Labor	Equipment	Total
Installed on balconies	H6@.092	LF	29.60	3.64	—	33.24
Installed on stairways	H6@.115	LF	33.50	4.55	—	38.05
Add for anodized finish, bronze or black	—	LF	10.80	—	—	10.80

Brass railings. Round tubular type, bright finish, standard 12' sections, .050" wall commercial quality with supports at 3' OC attached to existing structure using brass screws. Labor includes layout, measure, cut and end preparation

	Craft@Hrs	Unit	Material	Labor	Equipment	Total
1-1/2" tubing	H6@.058	LF	4.49	2.29	—	6.78
2" tubing	H6@.058	LF	5.71	2.29	—	8.00
Fittings, slip-on type, no soldering or brazing required						
1-1/2" elbows	H6@.092	Ea	13.00	3.64	—	16.64
1-1/2" tees	H6@.137	Ea	14.00	5.42	—	19.42
1-1/2" crosses	H6@.277	Ea	17.00	11.00	—	28.00
2" elbows	H6@.115	Ea	14.40	4.55	—	18.95
2" tees	H6@.171	Ea	19.30	6.76	—	26.06
2" crosses	H6@.346	Ea	21.70	13.70	—	35.40

Wrought iron railings. Standard 3'6" high, welded steel construction, shop prime painted, attached to existing structure

	Craft@Hrs	Unit	Material	Labor	Equipment	Total
Stock patterns	H6@.171	LF	13.70	6.76	—	20.46
Custom patterns	H6@.287	LF	20.50	11.40	—	31.90
Stairway handrails	H6@.343	LF	22.50	13.60	—	36.10

Space Frame System 10,000 SF or more. Equipment cost shown is $54.00 for a 15 ton truck crane at $42.50 per hour and a 2 ton truck at $11.50 per hour equipped for work of this type.

	Craft@Hrs	Unit	Material	Labor	Equipment	Total
5' module 4.5 lb live load	H5@.146	SF	15.30	5.72	1.75	22.77
4' module, add	H5@.025	SF	2.20	.98	.30	3.48

Steel Floor and Roof Decking These costs assume a 40,000 SF job not over six stories high. Weight and cost include 10% for waste and connections. Equipment cost shown is for a 15 ton truck crane at $42.50 per hour and a 2 ton truck at $11.50 per hour equipped for work of this type.

Floor and roof decking, 1-1/2" standard rib, simple non-composite section, non-cellular, shop prime painted finish

	Craft@Hrs	Unit	Material	Labor	Equipment	Total
22 gauge, 2.1 lbs per SF	H5@.013	SF	1.30	.51	.16	1.97
20 gauge, 2.5 lbs per SF	H5@.013	SF	1.37	.51	.16	2.04
18 gauge, 3.3 lbs per SF	H5@.013	SF	1.71	.51	.16	2.38
16 gauge, 4.1 lbs per SF	H5@.013	SF	1.89	.51	.16	2.56
Add for galvanized finish	—	SF	.22	—	—	.22

Steel floor deck system. Consists of 18 gauge, 1-1/2" standard rib, shop prime painted steel decking, installed over a structural steel beam and girder support system (cost of support structure not included) covered with 3" thick 6.0 sack mix, 3,000 PSI lightweight ready-mix concrete at $98.00 per CY before 3% allowance for waste, placed by pump, reinforced with 6" x 6", W1.4 x W1.4. welded wire mesh. Costs shown assume shoring is not required and include curing and finishing the concrete.

	Craft@Hrs	Unit	Material	Labor	Equipment	Total
18 gauge decking	H5@.013	SF	1.71	.51	.12	2.34
Welded wire mesh	RB@.004	SF	.11	.16	—	.27
Lightweight concrete	M2@.006	SF	1.13	.21	.09	1.43
Cure and finish concrete	P8@.011	SF	.11	.37	—	.48
Total floor deck system	**—@.032**	**SF**	**3.06**	**1.25**	**.21**	**4.52**
Add for 16 gauge decking	—	SF	.23	—	—	.23

	Craft@Hrs	Unit	Material	Labor	Equipment	Total

Steel roof deck system. Consists of 18 gauge, 1-1/2" standard rib, shop prime painted steel decking, installed over open web steel joists 4' OC (costs of joists not included) covered with 1-1/4" R-8.3 urethane insulation board overlaid with .045" loose laid membrane EPDM (ethylene propylene diene monomer) elastomeric roofing. Joist pricing assumes H or J series at 9.45 lbs per LF spanning 20'0", spaced 4' OC.

	Craft@Hrs	Unit	Material	Labor	Equipment	Total
18 gauge decking	H5@.013	SF	1.71	.51	.12	2.34
Insulation board	A1@.011	SF	1.18	.40	—	1.58
EPDM roofing	R3@.006	SF	.91	.21	—	1.12
Total roof deck system	**—@.030**	**SF**	**3.80**	**1.12**	**.12**	**5.04**
Add for 16 gauge decking	—	SF	.23	—	—	.23
Add open web steel joists	H5@.005	SF	.64	.20	.04	.88

Steel Handrails, Kickplates, Ladders, Landings, Railings, and Stairways Material costs assume shop fabrication and assembly prime painted and erection bolts. Equipment cost is $47.30 per hour and includes a 10 ton hydraulic truck-mounted crane, an electric powered welding machine, an oxygen/acetylene welding rig, a diesel powered 100 CFM air compressor and accessories.

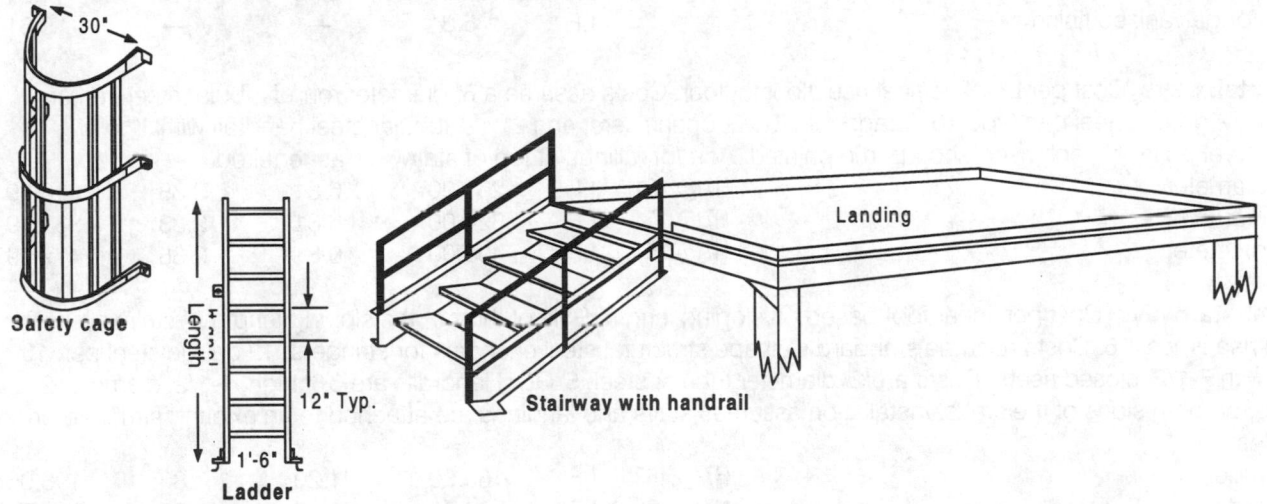

Safety cage

Ladder — Handrail Length, 12" Typ., 1'-6"

Stairway with handrail

Landing

Handrails. Wall mounted, with brackets 5' OC, based on 12' lengths, welded steel, shop prime painted

	Craft@Hrs	Unit	Material	Labor	Equipment	Total
1-1/4" diameter rail	H7@.092	LF	5.60	4.15	1.45	11.20
1-1/2" diameter rail	H7@.102	LF	6.55	4.60	1.61	12.76
Add for galvanized finish	—	LF	1.07	—	—	1.07

Kickplates, 1/4" thick flat steel plate, shop prime painted, attached to steel railings, stairways or landings

	Craft@Hrs	Unit	Material	Labor	Equipment	Total
4" high	H7@.058	LF	4.49	2.62	.91	8.02
6" high	H7@.069	LF	4.88	3.11	1.09	9.08
Add for galvanized finish	—	LF	.80	—	—	.80

Ladders, vertical type. 1'6" wide, attached to existing structure

	Craft@Hrs	Unit	Material	Labor	Equipment	Total
Without safety cage	H7@.100	LF	30.40	4.51	1.58	36.49
With steel safety cage	H7@.150	LF	36.10	6.77	2.36	45.23
Add for galvanized finish	—	LF	3.19	—	—	3.19

Landings. Costs assume standard C shape structural steel channels for joists and frame, 10 gauge steel pan deck areas, shop prime painted. Add for support legs from Structural Steel Section.

	Craft@Hrs	Unit	Material	Labor	Equipment	Total
Cost per square foot	H7@.250	SF	109.00	11.30	3.94	124.24
Add for galvanized finish	—	SF	11.20	—	—	11.20

	Craft@Hrs	Unit	Material	Labor	Equipment	Total

Pipe and chain railings. Welded steel pipe posts with U-brackets attached to accept 1/4" chain, shop prime painted. Installed in existing embedded sleeves. Posts are 5'0" long (1'6" in the sleeve and 3'6" above grade). Usual spacing is 4' OC. Add chain per lineal foot

	Craft@Hrs	Unit	Material	Labor	Equipment	Total
1-1/4" diameter post	H7@.287	Ea	12.40	12.90	—	25.30
1-1/2" diameter post	H7@.287	Ea	14.70	12.90	—	27.60
2" diameter post	H7@.343	Ea	20.00	15.50	—	35.50
2-1/2" diameter post	H7@.343	Ea	26.90	15.50	—	42.40
Add for galvanized finish	—	Ea	2.12	—	—	2.12
Chain, 1/4" diameter galvanized	H7@.058	LF	1.82	2.62	—	4.44

Railings. Floor mounted, 3'6" high with posts 5' OC, based on 12' length, welded steel, shop prime painted

	Craft@Hrs	Unit	Material	Labor	Equipment	Total
1-1/4" diameter rails and posts						
2-rail type	H7@.115	LF	23.10	5.19	1.82	30.11
3-rail type	H7@.125	LF	29.90	5.64	1.97	37.51
1-1/2" diameter rails and posts						
2-rail type	H7@.125	LF	27.70	5.64	1.97	35.31
3-rail type	H7@.137	LF	36.60	6.18	2.16	44.94
Add for galvanized finish	—	LF	5.31	—	—	5.31

Spiral stairways Cost per LF of vertical rise, floor to floor. Costs assume a 6" diameter round tubular steel center column, 12 gauge steel pan type 10" treads with 7-1/2" open risers, and 2" x 2" tubular steel handrail with 1" x 1" tubular steel posts at each riser, shop prime painted. Add for railings at top of stairways as required.

	Craft@Hrs	Unit	Material	Labor	Equipment	Total
4'0" diameter	H7@.151	LF	237.00	6.81	2.38	246.19
6'0" diameter	H7@.151	LF	324.00	6.81	2.38	333.19
8'0" diameter	H7@.151	LF	408.00	6.81	2.38	417.19

Straight stairways Cost per linear foot based on sloping length. Rule of thumb: the sloping length of stairways is the vertical rise times 1.6. Costs assume standard C shape structural steel channels for stringers, 12 gauge steel pan 10" treads with 7-1/2" closed risers. Posts are 2" diameter tubular steel, 5' OC. Handrails are 3'6" high, 1-1/2" diameter 2-rail type, on both sides of the stairs. Installation assumes stairs and landings are attached to an existing structure on two sides.

	Craft@Hrs	Unit	Material	Labor	Equipment	Total
3'0" wide	H7@.469	LF	169.00	21.20	7.86	198.06
4'0" wide	H7@.630	LF	223.00	28.40	10.60	262.00
5'0" wide	H7@.781	LF	279.00	35.20	13.10	327.30

Structural Steel Costs shown below are for 60 to 100 ton jobs up to six stories high where the fabricated structural steel is purchased from a steel fabricator located within 50 miles of the job site. Material costs are for ASTM A36 steel, fabricated according to plans, specifications and AISC recommendations and delivered to the job site with connections attached, piece marked and ready for erection. Labor includes time to unload and handle the steel, erect and plumb, install permanent bolts, weld and all other work usually associated with erecting the steel including field touch-up painting. Equipment cost is $65 per hour. Equipment includes a 2 CY crawler mounted crane at $50 per hour, plus $15 per hour for an electric powered welding machine, an oxygen/acetylene welding and cutting torch, a diesel powered 100 CFM air compressor with pneumatic grinders, ratchet wrenches, hoses, and other tools usually associated with work of this type. Structural steel is usually estimated by the tons of the steel in the project. Labor for erection will vary with the weight of steel per linear foot. For example a member weighing 25 pounds per foot will require more manhours per ton than one weighing 60 pounds per foot.

Open web steel joists Estimate the total weight of "bare" joists and then add 10% to allow for connections and accessories.

	Craft@Hrs	Unit	Material	Labor	Equipment	Total
"H" and "J" series, 8H3 thru 26H10 or 8J3 thru 30J8						
Under 10 lbs per LF	H8@10.0	Ton	1,030.00	441.00	110.00	1,581.00
Over 10 to 20 lbs per LF	H8@8.00	Ton	1,020.00	353.00	88.00	1,461.00

	Craft@Hrs	Unit	Material	Labor	Equipment	Total
"LH" and "LJ" series, 18LH02 thru 48LH10 or 18LJ02 thru 48LJ11						
Under 20 lbs per LF	H8@8.00	Ton	957.00	353.00	130.40	1,440.40
Over 20 to 50 lbs per LF	H8@7.00	Ton	923.00	309.00	114.00	1,346.00
Over 50 lbs per LF	H8@6.00	Ton	884.00	265.00	100.00	1,249.00
"DLH" or "DLJ" series, 52DLH10 thru 72DLH18 or 52DLJ12 thru 52DLJ20						
20 to 50 lbs per LF	H8@8.00	Ton	1,020.00	353.00	130.00	1,503.00
Over 50 lbs per LF	H8@6.00	Ton	969.00	265.00	100.00	1,334.00

Structural shapes Estimate the total weight of "bare" fabricated structural shapes and then add 15% to allow for the weight of connections.

	Craft@Hrs	Unit	Material	Labor	Equipment	Total
Beams, purlins, and girts						
Under 20 pounds per LF	H8@14.0	Ton	1,930.00	617.00	229.00	2,776.00
From 20 to 50 lbs per LF	H8@11.0	Ton	1,240.00	485.00	180.00	1,905.00
Over 50 to 75 lbs per LF	H8@9.40	Ton	1,170.00	415.00	154.00	1,739.00
Over 75 to 100 lbs per LF	H8@8.00	Ton	1,140.00	353.00	130.00	1,623.00
Columns						
Under 20 pounds per LF	H8@12.0	Ton	1,800.00	529.00	196.00	2,525.00
From 20 to 50 lbs per LF	H8@10.0	Ton	1,170.00	441.00	163.00	1,774.00
Over 50 to 75 lbs per LF	H8@9.00	Ton	1,120.00	397.00	147.00	1,664.00
Over 75 to 100 lbs per LF	H8@7.00	Ton	1,120.00	309.00	114.00	1,543.00
Sag rods and X type bracing						
Under 1 pound per LF	H8@82.1	Ton	2,930.00	3,620.00	1,340.00	7,890.00
From 1 to 3 lbs per LF	H8@48.0	Ton	2,710.00	2,120.00	785.00	5,615.00
Over 3 to 6 lbs per LF	H8@34.8	Ton	2,530.00	1,530.00	570.00	4,630.00
Over 6 to 10 lbs per LF	H8@28.0	Ton	2,310.00	1,230.00	457.00	3,997.00
Trusses and girders						
Under 20 pounds per LF	H8@10.0	Ton	1,790.00	441.00	163.00	2,394.00
From 20 to 50 lbs per LF	H8@7.50	Ton	1,170.00	331.00	123.00	1,624.00
Over 50 to 75 lbs per LF	H8@6.00	Ton	1,120.00	265.00	98.00	1,483.00
Over 75 to 100 lbs per LF	H8@5.00	Ton	1,080.00	220.00	81.50	1,381.50

Structural Steel, Subcontract, by Square Foot of Floor Use these figures only for preliminary estimates on buildings with 20,000 to 80,000 SF per story, 100 PSF live load and 14' to 16' story height. A "bay" is the center to center distance between columns. Costs assume steel at $2,000 to $2,100 per ton including installation labor and material and the subcontractor's overhead and profit. Include all area under the roof when calculating the total square footage.

	Craft@Hrs	Unit	Material	Labor	Equipment	Total
Single story						
20' x 20' bays [8 to 9 lbs. per SF]	—	SF	—	—	—	9.28
30' x 30' bays [10 to 11 lbs. per SF]	—	SF	—	—	—	10.86
40' x 40' bays [12 to 13 lbs. per SF]	—	SF	—	—	—	12.95
Two story to six story						
20' x 20' bays [9 to 10 lbs per SF]	—	SF	—	—	—	9.79
30' x 30' bays [11 to 12 lbs per SF]	—	SF	—	—	—	11.83
40' x 40' bays [13 to 14 lbs per SF]	—	SF	—	—	—	13.65

Tubular Columns Rectangular, round, and square structural columns. When estimating total weight of columns, figure the column weight and add 30% to allow for a shop attached U-bracket on the top, a square base plate on the bottom and four anchor bolts. This should be sufficient for columns 15' to 20' in length. Equipment cost is $45 per hour and includes a 10 ton hydraulic truck-mounted crane at $30 per hour plus $15 per hour for an electric powered welding machine, an oxygen/acetylene welding and cutting torch, a diesel powered 100 CFM air compressor with pneumatic grinders, ratchet wrenches, hoses, and other tools usually associated with work of this type.

	Craft@Hrs	Unit	Material	Labor	Equipment	Total
Rectangular tube columns, in pounds per LF						
3" x 2" to 12" x 4", to 20 lbs per LF	H7@9.00	Ton	2,140.00	406.00	135.00	2,681.00
6" x 4" to 16" x 8", 21 to 50 lbs per LF	H7@8.00	Ton	1,710.00	361.00	120.00	2,191.00
12 " x 6" to 20" x 12",						
51 to 75 lbs per LF	H7@6.00	Ton	1,670.00	271.00	90.00	2,031.00
14" x 10" to 20" x 12", 76 to 100 lbs/ LF	H7@5.00	Ton	1,630.00	226.00	75.00	1,931.00
Round columns, in pounds per LF						
3" to 6" diameter pipe, to 20 lbs per LF	H7@9.00	Ton	2,260.00	406.00	135.00	2,801.00
4" to 12" pipe, 20 to 50 lbs per LF	H7@8.00	Ton	1,810.00	361.00	120.00	2,291.00
6" to 12" pipe, 51 to 75 lbs per LF	H7@6.00	Ton	1,760.00	271.00	90.00	2,121.00
Square tube columns, in pounds per LF						
2" x 2" to 8" x 8", to 20 lbs per LF	H7@9.00	Ton	2,040.00	406.00	135.00	2,581.00
4" x 4" to 12" x 12", 21 to 50 lbs per LF	H7@8.00	Ton	1,670.00	361.00	120.00	2,151.00
8" x 8" to 16" x 16", 51 to 75 lbs per LF	H7@6.00	Ton	1,640.00	271.00	90.00	2,001.00
10" x 10" to 16" x 16", 76 to 100 lbs/LF	H7@5.00	Ton	1,610.00	226.00	75.00	1,911.00
Add for concrete fill in columns, 3,000 PSI design pump mix at $64.00 per CY before 10% allowance for waste						
Placed by pump	M2@.738	CY	70.40	25.30	14.80	110.50

	Craft@Hrs	Unit	Material	Labor	Total
Fabricated Metals					
Catch basin grating and frame					
24" x 24", Standard duty	CL@8.68	Ea	241.00	263.00	504.00
Manhole rings with cover					
24", 330 lbs	CL@6.80	Ea	232.00	206.00	438.00
30", 400 lbs	CL@6.80	Ea	241.00	206.00	447.00
38", 730 lbs	CL@6.80	Ea	663.00	206.00	869.00

Welded steel grating Weight shown is approximate. Costs shown are per square foot of grating, shop prime painted or galvanized after fabrication as shown.

	Craft@Hrs	Unit	Material	Labor	Total
Grating bars as shown, 1/2" cross bars at 4" OC					
1-1/4" x 3/16" at 1-1/4" OC, 9 lbs per SF, painted	IW@.019	SF	6.25	.93	7.18
1-1/4" x 3/16" at 1-1/4" OC, 9 lbs per SF, galvanized	IW@.019	SF	7.73	.93	8.66
Add for banding on edges	—	LF	3.27	—	3.27
2" x 1/4" at 1-1/4" OC, 18.7 lbs per SF, painted	IW@.026	SF	14.70	1.27	15.97
2" x 1/4" at 1-1/4" OC, 18.7 lbs per SF, galvanized	IW@.026	SF	19.80	1.27	21.07
Add for banding on edges	—	LF	4.89	—	4.89
Add for toe-plates, steel, shop prime painted					
1/4" thick x 4" high attached after grating is in place	IW@.048	LF	6.42	2.34	8.76

Checkered-steel floor plate Shop prime painted or galvanized after fabrication as shown

	Craft@Hrs	Unit	Material	Labor	Total
1/4", 11-1/4 lbs per SF, painted	IW@.025	SF	11.80	1.22	13.02
3/8", 16-1/2 lbs per SF, painted	IW@.025	SF	16.90	1.22	18.12
1/4", 11-1/4 lbs per SF, galvanized	IW@.025	SF	12.80	1.22	14.02
3/8", 16-1/2 lbs per SF, galvanized	IW@.025	SF	18.80	1.22	20.02

Checkered-steel trench cover plate Shop prime painted or galvanized after fabrication as shown

	Craft@Hrs	Unit	Material	Labor	Total
1/4", 11-1/4 lbs per SF, primed	IW@.045	SF	15.40	2.20	17.60
1/4", 11-1/4 lbs per SF, galvanized	IW@.045	SF	17.90	2.20	20.10
3/8", 16-1/2 lbs per SF, primed	IW@.045	SF	22.20	2.20	24.40
3/8", 16-1/2 lbs per SF, galvanized	IW@.045	SF	26.90	2.20	29.10

Metals 5

	Craft@Hrs	Unit	Material	Labor	Total
Fabrications					
Canopy framing, steel	IW@.009	Lb	.97	.44	1.41
Miscellaneous supports, steel	IW@.019	Lb	1.54	.93	2.47
Miscellaneous supports, aluminum	IW@.061	Lb	10.80	2.98	13.78
Embedded steel, (embedded in concrete)					
Light, to 20 lbs per LF	T3@.042	Lb	1.92	1.45	3.37
Medium, over 20 to 50 lbs per LF	T3@.035	Lb	1.74	1.21	2.95
Heavy, over 50 lbs per LF	T3@.025	Lb	1.46	.86	2.32
Channel sill, 1/8" steel, 1-1/2" x 8" wide	T3@.131	LF	15.00	4.53	19.53
Angle sill, 1/8" steel, 1-1/2" x 1-1/2"	T3@.095	LF	4.74	3.29	8.03
Fire escapes, ladder and balcony, per floor	IW@16.5	Ea	1,780.00	806.00	2,586.00
Grey iron foundry items	T3@.014	Lb	.87	.48	1.35
Stair nosing, 2-1/2" x 2-1/2", set in concrete	T3@.139	LF	7.44	4.81	12.25
Cast iron					
Lightweight sections	T3@.012	Lb	.92	.42	1.34
Heavy section	T3@.010	Lb	.74	.35	1.09
Cast column bases					
Iron, 16" x 16", custom	T3@.759	Ea	83.60	26.30	109.90
Iron, 32" x 32", custom	T3@1.54	Ea	347.00	53.30	400.30
Aluminum, 8" column, stock	T3@.245	Ea	19.20	8.48	27.68
Aluminum, 10" column, stock	T3@.245	Ea	24.20	8.48	32.68
Wheel type corner guards, per LF of height	T3@.502	LF	56.30	17.40	73.70
Angle corner guards, with anchors	T3@.018	Lb	1.69	.62	2.31
Channel door frames, with anchors	T3@.012	Lb	1.56	.42	1.98
Steel lintels, painted	T3@.013	Lb	.93	.45	1.38
Steel lintels, galvanized	T3@.013	Lb	1.23	.45	1.68
Anodizing Add to cost of fabrication					
Aluminum	—	%	15.0	—	—
Bronze	—	%	30.0	—	—
Black	—	%	50.0	—	—
Galvanizing, add to cost of fabrication	—	Lb	.41	—	.41
Ornamental Metal					
Ornamental screens, aluminum	H6@.135	SF	25.60	5.34	30.94
Ornamental screens, extruded metal	H6@.151	SF	21.50	5.97	27.47
Gates, wrought iron, 6' x 7', typical installed cost	—	Ea	—	—	1,250.00
Sun screens, aluminum stock design, manual	H6@.075	SF	27.80	2.97	30.77
Sun screens, aluminum stock design, motorized	H6@.102	SF	50.10	4.03	54.13
Brass and bronze work general fabrication,	H6@.079	Lb	19.80	3.12	22.92
Aluminum work, general fabrication	H6@.160	Lb	13.70	6.33	20.03
Stainless steel, general fabrication	H6@.152	Lb	14.80	6.01	20.81
Security Gratings					
Door grating, 1" tool steel vertical bars at 4" OC, 3/8" x 2-1/4" steel frame and horizontal bars at 18" OC,					
3' x 7' overall	H6@10.5	Ea	1,140.00	415.00	1,555.00
Add for butt hinges and locks, per door	H6@2.67	Ea	345.00	106.00	451.00
Add for field painting, per door	H6@1.15	Ea	5.93	45.50	51.43
Window grating, tool steel vertical bars at 4" OC, 3/8" x 2-1/4" steel frame, horizontal bars 6" OC,					
20" x 30" overall set with 1/2" bolts at 6" OC	H6@3.21	Ea	151.00	127.00	278.00
Louver vent grating, 1" x 3/16", "I" bearing bars 1-3/16" OC with 3/16" spaced bars 4" OC and banded ends.					
Set with 1/2" bolts at 1" OC, prime painted	H6@.114	SF	18.50	4.51	23.01
Steel plate window covers, hinged, 3/8" plate steel, 32" W x 48" H overall, 10 lbs per SF,					
including hardware	H6@.150	SF	8.39	5.93	14.32

Steel Fabrication These costs are for work occurring at the job site and assume engineering and design, including detail drawings sufficient to do the work, have been prepared by others at no cost to the contractor performing the work. Costs are based on ASTM A36 all-purpose carbon steel which is widely used in building construction. Equipment cost is $13 per hour and includes a 3/4-ton flatbed truck equipped with a diesel powered welding machine, an oxygen/acetylene welding and cutting torch, a diesel powered 100 CFM air compressor with pneumatic grinders, ratchet wrenches, hoses, and other tools usually associated with work of this type. Material costs for the steel is not included.

	Craft@Hrs	Unit	Material	Labor	Equipment	Total
Flat plate, cutting and drilling holes						
Flame cutting Labor shown includes layout, marking, cutting and grinding of finished cut						
Square cut, thickness of plate as shown						
1/4", 5/16" or 3/8"	IW@.020	LF	—	.98	.25	1.23
1/2", 5/8" or 3/4"	IW@.025	LF	—	1.22	.31	1.53
7/8", 1" or 1-1/4"	IW@.030	LF	—	1.47	.37	1.84
Bevel cut 30 to 45 degrees, thickness of plate as shown						
1/4", 5/16" or 3/8"	IW@.030	LF	—	1.47	.37	1.84
1/2", 5/8" or 3/4"	IW@.035	LF	—	1.71	.43	2.14
7/8", 1" or 1-1/4"	IW@.043	LF	—	2.10	.53	2.63

Drill holes. Using a hand-held drill motor. Labor shown includes layout, drill and ream holes with plate thickness and hole size as shown (note: hole size is usually 1/16" greater than bolt diameter)

	Craft@Hrs	Unit	Material	Labor	Equipment	Total
1/4", 5/16" or 3/8" plate thickness,						
5/16", 3/8" or 1/2" hole	IW@.084	Ea	—	4.10	1.04	5.14
1/2", 5/8" or 3/4" plate thickness,						
9/16", 3/4" or 13/16" hole	IW@.134	Ea	—	6.54	1.67	8.21
7/8", 1" or 1-1/4" plate thickness,						
1", 1-1/16" or 1-5/16" hole	IW@.200	Ea	—	9.77	2.48	12.25

Welding Using manual inert gas shielded arc electric welding process utilizing coated eletrodes (rod) of appropriate size and type with welding machine amperage suitable for the welding rod being used. Labor shown includes layout, fitup and weld.

Fillet welds for attaching base plates, clip angles, gussets. Welding in flat position.

	Craft@Hrs	Unit	Material	Labor	Equipment	Total
Size of continuous fillet weld.						
3/16"	IW@.300	LF	—	14.70	3.71	18.41
1/4"	IW@.400	LF	—	19.50	5.00	24.50
3/8"	IW@.585	LF	—	28.60	7.24	35.84
1/2"	IW@.810	LF	—	39.60	10.00	49.60
5/8"	IW@1.00	LF	—	48.80	12.40	61.20
3/4"	IW@1.20	LF	—	58.60	14.90	73.50

Buttweld joints, 100% penetration, flat plates.

	Craft@Hrs	Unit	Material	Labor	Equipment	Total
Thickness at joint						
3/16"	IW@.200	LF	—	9.77	2.50	12.27
1/4"	IW@.250	LF	—	12.20	3.10	15.30
3/8"	IW@.440	LF	—	21.50	5.45	26.95
1/2"	IW@.660	LF	—	32.20	8.17	40.37
5/8"	IW@.825	LF	—	40.30	10.20	50.50
3/4"	IW@1.05	LF	—	51.30	13.00	64.30

	Craft@Hrs	Unit	Material	Labor	Total

Rough Carpentry Industrial and commercial work. Price shown in parentheses is price before allowance for waste. Costs shown include 10% waste.

Stud walls, including plates, blocking, headers and diagonal bracing, per SF of wall, 16" OC

	Craft@Hrs	Unit	Material	Labor	Total
2" x 4", 8' high, 1.1 BF per SF (at $523 per MBF)	C8@.021	SF	.63	.76	1.39
2" x 4", 10' high, 1.0 BF per SF (at $523 per MBF)	C8@.021	SF	.58	.76	1.34
2" x 6", 8' high, 1.6 BF per SF (at $515 per MBF)	C8@.031	SF	.91	1.12	2.03
2" x 6", 10' high, 1.5 BF per SF (at $515 per MBF)	C8@.030	SF	.85	1.09	1.94
12" OC, 2" x 6", 8' high, 2.0 BF per SF (at $515 per MBF)	C8@.037	SF	1.13	1.34	2.47
8" OC, 2" x 4", staggered on 2" x 6" plates, 1.8 BF per SF (at $523 per MBF)	C8@.034	SF	1.04	1.23	2.27

Pier caps, .60 ACZA treated

	Craft@Hrs	Unit	Material	Labor	Total
2" x 6" (at $515 per MBF + $198 per MBF)	C8@.024	LF	.78	.87	1.65

Mud sills, treated material, add bolts or fastener

	Craft@Hrs	Unit	Material	Labor	Total
Bolted, 2" x 4" (at $523 + $198 per MBF)	C8@.023	LF	.53	.83	1.36
Bolted, 2" x 6" (at $515 + $198 per MBF)	C8@.024	LF	.78	.87	1.65
Shot, 2" x 4" (at $523 + $198 per MBF)	C8@.018	LF	.53	.65	1.18
Shot, 2" x 6" (at $515 + $198 per MBF)	C8@.020	LF	.78	.73	1.51

Plates, (add for .60 ACZA treatment at $198 per MBF, if required)

	Craft@Hrs	Unit	Material	Labor	Total
2" x 4" (at $523 per MBF)	C8@.013	LF	.38	.47	.85
2" x 6" (at $515 per MBF)	C8@.016	LF	.57	.58	1.15

Posts, #1 structural

	Craft@Hrs	Unit	Material	Labor	Total
4" x 4" (at $735 per MBF)	C8@.110	LF	1.08	3.99	5.07
6" x 6" (at $870 per MBF)	C8@.145	LF	2.87	5.26	8.13
Headers, 4" x 12" (at $735 per MBF)	C8@.067	LF	3.23	2.43	5.66
Diagonal bracing, 1" x 4", let in (at $573 per MBF)	C8@.021	LF	.21	.76	.97
Blocking and nailers, 2" x 6", (at $515 per MBF)	C8@.034	LF	.57	1.23	1.80
Blocking, solid diagonal bracing, 2" x 4" (at $523 per MBF)	C8@.023	LF	.38	.83	1.21

Beams and girders, #2 & Btr

	Craft@Hrs	Unit	Material	Labor	Total
4" x 8" to 4" x 12" (at $750 per MBF)	C8@.094	LF	3.30	3.41	6.71
6" x 8" to 6" x 12" (at $870 per MBF)	C8@.115	LF	5.74	4.17	9.91

Ceiling Joists and blocking, including rim joists, Std & Btr, per LF of joist

	Craft@Hrs	Unit	Material	Labor	Total
2" x 6" (at $515 per MBF)	C8@.028	LF	.57	1.02	1.59
2" x 8" (at $523 per MBF)	C8@.037	LF	.77	1.34	2.11
2" x 10" (at $571 per MBF)	C8@.046	LF	1.05	1.67	2.72
2" x 12" (at $587 per MBF)	C8@.057	LF	1.29	2.07	3.36

Floor joists and blocking including rim joists and 6% waste, per LF of joist

	Craft@Hrs	Unit	Material	Labor	Total
2" x 6", 1.02 BF per LF (at $515 per MBF)	C8@.017	LF	.56	.62	1.18
2" x 8", 1.36 BF per LF (at $523 per MBF)	C8@.018	LF	.75	.65	1.40
2" x 10", 1.71 BF per LF (at $571 per MBF)	C8@.019	LF	1.03	.69	1.72
2" x 12", 2.05 BF per LF (at $587 per MBF)	C8@.021	LF	1.28	.76	2.04

Purlins, struts, ridge boards, Std & Btr

	Craft@Hrs	Unit	Material	Labor	Total
4" x 10" (at $750 per MBF)	C8@.050	LF	2.75	1.81	4.56
4" x 12" (at $735 per MBF)	C8@.055	LF	3.23	1.99	5.22
6" x 10" (at $900 per MBF)	C8@.066	LF	4.95	2.39	7.34
6" x 12" (at $900 per MBF)	C8@.066	LF	5.94	2.39	8.33

Rafters and outriggers, to 4 in 12 pitch, Std & Btr

	Craft@Hrs	Unit	Material	Labor	Total
2" x 4" to 2" x 8" (at $523 per MBF)	C8@.033	LF	.77	1.20	1.97
2" x 10" to 2" x 12" (at $587 per MBF)	C8@.036	LF	1.25	1.31	2.56
Add for over 4 in 12 pitch roof	—	%	—	30.0	—

	Craft@Hrs	Unit	Material	Labor	Total

Rafters, including frieze blocks, ribbon, strongbacks, 24" OC, per SF of roof plan area, Std & Btr

	Craft@Hrs	Unit	Material	Labor	Total
2" x 4", .44 BF per SF (at $523 per MBF)	C8@.015	SF	.25	.54	.79
2" x 6", .65 BF per SF (at $515 per MBF)	C8@.021	SF	.37	.76	1.13
2" x 8", .86 BF per SF (at $523 per MBF)	C8@.027	SF	.49	.98	1.47
2" x 10", 1.10 BF per SF (at $571 per MBF)	C8@.031	SF	.69	1.12	1.81
2" x 12", 1.40 BF per SF (at $587 per MBF)	C8@.041	SF	.90	1.49	2.39

Sills and Ledgers Bolted to Concrete or Masonry Walls Costs are per linear foot of Std & Btr pressure treated sill or ledger with anchor bolt holes drilled each 2'6" center. Labor costs include handling, measuring, cutting to length and drilling holes at floor level and installation working on a scaffold 10' to 20' above floor level. Figures in parentheses show the board feet per linear foot of sill or ledger. Costs shown include 5% waste.

For scheduling purposes, estimate that a crew of 2 can install 150 to 200 LF of sills with a single row of anchor bolts in an 8-hour day when working from scaffolds. Estimate that a crew of 2 can install 150 to 175 LF of ledgers with a single row of anchor bolt holes or 50 to 60 LF of ledgers with a double row of anchor bolt holes in an 8-hour day working from scaffolds. Add the cost of anchor bolts and scaffold, if required.

Sills with a single row of anchor bolt holes

	Craft@Hrs	Unit	Material	Labor	Total
2" x 4" at $523 per MBF (0.70 BF per LF)	C8@.040	LF	.37	1.45	1.82
2" x 6" at $515 per MBF (1.05 BF per LF)	C8@.040	LF	.54	1.45	1.99
2" x 8" at $523 per MBF (1.40 BF per LF)	C8@.040	LF	.73	1.45	2.18
2" x 10" at $571 per MBF (1.75 BF per LF)	C8@.040	LF	1.00	1.45	2.45
2" x 12" at $587 per MBF (2.10 BF per LF)	C8@.040	LF	1.23	1.45	2.68
4" x 6" at $750 per MBF (2.10 BF per LF)	C8@.051	LF	1.58	1.85	3.43
4" x 8" at $740 per MBF (2.80 BF per LF)	C8@.051	LF	2.07	1.85	3.92
4" x 10" at $750 per MBF (3.33 BF per LF)	C8@.059	LF	2.50	2.14	4.64
4" x 12" at $735 per MBF (4.20 BF per LF)	C8@.059	LF	3.09	2.14	5.23

Ledgers with a single row of anchor bolt holes

	Craft@Hrs	Unit	Material	Labor	Total
2" x 4" at $523 per MBF (0.70 BF per LF)	C8@.059	LF	.37	2.14	2.51
4" x 6" at $750 per MBF (2.10 BF per LF)	C8@.079	LF	1.58	2.87	4.45
4" x 8" at $740 per MBF (2.80 BF per LF)	C8@.079	LF	2.07	2.87	4.94

Ledgers with a double row of anchor bolt holes

	Craft@Hrs	Unit	Material	Labor	Total
4" x 4" at $735 per MBF (1.40 BF per LF)	C8@.120	LF	1.03	4.35	5.38
4" x 6" at $750 per MBF (2.10 BF per LF)	C8@.160	LF	1.58	5.80	7.38
4" x 8" at $740 per MBF (2.80 BF per LF)	C8@.160	LF	2.07	5.80	7.87
4" x 12" at $735 per MBF (4.20 BF per LF)	C8@.250	LF	3.09	9.07	12.16

Plywood and Lumber Sheathing Price shown in parentheses is before waste allowance. Costs shown include nails and normal waste.

Floors, hand nailed, underlayment C-D grade

	Craft@Hrs	Unit	Material	Labor	Total
1/2" (at $.45 per SF)	C8@.012	SF	.47	.44	.91
5/8" (at $.56 per SF)	C8@.012	SF	.59	.44	1.03
3/4" (at $.67 per SF)	C8@.013	SF	.70	.47	1.17
Deduct for machine nailing	—	%	—	-25.0	—

Walls, B-D, sanded, standard, interior grade

	Craft@Hrs	Unit	Material	Labor	Total
1/4" (at $.50 per SF)	C8@.015	SF	.53	.54	1.07
3/8" (at $.58 per SF)	C8@.015	SF	.61	.54	1.15
1/2" (3 ply) (at $.71 per SF)	C8@.016	SF	.75	.58	1.33
5/8" (at $.85 per SF)	C8@.018	SF	.89	.65	1.54
3/4" (at $.95 per SF)	C8@.018	SF	1.00	.65	1.65
5/8" fire retardant, A-B, screwed to metal frame ($1.10)	C8@.026	SF	1.16	.94	2.10

	Craft@Hrs	Unit	Material	Labor	Total
Walls, exterior grade					
1/2" (at $.40 per SF)	C8@.016	SF	.43	.58	1.01
5/8" (at $.56 per SF)	C8@.018	SF	.59	.65	1.24
3/4" (at $.67 per SF)	C8@.020	SF	.70	.73	1.43
7/16" OSB (at $.30 per SF)	C8@.018	SF	.32	.65	.97
Roofs, CDX					
3/8" (at $.36 per SF)	C8@.009	SF	.38	.33	.71
1/2" (at $.45 per SF)	C8@.009	SF	.47	.33	.80
5/8" (at $.56 per SF)	C8@.009	SF	.59	.33	.92
3/4" (at $.67 per SF)	C8@.011	SF	.70	.40	1.10
Additional items to consider for plywood sheathing					
Add for T&G edge	C8@.001	SF	.02	.04	.06
Deduct for non-structural	—	%	-15.0	—	—
Add for nailing shear walls	—	%	—	20.0	—
Deduct for machine nailing	—	%	—	-25.0	—
Blocking for plywood joints,					
2" x 4" (at $523 per MBF)	C8@.010	LF	.38	.36	.74
Lumber floor sheathing					
Diagonal, 1" x 6" Std & Btr					
1.14 BF per SF (at $587 per MBF)	C8@.018	SF	.67	.65	1.32
Straight 2" x 6" commercial grade T&G					
fir, 2.28 BF per SF (at $515 per MBF)	C8@.032	SF	1.17	1.16	2.33
Straight 2" x 6" commercial grade T&G					
cedar, 2.28 BF per SF (at $1,070 per MBF)	C8@.032	SF	2.44	1.16	3.60
Timber deck, Southern Pine, 5/4" x 6", premium grade,					
1.25 BF per SF (at $1,400 per MBF)	C8@.012	SF	1.75	.44	2.19
Lumber wall sheathing, diagonal, 1" x 6"					
Std & Btr, 1.14 BF per SF (at $637 per MBF)	C8@.024	SF	.73	.87	1.60
Lumber roof sheathing, diagonal, 1" x 6"					
Std & Btr, 1.14 BF per SF (at $637 per MBF)	C8@.018	MSF	.73	.65	1.38

Panelized Wood Roof Systems The costs shown will apply on 20,000 to 30,000 SF jobs where roof panels are fabricated near the point of installation and moved into place with a forklift. Costs will be higher on smaller jobs and jobs with poor access. 4' x 8' and 4' x 10' roof panels are made of plywood nailed to 2" x 4" or 2" x 6" stiffeners spaced 24" on centers and running the long dimension of the panel. Based on fabricated panels lifted by forklift (with metal hangers attached) and nailed to 4" x 12" or 4" x 14" roof purlins spaced 8' on center. Equipment cost is $36.80 per hour for a 4-ton forklift (30' lift), a truck with hand tools, a portable air compressor with air hoses and two pneumatic-operated nailing guns. Costs are per 1,000 square feet (MSF) of roof surface and do not include roof supports except as noted. Lumber costs assume a lumber price of $523 per MBF for 2" x 4" stiffeners, $515 per MBF for 2" x 6" stiffeners and $750 per MBF for purlins and $405 per MSF for 1/2" CDX plywood.

For scheduling purposes, estimate that a crew of 5 can fabricate 6,000 to 6,400 SF of 4' x 8' or 4' x 10' panels in an 8-hour day. Estimate that the same crew of 5 can install and nail down 7,600 to 8,000 SF of panels in an 8-hour day.

	Craft@Hrs	Unit	Material	Labor	Equipment	Total
4' x 8' roof panels						
With 1/2" CDX structural #1 plywood attached to 2" x 4" const. grade fir stiffeners						
Panels fabricated, ready to install	F6@6.60	MSF	656.00	243.00	43.60	942.60
Equipment & labor						
to install and nail panels	F6@5.21	MSF	—	192.00	34.40	226.40
Total installed cost, 4' x 8' panels	**F6@11.8**	**MSF**	**656.00**	**435.00**	**78.00**	**1,169.00**

	Craft@Hrs	Unit	Material	Labor	Equipment	Total

4' x 10' roof panels
With 1/2" CDX structural #1 plywood attached to 2" x 6" construction grade fir stiffeners.

	Craft@Hrs	Unit	Material	Labor	Equipment	Total
Panels fabricated, ready to install	F6@6.22	MSF	763.00	229.00	41.00	1,033.00
Equipment & labor						
to install and nail panels	F6@4.91	MSF	—	181.00	32.40	213.40
Total installed cost, 4' x 10' panels	**F6@11.1**	**MSF**	**763.00**	**410.00**	**73.40**	**1,246.40**

8' x 20' and 8' x 32' roof panels
These panels can be fabricated on site using the same crew and equipment as shown on the preceding page. However, the stiffeners run across the short dimension of the panel and one purlin is attached to the outside edge of the long dimension of each panel before lifting into place. Figures below include the cost of purlins but not supporting beams.

For scheduling purposes, estimate that a crew of 5 can fabricate 6,400 to 7,000 SF of 8' x 20' or 8' x 32' panels in an 8-hour day. Estimate that the same crew of 5 can install and nail down 10,000 to 11,000 SF of panels in an 8-hour day.

8' x 20' roof panels with purlins attached
4' x 10' x 1/2" CDX structural #1 plywood attached
to 2" x 4" const. grade fir stiffeners and 4" x 12" #1 grade fir purlins

	Craft@Hrs	Unit	Material	Labor	Equipment	Total
Panels fabricated, ready to install	F6@6.22	MSF	1,350.00	229.00	41.10	1,620.10
Equipment & labor						
to install and nail panels	F6@3.96	MSF	—	146.00	26.15	172.15
Total installed cost, 8' x 20' panels	**F6@10.2**	**MSF**	**1,350.00**	**375.00**	**67.25**	**1,792.25**

8' x 32' panel with purlins attached
4' x 8' x 1/2" CDX structural #1 plywood attached
to 2" x 4" const. grade fir stiffeners and 4" x 12" #1 grade fir purlins

	Craft@Hrs	Unit	Material	Labor	Equipment	Total
Panels fabricated, ready to install	F6@5.81	MSF	1,460.00	214.00	38.50	1,712.50
Equipment & labor						
to install and nail panels	F6@3.55	MSF	—	131.00	23.50	154.50
Total installed cost, 8' x 32' panels	**F6@9.36**	**MSF**	**1,460.00**	**345.00**	**62.00**	**1,867.00**
Adjust to any of the panel costs for						
5/8" CDX #1 structural plywood, instead of 1/2"						
attached before panels are erected	—	MSF	105.00	—	—	105.00
Add for roof hatches, includes reinforced opening, curb & hatch,						
installed before panel is erected	F6@.224	SF	22.50	8.25	—	30.75

Panelized Wood Wall Panels These costs will apply on jobs with 20,000 to 30,000 SF of wall panels. Panel material costs are for prefabricated panels purchased from a modular plant located within 50 miles of the job site. Costs will be higher on smaller jobs and on jobs with poor access. Costs assume wall panels are fabricated according to job plans and specifications, delivered piece-marked, ready to install and finish. Labor costs for installation of panels are listed at the end of this section and include unloading and handling, sorting, moving the panels into position using a forklift, installing and nailing. Maximum panel height is usually 12'. Equipment cost is $40 per hour and includes a forklift, hand tools, a portable air compressor with air hoses and two pneumatic-operated nailing guns. Costs are per square foot of wall, measured on one face. For scheduling purposes, estimate that a crew of 4 can place and fasten 3,500 to 4,500 SF of panels in an 8-hour day.

2" x 4" studs 16" OC with gypsum wallboard on both sides

	Craft@Hrs	Unit	Material	Labor	Equipment	Total
1/2" wallboard	—	SF	2.16	—	—	2.16
5/8" wallboard	—	SF	1.71	—	—	1.71

2" x 4" studs 16" OC with gypsum wallboard inside and plywood outside

	Craft@Hrs	Unit	Material	Labor	Equipment	Total
1/2" wallboard and 1/2" CDX plywood	—	SF	2.08	—	—	2.08
1/2" wallboard and 5/8" CDX plywood	—	SF	2.55	—	—	2.55
5/8" wallboard and 1/2" CDX plywood	—	SF	2.41	—	—	2.41
5/8" wallboard and 5/8" CDX plywood	—	SF	2.64	—	—	2.64

	Craft@Hrs	Unit	Material	Labor	Equipment	Total
2" x 6" studs 12" OC with gypsum wallboard on both sides						
1/2" wallboard	—	SF	2.43	—	—	2.43
5/8" wallboard	—	SF	2.58	—	—	2.58
2" x 6" studs 12" OC with gypsum wallboard inside and plywood outside						
1/2" wallboard and 1/2" CDX plywood	—	SF	2.73	—	—	2.73
1/2" wallboard and 5/8" CDX plywood	—	SF	2.83	—	—	2.83
5/8" wallboard and 1/2" CDX plywood	—	SF	2.74	—	—	2.74
5/8" wallboard and 5/8" CDX plywood	—	SF	2.93	—	—	2.93
2" x 6" studs 16" OC with gypsum wallboard on both sides						
1/2" wallboard	—	SF	2.35	—	—	2.35
5/8" wallboard	—	SF	2.54	—	—	2.54
2" x 6" studs 16" OC with gypsum wallboard inside and plywood outside						
1/2" wallboard and 1/2" CDX plywood	—	SF	2.64	—	—	2.64
1/2" wallboard and 5/8" CDX plywood	—	SF	2.83	—	—	2.83
5/8" wallboard and 1/2" CDX plywood	—	SF	2.64	—	—	2.64
5/8" wallboard and 5/8" CDX plywood	—	SF	2.87	—	—	2.87
Add to any of the above for the following, installed by panel fabricator:						
R-19 insulation	—	SF	.52	—	—	.52
R-11 insulation	—	SF	.47	—	—	.47
Vapor barrier	—	SF	.05	—	—	.05
Door or window openings to 16 SF each	—	SF	1.29	—	—	1.29
Door or window openings over 16 SF each	—	SF	1.19	—	—	1.19
Electrical receptacles	—	Ea	36.20	—	—	36.20
Electrical switches	—	Ea	41.40	—	—	41.40
Labor and equipment cost to install panels						
Up to 8' high	F7@.003	SF	—	.12	.0	.12
Over 8' to 12' high	F7@.002	SF	—	.08	.02	.10

	Craft@Hrs	Unit	Material	Labor	Total
Miscellaneous Rough Carpentry					
Furring, Utility grade at $425 per MBF before 5% allowance for waste and nails					
2" x 4" on masonry walls	C8@.024	LF	.30	.87	1.17
1" x 4" machine nailed to ceiling	C8@.017	LF	.15	.62	.77
1" x 4" nailed on concrete	C8@.016	LF	.15	.58	.73
Skip sheathing, on roof rafters, machine nailed, #2 and Btr at $585 per MBF before 7% allowance for waste & nails					
1" x 6" at 9" centers (.67 BF per SF)					
Per SF of roof area	C8@.020	SF	.42	.73	1.15
1" x 4" at 7" centers (.50 BF per SF)					
Per SF of roof area	C8@.030	SF	.31	1.09	1.40
Hardboard, 4' x 8', used as underlayment					
1/8" tempered	C8@.011	SF	.38	.40	.78
1/4" tempered	C8@.011	SF	.48	.40	.88
Add for attaching to metal frame	C8@.005	SF	—	.18	.18
Glassweld panels					
1/8", finish 1 side	C8@.040	SF	4.50	1.45	5.95
1/4", finish 2 sides	C8@.055	SF	6.05	1.99	8.04
15 pound felt on walls	C8@.003	SF	.05	.11	.16
Kraft flashing with foil 2 sides, 30 pound	C8@.004	SF	.10	.15	.25

	Craft@Hrs	Unit	Material	Labor	Total
Insulation board on walls, 4' x 8'					
Sound control board, 1/2"	C8@.010	SF	.40	.36	.76
Building board, 1/2"	C8@.010	SF	.50	.36	.86
Asphalt impregnated 1/2" Firtex	C8@.012	SF	.40	.44	.84

	Craft@Hrs	Unit	Material	Labor	Equipment	Total

Wood Trusses Material costs are for prefabricated wood trusses shipped to job site fully assembled ready for installation. Equipment is a 6000 lb capacity forklift at $25.20 per hour. Labor includes unloading and handling and installing the trusses not over 20' above grade.

	Craft@Hrs	Unit	Material	Labor	Equipment	Total
Truss joists, TJI type at 50 PSF load design, at 16" OC						
9-1/2" TJI/15 ($2.12 LF)	C8@.017	SF	1.69	.62	.21	2.52
11-7/8" TJI/15 ($2.30 LF)	C8@.017	SF	1.84	.62	.21	2.67
14" TJI/35 ($3.32 LF)	C8@.018	SF	2.61	.65	.23	3.49
16" TJI/35 ($3.63 LF)	C8@.018	SF	2.84	.65	.23	3.72
Add for job under 3,000 SF	C8@.004	SF	.46	.15	.05	.66

Roof trusses, 24" OC, any slope from 3 in 12 to 12 in 12, total height not to exceed 12' high from bottom chord to highest point on truss. Prices for trusses over 12' high will be up to 100% higher. Square foot (SF) costs, where shown, are per square foot of roof area to be covered.

	Craft@Hrs	Unit	Material	Labor	Equipment	Total
Scissor trusses						
2" x 4" top and bottom chords						
Up to 38' span	F5@.022	SF	1.85	.82	.11	2.78
40' to 50' span	F5@.028	SF	2.33	1.05	.14	3.52
Fink truss "W" (conventional roof truss)						
2" x 4" top and bottom chords						
Up to 38' span	F5@.017	SF	1.61	.64	.09	2.34
40' to 50' span	F5@.022	SF	1.90	.82	.11	2.83
2" x 6" top and bottom chords						
Up to 38' span	F5@.020	SF	1.94	.75	.10	2.79
40' to 50' span	F5@.026	SF	2.34	.97	.13	3.44
Truss with gable fill at 16" OC						
28' span, 5 in 12 slope	F5@.958	Ea	132.00	35.90	4.82	172.72
32' span, 5 in 12 slope	F5@1.26	Ea	163.00	47.20	6.35	216.55
40' span, 5 in 12 slope	F5@1.73	Ea	233.00	64.80	8.72	306.52

Laminated Glued Beams (Glu-lam beams) The following gives typical costs for horizontal laminated glu-lam beams. Figures in parentheses give approximate weight per linear foot of beam. Equipment cost is based on using a 4,000 lb capacity forklift, with 30' maximum lift, at $21 per hour.

	Craft@Hrs	Unit	Material	Labor	Equipment	Total
Steel connectors and shoes for glu-lam beams, including bolts,						
Typical per end or splice connection	—	Ea	61.80	—	—	61.80

Drill holes in glu-lam beams and attach connectors at grade before installing beam, based on number of holes in the connector, with work performed at job site

	Craft@Hrs	Unit	Material	Labor	Equipment	Total
Labor per each hole	F6@.236	Ea	—	8.69	—	8.69
Install beams with connections attached	F6@.133	LF	—	4.90	.56	5.46
Stock glu-lam beams, typical costs						

West coast Douglas fir/hemlock, 24F-V5, H=155 PSI, exterior glue, camber 5000' R
Industrial grade, as is, no wrap. For other grades see additive adjustment data that follows below.

	Craft@Hrs	Unit	Material	Labor	Equipment	Total
2-1/2" x 6" (3.6 pounds per LF)	—	LF	3.63	—	—	3.63
2-1/2" x 9" (5.4 pounds per LF)	—	LF	5.17	—	—	5.17
2-1/2" x 12" (7.1 pounds per LF)	—	LF	7.23	—	—	7.23

West coast Douglas fir, 24F-V4, H=165 PSI, exterior glue, camber is 5000' R for beams through 15" deep and 2000' R for beams over 15" deep.

	Craft@Hrs	Unit	Material	Labor	Equipment	Total
Industrial grade, as is, no wrap. For other grades see additive adjustment data that follows.						
3-1/8" x 6" (4.4 pounds per LF)	—	LF	4.03	—	—	4.03
3-1/8" x 9" (6.6 pounds per LF)	—	LF	5.95	—	—	5.95
3-1/8" x 10-1/2" or 12"						
(8.0 pounds per LF)	—	LF	7.00	—	—	7.00
3-1/8" x 15" (10.0 pounds per LF)	—	LF	9.95	—	—	9.95
3-1/8" x 18" (13.2 pounds per LF)	—	LF	12.00	—	—	12.00
5-1/8" x 6" (7.3 pounds per LF)	—	LF	6.21	—	—	6.21
5-1/8" x 9" (10.7 pounds per LF)	—	LF	9.01	—	—	9.01
5-1/8" x 12" (14.5 pounds per LF)	—	LF	12.00	—	—	12.00
5-1/8" x 15" (18.1 pounds per LF)	—	LF	15.00	—	—	15.00
5-1/8" x 18" (21.7 pounds per LF)	—	LF	17.80	—	—	17.80
5-1/8" x 21" (25.4 pounds per LF)	—	LF	21.00	—	—	21.00
5-1/8" x 24" (29.0 pounds per LF)	—	LF	24.10	—	—	24.10
6-3/4" x 12" (19.1 pounds per LF)	—	LF	16.00	—	—	16.00
6-3/4" x 15" (23.8 pounds per LF)	—	LF	20.10	—	—	20.10
6-3/4" x 18" (28.6 pounds per LF)	—	LF	24.10	—	—	24.10
6-3/4" x 21" (33.3 pounds per LF)	—	LF	28.10	—	—	28.10
6-3/4" x 24" (38.1 pounds per LF)	—	LF	32.90	—	—	32.90
Additive adjustments, add to any of the beam costs shown in this section for:						
Industrial grade, clean, wrapped, add	—	%	12.5	—	—	—
Architectural grade, wrapped, add	—	%	20.0	—	—	—
One-coat rez-latex stain, add	—	%	10.0	—	—	—
Transportation to job site, per delivery,						
maximum 50 miles one way, add	—	LS	200.00	—	—	200.00

	Craft@Hrs	Unit	Material	Labor	Total
Stairs Job-built stairways, 3 stringers cut from 2" x 12" material (at $600 per MBF), treads and risers from 3/4" CDX plywood (at $612 per MSF).					
Cost per 7-1/2" rise, 36" wide.					
Straight run, 8'0" to 10'0" rise, per riser	C8@.530	Ea	9.69	19.20	28.89
"L" or "U" shape, per riser, add for landings	C8@.625	Ea	11.00	22.70	33.70
Landings, framed and 3/4" CDX plywood surfaced					
Per SF of landing surface	C8@.270	SF	2.29	9.79	12.08
Factory cut and assembled straight closed box stairs, unfinished, 36" width					
Oak treads with 7-1/2" risers, price per riser					
3'0" to 3'6" wide, pre riser	C8@.324	Ea	104.00	11.80	115.80
4'0" wide, per riser	C8@.357	Ea	125.00	12.90	137.90
Add for prefinished assembled stair rail with					
balusters and newel, per riser	C8@.257	Ea	45.80	9.32	55.12
Add for prefinished handrail, brackets and balusters	C8@.135	LF	13.50	4.90	18.40
Spiral stairs, 5' diameter. Steel, tubular handrail, composition board treads, prefabricated shipped unassembled,					
86" to 95" with 10 risers	C8@9.98	Ea	878.00	362.00	1,240.00
95" to 105" with 11 risers	C8@10.7	Ea	970.00	388.00	1,358.00
105" to 114" with 12 risers	C8@11.3	Ea	1,020.00	410.00	1,430.00
114" to 124" with 13 risers	C8@12.0	Ea	1,120.00	435.00	1,555.00
124" to 133" with 14 risers	C8@12.7	Ea	1,280.00	461.00	1,741.00
Add for 6' diameter	—	%	24.0	—	—
Add for oak treads	—	%	25.0	—	—
Add for oak handrail	—	%	45.0	—	—

	Craft@Hrs	Unit	Material	Labor	Total
Dampproofing					
Asphalt wall primer, per coat, gallon at $7.50					
covers 250 SF	CL@.010	SF	.03	.30	.33
Asphalt emulsion, wall, per coat, gallon at $2.00					
covers 33 SF, brush on	CL@.013	SF	.06	.39	.45
Hot mop concrete wall, 2 coats and glass fabric	CL@.033	SF	.30	1.00	1.30
Hot mop deck, 2 ply, felt, typical subcontract price	—	Sq	—	—	75.00
Hot mop deck, 3 ply, felt, typical subcontract price	—	Sq	—	—	100.00
Add for 1/2" asphalt fiberboard, subcontract price	—	Sq	—	—	50.00
Bituthene waterproofing membrane,					
1/16" plain surface	CL@.025	SF	2.45	.76	3.21
Waterproof baths in high rise	CL@.025	SF	1.27	.76	2.03
Iron compound, internal coating, 2 coats	CL@.012	SF	1.16	.36	1.52
Roof vapor barrier, .004" polyethylene	CL@.002	SF	.05	.06	.11

Exterior Wall Insulation and Finish System (EIFS) Polymer based exterior non-structural wall finish applied to concrete, masonry, stucco or exterior grade gypsum sheathing (such as Dryvit). System consists of a plaster base coat approximately 1/4" thick, and a glass fiber mesh that is embedded in the adhesive base coat. The base coat is applied over polystyrene insulation boards that are bonded to an existing flat vertical wall surface with adhesive. Costs shown are per SF of finished wall surface on structures not over three stories in height. These costs do not include surface preparation or wall costs. Cost for scaffolding is not included: add for scaffolding if required.

For scheduling purposes, estimate that a crew of three can field apply 400 SF of exterior wall insulation and finish systems in an 8-hour day.

	Craft@Hrs	Unit	Material	Labor	Total
Adhesive mixture, 2 coats					
2.1 SF per SF of wall area including waste	F8@.013	SF	.53	.47	1.00
Glass fiber mesh					
1.1 SF per SF of wall area including waste	F8@.006	SF	.30	.22	.52
Insulation board, 1" thick					
1.05 SF per SF of wall area including waste	F8@.013	SF	.19	.47	.66
Textured finish coat, with integral color.					
1.05 SF per SF of wall area including waste	F8@.024	SF	.53	.87	1.40
Total with 1" thick insulation	F8@.056	SF	1.54	2.02	3.56
Add for 2" thick insulation board	—	SF	.21	—	.21
Add for 3" thick insulation board	—	SF	.42	—	.42
Add for 4" thick insulation board	—	SF	.61	—	.61

Insulation Roof insulation. Due to regulations that may be imposed on manufacturers by government agencies, estimators should verify with material suppliers the cost of urethanes, isocyanurates, phenolics and polystyrene roof insulation board before preparing estimates on major projects.

	Craft@Hrs	Unit	Material	Labor	Total
Fiberglass board roof insulation					
3/4", R-2.80, C-0.36	R3@.570	Sq	40.50	20.20	60.70
1", R-4.20, C-0.24	R3@.636	Sq	48.60	22.50	71.10
1-3/8", R-5.30, C-0.19	R3@.684	Sq	62.50	24.20	86.70
1-5/8", R-6.70, C-0.15	R3@.706	Sq	70.70	25.00	95.70
2-1/4", R-8.30, C-0.12	R3@.797	Sq	85.80	28.20	114.00
Perlite-urethane composition board roof insulation					
1-1/2", R-6.70, C-0.15	R3@.478	Sq	80.40	16.90	97.30
1-5/8", R-7.70, C-0.13	R3@.478	Sq	85.80	16.90	102.70
2", R-10.0, C-0.10	R3@.684	Sq	102.00	24.20	126.20
2-1/4", R-12.5, C-0.08	R3@.728	Sq	124.00	25.80	149.80
2-1/2", R-14.3, C-0.07	R3@.728	Sq	139.00	25.80	164.80
2-3/4", R-16.7, C-0.06	R3@.820	Sq	164.00	29.00	193.00
3-1/4", R-20.0, C-0.05	R3@.889	Sq	197.00	31.50	228.50

	Craft@Hrs	Unit	Material	Labor	Total
Phenolic board roof insulation					
1-1/4", R-10.0, C-0.10	R3@.500	Sq	53.70	17.70	71.40
1-1/2", R-12.5, C-0.08	R3@.592	Sq	60.20	20.90	81.10
1-3/4", R-14.6, C-0.07	R3@.592	Sq	65.50	20.90	86.40
2", R-16.7, C-0.06	R3@.684	Sq	75.50	24.20	99.70
2-1/2", R-20.0, C-0.05	R3@.728	Sq	82.40	25.80	108.20
3", R-25.0, C-0.04	R3@.889	Sq	110.00	31.50	141.50
3-1/4", R-28.0, C-0.035	R3@1.05	Sq	142.00	37.20	179.20
Urethane board roof insulation					
3/4", R-5.30, C-0.19	R3@.478	Sq	45.40	16.90	62.30
1", R-6.70, C-0.15	R3@.478	Sq	59.20	16.90	76.10
1-1/4", R-8.30, C-0.12	R3@.478	Sq	65.50	16.90	82.40
1-1/2", R-11.1, C-0.09	R3@.570	Sq	78.90	20.20	99.10
2", R-14.3, C-0.07	R3@.684	Sq	104.00	24.20	128.20
2-1/4", R-16.7, C-0.06	R3@.684	Sq	112.00	24.20	136.20
Perlite board roof insulation					
1", R-2.80, C-0.36	R3@.639	Sq	45.40	22.60	68.00
1-1/2", R-4.20, C-0.24	R3@.728	Sq	66.50	25.80	92.30
2", R-5.30, C-0.19	R3@.820	Sq	81.60	29.00	110.60
2-1/2", R-6.70, C-0.15	R3@.956	Sq	102.00	33.80	135.80
3", R-8.30, C-0.12	R3@1.03	Sq	124.00	36.40	160.40
4", R-10.0, C-0.10	R3@1.23	Sq	164.00	43.50	207.50
5-1/4", R-14.3, C-0.07	R3@1.57	Sq	217.00	55.50	272.50
Cellular Foamglass board insulation, on roofs					
1-1/2"	R3@.775	Sq	164.00	27.40	191.40
Tapered 3/16" per foot	R3@.775	Sq	149.00	27.40	176.40
Polystyrene board roof insulation					
1", R-4.20, C-0.24	R3@.545	Sq	34.70	19.30	54.00
1-1/2", R-6.30, C-0.16	R3@.639	Sq	46.50	22.60	69.10
2", R-8.30, C-0.12	R3@.797	Sq	60.20	28.20	88.40
Fasteners for board type insulation on roofs					
Typical 2-1/2" to 3-1/2" screws with 3" metal disks					
I-60	R3@.053	Sq	2.59	1.88	4.47
I-90	R3@.080	Sq	3.92	2.83	6.75
Add for plastic disks	—	Sq	.65	—	.65
Fiberglass batts, placed between or over framing members					
3-1/2" unfaced, R-11, between ceiling joists	A1@.006	SF	.27	.22	.49
3-1/2" unfaced, R-11, on suspended ceiling; working off ladders from below	A1@.018	SF	.27	.66	.93
3-1/2" unfaced, R-11, in crawl space	A1@.020	SF	.27	.73	1.00
3-1/2" unfaced, R-19, between studs	A1@.006	SF	.28	.22	.50
6-1/4" unfaced, R-19, between ceiling joists	A1@.006	SF	.41	.22	.63
6-1/4" unfaced, R-19, on suspended ceiling, working off ladders from below	A1@.018	SF	.42	.66	1.08
6-1/4" unfaced, R-19, in crawl space	A1@.020	SF	.42	.73	1.15
6-1/4" unfaced, R-19, between studs	A1@.006	SF	.42	.22	.64
Add for supporting batts on wire rods, 1 per SF	A1@.001	SF	.03	.04	.07
Add for foil one side	—	SF	.03	—	.03
Add for foil two sides	—	SF	.06	—	.06
Add for Kraft paper one side	—	SF	.03	—	.03
Add for Kraft paper two sides	—	SF	.03	—	.03

	Craft@Hrs	Unit	Material	Labor	Total
Fasteners for batts placed under a ceiling deck, including flat washers					
2-1/2" long	A1@.025	Ea	.16	.92	1.08
4-1/2" long	A1@.025	Ea	.18	.92	1.10
6-1/2" long	A1@.025	Ea	.25	.92	1.17
Domed self-locking washers	A1@.005	Ea	.14	.18	.32
Blown fiberglass or mineral wool, over ceiling joists. Add blowing equipment cost at $300 per day					
R-11, 5"	A1@.008	SF	.16	.29	.45
R-13, 6"	A1@.010	SF	.21	.37	.58
R-19, 9"	A1@.014	SF	.28	.51	.79
Urethane board installed on perimeter walls					
1", R-6.70, C-0.15	A1@.010	SF	.60	.37	.97
1-1/2", R-11.0, C-0.09	A1@.010	SF	.77	.37	1.14
2", R-14.3, C-0.07	A1@.011	SF	1.05	.40	1.45
Polystyrene board installed on perimeter walls					
1", R-4.20, C-0.24	A1@.010	SF	.39	.37	.76
1-1/2", R-6.30, C-0.16	A1@.010	SF	.49	.37	.86
2", R-8.30, C-0.12	A1@.010	SF	.62	.37	.99
Perlite or vermiculite (at $2.29 per CF), poured in concrete block cores					
4" wall, 8.1 SF per CF	M1@.004	SF	.28	.13	.41
6" wall, 5.4 SF per CF	M1@.007	SF	.43	.24	.67
8" wall, 3.6 SF per CF	M1@.009	SF	.64	.30	.94
10" wall, 3.0 SF per CF	M1@.011	SF	.76	.37	1.13
12" wall, 2.1 SF per CF	M1@.016	SF	1.10	.54	1.64
Poured perlite or vermiculite in wall cavities	M1@.032	CF	2.26	1.08	3.34
Sound board, 1/2", on walls	C8@.010	SF	.39	.36	.75
Sound board, 1/2", on floors	C8@.009	SF	.39	.33	.72
Cold box insulation					
2" polystyrene	A1@.012	SF	1.11	.44	1.55
1" cork	A1@.013	SF	.90	.48	1.38
Add for overhead work or poor access areas	—	SF	—	.07	.07
Add for enclosed areas	—	SF	—	.07	.07
Add for 1 hour fire rating	—	SF	.20	—	.20

Aluminum Wall Cladding System, Exterior These costs assume installation by a manufacturer-trained contractor, on a framed and insulated exterior building surface, including track system, panels and trim. Alply, Inc.

	Craft@Hrs	Unit	Material	Labor	Total
Techwall solid aluminum panel, 1/8" thick, joints sealed with silicon sealer					
Kynar 500 paint finish	SM@.158	SF	18.50	6.53	25.03
Anodized aluminum finish	SM@.158	SF	18.20	6.53	24.73
Add for radius panels	SM@.032	SF	3.80	1.32	5.12

Cladding, Preformed Roofing and Siding Applied on metal framing.
Architectural insulated metal panels with smooth .040 aluminum exterior face with Kynar 500 finish, 2" thick isocyanurate insulation, .040 mill finish aluminum interior sheet, shop applied extrusion and dry seal gasketed joint system, shop fabricated. Alply, Inc.

	Craft@Hrs	Unit	Material	Labor	Total
Kynar 500 paint finish	SM@.136	SF	13.50	5.62	19.12
Add for radius panels	SM@.020	SF	1.90	.83	2.73
Corrugated or ribbed roofing, colored galvanized steel, 9/16" deep, uninsulated					
18 gauge	SM@.026	SF	2.29	1.07	3.36
20 gauge	SM@.026	SF	2.01	1.07	3.08
22 gauge	SM@.026	SF	1.79	1.07	2.86
24 gauge	SM@.026	SF	1.53	1.07	2.60

	Craft@Hrs	Unit	Material	Labor	Total
Corrugated siding, colored galvanized steel, 9/16" deep, uninsulated					
18 gauge	SM@.034	SF	2.29	1.40	3.69
20 gauge	SM@.034	SF	2.01	1.40	3.41
22 gauge	SM@.034	SF	1.79	1.40	3.19
Ribbed siding, colored galvanized steel, 1-3/4" deep, box rib					
18 gauge	SM@.034	SF	2.41	1.40	3.81
20 gauge	SM@.034	SF	2.11	1.40	3.51
22 gauge	SM@.034	SF	1.87	1.40	3.27
Corrugated aluminum roofing, natural finish					
.020" aluminum	SM@.026	SF	1.48	1.07	2.55
.032" aluminum	SM@.026	SF	1.82	1.07	2.89
Add for factory painted finish	—	SF	.19	—	.19
Corrugated aluminum siding, natural finish					
.020" aluminum	SM@.034	SF	1.48	1.40	2.88
.032" aluminum	SM@.034	SF	1.82	1.40	3.22
Add for factory painted finish	—	SF	.19	—	.19
Solid vinyl, .040", horizontal siding	SM@.026	SF	.77	1.07	1.84
Add for insulated siding backer panel	SM@.001	SF	.16	.04	.20
Vinyl window and door trim	SM@.040	LF	.27	1.65	1.92
Vinyl siding and fascia system	SM@.032	SF	.79	1.32	2.11
Laminated sandwich panel siding, enameled 22 gauge aluminum face one side, polystyrene core					
1" core	SM@.069	SF	4.08	2.85	6.93
1-1/2" core	SM@.069	SF	4.54	2.85	7.39

Membrane Roofing (One square "Sq" = 100 square feet.)

	Craft@Hrs	Unit	Material	Labor	Total
Built-up roofing, asphalt felt					
3 ply, smooth surface top sheet	R3@1.61	Sq	55.70	57.00	112.70
4 ply, base sheet, 3-ply felt, smooth surface top sheet	R3@1.85	Sq	65.40	65.50	130.90
5 ply, 20 year	R3@2.09	Sq	78.00	73.90	151.90
Built-up fiberglass felt roof					
Base and 3-ply felt	R3@1.93	Sq	73.20	68.30	141.50
Add for each additional ply	R3@.188	Sq	11.10	6.65	17.75
Add for light rock dress off (250 lbs per Sq)	R3@.448	Sq	10.70	15.90	26.60
Add for heavy rock dress off (400 lbs per Sq)	R3@.718	Sq	15.60	25.40	41.00
Remove and replace 400 lb per square rock					
Including new flood coat	R3@1.32	Sq	25.90	46.70	72.60
Aluminized coating for built-up roofing	R3@.599	Sq	15.30	21.20	36.50
Fire rated type	R3@.609	Sq	19.50	21.50	41.00
Modified APP, SBS flashing membrane	R3@.033	SF	.68	1.17	1.85
Cap sheet	R3@.277	Sq	19.50	9.80	29.30
Modified asphalt, APP, SBS base sheet and membrane	R3@1.97	Sq	73.20	69.70	142.90
Roll roofing					
90 lb mineral surface	R3@.399	Sq	22.20	14.10	36.30
19" wide selvage edge mineral surface (110 lb)	R3@.622	Sq	42.00	22.00	64.00
Double coverage selvage edge roll (140 lb)	R3@.703	Sq	42.00	24.90	66.90
Asphalt impregnated walkway for built-up roofing					
1/2"	R3@.016	SF	1.43	.57	2.00
3/4"	R3@.017	SF	1.81	.60	2.41
1"	R3@.020	SF	2.18	.71	2.89
Strip off existing 4-ply roof, no disposal included	R3@1.32	Sq	—	46.70	46.70

	Craft@Hrs	Unit	Material	Labor	Total
Elastomeric Roofing					
Butyl					
1/16"	R3@1.93	Sq	125.00	68.30	193.30
1/32"	R3@1.93	Sq	109.00	68.30	177.30
Neoprene, 1/16"	R3@2.23	Sq	188.00	78.90	266.90
Acrylic-urethane foam roof system					
Remove gravel and prepare existing built-up roof	R3@1.26	Sq	4.65	44.60	49.25
Apply base and top coats of acrylic (.030")	R3@.750	Sq	100.00	26.50	126.50
Urethane foam 1" thick (R-7.1)	R3@.280	Sq	88.90	9.91	98.81
Urethane foam 2" thick (R-14.3)	R3@1.10	Sq	150.00	38.90	188.90
Spray-on mineral granules, 40 lbs per CSF	R3@.600	Sq	14.50	21.20	35.70
EPDM roofing system (Ethylene Propylene Diene Monomer)					
.045" loose laid membrane	R3@.743	Sq	82.80	26.30	109.10
.060" adhered membrane (excluding adhesive)	R3@1.05	Sq	94.60	37.20	131.80
Bonding adhesive	R3@.305	Sq	38.40	10.80	49.20
Lap splice cement, per 100 LF	R3@.223	CLF	22.60	7.89	30.49
Ballast rock, 3/4" to 1-1/2" gravel	R3@.520	Sq	15.60	18.40	34.00
Neoprene sheet flashing, .060"	R3@.021	SF	2.06	.74	2.80
CSPE (Chloro-sulfonated Polyethylene)					
.045" single ply membrane	R3@.743	Sq	130.00	26.30	156.30
Add for mechanical fasteners and washers					
6" long, 12" OC at laps	—	Ea	.61	—	.61
12" long, 12" OC at laps	—	Ea	1.23	—	1.23
Cant Strips					
3" fiber	R3@.017	LF	.42	.60	1.02
6" fiber	R3@.017	LF	.75	.60	1.35
4" fiber or wood	R3@.017	LF	.73	.60	1.33
6" wood	R3@.017	LF	.75	.60	1.35
Flashing and Sheet Metal					
Prefinished sheet metal fascia and mansards, standing beam and batten					
Straight or simple	SM@.054	SF	3.36	2.23	5.59
Curved or complex	SM@.082	SF	5.06	3.39	8.45
Coping, gravel stop and flashing	SM@.037	SF	2.45	1.53	3.98
Sheet metal wainscot, galvanized	SM@.017	SF	.68	.70	1.38
Aluminum flashing, .032"					
Coping and wall cap, 16" girth	SM@.069	LF	2.79	2.85	5.64
Counter flash, 8"	SM@.054	LF	2.45	2.23	4.68
Reglet flashing, 8"	SM@.058	LF	1.01	2.40	3.41
Neoprene gasket for flashing	SM@.015	LF	.77	.62	1.39
Gravel stop and fascia, 10"	SM@.084	LF	2.91	3.47	6.38
Side wall flashing, 9"	SM@.027	LF	1.51	1.12	2.63
Gravel stop					
4"	SM@.058	LF	3.31	2.40	5.71
6"	SM@.063	LF	3.75	2.60	6.35
8"	SM@.068	LF	4.53	2.81	7.34
12"	SM@.075	LF	7.54	3.10	10.64
Valley, 24"	SM@.070	LF	1.95	2.89	4.84
Copper flashing, 16 oz					
Coping and wall cap, 16" girth	SM@.074	LF	6.13	3.06	9.19
Counter flash, 8"	SM@.057	LF	3.81	2.35	6.16

	Craft@Hrs	Unit	Material	Labor	Total
Gravel stop					
4"	SM@.058	LF	4.75	2.40	7.15
6"	SM@.063	LF	5.51	2.60	8.11
8"	SM@.068	LF	6.46	2.81	9.27
10"	SM@.075	LF	7.94	3.10	11.04
Reglet flashing, 8"	SM@.054	LF	2.96	2.23	5.19
Neoprene gasket for flashing	SM@.015	LF	.70	.62	1.32
Side wall flashing, 6"	SM@.027	LF	2.08	1.12	3.20
Base, 20 oz	SM@.072	LF	4.30	2.97	7.27
Valley, 24"	SM@.069	LF	3.85	2.85	6.70
Sheet lead flashing					
Plumbing vents, soil stacks, 4 lbs/SF material					
4" pipe size	P6@.709	Ea	19.40	25.60	45.00
6" pipe size	P6@.709	Ea	22.50	25.60	48.10
8" pipe size	P6@.709	Ea	27.70	25.60	53.30
Roof drains, flat pan type, 36" x 36"					
2.5 lbs/SF material	P6@1.49	Ea	23.80	53.80	77.60
4 lbs/SF material	P6@1.49	Ea	40.10	53.80	93.90
6 lbs/SF material	P6@1.49	Ea	57.00	53.80	110.80
Stainless steel flashing					
Fascia roof edge, .018"	SM@.054	LF	3.78	2.23	6.01
Base, .018"	SM@.054	LF	4.52	2.23	6.75
Counter flash, .015"	SM@.054	LF	3.56	2.23	5.79
Valley, .015", 24"	SM@.069	LF	3.54	2.85	6.39
Reglets, .020", 8"	SM@.035	LF	3.08	1.45	4.53
Galvanized sheet metal flashing					
Cap and counter flash, 8"	SM@.046	LF	1.11	1.90	3.01
Coping and wall cap, 16" girth	SM@.069	LF	1.45	2.85	4.30
Gravel stop, 6"	SM@.060	LF	.44	2.48	2.92
Neoprene gasket	SM@.005	LF	.63	.21	.84
Pitch pockets, 24 gauge, filled					
4" x 4"	SM@.584	Ea	16.50	24.10	40.60
6" x 6"	SM@.584	Ea	20.90	24.10	45.00
8" x 8"	SM@.584	Ea	24.70	24.10	48.80
8" x 10"	SM@.777	Ea	38.90	32.10	71.00
8" x 12"	SM@.777	Ea	34.40	32.10	66.50
Reglet flashing, 8"	SM@.054	LF	1.12	2.23	3.35
Shingles (chimney flash)	RF@.064	LF	.85	2.43	3.28
Side wall flashing, 9"	SM@.027	LF	.83	1.12	1.95
"W" valley, 24"	SM@.069	LF	1.64	2.85	4.49
Plumbers counter flash cone, 4" diameter	P8@.204	Ea	5.46	6.91	12.37
Roof safes and caps					
4" galvanized sheet metal	SM@.277	Ea	10.50	11.40	21.90
4" aluminum	SM@.277	Ea	11.90	11.40	23.30
Pitch pockets, 16 oz copper, filled					
4" x 4"	SM@.547	Ea	29.80	22.60	52.40
6" x 6"	SM@.547	Ea	38.00	22.60	60.60
8" x 8"	SM@.547	Ea	44.50	22.60	67.10
8" x 10"	SM@.679	Ea	52.20	28.00	80.20
8" x 12"	SM@.679	Ea	61.50	28.00	89.50

	Craft@Hrs	Unit	Material	Labor	Total
Galvanized corner guards, 4" x 4"	SM@.039	LF	11.80	1.61	13.41
Copper sheet metal roofing, 16 oz					
Batten seam	SM@.065	SF	3.22	2.68	5.90
Standing seam	SM@.056	SF	3.74	2.31	6.05

Gutters and Downspouts

	Craft@Hrs	Unit	Material	Labor	Total
Aluminum, .032", including hangers					
Fascia gutter, 5"	SM@.050	LF	1.63	2.07	3.70
Box gutter, 4"	SM@.050	LF	1.42	2.07	3.49
Dropouts, elbows, for either of above	SM@.101	Ea	3.16	4.17	7.33
Downspouts, to 24' height					
2" x 3"	SM@.037	LF	1.16	1.53	2.69
3" x 4"	SM@.047	LF	1.82	1.94	3.76
4" diameter, round, 12' to 24' high	SM@.047	LF	1.46	1.94	3.40
Add for height over 24'	SM@.017	LF	—	.70	.70
Copper, 16 oz, including hangers					
Box gutter, 4"	SM@.054	LF	3.96	2.23	6.19
Half round gutter					
4" gutter	SM@.039	LF	3.43	1.61	5.04
5" gutter	SM@.051	LF	4.29	2.11	6.40
6" gutter	SM@.056	LF	5.31	2.31	7.62
Dropouts, elbows, for either of above	SM@.154	Ea	8.59	6.36	14.95
Downspouts, to 24' height					
2" x 3"	SM@.036	LF	3.78	1.49	5.27
3" x 4"	SM@.047	LF	5.12	1.94	7.06
3" diameter	SM@.037	LF	2.52	1.53	4.05
4" diameter	SM@.047	LF	3.84	1.94	5.78
5" diameter	SM@.049	LF	4.65	2.02	6.67
Add for heights over 24'	SM@.017	LF	—	.70	.70
Scuppers, copper, 16 oz,					
8" x 8"	SM@.972	Ea	97.90	40.10	138.00
10" x 10"	SM@.972	Ea	116.00	40.10	156.10
Galvanized steel, 26 gauge, including hangers					
Box gutter 4"	SM@.054	LF	1.08	2.23	3.31
Dropouts, elbows	SM@.082	Ea	1.66	3.39	5.05
Fascia gutter					
5" face	SM@.054	LF	1.28	2.23	3.51
7" face	SM@.062	LF	1.72	2.56	4.28
Dropouts, elbows, for either of above	SM@.088	Ea	2.53	3.63	6.16
Downspouts, with all fittings, height to 24'					
2" x 3"	SM@.036	LF	.91	1.49	2.40
3" x 4"	SM@.047	LF	1.39	1.94	3.33
3" round	SM@.037	LF	1.39	1.53	2.92
6" round	SM@.055	LF	1.30	2.27	3.57
Add for heights over 24'	SM@.017	LF	—	.70	.70
Roof sump, 18" x 18" x 5"	SM@.779	Ea	7.07	32.18	39.25
Scupper, 6" x 6" x 8"	SM@.591	Ea	21.10	24.40	45.50
Stainless steel, .015" thick, including hangers					
Box gutter, 5" wide, 4-1/2" high	SM@.054	LF	6.79	2.23	9.02
Dropouts, elbows	SM@.076	Ea	10.00	3.14	13.14
Downspout, 4" x 5", to 24' high	SM@.051	LF	6.08	2.11	8.19
Add for heights over 24'	SM@.017	LF	—	.70	.70

	Craft@Hrs	Unit	Material	Labor	Total
Vents, Louvers and Screens					
Fixed louvers, with screen, typical	SM@.402	SF	10.20	16.60	26.80
Door louvers, typical	SM@.875	Ea	21.90	36.10	58.00
Manual operating louvers, typical	SM@.187	SF	16.20	7.72	23.92
Cooling tower screens	SM@.168	SF	9.00	6.94	15.94
Bird screens with frame	SM@.028	SF	1.35	1.16	2.51
Concrete block vents, aluminum	M1@.370	SF	12.00	12.40	24.40
Frieze vents, with screen, 14" x 4"	C8@.198	Ea	2.49	7.18	9.67
Foundation vents, 6" x 14", ornamental	C8@.317	Ea	6.73	11.50	18.23
Attic vents, with louvers, 14" x 24"	C8@.377	Ea	19.60	13.70	33.30
Architectural facade screen, aluminum	SM@.187	SF	18.80	7.72	26.52
Add for enamel or light anodized	—	SF	2.09	—	2.09
Add for porcelain or heavy anodized	—	SF	5.10	—	5.10
Roof Accessories					
Roof hatches, steel, not including ladder					
30" x 30"	C8@3.09	Ea	333.00	112.00	445.00
30" x 72"	C8@4.69	Ea	805.00	170.00	975.00
36" x 36"	C8@3.09	Ea	381.00	112.00	493.00
36" x 72"	C8@4.69	Ea	852.00	170.00	1,022.00
Ceiling access hatches					
30" x 30"	C8@1.59	Ea	201.00	57.70	258.70
42" x 42"	C8@1.80	Ea	240.00	68.50	308.50
Smoke vents					
Aluminum, 48" x 48"	C8@3.09	Ea	1,020.00	112.00	1,132.00
Galvanized, 48" x 48"	C8@3.09	Ea	886.00	112.00	998.00
Fusible "shrink-out" heat and smoke vent (PVC dome in aluminum frame),					
4' x 8'	C8@5.15	Ea	805.00	187.00	992.00
Roof scuttle, aluminum, 2'6" x 3'0"	C8@3.50	Ea	404.00	127.00	531.00
Ventilators					
Rotary, wind driven					
6" diameter	SM@.913	Ea	56.20	37.70	93.90
12" diameter	SM@.913	Ea	68.50	37.70	106.20
24" diameter	SM@1.36	Ea	195.00	56.20	251.20
Ventilators, mushroom type, motorized, single speed, including damper and bird screen but no electrical work					
8", 180 CFM	SM@2.93	Ea	276.00	121.00	397.00
12", 1,360 CFM	SM@2.93	Ea	457.00	121.00	578.00
18", 2,000 CFM	SM@4.00	Ea	657.00	165.00	822.00
24", 4,000 CFM	SM@4.86	Ea	1,010.00	201.00	1,211.00
Add for two-speed motor	—	%	35.0	—	—
Add for explosive proof units	—	Ea	377.00	—	377.00

Sealants and Caulking By joint size. Material prices are based on C.R. Laurence Co. products in 5 gallon quantities. Costs assume 10% waste. Twelve 10.6 ounce cartridges equals one gallon and yields 231 cubic inches of caulk or sealant. Productivity shown assumes one roofer using a bulk dispenser. Add the cost of joint cleaning, if required.

	Craft@Hrs	Unit	Material	Labor	Total
Oil base caulk, $16.00 per gallon					
1/4" x 1/4", 277 LF/gallon, 63 LF per hour	RF@.016	LF	.06	.61	.67
1/4" x 3/8", 185 LF/gallon, 60 LF per hour	RF@.017	LF	.10	.64	.74
1/4" x 1/2", 139 LF/gallon, 58 LF per hour	RF@.017	LF	.13	.64	.77
3/8" x 3/8", 123 LF/gallon, 58 LF per hour	RF@.017	LF	.14	.64	.78
3/8" x 1/2", 92 LF/gallon, 57 LF per hour	RF@.017	LF	.19	.64	.83

	Craft@Hrs	Unit	Material	Labor	Total
3/8" x 5/8", 74 LF/gallon, 56 LF per hour	RF@.018	LF	.24	.68	.92
3/8" x 3/4", 60 LF/gallon, 52 LF per hour	RF@.019	LF	.29	.72	1.01
1/2" x 1/2", 69 LF/gallon, 56 LF per hour	RF@.018	LF	.26	.68	.94
1/2" x 5/8", 55 LF/gallon, 50 LF per hour	RF@.020	LF	.32	.76	1.08
1/2" x 3/4", 46 LF/gallon, 48 LF per hour	RF@.021	LF	.38	.80	1.18
1/2" x 1", 35 LF/gallon, 45 LF per hour	RF@.022	LF	.50	.83	1.33
1" x 1", 17 LF/gallon, 30 LF per hour	RF@.033	LF	1.04	1.25	2.29
Butyl rubber sealant, $23.50 per gallon					
1/4" x 1/4", 277 LF/gallon, 50 LF per hour	RF@.020	LF	.08	.76	.84
1/2" x 1/2", 69 LF/gallon, 45 LF per hour	RF@.022	LF	.34	.83	1.17
1/2" x 3/4", 46 LF/gallon, 40 LF per hour	RF@.025	LF	.51	.95	1.46
3/4" x 3/4", 31 LF/gallon, 35 LF per hour	RF@.029	LF	.76	1.10	1.86
1" x 1", 17 LF/gallon, 24 LF per hour	RF@.042	LF	1.38	1.59	2.97
Polyurethane, 1 or 2 component sealant, $34.40 per gallon					
1/4" x 1/4", 277 LF/gallon, 63 LF per hour	RF@.016	LF	.14	.62	.76
1/4" x 3/8", 185 LF/gallon, 60 LF per hour	RF@.017	LF	.20	.64	.84
1/4" x 1/2", 139 LF/gallon, 58 LF per hour	RF@.017	LF	.27	.64	.91
3/8" x 3/8", 123 LF/gallon, 58 LF per hour	RF@.017	LF	.30	.64	.94
3/8" x 1/2", 92 LF/gallon, 57 LF per hour	RF@.017	LF	.41	.64	1.05
3/8" x 5/8", 74 LF/gallon, 56 LF per hour	RF@.018	LF	.51	.68	1.19
3/8" x 3/4", 60 LF/gallon, 52 LF per hour	RF@.019	LF	.62	.72	1.34
1/2" x 1/2", 69 LF/gallon, 56 LF per hour	RF@.018	LF	.54	.68	1.22
1/2" x 5/8", 55 LF/gallon, 50 LF per hour	RF@.020	LF	.68	.76	1.44
1/2" x 3/4", 46 LF/gallon, 48 LF per hour	RF@.021	LF	.78	.80	1.58
1/2" x 7/8", 40 LF/gallon, 46 LF per hour	RF@.022	LF	.94	.83	1.77
1/2" x 1", 35 LF/gallon, 45 LF per hour	RF@.022	LF	1.07	.83	1.90
3/4" x 3/4", 31 LF/gallon, 44 LF per hour	RF@.023	LF	1.21	.87	2.08
1" x 1", 17 LF/gallon, 30 LF per hour	RF@.033	LF	2.20	1.25	3.45
Silicone RTV 1 component sealant, $45.60 per gallon					
1/4" x 1/4", 277 LF/gallon, 92 LF per hour	RF@.011	LF	.16	.42	.58
1/2" x 1/2", 69 LF/gallon, 82 LF per hour	RF@.012	LF	.66	.46	1.12
1/2" x 3/4", 46 LF/gallon, 55 LF per hour	RF@.018	LF	.99	.68	1.67
3/4" x 3/4", 31 LF/gallon, 41 LF per hour	RF@.024	LF	1.47	.91	2.38
1/8" x 1", 138 LF/gallon, 87 LF per hour	RF@.012	LF	.33	.46	.79
Polysulfide or polyurethane sealant, 2 component, $35.40 per gallon					
1/4" x 1/4", 277 LF/gallon, 63 LF per hour	RF@.016	LF	.14	.61	.75
1/4" x 1/2", 139 LF/gallon, 58 LF per hour	RF@.017	LF	.28	.64	.92
3/8" x 1/2", 92 LF/gallon, 57 LF per hour	RF@.017	LF	.42	.64	1.06
1/2" x 1/2", 69 LF/gallon, 69 LF per hour	RF@.014	LF	.56	.53	1.09
3/8" x 3/4", 60 LF/gallon, 52 LF per hour	RF@.019	LF	.64	.72	1.36
Backing rods for sealant					
Open cell foam, 3/8" rod, 1/4" joint	RF@.010	LF	.05	.38	.43
Open cell foam, 5/8" rod, 1/2" joint	RF@.010	LF	.08	.38	.46
Open cell foam, 1" rod, 3/4" joint	RF@.010	LF	.15	.38	.53
Closed cell, 3/8" or 5/8" rod	RF@.011	LF	.12	.42	.54
Closed cell, 1-1/8" rod	RF@.012	LF	.20	.46	.66

Expansion Joints, Preformed

Wall and ceiling, aluminum cover

	Craft@Hrs	Unit	Material	Labor	Total
Drywall or panel type	CC@.067	LF	12.90	2.83	15.73
Plaster type	CC@.067	LF	12.90	2.83	15.73
Floor to wall, 3" x 3"	CC@.038	LF	7.16	1.61	8.77
Floor, 3"	CC@.038	LF	8.93	1.61	10.54
Neoprene joint with aluminum cover	CC@.079	LF	19.40	3.34	22.74
Wall to roof joint	CC@.094	LF	7.89	3.97	11.86

	Craft@Hrs	Unit	Material	Labor	Total
Bellows type expansion joints, butyl with neoprene backer					
16 oz copper, 4" wide	CC@.046	LF	12.10	1.94	14.04
28 gauge stainless steel, 4" wide	CC@.046	LF	12.00	1.94	13.94
26 gauge galvanized sheet metal, 4" wide	CC@.046	LF	7.02	1.94	8.96
16 oz copper, 6" wide	CC@.046	LF	13.80	1.94	15.74
28 gauge stainless steel, 6" wide	CC@.046	LF	13.60	1.94	15.54
26 gauge galvanized sheet metal, 6" wide	CC@.046	LF	8.48	1.94	10.42
Neoprene gaskets, closed cell					
1/8" x 2"	CC@.019	LF	.61	.80	1.41
1/8" x 6"	CC@.023	LF	1.41	.97	2.38
1/4" x 2"	CC@.022	LF	.71	.93	1.64
1/4" x 6"	CC@.024	LF	1.50	1.01	2.51
1/2" x 6"	CC@.025	LF	2.22	1.06	3.28
1/2" x 8"	CC@.028	LF	3.26	1.18	4.44
Acoustical caulking, drywall or plaster type	CC@.038	LF	.25	1.61	1.86
Polyisobutylene tapes, non-drying					
Polybutene	CC@.028	LF	.20	1.18	1.38
Polyisobutylene/butyl, preformed	CC@.020	LF	.09	.85	.94

Expansion Joint Covers

Surface type, aluminum

	Craft@Hrs	Unit	Material	Labor	Total
Floor, 1-1/2"	CC@.194	LF	21.00	8.20	29.20
Wall and ceiling, 1-1/2"	CC@.185	LF	11.90	7.82	19.72
Gymnasium base	CC@.099	LF	9.56	4.18	13.74
Roof, typical	CC@.145	LF	20.00	6.13	26.13

Cast-in-Place Concrete Fireproofing. Costs per linear foot (LF) of steel beam, column or girder. Use the figures below to estimate the average cubic feet (CF) of concrete per linear foot and square feet (SF) of form required per linear foot of each type of beam, column or girder. Quantities include concrete required to fill the void between the web and flange and provide 2" protection at the flanges. A small change in concrete thickness at flanges has very little effect on cost. Use $7,500.00 as a minimum a minimum job cost for work of this type. Concrete thickness for fireproofing will usually be:

Members at least 6" x 6" but less than 8" x 8", 3" for a 4 hour rating, 2" for a 3 hour rating, 1-1/2" for a 2 hour rating, and 1" for a 1 hour rating.

Members at least 8" x 8" but less than 12" x 12", 2-1/2" for a 4 hour rating, 2" for a 3 hour rating, 1" for a 1 or 2 hour rating.

Members 12" x 12" or greater, 2" for a 4 hour rating and 1" for a 1, 2, or 3 hour rating.

Labor costs listed here include the time needed to prepare formwork sketches at the job site, measure for the forms, fabricate, erect, align and brace the forms, cut, bend, place and tie the reinforcing steel, install embedded steel items, place and finish the concrete and strip, clean and stack the forms. Labor costs assume a concrete forming, placing and finishing crew of 1 carpenter, 1 laborer and 1 cement finisher. Concrete placing equipment assumes the use of a concrete pump.

For scheduling purposes, estimate that a crew of 3 will fireproof the following quantities of steel in an 8-hour day:

15 LF of W21 through W36 members,
25 LF of W8 through W18 members, 22 LF of S12 through S24 members, and 42 LF of M4 through W6 or S6 through S8 members.

Material costs assume reinforcing steel is #3 and #4 bars at 12" on center each way and includes waste and laps. Form cost assumes 3 uses without completely disassembling the form. Normal cleaning and repairs are included. No salvage value is assumed. Cost for forms, before allowance for waste, is based on Std & Btr lumber at $525 per MBF and 3/4" plyform at $880 per MSF.

	Craft@Hrs	Unit	Material	Labor	Equipment	Total

Concrete and accessories for fireproofing

Concrete, 3,000 PSI pump mix, at $71.00 CY before 5% allowance for waste

with pump cost at $15.00 per CY	P9@.600	CY	74.60	22.00	15.30	111.90

Grade A60 reinforcing bars, set and tied, material at $.30 lb

with 50 lbs per CY of concrete	P9@.790	CY	15.50	29.00	—	44.50

Hangers, snap-ties, misc. embedded steel material at $3.00 lb

with 10 lb per CY of concrete	P9@.664	CY	31.00	24.40	—	55.40

Concrete test cylinders including test reports, at $13.00 each

with 5 per 100 CY	—	CY	.65	—	—	.65
Total cost for concrete and accessories	**P9@2.05**	**CY**	**121.75**	**75.40**	**15.30**	**212.45**

Forms for fireproofing. These costs assume the following materials are used per square foot of contact area (SFCA): Nails, clamps and form oil costing $.12 per board foot of lumber, 1.2 SF of 3/4" plyform and 3.7 BF of lumber per SFCA.

Make, erect, align and strip forms, 3 uses	P9@.150	SF	1.72	5.50	—	7.22
Sack and patch concrete SF	P9@.010	SF	.05	.37	—	.42
Total cost for forms and finishing, per use	**P9@.160**	**SF**	**1.77**	**5.87**	**—**	**7.64**

See sample fireproofing estimate below

Sample fireproofing estimate for ten W24 x 55 beams 20' long. Note from the table below that this beam requires 2.7 cubic feet of concrete per linear foot and 6.9 square feet of form per linear foot. Using labor and material prices from above to estimate 200 linear feet of beam. (Total costs have been rounded)

Concrete (200 x 2.7 = 540 CF or 20 CY)	P9@41.0	LS	2,435.00	1,500.00	306.00	4,241.00
Form and finish (200 x 6.9 = 1,380 SF)	P9@221	LS	2,443.00	8,110.00	—	10,553.00
Total job cost shown above	**P9@262**	**LS**	**4,878.00**	**9,610.00**	**306.00**	**14,794.00**
Cost per linear foot of beam	P9@1.31	LF	24.40	48.10	1.53	74.03

Cast-in-Place Concrete Fireproofing Cost per LF, including concrete and forming

	Craft@Hrs	Unit	Material	Labor	Equipment	Total
W36 x 135 to W36 x 300						
5.13 CF and 10.07 SF per LF	P9@2.00	LF	41.00	73.40	2.91	117.31
W33 x 118 to W30 x 241						
4.59 CF and 9.35 SF per LF	P9@1.84	LF	37.30	67.50	2.60	107.40
W30 x 99 to W30 x 211						
3.78 CF and 8.40 SF per LF	P9@1.63	LF	31.90	59.80	2.14	93.84
W27 x 84 to W27 x 178						
3.51 CF and 8.19 SF per LF	P9@1.57	LF	30.30	57.60	1.99	89.89
W24 x 55 to W24 x 162						
2.70 CF and 6.90 SF per LF	P9@1.31	LF	24.40	48.10	1.53	74.03
W21 x 44 to W21 x 147						
2.43 CF and 6.75 SF per LF	P9@1.26	LF	22.90	46.20	1.38	70.48
W18 x 37 to W18 x 71						
1.62 CF and 5.34 SF per LF	P9@.974	LF	16.80	35.70	.92	53.42
W16 x 26 to W16 x 100						
1.62 CF and 5.40 SF per LF	P9@.987	LF	16.90	36.20	.92	54.02
W14 x 90 to W14 x 730						
2.97 CF and 7.15 SF per LF	P9@1.37	LF	26.10	50.30	1.68	78.08
W14 x 22 to W14 x 82						
1.35 CF and 4.70 SF per LF	P9@.854	LF	14.40	31.30	.77	46.47
W12 x 65 to W12 x 336						
2.16 CF and 6.08 SF per LF	P9@1.13	LF	20.50	41.50	1.22	63.22

	Craft@Hrs	Unit	Material	Labor	Equipment	Total
W12 x 40 to W12 x 58						
1.62 CF and 5.40 SF per LF	P9@.987	LF	16.90	36.20	.92	54.02
W12 x 14 to W12 x 35						
1.08 CF and 4.44 SF per LF	P9@.790	LF	12.70	29.00	.61	42.31
W10 x 33 to W10 x 112						
1.35 CF and 4.75 SF per LF	P9@.862	LF	14.50	31.60	.77	46.87
W10 x 12 to W10 x 30						
1.08 CF and 4.56 SF per LF	P9@.808	LF	12.90	29.60	.61	43.11
W8 x 24 to W8 x 67						
1.08 CF and 4.40 SF per LF	P9@.784	LF	12.70	28.80	.61	42.11
W8 x 10 to W8 x 21						
.81 CF and 3.87 SF per LF	P9@.678	LF	10.50	24.90	.46	35.86
W6 x 15 to W6 x 25						
.81 CF and 3.84 SF per LF	P9@.672	LF	10.50	24.70	.46	35.66
W6 x 9 to W6 x 16						
.54 CF and 2.92 SF per LF	P9@.507	LF	7.60	18.60	.31	26.51
W5 x 16 to W5 x 19						
.54 CF and 2.86 SF per LF	P9@.499	LF	7.50	18.30	.31	26.11
W4 x 13						
.54 CF and 3.20 SF per LF	P9@.552	LF	8.10	20.20	.31	28.61
M4 x 13 to M14 x 18						
.54 CF and 2.90 SF per LF	P9@.504	LF	7.57	18.50	.31	26.38
S24 x 80 to S24 x 121						
2.43 CF and 7.11 SF per LF	P9@1.32	LF	23.50	48.40	1.38	73.28
S20 x 66 to S20 x 96						
1.89 CF and 6.09 SF per LF	P9@1.12	LF	19.30	41.10	1.07	61.47
S15 x 43 to S18 x 70						
1.62 CF and 5.82 SF per LF	P9@1.05	LF	17.60	38.50	.92	57.02
S12 x 35 to S12 x 50						
1.08 CF and 4.44 SF per LF	P9@.790	LF	12.70	29.00	.61	42.31
S8 x 23 to S10 x 35						
0.81 CF and 3.81 SF per LF	P9@.670	LF	10.40	24.60	.46	35.46
S6 x 13 to S7 x 20						
0.54 CF and 2.91 SF per LF	P9@.504	LF	7.59	18.50	.31	26.40

	Craft@Hrs	Unit	Material	Labor	Total
Spray-Applied Fireproofing, Subcontract Fire endurance coating made from inorganic vermiculite and portland cement. Costs assume a 10,000 board foot job. (One BF is one square foot covered 1" thick.) For smaller jobs, increase the cost by 5% for each 500 BF less than 10,000 BF. Use $2,000 as a minimum subcontract price. For thickness other than 1", adjust these costs proportionately. For scheduling purposes, estimate that a crew of 2 plasterers and 1 helper can apply 200 to 250 board feet per hour.					
Structural steel columns	—	BF	—	—	1.37
Structural steel beams	—	BF	—	—	1.22
Purlins, girts, and miscellaneous members	—	BF	—	—	1.12
Decks, ceilings or walls	—	BF	—	—	1.02
Add for 18 gauge 2" hex mesh reinforcing	—	BF	—	—	.25
Add for key coat bonded on primed surfaces	—	SF	—	—	.28

Rule-of-thumb method for estimating spray-applied fireproofing on bare structural steel by member size:

Beams and Columns	Beams	Cost LF	Columns	Cost LF
W36 x 135 to 300	10 BF/LF	12.20	11 BF/LF	15.10
W33 x 118 to 241	9 BF/LF	11.00	10 BF/LF	13.70
W30 x 99 to 211	8 BF/LF	9.76	9 BF/LF	12.30
W27 x 84 to 178	7 BF/LF	8.54	8 BF/LF	11.00
W24 x 55 to 182	6.5 BF/LF	7.93	7.5 BF/LF	10.30
W21 x 44 to 147	6 BF/LF	7.32	7 BF/LF	9.59
W18 x 35 to 118	5 BF/LF	6.10	6 BF/LF	8.22
W16 x 28 to 57	3.5 BF/LF	4.27	4.5 BF/LF	6.17
W14 x 61 to 132	5.5 BF/LF	6.71	6.5 BF/LF	8.91
W14 x 22 to 63	4 BF/LF	4.88	5 BF/LF	6.85
W12 x 65 to 190	5 BF/LF	6.10	6 BF/LF	8.22
W12 x 40 to 58	4 BF/LF	4.88	5 BF/LF	6.85
W12 x 14 to 35	3 BF/LF	3.66	4 BF/LF	5.48
W10 x 49 to 112	4 BF/LF	4.88	5 BF/LF	6.85
W10 x 22 to 45	3 BF/LF	3.66	4 BF/LF	5.48
W10 x 12 to 19	2 BF/LF	2.44	3 BF/LF	4.11
W8 x 24 to 67	3 BF/LF	3.66	4 BF/LF	5.48
W6 x 9 to 25	2 BF/LF	2.44	3 BF/LF	4.11
W5 x 16 to 19	2 BF/LF	2.44	2.5 BF/LF	3.43
W4 x 13	1 BF/LF	1.22	2 BF/LF	2.74

Purlins and Girts			Purlins or girts	Cost LF
MC 18 members			4 BF/LF	4.48
MC 10 to MC 13 members			3 BF/LF	3.36
MC 8 members			2 BF/LF	2.24
C 10 to C 15 members			3 BF/LF	3.36
C 7 to C 9 members			2 BF/LF	2.24

	Craft@Hrs	Unit	Material	Labor	Total

Flexible Strip Doors Suitable for interior or exterior use. High strength USDA-approved clear transparent PVC plastic strips, with aluminum mounting hardware. Typical uses include truck dock door openings, railway car entrances, freezer doors and as positive draft protection to check the flow of unwanted air at +140° to -40° F. Costs shown per SF are per square foot of flexible strip area. To calculate the flexible strip area, add 2 linear feet to the opening width and height to allow for coverage at the sides and top. Costs shown per LF are per linear foot of mounting hardware. To calculate the length of the mounting hardware, add 2 LF to the opening width. Labor costs are based on work performed using ladders. Add for scaffolding costs if required. Based on Thermoscreen™ by Environmental Products.

	Craft@Hrs	Unit	Material	Labor	Total
Openings to 8' in height, 8" wide strips x .080" thick					
With 1/2 strip overlap	C8@.058	SF	2.88	2.10	4.98
With 3/4 strip overlap	C8@.076	SF	3.20	2.76	5.96
With full strip overlap	C8@.096	SF	4.47	3.48	7.95
Openings over 8' to 14' in height, 12" wide strips x .120" thick					
With 2/3 strip overlap	C8@.100	SF	5.11	3.63	8.74
With full strip overlap	C8@.115	SF	6.13	4.17	10.30
Openings over 14' to 20' in height, 16" wide strips x .120" thick					
With 1/2 strip overlap	C8@.075	SF	5.47	2.72	8.19
With 3/4 strip overlap	C8@.100	SF	7.16	3.63	10.79
With full strip overlap	C8@.125	SF	8.17	4.53	12.70

	Craft@Hrs	Unit	Material	Labor	Total
Add for mounting hardware, aluminum, pre-drilled, complete with nuts, bolts, washers and flex strip mounting pins					
Ceiling or header mounted (1.5 lbs per LF)	C8@.400	LF	7.42	14.50	21.92
Wall mounted (2.5 lbs per LF)	C8@.500	LF	27.30	18.10	26.70

Complete Hollow Metal Door Assembly These figures show the costs normally associated with installing an exterior hollow core steel door.

Hollow metal exterior 3' x 7' flush door, with frame, hardware and trim, complete

	Craft@Hrs	Unit	Material	Labor	Total
Total cost door, frame and trim as described below	**C8@7.62**	**Ea**	**784.00**	**276.00**	**1,060.00**
Stock hollow metal flush door,					
16 gauge, 3' x 7' x 1-3/4", non-rated	C8@.721	Ea	275.00	26.10	301.10
Closer plate reinforcing on door	—	Ea	11.25	—	11.25
Stock 18 gauge frame, 6" jamb, non-rated	C8@.944	Ea	85.70	34.20	119.90
Closer plate on frame	—	Ea	4.77	—	4.77
Three hinges, 4-1/2" x 4-1/2"	C8@.578	LS	51.30	21.00	72.30
Lockset, mortise type	C8@.949	Ea	213.00	34.40	247.40
Saddle type threshold, aluminum, 3'	C8@.276	Ea	15.30	10.00	25.30
Standard duty closer	C8@.787	Ea	79.80	28.50	108.30
Weatherstripping, bronze and neoprene	C8@2.58	LS	38.20	93.60	131.80
Paint with primer and 2 coats enamel	C8@.781	LS	9.54	28.30	37.84

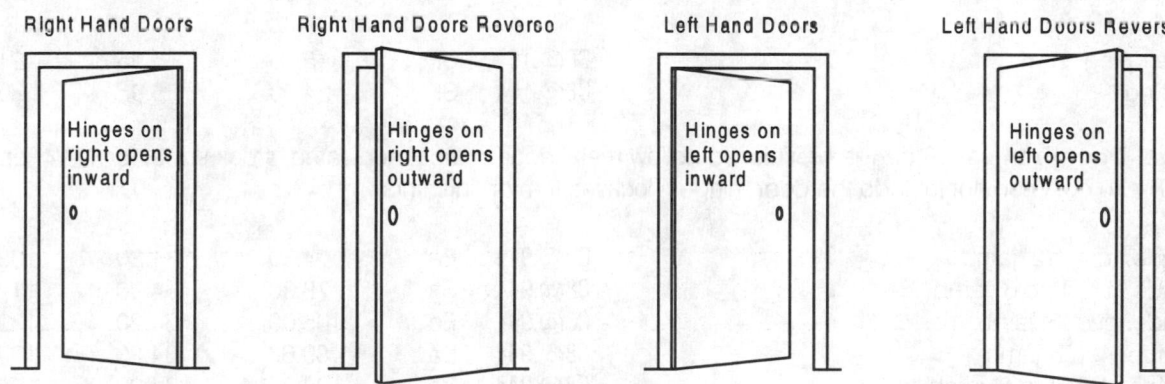

Right Hand Doors	Right Hand Doors Reverse	Left Hand Doors	Left Hand Doors Reverse
Hinges on right opens inward	Hinges on right opens outward	Hinges on left opens inward	Hinges on left opens outward

Hollow Metal Exterior Doors Commercial quality doors. See illustrations above. These costs do not include the frame, hinges, lockset, trim or finishing.

	Craft@Hrs	Unit	Material	Labor	Total
6'8" high hollow flush doors, 1-3/8" thick, non-rated					
2'6" wide, 20 gauge	C8@.681	Ea	156.00	24.70	180.70
2'8" wide, 20 gauge	C8@.683	Ea	161.00	24.80	185.80
3'0" wide, 20 gauge	C8@.721	Ea	173.00	26.10	199.10
6'8" high hollow flush doors, 1-3/4" thick, non-rated					
2'6" wide, 20 gauge	C8@.681	Ea	189.00	24.70	213.70
2'8" wide, 20 gauge	C8@.697	Ea	195.00	25.30	220.30
2'8" wide, 18 gauge	C8@.697	Ea	226.00	25.30	251.30
2'8" wide, 16 gauge	C8@.697	Ea	264.00	25.30	289.30
3'0" wide, 20 gauge	C8@.721	Ea	205.00	26.10	231.10
3'0" wide, 18 gauge	C8@.721	Ea	237.00	26.10	263.10
3'0" wide, 16 gauge	C8@.721	Ea	276.00	26.10	302.10
7'0" high hollow flush doors, 1-3/4" thick, non-rated					
2'6" or 2'8" wide, 20 gauge	C8@.757	Ea	200.00	27.50	227.50
2'8" or 3'0" wide, 18 gauge	C8@.757	Ea	241.00	27.50	268.50
2'8" or 3'0" wide, 16 gauge	C8@.757	Ea	281.00	27.50	308.50
3'0" wide, 20 gauge	C8@.816	Ea	215.00	29.60	244.60

	Craft@Hrs	Unit	Material	Labor	Total
Additional costs for hollow metal doors, cost per door					
Add for factory applied steel astragal	—	Ea	29.70	—	29.70
Add for steel astragal set on site	C8@.496	Ea	29.70	18.00	47.70
Add for 90 minute "B" fire label rating	—	Ea	33.10	—	33.10
Add for R-7 polyurethane foam core	—	Ea	79.80	—	79.80
Add for 10" x 10" wired glass panel	—	Ea	57.70	—	57.70
Add for closer reinforcing plate	—	Ea	11.20	—	11.20
Add for chain or bolt reinforcing plate	—	Ea	85.30	—	85.30
Add for rim exit latch reinforcing	—	Ea	24.40	—	24.40
Add for vertical exit latch reinforcing	—	Ea	44.20	—	44.20
Add for pull plate reinforcing	—	Ea	17.80	—	17.80
Add for galvanizing 1-3/8" doors	—	Ea	22.50	—	22.50
Add for galvanizing 1-3/4" door	—	Ea	27.50	—	27.50
Add for cutouts to 4 SF	—	Ea	46.20	—	46.20
Add for cutouts over 4 SF	—	SF	9.70	—	9.70
Add for stainless steel doors	—	Ea	776.00	—	776.00
Add for baked enamel finish	—	Ea	58.50	—	58.50
Add for porcelain enamel finish	—	Ea	140.00	—	140.00
Add for larger sizes	—	SF	3.28	—	3.28
Add for special dapping	—	Ea	25.00	—	25.00
Kalamein doors					
90 minute rating	C8@.164	SF	15.40	5.95	21.35
2 hour rating	C8@.164	SF	24.90	5.95	30.85

Hollow Metal Door Frames 18 gauge prefinished hollow metal door frames, non-rated, stock sizes, for 1-3/4" or 1-3/8" doors. These costs do not include the door, hinges, lockset, trim or finishing.

	Craft@Hrs	Unit	Material	Labor	Total
6'8" high					
To 3'6" wide, to 4-1/2" jamb	C8@.944	Ea	74.30	34.20	108.50
To 3'6" wide, 4-3/4" to 6" jamb	C8@.944	Ea	78.30	34.20	112.50
To 3'6" wide, over 6" jamb	C8@.944	Ea	105.00	34.20	139.20
Over 4' wide, 4-1/2" jamb	C8@.944	Ea	89.60	34.20	123.80
Over 4' wide, 4-3/4" to 6" jamb	C8@.944	Ea	94.30	34.20	128.50
Over 4' wide, over 6" jamb	C8@.944	Ea	122.00	34.20	156.20
7' high					
To 3'6" wide, to 4-1/2" jamb	C8@.944	Ea	80.90	34.20	115.10
To 3'6" wide, 4-3/4" to 6" jamb	C8@.944	Ea	85.70	34.20	119.90
To 3'6" wide, over 6" jamb	C8@.944	Ea	111.00	34.20	145.20
Over 4' wide, 4-1/2" jamb	C8@.944	Ea	96.10	34.20	130.30
Over 4' wide, 4-3/4" to 6" jamb	C8@.944	Ea	99.90	34.20	134.10
Over 4' wide, over 6" jamb	C8@.944	Ea	127.00	34.20	161.20
8' high					
To 3'6" wide, to 4-1/2" jamb	C8@1.05	Ea	93.30	38.10	131.40
To 3'6" wide, 4-3/4" to 6" jamb	C8@1.05	Ea	98.00	38.10	136.10
To 3'6" wide, over 6" jamb	C8@1.05	Ea	126.00	38.10	164.10
Over 4' wide, 4-1/2" jamb	C8@1.05	Ea	109.00	38.10	147.10
Over 4' wide, 4-3/4" to 6" jamb	C8@1.05	Ea	112.00	38.10	150.10
Over 4' wide, over 6" jamb	C8@1.05	Ea	143.00	38.10	181.10
Cost additions or deductions for hollow metal door frames, per frame					
Add for 90 minute UL label	—	Ea	24.10	—	24.10
Add for aluminum casing	—	Ea	20.20	—	20.20
Add for communicator frame (back-to-back doors)	—	Ea	14.80	—	14.80
Add for galvanized finish	—	%	12.0	—	—

	Craft@Hrs	Unit	Material	Labor	Total
Add for lengthening, 7' to 8'10"	—	Ea	13.00	—	13.00
Add for extra hinge reinforcing	—	Ea	4.44	—	4.44
Add for stainless steel frames	—	Ea	425.00	—	425.00
Add for porcelain enamel finish	—	Ea	138.00	—	138.00
Add for reinforcing for chain and bolt	—	Ea	5.65	—	5.65
Add for closer plate reinforcing	—	Ea	4.92	—	4.92
Add for exit latch reinforcing	—	Ea	9.17	—	9.17
Add for concrete filled frames	C8@.382	Ea	6.35	13.90	20.25
Add for borrowed lite	C8@.086	SF	6.06	3.12	9.18
Add for fixed transom lite	C8@.099	SF	7.99	3.59	11.58
Add for movable transom lite	C8@.107	SF	9.17	3.88	13.05
Add for window frame sections	C8@.071	LF	4.92	2.58	7.50
Add for wall frame sections	C8@.071	LF	5.61	2.58	8.19
Deduct for 22 gauge frames	—	Ea	-11.05	—	-11.05

Prehung Steel Doors 18 gauge primed insulated 1-3/4" thick entry doors with sweep and 18 gauge steel frame. Costs shown Include three 4" x 4" x 1/4" hinges but no lockset.

Flush doors

	Craft@Hrs	Unit	Material	Labor	Total
Jamb to 5-1/2" wide, 2'8" x 6'8", or 3'0" x 6'8"	C8@1.05	Ea	328.00	38.10	366.10
Jamb over 5-1/2" wide, 2'8" x 6'8", or 3'0" x 6'8"	C8@1.05	Ea	334.00	38.10	372.10

6 or 8 panel doors

	Craft@Hrs	Unit	Material	Labor	Total
Jamb to 5-1/2" wide, 2'8" x 6'8", or 3'0" x 6'8"	C8@1.05	Ea	338.00	38.10	376.10
Jamb over 5-1/2" wide, 2'8" x 6'8", or 3'0" x 6'8"	C8@1.05	Ea	343.00	38.10	381.10
Additional costs for prehung steel doors					
90 minute Factory Mutual fire rating	—	Ea	17.90	—	17.90
60 minute Factory Mutual fire rating	—	Ea	13.90	—	13.90
Add for 16 gauge door and frame	—	Ea	11.00	—	11.00
Deduct for 22 gauge and door	—	Ea	-3.10	—	-3.10
Add for galvanized steel finish	—	Ea	9.52	—	9.52
Deduct for unfinished door	—	Ea	-11.50	—	-11.50
Add for installed weatherstripping	—	Ea	27.60	—	27.60
Add for aluminum threshold	—	Ea	27.30	—	27.30

Wood Doors and Frames Commercial and institutional quality. Unfinished, no hardware included.

Hollow core 1-3/8" thick flush interior doors

	Craft@Hrs	Unit	Material	Labor	Total
2'6" x 6'8", hardboard face	C8@.936	Ea	30.20	33.90	64.10
2'8" x 6'8", hardboard face	C8@.936	Ea	33.40	33.90	67.30
3'0" x 6'8", hardboard face	C8@.982	Ea	34.40	35.60	70.00
2'6" x 6'8", hardwood face	C8@.936	Ea	53.10	33.90	87.00
2'8" x 6'8", hardwood face	C8@.936	Ea	55.30	33.90	89.20
3'0" x 6'8", hardwood face	C8@.982	Ea	59.40	35.60	95.00

Hollow core 1-3/4" thick flush exterior doors

	Craft@Hrs	Unit	Material	Labor	Total
2'6" x 6'8", hardwood face	C8@.982	Ea	58.20	35.60	93.80
2'8" x 6'8", hardwood face	C8@.982	Ea	60.40	35.60	96.00
3'0" x 6'8", hardwood face	C8@1.01	Ea	65.60	36.60	102.20
2'6" x 7'0", hardwood face	C8@.982	Ea	65.60	35.60	101.20
2'8" x 7'0", hardwood face	C8@.982	Ea	68.70	35.60	104.30
3'0" x 7'0", hardwood face	C8@1.01	Ea	71.90	36.60	108.50

Solid core 1-3/4" thick flush exterior doors, "B" label, 1 hour rating

	Craft@Hrs	Unit	Material	Labor	Total
2'6" x 6'8", hardwood face	C8@1.05	Ea	172.00	38.10	210.10
2'8" x 6'8", hardwood face	C8@1.05	Ea	172.00	38.10	210.10
3'0" x 6'8", hardwood face	C8@1.10	Ea	199.00	39.90	238.90

	Craft@Hrs	Unit	Material	Labor	Total
2'0" x 7'0", hardwood face	C8@1.05	Ea	169.00	38.10	207.10
2'4" x 7'0", hardwood face	C8@1.05	Ea	173.00	38.10	211.10
2'6" x 7'0", hardwood face	C8@1.05	Ea	182.00	38.10	220.10
2'8" x 7'0", hardwood face	C8@1.05	Ea	183.00	38.10	221.10
3'0" x 7'0", hardwood face	C8@1.10	Ea	204.00	39.90	243.90

Exterior door frames, fir, with oak sill

	Craft@Hrs	Unit	Material	Labor	Total
2'6" x 6'8"	C8@1.48	Ea	77.30	53.70	131.00
2'8" x 6'8"	C8@1.50	Ea	78.60	54.40	133.00
3'0" x 6'8"	C8@1.57	Ea	79.00	56.90	135.90
5'0" x 6'8", for two doors	C8@1.85	Ea	168.00	67.10	235.10
6'0" x 6'8", for two doors	C8@1.99	Ea	170.00	72.20	242.20
2'6" x 7'0"	C8@1.48	Ea	85.00	53.70	138.70
2'8" x 7'0"	C8@1.50	Ea	87.10	54.40	141.50
3'0" x 7'0"	C8@1.57	Ea	89.00	56.90	145.90
5'0" x 7'0", for two doors	C8@1.85	Ea	178.00	67.10	245.10
6'0" x 7'0", for two doors	C8@1.99	Ea	181.00	72.20	253.20

Interior door jambs, fir, without stops

	Craft@Hrs	Unit	Material	Labor	Total
2'6" x 6'8"	C8@.871	Ea	12.50	31.60	44.10
2'8" x 6'8"	C8@.879	Ea	12.90	31.90	44.80
3'0" x 6'8"	C8@.922	Ea	13.30	33.40	46.70
5'0" x 6'8", for two doors	C8@1.10	Ea	26.80	39.90	66.70
6'0" x 6'8", for two doors	C8@1.16	Ea	27.20	42.10	69.30
2'6" x 7'0"	C8@.871	Ea	13.90	31.60	45.50
2'8" x 7'0"	C8@.879	Ea	14.90	31.90	46.80
3'0" x 7'0"	C8@.922	Ea	15.30	33.40	48.70
5'0" x 7'0", for two doors	C8@1.10	Ea	28.80	39.90	68.70
6'0" x 7'0", for two doors	C8@1.16	Ea	29.60	42.10	71.70

Prehung exterior flush doors with hinge set, casing and jambs, but no lockset.

1-3/4", solid core, hardwood veneer, prefinished

	Craft@Hrs	Unit	Material	Labor	Total
To 3'6" x 7'0"	C8@1.29	Ea	300.00	46.80	346.80
To 3'0" x 8'0"	C8@1.63	Ea	342.00	59.10	401.10
3'6" x 8'0"	C8@1.72	Ea	360.00	62.40	422.40
Deduct for hollow core 1-3/8" prehung doors	—	%	-25.0	-10.0	—

	Craft@Hrs	Unit	Material	Labor	Equipment	Total
Special Doors These costs do not include electrical work. Equipment is a 4000 lb forklift at $16.30 per hour.						

Fire doors, overhead roll-up type, sectional steel, fusible link, motor operated, UL label

	Craft@Hrs	Unit	Material	Labor	Equipment	Total
6' x 7', manual operation	F7@13.4	Ea	1,470.00	515.00	54.40	2,039.40
5' x 8', manual operation	F7@13.4	Ea	1,000.00	515.00	54.40	1,569.40
14' x 14', chain hoist operated	F7@26.2	Ea	2,870.00	1,010.00	106.00	3,986.00
18' x 14', chain hoist operated	F7@32.9	Ea	3,970.00	1,270.00	134.00	5,374.00

Fire doors, sliding type, one-piece steel, including hardware, fusible link, motor operated, UL label

	Craft@Hrs	Unit	Material	Labor	Equipment	Total
4' x 7'	F7@13.3	Ea	1,040.00	512.00	54.10	1,606.10
6' x 7'	F7@14.8	Ea	1,380.00	569.00	60.10	2,009.10
10' x 10'	F7@32.1	Ea	2,640.00	1,230.00	130.00	4,000.00

Refrigerator doors, manually operated, with hardware and standard frame

	Craft@Hrs	Unit	Material	Labor	Equipment	Total
Galvanized, cooler, 3' x 7', hinged	F7@6.60	Ea	1,236.00	254.00	26.80	1,516.80
Stainless, cooler, 3' x 7', hinged	F7@6.60	Ea	1,423.00	254.00	26.80	1,703.80
Stainless, freezer, 3' x 7', hinged	F7@6.60	Ea	1,843.00	254.00	26.80	2,123.80
Stainless, cooler, 4' x 7', hinged	F7@6.60	Ea	1,601.00	254.00	26.80	1,881.80
Galvanized, cooler, 5' x 7', sliding	F7@7.93	Ea	1,811.00	305.00	32.20	2,148.20
Galvanized, freezer, 5' x 7' sliding	F7@7.93	Ea	2,130.00	305.00	32.20	2,467.20

	Craft@Hrs	Unit	Material	Labor	Equipment	Total
Revolving doors, 7' diameter, complete. Typical prices						
Aluminum	F7@50.3	Ea	23,600.00	1,930.00	204.00	25,734.00
Stainless steel	F7@50.3	Ea	26,800.00	1,930.00	204.00	28,934.00
Bronze	F7@69.0	Ea	34,300.00	2,650.00	280.00	37,230.00
Security grilles, aluminum, overhead roll-up type, horizontal rods at 2" OC, chain hoist operated						
8' x 8', clear anodized aluminum	F7@18.5	Ea	1,720.00	712.00	75.10	2,507.10
8' x 8', medium bronze aluminum	F7@18.5	Ea	2,440.00	712.00	75.10	3,227.10
10' x 10', anodized aluminum	F7@21.8	Ea	2,320.00	839.00	88.50	3,247.50
18' x 8', anodized aluminum	F7@26.3	Ea	3,550.00	1,010.00	107.00	4,667.00
18' x 18', anodized, motor operated	F7@28.7	Ea	4,170.00	1,100.00	117.00	5,387.00
Service doors, aluminum, overhead roll-up type						
4' x 4', manual counter shutter	F7@13.6	Ea	689.00	523.00	55.20	1,267.20
8' x 4', manual operation	F7@12.7	Ea	999.00	488.00	51.60	1,538.60
10' x 10', chain hoist operated	F7@12.7	Ea	1,520.00	488.00	51.60	2,059.60
Service doors, steel, overhead roll-up, chain hoist operated						
8' x 8'	F7@11.7	Ea	987.00	450.00	47.50	1,484.50
10' x 10'	F7@11.7	Ea	1,200.00	450.00	47.50	1,697.50
12' x 12'	F7@14.7	Ea	1,610.00	565.00	59.70	2,234.70
14' x 14'	F7@18.2	Ea	1,770.00	700.00	73.90	2,543.90
18' x 18'	F7@22.0	Ea	2,390.00	846.00	89.30	3,325.30
18' x 14', 240 volt motor operated	F7@25.0	Ea	3,130.00	962.00	102.00	4,194.00
Service doors, weather-stripped, steel, overhead roll-up type, chain hoist operated, bottom and side astragal, with hood baffles but no insulation						
10' x 10'	F7@12.8	Ea	1,180.00	492.00	52.00	1,724.00
14' x 14'	F7@26.9	Ea	1,610.00	1,030.00	109.00	2,749.00
Add for insulated weather-stripped doors		%	150.0	—	—	—

	Craft@Hrs	Unit	Material	Labor	Total
Steel access doors, with casing and ground					
12" x 12"	CC@.150	Ea	16.40	6.34	22.74
18" x 18"	CC@.155	Ea	23.00	6.55	29.55
24" x 24"	CC@.160	Ea	34.90	6.76	41.66

	Craft@Hrs	Unit	Material	Labor	Equipment	Total
Stock room doors, double-action .063" aluminum, with hardware and bumper strips both sides, two 7' high doors per opening, plastic laminate finish, 12" high baseplate in each door						
4' wide opening, 2 doors	F7@3.36	Pr	550.00	129.00	13.60	692.60
5' wide opening, 2 doors	F7@3.36	Pr	634.00	129.00	13.60	776.60
6' wide opening, 2 doors	F7@3.98	Pr	680.00	153.00	16.20	849.20
7' wide opening, 2 doors	F7@3.98	Pr	766.00	153.00	16.20	935.20
Vault doors, minimum security						
3' x 7', 2 hour fire rating	F7@22.0	Ea	2,040.00	846.00	89.30	2,975.30
4' x 7', 2 hour fire rating	F7@23.4	Ea	2,910.00	900.00	95.00	3,905.00
3' x 7', 4 hour fire rating	F7@27.2	Ea	2,230.00	1,050.00	110.60	3,390.60
4' x 7', 4 hour fire rating	F7@28.7	Ea	3,030.00	1,100.00	117.00	4,247.00

	Craft@Hrs	Unit	Material	Labor	Total
Commercial and Industrial Grade Steel Windows, glazed.					
Industrial grade, fixed 100%	G1@.086	SF	15.10	2.87	17.97
Industrial grade, vented 50%	G1@.086	SF	19.40	2.87	22.27
Projected, vented 50%	G1@.086	SF	21.80	2.87	24.67
Add for screen, SF of screen	—	SF	2.57	—	2.57

	Craft@Hrs	Unit	Material	Labor	Total
Hardware Commercial and industrial quality. Installation costs include drilling and routing of doors where required.					
Rule of thumb. Use these typical finish hardware costs for preliminary estimates only.					
Hardware for commercial buildings, per square foot of floor area					
Economy grade, for doors, cabinets, and toilet rooms	CC@.002	SF	.14	.08	.22
Standard grade, for doors, cabinets, and toilet rooms	CC@.004	SF	.33	.17	.50
Hospital, not including panic hardware, per door	CC@2.46	Ea	415.00	104.00	519.00
Office, not including panic hardware, per door	CC@1.57	Ea	199.00	66.40	265.40
School, not including panic hardware, per door	CC@1.94	Ea	277.00	82.00	359.00
Detail cost for individual hardware components					
Cabinet hardware, per LF of face, typical prices					
Commercial grade	CC@.145	LF	9.42	6.13	15.55
Institutional grade	CC@.181	LF	11.80	7.65	19.45
Closer, surface mounted. Based on Schalge Lock Company					
Interior doors	CC@.616	Ea	70.40	26.00	96.40
Exterior doors	CC@.709	Ea	88.00	30.00	118.00
Exterior, heavy duty	CC@.894	Ea	102.00	37.80	139.80
Floor mounted	CC@1.16	Ea	176.00	49.00	225.00
Deadbolts and deadlatches, chrome finish					
Standard deadbolt, single cylinder	CC@.745	Ea	26.50	31.50	58.00
Standard deadbolt, double cylinder	CC@.745	Ea	36.40	31.50	67.90
Standard turnbolt, no cylinder	CC@.745	Ea	20.30	31.50	51.80
Nightlatch, single cylinder	CC@.745	Ea	64.00	31.50	95.50
Nightlatch, double cylinder	CC@.745	Ea	70.70	31.50	102.20
Nightlatch, no cylinder, turnbolt	CC@.745	Ea	55.20	31.50	86.70
Heavy duty deadbolt, single cylinder	CC@.745	Ea	49.60	31.50	81.10
Heavy duty deadbolt, double cylinder	CC@.745	Ea	61.80	31.50	93.30
Heavy duty turnbolt, no cylinder	CC@.745	Ea	49.60	31.50	81.10
Extra heavy deadbolt, single cylinder	CC@.745	Ea	78.50	31.50	110.00
Extra heavy deadbolt, double cylinder	CC@.745	Ea	92.40	31.50	123.90
Add for brass or bronze finish	—	Ea	2.23	—	2.23
Door stops					
Rubber tip, screw base	CC@.060	Ea	3.61	2.54	6.15
Floor mounted, with holder	CC@.279	Ea	14.90	11.80	26.70
Wall mounted, with holder	CC@.279	Ea	14.60	11.80	26.40
Overhead mounted	CC@.528	Ea	86.00	22.30	108.30
Hinges, butt type					
3-1/2" x 3-1/2"	CC@.357	Pr	8.23	15.10	23.33
5" x 5", hospital swing clear	CC@.547	Pr	108.00	23.10	131.10
4", spring, single acting	CC@.243	Pr	26.80	10.30	37.10
4-1/2", heavy duty	CC@.347	Pr	50.30	14.70	65.00
7", spring, double acting	CC@.491	Pr	66.10	20.80	86.90
Floor-mounted, interior, per door	CC@.665	Ea	220.00	28.10	248.10
Floor-mounted, exterior, per 3' door	CC@.665	Ea	237.00	28.10	265.10
Floor-mounted, exterior, over 3' wide door	CC@.799	Ea	502.00	33.80	535.80
Invisible, blind doors, interior, per door	CC@.574	Ea	26.10	24.30	50.40
Invisible, for blind doors, exterior	CC@.574	Pr	33.30	24.30	57.60
Kick plates (see also Plates below)					
10" x 34" 16 gauge, bronze	CC@.452	Ea	36.10	19.10	55.20
10" x 34" 18 gauge, stainless	CC@.452	Ea	30.00	19.10	49.10
Letter drop, for wood doors					
brass plated	CC@.369	Ea	25.20	15.60	40.80

	Craft@Hrs	Unit	Material	Labor	Total
Locksets					
Heavy duty residential or light duty commercial locksets, chrome finish					
Classroom or storeroom, key lock	CC@.745	Ea	78.60	31.50	110.10
Dummy knob	CC@.247	Ea	14.70	10.40	25.10
Entrance lockset, key lock	CC@.745	Ea	72.00	31.50	103.50
Passage latch, no lock	CC@.689	Ea	34.10	29.10	63.20
Privacy latch, button lock	CC@.689	Ea	40.90	29.10	70.00
Add for lever handle (handicapped)	—	Ea	27.70	—	27.70
Add for brass or bronze finish	—	Ea	4.62	—	4.62
Heavy duty commercial, chrome finish					
Bored entrance lockset	CC@.745	Ea	181.00	31.50	212.50
Grip handle, with trim	CC@.855	Ea	199.00	36.10	235.10
Single deadbolt-lockset	CC@.745	Ea	87.80	31.50	119.30
Double deadbolt-lockset	CC@.745	Ea	115.00	31.50	146.50
Mortise deadbolt-lockset	CC@.855	Ea	199.00	36.10	235.10
Add for lever handle (handicapped)	—	Ea	18.50	—	18.50
Add for brass or bronze finish	—	Ea	7.39	—	7.39
Panic type exit door hardware, with trim					
Mortise type, for aluminum, steel or wood doors					
Satin aluminum finish	CC@2.41	Ea	364.00	102.00	466.00
Dark bronze finish	CC@2.41	Ea	375.00	102.00	477.00
Polished chrome finish	CC@2.41	Ea	443.00	102.00	545.00
Add for external lockset	CC@.401	Ea	49.60	16.90	66.50
Rim lock type, for aluminum, steel or wood doors					
Satin aluminum finish	CC@1.50	Ea	267.00	63.40	330.40
Dark bronze finish	CC@1.50	Ea	279.00	63.40	342.40
Polished chrome finish	CC@1.50	Ea	344.00	63.40	407.40
Add for external lockset	CC@.401	Ea	37.50	16.90	54.40
Vertical rod type, satin aluminum finish, with trim					
Aluminum doors, concealed rod	CC@2.14	Ea	354.00	90.50	444.50
Metal or wood doors, external rod	CC@1.79	Ea	369.00	75.70	444.70
Wood doors, concealed rod	CC@2.14	Ea	434.00	90.50	524.50
Add for external lockset	CC@.401	Ea	47.60	16.90	64.50
Add for bronze or black finish	—	%	37.5	—	—
Add for chrome finish	—	%	70.0	—	—
Plates (see also Kick Plates above)					
Pull plates, bronze, 4" x 16"	CC@.190	Ea	29.90	8.03	37.93
Push plates bronze, 4" x 16"	CC@.190	Ea	23.00	8.03	31.03
Surface bolts					
4" bolt	CC@.308	Ea	4.62	13.00	17.62
6" bolt	CC@.310	Ea	6.28	13.10	19.38
Threshold					
Aluminum, 36"	CC@.241	Ea	14.80	10.20	25.00
Bronze, 36"	CC@.241	Ea	60.70	10.20	70.90
Weatherstripping					
Bronze and neoprene, 3' x 7' door					
Wood door	CC@1.53	Ea	15.20	64.70	79.90
Steel door, adjustable	CC@2.32	Ea	36.60	98.10	134.70
Astragal, mortise mounted, bronze, adjustable	CC@.064	LF	9.38	2.71	12.09
Glue-back foam for 3' x 7' door	CC@.006	LF	.27	.25	.52

	Craft@Hrs	Unit	Material	Labor	Total
Glazing					
Sheet (window) glass					
Single strength "B"					
To 60" width plus length	G1@.057	SF	1.98	1.90	3.88
Over 60" to 70" width plus length	G1@.057	SF	2.03	1.90	3.93
Over 70" to 80" width plus length	G1@.057	SF	2.21	1.90	4.11
Over 80" width plus length	G1@.057	SF	2.25	1.90	4.15
Double strength window glass					
Double strength "B"					
To 60" width plus length	G1@.057	SF	2.57	1.90	4.47
Over 60" to 70" width plus length	G1@.057	SF	2.67	1.90	4.57
Over 70" to 80" width plus length	G1@.057	SF	2.76	1.90	4.66
Over 80" width plus length	G1@.057	SF	2.85	1.90	4.75
Tempered "B"	G1@.057	SF	4.19	1.90	6.09
Obscure	G1@.057	SF	3.96	1.90	5.86
3/16" glass					
Crystal (clear)	G1@.068	SF	3.43	2.27	5.70
Tempered	G1@.068	SF	5.66	2.27	7.93
1/4" glass					
Float, clear, quality 3	G1@.102	SF	3.27	3.40	6.67
Float, bronze or gray, quality 3	G1@.102	SF	3.44	3.40	6.84
Float, obscure	G1@.102	SF	4.97	3.40	8.37
Float, heat absorbing	G1@.102	SF	5.43	3.40	8.83
Float, safety, laminated, clear	G1@.102	SF	5.99	3.40	9.39
Float, tempered	G1@.102	SF	5.43	3.40	8.83
Spandrel, plain	G1@.102	SF	7.31	3.40	10.71
Spandrel, tinted gray or bronze	G1@.102	SF	8.04	3.40	11.44
Tempered, reflective	G1@.102	SF	8.36	3.40	11.76
3/8" glass					
Float, clear	G1@.128	SF	5.57	4.27	9.84
Float, tinted	G1@.128	SF	7.82	4.27	12.09
Obscure, tempered	G1@.128	SF	8.36	4.27	12.63
Corrugated glass	G1@.128	SF	7.17	4.27	11.44
1/2" float, tempered	G1@.157	SF	20.50	5.24	25.74
1" bullet resistant, 12" x 12" panels	G1@1.32	SF	79.80	44.00	123.80
2" bullet resistant, 15 to 20 SF	G1@1.62	SF	47.70	54.00	101.70
Insulating glass, 2 layers of 1/8", 1/2" overall, 10 to 15 square foot lites					
"B" quality sheet glass	G1@.131	SF	6.34	4.37	10.71
Float, polished bronze or gray	G1@.131	SF	8.83	4.37	13.20
Insulating glass, 2 layers of 1/4" float, 1" overall, 30 to 40 square foot lites					
Clear float glass	G1@.136	SF	9.73	4.54	14.27
Plexiglass					
Clear 1/8"	G1@.086	SF	2.39	2.87	5.26
Clear 1/4"	G1@.086	SF	3.27	2.87	6.14
Colors 1/4"	G1@.086	SF	4.50	2.87	7.37
Shatterproof 1/4"	G1@.086	SF	5.20	2.87	8.07
Wired glass, 1/4", type 3					
Clear	G1@.103	SF	6.90	3.43	10.33
Hammered	G1@.103	SF	6.45	3.43	9.88
Obscure	G1@.103	SF	6.20	3.43	9.63
Mirrors, unframed					
Sheet glass, 3/16"	G1@.103	SF	5.01	3.43	8.44
Float glass, 1/4"	G1@.106	SF	7.76	3.53	11.29

	Craft@Hrs	Unit	Material	Labor	Total
Reflective, 1 way, in wood stops	G1@.118	SF	14.30	3.94	18.24
Small lites, adhesive mount	G1@.057	SF	3.27	1.90	5.17

Curtain Walls Typical costs.

Structural sealant curtain wall systems with aluminum or steel frame

Regular weight, float glass	—	SF	—	—	31.00
Heavy duty, float glass	—	SF	—	—	33.70
Corning glasswall glass	—	SF	—	—	35.80
Add for spandrel glass	—	SF	—	—	4.40
Add for heat absorbing glass	—	SF	—	—	7.70
Add for low transmission glass	—	SF	—	—	5.10
Add for 1" insulating glass	—	SF	—	—	7.50
Add for bronze anodizing	—	%	—	—	15.0
Add for black anodizing	—	%	—	—	18.0

Finishes 9

Furring and Lathing For commercial applications. These costs do not include plaster.

Furring, no plaster included

Walls, steel hat channel at 16" OC galvanized

	Craft@Hrs	Unit	Material	Labor	Total
3/4" x 5/8"	C8@.243	SY	3.55	8.81	12.36
3/4" x 3/4"	C8@.243	SY	3.18	8.81	11.99
1-1/2" x 1-1/2"	C8@.269	SY	4.31	9.76	14.07

Beams and columns, steel hat channel at 16" OC galvanized

3/4" x 3/4" at 12" OC	C8@.423	SY	4.00	15.30	19.30
3/4" x 3/4" at 16" OC	C8@.382	SY	3.14	13.90	17.04

Ceilings, main runners at 4' OC, furring channels at 16" OC, 8 gauge wire at 4' OC

1-1/2" x 3/4", standard centering	C8@.344	SY	4.61	12.50	17.11
1-1/2" x 7/8", hat channel, standard centering	C8@.363	SY	4.45	13.20	17.65
3-1/4" x 1-1/2" x 3/4", triple hung	C8@.683	SY	8.45	24.80	33.25
1-1/2" x 3/4", coffered	C8@.887	SY	8.00	32.20	40.20
Add for resilient spring system	—	SY	11.10	—	11.10

Lathing, walls and ceilings, nailed or wired in place

2.5 lb, standard diamond galvanized	F8@.099	SY	2.26	3.58	5.84
2.5 lb, painted	F8@.099	SY	2.02	3.58	5.60
3.4 lb, standard diamond galvanized	F8@.099	SY	2.48	3.58	6.06
3.4 lb, painted copper alloy	F8@.099	SY	2.38	3.58	5.96
1/8" flat rib, 3.4 lb, galvanized	F8@.099	SY	2.73	3.58	6.31
1/8" flat rib, 2.75 lb, painted	F8@.099	SY	2.63	3.58	6.21
3/8" rib lath, galvanized, 3.4 lb	F8@.099	SY	3.18	3.58	6.76
3/8" painted copper alloy, 3.4 lb	F8@.099	SY	2.68	3.58	6.26
Wire mesh, 15 lb felt and line wire					
1" mesh, 18 gauge	F8@.079	SY	1.88	2.86	4.74
1-1/2" mesh, 17 gauge	F8@.079	SY	1.97	2.86	4.83
Paper back lath	F8@.063	SY	2.54	2.28	4.82
Stucco rite 2" corner bead, 16 gauge	F8@.028	CLF	2.00	1.01	3.01
Aqua lath, 2" x 16 gauge	F8@.076	SY	2.63	2.75	5.38

	Craft@Hrs	Unit	Material	Labor	Total

Plastering These costs do not include the furring, lath or drywall

Gypsum interior plaster, lime putty trowel finish

	Craft@Hrs	Unit	Material	Labor	Total
One coat application, on walls, extra smooth	F8@.330	SY	3.75	11.90	15.65
One coat application, on ceilings, extra smooth	F8@.381	SY	3.75	13.80	17.55
Two coat application, on ceilings, regular	F8@.381	SY	3.75	13.80	17.55
Two coat application, on walls regular	F8@.330	SY	3.75	11.90	15.65
Three coat application, on ceilings regular	F8@.448	SY	4.93	16.20	21.13
Three coat application, on walls regular	F8@.405	SY	4.93	14.60	19.53

Gypsum interior vermiculite plaster, trowel finish

	Craft@Hrs	Unit	Material	Labor	Total
Two coat application, on ceilings	F8@.381	SY	3.94	13.80	17.74
Two coat application, on walls	F8@.330	SY	3.94	11.90	15.84
Three coat application, on ceilings	F8@.448	SY	5.13	16.20	21.33
Three coat application, on walls	F8@.405	SY	5.13	14.60	19.73

Keene's cement plaster, troweled lime putty medium hard finish

	Craft@Hrs	Unit	Material	Labor	Total
Two coat application, on ceilings	F8@.435	SY	4.24	15.70	19.94
Two coat application, on walls	F8@.381	SY	4.24	13.80	18.04
Three coat application, on ceilings	F8@.521	SY	5.22	18.80	24.02
Three coat application, on walls	F8@.459	SY	5.22	16.60	21.82

Portland cement stucco on exterior walls, 3 coats totaling 1" thick

	Craft@Hrs	Unit	Material	Labor	Total
Natural gray, sand float finish	F8@.567	SY	3.75	20.50	24.25
Natural gray, trowel finish	F8@.640	SY	3.75	23.10	26.85
White cement, sand float finish	F8@.675	SY	4.43	24.40	28.83
White cement, trowel finish	F8@.735	SY	4.43	26.60	31.03

Portland cement stucco on soffits, 3 coats totaling 1" thick

	Craft@Hrs	Unit	Material	Labor	Total
Natural gray, sand float finish	F8@.678	SY	3.75	24.50	28.25
Natural gray, trowel finish	F8@.829	SY	3.75	30.00	33.75
White cement, sand float finish	F8@.878	SY	4.39	31.70	36.09
White cement, trowel finish	F8@1.17	SY	4.39	42.30	46.69

	Craft@Hrs	Unit	Material	Labor	Total
Brown and scratch coat base for tile	F8@.308	SY	2.96	11.10	14.06
Scratch coat only for tile	F8@.152	SY	1.39	5.50	6.89

Thin coat plaster, with board but no studs

	Craft@Hrs	Unit	Material	Labor	Total
1/2" board on wood studs	F8@.265	SY	3.35	9.58	12.93
5/8" board on wood studs	F8@.265	SY	5.76	9.58	15.34
1/2" board on metal studs	F8@.277	SY	4.74	10.00	14.74
5/8" board on metal studs	F8@.262	SY	6.02	9.47	15.49

Simulated acoustic texture, sometimes called "popcorn" or "cottage cheese"

	Craft@Hrs	Unit	Material	Labor	Total
On ceilings	F8@.052	SY	.81	1.88	2.69
Texture coat on exterior walls and soffits	F8@.045	SY	.81	1.63	2.44

Patching gypsum plaster, including lath repair but no studding

	Craft@Hrs	Unit	Material	Labor	Total
To 5 SF repairs	P5@.282	SF	.74	9.11	9.85
Over 5 SF repairs	P5@.229	SF	.74	7.39	8.13
Repair cracks only ($150 per job minimum typical)	P5@.055	LF	.14	1.78	1.92

Plastering bead and expansion joint, galvanized

	Craft@Hrs	Unit	Material	Labor	Total
Stop and casing, square nose	F8@.028	LF	.37	1.01	1.38
Corner bead, 3/4" radius	F8@.023	LF	.41	.83	1.24
Corner bead, expanded, zinc nose	F8@.024	LF	.79	.87	1.66
Expansion joint, 3/4", 26 gauge, 1 piece	F8@.025	LF	.79	.90	1.69
Expansion joint, 1-1/2", 2 piece	F8@.028	LF	1.36	1.01	2.37

	Craft@Hrs	Unit	Material	Labor	Total
Plastering accessories					
Base screed, 1/2", 26 gauge	F8@.032	LF	.49	1.16	1.65
Vents 1-1/2", galvanized	F8@.042	LF	1.36	1.52	2.88
Vents 4", galvanized	F8@.061	LF	1.61	2.21	3.82
Archbead, plastic nose	F8@.028	LF	.82	1.01	1.83
Plaster moldings, ornate designs					
2"	CC@.090	LF	3.49	3.80	7.29
4"	CC@.115	LF	7.12	4.86	11.98
6"	CC@.132	LF	10.40	5.58	15.98

Steel Studding (Hollow Metal Studs) Framing Systems

Hollow Metal Stud Partitions for plaster or gypsum wallboard. Galvanized metal studs, including studs, top and bottom track, screws and normal waste. Do not subtract for openings less than 16' wide. Add door and window opening framing, backing, let-in bracing, and sheathing for shear walls. These costs do not include the lath, plaster or gypsum wallboard. Costs per square foot of wall area, measured on one side. The figures in parentheses show production per day for a 2-man crew. Use one day as a minimum job charge

	Craft@Hrs	Unit	Material	Labor	Total
Non-load bearing partitions, 25 gauge, 16" on center spacing					
1-5/8" wide studs (615 SF per day)	C8@.012	SF	.17	.44	.61
2-1/2" wide studs (575 SF per day)	C8@.013	SF	.19	.47	.66
3-5/8" wide studs (535 SF per day)	C8@.016	SF	.22	.58	.80
4" wide studs (500 SF per day)	C8@.017	SF	.24	.62	.86
6" wide studs (470 SF per day)	C8@.018	SF	.31	.65	.96
Add for 20 gauge, 16" on center spacing	—	%	25.0	—	—
Deduct for 25 gauge, 24" on center spacing	—	%	-20.0	-15.0	—
Load bearing partitions. 16 gauge "C" section, 16" on center spacing					
2 1/2" wide studs (430 SF per day)	C8@.023	SF	.36	.83	1.19
3-5/8" wide studs (380 SF per day)	C8@.025	SF	.40	.91	1.31
4" wide studs (355 SF per day)	C8@.026	SF	.45	.94	1.39
6" wide studs (335 SF per day)	C8@.027	SF	.47	.98	1.45
Deduct for 20 gauge, 16" on center spacing	—	%	-30.0	-15.0	—
Deduct for 16 gauge, 24" on center spacing	—	%	-20.0	-10.0	—
Add for 16 gauge, 12" on center spacing	—	%	50.0	20.0	—

Stud adapters (internal slotted standards for supporting shelf brackets). Assembly includes two load-bearing 3-5/8" wide 16 gauge galvanized "C" section hollow metal studs spaced 1" apart, a steel or aluminum stud adapter, screws and normal waste. These costs assume that studs and adapter are installed at the same time. Capitol Hardware A-Line 1700. Costs per vertical linear foot of assembly height

	Craft@Hrs	Unit	Material	Labor	Total
.080" steel adapter, galvanized, 1/2" slots 1" OC	C8@.042	VLF	9.69	1.52	11.21
.080" aluminum adapter, mill finish, 1/2" slots 1" OC	C8@.042	VLF	8.99	1.52	10.51
Add for caps and foot pieces to meet fire code requirements, press-fit type					
Per set of top and bottom caps for each adapter	C8@.021	Set	1.14	.76	1.90

Door opening steel framing using 16 gauge load bearing steel studs (hollow metal studs), based on walls 8' high. Figures in parentheses indicate typical built-up header size. Costs shown are per door opening and include header (built up using steel studs), double vertical steel studs each side of the opening less than 8' wide (triple vertical steel studs each side of openings 8' wide or wider), with steel stud cripples and blocking installed using screws. These costs include the studs, screws and normal waste. Width shown is size of finished opening.

	Craft@Hrs	Unit	Material	Labor	Total
3-5/8" wall studs, opening size as shown					
Up to 3'0" wide (4" x 4" header)	C8@1.50	Ea	14.40	54.40	68.80
Up to 4'0" wide (4" x 6" header)	C8@1.60	Ea	16.30	58.00	74.30
Up to 5'0" wide (4" x 6" header)	C8@1.80	Ea	17.90	65.30	83.20
Up to 6'0" wide (4" x 8" header)	C8@2.00	Ea	21.60	72.50	94.10

	Craft@Hrs	Unit	Material	Labor	Total
Up to 8'0" wide (4" x 10" header)	C8@2.30	Ea	42.50	83.40	125.90
Up to 10'0" wide (4" x 12" header)	C8@2.40	Ea	55.20	87.00	142.20
Up to 12'0" wide (4" x 14" header)	C8@2.50	Ea	70.00	90.70	160.70
Add per foot of vertical height, walls over 8' high	C8@.250	VLF	1.15	9.07	10.22
Deduct for 2-1/2" wall studs	—	%	-15.0	-10.0	—
Add for 6" wall studs	—	%	30.0	10.0	—

Window opening steel framing using 16 gauge load bearing steel studs (hollow metal studs), based on walls 8' high. Figures in parentheses indicate typical built-up header size. Costs shown are per window opening and include header (built up using steel studs), sub-sill plate (double sub-sill if opening is 8' wide or wider), double vertical steel studs each side of openings less than 8' wide (triple vertical steel studs each side of openings 8' wide or wider), with steel stud top and bottom cripples and blocking installed using screws. These costs include the studs, screws and normal waste.

3-5/8" wall studs, opening size shown is width of finished opening

	Craft@Hrs	Unit	Material	Labor	Total
Up to 2'0" wide (4" x 4" header)	C8@1.90	Ea	18.50	68.90	87.40
Up to 3'0" wide (4" x 4" header)	C8@2.20	Ea	23.20	79.80	103.00
Up to 4'0" wide (4" x 6" header)	C8@2.50	Ea	24.50	90.70	115.20
Up to 5'0" wide (4" x 6" header)	C8@2.70	Ea	28.70	97.90	126.60
Up to 6'0" wide (4" x 8" header)	C8@3.00	Ea	36.20	109.00	145.20
Up to 7'0" wide (4" x 8" header)	C8@3.30	Ea	41.04	120.00	161.04
Up to 8'0" wide (4" x 10" header)	C8@3.50	Ea	66.40	127.00	193.40
Up to 10'0" wide (4" x 12" header)	C8@3.80	Ea	79.90	138.00	217.90
Up to 12'0" wide (4" x 14" header)	C8@4.00	Ea	99.90	145.00	244.90
Add per foot of vertical height, walls over 8' high	C8@.400	VLF	1.82	14.50	16.32
Deduct for 2-1/2" wall studs	—	%	-15.0	-10.0	—
Add for 6" wall studs	—	%	30.0	10.0	—

Bracing for steel framing See also sheathing for plywood bracing and shear panels.

Let-in wall bracing, using 16 gauge load bearing steel studs installed using screws. These costs include screws and normal waste

	Craft@Hrs	Unit	Material	Labor	Total
2-1/2" brace	C8@.027	LF	.31	.98	1.29
3-5/8" brace	C8@.035	LF	.35	1.27	1.62
6" brace	C8@.035	LF	.47	1.27	1.74
Steel strap bracing, 1-1/4" wide,					
9'6" or 11'6" lengths	C8@.010	LF	.42	.36	.78
Steel "V" bracing, 3/4" x 3/4"	C8@.010	LF	.58	.36	.94
Temporary steel stud frame wall bracing, assumes salvage at 50% and 3 uses					
2-1/2", 16 gauge load bearing steel studs	C8@.010	LF	.03	.36	.39
3-5/8", 16 gauge load bearing steel studs	C8@.012	LF	.05	.44	.49
6", 16 gauge load bearing steel studs	C8@.018	LF	.08	.65	.73

Gypsum Wallboard Commercial grade work. Material costs include 6% for waste.
Costs per square foot of area covered.
Gypsum wallboard nailed or screwed to wood framing or wood furring, no taping or finishing included

	Craft@Hrs	Unit	Material	Labor	Total
3/8" on walls	CD@.007	SF	.21	.26	.47
3/8" on ceilings	CD@.009	SF	.21	.33	.54
3/8" on furred columns and beams	CD@.013	SF	.21	.48	.69
1/2" on walls	CD@.007	SF	.21	.26	.47
1/2" on ceilings	CD@.009	SF	.21	.33	.54
1/2" on furred columns and beams	CD@.015	SF	.21	.55	.76
5/8" on walls	CD@.007	SF	.23	.26	.49
5/8" on ceilings	CD@.010	SF	.23	.37	.60

	Craft@Hrs	Unit	Material	Labor	Total
5/8" on furred columns and beams	CD@.015	SF	.23	.55	.78
Add for taping and finishing wall joints	CD@.007	SF	.03	.26	.29
Add for taping and finishing ceiling joints	CD@.009	SF	.03	.33	.36
Gypsum wallboard clipped to metal furring, no taping or finishing included					
3/8" on wall furring	CD@.008	SF	.22	.29	.51
3/8" on ceiling furring	CD@.011	SF	.22	.40	.62
1/2" on wall furring	CD@.009	SF	.22	.33	.55
1/2" on ceiling furring	CD@.011	SF	.22	.40	.62
1/2" on column or beam furring	CD@.018	SF	.22	.66	.88
5/8" on wall furring	CD@.009	SF	.23	.33	.56
5/8" on ceiling furring	CD@.011	SF	.23	.40	.63
5/8" on column or beam furring	CD@.018	SF	.23	.66	.89
1/2", two layers on ceiling furring	CD@.022	SF	.39	.81	1.20
1/2", two layers on furring	CD@.017	SF	.39	.62	1.01
1/4" sound board	CD@.008	SF	.23	.29	.52
Add for taping and finishing wall joints	CD@.007	SF	.03	.26	.29
Add for taping and finishing ceiling joints	CD@.009	SF	.03	.33	.36
Add for taping only, no joint finishing	CD@.003	SF	.03	.11	.14
Additional costs for gypsum wallboard					
Add for foil-backed board	—	SF	.05	—	.05
Add for fire resistant board	—	SF	.03	—	.03
Add for water resistant board	—	SF	.13	—	.13
Add for 10' or 12' wall heights	CD@.001	SF	.02	.04	.06
Add for school jobs	CD@.002	SF	—	.07	.07
Add for trowel textured finish	CD@.009	SF	.38	.33	.71
Add for adhesive application, 1/4" bead					
Studs 16" on center	CD@.002	SF	.05	.07	.12
Joists 16" on center	CD@.002	SF	.08	.07	.15
Bead, casing and channels					
Corner bead, 1-1/4" x 1-1/4"	CD@.017	LF	.16	.62	.78
Stop or casing	CD@.021	LF	.20	.77	.97
Jamb casing	CD@.023	LF	.20	.85	1.05
RC-1 channel or 7/8" hat channel	CD@.011	LF	.19	.40	.59
Vinyl clad gypsum board, adhesive or clip application on walls					
1/2", no mouldings included	CD@.010	SF	.69	.37	1.06
5/8", no mouldings included	CD@.011	SF	.73	.40	1.13

Metal Framed Shaft Walls Shaft walls are used to enclose vertical shafts that surround pipe chases, electrical conduit, elevators or stairwells. A 2-hour wall can be made from 1" gypsum wallboard screwed to a 2-1/2" metal stud partition on the shaft side with two layers of 5/8" gypsum wallboard screwed to the partition exterior face. The wall cavity is filled with fiberglass batt insulation and the drywall is taped and finished on both sides. This wall will be 4-3/4" thick. Labor includes installing metal studs, installing the insulation, hanging, taping and finishing the drywall, and cleanup. Metal studs and insulation include a 10% allowance for waste. Wallboard costs include corner and edge trim and a 15% allowance for waste. Costs shown are per square foot of wall measured on one side. For scheduling purposes, estimate that a crew of 4 can install metal studs and insulation, hang, tape and finish 320 SF of shaft wall per 8-hour day.

	Craft@Hrs	Unit	Material	Labor	Total
Metal studs, "C" section, 2-1/2" wide, 16 gauge, 24" on center,					
Complete with top runner and bottom plate	C8@.036	SF	.29	1.31	1.60
1" type X gypsum shaftboard	CD@.029	SF	.46	1.07	1.53
2 layers 5/8" type X gypsum wallboard	CD@.027	SF	.50	.99	1.49
Fiberglass insulation, foil faced,	A1@.006	SF	.28	.22	.50
Total for metal framed shaft wall as described	—@.098	**SF**	**1.53**	**3.59**	**5.12**

	Craft@Hrs	Unit	Material	Labor	Total

Ceramic Tile Set in adhesive and grouted. Add the cost of adhesive below. No scratch or brown coat included. Based on standard U.S. grades and stock colors. Custom colors, designs and imported tile will cost more.

	Craft@Hrs	Unit	Material	Labor	Total
4-1/4" x 4-1/4" glazed wall tile					
Smooth gloss glaze, minimum quality	T4@.131	SF	2.04	4.20	6.24
Smooth gloss glaze, standard quality	T4@.131	SF	4.50	4.20	8.70
Matte finish, better quality	T4@.131	SF	5.93	4.20	10.13
High gloss finish	T4@.131	SF	7.89	4.20	12.09
Commercial quality					
Group 1 (matte glaze)	T4@.131	SF	3.61	4.20	7.81
Group 2 (bright glaze)	T4@.131	SF	4.42	4.20	8.62
Group 3 (crystal glaze)	T4@.131	SF	5.08	4.20	9.28
6" wall tile, smooth glaze					
6" x 6", gray or brown	T4@.131	SF	2.42	4.20	6.62
6" x 8", quilted look	T4@.131	SF	4.91	4.20	9.11
6" x 8", fume look	T4@.131	SF	4.33	4.20	8.53
Trim pieces for glazed wall tile					
Surface bullnose	T4@.077	LF	1.65	2.47	4.12
Surface bullnose corner	T4@.039	Ea	1.07	1.25	2.32
Sink rail or cap	T4@.077	LF	4.78	2.47	7.25
Radius bullnose	T4@.077	LF	2.18	2.47	4.65
Quarter round or bead	T4@.077	LF	2.58	2.47	5.05
Outside corner or bead	T4@.039	Ea	.85	1.25	2.10
Base	T4@.077	LF	5.85	2.47	8.32
Soap holder, soap dish	T4@.265	Ea	11.60	8.49	20.09
Tissue holder, towel bar	T4@.268	Ea	17.00	8.59	25.59
Glazed floor tile, 1/4" thick					
Minimum quality, standard colors	T4@.210	SF	2.32	6.73	9.05
Better quality, patterns	T4@.210	SF	2.77	6.73	9.50
Commercial quality, 6" x 6" to 6" x 9"					
Standard colors and reds and browns	T4@.210	SF	2.68	6.73	9.41
Blues and greens	T4@.210	SF	3.27	6.73	10.00
Add for abrasive surface	—	SF	.28	—	.28
1" x 1" mosaic tile, back-mounted, natural clays					
Group I colors (browns, reds)	T4@.143	SF	2.32	4.58	6.90
Group II colors (tans, grays)	T4@.143	SF	2.54	4.58	7.12
Group III colors (charcoal, blues)	T4@.143	SF	2.77	4.58	7.35
Adhesive for ceramic tile					
Thinset mortar on concrete					
(25 lbs at $6.00 covers 25 SF)	—	SF	.05	—	.05
Latex floor and wall adhesive					
(1 gallon at $5.00 covers 50 SF)	—	SF	.09	—	.09
Epoxy adhesive on concrete or wood					
(12.5 lbs at $20.00 covers 40 SF)	—	SF	.22	—	.22
Acrylic adhesive on countertops or wood					
(25 lbs at $16.00 covers 150 SF)	—	SF	.18	—	.18
Decorator tile panels, institutional quality, 6" x 9" x 3/4", typical costs					
Unglazed domestic	T4@.102	SF	5.53	3.27	8.80
Wash-glazed domestic	T4@.102	SF	9.20	3.27	12.47
Full glazed domestic	T4@.102	SF	9.73	3.27	13.00
Glazed imported, decorative	T4@.102	SF	23.10	3.27	26.37
Deduct for imported tile	—	%	-20.00	—	—

	Craft@Hrs	Unit	Material	Labor	Total
Plastic tile					
4-1/4" x 4-1/4" x .110"	T4@.052	SF	1.40	1.67	3.07
4-1/4" x 4-1/4" x .050"	T4@.052	SF	1.08	1.67	2.75
Aluminum tile, 4-1/4"	T4@.073	SF	2.55	2.34	4.89
Copper on aluminum, 4-1/4" x 4-1/4"	T4@.076	SF	2.80	2.44	5.24
Stainless steel tile, 4-1/4" x 4-1/4"	T4@.076	SF	4.50	2.44	6.94
Quarry tile set in Portland cement					
5" high base tile	T4@.109	LF	2.77	3.49	6.26
4" x 4" x 1/2" floor tile	T4@.096	SF	3.26	3.08	6.34
6" x 6" x 1/2" floor tile	T4@.080	SF	2.68	2.56	5.24
Quarry tile set in furan resin					
5" high base tile	T4@.096	LF	1.92	3.08	5.00
6" x 6" x 3/4" floor tile	T4@.093	SF	4.60	2.98	7.58
Add for abrasive finish	—	SF	.42	—	.42

Terrazzo

Floors, 1/2" terrazzo topping on an underbed bonded to an existing concrete slab, #1 and #2 chips in gray portland cement. Add divider strips below

	Craft@Hrs	Unit	Material	Labor	Total
2", epoxy or polyester	T4@.128	SF	2.20	4.10	6.30
2" conductive	T4@.134	SF	2.85	4.29	7.14
2" with #3 and larger chips	T4@.134	SF	3.77	4.29	8.06
3" with 15 lb felt and sand cushion	T4@.137	SF	2.85	4.39	7.24
2-3/4" with mesh, felt and sand cushion	T4@.137	SF	2.76	4.39	7.15

Additional costs for terrazzo floors. Add for:

	Craft@Hrs	Unit	Material	Labor	Total
White portland cement	—	SF	.36	—	.36
Non-slip abrasive, light	—	SF	.92	—	.92
Non-slip abrasive, heavy	—	SF	1.38	—	1.38
Countertops, mud set	T4@.250	SF	7.55	8.01	15.56
Wainscot, precast, mud set	T4@.198	SF	6.62	6.35	12.97
Cove base, mud set	T4@.198	LF	1.60	6.35	7.95
Add for 2,000 SF job	—	%	10.0	15.0	—
Add for 1,000 SF job	—	%	40.0	60.0	—
Add for waterproof membrane	—	SF	1.18	—	1.18
Divider strips					
Brass, 12 gauge	T4@.009	LF	2.20	.29	2.49
White metal, 12 gauge	T4@.009	LF	1.02	.29	1.31
Brass 4' OC each way	T4@.004	SF	1.15	.13	1.28
Brass 2' OC each way	T4@.007	SF	2.24	.22	2.46

Acoustical Treatment

Tile board and panels, no suspension grid included, applied to ceilings with staples, not including furring

Decorative (non-acoustic) Armstrong ceiling tile, 12" x 12" x 1/2", T&G

	Craft@Hrs	Unit	Material	Labor	Total
Washable white or Grenoble	C8@.017	SF	.49	.62	1.11
Chaperone	C8@.017	SF	.54	.62	1.16
Conestoga	C8@.017	SF	.58	.62	1.20
Windstone	C8@.017	SF	.63	.62	1.25

Acoustic (Cushiontone) Armstrong ceiling tile, 12" x 12" x 1/2", T&G

	Craft@Hrs	Unit	Material	Labor	Total
Classic random perf or Verona	C8@.017	SF	.71	.62	1.33
Textured random perf	C8@.017	SF	.85	.62	1.47
Fissured or Plank n' Plaster	C8@.017	SF	1.11	.62	1.73
Rush Square	C8@.017	SF	1.21	.62	1.83
Colonial Sampler	C8@.017	SF	1.34	.62	1.96
Constitution	C8@.017	SF	1.70	.62	2.32

	Craft@Hrs	Unit	Material	Labor	Total
Fire-rated Armstrong acoustic tile, mineral fiber 12" x 12" x 5/8", beveled edge					
Fissured Minitone	C8@.017	SF	1.03	.62	1.65
Cortega Minitone	C8@.017	SF	1.03	.62	1.65
Add for adhesive applications	C8@.001	SF	.18	.04	.22
Deduct for wall applications	—	%	—	-10.0	—
Sound-absorbing fabric covered wall panels with splines for attaching to structural walls, Armstrong Soundsoak, 9' or 10' high					
Embossed 60, with internal splines, 30" wide, 3/4" thick	C8@.038	SF	5.81	1.38	7.19
Soundsoak 80, with internal splines, 24" wide, 1" thick	C8@.038	SF	7.95	1.38	9.33
Vinyl Soundsoak, 24" wide, 3/4" thick	C8@.038	SF	5.08	1.38	6.46
Encore, 30" wide, 3/4" thick	C8@.038	SF	8.03	1.38	9.41
Aluminum starter spline for panels	C8@.020	LF	.99	.73	1.72
Aluminum internal spline for panels	C8@.021	LF	1.03	.76	1.79

Suspended Ceiling Grid Systems 5,000 SF job. Add for Ceiling Tile below.

	Craft@Hrs	Unit	Material	Labor	Total
"T" bar suspension system, no tile included, suspended from ceiling joists with wires					
2' x 2' grid	C8@.011	SF	.47	.40	.87
2' x 4' grid	C8@.010	SF	.35	.36	.71
4' x 4' grid	C8@.009	SF	.23	.33	.56
Add for jobs under 5,000 SF					
Less than 2,500 SF	C8@.007	SF	.20	.25	.45
2,500 SF to 4,500 SF	C8@.005	SF	.15	.18	.33

Ceiling Tile Tile laid in suspended ceiling grid. No suspended ceiling grid system included, add for same from above. Labor column shows the cost of laying tile in a suspended ceiling grid.

	Craft@Hrs	Unit	Material	Labor	Total
1/2" thick non-acoustic (Armstrong Temlock) panels, 2' x 4'					
Plain white	C8@.004	SF	.28	.15	.43
Conestoga or Chaperone	C8@.004	SF	.36	.15	.51
Grenoble or Windstone vinyl	C8@.004	SF	.38	.15	.53
Glenwood vinyl	C8@.004	SF	.41	.15	.56
1/2" thick acoustic (Armstrong Cushiontone) panels, 2' x 4'					
Kingsley	C8@.004	SF	.28	.15	.43
Classic	C8@.004	SF	.36	.15	.51
Verona	C8@.004	SF	.37	.15	.52
Plaza	C8@.004	SF	.44	.15	.59
5/8" thick (Armstrong Fashiontone) panels, 2' x 4' mineral fiber panels					
Random perf or Cortega	C8@.004	SF	.31	.15	.46
Textured or fissured	C8@.004	SF	.36	.15	.51
Rock Castle	C8@.004	SF	.35	.15	.50
Royal Oak	C8@.004	SF	.67	.15	.82
Victoria	C8@.004	SF	.72	.15	.87
Cumberland	C8@.004	SF	.61	.15	.76
5/8" special purpose panels, 2' x 4'					
Panels with embossed aluminum foil face	C8@.004	SF	.60	.15	.75
Clean room rated panels	C8@.004	SF	.65	.15	.80
3/4" thick panels					
2' x 2' reveal edge panels	C8@.005	SF	.46	.18	.64
2' x 2' non-woven fabric covered with radius reveal	C8@.012	SF	2.95	.44	3.39

	Craft@Hrs	Unit	Material	Labor	Total
7/8" sound rated mineral fiber panels NRC (75-85), 2' x 4'					
White panels	C8@.004	SF	.53	.15	.68
Fire-rated white panels	C8@.004	SF	.59	.15	.74
Rabbeted reveal edge white panels	C8@.004	SF	.64	.15	.79
Translucent panels, 2' x 4'					
Clear or frosted prismatic	C8@.004	SF	.48	.15	.63
Arctic opal or cracked ice	C8@.004	SF	.48	.15	.63
Clear acrylic	C8@.004	SF	1.05	.15	1.20
White acrylic	C8@.004	SF	1.05	.15	1.20

Wood Flooring

	Craft@Hrs	Unit	Material	Labor	Total
Oak, 3/4" x 2-1/4", unfinished					
Number 1 common, red and white	F9@.062	SF	3.01	2.07	5.08
Select, plain red and white	F9@.062	SF	3.85	2.07	5.92
Plank natural oak, 3/4", 3" to 8" wide	F9@.062	SF	5.53	2.07	7.60
Parquet, prefinished, 5/16"					
Red oak, 9" x 9"	F9@.053	SF	6.00	1.77	7.77
Tropical walnut and cherry, 9" x 9"	F9@.053	SF	6.00	1.77	7.77
Imported teak, neutral finish, 12" x 12"	F9@.053	SF	7.95	1.77	9.72

Laminate wood floor (Pergo), plastic laminate on pressed wood base, 7-7/8" x 47-1/4" x 3/8" (20 cm x 120 cm x 8 mm), loose lay with glued tongue and groove joints. Many wood finishes available. 20.67 SF (1.9 square meters) per pack. Includes 10% waste and joint glue (bottle at $5.90 covers 150 SF)

	Craft@Hrs	Unit	Material	Labor	Total
Residential grade, on floors	BF@.051	SF	4.00	1.25	5.25
Commercial grade, on floors	BF@.051	SF	5.30	1.25	6.55
Glued on stairs (treads and risers)	BF@.115	SF	4.70	2.83	7.53
Add for 2' x 4' x 3/8" WhisperWalk underlay (110 SF per pack)	BF@.010	SF	.66	.25	.91
Add for 8 mil poly vapor barrier (over concrete)	BF@.002	LF	.08	.05	.13
T mould strip (joins same level surface)	BF@.040	LF	4.50	.98	5.48
Reducer strip (joins higher surface)	BF@.040	LF	4.12	.98	5.10
End molding (joins lower surface)	BF@.040	LF	2.85	.98	3.83
Wall base, 4" high	BF@.040	LF	2.72	.98	3.70
Stair nosing	BF@.050	LF	2.85	1.23	4.08
Gym floors, 3/4", maple, including finishing					
On sleepers and membrane	F9@.086	SF	5.65	2.88	8.53
On steel spring system	F9@.114	SF	6.71	3.81	10.52
Connected with steel channels	F9@.097	SF	5.89	3.24	9.13
Softwood floors, fir, unfinished					
"B" grade	F9@.031	SF	4.75	1.04	5.79
"C" grade	F9@.031	SF	4.07	1.04	5.11
Wood block flooring					
Natural, 1-1/2" thick	F9@.023	SF	4.60	.77	5.37
Creosoted, 2" thick	F9@.025	SF	3.48	.84	4.32
Creosoted, 2-1/2" thick	F9@.027	SF	3.91	.90	4.81
Creosoted, 3" thick	F9@.027	SF	4.82	.90	5.72
Finishing wood flooring					
Fill and stain	F9@.010	SF	.09	.33	.42
Two coats shellac	F9@.047	SF	.25	1.57	1.82
Wax coating	F9@.001	SF	.03	.03	.06

	Craft@Hrs	Unit	Material	Labor	Total
Sand, scrape and edge hardwood floor					
Large room or hall, using 12" drum sander and 7" disc sander					
Three cuts	F9@.009	SF	.05	.30	.35
Four cuts	F9@.011	SF	.06	.37	.43
Closet or 4' x 10' area, using 7" disc sander					
Three cuts	F9@.010	SF	.04	.33	.37
Four cuts	F9@.012	SF	.04	.40	.44

Resilient Flooring

	Craft@Hrs	Unit	Material	Labor	Total
Asphalt tile, 9" x 9" x 1/8"					
"B" grade	F9@.018	SF	1.35	.60	1.95
"C" grade	F9@.018	SF	1.51	.60	2.11
"D" grade	F9@.018	SF	1.51	.60	2.11
Vinyl composition and vinyl tile, 12" x 12"					
Minimum grade 1/16" vinyl composition tile	F9@.016	SF	1.68	.53	2.21
Most .045" no wax, self-stick tile	F9@.016	SF	.95	.53	1.48
Most .050" no wax, self-stick tile	F9@.016	SF	1.18	.53	1.71
Most .080" no wax, self-stick tile	F9@.016	SF	2.14	.53	2.67
Most 1/8" no wax vinyl tile	F9@.016	SF	2.44	.53	2.97
Vinyl composition and vinyl tile					
Imperial texture, 3/32", fire rated	F9@.016	SF	.95	.53	1.48
Imperial texture, 1/8", fire rated	F9@.016	SF	.84	.53	1.37
Solid vinyl tile, Azrock					
1/8" standard	F9@.021	SF	3.38	.70	4.08
.10" heavy duty	F9@.022	SF	3.44	.74	4.18

Armstrong Solarian sheet vinyl flooring, no wax, Armstrong Bonded with adhesive at perimeter and at seams. Ten year limited warranty. Performance Appearance Rating 4.1

	Craft@Hrs	Unit	Material	Labor	Total
Designer Solarian II	F9@.250	SY	40.80	8.36	49.16
Designer Solarian	F9@.250	SY	36.20	8.36	44.56
Etchings Solarian	F9@.250	SY	36.20	8.36	44.56
Rhythms Solarian	F9@.250	SY	32.94	8.36	41.30
Stencil Craft Solarian	F9@.250	SY	28.50	8.36	36.86
Starstep Solarian	F9@.250	SY	26.00	8.36	34.36
Traditions Solarian	F9@.250	SY	26.00	8.36	34.36

Armstrong sheet vinyl flooring. By Performance Appearance Rating (PAR). Bonded with adhesive at perimeter and at seams.

	Craft@Hrs	Unit	Material	Labor	Total
Fundamentals Successor, 5 year	F9@.250	SY	15.50	8.36	23.86
Caspian, PAR 3.8, urethane wear layer, 10 year	F9@.250	SY	14.40	8.36	22.76
Themes, PAR 2.9, urethane wear layer, 5 year	F9@.250	SY	12.10	8.36	20.46
Metro, PAR 2.3, vinyl wear layer, 3 year	F9@.250	SY	9.80	8.36	18.16
Cambray FHA, PAR 1.7, vinyl wear layer, 2 year	F9@.250	SY	8.50	8.36	16.86
Royelle, PAR 1.3, vinyl wear layer, 1 year	F9@.250	SY	5.90	8.36	14.26

Sheet vinyl, no wax, Congoleum Bonded with adhesive at perimeter and at seams.

	Craft@Hrs	Unit	Material	Labor	Total
Futura	F9@.250	SY	46.60	8.36	54.96
Celestial	F9@.250	SY	40.90	8.36	49.26
Forum Vinyl Plank, 3" or 4-1/2" x 36" x 1/8"	F9@.250	SY	38.50	8.36	46.86
Triumph	F9@.250	SY	34.00	8.36	42.36
Ovations Flex	F9@.250	SY	32.80	8.36	41.16
Ovations	F9@.250	SY	29.50	8.36	37.86
Starlight	F9@.250	SY	27.60	8.36	35.96
Endurance	F9@.250	SY	23.00	8.36	31.36
Exclamation	F9@.250	SY	23.60	8.36	31.96

	Craft@Hrs	Unit	Material	Labor	Total
Highlight Flex	F9@.250	SY	23.60	8.36	31.96
Highlight	F9@.250	SY	17.90	8.36	26.26

Sheet vinyl, no wax, Mannington. Bonded with adhesive at perimeter and at seams.

	Craft@Hrs	Unit	Material	Labor	Total
Gold sheet vinyl	F9@.250	SY	42.90	8.36	51.26
Aristocon or Classicon	F9@.250	SY	37.40	8.36	45.76
Lustrecon or Sterling	F9@.250	SY	29.40	8.36	37.76
Omnia, Resolution or Silverado	F9@.250	SY	23.60	8.36	31.96
Stardance	F9@.250	SY	21.40	8.36	29.76
Vega II	F9@.250	SY	13.90	8.36	22.26

Plastic Interlocking Safety Floor Tiles Based on Duragrid, for a job with 1,000 to 5,000 SF of floor tiles.

	Craft@Hrs	Unit	Material	Labor	Total
Firm, smooth tiles, for athletic courts, gym floors, industrial areas					
12" x 12" x 1/2"	F9@.003	SF	2.61	.10	2.71
2" x 12" x 1/2" line strips	F9@.003	LF	1.26	.10	1.36
Firm, non-slip (nippled) tile, for tennis courts, wet-weather surface					
12" x 12" x 1/2"	F9@.003	SF	3.34	.10	3.44
2" x 12" x 1/2" line strips	F9@.003	LF	1.26	.10	1.36
Soft tiles, for anti-fatigue matting, household or deck areas					
12" x 12" x 1/2"	F9@.003	SF	3.63	.10	3.73
2" x 12" x 1/2" line strips	F9@.003	LF	1.26	.10	1.36
Mitered edging strip, 2" x 12" x 1/2"	F9@.003	LF	1.21	.10	1.31
Monogramming/graphics pegs	F9@.085	C	2.51	2.84	5.35
Add for jobs under 1,000 SF	—	%	10.0	20.0	—
Deduct for jobs over 5,000 SF	—	%	-5.0	-5.0	—

Rubber Flooring and Accessories

	Craft@Hrs	Unit	Material	Labor	Total
Rubber tile flooring					
17-13/16" x 17-13/16" x 1/8" circular pattern	F9@.021	SF	6.12	.70	6.82
Base, top set rubber, 1/8" thick					
2-1/2" high	F9@.018	LF	.61	.60	1.21
4" high	F9@.018	LF	.77	.60	1.37
6" high	F9@.018	LF	1.01	.60	1.61
Base corners, top set rubber, 1/8" thick					
2-1/2" or 4" high	F9@.028	Ea	1.01	.94	1.95
6" high	F9@.028	Ea	1.28	.94	2.22
Stair treads, molded rubber, 12-1/2" tread width, per LF of tread length					
1/4" circle design, colors	F9@.067	LF	9.38	2.24	11.62
Rib design, light duty	F9@.067	LF	10.20	2.24	12.44
Tread with abrasive strip, light duty	F9@.071	LF	11.50	2.37	13.87
Tread with abrasive strip, heavy duty	F9@.071	LF	14.60	2.37	16.97
Stair risers, molded rubber, 7" x 1/8"	F9@.038	LF	2.96	1.27	4.23
Stringer cover, .100"					
10" high	F9@.106	LF	2.96	3.54	6.50
Add for project under 2,000 SF	F9@.010	LF	.21	.33	.54
Underlayment, plywood, 1/2"	C8@.009	SF	1.00	.33	1.33

Composition Flooring, 1/4" thick

	Craft@Hrs	Unit	Material	Labor	Total
Acrylic	F9@.075	SF	3.13	2.51	5.64
Epoxy terrazzo	F9@.105	SF	3.57	3.51	7.08
Conductive epoxy terrazzo	F9@.132	SF	4.91	4.41	9.32
Troweled neoprene	F9@.082	SF	4.02	2.74	6.76
Polyester	F9@.065	SF	2.68	2.17	4.85

	Craft@Hrs	Unit	Material	Labor	Total

Carpeting Glue-down carpet installation includes sweeping the floor and adhesive. Carpet over pad installation includes laying the pad and stapling or spot adhesive.

Nylon, cut pile, with 3/8" nova pad

	Craft@Hrs	Unit	Material	Labor	Total
30 oz	F9@.114	SY	11.90	3.81	15.71
36 oz	F9@.114	SY	13.80	3.81	17.61
40 oz	F9@.114	SY	15.20	3.81	19.01

Nylon, level loop, with 3/8" nova pad

	Craft@Hrs	Unit	Material	Labor	Total
20 oz	F9@.114	SY	11.90	3.81	15.71
24 oz	F9@.114	SY	13.80	3.81	17.61

Base grade, level loop, with 50 oz pad

	Craft@Hrs	Unit	Material	Labor	Total
18 oz	F9@.114	SY	6.04	3.81	9.85
20 oz	F9@.114	SY	6.48	3.81	10.29
26 oz	F9@.114	SY	7.15	3.81	10.96
Antron nylon, 20 oz, anti-static, no pad	F9@.068	SY	8.11	2.27	10.38
Lee's "Faculty IV" with unibond back, direct glue	F9@.114	SY	16.10	3.81	19.91

Carpet Tile

Modular carpet tiles

	Craft@Hrs	Unit	Material	Labor	Total
"Faculty 11" tiles, 26 oz face weight	F9@.170	SY	27.50	5.68	33.18
"Surfaces" tiles, 38 oz face weight	F9@.170	SY	30.20	5.68	35.88
ESD 26 oz loop pile for computer floors	F9@.170	SY	32.30	5.68	37.98

Velvet Tile Recycled tire material for entrances, stairs, wet areas, etc. Installation includes adhesive. Add for border strips below. El-Do Velvet Tile

	Craft@Hrs	Unit	Material	Labor	Total
12" x 12" tiles	F9@.044	SF	5.75	1.47	7.22
Aluminum border strips for velvet tile					
add to any tile above	F9@.031	LF	3.00	1.04	4.04

Decontamination and Surface Preparation Dustless removal of undesirable surface contamination and paint from steel and concrete substrates. Applications include PCBs, radioactivity, toxic chemicals and lead-based paints. The removal system is a mechanical process and no water, chemicals or abrasive grit are used. Containment or special ventilation is not required. Based on PENTEK Inc. For estimating purposes figure one 55-gallon drum of contaminated waste per 2,500 square feet of coatings removed from surfaces with an average coating thickness of 12 mils. Waste disposal costs are not included. Unit prices shown are based on a three-man crew. Use $4,500 as a minimum charge for work of this type.

Steel and concrete surfaces

Equipment is based on using three pneumatic needle scaler units attached to one air-powered vacuum-waste packing unit. Equipment cost is $45 per hour and includes one vacuum-waste packing unit (at $4,450 per month), three needle scalers (at $965 each per month) and one 150 CFM gas or diesel powered air compressor (at $621 per month) including all hoses and connectors. Add the one-time charge for move-on, move-off and refurbishing of equipment as shown.

	Craft@Hrs	Unit	Material	Labor	Equipment	Total

Steel surfaces Steel Structures Painting Council Surface Preparation (SSPC-SP) specification codes are shown.

Steel surfaces, SSPC-SP-11 (coating mechanically removed, bare metal exposed)

Overhead surfaces, box girders, obstructed areas

	Craft@Hrs	Unit	Material	Labor	Equipment	Total
30 SF of surface area per hour	PA@.100	SF	—	3.70	1.53	5.23

Large unobstructed structural steel members

	Craft@Hrs	Unit	Material	Labor	Equipment	Total
40 SF of surface area per hour	PA@.075	SF	—	2.78	1.15	3.93

Above ground storage tanks

	Craft@Hrs	Unit	Material	Labor	Equipment	Total
50 SF of surface area per hour	PA@.060	SF	—	2.22	.92	3.14

Steel surfaces, SSPC-SP-3 (coating mechanically removed, mill scale left in place)

Overhead surfaces, box girders, obstructed areas

	Craft@Hrs	Unit	Material	Labor	Equipment	Total
50 SF of surface area per hour	PA@.060	SF	—	2.22	.92	3.14

	Craft@Hrs	Unit	Material	Labor	Equipment	Total
Large unobstructed structural steel members						
70 SF of surface area per hour	PA@.042	SF	—	1.55	.64	2.19
Above ground storage tanks						
90 SF of surface area per hour	PA@.033	SF	—	1.22	.51	1.73
Door or window frames						
To 3'0" x 6'8"	PA@1.50	Ea	—	55.50	22.95	78.45
Over 3'0" x 6'8" to 6'0" x 8'0"	PA@2.00	Ea	—	74.00	30.60	104.60
Over 6'0" x 8'0" to 12'0 x 20'0"	PA@3.00	Ea	—	111.00	45.90	156.90

Concrete Surfaces

	Craft@Hrs	Unit	Material	Labor	Equipment	Total
Concrete lintels, per SF of area	PA@.033	SF	—	1.22	.56	1.78
Concrete walls. Measured on one face of wall						
80 SF of surface per hour	PA@.038	SF	—	1.41	.58	1.99
Concrete block walls measured on one face of wall						
70 SF of surface area per hour	PA@.042	SF	—	1.55	.64	2.19
Add to total estimated costs arrived at using above, one-time cost						
Move-on, move-off						
and refurbish equipment	—	LS	—	—	—	3,400.00

Concrete floor and slab surfaces

Equipment is based on using three pneumatic manually-operated wheel-mounted scabblers attached to one air-powered vacuum-waste packing unit to scarify concrete floors and slabs. Equipment cost is $89 per hour and includes one vacuum-waste packing unit (at $4,670 per month), three scabblers (at $3,520 each per month) and one 150 CFM gas or diesel powered air compressor (at $652 per month) including all hoses and connectors. Add the one-time charge for move-on, move-off and refurbishing of equipment as shown.

	Craft@Hrs	Unit	Material	Labor	Equipment	Total
Large unobstructed areas; warehouses and aircraft hanger floor slabs						
90 SF of surface area per hour	PA@.033	SF	—	1.22	1.01	2.23
Obstructed areas; narrow aisles, areas in pits						
60 SF of surface area per hour	PA@.050	SF	—	1.85	1.52	3.37
Small areas; around equipment, piping and conduit						
40 SF of surface area per hour	PA@.075	SF	—	2.78	2.27	5.05
Add to total estimated costs arrived at using above, one-time cost						
Move-on, move-off						
and refurbish equipment	—	LS	—	—	—	4,450.00

Painting These costs do not include surface preparation, scaffolding or masking unless noted.

	Craft@Hrs	Unit	Material	Labor	Total
Rule of thumb, in place costs, by SF of floor area, typical costs					
Commercial or industrial, 2 coats flat and enamel	—	SF	—	—	1.05
Institutional, 2 coats flat, 3 coats enamel	—	SF	—	—	1.27
Hospitals, public buildings	—	SF	—	—	1.32

Surface preparation for painting, typical costs, SF of area prepared for painting, including normal preparation on walls and ceilings. Hand work, no unusual conditions

	Craft@Hrs	Unit	Material	Labor	Total
Light cleaning (wipe down or hose down)	PA@.001	SF	.02	.04	.06
New cement asbestos siding	PA@.002	SF	.02	.07	.09
Painted cement asbestos siding	PA@.003	SF	.02	.11	.13
New masonry or concrete	PA@.002	SF	.02	.07	.09
Painted masonry or concrete	PA@.003	SF	.03	.11	.14
New plaster or wallboard	PA@.001	SF	.02	.04	.06
Painted plaster or wallboard	PA@.003	SF	.02	.11	.13
New wood	PA@.001	SF	.02	.04	.06
Painted wood	PA@.003	SF	.02	.11	.13
Painted steel (light brushing)	PA@.007	SF	.03	.26	.29

	Craft@Hrs	Unit	Material	Labor	Total
Cabinets and shelves, per square foot of surface painted, brush work					
Wood, 2 coats sealer, 2 coats enamel	PA@.026	SF	.26	.96	1.22
Wood, 2 coats enamel	PA@.014	SF	.15	.52	.67
Wood, seal and varnish, 3 coats	PA@.016	SF	.17	.59	.76
Wood, back priming	PA@.006	SF	.04	.22	.26
Metal, primer and 2 coats enamel	PA@.014	SF	.17	.52	.69
Cement asbestos siding, 2 coats acrylic latex					
Brush	PA@.016	SF	.09	.59	.68
Roller	PA@.009	SF	.11	.33	.44
Spray	PA@.001	SF	.13	.04	.17
Concrete walls, 2 coats acrylic latex					
Brush	PA@.012	SF	.08	.44	.52
Roller	PA@.006	SF	.13	.22	.35
Spray	PA@.003	SF	.14	.11	.25
Doors, brush application, per square foot painted					
Plywood, 2 coats primer, 1 coat alkyd	PA@.016	SF	.17	.59	.76
Wood, 2 coats alkyd oil base paint	PA@.010	SF	.12	.37	.49
Metal, 1 coat primer, 2 coats enamel	PA@.016	SF	.17	.59	.76
Metal, 2 coats enamel	PA@.012	SF	.13	.44	.57
Deduct for roller application	—	%	—	-25.0	—
Deduct for spray application	—	%	—	-65.0	—
Add for spray application	—	%	25.0	—	—
Add for panel or carved doors	—	%	10.0	35.0	—
Masonry walls, 1 coat primer, 2 coats acrylic latex					
Brush	PA@.019	SF	.19	.70	.89
Roller	PA@.010	SF	.20	.37	.57
Spray	PA@.003	SF	.21	.11	.32
Masonry walls, 2 coats acrylic latex					
Brush	PA@.014	SF	.15	.52	.67
Roller	PA@.006	SF	.16	.22	.38
Spray	PA@.002	SF	.17	.07	.24
Metal ladders and cages, per linear foot of height, widths to 4', brush work					
Wire brush ladder	PA@.027	LF	—	1.00	1.00
Wire brush ladder and cage	PA@.054	LF	—	2.00	2.00
Primer, 2 coats enamel, ladder or cage	PA@.019	LF	.60	.70	1.30
Metal trim with brush, figure trim less than 6" wide as being 1 square foot for each linear foot					
1 coat primer, 1 coat enamel	PA@.012	SF	.16	.44	.60
Metal walls, 1 coat primer, 2 coats enamel					
Brush	PA@.013	SF	.13	.48	.61
Roller	PA@.008	SF	.14	.30	.44
Spray	PA@.003	SF	.17	.11	.28
Metal walls, 2 coats enamel					
Brush	PA@.009	SF	.09	.33	.42
Roller	PA@.006	SF	.10	.22	.32
Spray	PA@.001	SF	.10	.04	.14
Most metals, primer and 2 coats enamel					
Brush	PA@.023	SF	.14	.85	.99
Roller	PA@.010	SF	.15	.37	.52
Spray	PA@.004	SF	.17	.15	.32

	Craft@Hrs	Unit	Material	Labor	Total
Miscellaneous metals, 2 coats enamel					
Brush	PA@.015	SF	.10	.56	.66
Roller	PA@.007	SF	.10	.26	.36
Spray	PA@.003	SF	.12	.11	.23
Pipe, figure 1 SF per LF for pipe 4" diameter or less					
Per coat, brush or glove	PA@.008	SF	.06	.30	.36
Per coat, spray	PA@.004	SF	.10	.15	.25
Pipe insulation, glue size, primer, enamel	PA@.013	SF	.12	.48	.60
Plaster or wallboard ceilings, 2 coats acrylic latex					
Brush	PA@.011	SF	.10	.41	.51
Roller	PA@.007	SF	.12	.26	.38
Spray	PA@.003	SF	.17	.11	.28
Plaster or wallboard walls, enamel coats, per coat					
Brush	PA@.005	SF	.05	.19	.24
Roller	PA@.003	SF	.06	.11	.17
Spray	PA@.001	SF	.07	.04	.11
Plaster or wallboard walls, primer coat					
Brush	PA@.005	SF	.05	.19	.24
Roller	PA@.003	SF	.05	.11	.16
Spray	PA@.001	SF	.06	.04	.10
Roofing, 2 coats aluminum paint					
Roller	PA@.005	SF	.11	.19	.30
Spray	PA@.001	SF	.16	.04	.20
Steel tank exteriors, 2 coats red lead and 1 coat enamel					
Brush	PA@.013	SF	.30	.48	.78
Roller	PA@.008	SF	.37	.30	.67
Spray	PA@.002	SF	.38	.07	.45
Structural steel, brush, medium or heavy members					
Primer and 2 coats enamel	PA@.012	SF	.15	.44	.59
2 coats enamel	PA@.008	SF	.10	.30	.40
2 coats primer	PA@.008	SF	.10	.30	.40
2 coats enamel, spray	PA@.004	SF	.16	.15	.31
Epoxy touch-up and 1 coat enamel	PA@.008	SF	.29	.30	.59
Structural steel, brush, light members					
Primer and 2 coats enamel	PA@.024	SF	.15	.89	1.04
2 coats enamel	PA@.018	SF	.10	.67	.77
2 coats primer	PA@.018	SF	.10	.67	.77
Stucco walls, 2 coats acrylic latex					
Brush	PA@.010	SF	.11	.37	.48
Roller	PA@.006	SF	.14	.22	.36
Spray	PA@.003	SF	.15	.11	.26
Window sash and trim, per linear foot of sash or trim painted each side, brush					
Wood, 1 coat primer, 2 coats oil base	PA@.022	LF	.15	.81	.96
Wood, 2 coats oil base	PA@.010	LF	.10	.37	.47
Metal, 1 coat primer, 2 coats enamel	PA@.018	LF	.14	.67	.81
Metal, 2 coats enamel	PA@.012	LF	.09	.44	.53
Window sash and trim, using Trimex™ primer to mask glass and prime, 2 coats oil base					
1 coat Trimex™ primer, 2 coats oil base, brush	PA@.013	SF	.33	.48	.81
1 coat Trimex™ primer, 2 coats oil base, spray	PA@.005	SF	.37	.19	.56
Score and peel Trimex™ primer from glass, per lite	PA@.020	Ea	—	.74	.74

	Craft@Hrs	Unit	Material	Labor	Total
Wood or plywood ceilings, brush application					
2 coats sealer, 2 coats enamel	PA@.023	SF	.25	.85	1.10
2 coats enamel	PA@.010	SF	.12	.37	.49
Wood or plywood walls, 2 coats primer, 1 coat alkyd					
Brush	PA@.013	SF	.10	.48	.58
Roller	PA@.007	SF	.12	.26	.38
Spray	PA@.005	SF	.15	.19	.34
Wood or plywood walls, 2 coats primer					
Brush	PA@.009	SF	.10	.33	.43
Roller	PA@.005	SF	.10	.19	.29
Spray	PA@.003	SF	.14	.11	.25
Wood or plywood walls, brush application					
2 coats sealer and varnish, 2 coats enamel	PA@.019	SF	.24	.70	.94
2 coats enamel	PA@.008	SF	.10	.30	.40
Sealer and varnish, 3 coats	PA@.010	SF	.22	.37	.59
Wood trim, brush, figure trim less than 6" wide at 1 SF per LF					
1 coat primer, 1 coat enamel	PA@.012	SF	.15	.44	.59
Sealer and varnish, 3 coats	PA@.013	SF	.20	.48	.68

Paper Wallcoverings Based on 2,000 SF job. These costs include 10% waste allowance. For estimating purposes figure an average roll of wallpaper will cover 40 SF.

	Craft@Hrs	Unit	Material	Labor	Total
Paperhanging, average quality, labor only	PA@.011	SF	—	.41	.41
Paperhanging, good quality, labor only	PA@.013	SF	—	.48	.48
Vinyl wallcovering, typical costs					
7 oz, light grade	PA@.012	SF	.45	.44	.89
14 oz, medium grade	PA@.013	SF	.80	.48	1.28
22 oz, heavy grade, premium quality	PA@.017	SF	1.05	.63	1.68
Vinyl, 14 oz, on aluminum backer	PA@.035	SF	.85	1.30	2.15
Flexwood, many veneers, typical job	PA@.053	SF	2.54	1.96	4.50
Cloth wallcovering					
Linen, acrylic backer, scotch guarded	PA@.021	SF	1.05	.78	1.83
Grasscloth	PA@.021	SF	1.67	.78	2.45
Felt	PA@.027	SF	2.10	1.00	3.10
Cork sheathing, 1/8", typical	PA@.027	SF	1.32	1.00	2.32
Blank stock (underliner)	PA@.011	SF	.30	.41	.71
Paper backed vinyl, prepasted, typical	PA@.012	SF	.74	.44	1.18
Untrimmed paper, hand prints, foils, custom	PA@.029	SF	2.99	1.07	4.06
Gypsum impregnated jute fabric Flexi-wall™, with #500 adhesive. Use 360 SF as minimum job size					
Medium weight plaster wall liner #605	PA@.010	SF	.50	.37	.87
Heavy duty plaster wall liner #609	PA@.010	SF	.58	.37	.95
Classics line	PA@.012	SF	.65	.44	1.09
Contemporary line	PA@.012	SF	.76	.44	1.20
Images line	PA@.012	SF	.79	.44	1.23
Scotland or Indian jute weave	PA@.013	SF	.65	.48	1.13
French weave	PA@.013	SF	.79	.48	1.27
Trimmed paper, hand prints, foils	PA@.012	SF	1.00	.44	1.44

	Craft@Hrs	Unit	Material	Labor	Total
Add for job less than 200 SF (5 rolls)	—	%	—	20.0	—
Add for small rooms (kitchen, bath)	—	%	10.0	20.0	—
Add for patterns	—	%	—	15.0	—
Deduct for job over 2,000 SF (50 rolls)	—	%	—	-5.0	—

Laminated Plastic Wallcovering Adhesive application, standard patterns and colors.

	Craft@Hrs	Unit	Material	Labor	Total
1/16", most colors and textures	PA@.028	SF	1.70	1.04	2.74
Vertical grade, .032", most finishes	PA@.028	SF	1.45	1.04	2.49
1/16" acid resisting	PA@.035	SF	2.15	1.30	3.45
.020" backing sheet	PA@.017	SF	.36	.63	.99
1/16" suede, leather, wood grain, solid colors	PA@.028	SF	2.36	1.04	3.40
Black, #840-7	PA@.028	SF	5.24	1.04	6.28
Metallic slate	PA@.028	SF	9.02	1.04	10.06
Add for post forming grade	PA@.004	SF	.15	.15	.30
Add for adhesive	—	SF	.29	—	.29

Hardboard Wallcovering Vinyl clad, printed. Add metal trim at $.35 per LF.

	Craft@Hrs	Unit	Material	Labor	Total
3/16" thick	C8@.025	SF	.71	.91	1.62
1/4" thick	C8@.025	SF	.98	.91	1.89
3/16" pegboard	C8@.025	SF	1.02	.91	1.93
1/4" pegboard	C8@.025	SF	1.11	.91	2.02
Add for adhesive, 1/4" bead, 16" OC	—	SF	.11	—	.11

Chalkboards and Tackboards

Typical subcontract prices. Use these prices for preliminary estimates only.

	Craft@Hrs	Unit	Material	Labor	Total
School and institutional grade chalkboards					
With trim, map and chalk rail	—	SF	—	—	12.10
Without trim, average	—	SF	—	—	6.60
Tackboards with trim, cork or vinyl	—	SF	—	—	11.10
Vertical sliding boards	—	SF	—	—	54.90
Horizontal sliding boards	—	SF	—	—	35.90
Swing leaf panels	—	SF	—	—	31.70
Chalkboards, no frame included					
Tempered hardboard, 1/4"	CC@.034	SF	2.55	1.44	3.99
Tempered hardboard, 1/2"	CC@.036	SF	4.19	1.52	5.71
Metal back, 24 gauge, 1/4", porcelain enamel	CC@.041	SF	7.65	1.73	9.38
Metal back, 24 gauge, 1/2", porcelain enamel	CC@.048	SF	7.85	2.03	9.88
Slate, 3/8"	CC@.055	SF	11.20	2.32	13.52
Tackboards, no frame or furring included					
Cork sheets					
Unbacked, 1/8"	CC@.032	SF	3.05	1.35	4.40
Unbacked, 1/4"	CC@.032	SF	3.30	1.35	4.65
Burlap backed, 1/8"	CC@.032	SF	4.54	1.35	5.89
Burlap backed, 1/4"	CC@.032	SF	5.08	1.35	6.43
Vinyl covered fiberboard, 1/2"	CC@.032	SF	4.35	1.35	5.70
1/4" vinyl cork on 1/4" hardboard	CC@.032	SF	5.94	1.35	7.29
Frames, trim, aluminum	CC@.018	LF	1.71	.76	2.47
Chalk tray, aluminum	CC@.040	LF	4.59	1.69	6.28
Map and display rail, aluminum	CC@.032	LF	3.08	1.35	4.43

	Craft@Hrs	Unit	Material	Labor	Total
Compartments and Cubicles					
Toilet partitions, floor or ceiling mounted, standard sizes, includes 1 wall and 1 door					
Baked enamel	T5@2.66	Ea	343.00	95.20	438.20
Porcelain enamel	T5@2.66	Ea	701.00	95.20	796.20
Laminated plastic	T5@2.66	Ea	458.00	95.20	553.20
Stainless steel	T5@2.66	Ea	827.00	95.20	922.20
Add for wheel chair units	—	Ea	53.10	—	53.10
Urinal screens, wall mounted with brackets, 18" wide					
Standard quality, baked enamel	T5@1.95	Ea	138.00	69.80	207.80
Porcelain enamel	T5@1.03	Ea	286.00	36.90	322.90
Laminated plastic	T5@1.03	Ea	202.00	36.90	238.90
Stainless steel	T5@1.03	Ea	297.00	36.90	333.90
Add for floor mounted, 24" wide	T5@.476	Ea	34.50	17.00	51.50
Add for wheel chair units	—	Ea	34.50	—	34.50
Sight screens, 3' x 7'					
Baked enamel, floor mounted	T5@1.28	Ea	189.00	45.80	234.80
Porcelain enamel, floor mounted	T5@1.28	Ea	308.00	45.80	353.80
Laminated plastic, floor mounted	T5@1.28	Ea	223.00	45.80	268.80
Stainless steel, floor mounted	T5@1.28	Ea	383.00	45.80	428.80
Add for ceiling-mounted units	—	Ea	39.80	—	39.80
Accessories mounted on partitions					
Coat hooks and door stop, chrome	T5@.166	Ea	9.99	5.94	15.93
Purse shelf, chrome, 5" x 14"	T5@.166	Ea	22.50	5.94	28.44
Dressing cubicles, 80" high, with curtain, floor standing, wall hung					
Baked enamel	T5@3.41	Ea	343.00	122.00	465.00
Porcelain enamel	T5@3.41	Ea	608.00	122.00	730.00
Laminated plastic	T5@3.41	Ea	440.00	122.00	562.00
Stainless steel	T5@3.41	Ea	734.00	122.00	856.00
Shower compartments, industrial type, with receptor but no door or plumbing					
Baked enamel	T5@2.68	Ea	927.00	95.90	1,022.90
Porcelain enamel	T5@2.68	Ea	1,650.00	95.90	1,745.90
Laminated plastic	T5@2.68	Ea	1,090.00	95.90	1,185.90
Stainless steel	T5@2.68	Ea	1,840.00	95.90	1,935.90
Add for soap dish	T5@.166	Ea	11.50	5.94	17.44
Add for curtain rod	T5@.166	Ea	7.97	5.94	13.91
Doors for industrial shower compartments					
Wire or tempered glass, 24" x 72"	G1@1.03	Ea	121.00	34.30	155.30
Plastic, 24" x 72"	G1@1.03	Ea	82.10	34.30	116.40
Plastic, 1 panel door and 1 side panel	G1@1.03	Ea	170.00	34.30	204.30
Shower stalls, with receptor, 32" x 32", with door but no plumbing					
Fiberglass	P6@2.65	Ea	319.00	95.70	414.70
Painted metal	P6@2.65	Ea	334.00	95.70	429.70
Hospital cubicle curtain track, ceiling mount	MW@.060	LF	5.67	2.62	8.29

Wall Protection Systems These costs include a .063" extruded aluminum retainer, with a .100" extruded plastic (Acrovyn) covering, installed with adhesives or mechanical fasteners over existing drywall, doors or door frames. No substrate costs are included.

	Craft@Hrs	Unit	Material	Labor	Total
Flush-mounted corner guards, 3" x 3" wide legs	CC@.194	LF	9.95	8.20	18.15
Surface-mounted corner guards, 3" x 3" wide legs	CC@.069	LF	5.55	2.92	8.47
Handrail	CC@.111	LF	11.00	4.69	15.69
Bumper guards, bed bumpers	CC@.138	LF	7.43	5.83	13.26
Door frame protectors	CC@.056	LF	11.20	2.37	13.57
Door protectors	CC@.083	SF	3.56	3.51	7.07
Protective wallcovering (vinyl only)	PA@.087	SF	2.81	3.22	6.03

	Craft@Hrs	Unit	Material	Labor	Total

Access Flooring Standard 24" x 24" steel flooring panels with 1/8" high pressure laminate surface. Costs shown include the floor panels and the supporting understructures listed. Add for accessories below. Raised Floor Installation, Inc.

Stringerless system, up to 10" finish floor height

	Craft@Hrs	Unit	Material	Labor	Total
Under 1,000 SF job	D3@.058	SF	11.30	2.37	13.67
1,000 to 5,000 SF job	D3@.046	SF	9.63	1.88	11.51
Over 5,000 SF job	D3@.039	SF	8.49	1.60	10.09

Snap-on stringer system, up to 10" finish floor height

	Craft@Hrs	Unit	Material	Labor	Total
Under 1,000 SF job	D3@.058	SF	12.20	2.37	14.57
1,000 to 5,000 SF job	D3@.046	SF	10.40	1.88	12.28
Over 5,000 SF job	D3@.039	SF	9.42	1.60	11.02

Rigid bolted stringer system, up to 18" finish floor height

	Craft@Hrs	Unit	Material	Labor	Total
Under 1,000 SF job	D3@.058	SF	13.00	2.37	15.37
1,000 to 5,000 SF job	D3@.046	SF	11.00	1.88	12.88
Over 5,000 SF job	D3@.039	SF	9.99	1.60	11.59

	Craft@Hrs	Unit	Material	Labor	Total
Corner lock system, under 2,500 SF office areas, 6" to 10" finish floor height, blank panels	D3@.036	SF	5.96	1.47	7.43

Used flooring, most floor system types

	Craft@Hrs	Unit	Material	Labor	Total
Unrefurbished, typical	D3@.058	SF	8.13	2.37	10.50
Refurbished, new panel surface and trim	D3@.058	SF	11.20	2.37	13.57

Extras for all floor systems, add to the costs above

	Craft@Hrs	Unit	Material	Labor	Total
Carpeted panels	—	SF	1.16	—	1.16
Concrete-filled steel panels	—	SF	.67	—	.67
Ramps, per SF of actual ramp area	D3@.104	SF	7.50	4.25	11.75
Non-skid tape on ramp surface	D3@.033	SF	4.59	1.35	5.94
Portable ramp for temporary use	—	SF	51.00	—	51.00
Steps, per SF of actual tread area	D3@.208	SF	11.90	8.51	20.41
Fascia at steps and ramps, to 18" high	D3@.117	LF	8.26	4.79	13.05
Cove base, 4" high, black or brown	D3@.026	LF	1.35	1.06	2.41
Guardrail	D3@.573	LF	49.00	23.40	72.40
Handrail	D3@.573	LF	22.40	23.40	45.80
Cable cutouts					
Standard "L" type trim	D3@.407	Ea	7.76	16.60	24.36
Type "F" trim, under floor plenum systems	D3@.407	Ea	8.96	16.60	25.56
Perforated air distribution panels	—	SF	3.24	—	3.24
Air distribution grilles, 6" x 18", with adjustable damper, including cutout	D3@.673	Ea	59.20	27.50	86.70

Fire extinguishing systems, smoke detectors, halon type, subcontract

	Craft@Hrs	Unit	Material	Labor	Total
With halon gas suppression	—	SF	—	—	5.10
Automatic fire alarm	—	SF	—	—	.77

Flagpoles Including erection cost but no foundation. See foundation costs below. By pole height above ground, except as noted. Buried dimension is usually 10% of pole height.

Fiberglass, semi-gloss white, tapered, external halyard, commercial grade

	Craft@Hrs	Unit	Material	Labor	Total
25' with 5.90" butt diameter	D4@8.19	Ea	880.00	303.00	1,183.00
30' with 6.50" butt diameter	D4@9.21	Ea	1,060.00	340.00	1,400.00
35' with 6.90" butt diameter	D4@9.96	Ea	1,370.00	368.00	1,738.00
39' with 7.10" butt diameter	D4@10.2	Ea	1,560.00	377.00	1,937.00
50' with 9.90" butt diameter	D4@15.1	Ea	3,730.00	558.00	4,288.00
60' with 10.50" butt diameter	D4@18.5	Ea	4,750.00	683.00	5,433.00
70' with 11.40" butt diameter	D4@21.7	Ea	6,720.00	802.00	7,522.00
80' with 16.00" butt diameter	D4@22.8	Ea	8,710.00	842.00	9,552.00

	Craft@Hrs	Unit	Material	Labor	Total
As above but with internal halyard, jam cleat type.					
25' with 5.90" butt diameter	D4@8.19	Ea	1088.00	303.00	1,391.00
30' with 6.50" butt diameter	D4@9.21	Ea	1,280.00	340.00	1,620.00
35' with 6.90" butt diameter	D4@9.96	Ea	1,650.00	368.00	2,018.00
39' with 7.10" butt diameter	D4@10.2	Ea	1,830.00	377.00	2,207.00
50' with 9.90" butt diameter	D4@15.1	Ea	5,200.00	558.00	5,758.00
60' with 10.50" butt diameter	D4@18.5	Ea	6,200.00	683.00	6,883.00
70' with 11.40" butt diameter	D4@21.7	Ea	8,110.00	802.00	8,912.00
80' with 16.00" butt diameter	D4@22.8	Ea	10,200.00	842.00	11,042.00
Add for nautical yardarm mast, based on height of flagpole, per foot of pole height	—	LF	28.00	—	28.00
Aluminum, satin finish, tapered, external halyard, heavy duty type					
20' x .188" wall, with 5" butt diam.	D4@8.19	Ea	790.00	303.00	1,093.00
25' x .188" wall, with 5" butt diam.	D4@8.19	Ea	925.00	303.00	1,228.00
30' x .188" wall, with 6" butt diam.	D4@9.21	Ea	1,140.00	340.00	1,480.00
35' x .188" wall, with 7" butt diam.	D4@9.96	Ea	1,490.00	368.00	1,858.00
40' x .188" wall, with 8" butt diam.	D4@10.2	Ea	2,010.00	377.00	2,387.00
45' x .188" wall, with 8" butt diam.	D4@10.5	Ea	2,300.00	388.00	2,688.00
50' x .188" wall, with 10" butt diam.	D4@15.1	Ea	3,170.00	558.00	3,728.00
60' x .250" wall, with 12" butt diam.	D4@18.5	Ea	5,840.00	683.00	6,523.00
70' x .250" wall, with 12" butt diam.	D4@21.7	Ea	6,560.00	802.00	7,362.00
80' x .375" wall, with 12" butt diam.	D4@25.2	Ea	9,910.00	931.00	10,841.00
Wall-mounted aluminum poles, including wall mount on concrete or masonry structure					
8' high	D4@7.95	Ea	835.00	294.00	1,129.00
10' high	D4@9.69	Ea	850.00	358.00	1,208.00
15' with wall support	D4@11.1	Ea	1,150.00	410.00	1,560.00
20' with wall support	D4@14.4	Ea	1,520.00	532.00	2,052.00
Steel poles, white enamel finish, lightweight					
12' high	D4@6.41	Ea	55.20	237.00	292.20
18' high	D4@6.41	Ea	73.60	237.00	310.60
22' high	D4@7.13	Ea	152.00	263.00	415.00
30' high	D4@8.97	Ea	317.00	331.00	648.00
Concrete poles, off-white or brown finish, with internal halyard, overall heights including buried portion					
20' high	D4@6.41	Ea	1,240.00	237.00	1,477.00
30' high	D4@8.83	Ea	1,660.00	326.00	1,986.00
40' high	D4@10.4	Ea	2,110.00	384.00	2,494.00
50' high	D4@14.4	Ea	4,790.00	532.00	5,322.00
60' high	D4@16.0	Ea	6,670.00	591.00	7,261.00
70' high	D4@19.2	Ea	8,550.00	709.00	9,259.00
80' high	D4@22.3	Ea	10,500.00	824.00	11,324.00
90' high	D4@27.2	Ea	12,700.00	1,000.00	13,700.00
100' high	D4@31.8	Ea	14,100.00	1,170.00	15,270.00
Add for yardarm nautical mast					
11' aluminum yardarm	—	Ea	928.00	—	928.00
13'6" aluminum yardarm	—	Ea	982.00	—	982.00
12' fiberglass yardarm	—	Ea	803.00	—	803.00
Add for locking cleat box	D4@2.42	Ea	107.00	89.40	196.40
Add for eagle mounted on spindle rod					
12" bronze	—	Ea	59.90	—	59.90
12" natural	—	Ea	74.00	—	74.00
16" bronze	—	Ea	101.00	—	101.00
16" natural	—	Ea	125.00	—	125.00

		Unit	Clear	Bronze	Black
Add for anodized finish, aluminum or steel pole					
Per foot of pole height, heights to 50' per LF		LF	10.00	13.00	15.00
Per foot of pole height, heights over 50' per LF		LF	15.00	20.00	25.00

	Craft@Hrs	Unit	Material	Labor	Total
Poured concrete flagpole foundations, including excavation, metal collar, steel casing, base plate, sand fill and ground spike, by pole height above ground, typical prices					
30' high or less	P9@10.5	Ea	50.20	385.00	435.20
35' high	P9@10.8	Ea	54.70	396.00	450.70
40' high	P9@12.5	Ea	59.20	459.00	518.20
45' high	P9@13.2	Ea	63.90	484.00	547.90
50' high	P9@15.8	Ea	77.50	580.00	657.50
60' high	P9@19.2	Ea	91.20	704.00	795.20
70' high	P9@22.5	Ea	118.00	825.00	943.00
80' high	P9@26.0	Ea	146.00	954.00	1,100.00
90' to 100' high	P9@30.1	Ea	182.00	1,100.00	1,282.00
Direct embedded pole base, augered in place, no concrete required, by pole height					
20' to 25' pole	D4@1.88	Ea	255.00	69.50	324.50
25' to 40' pole	D4@2.71	Ea	365.00	100.00	465.00
45' to 55' pole	D4@2.99	Ea	410.00	110.00	520.00
60' to 70' pole	D4@3.42	Ea	456.00	126.00	582.00

Identifying Devices

	Craft@Hrs	Unit	Material	Labor	Total
Architectural signage, self-adhesive plastic signs with high contrast white lettering					
4" x 12", stock wording	PA@.191	Ea	15.70	7.07	22.77
4" x 12", custom wording	PA@.191	Ea	23.20	7.07	30.27
3" x 8", stock wording	PA@.191	Ea	14.20	7.07	21.27
3" x 8", custom wording	PA@.191	Ea	20.00	7.07	27.07
Add for aluminum mounting frame					
4" x 12"	PA@.096	Ea	15.90	3.55	19.45
3" x 8"	PA@.096	Ea	9.55	3.55	13.10
Directory boards, felt-covered letter boards, indoor mount					
Aluminum frame					
18" x 2', 1 door	C8@.635	Ea	110.00	23.00	133.00
2' x 3', 1 door	C8@1.31	Ea	157.00	47.50	204.50
4' x 3', 2 doors	C8@1.91	Ea	246.00	69.30	315.30
Wood frame					
18" x 2', 1 door	C8@.635	Ea	94.20	23.00	117.20
2' x 3', 1 door	C8@1.31	Ea	130.00	47.50	177.50
4' x 3', 2 doors	C8@1.91	Ea	193.00	69.30	262.30
Acrylic door, aluminum frame					
18" x 2', 1 door	C8@.635	Ea	246.00	23.00	269.00
2' x 3', 1 door	C8@1.31	Ea	346.00	47.50	393.50
4' x 3', 2 doors	C8@1.91	Ea	645.00	69.30	714.30
Hand-painted lettering, typical subcontract prices					
Standard, per square foot	—	SF	—	—	13.90
Gold leaf, per square foot	—	SF	—	—	27.60
Illuminated letters, porcelain enamel finish, stock letters, aluminum with Plexiglass face, add electrical work					
12" high x 2-1/2" wide, 1 tube per letter	C8@.849	Ea	133.00	30.80	163.80
24" high x 5" wide, 2 tubes per letter	C8@1.61	Ea	271.00	58.40	329.40
36" high x 6-1/2" wide, 2 tubes per letter	C8@3.28	Ea	420.00	119.00	539.00
Fabricated back-lighted letters, aluminum with neon illumination, baked enamel finish, script, add electrical work					
12" high x 3" deep, 1 tube	G1@.834	Ea	71.40	27.80	99.20
24" high x 6" deep, 2 tubes	G1@1.41	Ea	157.00	47.00	204.00
36" high x 8" deep, 2 tubes	G1@2.53	Ea	285.00	84.40	369.40

	Craft@Hrs	Unit	Material	Labor	Total
Stock cast aluminum letters, various styles, on masonry wall					
2" high	C8@.215	Ea	13.50	7.80	21.30
6" high	C8@.320	Ea	57.00	11.60	68.60
12" high	C8@.320	Ea	118.00	11.60	129.60
14" high	C8@.456	Ea	180.00	16.50	196.50
Add for baked enamel finish	—	%	2.0	—	—
Add for satin bronze finish	—	%	75.0	—	—
Braille signage, meets ADA requirements, various colors, interior (See ADA section)					
8" x 8"	PA@.25	Ea	35.90	9.25	45.15
2" x 8"	PA@.25	Ea	21.70	9.25	30.95
Plaques, including standard lettering, expansion or toggle bolt applied					
Aluminum, 12" x 4"	G1@.900	Ea	166.00	30.00	196.00
Aluminum, 24" x 18"	G1@2.70	Ea	840.00	90.00	930.00
Aluminum, 24" x 36"	G1@2.91	Ea	1,530.00	97.00	1,627.00
Bronze, 12" x 4"	G1@.900	Ea	222.00	30.00	252.00
Bronze, 24" x 18"	G1@2.70	Ea	1,270.00	90.00	1,360.00
Bronze, 24" x 36"	G1@2.91	Ea	2,360.00	97.00	2,457.00
Safety and warning signs, stock signs ("High Voltage," "Fire Exit," etc.) Indoor, high performance plastic					
10" x 7"	PA@.099	Ea	12.50	3.66	16.16
14" x 10"	PA@.099	Ea	17.90	3.66	21.56
Add for aluminum	—	%	40.0		—
Indoor, pressure sensitive vinyl signs					
10" x 7"	PA@.099	Ea	9.80	3.66	13.46
14" x 10"	PA@.099	Ea	11.90	3.66	15.56
Turnstiles					
Non-register type, manual	D4@2.78	Ea	582.00	103.00	685.00
Register type, manual	D4@2.78	Ea	826.00	103.00	929.00
Attached ticket collection box, register type, add	—	Ea	196.00	—	196.00
Register type, portable	—	Ea	1,300.00	—	1,300.00
Lockers, Metal Baked enamel finish, key lock 12" wide x 60" high x 15" deep overall dimensions					
Single tier, cost per locker	D4@.571	Ea	122.00	21.10	143.10
Double tier, cost per 2 lockers	D4@.571	Ea	141.00	21.10	162.10
Triple tier, cost per 3 lockers	D4@.571	Ea	145.00	21.10	166.10
Quadruple tier, cost per 4 lockers	D4@.571	Ea	163.00	21.10	184.10
Add for 72" height	—	%	10.0	—	—
Add for 15" width	—	%	25.0	—	—
Benches, Locker Room Wooden, floor-mounted.					
Maple, 6' long	D4@.753	Ea	191.00	27.80	218.80
Postal Specialties					
Mail chutes, cost per floor, typical					
Aluminum and glass					
8-3/4" x 3-1/2"	D4@11.1	Ea	502.00	410.00	912.00
14-1/4" x 4-5/8"	D4@11.1	Ea	598.00	410.00	1,008.00
14-1/4" x 8-5/8"	D4@11.1	Ea	693.00	410.00	1,103.00
Stainless or bronze					
8-3/4" x 3-1/2"	D4@11.1	Ea	598.00	410.00	1,008.00
14-1/4" x 4-5/8"	D4@11.1	Ea	654.00	410.00	1,064.00
14-1/4" x 8-5/8"	D4@13.9	Ea	804.00	513.00	1,317.00

	Craft@Hrs	Unit	Material	Labor	Total
Collection boxes					
Single box, 36" H x 20" W x 12" deep					
Aluminum	D4@4.16	Ea	1,520.00	154.00	1,674.00
Stainless or bronze	D4@4.16	Ea	1,650.00	154.00	1,804.00
Double box, 36" H x 40" W x 12" deep					
Add to cost of single box	—	%	100.0	—	—

Movable Office Partitions Prefabricated units. Cost per SF of partition wall measured one side. Use 100 SF as minimum job size.

	Craft@Hrs	Unit	Material	Labor	Total
Gypsum board on 2-1/2" metal studs (includes finish on both sides)					
1/2" board, unpainted, STC 38	C8@.028	SF	3.57	1.02	4.59
5/8" board, unpainted, STC 40	C8@.028	SF	3.65	1.02	4.67
1/2" or 5/8", unpainted, STC 45	C8@.028	SF	4.03	1.02	5.05
1/2" vinyl-covered board, STC 38	C8@.028	SF	4.30	1.02	5.32
5/8" vinyl-covered board, STC 40	C8@.028	SF	4.52	1.02	5.54
Metal-covered, baked enamel finish	C8@.028	SF	9.55	1.02	10.57
Add for factory vinyl wrap, 15 to 22 oz	—	SF	.36	—	.36
Windows, including frame and glass, add to above, factory installed					
3'6" x 2'0"	—	Ea	84.80	—	84.80
3'6" x 4'0"	—	Ea	122.00	—	122.00
Wood doors, hollow core, prefinished, add to above, factory installed					
1-3/4", 3'0" x 7'0"	—	Ea	240.00	—	240.00
Metal door jambs, 3'6" x 7'0"	—	Ea	144.00	—	144.00
Metal door jambs, 6'0" x 7'0"	—	Ea	235.00	—	235.00
Passage hardware set	—	Ea	55.00	—	55.00
Lockset hardware	—	LS	73.70	—	73.70
Corner for gypsum partitions	—	Ea	90.90	—	90.90
Corner for metal-covered partitions	—	Ea	179.00	—	179.00
Starter, gypsum partitions	—	Ea	30.30	—	30.30
Starter, metal-covered partitions	—	Ea	64.60	—	64.60
Metal base	—	LF	1.82	—	1.82
Cubicles and welded booths, 5' x 5' x 5'	C8@.028	SF	10.40	1.02	11.42
Banker type divider partition, subcontract					
5'6" high including 18" glass top, 2" thick, vinyl finish, not including door or gates	—	SF	—	—	9.35

Demountable Partitions Prefinished units. Cost per SF of partition measured one side.
Use 100 SF as minimum job size. Labor shown applies to installing or removing

	Craft@Hrs	Unit	Material	Labor	Total
Modular panels, spring pressure, mounted to floor and ceiling with removable tracks					
Standard panel, non-rated	C8@.124	SF	12.30	4.50	16.80

Accordion Folding Partitions

	Craft@Hrs	Unit	Material	Labor	Total
Vinyl-covered particleboard core, top hung, not including structural supports or placing track in concrete					
Economy, to 8' high, STC 36	C8@.090	SF	9.33	3.26	12.59
Economy, to 30' wide x 17' high, STC 41	C8@.124	SF	12.60	4.50	17.10
Good quality, to 30' W x 17' H, STC 43	C8@.124	SF	14.80	4.50	19.30
Better quality, to 30' W x 17' H, STC 44	C8@.124	SF	15.30	4.50	19.80
Better quality, large openings, STC 45	C8@.124	SF	14.30	4.50	18.80
Better quality, extra large openings, STC 47	C8@.124	SF	15.30	4.50	19.80
Add for prefinished birch or ash wood slats accordion folding partitions					
With vinyl hinges, 15' W x 8' H maximum	—	SF	6.12	—	6.12

	Craft@Hrs	Unit	Material	Labor	Total

Folding fire partitions, accordion, with automatic closing system, baked enamel finish, with standard hardware and motor, cost per 5' x 7' module

	Craft@Hrs	Unit	Material	Labor	Total
20 minute rating	C8@22.6	Ea	3,980.00	820.00	4,800.00
60 minute rating	C8@22.6	Ea	5,050.00	820.00	5,870.00
90 minute rating	C8@22.6	Ea	5,430.00	820.00	6,250.00
2 hour rating	C8@22.6	Ea	6,380.00	820.00	7,200.00

Folding leaf partitions, typical subcontract prices. Not including header or floor track recess, vinyl covered

	Craft@Hrs	Unit	Material	Labor	Total
Metal panels, 7.5 lbs per SF, top hung, hinged, (STC 52), 16' high x 60' wide maximum	—	SF	—	—	46.50
Metal panels, 7.5 lbs per SF, top hung, single panel pivot, (STC 48), 18' high x 60' wide maximum	—	SF	—	—	47.50
Wood panels, hinged, floor or ceiling hung, 6 lbs per SF, (STC 40), 12' high x 36' wide maximum	—	SF	—	—	38.40
Add for laminated plastic	—	SF	—	—	2.15
Add for unfinished wood veneer	—	SF	—	—	2.65
Add for chalkboard, large area	—	SF	—	—	2.10

Woven Wire Partitions 10 gauge painted, 1-1/2" steel mesh.

Wall panels, 8' high, installed on a concrete slab

	Craft@Hrs	Unit	Material	Labor	Total
1' wide	H6@.437	Ea	81.30	17.30	98.60
2' wide	H6@.491	Ea	95.20	19.40	114.60
3' wide	H6@.545	Ea	111.00	21.60	132.60
4' wide	H6@.545	Ea	121.00	21.60	142.60
5' wide	H6@.598	Ea	127.00	23.70	150.70
Add for galvanized mesh	—	SF	.49	—	.49
Corner posts for 8' high walls	H6@.270	Ea	23.50	10.70	34.20
Stiffener channel posts for 8' walls	H6@.277	Ea	50.10	11.00	61.10
Service window panel, 8' H x 5' W panel	H6@.836	Ea	286.00	33.10	319.10

Sliding doors, 8' high

	Craft@Hrs	Unit	Material	Labor	Total
3' wide	H6@2.94	Ea	413.00	116.00	529.00
4' wide	H6@3.29	Ea	439.00	130.00	569.00
5' wide	H6@3.29	Ea	468.00	130.00	598.00
6' wide	H6@3.58	Ea	551.00	142.00	693.00

Hinged doors, 8' high

	Craft@Hrs	Unit	Material	Labor	Total
3' wide	H6@2.85	Ea	333.00	113.00	446.00
4' wide	H6@3.14	Ea	368.00	124.00	492.00

Dutch doors, 8' high

	Craft@Hrs	Unit	Material	Labor	Total
3' wide	H6@2.85	Ea	504.00	113.00	617.00
4' wide	H6@3.14	Ea	589.00	124.00	713.00

Complete wire partition installation including posts and typical doors, per SF of wall and ceiling panel

	Craft@Hrs	Unit	Material	Labor	Total
Painted wall panels	H6@.029	SF	3.76	1.15	4.91
Galvanized wall panels	H6@.029	SF	4.31	1.15	5.46
Painted ceiling panels	H6@.059	SF	3.43	2.33	5.76
Galvanized ceiling panels	H6@.059	SF	3.91	2.33	6.24

Truck Scales With dial and readout. Steel platform. These costs do not include concrete work or excavation.

	Craft@Hrs	Unit	Material	Labor	Total
5 ton, 6' x 8' platform	D4@45.9	Ea	7,100.00	1,700.00	8,800.00
20 ton, 10' x 24' platform	D4@70.2	Ea	10,400.00	2,590.00	12,990.00
50 ton, 10' x 60' platform	D4@112.	Ea	18,100.00	4,140.00	22,240.00
80 ton, 10' x 60' platform	D4@112.	Ea	21,000.00	4,140.00	25,140.00

	Craft@Hrs	Unit	Material	Labor	Total

Recessed Wall and Floor Safes Safes with 1" thick steel door, 1/4" hardplate protecting locking mechanism, welded deadbolts, and key or combination lock, internal relock device.

In-floor safes

	Craft@Hrs	Unit	Material	Labor	Total
15" L x 14" W x 15" D, 12" x 12" door	D4@2.33	Ea	541.00	86.10	627.10
15" L x 14" W x 20" D, 12" x 12" door	D4@2.33	Ea	554.00	86.10	640.10
24" L x 24" W x 17" D, 12" x 12" door	D4@2.33	Ea	773.00	86.10	859.10
Add for installation in existing building	D4@2.33	Ea	—	86.10	86.10

Storage Shelving

Retail store gondolas (supermarket shelving units), material prices vary widely, check local source

	Craft@Hrs	Unit	Material	Labor	Total
Install only	D4@.333	LF	—	12.30	12.30

Steel industrial shelving, enamel finish, 2 sides & back, 12" deep x 30" wide x 75" high

	Craft@Hrs	Unit	Material	Labor	Total
5 shelves high	D4@.758	Ea	87.50	28.00	115.50
8 shelves high	D4@1.21	Ea	120.00	44.70	164.70

Steel industrial shelving, enamel finish, open on sides, 12" deep x 36" wide x 75" high

	Craft@Hrs	Unit	Material	Labor	Total
5 shelves high	D4@.686	Ea	50.90	25.30	76.20
8 shelves high	D4@1.06	Ea	77.70	39.20	116.90

Stainless steel modular shelving, 20" deep x 36" wide x 63" high,

	Craft@Hrs	Unit	Material	Labor	Total
3 shelves, 16 gauge	D4@1.19	Ea	857.00	44.00	901.00
Stainless steel wall shelf, 12" deep, bracket mount	D4@.675	LF	39.30	24.90	64.20

Stainless steel work counters, with under-counter storage shelf, 16 gauge

	Craft@Hrs	Unit	Material	Labor	Total
4' wide x 2' deep x 3' high	D4@1.36	Ea	924.00	50.20	974.20
6' wide x 2' deep x 3' high	D4@2.01	Ea	1,400.00	74.30	1,474.30

Library type metal shelving, 90" high, 8" deep, 5 shelves

	Craft@Hrs	Unit	Material	Labor	Total
30" long units	D4@1.17	Ea	152.00	43.20	195.20
36" long units	D4@1.30	Ea	165.00	48.00	213.00

Telephone Enclosures Typical subcontract prices.

	Craft@Hrs	Unit	Material	Labor	Total
Shelf type, wall mounted, 15" D x 28" W x 30" H	—	Ea	—	—	610.00
Desk type	—	Ea	—	—	566.00
Full height booth	—	Ea	—	—	2,930.00

Wall mounted stainless steel directory shelf. Typical subcontract price.

	Craft@Hrs	Unit	Material	Labor	Total
3 directory	—	Ea	—	—	645.00

		Unit	Commercial/ Industrial Grade		Institutional Grade

Toilet and Bath Accessories Material cost only. See labor following this section

Toilet paper dispensers

	Unit	Commercial/ Industrial Grade	Institutional Grade
Single roll, surface mounted	Ea	19.40	31.60
Double roll, surface mounted	Ea	22.40	34.60
Single roll, recessed	Ea	24.80	31.60
Cabinet, double roll, recessed	Ea	—	185.00

Toilet seat cover dispensers

	Unit	Commercial/ Industrial Grade	Institutional Grade
Recessed	Ea	77.60	112.00
Surfaced mounted	Ea	40.80	47.50
With sanitary napkin disposal	Ea	226.00	325.00
Feminine napkin dispenser, recessed	Ea	298.00	440.00
Feminine napkin dispenser, surface mounted	Ea	44.50	61.20
Soap dish, chrome, surface mounted	Ea	8.59	15.30
Soap dish, chrome, recessed	Ea	9.39	15.50

		Commercial/ Industrial Grade	Institutional Grade
Soap dispensers			
Liquid, 20 oz, deck mounted	Ea	34.50	42.70
Liquid, 40 oz, deck mounted	Ea	42.50	51.60
Powder	Ea	30.50	42.80
System, liquid, with tank and 3 stations	Ea	—	849.00
Add for additional stations	Ea	—	73.00
Soap dishes			
With grab bar, recessed	Ea	17.20	20.80
With grab bar, surface mounted	Ea	27.20	34.30
Grab bars, stainless steel, wall mounted			
1" x 12"	Ea	32.60	37.40
1" x 18"	Ea	34.90	41.60
1" x 24"	Ea	39.30	42.70
1" x 30"	Ea	43.10	46.10
1" x 36"	Ea	45.80	48.20
1" x 48"	Ea	51.00	52.80
1-1/4" x 18"	Ea	45.40	59.60
1-1/4" x 24"	Ea	49.20	62.90
1-1/4" x 36"	Ea	59.70	74.90
1", 90 degree angle, 30"	Ea	70.60	80.00
1-1/4", 90 degree angle, 30"	Ea	86.80	117.00
Towel bars			
18" chrome	Ea	25.10	29.40
24" chrome	Ea	26.50	30.20
18" stainless steel	Ea	—	30.20
24" stainless steel	Ea	—	33.30
Paper towel dispensers			
Recessed	Ea	124.00	177.00
Semi-recessed or wall mounted	Ea	154.00	259.00
Recessed, with waste receptacle	Ea	337.00	476.00
Semi-recessed with waste receptacle	Ea	413.00	595.00
Waste receptacles, recessed	Ea	232.00	330.00
Waste receptacles, semi-recessed	Ea	320.00	447.00
Medicine cabinets, with mirror			
Swing door, 16" x 22", wall hung	Ea	55.10	82.80
Sliding doors, recessed	Ea	97.00	124.20
Swing door, 16" x 22", recessed	Ea	106.10	137.40
Mirrors, unframed, 1/4" float glass, polished edges			
To 5 SF	SF	7.30	7.30
Over 5 SF	SF	8.80	8.30
Mirrors, stainless steel frame			
18" x 24"	Ea	—	86.20
18" x 32"	Ea	—	95.60
18" x 36"	Ea	—	100.20
24" x 30"	Ea	—	124.20
24" x 48"	Ea	—	237.40
24" x 60"	Ea	—	292.90
30" x 72"	Ea	—	346.40
Shelves, stainless steel, 6" deep			
18" long	Ea	40.10	41.80
24" long	Ea	42.40	47.00
36" long	Ea	54.80	64.60
72" long	Ea	89.80	117.20

		Commercial/ Industrial Grade	Institutional Grade
Shower rod, end flanges, vinyl curtain, 6' long, straight	Ea	44.10	52.50
Robe hooks			
Single	Ea	6.46	9.21
Double	Ea	7.90	10.40
Single with door stop	Ea	7.27	11.60
Straddle bar, 24" x 24" x 20", stainless steel	Ea	94.40	107.00
Urinal bar, 24" long, stainless steel	Ea	76.80	90.40
Ash urns, wall mounted, aluminum	Ea	111.00	133.00
Hot air hand dryer, 110 volt, add electric connection	Ea	357.00	370.00

	Craft@Hrs	Unit	Material	Labor	Total
Labor to install toilet and bath accessories					
Toilet paper dispensers					
Surface mounted	CC@.182	Ea	—	7.69	7.69
Recessed	CC@.310	Ea	—	13.10	13.10
Toilet seat cover dispensers					
Surface mounted	CC@.413	Ea	—	17.50	17.50
Recessed	CC@.662	Ea	—	28.00	28.00
Feminine napkin dispenser, surface mounted	CC@.760	Ea	—	32.10	32.10
Feminine napkin disposer, recessed	CC@.882	Ea	—	37.30	37.30
Soap dishes	CC@.230	Ea	—	9.72	9.72
Soap dispensers, surface mounted	CC@.254	Ea	—	10.70	10.70
Soap dispensers, recessed	CC@.628	Ea	—	26.50	26.50
Soap dispenser system	CC@2.91	Ea	—	123.00	123.00
Paper towel dispensers and waste receptacles					
Semi-recessed	CC@1.96	Ea	—	82.80	82.80
Recessed	CC@2.57	Ea	—	109.00	109.00
Medicine cabinets with mirror, swing door wall hung	CC@.943	Ea	—	39.90	39.90
Mirrors					
To 5 SF	CG@.566	Ea	—	20.60	20.60
5 to 10 SF	CG@.999	Ea	—	36.40	36.40
Over 10 SF	CG@.125	SF	—	4.55	4.55
Towel bars and grab bars	CC@.286	Ea	—	12.10	12.10
Shelves, stainless steel, to 72"	CC@.569	Ea	—	24.10	24.10
Robe hooks	CC@.182	Ea	—	7.69	7.69
Straddle and urinal bars	CC@.484	Ea	—	20.50	20.50
Hot air hand dryer, add for electric circuit wiring	CE@.908	Ea	—	36.40	36.40

Coat and Hat Racks Assembly and installation of prefabricated coat and hat racks. Bevco.

Wall-mounted coat and hat rack with pilfer-resistant hangers, chrome (#200-SHB)

	Craft@Hrs	Unit	Material	Labor	Total
24" length, 9 hangers	CC@.665	Ea	59.00	28.10	87.10
36" length, 13 hangers	CC@.748	Ea	71.00	31.60	102.60
48" length, 17 hangers	CC@.833	Ea	83.00	35.20	118.20
60" length, 21 hangers	CC@.916	Ea	97.00	38.70	135.70
72" length, 25 hangers	CC@.997	Ea	113.00	42.10	155.10
Wall-mounted hook bar, chrome (#H-2)					
24" length, 5 hooks	CC@.252	Ea	12.00	10.70	22.70
36" length, 7 hooks	CC@.303	Ea	14.00	12.80	26.80
48" length, 9 hooks	CC@.352	Ea	17.00	14.90	31.90
60" length, 11 hooks	CC@.391	Ea	21.00	16.50	37.50
72" length, 13 hooks	CC@.457	Ea	24.00	19.30	43.30

	Craft@Hrs	Unit	Material	Labor	Total

Church Glass Material cost includes fabrication of the lite. Labor cost includes setting in a prepared opening only. Minimum labor cost is usually $150 per job. Add the cost of scaffold or a hoist, if needed.

	Craft@Hrs	Unit	Material	Labor	Total
Simple artwork stained glass	G1@.184	SF	62.20	6.14	68.34
Moderate artwork stained glass	G1@.318	SF	157.00	10.60	167.60
Elaborate artwork stained glass	G1@.633	SF	501.00	21.10	522.10
Faceted glass					
Simple artwork	G1@.184	SF	55.70	6.14	61.84
Moderate artwork	G1@.318	SF	112.00	10.60	122.60
Elaborate artwork	G1@.633	SF	152.00	21.10	173.10
Colored glass, no artwork					
Single pane	G1@.056	SF	6.19	1.87	8.06
Patterned	G1@.184	SF	13.10	6.14	19.24
Small pieces	G1@.257	SF	33.50	8.57	42.07

Equipment 11

Bank Equipment

	Craft@Hrs	Unit	Material	Labor	Total
Vault door and frame, 78" x 32", 10", single					
2 hour, 10 bolt	D4@11.9	Ea	2,990.00	440.00	3,430.00
4 hour, 10 bolt	D4@11.9	Ea	3,370.00	440.00	3,810.00
6 hour, 10 bolt	D4@11.9	Ea	3,940.00	440.00	4,380.00
Vault door and frame, 78" x 40", 10", double					
4 hour, 18 bolt	D4@20.8	Ea	4,380.00	768.00	5,148.00
6 hour, 18 bolt	D4@20.8	Ea	4,720.00	768.00	5,488.00
Teller windows					
Drive-up, motorized drawer, projected	D4@26.8	Ea	9,920.00	990.00	10,910.00
Walk-up, one teller	D4@17.6	Ea	5,060.00	650.00	5,710.00
Walk-up, two teller	D4@17.6	Ea	6,050.00	650.00	6,700.00
After hours depository doors					
Envelope and bag, whole chest, complete	D4@37.1	Ea	8,540.00	1,370.00	9,910.00
Envelope only	D4@4.72	Ea	1,020.00	174.00	1,194.00
Flush mounted, bag only	D4@5.44	Ea	1,520.00	201.00	1,721.00
Teller counter, modular components	C8@.532	LF	99.00	19.30	118.30
Square check desks, 48" x 48", 4 person	C8@2.00	Ea	1,130.00	72.50	1,202.50
Round check desks, 48" diameter, 4 person	C8@2.00	Ea	1,240.00	72.50	1,312.50
Rectangular check desks, 72" x 24", 4 person	C8@2.00	Ea	1,100.00	72.50	1,172.50
Rectangular check desks, 36" x 72", 8 person	C8@2.00	Ea	2,150.00	72.50	2,222.50
Safe deposit boxes, modular units					
42 openings, 2" x 5"	D4@3.42	Ea	1,830.00	126.00	1,956.00
30 openings, 2" x 5"	D4@3.42	Ea	1,440.00	126.00	1,566.00
18 openings, 2" x 5"	D4@3.42	Ea	1,140.00	126.00	1,266.00
Base, 32" x 24" x 3"	D4@1.11	Ea	109.00	41.00	150.00
Canopy top	D4@.857	Ea	62.30	31.70	94.00
Cash dispensing automatic teller, with phone	CE@84.3	Ea	41,400.00	3,380.00	44,780.00
Video camera surveillance system, 3 cameras	CE@15.6	Ea	2,670.00	625.00	3,295.00
Perimeter alarm system, typical price	CE@57.7	Ea	6,700.00	2,310.00	9,010.00

Ecclesiastical Equipment Labor includes unpacking, layout and complete installation of prefinished furniture and equipment.

	Craft@Hrs	Unit	Material	Labor	Total
Lecterns, 16" x 24"					
Plain	C8@4.48	Ea	541.00	162.00	703.00
Deluxe	C8@4.48	Ea	1,010.00	162.00	1,172.00
Premium	C8@4.48	Ea	1,890.00	162.00	2,052.00

Equipment 11

	Craft@Hrs	Unit	Material	Labor	Total
Pulpits					
Plain	C8@4.77	Ea	614.00	173.00	787.00
Deluxe	C8@4.77	Ea	974.00	173.00	1,147.00
Premium	C8@4.77	Ea	2,390.00	173.00	2,563.00
Arks, with curtain					
Plain	C8@5.97	Ea	636.00	217.00	853.00
Deluxe	C8@5.97	Ea	832.00	217.00	1,049.00
Premium	C8@5.97	Ea	1,390.00	217.00	1,607.00
Arks, with door					
Plain	C8@8.95	Ea	721.00	325.00	1,046.00
Deluxe	C8@8.95	Ea	1,220.00	325.00	1,545.00
Premium	C8@8.95	Ea	1,890.00	325.00	2,215.00
Pews, bench type					
Plain	C8@.382	LF	39.90	13.90	53.80
Deluxe	C8@.382	LF	43.50	13.90	57.40
Premium	C8@.382	LF	49.50	13.90	63.40
Pews, seat type					
Plain	C8@.464	LF	50.10	16.80	66.90
Deluxe	C8@.491	LF	65.10	17.80	82.90
Premium	C8@.523	LF	77.50	19.00	96.50
Kneelers					
Plain	C8@.185	LF	9.93	6.71	16.64
Deluxe	C8@.185	LF	12.60	6.71	19.31
Premium	C8@.185	LF	19.50	6.71	26.21
Cathedral chairs, shaped wood					
Without accessories	C8@.235	Ea	120.00	8.52	130.52
With book rack	C8@.235	Ea	139.00	8.52	147.52
With book rack and kneeler	C8@.235	Ea	153.00	8.52	161.52
Cathedral chairs, upholstered					
Without accessories	C8@.285	Ea	134.00	10.30	144.30
With book rack	C8@.285	Ea	144.00	10.30	154.30
With book rack and kneeler	C8@.285	Ea	172.00	10.30	182.30
Adaptable fabric upholstered chair seating (Sauder)					
Hardwood laminated frame, interlocking legs, book rack, stackable, "Modlok"					
Standard	C8@.066	Ea	109.00	2.39	111.39
Deluxe	C8@.066	Ea	114.00	2.39	116.39
Solid oak frame, stackable, "Oaklok"	C8@.066	Ea	88.90	2.39	91.29
Laminated beech frame, stackable, "Plylok"	C8@.066	Ea	96.60	2.39	98.99
Laminated beech frame, stacks & folds, "Plyfold"	C8@.066	Ea	84.00	2.39	86.39
Confessionals, single, with curtain					
Plain	C8@9.11	Ea	2,300.00	330.00	2,630.00
Deluxe	C8@9.11	Ea	2,750.00	330.00	3,080.00
Premium	C8@9.11	Ea	3,150.00	330.00	3,480.00
Add for door in place of curtain	—	Ea	370.00	—	370.00
Confessionals, double, with curtain					
Plain	C8@13.7	Ea	3,630.00	497.00	4,127.00
Deluxe	C8@13.7	Ea	4,220.00	497.00	4,717.00
Premium	C8@13.7	Ea	5,190.00	497.00	5,687.00
Communion rail, hardwood, with standards					
Plain	C8@.315	LF	42.10	11.40	53.50
Deluxe	C8@.315	LF	52.90	11.40	64.30
Premium	C8@.315	LF	63.00	11.40	74.40

	Craft@Hrs	Unit	Material	Labor	Total
Communion rail, carved oak, with standards					
Embellished	C8@.315	LF	68.50	11.40	79.90
Ornate	C8@.315	LF	91.60	11.40	103.00
Premium	C8@.315	LF	150.00	11.40	161.40
Communion rail, bronze or stainless steel,					
With standards	H6@.285	LF	99.30	11.30	110.60
Altars, hardwood, sawn					
Plain	C8@5.97	Ea	739.00	217.00	956.00
Deluxe	C8@5.97	Ea	872.00	217.00	1,089.00
Premium	C8@5.97	Ea	1,190.00	217.00	1,407.00
Altars, carved hardwood					
Plain	C8@5.97	Ea	2,210.00	217.00	2,427.00
Deluxe	C8@6.73	Ea	5,100.00	244.00	5,344.00
Premium	C8@7.30	Ea	5,790.00	265.00	6,055.00
Altars, with Nimmerillium fonts, marble	D4@27.8	Ea	3,770.00	1,030.00	4,800.00
Altars, marble or granite					
Plain	D4@29.1	Ea	3,920.00	1,080.00	5,000.00
Deluxe	D4@29.1	Ea	5,690.00	1,080.00	6,770.00
Premium	D4@29.1	Ea	8,430.00	1,080.00	9,510.00
Altars, marble legs and base					
Deluxe	D4@29.1	Ea	2,360.00	1,080.00	3,440.00
Premium	D4@29.1	Ea	3,380.00	1,080.00	4,460.00

Theater & Stage Curtains

Straight track for manual curtains, see lifting and drawing equipment on next page.

	Craft@Hrs	Unit	Material	Labor	Total
Standard duty					
24' wide stage	D4@.486	LF	18.25	18.00	36.25
40' wide stage	D4@.315	LF	18.25	11.60	29.85
60' wide stage	D4@.233	LF	18.25	8.61	26.86
Medium duty					
24' wide stage	D4@.486	LF	23.50	18.00	41.50
40' wide stage	D4@.315	LF	23.50	11.60	35.10
60' wide stage	D4@.233	LF	23.50	8.61	32.11
Heavy duty					
24' wide stage	D4@.486	LF	26.60	18.00	44.60
40' wide stage	D4@.315	LF	26.60	11.60	38.20
60' wide stage	D4@.233	LF	26.60	8.61	35.21
Curved track, manual operation, see lifting and drawing equipment following					
Standard duty					
24' wide stage	D4@.486	LF	44.90	18.00	62.90
40' wide stage	D4@.371	LF	35.00	13.70	48.70
60' wide stage	D4@.315	LF	32.20	11.60	43.80
Heavy duty					
24' wide stage	D4@.584	LF	100.00	21.60	121.60
40' wide stage	D4@.427	LF	88.75	15.80	104.55
60' wide stage	D4@.342	LF	88.75	12.60	101.35
Cyclorama track, manual, see lifting and drawing equipment following					
24' wide stage	D4@.486	LF	26.60	18.00	44.60
40' wide stage	D4@.315	LF	26.60	11.60	38.20
60' wide stage	D4@.233	LF	26.60	8.61	35.21

	Craft@Hrs	Unit	Material	Labor	Total

Stage curtains, Subcontract Complete costs for custom fabricated, flameproof stage curtains including labor to hang. No track or curtain lifting or drawing equipment included. By percent of added fullness

	Craft@Hrs	Unit	Material	Labor	Total
Heavyweight velour					
0% fullness	—	SF	—	—	3.20
50% fullness	—	SF	—	—	3.60
75% fullness	—	SF	—	—	4.05
100% fullness	—	SF	—	—	4.30
Lightweight velour					
0% fullness	—	SF	—	—	2.55
50% fullness	—	SF	—	—	2.85
75% fullness	—	SF	—	—	3.20
100% fullness	—	SF	—	—	3.50
Muslin, 0% fullness	—	SF	—	—	1.55
Repp cloth					
0% fullness	—	SF	—	—	2.05
50% fullness	—	SF	—	—	2.85
75% fullness	—	SF	—	—	3.10
100% fullness	—	SF	—	—	3.40

Stage curtain lifting and drawing equipment Labor is equipment installation, see electrical connection below.

	Craft@Hrs	Unit	Material	Labor	Total
Lift curtain equipment					
1/8 HP	D4@7.50	Ea	1,960.00	277.00	2,237.00
1/2 HP	D4@7.50	Ea	2,980.00	277.00	3,257.00
3/4 HP	D4@7.50	Ea	3,200.00	277.00	3,477.00
Draw curtain equipment					
1/8 HP	D4@7.50	Ea	1,910.00	277.00	2,187.00
1/2 HP	D4@7.50	Ea	2,660.00	277.00	2,937.00
3/4 HP	D4@7.50	Ea	2,805.00	277.00	3,082.00

Add for electrical connection. Typical subcontact price for 40' wire run with conduit, circuit breaker, junction box and starter switch

	Craft@Hrs	Unit	Material	Labor	Total
Per electrical connection	—	Ea	—	—	850.00

Educational Equipment Labor includes unpacking, layout and complete installation of prefinished furniture and equipment.

	Craft@Hrs	Unit	Material	Labor	Total
Wardrobes, 40" x 78" x 26"					
Teacher type	C8@2.00	Ea	1,140.00	72.50	1,212.50
Student type	C8@2.00	Ea	575.00	72.50	647.50
Seating, per seat					
Lecture room, pedestal, with folding arm	C8@2.00	Ea	145.00	72.50	217.50
Horizontal sections	C8@2.00	Ea	72.00	72.50	144.50
Tables, fixed pedestal					
48" x 16"	C8@2.00	Ea	321.00	72.50	393.50
With chairs, 48" x 16"	C8@2.00	Ea	467.00	72.50	539.50
Projection screen, ceiling mounted, pull down	D4@.019	SF	3.63	.70	4.33
Videotape recorder	CE@.500	Ea	550.00	20.00	570.00
T.V. camera, color	CE@.800	Ea	1,000.00	32.10	1,032.10
T.V. monitor, wall mounted, color	CE@2.00	Ea	495.00	80.10	575.10
Language laboratory study carrels, plastic laminated wood					
Individual, 48" x 30" x 54"	C8@1.43	Ea	202.00	51.90	253.90
Two position, 73" x 30" x 47"	C8@1.93	Ea	548.00	70.00	618.00
Island, 4 position, 66" x 66" x 47"	C8@3.17	Ea	619.00	115.00	734.00

	Craft@Hrs	Unit	Material	Labor	Total

Checkroom Equipment

Automatic electric checkroom conveyor, 350 coat capacity

	Craft@Hrs	Unit	Material	Labor	Total
Conveyor and controls	D4@15.2	Ea	2,870.00	562.00	3,432.00

Add for electrical connection for checkroom conveyor, typical subcontract price

40' wire run with conduit, circuit breaker, and j-box	—	LS	—	—	357.00

Food Service Equipment These costs do not include electrical work or plumbing. Add the cost of hookup when needed. Based on Superior Products Company. (Figure in parentheses is equipment weight.)

Beverage dispenser, non -carbonated, refrigerated, countertop model

	Craft@Hrs	Unit	Material	Labor	Total
5 gallon, 1 bowl	P6@2.00	Ea	690.00	72.20	762.20
5 gallon, 2 bowls	P6@2.00	Ea	960.00	72.20	1,032.20
2.4 gallon, 2 mini bowls	P6@2.00	Ea	820.00	72.20	892.20
2.4 gallon, 4 mini bowls	P6@2.00	Ea	1,180.00	72.20	1,252.20

Charbroiler, countertop model, burner and control every 12", natural or LP gas

	Craft@Hrs	Unit	Material	Labor	Total
25,000 Btu, 15" W (112 Lbs)	P6@2.00	Ea	390.00	72.20	462.20
50,000 Btu, 24" W (186 Lbs)	P6@2.00	Ea	510.00	72.20	582.20
75,000 Btu, 36" W (230 Lbs)	P6@2.00	Ea	700.00	72.20	772.20

Deep fat fryers, free standing floor model, natural or LP gas, 36" height, stainless steel front and top

	Craft@Hrs	Unit	Material	Labor	Total
35 Lbs capacity (200 Lbs)	P6@2.00	Ea	945.00	72.20	1,017.20
45 Lbs capacity (200 Lbs)	P6@2.00	Ea	1,160.00	72.20	1,232.20
65 Lbs capacity (225 Lbs)	P6@2.00	Ea	1,450.00	72.20	1,522.20

Dishwashers

Under counter, free standing, by rack capacity per hour

	Craft@Hrs	Unit	Material	Labor	Total
18 racks per hour (245 Lbs)	P6@4.00	Ea	3,330.00	144.00	3,474.00
30 racks per hour (245 Lbs)	P6@4.00	Ea	3,700.00	144.00	3,844.00
40 racks per hour (255 Lbs)	P6@4.00	Ea	4,540.00	144.00	4,684.00

Upright, heavy duty, semi-automatic, 53 racks per hour

	Craft@Hrs	Unit	Material	Labor	Total
Straight-thru model (450 Lbs)	P6@6.00	Ea	7,300.00	217.00	7,517.00
Corner model (450 Lbs)	P6@6.00	Ea	7,450.00	217.00	7,667.00

Exhaust hoods, 16 gauge type 304 stainless steel with #3 polished finish. Conforms to NFPA code #96 and the National Sanitation Foundation requirements. Add for exhaust fan from next section.

48" deep angled front hood, 24" high with length as shown

	Craft@Hrs	Unit	Material	Labor	Total
6' long (205 Lbs)	T5@2.00	Ea	840.00	71.60	911.60
8' long (250 Lbs)	T5@3.00	Ea	1,100.00	107.00	1,207.00
10' long (285 Lbs)	T5@4.00	Ea	1,360.00	143.00	1,503.00

Additives to any hood above

	Craft@Hrs	Unit	Material	Labor	Total
Installation kit	—	Ea	152.00	—	152.00
Grease filters, aluminum, cleanable, 2' deep	T5@.125	LF	15.50	4.47	19.97

Exhaust fans, 2,800 CFM, 1,265 RPM. Wall or roof mounted, for use with exhaust hoods. Heavy duty aluminum with 115 volt AC motor enclosed in weathertight enclosure.

These costs are based on installing the fan on an existing opening, electrical hook up not included

Fan for 6' long hood, 18-½" square curb size

	Craft@Hrs	Unit	Material	Labor	Total
1/2 HP (75 lbs)	T5@2.00	Ea	725.00	71.60	796.60

Fan for 8' or 10' long hood, 23-½" square curb size

	Craft@Hrs	Unit	Material	Labor	Total
1/2 HP (100 lbs)	T5@2.00	Ea	760.00	71.60	831.60

Additives to either fan above

	Craft@Hrs	Unit	Material	Labor	Total
Roof curb	T5@1.00	Ea	220.00	35.80	255.80
Variable speed motor controller, 115 volt AC	—	Ea	121.00	—	121.00
Lights, incandescent, pre-wired, 115 volt AC	—	Ea	49.00	—	49.00
Ductwork, aluminized steel	T5@.100	LF	23.00	3.58	26.58
90 degree elbow for ductwork, aluminized steel	T5@.500	Ea	170.00	17.90	187.90

	Craft@Hrs	Unit	Material	Labor	Total
Food mixing machines, floor mounted					
20 quart, (210 lbs)	D4@2.00	Ea	3,280.00	73.90	3,353.90
30 quart, (350 lbs)	D4@3.00	Ea	5,750.00	111.00	5,861.00
60 quart, (722 lbs)	D4@4.00	Ea	9,750.00	148.00	9,898.00
Food stations, vinyl clad steel, with stainless steel work area and clear acrylic breath-guard					
24" D x 35" H, mounted on 4" casters					
Hot food station, 60" long (182 lbs)	D4@1.00	Ea	1,390.00	36.90	1,426.90
Plate rest with mounting kit (10 lbs)	D4@.250	Ea	236.00	9.24	245.24
Incandescent light assembly (20 lbs)	D4@.250	Ea	430.00	9.24	439.24
Food warmers, stainless steel top, hardwood cutting board, 12" x 20" openings, waterless					
52-1/2", natural gas, 4 openings (135 lbs)	P6@1.00	Ea	750.00	36.10	786.10
52-1/2", LP gas, 4 openings (135 lbs)	P6@1.00	Ea	750.00	36.10	786.10
52-1/2", electric, 3000 watt, 4 openings, (135 lbs)	E4@1.00	Ea	772.00	35.20	807.20
Freezers, under counter type, stainless steel front and top, CFC free refrigeration system					
6.2 cu. ft. (162 lbs)	P6@.500	Ea	1030.00	18.10	1,048.10
11.8 cu ft (239 lbs)	P6@1.00	Ea	1,425.00	36.10	1,461.10
Garbage disposers, includes control switch, solenoid valve, siphon breaker and flow control valve					
1/2 HP, (56 lbs)	P6@1.00	Ea	803.00	36.10	839.10
3/4 HP, (57 lbs)	P6@1.00	Ea	998.00	36.10	1,034.10
1 HP, (60 lbs)	P6@1.00	Ea	1,070.00	36.10	1,106.10
1-1/2 HP, (76 lbs)	P6@2.00	Ea	1,350.00	72.20	1,422.20
2 HP, (81 lbs)	P6@2.00	Ea	1,480.00	72.20	1,552.20
3 HP, (129 lbs)	P6@3.00	Ea	1,750.00	108.00	1,858.00
Gas ranges, free standing, floor mounted, 29-3/4" D x 59" H x 62-3/4" W, with twin ovens and 24" W griddle					
6 burners, 17,500 Btu each (735 Lbs)	P6@4.00	Ea	2,400.00	144.00	2,544.00
Griddles, gas, 20,000 Btu, 1" thick x 24" deep griddle plate, countertop mounted					
24" wide (295 lbs)	P6@2.00	Ea	1,140.00	72.20	1,212.20
36" wide (385 lbs)	P6@2.00	Ea	1,460.00	72.20	1,532.20
48" wide (520 lbs)	P6@3.00	Ea	1,820.00	108.00	1,928.00
60" wide (650 lbs)	P6@3.00	Ea	2,230.00	108.00	2,338.00
Ice-maker, free standing, floor mounted model					
210 Lbs per day capacity (145 lbs)	P6@2.00	Ea	1,630.00	72.20	1,702.20
535 Lbs per day capacity (197 lbs)	P6@2.00	Ea	2,300.00	72.20	2,372.20
700 Lbs per day capacity (223 lbs)	P6@2.00	Ea	2,840.00	72.20	2,912.20
1270 Lbs per day capacity (337 lbs)	P6@2.00	Ea	3,380.00	72.20	3,452.20
Ice storage unit, self service for bagged ice, 40 cu. ft. Capacity, floor mounted. Includes refrigeration unit					
Indoor type	E4@3.00	Ea	1,610.00	106.00	1,716.00
Outdoor type	E4@3.00	Ea	1,180.00	106.00	1,286.00
Meat and deli case, self-contained with gravity cooling, stainless steel top, 54" H x 34" D					
45" long (482 lbs)	D4@3.00	Ea	2,550.00	111.00	2,661.00
69" long (610 lbs)	D4@4.00	Ea	2,960.00	148.00	3,108.00
93" long (805 lbs)	D4@4.00	Ea	3,700.00	148.00	3,848.00
117" long (960 lbs)	D4@4.00	Ea	3,980.00	148.00	4,128.00
Ovens, free standing, floor mounted					
Convection oven, natural gas. Stainless steel, porcelain interior, 38" W x 36" D x 54-1 /4" H					
40,000 BTU, (491 lbs)	D4@4.00	Ea	2,400.00	148.00	2,548.00
Conveyer oven, electric. Stainless steel body, 18" H x 50" W x 31-3/8" D, 208 or 240 V AC					
16" W conveyor, 6 KW, UL listed (190 lbs)	D4@4.00	Ea	3,100.00	148.00	3,248.00
Pot and pan racks, galvanized steel, pot hooks placed 4 per linear foot					
Ceiling mounted, 6'	C8@2.00	Ea	230.00	72.50	302.50
Wall mounted, 6'	C8@2.00	Ea	160.00	72.50	232.50

	Craft@Hrs	Unit	Material	Labor	Total
Refrigerators, CFC free refrigeration system 115 V AC					
Under counter, stainless steel front and top,					
6.2 cu. ft. (157 lbs)	P6@3.00	Ea	870.00	108.00	978.00
11.8 cu. ft. (226 lbs)	P6@4.00	Ea	1,245.00	144.00	1,389.00
Reach-in, bottom mounted, stainless steel doors, aluminum sides					
23 cu. ft.(290 lbs)	P6@4.00	Ea	1,880.00	144.00	2,024.00
49 cu. ft.(478 lbs)	P6@4.00	Ea	2,600.00	144.00	2,744.00

Sandwich preparation units. Free standing, stainless steel top, sides and back, 115 V AC, refrigerated 30" D x 43" H, full length 12" deep cutting board

	Craft@Hrs	Unit	Material	Labor	Total
6.3 cu. ft. 27-½" long	D4@2.00	Ea	1000.00	73.90	1,073.90
12 cu. ft. 48" long	D4@2.00	Ea	1,530.00	73.90	1,603.90

Shelving, aluminum, free standing, 2000 lb capacity lower shelf, 400 lb capacity uppers

	Craft@Hrs	Unit	Material	Labor	Total
18" x 36" x 68"H, 3 tier	C8@2.00	Ea	137.00	72.50	209.50
18" x 48" x 68"H, 3 tier	C8@2.00	Ea	152.00	72.50	224.50
18" x 36" x 76"H, 4 tier	C8@2.00	Ea	168.00	72.50	240.50
18" x 48" x 76"H, 4 tier	C8@2.00	Ea	189.00	72.50	261.50

Sinks, stainless steel, free standing, floor mounted, with 7" high backsplash and faucet holes drilled on 8" centers, no trim included

	Craft@Hrs	Unit	Material	Labor	Total
Scullery sink, 16" x 20" tubs, 12" deep					
40" long, 1 tub, 1 drainboard (65 lbs)	P6@2.00	Ea	660.00	72.20	732.20
58" long, 2 tubs, 1 drainboard (80 lbs)	P6@2.00	Ea	845.00	72.20	917.20
103" long, 3 tubs, 2 drainboards (110 lbs)	P6@2.00	Ea	890.00	72.20	962.20
122" long, 4 tubs, 2 drainboards (150 lbs)	P6@2.00	Ea	1,310.00	72.20	1,382.20
Mop sink					
21" x 21" x 41", single compartment	P6@2.00	Ea	188.00	72.20	260.20

Toaster, conveyor type, stainless steel, fused quartz sheathing, countertop model

	Craft@Hrs	Unit	Material	Labor	Total
300 slices per hour, 115 V AC, 1,400 watts	D4@3.00	Ea	656.00	111.00	767.00

Work tables, 16 gauge type 430 stainless steel, 30" high. Includes two 12"deep over-shelves, full depth under-shelf and pot and ladle rack

	Craft@Hrs	Unit	Material	Labor	Total
48" long (118 lbs)	D4@1.00	Ea	595.00	36.90	631.90
60" long (135 lbs)	D4@1.00	Ea	695.00	36.90	731.90
72" long(154 lbs)	D4@1.00	Ea	735.00	36.90	771.90
Add to any work table above					
20" x 20" x 5" pull flange drawer (13 lbs)	D4@.250	Ea	67.00	9.24	76.24
Stainless steel pot hooks (1/4 lbs)	D4@.250	Ea	4.50	9.24	13.74

Gymnasium and Athletic Equipment

	Craft@Hrs	Unit	Material	Labor	Total
Basketball backstops					
Fixed, wall mounted	D4@8.00	Ea	309.00	296.00	605.00
Swing-up, wall mounted	D4@10.0	Ea	815.00	369.00	1,184.00
Ceiling suspended, to 20', swing-up manual	D4@16.0	Ea	1,790.00	591.00	2,381.00
Add for glass backstop, fan shape	—	Ea	—	—	600.00
Add for glass backstop, rectangular	—	Ea	—	—	670.00
Add for power operation	—	Ea	—	—	590.00
Gym floors, typical subcontract prices, not including subfloor or base					
Synthetic gym floor, 3/16"	—	SF	—	—	5.45
Synthetic gym floor, 3/8"	—	SF	—	—	7.05
Built-up wood "spring system" maple floor	—	SF	—	—	9.80
Rubber cushioned maple floor	—	SF	—	—	8.50
Maple floor on sleepers and membrane					
Unfinished	—	SF	—	—	6.90
Apparatus inserts	—	Ea	—	—	37.50

	Craft@Hrs	Unit	Material	Labor	Total
Bleachers, telescoping, manual, per seat	D4@.187	Ea	40.60	6.91	47.51
Bleachers, hydraulic operated, per seat	D4@.235	Ea	55.40	8.68	64.08
Scoreboards, basketball, single face					
Economy	CE@7.01	Ea	1,200.00	281.00	1,481.00
Good	CE@13.0	Ea	1,650.00	521.00	2,171.00
Premium	CE@21.6	Ea	2,600.00	865.00	3,465.00

Rolling Ladders Rolling ladders for mercantile, library or industrial use. Putnam Rolling Ladder.
Oak rolling ladder with top and bottom rolling fixtures, non-slip treads, pre-finished. By floor to track height. Painted aluminum hardware. Add cost of track from below

	Craft@Hrs	Unit	Material	Labor	Total
7'6" high	D4@.819	Ea	380.00	30.30	410.30
9'0" high	D4@.819	Ea	348.00	30.30	378.30
11'8" high	D4@1.09	Ea	440.00	40.30	480.30
12'0" high	D4@1.09	Ea	461.00	40.30	501.30
13'9" high	D4@1.09	Ea	482.00	40.30	522.30
14'0" high	D4@1.09	Ea	504.00	40.30	544.30
Add for bend at ladder top or bottom	—	Ea	71.20	—	71.20
Add for brass or chrome hardware	—	Ea	91.10	—	91.10
Add for ladder work shelf	—	Ea	53.60	—	53.60
Add for handrail	—	Ea	40.20	—	40.20
Rolling ladder track, 7/8" slotted steel, per foot of length, 9' to 14' high					
Painted aluminum finish	D4@.089	LF	3.99	3.29	7.28
Painted black finish	D4@.089	LF	4.99	3.29	8.28
Brass or chrome plated finish	D4@.089	LF	11.50	3.29	14.79
Corner bend track, 90 degree, 30" radius					
Painted finish	D4@.363	Ea	39.90	13.40	53.30
Chrome or brass finish	D4@.363	Ea	64.60	13.40	78.00

Service Station Equipment These costs do not include excavation, concrete work, piping or electrical work

	Craft@Hrs	Unit	Material	Labor	Total
Air compressor, 1-1/2 HP with receiver	PF@5.54	Ea	1,570.00	237.00	1,807.00
Air compressor, 3 HP with receiver	PF@5.54	Ea	1,860.00	237.00	2,097.00
Hose reels with hose					
Air hose, 50', heavy duty	PF@8.99	Ea	785.00	385.00	1,170.00
Water hose, 50' and valve	PF@8.01	Ea	800.00	343.00	1,143.00
Lube rack, 5 hose, remote pump	PF@48.4	Ea	8,300.00	2,070.00	10,370.00
Air-operated pumps for hose reels, fit 55 gallon drum					
Motor oil or gear oil	PF@1.00	Ea	788.00	42.80	830.80
Lube oil	PF@1.00	Ea	1,680.00	42.80	1,722.80
Gasoline pumps, with cabinet, industrial type,					
15 gallons per minute	PF@5.93	Ea	1,590.00	254.00	1,844.00
Add for totalizer	—	Ea	280.00	—	280.00
Submerged turbine, 1/3 HP, serves 4 dispensers	PF@22.5	Ea	777.00	962.00	1,739.00
Submerged turbine, 3/4 HP, serves 8 dispensers	PF@22.5	Ea	1,080.00	962.00	2,042.00
Gasoline dispenser, computing, single hose, single product					
not including rough-in of pit box	PF@9.94	Ea	2,550.00	425.00	2,975.00
Gasoline dispenser, computing, dual hose, dual product,					
not including rough-in of pit box	PF@9.94	Ea	4,510.00	425.00	4,935.00
Pit box for submerged pump, 22" x 22",					
installed in existing pit	PF@1.26	Ea	124.00	53.90	177.90
Fill boxes, 12" diameter, cast iron	PF@.949	Ea	250.00	40.60	290.60
Air and water service, post type with auto-inflator and disappearing hoses,					
hoses included	PF@9.58	Ea	1,390.00	410.00	1,800.00
Expect hoists to vary in cost and rating depending on supplier.					
Auto hoist, single post frame contact					
8,000 lb semi hydraulic	D4@71.3	Ea	3,340.00	2,630.00	5,970.00

	Craft@Hrs	Unit	Material	Labor	Total
Hoist, single post frame contact					
8,000 lb fully hydraulic	D4@71.3	Ea	3,560.00	2,630.00	6,190.00
Hoist, two post, pneumatic, 11,000 lb	D4@114.	Ea	6,500.00	4,210.00	10,710.00
Hoist, two post, pneumatic, 24,000 lb	D4@188.	Ea	9,200.00	6,950.00	16,150.00
Two-post fully hydraulic hoists					
11,000 pound capacity	D4@188.	Ea	5,100.00	6,950.00	12,050.00
13,000 pound capacity	D4@188.	Ea	5,220.00	6,950.00	12,170.00
18,500 pound capacity	D4@188.	Ea	6,980.00	6,950.00	13,930.00
24,000 pound capacity	D4@188.	Ea	9,370.00	6,950.00	16,320.00
26,000 pound capacity	D4@188.	Ea	10,800.00	6,950.00	17,750.00
Cash box and pedestal stand	D4@2.55	Ea	250.00	94.20	344.20
Tire changer, 120 PSI pneumatic, auto tire	D4@10.6	Ea	2,350.00	392.00	2,742.00
Tire changer, hydraulic, truck tire	D4@21.4	Ea	7,300.00	791.00	8,091.00
Exhaust fume system, underground, complete					
Per station	D5@9.52	Ea	515.00	342.00	857.00

Parking Control Equipment Costs are for parking equipment only, no concrete work, paving, or electrical work included. Labor costs include installation and testing.

	Craft@Hrs	Unit	Material	Labor	Total
Parking gates, controllable by attendant, vehicle detector, ticket dispenser, card reader, or coin/token machine					
Gate with 10' arm	D4@9.13	Ea	2,780.00	337.00	3,117.00
Folding arm gate (low headroom), 10'	D4@9.13	Ea	3,090.00	337.00	3,427.00
Vehicle detector, for installation in approach lane					
including cutting and grouting	D4@4.96	Ea	387.00	183.00	570.00
Ticket dispenser (spitter), with date/time imprinter					
actuated by vehicle detector or hand button	D4@9.13	Ea	5,150.00	337.00	5,487.00
Card readers. Non-programmable, parking areas with long term parking privileges					
per station	D4@3.39	Ea	721.00	125.00	846.00
Programmable, parking areas for users paying a fee to renew parking privileges					
per reader	D4@3.39	Ea	4,270.00	125.00	4,395.00
Lane spikes, dragon teeth, installation includes cutting and grouting					
6' wide	D4@9.69	Ea	1,130.00	358.00	1,488.00
Lighted warning sign for lane spikes	D4@5.71	Ea	515.00	211.00	726.00

Loading Dock Equipment Costs shown assume that equipment is set in prepared locations. No excavation, concrete work or electrical work included. Equipment cost shown is based on using a 4000 lb forklift

	Craft@Hrs	Unit	Material	Labor	Equipment	Total
Move equipment on or off job						
Allow, per move	—	LS	—	—	60.00	60.00
Dock levelers, Systems, Inc.						
Edge-of-dock leveler, manual	D6@2.25	Ea	540.00	94.70	12.20	646.90
Mechanical platform dock leveler, recessed, 20,000 lb capacity						
6' wide x 6' long	D6@3.00	Ea	2,190.00	126.00	16.30	2,332.30
6' wide x 8' long	D6@3.00	Ea	2,600.00	126.00	16.30	2,742.30
7' wide x 6' long	D6@3.00	Ea	2,560.00	126.00	16.30	2,702.30
7' wide x 8' long	D6@3.00	Ea	2,560.00	126.00	16.30	2,702.30
Hydraulic, pit type						
6' wide x 6' long	D6@4.50	Ea	3,460.00	189.00	24.40	3,673.40
6' wide x 8' long	D6@4.50	Ea	3,690.00	189.00	24.40	3,903.40
Dock seals, foam pad type, 12" projection, with anchor brackets (bolts not included), Systems, Inc., door opening size as shown						
8' wide x 8' high	D6@3.75	Ea	550.00	158.00	20.40	728.40
8' wide x 10' high	D6@3.75	Ea	635.00	158.00	20.40	813.40
Dock lifts, portable, electro-hydraulic scissor lifts, Systems, Inc.						
4,000 lb capacity, 6' x 6' dock	D6@2.25	Ea	6,390.00	94.70	12.20	6,496.90

	Craft@Hrs	Unit	Material	Labor	Equipment	Total
Dock lifts, pit recessed, electro-hydraulic scissor lifts, Systems, Inc.						
5,000 lb capacity, 6' x 8' platform	D6@12.0	Ea	6,540.00	505.00	65.00	7,110.00
6,000 lb capacity, 6' x 8' platform	D6@12.0	Ea	6,670.00	505.00	65.00	7,240.00
8,000 lb capacity						
6' x 8' platform	D6@12.0	Ea	9,680.00	505.00	65.00	10,250.00
8' x 10' platform	D6@12.0	Ea	10,900.00	505.00	65.00	11,470.00
10,000 lb capacity						
6' x 8' platform	D6@12.0	Ea	11,000.00	505.00	65.00	11,570.00
8' x 10' platform	D6@12.0	Ea	12,100.00	505.00	65.00	12,670.00
12,000 lb capacity						
6' x 10' platform	D6@15.0	Ea	14,200.00	631.00	81.30	14,912.30
8' x 12' platform	D6@15.0	Ea	15,100.00	631.00	81.30	15,812.30
15,000 lb capacity						
6' x 10' platform	D6@15.0	Ea	15,800.00	631.00	81.30	16,512.30
8' x 12' platform	D6@15.0	Ea	17,200.00	631.00	81.30	17,912.30
18,000 lb capacity						
6' x 10' platform	D6@15.0	Ea	16,600.00	631.00	81.30	17,312.30
8' x 12' platform	D6@15.0	Ea	17,700.00	631.00	81.30	18,412.30
20,000 lb capacity						
6' x 12' platform	D6@15.0	Ea	18,400.00	631.00	81.30	19,112.30
8' x 12' platform	D6@15.0	Ea	20,100.00	631.00	81.30	20,812.30
Combination dock lift and dock leveler, electro-hydraulic scissor lifts, pit recessed, 5,000 lb capacity, Advance						
6' x 7'2" platform	D6@12.0	Ea	8,550.00	505.00	65.00	9,120.00
6' x 8' platform	D6@12.0	Ea	8,800.00	505.00	65.00	9,370.00
Dock bumpers, laminated rubber, not including masonry anchors						
10" high x 14" wide x 4-1/2" thick	D4@.248	Ea	51.50	9.16	—	60.66
10" high x 14" wide x 6" thick	D4@.299	Ea	70.00	11.00	—	81.00
10" high x 24" wide x 4-1/2" thick	D4@.269	Ea	59.80	9.94	—	69.74
10" high x 24" wide x 6" thick	D4@.358	Ea	84.20	13.20	—	97.40
10" high x 36" wide x 4-1/2" thick	D4@.416	Ea	70.70	15.40	—	86.10
10" high x 36" wide x 6" thick	D4@.499	Ea	102.00	18.40	—	120.40
12" high x 14" wide x 4-1/2" thick	D4@.259	Ea	61.20	9.57	—	70.77
12" high x 24" wide x 4-1/2" thick	D4@.328	Ea	71.00	12.10	—	83.10
12" high x 36" wide x 4-1/2" thick	D4@.432	Ea	87.60	16.00	—	103.60
22" high x 22" wide x 3" thick	D4@.251	Ea	77.00	9.27	—	86.27
Add for 4 masonry anchors	D4@1.19	LS	2.30	44.00	—	46.30
Barrier posts, 8' long concrete-filled steel posts set 4' underground in concrete						
3" diameter	P8@.498	Ea	73.80	16.90	—	90.70
4" diameter	P8@.498	Ea	105.00	16.90	—	121.90
6" diameter	P8@.752	Ea	131.00	25.50	—	156.50
8" diameter	P8@1.00	Ea	158.00	33.90	—	191.90
Loading dock shelter, fabric covered, steel frame,						
24" extension for 10' x 10' door	D4@9.91	Ea	866.00	366.00	—	1,232.00
Perimeter door seal,						
12" x 12", vinyl cover	D4@.400	LF	35.20	14.80	—	50.00

	Craft@Hrs	Unit	Material	Labor	Total
Medical Refrigeration Equipment Labor costs include unpacking, assembly and installation. Jewett Refrigerator. Blood bank refrigerators, upright enameled steel cabinets with locking glass doors, 7 day recording thermometer, LED temperature display, failure alarm, interior lights, and removable drawers. 2 to 4 degree C operating range					
5 drawer, 240 bag, 16.9 CF, 29" wide x 30" deep x 74" high, #BBR17	D4@2.25	Ea	5,520.00	83.10	5,603.10
6 drawer, 360 bag, 24.8 CF, 29" wide x 36" deep x 83" high, #BBR25	D4@2.25	Ea	5,760.00	83.10	5,843.10

	Craft@Hrs	Unit	Material	Labor	Total
10 drawer, 480 bag, 37.4 CF, 59" wide x 30" deep x 74" high, #BBR37	D4@3.42	Ea	8,000.00	126.00	8,126.00
12 drawer, 720 bag, 55.0 CF, 59" wide x 36" deep x 83" high, #BBR55	D4@3.42	Ea	8,220.00	126.00	8,346.00

Blood plasma storage freezers, upright enameled steel cabinets with locking steel doors, 7 day recording thermometer, LED temperature display, failure alarm, and removable drawers. -30 degree C operating temperature

	Craft@Hrs	Unit	Material	Labor	Total
6 drawer, 265 pack, 13.2 CF, 34" wide x 30" deep x 74" high, #BPL13	D4@2.25	Ea	9,400.00	83.10	9,483.10
8 drawer, 560 pack, 24.8 CF, 29" wide x 36" deep x 83" high, #BPL25	D4@2.25	Ea	8,220.00	83.10	8,303.10
16 drawer, 1,120 pack, 55 CF, 59" wide x 36" deep x 83" high, #BPL55	D4@3.42	Ea	12,940.00	126.00	13,066.00

Laboratory refrigerators, upright enameled steel cabinets with locking steel doors, enameled steel interior, and adjustable stainless steel shelves. 2 to 4 degree C operating temperature

	Craft@Hrs	Unit	Material	Labor	Total
4 shelf, 16.9 CF, 29" wide x 30" deep x 74" high #LR17B	D4@2.25	Ea	3,760.00	83.10	3,843.10
8 shelf, 37.4 CF, 59" wide x 30" deep x 74" high #LR37B	D4@3.42	Ea	5,760.00	126.00	5,886.00

Pharmacy refrigerators, upright enameled steel cabinets with locking steel doors, enameled steel interior, and adjustable stainless steel drawers and shelves. -2 degree C operating temperature

	Craft@Hrs	Unit	Material	Labor	Total
6 drawer, 1 shelf, 24.8 CF, 29" wide x 36" deep x 83" high, #PR25B	D4@2.25	Ea	4,220.00	83.10	4,303.10
9 drawer, 5 shelf, 55.0 CF, 59" wide x 36" deep x 83" high, #PR55B	D4@3.42	Ea	6,700.00	126.00	6,826.00

Hospital/Lab/Pharmacy undercounter refrigerators, stainless steel interior and exterior, locking door. 24" wide x 24" deep x 34-1/2" high. 2 degree C operating temperature

	Craft@Hrs	Unit	Material	Labor	Total
5.4 CF, with blower coil cooling system, stainless steel racks, and auto-defrost, #UC5B	D4@1.67	Ea	1,940.00	61.70	2,001.70
5.4 CF with cold wall cooling system, stainless steel racks, and auto-defrost, #UC5C	D4@1.69	Ea	2,460.00	62.40	2,522.40

Hospital/Lab/Pharmacy undercounter freezers, stainless steel interior and exterior, locking door. 24" wide x 24" deep x 34-1/2" high. -20 degree C operating temperature

	Craft@Hrs	Unit	Material	Labor	Total
4.6 CF, stainless steel racks, Auto-defrost system, #UCF5B	D4@1.69	Ea	2,720.00	62.40	2,782.40
4.6 CF, stainless steel racks, manual hot gas defrost system, #UCF5C	D4@1.69	Ea	3,220.00	62.40	3,282.40

Morgue refrigerators

	Craft@Hrs	Unit	Material	Labor	Total
1 or 2 body roll-in type, 39" wide x 96" deep x 76" high, with 1/2 HP, 4380 Btu/hour condenser, #1SPEC	D4@15.8	Ea	8,700.00	584.00	9,284.00
2 body sliding tray type, 39" wide x 96" deep x 76" high, with 1/2 HP, 4380 Btu/hour condenser, #2EC	D4@18.0	Ea	8,580.00	665.00	9,245.00
2 body side opening type, 96" wide x 39" deep x 76" high, with 1/2 HP, 4380 Btu/hour condenser, #2SC	D4@18.0	Ea	9,500.00	665.00	10,165.00
4 or 6 body with roll-in type lower compartments, sliding tray type upper compartments. 73" W x 96" D x 102" High, 6060 Btu/hr, 3/4 HP condenser, #4SPEC2W	D4@22.5	Ea	14,840.00	831.00	15,671.00

	Craft@Hrs	Unit	Material	Labor	Total

Cabinets Cost per linear foot of face or back dimension. Base cabinets are 34" high x 24" deep, wall cabinets are 42" high x 12" deep and full height cabinets are 94" high x 24" deep. Labor shown is for installing shop fabricated units. Add for countertops at the end of this section.

Metal cabinets, shop and commercial type, with hardware but no countertop

	Craft@Hrs	Unit	Material	Labor	Total
Base cabinet, drawer, door and shelf	D4@.379	LF	91.60	14.00	105.60
Wall cabinet with door and 2 shelves	D4@.462	LF	83.80	17.10	100.90
Full height cabinet, 5 shelves	D4@.595	LF	172.00	22.00	194.00
Library shelving, 48" high x 8" deep modules	D4@.336	LF	60.00	12.40	72.40
Wardrobe units, 72" high x 42" deep	D4@.398	LF	148.00	14.70	162.70

Classroom type wood cabinets, laminated plastic face, with hardware but no countertop

	Craft@Hrs	Unit	Material	Labor	Total
Base cabinet, drawer, door and shelf	C8@.391	LF	155.00	14.20	169.20
Wall cabinet with door and 2 shelves	C8@.475	LF	136.00	17.20	153.20
Full height cabinet with doors	C8@.608	LF	218.00	22.10	240.10

Hospital type wood cabinets, laminated plastic face, with hardware but no countertop

	Craft@Hrs	Unit	Material	Labor	Total
Base cabinet, drawer, door and shelf	C8@.982	LF	158.00	35.60	193.60
Wall, with 2 shelves and door	C8@1.15	LF	146.00	41.70	187.70
Full height with doors	C8@1.35	LF	215.00	49.00	264.00

Laboratory type metal cabinets, with hardware but no countertops

	Craft@Hrs	Unit	Material	Labor	Total
Base cabinets, doors	D4@.785	LF	85.00	29.00	114.00
Base cabinets, drawers	D4@.785	LF	158.00	29.00	187.00
Base cabinets, island base	D4@.785	LF	158.00	29.00	187.00
Wall cabinets, doors	D4@.785	LF	137.00	29.00	166.00
Wardrobe or storage cabinets, 80" high	D4@.939	LF	156.00	34.70	190.70
Fume hoods, steel, without duct, typical price	T5@3.88	Ea	635.00	139.00	774.00
Add for duct, without electric work, typical price	T5@16.6	LF	960.00	594.00	1,554.00

Custom made wood cabinets, no hardware or countertops included, unfinished, 36" wide modules

Standard grade, mahogany or laminated plastic face

	Craft@Hrs	Unit	Material	Labor	Total
Base	C8@.256	LF	67.20	9.28	76.48
Wall	C8@.380	LF	51.10	13.80	64.90
Full height	C8@.426	LF	115.00	15.50	130.50

Custom grade, birch

	Craft@Hrs	Unit	Material	Labor	Total
Base	C8@.361	LF	102.00	13.10	115.10
Wall	C8@.529	LF	87.40	19.20	106.60
Full height	C8@.629	LF	161.00	22.80	183.80
Sink fronts	C8@.361	LF	85.00	13.10	98.10
Drawer units, 4 drawers	C8@.361	LF	168.00	13.10	181.10
Apron (knee space)	C8@.148	LF	21.70	5.37	27.07

Additional costs for custom cabinets

	Craft@Hrs	Unit	Material	Labor	Total
Ash face	—	LF	3.75	—	3.75
Walnut face	—	LF	56.10	—	56.10
Edge banding	—	LF	6.24	—	6.24
Prefinished exterior	—	LF	5.62	—	5.62
Prefinished interior	—	LF	9.57	—	9.57
Standard hardware, installed	—	LF	4.52	—	4.52
Institutional grade hardware, installed	—	LF	6.59	—	6.59
Drawer roller guides, per drawer	—	Ea	7.68	—	7.68
Drawer roller guides, suspension	—	Ea	23.50	—	23.50

Extra drawers

	Craft@Hrs	Unit	Material	Labor	Total
12" wide	—	Ea	47.90	—	47.90
18" wide	—	Ea	56.10	—	56.10
24" wide	—	Ea	61.10	—	61.10

	Craft@Hrs	Unit	Material	Labor	Total
Countertops, shop fabricated, delivered ready for installation					
Laminated plastic tops 2'0" wide over plywood					
Custom work, square edge front, 4" splash	C8@.187	LF	24.50	6.78	31.28
Add for square edge 9" backsplash	—	LF	8.01	—	8.01
Add for solid colors	—	LF	2.36	—	2.36
Add for acid proof tops	—	LF	18.50	—	18.50
Komar or simulated marble tops					
Standard vanity unit	C8@.192	LF	23.60	6.96	30.56
Custom moulded top with splash	C8@.049	SF	13.20	1.78	14.98
Stainless steel, 2'0" wide with backsplash	C8@.366	LF	65.90	13.30	79.20

Window Treatment

	Craft@Hrs	Unit	Material	Labor	Total
Shades, custom made, cost per SF of glass covered					
Standard roll-up unit	D7@.013	SF	3.12	.44	3.56
Better quality roll-up unit	D7@.013	SF	4.35	.44	4.79
Decorator quality unit	D7@.013	SF	10.80	.44	11.24
Mini-blinds, custom made, cost per SF of glass covered					
Aluminum, horizontal, 1"	D7@.013	SF	4.06	.44	4.50
Cloth type, vertical	D7@.013	SF	6.46	.44	6.90
Steel, horizontal, 2"	D7@.013	SF	7.24	.44	7.68

Fabrics Measure glass size only. Includes allowance for overlap, pleating and all hardware.
Draperies, 8' height, fabric quality will cause major variation in prices, per LF of opening, typical installed prices

	Craft@Hrs	Unit	Material	Labor	Total
Standard quality, window or sliding door	—	LF	—	—	16.00
Better quality, window	—	LF	—	—	53.00
Better quality, sliding door	—	LF	—	—	50.50
Drapery lining	—	LF	—	—	6.00
Fiberglass fabric	—	LF	—	—	51.50
Filter light control type	—	LF	—	—	46.50
Flameproofed	—	LF	—	—	50.00
Velour, grand	—	LF	—	—	66.00

Recessed Open Link Entry Mats These costs assume that mats are mounted in an aluminum frame recessed into a concrete walkway and do not include concrete work, special designs or lettering.

	Craft@Hrs	Unit	Material	Labor	Total
Aluminum link, 3/8" thick	D4@.560	SF	13.30	20.70	34.00
Steel link, 3/8" thick	D4@.560	SF	5.99	20.70	26.69
Aluminum hinged strips, carpet inserts, 1/2"	D4@.560	SF	28.90	20.70	49.60
Black Koroseal, 1/2" thick, open link	D4@.560	SF	11.50	20.70	32.20
Aluminum frame for entry mats	D4@.086	LF	5.05	3.18	8.23

Entrance Mats and Recessed Foot Grilles These costs assume installation of an extruded aluminum tread foot grille on a level surface or over a catch basin with or without a drain. Conspec Systems

	Craft@Hrs	Unit	Material	Labor	Total
Pedigrid/Pedimat extruded aluminum entrance mat					
Grid mat with carpet treads	D4@.150	SF	30.00	5.54	35.54
Grid mat with vinyl treads	D4@.150	SF	32.80	5.54	38.34
Grid mat with pool or shower vinyl treads	D4@.150	SF	36.50	5.54	42.04
Grid mat with abrasive serrated aluminum treads	D4@.150	SF	36.50	5.54	42.04
Recessed catch basin with 2" drain	D4@.302	SF	42.50	11.20	53.70
Recessed catch basin without drain	D4@.242	SF	38.10	8.94	47.04
Level base recessed mount	D4@.150	SF	36.50	5.54	42.04
Surface-mount frame, vinyl	—	LF	3.65	—	3.65
Recessed-mount frame, bronze aluminum	D4@.092	LF	8.31	3.40	11.71

	Craft@Hrs	Unit	Material	Labor	Total
Seating These costs do not include concrete work.					
Theater style, economy	D4@.424	Ea	105.00	15.70	120.70
Theater style, lodge, rocking type	D4@.470	Ea	210.00	17.40	227.40
Bleachers, closed riser, extruded aluminum, cost per 18" wide seat					
Portable bleachers	D4@.090	Ea	45.90	3.32	49.22
Portable bleachers with guardrails	D4@.136	Ea	51.00	5.02	56.02
Team bleachers, stationary	D4@.090	Ea	39.40	3.32	42.72
Stadium seating, on concrete foundation by others, extruded aluminum, cost per 18" seat width					
10" flat bench seating, seatboard only	D4@.090	Ea	13.20	3.32	16.52
12" contour bench seating, seatboard only	D4@.090	Ea	14.20	3.32	17.52
Chair seating, molded plastic	D4@.310	Ea	76.50	11.50	88.00
Replacement bench seating, stadium or bleacher, extruded aluminum					
Cost per 18" seat width					
10" wide flat seatboard only	D4@.069	Ea	8.40	2.55	10.95
10" wide flat seatboard and footboard	D4@.136	Ea	17.90	5.02	22.92
12" contour seatboard only	D4@.069	Ea	9.99	2.55	12.54

Special Construction 13

Access for the Disabled

Use the following cost estimates when altering an existing building to meet requirements of the Americans with Disabilities Act (ADA). The ADA requires that existing public and private buildings be altered to accommodate the physically handicapped when that can be done without unreasonable difficulty or expense.

Ramps, flared aprons and landings Reinforced concrete, 2,000 PSI (at $57.50 per CY), placed directly from the chute of a ready-mix truck, vibrated and finished. Costs include typical wire mesh reinforcing and an allowance for waste but no excavation, backfill or foundations.

	Craft@Hrs	Unit	Material	Labor	Equipment	Total
Access ramps, based on ramps with 1" rise per 1' of run, 4" thick walls and 4" thick slab. Includes pipe sleeves for railings, sand fill in cavity and non-slip surface finish						
24" vertical rise (4' wide x 25' long)	C8@40.0	Ea	884.00	1,450.00	—	2,334.00
36" vertical rise (4' wide x 38' long)	C8@60.0	Ea	1,610.00	2,180.00	—	3,790.00
48" vertical rise (4' wide x 50' long)	C8@96.0	Ea	2,330.00	3,480.00	—	5,810.00
Access aprons at curbs, flared and inclined, 2' long x 8' wide, inclined						
4" thick (80 SF per CY)	P9@.024	Ea	1.03	.88	—	1.91
6" thick (54 SF per CY)	P9@.032	Ea	1.39	1.17	—	2.56
Landings, 4" thick walls on three sides and 4" thick top slab cast against an existing structure on one side. Costs include pipe sleeves for railings, sand fill in cavity and an allowance for waste.						
3'0" high with 6' x 6' square slab	C8@32.0	Ea	550.00	1,160.00	—	1,710.00
Add per vertical foot of height over 3'0"	—	VLF	93.00	—	—	93.00
Snow melting cable for ramps and landings. Includes 208-277 volt AC self-regulating heater cable designed to be encased in concrete. Based on cable connected to an existing power source adjacent to the installation						
Heater cable						
(1.1 LF per SF of pavement)	CE@.015	SF	7.56	.60	—	8.16
Power connection kit	CE@.500	Ea	27.50	20.00	—	47.50
Thermostat	CE@.500	Ea	231.00	20.00	—	251.00

	Craft@Hrs	Unit	Material	Labor	Equipment	Total

Railing for ramps and landings Welded steel. Equipment cost shown is for a flatbed truck and a welding machine at $43 per hour. Wall mounted handrails with brackets 5' OC, based on 12' lengths

1-1/4" diameter rail, shop prime painted	H6@.092	LF	5.44	3.64	1.30	10.38
1-1/2" diameter rail, shop prime painted	H6@.102	LF	6.35	4.03	1.45	11.83
Add for galvanized finish	—	%	20.0	—	—	—

Floor mounted railings 3'6" high with posts 5' OC, based on 12' length

1-1/4" diameter rails and posts, shop prime painted

2-rail type	H6@.115	LF	22.30	4.55	1.63	28.48
3-rail type	H6@.125	LF	29.00	4.94	1.77	35.71

1-1/2" diameter rails and posts, shop prime painted

2-rail type	H6@.125	LF	26.90	4.94	1.77	33.61
3-rail type	H6@.137	LF	35.50	5.42	1.91	42.83
Add for galvanized finish	—	%	20.0	—	—	—

Drinking fountains and water coolers for the disabled

Remove an existing drinking fountain or water cooler and disconnect waste and water piping at the wall adjacent to the fixture for use with the new fixture. No salvage value assumed

Floor or wall mounted fixture	P6@1.00	Ea	—	36.10	—	36.10

Install a wall-hung white vitreous china fountain with chrome plated spout, 14" wide, 13" high, non-electric, self-closing dispensing valve and automatic volume control

Drinking fountain, trim and valves	P6@2.80	Ea	600.00	101.00	—	701.00

Install floor or wall-mounted electric water cooler, 8 gallons per hour, cold water only, connected to existing electrical outlet adjacent to the cooler (no electrical work included)

Water cooler, trim and valves	P6@2.80	Ea	1,170.00	101.00	—	1,271.00

Parking stall marking for the disabled

Remove existing pavement marking by water blasting

4" wide stripes (allow 40 LF per stall)	CL@.032	LF	—	.97	.21	1.18

Markings, per square foot,

(minimum 10 SF)	CL@.098	SF	—	2.97	.64	3.61

Mark parking stall with handicapped symbol painted on, including layout

Reflectorized stripes and symbol, one color

per stall	PA@.425	Ea	6.64	15.70	8.40	30.74

Parking lot handicapped sign, 18" x 24" 10 gauge metal, reflective lettering and handicapped symbol

Sign on 2" galvanized steel pipe post 10' long, set 2' into the ground. Includes digging of hole

Using a manual auger and backfill	CL@1.25	Ea	81.00	37.80	—	118.80
On walls with mechanical fasteners	PA@.455	Ea	23.90	16.80	—	40.70

Paving and curb removal These costs include breaking out the paving or curb, loading and hauling debris to a legal dump within 6 miles but no dump fees. Equipment cost per hour is $45.00 and includes one 90 CFM air compressor, one paving breaker and jackhammer bits, one 55 HP wheel loader with integral backhoe and one 5 CY dump truck. The figures in parentheses show the approximate "loose" volume of the materials (volume after being demolished). Use $250 as a minimum charge for this type work and add dump charges as a separate item.

Asphalt paving, depths to 3"

(27 SF per CY)	C3@.004	SF	—	.13	.06	.19

Asphalt curbs, to 12" width

(18 LF per CY)	C3@.028	LF	—	.94	.44	1.38

Concrete paving and slabs on grade

4" concrete without rebars

(54 SF per CY)	C3@.006	SF	—	.20	.10	.30

	Craft@Hrs	Unit	Material	Labor	Equipment	Total
6" concrete without rebars						
(45 SF per CY)	C3@.011	SF	—	.37	.17	.54
Add for removal of bar reinforced paving	C3@.002	SF	—	.07	.03	.10

Curb, gutter and paving for the disabled These costs assume 3 uses of forms and 2,000 PSI concrete (at $58.50 per CY) placed directly from the chute of a ready-mix truck. Use $250 as a minimum charge for this type work. See also Snow melting cable on the preceding page.

Concrete cast-in-place vertical curb, including 2% waste, forms and finishing. These costs include typical excavation and backfill with excess excavation spread on site

	Craft@Hrs	Unit	Material	Labor	Equipment	Total
6" x 12" straight curb	P9@.136	LF	2.23	4.99	2.28	9.50
6" x 18" straight curb	P9@.149	LF	3.32	5.47	2.50	11.29
6" x 24" straight curb	P9@.154	LF	4.36	5.65	2.58	12.59
Add for curved sections	—	%	—	40.0	40.0	—
Curb and 24" monolithic gutter						
(7 LF per CY)	C3@.100	LF	—	3.36	1.61	4.97
Concrete paving and walkway						
4" thick	P8@.018	SF	1.03	.61	—	1.64
Add for integral colors, most pastels	—	SF	.90	—	—	.90

Plumbing fixtures for the disabled

Remove existing plumbing fixtures. Disconnect from existing waste and water piping at the wall adjacent to the fixture. Waste and water piping to remain in place for connection to a new fixture. No salvage value assumed

	Craft@Hrs	Unit	Material	Labor	Equipment	Total
Lavatories	P6@1.00	Ea	—	36.10	—	36.10
Urinals	P6@1.00	Ea	—	36.10	—	36.10
Water closets, floor or wall mounted	P6@1.00	Ea	—	36.10	—	36.10

Install new fixtures. White vitreous china fixtures with typical fittings, hangers, trim and valves. Connected to waste and supply piping from lines adjacent to the fixture

	Craft@Hrs	Unit	Material	Labor	Equipment	Total
Lavatories, wall hung	P6@1.86	Ea	728.00	67.20	—	795.20
Urinals, wall hung, with valve/breaker	P6@1.86	Ea	501.00	67.20	—	568.20
Water closets, tank type, with elongated bowl, toilet seat, trim and valves						
Floor mounted, with bowl ring	P6@1.86	Ea	259.00	67.20	—	326.20
Wall hung, with carrier	P6@2.80	Ea	547.00	101.00	—	648.00

Toilet partitions and grab bars for the disabled

Remove a standard toilet partition, no salvage value included. Floor or ceiling mounted

	Craft@Hrs	Unit	Material	Labor	Equipment	Total
Removal of one wall and one door	T5@1.00	Ea	—	35.80	—	35.80

Install toilet partition designed for wheel chair access. Floor or ceiling mounted. These costs include one wall and one door, mechanical fasteners and drilling of holes

	Craft@Hrs	Unit	Material	Labor	Equipment	Total
Baked enamel	T5@2.66	Ea	396.00	95.20	—	491.20
Porcelain enamel	T5@2.66	Ea	754.00	95.20	—	849.20
Laminated plastic	T5@2.66	Ea	511.00	95.20	—	606.20
Stainless steel	T5@2.66	Ea	880.00	95.20	—	975.20

Grab bars, stainless steel, wall mounted, commercial grade. Set of two required for each stall. Including mechanical fasteners and drilling of holes

	Craft@Hrs	Unit	Material	Labor	Equipment	Total
1-1/4" x 24" urinal bars	CC@1.00	Set	74.70	42.30	—	117.00
1-1/4" x 36" toilet partition bars	CC@1.00	Set	90.60	42.30	—	132.90

	Craft@Hrs	Unit	Material	Labor	Total

Signage in braille Standard signs (Rest Room, Elevator, Telephone, Stairway, etc.) mounted with self-stick foam, velcro or in frame. Add the cost of frames, if required, from below

	Craft@Hrs	Unit	Material	Labor	Total
5/8" lettering raised 1/32". Grade 2 braille raised 1/32"					
2" x 8" sign, raised letters and tactile braille	PA@0.25	Ea	21.70	9.25	30.95
8" x 8" sign with words, graphic symbols and tactile braille					
standard graphic symbols	PA@0.25	Ea	35.90	9.25	45.15
Elevator plates. Sign with graphic symbols and tactile braille, standard colors					
Door jamb plates, 3" x 4", two per opening	PA@0.75	Set	44.80	27.80	72.60
Operation plates, Open Door, Close Door, Emergency Stop, Alarm,					
1-1/4" x 1-1/4", set of four required per car	PA@1.00	Set	60.00	37.00	97.00
Operation plates, Floor Number, 1-1/4" x 1-1/4"					
One required for each floor per car	PA@0.25	Ea	15.00	9.25	24.25

Bus Stop Shelters Prefabricated structure systems, shipped knocked-down complete with all necessary hardware and foundation anchors and installation instructions, assembled at installation site. These costs do not include foundation or floor slab. Dimensions are approximate outside measurements. Based on Columbia Equipment Company.

Three-sided open front type. Clear satin silver anodized aluminum structure and fascia, 1/4" clear acrylic sheet glazing in panels and white baked enamel finish aluminum V-beam roof.

	Craft@Hrs	Unit	Material	Labor	Total
9'2" L x 5'3" W x 7'5" H					
Two rear panels, single panel each side	C8@8.00	Ea	2,820.00	290.00	3,110.00
Four rear panels, two panels each side	C8@12.0	Ea	2,960.00	435.00	3,395.00
11'3" L x 5'3" W x 7'5" H					
Four rear panels, two panels each side	C8@12.0	Ea	3,450.00	435.00	3,885.00
Add to three-sided open front shelters for the following:					
Wind break front entrance panel					
5'3" W x 7'5" H, with two glazed panels	C8@2.00	Ea	340.00	72.50	412.50
Bench, all aluminum vandal-resistant seat & backrest					
8'0" long	C8@1.00	Ea	282.00	36.30	318.30
10'0" long	C8@1.50	Ea	359.00	54.40	413.40
Light fixture with unbreakable Lexan diffuser, wiring not included					
100 watt incandescent, w/photoelectric cell control	C8@2.00	Ea	153.00	72.50	225.50
Radiant heater, with vandal-resistant electric heating element in heavy duty enclosure and metal mesh guard over opening, wiring not included					
2500 watt unit (no controls)	C8@1.50	Ea	610.00	54.40	664.40
5000 watt unit (no controls)	C8@1.50	Ea	716.00	54.40	770.40
Electronic timer control, add to above	C8@1.00	Ea	308.00	36.30	344.30
Integrated map/schedule display panel, anodized aluminum frame, tamper-proof fasteners and "spanner-head" tool for opening					
Full panel width x 30" high					
3/16" clear Plexiglass	C8@1.00	Ea	102.00	36.30	138.30
3/16" polycarbonate with mar-resistant coating	C8@1.00	Ea	143.00	36.30	179.30
Graphics, words "Bus Stop" in white letters, on fascia					
First side	—	LS	85.70	—	85.70
Each additional side	—	Ea	29.60	—	29.60

Modular Building Construction (Manufactured housing)
Relocatable structures and institutional type housing for remote areas. Costs are per square foot covered by the roof. Based on 2" x 4" wood studs 16" OC, 1/2" gypsum wallboard inside and 1/2" CDX plywood outside, with minimum plumbing, heating and electrical systems for the intended use. Costs include factory assembly, delivery by truck to 50 miles and setup on site. Add 1.5% for each additional 50 miles of delivery. No site preparation, earthwork, foundations, or furnishings included. Cost for smaller, more complex structures will be higher.

	Craft@Hrs	Unit	Material	Labor	Total
Portable structures (temporary offices, school classrooms, etc.)					
Medium quality	—	SF	—	—	38.00
Better quality	—	SF	—	—	47.00

	Craft@Hrs	Unit	Material	Labor	Total
Institutional housing, four-plex modules					
Medium quality	—	SF	—	—	49.00
Better quality	—	SF	—	—	49.50
Construction or mining camp barracks, mess halls, kitchens, etc.					
Single story structures	—	SF	—	—	44.00
Add for 2nd story (per SF of floor)	—	SF	—	—	4.40

Special Purpose Rooms and Buildings
Air-supported pool or tennis court enclosures, polyester reinforced vinyl dome, with zippered entry air lock doors, cable tiedowns and blower, typical cost per square foot of area covered.

	Craft@Hrs	Unit	Material	Labor	Total
Air-supported structures	D4@.030	SF	2.95	1.11	4.06
Add for 40" revolving door	D4@2.14	Ea	1,440.00	79.10	1,519.10
Add for 66" air lock door	D4@3.28	Ea	1,440.00	121.00	1,561.00

X-Ray Viewing Panels, Clear Lead-plastic Based on CLEAR-Pb, by Nuclear Associates. Based on SF of panel (rounded up to the next higher whole square foot). For panels 12 square feet or larger, add a crating charge of $70.00. Panels larger than 72" x 96" are special order items. Weights shown are approximate.

	Craft@Hrs	Unit	Material	Labor	Total
7mm thick, 0.3mm lead equivalence, 2.3 lbs/SF	G1@.115	SF	122.00	3.84	125.84
12mm thick, 0.5mm lead equivalence, 3.9 lbs/SF	G1@.195	SF	165.00	6.50	171.50
18mm thick, 0.8mm lead equivalence, 5.9 lbs/SF	G1@.294	SF	179.00	9.80	188.80
22mm thick, 1.0mm lead equivalence, 7.2 lbs/SF	G1@.361	SF	183.00	12.00	195.00
35mm thick, 1.5mm lead equivalence, 11.5 lbs/SF	G1@.574	SF	205.00	19.10	224.10
46mm thick, 2.0mm lead equivalence, 15.0 lbs/SF	G1@.751	SF	270.00	25.00	295.00

Mobile X-Ray Barriers Clear lead-plastic window panels on the upper portion and opaque panels on the lower portion, mounted within a framework with casters on the bottom. Based on Nuclear Associates. Labor shown is to uncrate factory assembled barrier and attach casters.

	Craft@Hrs	Unit	Material	Labor	Total
30" W x 75" H overall					
0.5mm lead equiv. window panel 30" x 24" and 0.8mm lead equiv.					
opaque panel 30" x 48"	MW@.501	Ea	1,870.00	21.90	1,891.90
48" W x 75" H overall					
0.5mm lead equiv. window panel 48" x 36" and 0.8mm lead equiv.					
opaque panel 48" x 36"	MW@.501	Ea	3,300.00	21.90	3,321.90
1.0mm lead equiv. window panel 48" x 36" and 1.5mm lead equiv.					
opaque panel 48" x 36"	MW@.501	Ea	3,990.00	21.90	4,011.90
72" W x 75" H, overall					
0.5mm lead equiv. window panel 72" x 36" and 0.8mm lead equiv.					
opaque panel 72" x 36"	MW@.751	Ea	3,910.00	32.80	3,942.80
1.0mm lead equiv. window panel 72" x 36" and 1.5mm lead equiv.					
opaque panel 72" x 36"	MW@.751	Ea	4,950.00	32.80	4,982.80

Modular X-Ray Barriers Panels are mounted within a framework for attaching to floor, wall or ceiling. Shipped unassembled. Based on Nuclear Associates. Structural supports not included. Costs shown are based on typical 36" wide x 84" high panel sections. Clear lead-plastic window panels 48" high are the upper portion and opaque leaded panels 36" high are the bottom portion.

	Craft@Hrs	Unit	Material	Labor	Total
1-section barrier, 36" W x 84" H overall					
0.5mm lead equiv. panels	G1@2.78	Ea	2,750.00	92.70	2,842.70
0.8mm lead equiv. panels	G1@2.78	Ea	2,910.00	92.70	3,002.70
1.0mm lead equiv. panels	G1@3.34	Ea	3,020.00	111.00	3,131.00
1.5mm lead equiv. panels	G1@4.46	Ea	3,210.00	149.00	3,359.00
2-section barrier, 72" W x 84" H overall					
0.5mm lead equiv. panels	G1@5.56	Ea	5,700.00	185.00	5,885.00
0.8mm lead equiv. panels	G1@5.56	Ea	6,020.00	185.00	6,205.00
1.0mm lead equiv. panels	G1@6.68	Ea	6,150.00	223.00	6,373.00
1.5mm lead equiv. panels	G1@8.90	Ea	6,660.00	297.00	6,957.00

	Craft@Hrs	Unit	Material	Labor	Total
3-section barrier, 108" W x 84" H overall					
0.5mm lead equiv. panels	G1@8.34	Ea	8,150.00	278.00	8,428.00
0.8mm lead equiv. panels	G1@8.34	Ea	8,560.00	278.00	8,838.00
1.0mm lead equiv. panels	G1@10.0	Ea	8,750.00	333.00	9,083.00
1.5mm lead equiv. panels	G1@13.4	Ea	9,500.00	447.00	9,947.00
Larger than 3-section barriers, add to the cost of 1-section barriers for each section over 3					
Add per 18" W x 84" H section	—	%	52.0	—	—
Add per 36" W x 84" H section	—	%	104.0	—	—

Swimming Pools Complete gunite pools including filter, chlorinator, ladder, diving board, basic electrical system, but excluding deck beyond edge of coping. Typical subcontract prices per SF of water surface. Use the prices for preliminary estimates only.

Apartment building	—	SF	—	—	37.20
Community	—	SF	—	—	47.80
Hotel or resort	—	SF	—	—	45.00
School, training (42' x 75')	—	SF	—	—	50.50
School, Olympic (42' x 165')	—	SF	—	—	76.80

Water Tanks, Elevated Typical subcontract prices for 100' high steel tanks, including design, fabrication and erection. Complete tank includes 36" pipe riser with access manhole, 8" overflow to ground, ladder with protective cage, vent, balcony catwalk, concrete foundation, painting and test. No well, pump, fencing, drainage or water distribution piping included. Costs are per tank with gallon capacity as shown. Tanks built in areas without earthquake or high wind risk may cost 15% to 25% less.

80,000 gallon	—	Ea	—	—	319,000.00
100,000 gallon	—	Ea	—	—	333,000.00
150,000 gallon	—	Ea	—	—	399,000.00
200,000 gallon	—	Ea	—	—	547,000.00
300,000 gallon	—	Ea	—	—	742,000.00
400,000 gallon	—	Ea	—	—	842,000.00
Add or subtract for each foot of height more or less than 100, per 100 gallons of capacity	—	Ea	—	—	.75
Cathodic protection system, per tank	—	LS	—	—	15,100.00

Electric service, including float switches, obstruction marker lighting system, and hookup of cathodic protection system

Per tank	—	LS	—	—	11,600.00

Solar Water Heating Systems

Solar collector panels, no piping included					
Typical 24 SF panel	P6@.747	Ea	281.00	27.00	308.00
Supports for solar panels, per panel	P6@2.25	Ea	30.10	81.20	111.30
Control panel for solar heating system	P6@1.56	Ea	104.00	56.30	160.30
Solar collector system pumps, no wiring or piping included					
5.4 GPM, 7.9' head	P6@1.89	Ea	198.00	68.20	266.20
10.5 GPM, 11.6' head or 21.6 GPM, 8.5' head	P6@2.25	Ea	357.00	81.20	438.20
12.6 GPM, 20.5' head or 22.5 GPM, 13.3' head	P6@3.26	Ea	549.00	118.00	667.00
46.5 GPM, 13.7' head	P6@4.59	Ea	821.00	166.00	987.00
60.9 GPM, 21.0' head	P6@5.25	Ea	939.00	190.00	1,129.00
Solar hot water storage tanks, no piping and no pad included					
40 gallon, horizontal	P6@1.12	Ea	220.00	40.40	260.40
110 gallon, horizontal	P6@3.54	Ea	901.00	128.00	1,029.00
180 gallon, vertical	P6@5.43	Ea	1,470.00	196.00	1,666.00
650 gallon, horizontal	P6@5.84	Ea	2,160.00	211.00	2,371.00

Special Construction 13

	Craft@Hrs	Unit	Material	Labor	Total
1,250 gallon, horizontal	P6@5.84	Ea	3,180.00	211.00	3,391.00
1,500 gallon, vertical	P6@11.4	Ea	5,020.00	412.00	5,432.00
2,200 gallon, horizontal	P6@17.1	Ea	4,560.00	617.00	5,177.00
2,700 gallon, vertical	P6@14.9	Ea	5,190.00	538.00	5,728.00

Conveying Systems 14

Dumbwaiters These costs do not include allowance for constructing the hoistway walls or electrical work.

Manual, 2 stop, 25 feet per minute, no doors, by rated capacity, 24" x 24" x 36" high car

	Craft@Hrs	Unit	Material	Labor	Total
25 pounds to 50 pounds	CV@39.8	Ea	1,690.00	1,490.00	3,180.00
75 pounds to 200 pounds	CV@42.7	Ea	2,300.00	1,600.00	3,900.00
Add for each additional stop	—	Ea	—	—	1,280.00

Electric, with machinery mounted above, floor loading, no security gates included

	Craft@Hrs	Unit	Material	Labor	Total
50 lbs, 25 FPM, 2 stop, no doors	CV@42.2	Ea	2,020.00	1,580.00	3,600.00
50 lbs, 25 FPM, 2 stop, manual doors	CV@106.	Ea	5,030.00	3,960.00	8,990.00
50 lbs, 50 FPM, 2 stop, manual doors	CV@107.	Ea	5,030.00	4,000.00	9,030.00
75 lbs, 25 FPM, 2 stop, no doors	CV@42.2	Ea	2,020.00	1,580.00	3,600.00
75 lbs, 25 FPM, 2 stop, manual doors	CV@107.	Ea	6,060.00	4,000.00	10,060.00
75 lbs, 50 FPM, 2 stop, manual doors	CV@111.	Ea	6,060.00	4,150.00	10,210.00
100 lbs, 25 FPM, 2 stop, no doors	CV@50.6	Ea	2,940.00	1,890.00	4,830.00
100 lbs, 25 FPM, 2 stop, manual doors	CV@107.	Ea	6,230.00	4,000.00	10,230.00
100 lbs, 50 FPM, 2 stop, manual doors	CV@111.	Ea	6,370.00	4,150.00	10,520.00
100 lbs, 100 FPM, 5 stop, manual doors	CV@162.	Ea	6,670.00	6,060.00	12,730.00
200 lbs, 25 FPM, 2 stop, no doors	CV@52.3	Ea	3,020.00	1,960.00	4,980.00
200 lbs, 25 FPM, 2 stop, manual doors	CV@111.	Ea	6,370.00	4,150.00	10,520.00
200 lbs, 100 FPM, 5 stop, manual doors	CV@162.	Ea	7,790.00	6,060.00	13,850.00
300 lbs, 50 FPM, 2 stop, manual doors	CV@113.	Ea	13,220.00	4,230.00	17,450.00
300 lbs, 100 FPM, 5 stop, manual doors	CV@171.	Ea	17,600.00	6,400.00	24,000.00
400 lbs, 50 FPM, 2 stop, manual doors	CV@114.	Ea	13,380.00	4,260.00	17,640.00
400 lbs, 100 FPM, 5 stop, manual doors	CV@174.	Ea	19,520.00	6,510.00	26,030.00
500 lbs, 50 FPM, 2 stop, manual doors	CV@117.	Ea	13,860.00	4,380.00	18,240.00
500 lbs, 100 FPM, 5 stop, manual doors	CV@178.	Ea	20,600.00	6,660.00	27,260.00

Elevators, Passenger Typical subcontract costs. These costs do not include allowance for the shaftwall.

Hydraulic, office type

100 FPM, 10 to 13 passenger, automatic exit door, 7' x 5' cab, 2,500 lb capacity

	Craft@Hrs	Unit	Material	Labor	Total
3 stop	—	LS	—	—	50,400.00
4 stop	—	LS	—	—	54,600.00
Each additional stop over 4	—	Ea	—	—	4,880.00

150 FPM, 10 to 13 passenger, center opening automatic door, 7' x 5' cab, 2,500 lb capacity

	Craft@Hrs	Unit	Material	Labor	Total
3 stop	—	LS	—	—	53,600.00
4 stop	—	LS	—	—	56,800.00
Each additional stop over 4	—	Ea	—	—	6,970.00

Geared, 350 FPM, 3,500 lb automatic, selective collective, 5' x 8' cab

	Craft@Hrs	Unit	Material	Labor	Total
5 stop	—	LS	—	—	91,600.00
6 stop	—	LS	—	—	96,800.00
10 stop	—	LS	—	—	117,000.00
15 stop	—	LS	—	—	157,000.00
Each additional stop	—	Ea	—	—	5,960.00
Add for 5,000 lb capacity	—	LS	—	—	3,160.00

	Craft@Hrs	Unit	Material	Labor	Total
Gearless, 500 FPM, 3,500 lb, 6' x 9' cab					
10 stop	—	LS	—	—	185,000.00
15 stop	—	LS	—	—	217,000.00
Add for each additional stop	—	Ea	—	—	5,250.00
Gearless, 700 FPM, 4,500 lb, 6' x 9' cab, passenger					
10 stop	—	LS	—	—	202,000.00
15 stop	—	LS	—	—	227,000.00
20 stop	—	LS	—	—	253,000.00
Add for each additional stop	—	Ea	—	—	6,290.00
Gearless, 1,000 FPM, 4,500 lb, 6' x 9' cab, passenger					
10 stop	—	LS	—	—	215,000.00
Add for each additional stop	—	Ea	—	—	8,260.00
Sidewalk elevators, 2 stops, 2,500 pounds	—	LS	—	—	38,600.00

Elevators, Freight Typical subcontract costs. These costs do not include allowance for the shaftwall.

	Craft@Hrs	Unit	Material	Labor	Total
Hydraulic, 2,500 pound capacity, 100 FPM, single entry cab					
Manual vertical doors, 2 stops	—	LS	—	—	41,100.00
Hydraulic, 6,000 pound capacity, 100 FPM, single entry cab					
Manual vertical doors, 5 stops	—	LS	—	—	60,400.00
Hydraulic, 6,000 pound capacity, 100 FPM, single entry cab					
Powered vertical doors, 5 stops	—	LS	—	—	71,200.00
Hydraulic, 10,000 pound capacity, 100 FPM, single entry cab					
Manual vertical doors, 5 stops	—	LS	—	—	67,700.00
Powered vertical doors, 5 stops	—	LS	—	—	76,500.00
Geared, 2,500 pound capacity, 100 FPM, single entry cab					
Manual doors, 2 stops	—	LS	—	—	65,500.00

Hoists and Cranes These costs do not include electrical work.

	Craft@Hrs	Unit	Material	Labor	Total
Electric hoists, manual trolley, 20 FPM lift. Add cost of monorail on next page					
1/2 ton, swivel mount, 25' lift	D4@9.34	Ea	3,480.00	345.00	3,825.00
1/2 ton, swivel mount, 45' lift	D4@13.7	Ea	3,690.00	506.00	4,196.00
1/2 ton, swivel mount, 75' lift	D4@18.0	Ea	4,100.00	665.00	4,765.00
1 ton, geared trolley, 20' lift	D4@11.9	Ea	4,150.00	440.00	4,590.00
1 ton, geared trolley, 30' lift	D4@15.8	Ea	4,390.00	584.00	4,974.00
1 ton, geared trolley, 55' lift	D4@19.2	Ea	4,810.00	709.00	5,519.00
2 ton, geared trolley, 15' lift	D4@15.8	Ea	5,080.00	584.00	5,664.00
2 ton, geared trolley, 35' lift	D4@15.8	Ea	5,400.00	584.00	5,984.00
2 ton, geared trolley, 55' lift	D4@15.8	Ea	5,890.00	584.00	6,474.00
2 ton, geared trolley, 70' lift	D4@15.8	Ea	6,180.00	584.00	6,764.00
Electric hoists, power trolley, 20 FPM trolley speed. Add cost of monorail from below					
1/2 ton, 50 FPM lift to 25'	D4@8.94	Ea	4,100.00	330.00	4,430.00
1/2 ton, 100 FPM lift to 25'	D4@15.8	Ea	6,030.00	584.00	6,614.00
1/2 ton, 50 FPM lift to 45'	D4@8.94	Ea	4,330.00	330.00	4,660.00
1/2 ton, 100 FPM lift to 45'	D4@15.8	Ea	6,310.00	584.00	6,894.00
1/2 ton, 50 FPM lift to 75'	D4@11.9	Ea	4,740.00	440.00	5,180.00
1/2 ton, 100 FPM lift to 75'	D4@18.0	Ea	6,640.00	665.00	7,305.00
1 ton, 50 FPM lift to 20'	D4@15.8	Ea	4,180.00	584.00	4,764.00
1 ton, 50 FPM lift to 30'	D4@15.8	Ea	4,430.00	584.00	5,014.00
1 ton, 50 FPM lift to 55'	D4@18.0	Ea	4,840.00	665.00	5,505.00
2 ton, 50 FPM lift to 15'	D4@17.9	Ea	5,030.00	661.00	5,691.00
2 ton, 50 FPM lift to 35'	D4@19.9	Ea	5,530.00	735.00	6,265.00
2 ton, 50 FPM lift to 55'	D4@19.9	Ea	5,930.00	735.00	6,665.00
2 ton, 50 FPM lift to 70'	D4@19.9	Ea	6,240.00	735.00	6,975.00

	Craft@Hrs	Unit	Material	Labor	Total
Monorail for electric hoists, channel type					
100 pounds per LF	D4@.443	LF	8.15	16.40	24.55
200 pounds per LF	D4@.555	LF	13.10	20.50	33.60
300 pounds per LF	D4@.761	LF	19.60	28.10	47.70
Jib cranes, self-supporting, swinging 8' boom, 220 degree rotation					
1,000 pounds	D4@7.90	Ea	1,290.00	292.00	1,582.00
2,000 pounds	D4@11.9	Ea	1,370.00	440.00	1,810.00
3,000 pounds	D4@13.5	Ea	1,560.00	499.00	2,059.00
4,000 pounds	D4@13.5	Ea	1,820.00	499.00	2,319.00
6,000 pounds	D4@15.8	Ea	1,960.00	584.00	2,544.00
10,000 pounds	D4@19.7	Ea	2,640.00	728.00	3,368.00
Jib cranes, wall mounted, swinging 8' boom, 180 degree rotation					
1,000 pounds	D4@8.19	Ea	695.00	303.00	998.00
2,000 pounds	D4@13.5	Ea	745.00	499.00	1,244.00
4,000 pounds	D4@13.5	Ea	1,140.00	499.00	1,639.00
6,000 pounds	D4@15.8	Ea	1,290.00	584.00	1,874.00
10,000 pounds	D4@19.7	Ea	2,450.00	728.00	3,178.00

Material Handling Systems

Conveyors, typical subcontract price. Foundations, support structures or electrical work not included.

	Craft@Hrs	Unit	Material	Labor	Total
Belt type, 24" width	—	LF	—	—	163.00
Mail conveyors, automatic, electronic					
Horizontal	—	LF	—	—	1,400.00
Vertical, per 12' floor	—	Ea	—	—	17,900.00

Chutes, linen or solid waste handling, prefabricated unit including roof vent, 1-1/2 hour "B" rated doors, discharge and sprinkler system, gravity feed. Costs shown are for each floor based on 8' to 10' floor to floor height.

	Craft@Hrs	Unit	Material	Labor	Total
Light duty, aluminum, 20" diameter	D4@5.85	Ea	649.00	216.00	865.00
Standard, 18 gauge steel, 24" diameter	D4@5.85	Ea	727.00	216.00	943.00
Standard, 18 gauge steel, 30" diameter	D4@5.85	Ea	872.00	216.00	1,088.00
Heavy duty, 18 gauge stainless steel					
24" diameter	D4@5.85	Ea	1,080.00	216.00	1,296.00
30" diameter	D4@5.85	Ea	1,150.00	216.00	1,366.00
Manual door with stainless steel rim	D4@2.17	Ea	310.00	80.20	390.20
Disinfecting and sanitizing unit	D4@2.17	Ea	145.00	80.20	225.20
Discharge storage unit					
Aluminum	D4@1.00	Ea	545.00	36.90	581.90
Stainless steel	D4@1.00	Ea	995.00	36.90	1,031.90

Turntables These costs do not include electrical work.

	Craft@Hrs	Unit	Material	Labor	Total
Baggage handling carousels					
Round, 20' diameter	D4@208.	Ea	27,100.00	7,680.00	34,780.00
Round, 25' diameter	D4@271.	Ea	36,500.00	10,000.00	46,500.00

Moving Stairs and Walks Typical subcontract prices.

Escalators, 90 FPM, steel trim, 32" step width, opaque balustrade

	Craft@Hrs	Unit	Material	Labor	Total
13' rise	—	Ea	—	—	96,000.00
15' rise	—	Ea	—	—	100,000.00
17' rise	—	Ea	—	—	106,000.00
19' rise	—	Ea	—	—	110,000.00
21' rise	—	Ea	—	—	127,000.00
Add for 48" width	—	%	—	—	12.0

	Craft@Hrs	Unit	Material	Labor	Total
Pneumatic Pipe Systems Typical subcontract prices.					
3" diameter, two station, single pipe 100' between stations	—	LS	—	—	3,500.00
3" diameter, two station, twin pipe, 100' between stations	—	LS	—	—	4,800.00

Mechanical 15

Black Steel Pipe and Fittings Pipe is welded seam or seamless. Installation heights indicate height above the working floor.

1/2" black steel (A53 or A120) threaded pipe and fittings installed in a building

	Craft@Hrs	Unit	Material	Labor	Total
Pipe, no fittings or supports included, 1/2", threaded ends					
Schedule 40 to 10' high	M5@.060	LF	.49	2.19	2.68
Schedule 40 over 10' to 20'	M5@.070	LF	.49	2.56	3.05
Schedule 80 to 10' high	M5@.070	LF	.65	2.56	3.21
Schedule 80 over 10' to 20'	M5@.080	LF	.65	2.92	3.57
90 degree ells, 1/2", threaded					
150 lb, malleable iron	M5@.120	Ea	.81	4.38	5.19
300 lb, malleable iron	M5@.132	Ea	4.62	4.82	9.44
45 degree ells, 1/2", threaded					
150 lb, malleable iron	M5@.120	Ea	1.30	4.38	5.68
300 lb, malleable iron	M5@.132	Ea	6.66	4.82	11.48
Tees, 1/2", threaded					
150 lb, malleable iron	M5@.180	Ea	1.08	6.57	7.65
300 lb, malleable iron	M5@.198	Ea	6.83	7.23	14.06
150 lb, reducing, malleable iron	M5@.170	Ea	1.79	6.21	8.00
300 lb, reducing, malleable iron	M5@.180	Ea	10.20	6.57	16.77
Caps, 1/2" threaded					
150 lb, malleable iron	M5@.090	Ea	.82	3.29	4.11
300 lb, malleable iron	M5@.100	Ea	3.70	3.65	7.35
Couplings, 1/2" threaded					
150 lb, malleable iron	M5@.120	Ea	1.09	4.38	5.47
300 lb, malleable iron	M5@.132	Ea	4.05	4.82	8.87
Flanges, forged steel, 1/2"					
150 lb, threaded flange	M5@.489	Ea	32.60	17.90	50.50
300 lb, threaded flange	M5@.781	Ea	36.50	28.50	65.00
600 lb, threaded flange	M5@.859	Ea	52.10	31.40	83.50
Unions, 1/2"					
150 lb, malleable iron	M5@.140	Ea	3.54	5.11	8.65
300 lb, malleable iron	M5@.150	Ea	6.72	5.48	12.20
For galvanized pipe, add	—	%	32.0	—	—
For galvanized 150 lb, malleable fittings, add	—	%	15.0	—	—
For galvanized 300 lb, malleable fittings, add	—	%	55.0	—	—
Pipe assembly, sch.40, fittings, hangers and supports	M5@.130	LF	1.27	4.75	6.02
For galvanized pipe assembly, sch 40, add	—	%	25.0	—	—

3/4" black steel (A53 or A120) threaded pipe and fittings installed in a building

	Craft@Hrs	Unit	Material	Labor	Total
Pipe, no fittings or supports included, 3/4", threaded					
Schedule 40 to 10' high	M5@.070	LF	.60	2.56	3.16
Schedule 40 over 10' to 20'	M5@.080	LF	.60	2.92	3.52

	Craft@Hrs	Unit	Material	Labor	Total
Schedule 80 to 10' high	M5@.080	LF	.84	2.92	3.76
Schedule 80 over 10' to 20'	M5@.095	LF	.84	3.47	4.31
90 degree ells, 3/4" threaded					
150 lb, malleable iron	M5@.130	Ea	.98	4.75	5.73
300 lb, malleable iron	M5@.143	Ea	5.34	5.22	10.56
45 degree ells, 3/4" threaded					
150 lb, malleable iron	M5@.130	Ea	1.59	4.75	6.34
300 lb, malleable iron	M5@.143	Ea	7.40	5.22	12.62
Tees, 3/4" threaded					
150 lb, malleable iron	M5@.190	Ea	1.55	6.94	8.49
300 lb, malleable iron	M5@.209	Ea	7.35	7.63	14.98
150 lb, reducing, malleable iron	M5@.180	Ea	2.29	6.57	8.86
300 lb, reducing, malleable iron	M5@.205	Ea	10.40	7.49	17.89
Caps, 3/4" threaded					
150 lb, malleable iron	M5@.100	Ea	1.09	3.65	4.74
300 lb, malleable iron	M5@.110	Ea	4.65	4.02	8.67
Couplings, 3/4" threaded					
150 lb, malleable iron	M5@.130	Ea	1.28	4.75	6.03
300 lb, malleable iron	M5@.143	Ea	4.65	5.22	9.87
Unions, 3/4"					
150 lb, malleable iron	M5@.150	Ea	4.07	5.48	9.55
300 lb, malleable iron	M5@.165	Ea	7.44	6.03	13.47
Flanges, forged steel, 3/4"					
150 lb, threaded flange	M5@.585	Ea	32.60	21.40	54.00
300 lb, threaded flange	M5@.781	Ea	36.50	28.50	65.00
600 lb, threaded flange	M5@.859	Ea	52.10	31.40	83.50
For galvanized pipe, add	—	%	32.0	—	—
Galvanized 150 lb, malleable fittings, add	—	%	25.0	—	—
Galvanized 300 lb, malleable fittings, add	—	%	55.0	—	—
Pipe assembly, sch.40, fittings, hangers and supports	M5@.140	LF	1.45	5.11	6.56
For galvanized pipe assembly, sch 40, add	—	%	30.0	—	—

1" black steel (A53 or A120) threaded pipe and fittings installed in a building

	Craft@Hrs	Unit	Material	Labor	Total
Pipe, no fittings or supports included, 1", threaded					
Schedule 40 to 10' high	M5@.090	LF	.75	3.29	4.04
Schedule 40 over 10' to 20'	M5@.110	LF	.75	4.02	4.77
Schedule 80 to 10' high	M5@.100	LF	1.01	3.65	4.66
Schedule 80 over 10' to 20'	M5@.120	LF	1.01	4.38	5.39
90 degree ells, 1" threaded					
150 lb, malleable iron	M5@.180	Ea	1.69	6.57	8.26
300 lb, malleable iron	M5@.198	Ea	6.83	7.23	14.06
45 degree ells, 1" threaded					
150 lb, malleable iron	M5@.180	Ea	2.03	6.57	8.60
300 lb, malleable iron	M5@.198	Ea	8.20	7.23	15.43
Tees, 1" threaded					
150 lb, malleable iron	M5@.230	Ea	2.63	8.40	11.03
300 lb, malleable iron	M5@.253	Ea	8.84	9.24	18.08
150 lb, reducing, malleable iron	M5@.205	Ea	2.92	7.49	10.41
300 lb, reducing, malleable iron	M5@.220	Ea	12.80	8.03	20.83

	Craft@Hrs	Unit	Material	Labor	Total
Caps, 1" threaded					
150 lb, malleable iron	M5@.140	Ea	1.34	5.11	6.45
300 lb, malleable iron	M5@.155	Ea	6.10	5.66	11.76
Couplings, 1" threaded					
150 lb, malleable iron	M5@.180	Ea	1.95	6.57	8.52
300 lb, malleable iron	M5@.198	Ea	5.31	7.23	12.54
Flanges, forged steel, 1"					
150 lb, threaded flange	M5@.731	Ea	32.60	26.70	59.30
300 lb, threaded flange	M5@1.05	Ea	36.50	38.30	74.80
600 lb, threaded flange	M5@1.15	Ea	52.10	42.00	94.10
Unions, 1"					
150 lb, malleable iron	M5@.210	Ea	5.31	7.67	12.98
300 lb, malleable iron	M5@.230	Ea	9.69	8.40	18.09
For galvanized pipe, add	—	%	32.0	—	—
Galvanized 150 lb, malleable fittings, add	—	%	20.0	—	—
Galvanized 300 lb, malleable fittings, add	—	%	60.0	—	—
Pipe assembly, sch.40, fittings, hangers and supports	M5@.160	LF	2.01	5.84	7.85
For galvanized pipe assembly, sch 40, add	—	%	30.0	—	—

1-1/4" black steel (A53 or A120) threaded pipe and fittings installed in a building

	Craft@Hrs	Unit	Material	Labor	Total
Pipe, no fittings or supports included, 1-1/4", threaded					
Schedule 40 to 10' high	M5@.100	LF	1.01	3.65	4.66
Schedule 40 over 10' to 20'	M5@.120	LF	1.01	4.38	5.39
Schedule 80 to 10' high	M5@.110	LF	1.40	4.02	5.42
Schedule 80 over 10' to 20'	M5@.130	LF	1.40	4.75	6.15
90 degree ells, 1-1/4" threaded					
150 lb, malleable iron	M5@.240	Ea	2.77	8.77	11.54
300 lb, malleable iron	M5@.264	Ea	9.78	9.64	19.42
45 degree ells, 1-1/4" threaded					
150 lb, malleable iron	M5@.240	Ea	3.58	8.77	12.35
300 lb, malleable iron	M5@.264	Ea	13.10	9.64	22.74
Tees, 1-1/4" threaded					
150 lb, malleable iron	M5@.310	Ea	4.32	11.30	15.62
300 lb, malleable iron	M5@.341	Ea	12.70	12.50	25.20
150 lb, reducing, malleable iron	M5@.290	Ea	5.04	10.60	15.64
300 lb, reducing, malleable iron	M5@.310	Ea	17.60	11.30	28.90
Caps, 1-1/4" threaded					
150 lb, malleable iron	M5@.180	Ea	1.73	6.57	8.30
300 lb, malleable iron	M5@.200	Ea	6.76	7.30	14.06
Couplings, 1-1/4" threaded					
150 lb, malleable iron	M5@.240	Ea	2.52	8.77	11.29
300 lb, malleable iron	M5@.264	Ea	7.44	9.64	17.08
Unions, 1-1/4"					
150 lb, malleable iron	M5@.280	Ea	7.61	10.20	17.81
300 lb, malleable iron	M5@.310	Ea	15.50	11.30	26.80
Flanges, forged steel, 1-1/4"					
150 lb, threaded flange	M5@.788	Ea	32.60	28.80	61.40
300 lb, threaded flange	M5@1.04	Ea	36.50	38.00	74.50
600 lb, threaded flange	M5@1.14	Ea	54.80	41.60	96.40

	Craft@Hrs	Unit	Material	Labor	Total
For galvanized pipe, add	—	%	32.0	—	—
For galvanized 150 lb, malleable fittings, add	—	%	20.0	—	—
For galvanized 300 lb, malleable fittings, add	—	%	65.0	—	—
Pipe assembly, sch.40, fittings, hangers and supports	M5@.200	LF	2.46	7.30	9.76
For galvanized pipe assembly, sch 40, add	—	%	30.0	—	—

1-1/2" black steel (A53 or A120) threaded pipe and fittings installed in a building

	Craft@Hrs	Unit	Material	Labor	Total
Pipe, no fittings or supports included, 1-1/2", threaded					
Schedule 40 to 10' high	M5@.110	LF	1.23	4.02	5.25
Schedule 40 over 10' to 20'	M5@.130	LF	1.23	4.75	5.98
Schedule 80 to 10' high	M5@.120	LF	1.74	4.38	6.12
Schedule 80 over 10' to 20'	M5@.140	LF	1.74	5.11	6.85
90 degree ells, 1-1/2" threaded					
150 lb, malleable iron	M5@.300	Ea	3.66	11.00	14.66
300 lb, malleable iron	M5@.333	Ea	11.50	12.20	23.70
45 degree ells, 1-1/2" threaded					
150 lb, malleable iron	M5@.390	Ea	4.41	14.20	18.61
300 lb, malleable iron	M5@.429	Ea	17.10	15.70	32.80
Tees, 1-1/2" threaded					
150 lb, malleable iron	M5@.390	Ea	5.51	14.20	19.71
300 lb, malleable iron	M5@.429	Ea	15.30	15.70	31.00
150 lb, reducing, malleable iron	M5@.360	Ea	6.08	13.10	19.18
300 lb, reducing, malleable iron	M5@.390	Ea	21.20	14.20	35.40
Caps, 1-1/2" threaded					
150 lb, malleable iron	M5@.230	Ea	2.35	8.40	10.75
300 lb, malleable iron	M5@.250	Ea	9.86	9.13	18.99
Couplings, 1-1/2" threaded					
150 lb, malleable iron	M5@.300	Ea	3.34	11.00	14.34
300 lb, malleable iron	M5@.330	Ea	9.44	12.10	21.54
Unions, 1-1/2"					
150 lb, malleable iron	M5@.360	Ea	9.52	13.10	22.62
300 lb, malleable iron	M5@.400	Ea	16.50	14.60	31.10
Flanges, forged steel, 1-1/2"					
150 lb, threaded flange	M5@.840	Ea	32.60	30.70	63.30
300 lb, threaded flange	M5@1.15	Ea	36.50	42.00	78.50
600 lb, threaded flange	M5@1.26	Ea	54.80	46.00	100.80
For 1-1/2" galvanized pipe, add	—	%	32.0	—	—
For 1-1/2" galvanized 150 lb, malleable fittings, add	—	%	20.0	—	—
For 1-1/2" galvanized 300 lb, malleable fittings, add	—	%	55.0	—	—
Pipe assembly, sch.40, fittings, hangers and supports	M5@.250	LF	1.27	9.13	10.40
For galvanized pipe assembly, sch 40, add	—	%	30.0	—	—

2" black steel (A53 or A120) threaded pipe and fittings installed in a building

	Craft@Hrs	Unit	Material	Labor	Total
Pipe, no fittings or supports included, 2", threaded					
Schedule 40 to 10' high	M5@.120	LF	1.66	4.38	6.04
Schedule 40 over 10' to 20'	M5@.140	LF	1.66	5.11	6.77
Schedule 80 to 10' high	M5@.130	LF	2.40	4.75	7.15
Schedule 80 over 10' to 20'	M5@.150	LF	2.40	5.48	7.88

	Craft@Hrs	Unit	Material	Labor	Total
90 degree ells, 2" threaded					
150 lb, malleable iron	M5@.380	Ea	6.23	13.90	20.13
300 lb, malleable iron	M5@.418	Ea	16.80	15.30	32.10
45 degree ells, 2" threaded					
150 lb, malleable iron	M5@.380	Ea	6.61	13.90	20.51
300 lb, malleable iron	M5@.418	Ea	25.80	15.30	41.10
Tees, 2" threaded					
150 lb, malleable iron	M5@.490	Ea	9.27	17.90	27.17
300 lb, malleable iron	M5@.539	Ea	22.60	19.70	42.30
150 lb, reducing, malleable iron	M5@.460	Ea	9.95	16.80	26.75
300 lb, reducing, malleable iron	M5@.500	Ea	30.30	18.30	48.60
Caps, 2" threaded					
150 lb, malleable iron	M5@.290	Ea	3.48	10.60	14.08
300 lb, malleable iron	M5@.320	Ea	13.70	11.70	25.40
Couplings, 2" threaded					
150 lb, malleable iron	M5@.380	Ea	4.86	13.90	18.76
300 lb, malleable iron	M5@.418	Ea	13.60	15.30	28.90
Flanges, forged steel, 2"					
150 lb, threaded flange	M5@.290	Ea	32.60	10.60	43.20
300 lb, threaded flange	M5@.310	Ea	36.50	11.30	47.80
600 lb, threaded flange	M5@.320	Ea	74.80	11.70	86.50
Unions, 2"					
150 lb, malleable iron	M5@.450	Ea	11.00	16.40	27.40
300 lb, malleable iron	M5@.500	Ea	20.50	18.30	38.80
For 2" galvanized pipe, add	—	%	32.0	—	—
For 2" galvanized 150 lb, malleable fittings, add	—	%	20.0	—	—
For 2" galvanized 300 lb, malleable fittings, add	—	%	55.0	—	—
Pipe assembly, sch.40, fittings, hangers and supports	M5@.300	LF	3.72	11.00	14.72
For galvanized pipe assembly, sch 40, add	—	%	30.0	—	—

3" black steel pipe (A53 or A120) and welded fittings installed in a building

	Craft@Hrs	Unit	Material	Labor	Total
Pipe, no fittings or supports included, 3", plain end					
Schedule 40 to 10' high	Ml@.190	LF	3.30	7.34	10.64
Schedule 40 over 10' to 20' high	Ml@.230	LF	3.30	8.88	12.18
Schedule 80 to 10' high	Ml@.280	LF	4.61	10.80	15.41
Schedule 80 over 10' to 20' high	Ml@.335	LF	4.61	12.90	17.51
90 degree ells, 3"					
Schedule 40 carbon steel, butt weld, long turn	Ml@1.33	Ea	9.27	51.30	60.57
Schedule 80 carbon steel, butt weld, long turn	Ml@1.77	Ea	13.90	68.30	82.20
45 degree ells, 3"					
Schedule 40, carbon steel, butt weld	Ml@1.33	Ea	8.11	51.30	59.41
Schedule 80, carbon steel, butt weld	Ml@1.77	Ea	10.70	68.30	79.00
Tees, 3"					
Schedule 40, carbon steel, butt weld	Ml@2.00	Ea	20.60	77.20	97.80
Schedule 80, carbon steel, butt weld	Ml@2.66	Ea	28.40	103.00	131.40
Reducing tee, Schedule 40, carbon steel, butt weld	Ml@1.89	Ea	26.60	73.00	99.60
Caps, 3"					
Schedule 40, carbon steel, butt weld	Ml@.930	Ea	5.51	35.90	41.41
Schedule 80, carbon steel, butt weld	Ml@1.11	Ea	6.66	42.90	49.56

	Craft@Hrs	Unit	Material	Labor	Total
Reducers, Schedule 40, carbon steel, butt weld	Ml@1.20	Ea	9.98	46.30	56.28
Bolt-ups, 3"	Ml@.750	Ea	3.75	29.00	32.75
Weldolets, Schedule 40, carbon steel, 3"	Ml@2.00	Ea	24.50	77.20	101.70
Welded pipe flanges, forged steel, 3" butt weld					
150 lb, slip-on flange	Ml@.730	Ea	15.20	28.20	43.40
300 lb, slip-on flange	Ml@.980	Ea	21.80	37.80	59.60
600 lb, slip-on flange	Ml@1.03	Ea	36.60	39.80	76.40
150 lb, weld neck flange	Ml@.730	Ea	20.00	28.20	48.20
300 lb, weld neck flange	Ml@.980	Ea	25.30	37.80	63.10
600 lb, weld neck flange	Ml@1.03	Ea	35.40	39.80	75.20
150 lb, threaded flange	Ml@.460	Ea	20.60	17.80	38.40
300 lb, threaded flange	Ml@.500	Ea	29.50	19.30	48.80
600 lb, threaded flange	Ml@.510	Ea	56.90	19.70	76.60
Butt welded joints, 3" pipe					
Schedule 40 pipe	Ml@.930	Ea	—	35.90	35.90
Schedule 80 pipe	Ml@1.11	Ea	—	42.90	42.90
Pipe assembly, sch.40, fittings, hangers and supports	Ml@.350	LF	7.54	13.50	21.04
For galvanized pipe assembly, sch 40, add	—	%	30.0	—	—

4" black steel pipe (A53 or A120) and welded fittings installed in a building

	Craft@Hrs	Unit	Material	Labor	Total
Pipe, no fittings or supports included, 4", plain end					
Schedule 40 to 10' high	Ml@.300	LF	4.71	11.60	16.31
Schedule 40 over 10' to 20' high	Ml@.360	LF	4.71	13.90	18.61
Schedule 80 to 10' high	Ml@.340	LF	8.84	13.10	21.94
Schedule 80 over 10' to 20' high	Ml@.400	LF	8.84	15.40	24.24
90 degree ells, 4"					
Schedule 40 carbon steel, butt weld, long turn	Ml@1.33	Ea	15.70	51.30	67.00
Schedule 80 carbon steel, butt weld, long turn	Ml@2.37	Ea	23.20	91.50	114.70
45 degree ells, 4"					
Schedule 40, carbon steel, butt weld	Ml@1.33	Ea	13.60	51.30	64.90
Schedule 80, carbon steel, butt weld	Ml@2.37	Ea	16.20	91.50	107.70
Tees, 4"					
Schedule 40, carbon steel, butt weld	Ml@2.67	Ea	28.60	103.00	131.60
Schedule 80, carbon steel, butt weld	Ml@3.55	Ea	46.10	137.00	183.10
Reducing, Schedule 40 carbon steel, butt weld	Ml@2.44	Ea	33.90	94.20	128.10
Caps, 4"					
Schedule 40, carbon steel, butt weld	Ml@1.25	Ea	7.24	48.30	55.54
Schedule 80, carbon steel, butt weld	Ml@1.50	Ea	9.61	57.90	67.51
Reducers, Schedule 40, carbon steel, butt weld	Ml@1.60	Ea	11.05	61.80	72.85
Bolt-ups, 4"	Ml@1.00	Ea	6.75	38.60	45.35
Weldolets, Schedule 40, carbon steel, 4"	Ml@2.67	Ea	31.20	103.00	134.20
Welded pipe flanges, forged steel, 4"					
150 lb, slip-on flange	Ml@.980	Ea	19.00	37.80	56.80
300 lb, slip-on flange	Ml@1.30	Ea	31.50	50.20	81.70
600 lb, slip-on flange	Ml@1.80	Ea	70.80	69.50	140.30
150 lb, weld neck flange	Ml@.980	Ea	23.80	37.80	61.60
300 lb, weld neck flange	Ml@1.30	Ea	41.70	50.20	91.90
600 lb, weld neck flange	Ml@1.80	Ea	63.40	69.50	132.90
150 lb, threaded flange	Ml@.600	Ea	24.10	23.20	47.30
300 lb, threaded flange	Ml@.640	Ea	41.70	24.70	66.40
600 lb, threaded flange	Ml@.650	Ea	82.50	25.10	107.60

	Craft@Hrs	Unit	Material	Labor	Total
Butt welded joints, 4" pipe					
Schedule 40 pipe	Ml@1.25	Ea	—	48.30	48.30
Schedule 80 pipe	Ml@1.50	Ea	—	57.90	57.90
Pipe assembly, sch.40, fittings, hangers and supports	Ml@.400	LF	10.96	15.44	26.40
For galvanized pipe assembly, sch 40, add	—	%	30.0	—	—

6" black steel pipe (A53 or A106) and welded fittings installed in a building

Pipe, no fittings or supports included, 6", plain end	Craft@Hrs	Unit	Material	Labor	Total
Schedule 40 to 10' high	M8@.420	LF	7.50	16.70	24.20
Schedule 40 over 10' to 20'	M8@.500	LF	7.50	19.80	27.30
Schedule 80 to 10' high	M8@.460	LF	11.30	18.20	29.50
Schedule 80 over 10' to 20'	M8@.550	LF	11.30	21.80	33.10
90 degree ells, 6"					
Schedule 40 carbon steel, butt weld, long turn	M8@2.67	Ea	38.00	106.00	144.00
Schedule 80 carbon steel, butt weld, long turn	M8@3.55	Ea	56.40	141.00	197.40
45 degree ells, 6"					
Schedule 40, carbon steel, butt weld	M8@2.67	Ea	29.00	106.00	135.00
Schedule 80, carbon steel, butt weld	M8@3.55	Ea	43.10	141.00	184.10
Tees, 6"					
Schedule 40, carbon steel, butt weld	M8@4.00	Ea	52.40	159.00	211.40
Schedule 80, carbon steel, butt weld	M8@5.32	Ea	73.30	211.00	284.30
Reducing, Schedule 40 carbon steel, butt weld	M8@3.78	Ea	67.20	150.00	217.20
Caps, 6"					
Schedule 40, carbon steel, butt weld	M8@1.87	Ea	13.30	74.10	87.40
Schedule 80, carbon steel, butt weld	M8@2.24	Ea	19.00	88.80	107.80
Schedule 40, carbon steel, butt weld	M8@2.40	Ea	20.80	95.20	116.00
Bolt-ups, 6"	M8@1.20	Ea	10.50	47.60	58.10
Weldolets, Schedule 40, carbon steel, 6"	Ml@3.65	Ea	86.70	141.00	227.70
Welded pipe flanges, forged steel, 6"					
150 lb, slip-on flange	M8@1.47	Ea	30.90	58.30	89.20
300 lb, slip-on flange	M8@1.95	Ea	51.80	77.30	129.10
150 lb, weld neck flange	M8@1.47	Ea	36.60	58.30	94.90
300 lb, weld neck flange	M8@1.95	Ea	60.70	77.30	138.00
Butt welded joints, 6" pipe					
Schedule 40 pipe	M8@1.87	Ea	—	74.10	74.10
Schedule 80 pipe	M8@2.24	Ea	—	88.80	88.80
Pipe assembly, sch.40, fittings, hangers and supports	Ml@.800	LF	18.84	30.89	49.73
For galvanized pipe assembly, sch 40, add	—	%	30.0	—	—

8" black steel pipe (A53 or A106) and welded fittings installed in a building

Pipe, no fittings or supports included, 8" plain end	Craft@Hrs	Unit	Material	Labor	Total
Schedule 40 to 10' high	M8@.510	LF	11.30	20.20	31.50
Schedule 40 over 10' to 20'	M8@.600	LF	11.30	23.80	35.10
Schedule 80 to 10' high	M8@.560	LF	17.20	22.20	39.40
Schedule 80 over 10' to 20' high	M8@.670	LF	17.20	26.60	43.80
90 degree ells, 8"					
Schedule 40 carbon steel, butt weld, long turn	M8@3.20	Ea	70.60	127.00	197.60
Schedule 80 carbon steel, butt weld, long turn	M8@4.26	Ea	105.00	169.00	274.00
45 degree ells, 8"					
Schedule 40, carbon steel, butt weld	M8@3.20	Ea	50.40	127.00	177.40
Schedule 80, carbon steel, butt weld	M8@4.26	Ea	70.00	169.00	239.00

	Craft@Hrs	Unit	Material	Labor	Total
Tees, 8"					
Schedule 40, carbon steel, butt weld	M8@4.80	Ea	96.90	190.00	286.90
Schedule 80, carbon steel, butt weld	M8@6.38	Ea	149.00	253.00	402.00
Reducing, Schedule 40, carbon steel, butt weld	M8@4.40	Ea	130.00	174.00	304.00
Caps, 8"					
Schedule 40, carbon steel, butt weld	M8@2.24	Ea	20.60	88.80	109.40
Schedule 80, carbon steel, butt weld	M8@2.68	Ea	33.70	106.00	139.70
Reducers, Schedule 40, carbon steel, butt weld	M8@2.88	Ea	30.77	114.00	144.77
Bolt-ups, 8"	M8@1.25	Ea	11.00	49.60	60.60
Weldolets, Schedule 40, carbon steel, 8"	Ml@4.76	Ea	153.00	184.00	337.00
Welded pipe flanges, forged steel, 8"					
150 lb, slip-on flange	M8@1.77	Ea	46.80	70.20	117.00
300 lb, slip-on flange	M8@2.35	Ea	86.70	93.20	179.90
150 lb, weld neck flange	M8@1.77	Ea	64.00	70.20	134.20
300 lb, weld neck flange	M8@2.35	Ea	105.00	93.20	198.20
Butt welded joints, 8" pipe					
Schedule 40 pipe	M8@2.24	Ea	—	88.80	88.80
Schedule 80 pipe	M8@2.68	Ea	—	106.00	106.00

10" black steel pipe (A53 or A106) and welded fittings installed in a building

	Craft@Hrs	Unit	Material	Labor	Total
Pipe, no fittings or supports included, 10", plain end					
Schedule 40 to 20' high	M8@.720	LF	16.50	28.50	45.00
Schedule 80 to 20' high	M8@.830	LF	21.60	32.90	54.50
90 degree ells, 10"					
Schedule 40 carbon steel, butt weld, long turn	M8@4.00	Ea	130.00	159.00	289.00
Schedule 80 carbon steel, butt weld, long turn	M8@5.75	Ea	188.00	228.00	416.00
45 degree ells, 10"					
Schedule 40, carbon steel, butt weld	M8@4.00	Ea	122.00	159.00	281.00
Schedule 80, carbon steel, butt weld	M8@5.75	Ea	113.00	228.00	341.00
Tees, 10"					
Schedule 40, carbon steel, butt weld	M8@6.00	Ea	162.00	238.00	400.00
Schedule 80, carbon steel, butt weld	M8@8.61	Ea	241.00	341.00	582.00
Reducing tees, Schedule 40, carbon steel, butt weld	M8@5.60	Ea	214.00	222.00	436.00
Caps, 10"					
Schedule 40, carbon steel, butt weld	M8@2.80	Ea	36.20	111.00	147.20
Schedule 80, carbon steel, butt weld	M8@3.62	Ea	52.40	144.00	196.40
Reducers, Schedule 40, carbon steel, butt weld	M8@3.60	Ea	46.70	143.00	189.70
Bolt-ups, 10"	M8@1.70	Ea	22.70	67.40	90.10
Weldolets, Schedule 40, carbon steel, 10"	Ml@6.35	Ea	222.00	245.00	467.00
Welded pipe flanges, forged steel, 10"					
150 lb, slip-on flange	M8@2.20	Ea	83.30	87.20	170.50
150 lb, weld neck flange	M8@2.20	Ea	105.00	87.20	192.20
300 lb, weld neck flange	M8@2.90	Ea	207.00	115.00	322.00

12" black steel pipe (A53 or A106) and welded fittings installed in a building

	Craft@Hrs	Unit	Material	Labor	Total
Pipe, no fittings or supports included, 12", plain end					
Schedule 40 to 20' high	M8@.910	LF	23.20	36.10	59.30
Schedule 80 to 20' high	M8@1.05	LF	38.30	41.60	79.90
90 degree ells, 12"					
Schedule 40 carbon steel, butt weld, long turn	M8@4.80	Ea	179.00	190.00	369.00
Schedule 80 carbon steel, butt weld, long turn	M8@7.76	Ea	266.00	308.00	574.00

	Craft@Hrs	Unit	Material	Labor	Total
45 degree ells, 12"					
Schedule 40, carbon steel, butt weld	M8@4.80	Ea	129.00	190.00	319.00
Schedule 80, carbon steel, butt weld	M8@7.76	Ea	170.00	308.00	478.00
Tees, 12"					
Schedule 40, carbon steel, butt weld	M8@7.20	Ea	247.00	285.00	532.00
Schedule 80, carbon steel, butt weld	M8@11.6	Ea	349.00	460.00	809.00
Reducing tee, Schedule 40, carbon steel, butt weld	M8@6.80	Ea	322.00	270.00	592.00
Caps, 12"					
Schedule 40, carbon steel, butt weld	M8@3.36	Ea	44.90	133.00	177.90
Schedule 80, carbon steel, butt weld	Ml@4.88	Ea	75.70	188.00	263.70
Reducers, Schedule 40, carbon steel, butt weld	M8@4.30	Ea	74.10	170.00	244.10
Bolt-ups, 12"	M8@2.20	Ea	24.90	87.20	112.10
Weldolets, Schedule 40, carbon steel, 12"	Ml@8.48	Ea	357.00	327.00	684.00
Welded pipe flanges, forged steel, 12"					
150 lb, slip-on flange	M8@2.64	Ea	123.00	105.00	228.00
300 lb, slip-on flange	M8@3.50	Ea	200.00	139.00	339.00
150 lb, weld neck flange	M8@2.64	Ea	153.00	105.00	258.00
300 lb, weld neck flange	M8@3.50	Ea	265.00	139.00	404.00

Copper Pressure Pipe and Soldered Copper Fittings Note that copper pipe and fitting prices can change very quickly as the price of copper changes.

Type K, L and M copper pipe No fittings or hangers included (See fittings and hangers in the sections that follow). Pipe installed either horizontally or vertically in a building and up to 10' above floor level. Add 25% to the labor cost for heights over 10' above floor level.

Type M hard copper pipe. Type M is acceptable for above ground inside building applications only. Based on commercial quantity purchases. Smaller quantities will cost up to 25% more. Add the cost of soldered wrought copper fittings in the section that follows.

	Craft@Hrs	Unit	Material	Labor	Total
1/2" pipe	P6@.032	LF	.32	1.16	1.48
3/4" pipe	P6@.035	LF	.52	1.26	1.78
1" pipe	P6@.038	LF	.82	1.37	2.19
1-1/4" pipe	P6@.042	LF	1.22	1.52	2.74
1-1/2" pipe	P6@.046	LF	1.72	1.66	3.38
2" pipe	P6@.053	LF	3.20	1.91	5.11
2-1/2" pipe	P6@.060	LF	4.35	2.17	6.52
3" pipe	P6@.066	LF	5.53	2.38	7.91
4" pipe	P6@.080	LF	9.36	2.89	12.25

Type L hard copper pipe. Type L is acceptable for above and below ground, inside of building applications only. Based on commercial quantity puchases. Smaller quantities will cost up to 35% more. Add the cost of soldered wrought copper fittings in the section that follows

	Craft@Hrs	Unit	Material	Labor	Total
1/2" pipe	P6@.032	LF	.50	1.16	1.66
3/4" pipe	P6@.035	LF	.78	1.26	2.04
1" pipe	P6@.038	LF	1.10	1.37	2.47
1-1/4" pipe	P6@.042	LF	1.55	1.52	3.07
1-1/2" pipe	P6@.046	LF	1.98	1.66	3.64
2" pipe	P6@.053	LF	3.45	1.91	5.36
2-1/2" pipe	P6@.060	LF	5.00	2.17	7.17
3" pipe	P6@.066	LF	6.78	2.38	9.16
4" pipe	P6@.080	LF	10.71	2.89	13.60
Add for soft Type L copper pipe	—	%	12.0	—	—

	Craft@Hrs	Unit	Material	Labor	Total

Type K hard copper pipe. Installed in an open trench to 5' deep. No excavation or backfill included. Type K is acceptable for above and below ground, inside and outside of building applications. Do not use soldered jointing methods in underground applications. Use corporation (compression) fittings. Based on commercial quantity purchases. Smaller quantities will cost up to 35% more. Add the cost of fittings and hangers when required.

	Craft@Hrs	Unit	Material	Labor	Total
3/4" pipe	P6@.024	LF	1.28	.87	2.15
1" pipe	P6@.024	LF	1.65	.87	2.52
1-1/4" pipe	P6@.028	LF	2.05	1.01	3.06
1-1/2" pipe	P6@.029	LF	2.65	1.05	3.70
2" pipe	P6@.032	LF	4.20	1.16	5.36
2-1/2" pipe	P6@.060	LF	6.05	2.17	8.22
3" pipe	P6@.066	LF	8.40	2.38	10.78
4" pipe	P6@.080	LF	13.05	2.89	15.94
Add for soft copper Type K pipe	—	%	10.0	—	—

Copper pressure fittings Soldered wrought copper fittings. Solder is 95% tin and 5% antimony (lead free). Installed either horizontally or vertically in a building up to 10' above floor level.

	Craft@Hrs	Unit	Material	Labor	Total
1/2" copper pressure fittings					
1/2" 90 degree ell	P6@.107	Ea	.26	3.86	4.12
1/2" 45 degree ell	P6@.107	Ea	.45	3.86	4.31
1/2" tee	P6@.129	Ea	.56	4.66	5.22
1/2" x 1/2" x 3/8" reducing tee	P6@.121	Ea	3.11	4.37	7.48
1/2" cap	P6@.069	Ea	.20	2.49	2.69
1/2" coupling	P6@.107	Ea	.18	3.86	4.04
1/2" copper by male pipe thread adapter	P6@.075	Ea	.66	2.71	3.37
3/4" copper pressure fittings					
3/4" 90 degree ell	P6@.150	Ea	.57	5.42	5.99
3/4" 45 degree ell	P6@.150	Ea	.76	5.42	6.18
3/4" tee	P6@.181	Ea	1.05	6.54	7.59
3/4" x 3/4" x 1/2" reducing tee	P6@.170	Ea	1.04	6.14	7.18
3/4" cap	P6@.096	Ea	.31	3.47	3.78
3/4" coupling	P6@.150	Ea	.32	5.42	5.74
3/4" copper by male pipe thread adapter	P6@.105	Ea	.72	3.79	4.51
1" copper pressure fittings					
1" 90 degree ell	P6@.193	Ea	1.72	6.97	8.69
1" 45 degree ell	P6@.193	Ea	2.50	6.97	9.47
1" tee	P6@.233	Ea	4.05	8.41	12.46
1" x 1" x 3/4" reducing tee	P6@.219	Ea	4.35	7.91	12.26
1" cap	P6@.124	Ea	1.10	4.48	5.58
1" coupling	P6@.193	Ea	1.05	6.97	8.02
1" copper by male pipe thread adapter	P6@.134	Ea	3.25	4.84	8.09
1-1/4" copper pressure fittings					
1-1/4" 90 degree ell	P6@.236	Ea	2.79	8.52	11.31
1-1/4" 45 degree ell	P6@.236	Ea	3.79	8.52	12.31
1-1/4" tee	P6@.285	Ea	6.47	10.30	16.77
1-1/4" x 1-1/4" x 1" reducing tee	P6@.268	Ea	7.35	9.68	17.03
1-1/4" cap	P6@.151	Ea	1.57	5.45	7.02
1-1/4" coupling	P6@.236	Ea	1.82	8.52	10.34
1-1/4" copper by male pipe thread adapter	P6@.163	Ea	6.06	5.89	11.95
1-1/2" copper pressure fittings					
1-1/2" 90 degree ell	P6@.278	Ea	4.25	10.00	14.25
1-1/2" 45 degree ell	P6@.278	Ea	4.55	10.00	14.55
1-1/2" tee	P6@.337	Ea	9.40	12.20	21.60
1-1/2" x 1-1/2" x 1-1/4" reducing tee	P6@.302	Ea	8.80	10.90	19.70

	Craft@Hrs	Unit	Material	Labor	Total
1-1/2" cap	P6@.178	Ea	2.30	6.43	8.73
1-1/2" coupling	P6@.278	Ea	2.65	10.00	12.65
1-1/2" copper by male pipe thread adapter	P6@.194	Ea	6.27	7.00	13.27
2" copper pressure fittings					
2" 90 degree ell	P6@.371	Ea	7.56	13.40	20.96
2" 45 degree ell	P6@.371	Ea	7.42	13.40	20.82
2" tee	P6@.449	Ea	15.05	16.20	31.25
2" x 2" x 1-1/2" reducing tee	P6@.422	Ea	11.85	15.20	27.05
2" cap	P6@.238	Ea	4.18	8.59	12.77
2" coupling	P6@.371	Ea	4.05	13.40	17.45
2" copper by male pipe thread adapter	P6@.259	Ea	8.95	9.35	18.30
2-1/2" copper pressure fittings					
2-1/2" 90 degree ell	P6@.457	Ea	15.66	16.50	32.16
2-1/2" 45 degree ell	P6@.457	Ea	19.15	16.50	35.65
2-1/2" tee	P6@.552	Ea	30.50	19.90	50.40
2-1/2" x 2 1/2" x 2" reducing tee	P6@.519	Ea	36.00	18.70	54.70
2-1/2" cap	P6@.292	Ea	13.50	10.50	24.00
2-1/2" coupling	P6@.457	Ea	9.35	16.50	25.85
2-1/2" copper by male pipe thread adapter	P6@.319	Ea	28.50	11.50	40.00
3" copper pressure fittings					
3" 90 degree ell	P6@.543	Ea	19.85	19.60	39.45
3" 45 degree ell	P6@.543	Ea	25.00	19.60	44.60
3" tee	P6@.656	Ea	46.75	23.70	70.45
3" x 3" x 2" reducing tee	P6@.585	Ea	41.00	21.10	62.10
3" cap	P6@.347	Ea	16.50	12.50	29.00
3" coupling	P6@.543	Ea	14.30	19.60	33.90
3" copper by male pipe thread adapter	P6@.375	Ea	48.00	13.50	61.50
4" copper pressure fittings					
4" 90 degree ell	P6@.714	Ea	48.00	25.80	73.80
4" 45 degree ell	P6@.714	Ea	52.50	25.80	78.30
4" tee	P6@.863	Ea	101.25	31.20	132.45
4" x 4" x 3" reducing tee	P6@.811	Ea	76.75	29.30	106.05
4" cap	P6@.457	Ea	34.75	16.50	51.25
4" coupling	P6@.714	Ea	33.50	25.80	59.30
4" copper by male pipe thread adapter	P6@.485	Ea	102.50	17.50	120.00

Schedule 40 PVC Pressure Pipe and Socket-weld Fittings Pipe installed in a building vertically or horizontally up to 10' above floor level.

	Craft@Hrs	Unit	Material	Labor	Total
1/2" Schedule 40 PVC pressure pipe and fittings					
1/2" pipe	P6@.020	LF	.17	.72	.89
1/2" 90 degree ell	P6@.100	Ea	.22	3.61	3.83
1/2" tee	P6@.130	Ea	.28	4.69	4.97
3/4" Schedule 40 PVC pressure pipe and fittings					
3/4" pipe	P6@.025	LF	.24	.90	1.14
3/4" 90 degree ell	P6@.115	Ea	.24	4.15	4.39
3/4" tee	P6@.140	Ea	.31	5.06	5.37
1" Schedule 40 PVC pressure pipe and fittings					
1" pipe	P6@.030	LF	.32	1.08	1.40
1" 90 degree ell	P6@.120	Ea	.44	4.33	4.77
1" tee	P6@.170	Ea	.58	6.14	6.72

	Craft@Hrs	Unit	Material	Labor	Total
1-1/4" Schedule 40 PVC pressure pipe and fittings					
1-1/4" pipe	P6@.035	LF	.39	1.26	1.65
1-1/4" 90 degree ell	P6@.160	Ea	.77	5.78	6.55
1-1/4" tee	P6@.220	Ea	.91	7.94	8.85
1-1/2" Schedule 40 PVC pressure pipe and fittings					
1-1/2" pipe	P6@.040	LF	.46	1.44	1.90
1-1/2" 90 degree ell	P6@.180	Ea	.83	6.50	7.33
1-1/2" tee	P6@.250	Ea	1.01	9.03	10.04
2" Schedule 40 PVC pressure pipe and fittings					
2" pipe	P6@.046	LF	.62	1.66	2.28
2" 90 degree ell	P6@.200	Ea	1.30	7.22	8.52
2" tee	P6@.280	Ea	1.30	10.10	11.40
2-1/2" Schedule 40 PVC pressure pipe and fittings					
2-1/2" pipe	P6@.053	LF	.98	1.91	2.89
2-1/2" 90 degree ell	P6@.250	Ea	3.96	9.03	12.99
2-1/2" tee	P6@.350	Ea	5.30	12.60	17.90
3" Schedule 40 PVC pressure pipe and fittings					
3" pipe	P6@.055	LF	1.24	1.99	3.23
3" 90 degree ell	P6@.300	Ea	4.73	10.80	15.53
3" tee	P6@.420	Ea	6.89	15.20	22.09
4" Schedule 40 PVC pressure pipe and fittings					
4" pipe	P6@.070	LF	1.80	2.53	4.33
4" 90 degree ell	P6@.500	Ea	8.48	18.10	26.58
4" tee	P6@.560	Ea	12.60	20.20	32.80

Schedule 80 CPVC Chlorinated Pressure Pipe and socket weld type fittings installed in a building vertically or horizontally up to 10' above floor level.

	Craft@Hrs	Unit	Material	Labor	Total
1/2" Schedule 80 CPVC chlorinated pressure pipe and fittings					
1/2" pipe	P6@.021	LF	.21	.76	.97
1/2" 90 degree ell	P6@.105	Ea	.60	3.79	4.39
1/2" tee	P6@.135	Ea	1.68	4.87	6.55
3/4" Schedule 80 CPVC chlorinated pressure pipe and fittings					
3/4" pipe	P6@.026	LF	.29	.94	1.23
3/4" 90 degree ell	P6@.120	Ea	.77	4.33	5.10
3/4" tee	P6@.145	Ea	1.76	5.24	7.00
1" Schedule 80 CPVC chlorinated pipe and fittings					
1" pipe	P6@.032	LF	.42	1.16	1.58
1" 90 degree ell	P6@.125	Ea	1.23	4.51	5.74
1" tee	P6@.180	Ea	2.20	6.50	8.70
1-1/4" Schedule 80 CPVC chlorinated pipe and fittings					
1-1/4" pipe	P6@.037	LF	.50	1.34	1.84
1-1/4" 90 degree ell	P6@.170	Ea	1.64	6.14	7.78
1-1/4" tee	P6@.230	Ea	6.05	8.30	14.35
1-1/2" Schedule 80 CPVC chlorinated pipe and fittings					
1-1/2" pipe	P6@.042	LF	.59	1.52	2.11
1-1/2" 90 degree ell	P6@.190	Ea	1.76	6.86	8.62
1-1/2" tee	P6@.265	Ea	6.05	9.57	15.62
2" Schedule 80 CPVC chlorinated pipe and fittings					
2" pipe	P6@.048	LF	.79	1.73	2.52
2" 90 degree ell	P6@.210	Ea	2.13	7.58	9.71
2" tee	P6@.295	Ea	7.57	10.70	18.27

	Craft@Hrs	Unit	Material	Labor	Total

Copper drain, waste and vent pipe (DWV)
Includes a clevis hanger every 6' and fitting installed every 10'. Use these figures for preliminary estimates.

	Craft@Hrs	Unit	Material	Labor	Total
1-1/4" pipe assembly, horizontal	P6@.110	LF	3.28	3.97	7.25
1-1/2" pipe assembly, horizontal	P6@.110	LF	3.89	3.97	7.86
2" pipe assembly, horizontal	P6@.120	LF	4.80	4.33	9.13
3" pipe assembly, horizontal	P6@.170	LF	10.20	6.14	16.34
4" pipe assembly, horizontal	P6@.190	LF	20.30	6.86	27.16

PVC SDR 35 DWV drainage pipe, Solvent weld, installed in an open trench inside a building, (main building drain) including typical fittings (One wye and 45 degree elbow every 20') SDR 35 is thin wall pipe suitable for below grade applications only. Add for excavation and backfill. Use these figures for preliminary estimates.

	Craft@Hrs	Unit	Material	Labor	Total
3" pipe assembly below grade	P6@.160	LF	2.55	5.78	8.33
4" pipe assembly below grade	P6@.175	LF	3.85	6.32	10.17
6" pipe assembly below grade	P6@.220	LF	8.05	7.94	15.99
Add for interior excavation and backfill (shallow)	—	LF	7.00	—	7.00

ABS "DWV" Drain waste and vent pipe, Solvent weld, installed in a building including typical fittings, clevis hangers and supports. (One fitting each 10' and typical supports.) Use these figures for preliminary estimates.

	Craft@Hrs	Unit	Material	Labor	Total
1-1/2" pipe assembly, horizontal	P6@.200	LF	1.05	7.22	8.27
2" pipe assembly, horizontal	P6@.220	LF	1.65	7.94	9.59
3" pipe assembly,	P6@.270	LF	2.71	9.75	12.46
4" pipe assembly, horizontal	P6@.310	LF	4.05	11.20	15.25
6" pipe assembly, horizontal	P6@.370	LF	8.65	13.40	22.05
Deduct for below grade applications	—	%	-15.0	—	—
Add for interior excavation and backfill (shallow)	—	LF	7.00	—	7.00

Stainless Steel Pipe

1/2" stainless steel Schedule 40 pipe and screwed 150 lb fittings installed in a building

Pipe, screwed, no fittings or supports included, 1/2"

	Craft@Hrs	Unit	Material	Labor	Total
Type 304L pipe installed to 10' high	M5@.094	LF	3.63	3.43	7.06
Type 304L pipe over 10' to 20' high	M5@.105	LF	3.63	3.83	7.46
Type 316L pipe installed to 10' high	M5@.094	LF	4.80	3.43	8.23
Type 316L pipe installed over 10' to 20' high	M5@.105	LF	4.80	3.83	8.63

90 degree ells, screwed, long radius, 1/2"

	Craft@Hrs	Unit	Material	Labor	Total
Type 304	M5@.703	Ea	5.85	25.70	31.55
Type 316	M5@.703	Ea	7.60	25.70	33.30

45 degree ells, screwed, long radius, 1/2"

	Craft@Hrs	Unit	Material	Labor	Total
Type 304	M5@.703	Ea	7.22	25.70	32.92
Type 316	M5@.703	Ea	9.05	25.70	34.75

Tees, screwed, 1/2"

	Craft@Hrs	Unit	Material	Labor	Total
Type 304	M5@1.05	Ea	7.93	38.30	46.23
Type 316	M5@1.05	Ea	9.85	38.30	48.15

Caps, screwed, 1/2"

	Craft@Hrs	Unit	Material	Labor	Total
Type 304	M5@.354	Ea	3.89	12.90	16.79
Type 316	M5@.354	Ea	4.95	12.90	17.85

3/4" stainless steel Schedule 40 pipe and screwed 150 lb fittings installed in a building

Pipe, screwed, no fittings or supports included, 3/4"

	Craft@Hrs	Unit	Material	Labor	Total
Type 304L pipe to 10' high	M5@.096	LF	4.48	3.51	7.99
Type 304L pipe over 10' to 20' high	M5@.105	LF	4.48	3.83	8.31
Type 316L pipe to 10' high	M5@.094	LF	5.42	3.43	8.85
Type 316L pipe over 10' to 20' high	M5@.105	LF	5.42	3.83	9.25

90 degree ells, screwed, long radius, 3/4"

	Craft@Hrs	Unit	Material	Labor	Total
Type 304	M5@.703	Ea	8.39	25.70	34.09
Type 316	M5@.703	Ea	11.00	25.70	36.70

	Craft@Hrs	Unit	Material	Labor	Total
45 degree ells, screwed, long radius, 3/4"					
Type 304	M5@.703	Ea	10.50	25.70	36.20
Type 316	M5@.703	Ea	11.90	25.70	37.60
Tees, screwed, 3/4"					
Type 304	M5@1.05	Ea	12.60	38.30	50.90
Type 316	M5@1.05	Ea	14.50	38.30	52.80
Caps, screwed, 3/4"					
Type 304	M5@.354	Ea	4.87	12.90	17.77
Type 316	M5@.354	Ea	5.69	12.90	18.59

1" stainless steel Schedule 40 pipe and screwed 150 lb fittings installed in a building

	Craft@Hrs	Unit	Material	Labor	Total
Pipe, screwed, no fittings or supports included, 1"					
Type 304L pipe to 10' high	M5@.103	LF	5.65	3.76	9.41
Type 304L pipe over 10' to 20' high	M5@.123	LF	5.65	4.49	10.14
Type 316L pipe to 10' high	M5@.103	LF	7.21	3.76	10.97
Type 316L pipe over 10' to 20' high	M5@.123	LF	7.21	4.49	11.70
90 degree ells, screwed, long radius, 1"					
Type 304	M5@.846	Ea	11.90	30.90	42.80
Type 316	M5@.846	Ea	13.80	30.90	44.70
45 degree ells, screwed, long radius, 1"					
Type 304	M5@.846	Ea	12.90	30.90	43.80
Type 316	M5@.846	Ea	14.90	30.90	45.80
Tees, screwed, 1"					
Type 304	M5@1.26	Ea	16.10	46.00	62.10
Type 316	M5@1.26	Ea	18.20	46.00	64.20
Caps, screwed, 1"					
Type 304	M5@.422	Ea	7.32	15.40	22.72
Type 316	M5@.422	Ea	8.72	15.40	24.12

1-1/4" stainless steel Schedule 40 pipe and screwed 150 lb fittings installed in a building

	Craft@Hrs	Unit	Material	Labor	Total
Pipe, screwed, no fittings or supports included, 1-1/4"					
Type 304L pipe to 10', high	M5@.111	LF	7.10	4.05	11.15
Type 304L pipe over 10' to 20' high	M5@.131	LF	7.10	4.78	11.88
Type 316L pipe to 10' high	M5@.111	LF	9.10	4.05	13.15
Type 316L pipe over 10' to 20' high	M5@.131	LF	9.10	4.78	13.88
90 degree ells, screwed, long radius, 1-1/4"					
Type 304	M5@.947	Ea	20.90	34.60	55.50
Type 316	M5@.947	Ea	23.70	34.60	58.30
45 degree ells, screwed, long radius, 1-1/4"					
Type 304	M5@.947	Ea	20.60	34.60	55.20
Type 316	M5@.880	Ea	23.40	32.10	55.50
Tees, screwed, 1-1/4"					
Type 304	M5@1.41	Ea	32.00	51.50	83.50
Type 316	M5@1.41	Ea	36.30	51.50	87.80
Caps, screwed, 1-1/4"					
Type 304	M5@.466	Ea	13.30	17.00	30.30
Type 316	M5@.466	Ea	17.10	17.00	34.10

1-1/2" stainless steel Schedule 40 pipe and screwed, 150 lb fittings installed in a building

	Craft@Hrs	Unit	Material	Labor	Total
Pipe, screwed, no fittings or supports included, 1-1/2"					
Type 304L pipe to 10' high	M5@.116	LF	8.33	4.24	12.57
Type 304L pipe over 10' to 20' high	M5@.140	LF	8.33	5.11	13.44
Type 316L pipe to 10' high	M5@.116	LF	10.80	4.24	15.04
Type 316L pipe over 10' to 20' high	M5@.140	LF	10.80	5.11	15.91

	Craft@Hrs	Unit	Material	Labor	Total
90 degree ells, screwed, long radius, 1-1/2"					
Type 304	M5@1.02	Ea	27.80	37.30	65.10
Type 316	M5@1.02	Ea	31.60	37.30	68.90
45 degree ells, screwed, long radius, 1-1/2"					
Type 304	M5@1.02	Ea	24.50	37.30	61.80
Type 316	M5@1.02	Ea	27.90	37.30	65.20
Tees, screwed, 1-1/2"					
Type 304	M5@1.54	Ea	39.70	56.20	95.90
Type 316	M5@1.54	Ea	44.00	56.20	100.20
Caps, screwed, 1-1/2"					
Type 304	M5@.515	Ea	17.40	18.80	36.20
Type 316	M5@.515	Ea	22.00	18.80	40.80

2" stainless steel Schedule 40 pipe and screwed, 150 lb fittings installed in a building

	Craft@Hrs	Unit	Material	Labor	Total
Pipe, screwed, no fittings or supports included, 2"					
Type 304L pipe to 10' high	Ml@.118	LF	10.80	4.56	15.36
Type 304L pipe over 10' to 20' high	Ml@.141	LF	10.80	5.44	16.24
Type 316L pipe to 10' high	Ml@.118	LF	14.20	4.56	18.76
Type 316L pipe over 10' to 20' high	Ml@.141	LF	14.20	5.44	19.64
90 degree ells, screwed, long radius, 2"					
Type 304	Ml@1.25	Ea	35.20	48.30	83.50
Type 316	Ml@1.25	Ea	40.00	48.30	88.30
45 degree ells, screwed, long radius, 2"					
Type 304	Ml@1.25	Ea	34.50	48.30	82.80
Type 316	Ml@1.25	Ea	38.90	48.30	87.20
Tees, screwed, 2"					
Type 304	Ml@1.91	Ea	51.00	73.70	124.70
Type 316	Ml@1.91	Ea	56.70	73.70	130.40
Caps, screwed, 2"					
Type 304	Ml@.625	Ea	22.70	24.10	46.80
Type 316	Ml@.625	Ea	28.10	24.10	52.20

Hangers and Supports

	Craft@Hrs	Unit	Material	Labor	Total
Angle bracket hangers, steel, by rod size					
3/8" angle bracket hanger	M5@.070	Ea	1.41	2.56	3.97
1/2" angle bracket hanger	M5@.070	Ea	1.58	2.56	4.14
5/8" angle bracket hanger	M5@.070	Ea	3.48	2.56	6.04
3/4" angle bracket hanger	M5@.070	Ea	5.35	2.56	7.91
7/8" angle bracket hanger	M5@.070	Ea	9.66	2.56	12.22
Angle supports, wall mounted, welded steel					
12" x 18", medium weight	M5@.950	Ea	65.00	34.70	99.70
18" x 24", medium weight	M5@1.10	Ea	81.30	40.20	121.50
24" x 30", medium weight	M5@1.60	Ea	103.00	58.40	161.40
12" x 30", heavy weight	M5@1.10	Ea	92.10	40.20	132.30
18" x 24", heavy weight	M5@1.10	Ea	130.00	40.20	170.20
24" x 30", heavy weight	M5@1.60	Ea	146.00	58.40	204.40
C-clamps with lock nut, steel, by rod size					
3/8" C-clamp	M5@.070	Ea	1.56	2.56	4.12
1/2" C-clamp	M5@.070	Ea	1.89	2.56	4.45
5/8" C-clamp	M5@.070	Ea	3.83	2.56	6.39
3/4" C-clamp	M5@.070	Ea	5.54	2.56	8.10
7/8" C-clamp	M5@.070	Ea	14.90	2.56	17.46

	Craft@Hrs	Unit	Material	Labor	Total
Top beam clamps, steel, by rod size					
3/8" top beam clamp	M5@.070	Ea	1.83	2.56	4.39
1/2" top beam clamp	M5@.070	Ea	2.98	2.56	5.54
5/8" top beam clamp	M5@.070	Ea	3.65	2.56	6.21
3/4" top beam clamp	M5@.070	Ea	5.34	2.56	7.90
7/8" top beam clamp	M5@.070	Ea	7.00	2.56	9.56
Clevis hangers, standard duty, by intended pipe size, no threaded rod or beam clamp included					
1/2" clevis hanger	M5@.156	Ea	1.44	5.70	7.14
3/4" clevis hanger	M5@.156	Ea	1.44	5.70	7.14
1" clevis hanger	M5@.156	Ea	1.53	5.70	7.23
1-1/4" clevis hanger	M5@.173	Ea	1.74	6.32	8.06
1-1/2" clevis hanger	M5@.173	Ea	1.91	6.32	8.23
2" clevis hanger	M5@.173	Ea	2.13	6.32	8.45
2-1/2" clevis hanger	M5@.189	Ea	2.38	6.90	9.28
3" clevis hanger	M5@.189	Ea	2.98	6.90	9.88
4" clevis hanger	M5@.189	Ea	3.68	6.90	10.58
6" clevis hanger	M5@.202	Ea	6.18	7.38	13.56
8" clevis hanger	M5@.202	Ea	9.10	7.38	16.48
Add for epoxy coated	—	%	25.0	—	—
Add for copper plated	—	%	35.0	—	—
Deduct for light duty	—	%	-20.0	—	—
Swivel ring hangers, light duty, by intended pipe size, band hanger, galvanized steel, no hanger rod, beam clamp or insert included					
1/2" swivel ring hanger	M5@.125	Ea	.35	4.57	4.92
3/4" swivel ring hanger	M5@.125	Ea	.35	4.57	4.92
1" swivel ring hanger	M5@.125	Ea	.35	4.57	4.92
1-1/4" swivel ring hanger	M5@.125	Ea	.42	4.57	4.99
1-1/2" swivel ring hanger	M5@.125	Ea	.48	4.57	5.05
2" swivel ring hanger	M5@.125	Ea	.59	4.57	5.16
2-1/2" swivel ring hanger	M5@.135	Ea	.78	4.93	5.71
3" swivel ring hanger	M5@.135	Ea	.95	4.93	5.88
4" swivel ring hanger	M5@.135	Ea	1.30	4.93	6.23
6" swivel ring hanger	M5@.145	Ea	2.45	5.30	7.75
8" swivel ring hanger	M5@.145	Ea	2.95	5.30	8.25
Add for epoxy coating	—	%	45.0	—	—
Add for copper plated	—	%	55.0	—	—
Riser clamps, black steel by intended pipe size					
1/2" riser clamp	M5@.100	Ea	1.25	3.65	4.90
3/4" riser clamp	M5@.100	Ea	1.85	3.65	5.50
1" riser clamp	M5@.100	Ea	1.90	3.65	5.55
1-1/4" riser clamp	M5@.115	Ea	2.25	4.20	6.45
1-1/2" riser clamp	M5@.115	Ea	2.40	4.20	6.60
2" riser clamp	M5@.125	Ea	2.50	4.57	7.07
2-1/2" riser clamp	M5@.135	Ea	2.65	4.93	7.58
3" riser clamp	M5@.140	Ea	2.90	5.11	8.01
4" riser clamp	M5@.145	Ea	3.65	5.30	8.95
6" riser clamp	M5@.155	Ea	6.35	5.66	12.01
8" riser clamp	M5@.165	Ea	10.30	6.03	16.33
Add for epoxy coating	—	%	25.0	—	—
Add for copper plated	—	%	35.0	—	—

	Craft@Hrs	Unit	Material	Labor	Total
Pipe clamps, medium duty, by intended pipe size					
3" pipe clamp	M5@.189	Ea	3.60	6.90	10.50
4" pipe clamp	M5@.189	Ea	5.10	6.90	12.00
6" pipe clamp	M5@.202	Ea	12.45	7.38	19.83
Pipe rolls					
3" pipe roll, complete	M5@.403	Ea	11.60	14.70	26.30
4" pipe roll, complete	M5@.518	Ea	12.10	18.90	31.00
6" pipe roll, complete	M5@.518	Ea	12.10	18.90	31.00
8" pipe roll, complete	M5@.578	Ea	22.50	21.10	43.60
3" pipe roll stand, complete	M5@.323	Ea	34.10	11.80	45.90
4" pipe roll stand, complete	M5@.604	Ea	46.40	22.10	68.50
6" pipe roll stand, complete	M5@.775	Ea	46.40	28.30	74.70
8" pipe roll stand, complete	M5@.775	Ea	81.00	28.30	109.30
Pipe straps, galvanized, two hole, lightweight					
2-1/2" pipe strap	M5@.119	Ea	1.48	4.35	5.83
3" pipe strap	M5@.128	Ea	1.91	4.67	6.58
3-1/2" pipe strap	M5@.135	Ea	2.42	4.93	7.35
4" pipe strap	M5@.144	Ea	2.70	5.26	7.96
Plumber's tape, galvanized, 3/4" x 10' roll					
26 gauge	—	LF	1.30	—	1.30
22 gauge	—	LF	1.75	—	1.75
Sliding pipe guides, welded					
3" pipe guide	M5@.354	Ea	13.30	12.90	26.20
4" pipe guide	M5@.354	Ea	17.90	12.90	30.80
6" pipe guide	M5@.354	Ea	26.00	12.90	38.90
8" pipe guide	M5@.471	Ea	38.90	17.20	56.10
Threaded hanger rod, by rod size, in 6' to 10' lengths, per LF					
3/8" rod	—	LF	.31	—	.31
1/2" rod	—	LF	.49	—	.49
5/8" rod	—	LF	.69	—	.69
3/4" rod	—	LF	1.21	—	1.21
U bolts, with hex nuts, light duty					
4" U-bolt	M5@.089	Ea	2.48	3.25	5.73
6" U-bolt	M5@.135	Ea	4.57	4.93	9.50
Wall sleeves, cast iron					
4" sleeve	M5@.241	Ea	5.75	8.80	14.55
6" sleeve	M5@.284	Ea	7.15	10.40	17.55
8" sleeve	M5@.377	Ea	9.45	13.80	23.25
Gate Valves bronze trim, non-rising stem					
2-1/2" 125 lb, cast iron, flanged	Ml@.600	Ea	137.00	23.20	160.20
2-1/2" 250 lb, cast iron, flanged	Ml@.600	Ea	269.00	23.20	292.20
3" 125 lb, cast iron, flanged	Ml@.750	Ea	148.00	29.00	177.00
4" 125 lb, cast iron, flanged	Ml@1.35	Ea	218.00	52.10	270.10
4" 250 lb, cast iron, flanged	Ml@1.35	Ea	560.00	52.10	612.10
Steel Gate, Globe and Check Valves					
Cast steel Class 150 gate valves, OS&Y, flanged ends					
2-1/2" valve	Ml@.600	Ea	556.00	23.20	579.20
3" valve	Ml@.750	Ea	556.00	29.00	585.00
4" valve	Ml@1.35	Ea	677.00	52.10	729.10

Mechanical 15

	Craft@Hrs	Unit	Material	Labor	Total
Cast steel Class 300 gate valves, OS&Y, flanged ends					
2-1/2" valve	Ml@.600	Ea	725.00	23.20	748.20
3" valve	Ml@.750	Ea	725.00	29.00	754.00
4" valve	Ml@1.35	Ea	1009.00	52.10	1,061.10

Outside Stem and Yoke (OS&Y) All Iron Gate Valves

	Craft@Hrs	Unit	Material	Labor	Total
2-1/2" flanged iron OS&Y gate valves					
125 lb	Ml@.600	Ea	214.00	23.20	237.20
250 lb	Ml@.600	Ea	480.00	23.20	503.20
3" flanged iron OS&Y gate valves					
125 lb	Ml@.750	Ea	226.00	29.00	255.00
250 lb	Ml@.750	Ea	546.00	29.00	575.00
4" flanged iron OS&Y gate valves					
125 lb	Ml@1.35	Ea	314.00	52.10	366.10
250 lb	Ml@1.35	Ea	790.00	52.10	842.10
6" flanged iron OS&Y gate valves					
125 lb	Ml@2.50	Ea	530.00	96.50	626.50
250 lb	Ml@2.50	Ea	1,340.00	96.50	1,436.50

Butterfly Valves, Ductile Iron
Wafer body, flanged, with lever lock, nickel plated ductile iron disk and Buna-N seals.

	Craft@Hrs	Unit	Material	Labor	Total
2-1/2" butterfly valves					
200 lb	Ml@.400	Ea	87.00	15.40	102.40
3" butterfly valves					
200 lb	Ml@.450	Ea	92.00	17.40	109.40
4" butterfly valves					
200 lb	Ml@.550	Ea	115.00	21.20	136.20
6" butterfly valves					
200 lb	Ml@.750	Ea	185.00	29.00	214.00
8" butterfly valves, with gear operator					
200 lb	Ml@.950	Ea	345.00	36.70	381.70
10" butterfly valves, with gear operator					
200 lb	Ml@1.40	Ea	445.00	54.00	499.00

Globe Valves
Flanged, iron body, OS&Y, (outside stem & yoke) 125 psi steam pressure and 200 psi cold water pressure rated.

	Craft@Hrs	Unit	Material	Labor	Total
3" flanged O.S&Y, globe valve	Ml@.750	Ea	330.00	29.00	359.00
4" flanged O.S&Y, globe valve	Ml@1.35	Ea	460.00	52.10	512.10
6" flanged O.S&Y, globe valve	Ml@2.50	Ea	854.00	96.50	950.50
8" flanged O.S&Y, globe valve	Ml@3.00	Ea	1,500.00	116.00	1,616.00

Swing Check Valves
Flanged, Iron body, bolted cap, 125 psi steam/200 psi cold water pressure rated or 250 psi steam/500 psi cold water pressure rated

	Craft@Hrs	Unit	Material	Labor	Total
2-1/2" flanged iron body check valve					
Swing check, bronze trim, 125 psi steam	Ml@.600	Ea	115.00	23.20	138.20
Swing check, iron trim, 125 psi steam	Ml@.600	Ea	135.00	23.20	158.20
Swing check, bronze trim, 250 psi steam	Ml@.750	Ea	350.00	29.00	379.00
3" flanged iron body check valve					
Swing check, bronze trim, 125 psi steam	Ml@.750	Ea	140.00	29.00	169.00
Swing check, iron trim, 125 psi steam	Ml@.750	Ea	185.00	29.00	214.00
Swing check, bronze trim, 250 psi steam	Ml@1.25	Ea	435.00	48.30	483.30

	Craft@Hrs	Unit	Material	Labor	Total
4" flanged iron body check valve					
Swing check, bronze trim, 125 psi steam	Ml@1.35	Ea	225.00	52.10	277.10
Swing check, iron trim, 125 psi steam	Ml@1.35	Ea	305.00	52.10	357.10
Swing check, bronze trim, 250 psi steam	Ml@1.85	Ea	555.00	71.40	626.40
6" flanged iron body check valve					
Swing check, bronze trim, 125 psi steam	Ml@2.50	Ea	365.00	96.50	461.50
Swing check, iron trim, 125 psi steam	Ml@2.50	Ea	485.00	96.50	581.50
Swing check, bronze trim, 250 psi steam	Ml@3.00	Ea	1,025.00	116.00	1,141.00
8" flanged iron body check valve					
Swing check, bronze trim, 125 psi steam	Ml@3.00	Ea	675.00	116.00	791.00
Swing check, iron trim, 125 psi steam	Ml@3.00	Ea	900.00	116.00	1,016.00
10" flanged iron body check valve					
Swing check, bronze trim, 125 psi steam	Ml@4.00	Ea	1,375.00	154.00	1,529.00
Swing check, iron trim, 125 psi steam	Ml@4.00	Ea	1,830.00	154.00	1,984.00
12" flanged iron body check valve					
Swing check, bronze trim, 125 psi steam	Ml@4.50	Ea	1,805.00	174.00	1,979.00
Swing check, iron trim, 125 psi steam	Ml@4.50	Ea	2,390.00	174.00	2,564.00

Miscellaneous Valves and Regulators

	Craft@Hrs	Unit	Material	Labor	Total
Backflow preventers reduced pressure					
2-1/2", with 2 gate valves, flanged, iron body	Ml@2.00	Ea	1,400.00	77.20	1,477.20
3", with 2 gate valves, flanged, iron body	Ml@3.00	Ea	1,675.00	116.00	1,791.00
4", with 2 gate valves, flanged, iron body	Ml@4.50	Ea	2,100.00	174.00	2,274.00
6", without gate valves, flanged, iron body	Ml@8.00	Ea	3,490.00	309.00	3,799.00
6", with 2 gate valves, flanged, iron body	Ml@9.00	Ea	4,000.00	347.00	4,347.00
8", without gate valves, flanged, iron body	Ml@11.5	Ea	6,050.00	444.00	6,494.00
8", with 2 gate valves, flanged, iron body	Ml@12.5	Ea	8,780.00	483.00	9,263.00
Ball valves, bronze, 150 lb steam working pressure, threaded, Teflon seat					
2-1/2" valve	Ml@.750	Ea	134.00	29.00	163.00
3" valve	Ml@.950	Ea	155.00	36.70	191.70
Hose gate valves, with cap, 200 lb					
2-1/2", Non-rising stem	P6@1.00	Ea	58.40	36.10	94.50
Pressure regulator valves, bronze, threaded, Class 300 PSI					
2-1/2" valve	P6@.750	Ea	701.00	27.10	728.10
3" valve	P6@.950	Ea	814.00	34.30	848.30
Water control valves, threaded, 3-way with actuator					
2-1/2", CV 54	Ml@.750	Ea	1,220.00	29.00	1,249.00
3", CV 80	Ml@.950	Ea	1,460.00	36.70	1,496.70
4", CV 157	Ml@1.30	Ea	2,780.00	50.20	2,830.20

Piping Specialties

	Craft@Hrs	Unit	Material	Labor	Total
Stainless steel bellows type expansion joints, 150 lb. Flanged units include bolt and gasket sets					
3" pipe, 4" travel, welded	Ml@1.33	Ea	788.00	51.30	839.30
3" pipe, 6" travel, welded	Ml@1.33	Ea	860.00	51.30	911.30
4" pipe, 3" travel, welded	Ml@1.78	Ea	823.00	68.70	891.70
4" pipe, 6" travel, welded	Ml@1.78	Ea	1,030.00	68.70	1,098.70
5" pipe, 4" travel, flanged	Ml@2.20	Ea	2,180.00	84.90	2,264.90
5" pipe, 7" travel, flanged	Ml@2.20	Ea	2,250.00	84.90	2,334.90
6" pipe, 4" travel, flanged	Ml@2.67	Ea	2,320.00	103.00	2,423.00
8" pipe, 8" travel, flanged	Ml@3.20	Ea	3,440.00	124.00	3,564.00

	Craft@Hrs	Unit	Material	Labor	Total
Ball joints, welded ends, 150 lb					
3" ball joint	MI@2.40	Ea	349.00	92.70	441.70
4" ball joint	MI@3.35	Ea	512.00	129.00	641.00
5" ball joint	MI@4.09	Ea	866.00	158.00	1,024.00
6" ball joint	MI@4.24	Ea	928.00	164.00	1,092.00
8" ball joint	MI@4.79	Ea	1,630.00	185.00	1,815.00
10" ball joint	MI@5.81	Ea	2,000.00	224.00	2,224.00
14" ball joint	MI@7.74	Ea	2,790.00	299.00	3,089.00

In-Line Circulating Pumps Flanged, for hot or cold water circulation. By flange size and pump horsepower rating. See flange costs below. No electrical work included.

	Craft@Hrs	Unit	Material	Labor	Total
Horizontal drive, iron body					
3/4" to 1-1/2", 1/12 HP	MI@1.25	Ea	144.00	48.30	192.30
2", 1/4 HP	MI@1.40	Ea	248.00	54.00	302.00
2-1/2" or 3", 1/4 HP	MI@1.40	Ea	356.00	54.00	410.00
3/4" to 1-1/2", 1/6 HP	MI@1.25	Ea	212.00	48.30	260.30
3/4" to 1-1/2", 1/4 HP	MI@1.40	Ea	282.00	54.00	336.00
3", 1/3 HP	MI@1.50	Ea	481.00	57.90	538.90
3", 1/2 HP	MI@1.70	Ea	507.00	65.60	572.60
3", 3/4 HP	MI@1.80	Ea	583.00	69.50	652.50
Horizontal drive, all bronze					
3/4" to 1-1/2", 1/12 HP	MI@1.25	Ea	220.00	48.30	268.30
3/4" to 1-1/2", 1/6 HP	MI@1.25	Ea	329.00	48.30	377.30
3/4" to 2", 1/4 HP	MI@1.40	Ea	389.00	54.00	443.00
2-1/2" or 3", 1/4 HP	MI@1.50	Ea	645.00	57.90	702.90
3", 1/3 HP	MI@1.50	Ea	855.00	57.90	912.90
3", 1/2 HP	MI@1.70	Ea	880.00	65.60	945.60
3", 3/4 HP	MI@1.80	Ea	917.00	69.50	986.50
Vertical drive, iron body					
3/4" to 1/2", 1/6 HP	MI@1.25	Ea	290.00	48.30	338.30
2", 1/4 HP	MI@1.40	Ea	344.00	54.00	398.00
2-1/2", 1/4 HP	MI@1.40	Ea	406.00	54.00	460.00
3", 1/4 HP	MI@1.40	Ea	406.00	54.00	460.00

Pump connectors, vibration isolators In line, stainless steel braided connections, threaded

	Craft@Hrs	Unit	Material	Labor	Total
1/2" x 9"	M5@.210	Ea	4.44	7.67	12.11
1/2" x 12"	M5@.300	Ea	4.72	11.00	15.72
1/2" x 16"	M5@.400	Ea	5.10	14.60	19.70
1/2" x 20"	M5@.450	Ea	5.22	16.40	21.62
1/2" x 30"	MI@.500	Ea	6.28	19.30	25.58
1/2" x 36"	MI@.750	Ea	7.74	29.00	36.74

Flanges for circulating pumps Per set of two, including bolts and gaskets

	Craft@Hrs	Unit	Material	Labor	Total
3/4" to 1-1/2", iron	M5@.290	Ea	6.50	10.60	17.10
2", iron	M5@.290	Ea	9.25	10.60	19.85
2-1/2" to 3", iron	M5@.420	Ea	29.50	15.30	44.80
3/4" to 1", bronze	M5@.290	Ea	14.50	10.60	25.10
2", bronze	M5@.290	Ea	58.00	10.60	68.60
2-1/2" to 3", iron	M5@.420	Ea	30.00	15.30	45.30
3/4" copper sweat flanges	M5@.290	Ea	9.05	10.60	19.65
1" copper sweat flanges	M5@.290	Ea	13.50	10.60	24.10
1-1/2" copper sweat flanges	M5@.290	Ea	15.50	10.60	26.10

	Craft@Hrs	Unit	Material	Labor	Total
Base Mounted Centrifugal Pumps Single stage cast iron pumps with flanges but no electrical connection.					
50 GPM at 100' head	MI@3.15	Ea	1,920.00	122.00	2,042.00
50 GPM at 200' head	MI@4.15	Ea	2,190.00	160.00	2,350.00
50 GPM at 300' head	MI@5.83	Ea	2,440.00	225.00	2,665.00
200 GPM at 100' head	MI@6.70	Ea	2,470.00	259.00	2,729.00
200 GPM at 200' head	MI@7.15	Ea	2,550.00	276.00	2,826.00
200 GPM at 300' head	MI@10.2	Ea	2,770.00	394.00	3,164.00
240 GPM, 7-1/2 HP	MI@6.20	Ea	1,630.00	239.00	1,869.00
300 GPM, 5 HP	MI@3.91	Ea	1,840.00	151.00	1,991.00
340 GPM, 7-1/2 HP	MI@6.72	Ea	2,010.00	259.00	2,269.00
370 GPM, 5 HP	MI@4.02	Ea	1,880.00	155.00	2,035.00
375 GPM, 7-1/2 HP	MI@7.14	Ea	2,110.00	276.00	2,386.00
465 GPM, 10 HP	MI@7.63	Ea	2,200.00	295.00	2,495.00
Vertical Turbine Pumps 3,550 RPM These costs do not include electrical work.					
Cast iron single stage vertical turbine pumps					
50 GPM at 50' head	MI@9.9	Ea	3,190.00	382.00	3,572.00
50 GPM at 100' head	MI@9.9	Ea	3,670.00	382.00	4,052.00
100 GPM at 300' head	MI@9.9	Ea	4,750.00	382.00	5,132.00
200 GPM at 50' head	MI@9.9	Ea	4,000.00	382.00	4,382.00
Cast iron multi-stage vertical turbine pumps					
50 GPM at 100' head	D9@15.6	Ea	3,630.00	611.00	4,241.00
50 GPM at 200' head	D9@15.6	Ea	4,180.00	611.00	4,791.00
50 GPM at 300' head	D9@15.6	Ea	4,740.00	611.00	5,351.00
100 GPM at 100' head	D9@15.6	Ea	3,910.00	611.00	4,521.00
100 GPM at 200' head	D9@15.6	Ea	4,380.00	611.00	4,991.00
100 GPM at 300' head	D9@15.6	Ea	5,060.00	611.00	5,671.00
200 GPM at 100' head	D9@15.6	Ea	4,270.00	611.00	4,881.00
200 GPM at 200' head	D9@20.5	Ea	5,060.00	803.00	5,863.00
200 GPM at 300' head	D9@20.5	Ea	5,390.00	803.00	6,193.00
Bronze single stage vertical turbine pumps					
50 GPM at 50' head	D9@15.6	Ea	3,170.00	611.00	3,781.00
100 GPM at 50' head	D9@15.6	Ea	3,400.00	611.00	4,011.00
200 GPM at 50' head	D9@15.6	Ea	3,670.00	611.00	4,281.00
Bronze multi-stage vertical turbine pumps					
50 GPM at 50' head	D9@15.6	Ea	3,630.00	611.00	4,241.00
50 GPM at 100' head	D9@15.6	Ea	3,910.00	611.00	4,521.00
50 GPM at 150' head	D9@15.6	Ea	4,010.00	611.00	4,621.00
100 GPM at 50' head	D9@15.6	Ea	3,630.00	611.00	4,241.00
100 GPM at 100' head	D9@15.6	Ea	3,910.00	611.00	4,521.00
Bronze multi-stage vertical turbine pumps					
100 GPM at 150' head	D9@15.6	Ea	4,110.00	611.00	4,721.00
200 GPM at 50' head	D9@15.6	Ea	3,910.00	611.00	4,521.00
200 GPM at 100' head	D9@15.6	Ea	4,270.00	611.00	4,881.00
200 GPM at 150' head	D9@15.6	Ea	4,700.00	611.00	5,311.00
Sump Pumps Bronze 1,750 RPM sump pumps. These costs do not include electrical work.					
25 GPM at 25' head	MI@17.1	Ea	2,550.00	660.00	3,210.00
25 GPM at 50' head	MI@17.1	Ea	2,670.00	660.00	3,330.00
25 GPM at 100' head	MI@17.1	Ea	2,800.00	660.00	3,460.00
25 GPM at 150' head	MI@17.1	Ea	3,700.00	660.00	4,360.00
50 GPM at 25' head	MI@20.1	Ea	2,460.00	776.00	3,236.00
50 GPM at 50' head	MI@20.1	Ea	2,580.00	776.00	3,356.00

	Craft@Hrs	Unit	Material	Labor	Total
50 GPM at 100' head	Ml@20.1	Ea	2,700.00	776.00	3,476.00
50 GPM at 150' head	Ml@20.1	Ea	3,790.00	776.00	4,566.00
100 GPM at 25' head	Ml@22.8	Ea	2,610.00	880.00	3,490.00
100 GPM at 50' head	Ml@22.8	Ea	2,580.00	880.00	3,460.00
100 GPM at 100' head	Ml@22.8	Ea	3,310.00	880.00	4,190.00
100 GPM at 150' head	Ml@22.8	Ea	3,770.00	880.00	4,650.00

Miscellaneous Pumps 1,750 RPM. These costs do not include electrical work.

	Craft@Hrs	Unit	Material	Labor	Total
Salt water pump, 135 GPM at 300' head	D9@19.9	Ea	10,300.00	779.00	11,079.00
Sewage pumps, vertical mount, centrifugal					
1/2 HP, 25 GPM at 25' head	P6@8.00	Ea	800.00	289.00	1,089.00
1-1/2 HP, 100 GPM at 70' head	P6@14.6	Ea	3,480.00	527.00	4,007.00
10 HP, 300 GPM at 70' head	P6@18.3	Ea	5,700.00	661.00	6,361.00
Medical vacuum pump, duplex, tank mount with 30 gallon tank,					
1/2 HP, 20" vacuum	D4@20.6	Ea	5,990.00	761.00	6,751.00
Oil pump, base mounted, 1-1/2 GPM at 15' head,					
1/3 HP, 120 volt	D4@1.50	Ea	174.00	55.40	229.40

Fiberglass Pipe Insulation All Service Jacket, (ASJ) Self-Sealing Lap joint (SSL) By nominal pipe diameter. R factor equals 2.56 at 300 degrees F. Make additional allowances for scaffolding if required Also see fittings, flanges and valves at the end of this section.

	Craft@Hrs	Unit	Material	Labor	Total
2-1/2" diameter pipe					
1" thick insulation	A1@.047	LF	1.21	1.72	2.93
1-1/2" thick insulation	A1@.048	LF	2.13	1.76	3.89
2" thick insulation	A1@.050	LF	3.47	1.83	5.30
3" diameter pipe					
1" thick insulation	A1@.051	LF	.84	1.87	2.71
1-1/2" thick insulation	A1@.054	LF	2.31	1.98	4.29
2" thick insulation	A1@.056	LF	3.80	2.05	5.85
4" diameter pipe					
1" thick insulation	A1@.063	LF	1.04	2.31	3.35
1-1/2" thick insulation	A1@.066	LF	2.66	2.42	5.08
2" thick insulation	A1@.069	LF	4.55	2.53	7.08
6" diameter pipe					
1" thick insulation	A1@.077	LF	2.43	2.82	5.25
1-1/2" thick insulation	A1@.080	LF	3.34	2.93	6.27
2" thick insulation	A1@.085	LF	5.61	3.12	8.73
8" diameter pipe					
1" thick insulation	A1@.117	LF	3.34	4.29	7.63
1-1/2" thick insulation	A1@.123	LF	4.09	4.51	8.60
2" thick insulation	A1@.129	LF	6.60	4.73	11.33
10" diameter pipe					
1" thick insulation	A1@.140	LF	3.94	5.13	9.07
1-1/2" thick insulation	A1@.147	LF	5.01	5.39	10.40
2" thick insulation	A1@.154	LF	7.92	5.64	13.56
12" diameter pipe					
1" thick insulation	A1@.155	LF	4.55	5.68	10.23
1-1/2" thick insulation	A1@.163	LF	5.61	5.97	11.58
2" thick insulation	A1@.171	LF	8.72	6.27	14.99

	Craft@Hrs	Unit	Material	Labor	Total
Aluminum pipe insulation cover					
Add for .016" aluminum jacket, SF of surface	A1@.018	SF	.56	.66	1.22

Additional cost for insulating pipe fittings and flanges

For each fitting or flange, add the cost of insulating 3 LF of pipe of the same size.

Additional cost for insulating valves

Valve body only. Add the cost of insulating 5 LF of pipe of the same size.

Body and bonnet or yoke valves. Add the cost of insulating 10 LF of pipe of the same size.

Calcium Silicate Pipe Insulation   Without cover. Make additional allowances for canvas, weather, aluminum or stainless steel jacket. By nominal pipe diameter, R factor equals 2.23 at 300 degrees F. Add the cost of scaffolding or lifts if required Also see fittings, flanges and valves at the end of this section.

	Craft@Hrs	Unit	Material	Labor	Total
2-1/2" diameter pipe					
2" thick insulation	A1@.09	LF	5.29	3.30	8.59
4" thick insulation	A1@.10	LF	12.60	3.67	16.27
3" diameter pipe					
2" thick insulation	A1@.10	LF	5.49	3.67	9.16
4" thick insulation	A1@.11	LF	13.70	4.03	17.73
4" diameter pipe					
2" thick insulation	A1@.105	LF	6.44	3.85	10.29
4" thick insulation	A1@.125	LF	15.80	4.58	20.38
6" diameter pipe					
2" thick insulation	A1@.109	LF	8.30	4.00	12.30
4" thick insulation	A1@.133	LF	19.40	4.87	24.27
6" thick insulation	A1@.178	LF	30.40	6.52	36.92
8" diameter pipe					
2" thick insulation	A1@.115	LF	9.96	4.21	14.17
4" thick insulation	A1@.143	LF	22.60	5.24	27.84
6" thick insulation	A1@.200	LF	34.80	7.33	42.13
10" diameter pipe					
2" thick insulation	A1@.122	LF	11.70	4.47	16.17
4" thick insulation	A1@.143	LF	26.40	5.24	31.64
6" thick insulation	A1@.230	LF	41.90	8.43	50.33
12" diameter pipe					
2" thick insulation	A1@.127	LF	13.50	4.65	18.15
4" thick insulation	A1@.158	LF	29.70	5.79	35.49
6" thick insulation	A1@.242	LF	47.50	8.87	56.37

Additional cost for insulating pipe fittings and flanges

For each fitting or flange, add the cost of insulating 3 LF of pipe of the same size.

Additional cost for insulating valves

Valve body only. Add the cost of insulating 5 LF of pipe of the same size.

Body and bonnet or yoke valves. Add the cost of insulating 10 LF of pipe of the same size.

Calcium Silicate Pipe Insulation with .016" Aluminum Jacket By nominal pipe diameter. R factor equals 2.20 at 300 degrees F. These costs do not include scaffolding, add for same if required. Also see fittings, flanges and valves at the end of this section.

	Craft@Hrs	Unit	Material	Labor	Total
6" diameter pipe					
2" thick insulation	A1@.183	LF	9.77	6.71	16.48
4" thick insulation	A1@.218	LF	21.40	7.99	29.39
6" thick insulation	A1@.288	LF	30.00	10.60	40.60
8" diameter pipe					
2" thick insulation	A1@.194	LF	11.70	7.11	18.81
4" thick insulation	A1@.228	LF	24.90	8.36	33.26
6" thick insulation	A1@.308	LF	37.70	11.30	49.00

	Craft@Hrs	Unit	Material	Labor	Total
10" diameter pipe					
2" thick insulation	A1@.202	LF	13.70	7.40	21.10
4" thick insulation	A1@.239	LF	29.00	8.76	37.76
6" thick insulation	A1@.337	LF	45.10	12.40	57.50
12" diameter pipe					
2" thick insulation	A1@.208	LF	15.90	7.62	23.52
4" thick insulation	A1@.261	LF	32.30	9.57	41.87
6" thick insulation	A1@.388	LF	51.10	14.20	65.30

Additional cost for insulating pipe fittings and flanges

For each fitting or flange, add the cost of insulating 3 LF of pipe of the same size.

Additional cost for insulating valves

Valve body only. Add the cost of insulating 5 LF of pipe of the same size.

Body and bonnet or yoke valves. Add the cost of insulating 10 LF of pipe of the same size.

Closed Cell Elastomeric Pipe Semi split, thickness as shown, by nominal pipe diameter. R factor equals 3.58 at 220 degrees F. Make additional allowances for scaffolding if required. Also see fittings, flanges and valves at the end of this section.

	Craft@Hrs	Unit	Material	Labor	Total
2-1/2", 3/8" thick	A1@.056	LF	.99	2.05	3.04
2-1/2", 1/2" thick	A1@.056	LF	1.33	2.05	3.38
2-1/2", 3/4" thick	A1@.056	LF	1.73	2.05	3.78
3", 3/8" thick	A1@.082	LF	1.14	3.01	4.15
3", 1/2" thick	A1@.082	LF	1.50	3.01	4.51
3", 3/4" thick	A1@.082	LF	1.95	3.01	4.96
4", 3/8" thick	A1@.095	LF	1.33	3.48	4.81
4", 1/2" thick	A1@.095	LF	1.76	3.48	5.24
4", 3/4" thick	A1@.095	LF	2.29	3.48	5.77

Additional cost for insulating pipe fittings and flanges

For each fitting or flange, add the cost of insulating 3 LF of pipe of the same size.

Additional cost for insulating valves

Valve body only. Add the cost of insulating 5 LF of pipe of the same size.

Body and bonnet or yoke valves. Add the cost of insulating 10 LF of pipe of the same size.

Meters, Gauges and Indicators

	Craft@Hrs	Unit	Material	Labor	Total
Water meters					
Disc type, 1-1/2"	P6@1.00	Ea	136.00	36.10	172.10
Turbine type, 1-1/2"	P6@1.00	Ea	206.00	36.10	242.10
Displacement type, AWWA C7000, 1"	P6@.941	Ea	211.00	34.00	245.00
Displacement type, AWWA C7000, 1-1/2"	P6@1.23	Ea	725.00	44.40	769.40
Displacement type, AWWA C7000, 2"	P6@1.57	Ea	1,090.00	56.70	1,146.70
Curb stop and waste valve, 1"	P6@1.00	Ea	76.40	36.10	112.50
Thermoflow indicators, soldered					
1-1/4" indicator	M5@2.01	Ea	315.00	73.40	388.40
2" indicator	M5@2.01	Ea	359.00	73.40	432.40
2-1/2" indicator	Ml@1.91	Ea	526.00	73.70	599.70
Cast steel steam meters, in-line, flanged, 300 pound					
1" pipe, threaded	M5@1.20	Ea	3,010.00	43.80	3,053.80
2" pipe	Ml@1.14	Ea	3,400.00	44.00	3,444.00
3" pipe	Ml@2.31	Ea	3,790.00	89.20	3,879.20
4" pipe	Ml@2.71	Ea	4,210.00	105.00	4,315.00
Cast steel by-pass steam meters, 2", by line pipe size					
6" line	Ml@16.7	Ea	7,300.00	645.00	7,945.00
8" line	Ml@17.7	Ea	7,500.00	683.00	8,183.00
10" line	Ml@18.5	Ea	7,800.00	714.00	8,514.00
12" line	Ml@19.7	Ea	8,100.00	761.00	8,861.00

	Craft@Hrs	Unit	Material	Labor	Total
14" line	MI@21.4	Ea	8,570.00	826.00	9,396.00
16" line	MI@22.2	Ea	8,830.00	857.00	9,687.00
Add for steam meter pressure-compensated counter	—	Ea	1,300.00	—	1,300.00
Add for contactor used with dial counter	—	Ea	604.00	—	604.00
Add for wall or panel-mounted remote totalizer with contactor	—	Ea	964.00	—	964.00
Add for direct reading pressure gauge	M5@.250	Ea	71.80	9.13	80.93
Add for direct reading thermometer, stainless steel case, with trim, 2% accuracy					
3" dial	M5@.250	Ea	56.20	9.13	65.33

Plumbing Specialties

Access doors, painted steel, with screwdriver catch

	Craft@Hrs	Unit	Material	Labor	Total
8" x 8"	SM@.400	Ea	28.40	16.50	44.90
12" x 12"	SM@.500	Ea	32.90	20.70	53.60
18" x 18"	SM@.800	Ea	46.40	33.00	79.40
24" x 24"	SM@1.25	Ea	68.80	51.60	120.40
36" x 36"	SM@1.60	Ea	127.00	66.10	193.10
Add for fire rating	—	%	400.0	—	—
Add for stainless steel	—	%	300.0	—	—
Add for cylinder lock	—	Ea	11.10	—	11.10

Cleanouts For drainage piping. A cleanout may be required: at the end of each waste line, or where a main building drain exits the building, or at regular 100' intervals for straight runs longer than 100', or at the base of soil or waste stacks, or where a 90 degree change in direction of a main building drain takes place. Consult your plumbing code for the regulations that apply to your area.

Finished grade end cleanouts, (Malcolm) with cast iron top and body

	Craft@Hrs	Unit	Material	Labor	Total
2" pipe, 4-1/8" round top	P6@.500	Ea	10.70	18.10	28.80
3" pipe, 5-1/8" round top	P6@.500	Ea	18.40	18.10	36.50
4" pipe, 6-1/8" round top	P6@.500	Ea	33.00	18.10	51.10

Finished grade end cleanouts, (Malcolm) with cast iron body and nickel bronze top

	Craft@Hrs	Unit	Material	Labor	Total
2" pipe, 4-1/8" round top	P6@.500	Ea	53.50	18.10	71.60
3" pipe, 5-1/8" round top	P6@.500	Ea	53.50	18.10	71.60
4" pipe, 6-1/8" round top	P6@.500	Ea	60.60	18.10	78.70

Cast iron in-line cleanouts (Barrett) make additional allowances for access doors

	Craft@Hrs	Unit	Material	Labor	Total
2" pipe	P6@.750	Ea	20.80	27.10	47.90
4" pipe	P6@.750	Ea	34.70	27.10	61.80
6" pipe	P6@.750	Ea	74.00	27.10	101.10

Cleanout access cover accessories

	Craft@Hrs	Unit	Material	Labor	Total
Round polished bronze cover	—	Ea	15.00	—	15.00
Square chrome plated cover	—	Ea	140.00	—	140.00
Square bronze hinged cover box	—	Ea	100.00	—	100.00

Floor drains, body and grate only, add rough-in

	Craft@Hrs	Unit	Material	Labor	Total
3" x 2" plastic grate and body	P6@.25	Ea	10.00	9.03	19.03
5" x 3" plastic grate and body	P6@.25	Ea	14.00	9.03	23.03
3" x 2" cast iron body, nickel-bronze grate	P6@.50	Ea	25.00	18.10	43.10
5" x 3" cast iron body, nickel-bronze grate	P6@.50	Ea	30.00	18.10	48.10
6" x 4" cast iron body, nickel-bronze grate	P6@.50	Ea	35.00	18.10	53.10
Add for a sediment bucket	—	Ea	6.00	—	6.00
Add for plastic backwater valve	—	Ea	22.00	—	22.00
Add for cast iron backwater valve	—	Ea	98.00	—	98.00
Add for trap seal primer fitting	—	Ea	5.00	—	5.00

	Craft@Hrs	Unit	Material	Labor	Total
Roof (jack) flashing, round					
4" jack	SM@.311	Ea	6.04	12.80	18.84
6" jack	SM@.416	Ea	6.56	17.20	23.76
8" jack	SM@.516	Ea	8.93	21.30	30.23
Drain, general use, medium duty, PVC plastic, 3" or 4" pipe					
3" x 4", stainless steel strainer	P6@1.16	Ea	6.35	41.90	48.25
3" x 4", brass strainer	P6@1.32	Ea	15.60	47.70	63.30
4" with 5" brass strainer & spigot outlet	P6@1.55	Ea	30.50	56.00	86.50
Roof drains, cast iron mushroom strainer, bottom outlet					
2" to 4" drain	P6@1.75	Ea	53.50	63.20	116.70
5" or 6" drain	P6@1.98	Ea	60.60	71.50	132.10
8" drain	P6@2.36	Ea	110.00	85.20	195.20
Deduct for PVC body and dome	—	%	-40.0	—	—

Plumbing Equipment Valves, supports, vents, gas or electric hookup and related equipment are not included except as noted.

Garbage disposers commercial, stainless steel, with water flow interlock switch, (Insinkerator)

	Craft@Hrs	Unit	Material	Labor	Total
1/2 HP, 1-1/2" drain connection	PL@1.47	Ea	418.00	61.70	479.70
3/4 HP, 1-1/2" drain connection	PL@2.09	Ea	575.00	87.70	662.70
1 HP, 1 1/2" drain connection	PL@3.92	Ea	603.00	164.00	767.00
1-1/2 HP, 2" drain connection	PL@5.24	Ea	896.00	220.00	1,116.00

Water heaters, commercial, gas fired 3 year warranty, Glasslined. Set in place only. Make additional allowances for pipe, recirculating pump, gas and flue connections.

Standard efficiency

	Craft@Hrs	Unit	Material	Labor	Total
50 gallons, 98 MBtu, 95 GPH recovery	P6@2.00	Ea	797.00	72.20	869.20
60 gallons, 55 MBtu, 50 GPH recovery	P6@2.00	Ea	447.00	72.20	519.20
67 gallons, 114 MBtu, 108 GPH recovery	P6@2.00	Ea	939.00	72.20	1,011.20
76 gallons, 180 MBtu, 175 GPH recovery	P6@2.25	Ea	1,270.00	81.20	1,351.20
82 gallons, 156 MBtu, 151 GPH recovery	P6@2.25	Ea	1,210.00	81.20	1,291.20
91 gallons, 300 MBtu, 291 GPH recovery	P6@2.50	Ea	1,500.00	90.30	1,590.30

Energy Miser. Meets ASHRAE 90.1b-1992 standards

	Craft@Hrs	Unit	Material	Labor	Total
50 gallons, 98 MBtu, 95 GPH recovery	P6@2.00	Ea	1,150.00	72.20	1,222.20
65 gallons, 360 MBtu, 350 GPH recovery	P6@2.00	Ea	1,710.00	72.20	1,782.20
67 gallons, 114 MBtu, 108 GPH recovery	P6@2.00	Ea	1,190.00	72.20	1,262.20
72 gallons, 250 MBtu, 242 GPH recovery	P6@2.25	Ea	1,580.00	81.20	1,661.20
76 gallons, 180 MBtu, 175 GPH recovery	P6@2.25	Ea	1,470.00	81.20	1,551.20
82 gallons, 156 MBtu, 151 GPH recovery	P6@2.25	Ea	1,440.00	81.20	1,521.20
85 gallons, 400 MBtu, 388 GPH recovery	P6@2.25	Ea	1,770.00	81.20	1,851.20
91 gallons, 200 MBtu, 194 GPH recovery	P6@2.50	Ea	1,600.00	90.30	1,690.30
91 gallons, 300 MBtu, 291 GPH recovery	P6@2.50	Ea	1,750.00	90.30	1,840.30
Power vent kit	P6@2.00	Ea	386.00	72.20	458.20
Equal flow manifold kit, duplex	—	Ea	254.00	—	254.00
Equal flow manifold kit, triplex	—	Ea	508.00	—	508.00
Add for propane fired units, any of above	—	Ea	64.20	—	64.20

Water heaters, commercial, electric 208/240 volts, with surface mounted thermostat, Rheem Glasslined Energy Miser. Set in place only. Make additional allowances for pipe, recirculating pump, and electrical connections.

	Craft@Hrs	Unit	Material	Labor	Total
50 gallons, 9 kw	P6@1.75	Ea	1,020.00	63.20	1,083.20
50 gallons, 27 kw	P6@1.75	Ea	1,270.00	63.20	1,333.20
85 gallons, 18 kw	P6@2.00	Ea	1,220.00	72.20	1,292.20
120 gallons, 45 kw	P6@3.00	Ea	1,830.00	108.00	1,938.00

	Craft@Hrs	Unit	Material	Labor	Total
Water heater stands 16 gauge steel, installed at same time as water heater					
18" x 18" square x 18" high	—	Ea	45.00	—	45.00
24" x 24" square x 18" high	—	Ea	62.50	—	62.50
Water heater safety pans with side drain, 16 gauge steel, installed at same time as water heater					
20" diameter	—	Ea	12.10	—	12.10
26" diameter	—	Ea	19.70	—	19.70
Water heater connection assembly Includes typical pipe, fittings, valves, gages, and hangers					
Copper pipe connection, 3/4" supply	P6@2.75	Ea	203.00	99.30	302.30
Copper pipe connection, 1" supply	P6@3.20	Ea	269.00	116.00	385.00
Copper pipe connection, 1-1/4" supply	P6@3.85	Ea	330.00	139.00	469.00
Copper pipe connection, 1-1/2" supply	P6@4.00	Ea	340.00	144.00	484.00
Copper pipe connection, 2" supply	P6@4.40	Ea	355.00	159.00	514.00
Copper pipe connection, 2-1/2" supply	P6@5.82	Ea	787.00	210.00	997.00
Copper pipe connection, 3" supply	P6@6.70	Ea	1,170.00	242.00	1,412.00
Black steel gas pipe connection, 3/4"	P6@1.40	Ea	82.00	50.60	132.60
Black steel gas pipe connection, 1"	P6@1.50	Ea	107.00	54.20	161.20
Black steel gas pipe connection, 1-1/2"	P6@2.00	Ea	157.00	72.20	229.20
Double wall vent pipe connection, 4"	P6@1.40	Ea	102.00	50.60	152.60
Grease interceptors					
2" cast iron, 4 GPM, 8 lb capacity	P6@3.24	Ea	309.00	117.00	426.00
2" cast iron, 10 GPM, 20 lb capacity	P6@4.13	Ea	503.00	149.00	652.00
2" steel, 15 GPM, 30 lb capacity	P6@5.63	Ea	514.00	203.00	717.00
3" steel, 20 GPM, 40 lb capacity	P6@16.8	Ea	763.00	607.00	1,370.00
Hair interceptors 1-1/2" cast iron					
Small	P6@3.24	Ea	116.00	117.00	233.00
Large	P6@4.72	Ea	162.00	170.00	332.00
Add for nickel bronze construction	—	Ea	100.00	—	100.00
Sewage ejector packaged systems with vertical pumps, float switch controls, poly tank, cover and fittings, no electrical work or pipe connections included.					
Simplex sewage ejector systems (single pump)					
1/2 HP, 2" line, 20" x 30" tank	P6@4.75	Ea	385.00	172.00	557.00
3/4 HP, 2" line, 24" x 36" tank	P6@4.75	Ea	919.00	172.00	1,091.00
1 HP, 3" line, 24" x 36" tank	P6@6.00	Ea	995.00	217.00	1,212.00
Duplex sewage ejector systems (two pumps)					
1/2 HP, 2" line, 30" x 36" tank	P6@6.50	Ea	1,480.00	235.00	1,715.00
3/4 HP, 2" line, 30" x 36" tank	P6@6.50	Ea	1,930.00	235.00	2,165.00
1 HP, 3" line, 30" x 36" tank	P6@8.00	Ea	2,170.00	289.00	2,459.00
Sump pumps with float switch, no electrical work included					
Submersible type					
1/4 HP, 1-1/2" outlet, ABS plastic	P6@2.00	Ea	96.00	72.20	168.20
1/3 HP, 1-1/4" outlet, Cast iron	P6@2.00	Ea	114.00	72.20	186.20
1/3 HP, 1-1/4" outlet, ABS plastic	P6@2.00	Ea	102.00	72.20	174.20
Upright type					
1/3 HP, 1-1/4" outlet, cast iron	P6@2.00	Ea	110.00	72.20	182.20
1/3 HP, 1-1/2" outlet, ABS plastic	P6@2.00	Ea	82.00	72.20	154.20

	Craft@Hrs	Unit	Material	Labor	Total

Tanks See also ASME tanks

Hot water storage tanks insulated, floor mounted, includes pipe connection

	Craft@Hrs	Unit	Material	Labor	Total
80 gallons	P6@4.00	Ea	381.00	144.00	525.00
120 gallons	P6@4.50	Ea	584.00	162.00	746.00
200 gallons	P6@5.25	Ea	1,570.00	190.00	1,760.00
980 gallons	P6@20.0	Ea	5,390.00	722.00	6,112.00
1,230 gallons	P6@22.0	Ea	6,420.00	794.00	7,214.00
1,615 gallons	P6@26.0	Ea	8,910.00	939.00	9,849.00

Septic tanks including typical excavation and piping

	Craft@Hrs	Unit	Material	Labor	Total
500 gallons, polyethylene	U2@15.2	Ea	442.00	519.00	961.00
1,000 gallons, fiberglass	U2@16.1	Ea	629.00	550.00	1,179.00
2,000 gallons, concrete	U2@17.0	Ea	949.00	581.00	1,530.00
5,000 gallons, concrete	U2@22.3	Ea	3,530.00	762.00	4,292.00

Water softeners automatic two tank unit, 3/4" valve

	Craft@Hrs	Unit	Material	Labor	Total
32,000 grain capacity	P6@3.59	Ea	518.00	130.00	648.00
60,000 grain capacity	P6@3.59	Ea	619.00	130.00	749.00

Water filters in-line, filters for taste, odor, chlorine and sediment to 20 microns, with by-pass assembly. Includes pipe connection

	Craft@Hrs	Unit	Material	Labor	Total
Housing, by-pass and carbon filter, complete	P6@2.50	Ea	122.00	90.30	212.30
Water filter housing only	P6@1.00	Ea	30.50	36.10	66.60
By-pass assembly only, 3/4" copper	P6@1.25	Ea	66.00	45.10	111.10
Carbon core cartridge, 20 microns	P6@.250	Ea	15.20	9.03	24.23
Ceramic disinfection and carbon cartridge	P6@.250	Ea	40.60	9.03	49.63
Lead, chlorine, taste and odor cartridge	P6@.250	Ea	30.50	9.03	39.53

Water sterilizers ultraviolet, point of use water. Add the cost of electrical and pipe connections. In gallons per minute (GPM) of rated capacity. 2 GPM and larger units include light-emitting diode (LED)

	Craft@Hrs	Unit	Material	Labor	Total
1/2 GPM point of use filter, 1/2" connection	P6@2.00	Ea	173.00	72.20	245.20
3/4 GPM point of use filter, 1/2" connection	P6@2.00	Ea	183.00	72.20	255.20
2 GPM, 1/2" connection	P6@2.50	Ea	203.00	90.30	293.30
2 GPM, with alarm, 1/2" connection	P6@2.50	Ea	264.00	90.30	354.30
2 GPM, with alarm and monitor, 1/2" connection	P6@3.00	Ea	462.00	108.00	570.00
5 GPM, 3/4" connection	P6@2.50	Ea	254.00	90.30	344.30
5 GPM, with alarm, 3/4" connection	P6@2.50	Ea	335.00	90.30	425.30
5 GPM, with alarm and monitor, 3/4" connection	P6@3.00	Ea	533.00	108.00	641.00
8 GPM, with alarm, 3/4" connection	P6@2.50	Ea	406.00	90.30	496.30
8 GPM, with alarm and monitor, 3/4" connection	P6@3.00	Ea	604.00	108.00	712.00
12 GPM, with alarm, 3/4" connection	P6@2.50	Ea	477.00	90.30	567.30
12 GPM, with alarm and monitor, 3/4" connection	P6@3.00	Ea	680.00	108.00	788.00

Reverse osmosis water filters Economy system. Add the cost of electrical and pipe connections. In gallons per day of rated capacity

	Craft@Hrs	Unit	Material	Labor	Total
12 gallons per day	P6@3.50	Ea	400.00	126.00	526.00
16 gallons per day	P6@3.50	Ea	475.00	126.00	601.00
25 gallons per day	P6@3.50	Ea	700.00	126.00	826.00

Plumbing Assemblies

Rough-in Assemblies: Includes rough-ins for commercial industrial and residential buildings. This section shows costs for typical rough-ins of drain, waste and vent (DWV) and water supply piping using either plastic or metal pipe and fittings. Plastic ABS or PVC DWV pipe can be combined with plastic "poly b" supply pipe for a complete residential rough-in assembly. Cast iron or copper drainage DWV pipe can be combined with copper supply pipe for a complete commercial rough-in assembly. (Check local plumbing and building codes to determine acceptable materials for above and below grade installations in your area.) Detailed estimates of the piping requirements for each type plumbing fixture rough-in were made using the Uniform Plumbing Code as a guideline. For scheduling purposes, estimate that a plumber and an apprentice can install rough-in piping for 3 to 4 plumbing fixtures in an 8-hour day. The same plumber and apprentice will install, connect and test 3 to 4 fixtures a day. When roughing-in less than 4 fixtures in a building, increase the rough-in cost 25% for each fixture less than 4. Add the cost of the fixture and final connection assembly (as shown in Fixtures and Final Connection Assemblies) to the rough-in assembly cost.

Rule of thumb: Use plastic drainage and supply pipe in economy residential applications. Use plastic drainage and copper supply in quality residential applications. Use cast iron or copper DWV drainage and copper supply pipe in quality commercial applications.

Tricks of the trade: In some jurisdictions, plastic drainage pipe is allowed below grade for commercial applications only. In residential applications, plastic drainage pipe may be allowed in above and below grade applications. The difference is the liability of the owner to the patrons of a public building vs. the liability of the owner of a private residence in case of fire. ABS pipe is toxic when it burns.

	Craft@Hrs	Unit	Material	Labor	Total
Water closets					

Floor mounted, tank type, rough-in includes 7' of 3", 8' of 1 1/2" DWV pipe and 8' of 1/2" poly b supply pipe with typical fittings, hangers, and strapping.

	Craft@Hrs	Unit	Material	Labor	Total
Plastic DWV and poly b supply rough-in	P6@1.90	Ea	27.00	68.60	95.60
Plastic DWV and copper supply rough-in	P6@2.00	Ea	29.00	72.20	101.20
Cast iron, copper DWV and copper supply rough-in	P6@2.25	Ea	117.00	81.20	198.20

Floor mounted flush valve type, rough-in includes 7' of 3", 8' of 1 1/2" DWV pipe and 8' of 1" supply pipe with typical fittings, hangers, and strapping.

	Craft@Hrs	Unit	Material	Labor	Total
Plastic DWV and PVC schedule 40 supply rough-in	P6@1.90	Ea	27.00	68.60	95.60
Plastic DWV and copper supply rough-in	P6@2.00	Ea	29.00	72.20	101.20
Cast iron, copper DWV and copper supply rough-in	P6@2.25	Ea	117.00	81.20	198.20

Wall hung, tank type, rough-in includes 5' of 4", 8' of 1 1/2" DWV pipe and 8' of 1/2" supply pipe with typical fittings, hangers, and strapping.

	Craft@Hrs	Unit	Material	Labor	Total
Plastic DWV and poly b supply rough-in	P6@1.80	Ea	19.00	65.00	84.00
Plastic DWV and copper supply rough-in	P6@1.90	Ea	24.00	68.60	92.60
Cast iron, copper DWV and copper supply rough-in	P6@2.35	Ea	135.00	84.90	219.90
Add for fixture carrier	P6@.750	Ea	120.00	27.10	147.10

Wall hung flush valve type, rough-in includes 5' of 4", 6' of 1 1/2" DWV pipe and 8' of 1" supply pipe with typical fittings, hangers, and strapping.

	Craft@Hrs	Unit	Material	Labor	Total
Plastic DWV and PVC schedule 40 supply rough-in	P6@1.70	Ea	26.00	61.40	87.40
Plastic DWV and copper supply rough-in	P6@1.00	Ea	28.00	36.10	64.10
Cast iron, copper DWV and copper supply rough-in	P6@1.95	Ea	69.00	70.40	139.40
Add for fixture carrier	P6@.750	Ea	120.00	27.10	147.10

Urinals

Wall hung urinal, rough-in includes 15' of DWV pipe and 8' of supply pipe with typical fittings, hangers and strapping.

	Craft@Hrs	Unit	Material	Labor	Total
Plastic DWV and supply rough-in	P6@2.00	Ea	50.00	72.20	122.20
Plastic DWV and copper supply rough-in	P6@2.10	Ea	55.00	75.80	130.80
Copper DWV and copper supply rough-in	P6@2.35	Ea	117.00	84.90	201.90
Add for fixture carrier	P6@.850	Ea	90.00	30.70	120.70

	Craft@Hrs	Unit	Material	Labor	Total
Floor mounted urinal, rough-in includes 15' of DWV pipe and 8' of supply pipe with typical fittings, hangers and strapping.					
Plastic DWV and supply rough-in	P6@2.15	Ea	65.00	77.60	142.60
Plastic DWV and copper supply rough-in	P6@2.25	Ea	70.00	81.20	151.20
Copper DWV and copper supply rough-in	P6@2.50	Ea	145.00	90.30	235.30
Lavatory rough-in includes 15' of 1 1/2" DWV pipe and 15' of 1/2" supply pipe with typical fittings, hangers and strapping.					
Plastic DWV and poly b supply rough-in	P6@1.70	Ea	20.00	61.40	81.40
Plastic DWV and copper supply rough-in	P6@1.75	Ea	22.00	63.20	85.20
Copper DWV and copper supply rough-in	P6@1.90	Ea	82.00	68.60	150.60
Add for carrier	P6@.75	Ea	90.00	27.10	117.10
Bathtub and Shower Rough-in includes 15' of 1 1/2" DWV pipe and 15' of 1/2" supply pipe with typical fittings, hangers, and strapping.					
Plastic DWV and poly b supply rough-in	P6@1.95	Ea	20.00	70.41	90.41
Plastic DWV and copper supply rough-in	P6@2.10	Ea	22.00	75.80	97.80
Copper DWV and copper supply rough-in	P6@2.35	Ea	126.00	84.90	210.90
Kitchen sink Rough-in includes 15' of 1 1/2" DWV pipe and 15' of 1/2" supply pipe with typical, fittings, hangers and strapping					
Plastic DWV and poly b supply rough-in	P6@1.70	Ea	22.00	61.40	83.40
Plastic DWV and copper supply rough-in	P6@1.75	Ea	24.00	63.20	87.20
Copper DWV and copper supply rough-in	P6@1.90	Ea	88.00	68.60	156.60
Drinking fountain and Water cooler rough-in includes 15' of 1 1/2" DWV pipe and 10' of 1/2" supply pipe with typical, fittings, strapping and hangers.					
Plastic DWV and poly b supply rough-in	P6@1.80	Ea	18.00	65.00	83.00
Plastic DWV and copper supply rough-in	P6@2.00	Ea	20.00	72.20	92.20
Copper DWV and copper supply rough-in	P6@2.20	Ea	65.00	79.40	144.40
Wash fountain rough-in includes 20' of 1 1/2" DWV pipe and 20' of 1/2" supply pipe with typical fittings, hangers and strapping.					
Plastic DWV and supply rough-in	P6@1.90	Ea	34.00	68.60	102.60
Plastic DWV and copper supply rough-in	P6@2.00	Ea	38.00	72.20	110.20
Copper DWV and copper supply rough-in	P6@2.10	Ea	96.00	75.80	171.80
Service sink rough-in includes 15' of DWV pipe and 15' of supply pipe with typical fittings, hangers, and strapping.					
Wall hung slop sink					
Plastic DWV and supply rough-in	P6@2.25	Ea	42.00	81.20	123.20
Plastic DWV and copper supply rough-in	P6@2.35	Ea	45.00	84.90	129.90
Cast iron DWV and copper supply rough-in	P6@2.45	Ea	85.00	88.50	173.50
Floor mounted mop sink					
Plastic DWV and supply rough-in	P6@2.35	Ea	55.00	84.90	139.90
Plastic DWV and copper supply rough-in	P6@2.45	Ea	60.00	88.50	148.50
Cast iron DWV and copper supply rough-in	P6@2.60	Ea	105.00	93.90	198.90
Emergency drench shower and eye wash station rough-in including a valved water supply with 15' of supply pipe, typical fittings and hangers. Add the cost of a floor drain if required.					
Plastic poly b pipe	P6@1.25	Ea	15.00	45.10	60.10
Copper pipe	P6@1.50	Ea	20.00	54.20	74.20
Add for floor drain	P6@3.25	Ea	95.00	117.00	212.00

	Craft@Hrs	Unit	Material	Labor	Total

Surgeons' scrub sink rough-in includes 7' of 2" and 8' of 1 1/2" DWV pipe and 15' of 3/4" supply pipe with typical fittings and hangers.

	Craft@Hrs	Unit	Material	Labor	Total
Plastic DWV and supply rough-in	P6@2.25	Ea	55.00	81.20	136.20
Plastic DWV and copper supply rough-in	P6@2.35	Ea	65.00	84.90	149.90
Cast iron DWV and copper supply rough-in	P6@2.75	Ea	120.00	99.30	219.30

Floor sink, floor drain rough-in includes 15' of drain and 15' of vent pipe, P trap, typical fittings and hangers

	Craft@Hrs	Unit	Material	Labor	Total
2" plastic DWV rough-in	P6@1.50	Ea	35.00	54.20	89.20
2" cast iron DWV rough-in	P6@2.00	Ea	75.00	72.20	147.20
3" plastic DWV rough-in	P6@1.75	Ea	50.00	63.20	113.20
3" cast iron DWV rough-in	P6@2.50	Ea	90.00	90.30	180.30
4" cast iron DWV rough-in	P6@2.50	Ea	120.00	90.30	210.30

Fixtures and final connection assemblies

Fixture costs are based on good to better quality white fixtures of standard dimensions and include all trim and valves. Add 30% for colored fixtures. Detailed estimates of the piping requirements for each type of plumbing fixture were made using the Uniform Plumbing Code as a guideline.

Water closets

Floor mounted, tank type, elongated vitreous china bowl

	Craft@Hrs	Unit	Material	Labor	Total
Water closet, trim and valves (add rough-in)	P6@2.10	Ea	140.00	75.80	215.80
Deduct for round bowl	—	Ea	-60.00	—	-60.00

Floor mounted flush valve type, elongated vitreous china bowl

	Craft@Hrs	Unit	Material	Labor	Total
Water closet, trim and valves (add rough-in)	P6@2.60	Ea	198.00	93.90	291.90

Floor mounted flush valve type, elongated vitreous china bowl, meets ADA requirements (18" high)

	Craft@Hrs	Unit	Material	Labor	Total
Water closet, trim and valves (add rough-in)	P6@2.60	Ea	327.00	93.90	420.90

Wall hung, tank type, elongated vitreous china bowl, cast iron carrier

	Craft@Hrs	Unit	Material	Labor	Total
Water closet, trim and valves (add rough-in)	P6@3.00	Ea	540.00	108.00	648.00
Deduct for round bowl	—	Ea	-60.00	—	-60.00

Wall hung flush valve type elongated vitreous china bowl, cast iron carrier

	Craft@Hrs	Unit	Material	Labor	Total
Water closet, trim and valves (add rough-in)	P6@3.55	Ea	337.00	128.00	465.00
Add for bed pan cleanser (flush valve and rim lugs)	P6@1.50	Ea	167.00	54.20	221.20

Urinals

Wall hung urinal, vitreous china, washout flush action, 3/4" top spud with hand operated flush valve/vacuum breaker

	Craft@Hrs	Unit	Material	Labor	Total
Urinal, trim and valves (add rough-in)	P6@2.35	Ea	284.00	84.90	368.90
Add for fixture carrier	P6@1.25	Ea	90.00	45.10	135.10

Floor mounted urinal, 18" wide, sloping front vitreous china, washout flush action, 3/4" top spud with hand operated flush valve/vacuum breaker

	Craft@Hrs	Unit	Material	Labor	Total
Urinal, trim and valves (add rough-in)	P6@5.25	Ea	750.00	190.00	940.00

Lavatories

Wall hung vitreous china lavatory with single handle faucet and pop-up waste

	Craft@Hrs	Unit	Material	Labor	Total
Lavatory, trim and valves (add rough-in)	P6@2.50	Ea	260.00	90.30	350.30
Add for wheelchair type (ADA standards)	—	Ea	218.00	—	218.00

Countertop mounted vitreous china lavatory with single handle water and pop-up waste

	Craft@Hrs	Unit	Material	Labor	Total
Lavatory, trim and valves (add rough-in)	P6@2.00	Ea	240.00	72.20	312.20

Countertop mounted enameled steel lavatory with single handle water faucet and pop-up waste

	Craft@Hrs	Unit	Material	Labor	Total
Lavatory, trim and valves (add rough-in)	P6@2.00	Ea	185.00	72.20	257.20

	Craft@Hrs	Unit	Material	Labor	Total
Bathtubs and Showers					
Recessed enameled steel tub with shower					
Tub, trim and valves (add rough-in)	P6@2.50	Ea	340.00	90.30	430.30
Recessed enameled cast iron tub with shower					
Tub, trim and valves (add rough-in)	P6@4.50	Ea	640.00	162.00	802.00
Acrylic one-piece tub and shower enclosure					
Enclosure, trim and valves (add rough-in)	P6@4.75	Ea	610.00	172.00	782.00
Acrylic three-piece tub and shower enclosure (renovation)					
Enclosure, trim and valves (add rough-in)	P6@5.75	Ea	820.00	208.00	1,028.00
Fiberglass one-piece tub and shower enclosure					
Enclosure, trim and valves (add rough-in)	P6@4.75	Ea	450.00	172.00	622.00
Fiberglass three-piece tub and shower enclosure					
Enclosure, trim and valves (add rough-in)	P6@5.75	Ea	680.00	208.00	888.00
Acrylic one-piece shower stall					
Shower stall, trim and valves (add rough-in)	P6@3.50	Ea	580.00	126.00	706.00
Add for larger stall	—	Ea	100.00	—	100.00
Acrylic three-piece shower stall (renovation)					
Shower stall, trim and valves (add rough-in)	P6@4.25	Ea	760.00	153.00	913.00
Add for larger stall	—	Ea	120.00	—	120.00
Fiberglass one-piece shower stall					
Shower stall, trim and valves (add rough-in)	P6@3.50	Ea	390.00	126.00	516.00
Add for larger stall	—	Fa	80.00	—	80.00
Add for corner entry	—	Ea	65.00	—	65.00
Fiberglass three-piece shower stall (renovation)					
Shower stall, trim and valves (add rough-in)	P6@4.25	Ea	500.00	153.00	653.00
Add for larger stall	—	Ea	110.00	—	110.00
Add for corner entry	—	Ea	90.00	—	90.00
Acrylic shower base					
Base , trim and valves (add rough-in)	P6@2.50	Ea	245.00	90.30	335.30
Add for larger base	—	Ea	10.00	—	10.00
Add for corner entry	—	Ea	10.00	—	10.00
Fiberglass shower base					
Base , trim and valves (add rough-in)	P6@2.50	Ea	275.00	90.30	365.30
Add for larger base	—	Ea	10.00	—	10.00
Add for corner entry	—	Ea	10.00	—	10.00
Tub and shower doors					
Three panel sliding tub doors	P6@.50	Ea	250.00	18.10	268.10
Three panel sliding shower doors	P6@.50	Ea	220.00	18.10	238.10
Two panel sliding shower doors	P6@.50	Ea	185.00	18.10	203.10
One panel swinging shower door	P6@.50	Ea	90.00	18.10	108.10

Kitchen sinks

Single compartment kitchen sink, 20" long, 21" wide, 8" deep, self rimming, countertop mounted, stainless steel with single handle faucet and deck spray.

	Craft@Hrs	Unit	Material	Labor	Total
Sink, trim and valves (add rough-in)	P6@2.00	Ea	180.00	72.20	252.20

Double compartment kitchen sink, 32" long, 21" wide, 8" deep, self rimming, countertop mounted, stainless steel with single handle faucet and deck spray.

	Craft@Hrs	Unit	Material	Labor	Total
Sink, trim and valves (add rough-in)	P6@2.15	Ea	230.00	77.60	307.60

Drinking fountain and water cooler

Wall hung drinking fountains, 10" wide, 10" high , non-electric, white vitreous china

	Craft@Hrs	Unit	Material	Labor	Total
Wall hung fountain, and trim (add rough-in)	P6@2.00	Ea	275.00	72.22	347.22

	Craft@Hrs	Unit	Material	Labor	Total
Recessed drinking fountains, 14" wide, 26" high, non-electric, white vitreous china,					
Recessed fountain, and trim (add rough-in)	P6@2.00	Ea	500.00	72.22	572.22
Semi-recessed water cooler, vitreous china, refrigerated, (no electrical work included)					
Semi-recessed cooler, and trim (add rough-in)	P6@2.50	Ea	850.00	90.30	940.30
Add for fully recessed	P6@.750	Ea	650.00	27.10	677.10
Wheelchair wall hung water cooler, vitreous china, refrigerated, (no electrical work included)					
Wall hung water cooler, and trim (add rough-in)	P6@2.50	Ea	675.00	90.30	765.30
Free standing water cooler, refrigerated, (no electrical work included)					
Wall hung water cooler, and trim (add rough-in)	P6@2.50	Ea	650.00	90.30	740.30

Wash fountains

	Craft@Hrs	Unit	Material	Labor	Total
54" circular wash fountain, pre-cast stone, (terrazzo) with foot operated water spray control					
Wash fountain, trim and valves (add rough-in)	P6@3.65	Ea	1,940.00	132.00	2,072.00
Add for stainless steel in lieu of pre-cast stone	—	Ea	170.00	—	170.00
54" semi-circular wash fountain, pre-cast stone, (terrazzo), with foot operated water spray control					
Wash fountain, trim and valves (add rough-in)	P6@3.65	Ea	1,740.00	132.00	1,872.00
Add for stainless steel in lieu of pre-cast stone	—	Ea	90.00	—	90.00
36" circular wash fountain, pre-cast stone, (terrazzo) with foot operated water spray control					
Wash fountain, trim and valves (add rough-in)	P6@3.50	Ea	1,560.00	126.00	1,686.00
Add for stainless steel in lieu of pre-cast stone	—	Ea	80.00	—	80.00
36" semi-circular wash fountain, pre-cast stone, (terrazzo), with foot operated water spray control					
Wash fountain, trim and valves (add rough-in)	P6@3.50	Ea	1,450.00	126.00	1,576.00
Add for stainless steel in lieu of pre-cast stone	—	Ea	40.00	—	40.00

Service sinks

	Craft@Hrs	Unit	Material	Labor	Total
Wall hung enameled cast iron slop sink, with wall-mounted faucet, hose and P strap standard					
Slop sink, trim and valves (add rough-in)	P6@3.30	Ea	620.00	119.00	739.00
Floor mounted mop sink pre-cast molded stone, (terrazzo) with wall mounted faucet and hose					
Mop sink, 24" x 24", 6" curbs (add rough-in)	P6@2.60	Ea	325.00	93.90	418.90
Mop sink, 32" x 32", 6" curbs (add rough-in)	P6@2.60	Ea	360.00	93.90	453.90
Mop sink, 36" x 36", 6" curbs (add rough-in)	P6@2.60	Ea	380.00	93.90	473.90
Add for 12" curbs, 6" drop front and stainless steel caps	—	Ea	230.00	—	230.00

Drench shower and emergency eye wash stations

	Craft@Hrs	Unit	Material	Labor	Total
Drench shower, plastic pipe, plastic head, wall mounted	P6@2.00	Ea	180.00	72.20	252.20
Drench shower galvanized pipe, free standing	P6@2.50	Ea	455.00	90.30	545.30
Drench shower, emergency eyewash combination	P6@2.75	Ea	685.00	99.30	784.30
Emergency eyewash station,	P6@2.00	Ea	200.00	72.20	272.20
Walk-thru station, 18 heads with plastic pipe	P6@12.5	Ea	2,480.00	451.00	2,931.00
Walk-thru station, 18 heads with copper pipe	P6@14.9	Ea	2,650.00	538.00	3,188.00

Surgeons' scrub sink

	Craft@Hrs	Unit	Material	Labor	Total
Wall hung wash sink, white enameled cast iron, 48" long, 18" deep, trough-type with 2 twin-handled faucets and two soap dishes					
Wash sink, trim and valves (add rough-in)	P6@2.50	Ea	575.00	90.30	665.30
Add for foot or knee pedal operation	P6@.500	Ea	85.00	18.10	103.10
Add for electronic infrared hands free operation	P6@.500	Ea	275.00	18.10	293.10
Add for stainless steel in lieu of enameled cast iron	—	Ea	80.00	—	80.00

Floor sinks

	Craft@Hrs	Unit	Material	Labor	Total
9" square sink, cast iron top and body (add rough-in)	P6@1.00	Ea	150.00	36.10	186.10
Add for an acid-resistant finish to the sink	—	Ea	25.00	—	25.00

	Craft@Hrs	Unit	Material	Labor	Total

Fire Protection

Sprinkler systems Typical subcontract prices including the subcontractor's overhead and profit. These prices include design drawings, valves and connection to piping to within 5'0" outside the building. Costs will be higher where room sizes are smaller or where coverage per head averages less than 110 SF. Make additional allowances if a booster pump is required.

	Craft@Hrs	Unit	Material	Labor	Total
Exposed systems, wet, complete, cost per SF of floor protected					
5,000 SF	—	SF	—	—	1.79
Over 5,000 to 15,000 SF	—	SF	—	—	1.52
15,000 SF or more	—	SF	—	—	1.34
Concealed systems, wet, complete, cost per SF of floor protected					
5,000 SF	—	SF	—	—	2.00
Over 5,000 to 15,000 SF	—	SF	—	—	1.61
15,000 SF or more	—	SF	—	—	1.52
Add for dry systems	—	%	—	—	15.0

Fire sprinkler system components Make additional allowances to shut down and drain the system when required.

	Craft@Hrs	Unit	Material	Labor	Total
Sprinkler heads only, (remove and replace)					
Brass pendent or upright head, 155-200 degree	SP@.350	Ea	4.60	15.50	20.10
Brass pendent or upright head, 286 degree	SP@.350	Ea	4.90	15.50	20.40
Brass pendent or upright head, 360 degree	SP@.350	Ea	5.30	15.50	20.80
Brass pendent or upright head, 400-500 degree	SP@.350	Ea	14.00	15.50	29.50
Chrome pendent or upright head, 155-200 degree	SP@.350	Ea	5.25	15.50	20.75
Chrome pendent or upright head, 286 degree	SP@.350	Ea	5.45	15.50	20.95
Dry pendent or upright head	SP@.350	Ea	42.00	15.50	57.50
Relocate wet sprinkler head and branch drop	SP@1.50	Ea	20.00	66.60	86.60
Add for heads more than 14' above floor	—	%	—	20.0	—

Wet system components (where freezing is not a hazard)

	Craft@Hrs	Unit	Material	Labor	Total
Zone valves					
2" O.S.& Y. gate valve, flanged or grooved	SP@1.75	Ea	145.00	77.70	222.70
3" O.S.& Y. gate valve, flanged or grooved	SP@2.00	Ea	175.00	88.80	263.80
4" O.S.& Y. gate valve, flanged or grooved	SP@2.25	Ea	245.00	99.80	344.80
6" O.S.& Y. gate valve, flanged or grooved	SP@3.00	Ea	385.00	133.00	518.00
8" O.S.& Y. gate valve, flanged or grooved	SP@3.50	Ea	790.00	155.00	945.00
4" alarm valve, flanged or grooved	SP@2.65	Ea	330.00	118.00	448.00
6" alarm valve, flanged or grooved	SP@3.50	Ea	415.00	155.00	570.00
8" alarm valve, flanged or grooved	SP@4.25	Ea	595.00	189.00	784.00
Alarm valve trim only (retard chamber and gauges)	SP@1.25	Ea	185.00	55.50	240.50
4"Alarm valve package, complete	SP@4.00	Ea	1,500.00	178.00	1,678.00
6"Alarm valve package, complete	SP@4.50	Ea	1,640.00	200.00	1,840.00
8"Alarm valve package, complete	SP@5.25	Ea	2,150.00	233.00	2,383.00
Retard pressure switch	SP@1.75	Ea	175.00	77.70	252.70
Check valves					
3" swing check valve	SP@2.00	Ea	135.00	88.80	223.80
4" swing check valve	SP@2.25	Ea	145.00	99.80	244.80
6" swing check valve	SP@3.00	Ea	275.00	133.00	408.00
3" wafer check valve	SP@2.00	Ea	115.00	88.80	203.80
4" wafer check valve	SP@2.25	Ea	125.00	99.80	224.80
6" wafer check valve	SP@3.00	Ea	200.00	133.00	333.00
8" wafer check valve	SP@3.50	Ea	265.00	155.00	420.00
10" wafer check valve	SP@4.25	Ea	510.00	189.00	699.00
3" double check detector assembly	SP@3.75	Ea	1,190.00	166.00	1,356.00
4" double check detector assembly	SP@4.50	Ea	1,600.00	200.00	1,800.00
6" double check detector assembly	SP@5.75	Ea	2,450.00	255.00	2,705.00
8" double check detector assembly	SP@6.25	Ea	4,550.00	277.00	4,827.00

	Craft@Hrs	Unit	Material	Labor	Total
Dry system components (distribution piping holds no water)					
Dry valves (Deluge)					
3" dry pipe valve	SP@2.25	Ea	650.00	99.80	749.80
4" dry pipe valve	SP@2.65	Ea	785.00	118.00	903.00
6" dry pipe valve	SP@3.50	Ea	1,050.00	155.00	1,205.00
Dry valve trim and gauges	SP@1.00	Ea	260.00	44.40	304.40
Wall hydrants for sprinkler systems, brass					
Single outlet, 2-1/2" x 2-1/2"	SP@2.00	Ea	234.00	88.80	322.80
2-way outlet, 2-1/2" x 2-1/2" x 4"	SP@2.50	Ea	375.00	111.00	486.00
3-way outlet, 2-1/2" x 2-1/2" x 3-1/3" x 4"	SP@2.75	Ea	685.00	122.00	807.00
Fire dept. connection, 4" (Siamese)	SP@2.50	Ea	375.00	111.00	486.00
Pumper connection, 6"	SP@3.00	Ea	742.00	133.00	875.00
Roof manifold, with valves	SP@3.90	Ea	439.00	173.00	612.00
Fire hose cabinets recessed, 24" x 30" x 5 1/2" with rack, full glass door, 1-1/2" hose, valve and nozzle, Standard					
Painted steel, 75' hose, 24" x 30"	SP@1.55	Ea	300.00	68.80	368.80
Painted steel, 100' hose, 24" x 36"	SP@1.55	Ea	335.00	68.80	403.80
Aluminum, 75' hose	SP@1.55	Ea	465.00	68.80	533.80
Stainless steel, 75' hose	SP@1.55	Ea	600.00	68.80	668.80
Fire extinguishers factory charged, complete with hose and horn and wall mounting bracket					
5 lb carbon dioxide	C8@.488	Ea	179.00	17.70	196.70
10 lb carbon dioxide	C8@.488	Ea	260.00	17.70	277.70
15 lb carbon dioxide	C8@.488	Ea	354.00	17.70	371.70
5 lb dry chemical	C8@.488	Ea	56.70	17.70	74.40
10 lb dry chemical	C8@.488	Ea	87.10	17.70	104.80
20 lb dry chemical	C8@.488	Ea	142.00	17.70	159.70
Extinguisher cabinets no extinguishers included, painted steel, recessed					
9" x 24" x 5" full glass door	C8@1.67	Ea	70.90	60.60	131.50
9" x 24" x 5" break glass door	C8@1.67	Ea	85.80	60.60	146.40
12" x 27" x 8" full glass door	C8@1.67	Ea	101.00	60.60	161.60
12" x 27" x 8" break glass door	C8@1.67	Ea	115.00	60.60	175.60
Add for semi-recessed or surface mount	—	%	10.0	—	—
Add for chrome cabinets	—	%	15.0	—	—
Add for stainless steel cabinets	—	Ea	160.00	—	160.00
Miscellaneous sprinkler system components					
Air maintenance device	SP@.781	Ea	136.00	34.70	170.70
Low air supervisory unit	SP@1.11	Ea	424.00	49.30	473.30
Low air pressure switch	SP@.784	Ea	136.00	34.80	170.80
Pressure switch (double circuit, open and close)	SP@1.92	Ea	132.00	85.20	217.20
1/2" ball drip at check valve and fire dept. connection	SP@.193	Ea	15.30	8.56	23.86
Water powered gong (local alarm)	SP@1.94	Ea	166.00	86.10	252.10
Cabinet with 6 spare heads and wrench	SP@.291	LS	63.70	12.90	76.60
Inspector's test connection	SP@.973	Ea	53.00	43.20	96.20
Escutcheon for sprinkler heads, chrome	SP@.071	Ea	.90	3.15	4.05
Field testing and flushing, subcontract	—	LS	—	—	150.00
Disinfection of distribution system, subcontract	—	LS	—	—	150.00
Hydraulic design fee, subcontract					
Sprinklers in a high density or high risk area, typical	—	LS	—	—	2,300.00
Underground valve vault (if required), subcontract	—	LS	—	—	2,770.00

	Craft@Hrs	Unit	Material	Labor	Total

Tanks Including fittings but no excavation, dewatering, shoring, supports, piping, backfill, paving, or concrete.

Underground steel oil storage tanks

550 gallon	Ml@4.86	Ea	979.00	188.00	1,167.00
2,500 gallon	M8@18.5	Ea	2,280.00	733.00	3,013.00
5,000 gallon	M8@24.0	Ea	4,100.00	952.00	5,052.00
10,000 gallon	M8@37.3	Ea	7,010.00	1,480.00	8,490.00

Above ground steel oil storage tanks

275 gallon	Ml@2.98	Ea	342.00	115.00	457.00
500 gallon	Ml@3.80	Ea	729.00	147.00	876.00
1,000 gallon	M8@7.64	Ea	1,230.00	303.00	1,533.00
1,500 gallon	M8@8.29	Ea	1,340.00	329.00	1,669.00
2,000 gallon	M8@10.5	Ea	1,740.00	416.00	2,156.00
5,000 gallon	M8@29.6	Ea	4,810.00	1,170.00	5,980.00
Add for vent caps	M8@.155	Ea	1.14	6.15	7.29
Add for filler cap	M8@.155	Ea	24.90	6.15	31.05

Underground fiberglass oil storage tanks

550 gallon	Ml@8.36	Ea	2,220.00	323.00	2,543.00
1,000 gallon	M8@16.8	Ea	4,010.00	666.00	4,676.00
2,000 gallon	M8@16.8	Ea	4,530.00	666.00	5,196.00
4,000 gallon	M8@25.9	Ea	5,500.00	1,030.00	6,530.00
6,000 gallon	M8@33.5	Ea	6,130.00	1,330.00	7,460.00
8,000 gallon	M8@33.5	Ea	6,900.00	1,330.00	8,230.00
10,000 gallon	M8@33.5	Ea	7,570.00	1,330.00	8,900.00
12,000 gallon	M8@84.4	Ea	9,330.00	3,350.00	12,680.00
15,000 gallon	M8@89.0	Ea	10,900.00	3,530.00	14,430.00
20,000 gallon	M8@89.0	Ea	12,900.00	3,530.00	16,430.00

Propane tanks with valves, underground

1,000 gallon	M8@10.0	Ea	1,930.00	396.00	2,326.00

Heat Generation

Gas fired heating equipment No gas pipe or electric runs included
Unit heaters, gas fired, ceiling suspended, with flue and gas valve, propeller fan with aluminized heat exchanger, Modine PAE series

50 MBtu input	M5@4.75	Ea	550.00	173.00	723.00
75 MBtu input	M5@4.75	Ea	600.00	173.00	773.00
125 MBtu input	M5@5.25	Ea	755.00	192.00	947.00
175 MBtu input	M5@5.50	Ea	845.00	201.00	1,046.00
225 MBtu input	Ml@6.75	Ea	960.00	261.00	1,221.00
300 MBtu input	Ml@7.50	Ea	1,200.00	290.00	1,490.00
400 MBtu input	Ml@8.75	Ea	1,575.00	338.00	1,913.00
Add for single stage intermittent ignition pilot	—	LS	131.00	—	131.00
Add for high efficiency units (PV series)	—	%	35.0	—	—
Add for stainless steel burner	—	%	20.0	—	—
Add for air blower style fan	—	%	10.0	—	—
Add for switch and thermostat	CE@1.00	Ea	33.80	40.10	73.90
Add for gas supply pipe connection	M5@1.75	Ea	55.00	63.90	118.90
Add for typical electric run, BX	E4@2.50	Ea	70.70	87.90	158.60

Furnaces, wall type, upflow, gas fired with gas valve, flue vent and electronic ignition

35 MBtu input	M5@4.25	Ea	576.00	155.00	731.00
50 MBtu input	M5@4.25	Ea	651.00	155.00	806.00
65 MBtu input	M5@4.25	Ea	657.00	155.00	812.00

	Craft@Hrs	Unit	Material	Labor	Total
Add for gas supply pipe connection	M5@1.75	Ea	55.00	63.90	118.90
Add for power vent kit (direct sidewall vent)	M5@2.50	Ea	75.00	91.30	166.30
Add for thermostat	CE@1.00	Ea	30.00	40.10	70.10
Furnaces, non condensing gas fired, multi-positional, mid-efficiency with gas valve, flue vent and electronic ignition					
40 MBtu input, 80% AFUE	M5@4.25	Ea	675.00	155.00	830.00
60 MBtu input, 80% AFUE	M5@4.25	Ea	725.00	155.00	880.00
80 MBtu input, 80% AFUE	M5@4.25	Ea	750.00	155.00	905.00
100 MBtu input, 80% AFUE	M5@4.50	Ea	775.00	164.00	939.00
120 MBtu input, 80% AFUE	M5@4.50	Ea	825.00	164.00	989.00
140 MBtu input, 80% AFUE	M5@4.50	Ea	920.00	164.00	1,084.00
Add for gas supply pipe connection	M5@1.75	Ea	55.00	63.90	118.90
Add for power vent kit (direct sidewall vent)	M5@2.50	Ea	75.00	91.30	166.30
Add for thermostat	CE@1.00	Ea	30.00	40.10	70.10
Furnaces, condensing gas fired, upflow, high-efficiency with gas valve, flue vent and electronic ignition					
50 MBtu input, 90% AFUE	M5@4.25	Ea	1,200.00	155.00	1,355.00
70 MBtu input, 90% AFUE	M5@4.25	Ea	1,320.00	155.00	1,475.00
90 MBtu input, 90% AFUE	M5@4.25	Ea	1,330.00	155.00	1,485.00
110 MBtu input, 90% AFUE	M5@4.50	Ea	1,425.00	164.00	1,589.00
130 MBtu input, 90% AFUE	M5@4.50	Ea	1,425.00	164.00	1,589.00
Add for gas supply pipe connection	M5@1.75	Ea	55.00	63.90	118.90
Add for power vent kit (direct sidewall vent)	M5@2.50	Ea	75.00	91.30	166.30
Air curtain, gas fired, ceiling suspended, includes flue, vent, and electrical ignition					
256 MBtu, 2,840 CFM	MI@7.50	Ea	1,310.00	290.00	1,600.00
290 MBtu, 4,740 CFM	MI@8.25	Ea	1,420.00	318.00	1,738.00
400 MBtu, 5,660 CFM	MI@9.75	Ea	1,590.00	376.00	1,966.00
525 MBtu, 7,580 CFM	MI@11.0	Ea	1,680.00	425.00	2,105.00
630 MBtu, 9,480 CFM	MI@12.5	Ea	1,840.00	483.00	2,323.00
850 MBtu, 11,400 CFM	MI@16.0	Ea	2,630.00	618.00	3,248.00
Add for gas supply pipe connection	M5@1.75	Ea	55.00	63.90	118.90

Electric Heat

	Craft@Hrs	Unit	Material	Labor	Total
Air curtain electric heaters, wall mounted, includes electrical connection only					
37" long, 9.5 kW	H9@6.33	Ea	542.00	258.00	800.00
49" long, 12.5 kW	H9@8.05	Ea	767.00	328.00	1,095.00
61" long, 16.0 kW	H9@11.2	Ea	921.00	456.00	1,377.00
Baseboard electric heaters, with remote thermostat, 185 watts per LF	CE@.175	LF	18.20	7.01	25.21

	Craft@Hrs	Unit	Material	Labor	Equipment	Total
Hydronic hot water generators 30 PSI, with cast iron burners, factory installed boiler trim and circulating pump. For commercial and light industrial applications. Equipment cost is $11.00 per hour for a 2 ton capacity forklift						
4 BHP, 133 MBtu/Hr	M5@4.00	Ea	1,550.00	146.00	32.60	1,728.60
8 BHP, 263 MBtu/Hr	M5@4.00	Ea	2,070.00	146.00	32.60	2,248.60
12 BHP, 404 MBtu/Hr	M5@4.00	Ea	2,660.00	146.00	32.60	2,838.60
15 BHP, 514 MBtu/Hr	M5@5.00	Ea	4,030.00	183.00	40.75	4,253.75
19 BHP, 624 MBtu/Hr	M5@5.00	Ea	5,340.00	183.00	40.75	5,563.75
22 BHP, 724 MBtu/Hr	M5@6.00	Ea	5,560.00	219.00	48.90	5,827.90
29 BHP, 962 MBtu/Hr	M5@6.00	Ea	6,630.00	219.00	48.90	6,897.90
34 BHP, 1,125 MBtu/Hr	MI@8.00	Ea	7,390.00	309.00	43.60	7,742.60
40 BHP, 1,336 MBtu/Hr	MI@8.00	Ea	8,530.00	309.00	43.60	8,882.60
47 BHP, 1,571 MBtu/Hr	MI@9.00	Ea	8,840.00	347.00	48.90	9,235.90
53 BHP, 1,758 MBtu/Hr	MI@10.0	Ea	9,630.00	386.00	54.30	10,070.30
63 BHP, 2,100 MBtu/Hr	MI@12.0	Ea	9,990.00	463.00	65.20	10,518.20
89 BHP, 3,001 MBtu/Hr	MI@18.0	Ea	12,800.00	695.00	97.80	13,592.80
118 BHP, 4,001 MBtu/Hr	MI@22.0	Ea	13,200.00	849.00	120.00	14,169.00

	Craft@Hrs	Unit	Material	Labor	Total

Hydronic Heating

Hot water baseboard fin tube radiation, copper tube and aluminum fin element, commercial quality, 18 gauge enclosure. Make additional allowances for control wiring

	Craft@Hrs	Unit	Material	Labor	Total
9-7/8" enclosure and 1-1/4" element	P6@.280	LF	16.70	10.10	26.80
13-7/8" enclosure and 1-1/4" element, 2 rows	P6@.320	LF	28.90	11.60	40.50
9-7/8" enclosure only	P6@.130	LF	5.24	4.69	9.93
13-7/8" enclosure only	P6@.150	LF	8.06	5.42	13.48
1-1/4" element only	P6@.150	LF	11.74	5.42	17.16
1" element only	P6@.150	LF	5.96	5.42	11.38
3/4" element only	P6@.140	LF	4.78	5.06	9.84
Add for pipe connection and control valve	P6@2.25	Ea	110.00	81.20	191.20
Add for corners, fillers and caps, average per foot	—	%	10.0	—	—
End cap, 4"	P6@.090	Ea	3.60	3.25	6.85
End cap, 2"	P6@.090	Ea	1.80	3.25	5.05
Valve access door	P6@.250	Ea	9.08	9.03	18.11
Inside corner	P6@.120	Ea	3.58	4.33	7.91
Outside corner	P6@.120	Ea	5.10	4.33	9.43
Filler sleeve	P6@.090	Ea	2.25	3.25	5.50

Hot water baseboard fin tube radiation, steel element, commercial quality, 16 gauge, sloping top cover. Make additional allowances for control wiring

	Craft@Hrs	Unit	Material	Labor	Total
9-7/8" enclosure and 1-1/4" steel element	P6@.330	LF	19.30	11.90	31.20
9-7/8" enclosure only	P6@.130	LF	5.80	4.69	10.49
1-1/4" steel element only	P6@.160	LF	13.50	5.78	19.28
Add for pipe connection and control valve	P6@2.25	Ea	110.00	81.20	191.20
Add for corners, fillers and caps, average per foot	—	%	10.0	—	—
End cap, 4"	P6@.090	Ea	6.92	3.25	10.17
End cap, 2"	P6@.090	Ea	3.31	3.25	6.56
Valve access door	P6@.250	Ea	9.08	9.03	18.11
Inside corner	P6@.120	Ea	6.06	4.33	10.39
Outside corner	P6@.120	Ea	8.16	4.33	12.49
Filler sleeve	P6@.090	Ea	4.40	3.25	7.65

Unit heaters, hot water, horizontal, fan driven

	Craft@Hrs	Unit	Material	Labor	Total
12.5 MBtu, 200 CFM	M5@2.00	Ea	242.00	73.00	315.00
17 MBtu, 300 CFM	M5@2.00	Ea	268.00	73.00	341.00
25 MBtu, 500 CFM	M5@2.50	Ea	305.00	91.30	396.30
30 MBtu, 700 CFM	M5@3.00	Ea	322.00	110.00	432.00
50 MBtu, 1,000 CFM	M5@3.50	Ea	373.00	128.00	501.00
60 MBtu, 1,300 CFM	M5@4.00	Ea	443.00	146.00	589.00

Unit heaters, hot water, vertical, fan driven

	Craft@Hrs	Unit	Material	Labor	Total
12.5 MBtu, 200 CFM	M5@2.00	Ea	271.00	73.00	344.00
17 MBtu, 300 CFM	M5@2.00	Ea	276.00	73.00	349.00
25 MBtu, 500 CFM	M5@2.50	Ea	357.00	91.30	448.30
30 MBtu, 700 CFM	M5@3.00	Ea	374.00	110.00	484.00
50 MBtu, 1,000 CFM	M5@3.50	Ea	449.00	128.00	577.00
60 MBtu, 1,300 CFM	M5@4.00	Ea	495.00	146.00	641.00

Steam Heating

Steam baseboard radiation, per linear foot, wall mounted

	Craft@Hrs	Unit	Material	Labor	Total
1/2" tube and cover	P6@.456	LF	6.42	16.50	22.92
3/4" tube and cover	P6@.493	LF	6.90	17.80	24.70
3/4" tube and cover, high capacity	P6@.493	LF	8.17	17.80	25.97
1/2" tube alone	P6@.350	LF	3.50	12.60	16.10
3/4" tube alone	P6@.377	LF	3.98	13.60	17.58
Cover only	P6@.137	LF	4.19	4.95	9.14

	Craft@Hrs	Unit	Material	Labor	Total
Unit heater, steam, horizontal, fan driven					
18 MBtu, 300 CFM	M5@2.00	Ea	222.00	73.00	295.00
45 MBtu, 500 CFM	M5@2.50	Ea	367.00	91.30	458.30
60 MBtu, 700 CFM	M5@3.00	Ea	414.00	110.00	524.00
85 MBtu, 1,000 CFM	M5@3.50	Ea	437.00	128.00	565.00
Unit heater, steam, vertical, fan driven					
12.5 MBtu, 200 CFM	M5@2.00	Ea	202.00	73.00	275.00
17 MBtu, 300 CFM	M5@2.00	Ea	205.00	73.00	278.00
40 MBtu, 500 CFM	M5@2.50	Ea	397.00	91.30	488.30
60 MBtu, 700 CFM	M5@3.00	Ea	460.00	110.00	570.00
70 MBtu, 1,000 CFM	M5@3.50	Ea	580.00	128.00	708.00

	Craft@Hrs	Unit	Material	Labor	Equipment	Total
ASME Tanks Add the cost of a concrete pad and an insulation jacket						
Expansion tanks, galvanized steel, ASME code, 125 PSI						
15 gallons	P6@1.75	Ea	384.00	63.20	—	447.20
30 gallons	P6@2.00	Ea	415.00	72.20	—	487.20
60 gallons	P6@2.70	Ea	510.00	97.50	—	607.50
80 gallons	P6@2.95	Ea	549.00	107.00	—	656.00

Vertical cylindrical, epoxy-lined, with pipe legs, two 1/2" openings, a 1" temperature and pressure relief valve opening and a 1" drain opening. Equipment cost is $14.00 per hour for a 2 ton forklift. Make additional allowances for pipe connections.

	Craft@Hrs	Unit	Material	Labor	Equipment	Total
115 gallons, 24" dia	Ml@4.00	Ea	1,570.00	154.00	18.70	1,742.70
261 gallons, 30" dia	Ml@4.00	Ea	2,150.00	154.00	18.70	2,322.70
327 gallons, 36" dia	Ml@4.00	Ea	2,510.00	154.00	18.70	2,682.70
462 gallons, 42" dia	Ml@4.00	Ea	3,790.00	154.00	18.70	3,962.70
534 gallons, 42" dia	Ml@4.00	Ea	3,990.00	154.00	18.70	4,162.70
606 gallons, 42" dia	Ml@5.00	Ea	4,340.00	193.00	23.30	4,556.30
712 gallons, 48" dia	Ml@5.00	Ea	5,160.00	193.00	23.30	5,376.30
803 gallons, 54" dia	Ml@5.00	Ea	6,170.00	193.00	23.30	6,386.30
922 gallons, 54" dia	Ml@6.00	Ea	6,470.00	232.00	28.00	6,730.00
1,017 gallons, 60" dia	Ml@6.00	Ea	7,260.00	232.00	28.00	7,520.00
1,164 gallons, 60" dia	Ml@6.00	Ea	7,610.00	232.00	28.00	7,870.00
1,311 gallons, 60" dia	Ml@8.00	Ea	7,990.00	309.00	37.30	8,336.30
1,486 gallons, 60" dia	Ml@8.00	Ea	8,320.00	309.00	37.30	8,666.30
1,616 gallons, 66" dia	Ml@8.00	Ea	9,520.00	309.00	37.30	9,866.30
1,953 gallons, 72" dia	Ml@8.00	Ea	10,900.00	309.00	37.30	11,246.30
2,378 gallons, 72" dia	Ml@8.00	Ea	12,700.00	309.00	37.30	13,046.30
2,590 gallons, 72" dia	Ml@8.00	Ea	13,100.00	309.00	37.30	13,446.30
3,042 gallons, 84" dia	Ml@8.00	Ea	15,500.00	309.00	37.30	15,846.30
4,500 gallons, 96" dia	Ml@8.00	Ea	21,200.00	309.00	37.30	21,546.30
5,252 gallons, 96" dia	Ml@8.00	Ea	28,100.00	309.00	37.30	28,446.30

Compressors Costs shown include the concrete mounting pad, vibration isolation and anchor bolting, steel stand and structural supports, rigging, bolts and gasket sets for flanges, integrated intercooling and lubrication systems, typical vendor engineering and calibration, and inlet air cleaners. Add the cost of electrical connections, control valves, service bypass/isolation piping and the cost of piping to load and control valves. Equipment cost is $14.00 per hour for a 2 ton forklift.

	Craft@Hrs	Unit	Material	Labor	Equipment	Total
16.5 CFM at 175 PSI, 5 HP,						
80 gallon tank	Ml@3.50	Ea	2,050.00	135.00	16.40	2,201.40
34.4 CFM at 175 PSI, 10 HP,						
120 gallon tank	Ml@5.00	Ea	3,500.00	193.00	23.30	3,716.30

	Craft@Hrs	Unit	Material	Labor	Equipment	Total
53.7 CFM at 175 PSI, 15 HP, 120 gallon tank	Ml@8.00	Ea	4,970.00	309.00	37.30	5,316.30
91.0 CFM at 175 PSI, 25 HP, 120 gallon tank	Ml@10.0	Ea	6,190.00	386.00	46.60	6,622.60

	Craft@Hrs	Unit	Material	Labor	Total
Hydronic Heating Specialties					
Air eliminator-purger, screwed connections					
3/4", cast iron, steam & water, 15 PSI	PF@.250	Ea	285.00	10.70	295.70
3/4", cast iron, 150 PSI water	PF@.250	Ea	92.00	10.70	102.70
1", cast iron, 150 PSI water	PF@.250	Ea	105.00	10.70	115.70
3/8", brass, 125 PSI steam	PF@.250	Ea	75.00	10.70	85.70
1/2", cast iron, 250 PSI steam	PF@.250	Ea	155.00	10.70	165.70
3/4", cast iron, 250 PSI steam	PF@.250	Ea	195.00	10.70	205.70
Airtrol fitting, 3/4"	PF@.300	Ea	33.00	12.80	45.80
Air eliminator vents, 1/4"	P1@.150	Ea	11.60	3.67	15.27
Atmospheric vacuum breakers, screwed					
1/2" vacuum breaker	PF@.210	Ea	16.00	8.98	24.98
3/4" vacuum breaker	PF@.250	Ea	17.00	10.70	27.70
1" vacuum breaker	PF@.300	Ea	27.00	12.80	39.80
1-1/4" vacuum breaker	PF@.400	Ea	45.00	17.10	62.10
1-1/2" vacuum breaker	PF@.450	Ea	53.00	19.20	72.20
2" vacuum breaker	PF@.500	Ea	79.00	21.40	100.40
Circuit Balancing valves, brass, threaded					
1/2" circuit balancing valve	PF@.210	Ea	19.50	8.98	28.48
3/4" circuit balancing valve	PF@.250	Ea	25.00	10.70	35.70
1" circuit balancing valve	PF@.300	Ea	32.00	12.80	44.80
1-1/2" circuit balancing valve	PF@.450	Ea	42.00	19.20	61.20
Flow check valves, brass, threaded, horizontal type					
3/4" check valve	PF@.250	Ea	16.00	10.70	26.70
1" check valve	PF@.300	Ea	23.00	12.80	35.80
1-1/2" valve	PF@.350	Ea	50.80	15.00	65.80
2" valve	PF@.400	Ea	51.00	17.10	68.10
Pressure reducing valves, bronze body					
1/2", 25 – 75 PSI	PF@.210	Ea	68.00	8.98	76.98
1/2", 10 – 35 PSI	PF@.210	Ea	79.00	8.98	87.98
3/4", 25 – 75 PSI	PF@.250	Ea	81.50	10.70	92.20
3/4", 10 – 35 PSI	PF@.250	Ea	92.00	10.70	102.70
1", - 25 – 75 PSI	PF@.300	Ea	128.00	12.80	140.80
1", - 10 – 35 PSI	PF@.300	Ea	140.00	12.80	152.80
1-1/4", 25 – 75 PSI	PF@.350	Ea	258.00	15.00	273.00
1-1/4", 10 – 35 PSI	PF@.350	Ea	280.00	15.00	295.00
1-1/2", 25 – 75 PSI	PF@.250	Ea	277.00	10.70	287.70
1-1/2", 10 – 35 PSI	PF@.250	Ea	317.00	10.70	327.70
2", 25 – 75 PSI	PF@.250	Ea	414.00	10.70	424.70
2", 10 – 35 PSI	PF@.250	Ea	437.00	10.70	447.70
Pressure regulators, feed water 25 – 75 PSI, cast iron, screwed connections					
3/4" regulator	PF@.250	Ea	49.75	10.70	60.45
1" regulator	PF@.300	Ea	77.00	12.80	89.80
1-1/4" regulator	PF@.400	Ea	105.00	17.10	122.10

	Craft@Hrs	Unit	Material	Labor	Total
1-1/2" regulator	PF@.450	Ea	150.00	19.20	169.20
2" regulator	PF@.600	Ea	240.00	25.70	265.70
2-1/2" regulator	PF@.700	Ea	360.00	29.90	389.90
Combination pressure relief and reducing valves					
1/2", brass body	PF@.210	Ea	61.50	8.98	70.48
3/4", cast iron body	PF@.250	Ea	95.50	10.70	106.20
Wye pattern strainers, screwed connections, 250 PSI, cast iron body					
3/4" strainer	PF@.260	Ea	7.50	11.10	18.60
1" strainer	PF@.330	Ea	9.50	14.10	23.60
1-1/4" strainer	PF@.440	Ea	13.75	18.80	32.55
1-1/2" strainer	PF@.495	Ea	16.00	21.20	37.20
2" strainer	PF@.550	Ea	25.00	23.50	48.50
Wye pattern strainers, screwed connections, 250 PSI, cast bronze body					
1/2"	PF@.210	Ea	16.50	8.98	25.48
3/4"	PF@.250	Ea	22.45	10.70	33.15
1"	PF@.300	Ea	28.50	12.80	41.30
1-1/4"	PF@.400	Ea	40.25	17.10	57.35
1-1/2"	PF@.450	Ea	52.25	19.20	71.45
2"	PF@.500	Ea	90.00	21.40	111.40
Wye pattern strainers, screwed connections, 600 PSI, cast steel body					
3/4" strainer	PF@.260	Ea	162.00	11.10	173.10
1" strainer	PF@.330	Ea	227.00	14.10	241.10
1-1/4" strainer	PF@.440	Ea	275.00	18.80	293.80
1-1/2" strainer	PF@.495	Ea	375.00	21.20	396.20
2" strainer	PF@.550	Ea	522.00	23.50	545.50
Wye pattern strainers, flanged connections, 125 PSI, cast iron body					
2"	PF@.500	Ea	67.00	21.40	88.40
2-1/2"	PF@.600	Ea	69.00	25.70	94.70
3"	PF@.750	Ea	80.00	32.10	112.10
4"	PF@1.35	Ea	137.00	57.70	194.70
6"	PF@2.50	Ea	277.00	107.00	384.00
8"	PF@3.00	Ea	470.00	128.00	598.00
Tempering valves, screwed connections, high temperature					
3/4", screwed	PF@.250	Ea	162.00	10.70	172.70
1", screwed	PF@.300	Ea	178.00	12.80	190.80
1-1/4", screwed	PF@.400	Ea	283.00	17.10	300.10
1-1/2", screwed	PF@.450	Ea	309.00	19.20	328.20
2", screwed	PF@.500	Ea	465.00	21.40	486.40
Thermostatic mixing valves 110 to 150 degrees F					
1/2", soldered	PF@.240	Ea	35.00	10.30	45.30
3/4", threaded	PF@.250	Ea	45.00	10.70	55.70
3/4", soldered	PF@.300	Ea	38.50	12.80	51.30
1", threaded	PF@.300	Ea	175.00	12.80	187.80
Float and thermostatic traps, cast iron body, parallel connection					
3/4" (15 to 75 PSI)	PF@.250	Ea	102.00	10.70	112.70
1" (15 to 30 PSI)	PF@.300	Ea	102.00	12.80	114.80
3/4" – 1" (125 PSI)	PF@.250	Ea	108.00	10.70	118.70
3/4" – 1" (150 PSI)	PF@.300	Ea	330.00	10.40	166.40
3/4" – 1" (250 PSI)	PF@.300	Ea	345.00	12.80	357.80
1-1/4" (15 to 30 PSI)	PF@.450	Ea	122.00	19.20	141.20
1-1/4" – 1-1/2" (75 PSI)	PF@.450	Ea	233.00	19.20	252.20
1-1/4" – 1-1/2" (125 PSI)	PF@.500	Ea	240.00	21.40	261.40

Mechanical 15

	Craft@Hrs	Unit	Material	Labor	Total
1-1/4" – 1-1/2" (150 PSI)	PF@.500	Ea	545.00	21.40	566.40
1-1/4" – 1-1/2" (250 PSI)	PF@.500	Ea	555.00	21.40	576.40
1-1/2" (15 to 30 PSI)	PF@.450	Ea	230.00	19.20	249.20
2" (15 to 30 PSI)	PF@.450	Ea	297.00	19.20	316.20
2" (75 PSI)	PF@.550	Ea	320.00	23.50	343.50
2" (125 PSI)	PF@.600	Ea	330.00	25.70	355.70
2-1/2" (15 to 250 PSI)	PF@.800	Ea	1,540.00	34.20	1,574.20
Liquid level gauges					
3/4", aluminum	PF@.250	Ea	227.00	10.70	237.70
3/4", 125 PSI PVC	PF@.250	Ea	238.00	10.70	248.70
1/2", 175 PSI bronze	PF@.210	Ea	45.70	8.98	54.68
3/4", 150 PSI stainless steel	PF@.250	Ea	218.00	10.70	228.70
1", 150 PSI stainless steel	PF@.300	Ea	244.00	12.80	256.80

Electrical 16

	Craft@Hrs	Unit	Material	Labor	Total

Typical In-Place Costs Preliminary estimates per square foot of floor area for all electrical work, including service entrance, distribution and lighting fixtures, by building type. These costs include the subcontractor's overhead and profit

	Craft@Hrs	Unit	Material	Labor	Total
Commercial stores	—	SF	—	—	6.27
Market buildings	—	SF	—	—	7.04
Recreation facilities	—	SF	—	—	7.60
Schools	—	SF	—	—	8.87
Colleges	—	SF	—	—	12.00
Hospitals	—	SF	—	—	19.40
Office buildings	—	SF	—	—	7.65
Warehouses	—	SF	—	—	3.47
Parking lots	—	SF	—	—	1.27
Garages	—	SF	—	—	5.66

Concrete Envelopes for Duct No wire, duct, excavation, pavement removal or replacement included. Based on ready-mix concrete at $60.00 per CY before allowance for waste

	Craft@Hrs	Unit	Material	Labor	Total
9" x 9" (.56 CF per linear foot)	E2@.020	LF	1.29	.70	1.99
12" x 12" (1.00 CF per linear foot)	E2@.036	LF	2.29	1.27	3.56
12" x 24" (2.00 CF per linear foot)	E2@.072	LF	4.57	2.53	7.10
Per CY	E2@1.21	CY	61.80	42.60	104.40

Precast Concrete Pull and Junction Boxes Reinforced concrete with appropriate inserts, cast iron frame, cover and pulling hooks. Overall dimensions as shown.

	Craft@Hrs	Unit	Material	Labor	Total
Precast handholes, 4' deep					
2' wide, 3' long	E1@4.24	Ea	684.00	152.00	836.00
3' wide, 3' long	E1@5.17	Ea	1,010.00	186.00	1,196.00
4' wide, 4' long	E1@13.2	Ea	1,430.00	474.00	1,904.00
Precast power manholes, 7' deep					
4' wide, 6' long	E1@24.6	Ea	2,110.00	884.00	2,994.00
6' wide, 8' long	E1@26.6	Ea	2,320.00	956.00	3,276.00
8' wide, 10' long	E1@26.6	Ea	2,470.00	956.00	3,426.00

	Craft@Hrs	Unit	Material	Labor	Total

Light and Power Circuits Costs per circuit including stranded THHN copper wire (except where noted), conduit, handy box, cover, straps, fasteners and connectors, but without fixture, receptacle or switch, except as noted.

Circuits for an exit light, fire alarm bell or suspended ceiling lighting system, 20 amp, with 4" junction box and 30' of 1/2" EMT conduit and copper THHN wire.

3 #12 solid wire	E4@2.68	Ea	23.60	94.20	117.80
4 #12 solid wire	E4@2.90	Ea	25.60	102.00	127.60
5 #12 solid wire	E4@3.14	Ea	27.60	110.00	137.60

Under slab circuit, with 4" square junction box, 10' of 1/2" RSC and 20' of 1/2" PVC conduit

3 #12 wire, 20 amp	E4@1.75	Ea	18.20	61.50	79.70
4 #12 wire, 20 amp	E4@1.97	Ea	20.20	69.30	89.50
5 #12 wire, 20 amp	E4@2.21	Ea	22.20	77.70	99.90
3 #10 wire, 30 amp	E4@2.00	Ea	21.60	70.30	91.90
4 #10 wire, 30 amp	E4@2.31	Ea	24.70	81.20	105.90

Under slab circuit, with 4" square junction box, 10' of 1/2" RSC conduit, 20' of 1/2" PVC conduit and 6' of liquid tight flexible conduit

3 #12 wire, 20 amp	E4@2.41	Ea	34.70	84.80	119.50
4 #12 wire, 20 amp	E4@2.69	Ea	36.70	94.60	131.30
3 #10 wire, 30 amp	E4@2.75	Ea	38.10	96.70	134.80
4 #10 wire, 30 amp	E4@3.81	Ea	41.20	134.00	175.20

20 amp switch circuit, 277 volt, with 30' of 1/2" EMT conduit, switch, cover and 2 indenter connectors

One gang box

Single pole switch	E4@2.82	Ea	40.80	99.20	140.00
Three-way switch	E4@2.93	Ea	42.30	103.00	145.30
Four-way switch	E4@3.14	Ea	68.40	110.00	178.40

Two gang box

Single pole switch	E4@2.95	Ea	43.40	104.00	147.40
Three-way switch	E4@3.08	Ea	44.80	108.00	152.80
Four-way switch	E4@3.24	Ea	70.90	114.00	184.90

15 amp switch circuit, 125 volt, with 30' of #14 two conductor Romex non-metallic cable (NM) with ground, switch, cover and switch connection

One gang box

Single pole switch	E4@1.05	Ea	12.20	36.90	49.10
Three-way switch	E4@1.18	Ea	18.40	41.50	59.90

Two gang box

Single pole switch	E4@1.17	Ea	18.40	41.10	59.50
Single pole switch & receptacle	E4@1.36	Ea	23.40	47.80	71.20
Single pole & 3-way switch	E4@1.43	Ea	22.40	50.30	72.70
One two pole & 3-way switch	E4@1.66	Ea	32.30	58.40	90.70

20 amp 2 pole duplex receptacle outlet circuit, 125 volt, 30' of 1/2" EMT conduit, wire and specification grade receptacle, connectors and box

3 #12 wire	E4@2.80	Ea	40.00	98.50	138.50
4 #12 wire	E4@3.06	Ea	42.10	108.00	150.10
5 #12 wire	E4@3.26	Ea	44.20	115.00	159.20

Duplex receptacle outlet circuit with 30' of Romex non-metallic cable (NM) with ground, receptacle, cover and connection

15 amp, 125 volt, #14/2 wire	E4@1.11	Ea	11.00	39.00	50.00
15 amp, 125 volt, #14/2 wire with ground fault circuit interrupter (GFCI)	E4@1.30	Ea	14.90	45.70	60.60
20 amp, 125 volt, #12/2 wire	E4@1.36	Ea	17.00	47.80	64.80
30 amp, 250 volt, #10/3 wire, dryer circuit	E4@1.77	Ea	23.40	62.20	85.60
50 amp, 250 volt, #8/3 wire, range circuit	E4@2.06	Ea	38.40	72.40	110.80

	Craft@Hrs	Unit	Material	Labor	Total

Rigid Galvanized Steel Conduit Installed exposed in a building either vertically or horizontally up to 10' above floor level. No wire, fittings or supports included except as noted.

Rigid galvanized steel conduit, standard wall. Add the cost of couplings, fittings, boxes supports and wire.

	Craft@Hrs	Unit	Material	Labor	Total
1/2" conduit	E4@.043	LF	.63	1.51	2.14
3/4" conduit	E4@.054	LF	.79	1.90	2.69
1" conduit	E4@.066	LF	1.04	2.32	3.36
1-1/4" conduit	E4@.077	LF	1.33	2.71	4.04
1-1/2" conduit	E4@.100	LF	1.62	3.52	5.14
2" conduit	E4@.116	LF	2.18	4.08	6.26
2-1/2" conduit	E4@.143	LF	3.58	5.03	8.61
3" conduit	E4@.164	LF	4.59	5.77	10.36
3-1/2" conduit	E4@.187	LF	5.59	6.58	12.17
4" conduit	E4@.221	LF	6.54	7.77	14.31
6" conduit	E4@.383	LF	18.60	13.50	32.10
Add for red plastic coating	—	%	80.0	—	—
Deduct for rigid steel conduit installed in a concrete slab or open trench	—	%	-5.0	-40.0	—

90 or 45 degree ells, standard weight rigid galvanized steel, threaded both ends

	Craft@Hrs	Unit	Material	Labor	Total
1/2"	E4@.170	Ea	2.16	5.98	8.14
3/4"	E4@.190	Ea	2.60	6.68	9.28
1"	E4@.210	Ea	3.85	7.39	11.24
1-1/4"	E4@.311	Ea	5.52	10.90	16.42
1-1/2"	E4@.378	Ea	7.17	13.30	20.47
2"	E4@.448	Ea	10.40	15.80	26.20
2-1/2"	E4@.756	Ea	18.90	26.60	45.50
3"	E4@1.10	Ea	27.50	38.70	66.20
3-1/2"	E4@1.37	Ea	43.10	48.20	91.30
4"	E4@1.73	Ea	49.30	60.80	110.10
Add for PVC coated ells	—	%	40.0	—	—

Rigid steel threaded couplings

	Craft@Hrs	Unit	Material	Labor	Total
1/2"	E4@.062	Ea	.91	2.18	3.09
3/4"	E4@.080	Ea	1.12	2.81	3.93
1"	E4@.092	Ea	1.66	3.24	4.90
1-1/4"	E4@.102	Ea	2.07	3.59	5.66
1-1/2"	E4@.140	Ea	2.62	4.92	7.54
2"	E4@.170	Ea	3.46	5.98	9.44
2-1/2"	E4@.202	Ea	8.09	7.10	15.19
3"	E4@.230	Ea	10.50	8.09	18.59
3-1/2"	E4@.275	Ea	14.10	9.67	23.77
4"	E4@.425	Ea	14.11	14.90	29.01

Three-piece "Erickson" unions, by pipe size

	Craft@Hrs	Unit	Material	Labor	Total
1/2"	E4@.285	Ea	2.55	10.00	12.55
3/4"	E4@.324	Ea	3.28	11.40	14.68
1"	E4@.394	Ea	6.30	13.90	20.20
1-1/4"	E4@.541	Ea	12.70	19.00	31.70
1-1/2"	E4@.645	Ea	16.00	22.70	38.70
2"	E4@.860	Ea	32.10	30.20	62.30
2-1/2"	E4@1.08	Ea	73.90	38.00	111.90
3"	E4@1.15	Ea	112.00	40.40	152.40
3-1/2"	E4@1.40	Ea	181.00	49.20	230.20
4"	E4@1.62	Ea	211.00	57.00	268.00

	Craft@Hrs	Unit	Material	Labor	Total
Type A, C, LB, E, LL or LR two hub malleable threaded conduit bodies, with covers					
1/2"	E4@.191	Ea	6.20	6.72	12.92
3/4"	E4@.238	Ea	7.41	8.37	15.78
1"	E4@.285	Ea	11.10	10.00	21.10
1-1/4"	E4@.358	Ea	19.20	12.60	31.80
1-1/2"	E4@.456	Ea	25.00	16.00	41.00
2"	E4@.598	Ea	40.00	21.00	61.00
2-1/2"	E4@1.43	Ea	88.10	50.30	138.40
3"	E4@1.77	Ea	117.00	62.20	179.20
3-1/2"	E4@2.38	Ea	161.00	83.70	244.70
4"	E4@2.80	Ea	188.00	98.50	286.50
Add for explosion-proof 2 hub junction boxes	—	%	300.0	100.0	—
Type T malleable threaded conduit boxes, with covers					
1/2"	E4@.285	Ea	7.73	10.00	17.73
3/4"	E4@.358	Ea	9.28	12.60	21.88
1"	E4@.433	Ea	13.91	15.20	29.11
1-1/4"	E4@.534	Ea	20.43	18.80	39.23
1-1/2"	E4@.679	Ea	27.20	23.90	51.10
2"	E4@.889	Ea	42.10	31.30	73.40
2-1/2"	E4@2.15	Ea	92.10	75.60	167.70
3"	E4@2.72	Ea	122.00	95.70	217.70
3-1/2"	E4@3.63	Ea	226.00	128.00	354.00
4"	E4@4.22	Ea	253.00	148.00	401.00
Add for explosion-proof 3 hub junction boxes	—	%	230.0	100.0	—
Conduit seal fittings, explosion proof, malleable iron or ferrous alloy, horizontal or vertical, male or female					
1/2"	E4@.689	Ea	11.50	24.20	35.70
3/4"	E4@.798	Ea	13.70	28.10	41.80
1"	E4@.990	Ea	17.00	34.80	51.80
1-1/4"	E4@1.19	Ea	22.40	41.80	64.20
1-1/2"	E4@1.43	Ea	32.70	50.30	83.00
2"	E4@1.76	Ea	43.00	61.90	104.90
2-1/2"	E4@1.85	Ea	73.50	65.10	138.60
3"	E4@1.97	Ea	84.00	69.30	153.30
3-1/2"	E4@2.15	Ea	177.00	75.60	252.60
4"	E4@2.33	Ea	296.00	81.90	377.90
Entrance caps, threaded					
1/2"	E4@.319	Ea	4.40	11.20	15.60
3/4"	E4@.373	Ea	5.30	13.10	18.40
1"	E4@.376	Ea	6.30	13.20	19.50
1-1/4"	E4@.482	Ea	7.86	17.00	24.86
1-1/2"	E4@.536	Ea	13.40	18.80	32.20
2"	E4@.591	Ea	20.10	20.80	40.90
2-1/2"	E4@.671	Ea	68.40	23.60	92.00
3"	E4@.749	Ea	112.00	26.30	138.30
3-1/2"	E4@.834	Ea	149.00	29.30	178.30
4"	E4@.912	Ea	156.00	32.10	188.10
Add for heights over 10' to 20'	—	%	—	20.0	—

	Craft@Hrs	Unit	Material	Labor	Total

Electric Metallic Tubing (EMT) Installed exposed in a building either vertically or horizontally up to 10' above floor level. No wire, fittings or supports included except as noted.

EMT with fittings and hangers. These figures assume a typical installation with tube bent once each 10', one set-screw coupling each 10', one-set screw connector each 50', and one stamped steel strap with lag screw each 6'. Use these costs for preliminary estimates. Typical retail prices for EMT in commercial quantities. Expect specialized supplier's prices to be somewhat higher.

	Craft@Hrs	Unit	Material	Labor	Total
1/2" EMT with fittings and hangers	E4@.059	LF	.26	2.07	2.33
3/4" EMT with fittings and hangers	E4@.067	LF	.40	2.36	2.76
1" EMT with fittings and hangers	E4@.080	LF	.65	2.81	3.46
1-1/4" EMT with fittings and hangers	E4@.092	LF	1.10	3.24	4.34
1-1/2" EMT with fittings and hangers	E4@.100	LF	1.47	3.52	4.99
2" EMT with fittings and hangers	E4@.119	LF	1.93	4.18	6.11
2-1/2" EMT with fittings and hangers	E4@.143	LF	5.62	5.03	10.65
3" EMT with fittings and hangers	E4@.150	LF	6.54	5.28	11.82
4" EMT with fittings and hangers	E4@.175	LF	8.38	6.15	14.53

EMT conduit with one-set screw coupling each 10' (no hangers included)

	Craft@Hrs	Unit	Material	Labor	Total
1/2" EMT conduit	E4@.036	LF	.24	1.27	1.51
3/4" EMT conduit	E4@.043	LF	.38	1.51	1.89
1" EMT conduit	E4@.055	LF	.63	1.93	2.56
1-1/4" EMT conduit	E4@.067	LF	1.06	2.36	3.42
1-1/2" EMT conduit	F4@.070	LF	1.41	2.46	3.87
2" EMT conduit	E4@.086	LF	1.84	3.02	4.86
2-1/2" EMT conduit	E4@.106	LF	5.36	3.73	9.09
3" EMT conduit	E4@.127	LF	6.22	4.47	10.69
4" EMT conduit	E4@.170	LF	8.25	5.98	14.23

EMT 90 or 45 degree ells

	Craft@Hrs	Unit	Material	Labor	Total
1/2" EMT ell	E4@.143	Ea	1.15	5.03	6.18
3/4" EMT ell	E4@.170	Ea	1.50	5.98	7.48
1" EMT ell	E4@.190	Ea	1.73	6.68	8.41
1-1/4" EMT ell	E4@.216	Ea	2.50	7.60	10.10
1-1/2" EMT ell	E4@.288	Ea	3.27	10.10	13.37
2" EMT ell	E4@.358	Ea	4.98	12.60	17.58
2-1/2" EMT ell	E4@.648	Ea	12.10	22.80	34.90
3" EMT ell	E4@.858	Ea	18.10	30.20	48.30
4" EMT ell	E4@1.33	Ea	28.50	46.80	75.30

EMT set screw die cast connectors

	Craft@Hrs	Unit	Material	Labor	Total
1/2" EMT connector	E4@.085	Ea	.35	2.99	3.34
3/4" EMT connector	E4@.096	Ea	.54	3.38	3.92
1" EMT connector	E4@.124	Ea	.83	4.36	5.19
1-1/4" EMT connector	E4@.143	Ea	1.75	5.03	6.78
1-1/2" EMT connector	E4@.162	Ea	2.26	5.70	7.96
2" EMT connector	E4@.237	Ea	3.42	8.33	11.75
2-1/2" EMT connector	E4@.363	Ea	12.00	12.80	24.80
3" EMT connector	E4@.479	Ea	15.10	16.80	31.90
4" EMT connector	E4@.655	Ea	22.30	23.00	45.30

Add for similar sizes of connectors

Regular grade connectors

	Craft@Hrs	Unit	Material	Labor	Total
Set screw insulated throat	—	%	30.0	—	—
Compression connectors	—	%	50.0	—	—
Insulated throat connectors	—	%	80.0	—	—
Add for heights over 10' to 20'	—	%	—	20.0	—

	Craft@Hrs	Unit	Material	Labor	Total

EMT Conduit with Wire THHN wire pulled in EMT conduit bent each 10', with one tap-on coupling each 10', one-set screw connector each 50' and a strap and lag screw each 6'. Installed exposed on walls or ceilings. Add 15% to the labor cost for heights over 10' to 20'. Use these figures for preliminary estimates.

1-1/2" EMT conduit and wire as shown below

	Craft@Hrs	Unit	Material	Labor	Total
2 #1 aluminum, 1 #8 copper	E4@.140	LF	2.33	4.92	7.25
3 #1 aluminum, 1 #8 copper	E4@.156	LF	2.87	5.49	8.36
4 #1 aluminum, 1 #8 copper	E4@.173	LF	3.41	6.08	9.49
4 #4 copper, 1 #8 copper	E4@.162	LF	2.97	5.70	8.67
2 #3 copper, 1 #8 copper	E4@.140	LF	2.27	4.92	7.19
3 #3 copper, 1 #8 copper	E4@.155	LF	2.77	5.45	8.22
4 #3 copper, 1 #8 copper	E4@.170	LF	3.28	5.98	9.26
2 #1 copper, 1 #8 copper	E4@.146	LF	2.97	5.13	8.10
3 #1 copper, 1 #8 copper	E4@.164	LF	3.83	5.77	9.60
4 #1 copper, 1 #8 copper	E4@.183	LF	4.69	6.44	11.13

2" EMT conduit and wire as shown below

	Craft@Hrs	Unit	Material	Labor	Total
2 #2/0 aluminum, 1 #6 copper	E4@.170	LF	3.18	5.98	9.16
3 #2/0 aluminum, 1 #6 copper	E4@.190	LF	3.93	6.68	10.61
4 #2/0 aluminum, 1 #6 copper	E4@.210	LF	4.68	7.39	12.07
2 #3/0 aluminum, 1 #6 copper	E4@.174	LF	3.54	6.12	9.66
3 #3/0 aluminum, 1 #6 copper	E4@.195	LF	4.46	6.86	11.32
4 #3/0 aluminum, 1 #6 copper	E4@.218	LF	5.38	7.67	13.05
2 #1/0 copper, 1 #6 copper	E4@.171	LF	3.78	6.01	9.79
3 #1/0 copper, 1 #6 copper	E4@.193	LF	4.82	6.79	11.61
4 #1/0 copper, 1 #6 copper	E4@.213	LF	5.87	7.49	13.36
2 #3/0 copper, 1 #4 copper	E4@.175	LF	5.05	6.15	11.20
3 #3/0 copper, 1 #4 copper	E4@.197	LF	6.66	6.93	13.59

2-1/2" EMT conduit and wire as shown below

	Craft@Hrs	Unit	Material	Labor	Total
2 250 MCM alum, 1 #4 copper	E4@.210	LF	4.73	7.39	12.12
3 250 MCM alum, 1 #4 copper	E4@.237	LF	5.97	8.33	14.30
4 #3 copper, 1 #4 copper	E4@.216	LF	4.29	7.60	11.89
2 #4 copper, 1 #2 copper	E4@.186	LF	3.36	6.54	9.90
3 #4 copper, 1 #2 copper	E4@.198	LF	3.79	6.96	10.75
4 #4 copper, 1 #2 copper	E4@.210	LF	4.22	7.39	11.61
3 250 MCM copper, 1 #2 copper	E4@.253	LF	9.62	8.90	18.52

3" EMT conduit and wire

	Craft@Hrs	Unit	Material	Labor	Total
4 250 MCM alum, 1 #4 copper	E4@.288	LF	10.72	10.10	20.82
4 250 MCM copper, 1 #2 copper	E4@.290	LF	15.50	10.20	25.70
5 250 MCM copper, 1 #2 copper	E4@.319	LF	17.90	11.20	29.10

EMT Conduit Circuits Cost per circuit based on 30' run from the panel. Includes THHN copper wire pulled in conduit, 2 compression connectors, 2 couplings, conduit bending, straps, bolts and washers. Three-wire circuits are 120, 240 or 277 volt 1 phase, with 2 conductors and ground. Four-wire circuits are 120/208 or 480 volt 3 phase, with 3 conductors and ground. Five-wire circuits are 120/208 or 480 volt 3 phase, with 3 conductors, neutral and ground. Installed exposed in vertical or horizontal runs to 10' above floor level in a building. Add 15% to the manhours and labor costs for heights over 10' to 20'. Use these figures for preliminary estimates. Based on commercial quantity discounts.

15 amp circuits

	Craft@Hrs	Unit	Material	Labor	Total
3 #14 wire, 1/2" conduit	E4@2.39	Ea	11.70	84.00	95.70
4 #14 wire, 1/2" conduit	E4@2.58	Ea	13.30	90.70	104.00

20 amp circuits

	Craft@Hrs	Unit	Material	Labor	Total
3 #12 wire, 1/2" conduit	E4@2.49	Ea	14.00	87.60	101.60
4 #12 wire, 1/2" conduit	E4@2.72	Ea	16.30	95.70	112.00
5 #12 wire, 1/2" conduit	E4@2.95	Ea	18.60	104.00	122.60

	Craft@Hrs	Unit	Material	Labor	Total
30 amp circuits					
3 #10 wire, 1/2" conduit	E4@2.75	Ea	17.50	96.70	114.20
4 #10 wire, 1/2" conduit	E4@3.08	Ea	20.90	108.00	128.90
5 #10 wire, 3/4" conduit	E4@3.58	Ea	21.70	126.00	147.70
50 amp circuits					
3 #8 wire, 3/4" conduit	E4@2.98	Ea	29.00	105.00	134.00
4 #8 wire, 3/4" conduit	E4@3.32	Ea	34.90	117.00	151.90
5 #8 wire, 1" conduit	E4@4.02	Ea	48.20	141.00	189.20
65 amp circuits					
2 #6, 1 #8 wire, 1" conduit	E4@3.42	Ea	41.40	120.00	161.40
3 #6, 1 #8 wire, 1" conduit	E4@3.76	Ea	49.70	132.00	181.70
4 #6, 1 #8 wire, 1-1/4" conduit	E4@4.48	Ea	70.90	158.00	228.90
85 amp circuits in 1-1/4" conduit					
2 #4, 1 #8 wire	E4@3.89	Ea	63.40	137.00	200.40
3 #4, 1 #8 wire	E4@4.25	Ea	76.30	149.00	225.30
4 #4, 1 #8 wire	E4@4.48	Ea	89.20	158.00	247.20
100 amp circuits in 1-1/2" conduit					
2 #3, 1 #8 wire	E4@4.20	Ea	78.50	148.00	226.50
3 #3, 1 #8 wire	E4@4.66	Ea	93.60	164.00	257.60
4 #3, 1 #8 wire	E4@5.10	Ea	109.00	179.00	288.00
125 amp circuits in 1-1/2" conduit					
2 #1, 1 #8 wire	E4@4.40	Ea	99.70	155.00	254.70
3 #1, 1 #8 wire	E4@4.92	Ea	125.00	173.00	298.00
4 #1, 1 #8 wire	E4@5.47	Ea	151.00	192.00	343.00
150 amp circuits in 2" conduit					
2 #1/0, 1 #6 wire	E4@5.16	Ea	126.00	181.00	307.00
3 #1/0, 1 #6 wire	E4@5.78	Ea	158.00	203.00	361.00
4 #1/0, 1 #6 wire	E4@6.43	Ea	189.00	226.00	415.00
200 amp circuits in 2" conduit					
2 #3/0, 1 #4 wire	E4@5.49	Ea	164.00	193.00	357.00
3 #3/0, 1 #4 wire	E4@6.27	Ea	213.00	220.00	433.00
4 #3/0, 1 #4 wire	E4@7.80	Ea	261.00	274.00	535.00
225 amp circuits in 2-1/2" conduit					
2 #4/0, 1 #2 wire	E4@6.48	Ea	300.00	228.00	528.00
3 #4/0, 1 #2 wire	E4@7.33	Ea	360.00	258.00	618.00
4 #4/0, 1 #2 wire	E4@8.19	Ea	419.00	288.00	707.00
250 amp circuits in 2-1/2" conduit					
2 250 MCM, 1 #2 wire	E4@6.63	Ea	323.00	233.00	556.00
3 250 MCM, 1 #2 wire	E4@8.24	Ea	394.00	290.00	684.00
4 250 MCM, 1 #2 wire	E4@9.56	Ea	465.00	336.00	801.00

Intermediate Metal Conduit (IMC) Installed exposed in a building either vertically or horizontally up to 10' above floor level. No wire, fittings or supports included except as noted. Based on commercial quantity discounts.
IMC conduit and one coupling each 10'

	Craft@Hrs	Unit	Material	Labor	Total
1/2" IMC conduit and coupling	E4@.038	LF	.71	1.34	2.05
3/4" IMC conduit and coupling	E4@.048	LF	.82	1.69	2.51
1" IMC conduit and coupling	E4@.056	LF	1.25	1.97	3.22
1-1/4" IMC conduit and coupling	E4@.073	LF	1.57	2.57	4.14
1-1/2" IMC conduit and coupling	E4@.086	LF	1.86	3.02	4.88
2" IMC conduit and coupling	E4@.101	LF	2.55	3.55	6.10
2-1/2" IMC conduit and coupling	E4@.127	LF	5.48	4.47	9.95
3" IMC conduit and coupling	E4@.146	LF	6.77	5.13	11.90
3-1/2" IMC conduit and coupling	E4@.164	LF	7.88	5.77	13.65
4" IMC conduit and coupling	E4@.194	LF	9.25	6.82	16.07

	Craft@Hrs	Unit	Material	Labor	Total
90 degree ells					
1/2" IMC ell	E4@.160	Ea	2.15	5.63	7.78
3/4" IMC ell	E4@.191	Ea	2.63	6.72	9.35
1" IMC ell	E4@.225	Ea	3.79	7.91	11.70
1-1/4" IMC ell	E4@.282	Ea	6.14	9.92	16.06
1-1/2" IMC ell	E4@.347	Ea	6.89	12.20	19.09
2" IMC ell	E4@.407	Ea	9.92	14.30	24.22
2-1/2" IMC ell	E4@.692	Ea	17.60	24.30	41.90
3" IMC ell	E4@.995	Ea	27.00	35.00	62.00
3-1/2" IMC ell	E4@1.25	Ea	41.10	44.00	85.10
4" IMC ell	E4@1.58	Ea	47.70	55.60	103.30
Add for heights over 10' to 20'	—	%	—	20.0	—
Deduct for IMC conduit					
installed in a concrete slab or open trench	—	%	-5.0	-40.0	—

PVC Schedule 40 Conduit Installed in or under a building slab. No wire, fittings or supports included except as noted.

PVC conduit with one coupling each 10', regular Type 40 heavy wall conduit. Based on commercial quantity discounts.

	Craft@Hrs	Unit	Material	Labor	Total
1/2" PVC conduit and coupling	E4@.030	LF	.17	1.06	1.23
3/4" PVC conduit and coupling	E4@.036	LF	.23	1.27	1.50
1" PVC conduit and coupling	E4@.046	LF	.29	1.62	1.91
1-1/4" PVC conduit and coupling	E4@.056	LF	.37	1.97	2.34
1-1/2" PVC conduit and coupling	E4@.061	LF	.55	2.15	2.70
2" PVC conduit and coupling	E4@.072	LF	.73	2.53	3.26
2-1/2" PVC conduit and coupling	E4@.089	LF	1.18	3.13	4.31
3" PVC conduit and coupling	E4@.105	LF	1.52	3.69	5.21
3-1/2" PVC conduit and coupling	E4@.121	LF	1.90	4.26	6.16
4" PVC conduit and coupling	E4@.151	LF	2.17	5.31	7.48
5" PVC conduit and coupling	E4@.176	LF	3.80	6.19	9.99
6" PVC conduit and coupling	E4@.208	LF	5.17	7.31	12.48
Add for Type 80 extra heavy wall	—	%	30.0	10.0	—
Deduct for Type A thin wall	—	%	-35.0	—	—
Rigid straight couplings					
1/2" rigid coupling	E4@.040	Ea	.29	1.41	1.70
3/4" rigid coupling	E4@.051	Ea	.35	1.79	2.14
1" rigid coupling	E4@.062	Ea	.54	2.18	2.72
1-1/4" rigid coupling	E4@.078	Ea	.72	2.74	3.46
1-1/2" rigid coupling	E4@.092	Ea	1.00	3.24	4.24
2" rigid coupling	E4@.102	Ea	1.31	3.59	4.90
2-1/2" rigid coupling	E4@.113	Ea	2.32	3.97	6.29
3" rigid coupling	E4@.129	Ea	3.68	4.54	8.22
3-1/2" rigid coupling	E4@.144	Ea	4.15	5.06	9.21
4" rigid coupling	E4@.156	Ea	5.69	5.49	11.18
5" rigid coupling	E4@.191	Ea	15.00	6.72	21.72
6" rigid coupling	E4@.226	Ea	19.20	7.95	27.15
90 degree regular weight ells					
1/2" ell	E4@.082	Ea	.95	2.88	3.83
3/4" ell	E4@.102	Ea	1.04	3.59	4.63
1" ell	E4@.124	Ea	1.61	4.36	5.97
1-1/4" ell	E4@.154	Ea	2.30	5.42	7.72
1-1/2" ell	E4@.195	Ea	3.11	6.86	9.97
2" ell	E4@.205	Ea	4.52	7.21	11.73
2-1/2" ell	E4@.226	Ea	8.20	7.95	16.15

	Craft@Hrs	Unit	Material	Labor	Total
3" ell	E4@.259	Ea	14.40	9.11	23.51
3-1/2" ell	E4@.288	Ea	19.90	10.10	30.00
4" ell	E4@.313	Ea	24.90	11.00	35.90
5" ell	E4@.381	Ea	43.80	13.40	57.20
6" ell	E4@.453	Ea	74.20	15.90	90.10
Deduct for 45 or 30 degree ells	—	%	-30.0	—	—
Add for					
Extra heavy weight ells	—	%	70.0	—	—
24" radius sweep ells	E4@.187	Ea	5.00	6.58	11.58
36" radius sweep ells	E4@.187	Ea	10.00	6.58	16.58
Type TA terminal adapters					
1/2" adapter	E4@.065	Ea	.31	2.29	2.60
3/4" adapter	E4@.081	Ea	.57	2.85	3.42
1" adapter	E4@.097	Ea	.71	3.41	4.12
1-1/4" adapter	E4@.121	Ea	.90	4.26	5.16
1-1/2" adapter	E4@.144	Ea	1.11	5.06	6.17
2" adapter	E4@.162	Ea	1.59	5.70	7.29
2-1/2" adapter	E4@.178	Ea	2.69	6.26	8.95
3" adapter	E4@.203	Ea	3.91	7.14	11.05
3-1/2" adapter	E4@.226	Ea	5.13	7.95	13.08
4" adapter	E4@.244	Ea	6.72	8.58	15.30
5" adapter	E4@.301	Ea	13.20	10.60	23.80
6" adapter	E4@.358	Ea	15.90	12.60	28.50
Bell ends					
1" bell end	E4@.105	Ea	1.54	3.69	5.23
1-1/4" bell end	E4@.115	Ea	1.86	4.04	5.90
1-1/2" bell end	E4@.124	Ea	1.86	4.36	6.22
2" bell end	E4@.143	Ea	2.89	5.03	7.92
2-1/2" bell end	E4@.221	Ea	3.03	7.77	10.80
3" bell end	E4@.285	Ea	3.38	10.00	13.38
3-1/2" bell end	E4@.381	Ea	3.64	13.40	17.04
4" bell end	E4@.479	Ea	4.04	16.80	20.84
5" bell end	E4@.578	Ea	6.21	20.30	26.51
6" bell end	E4@.741	Ea	6.81	26.10	32.91
PVC conduit access fittings (conduit bodies), Type LL, LB, LR, C or E					
1/2" fitting	E4@.100	Ea	1.72	3.52	5.24
3/4" fitting	E4@.132	Ea	2.44	4.64	7.08
1" fitting	E4@.164	Ea	2.68	5.77	8.45
1-1/4" fitting	E4@.197	Ea	3.83	6.93	10.76
1-1/2" fitting	E4@.234	Ea	4.47	8.23	12.70
2" fitting	E4@.263	Ea	7.92	9.25	17.17
Add for Type T conduit access fittings	—	%	20.0	50.0	—
Add for PVC conduit installed exposed either vertically or horizontally in building walls					
To 10' above floor level	—	%	5.0	35.0	—
Over 10' to 20'	—	%	—	20.0	—

PVC Conduit with Wire Includes THHN wire pulled in Type 40 rigid PVC heavy wall conduit placed in a slab. Use these figures for preliminary estimates. One rigid coupling every 10 ft. Add for fittings as required. Based on commercial quantity discounts.

	Craft@Hrs	Unit	Material	Labor	Total
1/2" PVC conduit with wire					
3 #12 solid copper	E4@.054	LF	.37	1.90	2.27
4 #12 solid copper	E4@.061	LF	.43	2.15	2.58
5 #12 solid copper	E4@.069	LF	.50	2.43	2.93

	Craft@Hrs	Unit	Material	Labor	Total
3 #10 solid copper	E4@.062	LF	.48	2.18	2.66
4 #10 solid copper	E4@.073	LF	.58	2.57	3.15
3/4" PVC conduit with wire					
5 #10 solid copper	E4@.089	LF	.74	3.13	3.87
3 #8 copper	E4@.069	LF	.81	2.43	3.24
4 #8 copper	E4@.080	LF	1.01	2.81	3.82
1" PVC conduit with wire					
2 #6 copper, 1 #8 copper	E4@.081	LF	1.04	2.85	3.89
3 #6 copper, 1 #8 copper	E4@.092	LF	1.32	3.24	4.56
5 #8 copper	E4@.100	LF	1.27	3.52	4.79
1-1/2" PVC conduit with wire					
4 #6 copper, 1 #8 copper	E4@.117	LF	1.85	4.11	5.96
2 #4 copper, 1 #8 copper	E4@.097	LF	1.60	3.41	5.01
2 #1 alum, 1 #8 copper	E4@.102	LF	1.82	3.59	5.41
3 #1 alum, 1 #8 copper	E4@.119	LF	2.36	4.18	6.54
4 #4 copper, 1 #8 copper	E4@.124	LF	2.46	4.36	6.82
3 #3 copper, 1 #8 copper	E4@.116	LF	2.26	4.08	6.34
4 #3 copper, 1 #8 copper	E4@.132	LF	2.77	4.64	7.41
2 #1 copper, 1 #8 copper	E4@.108	LF	2.46	3.80	6.26
3 #1 copper, 1 #8 copper	E4@.125	LF	3.32	4.40	7.72
4 #1 copper, 1 #8 copper	E4@.143	LF	4.18	5.03	9.21
2" PVC conduit with wire					
4 #1 alum, 1 #8 copper	E4@.154	LF	3.08	5.42	8.50
2 2/0 alum, 1 #6 copper	E4@.123	LF	2.50	4.33	6.83
3 2/0 alum, 1 #6 copper	E4@.143	LF	3.25	5.03	8.28
4 2/0 alum, 1 #6 copper	E4@.163	LF	4.00	5.73	9.73
2 3/0 alum, 1 #6 copper	E4@.127	LF	2.86	4.47	7.33
3 3/0 alum, 1 #6 copper	E4@.148	LF	3.78	5.20	8.98
2" PVC conduit with wire					
4 3/0 alum, 1 #6 copper	E4@.171	LF	4.70	6.01	10.71
2 1/0 & 1 #6 copper	E4@.124	LF	2.26	4.36	6.62
3 1/0 & 1 #6 copper	E4@.146	LF	2.89	5.13	8.02
4 1/0 & 1 #6 copper	E4@.167	LF	3.52	5.87	9.39
2 3/0 & 1 #4 copper	E4@.136	LF	3.01	4.78	7.79
3 3/0 & 1 #4 copper	E4@.162	LF	3.93	5.70	9.63
2-1/2" PVC conduit with wire					
2 250 MCM alum, 1 #4 copper	E4@.156	LF	4.08	5.49	9.57
3 250 MCM alum, 1 #4 copper	E4@.183	LF	5.31	6.44	11.75
4 3/0 alum, 1 #4 copper	E4@.206	LF	4.38	7.24	11.62
2 4/0 alum, 1 #2 copper	E4@.164	LF	3.92	5.77	9.69
3 4/0 alum, 1 #2 copper	E4@.193	LF	4.96	6.79	11.75
4 4/0 alum, 1 #2 copper	E4@.221	LF	6.00	7.77	13.77
2 250 MCM copper, 1 #2 copper	E4@.170	LF	6.59	5.98	12.57
3" PVC conduit with wire					
4 250 MCM alum, 1 #4 copper	E4@.226	LF	6.88	7.95	14.83
2 350 MCM alum, 1 #2 copper	E4@.186	LF	5.65	6.54	12.19
3 350 MCM alum, 1 #2 copper	E4@.217	LF	7.38	7.63	15.01
3 250 MCM alum, 1 #2 copper	E4@.217	LF	5.88	7.63	13.51
4 250 MCM alum, 1 #2 copper	E4@.248	LF	7.11	8.72	15.83
3-1/2" PVC conduit with wire					
4 350 MCM alum, 1 #2 copper	E4@.264	LF	9.49	9.28	18.77

	Craft@Hrs	Unit	Material	Labor	Total
4" PVC conduit with wire					
6 350 MCM alum, 1 #2 copper	E4@.293	LF	13.20	10.30	23.50
7 350 MCM alum, 1 #2 copper	E4@.391	LF	15.00	13.80	28.80
3 500 MCM alum, 1 1/0 copper	E4@.311	LF	9.92	10.90	20.82
4 500 MCM alum, 1 1/0 copper	E4@.355	LF	12.20	12.50	24.70

PVC-RSC Conduit Circuits Cost per circuit for a 30' run from the panel. Based on THHN wire pulled in 20' of Type 40 rigid PVC heavy wall conduit placed in a slab and 10' of galvanized rigid steel conduit installed exposed in a building. Includes two 90 degree RSC ells, 2 PVC terminal adapters, hangers and bolts. Use these figures for preliminary estimates. Based on commercial quantity discounts.

	Craft@Hrs	Unit	Material	Labor	Total
20 amp circuits, in 1/2" conduit					
3 #12 copper wire	E4@2.23	Ea	20.00	78.40	98.40
4 #12 copper wire	E4@2.47	Ea	22.00	86.90	108.90
5 #12 copper wire	E4@2.56	Ea	24.00	90.00	114.00
30 amp circuits in 1/2" conduit					
3 #10 copper wire	E4@2.50	Ea	23.40	87.90	111.30
4 #10 copper wire	E4@2.82	Ea	26.50	99.20	125.70
5 #10 copper wire	E4@3.47	Ea	29.60	122.00	151.60
50 amp circuits in 3/4" conduit					
3 #8 copper wire	E4@2.85	Ea	35.80	100.00	135.80
4 #8 copper wire	E4@3.21	Ea	41.70	113.00	154.70
5 #8 copper wire	E4@3.96	Ea	47.60	139.00	186.60
65 amp circuits in 1" conduit					
2 #6 copper, 1 #8 copper	E4@3.26	Ea	45.60	115.00	160.60
3 #6 copper, 1 #8 copper	E4@3.60	Ea	53.90	127.00	180.90
4 #6 copper, 1 #8 copper	E4@4.97	Ea	62.20	175.00	237.20
85 amp circuits in 1-1/2" conduit					
2 #4 copper, 1 #8 copper	E4@4.35	Ea	67.10	153.00	220.10
3 #4 copper, 1 #8 copper	E4@4.77	Ea	80.00	168.00	248.00
4 #4 copper, 1 #8 copper	E4@5.18	Ea	92.90	182.00	274.90
100 amp circuits in 1-1/2" conduit					
2 #3 copper, 1 #8 copper	E4@4.40	Ea	71.70	155.00	226.70
3 #3 copper, 1 #8 copper	E4@5.10	Ea	86.80	179.00	265.80
4 #3 copper, 1 #8 copper	E4@5.42	Ea	102.00	191.00	293.00
2 #1 aluminum, 1 #8 copper	E4@4.53	Ea	73.70	159.00	232.70
3 #1 aluminum, 1 #8 copper	E4@5.00	Ea	89.90	176.00	265.90
125 amp circuits in 1-1/2" conduit					
2 #1 copper, 1 #8 copper	E4@4.69	Ea	92.90	165.00	257.90
3 #1 copper, 1 #8 copper	E4@5.23	Ea	119.00	184.00	303.00
4 #1 copper, 1 #8 copper	E4@5.80	Ea	144.00	204.00	348.00
125 amp circuits in 2" conduit					
2 2/0 aluminum, 1 #6 copper	E4@5.54	Ea	101.00	195.00	296.00
3 2/0 aluminum, 1 #6 copper	E4@6.14	Ea	123.00	216.00	339.00
4 2/0 aluminum, 1 #6 copper	E4@6.76	Ea	145.00	238.00	383.00
150 amp circuits in 2" conduit					
2 1/0 copper, 1 #6 copper	E4@5.21	Ea	119.00	183.00	302.00
3 1/0 copper, 1 #6 copper	E4@6.30	Ea	150.00	222.00	372.00
4 1/0 copper, 1 #6 copper	E4@7.44	Ea	182.00	262.00	444.00
2 3/0 aluminum, 1 #6 copper	E4@5.86	Ea	112.00	206.00	318.00
3 3/0 aluminum, 1 #6 copper	E4@6.61	Ea	139.00	232.00	371.00
4 3/0 aluminum, 1 #6 copper	E4@10.1	Ea	167.00	355.00	522.00

	Craft@Hrs	Unit	Material	Labor	Total
200 amp circuits in 2" conduit					
2 3/0 copper, 1 #4 copper	E4@6.11	Ea	157.00	215.00	372.00
3 3/0 copper, 1 #4 copper	E4@6.71	Ea	205.00	236.00	441.00
200 amp circuits in 2-1/2" conduit					
4 3/0 copper, 1 #4 copper	E4@10.3	Ea	279.00	362.00	641.00
2 250 MCM alum., 1 #4 copper	E4@8.76	Ea	135.00	308.00	443.00
3 250 MCM alum., 1 #4 copper	E4@9.59	Ea	172.00	337.00	509.00
4 250 MCM alum., 1 #4 copper	E4@11.3	Ea	209.00	397.00	606.00
225 amp circuits in 2-1/2" conduit					
2 4/0 copper, 1 #2 copper	E4@8.94	Ea	187.00	314.00	501.00
3 4/0 copper, 1 #2 copper	E4@9.85	Ea	247.00	346.00	593.00
4 4/0 copper, 1 #2 copper	E4@9.64	Ea	306.00	339.00	645.00
250 amp circuits in 3" conduit					
2 250 MCM copper, 1 #2 copper	E4@9.15	Ea	252.00	322.00	574.00
3 250 MCM copper, 1 #2 copper	E4@11.0	Ea	324.00	387.00	711.00
4 250 MCM copper, 1 #2 copper	E4@12.0	Ea	395.00	422.00	817.00
2 350 MCM aluminum, 1 #2 copper	E4@10.6	Ea	308.00	373.00	681.00
3 350 MCM aluminum, 1 #2 copper	E4@11.9	Ea	407.00	418.00	825.00
4 350 MCM aluminum, 1 #2 copper	E4@14.9	Ea	506.00	524.00	1,030.00
400 amp circuits in 4" conduit					
3 500 MCM copper, 1 1/0 copper	E4@16.7	Ea	579.00	587.00	1,166.00
4 500 MCM copper, 1 1/0 copper	E4@18.1	Ea	719.00	637.00	1,356.00
6 350 MCM aluminum, 1 2/0 copper	E4@18.6	Ea	478.00	654.00	1,132.00
7 350 MCM aluminum, 1 2/0 copper	E4@20.5	Ea	530.00	721.00	1,251.00

Rigid Aluminum Conduit Installed exposed in a building either vertically or horizontally up to 10' above floor level. No wire, fittings or supports included.

	Craft@Hrs	Unit	Material	Labor	Total
Rigid aluminum conduit with one coupling each 10'					
1/2" conduit with coupling	E4@.040	LF	.79	1.41	2.20
3/4" conduit with coupling	E4@.048	LF	1.04	1.69	2.73
1" conduit with coupling	E4@.059	LF	1.49	2.07	3.56
1-1/4" conduit with coupling	E4@.075	LF	1.96	2.64	4.60
1-1/2" conduit with coupling	E4@.093	LF	2.42	3.27	5.69
2" conduit with coupling	E4@.108	LF	3.24	3.80	7.04
2-1/2" conduit with coupling	E4@.081	LF	5.13	2.85	7.98
3" conduit with coupling	E4@.097	LF	6.74	3.41	10.15
3-1/2" conduit with coupling	E4@.129	LF	8.09	4.54	12.63
4" conduit with coupling	E4@.162	LF	9.59	5.70	15.29
6" conduit with coupling	E4@.242	LF	18.00	8.51	26.51
Type LL, LR, or LB cast metal conduit bodies, with aluminum covers and gaskets					
1/2" conduit body	E4@.313	Ea	7.02	11.00	18.02
3/4" conduit body	E4@.345	Ea	8.46	12.10	20.56
1" conduit body	E4@.464	Ea	12.50	16.30	28.80
1-1/4" conduit body	E4@.648	Ea	18.80	22.80	41.60
1-1/2" conduit body	E4@.777	Ea	23.30	27.30	50.60
2" conduit body	E4@.902	Ea	38.80	31.70	70.50
2-1/2" conduit body	E4@1.42	Ea	78.10	49.90	128.00
3" conduit body	E4@1.97	Ea	101.00	69.30	170.30
3-1/2" conduit body	E4@2.60	Ea	161.00	91.40	252.40
4" conduit body	E4@3.06	Ea	188.00	108.00	296.00
Add for Type T aluminum conduit bodies	—	%	20.0	30.0	—
Add for Heights over 10' to 20'	—	%	—	20.0	—

	Craft@Hrs	Unit	Material	Labor	Total

Lightweight Flexible Metal Conduit Installed exposed in a building either vertically or horizontally up to 10' above floor level. No wire, fittings or supports included except as noted. Typical flexible conduit connection for electrical equipment consisting of a 4" square junction box, 6' of aluminum reduced wall conduit, two screw in flex connectors and solid copper wire pulled in conduit as noted. Based on commercial quantity discounts. Per 6' connection

3/8" conduit, 3 #14 THHN copper wire	E4@.630	Ea	4.35	22.20	26.55
1/2" conduit, 4 #12 THHN copper wire	E4@.733	Ea	5.91	25.80	31.71
1/2" conduit, 4 #10 THHN copper wire	E4@.785	Ea	6.63	27.60	34.23
3/4" conduit, 4 #8 THHN copper wire	E4@.883	Ea	9.97	31.10	41.07
1" conduit, 4 #4 THHN copper wire	E4@1.12	Ea	19.80	39.40	59.20

Flexible aluminum reduced wall conduit installed in a building to 10' above floor level. Based on commercial quantity discounts. No wire, fittings or supports included

3/8" conduit	E4@.030	LF	.21	1.06	1.27
1/2" conduit	E4@.039	LF	.30	1.37	1.67
3/4" conduit	E4@.051	LF	.42	1.79	2.21
1" conduit	E4@.078	LF	.86	2.74	3.60
1-1/4" conduit	E4@.113	LF	1.10	3.97	5.07
1-1/2" conduit	E4@.147	LF	1.41	5.17	6.58
2" conduit	E4@.191	LF	2.05	6.72	8.77
2-1/2" conduit	E4@.282	LF	2.61	9.92	12.53
3" conduit	E4@.394	LF	3.02	13.90	16.92
Add for standard wall flex steel conduit	—	%	10.0	10.0	—
Deduct for reduced wall flex steel conduit	—	%	-15.0	—	—

Flexible metal one screw tite-bite straight connectors

3/8" connector	E4@.070	Ea	1.09	2.46	3.55
1/2" connector	E4@.086	Ea	2.06	3.02	5.08
3/4" connector	E4@.100	Ea	2.76	3.52	6.28
1" connector	E4@.113	Ea	4.80	3.97	8.77
1-1/4" connector	E4@.148	Ea	7.86	5.20	13.06
1-1/2" connector	E4@.191	Ea	12.50	6.72	19.22
2" connector	E4@.260	Ea	17.20	9.14	26.34
2-1/2" connector	E4@.326	Ea	30.20	11.50	41.70
3" connector	E4@.435	Ea	42.20	15.30	57.50
Add for 90 or 45 degree tite-bite connectors	—	%	100.0	—	—
Add for height over 10' to 20'	—	%	—	20.0	—

Liquid-Tight Flexible Metal Conduit (Sealtite) Installed exposed in a building either vertically or horizontally up to 10' above floor level. No wire, fittings or supports included except as noted.

Typical liquid-tight flexible conduit connection for electrical equipment consisting of 6' of conduit, 2 connectors and solid copper wire pulled in conduit as noted. Per 6' connection. Prices are typical retail costs and based on commercial quantities. Expect specialized supplier's prices to be somewhat higher.

3/8" conduit, 3 #12 THHN copper wire	E4@.622	Ea	10.80	21.90	32.70
1/2" conduit, 4 #12 THHN copper wire	E4@.746	Ea	11.80	26.20	38.00
1/2" conduit, 4 #10 THHN copper wire	E4@.813	Ea	12.50	28.60	41.10
3/4" conduit, 3 #8, 1 #10 THHN copper wire	E4@.891	Ea	17.70	31.30	49.00
1" conduit, 3 #6, 1 #8 THHN copper wire	E4@1.23	Ea	26.40	43.30	69.70

Liquid-tight flexible metal conduit installed in a building to 10' above floor level. No wire, supports or fittings included (Type E.F. or L.T. extra flex)

3/8" conduit	E4@.045	LF	.91	1.58	2.49
1/2" conduit	E4@.051	LF	1.00	1.79	2.79
3/4" conduit	E4@.067	LF	1.33	2.36	3.69
1" conduit	E4@.094	LF	2.02	3.31	5.33
1-1/4" conduit	E4@.116	LF	2.74	4.08	6.82
1-1/2" conduit	E4@.132	LF	3.73	4.64	8.37

	Craft@Hrs	Unit	Material	Labor	Total
2" conduit	E4@.156	LF	4.71	5.49	10.20
2-1/2" conduit	E4@.234	LF	8.70	8.23	16.93
3" conduit	E4@.313	LF	12.10	11.00	23.10
4" conduit	E4@.624	LF	17.50	21.90	39.40
Add for other types of liquid-tight flexible metal conduit					
Type U.A. or L.A.	—	%	50.0	—	—
Type O.R. or L.O.R.	—	%	20.0	—	—
Type H.C. or A.T.	—	%	28.0	—	—
Straight sealtite connectors for flexible metal conduit					
3/8" connector	E4@.110	Ea	1.93	3.87	5.80
1/2" connector	E4@.129	Ea	1.93	4.54	6.47
3/4" connector	E4@.140	Ea	2.75	4.92	7.67
1" connector	E4@.221	Ea	4.03	7.77	11.80
1-1/4" connector	E4@.244	Ea	6.92	8.58	15.50
1-1/2" connector	E4@.293	Ea	9.86	10.30	20.16
2" connector	E4@.492	Ea	18.10	17.30	35.40
2-1/2" connector	E4@.741	Ea	83.60	26.10	109.70
3" connector	E4@.982	Ea	94.00	34.50	128.50
4" connector	E4@1.47	Ea	119.00	51.70	170.70
Add for					
Height over 10' to 20'	—	%	—	20.0	—
Insulated throat sealtite connectors	—	%	20.0	—	—
For 45 or 90 degree sealtite connectors	—	%	30.0	—	—

Conduit Hangers and Supports By conduit size. Installed exposed in a building to 10' above floor level. Hangers and supports installed at heights over 10' will add about 20% to the labor cost.

	Craft@Hrs	Unit	Material	Labor	Total
Right angle conduit supports					
3/8" support	E4@.242	Ea	2.94	8.51	11.45
1/2" support or 3/4" support	E4@.242	Ea	2.39	8.51	10.90
1" support	E4@.269	Ea	2.76	9.46	12.22
1-1/4" support or 1-1/2" support	E4@.269	Ea	3.48	9.46	12.94
2" support	E4@.295	Ea	5.06	10.40	15.46
2-1/2" support	E4@.295	Ea	7.00	10.40	17.40
3" support	E4@.295	Ea	7.55	10.40	17.95
3-1/2" support	E4@.324	Ea	11.40	11.40	22.80
4" support	E4@.324	Ea	11.10	11.40	22.50
Add for					
Parallel conduit supports	—	%	50.0	—	—
Edge conduit supports	—	%	120.0	—	—

Conduit hangers, rigid with bolt, labor includes the cost of cutting threaded rod to length and attaching the rod but not the rod itself. Add the material cost of threaded rod from below.

	Craft@Hrs	Unit	Material	Labor	Total
1/2" hanger	E4@.097	Ea	.45	3.41	3.86
3/4" hanger	E4@.097	Ea	.47	3.41	3.88
1" hanger	E4@.108	Ea	.75	3.80	4.55
1-1/4" hanger	E4@.108	Ea	.99	3.80	4.79
1-1/2" hanger	E4@.108	Ea	1.27	3.80	5.07
2" hanger	E4@.119	Ea	1.39	4.18	5.57
2-1/2" hanger	E4@.119	Ea	1.78	4.18	5.96
3" hanger	E4@.119	Ea	2.13	4.18	6.31
3-1/2" hanger	E4@.129	Ea	3.03	4.54	7.57
4" hanger	E4@.129	Ea	7.16	4.54	11.70

	Craft@Hrs	Unit	Material	Labor	Total
All threaded rod, plated steel, per linear foot of rod					
1/4", 20 thread	—	LF	.74	—	.74
5/16", 18 thread	—	LF	1.04	—	1.04
3/8", 16 thread	—	LF	1.23	—	1.23
1/2", 13 thread	—	LF	1.96	—	1.96
5/8", 11 thread	—	LF	2.92	—	2.92
Deduct for plain steel rod	—	%	-10.0	—	—
Rod beam clamps, for 1/4" or 3/8" rod					
1" flange	E4@.234	Ea	1.58	8.23	9.81
1-1/2" flange	E4@.234	Ea	4.10	8.23	12.33
2" flange	E4@.234	Ea	5.78	8.23	14.01
2-1/2" flange	E4@.234	Ea	8.54	8.23	16.77
Conduit clamps for EMT conduit					
1/2" or 3/4" clamp	E4@.062	Ea	.18	2.18	2.36
1" clamp	E4@.069	Ea	.25	2.43	2.68
1-1/4" clamp	E4@.069	Ea	.37	2.43	2.80
1-1/2" clamp	E4@.069	Ea	.39	2.43	2.82
2" clamp	E4@.075	Ea	.44	2.64	3.08
One hole heavy duty stamped steel conduit straps					
1/2" strap	E4@.039	Ea	.28	1.37	1.65
3/4" strap	E4@.039	Ea	.31	1.37	1.68
1" strap	E4@.039	Ea	.67	1.37	2.04
1-1/4" strap	E4@.039	Ea	.93	1.37	2.30
1-1/2" strap	E4@.039	Ea	1.26	1.37	2.63
2" strap	E4@.039	Ea	2.09	1.37	3.46
2-1/2" strap	E4@.053	Ea	3.57	1.86	5.43
3" strap	E4@.053	Ea	4.48	1.86	6.34
3-1/2" strap	E4@.059	Ea	7.07	2.07	9.14
4" strap	E4@.059	Ea	9.14	2.07	11.21
Add for malleable iron one hole straps	—	%	30.0	—	—
Deduct for nail drive straps	—	%	-20.0	-30.0	—
Deduct for two hole stamped steel EMT straps	—	%	-70.0	—	—
Hanger channel, 1-5/8" x 1-5/8", based on 12" length					
No holes, solid back,					
12 gauge steel	E4@.119	Ea	3.23	4.18	7.41
Holes 1-1/2" on center,					
12 gauge aluminum	E4@.119	Ea	3.71	4.18	7.89
Channel strap for rigid steel conduit or EMT conduit, by conduit size, with bolts					
1/2" or 3/4" strap	E4@.020	Ea	.91	.70	1.61
1" or 1-1/4" strap	E4@.023	Ea	1.12	.81	1.93
1-1/2"or 2" strap	E4@.023	Ea	1.45	.81	2.26
2-1/2" or 3" strap	E4@.039	Ea	1.76	1.37	3.13
3-1/2" strap	E4@.039	Ea	2.25	1.37	3.62
4" strap	E4@.062	Ea	2.51	2.18	4.69
5" strap	E4@.062	Ea	3.10	2.18	5.28
6" strap	E4@.062	Ea	4.15	2.18	6.33
Lag screws					
1/4" x 3" or 1/4" x 4"	E4@.092	Ea	.85	3.24	4.09
5/16" x 3-1/2"	E4@.092	Ea	.99	3.24	4.23
3/8" x 4"	E4@.119	Ea	1.15	4.18	5.33
1/2" x 5"	E4@.129	Ea	1.73	4.54	6.27

	Craft@Hrs	Unit	Material	Labor	Total
Lag screw short expansion shields, without screws. Labor is for drilling hole					
1/4" or 5/16" shield	E4@.175	Ea	.19	6.15	6.34
3/8" shield	E4@.272	Ea	.32	9.57	9.89
1/2" shield	E4@.321	Ea	.51	11.30	11.81
5/8" shield	E4@.321	Ea	.87	11.30	12.17
3/4" shield	E4@.345	Ea	1.21	12.10	13.31
Self drilling masonry anchors					
1/4" anchor	E4@.127	Ea	1.15	4.47	5.62
5/16" anchor	E4@.127	Ea	1.54	4.47	6.01
3/8" anchor	E4@.191	Ea	1.70	6.72	8.42
1/2" anchor	E4@.191	Ea	2.55	6.72	9.27
5/8" anchor	E4@.259	Ea	4.61	9.11	13.72
3/4" anchor	E4@.259	Ea	8.05	9.11	17.16
7/8" anchor	E4@.332	Ea	12.10	11.70	23.80

Electrical Wireway (Surface Duct) Wall mounted flangeless hinged cover or screw cover enameled steel wiring raceway. No wire included. Section prices include 1 connector or end plate with each section.

	Craft@Hrs	Unit	Material	Labor	Total
2-1/2" x 2-1/2" wireway					
1' section	E4@.265	Ea	10.40	9.32	19.72
2' sections	E4@.391	Ea	14.80	13.80	28.60
3' sections	E4@.671	Ea	19.90	23.60	43.50
4' sections	E4@.702	Ea	27.70	24.70	52.40
5' sections	E4@.865	Ea	33.70	30.40	64.10
10' sections	E4@1.76	Ea	68.50	61.90	130.40
Hangers and brackets.	E4@.222	Ea	6.36	7.81	14.17
Tee or cross pull boxes	E4@1.17	Ea	31.30	41.10	72.40
Internal or external corners	E4@.894	Ea	24.40	31.40	55.80
4" x 4" wireway					
1' section	E4@.272	Ea	11.30	9.57	20.87
2' sections	E4@.407	Ea	16.50	14.30	30.80
3' sections	E4@.609	Ea	24.40	21.40	45.80
4' sections	E4@.811	Ea	33.70	28.50	62.20
5' sections	E4@.902	Ea	36.30	31.70	68.00
10' sections	E4@1.85	Ea	75.50	65.10	140.60
Hangers and brackets	E4@.334	Ea	7.42	11.70	19.12
Tee or cross pull boxes	E4@1.79	Ea	37.70	62.90	100.60
Internal or external corners	E4@1.35	Ea	27.70	47.50	75.20
6" x 6" wireway					
1' section	E4@.326	Ea	21.70	11.50	33.20
2' sections	E4@.490	Ea	26.70	17.20	43.90
3' sections	E4@.733	Ea	37.70	25.80	63.50
4' sections	E4@.972	Ea	50.40	34.20	84.60
5' sections	E4@1.08	Ea	54.40	38.00	92.40
10' sections	E4@2.21	Ea	128.00	77.70	205.70
Hangers and brackets	E4@.391	Ea	10.60	13.80	24.40
Tee or cross pull boxes	E4@2.06	Ea	46.90	72.40	119.30
Internal or external corners	E4@1.58	Ea	31.30	55.60	86.90
8" x 8" wireway					
1' section	E4@.389	Ea	34.20	13.70	47.90
2' sections	E4@.591	Ea	52.40	20.80	73.20
3' sections	E4@.878	Ea	80.00	30.90	110.90
4' sections	E4@1.17	Ea	97.20	41.10	138.30

	Craft@Hrs	Unit	Material	Labor	Total
5' sections	E4@1.30	Ea	108.00	45.70	153.70
10' sections	E4@2.66	Ea	160.00	93.50	253.50
Hangers and brackets	E4@.451	Ea	31.80	15.90	47.70
Tee or cross pull boxes	E4@2.39	Ea	74.00	84.00	158.00
Internal or external corners	E4@1.82	Ea	50.70	64.00	114.70
12" x 12" wireway					
1' section	E4@.547	Ea	48.30	19.20	67.50
2' sections	E4@.819	Ea	93.20	28.80	122.00
3' sections	E4@1.23	Ea	139.00	43.30	182.30
4' sections	E4@1.64	Ea	167.00	57.70	224.70
5' sections	E4@1.82	Ea	193.00	64.00	257.00
Hangers and brackets	E4@.541	Ea	56.00	19.00	75.00
Tee or cross pull boxes	E4@2.85	Ea	140.00	100.00	240.00
Internal or external corners	E4@2.15	Ea	91.20	75.60	166.80
Add for					
Raintight (exterior) wireway	—	%	75.0	—	—
Flanged latch cover wireway	—	%	125.0	—	—

Wiremold Raceway Installed on a finished wall in a building
Raceway including one coupling each 10 feet

	Craft@Hrs	Unit	Material	Labor	Total
#200 2-piece midget raceway	F4@.046	LF	.60	1.62	2.22
#500 2 wire surface mounted raceway	E4@.059	LF	.61	2.07	2.68
#700 2 wire surface mounted raceway	E4@.067	LF	.69	2.36	3.05
#1500 "pancake" surface mounted raceway	E4@.080	LF	1.27	2.81	4.08
Wiremold fittings					
#200 90 degree flat ell	E4@.100	Ea	1.18	3.52	5.00
#200 internal ell	E4@.100	Ea	4.11	3.52	7.63
#200 extension adapter	E4@.199	Ea	5.26	7.00	12.26
#200, #500, #700 1-pole switch and box	E4@.253	Ea	8.12	8.90	17.02
#200, #500, #700 1-pole switch box only	E4@.360	Ea	8.00	12.70	20.70
#200, #500, #700 duplex receptacle and box	E4@.360	Ea	9.59	12.70	22.29
#500, #700 90 degree flat ell	E4@.100	Ea	1.01	3.52	4.53
#500, #700 internal twisted ell	E4@.100	Ea	2.99	3.52	6.51
#500, #700 utility box	E4@.360	Ea	5.84	12.70	18.54
#500, #700 corner box	E4@.360	Ea	9.12	12.70	21.82
#500, #700 fixture box	E4@.360	Ea	7.96	12.70	20.66
#1500 90 degree flat ell	E4@.151	Ea	3.22	5.31	8.53
#1500 internal ell	E4@.151	Ea	3.22	5.31	8.53
#1500 junction box	E4@.360	Ea	5.75	12.70	18.45

Cable Tray and Ducts

Cable tray, steel, ladder type

	Craft@Hrs	Unit	Material	Labor	Total
4" wide	E4@.144	LF	10.50	5.06	15.56
6" wide	E4@.166	LF	20.40	5.84	26.24
8" wide	E4@.187	LF	32.00	6.58	38.58
10" wide	E4@.211	LF	59.10	7.42	66.52
12" wide	E4@.238	LF	45.00	8.37	53.37

Steel underfloor duct, including typical supports, fittings and accessories

	Craft@Hrs	Unit	Material	Labor	Total
3-1/4" wide, 1 cell	E4@.097	LF	4.48	3.41	7.89
3-1/4" wide, 2 cell	E4@.104	LF	9.12	3.66	12.78
7-1/4" wide, 1 cell	E4@.148	LF	10.50	5.20	15.70
7-1/4" wide, 2 cell	E4@.148	LF	16.10	5.20	21.30

	Craft@Hrs	Unit	Material	Labor	Total

Wire and Cable

Copper wire and cable prices can change very quickly, sometimes as much as 10% in a week. The prices that follow were representative at the time of publication. Your material supplier will be able to quote more current wire and cable costs.

Copper Service Entrance Wire Type USE-RHH-RHW (XLPE) crosslinked polyethylene 600 volt stranded copper service entrance wire pulled in conduit. No excavation or conduit included. By American Wire Gauge. Note that wire prices can change very quickly.

	Craft@Hrs	Unit	Material	Labor	Total
#12 gauge	E4@.008	LF	.14	.28	.42
#10 gauge	E4@.009	LF	.20	.32	.52
#8 gauge	E4@.011	LF	.28	.39	.67
#6 gauge	E4@.013	LF	.31	.46	.77
#4 gauge	E4@.016	LF	.47	.56	1.03
#2 gauge	E4@.018	LF	.72	.63	1.35
#1 gauge	E4@.019	LF	.88	.67	1.55
#1/0 gauge	E4@.024	LF	1.11	.84	1.95
#2/0 gauge	E4@.028	LF	1.68	.98	2.66
#3/0 gauge	E4@.032	LF	1.71	1.13	2.84
#4/0 gauge	E4@.036	LF	2.07	1.27	3.34
250 MCM	E4@.039	LF	2.58	1.37	3.95
300 MCM	E4@.040	LF	3.03	1.41	4.44
350 MCM	E4@.045	LF	3.52	1.58	5.10
400 MCM	E4@.046	LF	3.90	1.62	5.52
500 MCM	E4@.047	LF	4.76	1.65	6.41
600 MCM	E4@.054	LF	6.47	1.90	8.37
750 MCM	E4@.066	LF	7.86	2.32	10.18
Deduct for aerial distribution	—	%	—	-25.0	—

Bare Copper Groundwire Stranded, pulled in conduit with conductor wires. No excavation or duct included. By American Wire Gauge. Note that wire prices can change very quickly.

	Craft@Hrs	Unit	Material	Labor	Total
#8 gauge	E4@.008	LF	.16	.28	.44
#6 gauge	E4@.009	LF	.25	.32	.57
#4 gauge	E4@.012	LF	.38	.42	.80
#2 gauge	E4@.014	LF	.60	.49	1.09
#1 gauge	E4@.015	LF	.77	.53	1.30
#1/0 gauge	E4@.018	LF	.99	.63	1.62
#2/0 gauge	E4@.021	LF	1.23	.74	1.97
#3/0 gauge	E4@.025	LF	1.57	.88	2.45
#4/0 gauge	E4@.028	LF	1.96	.98	2.94

Aluminum Service Entrance Wire Type USE-RHH-RHW (XLP) crosslinked polyethylene 600 volt stranded aluminum service entrance wire pulled in conduit. No excavation or conduit included. By American Wire Gauge. Note that wire prices can change very quickly.

	Craft@Hrs	Unit	Material	Labor	Total
#6 gauge	E4@.012	LF	.28	.42	.70
#4 gauge	E4@.014	LF	.34	.49	.83
#2 gauge	E4@.017	LF	.46	.60	1.06
#1 gauge	E4@.017	LF	.66	.60	1.26
#1/0 gauge	E4@.022	LF	.76	.77	1.53
#2/0 gauge	E4@.026	LF	.88	.91	1.79
#3/0 gauge	E4@.030	LF	1.09	1.06	2.15
#4/0 gauge	E4@.034	LF	1.21	1.20	2.41
250 MCM	E4@.035	LF	1.54	1.23	2.77
300 MCM	E4@.038	LF	2.10	1.34	3.44
350 MCM	E4@.040	LF	2.15	1.41	3.56
400 MCM	E4@.042	LF	2.50	1.48	3.98
500 MCM	E4@.043	LF	2.75	1.51	4.26

	Craft@Hrs	Unit	Material	Labor	Total
600 MCM	E4@.050	LF	3.51	1.76	5.27
700 MCM	E4@.054	LF	4.04	1.90	5.94
750 MCM	E4@.059	LF	4.10	2.07	6.17
1000 MCM	E4@.074	LF	6.01	2.60	8.61
Deduct for aerial distribution	—	%	—	-25.0	—

Bare Aluminum Groundwire Stranded, pulled in conduit with conductor wires. No excavation or conduit included. By American Wire Gauge. Note that wire prices can change very quickly.

	Craft@Hrs	Unit	Material	Labor	Total
#4 gauge	E4@.011	LF	.34	.39	.73
#2 gauge	E4@.012	LF	.31	.42	.73
#1/0 gauge	E4@.016	LF	.30	.56	.86
#2/0 gauge	E4@.019	LF	.29	.67	.96
#3/0 gauge	E4@.022	LF	.29	.77	1.06
#4/0 gauge	E4@.025	LF	.29	.88	1.17

Copper Building Wire Type THHN and single conductor flame-retardant, moisture and heat resistant thermoplastic insulated 600 volt stranded copper building wire pulled in conduit. No conduit included. Listed by American Wire Gauge. Note that wire prices can change very quickly.

	Craft@Hrs	Unit	Material	Labor	Total
#14 stranded	E4@.006	LF	.05	.21	.26
#14 solid	E4@.006	LF	.04	.21	.25
#12 stranded	E4@.007	LF	.08	.25	.33
#12 solid	E4@.008	LF	.07	.28	.35
#10 stranded	E4@.010	LF	.11	.35	.46
#10 solid	E4@.011	LF	.10	.39	.49
#8 gauge	E4@.011	LF	.20	.39	.59
#6 gauge	E4@.012	LF	.28	.42	.70
#4 gauge	E4@.013	LF	.43	.46	.89
#3 gauge	E4@.015	LF	.51	.53	1.04
#2 gauge	E4@.017	LF	.66	.60	1.26
#1 gauge	E4@.018	LF	.86	.63	1.49
#1/0 gauge solid	E4@.021	LF	1.05	.74	1.79
#2/0 gauge	E4@.023	LF	1.28	.81	2.09
#3/0 gauge	E4@.026	LF	1.61	.91	2.52
#4/0 gauge	E4@.029	LF	1.99	1.02	3.01
250 MCM	E4@.032	LF	2.37	1.13	3.50
300 MCM	E4@.035	LF	2.82	1.23	4.05
350 MCM	E4@.038	LF	3.30	1.34	4.64
400 MCM	E4@.042	LF	3.70	1.48	5.18
500 MCM	E4@.046	LF	4.67	1.62	6.29
600 MCM	E4@.051	LF	6.19	1.79	7.98
750 MCM	E4@.055	LF	7.68	1.93	9.61
1000 MCM	E4@.066	LF	11.50	2.32	13.82
Add for type XHHW (XLP) crosslinked polyethylene thermosetting moisture and heat-resistant insulated copper building wire	—	%	5.0	—	—
Add for type THHN flame-retardant, moisture and heat resistant thermoplastic insulated copper wire with extruded nylon jacket, 8 gauge and over	—	%	10.0	—	—

Aluminum Building Wire Type THHN single conductor flame-retardant, moisture and heat resistant thermoplastic insulated 600 volt stranded aluminum building wire pulled in conduit. No conduit included. Listed by American Wire Gauge. Note that wire prices can change very quickly.

	Craft@Hrs	Unit	Material	Labor	Total
#6 gauge	E4@.009	LF	.22	.32	.54
#4 gauge	E4@.011	LF	.27	.39	.66
#2 gauge	E4@.014	LF	.37	.49	.86

	Craft@Hrs	Unit	Material	Labor	Total
#1 gauge	E4@.016	LF	.54	.56	1.10
#1/0 gauge	E4@.018	LF	.63	.63	1.26
#2/0 gauge	E4@.020	LF	.75	.70	1.45
#3/0 gauge	E4@.022	LF	.92	.77	1.69
#4/0 gauge	E4@.025	LF	1.04	.88	1.92
250 MCM	E4@.027	LF	1.23	.95	2.18
300 MCM	E4@.030	LF	1.71	1.06	2.77
350 MCM	E4@.032	LF	1.73	1.13	2.86
400 MCM	E4@.035	LF	2.03	1.23	3.26
500 MCM	E4@.039	LF	2.24	1.37	3.61
600 MCM	E4@.043	LF	2.83	1.51	4.34
700 MCM	E4@.046	LF	3.27	1.62	4.89
750 MCM	E4@.047	LF	3.30	1.65	4.95
1000 MCM	E4@.055	LF	4.92	1.93	6.85
Add for XHHW crosslinked polyethylene thermosetting moisture and heat resistant insulated aluminum building wire	—	%	10.0	—	—

UF Direct Burial Cable Copper Type UF and UF-NMC thermoplastic jacketed "A-Z" cable, 600 volt. No excavation or backfill included. Note that wire prices can change very quickly.

	Craft@Hrs	Unit	Material	Labor	Total
With no ground wire					
#14, 2 conductors	E4@.005	LF	.15	.18	.33
#12, 2 conductors	E4@.005	LF	.20	.18	.38
#10, 2 conductors	E4@.005	LF	.30	.18	.48
#14, 3 conductors	E4@.007	LF	.21	.25	.46
#12, 3 conductors	E4@.008	LF	.31	.28	.59
#10, 3 conductors	E4@.009	LF	.44	.32	.76
# 8, 3 conductors	E4@.010	LF	1.20	.35	1.55
# 6, 3 conductors	E4@.011	LF	1.76	.39	2.15
With ground wire					
#14, 2 conductors	E4@.006	LF	.16	.21	.37
#12, 2 conductors	E4@.006	LF	.23	.21	.44
#10, 2 conductors	E4@.006	LF	.35	.21	.56
# 8, 2 conductors	E4@.007	LF	1.00	.25	1.25
# 6, 2 conductors	E4@.009	LF	1.51	.32	1.83
#14, 3 conductors	E4@.010	LF	.24	.35	.59
#12, 3 conductors	E4@.010	LF	.34	.35	.69
#10, 3 conductors	E4@.011	LF	.49	.39	.88
# 8, 3 conductors	E4@.012	LF	1.32	.42	1.74
# 6, 3 conductors	E4@.013	LF	1.90	.46	2.36

Romex Cable Non-metallic (NM) sheathed copper cable, 600 volt, with full size ground wire, installed in frame building, including boring out and pulling cable. Note that wire prices can change very quickly.

	Craft@Hrs	Unit	Material	Labor	Total
#14 wire, 2 conductor	E4@.027	LF	.12	.95	1.07
#12 wire, 2 conductor	E4@.030	LF	.15	1.06	1.21
#10 wire, 2 conductor	E4@.031	LF	.25	1.09	1.34
# 8 wire, 2 conductor	E4@.040	LF	.50	1.41	1.91
# 6 wire, 2 conductor	E4@.045	LF	.87	1.58	2.45
#14 wire, 3 conductor	E4@.030	LF	.20	1.06	1.26
#12 wire, 3 conductor	E4@.031	LF	.28	1.09	1.37
#10 wire, 3 conductor	E4@.035	LF	.41	1.23	1.64
# 8 wire, 3 conductor	E4@.043	LF	.82	1.51	2.33
# 6 wire, 3 conductor	E4@.059	LF	1.21	2.07	3.28
# 4 wire, 3 conductor	E4@.067	LF	1.97	2.36	4.33

	Craft@Hrs	Unit	Material	Labor	Total
Deduct for no ground wire					
#14 and #12 wire	—	%	-5.0	-5.0	—
#10, #8 and #6 wire	—	%	-15.0	-5.0	—

Armored Cable Type MC copper conductors with moisture-resistant and flame-retardant cover wrapped in flexible steel cover, 600 volt, with bonding strip, installed with clamps or staples in a frame building or embedded in masonry, including approved bushings at terminations. Note that wire prices can change very quickly.

	Craft@Hrs	Unit	Material	Labor	Total
#14 wire, 2 conductor, solid	E4@.030	LF	.35	1.06	1.41
#12 wire, 2 conductor, solid	E4@.030	LF	.36	1.06	1.42
#10 wire, 2 conductor, solid	E4@.031	LF	.64	1.09	1.73
# 8 wire, 2 conductor, stranded	E4@.035	LF	1.21	1.23	2.44
# 6 wire, 2 conductor, stranded	E4@.039	LF	1.62	1.37	2.99
#14 wire, 3 conductor, solid	E4@.031	LF	.45	1.09	1.54
#12 wire, 3 conductor, solid	E4@.035	LF	.53	1.23	1.76
#10 wire, 3 conductor, solid	E4@.039	LF	.83	1.37	2.20
# 8 wire, 3 conductor, stranded	E4@.043	LF	1.53	1.51	3.04
# 6 wire, 3 conductor, stranded	E4@.047	LF	2.21	1.65	3.86
# 4 wire, 3 conductor, stranded	E4@.051	LF	3.19	1.79	4.98
# 2 wire, 3 conductor, stranded	E4@.056	LF	4.54	1.97	6.51
#14 wire, 4 conductor, solid	E4@.035	LF	.61	1.23	1.84
#12 wire, 4 conductor, solid	E4@.038	LF	.74	1.34	2.08
#10 wire, 4 conductor, solid	E4@.043	LF	1.25	1.51	2.76
# 8 wire, 4 conductor, stranded	E4@.048	LF	2.12	1.69	3.81
# 6 wire, 4 conductor, stranded	E4@.054	LF	2.85	1.90	4.75
# 4 wire, 4 conductor, stranded	E4@.059	LF	4.24	2.07	6.31

Power Cable Single conductor medium voltage ozone resistant stranded copper cable pulled in conduit. No conduit included. Labor cost assumes three bundled conductors are pulled at one time on runs up to 100 feet. Labor cost will be higher on longer cable pulls and about 50% lower when cable is laid in an open trench. No splicing included. Listed by American Wire Gauge. Note that wire prices can change very quickly. Check with supplier.

5,000 volt, tape shielded, crosslinked polyethylene (XLP) insulated, with PVC jacket

	Craft@Hrs	Unit	Material	Labor	Total
#6 gauge	E4@.014	LF	.80	.49	1.29
#4 gauge	E4@.016	LF	.96	.56	1.52
#2 gauge	E4@.021	LF	1.20	.74	1.94
#1/0 gauge	E4@.026	LF	1.80	.91	2.71
#2/0 gauge	E4@.028	LF	2.10	.98	3.08
#4/0 gauge	E4@.038	LF	3.10	1.34	4.44
250 MCM	E4@.042	LF	3.50	1.48	4.98
350 MCM	E4@.047	LF	3.90	1.65	5.55
500 MCM	E4@.053	LF	4.60	1.86	6.46

15,000 volt, tape shielded, ethylene propylene rubber (EPR) insulated, with PVC jacket

	Craft@Hrs	Unit	Material	Labor	Total
#2 gauge	E4@.024	LF	1.80	.84	2.64
#1/0	E4@.030	LF	2.00	1.06	3.06
#2/0	E4@.032	LF	2.10	1.13	3.23
250 MCM	E4@.046	LF	3.30	1.62	4.92
350 MCM	E4@.055	LF	4.40	1.93	6.33
500 MCM	E4@.062	LF	5.30	2.18	7.48

	Craft@Hrs	Unit	Material	Labor	Total

Snow Melting Cable For concrete pavement. Self-regulating heater cable 208-277 volt AC, encased in concrete pavement for walkways, steps, loading ramps or parking garages. Based on Raychem ElectroMelt™ System for concrete pavement. The heating cable is cut to length at the installation site and typically laid in a serpentine pattern using 12-inch center to center spacing fastened to the top of the reinforcing using nylon cable ties before the concrete is placed. For scheduling purposes estimate that one man can lay out, cut to length, install and tie 600 LF of cable in an 8-hour day. Based on 1.1 LF of cable per SF of pavement including ties, "return bends" and waste, this yields 545 SF of pavement. These costs do not include reinforcing, concrete or concrete placing. A rule of thumb for sizing circuit breakers required for startup at 0 degrees F with heating cable voltage at 220 volts AC is to allow 0.20 amps per LF of cable.

	Craft@Hrs	Unit	Material	Labor	Total
Heater cable (1.1 LF per SF of pavement)	CE@.015	SF	7.56	.60	8.16
Power connection kit, including end seals	CE@.500	Ea	27.50	20.00	47.50
Cable splice kit	CE@.250	Ea	24.70	10.00	34.70
Cable expansion joint kit	CE@.250	Ea	19.70	10.00	29.70
ElectroMelt™ junction box	CE@.500	Ea	107.00	20.00	127.00
System controller, automatic	CE@1.50	Ea	1,790.00	60.10	1,850.10
Thermostat, line sensing	CE@.500	Ea	231.00	20.00	251.00
Ground fault protection device	CE@.969	Ea	341.00	38.80	379.80

Ice Melting Cable For roofs, gutters and downspouts. Self-regulating heater cable 120 or 208-277 volts AC, run exposed on the surface of the roof or within gutters. Based on Raychem IceStop™ System for roofs and gutters. The heating cable is cut to length at the installation site and typically laid in a serpentine pattern using 24-inch center to center spacing and 32-inch loop height fastened to the roof using clips supplied by Raychem. For scheduling purposes estimate that one man can lay out, cut to length and install 400 LF of cable in an 8-hour day. Based on 1.2 LF of cable per SF of protected roof area including clips, "return bends" and waste, this yields 330 SF of protected area. For each LF of gutter add costs equal to one SF of roof; for each downspout add costs equal to 15 SF of roof. These costs do not include the roofing, gutters, or downspouts. A rule of thumb for sizing circuit breakers required for startup at 0 degree F with heating cable voltage at 120 volts AC is to allow 0.17 amps per LF of cable, at 208-277 volts AC allow 0.11 amps per LF of cable.

	Craft@Hrs	Unit	Material	Labor	Total
Heater cable (1.2 LF per SF of protected area)	CE@.033	LF	6.41	1.32	7.73
Power connection kit	CE@.500	Ea	26.80	20.00	46.80
Cable splice kit	CE@.374	Ea	24.80	15.00	39.80
End seal kit	CE@.374	Ea	10.20	15.00	25.20
Downspout hanger	CE@.374	Ea	16.20	15.00	31.20
Metal roof mounting kit	CE@.500	Ea	128.00	20.00	148.00

Electrical Outlet Boxes Steel boxes installed on an exposed wall or ceiling. The material cost for phenolic, PVC and fiberglass boxes will be 50% to 70% less. These costs include fasteners but no switch or receptacle.

	Craft@Hrs	Unit	Material	Labor	Total
Square outlet boxes					
4" x 4" x 1-1/2" deep with					
1/2" and 3/4" knockouts	E4@.187	Ea	1.48	6.58	8.06
Side mounting bracket	E4@.187	Ea	2.02	6.58	8.60
4" x 4" x 2-1/8" deep with					
1/2" and 3/4" knockouts	E4@.245	Ea	2.43	8.62	11.05
Side mounting bracket	E4@.245	Ea	3.51	8.62	12.13
Add for					
1-1/2" deep extension rings	E4@.120	Ea	3.01	4.22	7.23
Romex or BX clamps	—	Ea	1.85	—	1.85
Steel flush cover blanks	E4@.039	Ea	.84	1.37	2.21
Plaster rings to 3/4" deep	E4@.080	Ea	2.45	2.81	5.26

	Craft@Hrs	Unit	Material	Labor	Total
Octagon outlet boxes					
4" x 1-1/2" deep with					
1/2" and 3/4" knockouts	E4@.187	Ea	1.43	6.58	8.01
Romex clamps	E4@.187	Ea	2.45	6.58	9.03
Mounting bracket, horizontal	E4@.187	Ea	2.15	6.58	8.73
4" x 2-1/8" deep with					
1/2" and 3/4" knockouts	E4@.245	Ea	2.12	8.62	10.74
Romex clamps	E4@.245	Ea	2.69	8.62	11.31
Add for					
1-1/2" deep extension rings	E4@.120	Ea	1.78	4.22	6.00
1-1/2" with 21" bar set	E4@.282	Ea	6.78	9.92	16.70
3" deep concrete ring	E4@.516	Ea	5.43	18.10	23.53
4" steel flush cover blanks	E4@.039	Ea	.78	1.37	2.15
Plaster rings to 3/4" deep	E4@.080	Ea	1.85	2.81	4.66
Handy boxes, 4" x 2-1/8"					
1-1/2" deep, 1/2" knockouts	E4@.187	Ea	1.13	6.58	7.71
1-7/8" deep, 3/4" knockouts	E4@.187	Ea	1.15	6.58	7.73
2-1/8" deep, 3/4" knockouts	E4@.245	Ea	2.11	8.62	10.73
2-1/8" deep, with side mounting bracket	E4@.245	Ea	2.50	8.62	11.12
1-1/2" deep extension rings	E4@.120	Ea	2.73	4.22	6.95
Blank or switch cover	E4@.039	Ea	.40	1.37	1.77
Weatherproof box and cover	E4@.059	Ea	3.30	2.07	5.37
Switch boxes, 3" x 2", square corner gangable boxes with mounting ears					
2" deep, 1/2" knockouts	E4@.245	Ea	1.92	8.62	10.54
2-1/2" deep, 3/4" knockouts	E4@.245	Ea	2.41	8.62	11.03
3-1/2" deep, 3/4" knockouts	E4@.324	Ea	2.64	11.40	14.04
Gang switch boxes, 2" x 3" x 1-5/8" deep, with cover					
2 gang	E4@.242	Ea	8.62	8.51	17.13
3 gang	E4@.342	Ea	9.72	12.00	21.72
4 gang	E4@.342	Ea	14.00	12.00	26.00
5 gang	E4@.482	Ea	21.80	17.00	38.80
6 gang	E4@.482	Ea	42.40	17.00	59.40
Floor boxes, watertight, cast iron, round					
4-3/16" x 3-3/4" deep, non-adjustable	E4@1.54	Ea	32.10	54.20	86.30
3-3/4" x 2" deep, semi-adjustable	E4@1.54	Ea	46.50	54.20	100.70
4-3/16" x 3-3/4" deep, adjustable	E4@1.54	Ea	55.20	54.20	109.40
2-1/8" single, round floor cover plates	E4@.156	Ea	26.20	5.49	31.69
Galvanized or gray enamel NEMA class 1 (indoor) pull boxes, with screw cover, 4" deep					
4" x 4"	E4@.373	Ea	4.86	13.10	17.96
4" x 6"	E4@.373	Ea	5.81	13.10	18.91
6" x 6"	E4@.373	Ea	6.89	13.10	19.99
6" x 8"	E4@.391	Ea	7.97	13.80	21.77
8" x 8"	E4@.391	Ea	9.59	13.80	23.39
8" x 10"	E4@.444	Ea	10.60	15.60	26.20
8" x 12"	E4@.483	Ea	12.10	17.00	29.10
10" x 10"	E4@.492	Ea	15.90	17.30	33.20
10" x 12"	E4@.539	Ea	18.40	19.00	37.40
12" x 12"	E4@.593	Ea	20.70	20.90	41.60
12" x 16"	E4@.663	Ea	28.90	23.30	52.20
12" x 18"	E4@.785	Ea	30.10	27.60	57.70

	Craft@Hrs	Unit	Material	Labor	Total
Galvanized or gray enamel NEMA class 1 (indoor) pull boxes, with screw cover, 6" deep					
6" x 6"	E4@.373	Ea	8.20	13.10	21.30
8" x 8"	E4@.391	Ea	11.40	13.80	25.20
8" x 10"	E4@.444	Ea	12.70	15.60	28.30
10" x 10"	E4@.492	Ea	15.00	17.30	32.30
10" x 12"	E4@.539	Ea	16.60	19.00	35.60
12" x 12"	E4@.593	Ea	18.70	20.90	39.60
12" x 16"	E4@.663	Ea	25.50	23.30	48.80
12" x 18"	E4@.785	Ea	27.50	27.60	55.10
16" x 16"	E4@.929	Ea	31.30	32.70	64.00
18" x 18"	E4@1.10	Ea	33.90	38.70	72.60
18" x 24"	E4@1.30	Ea	61.90	45.70	107.60
24" x 24"	E4@1.54	Ea	77.50	54.20	131.70
Cast aluminum NEMA class 3R (weatherproof) screw cover junction boxes, recessed, flush mounted, with cover					
6" x 6" x 4" deep	E4@.930	Ea	128.00	32.70	160.70
6" x 8" x 4" deep	E4@.930	Ea	156.00	32.70	188.70
6" x 6" x 6" deep	E4@.930	Ea	156.00	32.70	188.70
6" x 12" x 6" deep	E4@1.42	Ea	233.00	49.90	282.90
8" x 8" x 4" deep	E4@.982	Ea	179.00	34.50	213.50
8" x 8" x 6" deep	E4@1.14	Ea	192.00	40.10	232.10
8" x 12" x 6" deep	E4@1.30	Ea	292.00	45.70	337.70
10" x 10" x 6" deep	E4@1.33	Ea	165.00	46.80	211.80
12" x 12" x 6" deep	E4@1.35	Ea	345.00	47.50	392.50
12" x 12" x 8" deep	E4@1.73	Ea	405.00	60.80	465.80
12" x 24" x 6" deep	E4@1.97	Ea	773.00	69.30	842.30
18" x 36" x 8" deep	E4@2.56	Ea	1,780.00	90.00	1,870.00
Hinged cover panel enclosures, NEMA 1, enamel, with cover					
16" x 12" x 7" deep	E4@1.30	Ea	69.60	45.70	115.30
20" x 20" x 7" deep	E4@1.30	Ea	94.80	45.70	140.50
30" x 20" x 7" deep	E4@1.77	Ea	88.20	62.20	150.40
24" x 20" x 9" deep	E4@1.30	Ea	115.00	45.70	160.70
30" x 24" x 9" deep	E4@1.77	Ea	130.00	62.20	192.20
36" x 30" x 9" deep	E4@1.77	Ea	166.00	62.20	228.20

Electrical Receptacles Standard commercial grade except where noted. Ivory or brown receptacles with cover and screws. White or gray receptacles will cost about 10% more. No outlet boxes included. Labor includes connecting wire, securing device in the box, and attaching cover.

	Craft@Hrs	Unit	Material	Labor	Total
15 amp, 125 volt self-grounding, 2 pole, 3 wire duplex receptacles					
Residential quality grounded duplex	CE@.162	Ea	.43	6.49	6.92
Screwless, not self-grounding	CE@.162	Ea	5.60	6.49	12.09
Side terminals	CE@.162	Ea	5.90	6.49	12.39
Back and side terminals	CE@.162	Ea	8.90	6.49	15.39
Feed thru wiring, back and side terminals	CE@.162	Ea	14.50	6.49	20.99
Safety ground, side terminals	CE@.162	Ea	21.30	6.49	27.79
Ground fault circuit interrupter receptacle	CE@.324	Ea	32.90	13.00	45.90
Add for NEMA 5 single receptacles	—	%	15.0	—	—
20 amp, 125 volt self-grounding, 2 pole, 3 wire duplex receptacles					
Side terminals	CE@.162	Ea	8.90	6.49	15.39
Back and side terminals	CE@.162	Ea	12.60	6.49	19.09
Feed thru wiring, back and side terminals	CE@.162	Ea	17.50	6.49	23.99
Hospital grade	CE@.162	Ea	21.10	6.49	27.59
Ground fault circuit interrupter receptacle	CE@.324	Ea	34.80	13.00	47.80
Add for NEMA 5 single receptacles	—	%	15.0	—	—

	Craft@Hrs	Unit	Material	Labor	Total
250 volt receptacles, self-grounding, 2 pole, 3 wire, back & side terminals					
15 amp, duplex	CE@.187	Ea	10.00	7.49	17.49
15 amp, duplex, feed thru wiring	CE@.187	Ea	16.90	7.49	24.39
15 amp, single	CE@.187	Ea	11.70	7.46	19.16
20 amp, duplex	CE@.187	Ea	13.10	7.49	20.59
20 amp, duplex, feed thru wiring	CE@.187	Ea	20.60	7.49	28.09
20 amp, single	CE@.187	Ea	14.10	7.49	21.59
Clock receptacle, 2 pole, 15 amp, 125 volt	CE@.329	Ea	18.80	13.20	32.00
120/208 volt 20 amp 4 pole, duplex receptacle	CE@.795	Ea	24.20	31.90	56.10
125/250 volt 3 pole receptacles, flush mount					
15 amp/10 amp	CE@.714	Ea	13.70	28.60	42.30
20 amp	CE@.820	Ea	21.50	32.90	54.40
277 volt 30 amp 2 pole receptacle	CE@.268	Ea	51.20	10.70	61.90
Dryer receptacles, 250 volt, 30/50 amp, 3 wire	CE@.536	Ea	37.30	21.50	58.80
Accessories for 50 amp dryer receptacles					
Flush mounted plug	CE@.288	Ea	65.00	11.50	76.50
Surface mounted plug	CE@.288	Ea	44.50	11.50	56.00
Cord sets for dryer receptacles					
36 inch, three #10 wires	CE@.288	Ea	12.20	11.50	23.70
48 inch, three #10 wires	CE@.288	Ea	13.80	11.50	25.30
60 inch, three #10 wires	CE@.288	Ea	15.90	11.50	27.40

Electrical Switches Commercial grade, 120 to 277 volt rating, except where noted. Ivory or brown. Add 10% for white or gray. Includes cover plate but no fixture boxes except as noted. Labor includes connecting wire to switch, securing switch in box and attaching cover plate.

	Craft@Hrs	Unit	Material	Labor	Total
15 amp switches, back and side wired					
One pole quiet switch, residential (side wired only)	CE@.112	Ea	.59	4.49	5.08
One pole switch, commercial, minimum quality	CE@.112	Ea	2.95	4.49	7.44
Two pole switch	CE@.309	Ea	9.90	12.40	22.30
Three-way switch	CE@.227	Ea	10.20	9.09	19.29
Four-way switch	CE@.309	Ea	26.60	12.40	39.00
20 amp switches, back and side wired					
One pole switch	CE@.187	Ea	9.18	7.49	16.67
Two pole switch	CE@.433	Ea	10.40	17.30	27.70
Three-way switch	CE@.291	Ea	10.00	11.70	21.70
Four-way switch	CE@.435	Ea	26.90	17.40	44.30
30 amp switches, side wired					
One pole switch	CE@.246	Ea	11.60	9.86	21.46
Two pole switch	CE@.475	Ea	17.20	19.00	36.20
Three-way switch	CE@.372	Ea	16.80	14.90	31.70
Four-way switch	CE@.475	Ea	25.00	19.00	44.00
20 amp weatherproof switches, lever handle, with cover					
One pole switch	CE@.187	Ea	4.20	7.49	11.69
Two pole switch	CE@.433	Ea	7.08	17.30	24.38
Three-way switch	CE@.291	Ea	4.80	11.70	16.50
Single pole, two gang, 10 amp	CE@.358	Ea	9.54	14.30	23.84
Dimmer switches, push for off					
600 watt, one pole	CE@.417	Ea	3.84	16.70	20.54
600 watt, three way	CE@.626	Ea	5.94	25.10	31.04
Fluorescent dimmer,					
10 lamp load	CE@.626	Ea	30.00	25.10	55.10

	Craft@Hrs	Unit	Material	Labor	Total
Astro dial time switch, 40 amp, with box	CE@3.00	Ea	106.00	120.00	226.00
15 minute timer switch wall box mounted	CE@.426	Ea	8.82	17.10	25.92
Single pole, 1 throw time switch, 277 volt	CE@.890	Ea	46.80	35.70	82.50
Float switches for sump pumps, automatic 10A, 125/250/480 VAC	CE@1.03	Ea	22.70	41.30	64.00
One way 15 amp toggle switch with neon pilot	CE@.327	Ea	7.23	13.10	20.33
Three way 15 amp toggle switch with neon pilot	CE@.372	Ea	12.10	14.90	27.00
Photoelectric switches, flush, with wall plate					
120 volt, 1,000 watt	CE@.426	Ea	6.30	17.10	23.40
208 volt, 1,800 watt	CE@.626	Ea	7.56	25.10	32.66
480 volt, 3,000 watt	CE@.795	Ea	11.80	31.90	43.70
Button control stations, surface mounted, NEMA class 1, standard duty, 120/240 volt, with enclosure					
Start-stop switch, 2 button	CE@.624	Ea	17.30	25.00	42.30
Start-stop switch with lockout, 2 button	CE@.624	Ea	25.90	25.00	50.90
Forward, reverse and stop buttons, 3 button	CE@.624	Ea	34.60	25.00	59.60
Forward, reverse, stop and lockout, 3 button	CE@.624	Ea	34.60	25.00	59.60
Manual toggle starter switches, surface mounted, NEMA class 1, 120/240 volt, non-reversing, with enclosure					
Size 0 motors, 2 pole	CE@.613	Ea	91.80	24.60	116.40
Size 1 motors, 3 pole	CE@.698	Ea	119.00	28.00	147.00
Size 1-1/2 motors, 2 pole	CE@.901	Ea	137.00	36.10	173.10
Manual button starter switches, surface mounted, NEMA class 1, 110 to 240 volt, 2 pole, 1 phase, with enclosure, start, stop, reset, with relay					
Size 00 motors	CE@.712	Ea	123.00	28.50	151.50
Size 0 motors	CE@.712	Ea	133.00	28.50	161.50
Size 1 motors	CE@.820	Ea	147.00	32.90	179.90
Size 1-1/2 motors	CE@.975	Ea	178.00	39.10	217.10

Grounding Devices No excavation or concrete included.

	Craft@Hrs	Unit	Material	Labor	Total
Copperweld grounding rods, driven					
5/8" x 8'	E4@.365	Ea	23.30	12.80	36.10
5/8" x 10'	E4@.420	Ea	30.40	14.80	45.20
3/4" x 8'	E4@.365	Ea	37.90	12.80	50.70
3/4" x 10'	E4@.420	Ea	46.80	14.80	61.60
Grounding locknuts					
1/2" locknuts	E4@.050	Ea	1.93	1.76	3.69
3/4" locknuts	E4@.060	Ea	2.34	2.11	4.45
1" locknuts	E4@.080	Ea	3.51	2.81	6.32
1-1/4" locknuts	E4@.100	Ea	3.66	3.52	7.18
1-1/2" locknuts	E4@.100	Ea	3.94	3.52	7.46
2" locknuts	E4@.150	Ea	5.30	5.28	10.58
2-1/2" locknuts	E4@.200	Ea	9.75	7.03	16.78
3" locknuts	E4@.200	Ea	12.30	7.03	19.33
4" locknuts	E4@.300	Ea	26.10	10.60	36.70
Ground rod clamps					
5/8" clamp	E4@.249	Ea	3.49	8.76	12.25
3/4" clamp	E4@.249	Ea	4.24	8.76	13.00

	Craft@Hrs	Unit	Material	Labor	Total
Coupling for threaded ground rod					
5/8" clamp	E4@.249	Ea	6.43	8.76	15.19
3/4" clamp	E4@.249	Ea	9.54	8.76	18.30
Copper bonding connector straps, 3/4" x .050"					
1" strap	E4@.286	Ea	3.64	10.10	13.74
2" strap	E4@.300	Ea	4.67	10.60	15.27
3" strap	E4@.315	Ea	6.26	11.10	17.36
4" strap	E4@.330	Ea	7.43	11.60	19.03
Brazed connections for wire					
#6 wire	E4@.194	Ea	.95	6.82	7.77
#2 wire	E4@.194	Ea	1.19	6.82	8.01
#1/0 wire	E4@.295	Ea	3.00	10.40	13.40
#4/0 wire	E4@.391	Ea	7.40	13.80	21.20

Grounding using a 5/8" x 8' copper clad ground rod, includes 10' grounding conductor, clamps, connectors, locknut, and rod.

	Craft@Hrs	Unit	Material	Labor	Total
100 amp service, #8 copper wire	E4@1.13	Ea	32.00	39.70	71.70
150 amp service, #6 copper wire	E4@1.16	Ea	34.10	40.80	74.90

Electric Motors General purpose, 3 phase, open drip-proof, industrial duty, 1,700 - 1,780 RPM, 208-230/460 volt, rigid welded or solid base, furnished and placed, no hookup included. Based on Dayton motors.

	Craft@Hrs	Unit	Material	Labor	Total
1/2 HP	E4@1.99	Ea	203.00	70.00	273.00
3/4 HP	E4@1.99	Ea	300.00	70.00	370.00
1 HP	E4@1.99	Ea	351.00	70.00	421.00
1-1/2 HP	E4@1.99	Ea	386.00	70.00	456.00
2 HP	E4@1.99	Ea	367.00	70.00	437.00
3 HP	E4@1.99	Ea	476.00	70.00	546.00
5 HP	E4@2.22	Ea	595.00	78.10	673.10
7-1/2 HP	E4@2.40	Ea	784.00	84.40	868.40
10 HP	E4@2.51	Ea	973.00	88.30	1,061.30

Electric Motor Connection Includes a NEMA-1 fusible heavy duty safety switch, connectors, fittings, ells, adapters, supports, anchors and conduit with wire appropriate for the load connected but no starter or motor.

Single phase, using 10' RSC, 20' PVC and 6' flex conduit, NEMA class 1 (indoor) general duty switch, 240 V

	Craft@Hrs	Unit	Material	Labor	Total
For motor to 2 HP, 20 amps	E4@6.35	Ea	64.00	223.00	287.00
For 2.5 to 3 HP motor, 30 amps	E4@6.63	Ea	70.50	233.00	303.50
For 3 HP motor, 45 amps	E4@6.68	Ea	112.00	235.00	347.00
For 8 to 15 HP motor, 90 amps	E4@13.4	Ea	238.00	471.00	709.00

Three phase wiring using 10' RSC, 20' PVC and 6' flex conduit, NEMA class 1 (indoor) switch, 600 V

	Craft@Hrs	Unit	Material	Labor	Total
For motor to 2 HP, 20 amps	E4@7.18	Ea	209.00	252.00	461.00
For 3 to 5 HP motor, 30 amps	E4@7.90	Ea	216.00	278.00	494.00
For 7 to 10 HP motor, 45 amps	E4@8.30	Ea	274.00	292.00	566.00
For 25 to 50 HP motor, 90 amps	E4@9.30	Ea	520.00	327.00	847.00
For 50 to 100 HP motor, 135 amps	E4@14.1	Ea	785.00	496.00	1,281.00
For 100 to 200 HP motor, 270 amps	E4@21.9	Ea	1,770.00	770.00	2,540.00

Single phase, using 10' RSC, 20' PVC and 6' flex conduit, NEMA class 3R (weatherproof) switch, 240 V

	Craft@Hrs	Unit	Material	Labor	Total
For motor to 2 HP, 20 amps	E4@6.63	Ea	128.00	233.00	361.00
For 2 to 5 HP motor, 30 amps	E4@6.97	Ea	132.00	245.00	377.00
For 5 to 7.5 HP motor, 45 amps	E4@6.94	Ea	198.00	244.00	442.00

Three phase using 10' RSC, 20' PVC and 6' flex conduit, NEMA class 3R (weatherproof) switch, 600 V

	Craft@Hrs	Unit	Material	Labor	Total
For motor to 5 HP motor, 20 amps	E4@7.10	Ea	328.00	250.00	578.00
For 1 to 10 HP motor, 30 amps	E4@7.25	Ea	334.00	255.00	589.00
For 10 to 25 HP motor, 45 amps	E4@9.51	Ea	408.00	334.00	742.00

	Craft@Hrs	Unit	Material	Labor	Total
For 25 to 50 HP motor, 90 amps	E4@9.51	Ea	666.00	334.00	1,000.00
For 60 to 100 HP motor, 135 amps	E4@12.1	Ea	961.00	426.00	1,387.00
For 125 to 200 HP motor, 270 amps	E4@21.6	Ea	2,160.00	760.00	2,920.00

Electric Motor Connection complete installation. Including magnetic motor starter, overload relay, all wiring, a heavy duty fusible safety switch, junction box and cover, connectors, fittings, adapters, supports, anchors and conduit appropriate for the load connected. No motor included.

Single phase, using 10' RSC, 20' PVC and 6' flex conduit, NEMA class 1 (indoor) starter, 230 V

	Craft@Hrs	Unit	Material	Labor	Total
For motor to 2 HP, 20 amps	E4@8.11	Ea	264.00	285.00	549.00
For 2 to 3 HP motor, 30 amps	E4@8.32	Ea	328.00	293.00	621.00
For 3 to 8 HP motor, 45 amps	E4@8.45	Ea	507.00	297.00	804.00
For 8 to 15 HP motor, 90 amps	E4@15.5	Ea	633.00	545.00	1,178.00

Three phase wiring using 10' RSC, 20' PVC and 6' flex conduit, NEMA class 1 (indoor) starter, 480 V

	Craft@Hrs	Unit	Material	Labor	Total
For motor to 2 HP, 20 amps	E4@9.90	Ea	692.00	348.00	1,040.00
For 2 to 5 HP motor, 30 amps	E4@10.6	Ea	725.00	373.00	1,098.00
For 7 to 25 HP motor, 45 amps	E4@20.5	Ea	1,080.00	721.00	1,801.00
For 25 to 50 HP motor, 90 amps	E4@23.1	Ea	1,860.00	812.00	2,672.00

Single phase, using 10' RSC, 20' PVC and 6' flex conduit, NEMA Type 3R (weatherproof) starter, 230 V

	Craft@Hrs	Unit	Material	Labor	Total
For motor to 2 HP, 20 amps	E4@8.91	Ea	804.00	313.00	1,117.00
For 2 to 5 HP motor, 30 amps	E4@9.15	Ea	844.00	322.00	1,166.00
For 5 to 7.5 HP motor, 45 amps	E4@9.25	Ea	910.00	325.00	1,235.00

Three phase using 10' RSC, 20' PVC and 6' flex conduit, NEMA Type 3R (weatherproof) starter, 480 V

	Craft@Hrs	Unit	Material	Labor	Total
For motor to 5 HP, 20 amps	E4@10.9	Ea	1,040.00	383.00	1,423.00
For 5 to 10 HP motor, 30 amps	E4@11.7	Ea	1,040.00	411.00	1,451.00
For 10 to 25 HP motor, 45 amps	E4@22.5	Ea	1,538.00	791.00	2,329.00
For 25 to 50 HP motor, 90 amps	E4@25.4	Ea	1,800.00	893.00	2,693.00

Motor Starters Magnetic operated full voltage non-reversing motor controllers with thermal overload relays and enclosure.

Two pole, 1 phase contactors, NEMA class 1 (indoor), electrically held with holding interlock, one reset

	Craft@Hrs	Unit	Material	Labor	Total
Size 00, 1/3 HP, 9 amp	E4@1.47	Ea	154.00	51.70	205.70
Size 0, 1 HP, 18 amp	E4@1.62	Ea	176.00	57.00	233.00
Size 1, 2 HP, 27 amp	E4@1.73	Ea	200.00	60.80	260.80
Size 2, 3 HP, 45 amp	E4@1.73	Ea	257.00	60.80	317.80
Size 3, 7 1/2 HP, 90 amp	E4@2.02	Ea	395.00	71.00	466.00

Three pole polyphase NEMA class 1 (indoor) AC magnetic combination starters with fusible disconnect and overload relays but no heaters, 208 to 240 volts

	Craft@Hrs	Unit	Material	Labor	Total
size 0, 3 HP, 30 amps	E4@2.75	Ea	483.00	96.70	579.70
size 1, 7.5 HP, 30 amps	E4@3.89	Ea	509.00	137.00	646.00
size 1, 10 HP, 60 amps	E4@3.89	Ea	509.00	137.00	646.00
size 2, 10 HP, 60 amps	E4@5.34	Ea	806.00	188.00	994.00
size 2, 25 HP, 100 amps	E4@5.34	Ea	806.00	188.00	994.00
size 3, 25 HP, 100 amps	E4@6.76	Ea	1,340.00	238.00	1,578.00
size 3, 50 HP, 200 amps	E4@6.76	Ea	1,340.00	238.00	1,578.00
size 4, 40 HP, 200 amps	E4@12.0	Ea	2,570.00	422.00	2,992.00
size 4, 75 HP, 400 amps	E4@12.0	Ea	2,570.00	422.00	2,992.00

On and off switches for starter enclosures, any motor starter above

	Craft@Hrs	Unit	Material	Labor	Total
Switch kit without pilot light	E4@.744	Ea	30.20	26.20	56.40
Switch kit with pilot light	E4@.744	Ea	45.30	26.20	71.50

Accessories for any motor starter

	Craft@Hrs	Unit	Material	Labor	Total
Add for 440 to 600 volt starters	—	%	2.0	—	—
Add for NEMA Type 3R (rainproof) enclosure	—	%	40.0	10.0	—
Add for NEMA Type 4 (waterproof) enclosure	—	%	40.0	20.0	—

	Craft@Hrs	Unit	Material	Labor	Total
Add for NEMA Type 12 (dust-tight) enclosure	—	%	15.0	15.0	—
Add for starters in an oversize enclosure	—	%	21.0	—	—
Deduct for starters with circuit breakers	—	%	-4.0	—	—

Safety Switches Wall mounted switches with enclosures as noted. No fuses or hubs included.

Heavy duty (NEMA-1) 600 volt 2, 3 or 4 pole fusible safety switches

	Craft@Hrs	Unit	Material	Labor	Total
30 amp	E4@3.06	Ea	154.00	108.00	262.00
60 amp	E4@3.89	Ea	186.00	137.00	323.00
100 amp	E4@4.20	Ea	347.00	148.00	495.00
200 amp	E4@6.92	Ea	500.00	243.00	743.00
400 amp	E4@11.4	Ea	1,230.00	401.00	1,631.00
600 amp	E4@14.3	Ea	1,940.00	503.00	2,443.00
800 amp	E4@19.6	Ea	3,020.00	689.00	3,709.00
1200 amp	E4@22.7	Ea	4,250.00	798.00	5,048.00

Heavy duty rainproof (NEMA-3R) 600 volt 2, 3 or 4 pole fusible safety switches

	Craft@Hrs	Unit	Material	Labor	Total
30 amp	E4@3.32	Ea	261.00	117.00	378.00
60 amp	E4@4.61	Ea	308.00	162.00	470.00
100 amp	E4@4.77	Ea	480.00	168.00	648.00
200 amp	E4@7.51	Ea	660.00	264.00	924.00
400 amp	E4@12.4	Ea	1,590.00	436.00	2,026.00
600 amp	E4@15.6	Ea	3,180.00	549.00	3,729.00
800 amp	E4@21.4	Ea	4,790.00	753.00	5,543.00
1200 amp	E4@25.8	Ea	5,250.00	907.00	6,157.00

Heavy duty watertight (NEMA-4) 600 volt 3 pole, 4 wire fusible safety switches

	Craft@Hrs	Unit	Material	Labor	Total
30 amp	E4@3.76	Ea	700.00	132.00	832.00
60 amp	E4@5.21	Ea	770.00	183.00	953.00
100 amp	E4@5.34	Ea	1,520.00	188.00	1,708.00
200 amp	E4@8.52	Ea	2,140.00	300.00	2,440.00
400 amp	E4@13.7	Ea	4,240.00	482.00	4,722.00
600 amp	E4@17.0	Ea	6,080.00	598.00	6,678.00

Heavy duty dust-tight (NEMA-12) 600 volt 3 pole or 4 pole solid neutral fusible safety switches

	Craft@Hrs	Unit	Material	Labor	Total
30 amp	E4@3.32	Ea	273.00	117.00	390.00
60 amp	E4@4.20	Ea	274.00	148.00	422.00
100 amp	E4@4.48	Ea	426.00	158.00	584.00
200 amp	E4@7.51	Ea	666.00	264.00	930.00
400 amp	E4@12.7	Ea	1,520.00	447.00	1,967.00
600 amp	E4@15.9	Ea	2,560.00	559.00	3,119.00

General duty (NEMA-1) 240 volt 3 pole, 4 wire non-fusible safety switches

	Craft@Hrs	Unit	Material	Labor	Total
30 amp	E4@2.58	Ea	52.20	90.70	142.90
100 amp	E4@3.60	Ea	156.00	127.00	283.00
200 amp	E4@6.50	Ea	333.00	229.00	562.00
400 amp	E4@10.8	Ea	864.00	380.00	1,244.00
600 amp	E4@12.7	Ea	1,620.00	447.00	2,067.00

General duty rainproof (NEMA-3R) 240 volt 3 pole, 4 wire non-fusible safety switches

	Craft@Hrs	Unit	Material	Labor	Total
30 amp	E4@2.88	Ea	81.10	101.00	182.10
60 amp	E4@3.76	Ea	123.00	132.00	255.00
100 amp	E4@3.89	Ea	227.00	137.00	364.00
200 amp	E4@7.23	Ea	406.00	254.00	660.00
400 amp	E4@11.8	Ea	1,020.00	415.00	1,435.00
600 amp	E4@14.0	Ea	2,200.00	492.00	2,692.00

	Craft@Hrs	Unit	Material	Labor	Total
Heavy duty watertight (NEMA-4) 240 volt 3 pole, 4 wire fusible safety switches					
30 amp	E4@3.19	Ea	589.00	112.00	701.00
60 amp	E4@3.76	Ea	700.00	132.00	832.00
100 amp	E4@4.33	Ea	1,420.00	152.00	1,572.00
200 amp	E4@7.80	Ea	1,930.00	274.00	2,204.00
400 amp	E4@13.0	Ea	4,250.00	457.00	4,707.00
600 amp	E4@15.4	Ea	6,000.00	542.00	6,542.00
Heavy duty dust-tight (NEMA-12) 240 volt 3 pole, 4 wire fusible safety switches					
30 amp	E4@2.88	Ea	180.00	101.00	281.00
60 amp	E4@3.76	Ea	232.00	132.00	364.00
100 amp	E4@4.04	Ea	332.00	142.00	474.00
200 amp	E4@7.23	Ea	445.00	254.00	699.00
400 amp	E4@11.8	Ea	1,110.00	415.00	1,525.00
600 amp	E4@14.0	Ea	1,860.00	492.00	2,352.00
Add for conduit hubs					
3/4" to 1-1/2", to 100 amp	—	Ea	6.45	—	6.45
2", 200 amp	—	Ea	11.30	—	11.30
2-1/2", 200 amp	—	Ea	18.90	—	18.90

Service Entrance and Distribution Switchboards

NEMA Class 1 indoor, 600 volt, 3 phase, 4 wire, for 240/480 volt insulated case main breakers. Basic structure is 90" high by 21" deep. Width varies with equipment capacity. Labor cost includes setting and leveling on a prepared pad but excludes the pad cost. Add breaker, instrumentation and accessory costs for a complete installation. Based on Westinghouse Pow-R-Gear. These switchboards are custom designed for each installation. Costs can vary widely. Multiple units ordered at the same time may reduce costs per unit by 25% or more.

	Craft@Hrs	Unit	Material	Labor	Total
600 amp bus	E4@22.9	Ea	2,740.00	805.00	3,545.00
1,000 amp bus	E4@22.9	Ea	2,830.00	805.00	3,635.00
1,200 amp bus	E4@22.9	Ea	3,440.00	805.00	4,245.00
1,600 amp bus	E4@22.9	Ea	3,970.00	805.00	4,775.00
2,000 amp bus	E4@22.9	Ea	4,360.00	805.00	5,165.00
2,500 amp bus	E4@22.9	Ea	5,310.00	805.00	6,115.00
3,000 amp bus	E4@22.9	Ea	6,080.00	805.00	6,885.00
4,000 amp bus	E4@22.9	Ea	7,820.00	805.00	8,625.00

240/480 volt draw-out main breakers, 100K amp interrupt capacity. Includes connecting and testing breakers only. Based on Pow-R-Trip

	Craft@Hrs	Unit	Material	Labor	Total
100 to 800 amp, manual operation	E4@10.2	Ea	8,100.00	359.00	8,459.00
1,000 to 1,600 amp, manual operation	E4@17.5	Ea	17,000.00	615.00	17,615.00
2,000 amp, manual operation	E4@23.2	Ea	20,800.00	816.00	21,616.00
2,500 amp, manual operation	E4@23.2	Ea	35,300.00	816.00	36,116.00
3,000 amp, manual operation	E4@23.2	Ea	39,600.00	816.00	40,416.00
4,000 amp, manual operation	E4@23.2	Ea	47,500.00	816.00	48,316.00
100 to 800 amp electric operation	E4@10.2	Ea	11,100.00	359.00	11,459.00
1,000 to 1,600 amp electric operation	E4@17.5	Ea	18,600.00	615.00	19,215.00
2,000 amp electric operation	E4@23.2	Ea	24,700.00	816.00	25,516.00
2,500 amp electric operation	E4@23.2	Ea	40,400.00	816.00	41,216.00
3,000 amp electric operation	E4@23.2	Ea	43,300.00	816.00	44,116.00
Additional space for future draw-out breakers					
100 to 800 amp	—	Ea	2,160.00	—	2,160.00
1,000 to 1,600 amp	—	Ea	3,160.00	—	3,160.00
2,000 amp	—	Ea	3,550.00	—	3,550.00
2,500 amp	—	Ea	3,790.00	—	3,790.00
3,000 amp	—	Ea	6,910.00	—	6,910.00

	Craft@Hrs	Unit	Material	Labor	Total

Service Entrance and Distribution Switchboards, Weatherproof

600 volt, 3 phase, 4 wire, for 480 volt insulated case main breakers. Basic structure is 90" high by 21" deep. Width varies with equipment capacity. Labor cost includes setting and leveling on a prepared pad but excludes the pad cost. Add breaker, instrumentation and accessory costs from below for a complete installation. These switchboards are custom designed for each installation. Costs can vary widely. Multiple units ordered at the same time can reduce costs per unit by 25% or more.

	Craft@Hrs	Unit	Material	Labor	Total
600 amp bus	E4@23.2	Ea	3,250.00	816.00	4,066.00
1,000 amp bus	E4@23.2	Ea	3,800.00	816.00	4,616.00
1,200 amp bus	E4@23.2	Ea	4,060.00	816.00	4,876.00
1,600 amp bus	E4@23.2	Ea	4,680.00	816.00	5,496.00
2,000 amp bus	E4@23.2	Ea	5,190.00	816.00	6,006.00
2,500 amp bus	E4@23.2	Ea	6,240.00	816.00	7,056.00
3,000 amp bus	E4@23.2	Ea	6,360.00	816.00	7,176.00
4,000 amp bus	E4@23.2	Ea	9,260.00	816.00	10,076.00

480 volt manual operated draw-out breakers. By amp interrupt capacity (AIC) as noted. Includes connecting and testing factory installed breakers only

	Craft@Hrs	Unit	Material	Labor	Total
50 to 800 amp, 30K AIC	E4@10.2	Ea	11,100.00	359.00	11,459.00
50 to 1,600 amp, 50K AIC	E4@17.5	Ea	23,300.00	615.00	23,915.00
2,000 amp	E4@23.2	Ea	29,300.00	816.00	30,116.00
1,200 to 3,200 amp	E4@23.2	Ea	51,600.00	816.00	52,416.00
4,000 amp	E4@23.2	Ea	82,100.00	816.00	82,916.00

480 volt manual operated draw-out breakers. 200K amp interrupt capacity (AIC)
Includes connecting and testing factory installed breakers only

	Craft@Hrs	Unit	Material	Labor	Total
50 to 800 amp	E4@10.2	Ea	16,500.00	359.00	16,859.00
50 to 1,600 amp	E4@17.5	Ea	29,500.00	615.00	30,115.00
1,200 to 3,200 amp	E4@23.2	Ea	74,700.00	816.00	75,516.00
4,000 amp	E4@23.2	Ea	121,000.00	816.00	121,816.00

480 volt electrically operated draw-out breakers. By amp interrupt capacity (AIC) as noted. Includes connecting and testing factory installed breakers only

	Craft@Hrs	Unit	Material	Labor	Total
50 to 800 amp, 30K AIC	E4@10.2	Ea	14,700.00	359.00	15,059.00
50 to 1,600 amp, 50K AIC	E4@17.5	Ea	30,400.00	615.00	31,015.00
2,000 amp	E4@23.2	Ea	38,700.00	816.00	39,516.00
1,200 to 3,200 amp	E4@23.2	Ea	58,000.00	816.00	58,816.00
4,000 amp	E4@23.2	Ea	89,600.00	816.00	90,416.00

480 volt electrically operated draw-out breakers. 200K amp interrupt capacity (AIC)
Includes connecting and testing factory installed breakers only

	Craft@Hrs	Unit	Material	Labor	Total
50 to 800 amp	E4@10.2	Ea	20,100.00	359.00	20,459.00
50 to 1,600 amp	E4@17.5	Ea	40,500.00	615.00	41,115.00
1,200 to 3,200 amp	E4@23.2	Ea	83,600.00	816.00	84,416.00
4,000 amp	E4@23.2	Ea	121,000.00	816.00	121,816.00

Additional space for future draw-out breakers

	Craft@Hrs	Unit	Material	Labor	Total
50 to 800 amp	—	Ea	2,540.00	—	2,540.00
50 to 1,600 amp	—	Ea	4,120.00	—	4,120.00
2,000 amp	—	Ea	5,640.00	—	5,640.00
1,200 to 3,200 amp	—	Ea	8,760.00	—	8,760.00
4,000 amp	—	Ea	16,200.00	—	16,200.00

	Craft@Hrs	Unit	Material	Labor	Total

Switchboard Instrumentation and Accessories Add these costs to the cost of the basic enclosure and breakers to find the complete switchboard cost. These figures are for factory-installed instrumentation and accessories and include the cost of connecting and testing only.

Bus duct connection, 3 phase, 4 wire

	Craft@Hrs	Unit	Material	Labor	Total
225 amp	CE@.516	Ea	483.00	20.70	503.70
400 amp	CE@.516	Ea	563.00	20.70	583.70
600 amp	CE@.516	Ea	681.00	20.70	701.70
800 amp	CE@.516	Ea	1,090.00	20.70	1,110.70
1,000 amp	CE@.516	Ea	1,240.00	20.70	1,260.70
1,350 amp	CE@1.01	Ea	1,610.00	40.50	1,650.50
1,600 amp	CE@1.01	Ea	1,860.00	40.50	1,900.50
2,000 amp	CE@1.01	Ea	2,290.00	40.50	2,330.50
2,500 amp	CE@1.01	Ea	2,720.00	40.50	2,760.50
3,000 amp	CE@1.01	Ea	3,110.00	40.50	3,150.50
4,000 amp	CE@1.01	Ea	3,450.00	40.50	3,490.50

Recording meters

	Craft@Hrs	Unit	Material	Labor	Total
Voltmeter	CE@1.01	Ea	7,010.00	40.50	7,050.50
Ammeter	CE@1.01	Ea	7,150.00	40.50	7,190.50
Wattmeter	CE@1.01	Ea	8,950.00	40.50	8,990.50
Varmeter	CE@1.01	Ea	9,570.00	40.50	9,610.50
Power factor meter	CE@1.01	Ea	9,370.00	40.50	9,410.50
Frequency meter	CE@1.01	Ea	9,370.00	40.50	9,410.50
Watt-hour meter, 3 element	CE@1.01	Ea	12,100.00	40.50	12,140.50

Non-recording meters

	Craft@Hrs	Unit	Material	Labor	Total
Voltmeter	CE@.516	Ea	1,190.00	20.70	1,210.70
Ammeter for incoming line	CE@.516	Ea	1,210.00	20.70	1,230.70
Ammeter for feeder circuits	CE@.516	Ea	505.00	20.70	525.70
Wattmeter	CE@1.01	Ea	3,920.00	40.50	3,960.50
Varmeter	CE@1.01	Ea	3,540.00	40.50	3,580.50
Power factor meter	CE@1.01	Ea	2,870.00	40.50	2,910.50
Frequency meter	CE@1.01	Ea	3,390.00	40.50	3,430.50
Watt-hour meter, 3 element	CE@1.01	Ea	3,540.00	40.50	3,580.50
Instrument phase select switch	CE@.516	Ea	611.00	20.70	631.70
Adjustable short time pickup and delay	CE@.516	Ea	920.00	20.70	940.70
Ground fault trip pickup and delay, 4 wire	CE@.525	Ea	967.00	21.00	988.00
Ground fault test panel	CE@.516	Ea	1,060.00	20.70	1,080.70
Shunt trip for breakers	CE@.516	Ea	707.00	20.70	727.70
Key interlock for breakers	CE@.516	Ea	707.00	20.70	727.70
Breaker lifting and transport truck	—	Ea	4,800.00	—	4,800.00
Breaker lifting device	—	Ea	4,180.00	—	4,180.00

Current transformers, by primary capacity

	Craft@Hrs	Unit	Material	Labor	Total
800 amps and less	CE@1.01	Ea	850.00	40.50	890.50
1,000 to 1,500 amps	CE@1.01	Ea	1,250.00	40.50	1,290.50
2,000 to 6,000 amps	CE@1.01	Ea	1,490.00	40.50	1,530.50
Current transformer mount	—	Ea	415.00	—	415.00
Potential transformer	CE@1.01	Ea	1,140.00	40.50	1,180.50
Potential transformer mount	—	Ea	261.00	—	261.00

Feeder Section Breaker Panels

Flush or surface mounted 277/480 volt, 3 phase, 4 wire, FA frame panels.

	Craft@Hrs	Unit	Material	Labor	Total
225 amp, with 22K amp interrupt capacity breaker	E4@6.76	Ea	1,110.00	238.00	1,348.00
225 amp, with 25K amp interrupt capacity breaker	E4@6.76	Ea	2,020.00	238.00	2,258.00

	Craft@Hrs	Unit	Material	Labor	Total
400 amp, with 30K amp interrupt capacity breaker	E4@10.8	Ea	2,430.00	380.00	2,810.00
400 amp, with 35K amp interrupt capacity breaker	E4@10.8	Ea	2,890.00	380.00	3,270.00
225 amp, with main lugs only	E4@6.76	Ea	367.00	238.00	605.00
400 amp, with main lugs only	E4@10.8	Ea	516.00	380.00	896.00
600 amp, with main lugs only	E4@19.3	Ea	687.00	679.00	1,366.00

Transformers Indoor dry type transformer, floor mounted, including connections.

Single phase light and power circuit transformer, 240/480 volt primary, 120/240 secondary, no taps for voltage change

	Craft@Hrs	Unit	Material	Labor	Total
1 KVA	E4@3.76	Ea	213.00	132.00	345.00
2 KVA	E4@4.20	Ea	286.00	148.00	434.00
3 KVA	E4@4.33	Ea	308.00	152.00	460.00
5 KVA	E4@5.47	Ea	422.00	192.00	614.00
7.5 KVA	E4@8.34	Ea	543.00	293.00	836.00
10 KVA	E4@9.38	Ea	660.00	330.00	990.00
15 KVA	E4@12.6	Ea	826.00	443.00	1,269.00
25 KVA	E4@12.6	Ea	1,300.00	443.00	1,743.00

Three phase light and power circuit transformers, 480 volt primary, 120/208 volt secondary, with 6 taps for voltage change

	Craft@Hrs	Unit	Material	Labor	Total
30 KVA	E4@10.5	Ea	618.00	369.00	987.00
45 KVA	E4@12.6	Ea	774.00	443.00	1,217.00
50 KVA	E4@12.6	Ea	1,070.00	443.00	1,513.00
75 KVA	E4@14.8	Ea	1,090.00	520.00	1,610.00
112.5 KVA	E4@15.7	Ea	1,680.00	552.00	2,232.00
150 KVA	E4@16.7	Ea	2,130.00	587.00	2,717.00
225 KVA	E4@18.9	Ea	3,040.00	665.00	3,705.00
300 KVA	E4@21.2	Ea	3,880.00	746.00	4,626.00

Three phase 5000 volt load center transformers, 4160 volt primary, 480/277 volt secondary, with voltage change taps, insulation class H, 80 degree system

	Craft@Hrs	Unit	Material	Labor	Total
225 KVA	E4@29.5	Ea	10,100.00	1,040.00	11,140.00
300 KVA	E4@33.7	Ea	12,000.00	1,190.00	13,190.00
500 KVA	E4@42.5	Ea	18,200.00	1,490.00	19,690.00
750 KVA	E4@53.1	Ea	23,800.00	1,870.00	25,670.00
1,000 KVA	E4@63.2	Ea	28,000.00	2,220.00	30,220.00

Panelboards Surface mounted lighting and appliance distribution panelboards for bolt-on circuit breakers. No breakers included, add for same as required.

3 wire 120/240 volt, 250 amp, single phase AC panelboards with main lugs only for bolt-on breakers

	Craft@Hrs	Unit	Material	Labor	Total
18 circuits	E4@10.7	Ea	181.00	376.00	557.00
30 circuits	E4@15.3	Ea	185.00	538.00	723.00
42 circuits	E4@20.9	Ea	244.00	735.00	979.00
Deduct for 3 wire 120/240 panelboards					
With main lugs only	—	%	-10.0	-15.0	—

4 wire 120/208 volt, 250 amp, 3 phase AC panelboards for bolt-on breakers, with main circuit breaker only

	Craft@Hrs	Unit	Material	Labor	Total
18 circuits	E4@12.9	Ea	148.00	454.00	602.00
30 circuits	E4@21.5	Ea	209.00	756.00	965.00
42 circuits	E4@30.1	Ea	271.00	1,060.00	1,331.00
Deduct for 3 wire panelboards					
With circuit breaker mains only	—	%	-10.0	-15.0	—

	Craft@Hrs	Unit	Material	Labor	Total

Loadcenters Surface mounted 240 volt NEMA 1 (indoor) distribution panels for plug-in branch breakers. Add the cost of distribution breakers and conduit hubs.

Single phase 3 wire 120/240 volt loadcenters with main circuit breaker only

	Craft@Hrs	Unit	Material	Labor	Total
100 amps, 14 circuits	E4@1.20	Ea	150.00	42.20	192.20
100 amps, 18 circuits	E4@1.20	Ea	163.00	42.20	205.20
125 amps, 22 circuits	E4@1.52	Ea	305.00	53.50	358.50
150 amps, 24 circuits	E4@2.30	Ea	312.00	80.90	392.90
150 amps, 32 circuits	E4@2.30	Ea	335.00	80.90	415.90
200 amps, 24 circuits	E4@2.60	Ea	312.00	91.40	403.40
200 amps, 32 circuits	E4@2.60	Ea	345.00	91.40	436.40
200 amps, 42 circuits	E4@2.93	Ea	395.00	103.00	498.00
225 amps, 32 circuits	E4@2.93	Ea	452.00	103.00	555.00
225 amps, 42 circuits	E4@2.93	Ea	472.00	103.00	575.00
Add for NEMA-4 raintight loadcenters	—	%	25.0	—	—

Three phase 4 wire 120/208 volt loadcenters with main circuit breaker only

	Craft@Hrs	Unit	Material	Labor	Total
100 amps, 12 circuits	E4@1.21	Ea	249.00	42.60	291.60
100 amps, 18 circuits	E4@1.60	Ea	335.00	56.30	391.30
125 amps, 30 circuits	E4@1.91	Ea	662.00	67.20	729.20
150 amps, 24 circuits	E4@2.18	Ea	640.00	76.70	716.70
150 amps, 42 circuits	E4@2.82	Ea	760.00	99.20	859.20
200 amps, 42 circuits	E4@3.63	Ea	760.00	128.00	888.00
225 amps, 42 circuits	E4@3.78	Ea	799.00	133.00	932.00
400 amps, 24 circuits	E4@3.91	Ea	1,610.00	138.00	1,748.00
400 amps, 42 circuits	E4@4.51	Ea	1,870.00	159.00	2,029.00
Add for NEMA-4 raintight loadcenters	—	%	25.0	—	—

Single phase 3 wire, 120/240 volt loadcenters with main lugs only

	Craft@Hrs	Unit	Material	Labor	Total
150 amps, 12 circuits	E4@1.20	Ea	86.50	42.20	128.70
150 amps, 16 circuits	E4@1.20	Ea	110.00	42.20	152.20
150 amps, 20 circuits	E4@1.52	Ea	115.00	53.50	168.50
200 amps, 30 circuits	E4@1.70	Ea	158.00	59.80	217.80
200 amps, 40 circuits	E4@2.03	Ea	220.00	71.40	291.40
225 amps, 24 circuits	E4@2.50	Ea	256.00	87.90	343.90
225 amps, 42 circuits	E4@2.77	Ea	315.00	97.40	412.40

Three phase 4 wire 120/208 volt loadcenters with main lugs only

	Craft@Hrs	Unit	Material	Labor	Total
125 amps, 12 circuits	E4@1.19	Ea	111.00	41.80	152.80
150 amps, 18 circuits	E4@1.47	Ea	165.00	51.70	216.70
200 amps, 30 circuits	E4@1.90	Ea	225.00	66.80	291.80
200 amps, 42 circuits	E4@2.18	Ea	299.00	76.70	375.70
225 amps, 42 circuits	E4@2.75	Ea	315.00	96.70	411.70
400 amps, 24 circuits	E4@3.63	Ea	575.00	128.00	703.00
400 amps, 42 circuits	E4@3.78	Ea	640.00	133.00	773.00
600 amps, 24 circuits	E4@3.91	Ea	700.00	138.00	838.00
600 amps, 42 circuits	E4@6.53	Ea	855.00	230.00	1,085.00

Circuit Breakers No enclosures included. 10,000 amp interrupt capacity except as noted.

Plug-in molded case 120 volt, 100 amp frame circuit breakers, 1" or 1/2" module

	Craft@Hrs	Unit	Material	Labor	Total
15 to 50 amps, single pole	CE@.270	Ea	7.60	10.80	18.40
15 to 60 amps, two pole	CE@.374	Ea	17.50	15.00	32.50
70 amps, two pole	CE@.563	Ea	33.90	22.60	56.50
80 to 100 amps, two pole	CE@.680	Ea	47.40	27.20	74.60

	Craft@Hrs	Unit	Material	Labor	Total
Plug-in molded case 240 volt, 100 amp frame circuit breakers					
15 thru 60 amps, single pole	CE@.417	Ea	10.10	16.70	26.80
15 thru 60 amps, two pole	CE@.577	Ea	66.30	23.10	89.40
15 thru 60 amps, three pole	CE@.863	Ea	82.30	34.60	116.90
60 thru 100 amps, two pole	CE@.577	Ea	110.00	23.10	133.10
60 thru 100 amps, three pole	CE@.897	Ea	123.00	35.90	158.90
Plug-in molded case 600 volt, 100 amp frame circuit breakers					
15 thru 60 amps, two pole	CE@.577	Ea	210.00	23.10	233.10
15 thru 60 amps, three pole	CE@.897	Ea	269.00	35.90	304.90
Plug-in molded case 480 volt, 100 amp frame circuit breakers					
70 thru 100 amps, two pole	CE@1.22	Ea	250.00	48.90	298.90
70 thru 100 amps, three pole	CE@1.71	Ea	293.00	68.50	361.50
Plug-in molded case 600 volt, 100 amp frame circuit breakers					
70 thru 100 amps, two pole	CE@1.22	Ea	281.00	48.90	329.90
70 thru 100 amps, three pole	CE@1.71	Ea	353.00	68.50	421.50
Plug-in molded case 250/600 volt, 225 amp frame circuit breakers					
125 thru 225 amps, two pole	CE@2.06	Ea	655.00	82.50	737.50
125 thru 225 amps, three pole	CE@2.57	Ea	823.00	103.00	926.00
Plug-in molded case 250/600 volt, 400 amp frame circuit breakers					
250 thru 400 amps, two pole	CE@2.88	Ea	1,030.00	115.00	1,145.00
250 thru 400 amps, three pole	CE@4.33	Ea	1,240.00	173.00	1,413.00
Plug-in molded case 250/600 volt, 800 amp frame circuit breakers					
400 thru 600 amps, three pole	CE@6.04	Ea	2,270.00	242.00	2,512.00
700 thru 800 amps, two pole	CE@4.73	Ea	2,300.00	190.00	2,490.00
700 thru 800 amps, three pole	CE@6.71	Ea	2,950.00	269.00	3,219.00
Plug-in molded case 600 volt, 1,200 amp frame circuit breakers					
600 thru 1,200 amps, two pole	CE@9.60	Ea	2,150.00	385.00	2,535.00
600 thru 1,200 amps, three pole	CE@11.9	Ea	2,730.00	477.00	3,207.00
Space only for factory assembled panels					
Single pole space	—	Ea	17.00	—	17.00
Two pole space	—	Ea	15.60	—	15.60
Three pole space	—	Ea	21.10	—	21.10
Bolt-on molded case 120/240 volt circuit breakers					
15 thru 60 amps, one pole	CE@.320	Ea	13.30	12.80	26.10
15 thru 60 amps, two pole	CE@.484	Ea	29.00	19.40	48.40
15 thru 60 amps, three pole	CE@.638	Ea	90.40	25.60	116.00
70 thru 100 amps, two pole	CE@.712	Ea	110.00	28.50	138.50
70 thru 100 amps, three pole	CE@.946	Ea	130.00	37.90	167.90
Bolt-on molded case 480 volt circuit breakers, 65,000 amp interrupt capacity					
15 thru 60 amps, two pole	CE@.484	Ea	109.00	19.40	128.40
15 thru 60 amps, three pole	CE@.712	Ea	190.00	28.50	218.50
70 thru 100 amps, three pole	CE@.804	Ea	289.00	32.20	321.20
Circuit breaker panels, 4 wire 3 phase 240 volts, flush-mounted, with main breaker, assembled.					
100 amp service, 12 circuits	E4@4.60	Ea	388.00	162.00	550.00
100 amp service, 18 circuits	E4@6.90	Ea	408.00	243.00	651.00
100 amp service, 24 circuits	E4@8.10	Ea	442.00	285.00	727.00
100 amp service, 30 circuits	E4@8.20	Ea	475.00	288.00	763.00
100 amp service, 36 circuits	E4@8.30	Ea	508.00	292.00	800.00
100 amp service, 42 circuits	E4@8.40	Ea	542.00	295.00	837.00
225 amp service, 12 circuits	E4@8.50	Ea	840.00	299.00	1,139.00
225 amp service, 18 circuits	E4@8.60	Ea	860.00	302.00	1,162.00

	Craft@Hrs	Unit	Material	Labor	Total
225 amp service, 24 circuits	E4@8.70	Ea	890.00	306.00	1,196.00
225 amp service, 30 circuits	E4@10.0	Ea	926.00	352.00	1,278.00
225 amp service, 36 circuits	E4@12.8	Ea	960.00	450.00	1,410.00
225 amp service, 42 circuits	E4@15.1	Ea	994.00	531.00	1,525.00

Fuses No enclosures included.
250 volt cartridge fuses, one time non-renewable

	Craft@Hrs	Unit	Material	Labor	Total
To 30 amps	CE@.077	Ea	.85	3.09	3.94
35 to 60 amps	CE@.077	Ea	1.29	3.09	4.38
70 to 100 amps	CE@.077	Ea	5.80	3.09	8.89
110 to 200 amps	CE@.077	Ea	15.20	3.09	18.29
225 to 400 amps	CE@.077	Ea	26.00	3.09	29.09
450 to 600 amps	CE@.077	Ea	40.30	3.09	43.39

600 volt cartridge fuses, one time non-renewable

	Craft@Hrs	Unit	Material	Labor	Total
To 30 amps	CE@.077	Ea	5.06	3.09	8.15
35 to 60 amps	CE@.077	Ea	7.52	3.09	10.61
70 to 100 amps	CE@.077	Ea	15.80	3.09	18.89
110 to 200 amps	CE@.077	Ea	31.70	3.09	34.79
225 to 400 amps	CE@.087	Ea	63.20	3.49	66.69
450 to 600 amps	CE@.124	Ea	95.00	4.97	99.97

250 volt class H renewable fuses

	Craft@Hrs	Unit	Material	Labor	Total
To 10 amps	CE@.077	Ea	6.64	3.09	9.73
12 to 30 amps	CE@.077	Ea	5.24	3.09	8.33
35 to 60 amps	CE@.077	Ea	12.40	3.09	15.49
70 to 100 amps	CE@.077	Ea	28.00	3.09	31.09
110 to 200 amps	CE@.077	Ea	64.40	3.09	67.49
225 to 400 amps	CE@.080	Ea	117.00	3.21	120.21
450 to 600 amps	CE@.122	Ea	188.00	4.89	192.89

600 volt class H renewable fuses

	Craft@Hrs	Unit	Material	Labor	Total
To 10 amps	CE@.077	Ea	18.60	3.09	21.69
12 to 30 amps	CE@.077	Ea	14.90	3.09	17.99
35 to 60 amps	CE@.077	Ea	24.80	3.09	27.89
70 to 100 amps	CE@.077	Ea	55.70	3.09	58.79
110 to 200 amps	CE@.077	Ea	110.00	3.09	113.09
225 to 400 amps	CE@.153	Ea	231.00	6.13	237.13
450 to 600 amps	CE@.223	Ea	321.00	8.93	329.93

250 volt class RK5 non-renewable cartridge fuses

	Craft@Hrs	Unit	Material	Labor	Total
10 to 30 amps	CE@.077	Ea	5.21	3.09	8.30
35 to 60 amps	CE@.077	Ea	5.46	3.09	8.55
70 to 100 amps	CE@.077	Ea	13.90	3.09	16.99
110 to 200 amps	CE@.077	Ea	32.10	3.09	35.19
225 to 400 amps	CE@.230	Ea	63.10	9.22	72.32
450 to 600 amps	CE@.297	Ea	99.00	11.90	110.90
650 to 1,200 amps	CE@.372	Ea	121.00	14.90	135.90
1,400 to 1,600 amps	CE@.804	Ea	165.00	32.20	197.20
1,800 to 2,000 amps	CE@1.00	Ea	211.00	40.10	251.10

600 volt class RK5 non-renewable cartridge fuses

	Craft@Hrs	Unit	Material	Labor	Total
10 to 30 amps	CE@.077	Ea	7.20	3.09	10.29
35 to 60 amps	CE@.077	Ea	12.30	3.09	15.39
70 to 100 amps	CE@.077	Ea	25.40	3.09	28.49
110 to 200 amps	CE@.129	Ea	50.80	5.17	55.97
225 to 400 amps	CE@.257	Ea	101.00	10.30	111.30
450 to 600 amps	CE@.354	Ea	147.00	14.20	161.20

	Craft@Hrs	Unit	Material	Labor	Total
650 to 1,200 amps	CE@.372	Ea	232.00	14.90	246.90
1,400 to 1,600 amps	CE@.797	Ea	323.00	31.90	354.90
1,800 to 2,000 amps	CE@1.00	Ea	401.00	40.10	441.10
Add for type K5 600 volt fuses	—	%	20.0	—	—
Deduct for type J 600 volt fuses	—	%	-50.0	—	—
600 volt class L fuses, bolt-on					
600 to 1,200 amps	CE@.333	Ea	357.00	13.30	370.30
1,500 to 1,600 amps	CE@.615	Ea	481.00	24.60	505.60
1,400 to 2,000 amps	CE@.809	Ea	638.00	32.40	670.40
2,500 amps	CE@1.10	Ea	638.00	44.10	682.10
3,000 amps	CE@1.25	Ea	735.00	50.10	785.10
3,500 to 4,000 amps	CE@1.74	Ea	1,210.00	69.70	1,279.70
5,000 amps	CE@2.33	Ea	1,550.00	93.40	1,643.40

	Craft@Hrs	Unit	Material	Labor	Equipment	Total

Overhead Electrical Distribution

Poles, pressure treated wood, class 4, type C, set with crane or boom, including augured holes. Equipment is a flatbed truck with a boom and auger at $21 per hour.

	Craft@Hrs	Unit	Material	Labor	Equipment	Total
25' high	E1@7.62	Ea	248.00	274.00	33.60	555.60
30' high	E1@7.62	Ea	318.00	274.00	33.60	625.60
35' high	E1@9.72	Ea	413.00	349.00	42.80	804.80
40' high	E1@11.5	Ea	489.00	413.00	92.70	994.70

Cross arms, wood, with typical hardware

	Craft@Hrs	Unit	Material	Labor	Equipment	Total
4' high	E1@2.86	Ea	58.60	103.00	12.60	174.20
5' high	E1@2.86	Ea	73.50	103.00	12.60	189.10
6' high	E1@4.19	Ea	107.00	151.00	18.50	276.50
8' high	E1@4.91	Ea	141.00	176.00	21.60	338.60

Transformers, pole mounted, single phase 7620/13200 Y primary, 120/240 secondary, with two 2-1/2% taps above and below

	Craft@Hrs	Unit	Material	Labor	Equipment	Total
10 KVA	E1@4.14	Ea	723.00	149.00	18.30	890.30
15 KVA	E1@4.14	Ea	833.00	149.00	18.30	1,000.30
25 KVA	E1@4.14	Ea	1,020.00	149.00	18.30	1,187.30
36.5 KVA	E1@4.14	Ea	1,360.00	149.00	18.30	1,527.30
50 KVA	E1@4.14	Ea	1,550.00	149.00	18.30	1,717.30
75 KVA	E1@6.16	Ea	2,300.00	221.00	26.70	2,547.70
100 KVA	E1@6.16	Ea	2,630.00	221.00	26.70	2,877.70
167 KVA	E1@6.16	Ea	4,040.00	221.00	26.70	4,287.70
250 KVA	E1@6.16	Ea	4,120.00	221.00	26.70	4,367.70
333 KVA	E1@6.16	Ea	4,460.00	221.00	26.70	4,707.70
500 KVA	E1@6.16	Ea	6,210.00	221.00	26.70	6,457.70

Fused cutouts, pole mounted, 5 KV

	Craft@Hrs	Unit	Material	Labor	Equipment	Total
50 amp	E1@2.56	Ea	83.80	92.00	11.30	187.10
100 amp	E1@2.56	Ea	84.90	92.00	11.30	188.20
250 amp	E1@2.56	Ea	96.20	92.00	11.30	199.50

Switches, disconnect, pole mounted, pole arm throw, 5 KV, set of 3 with throw and lock

	Craft@Hrs	Unit	Material	Labor	Equipment	Total
400 amp	E1@45.8	Ea	1,510.00	1,650.00	202.00	3,362.00
600 amp	E1@45.8	Ea	2,870.00	1,650.00	202.00	4,722.00
1,200 amp	E1@51.4	Ea	5,700.00	1,850.00	227.00	7,777.00

	Craft@Hrs	Unit	Material	Labor	Total

Fluorescent Lighting Commercial, industrial and architectural grade. Rapid start. With ballasts but no lamps, wire or conduit. Includes setting and connecting prewired fixtures only.

Surface mounted fluorescent fixtures, better quality, baked enamel finish, with hinged wraparound acrylic prismatic lens suitable for office or classroom use, 4' long, Lithonia LB series

	Craft@Hrs	Unit	Material	Labor	Total
Two 40 watt lamp fixture (#240-A)	CE@.773	Ea	66.70	31.00	97.70
Four 40 watt lamp fixture (#440-A)	CE@.773	Ea	127.00	31.00	158.00
Eight 40 watt lamp fixture, 8' long (#8T440-A)	CE@.960	Ea	191.00	38.50	229.50

Surface mounted fluorescent fixtures, baked enamel finish, with wraparound acrylic prismatic lens suitable for industrial use, 4' long, Lithonia series

	Craft@Hrs	Unit	Material	Labor	Total
One 40 watt lamp fixture (CB140A)	CE@.806	Ea	59.70	32.30	92.00
Two 40 watt lamp fixture (LB240A)	CE@.858	Ea	51.10	34.40	85.50
Four 40 watt lamp fixture (LB440A)	CE@1.07	Ea	87.10	42.90	130.00
Add for 40 watt lamps	—	Ea	3.25	—	3.25

Surface mounted steel sided fluorescent fixtures with acrylic prismatic diffuser, 7-1/16" x 4-1/2"

	Craft@Hrs	Unit	Material	Labor	Total
Two 40 watt lamps	CE@.955	Ea	73.10	38.30	111.40
Four 40 watt lamps	CE@1.05	Ea	102.00	42.10	144.10
Add for 40 watt straight lamps	—	Ea	3.25	—	3.25

Wall surface mounted fluorescent fixtures, baked enamel finish, 2 lamps, 8" wide 5" high, hinged acrylic prismatic lens, Lightolier

	Craft@Hrs	Unit	Material	Labor	Total
26" long fixture	CE@.737	Ea	153.00	29.50	182.50
38" long fixture	CE@.795	Ea	177.00	31.90	208.90
50" long fixture	CE@.858	Ea	172.00	34.40	206.40
Add for 24" 20 watt lamps	—	Ea	2.98	—	2.98
Add for 36" 30 watt lamps	—	Ea	4.62	—	4.62
Add for 48" 40 watt lamps	—	Ea	2.11	—	2.11

Industrial pendant or surface mounted open striplights, baked enamel finish, Lithonia C series

	Craft@Hrs	Unit	Material	Labor	Total
One or two 40 watt lamps, 4' long (C240)	CE@.865	Ea	34.60	34.70	69.30
One or two 75 watt lamps, 8' long (C296)	CE@1.02	Ea	51.10	40.90	92.00
One or two 55 watt lamps, 6' long (C272)	CE@1.02	Ea	51.10	40.90	92.00
Add for 40 watt lamps	—	Ea	2.28	—	2.28
Add for 75 or 55 watt lamps	—	Ea	5.43	—	5.43
Add for 24" stem and canopy set, per fixture	—	Ea	5.43	—	5.43

Industrial pendant or surface mounted fluorescent fixtures, 12" wide baked enamel, steel reflector-shield, Lithonia L/LA series

	Craft@Hrs	Unit	Material	Labor	Total
Two 40 watt lamps, 4' long (L-240-120)	CE@.903	Ea	37.20	36.20	73.40
Two 75 watt lamps, 8' long (L-296-120)	CE@1.21	Ea	73.10	48.50	121.60
Add for 40 watt lamps	—	Ea	2.28	—	2.28
Add for 75 watt lamps	—	Ea	5.43	—	5.43

Ceiling grid mounted fluorescent troffer fixtures, baked enamel finish, with hinged acrylic prismatic lens with energy saving ballast, Day-Brite

	Craft@Hrs	Unit	Material	Labor	Total
Two 40 watt "U" lamp fixture, 2' x 2'	CE@.908	Ea	54.00	36.40	90.40
Two 40 watt lamp fixture, 2' x 4'	CE@.998	Ea	61.20	40.00	101.20
Four 40 watt lamp fixture, 2' x 4'	CE@1.05	Ea	69.50	42.10	111.60
Add for plaster frames, per fixture	—	Ea	7.54	—	7.54

Air-handling grid mounted floating door, heat transfer troffer fixtures with acrylic lens, no ductwork included

	Craft@Hrs	Unit	Material	Labor	Total
1' x 4', two 40 watt lamps	CE@.908	Ea	103.00	36.40	139.40
2' x 2', one U-shape 40 watt lamp	CE@.809	Ea	108.00	32.40	140.40
2' x 4', two 40 watt lamps	CE@1.00	Ea	135.00	40.10	175.10
2' x 4', three 40 watt lamps	CE@1.05	Ea	168.00	42.10	210.10
2' x 4', four 40 watt lamps	CE@1.05	Ea	168.00	42.10	210.10
Add for 40 watt 4' lamps	—	Ea	2.28	—	2.28
Add for 40 watt "U" lamps	—	Ea	9.49	—	9.49

	Craft@Hrs	Unit	Material	Labor	Total
Surface mounted vandal resistant luminaire with tamper-resistant screw-type latching, Sylvania					
1 lamp	CE@.761	Ea	112.00	30.50	142.50
2 lamps	CE@.761	Ea	114.00	30.50	144.50
4 lamps	CE@.761	Ea	195.00	30.50	225.50
Parabolic fluorescent luminaire, 6" x 24" x 24" with semi-specular anodized aluminum reflector and parabolic louvers, I.C.E. AR series					
Two 40 watt lamp, bracket or pendant mount	CE@.955	Ea	67.70	38.30	106.00
Four 40 watt lamps, bracket or pendant mount	CE@.955	Ea	107.00	38.30	145.30
Round surface mounted fluorescent fixtures, aluminum housing, matte black finish, polycarbonate or acrylic opal globe, Circleline					
22 to 32 watt, 14" diameter	CE@.478	Ea	44.80	19.20	64.00
32 to 40 watt, 18" diameter	CE@.478	Ea	61.80	19.20	81.00
Add for circline tubes	—	Ea	6.76	—	6.76
Outdoor sign fluorescent fixtures, anodized aluminum housing with parabolic specular reflector and acrylic lens, with high output lamps and remote ballast					
One 60 watt lamp. 48" long fixture	CE@2.39	Ea	214.00	95.80	309.80
One 85 watt lamp, 72" long fixture	CE@2.39	Ea	228.00	95.80	323.80
One 110 watt lamp, 96" long fixture	CE@2.39	Ea	251.00	95.80	346.80
Two 85 watt lamps, 144" long fixture	CE@2.39	Ea	338.00	95.80	433.80
Add for high output T12 800 MA lamps, sign white					
60 watt	—	Ea	5.97	—	5.97
85 watt	—	Ea	5.97	—	5.97
110 watt	—	Ea	6.76	—	6.76
Cost per linear foot of fixture	CE@.179	LF	57.50	7.17	64.67
Add for 90 degree elbows	CE@.475	Ea	148.00	19.00	167.00
Add for in-line connectors	CE@.475	Ea	74.40	19.00	93.40
Wet or damp location surface mounted fluorescent fixtures, hinge mounted acrylic diffuse lens, fiberglass or plastic housing with gaskets, Hubbell					
Two 40 watt lamps, 4' long fixture	CE@1.21	Ea	130.00	48.50	178.50
Two 40 watt lamps, 8' long fixture	CE@1.86	Ea	278.00	74.50	352.50
Add for 40 watt lamps	—	Ea	2.28	—	2.28

High Intensity Discharge Lighting

Commercial and architectural grade fixtures. With ballasts but no lamps, wire, conduit or concrete except as noted. See lamp costs below. Labor includes setting and connecting prewired fixtures only.

	Craft@Hrs	Unit	Material	Labor	Total
Recess mounted commercial luminaire, 2' x 2' x 13", aluminum reflector glass lens, Wide-Lite series					
175 watt, metal halide	CE@1.15	Ea	460.00	46.10	506.10
250 watt, high pressure sodium	CE@1.15	Ea	528.00	46.10	574.10
400 watt, metal halide	CE@1.15	Ea	464.00	46.10	510.10
400 watt, high pressure sodium	CE@1.15	Ea	578.00	46.10	624.10
Surface mounted commercial luminaire, 2' x 2' x 15", baked enamel finish with tempered glass lens, Holophane series					
70 watt, high pressure sodium	CE@1.42	Ea	277.00	56.90	333.90
100 watt, high pressure sodium	CE@1.42	Ea	298.00	56.90	354.90
150 watt, high pressure sodium	CE@1.42	Ea	305.00	56.90	361.90
175 watt, metal halide	CE@1.42	Ea	292.00	56.90	348.90
400 watt, metal halide	CE@1.42	Ea	315.00	56.90	371.90
Recess mounted indirect metal halide luminaire, aluminum reflector with glass lens					
175 watt, square	CE@1.71	Ea	460.00	68.50	528.50
250 watt, square	CE@1.71	Ea	460.00	68.50	528.50
400 watt, square	CE@1.71	Ea	464.00	68.50	532.50
175 watt, round	CE@1.71	Ea	441.00	68.50	509.50
250 watt, round	CE@1.71	Ea	441.00	68.50	509.50
400 watt, round	CE@1.71	Ea	446.00	68.50	514.50

	Craft@Hrs	Unit	Material	Labor	Total
Surface mounted indirect high pressure sodium luminaire, aluminum reflector with glass lens					
250 watt, square	CE@1.71	Ea	508.00	68.50	576.50
400 watt, square	CE@1.71	Ea	557.00	68.50	625.50
250 watt, round	CE@1.71	Ea	489.00	68.50	557.50
400 watt, round	CE@1.71	Ea	538.00	68.50	606.50
Outdoor wall pack, aluminum housing with hinged polycarbonate access door and lens, Sylvania					
100 watt, high pressure sodium	CE@.955	Ea	165.00	38.30	203.30
150 watt, high pressure sodium	CE@.955	Ea	145.00	38.30	183.30
175 watt, mercury vapor	CE@.955	Ea	141.00	38.30	179.30
175 watt, metal halide	CE@.955	Ea	180.00	38.30	218.30
Utility wall pack, aluminum housing with hinged polycarbonate or tempered glass lens, Holophane Wallpack 2					
70 watt, high pressure sodium	CE@.955	Ea	277.00	38.30	315.30
100 watt, high pressure sodium	CE@.955	Ea	298.00	38.30	336.30
150 watt, high pressure sodium	CE@.955	Ea	305.00	38.30	343.30
175 watt, metal halide	CE@.955	Ea	292.00	38.30	330.30
250 watt, mercury vapor	CE@.955	Ea	275.00	38.30	313.30
Add for photo cell automatic control, factory installed	—	Ea	19.50	—	19.50
High bay metal halide industrial fixture, cast aluminum housing, spun aluminum reflector, lens is tempered glass					
400 watt	CE@2.61	Ea	166.00	105.00	271.00
1,000 watt	CE@2.61	Ea	281.00	105.00	386.00
Add for power hook and cord	CE@.250	Ea	51.50	10.00	61.50
Add for thru wire power hook receptacle	CE@.199	Ea	43.80	7.97	51.77
Add for wire guard on lens	CE@.199	Ea	51.50	7.97	59.47
Low bay high pressure sodium industrial fixture, cast aluminum with epoxy finish, acrylic lens					
100 watt fixture	CE@1.42	Ea	243.00	56.90	299.90
150 watt fixture	CE@1.39	Ea	252.00	55.70	307.70
250 watt fixture	CE@1.39	Ea	361.00	55.70	416.70
Add for power hook and cord	CE@.250	Ea	47.50	10.00	57.50
Add for thru wire power hook receptacle	CE@.199	Ea	41.70	7.97	49.67
Add for wire guard on lens	CE@.199	Ea	57.70	7.97	65.67
Low bay metal halide industrial downlight, cast aluminum with epoxy finish, shock resistant glass lens					
100 watt fixture	CE@1.07	Ea	303.00	42.90	345.90
400 watt fixture	CE@1.39	Ea	380.00	55.70	435.70
Add for power hook and cord	CE@.250	Ea	47.50	10.00	57.50
Add for thru wire power hook receptacle	CE@.199	Ea	41.70	7.97	49.67
Add for wire guard on lens	CE@.199	Ea	57.70	7.97	65.67
Lamps for high intensity discharge luminaires					
400 watt, mercury vapor	—	Ea	49.10	—	49.10
1,000 watt, mercury vapor	—	Ea	89.60	—	89.60
35 watt, high pressure sodium	—	Ea	32.70	—	32.70
70 watt, high pressure sodium	—	Ea	32.70	—	32.70
100 watt, high pressure sodium	—	Ea	36.90	—	36.90
150 watt, high pressure sodium	—	Ea	32.00	—	32.00
250 watt, high pressure sodium	—	Ea	32.50	—	32.50
400 watt, high pressure sodium	—	Ea	27.80	—	27.80
1,000 watt, high pressure sodium	—	Ea	101.00	—	101.00
175 watt, metal halide	—	Ea	60.30	—	60.30
250 watt, metal halide	—	Ea	35.90	—	35.90
400 watt, metal halide	—	Ea	33.30	—	33.30
1,000 watt, metal halide	—	Ea	86.90	—	86.90
Add for diffuse coated lamps	—	Ea	5.00	—	5.00

	Craft@Hrs	Unit	Material	Labor	Total

Incandescent Lighting Commercial and architectural grade. No wire, conduit or concrete included. Includes setting and connecting prewired fixtures only.

Step illumination light fixtures, 16 gauge steel housing with white enamel finish, 5" high, 4" deep, 11" long

	Craft@Hrs	Unit	Material	Labor	Total
Polycarbonate or tempered glass lens fixture	CE@1.07	Ea	140.00	42.90	182.90
Louvered sheet metal front fixture	CE@1.07	Ea	120.00	42.90	162.90
Add for brushed aluminum trim, 2 lamps	—	Ea	32.50	—	32.50

Adjustable recessed spotlight, 20 gauge steel housing with matt finish, 30 degree adjustable with 350 degree rotation

	Craft@Hrs	Unit	Material	Labor	Total
150 watt fixture	CE@1.07	Ea	115.00	42.90	157.90

Recessed round eyelid wall wash, 6" milligroove baffle, Prescolite

	Craft@Hrs	Unit	Material	Labor	Total
150 watt fixture (1220)	CE@1.07	Ea	117.00	42.90	159.90

Exterior wall fixtures, cast anodized aluminum with white polycarbonate globe, gaskets between globe and housing and between housing and base, Prescolite

	Craft@Hrs	Unit	Material	Labor	Total
Cylinder globe fixture	CE@.320	Ea	20.00	12.80	32.80
Round globe fixture	CE@.430	Ea	23.80	17.20	41.00
Add for 100 watt lamp	—	Ea	1.50	—	1.50
Add for 150 watt lamp	—	Ea	2.20	—	2.20

Obstruction light, meets Federal Aviation Administration Specification L-810, cast aluminum housing, two one-piece red, heat-resistant fresnel globes, with photoelectric control, mounted on 1" rigid steel conduit, including junction box and mounting plate

	Craft@Hrs	Unit	Material	Labor	Total
Two 100 watt lamp fixture	CE@4.91	Ea	99.20	197.00	296.20

Vandal Resistant Lighting Commercial and architectural grade. Includes setting and connecting prewired fixtures but no wire, conduit or lamps.

Ceiling mounted 12" x 12" x 6" fixture, polycarbonate prismatic lens

	Craft@Hrs	Unit	Material	Labor	Total
Two 100 watt Incandescent lamps	CE@.764	Ea	72.00	30.60	102.60
35 watt, high pressure sodium lamp	CE@.764	Ea	210.00	30.60	240.60
50 watt, high pressure sodium lamp	CE@.764	Ea	215.00	30.60	245.60
70 watt, high pressure sodium lamp	CE@.764	Ea	225.00	30.60	255.60
22 watt fluorescent circline lamp	CE@.955	Ea	48.20	38.30	86.50

Wall mounted 6" wide, 9" high, 7" deep fixture, polycarbonate prismatic diffuser

	Craft@Hrs	Unit	Material	Labor	Total
One 100 watt incandescent lamp	CE@.572	Ea	51.00	22.90	73.90
13 watt fluorescent lamp	CE@.764	Ea	79.90	30.60	110.50
35 watt, high pressure sodium lamp	CE@.475	Ea	215.00	19.00	234.00
50 watt, high pressure sodium lamp	CE@.475	Ea	215.00	19.00	234.00
70 watt, high pressure sodium lamp	CE@.475	Ea	205.00	19.00	224.00

Lamps for vandal resistant fixtures

	Craft@Hrs	Unit	Material	Labor	Total
100 watt incandescent lamp	—	Ea	1.35	—	1.35
13 watt fluorescent lamp	—	Ea	7.00	—	7.00
35 to 70 watt high pressure sodium lamp	—	Ea	50.00	—	50.00

Lighted Exit Signs Includes setting and connecting prewired fixtures but no wire or conduit.

Universal mount (wall or ceiling), Thinline series, 6" letters on one side, stencil face,

	Craft@Hrs	Unit	Material	Labor	Total
single	CE@.764	Ea	55.80	30.60	86.40
twin	CE@.764	Ea	55.90	30.50	86.40

Universal mount, 6" letters on two sides of polycarbonate

	Craft@Hrs	Unit	Material	Labor	Total
housing, two 15 watt lamps	CE@.764	Ea	62.60	30.60	93.20
Add for battery backup for exit signs	CE@.381	Ea	120.00	15.30	135.30
Add for two circuit supply	—	Ea	8.13	—	8.13
Add for 24" stem hanger	—	Ea	11.10	—	11.10

	Craft@Hrs	Unit	Material	Labor	Total

Explosion Proof Lighting

No wire, conduit or concrete included. Includes setting and connecting prewired fixtures only.

Explosion proof fixtures, cast aluminum housing, heat-resistant prestressed globe, with fiberglass reinforced polyester reflector

	Craft@Hrs	Unit	Material	Labor	Total
60-200 watt incandescent fixture	CE@1.74	Ea	65.50	69.70	135.20
70 watt HID fixture	CE@1.74	Ea	254.00	69.70	323.70
150 watt HID fixture	CE@1.74	Ea	360.00	69.40	429.40

Explosion proof (Class 1 Division 1) fixtures, cast aluminum housing, heat-resistant prestressed globe, with fiberglass reinforced polyester reflector

	Craft@Hrs	Unit	Material	Labor	Total
250 watt, high pressure sodium fixture	CE@2.57	Ea	705.00	103.00	808.00
175 watt, metal halide fixture	CE@2.57	Ea	542.00	103.00	645.00
400 watt, metal halide fixture	CE@2.57	Ea	677.00	103.00	780.00

Explosion proof (Class 1 Division 2) fixtures, cast aluminum housing, heat-resistant prestressed globe, with fiberglass reinforced polyester reflector

	Craft@Hrs	Unit	Material	Labor	Total
70 watt, high pressure sodium fixture	CE@2.57	Ea	365.00	103.00	468.00
250 watt, high pressure sodium fixture	CE@2.57	Ea	499.00	103.00	602.00
175 watt, metal halide fixture	CE@2.57	Ea	338.00	103.00	441.00
400 watt, metal halide fixture	CE@2.57	Ea	478.00	103.00	581.00

Emergency Lighting

Includes setting and connecting prewired fixtures but no wire or conduit. Shelf or wall mounted solid state battery pack with charger, lead-acid battery, baked enamel finish, with 2 halogen lamps, 90 minute operation, indoor use only

	Craft@Hrs	Unit	Material	Labor	Total
Six volt, 15 watt	CE@1.74	Ea	88.00	69.70	157.70
Six volt, 30 watt	CE@1.74	Ea	135.00	69.40	247.40
Six volt, 50 watt	CE@1.74	Ea	160.00	69.70	229.70
Six volt, 100 watt	CE@1.74	Ea	215.00	69.70	284.70
Twelve volt, 100 watt	CE@1.74	Ea	215.00	69.70	284.70

Self diagnostic emergency lighting fixtures, microprocessor based, continuously monitors battery/lamps

	Craft@Hrs	Unit	Material	Labor	Total
Emergency fixture, two lamps, wall mount	CE@.825	Ea	164.00	33.10	197.10
LED Exit sign with emergency fixture, wall mount	CE@.825	Ea	288.30	33.10	321.40
LED Exit sign only, wall mount	CE@.825	Ea	246.00	19.10	279.10

Yard and Street Lighting

Includes setting and connecting prewired fixtures with ballast but no pole, concrete work, excavation, wire, lamps or conduit. See pole costs below. Equipment is a flatbed truck with a boom & auger at $21.00 per hour.

Mast arm mounted rectangular high pressure sodium flood, aluminum housing with baked enamel finish, anodized aluminum reflector, tempered glass lens, slip-fitting for mast mount, photoelectric cell

	Craft@Hrs	Unit	Material	Labor	Equipment	Total
70 watt	E1@1.55	Ea	365.00	55.70	6.84	427.54
100 watt	E1@1.55	Ea	360.00	55.70	6.84	422.54
150 watt	E1@1.55	Ea	375.00	55.70	6.84	437.54
250 watt	E1@1.55	Ea	403.00	55.70	6.84	465.54
400 watt	E1@1.55	Ea	478.00	55.70	6.84	540.54
Add for 6' x 2" mounting arm	E1@.547	Ea	52.20	19.70	2.42	74.32

Yoke mounted high intensity discharge flood, die-cast anodized aluminum housing with tempered glass lens

	Craft@Hrs	Unit	Material	Labor	Equipment	Total
250 watt, high pressure sodium	E1@2.71	Ea	332.00	97.40	12.00	441.40
250 watt, metal halide	E1@2.71	Ea	301.00	97.40	12.00	410.40
400 watt, high pressure sodium	E1@2.71	Ea	341.00	97.40	12.00	450.40
400 watt, metal halide	E1@2.71	Ea	308.00	97.40	12.00	417.40
1,000 watt, high pressure sodium	E1@2.71	Ea	580.00	97.40	12.00	689.40
1,000 watt, metal halide	E1@2.71	Ea	464.00	97.40	12.00	573.40
1,000 watt, mercury vapor	E1@2.71	Ea	426.00	97.40	12.00	535.40

	Craft@Hrs	Unit	Material	Labor	Equipment	Total
Round flood, aluminum reflector, impact-resistant glass lens						
400 watt, high pressure sodium	E1@2.71	Ea	348.00	97.40	12.00	457.40
1,000 watt, high pressure sodium	E1@2.71	Ea	558.00	97.40	12.00	667.40
400 watt, metal halide	E1@2.71	Ea	321.00	97.40	12.00	430.40
1,000 watt, metal halide	E1@2.71	Ea	438.00	97.40	12.00	547.40

Cobra head high intensity flood, anodized aluminum reflector, glass lens, die cast aluminum housing, with photoelectric control but no bulb or mounting arm

	Craft@Hrs	Unit	Material	Labor	Equipment	Total
1,000 watt, high pressure sodium	E1@2.71	Ea	564.00	97.40	12.00	673.40
1,000 watt, mercury vapor	E1@2.71	Ea	412.00	97.40	12.00	521.40
1,000 watt, metal halide	E1@2.71	Ea	446.00	97.40	12.00	555.40

Square metal halide flood, anodized aluminum housing, tempered glass door, trunnion or yoke mount

	Craft@Hrs	Unit	Material	Labor	Equipment	Total
400 watt	E1@2.71	Ea	431.00	97.40	12.00	540.40
1,000 watt	E1@2.71	Ea	512.00	97.40	12.00	621.40

Yard and Street Lighting Poles

Hole for a pole foundation dug with a truck-mounted auger at $21.00 per hour, no soil disposal included

	Craft@Hrs	Unit	Material	Labor	Equipment	Total
Per CF of undisturbed soil	E1@.064	CF	—	2.30	.27	2.57

Concrete pole foundations, formed, poured and finished, with anchor bolts. Material costs shown assume concrete, forms and anchor bolts at $160.00 per CY.

	Craft@Hrs	Unit	Material	Labor	Equipment	Total
12" diameter, 36" deep, for 12' pole	E1@.746	Ea	18.30	26.80	3.13	48.23
24" diameter, 72" deep, for 30' pole	E1@1.14	Ea	73.10	41.00	4.79	118.89
30" diameter, 76" deep, for 40' pole	E1@1.54	Ea	96.60	55.40	6.47	158.47

Yard and street light poles, including base plate but no top brackets. See labor costs below. Crouse-Hinds

	Unit	Round steel	Square steel	Aluminum
12' high	Ea	461.00	300.00	215.00
16' high	Ea	538.00	238.00	265.00
20' high, 6" wide	Ea	735.00	440.00	377.00
25' high, 6" wide	Ea	883.00	538.00	835.00
30' high, 8" wide	Ea	1,010.00	820.00	944.00
35' high, 8" wide	Ea	1,300.00	1,020.00	1,460.00
40' high, 9" wide	Ea	1,530.00	1,160.00	1,650.00
50' high, 10" wide	Ea	1,690.00	1,660.00	—

	Craft@Hrs	Unit	Material	Labor	Equipment	Total
Labor setting metal light poles Equipment is a flatbed truck with a boom & auger at $21.00 per hour						
To 20' high	E1@2.89	Ea	—	104.00	12.70	116.70
25' to 30' high	E1@4.35	Ea	—	156.00	19.20	175.20
35' to 40' high	E1@5.73	Ea	—	206.00	25.30	231.30
50' high	E1@8.57	Ea	—	308.00	37.80	345.80
Deduct for aluminum poles	—	%	—	-20.0	-20.0	—
Top brackets for steel and aluminum light poles, installed prior to setting pole						
Single arm	CE@.500	Ea	73.20	20.00	2.21	95.41
Two arm	CE@.500	Ea	86.70	20.00	2.21	108.91
Three arm	CE@.991	Ea	111.00	39.70	4.37	155.07
Four arm	CE@.991	Ea	139.00	39.70	4.37	183.07

	Craft@Hrs	Unit	Material	Labor	Equipment	Total

Area Lighting, Installed on Poles

These costs include excavation for pole, pole foundation (2' x 2' x 3' deep at $100), 150' of 2" PVC conduit run with 40' of RSC, 600' of #8 copper wire and terminations, 35' x 8" square steel pole and one arm and 480 volt mercury vapor 1,000 watt luminaire with lamp for each fixture. Equipment is a flatbed truck with a boom & auger at $21.00 per hour.

Cost per pole including fixtures as shown

	Craft@Hrs	Unit	Material	Labor	Equipment	Total
1 fixture per pole	E1@30.4	Ea	2,290.00	1,090.00	134.00	3,514.00
2 fixtures per pole	E1@32.5	Ea	2,850.00	1,170.00	149.00	4,169.00
3 fixtures per pole	E1@34.8	Ea	3,420.00	1,250.00	153.00	4,823.00
4 fixtures per pole	E1@36.8	Ea	3,980.00	1,320.00	163.00	5,463.00
Deduct for aluminum wire per pole	—	LS	-30.00	—	—	-30.00

Standby Power Generators

Typical subcontract prices including the subcontractor's overhead and profit. Use these prices for preliminary estimates only.

Diesel generators including exhaust system, fuel tank and pump, automatic transfer switch, equipment pad, vibration isolators, associated piping, conduit, and lifting into position

	Craft@Hrs	Unit	Material	Labor	Equipment	Total
10 kw	—	Ea	—	—	—	11,300.00
30 kw	—	Ea	—	—	—	20,100.00
45 kw	—	Ea	—	—	—	24,200.00
50 kw	—	Ea	—	—	—	25,200.00
60 kw	—	Ea	—	—	—	29,900.00
75 kw	—	Ea	—	—	—	30,400.00
90 kw	—	Ea	—	—	—	36,900.00
100 kw	—	Ea	—	—	—	40,900.00
125 kw	—	Ea	—	—	—	41,800.00
150 kw	—	Ea	—	—	—	51,300.00
175 kw	—	Ea	—	—	—	55,300.00
200 kw	—	Ea	—	—	—	55,800.00
300 kw	—	Ea	—	—	—	81,200.00
350 kw	—	Ea	—	—	—	95,900.00
400 kw	—	Ea	—	—	—	97,300.00
450 kw	—	Ea	—	—	—	135,000.00
500 kw	—	Ea	—	—	—	140,000.00
600 kw	—	Ea	—	—	—	170,000.00
750 kw	—	Ea	—	—	—	234,000.00
900 kw	—	Ea	—	—	—	283,000.00
1,000 kw	—	Ea	—	—	—	288,000.00
1,100 kw	—	Ea	—	—	—	353,000.00

Automatic Transfer Switches

for standby power systems.

Single phase, 3 wire, solid neutral, 50 to 60 cycle. NEMA Class 1 (indoor)

	Craft@Hrs	Unit	Material	Labor	Equipment	Total
100 amp, 250 volt	E4@4.97	Ea	460.00	175.00	—	635.00
150 amp, 250 volt	E4@4.97	Ea	2,430.00	175.00	—	2,605.00
200 amp, 250 volt	E4@4.97	Ea	1,640.00	175.00	—	1,815.00
100 amp, 600 volt	E4@4.97	Ea	2,560.00	175.00	—	2,735.00
200 amp, 600 volt	E4@8.76	Ea	3,500.00	308.00	—	3,808.00
400 amp, 600 volt	E4@15.1	Ea	4,810.00	531.00	—	5,341.00

	Craft@Hrs	Unit	Material	Labor	Total

Security and Alarm Systems

Sensor costs include connection but no wire runs.

Monitor panels, including accessory section and connection to signal and power supply wiring, cost per panel.

	Craft@Hrs	Unit	Material	Labor	Total
1 zone wall-mounted cabinet with 1 monitor panel, tone, standard line supervision, and 115 volt power supply	E4@2.02	Ea	902.00	71.00	973.00

	Craft@Hrs	Unit	Material	Labor	Total
5 zone wall-mounted cabinet with 5 monitor panels, 10 zone monitor rack, 5 monitor panel blanks, tone, standard line supervision and 115 volt power supply	E4@8.21	Ea	5,110.00	289.00	5,399.00
10 zone wall-mounted cabinet with 10 monitor panels, 10 zone monitor rack, tone, standard line supervision and 115 volt power supply	E4@12.1	Ea	5,880.00	426.00	6,306.00
10 zone monitor rack and 10 monitor panels with tone and standard line supervision but no power supply and no cabinet	E4@11.1	Ea	4,860.00	390.00	5,250.00
1 zone wall-mounted cabinet with 1 monitor panel, tone, high security line supervision and 115 volt power supply	E4@2.02	Ea	1,200.00	71.00	1,271.00
5 zone wall-mounted cabinet with 10 zone monitor rack, 5 monitor panel blanks, tone, high security line supervision but no power supply	E4@7.82	Ea	6,410.00	275.00	6,685.00
10 zone wall-mounted cabinet with 10 monitor panels, 10 zone monitor rack, tone, high security line supervision and 115 volt power supply	E4@12.0	Ea	7,510.00	422.00	7,932.00
Emergency power indicator for monitor panel	E4@1.14	Ea	240.00	40.10	280.10
Monitor panels					
Panel with accessory section, tone and standard Line supervision	E4@.990	Ea	396.00	34.80	430.80
Panel with accessory section, tone and high security Line supervision	E4@.904	Ea	596.00	31.80	627.80
Monitor racks					
1 zone with 115 volt power supply	E4@.904	Ea	326.00	31.80	357.80
10 zone with 115 volt power supply	E4@.904	Ea	1,680.00	31.80	1,711.80
10 zone with no power supply	E4@.904	Ea	1,290.00	31.80	1,321.80
Monitor cabinets					
1 zone monitor for wall mounted cabinet	E4@.990	Ea	511.00	34.80	545.80
5 zone monitor for wall mounted cabinet	E4@.990	Ea	656.00	34.80	690.80
10 zone monitor for wall mounted cabinet	E4@.990	Ea	800.00	34.80	834.80
20 zone monitor for wall mounted cabinet	E4@.990	Ea	1,080.00	34.80	1,114.80
50 zone monitor for wall mounted cabinet	E4@.990	Ea	2,420.00	34.80	2,454.80
50 zone monitor for floor mounted cabinet	E4@1.04	Ea	2,640.00	36.60	2,676.60
Balanced magnetic door switches					
Surface mounted	E4@2.02	Ea	100.00	71.00	171.00
Surface mounted with remote test	E4@2.02	Ea	140.00	71.00	211.00
Flush mounted	E4@2.02	Ea	96.60	71.00	167.60
Mounted bracket or spacer	E4@.249	Ea	5.79	8.76	14.55
Photoelectric sensors					
500' range, 12 volt DC	E4@2.02	Ea	274.00	71.00	345.00
800' range, 12 volt DC	E4@2.02	Ea	308.00	71.00	379.00
Fence type, 6 beam, without wire or trench	E4@3.89	Ea	9,960.00	137.00	10,097.00
Fence type, 9 beam, without wire or trench	E4@3.89	Ea	12,500.00	137.00	12,637.00
Capacitance wire grid systems					
Surface type	E4@.990	Ea	70.20	34.80	105.00
Duct type	E4@.990	Ea	57.10	34.80	91.90
Tube grid kit	E4@.990	Ea	104.00	34.80	138.80
Vibration sensors, to 30 per zone	E4@2.02	Ea	123.00	71.00	194.00
Audio sensors, to 30 per zone	E4@2.02	Ea	145.00	71.00	216.00
Inertia sensors, outdoor, without trenching	E4@2.02	Ea	96.40	71.00	167.40
Inertia sensors, indoor	E4@2.02	Ea	64.40	71.00	135.40
Electric sensor cable, 300 meters, no trenching	E4@3.89	Ea	1,460.00	137.00	1,597.00
Ultrasonic transmitters, to 20 per zone					
Omni-directional	E4@2.02	Ea	62.70	71.00	133.70
Directional	E4@2.02	Ea	70.20	71.00	141.20

	Craft@Hrs	Unit	Material	Labor	Total
Ultrasonic transceivers, to 20 per zone					
Omni-directional	E4@2.02	Ea	70.20	71.00	141.20
Directional	E4@2.02	Ea	75.10	71.00	146.10
High security	E4@2.02	Ea	255.00	71.00	326.00
Standard security	E4@2.02	Ea	104.00	71.00	175.00
Microwave perimeter sensor, to 4 per zone	E4@3.89	Ea	4,100.00	137.00	4,237.00
Passive interior infrared sensors, to 20 per zone					
Wide pattern	E4@2.02	Ea	568.00	71.00	639.00
Narrow pattern	E4@2.02	Ea	539.00	71.00	610.00
Access and secure control units					
For balanced magnetic door switches	E4@3.06	Ea	310.00	108.00	418.00
For photoelectric sensors	E4@3.06	Ea	568.00	108.00	676.00
For capacitance sensors	E4@3.06	Ea	576.00	108.00	684.00
For audio and vibration sensors	E4@3.06	Ea	665.00	108.00	773.00
For inertia sensors	E4@3.06	Ea	596.00	108.00	704.00
For nimmer (m.b.) detector	E4@3.06	Ea	486.00	108.00	594.00
For electric cable sensors	E4@3.06	Ea	2,990.00	108.00	3,098.00
For ultrasonic sensors	E4@3.06	Ea	912.00	108.00	1,020.00
For microwave sensors	E4@3.06	Ea	561.00	108.00	669.00
Accessories					
Tamper assembly for monitor cabinet	E4@.990	Ea	62.70	34.80	97.50
Monitor panel blank	E4@.249	Ea	7.00	8.76	15.76
Audible alarm	E4@.990	Ea	73.20	34.80	108.00
Audible alarm control	E4@.990	Ea	290.00	34.80	324.80
Termination screw terminal cabinet, 25 pair	E4@.990	Ea	210.00	34.80	244.80
Termination screw terminal cabinet, 50 pair	E4@1.36	Ea	342.00	47.80	389.80
Termination screw terminal cabinet, 150 pair	E4@3.45	Ea	564.00	121.00	685.00
Universal termination with remote test for cabinets and control panels	E4@1.47	Ea	51.50	51.70	103.20
Universal termination without remote test	E4@.990	Ea	30.10	34.80	64.90
High security line supervision termination with remote test and tone	E4@1.47	Ea	261.00	51.70	312.70
High security line supervision termination with tone only	E4@1.47	Ea	240.00	51.70	291.70
12" door cord for capacitance sensor	E4@.249	Ea	8.03	8.76	16.79
Insulation block kit for capacitance sensor	E4@.990	Ea	40.00	34.80	74.80
Termination block for capacitance sensor	E4@.249	Ea	8.50	8.76	17.26
200 zone graphic display	E4@3.06	Ea	4,000.00	108.00	4,108.00
210 zone multiplexer event recorder unit	E4@3.89	Ea	3,810.00	137.00	3,947.00
Guard alert display	E4@3.06	Ea	897.00	108.00	1,005.00
Uninterruptable power supply (battery)	E4@2.02	Ea	1,840.00	71.00	1,911.00
12 volt, 40VA plug-in transformer	E4@.495	Ea	34.50	17.40	51.90
18 volt, 40VA plug-in transformer	E4@.495	Ea	23.60	17.40	41.00
24 volt, 40VA plug-in transformer	E4@.495	Ea	17.10	17.40	34.50
Test relay	E4@.332	Ea	53.10	11.70	64.80
Sensor test control (average for sensors)	E4@.990	Ea	69.60	34.80	104.40
Cable, 2 pair 22 gauge, twisted, 1 pair shielded, run in exposed walls or per LF per pull, pulled in raceway and connected	E4@.008	LF	.52	.28	.80

Fire Alarm and Detection Systems No wiring included except as noted.

	Craft@Hrs	Unit	Material	Labor	Total
Pedestal mounted master fire alarm box	E4@3.89	Ea	1,240.00	137.00	1,377.00
Automatic coded signal transmitter	E4@3.89	Ea	651.00	137.00	788.00
Fire alarm control panel, simplex	E4@3.89	Ea	610.00	137.00	747.00

	Craft@Hrs	Unit	Material	Labor	Total
Fire detection annunciator panel					
8 zone drop	E4@3.89	Ea	404.00	137.00	541.00
12 zone drop	E4@5.03	Ea	574.00	177.00	751.00
16 zone drop	E4@5.93	Ea	825.00	209.00	1,034.00
8 zone lamp panel only	E4@4.33	Ea	134.00	152.00	286.00
12 zone lamp panel only	E4@5.93	Ea	197.00	209.00	406.00
16 zone lamp panel only	E4@7.23	Ea	261.00	254.00	515.00
Battery charger and cabinet, simplex	E4@6.19	Ea	351.00	218.00	569.00
Add for nickel cadmium batteries	E4@3.89	Ea	381.00	137.00	518.00
Fire alarm pull stations, manual operation	E4@.497	Ea	19.70	17.50	37.20
Fire alarm 10" bell with outlet box	E4@.497	Ea	64.10	17.50	81.60
Magnetic door holder	E4@1.33	Ea	60.10	46.80	106.90
Combination door holder and closer	E4@2.48	Ea	243.00	87.20	330.20
Detex door lock	E4@2.72	Ea	28.80	95.70	124.50
Thermodetector	E4@.497	Ea	8.69	17.50	26.19
Ionization AC smoke detector, with wiring	E4@.710	Ea	149.00	25.00	174.00
Fixed temp. and rate of rise smoke detector	E4@.746	Ea	13.40	26.20	39.60
Carbon dioxide pressure switch	E4@.746	Ea	4.58	26.20	30.78

Telephone Wiring

	Craft@Hrs	Unit	Material	Labor	Total
Plywood backboard mount for circuit wiring, set on masonry wall					
4' x 4' x 3/4"	E4@.938	Ea	17.50	33.00	50.50
4' x 8' x 3/4"	E4@1.10	Ea	34.90	38.70	73.60
Cable tap in a manhole or junction box					
25 to 50 pair cable	E4@7.80	Ea	103.00	274.00	377.00
100 to 200 pair cable	E4@10.1	Ea	118.00	355.00	473.00
300 pair cable	E4@12.0	Ea	120.00	422.00	542.00
400 pair cable	E4@13.0	Ea	130.00	457.00	587.00
Cable terminations in a manhole or junction box					
25 pair cable	E4@1.36	Ea	9.34	47.80	57.14
50 pair cable	E4@1.36	Ea	10.60	47.80	58.40
100 pair cable	E4@2.72	Ea	21.20	95.70	116.90
150 pair cable	E4@3.45	Ea	32.00	121.00	153.00
200 pair cable	E4@4.61	Ea	48.20	162.00	210.20
300 pair cable	E4@5.03	Ea	63.90	177.00	240.90
400 pair cable	E4@5.34	Ea	84.80	188.00	272.80
Communications cable, pulled in conduit or walls and connected					
2 pair cable	E4@.008	LF	.06	.28	.34
25 pair cable	E4@.030	LF	.59	1.06	1.65
50 pair cable	E4@.043	LF	1.28	1.51	2.79
75 pair cable	E4@.043	LF	1.59	1.51	3.10
100 pair cable	E4@.043	LF	1.85	1.51	3.36
150 pair cable	E4@.045	LF	2.58	1.58	4.16
200 pair cable	E4@.051	LF	3.37	1.79	5.16
300 pair cable	E4@.051	LF	4.97	1.79	6.76
Telephone outlets (no junction box included)					
Wall outlet	E4@.396	Ea	5.62	13.90	19.52
Floor outlet, flush mounted	E4@.495	Ea	6.99	17.40	24.39

Closed Circuit Television Systems

	Craft@Hrs	Unit	Material	Labor	Total
Outdoor television camera (16 mm f11.6 lens, black and white) in weatherproof housing					
Wall or roof mount	E4@1.67	Ea	1,680.00	58.70	1,738.70
Self-terminating outlet box, cover plate	E4@.689	Ea	5.06	24.20	29.26

	Craft@Hrs	Unit	Material	Labor	Total
Coaxial cable (RG59/J), 75 ohms, no raceway included	E4@.006	LF	.13	.21	.34
In-line cable taps (PTU) for 36 TV system	E4@.689	Ea	8.41	24.20	32.61
Cable blocks for in-line taps	—	Ea	4.37	—	4.37
Desk type black and white TV monitor, 12" diagonal screen, 75 ohms,					
0.5 to 2.0 Vp-p external sync input	E4@2.77	Ea	378.00	97.40	475.40

Public Address Paging Systems

	Craft@Hrs	Unit	Material	Labor	Total
Signal cable, #22 solid copper polyethylene insulated, PVC jacketed wire					
Pulled in conduit, no conduit included	E4@.008	LF	.07	.28	.35
Microphone, desk type, push to talk, with					
20' of three conductor cable	E4@.839	Ea	95.00	29.50	124.50
Wall mounted enameled steel amplifier cabinet (18" high, 24" wide, 6" deep) with electrical receptacle					
and lockable door	E4@1.67	Ea	190.00	58.70	248.70
Amplifier, 40 watts, 50 to 15,000 Hertz, 4, 8, 16 ohm outputs, built in AM/FM tuner					
Complete unit	E4@.801	Ea	650.00	28.20	678.20
Fanon, wall mounted paging horns, 250 to 13,000 Hertz, 8 ohms, 30 watts, 10" diameter, 12" deep, gray baked enamel					
Interior horns	E4@1.67	Ea	83.00	58.70	141.70
Exterior horns, weatherproof	E4@2.54	Ea	143.00	89.30	232.30

Computer Network Cable Pulled in conduit or walls. Add for connectors and cable testing. Prices are for cable purchased in bulk on 500' spools.

	Craft@Hrs	Unit	Material	Labor	Total
Coaxial cable					
Arcnet, 75 ohm (RG62)					
PVC cable	E4@.008	LF	.20	.28	.48
Plenum cable	E4@.008	LF	.53	.28	.81
Thick Ethernet, 50 ohm, 10BASE5					
PVC cable	E4@.043	LF	1.00	1.51	2.51
Plenum cable	E4@.043	LF	2.04	1.51	3.55
Thin Ethernet, 50 ohm, 10BASE2 (RG58)					
PVC cable	E4@.008	LF	.29	.28	.57
Plenum cable	E4@.008	LF	.57	.28	.85
Fiber optic duplex cable					
62.5-micron core	E4@.012	LF	.66	.42	1.08
Plenum duplex fiber optic cable					
62.5-micron core	E4@.012	LF	.85	.42	1.27
Shielded twisted pair (STP) IBM Type 1					
PVC, 2 pair	E4@.030	LF	.45	1.06	1.51
Plenum, 2 pair	E4@.030	LF	.98	1.06	2.04
IBM Type 2					
PVC, 6 pair	E4@.043	LF	.85	1.51	2.36
Plenum, 6 pair	E4@.043	LF	1.61	1.51	3.12
IBM Type 6, 2 pair	E4@.035	LF	.50	1.23	1.73
IBM Type 9, 2 pair	E4@.030	LF	.71	1.06	1.77
Twinaxial cable					
PVC cable	E4@.030	LF	.44	1.06	1.50
Plenum cable	E4@.030	LF	1.55	1.06	2.61
Unshielded twisted pair (UTP). EIA/TIA 568 standard categories shown. Labor shown is for single cable pulls					
Category 3, 10 Mbps 10BASE-T					
2 pair UTP, PVC	E4@.008	LF	.07	.28	.35
2 pair UTP, plenum	E4@.008	LF	.15	.28	.43
4 pair UTP, PVC	E4@.008	LF	.11	.28	.39
4 pair UTP, plenum	E4@.008	LF	.21	.28	.49

	Craft@Hrs	Unit	Material	Labor	Total
Category 4, 20 Mbps					
2 pair UTP, PVC	E4@.008	LF	.12	.28	.40
2 pair UTP, plenum	E4@.008	LF	.21	.28	.49
4 pair UTP, PVC	E4@.008	LF	.13	.28	.41
4 pair UTP, plenum	E4@.008	LF	.35	.28	.63
25 pair UTP, PVC	E4@.030	LF	.91	1.06	1.97
25 pair UTP, plenum	E4@.030	LF	2.16	1.06	3.22
Category 5, 100 Mbps					
2 pair UTP, PVC	E4@.008	LF	.15	.28	.43
2 pair UTP, plenum	E4@.008	LF	.25	.28	.53
4 pair UTP, PVC	E4@.008	LF	.17	.28	.45
4 pair UTP, plenum	E4@.008	LF	.39	.28	.67

Add for multiple cable pulls of thin Ethernet or unshielded twisted pair cables when cables are pulled simultaneously in plenum or PVC, based on quantity of cables pulled

	Craft@Hrs	Unit	Material	Labor	Total
Add for each cable pulled simultaneously, over one	E4@.250	Ea	—	8.79	8.79

Punch-down blocks

Plywood backboard mount for patch panel or punch-down blocks, set on masonry wall

	Craft@Hrs	Unit	Material	Labor	Total
4' x 4' x 3/4"	E4@.938	Ea	16.90	33.00	49.90
4' x 8' x 3/4"	E4@1.10	Ea	33.80	38.70	72.50
Distribution frame for punch-down blocks					
Holds 5 punch-down blocks	E4@.250	Ea	32.40	8.79	41.19
Holds 10 punch-down blocks	E4@.250	Ea	38.60	8.79	47.39

Punch-down blocks, labor is to install block in distribution frame only, add labor to punch-down wire pairs as shown below

	Craft@Hrs	Unit	Material	Labor	Total
25 pair	E4@.250	Ea	10.10	8.79	18.89
50 pair	E4@.250	Ea	10.30	8.79	19.09
Punch-down wire pairs, labor only, per pair	E4@.020	Ea	—	.70	.70

Computer network cable connectors

	Craft@Hrs	Unit	Material	Labor	Total
Coax					
Thick Ethernet					
N series, crimp on, female	E4@.396	Ea	12.00	13.90	25.90
N series, crimp on, male	E4@.396	Ea	5.95	13.90	19.85
In-line transceiver connector	E4@.396	Ea	15.00	13.90	28.90
Line taps, N series, BNC, or non-intrusive	E4@.396	Ea	50.00	13.90	63.90
Thin Ethernet or Arcnet					
BNC type, crimp on, female	E4@.450	Ea	3.36	15.80	19.16
Fiber optic cable, terminated, 1 meter length					
ST-ST duplex riser	E4@.100	Ea	70.00	3.52	73.52
ST-SC duplex riser	E4@.100	Ea	96.00	3.52	99.52
SC-SC duplex riser	E4@.100	Ea	110.00	3.52	113.52
FDDI OFNP riser	E4@.100	Ea	90.00	3.52	93.52
FDDI OFNP plenum	E4@.100	Ea	100.00	3.52	103.52
Shielded twisted pair (STP)					
IBM type 1, 2 pair data connector	E4@.396	Ea	5.56	13.90	19.46
IBM type 2, 6 pair data connector	E4@.396	Ea	6.72	13.90	20.62
IBM type 6, 2 pair data connector	E4@.450	Ea	17.50	15.80	33.30
IBM type 9, 2 pair data connector	E4@.450	Ea	17.40	15.80	33.20
Twinaxial, female connector	E4@.396	Ea	6.30	13.90	20.20

	Craft@Hrs	Unit	Material	Labor	Total
Unshielded twisted pair (UTP)					
RJ11	E4@.396	Ea	.79	13.90	14.69
RJ45	E4@.396	Ea	1.10	13.90	15.00
Cable testing Testing, with dedicated cable tester					
Coax cable, per conductor	E4@.125	Ea	—	4.40	4.40
Fiber optic cable, per segment	E4@.160	Ea	—	5.63	5.63
Shielded twisted pair (STP), per pair	E4@.125	Ea	—	4.40	4.40
Twinaxial, per pair	E4@.160	Ea	—	5.63	5.63
Unshielded twisted pair (UTP), per pair	E4@.125	Ea	—	4.40	4.40
Wall plates					
Coax wall plates with jacks (no junction box included)					
One or two jacks, including termination	E4@.396	Ea	7.95	13.90	21.85
Modular wall plates with jacks for UTP (no junction box included), including termination					
One 4 or 6 wire jack (RJ11)	E4@.396	Ea	4.95	13.90	18.85
One 8 wire jack (RJ45)	E4@.396	Ea	4.95	13.90	18.85
Two 8 wire jacks (RJ45)	E4@.450	Ea	10.95	15.80	26.75
Twinaxial wall plates with jacks (no junction box included), including termination					
One twinaxial feedthrough	E4@.396	Ea	8.00	13.90	21.90
Two twinaxial feedthroughs	E4@.396	Ea	9.00	13.90	22.90

National Estimator and Job Cost Wizard Quick Start

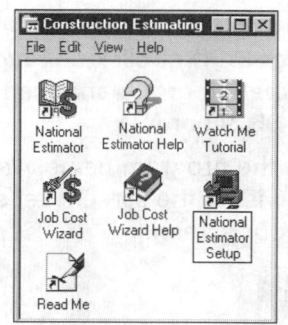

Construction Estimating program group.

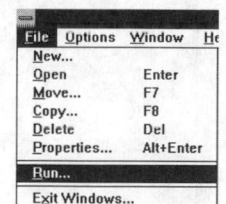

Click on File.
Click on Run in Windows 3.1.

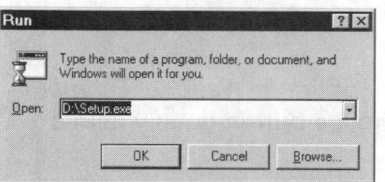

Type D:\SETUP
in Windows 95, 98 or NT.

National Estimator is a construction cost estimating program with all the cost estimates in this book. You can page through these costs one screen at a time or use the electronic index to search by keyword for exactly the information you need. When you find what you're looking for, split the screen in two so your estimate is on the bottom half of the screen and the database is on the top half. Then copy and paste into your estimate. Type the estimated quantity. The program extends prices and totals columns automatically. Add your overhead and profit as a percentage and then print the estimate.

Job Cost Wizard takes your estimates to the next level:

1. Turning estimates into invoices you can send out in a window envelope.

2. Exporting to *QuickBooks* where you can track actual costs against estimates.

The installation program creates a Construction Estimating program group and puts several icons in that group. If you have trouble installing *National Estimator*, call Craftsman tech support at 760-438-7828. If your computer can't use the CD, exchange it for *National Estimator* on 3½" disk. Send the CD to Craftsman, Box 6500, Carlsbad, CA 92018. There's no charge for this exchange.

Installing National Estimator and Job Cost Wizard

Put the *National Estimator* disk in your CD drive (such as D:). Start Windows.

In Windows 3.1 or 3.11
go to the Program Manager.

1. Click on **File**
2. Click on **Run** . . .
3. Type D:\SETUP
4. Press Enter ←

In Windows 95, 98 or NT

1. Click on Start
2. Click on **Run** . . .
3. Type D:\SETUP
4. Press Enter ←

We recommend accepting the installation defaults:

1. National directory

2. "Typical" setup

When installation is complete, click on **Finish**. Note that *Job Cost Wizard* won't be installed if you use Windows 3.1 or 3.11.

Uninstalling National Estimator and Job Cost Wizard

In Windows 3.1 or 3.11: Select the Construction Estimating program group. Then click on the Uninstall icon.

In Windows 95, 98 or NT: Click on **Start, Settings, Control Panel**. Double click on **Add/Remove Programs**. Then click on the name of the program to remove and **Add/Remove**.

Learn the Easy Way!

Estimating with Jay and Julie.

 Watch Me icon

The **Watch Me** tutorial is an interactive video guide to *National Estimator*. Let Jay and Julie help you get started using the *National Estimator* program. Sit back and relax and let the video run — or click on buttons and menus to jump from topic to topic inside the video. Exit any time you want. Then go back to Watch Me later to brush up on some topic. Watch Me takes up very little space on your hard drive and is part of the "Typical" installation. Watch Me requires 8Mb of RAM, a Windows-compatible sound system and at least a 100 MHz 486 computer. Click on the Watch Me icon to begin Watch Me.

Using National Estimator

 *National Estimator icon*

National Estimator begins when you click on the *National Estimator* icon (in Windows 3.1 or 3.11) or click on *Start*, *Programs*, the *Construction Estimating* group and then *National Estimator* (in Windows 95, 98 or NT).

On the title bar at the top of the screen you see the program name, *National Estimator*, and [Estimate1], showing that Estimate1 is on the screen. Let's take a closer look at the other information at the top of your screen.

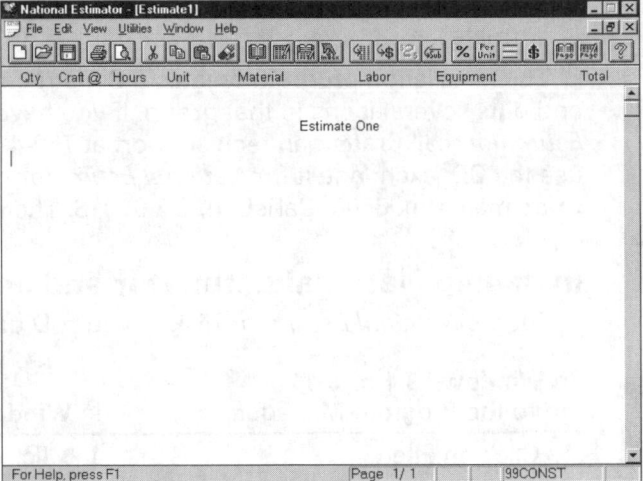

Your estimating form ready to begin the first estimate.

The Menu Bar

Below the title bar you see the menu bar: Every option in *National Estimator* is available on the menu bar. Click with your left mouse button on any item on the menu to open a list of available commands.

Title Bar —
Menu Bar —
Toolbar —
Column Heads —

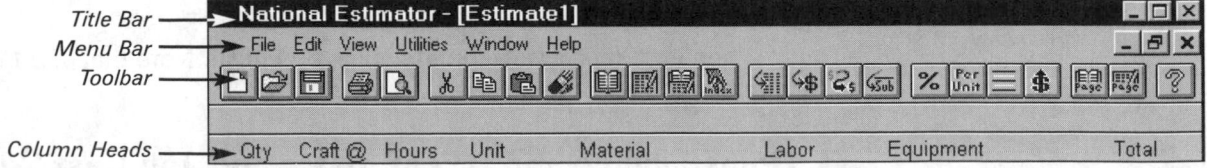

Buttons on the Toolbar

Below the menu bar you see 24 buttons that make up the toolbar. The options you use most in *National Estimator* are only a mouse click away on the toolbar.

Column Headings

Below the toolbar you'll see column headings for your estimate form:

Qty for quantity

Craft@Hours for craft (the crew doing the work) and manhours (to complete the task)

Unit for unit of measure, such as per linear foot or per square foot

Material for material cost

Labor for labor cost

Equipment for equipment cost

Total for the total of all cost columns

The Status Bar

The bottom line on your screen is the status bar. Here you'll find helpful information about the choices available. Notice "Page 1/1" near the right end of the status line. That's a clue that you're looking at page 1 of a one-page estimate. When you see PICT at the right of the status bar, a picture is available in the costbook. Click on PICT to show the picture. Click on PICT again to hide the picture.

Check the status bar occasionally for helpful tips and explanations of what you see on screen.

| For Help, press F1 | Page 1/ 1 | 99CONST | PICT |

Beginning an Estimate

Let's start by putting a heading on this estimate.

| The Blinking Cursor (insert point)

1. Press [Enter ↵] once to space down one line.

⇗ Mouse Pointer

2. Press [Tab ⇆] four times (or hold the space bar down) to move the Blinking Cursor (the insert point) near the middle of the line.

I Mouse Pointer

3. Type "Estimate One" and press [Enter ↵]. That's the title of this estimate, "Estimate One."

4. Press [Enter ↵] again to move the cursor down a line. That opens up a little space below the title.

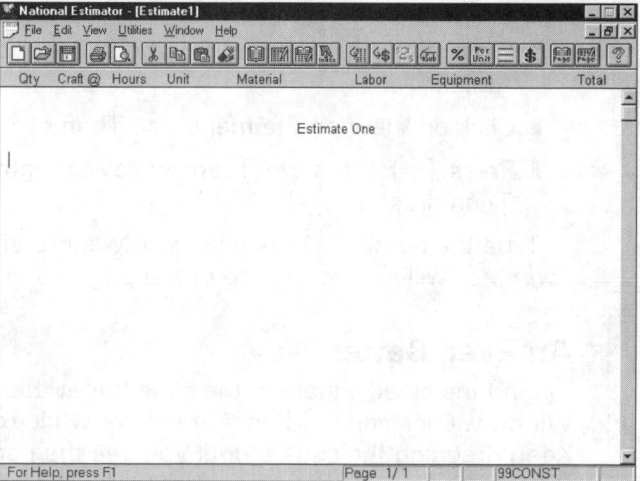

Begin by putting a title on your estimate, such as "Estimate One."

The Costbook

Let's leave your estimating form for a moment to take a look at estimates stored in the costbook. To switch to the costbook, either:

▮ Click on the 📖 (Costbook Window) button, -or-

▮ Click on **View** on the menu bar. Then click on **Costbook Window**, -or-

■ Press ⟨Esc⟩. Press the ⟨↓⟩ arrow key to highlight Costbook Window. Then press ⟨Enter ⏎⟩.

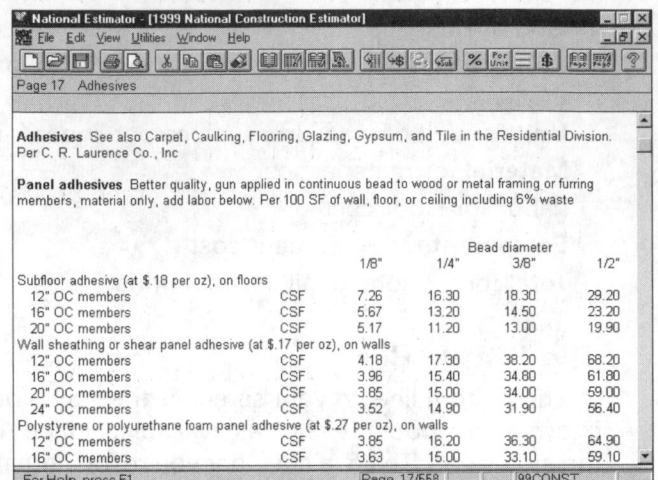

The costbook window has the entire National Construction Estimator.

The entire *1999 National Construction Estimator* is available in the Costbook Window. Notice the words *Page 17 Adhesives* at the left side of the screen just below the toolbar. That's your clue that the adhesives section of page 17 is on the screen. To turn to the next page, either:

■ Press ⟨PgDn⟩ (with Num Lock off), -or-

■ Click on the lower half of the scroll bar at the right edge of the screen.

To move down one line at a time, either:

■ Press the ⟨↓⟩ arrow key (with Num Lock off), -or-

■ Click on the arrow on the down scroll bar at the lower right corner of the screen.

Press ⟨PgDn⟩ about 1,400 times and you'll page through the entire *National Construction Estimator*. Obviously, there's a better way. To turn quickly to any page, either:

■ Click on the 🔖 (Turn to Costbook Page) button near the right end of the toolbar, -or-

■ Click on **View** on the menu bar. Then click on **Turn to Costbook Page**, -or-

■ Press ⟨Esc⟩. Press the ⟨↓⟩ arrow key to highlight Turn to Costbook Page. Then press ⟨Enter ⏎⟩.

Type the page number you want to see.

Type the number of the page you want to see and press Enter. *National Estimator* will turn to the top of the page you requested.

An Even Better Way

Find the small square in the slide bar at the right side of the Costbook Window. Click and hold on that square while rolling the mouse up or down. Keep dragging the square until you see the page you want in the Page: box. Release the mouse button to turn to the top of that page.

Drag the square to see any page.

*Use the electronic index
to find cost estimates
for any item.*

Enter Keyword to Locate:

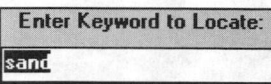

sand

Enter keyword to locate.

The index jumps to Sand.

A Still Better Way: Keyword Search

To find any cost estimate in seconds, search by keyword in the index. To go to the index, either:

▌ Click on the ▨ (Index) button near the center of the toolbar, -or-

▌ Press Esc. Press Enter↵, -or-

▌ Click on **View** on the menu bar. Then press Enter↵.

Notice that the cursor is blinking in the Enter Keyword to Locate box at the right of the screen. Obviously, the index is ready to begin a search.

Your First Estimate

Suppose we're estimating the cost of 4" sand fill under a 10' x 10' (100 square foot) concrete slab. Let's put the index to work with a search for sand fill. In the box under Enter Keyword to Locate, type *sand*. The index jumps to the heading *Sand*.

The fourth item under sand is *fill: 68, 95, 324*. Either:

▌ Click once on that line and press Enter↵, -or-

▌ Double click on that line, -or-

▌ Press Tab⇆ and the ↓ arrow key to move the highlight to *fill: 68, 95, 324*. Then press Enter↵.

If costs appear on several pages, click on the page you prefer.

To select the page you want to see (page 68 in this case), either:

▌ Click on the number 68, -or-

▌ Press Tab⇆ to highlight 68. Then press Enter↵.

National Estimator turns to the top of page 68. See the example below. Notice that estimates at the top of this page are for footings and grade beams. Press the ↓ arrow key (or click on the down scroll arrow) until "Sand fill base" moves to the top of your screen.

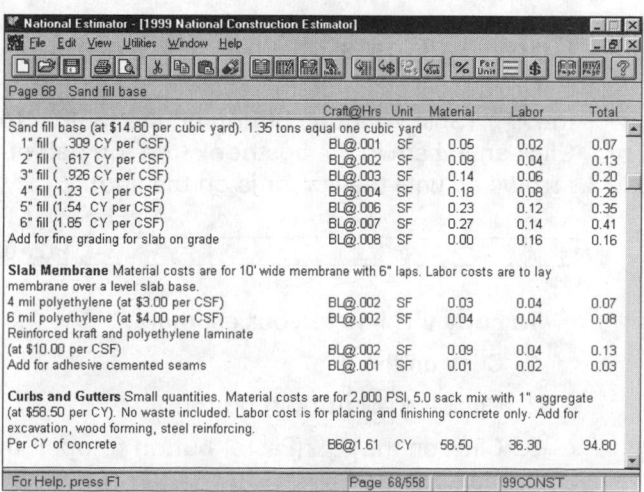

*Costs for sand fill base
on page 68.*

Splitting the Screen

Most of the time you'll want to see what's in both the costbook and your estimate. To split the screen into two halves, either:

▌ Click on the 📇 (Split Window) button near the center of the toolbar, -or-

▌ Press [Esc] and the [↓] arrow key to move the selection bar to Split Window, -or-

▌ Click on **View** on the menu bar. Then click on **Split Window** and your screen should look like the example below.

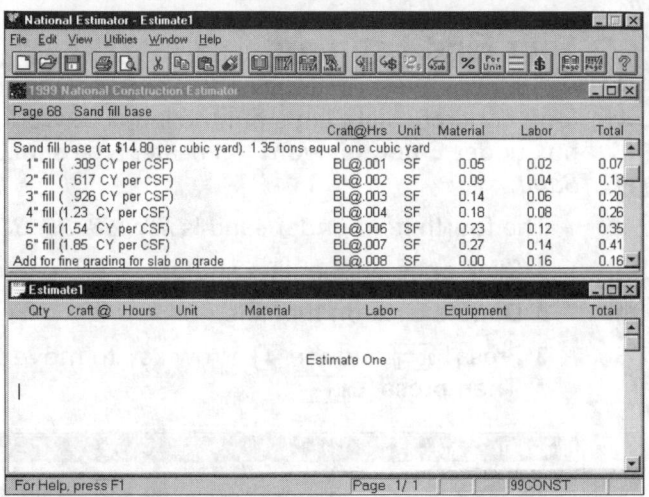

The split window: Costbook above and estimate below.

Notice the eight lines of the costbook are at the top of the screen and your estimate is at the bottom. You should recognize "Estimate One." It's your title for the estimate. Column headings are at the top of the costbook and across the middle of the screen (for your estimate).

To Switch from Window to Window

▌ Click in the window of your choice, -or-

▌ Hold the [Ctrl] key down and press [Tab⇆].

Notice that a window title bar turns dark when that window is selected. The selected window is where keystrokes appear as you type.

Copying Costs to Your Estimate

Next, we'll estimate the cost of 100 square feet of sand fill base. Click the 📇 (Split Window) button on the toolbar to be sure you're in the split window. Click anywhere in the costbook (the top half of your screen). Then press the [↓] arrow key until the cursor is on the line:

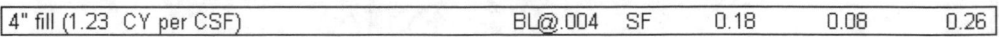

| 4" fill (1.23 CY per CSF) | BL@.004 | SF | 0.18 | 0.08 | 0.26 |

To copy this line to your estimate:

1. Click on the line.

2. Click the 📋 (Copy) button.

3. Click on the 📋 (Paste) button to open the Enter Cost Information dialog box.

Notice that the blinking cursor is in the Quantity box:

1. Type a quantity of 100 because the slab base is 100 square feet (10' x 10').
2. Press ⎀Tab⎀ and check the estimate for accuracy.

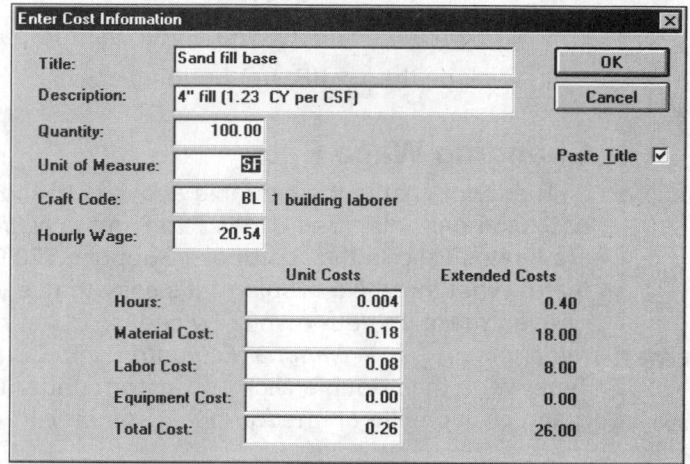

Costs for the 100 SF job (extended costs) are on the right.

▮ Notice that the column headed Unit Costs shows costs per unit, per "SF" (square foot) in this case.

▮ The column headed Extended Costs shows costs for 100 square feet.

▮ The lines opposite Title and Description show what's getting installed. You can change the words in either of these boxes. Just click on what you want to change and start typing or deleting.

▮ You can also change any numbers in the Unit Cost column. Just click and start typing.

▮ When the words and costs are exactly right, press ⎀Enter↵⎀ or click on OK to copy these figures to the end of your estimate.

The new line at the bottom of your estimate shows:

Extended costs for sand fill base as they appear on your estimate form.

Sand fill base						
4" fill (1.23 CY per CSF)						
100.00	BL@.4000	SF	18.00	8.00	0.00	26.00

100 is the quantity of sand in square feet

BL is the recommended crew, a building laborer

@.4000 shows the manhours required for the work

SF is the unit of measure, square feet in this case

18.00 is the material cost (the sand fill)

8.00 is labor cost for the job

0.00 shows there is no equipment cost

26.00 is the total of material, labor and equipment columns

Copy Anything to Anywhere in Your Estimate

Anything in the costbook can be copied to your estimate. Just click on the line (or select the words) you want to copy and press the ⎀F8⎀ key. It's copied to the last line of your estimating form. If your selection includes costs, you'll have a chance to enter the quantity. To copy to the *middle* of your estimate:

1. Select what you want to copy.
2. Click on the 📋 (Copy) button.
3. Click in the estimate where you want to paste.
4. Click on the 📋 (Paste) button.

Changing Wage Rates

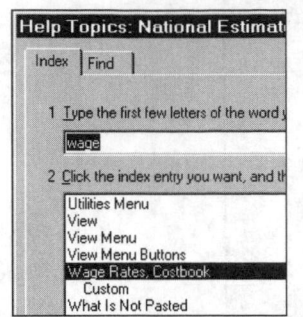

Search for information on setting wage rates.

The labor cost in the example above is based on a laborer working at a cost of $20.54 per hour. (See pages 7 to 9 in the *National Construction Estimator* for crew rates used in the costbook.) Suppose $20.54 per hour isn't right for your estimate. What then? No problem! It's easy to use your own wage rate for any crew or even make up your own crew codes. To get more information on setting wage rates, press F1. At *National Estimator* Help Contents, click on the **Search** button. Type *wage* then double click on **Custom** under Wage Rates. To return to your estimate, click on **File** on the *National Estimator* Help menu bar. Then click on **Exit**.

Changing Cost Estimates

Unit Costs	
Hours:	0.004
Material Cost:	0.25
Labor Cost:	0.08
Equipment Cost:	0.00
Total Cost:	0.33

Change the material cost to $.25.

With Num Lock off, use the ⬆ or ⬇ arrow key to move the cursor to the line you want to change (or click on that line). In this case, move to the line that begins with a quantity of 100. To open the Enter Cost Information Dialog box, either:

▪ Press Enter ↵, -or-

▪ Click on the 💲 (Change Cost) button on the toolbar.

To make a change, either:

▪ Click on what you want to change, -or-

▪ Press Tab⇆ until the cursor advances to what you want to change.

Then type the correct figure. In this case, change the material cost to $.25 (twenty-five cents).

Press Tab⇆ and check the Extended Costs column. If it looks OK, press Enter ↵ and the change is made on your estimating form.

Changing Text (Descriptions)

4" fill (1.23 CY per
100.00   BL@.4

To select, click and hold the mouse button while dragging the mouse.

Click on the 🖼 (Estimate Window) button on the toolbar to be sure you're in the estimate. With Num Lock off, use the ⬆ or ⬇ arrow key or click the mouse button to put the cursor where you want to make a change. In this case, we're going to make a change on the line that begins "Sand fill base"

To make a change, click where the change is needed. Then either:

▪ Press the Del or ←Bksp key to erase what needs deleting, -or-

▪ Select what needs deleting and click on the ✂ (Cut) button on the toolbar.

▪ Type what needs to be added.

In this case, click just before "base." Then hold the left mouse button down and drag the mouse to the right until you've put a dark background behind the word "base." The dark background shows that this word is selected and ready for editing.

Press the Del key, or click on the ✂ (Cut) button on the toolbar, and the selection is cut from the estimate. If that's not what you wanted, click on the (Undo) button and "base" is back again.

Adding Text (Descriptions)

Some of your estimates will require descriptions (text) and costs that can't be found in the *National Construction Estimator*. What then? With *National Estimator* it's easy to add descriptions and costs of your choice anywhere in the estimate. For practice, let's add an estimate for four reinforced corners to Estimate One.

Click on the ▓ (Estimate Window) button to be sure the estimate window is maximized. We can add lines anywhere on the estimate. But in this case, let's make the addition at the end. Press the ⬇ arrow key to move the cursor down until it's just above the horizontal line that separates estimate detail lines from estimate totals. To open a blank line, either:

▌ Press [Enter ↵], -or-

▌ Click on the ▤ (Insert Text) button on the toolbar, -or-

▌ Click on **Edit** on the menu bar. Then click on **Insert a Text Line**.

Type "Reinforced corners" and press [Enter ↵].

Adding a Cost Estimate Line

Now let's add a cost for "Reinforced corners" to your estimate. Begin by opening the Enter Cost Information dialog box. Either:

▌ Click on the ▤$ (Insert Cost) button on the toolbar, -or-

▌ Click on **Edit** on the menu bar. Then click on **Insert a Cost Line**.

1. The cursor is in the Quantity box. Type the number of units (4 in this case) and press [Tab⇆].

2. The cursor moves to the next box, Unit of Measure.

3. In the Unit of Measure box, type *Each* and press [Tab⇆].

4. Press [Tab⇆] twice to leave the Craft Code blank and Hourly Wage at zero.

5. Since these corners will be installed by the supplier, there's no material, labor or equipment cost. So press [Tab⇆] four times to skip over the Hours, Material Cost, Labor Cost and Equipment Cost boxes.

6. In the Total Cost box, type 20.00. That's the cost per corner quoted by your supplier.

7. Press [Tab⇆] once more to advance to OK.

8. Press [Enter ↵] and the cost of four reinforced corners is written to your estimate.

| Sand fill base |
| 4" fill (1.23 CY per CS |
| 100.00   BL@ .4000 |
| Reinforced corners |

Adding "Reinforced corners."

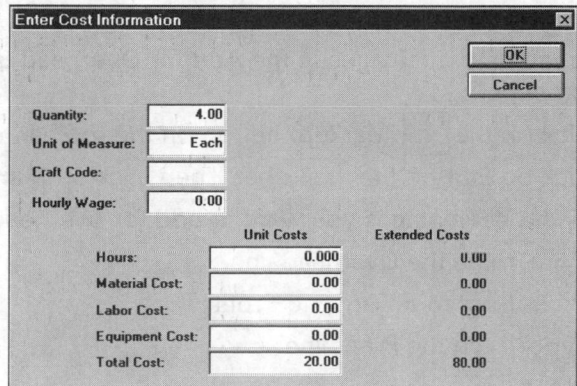

Unit and extended costs for four reinforced corners.

Note: The sum of material, labor and equipment costs appears automatically in the Total Cost box. If there's no cost entered in the Material Cost, Labor Cost or Equipment Cost boxes (such as for a subcontracted item), you can enter any figure in the Total Cost box.

Adding Lines to the Costbook

Add lines or make changes in the costbook the same way you add lines or make changes in an estimate. The additions and changes you make become part of the user costbook. For more information on user costbooks, press F1. Click on **Search**. Type "user" and press Enter⏎.

Subtotals Become QuickBooks Cost Categories

At the end of each section in your estimate, insert a subtotal. For a general contractor, estimate sections might be Demolition, Excavation, Foundation, Framing, etc. Estimate sections for a remodeling contractor might include Kitchen, Bathroom and Basement. Section subtotals help organize your estimates and make them easier to read and understand. Insert section subtotals wherever they make the most sense to you. These subtotals become cost categories when printing bids and invoices. The first ten characters of subtotal names become *QuickBooks* cost category names when exporting to *QuickBooks*.

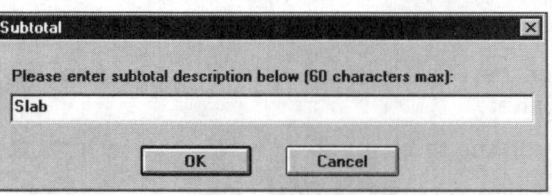

Insert a subtotal.

To insert a subtotal:

1. Click on the last cost line of the section (or on any blank line below the section).
2. Click on the 🔲 (Subtotal) button on the toolbar (or click on **Edit** and **Insert Subtotal**).
3. Type a name or description for the section (such as "Slab").
4. Press Enter⏎.

Adding Tax

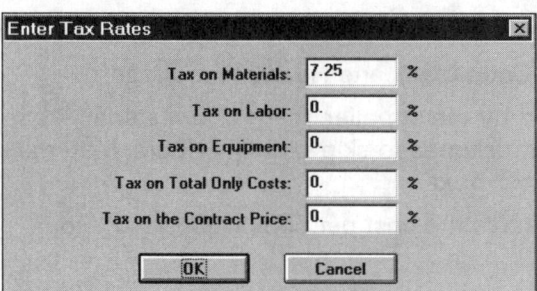

Type the tax rate that applies.

To include sales tax in your estimate:

1. Click on **Edit**.
2. Click on **Tax Rates**.
3. Type the tax rate in the appropriate box.
4. Press Tab⇆ to advance to the next box.
5. Press Enter⏎ or click on **OK** when done.

In this case, the tax rate is 7.25% on materials only. Tax will appear as the last line of the estimate.

Adding Overhead and Profit

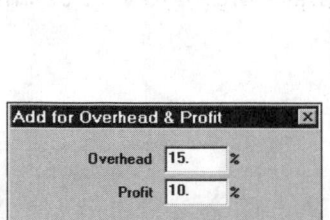

Adding overhead & profit.

Set markup percentages in the Add for Overhead & Profit dialog box. To open the box, either:

■ Click on the 💲 (Markup) button on the toolbar, -or-

■ Click on **Edit** on the menu bar. Then click on **Markup**.

Type the percentages you want to add for overhead. For this estimate:

1. Type 15 on the Overhead line.
2. Press Tab⇆ to advance to Profit.
3. Type 10 on the Profit line.
4. Press Enter⏎.

Markup percentages can be changed at any time. Just reopen the Add for Overhead & Profit dialog box and type the correct figure.

Preview Your Estimate

You can display an estimate on screen just the way it will look when printed on paper. To preview your estimate, either:

▌ Click on the [🔍] (Print Preview) button on the toolbar, -or-

▌ Click on **File** on the menu bar. Then click on **Print Preview**.

Use buttons on Print Preview to see your estimate as it will look when printed.

In Print Preview:

▌ Click on **Next Page** or **Prev Page** to turn pages.

▌ Click on **Two Page** to see two estimate pages side by side.

▌ Click on **Zoom In** to get a closer look.

▌ Click on **Close** when you've seen enough.

Printing Your Estimate

When you're ready to print the estimate, either:

▌ Click on the [🖨] (Print) button on the toolbar, -or-

▌ Click on File on the menu bar. Then click on **Print**, -or-

▌ Hold the Ctrl key down and type the letter P.

▌ Press Enter ↵ or click on **OK** to begin printing.

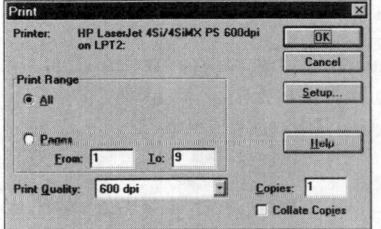

Options available depend on the printer you're using.

Save Your Estimate to Disk

To store your estimate on the hard disk where it can be re-opened and changed at any time, either:

▌ Click on the [💾] (Save) button on the toolbar, -or-

▌ Click on **File** on the menu bar. Then click on **Save**, -or-

▌ Hold the Ctrl key down and type the letter S.

The cursor is in the File Name box. Type the name you want to give this estimate, such as *First*. The name can be up to eight letters and numbers, but don't use symbols or spaces. Press Enter ↵ or click on **OK** and the estimate is written to disk.

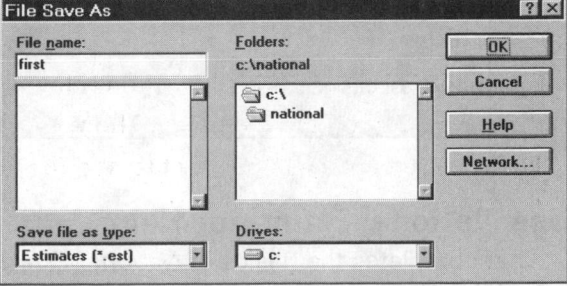

Type the estimate name in the File Name box to assign a file name.

Opening Other Costbooks

Many construction cost estimating databases are available for the *National Estimator* program. The order form at the back of this manual has more information on these costbooks.

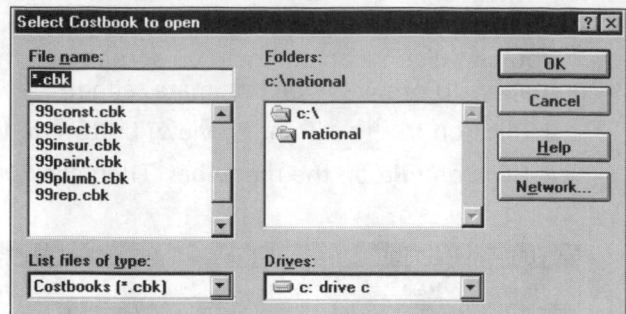

Open the costbook of your choice.

If you own several of these manuals and have installed the database that comes with each, you can open several costbooks at the same time. To open another costbook:

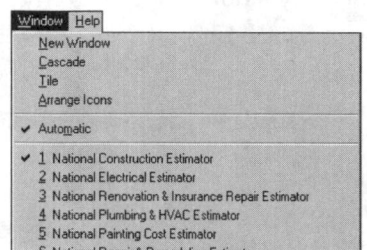

Click to switch costbooks.

▌ Click on **File**.

▌ Click on **Open Costbook**.

▌ Be sure the drive and directory are correct, usually *c:\national*.

▌ Double click on the costbook of your choice.

To see a list of the costbooks open, click on **Window**. The name of the current estimate or costbook will be checked. Click on any other costbook name to display that costbook. Click on **Window**, then click on **Tile** to display all open costbooks and estimates.

Select Your Default Costbook

Your default costbook is the last costbook installed. It opens automatically every time you begin using *National Estimator*. Save time by making the default costbook the one you use most.

To change your default costbook, click on **Utilities** on the menu bar. Then click on **Options**. Next, click on **Select Default Costbook**. Click on the costbook of your choice. Click on **OK**. Then click on **OK** again.

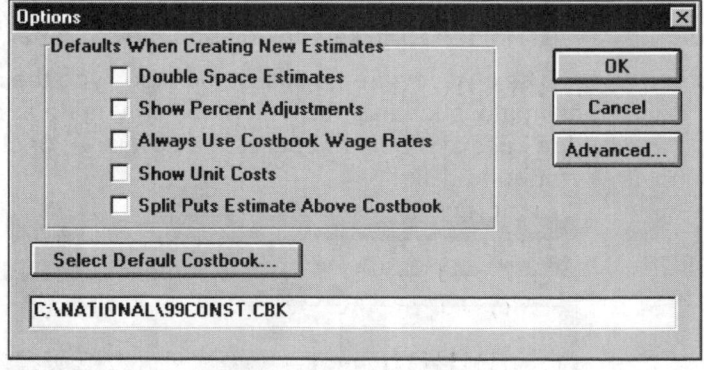

Select the default costbook.

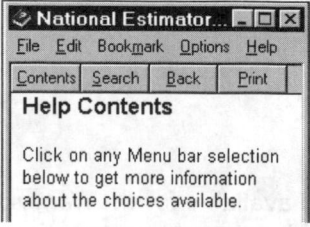

Click on File and then Print Topic.

Use National Estimator Help

That completes the basics of *National Estimator*. You've learned enough to complete most estimates. When you need more information about the fine points, use *National Estimator* Help. Click on the [?] (Help) button to see Help Contents. Then click on the menu selection of your choice. To print 28 pages of instructions for *National Estimator*, go to **Help Contents**. Click on **Print All Topics** (at the bottom of Help Contents). Click on **File** on the Help menu bar. Then click on **Print Topic**. Click on **OK**.

Converting Estimates with Job Cost Wizard

*Job Cost
Wizard icon*

To start *Job Cost Wizard*, click on the Job Cost Wizard icon. (Be sure to close the estimate in *National Estimator* before trying to open it in *Job Cost Wizard*.)

Job Cost Wizard starts by asking what estimate you want to open. The blinking cursor will be in the File Name box. To open the estimate of your choice:

▌ Double click on the estimate name, -or-

▌ Click on the estimate name and click on **Open**.

For practice, try opening Stillwel.est in the National folder.

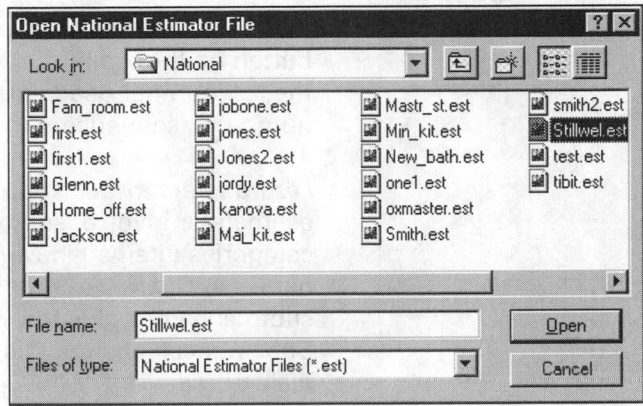

Open a National Estimator file.

The Company Information dialog box will open the first time you use *Job Cost Wizard*. Type your company name and address. This will appear at the top of every estimate and invoice.

Zoom, Scroll and Turn Pages

Zoom window

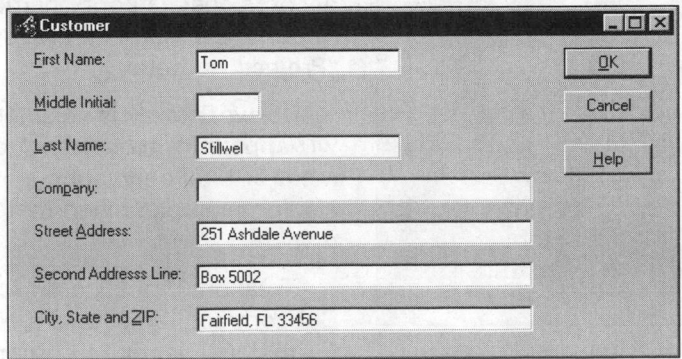

Previous Page

Next Page

If the estimate doesn't fit your screen, set the percentage of zoom. For 640x480 resolution, type 81% in the zoom window and press Enter↵.

Click and drag the vertical slide bar at the right of your screen to scroll down the page. Turn pages by clicking on the Previous Page or Next Page buttons.

Enter Job Information

Job Cost Wizard needs some information about the job to create a nice-looking bid or invoice. For practice, enter job information for the Stillwel estimate:

1. **Customer name and address.** Click on the 🔲 (Customer Info) button on the toolbar to enter information about the customer. Only the customer first name and last name are required. All other information is optional. When done, click on **OK**.

Fill in Customer Information.

Enter a Job Name.

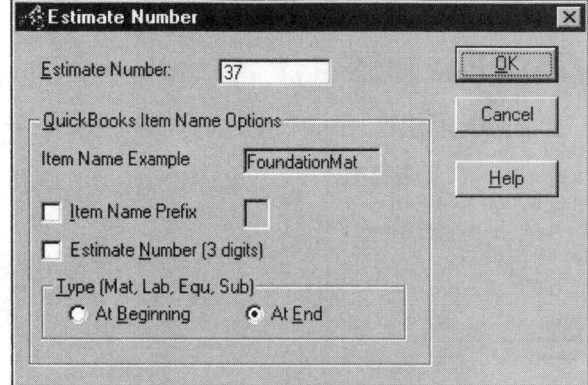

Check the Estimate Number.

2. Job Name. Click on the 🪖 (Job Name) button on the toolbar. Then type the name of the job, such as "Room Addition." When done, click on OK.

For transfers to QuickBooks You can change the customer or job name after the file has been imported into *QuickBooks*. In *QuickBooks*, click on **Lists**. Click on **Customers:Jobs**. Right-click on customer name or job name. Click on **Edit**. Then click on the tab of your choice.

3. Estimate Number. Click on the 📄 (Estimate Number) button on the toolbar. *Job Cost Wizard* keeps track of the last number used and recommends using the next number in sequence.

For transfers to QuickBooks When *QuickBooks* imports an estimate or invoice, subtotals in your estimate become cost categories ("items") in *QuickBooks*. By default, cost category names in *QuickBooks* are the first 10 characters of estimate subtotal names plus the work type, either Mat, Lab, Equ, or Sub. You can change this default in the Estimate Number dialog box.

Job Cost Wizard Prints Invoices Your Way

Your estimates should cover every cost in a job. But your bids and invoices don't have to show all the details and reveal your markup. So *Job Cost Wizard* gives you choices about showing or hiding the details and markup.

Amount of Detail

To set the amount of detail, click on the 🔍 (Detail) button.

If *Show All* is selected, every item in your bid or invoice will show a cost and each will become a cost category on the QuickBooks Items list. *Subtotals Only* is the default and will usually be a better choice.

If *Subtotals Only* is selected, subtotals will be the only costs in your bid or invoice. Each subtotal in your estimate becomes a cost category on the QuickBooks Items list. That's usually the best choice. If Subtotals Only is selected for all four cost categories, click on Omit Work Descriptions to show subtotal categories but hide all work descriptions.

The names you give to subtotals in *National Estimator* become cost category names in *QuickBooks*. Cost lines in *National Estimator* not followed by a subtotal become the "Project" subtotal.

If *Omit Entirely* is selected, neither costs nor descriptions will appear for that type of cost — either material, labor, equipment or total only (subcontract). Use Omit Entirely for materials, for example, when materials are being furnished by the owner.

"Total only" costs are assumed to be subcontract items. Subcontract items have a cost in the total column but no cost for material, labor, or equipment.

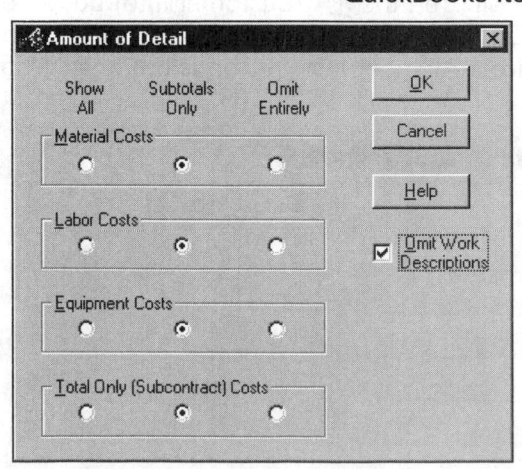

How much detail do you want to show?

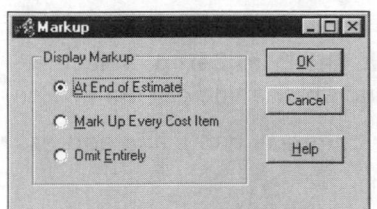

Where do you show markup?

Markup

To control how markup appears, click on the 💲 (Markup) button.

At End of Estimate puts overhead and profit at the end of the estimate, as in *National Estimator*.

Mark Up Every Cost Item distributes overhead and profit proportionately throughout the estimate. There's no mention of overhead, profit or markup anywhere in the estimate or invoice.

Omit Entirely omits overhead and profit from the *QuickBooks* estimate or invoice. Use this option if you prefer to add markup to each cost item in *QuickBooks*.

QuickBooks Account Names

Estimates and invoices imported into *QuickBooks* include expense and income account names. If the imported accounts do not exist already in your *QuickBooks* company, *QuickBooks* will create new accounts. You can control the names of these accounts by making changes in the QuickBooks Account Names box.

▌ Click on the 📄 (Account Names) button on the toolbar, -or-

▌ Click on **Options** on the menu bar. Then click on **QuickBooks Account Names**.

Enter the names you prefer for construction income and construction expense accounts. Change "ST Sales Tax" to "FL Sales Tax", for example, if the job is taxable under Florida law. *QuickBooks* will keep track of tax due in each state where you do business.

Check the QuickBooks account names.

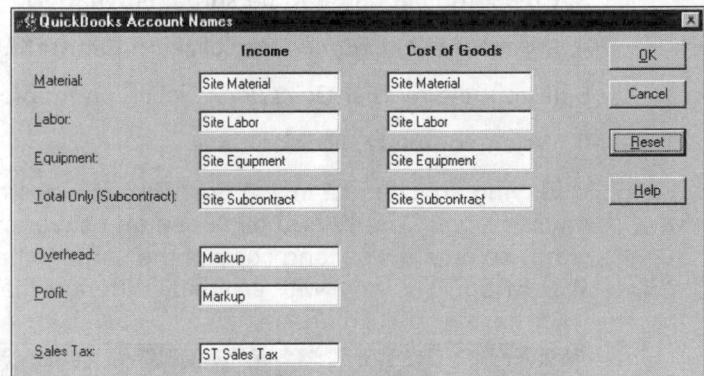

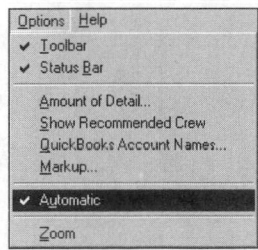

Set up for automatic operation.

Automatic Mode

Job Cost Wizard always requires customer information, a job name and a job number before saving an estimate. In automatic mode, *Job Cost Wizard* opens the Customer Info, Job Name and Estimate Number dialog boxes automatically after opening any estimate.

Job Cost Wizard runs in automatic mode when there is a check mark beside Automatic on the Options menu. To change to automatic mode, click on **Options** on the menu bar. Then click on **Automatic**.

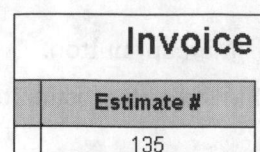

"Invoice" is the document title

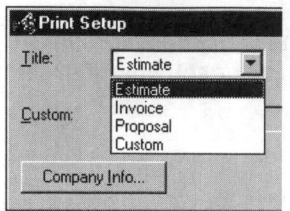

Change the title of a document

Printing a Bid or Invoice

To print one copy of the entire document, click on the 🖨 button on the toolbar. Then press `Enter ↵` or click on **OK**.

To change the title of the document, click on **File** and **Print Setup**. Select **Estimate, Invoice, Proposal,** or click on **Custom** and enter a title of your choice.

Click on **Company Info** to enter or change your company name and address to be printed on documents.

Exporting to QuickBooks

QuickBooks can not use estimates created by *National Estimator* until *Job Cost Wizard* converts them into the *iif* (Intuit Interface Format) type. This conversion is done automatically in *Job Cost Wizard* when you open and then save any estimate.

You can save to *QuickBooks* any estimate that is open and has been given a customer name and job name.

1. Open the *Save QuickBooks File* dialog box:

   ▌ Click on the 📲 (Save) button on the toolbar, -or-

   ▌ Click on **File** on the menu bar. Then click on **Save**, -or-

   ▌ Hold the `Ctrl` key down and tap the letter S.

2. Change the drive or the folder if the *QuickBooks* folder is not listed at the right of *Save in*.

3. Check the file name to be sure it is what you want.

4. If you use *QuickBooks Pro*, click on **Estimate**.

5. If you use regular *QuickBooks*, click on **Invoice**.

6. When complete, click on **Save**.

Saving an estimate in *Job Cost Wizard* does not affect the original estimate in any way. *Job Cost Wizard* can open and save an estimate as many times as you want. To create a second copy of the same estimate with different *Job Cost Wizard* options, save with a slightly different file name.

Saving as an estimate to QuickBooks Pro.

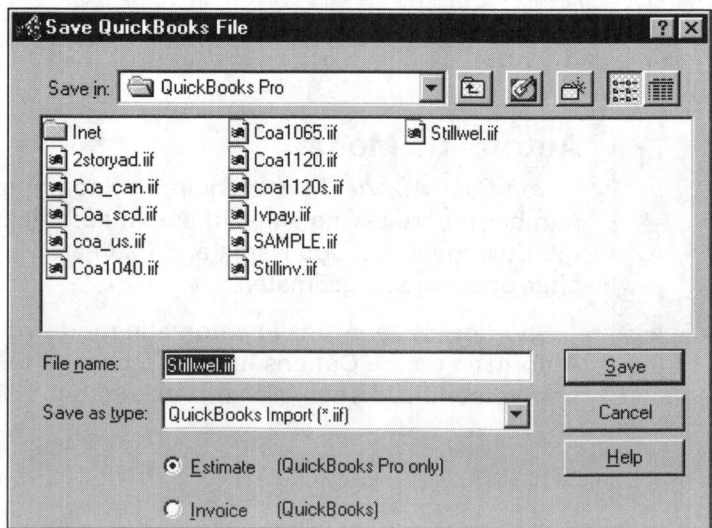

Job Cost Tracking with QuickBooks

If you're using *QuickBooks* for the first time, begin by setting up a new *QuickBooks* company. Click on **File** and **EasyStep Interview**. Allow a day or two to set up income and expense accounts, enter existing account balances, format reports to your liking and set up payroll processing. *QuickBooks Pro* "Qcards" will coach you through the setup process. A book titled *Contractor's Guide to QuickBooks Pro* (available at http://costbook.com or 800-829-8123) may also be helpful.

To track job costs, you'll want to have *QuickBooks* write company checks and prepare the company payroll. To turn on *QuickBooks* payroll, click on **File**, **Preferences**, **Payroll & Employees**, and **Full payroll features**. You'll need the current tax table for your state. Click on **Help** and **About Tax Table**. Your first tax table is free on the Internet. Thereafter the cost is about $60 per year.

When company files have been created in *QuickBooks*, click on **File** and **Back Up** to preserve the work you've done. *Backups are important.* Intuit, the developer of *QuickBooks*, recommends making a backup before opening import files.

Opening Import Files in QuickBooks

Importing into QuickBooks.

Once an estimate or invoice has been saved with *Job Cost Wizard*, start *QuickBooks*:

1. Click on **File**.
2. Click on **Import**.
3. Double click on the name of the estimate or invoice you want.
4. Click on **OK** when the import is complete.
5. Click on **Activities**.
6. Click on **Create Estimates** if you saved the file as an estimate.
7. Click on **Create Invoices** if you saved the file as an invoice.
8. Click on **Prev** to see the file just imported.

Entering Markup in QuickBooks Estimates

Entering zero markup in an estimate.

An imported estimate is not complete until some figure, either a percentage or number, is typed in the Markup column of each cost line. So long as the Markup column is blank, *QuickBooks* will consider the estimated cost for that line to be zero (even when numbers appear in the Total column). If the Markup column for an entire estimate is left blank, *QuickBooks* reports will show the estimated cost for that job to be zero.

The fastest way to fill in the Markup column for a *QuickBooks* estimate is to click in the Markup column on a cost line. Type a zero. Then click on the next cost line in the Markup column and type another zero. Continue clicking and entering zeros until zero markup has been entered on each cost line.

Turn an Estimate into an Invoice in QuickBooks

First, decide if you want an invoice for the whole job or for just part of the job (progress billing). If you prefer progress billing:

1. Click on **File**.
2. Click on **Preferences**.
3. Click on **Jobs & Estimates**.

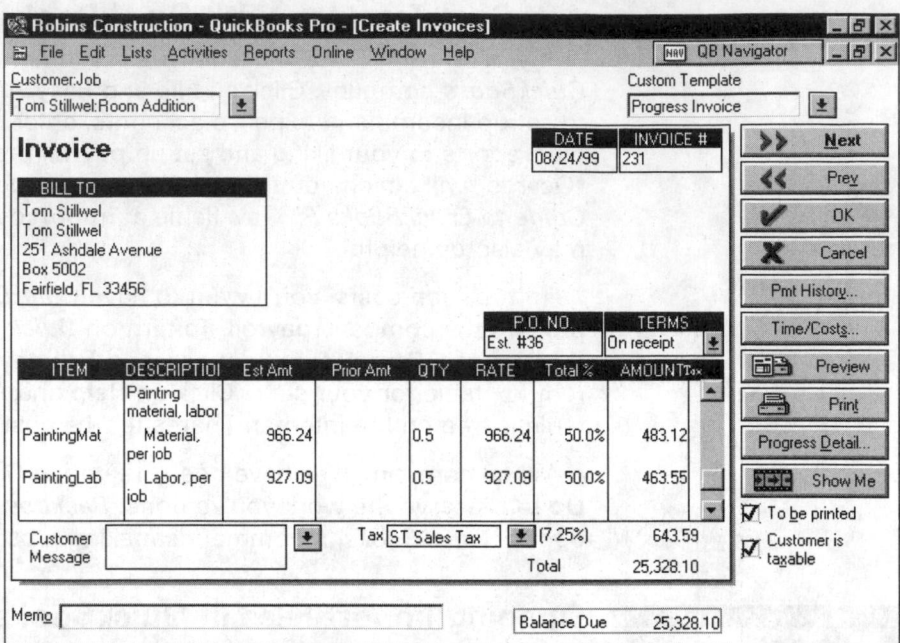

An invoice for the first half of the painting.

4. Click on the **Company Preferences** tab.

5. Click on **Yes** under **Do You Do Progress Invoicing?**

6. Click on **OK**.

To create the invoice:

1. Click on the **Create Invoice** button at the bottom of the *QuickBooks* estimate screen.

2. Click on **Yes** to record changes to the estimate.

3. If you selected progress billing, enter a percentage or select items to be invoiced.

4. Click on **OK** and *QuickBooks* creates the invoice.

5. Make changes to the invoice if you want.

6. Click to print one copy for your file and another for your customer.

7. Click on **OK** when done. (Processing the file may take a little time.) *QuickBooks* reports will now include totals from the job just invoiced.

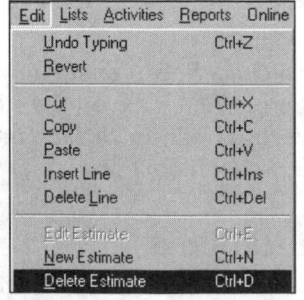

It's easy to delete an estimate.

Don't Worry About Making a Mistake

QuickBooks is very forgiving. Practice all you want. Experiment any way you want. Then delete any estimate or invoice to remove every trace of it from *QuickBooks*. With the offending estimate or invoice displayed:

1. Click on **Edit**.

2. Click on **Delete Estimate** (or Invoice).

3. Click on **OK**. The estimate (or invoice) is deleted.

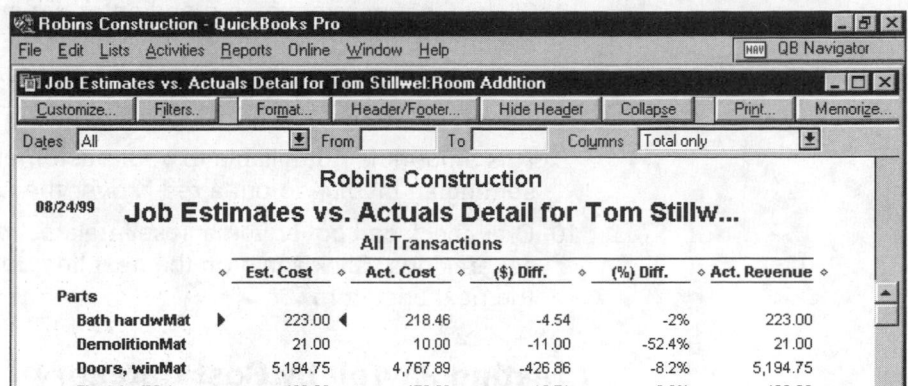

Estimates vs. actuals detail.

Your Jobs in QuickBooks

▌ Click on **Reports, Profit & Loss, Standard** to see job income and expense.

▌ Click on **Reports, Balance Sheet, Standard** to see the new receivables total.

▌ Click on **Reports, Project Reports, Job Estimates vs. Actuals Detail**, select the customer and job to see a detailed cost comparison for the job. Until you start paying bills, the Actual Cost column will be all zeros.

Select a cost category from the Item list.

Paying Bills by Cost Category

When you pay vendors and subcontractors, the amount paid is charged to the job and deducted from your bank balance.

1. Click on **Activities.**

2. Click on **Write Checks.**

3. Fill in the **Pay to the Order of** line.

4. Click on the **Items** tab.

5. Click on the down triangle in the Item column of the check stub to open the Item list.

6. Select a *Mat, Sub or Equ* cost category from the Item list. These are the material, subcontract and equipment subtotal costs from your estimate.

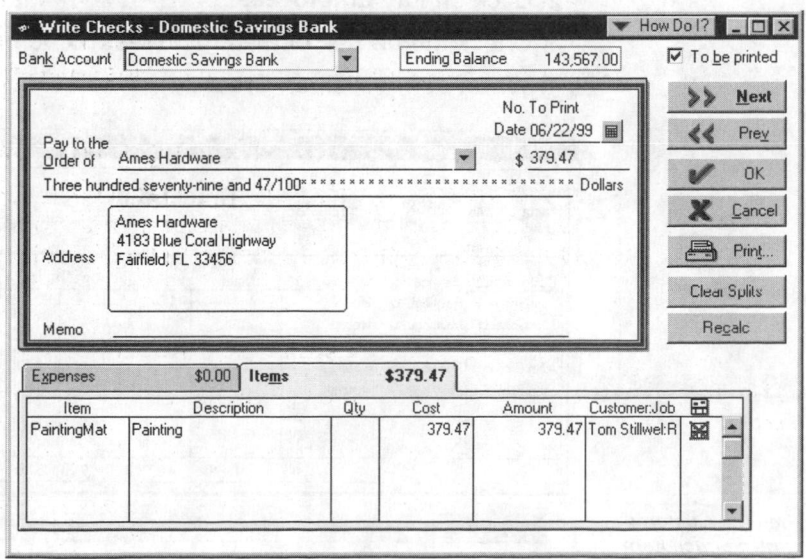

Paying a bill for the paint on the Stilwell job.

7. Click in the amount column and enter the amount paid for the item described.

8. Click on the down triangle under **Customer:Job** and select the correct customer and job.

9. This amount is not billable to your customer. So click on the icon representing an invoice to put a red X over the icon.

10. One check can cover items in several cost categories and even costs on several jobs. Click again on the next line down in the Item column and find the next cost item.

Creating Payroll by Cost Category

When you write payroll checks, the amount paid is charged to the job and deducted from your bank balance.

1. Click on **Activities**.

2. Click on **Time Tracking**.

3. Click on **Use Weekly Time Sheets**.

4. Click on the down triangle opposite **Name** and select the employee to be paid.

5. On the **Timesheet,** click on the down arrow under **Customer:Job** and select the first job where the employee worked.

6. Under **Service Item**, select a *Lab* cost category from the list. These are labor subtotal costs from your estimate.

7. Under **Payroll Item**, select the type of pay, such as hourly.

8. Enter the number of hours worked for that cost category.

9. Any timesheet can cover work done on several jobs and many service items.

10. When finished recording time for an employee, click on **New Sheet**.

11. When finished recording time for all employees, click on **OK**.

To actually produce paychecks:

1. Click on **Activities** and then **Payroll**.

2. Click on **Pay Employees**.

3. Check the names of the employees to be paid.

4. When done, click on **Create**.

Bath hardwLab	Bath hardware
DemolitionLab	Demolition
Doors, winLab	Doors, windows and...
ElectricalLab	Electrical
FireplaceLab	Fireplace
FlooringLab	Flooring
FoundationLab	Foundation
Heating anLab	Heating and coolin...
PaintingLab	Painting
Plumbing fLab	Plumbing fixtures
ProjectLab	Project
RoofingLab	Roofing
Rough framLab	Rough framing, ins...
Rough plumLab	Rough plumbing

Under Service Item, select a labor cost category from the list.

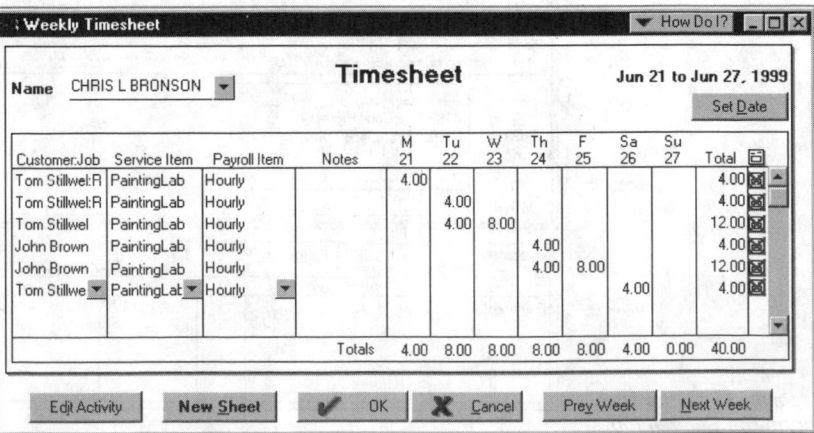

Fill out the timesheet by Customer and Service Item.

Index

559

Practical References for Builders

Rough Framing Carpentry

If you'd like to make good money working outdoors as a framer, this is the book for you. Here you'll find short-cuts to laying out studs; speed cutting blocks, trimmers and plates by eye; quickly building and blocking rake walls; installing ceiling backing, ceiling joists, and truss joists; cutting and assembling hip trusses and California fills; arches and drop ceilings — all with production line procedures that save you time and help you make more money. Over 100 on-the-job photos of how to do it right and what can go wrong. **304 pages, 8¹/₂ x 11, $26.50**

Builder's Guide to Room Additions

How to tackle problems that are unique to additions, such as requirements for basement conversions, reinforcing ceiling joists for second-story conversions, handling problems in attic conversions, what's required for footings, foundations, and slabs, how to design the best bathroom for the space, and much more. Besides actual construction, you'll even find help in designing, planning, and estimating your room addition jobs. **352 pages, 8¹/₂ x 11, $27.25**

Basic Engineering for Builders

If you've ever been stumped by an engineering problem on the job, yet wanted to avoid the expense of hiring a qualified engineer, you should have this book. Here you'll find engineering principles explained in non-technical language and practical methods for applying them on the job. With the help of this book you'll be able to understand engineering functions in the plans and how to meet the requirements, how to get permits issued without the help of an engineer, and anticipate requirements for concrete, steel, wood and masonry. See why you sometimes have to hire an engineer and what you can undertake yourself: surveying, concrete, lumber loads and stresses, steel, masonry, plumbing, and HVAC systems. This book is designed to help the builder save money by understanding engineering principles that you can incorporate into the jobs you bid. **400 pages, 8¹/₂ x 11, $34.00**

National Painting Cost Estimator

A complete guide to estimating painting costs for just about any type of residential, commercial, or industrial painting, whether by brush, spray, or roller. Shows typical costs and bid prices for fast, medium, and slow work, including material costs per gallon; square feet covered per gallon; square feet covered per manhour; labor, material, overhead, and taxes per 100 square feet; and how much to add for profit. Includes a CD-ROM with an electronic version of the book with *National Estimator*, a stand-alone Windows™ estimating program, plus an interactive multimedia video that shows how to use the disk to compile construction cost estimates. **456 pages, 8¹/₂ x 11, $48.00. Revised annually**

National Building Cost Manual

Square foot costs for residential, commercial, industrial, and farm buildings. Quickly work up a reliable budget estimate based on actual materials and design features, area, shape, wall height, number of floors, and support requirements. Includes all the important variables that can make any building unique from a cost standpoint. **240 pages, 8¹/₂ x 11, $23.00. Revised annually**

Excavation & Grading Handbook Revised

Explains how to handle all excavation, grading, compaction, paving and pipeline work: setting cut and fill stakes (with bubble and laser levels), working in rock, unsuitable material or mud, passing compaction tests, trenching around utility lines, setting grade pins and string line, removing or laying asphaltic concrete, widening roads, cutting channels, installing water, sewer, and drainage pipe. This is the completely revised edition of the popular guide used by over 25,000 excavation contractors. **384 pages, 5¹/₂ x 8¹/₂, $22.75**

Roofing Construction & Estimating

Installation, repair and estimating for nearly every type of roof covering available today in residential and commercial structures: asphalt shingles, roll roofing, wood shingles and shakes, clay tile, slate, metal, built-up, and elastomeric. Covers sheathing and underlayment techniques, as well as secrets for installing leakproof valleys. Many estimating tips help you minimize waste, as well as insure a profit on every job. Troubleshooting techniques help you identify the true source of most leaks. Over 300 large, clear illustrations help you find the answer to just about all your roofing questions. **432 pages, 8¹/₂ x 11, $35.00**

National Renovation & Insurance Repair Estimator

Current prices in dollars and cents for hard-to-find items needed on most insurance, repair, remodeling, and renovation jobs. All price items include labor, material, and equipment breakouts, plus special charts that tell you exactly how these costs are calculated. Includes a CD-ROM with an electronic version of the book with *National Estimator*, a stand-alone Windows™ estimating program, plus an interactive multimedia video that shows how to use the disk to compile construction cost estimates. **568 pages, 8¹/₂ x 11, $49.50. Revised annually**

Finish Carpenter's Manual

Everything you need to know to be a finish carpenter: assessing a job before you begin, and tricks of the trade from a master finish carpenter. Easy-to-follow instructions for installing doors and windows, ceiling treatments (including fancy beams, corbels, cornices and moldings), wall treatments (including wainscoting and sheet paneling), and the finishing touches of chair, picture, and plate rails. Specialized interior work includes cabinetry and built-ins, stair finish work, and closets. Also covers exterior trims and porches. Includes manhour tables for finish work, and hundreds of illustrations and photos. **208 pages, 8¹/₂ x 11, $22.50**

Drafting House Plans

Here you'll find step-by-step instructions for drawing a complete set of home plans for a one-story house, an addition to an existing house, or a remodeling project. This book shows how to visualize spatial relationships, use architectural scales and symbols, sketch preliminary drawings, develop detailed floor plans and exterior elevations, and prepare a final plot plan. It even includes code-approved joist and rafter spans and how to make sure that drawings meet code requirements. **192 pages, 8¹/₂ x 11, $27.50**

Contractor's Guide to the Building Code Revised

This new edition was written in collaboration with the International Conference of Building Officials, writers of the code. It explains in plain English exactly what the latest edition of the *Uniform Building Code* requires. Based on the 1997 code, it explains the changes and what they mean for the builder. Also covers the *Uniform Mechanical Code* and the *Uniform Plumbing Code*. Shows how to design and construct residential and light commercial buildings that'll pass inspection the first time. Suggests how to work with an inspector to minimize construction costs, what common building shortcuts are likely to be cited, and where exceptions may be granted. **320 pages, 8¹/₂ x 11, $39.00**

Basic Lumber Engineering for Builders

Beam and lumber requirements for many jobs aren't always clear, especially with changing building codes and lumber products. Most of the time you rely on your own "rules of thumb" when figuring spans or lumber engineering. This book can help you fill the gap between what you can find in the building code span tables and what you need to pay a certified engineer to do. With its large, clear illustrations and examples, this book shows you how to figure stresses for pre-engineered wood or wood structural members, how to calculate loads, and how to design your own girders, joists and beams. Included FREE with the book — an easy-to-use version of NorthBridge Software's *Wood Beam Sizing* program. **272 pages, 8¹/₂ x 11, $38.00**

Construction Forms & Contracts

125 forms you can copy and use — or load into your computer (from the FREE disk enclosed). Then you can customize the forms to fit your company, fill them out, and print. Loads into *Word* for *Windows*™, *Lotus 1-2-3*, *WordPerfect*, *Works*, or *Excel* programs. You'll find forms covering accounting, estimating, fieldwork, contracts, and general office. Each form comes with complete instructions on when to use it and how to fill it out. These forms were designed, tested and used by contractors, and will help keep your business organized, profitable and out of legal, accounting and collection troubles. Includes a CD-ROM for *Windows*™ and Macintosh. **400 pages, 8¹/₂ x 11, $39.75**

Building Contractor's Exam Preparation Guide

Passing today's contractor's exams can be a major task. This book shows you how to study, how questions are likely to be worded, and the kinds of choices usually given for answers. Includes sample questions from actual state, county, and city examinations, plus a sample exam to practice on. This book isn't a substitute for the study material that your testing board recommends, but it will help prepare you for the types of questions — and their correct answers — that are likely to appear on the actual exam. Knowing how to answer these questions, as well as what to expect from the exam, can greatly increase your chances of passing.
320 pages, 8¹/₂ x 11, $35.00

CD Estimator

If your computer has *Windows*™ and a CD-ROM drive, CD Estimator puts at your fingertips 85,000 construction costs for new construction, remodeling, renovation & insurance repair, electrical, plumbing, HVAC and painting. You'll also have the *National Estimator* program — a stand-alone estimating program for *Windows*™ that *Remodeling* magazine called a "computer wiz." Quarterly cost updates are available at no charge on the Internet. To help you create professional-looking estimates, the disk includes over 40 construction estimating and bidding forms in a format that's perfect for nearly any word processing or spreadsheet program for *Windows*™. And to top it off, a 70-minute interactive video teaches you how to use this CD-ROM to estimate construction costs. **CD Estimator is $68.50**

CD Estimator — Heavy

CD Estimator — Heavy has a complete 780-page heavy construction cost estimating volume for each of the 50 states. Select the cost database for the state where the work will be done. Includes thousands of cost estimates you won't find anywhere else, and in-depth coverage of demolition, hazardous materials remediation, tunneling, site utilities, precast concrete, structural framing, heavy timber construction, membrane waterproofing, industrial windows and doors, specialty finishes, built-in commercial and industrial equipment, and HVAC and electrical systems for commercial and industrial buildings. **CD Estimator — Heavy is $69.00**

Estimating Home Building Costs

Estimate every phase of residential construction from site costs to the profit margin you include in your bid. Shows how to keep track of manhours and make accurate labor cost estimates for footings, foundations, framing and sheathing finishes, electrical, plumbing, and more. Provides and explains sample cost estimate worksheets with complete instructions for each job phase.
320 pages, 5¹/₂ x 8¹/₂, $17.00

National Plumbing & HVAC Estimator

Manhours, labor and material costs for all common plumbing and HVAC work in residential, commercial, and industrial buildings. You can quickly work up a reliable estimate based on the pipe, fittings and equipment required. Every plumbing and HVAC estimator can use the cost estimates in this practical manual. Sample estimating and bidding forms and contracts also included. Explains how to handle change orders, letters of intent, and warranties. Describes the right way to process submittals, deal with suppliers and subcontract specialty work. Includes a CD-ROM with an electronic version of the book with *National Estimator*, a stand-alone *Windows*™ estimating program, plus an interactive multimedia video that shows you how to use the disk to compile construction cost estimates.
392 pages, 8¹/₂ x 11, $48.25. Revised annually

Residential Steel Framing Guide

Steel is stronger and lighter than wood — straight walls are guaranteed — steel framing will not wrap, shrink, split, swell, bow, or rot. Here you'll find full page schematics and details that show how steel is connected in just about all residential framing work. You won't find lengthy explanations here on how to run your business, or even how to do the work. What you will find are over 150 easy-to-ready full-page details on how to construct steel-framed floors, roofs, interior and exterior walls, bridging, blocking, and reinforcing for all residential construction. Also includes recommended fasteners and their applications, and fastening schedules for attaching every type of steel framing member to steel as well as wood.
170 pages, 8¹/₂ x 11, $38.80

Contractor's Survival Manual

How to survive hard times and succeed during the up cycles. Shows what to do when the bills can't be paid, finding money and buying time, transferring debt, and all the alternatives to bankruptcy. Explains how to build profits, avoid problems in zoning and permits, taxes, time-keeping, and payroll. Unconventional advice on how to invest in inflation, get high appraisals, trade and postpone income, and stay hip-deep in profitable work. **160 pages, 8¹/₂ x 11, $22.25**

The Contractor's Legal Kit

Stop "eating" the costs of bad designs, hidden conditions, and job surprises. Set ground rules that assign those costs to the rightful party ahead of time. And it's all in plain English, not "legalese." For less than the cost of an hour with a lawyer you'll learn the exclusions to put in your agreements, why your insurance company may pay for your legal defense, how to avoid liability for injuries to your sub and his employees or damages they cause, how to collect on lawsuits you win, and much more. It also includes a FREE computer disk with contracts and forms you can customize for your own use. **352 pages, 8¹/₂ x 11, $59.95**

Estimating Excavation

How to calculate the amount of dirt you'll have to move and the cost of owning and operating the machines you'll do it with. Detailed, step-by-step instructions on how to assign bid prices to each part of the job, including labor and equipment costs. Also, the best ways to set up an organized and logical estimating system, take off from contour maps, estimate quantities in irregular areas, and figure your overhead.
448 pages, 8¹/₂ x 11, $39.50

Residential Electrical Estimating

A fast, accurate pricing system proven on over 1000 residential jobs. Using the manhours provided, combined with material prices from your wholesaler, you quickly work up estimates based on degree of difficulty. These manhours come from a working electrical contractor's records — not some pricing agency. You'll find prices for every type of electrical job you're likely to estimate — from service entrances to ceiling fans.
320 pages, 8¹/₂ x 11, $29.00

Profits in Building Spec Homes

If you've ever wanted to make big profits in building spec homes yet were held back by the risks involved, you should have this book. Here you'll learn how to do a market study and feasibility analysis to make sure your finished home will sell quickly, and for a good profit. You'll find tips that can save you thousands in negotiating for land, learn how to impress bankers and get the financing package you want, how to nail down cost estimating, schedule realistically, work effectively yet harmoniously with subcontractors so they'll come back for your next home, and finally, what to look for in the agent you choose to sell your finished home. Includes forms, checklists, worksheets, and step-by-step instructions.
208 pages, 8¹/₂ x 11, $27.25

Concrete Construction & Estimating

Explains how to estimate the quantity of labor and materials needed, plan the job, erect fiberglass, steel, or prefabricated forms, install shores and scaffolding, handle the concrete into place, set joints, finish and cure the concrete. Full of practical reference data, cost estimates, and examples.
571 pages, 5¹/₂ x 8¹/₂, $25.00

Craftsman's Illustrated Dictionary of Construction Terms

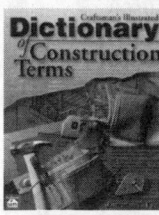

Almost everything you could possibly want to know about any word or technique in construction. Hundreds of up-to-date construction terms, materials, drawings and pictures with detailed, illustrated articles describing equipment and methods. Terms and techniques are explained or illustrated in vivid detail. Use this valuable reference to check spelling, find clear, concise definitions of construction terms used on plans and construction documents, or learn about little-known tools, equipment, tests and methods used in the building industry. It's all here. **416 pages, 8¹/₂ x 11, $36.00**

National Repair & Remodeling Estimator

The complete pricing guide for dwelling reconstruction costs. Reliable, specific data you can apply on every repair and remodeling job. Up-to-date material costs and labor figures based on thousands of jobs across the country. Provides recommended crew sizes; average production rates; exact material, equipment, and labor costs; a total unit cost and a total price including overhead and profit. Separate listings for high- and low-volume builders, so prices shown are specific for any size business. Estimating tips specific to repair and remodeling work to make your bids complete, realistic, and profitable. Includes a CD-ROM with an electronic version of the book with *National Estimator*, a stand-alone *Windows*™ estimating program, plus an interactive multi-media video that shows how to use the disk to compile construction cost estimates. **256 pages, 8¹/₂ x 11, $48.50. Revised annually**

Construction Estimating Reference Data

Provides the 300 most useful manhour tables for practically every item of construction. Labor requirements are listed for sitework, concrete work, masonry, steel, carpentry, thermal and moisture protection, door and windows, finishes, mechanical and electrical. Each section details the work being estimated and gives appropriate crew size and equipment needed. Includes a CD-ROM with an electronic version of the book with *National Estimator*, a stand alone *Windows*™ estimating program, plus an interactive multimedia video that shows how to use the disk to compile construction cost estimates. **432 pages, 11 x 8¹/₂, $39.50**

National Electrical Estimator

This year's prices for installation of all common electrical work: conduit, wire, boxes, fixtures, switches, outlets, loadcenters, panelboards, raceway, duct, signal systems, and more. Provides material costs, manhours per unit, and total installed cost. Explains what you should know to estimate each part of an electrical system. Includes a CD-ROM with an electronic version of the book with *National Estimator*, a stand-alone *Windows*™ estimating program, plus an interactive multimedia video that shows how to use the disk to compile construction cost estimates. **512 pages, 8¹/₂ x 11, $47.75. Revised annually**

Blueprint Reading for the Building Trades

How to read and understand construction documents, blueprints, and schedules. Includes layouts of structural, mechanical, HVAC and electrical drawings. Shows how to interpret sectional views, follow diagrams and schematics, and covers common problems with construction specifications. **192 pages, 5¹/₂ x 8¹/₂, $14.75**

Roof Framing

Shows how to frame any type of roof in common use today, even if you've never framed a roof before. Includes using a pocket calculator to figure any common, hip, valley, or jack rafter length in seconds. Over 400 illustrations cover every measurement and every cut on each type of roof: gable, hip, Dutch, Tudor, gambrel, shed, gazebo, and more. **480 pages, 5¹/₂ x 8¹/₂, $22.00**

Plumber's Handbook Revised

This new edition shows what will and won't pass inspection in drainage, vent, and waste piping, septic tanks, water supply, fire protection, and gas piping systems. All tables, standards, and specifications completely up-to-date with recent plumbing code changes. Covers common layouts for residential work, how to size piping, selecting and hanging fixtures, practical recommendations, and trade tips. The approved reference for the plumbing contractor's exam in many states. **240 pages, 8¹/₂ x 11, $32.00**

Home Inspection Handbook

Every area you need to check in a home inspection — especially in older homes. Twenty complete inspection checklists: building site, foundation and basement, structural, bathrooms, chimneys and flues, ceilings, interior & exterior finishes, electrical, plumbing, HVAC, insects, vermin and decay, and more. Also includes information on starting and running your own home inspection business. **324 pages, 5¹/₂ x 8¹/₂, $24.95**

Estimating Framing Quantities

Gives you hundreds of time-saving estimating tips. Shows how to make thorough step-by-step estimates of all rough carpentry in residential and light commercial construction: ceilings, walls, floors, and roofs. Lots of illustrations showing lumber requirements, nail quantities, and practical estimating procedures. **285 pages, 5¹/₂ x 8¹/₂, $34.95**

Carpentry Estimating

Simple, clear instructions to help you take off quantities and figure costs for all rough and finish carpentry. Provides checklists, manhour tables, and worksheets with calculation factors built in. Take the guesswork out of estimating carpentry so you get accurate labor and material costs every time. Includes FREE *National Estimator* software, with all the manhours in the book, plus the carpentry estimating worksheets in formats that work with *Word*, *Excel*, *WordPerfect*, and *MS Works for Windows*. **336 pages, 8¹/₂ x 11, $35.50**

Builder's Guide to Accounting Revised

Step-by-step, easy-to-follow guidelines for setting up and maintaining records for your building business. This practical, newly-revised guide to all accounting methods shows how to meet state and federal accounting requirements, explains the new depreciation rules, and describes how the Tax Reform Act can affect the way you keep records. Full of charts, diagrams, simple directions and examples, to help you keep track of where your money is going. Recommended reading for many state contractor's exams. **320 pages, 8¹/₂ x 11, $26.50**

Estimating Tables for Home Building

Produce accurate estimates for nearly any residence in just minutes. This handy manual has tables you need to find the quantity of materials and labor for most residential construction. Includes overhead and profit, how to develop unit costs for labor and materials, and how to be sure you've considered every cost in the job. **336 pages, 8¹/₂ x 11, $21.50**

Cost Records for Construction Estimating

How to organize and use cost information from jobs just completed to make more accurate estimates in the future. Explains how to keep the records you need to track costs for sitework, footings, foundations, framing, interior finish, siding and trim, masonry, and subcontract expense. Provides sample forms. **208 pages, 8¹/₂ x 11, $15.75**

Troubleshooting Guide to Residential Construction

How to solve practically every construction problem - before it happens to you! With this book you'll learn from the mistakes other builders made as they faced 63 typical residential construction problems. Filled with clear photos and drawings that explain how to enhance your reputation as well as your bottom line by avoiding problems that plague most builders. Shows how to avoid, or fix, problems ranging from defective slabs, walls and ceilings, through roofing, plumbing & HVAC, to paint. **304 pages, 8¹/₂ x 11, $32.50**

Renovating & Restyling Older Homes

Any builder can turn a run-down old house into a showcase of perfection — if the customer has unlimited funds to spend. Unfortunately, most customers are on a tight budget. They usually want more improvements than they can afford — and they expect you to deliver. This book shows how to add economical improvements that can increase the property value by two, five or even ten times the cost of the remodel. Sound impossible? Here you'll find the secrets of a builder who has been putting these techniques to work on Victorian and Craftsman-style houses for twenty years. You'll see what to repair, what to replace and what to leave, so you can remodel or restyle older homes for the least amount of money and the greatest increase in value. **416 pages, 8¹/₂ x 11, $33.50**

Professional Kitchen Design

Remodeling kitchens requires a "special" touch — one that blends artistic flair with function to create a kitchen with charm and personality as well as one that is easy to work in. Here you'll find how to make the best use of the space available in any kitchen design job, as well as tips and lessons on how to design one-wall, two-wall, L-shaped, U-shaped, peninsula and island kitchens. Also includes what you need to know to run a profitable kitchen design business. **176 pages, 8¹/₂ x 11, $24.50**

Paint Contractor's Manual

How to start and run a profitable paint contracting company: getting set up and organized to handle volume work, avoiding mistakes, squeezing top production from your crews and the most value from your advertising dollar. Shows how to estimate all prep and painting. Loaded with manhour estimates, sample forms, contracts, charts, tables and examples you can use. **224 pages, 8¹/₂ x 11, $24.00**

Land Development

The industry's bible. Nine chapters cover everything you need to know about land development from initial market studies to site selection and analysis. New and innovative design ideas for streets, houses, and neighborhoods are included. Whether you're developing a whole neighborhood or just one site, you shouldn't be without this essential reference. **360 pages, 5¹/₂ x 8¹/₂, $40.70**

Profits in Buying & Renovating Homes

Step-by-step instructions for selecting, repairing, improving, and selling highly profitable "fixer-uppers." Shows which price ranges offer the highest profit-to-investment ratios, which neighborhoods offer the best return, practical directions for repairs, and tips on dealing with buyers, sellers, and real estate agents. Shows you how to determine your profit before you buy, what "bargains" to avoid, and how to make simple, profitable, inexpensive upgrades. **304 pages, 8¹/₂ x 11, $19.75**

BNI Public Works Costbook

Labor and material prices for most public works and infrastructure projects: roads and streets, utilities, street lighting, manholes, and much more. Includes manhour data and a 200-city geographic modifier chart. Includes FREE estimating software and data. **450 pages, 8¹/₂ x 11, $64.95**

Plumbing & HVAC Manhour Estimates

Hundreds of tested and proven manhours for installing just about any plumbing and HVAC component you're likely to use in residential, commercial, and industrial work. You'll find manhours for installing piping systems, specialties, fixtures and accessories, ducting systems, and HVAC equipment. If you estimate the price of plumbing, you shouldn't be without the reliable, proven manhours in this unique book. **224 pages, 5¹/₂ x 8¹/₂, $28.25**

Planning Drain, Waste & Vent Systems

How to design plumbing systems in residential, commercial, and industrial buildings. Covers designing systems that meet code requirements for homes, commercial buildings, private sewage disposal systems, and even mobile home parks. Includes relevant code sections and many illustrations to guide you through what the code requires in designing drainage, waste, and vent systems. **192 pages, 8¹/₂ x 11, $19.25**

Commercial Electrical Wiring

Make the transition from residential to commercial electrical work. Here are wiring methods, spec reading tips, load calculations and everything you need for making the transition to commercial work: commercial construction documents, load calculations, electric services, transformers, overcurrent protection, wiring methods, raceway, boxes and fittings, wiring devices, conductors, electric motors, relays and motor controllers, special occupancies, and safety requirements. This book is written to help any electrician break into the lucrative field of commercial electrical work. **320 pages, 8¹/₂ x 11, $27.50**

Simplified Guide To Construction Law

Easy-to-read, paragraphed-sized samples of how the courts have viewed common areas of disagreement - and litigation - in the building industry. This book will tell you what you need to know about contracts, torts, fraud, misrepresentation, warranty and strict liability, construction defects, indemnity, insurance, mechanics liens, bonds and bonding, statutes of limitation, arbitration, and more. These are simplified examples that illustrate who the law has sided with in each case. **298 pages, 5¹/₂ x 8¹/₂, $29.95**

How to Succeed With Your Own Construction Business

Everything you need to start your own construction business: setting up the paperwork, finding the work, advertising, using contracts, dealing with lenders, estimating, scheduling, finding and keeping good employees, keeping the books, and coping with success. If you're considering starting your own construction business, all the knowledge, tips, and blank forms you need are here. **336 pages, 8¹/₂ x 11, $24.25**

Handbook of Construction Contracting

Volume 1: Everything you need to know to start and run your construction business; the pros and cons of each type of contracting, the records you'll need to keep, and how to read and understand house plans and specs so you find any problems before the actual work begins. All aspects of construction are covered in detail, including all-weather wood foundations, practical math for the job site, and elementary surveying. **416 pages, 8¹/₂ x 11, $28.75**

Volume 2: Everything you need to know to keep your construction business profitable; different methods of estimating, keeping and controlling costs, estimating excavation, concrete, masonry, rough carpentry, roof covering, insulation, doors and windows, exterior finishes, specialty finishes, scheduling work flow, managing workers, advertising and sales, spec building and land development, and selecting the best legal structure for your business. **320 pages, 8¹/₂ x 11, $30.75**

Basic Plumbing with Illustrations, Revised

This completely-revised edition brings this comprehensive manual fully up-to-date with all the latest plumbing codes. It is the journeyman's and apprentice's guide to installing plumbing, piping, and fixtures in residential and light commercial buildings: how to select the right materials, lay out the job and do professional-quality plumbing work, use essential tools and materials, make repairs, maintain plumbing systems, install fixtures, and add to existing systems. Includes extensive study questions at the end of each chapter, and a section with all the correct answers. **384 pages, 8¹/₂ x 11, $33.00**

Wood-Frame House Construction

Step-by-step construction details, from the layout of the outer walls, excavation and formwork, to finish carpentry and painting. Contains all new, clear illustrations and explanations updated for construction in the '90s. Everything you need to know about framing, roofing, siding, interior finishings, floor covering and stairs — your complete book of wood-frame homebuilding. **320 pages, 8¹/₂ x 11, $25.50. Revised edition**

Rafter Length Manual

Complete rafter length tables and the "how to" of roof framing. Shows how to use the tables to find the actual length of common, hip, valley, and jack rafters. Explains how to measure, mark, cut and erect the rafters; find the drop of the hip; shorten jack rafters; mark the ridge and much more. Loaded with explanations and illustrations. **369 pages, 5¹/₂ x 8¹/₂, $15.75**

Fences & Retaining Walls

Everything you need to know to run a profitable business in fence and retaining wall contracting. Takes you through layout and design, construction techniques for wood, masonry, and chain link fences, gates and entries, including finishing and electrical details. How to build retaining and rock walls. How to get your business off to the right start, keep the books, and estimate accurately. The book even includes a chapter on contractor's math. **400 pages, 8¹/₂ x 11, $23.25**

Residential Electrical Design Revised

If you've ever had to draw up an electrical plan for an addition, or add corrections to an existing plan, you know how complicated it can get. And how many electrical plans - no matter how well designed - fit the reality of what the homeowner wants? Here you'll find everything you need to know about blueprints, what the NEC requires, how to size electric service, calculate and size loads and conductors, install ground-fault circuit interrupters, ground service entrances, and recommended wiring methods, It covers branch circuite layout, how to analyze existing lighting layouts and install outdoor lighting, methods for remote-control switching, residential HVAC systems and controls, and more. **256 pages, 8¹/₂ x 11, $22.50**

Quicken for Contractors

Most builders, contractors, and remodelers came up through the ranks, learning their craft "in the trenches." They know how buildings go together, but office management skills like bookkeeping, accounting, payroll and job costing are often a new, and dangerous, challenge. Quicken for Contractors explains how to use Quicken, an affordable, easy-to-understand computer program published by Intuit. It shows step-by-step how Quicken can work in the builder's office, putting accurate bookkeeping within easy reach. This manual does not include the program, but has instructions and examples, with onscreen "pictures" of familiar forms like checks, deposit slips and time cards that you can create with Quicken. Even if you only have a few minutes a week, this book will help you use Quicken to set up a basic checkbook system that will quickly tell you your cash position and job costs. You'll learn how to set up your financial files and create valuable reports, including profit & loss statements, payroll reports and job cost reports. The companion diskette included with the book consists of a sample construction company that you can use as a tutorial, or as a template for you to plug in your own data. **240 pages, 8¹/₂ x 11, $32.50**

Contractor's Guide to QuickBooks Pro

This user-friendly manual walks you through QuickBooks Pro's detailed setup procedure and explains step-by-step how to create a first-rate accounting system. You'll learn in days, rather than weeks, how to use QuickBooks Pro to get your contracting business organized, with simple, fast accounting procedures. On the CD included with the book you'll find a full version of QuickBooks Pro, good for 25 uses, with a QuickBooks Pro file preconfigured for a construction company (you drag it over onto your computer and plug in your own company's data). You'll also get a complete estimating program, including a database, and a job costing program that lets you export your estimates to QuickBooks Pro. It even includes many useful construction forms to use in your business. **288 pages, 8¹/₂ x 11, $39.75**

Contracting in All 50 States

Every state has its own licensing requirements that you must meet to do business there. These are usually written exams, financial requirements, and letters of reference. This book shows how to get a building, mechanical or specialty contractor's license, qualify for DOT work, and register as an out-of-state corporation, for every state in the U.S. It lists addresses, phone numbers, application fees, requirements, where an exam is required, what's covered on the exam and how much weight each area of construction is given on the exam. You'll find just about everything you need to know in order to apply for your out-of-state license. **416 pages, 8¹/₂ x 11, $36.00**

Contractor's Index to the 1997 Uniform Building Code

Finally, there's a common-sense index that helps you quickly and easily find the section you're looking for in the UBC. It lists topics under the names builders actually use in construction. Best of all, it gives the full section number and the actual page in the UBC where you'll find it. If you need to know the requirements for windows in exit access corridor walls, just look under Windows™. You'll find the requirements you need are in Section 1004.3.4.3.2.2 in the UBC — on page 115. This practical index was written by a former builder and building inspector who knows the UBC from both perspectives. If you hate to spend valuable time hunting through pages of fine print for the information you need, this is the book for you. **192 pages, 8¹/₂ x 11, paperback edition, $26.00**
192 pages, 8¹/₂ x 11, loose-leaf edition, $29.00

Electrician's Exam Preparation Guide

Need help in passing the apprentice, journeyman, or master electrician's exam? This is a book of questions and answers based on actual electrician's exams over the last few years. Almost a thousand multiple-choice questions — exactly the type you'll find on the exam — cover every area of electrical installation: electrical drawings, services and systems, transformers, capacitors, distribution equipment, branch circuits, feeders, calculations, measuring and testing, and more. It gives you the correct answer, an explanation, and where to find it in the latest NEC. Also tells how to apply for the test, how best to study, and what to expect on examination day. **352 pages, 8¹/₂ x 11, $28.00**

Masonry & Concrete Construction Revised

This is the revised edition of the popular manual, with updated information on everything from on-site preplanning and layout through the construction of footings, foundations, walls, fireplaces and chimneys. There's an added appendix on safety regulations, with all the applicable OSHA sections pulled together into one handy condensed reference. There's new information on concrete, masonry and seismic reinforcement. Plus improved estimating techniques to help you win more construction bids. The emphasis is on integrating new techniques and improved materials with the tried-and-true methods. Includes information on cement and mortar types, mixes, coloring agents and additives, and suggestions on when, where and how to use them; calculating footing and foundation loads, with tables and formulas to use as references; forming materials and forming systems; pouring and reinforcing concrete slabs and flatwork; block and brick wall construction, including seismic requirements; crack control, masonry veneer construction, brick floors and pavements, including design considerations and materials; and cleaning, painting and repairing all types of masonry. **304 pages, 8¹/₂ x 11, $28.50**

ADA Accessibility Guidelines for Buildings and Facilities With ADA Technical Assistance Manuals

All buildings, both old and new, must comply with the American with Disabilities Act. This book provides a practical, step-by-step checklist to assure compliance for all types of buildings, from offices and retail to hospitals and public facilities. Here you'll find diagrams, forms and 150 pages of checklists to make sure that what you build or remodel will pass ADA guidelines. Also includes the Technical Assistance Manuals for the ADA Titles II and III. **350 pages, 8¹/₂ x 11, $69.95**

Greenbook Standard Specifications For Public Works Construction

Since 1967, ten previous editions of the popular "Greenbook" have been used as the official specification, bidding and contract document for many cities, counties and public agencies throughout the West. New federal regulations mandate that all public construction use metric documentation. This complete reference, which meets this new requirement, provides uniform standards of quality and sound construction practice easily understood and used by engineers, public works officials, and contractors across the U.S. Includes hundreds of charts and tables. **730 pages, 5 x 8, $49.95**

Pipe & Excavation Contracting

Shows how to read plans and compute quantities for both trench and surface excavation, figure crew and equipment productivity rates, estimate unit costs, bid the work, and get the bonds you need. Explains what equipment will deliver maximum productivity for a job, how to lay all types of water and sewer pipe, and how to switch your business to excavation work when you don't have pipe contracts. Covers asphalt and rock removal, working on steep slopes or in high groundwater, and how to avoid the pitfalls that can wipe out your profits on any job. **400 pages, 5¹/₂ x 8¹/₂, $29.00**

Circular Work in Carpentry and Joinery

Simple, illustrated details reveal the secrets of curved construction in wood. Over 240 illustrations covering doors, windows, stairs, moldings, cased frames, elliptical heads, conical roofs, skylights and much more. This book gives you the knowledge you need to do real curved work instead of faceted work. **Hardcover, 120 pages, 8 x 10, $26.95**

Basic Construction Management: The Superintendent's Job, 3rd Edition

Today's construction projects are more complex than ever. Managing these projects has also become more complex. This perennial NAHB best-seller, now in its third edition, addresses the issues facing today's construction manager. New managers can use this as a great training tool. More experienced superintendents can brush up on the latest techniques and technologies. **87 pages, 8½ x 11, $30.25**

A Treatise on Stairbuilding and Handrailing

A complete course on stairbuilding and handrailing contained in one book. Here you'll find every step in handrailing and building newelled stairs — even stone stairs. This classic reference includes many examples of designs and joinery and even describes the development of face molds. This is probably the most complete book available on stairbuilding and handrailings. **390 pages, 7½ x 11, $22.95**

Building Layout

Shows how to use a transit to locate a building correctly on the lot, plan proper grades with minimum excavation, find utility lines and easements, establish correct elevations, lay out accurate foundations, and set correct floor heights. Explains how to plan sewer connections, level a foundation that's out of level, use a story pole and batterboards, work on steep sites, and minimize excavation costs. **240 pages, 5½ x 8½, $15.00**

Managing the Small Construction Business

Overcome your share of business hassles by learning how 50 small contractors handled similar problems in their businesses. Here you'll learn how they handle bidding, unit pricing, contract clauses, change orders, job-site safety, quality control, overhead and markup, managing subs, scheduling systems, cost-plus contracts, pricing small jobs, insurance repair, finding solutions to conflicts, and much more. **243 pages, 8½ x 11, $27.95**

Craftsman Book Company
6058 Corte del Cedro
P.O. Box 6500
Carlsbad, CA 92018

☎ 24 hour order line
1-800-829-8123
Fax (760) 438-0398

In A Hurry?
We accept phone orders charged to your

○ Visa, ○ MasterCard, ○ Discover or ○ American Express

Card#_____

Exp.date_____ Initials_____

Order online http://www.craftsman-book.com
Free on the Internet! Download any of Craftsman's estimating costbooks for a 30-day free trial! http://costbook.com

Name_____

Company_____

Address_____

City/State/Zip
○ This is a residence

Total enclosed_____(In California add 7.25% tax)

We pay shipping when your check covers your order in full.

Tax Deductible: Treasury regulations make these references tax deductible when used in your work. Save the canceled check or charge card statement as your receipt.

10-Day Money Back Guarantee

- ○ 69.95 ADA Accessibility Guidelines for Buildings and Facilities With ADA Technical Assistance Manuals
- ○ 30.25 Basic Construction Management: The Superintendent's Job, 3rd Edition
- ○ 34.00 Basic Engineering for Builders
- ○ 38.00 Basic Lumber Engineering for Builders
- ○ 33.00 Basic Plumbing with Illustrations
- ○ 14.75 Blueprint Reading for Building Trades
- ○ 64.95 BNI Public Works Costbook
- ○ 26.50 Builder's Guide to Accounting Revised
- ○ 27.25 Builder's Guide to Room Additions
- ○ 35.00 Building Contractor's Exam Preparation Guide
- ○ 15.00 Building Layout
- ○ 35.50 Carpentry Estimating with FREE *National Estimator* carpentry estimating program on a 3½" disk.
- ○ 68.50 CD Estimator
- ○ 69.00 CD Estimator — Heavy
- ○ 26.95 Circular Work in Carpentry and Joinery
- ○ 27.50 Commercial Electrical Wiring
- ○ 25.00 Concrete Construction & Estimating
- ○ 39.50 Construction Estimating Reference Data with FREE *National Estimator* on a CD-ROM.
- ○ 39.75 Construction Forms & Contracts with a CD-ROM for *Windows*™ and Macintosh.
- ○ 36.00 Contracting in All 50 States
- ○ 39.75 Contractor's Guide to QuickBooks Pro
- ○ 39.00 Contractor's Guide to the Building Code Revised
- ○ 26.00 Contractor's Index to the *UBC* - Paperback
- ○ 29.00 Contractor's Index to the *UBC* - Loose-leaf

- ○ 59.95 Contractor's Legal Kit
- ○ 22.25 Contractor's Survival Manual
- ○ 15.75 Cost Records for Construction Estimating
- ○ 36.00 Craftsman's Illustrated Dictionary of Construction Terms
- ○ 27.50 Drafting House Plans
- ○ 28.00 Electrician's Exam Preparation Guide
- ○ 39.50 Estimating Excavation
- ○ 34.95 Estimating Framing Quantities
- ○ 17.00 Estimating Home Building Costs
- ○ 21.50 Estimating Tables for Home Building
- ○ 22.75 Excavation & Grading Handbook Revised
- ○ 23.25 Fences and Retaining Walls
- ○ 22.50 Finish Carpenter's Manual
- ○ 49.95 Greenbook Standard Specifications for Public Works Construction
- ○ 28.75 Handbook of Construction Contracting Volume 1
- ○ 30.75 Handbook of Construction Contracting Volume 2
- ○ 24.95 Home Inspection Handbook
- ○ 24.25 How to Succeed w/Your Own Construction Business
- ○ 40.70 Land Development
- ○ 27.95 Managing the Small Construction Business
- ○ 28.50 Masonry & Concrete Construction Revised
- ○ 23.00 National Building Cost Manual
- ○ 47.75 National Electrical Estimator with FREE *National Estimator* on a CD-ROM.
- ○ 48.00 National Painting Cost Estimator with FREE *National Estimator* on a CD-ROM.

- ○ 48.25 National Plumbing & HVAC Estimator with FREE *National Estimator* on a CD-ROM.
- ○ 49.50 National Renovation & Insurance Repair Estimator with FREE *National Estimator* on a CD-ROM.
- ○ 48.50 National Repair & Remodeling Estimator with FREE *National Estimator* on a CD-ROM.
- ○ 24.00 Paint Contractor's Manual
- ○ 29.00 Pipe & Excavation Contracting
- ○ 19.25 Planning Drain, Waste & Vent Systems
- ○ 32.00 Plumber's Handbook Revised
- ○ 28.25 Plumbing & HVAC Manhour Estimates
- ○ 24.50 Professional Kitchen Design
- ○ 27.25 Profits in Building Spec Homes
- ○ 19.75 Profits in Buying & Renovating Homes
- ○ 15.75 Rafter Length Manual
- ○ 32.50 *Quicken* for Contractors
- ○ 33.50 Renovating & Restyling Older Homes
- ○ 22.50 Residential Electrical Design Revised
- ○ 29.00 Residential Electrical Estimating
- ○ 38.80 Residential Steel Framing Guide
- ○ 22.00 Roof Framing
- ○ 35.00 Roofing Construction & Estimating
- ○ 26.50 Rough Framing Carpentry
- ○ 29.95 Simplified Guide to Construction Law
- ○ 22.95 A Treatise on Stairbuilding & Handrailing
- ○ 32.50 Troubleshooting Guide to Residential Construction
- ○ 25.50 Wood-Frame House Construction
- ○ 47.50 National Construction Estimator with FREE *National Estimator* on a CD-ROM.
- ○ FREE Full Color Catalog

Prices subject to change without notice

      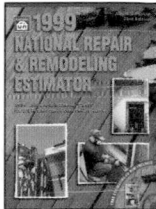
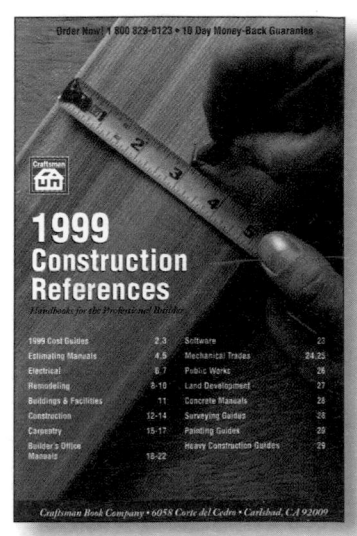

BUSINESS REPLY MAIL

FIRST CLASS MAIL PERMIT NO. 271 CARLSBAD, CA

POSTAGE WILL BE PAID BY ADDRESSEE

Craftsman Book Company

6058 Corte del Cedro
P.O. Box 6500
Carlsbad, CA 92018-9974

BUSINESS REPLY MAIL

FIRST CLASS MAIL PERMIT NO. 271 CARLSBAD, CA

POSTAGE WILL BE PAID BY ADDRESSEE

Craftsman Book Company

6058 Corte del Cedro
P.O. Box 6500
Carlsbad, CA 92018-9974

BUSINESS REPLY MAIL

FIRST CLASS MAIL PERMIT NO. 271 CARLSBAD, CA

POSTAGE WILL BE PAID BY ADDRESSEE

Craftsman Book Company

6058 Corte del Cedro
P.O. Box 6500
Carlsbad, CA 92018-9974